Student & Parent

One-Stop Int[...]s

M000279702

Log on to

www.geometryonline.com

Online Book

- Complete Student Edition
- Links to Online Study Tools

Online Study Tools

- Extra Examples
- Self-Check Quizzes
- Vocabulary Review
- Chapter Test Practice
- Standardized Test Practice

Online Activities

- WebQuest Projects
- USA TODAY Activities
- Career Links
- Data Updates

Graphing Calculator Keystrokes

- Calculator Keystrokes for other calculators

GLENCOE MATHEMATICS

Geometry

Boyd   Cummins   Malloy

Carter   Flores

Mc Graw Hill **Glencoe**

New York, New York
Columbus, Ohio
Chicago, Illinois
Peoria, Illinois
Woodland Hills, California

Send all inquiries to:
Glencoe/McGraw-Hill
8787 Orion Place
Columbus, OH 43240

ISBN: 0-07-865106-9

3 4 5 6 7 8 9 10   055/071   12 11 10 09 08 07 06 05

Contents in Brief

Authors

Cindy J. Boyd
Mathematics Teacher
Abilene High School
Abilene, Texas

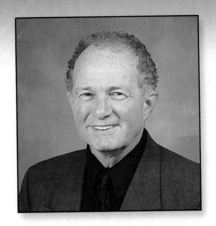

Jerry Cummins
Past President
National Council of
    Supervisors of
    Mathematics (NCSM)
Western Springs, Illinois

Carol Malloy, Ph.D.
Associate Professor, Math
    Education
The University of North
    Carolina at Chapel Hill
Chapel Hill, North Carolina

John Carter, Ph.D.
Director of Mathematics
Adlai E. Stevenson High
    School
Lincolnshire, Illinois

Alfinio Flores, Ph.D.
Professor
Arizona State University
Tempe, Arizona

Contributing Authors

USA TODAY
The USA TODAY Snapshots®, created by
USA TODAY®, help students make the connection
between real life and mathematics.

Dinah Zike
Educational Consultant
Dinah-Might Activities, Inc.
San Antonio, TX

Content Consultants

Each of the Content Consultants reviewed every chapter and gave suggestions for improving the effectiveness of the mathematics instruction.

Content Consultants

Ruth M. Casey
Mathematics Teacher/
    Department Chair
Anderson County High School
Lawrenceburg, KY

Gilbert Cuevas
Professor of Mathematics Education
University of Miami
Coral Gables, FL

Alan G. Foster
Former Mathematics
    Department Chair
Addison Trail High School
Addison, IL

Linda M. Hayek
Curriculum Facilitator
Ralston Public Schools
Omaha, NE

Berchie Holliday
Educational Consultant
Silver Spring, MD

Joseph Kavanaugh
Academic Head for Mathematics
Scotia-Glenville Senior High School
Scotia, NY

Yvonne Medina Mojica
Mathematics Coach
Verdugo Hills High School, Los Angeles
    Unified School District
Tujunga, CA

Reading Consultant

Lynn T. Havens
Director
Project CRISS
Kalispell, MT

ELL Consultant

Idania Dorta
Mathematics Educational Specialist
Miami-Dade County Public Schools
Miami, FL

Teacher Reviewers

Each Teacher Reviewer reviewed at least two chapters of the Student Edition, giving feedback and suggestions for improving the effectiveness of the mathematics instruction.

Liza Allen
Math Teacher
Conway High School West
Conway, AR

Molly M. Andaya
Mathematics Teacher
Ellensburg H.S.
Ellensburg, WA

David J. Armstrong
Mathematics Facilitator
Huntington Beach UHSD
Huntington Beach, CA

Jerry C. Bencivenga
Consultant
Connecticut VTSS
Connecticut SDE
Middletown, CT

Patrick M. Blake
Math Department Chairperson
Ritenour H.S.
St. Louis, MO

Donna L. Burns
Math Department Chairperson
Los Angeles H.S.
Los Angeles, CA

Nita Carpenter
Mathematics Teacher
Okemos H.S.
Okemos, MI

Vincent Ciraulo
Supervisor of Mathematics
J.P. Stevens H.S.
Edison, NJ

Keitha Cleveland
Mathematics Teacher
E.S. Aiken Optional School
Alexandria, LA

Janice Garner Coffer
Math Department Chairperson
Page H.S.
Greensboro, NC

Karyn S. Cummins
Mathematics Department Head
Franklin Central H.S.
Indianapolis, IN

Christine B. Denardo
Math Teacher and Department Chair
Blacksburg H.S.
Blacksburg, VA

Joseph N. Di Cioccio
Mathematics Teacher/Department
    Chair
Mahopac H.S.
Mahopac, NY

Robert A. Di Dio, M.S.Ed., P.D.
Assistant Principal
IS 192X
Bronx, NY

James S. Emery
Geometry Teacher
Beaverton H.S.
Beaverton, MI

Teacher Reviewers

Nancy S. Falls
Math Department Chair/
   Geometry Teacher
Northern York County H.S.
Dillsburg, PA

Jane Fasullo
Retired High School Teacher
Ward Melville H.S.
E. Setauket, NY

Susan Fischbein
Math Department Chair
Desert Mountain H.S.
Scottsdale, AZ

Dolores Fischer
Mathematics Teacher
Adlai E. Stevenson H.S.
Lincolnshire, IL

Joyce E. Fisher
Math Teacher
La Cueva H.S.
Albuquerque, NM

Candace Frewin
Teacher on Special Assignment
Pinellas County Schools
Largo, FL

Karen L. George
Math Teacher
Taunton H.S.
Taunton, MA

Douglas E. Hall
Geometry Teacher
Chaparral H.S.
Las Vegas, NV

Cynthia D. Hodges
Teacher
Shoemaker H.S.
Killeen, TX

Brian J. Johnson
K–12 Curriculum Specialist
Bay City Public Schools
Bay City, MI

Melissa L. Jones
Mathematics Teacher
Bexley H.S.
Bexley, OH

Nancy Lee Keen
Geometry Teacher
Martinsville H.S.
Martinsville, IN

John R. Kennedy
Mathematics Department Chairman
Derby H.S.
Derby, KS

Sharon Kenner
Algebra/Geometry Teacher
Coconut Creek H.S.
Coconut Creek, FL

Julie Kolb
Mathematics Teacher
Leesville Road H.S.
Raleigh, NC

Gary Kubina
Math Department Head
Citronelle H.S.
Citronelle, AL

Jenita Lyons
Mathematics Teacher
William M. Raines H.S.
Jacksonville, FL

Debra D. McCoy
Math Teacher
Hixson H.S.
Chattanooga, TN

Delia Dee Miller
Geometry/Algebra II Teacher
Caddo Parish Magnet H.S.
Shreveport, LA

Kenneth E. Montgomery
Mathematics Teacher
Tri-Cities H.S.
East Point, GA

Cynthia Orkin
Teacher/Department Chairperson
Brookville H.S.
Lynchburg, VA

Susan M. Parece
Mathematics Teacher
Plymouth South H.S.
Plymouth, MA

Mike Patterson
High School Mathematics Teacher
Advanced Technologies Academy
Las Vegas, NV

Cynthia W. Poché
Math Teacher
Salmen H.S.
Slidell, LA

David E. Rader
K–12 Math/Science Supervisor
Wissahickon School District
Ambler, PA

Monique Siedschlag
Math Teacher
Thoreau H.S.
Thoreau, NM

Frank Louis Sparks
Curriculum Design and Support
   Specialist—Mathematics
New Orleans Public Schools
New Orleans, LA

Alice Brady Sprinkle
Math Teacher
Broughton H.S.
Raleigh, NC

Joy F. Stanford
Mathematics Chair
Booker T. Washington Magnet H.S.
Montgomery, AL

Dora Swart
Math Teacher & Department Chair
W.F. West H.S.
Chehalis, WA

David Tate
Mathematics Teacher
Prairie Grove H.S.
Prairie Grove, AR

Kathy Thirkell
Math Curriculum Coordinator
Lewis-Palmer H.S.
Monument, CO

Marylu Tyndell
Teacher of Mathematics
Colts Neck H.S.
Colts Neck, NJ

Sarah L. Waldrop
Math Department Chairperson
Forestview H.S.
Gastonia, NC

Christine A. Watts
Mathematics Department Chairman
Monmouth H.S.
Monmouth, IL

René Wilkins
Mathematics Teacher
Fairhope H.S.
Fairhope, AL

Rosalyn Zeid
Mathematics Supervisor
Union Township Public Schools
Union, NJ

UNIT 1

Table of Contents

Lines and Angles 2

Web_Quest_ **Internet Project**

- Introduction 3
- Follow-Ups 23, 65, 155
- Culmination 164

Lesson 1-5, p. 37

Prerequisite Skills
- Getting Started 5
- Getting Ready for the Next Lesson 11, 19, 27, 36, 43

FOLDABLES™ Study Organizer 5

Reading and Writing Mathematics
- Describing What You See 12
- Reading Math Tips 6, 29, 45, 46
- Writing in Math 11, 19, 27, 35, 43, 50

Standardized Test Practice
- Multiple Choice 11, 19, 23, 25, 27, 35, 43, 50, 57, 58
- Short Response/Grid In 43, 50, 59
- Extended Response 59

USA TODAY Snapshots 16

Constructions 15, 18, 24, 31, 33, 44

Chapter 2   Reasoning and Proof   60

Prerequisite Skills
- Getting Started 61
- Getting Ready for the Next Lesson 66, 74, 80, 87, 93, 100, 106

Study Organizer 61

Reading and Writing Mathematics
- Biconditional Statements 81
- Reading Math Tips 75
- Writing in Math 66, 74, 79, 86, 93, 99, 106, 114

Standardized Test Practice
- Multiple Choice 66, 74, 80, 86, 87, 93, 96, 97, 99, 106, 114, 121, 122
- Short Response/Grid In 106, 123
- Extended Response 123

USA TODAY Snapshots 63

Lesson 2-3, p. 79

Chapter ③ **Parallel and Perpendicular Lines**

124

Lesson 3-6, p. 163

Prerequisite Skills
- Getting Started **125**
- Getting Ready for the Next Lesson **131, 138, 144, 150, 157**

 Study Organizer 125

Reading and Writing Mathematics
- Writing in Math **130, 138, 144, 149, 157, 164**

Standardized Test Practice
- Multiple Choice **131, 138, 144, 149, 157, 164, 171, 172**
- Short Response/Grid In **131, 135, 136, 164, 173**
- Extended Response **173**

 Snapshots 143

Constructions 151, 160

Chapter 4   Congruent Triangles   176

WebQuest Internet Project

- Introduction **175**
- Follow-Ups **218, 241, 325, 347**
- Culmination **390**

Lesson 4-4, p. 204

Prerequisite Skills
- Getting Started **177**
- Getting Ready for the Next Lesson **183, 191, 198, 206, 213, 221**

FOLDABLES Study Organizer **177**

Reading and Writing Mathematics
- Making Concept Maps **199**
- Reading Math Tips **186, 207**
- Writing in Math **183, 191, 198, 205, 213, 221, 226**

Standardized Test Practice
- Multiple Choice **183, 191, 198, 206, 213, 217, 219, 221, 226, 231, 232**
- Short Response/Grid In **233**
- Extended Response **233**

 Snapshots 206

Constructions 200, 202, 207, 214

Chapter ⑤ Relationships in Triangles 234

Lesson 5-5, p. 267

Chapter ⑥   Proportions and Similarity   280

Prerequisite Skills

- Getting Started 281
- Getting Ready for the Next Lesson
  287, 297, 306, 315, 323

Study Organizer 281

Reading and Writing Mathematics

- Reading Math Tips 283
- Writing in Math 286, 296, 305, 314, 322, 330

Standardized Test Practice

- Multiple Choice 282, 287, 297, 305, 314, 322, 331, 337, 338
- Short Response/Grid In 285, 287, 314, 322, 331, 339
- Extended Response 339

USA TODAY Snapshots 296

Constructions 311, 314

Lesson 6-3, p. 305

Chapter 7   Right Triangles and Trigonometry   340

Lesson 7-2, p. 350

Prerequisite Skills

FOLDABLES

Study Organizer 341

Reading and Writing Mathematics

Standardized Test Practice

USA TODAY Snapshots 347

Chapter 8   Quadrilaterals   **402**

WebQuest Internet Project

- Introduction **401**
- Follow-Ups **444, 469, 527**
- Culmination **580**

Lesson 8-4, p. 429

Prerequisite Skills
- Getting Started **403**
- Getting Ready for the Next Lesson **409, 416, 423, 430, 437, 445**

FOLDABLES Study Organizer **403**

Reading and Writing Mathematics
- Hierarchy of Polygons **446**
- Reading Math Tips **411, 432**
- Writing in Math **409, 416, 422, 430, 436, 444, 451**

Standardized Test Practice
- Multiple Choice **409, 413, 414, 416, 423, 430, 437, 445, 451, 457, 458**
- Short Response/Grid In **409, 416, 444, 459**
- Extended Response **459**

USA TODAY Snapshots **411**

Constructions **425, 433, 435, 438, 441, 444**

Chapter ⑨ Transformations   460

Prerequisite Skills

- Getting Started 461
- Getting Ready for the Next Lesson
  469, 475, 482, 488, 497, 505

FOLDABLES™

Study Organizer 461

Reading and Writing Mathematics

- Reading Math Tips 464, 470, 483
- Writing in Math 469, 474, 481, 487, 496, 505, 511

Standardized Test Practice

- Multiple Choice 469, 475, 481, 487, 493, 494, 496, 505, 511, 517, 518
- Short Response/Grid In 511, 519
- Extended Response 519

 Snapshots 474

Lesson 9-5, p. 495

Chapter ⑩   Circles   520

Lesson 10-3, p. 541

Prerequisite Skills
- Getting Started **521**
- Getting Ready for the Next Lesson **528, 535, 543, 551, 558, 568, 574**

FOLDABLES™ Study Organizer **521**

Reading and Writing Mathematics
- Reading Math Tips **522, 536**
- Writing in Math **527, 534, 542, 551, 558, 567, 574, 579**

Standardized Test Practice
- Multiple Choice **525, 526, 528, 535, 543, 551, 558, 567, 574, 580, 587, 588**
- Short Response/Grid In **528, 535, 543, 551, 558, 589**
- Extended Response **589**

 Snapshots 531

Constructions 542, 554, 556, 559, 560

Area and Volume 590

Chapter ⑪ Areas of Polygons and Circles 592

 **Internet Project**

- Introduction **591**
- Follow-Ups **618, 662, 703**
- Culmination **719**

Prerequisite Skills
- Getting Started **593**
- Getting Ready for the Next Lesson **600, 609, 616, 621**

Study Organizer 593

Reading and Writing Mathematics
- Prefixes **594**
- Reading Math Tips **617**
- Writing in Math **600, 608, 616, 620, 627**

Standardized Test Practice
- Multiple Choice **600, 608, 616, 621, 627, 631, 632**
- Short Response/Grid In **622, 625, 633**
- Extended Response **633**

USA TODAY Snapshots 614

Lesson 11-4, p. 617

Chapter 12 Surface Area   634

Prerequisite Skills

- Getting Started 635
- Getting Ready for the Next Lesson
  642, 648, 654, 659, 665, 670

FOLDABLES™

Study Organizer 635

Reading and Writing Mathematics

- Reading Math Tips 637, 649, 666
- Writing in Math 641, 648, 653, 658, 664, 669, 676

Standardized Test Practice

- Multiple Choice 642, 644, 645, 648, 653, 658, 664, 665, 670, 676, 683, 684
- Short Response/Grid In 685
- Extended Response 685

USA TODAY Snapshots 653

Lesson 12-5, p. 660

Chapter ⑬   Volume   **686**

Lesson 13-1, p. 693

Prerequisite Skills

- Getting Started **687**
- Getting Ready for the Next Lesson **694, 701, 706, 713**

 Study Organizer 687

Reading and Writing Mathematics

- Reading Math Tips **714**
- Writing in Math **693, 701, 706, 712, 719**

Standardized Test Practice

- Multiple Choice **694, 701, 706, 713, 719, 723, 724**
- Short Response/Grid In **703, 704, 725**
- Extended Response **725**

USA TODAY Snapshots 705

Student Handbook

Need extra help or information? Log on to math.glencoe.com or any of the web addresses below to learn more.

Online Study Tools

- www.geometryonline.com/extra_examples shows you additional worked-out examples that mimic the ones in your book.

- www.geometryonline.com/self_check_quiz provides you with a practice quiz for each lesson that grades itself.

- www.geometryonline.com/vocabulary_review lets you check your understanding of the terms and definitions used in each chapter.

- www.geometryonline.com/chapter_test allows you to take a self-checking test before the actual test.

- www.geometryonline.com/standardized_test is another way to brush up on your standardized test-taking skills.

Research Options

- www.geometryonline.com/webquest walks you step-by-step through a long-term project using the web. One WebQuest for each unit is explored using the mathematics from that unit.

- www.geometryonline.com/usa_today provides activities related to the concept of the lesson as well as up-to-date Snapshot data.

- www.geometryonline.com/careers links you to additional information about interesting careers.

- www.geometryonline.com/data_update links you to the most current data available for subjects such as basketball and family.

Calculator Help

- www.geometryonline.com/other_calculator_keystrokes provides you with instructions for using Cabri Jr. on the TI-83 Plus and keystrokes other than the TI-83 Plus used in your textbook.

UNIT 1

Lines and Angles

Lines and angles are all around us and can be used to model and describe real-world situations. In this unit, you will learn about lines, planes, and angles and how they can be used to prove theorems.

Chapter 1
Points, Lines, Planes, and Angles

Chapter 2
Reasoning and Proof

Chapter 3
Parallel and Perpendicular Lines

WebQuest Internet Project

When Is Weather Normal?

Source: *USA TODAY,* October 8, 2000

"Climate normals are a useful way to describe the average weather of a location. Several statistical measures are computed as part of the normals, including measures of central tendency, such as mean or median, of dispersion or how spread out the values are, such as the standard deviation or inter-quartile range, and of frequency or probability of occurrence." In this project, you will explore how latitude, longitude, and *degree distance* relate to differences in temperature for pairs of U.S. cities.

 Log on to www.geometryonline.com/webquest. Begin your WebQuest by reading the Task.

Continue working on your WebQuest as you study Unit 1.

Lesson	1-3	2-1	3-5
Page	23	65	155

USA TODAY Snapshots®

Coldest cities in the USA

City	Mean temperature
International Falls, Minn.	**36.4**
Duluth, Minn.	**38.2**
Caribou, Maine	**38.9**
Marquette, Mich.	**39.2**
Sault Ste. Marie, Mich.	**39.7**
Williston, N.D.	**40.1**
Fargo, N.D.	**40.5**
Alamosa, Colo.	**41.2**
Bismarck, N.D.	**41.3**
St. Cloud, Minn.	**41.4**

Source: Planet101.com   By Lori Joseph and Keith Simmons, USA TODAY

Points, Lines, Planes, and Angles

What You'll Learn

- **Lesson 1-1** Identify and model points, lines, and planes.
- **Lesson 1-2** Measure segments and determine accuracy of measurements.
- **Lesson 1-3** Calculate the distance between points and find the midpoint of a segment.
- **Lessons 1-4 and 1-5** Measure and classify angles and identify angle relationships.
- **Lesson 1-6** Identify polygons and find their perimeters.

Key Vocabulary

- line segment (p. 13)
- congruent (p. 15)
- segment bisector (p. 24)
- angle bisector (p. 32)
- perpendicular (p. 40)

Why It's Important

Points, lines, and planes are the basic building blocks used in geometry. They can be used to describe real-world objects. For example, a kite can model lines, angles, and planes in two and three dimensions. *You will explore the angles formed by the structure of a kite in Lesson 1-2.*

Getting Started

Prerequisite Skills To be successful in this chapter, you'll need to master these skills and be able to apply them in problem-solving situations. Review these skills before beginning Chapter 1.

For Lesson 1-1
Graph Points

Graph and label each point in the coordinate plane. *(For review, see pages 728 and 729.)*

1. $A(3, -2)$ **2.** $B(4, 0)$ **3.** $C(-4, -4)$ **4.** $D(-1, 2)$

For Lesson 1-2
Add and Subtract Fractions

Find each sum or difference.

5. $\frac{3}{4} + \frac{3}{8}$ **6.** $2\frac{5}{16} + 5\frac{1}{8}$ **7.** $\frac{7}{8} - \frac{9}{16}$ **8.** $11\frac{1}{2} - 9\frac{7}{16}$

For Lessons 1-3 through 1-5
Operations With Integers

Evaluate each expression. *(For review, see pages 734 and 735.)*

9. $2 - 17$ **10.** $23 - (-14)$ **11.** $[-7 - (-2)]^2$ **12.** $9^2 + 13^2$

For Lesson 1-6
Find Perimeter

Find the perimeter of each figure. *(For review, see pages 732 and 733.)*

13.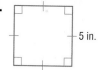
5 in.

14.
$2\frac{1}{2}$ ft

6 ft

15.
7.5 m

4.8 m

FOLDABLES™
Study Organizer

Lines and Angles Make this Foldable to help you organize your notes. Begin with a sheet of 11″ by 17″ paper.

Step 1 Fold

Fold the short sides to meet in the middle.

Step 2 Fold Again

Fold the top to the bottom.

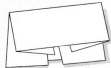

Step 3 Cut

Open. Cut flaps along the second fold to make four tabs.

Step 4 Label

Label the tabs as shown.

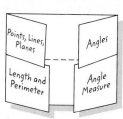

Points, Lines, Planes

Angles

Length and Perimeter

Angle Measure

Reading and Writing As you read and study the chapter, record examples and notes from each lesson under the appropriate tab.

1-1 Points, Lines, and Planes

What You'll Learn

- Identify and model points, lines, and planes.
- Identify collinear and coplanar points and intersecting lines and planes in space.

Vocabulary

- point
- line
- collinear
- plane
- coplanar
- undefined term
- space
- locus

Why do chairs sometimes wobble?

Have you ever noticed that a four-legged chair sometimes wobbles, but a three-legged stool never wobbles? This is an example of points and how they lie in a plane. All geometric shapes are made of points. In this book, you will learn about those shapes and their characteristics.

NAME POINTS, LINES, AND PLANES You are familiar with the terms *plane*, *line*, and *point* from algebra. You graph on a coordinate *plane*, and ordered pairs represent *points* on *lines*. In geometry, these terms have similar meanings.

Unlike objects in the real world that model these shapes, points, lines, and planes do not have any actual size.

- A **point** is simply a location.
- A **line** is made up of points and has no thickness or width. Points on the same line are said to be **collinear**.
- A **plane** is a flat surface made up of points. Points that lie on the same plane are said to be **coplanar**. A plane has no depth and extends infinitely in all directions.

Points are often used to name lines and planes. The letters of the points can be in any order.

Study Tip

Reading Math
The word *noncollinear* means not collinear or not lying on the same line. Likewise, *noncoplanar* means not lying in the same plane.

Key Concept   Points, Lines, and Planes

	Point	Line	Plane
Model	•P	A •———• B *n*	•X •Y •Z *T*
Drawn:	as a dot	with an arrowhead at each end	as a shaded, slanted 4-sided figure
Named by:	a capital letter	the letters representing two points on the line or a lowercase script letter	a capital script letter or by the letters naming three noncollinear points
Facts	A point has neither shape nor size.	There is exactly one line through any two points.	There is exactly one plane through any three noncollinear points.
Words/ Symbols	point *P*	line *n*, line *AB* or $\overleftrightarrow{AB}$, line *BA* or $\overleftrightarrow{BA}$	plane *T*, plane *XYZ*, plane *XZY*, plane *YXZ*, plane *YZX*, plane *ZXY*, plane *ZYX*

Example 1 · Name Lines and Planes

Use the figure to name each of the following.

a. a line containing point A

The line can be named as line ℓ.

There are four points on the line. Any two of the points can be used to name the line.

$\overleftrightarrow{AB}$ $\overleftrightarrow{BA}$ $\overleftrightarrow{AC}$ $\overleftrightarrow{CA}$ $\overleftrightarrow{AD}$ $\overleftrightarrow{DA}$ $\overleftrightarrow{BC}$ $\overleftrightarrow{CB}$ $\overleftrightarrow{BD}$ $\overleftrightarrow{DB}$ $\overleftrightarrow{CD}$ $\overleftrightarrow{DC}$

b. a plane containing point C

The plane can be named as plane $\mathcal{N}$.

You can also use the letters of any three *noncollinear* points to name the plane.
plane *ABE*   plane *ACE*   plane *ADE*   plane *BCE*   plane *BDE*   plane *CDE*

The letters of each of these names can be reordered to create other acceptable names for this plane. For example, *ABE* can also be written as *AEB*, *BEA*, *BAE*, *EBA*, and *EAB*. In all, there are 36 different three-letter names for this plane.

Example 2 · Model Points, Lines, and Planes

VISUALIZATION **Name the geometric shapes modeled by the picture.**

The pencil point models point *A*.

The blue rule on the paper models line *BC*.

The edge of the paper models line *BD*.

The sheet of paper models plane *ADC*.

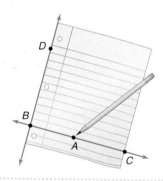

In geometry, *point*, *line*, and *plane* are considered **undefined terms** because they are only explained using examples and descriptions. Even though they are undefined, these terms can still be used to define other geometric terms and properties. For example, two lines intersect in a point. In the figure at the right, point *P* represents the intersection of $\overleftrightarrow{AB}$ and $\overleftrightarrow{CD}$. Lines can intersect planes, and planes can intersect each other.

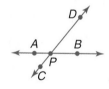

Example 3 · Draw Geometric Figures

Draw and label a figure for each relationship.

a. ALGEBRA Lines *GH* and *JK* intersect at *L* for *G*(−1, −3), *H*(2, 3), *J*(−3, 2), and *K*(2, −3) on a coordinate plane. Point *M* is coplanar with these points, but not collinear with $\overleftrightarrow{GH}$ or $\overleftrightarrow{JK}$.

Graph each point and draw $\overleftrightarrow{GH}$ and $\overleftrightarrow{JK}$.

Label the intersection point as *L*.

There are an infinite number of points that are coplanar with *G*, *H*, *J*, *K*, and *L*, but are not collinear with $\overleftrightarrow{GH}$ or $\overleftrightarrow{JK}$. In the graph, one such point is *M*(−4, 0).

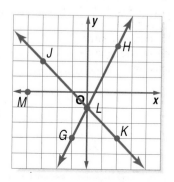

 www.geometryonline.com/extra_examples

b. $\overleftrightarrow{TU}$ lies in plane Q and contains point R.

Draw a surface to represent plane Q and label it.

Draw a line anywhere on the plane.

Draw dots on the line for points T and U.

Since $\overleftrightarrow{TU}$ contains R, point R lies on $\overleftrightarrow{TU}$.

Draw a dot on $\overleftrightarrow{TU}$ and label it R.

The locations of points T, R, and U are totally arbitrary.

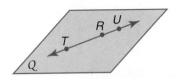

POINTS, LINES, AND PLANES IN SPACE   **Space** is a boundless, three-dimensional set of all points. Space can contain lines and planes.

Study Tip

Three-Dimensional Drawings
Because it is impossible to show space or an entire plane in a figure, edged shapes with different shades of color are used to represent planes. If the lines are hidden from view, the lines or segments are shown as dashed lines or segments.

Example 4 *Interpret Drawings*

a. How many planes appear in this figure?

There are four planes: plane P, plane ADB, plane BCD, plane ACD

b. Name three points that are collinear.

Points D, B, and G are collinear.

c. Are points G, A, B, and E coplanar? Explain.

Points A, B, and E lie in plane P, but point G does not lie in plane P. Thus, they are not coplanar. Points A, G, and B lie in a plane, but point E does not lie in plane AGB.

d. At what point do $\overleftrightarrow{EF}$ and $\overleftrightarrow{AB}$ intersect?

$\overleftrightarrow{EF}$ and $\overleftrightarrow{AB}$ do not intersect. $\overleftrightarrow{AB}$ lies in plane P, but only point E of $\overleftrightarrow{EP}$ lies in P.

Sometimes it is difficult to identify collinear or coplanar points in space unless you understand what a drawing represents. In geometry, a model is often helpful in understanding what a drawing is portraying.

Geometry Activity

Modeling Intersecting Planes

- Label one index card as Q and another as R.
- Hold the two index cards together and cut a slit halfway through both cards.

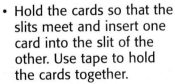

- Where the two cards meet models a line. Draw the line and label two points, C and D, on the line.

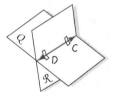

- Hold the cards so that the slits meet and insert one card into the slit of the other. Use tape to hold the cards together.

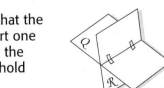

Analyze

1. Draw a point F on your model so that it lies in Q but not in R. Can F lie on $\overleftrightarrow{DC}$?

2. Draw point G so that it lies in R, but not in Q. Can G lie on $\overleftrightarrow{DC}$?

3. If point H lies in both Q and R, where would it lie? Draw point H on your model.

4. Draw a sketch of your model on paper. Label all points, lines, and planes appropriately.

Concept Check

1. **Name** three undefined terms from this lesson.

2. **OPEN ENDED** Fold a sheet of paper. Open the paper and fold it again in a different way. Open the paper and label the geometric figures you observe. Describe the figures.

3. **FIND THE ERROR** Raymond and Micha were looking for patterns to determine how many ways there are to name a plane given a certain number of points.

Raymond

If there are 4 points, then there are $4 \cdot 3 \cdot 2$ ways to name the plane.

Micha

If there are 5 noncollinear points, then there are $5 \cdot 4 \cdot 3$ ways to name the plane.

Who is correct? Explain your reasoning.

Guided Practice

4. Use the figure at the right to name a line containing point B and a plane containing points D and C.

Draw and label a figure for each relationship.

5. A line in a coordinate plane contains $X(3, -1)$, $Y(-3, -4)$, and $Z(-1, -3)$ and a point W that does not lie on $\overleftrightarrow{XY}$.

6. Plane Q contains lines r and s that intersect in P.

Refer to the figure.

7. How many planes are shown in the figure?

8. Name three points that are collinear.

9. Are points A, C, D, and J coplanar? Explain.

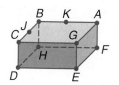

Application **VISUALIZATION** **Name the geometric term modeled by each object.**

10.

11. a pixel on a computer screen

12. a ceiling

Homework Help	
For Exercises	**See Examples**
13–18	1
21–28	3
30–37	4
38–46	2

Extra Practice
See page 754.

Refer to the figure.

13. Name a line that contains point P.

14. Name the plane containing lines n and m.

15. Name the intersection of lines n and m.

16. Name a point not contained in lines ℓ, m, or n.

17. What is another name for line n?

18. Does line ℓ intersect line m or line n? Explain.

MAPS **For Exercises 19 and 20, refer to the map, and use the following information.**
A map represents a plane. Points on this plane are named using a letter/number combination.

19. Name the point where Raleigh is located.

20. What city is located at (F, 5)?

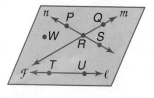

Draw and label a figure for each relationship.

21. Line AB intersects plane Q at W.

22. Point T lies on $\overleftrightarrow{WR}$.

23. Points $Z(4, 2)$, $R(-4, 2)$, and S are collinear, but points Q, Z, R, and S are not.

24. The coordinates for points C and R are $(-1, 4)$ and $(6, 4)$, respectively. $\overleftrightarrow{RS}$ and $\overleftrightarrow{CD}$ intersect at $P(3, 2)$.

25. Lines a, b, and c are coplanar, but do not intersect.

26. Lines a, b, and c are coplanar and meet at point F.

27. Point C and line r lie in M. Line r intersects line s at D. Point C, line r, and line s are not coplanar.

28. Planes $\mathcal{A}$ and $\mathcal{B}$ intersect in line s. Plane C intersects $\mathcal{A}$ and $\mathcal{B}$, but does not contain s.

29. **ALGEBRA** Name at least four ordered pairs for which the sum of coordinates is -2. Graph them and describe the graph.

Refer to the figure.

30. How many planes are shown in the figure?

31. How many planes contain points B, C, and E?

32. Name three collinear points.

33. Where could you add point G on plane $\mathcal{N}$ so that A, B, and G would be collinear?

34. Name a point that is not coplanar with A, B, and C.

35. Name four points that are coplanar.

36. Name an example that shows that three points are always coplanar, but four points are not always coplanar.

37. Name the intersection of plane $\mathcal{N}$ and the plane that contains points A, E, and C.

VISUALIZATION Name the geometric term(s) modeled by each object.

38.

39.

40.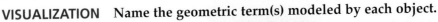

41. a table cloth

42. a partially-opened newspaper

43. a star in the sky

44. woven threads in a piece of cloth

45. a knot in a string

46. satellite dish signal

ONE-POINT PERSPECTIVE One-point perspective drawings use lines to convey depth in a picture. Lines representing horizontal lines in the real object can be extended to meet at a single point called the *vanishing point*.

47. Trace the figure at the right. Draw all of the vertical lines. Several are already drawn for you.

48. Draw and extend the horizontal lines to locate the vanishing point and label it.

49. Draw a one-point perspective of your classroom or a room in your house.

50. RESEARCH Use the Internet or other research resources to investigate one-point perspective drawings in which the vanishing point is in the center of the picture. How do they differ from the drawing for Exercises 47–49?

TWO-POINT PERSPECTIVE Two-point perspective drawings also use lines to convey depth, but two sets of lines can be drawn to meet at two vanishing points.

51. Trace the outline of the house. Draw all of the vertical lines.

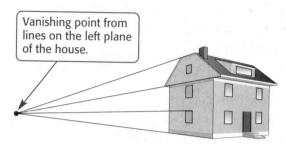

Vanishing point from lines on the left plane of the house.

52. Draw and extend the lines on your sketch representing horizontal lines in the real house to identify the vanishing point on the right plane in this figure.

53. Which types of lines seem unaffected by any type of perspective drawing?

54. CRITICAL THINKING Describe a real-life example of three lines in space that do not intersect each other and no two lines lie in the same plane.

55. WRITING IN MATH   Answer the question that was posed at the beginning of the lesson.

Why do chairs sometimes wobble?

Include the following in your answer:
- an explanation of how the chair legs relate to points in a plane, and
- how many legs would create a chair that does not wobble.

Standardized Test Practice
(A) (B) (C) (D)

56. Four lines are coplanar. What is the greatest number of intersection points that can exist?
(A) 4   (B) 5   (C) 6   (D) 7

57. ALGEBRA   If $2 + x = 2 - x$, then $x = ?$
(A) -1 (B) 0   (C) 1   (D) 2

Extending the Lesson
Another way to describe a group of points is called a locus. A locus is a set of points that satisfy a particular condition.

58. Find five points that satisfy the equation $4 - x = y$. Graph them on a coordinate plane and describe the geometric figure they suggest.

59. Find ten points that satisfy the inequality $y > -2x + 1$. Graph them on a coordinate plane and describe the geometric figure they suggest.

Getting Ready for the Next Lesson
BASIC SKILL   Replace each ● with >, <, or = to make a true statement.

60. $\frac{1}{2}$ in. ● $\frac{3}{8}$ in.   **61.** $\frac{4}{16}$ in. ● $\frac{1}{4}$ in.   **62.** $\frac{4}{5}$ in. ● $\frac{6}{10}$ in.

63. 10 mm ● 1 cm   **64.** 2.5 cm ● 28 mm   **65.** 0.025 cm ● 25 mm

Describing What You See

Figures play an important role in understanding geometric concepts. It is helpful to know what words and phrases can be used to describe figures. Likewise, it is important to know how to read a geometric description and be able to draw the figure it describes.

The figures and descriptions below help you visualize and write about points, lines, and planes.

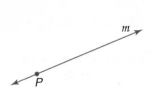

Point P is on m.
Line m contains P.
Line m passes through P.

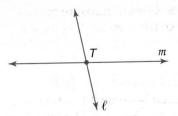

Lines ℓ and m intersect in T.
Point T is the intersection of ℓ and m.
Point T is on m. Point T is on ℓ.

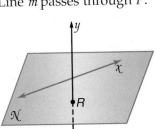

Line x and point R are in $\mathcal{N}$.
Point R lies in $\mathcal{N}$.
Plane $\mathcal{N}$ contains R and x.
Line y intersects $\mathcal{N}$ at R.
Point R is the intersection of y with $\mathcal{N}$.
Lines y and x do not intersect.

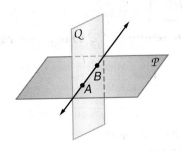

$\overleftrightarrow{AB}$ is in $\mathcal{P}$ and $\mathcal{Q}$.
Points A and B lie in both $\mathcal{P}$ and $\mathcal{Q}$.
Planes $\mathcal{P}$ and $\mathcal{Q}$ both contain $\overleftrightarrow{AB}$.
Planes $\mathcal{P}$ and $\mathcal{Q}$ intersect in $\overleftrightarrow{AB}$.
$\overleftrightarrow{AB}$ is the intersection of $\mathcal{P}$ and $\mathcal{Q}$.

Reading to Learn

Write a description for each figure.

1.

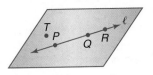

2.

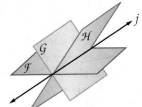

3.

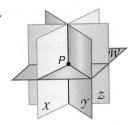

4. Draw and label a figure for the statement *Planes A, B, and C do not intersect.*

1-2 Linear Measure and Precision

What You'll Learn

- Measure segments and determine accuracy of measurement.
- Compute with measures.

Vocabulary

- line segment
- precision
- betweenness of points
- between
- congruent
- construction
- relative error

Why are units of measure important?

When you look at the sign, you probably assume that the unit of measure is miles. However, if you were in France, this would be 17 kilometers, which is a shorter distance than 17 miles. Units of measure give us points of reference when evaluating the sizes of objects.

Paris   17

MEASURE LINE SEGMENTS Unlike a line, a **line segment**, or *segment*, can be measured because it has two endpoints. A segment with endpoints A and B can be named as $\overline{AB}$ or $\overline{BA}$. The length or measure of $\overline{AB}$ is written as AB. The length of a segment is only as precise as the smallest unit on the measuring device.

Example 1 Length in Metric Units

Find the length of $\overline{CD}$ using each ruler.

a.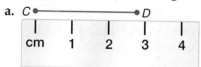

The ruler is marked in centimeters. Point D is closer to the 3-centimeter mark than to 2 centimeters. Thus, $\overline{CD}$ is about 3 centimeters long.

b.

The long marks are centimeters, and the shorter marks are millimeters. There are 10 millimeters for each centimeter. Thus, $\overline{CD}$ is about 28 millimeters long.

Example 2 Length in Customary Units

Find the length of $\overline{AB}$ using each ruler.

a.

Each inch is divided into fourths. The long marks are half-inch increments. Point B is closer to the $1\frac{2}{4}$-inch mark. Thus, $\overline{AB}$ is about $1\frac{2}{4}$ or $1\frac{1}{2}$ inches long.

b.

Each inch is divided into sixteenths. Point B is closer to the $1\frac{8}{16}$-inch mark. Thus, $\overline{AB}$ is about $1\frac{8}{16}$ or $1\frac{1}{2}$ inches long.

A measurement of
38.0 centimeters on a
ruler with millimeter
marks means a
measurement of
380 millimeters. So the
actual measurement is
between 379.5 millimeters
and 380.5 millimeters, not
37.5 centimeters and
38.5 centimeters. The
range of error in the
measurement is called the
tolerance and can be
expressed as ±0.5.

The **precision** of any measurement depends on the smallest unit available on the measuring tool. The measurement should be precise to within 0.5 unit of measure. For example, in part **a** of Example 1, 3 centimeters means that the actual length is no less than 2.5 centimeters, but no more than 3.5 centimeters.

Measurements of 28 centimeters and 28.0 centimeters indicate different precision in measurement. A measurement of 28 centimeters means that the ruler is divided into centimeters. However, a measurement of 28.0 centimeters indicates that the ruler is divided into millimeters.

Example 3 Precision

Find the precision for each measurement. Explain its meaning.

a. 5 millimeters

The measurement is precise to within 0.5 millimeter. So, a measurement of 5 millimeters could be 4.5 to 5.5 millimeters.

b. $8\frac{1}{2}$ inches

The measuring tool is divided into $\frac{1}{2}$-inch increments. Thus, the measurement is precise to within $\frac{1}{2}\left(\frac{1}{2}\right)$ or $\frac{1}{4}$ inch. Therefore, the measurement could be between $8\frac{1}{4}$ inches and $8\frac{3}{4}$ inches.

CALCULATE MEASURES Measures are real numbers, so all arithmetic operations can be used with them. You know that the whole usually equals the sum of its parts. That is also true of line segments in geometry.

Recall that for any two real numbers a and b, there is a real number n between a and b such that $a < n < b$. This relationship also applies to points on a line and is called **betweenness of points**. Point M is **between** points P and Q if and only if P, Q, and M are collinear and $PM + MQ = PQ$.

Because measures are
real numbers, you can
compare measures. If X, Y,
and Z are collinear in that
order, then one of these
statements is true.
$XY = YZ$, $XY > YZ$, or
$XY < YZ$.

Example 4 Find Measurements

a. Find AC.

AC is the measure of $\overline{AC}$.

Point B is between A and C. AC can be found by adding AB and BC.

$AB + BC = AC$ Sum of parts = whole

$3.3 + 3.3 = AC$ Substitution

$6.6 = AC$ Add.

So, $\overline{AC}$ is 6.6 centimeters long.

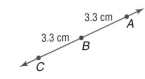

b. Find DE.

DE is the measure of $\overline{DE}$.

$DE + EF = DF$ Sum of parts = whole

$DE + 2\frac{3}{4} = 12$ Substitution

$DE + 2\frac{3}{4} - 2\frac{3}{4} = 12 - 2\frac{3}{4}$ Subtract $2\frac{3}{4}$ from each side.

$DE = 9\frac{1}{4}$ Simplify.

So, $\overline{DE}$ is $9\frac{1}{4}$ inches long.

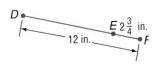

c. **Find y and PQ if P is between Q and R, $PQ = 2y$, $QR = 3y + 1$, and $PR = 21$.**

Draw a figure to represent this information.

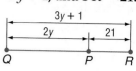

$QR = QP + PR$	
$3y + 1 = 2y + 21$	Substitute known values.
$3y + 1 - 1 = 2y + 21 - 1$	Subtract 1 from each side.
$3y = 2y + 20$	Simplify.
$3y - 2y = 2y + 20 - 2y$	Subtract 2y from each side.
$y = 20$	Simplify.
$PQ = 2y$	Given
$PQ = 2(20)$	$y = 20$
$PQ = 40$	Multiply.

Look at the figure in part **a** of Example 4. Notice that $\overline{AB}$ and $\overline{BC}$ have the same measure. When segments have the same measure, they are said to be **congruent**.

Key Concept   *Congruent Segments*

- **Words** Two segments having the same measure are congruent.

- **Symbol** $\cong$ is read *is congruent to*.
  Red slashes on the figure also indicate that segments are congruent.

- **Model** $\overline{XY} \cong \overline{PQ}$

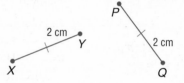

Constructions are methods of creating geometric figures without the benefit of measuring tools. Generally, only a pencil, straightedge, and compass are used in constructions. You can construct a segment that is congruent to a given segment by using a compass and straightedge.

Construction
Copy a Segment

① Draw a segment $\overline{XY}$. Elsewhere on your paper, draw a line and a point on the line. Label the point P.

② Place the compass at point X and adjust the compass setting so that the pencil is at point Y.

③ Using that setting, place the compass point at P and draw an arc that intersects the line. Label the point of intersection Q. Because of identical compass settings, $\overline{PQ} \cong \overline{XY}$.

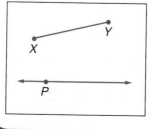

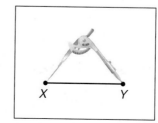

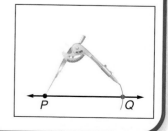

Example 5 Congruent Segments

TIME MANAGEMENT In the graph at the right, suppose a segment was drawn along the top of each bar. Which categories would have segments that are congruent? Explain.

The segments on the bars for grocery shopping and medical research would be congruent because they both have the same length, representing 12%.

The segments on bars for making appointments and personal shopping would be congruent because they have the same length, representing 7%.

USA TODAY Snapshots®

Taking care of business

Seventy-five percent of people who work outside the home take care of personal responsibilities at least once a month while on the job. The top responsibilities:

Banking, bill paying — 34%
Child care — 16%
Grocery shopping — 12%
Medical research — 12%
Making appointments — 7%
Personal shopping — 7%

Source: Xylo Report: shifts in Work and Home Life Boundaries, Nov. 2000 survey of 1,000 adults nationally

By Lori Joseph and Marcy E. Mullins, USA TODAY

Check for Understanding

Concept Check

1. **Describe** how to measure a segment with a ruler that is divided into eighths of an inch.

2. **OPEN ENDED** Name or draw some geometric figures that have congruent segments.

Guided Practice

Find the length of each line segment or object.

3.

4.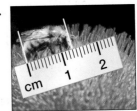

5. Find the precision for a measurement of 14 meters. Explain its meaning.

6. Find the precision for a measurement of $3\frac{1}{4}$ inches. Explain its meaning.

Find the measurement of each segment. Assume that each figure is not drawn to scale.

7. $\overline{EG}$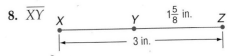

8. $\overline{XY}$

Find the value of the variable and *LM* if *L* is between *N* and *M*.

9. $NL = 5x$, $LM = 3x$, and $NL = 15$

10. $NL = 6x - 5$, $LM = 2x + 3$, and $NM = 30$

Application

11. **KITES** Kite making has become an art form using numerous shapes and designs for flight. The figure at the right is known as a *diamond kite*. The measures are in inches. Name all of the congruent segments in the figure.

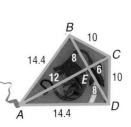

Homework Help

For Exercises	See Examples
12–15	1, 2
16–21	3
22–33	4
34–39	5

Extra Practice
See page 754.

Find the length of each line segment or object.

12.

13.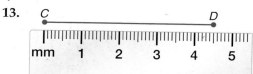

14.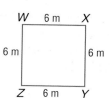

15.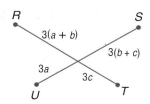

Find the precision for each measurement. Explain its meaning.

16. 80 in.

17. 22 mm

18. $16\frac{1}{2}$ in.

19. 308 cm

20. 3.75 meters

21. $3\frac{1}{4}$ ft

Find the measurement of each segment.

22. AC

A 16.7 mm B 12.8 mm C

23. XZ

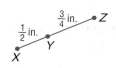

24. QR

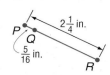

25. ST

R 1.2 cm S   T
|← 4.0 cm →|

26. WX

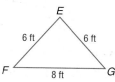

27. BC

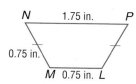

Find the value of the variable and ST if S is between R and T.

28. $RS = 7a$, $ST = 12a$, $RS = 28$

29. $RS = 12$, $ST = 2x$, $RT = 34$

30. $RS = 2x$, $ST = 3x$, $RT = 25$

31. $RS = 16$, $ST = 2x$, $RT = 5x + 10$

32. $RS = 3y + 1$, $ST = 2y$, $RT = 21$

33. $RS = 4y - 1$, $ST = 2y - 1$, $RT = 5y$

Use the figures to determine whether each pair of segments is congruent.

34. $\overline{AB}, \overline{CD}$

A 3 cm B
2 cm   2 cm
D 3 cm C

35. $\overline{EF}, \overline{FG}$

E
6 ft   6 ft
F   8 ft   G

36. $\overline{NP}, \overline{LM}$

N   1.75 in.   P
0.75 in.
M 0.75 in. L

37. $\overline{WX}, \overline{XY}$

W 6 m X
6 m   6 m
Z 6 m Y

38. $\overline{CH}, \overline{CM}$

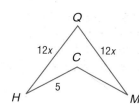

39. $\overline{TR}, \overline{SU}$

R   S
3(a + b)
3(b + c)
3a   3c
U   T

40. MUSIC A CD has a single spiral track of data, circling from the inside of the disc to the outside. Use a metric ruler to determine the full width of a music CD.

41. DOORS Name all segments in the crossbuck pattern in the picture that appear to be congruent.

42. CRAFTS Martin makes pewter figurines and wants to know how much molten pewter he needs for each mold. He knows that when a solid object with a volume of 1 cubic centimeter is submerged in water, the water level rises 1 milliliter. Martin pours 200 mL of water in a measuring cup, completely submerges a figurine in it, and watches it rise to 343 mL. What is the maximum amount of molten pewter, in cubic centimeters, Martin would need to make a figurine? Explain.

More About. . .

Recreation

There are more than 3300 state parks, historic sites, and natural areas in the United States. Most of the parks are open year round to visitors.

Source: *Parks Directory of the United States*

RECREATION For Exercises 43–45, refer to the graph that shows the states with the greatest number of visitors to state parks in a recent year.

43. To what number can the precision of the data be measured?

44. Find the precision for the California data.

45. Can you be sure that 1.9 million more people visited Washington state parks than Illinois state parks? Explain.

 Online Research Data Update Find the current park data for your state and determine the precision of its measure. Visit www.geometryonline.com/data_update to learn more.

Visitors to U.S. State Parks

CA	98.5
NY	59.1
OH	55.3
WA	46.4
IL	44.5
OR	38.6

Visitors (millions)

Source: National Association of Park Directors

PERIMETER For Exercises 46 and 47, use the following information.
The **perimeter** of a geometric figure is the sum of the lengths of its sides. Pablo used a ruler divided into centimeters and measured the sides of a triangle as 3 centimeters, 5 centimeters, and 6 centimeters. Use what you know about the accuracy of any measurement to answer each question.

46. What is the least possible perimeter of the triangle? Explain.

47. What is the greatest possible perimeter of the triangle? Explain.

CONSTRUCTION For Exercises 48 and 49, refer to the figure.

48. Construct a segment whose measure is $4(CD)$.

49. Construct a segment that has length $3(AB) - 2(CD)$.

A———B

C——D

50. CRITICAL THINKING **Significant digits** represent the accuracy of a measurement.
- Nonzero digits are always significant.
- In whole numbers, zeros are significant if they fall between nonzero digits.
- In decimal numbers greater than or equal to 1, every digit is significant.
- In decimal numbers less than 1, the first nonzero digit and every digit to its right are significant.

For example, 600.070 has six significant digits, but 0.0210 has only three. How many significant digits are there in each measurement below?

a. 83,000 miles   **b.** 33,002 miles   **c.** 450.0200 liters

51. WRITING IN MATH Answer the question that was posed at the beginning of the lesson.

Why are units of measure important?

Include the following in your answer.
- an example of how measurements might be misinterpreted, and
- what measurements you can assume from a figure.

Extending the Lesson

ERROR Accuracy is an indication of error. The absolute value of the difference between the actual measure of an object and the allowable measure is the **absolute error**. The **relative error** is the ratio of the absolute error to the actual measure. The relative error is expressed as a percent. For a length of 11 inches and an allowable error of 0.5 inches, the absolute error and relative error can be found as follows.

$$\frac{\text{absolute error}}{\text{measure}} = \frac{|\,11 \text{ in.} - 11.5 \text{ in.}\,|}{11 \text{ in.}} = \frac{0.5 \text{ in.}}{11 \text{ in.}} \approx 0.045 \text{ or } 4.5\%$$

Determine the relative error for each measurement.

52. 27 ft　　　　**53.** $14\frac{1}{2}$ in.　　　　**54.** 42.3 cm　　　　**55.** 63.7 km

Standardized Test Practice
Ⓐ Ⓑ Ⓒ Ⓓ

56. The pipe shown is divided into five equal sections. How many feet long is the pipe?

　Ⓐ 2.4 ft　　　　　Ⓑ 5 ft
　Ⓒ 28.8 ft　　　　Ⓓ 60 ft

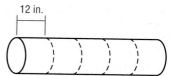

57. ALGEBRA Forty percent of a collection of 80 tapes are jazz tapes, and the rest are blues tapes. How many blues tapes are in the collection?

　Ⓐ 32　　　　Ⓑ 40　　　　Ⓒ 42　　　　Ⓓ 48

Maintain Your Skills

Mixed Review

Refer to the figure at the right. *(Lesson 1-1)*

58. Name three collinear points.

59. Name two planes that contain points B and C.

60. Name another point in plane DFA.

61. How many planes are shown?

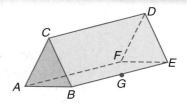

Getting Ready for the Next Lesson

PREREQUISITE SKILL Evaluate each expression if $a = 3$, $b = 8$, and $c = 2$.
*(To review **evaluating expressions**, see page 736.)*

62. $2a + 2b$　　　**63.** $ac + bc$　　　**64.** $\dfrac{a - c}{2}$　　　**65.** $\sqrt{(c - a)^2}$

Practice Quiz 1
Lessons 1-1 and 1-2

For Exercises 1–3, refer to the figure. *(Lesson 1-1)*

1. Name the intersection of planes $\mathcal{A}$ and $\mathcal{B}$.

2. Name another point that is collinear with points S and Q.

3. Name a line that is coplanar with $\overleftrightarrow{VU}$ and point W.

Given that R is between S and T, find each measure. *(Lesson 1-2)*

4. $RS = 6$, $TR = 4.5$, $TS = $ ___?___ .

5. $TS = 11.75$, $TR = 3.4$, $RS = $ ___?___ .

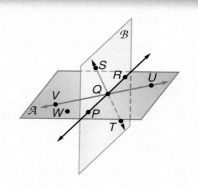

Probability and Segment Measure

You may remember that probability is often expressed as a fraction.

Probability (P) of an event $= \dfrac{\text{number of favorable outcomes}}{\text{total number of possible outcomes}}$

To find the probability that a point lies on a segment, you need to calculate the length of the segment.

Activity

Assume that point Q is contained in $\overline{RT}$. Find the probability that Q is contained in $\overline{RS}$.

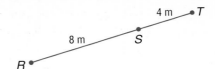

Collect Data

- Find the measures of all segments in the figure.

- $RS = 8$ and $ST = 4$, so $RT = RS + ST$ or 12.

- While a point has no dimension, the segment that contains it does have one dimension, length. To calculate the probability that a point, randomly selected, is in a segment contained by another segment, you must compare their lengths.

$$P(Q \text{ lies in } \overline{RS}) = \frac{RS}{RT}$$

$$= \frac{8}{12} \quad RS = 8 \text{ and } RT = 12$$

$$= \frac{2}{3} \quad \text{Simplify.}$$

The probability that Q is contained in $\overline{RS}$ is $\frac{2}{3}$.

Analyze

For Exercises 1–3, refer to the figure at the right.

1. Point J is contained in $\overline{WZ}$. What is the probability that J is contained in $\overline{XY}$?

2. Point R is contained in $\overline{WZ}$. What is the probability that R is contained in $\overline{YZ}$?

3. Point S is contained in $\overline{WY}$. What is the probability that S is contained in $\overline{XY}$?

Make a Conjecture

For Exercises 4–5, refer to the figure for Exercises 1–3.

4. Point T is contained in both $\overline{WY}$ and $\overline{XZ}$. What do you think is the probability that T is contained in $\overline{XY}$? Explain.

5. Point U is contained in $\overline{WX}$. What do you think is the probability that U is contained in $\overline{YZ}$? Explain.

1-3 Distance and Midpoints

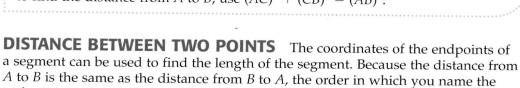

What You'll Learn

- Find the distance between two points.
- Find the midpoint of a segment.

Vocabulary

- midpoint
- segment bisector

How can you find the distance between two points without a ruler?

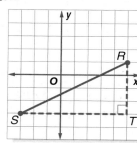

Whenever you connect two points on a number line or on a plane, you have graphed a line segment. Distance on a number line is determined by counting the units between the two points. On a coordinate plane, you can use the Pythagorean Theorem to find the distance between two points. In the figure, to find the distance from A to B, use $(AC)^2 + (CB)^2 = (AB)^2$.

DISTANCE BETWEEN TWO POINTS The coordinates of the endpoints of a segment can be used to find the length of the segment. Because the distance from A to B is the same as the distance from B to A, the order in which you name the endpoints makes no difference.

Key Concept   Distance Formulas

- **Number line**

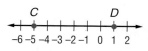

$$PQ = |b - a| \text{ or } |a - b|$$

- **Coordinate Plane**

The distance d between two points with coordinates (x_1, y_1) and (x_2, y_2) is given by $d = \sqrt{(x_2 - x_1)^2 + (y_2 - y_1)^2}$.

Example 1 Find Distance on a Number Line

Use the number line to find CD.

The coordinates of C and D are -5 and 1.

$CD = |-5 - 1|$ Distance Formula

    $= |-6|$ or 6 Simplify.

Study Tip

Pythagorean Theorem

Recall that the Pythagorean Theorem is often expressed as $a^2 + b^2 = c^2$, where a and b are the measures of the shorter sides (legs) of a right triangle, and c is the measure of the longest side (hypotenuse) of a right triangle.

Example 2 Find Distance on a Coordinate Plane

Find the distance between $R(5, 1)$ and $S(-3, -3)$.

Method 1 Pythagorean Theorem

Use the gridlines to form a triangle so you can use the Pythagorean Theorem.

$(RS)^2 = (RT)^2 + (ST)^2$ Pythagorean Theorem

$(RS)^2 = 4^2 + 8^2$ $RT = 4$ units, $ST = 8$ units

$(RS)^2 = 80$ Simplify.

$RS = \sqrt{80}$ Take the square root of each side.

Study Tip

Distance Formula
The Pythagorean Theorem is used to develop the Distance Formula. You will learn more about the Pythagorean Theorem in Lesson 7-2.

Method 2 Distance Formula

$$d = \sqrt{(x_2 - x_1)^2 + (y_2 - y_1)^2} \quad \text{Distance Formula}$$

$$RS = \sqrt{(-3 - 5)^2 + (-3 - 1)^2} \quad (x_1, y_1) = (5, 1) \text{ and } (x_2, y_2) = (-3, -3)$$

$$RS = \sqrt{(-8)^2 + (-4)^2} \quad \text{Simplify.}$$

$$RS = \sqrt{80} \quad \text{Simplify.}$$

The distance from R to S is $\sqrt{80}$ units. You can use a calculator to find that $\sqrt{80}$ is approximately 8.94.

MIDPOINT OF A SEGMENT The **midpoint** of a segment is the point halfway between the endpoints of the segment. If X is the midpoint of $\overline{AB}$, then $AX = XB$.

Geometry Activity

Midpoint of a Segment

Model
- Graph points $A(5, 5)$ and $B(-1, 5)$ on grid paper. Draw $\overline{AB}$.
- Hold the paper up to the light and fold the paper so that points A and B match exactly. Crease the paper slightly.
- Open the paper and put a point where the crease intersects $\overline{AB}$. Label this midpoint as C.
- Repeat the first three steps using endpoints $X(-4, 3)$ and $Y(2, 7)$. Label the midpoint Z.

Make a Conjecture
1. What are the coordinates of point C?
2. What are the lengths of $\overline{AC}$ and $\overline{CB}$?
3. What are the coordinates of point Z?
4. What are the lengths of $\overline{XZ}$ and $\overline{ZY}$?
5. Study the coordinates of points A, B, and C. Write a rule that relates these coordinates. Then use points X, Y, and Z to verify your conjecture.

The points found in the activity are both midpoints of their respective segments.

Study Tip

Common Misconception
The Distance Formula and the Midpoint Formula do not use the same relationship among the coordinates.

Key Concept   Midpoint

Words	The midpoint M of $\overline{PQ}$ is the point between P and Q such that $PM = MQ$.	
	Number Line	**Coordinate Plane**
Symbols	The coordinate of the midpoint of a segment whose endpoints have coordinates a and b is $\dfrac{a + b}{2}$.	The coordinates of the midpoint of a segment whose endpoints have coordinates (x_1, y_1) and (x_2, y_2) are $\left(\dfrac{x_1 + x_2}{2}, \dfrac{y_1 + y_2}{2}\right)$.
Models	 $P \qquad M \qquad Q$ $a \qquad \boxed{\frac{a+b}{2}} \qquad b$	$Q(x_2, y_2)$ $P(x_1, y_1) \quad M\left(\frac{x_1 + x_2}{2}, \frac{y_1 + y_2}{2}\right)$

Example 3 Find Coordinates of Midpoint

a. **TEMPERATURE** Find the coordinate of the midpoint of $\overline{PQ}$.

The coordinates of P and Q are -20 and 40.

Let M be the midpoint of $\overline{PQ}$.

$$M = \frac{-20 + 40}{2} \quad a = -20, b = 40$$

$$= \frac{20}{2} \text{ or } 10 \quad \text{Simplify.}$$

b. **Find the coordinates of M, the midpoint of $\overline{PQ}$, for $P(-1, 2)$ and $Q(6, 1)$.**

Let P be (x_1, y_1) and Q be (x_2, y_2).

$$M\left(\frac{x_1 + x_2}{2}, \frac{y_1 + y_2}{2}\right) = M\left(\frac{-1 + 6}{2}, \frac{2 + 1}{2}\right) \quad (x_1, y_1) = (-1, 2), (x_2, y_2) = (6, 1)$$

$$= M\left(\frac{5}{2}, \frac{3}{2}\right) \text{ or } M\left(2\frac{1}{2}, 1\frac{1}{2}\right) \quad \text{Simplify.}$$

You can also find the coordinates of the endpoint of a segment if you know the coordinates of its other endpoint and its midpoint.

Example 4 Find Coordinates of Endpoint

Find the coordinates of X if $Y(-2, 2)$ is the midpoint of $\overline{XZ}$ and Z has coordinates $(2, 8)$.

Let Z be (x_2, y_2) in the Midpoint Formula.

$$Y(-2, 2) = Y\left(\frac{x_1 + 2}{2}, \frac{y_1 + 8}{2}\right) \quad (x_2, y_2) = (2, 8)$$

Write two equations to find the coordinates of X.

$$-2 = \frac{x_1 + 2}{2}$$
$$-4 = x_1 + 2 \quad \text{Multiply each side by 2.}$$
$$-6 = x_1 \quad \text{Subtract 2 from each side.}$$

$$2 = \frac{y_1 + 8}{2}$$
$$4 = y_1 + 8 \quad \text{Multiply each side by 2.}$$
$$-4 = y_1 \quad \text{Subtract 8 from each side.}$$

The coordinates of X are $(-6, -4)$.

Standardized Test Practice
Ⓐ Ⓑ Ⓒ Ⓓ

Example 5 Use Algebra to Find Measures

Multiple-Choice Test Item

What is the measure of $\overline{BC}$ if B is the midpoint of $\overline{AC}$?

Ⓐ -5 Ⓑ 8
Ⓒ 17 Ⓓ 27

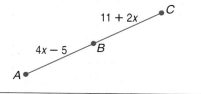

Test-Taking Tip
Eliminate Possibilities You can sometimes eliminate choices by looking at the reasonableness of the answer. In this test item, you can eliminate choice A because measures cannot be negative.

Read the Test Item

You know that B is the midpoint of $\overline{AC}$, and the figure gives algebraic measures for $\overline{AB}$ and $\overline{BC}$. You are asked to find the measure of $\overline{BC}$.

(continued on the next page)

 www.geometryonline.com/extra_examples

Solve the Test Item

Because B is the midpoint, you know that $AB = BC$. Use this equation and the algebraic measures to find a value for x.

$AB = BC$	Definition of midpoint
$4x - 5 = 11 + 2x$	$AB = 4x - 5$, $BC = 11 + 2x$
$4x = 16 + 2x$	Add 5 to each side.
$2x = 16$	Subtract $2x$ from each side.
$x = 8$	Divide each side by 2.

Now substitute 8 for x in the expression for BC.

$BC = 11 + 2x$	Original measure
$BC = 11 + 2(8)$	$x = 8$
$BC = 11 + 16$ or 27	Simplify.

The answer is D.

Any segment, line, or plane that intersects a segment at its midpoint is called a **segment bisector**. In the figure at the right, M is the midpoint of $\overline{AB}$. Plane $\mathcal{N}$, $\overline{MD}$, $\overleftrightarrow{RM}$, and point M are all bisectors of $\overline{AB}$. We say that they *bisect* $\overline{AB}$.

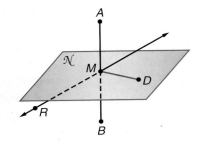

You can construct a line that bisects a segment without measuring to find the midpoint of the given segment.

Construction

Bisect a Segment

1 Draw a segment and name it $\overline{XY}$. Place the compass at point X. Adjust the compass so that its width is greater than $\frac{1}{2}XY$. Draw arcs above and below $\overline{XY}$.

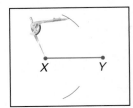

2 Using the same compass setting, place the compass at point Y and draw arcs above and below $\overline{XY}$ intersect the two arcs previously drawn. Label the points of the intersection of the arcs as P and Q.

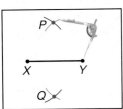

3 Use a straightedge to draw $\overline{PQ}$. Label the point where it intersects $\overline{XY}$ as M. Point M is the midpoint of $\overline{XY}$, and $\overline{PQ}$ is a bisector of $\overline{XY}$. Also $XM = MY = \frac{1}{2}XY$.

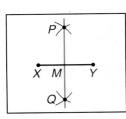

Concept Check

1. **Explain** three ways to find the midpoint of a segment.

2. **OPEN ENDED** Draw a segment. Construct the bisector of the segment and use a millimeter ruler to check the accuracy of your construction.

Guided Practice

Use the number line to find each measure.

3. *AB*

4. *CD*

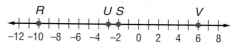

5. Use the Pythagorean Theorem to find the distance between $X(7, 11)$ and $Y(-1, 5)$.

6. Use the Distance Formula to find the distance between $D(2, 0)$ and $E(8, 6)$.

Use the number line to find the coordinate of the midpoint of each segment.

7. $\overline{RS}$

8. $\overline{UV}$

Find the coordinates of the midpoint of a segment having the given endpoints.

9. $X(-4, 3)$, $Y(-1, 5)$

10. $A(2, 8)$, $B(-2, 2)$

11. Find the coordinates of A if $B(0, 5.5)$ is the midpoint of $\overline{AC}$ and C has coordinates $(-3, 6)$.

Standardized Test Practice
Ⓐ Ⓑ Ⓒ Ⓓ

12. Point M is the midpoint of $\overline{AB}$. What is the value of x in the figure?

  Ⓐ 1.5   Ⓑ 5

  Ⓒ 5.5   Ⓓ 11

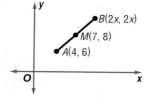

Homework Help

For Exercises	See Examples
13–18	1
19–28	2
29, 30	5
31–42	3
43–45	4

Extra Practice
See page 754.

Use the number line to find each measure.

13. *DE*

14. *CF*

15. *AB*

16. *AC*

17. *AF*

18. *BE*

Use the Pythagorean Theorem to find the distance between each pair of points.

19. $A(0, 0)$, $B(8, 6)$

20. $C(-10, 2)$, $D(-7, 6)$

21. $E(-2, -1)$, $F(3, 11)$

22. $G(-2, -6)$, $H(6, 9)$

Use the Distance Formula to find the distance between each pair of points.

23. $J(0, 0)$, $K(12, 9)$

24. $L(3, 5)$, $M(7, 9)$

25. $S(-3, 2)$, $T(6, 5)$

26. $U(2, 3)$, $V(5, 7)$

27.
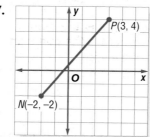

28.
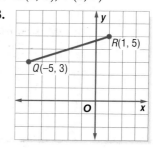

PERIMETER For Exercises 29 and 30, use the following information.
The perimeter of a figure is the sum of the lengths of its sides.

29. The vertices of a triangle are located at $X(-2, -1)$, $Y(2, 5)$, and $Z(4, 3)$. What is the perimeter of this triangle? Round to the nearest tenth.

30. What is the perimeter of a square whose vertices are $A(-4, -3)$, $B(-5, 1)$, $C(-1, 2)$, and $D(0, -2)$?

Use the number line to find the coordinate of the midpoint of each segment.

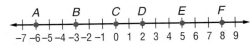

31. $\overline{AC}$　　　　　　　　　**32.** $\overline{DF}$　　　　　　　　　**33.** $\overline{CE}$

34. $\overline{BD}$　　　　　　　　　**35.** $\overline{AF}$　　　　　　　　　**36.** $\overline{BE}$

Find the coordinates of the midpoint of a segment having the given endpoints.

37. $A(8, 4)$, $B(12, 2)$　　　　　　　　　**38.** $C(9, 5)$, $D(17, 4)$

39. $E(-11, -4)$, $F(-9, -2)$　　　　　　**40.** $G(4, 2)$, $H(8, -6)$

41. $J(3.4, 2.1)$, $K(7.8, 3.6)$　　　　　　**42.** $L(-1.4, 3.2)$, $M(2.6, -5.4)$

Find the coordinates of the missing endpoint given that S is the midpoint of $\overline{RT}$.

43. $T(-4, 3)$, $S(-1, 5)$　　　**44.** $T(2, 8)$, $S(-2, 2)$　　　**45.** $R\left(\frac{2}{3}, -5\right)$, $S\left(\frac{5}{3}, 3\right)$

GEOGRAPHY For Exercises 46 and 47, use the following information.
The geographic center of Texas is located northeast of Brady at $(31.1°, 99.3°)$, which represent north latitude and west longitude. El Paso is located near the western border of Texas at $(31.8°, 106.4°)$.

46. If El Paso is one endpoint of a segment and the geographic center is its midpoint, find the latitude and longitude of the other endpoint.

47. Use an atlas or the Internet to find a city near this location.

SPREADSHEETS For Exercises 48 and 49, refer to the information at the left and use the following information.
Spreadsheets can be used to perform calculations quickly. Values are used in formulas by using a specific cell name. For example, the value of x_1 below is used in a formula using its cell name, A2. The spreadsheet below can be used to calculate the distance between two points.

Study Tip

Spreadsheets
Spreadsheets often use special commands to perform operations. For example, $\sqrt{x_1 - x_2}$ would be written as $= \text{SQRT(A2 } - \text{ C2)}$. To raise a number to a power, x^2 for example, write it as $x \char`\^ 2$.

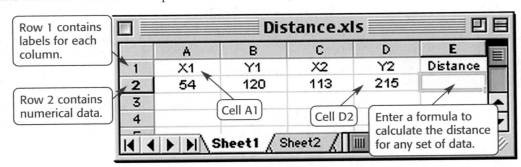

48. Write a formula for cell E2 that could be used to calculate the distance between (x_1, y_1) and (x_2, y_2).

49. Find the distance between each pair of points to the nearest tenth.

　　a. $(54, 120)$, $(113, 215)$　　　　　　　**b.** $(68, 153)$, $(175, 336)$

　　c. $(421, 454)$, $(502, 798)$　　　　　　**d.** $(837, 980)$, $(612, 625)$

　　e. $(1967, 3)$, $(1998, 24)$　　　　　　　**f.** $(4173.5, 34.9)$, $(2080.6, 22.4)$

ENLARGEMENT   For Exercises 50–53, use the following information.
The coordinates of the vertices of a triangle are $A(1, 3)$, $B(9, 10)$, and $C(11, 18)$.

50. Find the perimeter of $\triangle ABC$.

51. Suppose each coordinate is multiplied by 2. What is the perimeter of this triangle?

52. Find the perimeter of the triangle when the coordinates are multiplied by 3.

53. Make a conjecture about the perimeter of a triangle when the coordinates of its vertices are multiplied by the same positive factor.

54. CRITICAL THINKING   In the figure, $\overline{GE}$ bisects $\overline{BC}$, and $\overline{GF}$ bisects $\overline{AB}$. $\overline{GE}$ is a horizontal segment, and $\overline{GF}$ is a vertical segment.

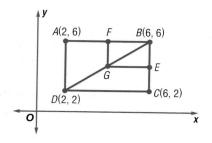

    a. Find the coordinates of points F and E.

    b. Name the coordinates of G and explain how you calculated them.

    c. Describe what relationship, if any, exists between $\overline{DG}$ and $\overline{GB}$. Explain.

55. CRITICAL THINKING $\overline{WZ}$ has endpoints $W(-3, -8)$ and $Z(5, 12)$. Point X lies between W and Z, such that $WX = \frac{1}{4}WZ$. Find the coordinates of X.

56. ▐ **WRITING IN MATH** ▐   Answer the question that was posed at the beginning of the lesson.

How can you find the distance between two points without a ruler?

Include the following in your answer:

- how to use the Pythagorean Theorem and the Distance Formula to find the distance between two points, and
- the length of $\overline{AB}$ from the figure on page 21.

57. Find the distance between points at $(6, 11)$ and $(-2, -4)$.

    Ⓐ 16 units   Ⓑ 17 units   Ⓒ 18 units   Ⓓ 19 units

58. ALGEBRA   Which equation represents the following problem?

Fifteen minus three times a number equals negative twenty-two. Find the number.

    Ⓐ $15 - 3n = -22$ Ⓑ $3n - 15 = -22$

    Ⓒ $3(15 - n) = -22$ Ⓓ $3(n - 15) = -22$

Maintain Your Skills

Mixed Review   **Find the measurement of each segment.**   *(Lesson 1-2)*

59. $\overline{WY}$

60. $\overline{BC}$

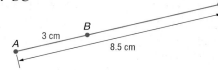

Draw and label a figure for each relationship.   *(Lesson 1-1)*

61. four noncollinear points A, B, C, and D that are coplanar

62. line m that intersects plane $\mathcal{A}$ and line n in plane $\mathcal{A}$

Getting Ready for the Next Lesson   **PREREQUISITE SKILL   Solve each equation.**   *(To review solving equations, see page 737.)*

63. $2k = 5k - 30$ **64.** $14x - 31 = 12x + 8$ **65.** $180 - 8t = 90 + 2t$

66. $12m + 7 = 3m + 52$ **67.** $8x + 7 = 5x + 20$ **68.** $13n - 18 = 5n + 32$

Geometry Activity

Modeling the Pythagorean Theorem

In Chapter 7, you will formally write a verification of the Pythagorean Theorem, but this activity will suggest that the Pythagorean Theorem holds for any right triangle. Remember that a right triangle is a triangle with a right angle, and that a right angle measures 90°.

Make a Model

- Draw right triangle ABC in the center of a piece of grid paper.

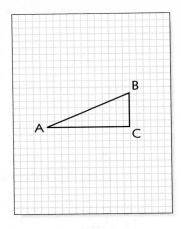

- Use another piece of grid paper to draw a square that is 5 units on each side, a square that is 12 units on each side, and a square that is 13 units on each side. Use colored pencils to shade each of these squares. Cut out the squares. Label them as 5×5, 12×12, and 13×13 respectively.

- Place the squares so that a side of the square matches up with a side of the right triangle.

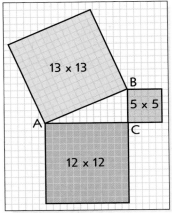

Analyze

1. Determine the number of grid squares in each square you drew.
2. How do the numbers of grid squares relate?
3. If $AB = c$, $BC = a$, and $AC = b$, write an expression to describe each of the squares.
4. How does this expression compare with what you know about the Pythagorean Theorem?

Make a Conjecture

5. Repeat the activity for triangles with each of the side measures listed below. What do you find is true of the relationship of the squares on the sides of the triangle?

   a. 3, 4, 5   **b.** 8, 15, 17   **c.** 6, 8, 10

6. Repeat the activity with a right triangle whose shorter sides are both 5 units long. How could you determine the number of grid squares in the larger square?

CONGRUENT ANGLES Just as segments that have the same measure are congruent, angles that have the same measure are congruent.

Key Concept
Congruent Angles

- **Words**   Angles that have the same measure are congruent angles. Arcs on the figure also indicate which angles are congruent.

- **Symbols** $\angle NMP \cong \angle QMR$

- **Model**

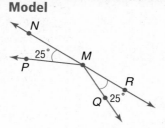

You can construct an angle congruent to a given angle without knowing the measure of the angle.

Construction

Copy an Angle

① Draw an angle like $\angle P$ on your paper. Use a straightedge to draw a ray on your paper. Label its endpoint T.

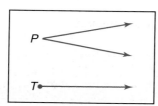

② Place the tip of the compass at point P and draw a large arc that intersects both sides of $\angle P$. Label the points of intersection Q and R.

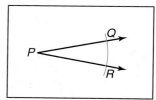

③ Using the same compass setting, put the compass at T and draw a large arc that intersects the ray. Label the point of intersection S.

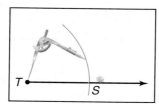

④ Place the point of your compass on R and adjust so that the pencil tip is on Q.

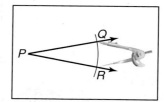

⑤ Without changing the setting, place the compass at S and draw an arc to intersect the larger arc you drew in Step 3. Label the point of intersection U.

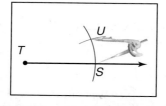

⑥ Use a straightedge to draw $\overrightarrow{TU}$.

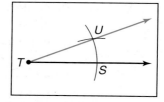

Example 3 **Use Algebra to Find Angle Measures**

GARDENING A trellis is often used to provide a frame for vining plants. Some of the angles formed by the slats of the trellis are congruent angles. In the figure, $\angle ABC \cong \angle DBF$. If $m\angle ABC = 6x + 2$ and $m\angle DBF = 8x - 14$, find the actual measurements of $\angle ABC$ and $\angle DBF$.

$\angle ABC \cong \angle DBF$	Given
$m\angle ABC = m\angle DBF$	Definition of congruent angles
$6x + 2 = 8x - 14$	Substitution
$6x + 16 = 8x$	Add 14 to each side.
$16 = 2x$	Subtract $6x$ from each side.
$8 = x$	Divide each side by 2.

Use the value of x to find the measure of one angle.

$$
\begin{aligned}
m\angle ABC &= 6x + 2 && \text{Given} \\
&= 6(8) + 2 && x = 8 \\
&= 48 + 2 \text{ or } 50 && \text{Simplify.}
\end{aligned}
$$

Since $m\angle ABC = m\angle DBF$, $m\angle DBF = 50$.

Both $\angle ABC$ and $\angle DBF$ measure 50°.

Geometry Activity

Bisect an Angle

Make a Model

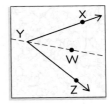

- Draw any $\angle XYZ$ on patty paper or tracing paper.
- Fold the paper through point Y so that $\overrightarrow{YX}$ and $\overrightarrow{YZ}$ are aligned together.
- Open the paper and label a point on the crease in the interior of $\angle XYZ$ as point W.

Analyze the Model

1. What seems to be true about $\angle XYW$ and $\angle WYZ$?
2. Measure $\angle XYZ$, $\angle XYW$, and $\angle WYZ$.
3. You learned about a segment bisector in Lesson 1-3. Write a sentence to explain the term *angle bisector*.

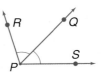

A ray that divides an angle into two congruent angles is called an **angle bisector**. If $\overrightarrow{PQ}$ is the angle bisector of $\angle RPS$, then point Q lies in the interior of $\angle RPS$ and $\angle RPQ \cong \angle QPS$.

You can construct the angle bisector of any angle without knowing the measure of the angle.

Construction

Bisect an Angle

1 Draw an angle on your paper. Label the vertex as *A*. Put your compass at point *A* and draw a large arc that intersects both sides of ∠*A*. Label the points of intersection *B* and *C*.

2 With the compass at point *B*, draw an arc in the interior of the angle.

3 Keeping the same compass setting, place the compass at point *C* and draw an arc that intersects the arc drawn in Step 2.

4 Label the point of intersection *D*. Draw $\overrightarrow{AD}$. $\overrightarrow{AD}$ is the bisector of ∠*A*. Thus, $m\angle BAD = m\angle DAC$ and ∠*BAD* ≅ ∠*DAC*.

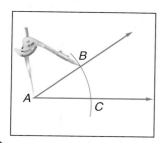
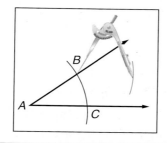
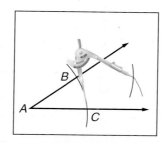
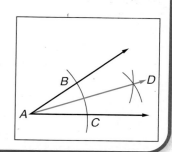

Check for Understanding

Concept Check

1. **Determine** whether all right angles are congruent.

2. **OPEN ENDED** Draw and label a figure to show $\overrightarrow{PR}$ that bisects ∠*SPQ* and $\overrightarrow{PT}$ that bisects ∠*SPR*. Use a protractor to measure each angle.

3. **Write** a statement about the measures of congruent angles *A* and *Z*.

Guided Practice For Exercises 4 and 5, use the figure at the right.

4. Name the vertex of ∠2.

5. Name the sides of ∠4.

6. Write another name for ∠*BDC*.

Measure each angle and classify as *right, acute,* or *obtuse.*

7. ∠*WXY*

8. ∠*WXZ*

ALGEBRA In the figure, $\overrightarrow{QP}$ and $\overrightarrow{QR}$ are opposite rays, and $\overrightarrow{QT}$ bisects ∠*RQS*.

9. If $m\angle RQT = 6x + 5$ and $m\angle SQT = 7x - 2$, find $m\angle RQT$.

10. Find $m\angle TQS$ if $m\angle RQS = 22a - 11$ and $m\angle RQT = 12a - 8$.

Application

11. **ORIGAMI** The art of origami involves folding paper at different angles to create designs and three-dimensional figures. One of the folds in origami involves folding a strip of paper so that the lower edge of the strip forms a right angle with itself. Identify each numbered angle as *right, acute,* or *obtuse.*

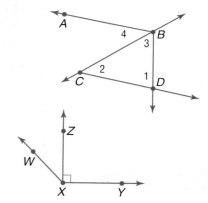
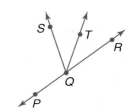
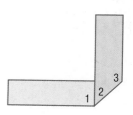

Homework Help

For Exercises	See Examples
12–27	1
28–33	2
34–39	3

Extra Practice
See page 755.

Name the vertex of each angle.

12. $\angle 1$

13. $\angle 2$

14. $\angle 6$

15. $\angle 5$

Name the sides of each angle.

16. $\angle ADB$

17. $\angle 6$

18. $\angle 3$

19. $\angle 5$

Write another name for each angle.

20. $\angle 7$

21. $\angle AEF$

22. $\angle ABD$

23. $\angle 1$

24. Name a point in the interior of $\angle GAB$.

25. Name an angle with vertex B that appears to be acute.

26. Name a pair of angles that share exactly one point.

27. If $\overrightarrow{AD}$ bisects $\angle EAB$ and $m\angle EAB = 60$, find $m\angle 5$ and $m\angle 6$.

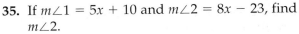

Measure each angle and classify it as *right*, *acute*, or *obtuse*.

28. $\angle BFD$

29. $\angle AFB$

30. $\angle DFE$

31. $\angle EFC$

32. $\angle AFD$

33. $\angle EFB$

ALGEBRA In the figure, $\overrightarrow{YX}$ and $\overrightarrow{YZ}$ are opposite rays. $\overrightarrow{YU}$ bisects $\angle ZYW$, and $\overrightarrow{YT}$ bisects $\angle XYW$.

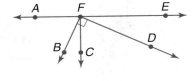

34. If $m\angle ZYU = 8p - 10$ and $m\angle UYW = 10p - 20$, find $m\angle ZYU$.

35. If $m\angle 1 = 5x + 10$ and $m\angle 2 = 8x - 23$, find $m\angle 2$.

36. If $m\angle 1 = y$ and $m\angle XYW = 6y - 24$, find y.

37. If $m\angle WYZ = 82$ and $m\angle ZYU = 4r + 25$, find r.

38. If $m\angle WYX = 2(12b + 7)$ and $m\angle ZYU = 9b - 1$, find $m\angle UYW$.

39. If $\angle ZYW$ is a right angle and $m\angle ZYU = 13a - 7$, find a.

40. DOG TRACKING A dog is *tracking* when it is following the scent trail left by a human being or other animal that has passed along a certain route. One of the training exercises for these dogs is a tracking trail. The one shown is called an acute tracking trail. Explain why it might be called this.

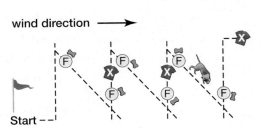

wind direction

Start

F = food drop   X = article

41. LANGUAGE The words *obtuse* and *acute* have other meanings in the English language. Look these words up and write how the everyday meaning relates to the mathematical meaning.

42. PATTERN BLOCKS   Pattern blocks can be arranged to fit in a circular pattern without leaving spaces. Remember that the measurement around a full circle is 360°. Determine the angle measure of the numbered angles shown below.

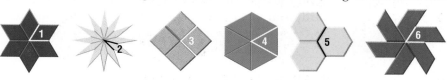

43. PHYSICS   A ripple tank can be used to study the behavior of waves in two dimensions. As a wave strikes a barrier, it is reflected. The angle of incidence and the angle of reflection are congruent. In the diagram at the right, if $m\angle IBR = 62$, find the angle of reflection and $m\angle IBA$.

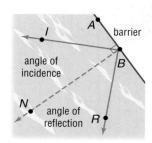

44. CRITICAL THINKING   How would you compare the size of $\angle P$ and $\angle Q$? Explain.

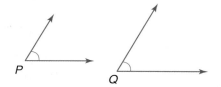

CRITICAL THINKING   **For Exercises 45–48, use the following information.**
Each figure below shows noncollinear rays with a common endpoint.

2 rays   3 rays   4 rays   5 rays   6 rays

45. Count the number of angles in each figure.

46. Describe the pattern between the number of rays and the number of angles.

47. **Make a conjecture** of the number of angles that are formed by 7 noncollinear rays and by 10 noncollinear rays.

48. Write a formula for the number of angles formed by n noncollinear rays with a common endpoint.

49.   WRITING IN MATH   Answer the question that was posed at the beginning of the lesson.

How big is a degree?

Include the following in your answer:
• how to find degree measure with a protractor, and
• drawings of several angles and their degree measures.

50. If $\overrightarrow{BX}$ bisects $\angle ABC$, which of the following are true?

Ⓐ $m\angle ABX = m\angle XBC$ Ⓑ $m\angle ABX = \frac{1}{2}m\angle ABC$

Ⓒ $\frac{1}{2}m\angle ABC = m\angle XBC$ Ⓓ all of these

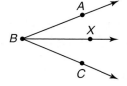

51. ALGEBRA   Solve $5n + 4 = 7(n + 1) - 2n$.

Ⓐ 0   Ⓑ −1   Ⓒ no solution   Ⓓ all numbers

Mixed Review   **Find the distance between each pair of points. Then find the coordinates of the midpoint of the line segment between the points.**   *(Lesson 1-3)*

52. $A(2, 3)$, $B(5, 7)$ **53.** $C(-2, 0)$, $D(6, 4)$ **54.** $E(-3, -2)$, $F(5, 8)$

Find the measurement of each segment.   *(Lesson 1-2)*

55. $\overline{WX}$

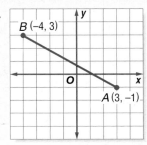

56. $\overline{YZ}$

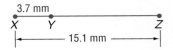

57. Find PQ if Q lies between P and R, $PQ = 6x - 5$, $QR = 2x + 7$, and $PQ = QR$.   *(Lesson 1-2)*

Refer to the figure at the right.   *(Lesson 1-1)*

58. How many planes are shown?

59. Name three collinear points.

60. Name a point coplanar with J, H, and F.

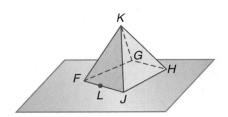

Getting Ready for the Next Lesson   **PREREQUISITE SKILL   Solve each equation.**
*(To review **solving equations**, see pages 737 and 738.)*

61. $14x + (6x - 10) = 90$ **62.** $2k + 30 = 180$

63. $180 - 5y = 90 - 7y$ **64.** $90 - 4t = \frac{1}{4}(180 - t)$

65. $(6m + 8) + (3m + 10) = 90$ **66.** $(7n - 9) + (5n + 45) = 180$

Practice Quiz 2   *Lessons 1-3 and 1-4*

Find the coordinates of the midpoint of each segment. Then find the distance between the endpoints.   *(Lesson 1-3)*

1.

2.

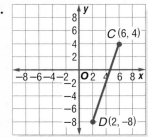

3.

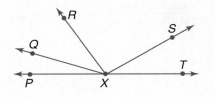

In the figure, $\overrightarrow{XP}$ and $\overrightarrow{XT}$ are opposite rays. Given the following conditions, find the value of a and the measure of the indicated angle.   *(Lesson 1-4)*

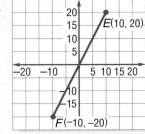

4. $m\angle SXT = 3a - 4$, $m\angle RXS = 2a + 5$, $m\angle RXT = 111$; $m\angle RXS$

5. $m\angle QXR = a + 10$, $m\angle QXS = 4a - 1$, $m\angle RXS = 91$; $m\angle QXS$

1-5 Angle Relationships

What You'll Learn

- Identify and use special pairs of angles.
- Identify perpendicular lines.

Vocabulary

- adjacent angles
- vertical angles
- linear pair
- complementary angles
- supplementary angles
- perpendicular

What kinds of angles are formed when streets intersect?

When two lines intersect, four angles are formed. In some cities, more than two streets might intersect to form even more angles. All of these angles are related in special ways.

PAIRS OF ANGLES Certain pairs of angles have special names.

Key Concept · Angle Pairs

- **Words**   **Adjacent angles** are two angles that lie in the same plane, have a common vertex, and a common side, but no common interior points.

- **Examples**
  ∠ABC and ∠CBD

  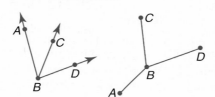

- **Nonexamples**
  ∠ABC and ∠ABD

  shared interior

  ∠ABC and ∠BCD

  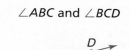

  no common vertex

- **Words**   **Vertical angles** are two nonadjacent angles formed by two intersecting lines.

- **Examples**
  ∠AEB and ∠CED
  ∠AED and ∠BEC

- **Nonexample**
  ∠AED and ∠BEC

  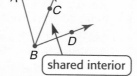

  D, E, and C are noncollinear.

- **Words**   A **linear pair** is a pair of adjacent angles whose noncommon sides are opposite rays.

- **Example**
  ∠BED and ∠BEC

  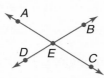

- **Nonexample**

  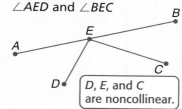

  D, E, and C are noncollinear.

Example 1 *Identify Angle Pairs*

Name an angle pair that satisfies each condition.

a. two obtuse vertical angles

∠VZX and ∠YZW are vertical angles.

They each have measures greater than 90°, so they are obtuse.

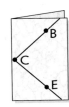

b. two acute adjacent angles

There are four acute angles shown.

Adjacent acute angles are ∠VZY and ∠YZT, ∠YZT and ∠TZW, and ∠TZW and ∠WZX.

The measures of angles formed by intersecting lines also have a special relationship.

Geometry Activity

Angle Relationships

Make a Model

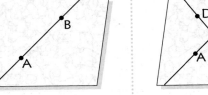

Step 1

Fold a piece of patty paper so that it makes a crease across the paper. Open the paper, trace the crease with a pencil, and name two points on the crease *A* and *B*.

Step 2

Fold the paper again so that the crease intersects $\overleftrightarrow{AB}$ between the two labeled points. Open the paper, trace this crease, and label the intersection *C*. Label two other points, *D* and *E*, on the second crease so that *C* is between *D* and *E*.

Step 3

Fold the paper again through point *C* so that $\overrightarrow{CB}$ aligns with $\overrightarrow{CD}$.

Analyze the Model

1. What do you notice about ∠BCE and ∠DCA when you made the last fold?
2. Fold the paper again through *C* so that $\overrightarrow{CB}$ aligns with $\overrightarrow{CE}$. What do you notice?
3. Use a protractor to measure each angle. Label the measure on your model.
4. Name pairs of vertical angles and their measures.
5. Name linear pairs of angles and their measures.
6. Compare your results with those of your classmates. Write a "rule" about the measures of vertical angles and another about the measures of linear pairs.

The Geometry Activity suggests that all vertical angles are congruent. It also supports the concept that the sum of the measures of a linear pair is 180.

There are other angle relationships that you may remember from previous math courses. These are complementary angles and supplementary angles.

- **Words** **Complementary angles** are two angles whose measures have a sum of 90.

- **Examples**
  $\angle 1$ and $\angle 2$ are complementary.
  $\angle PQR$ and $\angle XYZ$ are complementary.

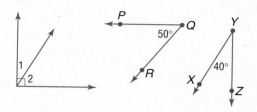

Study Tip

Complementary and Supplementary Angles
While the other angle pairs in this lesson share at least one point, complementary and supplementary angles need not share any points.

- **Words** **Supplementary angles** are two angles whose measures have a sum of 180.

- **Examples**
  $\angle EFH$ and $\angle HFG$ are supplementary.
  $\angle M$ and $\angle N$ are supplementary.

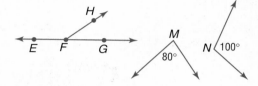

Remember that angle measures are real numbers. So, the operations for real numbers and algebra can be used with angle measures.

Example 2 Angle Measure

ALGEBRA Find the measures of two complementary angles if the difference in the measures of the two angles is 12.

Explore The problem relates the measures of two complementary angles. You know that the sum of the measures of complementary angles is 90.

Plan Draw two figures to represent the angles.

Let the measure of one angle be x.
If $m\angle A = x$, then because $\angle A$ and $\angle B$ are complementary, $m\angle B + x = 90$ or $m\angle B = 90 - x$.

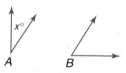

The problem states that the difference of the two angle measures is 12, or $m\angle B - m\angle A = 12$.

Solve

$m\angle B - m\angle A = 12$	Given
$(90 - x) - x = 12$	$m\angle A = x, m\angle B = 90 - x$
$90 - 2x = 12$	Simplify.
$-2x = -78$	Subtract 90 from each side.
$x = 39$	Divide each side by -2.

Use the value of x to find each angle measure.

$m\angle A = x$ $m\angle B = 90 - x$
$m\angle A = 39$ $m\angle B = 90 - 39$ or 51

Examine Add the angle measures to verify that the angles are complementary.

$m\angle A + m\angle B = 90$
$39 + 51 = 90$
$90 = 90$

PERPENDICULAR LINES   Lines that form right angles are **perpendicular**.
The following statements are also true when two lines are perpendicular.

Key Concept — Perpendicular Lines

- Perpendicular lines intersect to form four right angles.

- Perpendicular lines intersect to form congruent adjacent angles.

- Segments and rays can be perpendicular to lines or to other line segments and rays.

- The right angle symbol in the figure indicates that the lines are perpendicular.

- **Symbol** $\perp$ is read *is perpendicular to*.

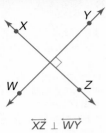

$$\overleftrightarrow{XZ} \perp \overleftrightarrow{WY}$$

Example 3   *Perpendicular Lines*

ALGEBRA   Find x and y so that $\overrightarrow{BE}$ and $\overrightarrow{AD}$ are perpendicular.

If $\overrightarrow{BE} \perp \overrightarrow{AD}$, then $m\angle BFD = 90$ and $m\angle AFE = 90$.

To find x, use $\angle BFC$ and $\angle CFD$.

$m\angle BFD = m\angle BFC + m\angle CFD$	Sum of parts = whole
$90 = 6x + 3x$	Substitution
$90 = 9x$	Add.
$10 = x$	Divide each side by 9.

To find y, use $\angle AFE$.

$m\angle AFE = 12y - 10$	Given
$90 = 12y - 10$	Substitution
$100 = 12y$	Add 10 to each side.
$\dfrac{25}{3} = y$	Divide each side by 12, and simplify.

While two lines may appear to be perpendicular in a figure, you cannot assume this is true unless other information is given. In geometry, figures are used to depict a situation. They are not drawn to reflect total accuracy of the situation. There are certain relationships you can assume to be true, but others that you cannot.

Study the figure at the right and then compare the lists below.

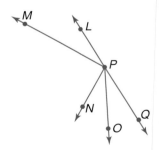

Can Be Assumed	Cannot Be Assumed
All points shown are coplanar.	Perpendicular lines: $\overrightarrow{PN} \perp \overrightarrow{PM}$
L, P, and Q are collinear.	Congruent angles: $\angle QPO \cong \angle LPM$
$\overrightarrow{PM}$, $\overrightarrow{PN}$, $\overrightarrow{PO}$, and $\overleftrightarrow{LQ}$ intersect at P.	$\angle QPO \cong \angle OPN$
	$\angle OPN \cong \angle LPM$
P is between L and Q.	
N is in the interior of MPO.	Congruent segments: $\overline{LP} \cong \overline{PQ}$
$\angle LPM$ and $\angle MPN$ are adjacent angles.	$\overline{PQ} \cong \overline{PO}$
	$\overline{PO} \cong \overline{PN}$
$\angle LPN$ and $\angle NPQ$ are a linear pair.	$\overline{PN} \cong \overline{PL}$
$\angle QPO$ and $\angle OPL$ are supplementary.	

Example 4 *Interpret Figures*

Determine whether each statement can be assumed from the figure below.

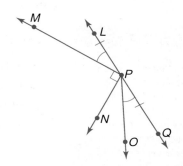

a. ∠LPM and ∠MPO are adjacent angles.

Yes; they share a common side and vertex and have no interior points in common.

b. ∠OPQ and ∠LPM are complementary.

No; they are congruent, but we do not know anything about their exact measures.

c. ∠LPO and ∠QPO are a linear pair.

Yes; they are adjacent angles whose noncommon sides are opposite rays.

Check for Understanding

Concept Check

1. **OPEN ENDED** Draw two angles that are supplementary, but not adjacent.

2. **Explain** the statement *If two adjacent angles form a linear pair, they must be supplementary.*

3. **Write** a sentence to explain why a linear pair of angles is called *linear*.

Guided Practice

For Exercises 4 and 5, use the figure at the right and a protractor.

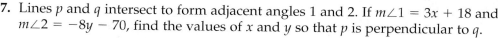

4. Name two acute vertical angles.

5. Name two obtuse adjacent angles.

6. The measure of the supplement of an angle is 60 less than three times the measure of the complement of the angle. Find the measure of the angle.

7. Lines p and q intersect to form adjacent angles 1 and 2. If $m\angle 1 = 3x + 18$ and $m\angle 2 = -8y - 70$, find the values of x and y so that p is perpendicular to q.

Determine whether each statement can be assumed from the figure. Explain.

8. ∠SRP and ∠PRT are complementary.

9. ∠QPT and ∠TPR are adjacent, but neither complementary or supplementary.

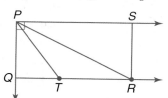

Application

10. **SKIING** Alisa Camplin won the gold medal in the 2002 Winter Olympics with a triple-twisting, double backflip jump in the women's freestyle skiing event. While she is in the air, her skis are positioned like intersecting lines. If ∠4 measures 60°, find the measures of the other angles.

Practice and Apply

Homework Help

For Exercises	See Examples
11–16	1
17–22	2
27–30	3
31–35	4

Extra Practice
See page 755.

For Exercises 11–16, use the figure at the right and a protractor.

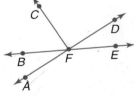

11. Name two acute vertical angles.

12. Name two obtuse vertical angles.

13. Name a pair of complementary adjacent angles.

14. Name a pair of complementary nonadjacent angles.

15. Name a linear pair whose vertex is T.

16. Name an angle supplementary to $\angle UVZ$.

17. Rays PQ and QR are perpendicular. Point S lies in the interior of $\angle PQR$. If $m\angle PQS = 4 + 7a$ and $m\angle SQR = 9 + 4a$, find $m\angle PQS$ and $m\angle SQR$.

18. The measures of two complementary angles are $16z - 9$ and $4z + 3$. Find the measures of the angles.

19. Find $m\angle T$ if $m\angle T$ is 20 more than four times its supplement.

20. The measure of an angle's supplement is 44 less than the measure of the angle. Find the measure of the angle and its supplement.

21. Two angles are supplementary. One angle measures 12° more than the other. Find the measures of the angles.

22. The measure of $\angle 1$ is five less than four times the measure of $\angle 2$. If $\angle 1$ and $\angle 2$ form a linear pair, what are their measures?

Determine whether each statement is *sometimes*, *always*, or *never* true.

23. If two angles are supplementary and one is acute, the other is obtuse.

24. If two angles are complementary, they are both acute angles.

25. If $\angle A$ is supplementary to $\angle B$ and $\angle B$ is supplementary to $\angle C$, then $\angle A$ is supplementary to $\angle C$.

26. If $\overline{PN} \perp \overline{PQ}$, then $\angle NPQ$ is acute.

ALGEBRA For Exercises 27–29, use the figure at the right.

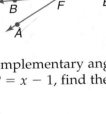

27. If $m\angle CFD = 12a + 45$, find a so that $\overrightarrow{FC} \perp \overrightarrow{FD}$.

28. If $m\angle AFB = 8x - 6$ and $m\angle BFC = 14x + 8$, find the value of x so that $\angle AFC$ is a right angle.

29. If $\angle BFA = 3r + 12$ and $m\angle DFE = -8r + 210$, find $m\angle AFE$.

30. $\angle L$ and $\angle M$ are complementary angles. $\angle N$ and $\angle P$ are complementary angles. If $m\angle L = y - 2$, $m\angle M = 2x + 3$, $m\angle N = 2x - y$, and $m\angle P = x - 1$, find the values of x, y, $m\angle L$, $m\angle M$, $m\angle N$, and $m\angle P$.

Determine whether each statement can be assumed from the figure. Explain.

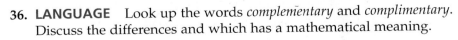

31. $\angle DAB$ is a right angle.

32. $\angle AEB \cong \angle DEC$

33. $\angle ADB$ and $\angle BDC$ are complementary.

34. $\angle DAE \cong \angle ADE$

35. $\overline{AB} \perp \overline{BC}$

36. **LANGUAGE** Look up the words *complementary* and *complimentary*. Discuss the differences and which has a mathematical meaning.

37. **CRITICAL THINKING** A counterexample is used to show that a statement is not necessarily true. Find a counterexample for the statement *Supplementary angles form linear pairs.*

38. STAINED GLASS In the stained glass pattern at the right, determine which segments are perpendicular.

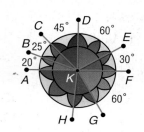

39. CRITICAL THINKING In the figure below, $\angle WUT$ and $\angle XUV$ are vertical angles, $\overline{YU}$ is the bisector of $\angle WUT$, and $\overline{UZ}$ is the bisector of $\angle TUV$. Write a convincing argument that $\overline{YU} \perp \overline{UZ}$.

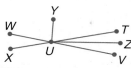

40. WRITING IN MATH Answer the question that was posed at the beginning of the lesson.

What kinds of angles are formed when streets intersect?

Include the following in your answer.
- the types of angles that might be formed by two intersecting lines, and
- a sketch of intersecting streets with angle measures and angle pairs identified.

Standardized Test Practice
Ⓐ Ⓑ Ⓒ Ⓓ

41. Which statement is true of the figure?

Ⓐ $x > y$ Ⓑ $x < y$

Ⓒ $x = y$ Ⓓ cannot be determined

42. SHORT RESPONSE The product of 4, 5, and 6 is equal to twice the sum of 10 and what number?

Extending the Lesson **43.** The concept of perpendicularity can be extended to include planes. If a line, line segment, or ray is perpendicular to a plane, it is perpendicular to every line, line segment, or ray in that plane that intersects it. In the figure at the right, $\overleftrightarrow{AB} \perp \mathcal{E}$. Name all pairs of perpendicular lines.

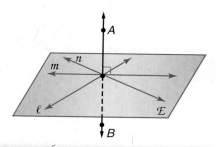

Maintain Your Skills

Mixed Review **Measure each angle and classify it as right, acute, or obtuse.** *(Lesson 1-4)*

44. $\angle KFG$ **45.** $\angle HFG$

46. $\angle HFK$ **47.** $\angle JFE$

48. $\angle HFJ$ **49.** $\angle EFK$

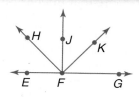

Find the distance between each pair of points. *(Lesson 1-3)*

50. $A(3, 5)$, $B(0, 1)$ **51.** $C(5, 1)$, $D(5, 9)$ **52.** $E(-2, -10)$, $F(-4, 10)$

53. $G(7, 2)$, $H(-6, 0)$ **54.** $J(-8, 9)$, $K(4, 7)$ **55.** $L(1, 3)$, $M(3, -1)$

Find the value of the variable and QR if Q is between P and R. *(Lesson 1-2)*

56. $PQ = 1 - x$, $QR = 4x + 17$, $PR = -3x$

57. $PR = 7n + 8$, $PQ = 4n - 3$, $QR = 6n + 2$

Getting Ready for the Next Lesson **PREREQUISITE SKILL** Evaluate each expression if $\ell = 3$, $w = 8$, and $s = 2$.
*(To review **evaluating expressions**, see page 736.)*

58. $2\ell + 2w$ **59.** ℓw **60.** $4s$ **61.** $\ell w + ws$ **62.** $s(\ell + w)$

Constructing Perpendiculars

You can use a compass and a straightedge to construct a line perpendicular to a given line through a point on the line, or through a point *not* on the line.

Activity 1 Perpendicular Through a Point on the Line

Construct a line perpendicular to line *n* and passing through point *C* on *n*.

1 Place the compass at point *C*. Using the same compass setting, draw arcs to the right and left of *C*, intersecting line *n*. Label the points of intersection *A* and *B*.

2 Open the compass to a setting greater than $\overline{AC}$. Put the compass at point *A* and draw an arc above line *n*.

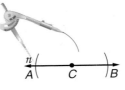

3 Using the same compass setting as in Step 2, place the compass at point *B* and draw an arc intersecting the arc drawn in Step 2. Label the point of intersection *D*.

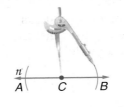

4 Use a straightedge to draw $\overleftrightarrow{CD}$.

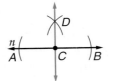

Activity 2 Perpendicular Through a Point not on the Line

Construct a line perpendicular to line *m* and passing through point *Z* not on *m*.

1 Place the compass at point *Z*. Draw an arc that intersects line *m* in two different places. Label the points of intersection *X* and *Y*.

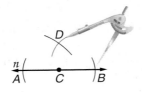

2 Open the compass to a setting greater than $\frac{1}{2}$ *XY*. Put the compass at point *X* and draw an arc below line *m*.

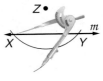

3 Using the same compass setting, place the compass at point *Y* and draw an arc intersecting the arc drawn in Step 2. Label the point of intersection *A*.

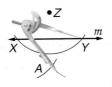

4 Use a straightedge to draw $\overleftrightarrow{ZA}$.

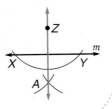

Model and Analyze

1. Draw a line and construct a line perpendicular to it through a point on the line. Repeat with a point not on the line.

2. How is the second construction similar to the first one?

1-6 Polygons

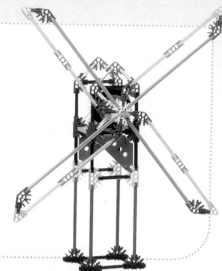

What You'll Learn

- Identify and name polygons.
- Find perimeters of polygons.

How are polygons related to toys?

There are numerous types of building sets that connect sticks to form various shapes. Whether they are made of plastic, wood, or metal, the sticks represent segments. When the segments are connected, they form angles. The sticks are connected to form closed figures that in turn are connected to make a model of a real-world object.

Vocabulary

- polygon
- concave
- convex
- *n*-gon
- regular polygon
- perimeter

Study Tip

Reading Math
The plural of vertex is *vertices*.

POLYGONS Each closed figure shown in the toy is a **polygon**. A polygon is a closed figure whose sides are all segments. The sides of each angle in a polygon are called *sides* of the polygon, and the vertex of each angle is a *vertex* of the polygon.

Key Concept
Polygon

- **Words** A polygon is a closed figure formed by a finite number of coplanar segments such that
  (1) the sides that have a common endpoint are noncollinear, and
  (2) each side intersects exactly two other sides, but only at their endpoints.

- **Symbol** A polygon is named by the letters of its vertices, written in consecutive order.

- **Examples**   • **Nonexamples**

polygons *ABC*, *WXYZ*, *EFGHJK*

Polygons can be **concave** or **convex**. Suppose the line containing each side is drawn. If any of the lines contain any point in the interior of the polygon, then it is concave. Otherwise it is convex.

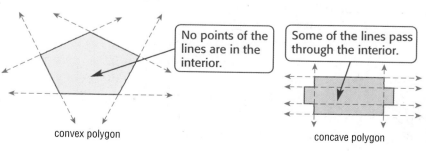

No points of the lines are in the interior.

Some of the lines pass through the interior.

convex polygon   concave polygon

You are already familiar with many polygon names, such as triangle, square, and rectangle. In general, polygons can be classified by the number of sides they have. A polygon with *n* sides is an **n-gon**. The table lists some common names for various categories of polygon.

A convex polygon in which all the sides are congruent and all the angles are congruent is called a **regular polygon**. Octagon *PQRSTUVW* below is a regular octagon.

Number of Sides	Polygon
3	triangle
4	quadrilateral
5	pentagon
6	hexagon
7	heptagon
8	octagon
9	nonagon
10	decagon
12	dodecagon
n	*n*-gon

Polygons and circles are examples of *simple closed curves.*

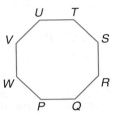

Example 1 Identify Polygons

Name each polygon by its number of sides. Then classify it as *convex* or *concave* and *regular* or *irregular*.

a.

b.

There are 5 sides, so this is a pentagon. No line containing any of the sides will pass through the interior of the pentagon, so it is convex.

The sides are congruent, and the angles are congruent. It is regular.

There are 8 sides, so this is an octagon. A line containing any of the sides will pass through the interior of the octagon, so it is concave.

The sides are congruent. However, since it is concave, it cannot be regular.

PERIMETER   The **perimeter** of a polygon is the sum of the lengths of its sides, which are segments. Some shapes have special formulas, but they are all derived from the basic definition of perimeter.

Key Concept   Perimeter

- **Words**   The perimeter P of a polygon is the sum of the lengths of the sides of the polygon.

- **Examples**   triangle
$P = a + b + c$

square
$P = s + s + s + s$
$P = 4s$

rectangle
$P = \ell + w + \ell + w$
$P = 2\ell + 2w$

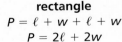

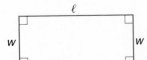

Example 2 *Find Perimeter*

GARDENING A landscape designer is putting black plastic edging around a rectangular flower garden that has length 5.7 meters and width 3.8 meters. The edging is sold in 5-meter lengths.

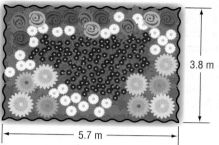

a. Find the perimeter of the garden and determine how much edging the designer should buy.

$P = 2\ell + 2w$

$\quad = 2(5.7) + 2(3.8)$ $\ell = 5.7, w = 3.8$

$\quad = 11.4 + 7.6 \text{ or } 19$

3.8 m

5.7 m

The perimeter of the garden is 19 meters.

The designer needs to buy 20 meters of edging.

b. Suppose the length and width of the garden are tripled. What is the effect on the perimeter and how much edging should the designer buy?

The new length would be 3(5.7) or 17.1 meters.

The new width would be 3(3.8) or 11.4 meters.

$P = 2\ell + 2w$

$\quad = 2(17.1) + 2(11.4) \text{ or } 57$

Compare the original perimeter to this measurement.

$57 = 3(19)$ meters

So, when the lengths of the sides of the rectangle are tripled, the perimeter also triples. The designer needs to buy 60 meters of edging.

You can use the Distance Formula to find the perimeter of a polygon graphed on a coordinate plane.

Example 3 *Perimeter on the Coordinate Plane*

COORDINATE GEOMETRY Find the perimeter of triangle *PQR* if *P*(−5, 1), *Q*(−1, 4), and *R*(−6, −8).

Study Tip

Look Back
To review the **Distance Formula**, see Lesson 1-3.

Use the Distance Formula,

$d = \sqrt{(x_2 - x_1)^2 + (y_2 - y_1)^2}$,

to find *PQ*, *QR*, and *PR*.

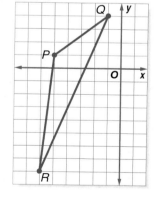

$PQ = \sqrt{[-1 - (-5)]^2 + (4 - 1)^2}$

$\quad\;\; = \sqrt{4^2 + 3^2}$

$\quad\;\; = \sqrt{25} \text{ or } 5$

$QR = \sqrt{[-6 - (-1)]^2 + (-8 - 4)^2}$ $PR = \sqrt{[-6 - (-5)]^2 + (-8 - 1)^2}$

$\quad\;\; = \sqrt{(-5)^2 + (-12)^2}$ $\quad\;\; = \sqrt{(-1)^2 + (-9)^2}$

$\quad\;\; = \sqrt{169} \text{ or } 13$ $\quad\;\; = \sqrt{82} \approx 9.1$

The perimeter of triangle *PQR* is $5 + 13 + \sqrt{82}$ or about 27.1 units.

You can also use algebra to find the lengths of the sides if the perimeter is known.

Example 4 **Use Perimeter to Find Sides**

ALGEBRA   The length of a rectangle is three times the width. The perimeter is 2 feet. Find the length of each side.

Let w represent the width. Then the length is $3w$.

$$P = 2\ell + 2w \qquad \text{Perimeter formula for rectangle}$$

$$2 = 2(3w) + 2w \qquad \ell = 3w$$

$$2 = 8w \qquad \text{Simplify.}$$

$$\frac{1}{4} = w \qquad \text{Divide each side by 8.}$$

The width is $\frac{1}{4}$ foot. By substituting $\frac{1}{4}$ for w, the length $3w$ becomes $3\left(\frac{1}{4}\right)$ or $\frac{3}{4}$ foot.

Check for Understanding

Concept Check

1. **OPEN ENDED**   Explain how you would find the length of a side of a regular decagon if the perimeter is 120 centimeters.

2. **FIND THE ERROR**   Saul and Tiki were asked to draw quadrilateral $WXYZ$ with $m\angle Z = 30$.

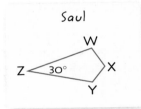

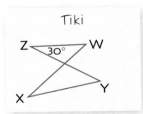

Who is correct? Explain your reasoning.

3. **Write** a formula for the perimeter of a triangle with congruent sides of length s.

4. **Draw** a concave pentagon and explain why it is concave.

Guided Practice

Name each polygon by its number of sides. Then classify it as *convex* or *concave* and *regular* or *irregular*.

5.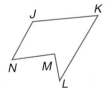

6.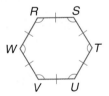

For Exercises 7 and 8, use pentagon $LMNOP$.

7. Find the perimeter of pentagon $LMNOP$.

8. Suppose the length of each side of pentagon $LMNOP$ is doubled. What effect does this have on the perimeter?

9. **COORDINATE GEOMETRY**   A polygon has vertices $P(-3, 4)$, $Q(0, 8)$, $R(3, 8)$, and $S(0, 4)$. Find the perimeter of $PQRS$.

10. **ALGEBRA**   Quadrilateral $ABCD$ has a perimeter of 95 centimeters. Find the length of each side if $AB = 3a + 2$, $BC = 2(a - 1)$, $CD = 6a + 4$, and $AD = 5a - 5$.

Application

11. **HISTORIC LANDMARKS**   The Pentagon building in Arlington, Virginia, is so named because of its five congruent sides. Find the perimeter of the outside of the Pentagon if one side is 921 feet long.

Homework Help

For Exercises	See Examples
12–18	1
19–25	2
26–28	3
29–34	4

Extra Practice
See page 755.

Name each polygon by its number of sides. Then classify it as *convex* **or** *concave* **and** *regular* **or** *irregular***.**

12.

13.

14.

TRAFFIC SIGNS **Identify the shape of each traffic sign.**

15. school zone

16. caution or warning

17. yield

18. railroad

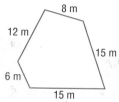

Find the perimeter of each figure.

19.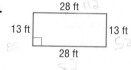

28 ft
13 ft   13 ft
28 ft

20.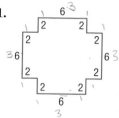

8 m
12 m
15 m
6 m
15 m

21.

6
2   2
2   2
2   2
2   2
6

22. What is the effect on the perimeter of the figure in Exercise 19 if each measure is multiplied by 4?

23. What is the effect on the perimeter of the figure in Exercise 20 if each measure is tripled?

24. What is the effect on the perimeter of the figure in Exercise 21 if each measure is divided by 2?

25. The perimeter of an *n*-gon is 12.5 meters. Find the perimeter of the *n*-gon if the length of each of its *n* sides is multiplied by 10.

COORDINATE GEOMETRY **Find the perimeter of each polygon.**

26. rectangle with vertices $A(-1, 1)$, $B(3, 4)$, $C(6, 0)$, and $D(2, -3)$

27. hexagon with vertices $P(-2, 3)$, $Q(3, 3)$, $R(7, 0)$, $S(3, -3)$, $T(-2, -3)$, and $U(-6, 0)$

28. pentagon with vertices $V(3, 0)$, $W(-2, 12)$, $X(-10, -3)$, $Y(-8, -12)$, and $Z(-2, -12)$

ALGEBRA **Find the length of each side of the polygon for the given perimeter.**

29. $P = 90$ centimeters

G   H
M   J
L   K

30. $P = 14$ miles

W   X
Z   Y

31. $P = 31$ units

S
x + 7
x − 1   R
T
3x − 5

More About. . .

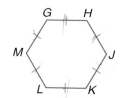

NEXT 5 MILES

Traffic Signs

The shape of a traffic sign is determined by the use of the sign. For example, wide rectangle-shaped signs are used to provide drivers with information.

Source: U.S. Federal Highway Administration

ALGEBRA Find the length of each side of the polygon for the given perimeter.

32. $P = 84$ meters

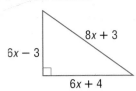

33. $P = 42$ inches

$3n + 2$
$n - 1$

34. $P = 41$ yards

35. NETS *Nets* are patterns that form a three-dimensional figure when cut out and folded. The net at the right makes a rectangular box. What is the perimeter of the net?

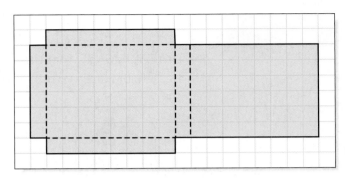

36. CRITICAL THINKING Use grid paper to draw all possible rectangles with length and width that are whole numbers and with a perimeter of 12. Record the number of grid squares contained in each rectangle.

a. What do you notice about the rectangle with the greatest number of squares?

b. The perimeter of another rectangle is 36. What would be the dimensions of the rectangle with the greatest number of squares?

37. WRITING IN MATH Answer the question that was posed at the beginning of the lesson.

How are polygons related to toys?

Include the following in your answer:

• names of the polygons shown in the picture of the toy structure, and
• sketches of other polygons that could be formed with construction toys with which you are familiar.

Standardized Test Practice
Ⓐ Ⓑ Ⓒ Ⓓ

38. SHORT RESPONSE A farmer fenced all but one side of a square field. If he has already used $3x$ meters of fence, how many meters will he need for the last side?

39. ALGEBRA If $5n + 5 = 10$, what is the value of $11 - n$?

Ⓐ -10 Ⓑ 0 Ⓒ 5 Ⓓ 10

Maintain Your Skills

Mixed Review **Determine whether each statement is *always*, *sometimes*, or *never* true.** *(Lesson 1-5)*

40. Two angles that form a linear pair are supplementary.

41. If two angles are supplementary, then one of the angles is obtuse.

In the figure, $\overrightarrow{AM}$ bisects $\angle LAR$, and $\overrightarrow{AS}$ bisects $\angle MAR$.
(Lesson 1-4)

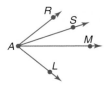

42. If $m\angle MAR = 2x + 13$ and $m\angle MAL = 4x - 3$, find $m\angle RAL$.

43. If $m\angle RAL = x + 32$ and $m\angle MAR = x - 31$, find $m\angle LAM$.

44. Find $m\angle LAR$ if $m\angle RAS = 25 - 2x$ and $m\angle SAM = 3x + 5$.

Measuring Polygons

You can use The Geometer's Sketchpad® to draw and investigate polygons. It can be used to find the measures of the sides and the perimeter of a polygon. You can also find the measures of the angles in a polygon.

Step 1 Draw △ABC.

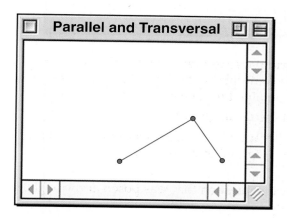

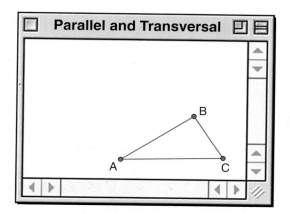

- Select the segment tool from the toolbar, and click to set the first endpoint A of side $\overline{AB}$. Then drag the cursor and click again to set the other endpoint B.

- Click on point B to set the endpoint of $\overline{BC}$. Drag the cursor and click to set point C.

- Click on point C to set the endpoint of $\overline{CA}$. Then move the cursor to highlight point A. Click on A to draw $\overline{CA}$.

- Use the pointer tool to click on points A, B, and C. Under the **Display** menu, select **Show Labels** to label the vertices of your triangle.

Step 2 Find AB, BC, and CA.

- Use the pointer tool to select $\overline{AB}$, $\overline{BC}$, and $\overline{CA}$.

- Select the **Length** command under the **Measure** menu to display the lengths of $\overline{AB}$, $\overline{BC}$, and $\overline{CA}$.

$AB = 5.30$
$BC = 3.80$
$CA = 6.54$

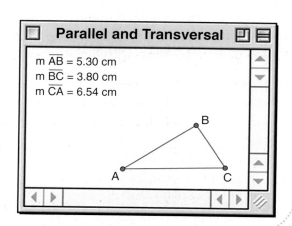

Geometry Software Investigation

Step 3 *Find the perimeter of △ABC.*

- Use the pointer tool to select points *A*, *B*, and *C*.

- Under the **Construct** menu, select **Triangle Interior.** The triangle will now be shaded.

- Select the triangle interior using the pointer.

- Choose the **Perimeter** command under the **Measure** menu to find the perimeter of △*ABC*. The perimeter of △*ABC* is 15.64 centimeters.

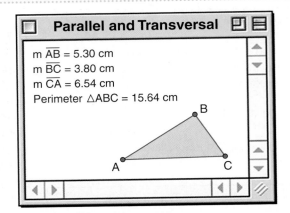

Step 4 *Find m∠A, m∠B, and m∠C.*

- Recall that ∠*A* can also be named ∠*BAC* or ∠*CAB*. Use the pointer to select points *B*, *A*, and *C* in order.

- Select the **Angle** command from the **Measure** menu to find *m∠A*.

- Select points *A*, *B*, and *C*. Find *m∠B*.

- Select points *A*, *C*, and *B*. Find *m∠C*.

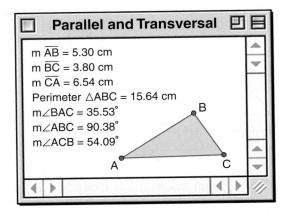

Analyze

1. Add the side measures you found in Step 2. Compare this sum to the result of Step 3. How do these compare?

2. What is the sum of the angle measures of △*ABC*?

3. Repeat the activities for each convex polygon.
   a. irregular quadrilateral   **b.** square   **c.** pentagon   **d.** hexagon

4. Draw another quadrilateral and find its perimeter. Then enlarge your figure using the **Dilate** command under the **Transform** menu. How does changing the sides affect the perimeter?

5. Compare your results with those of your classmates.

Make a Conjecture

6. Make a conjecture about the sum of the measures of the angles in any triangle.

7. What is the sum of the measures of the angles of a quadrilateral? pentagon? hexagon?

8. Make a conjecture about how the sums of the measures of the angles of polygons are related to the number of sides.

9. Test your conjecture on other polygons. Does your conjecture hold for these polygons? Explain.

10. When the sides of a polygon are changed by a common factor, does the perimeter of the polygon change by the same factor as the sides? Explain.

Vocabulary and Concept Check

acute angle (p. 30)
adjacent angles (p. 37)
angle (p. 29)
angle bisector (p. 32)
between (p. 14)
betweenness of points (p. 14)
collinear (p. 6)
complementary angles (p. 39)
concave (p. 45)
congruent (p. 15)
construction (p. 15)

convex (p. 45)
coplanar (p. 6)
degree (p. 29)
exterior (p. 29)
interior (p. 29)
line (p. 6)
line segment (p. 13)
linear pair (p. 37)
locus (p. 11)
midpoint (p. 22)

n-gon (p. 46)
obtuse angle (p. 30)
opposite rays (p. 29)
perimeter (p. 46)
perpendicular (p. 40)
plane (p. 6)
point (p. 6)
polygon (p. 45)
precision (p. 14)
ray (p. 29)

regular polygon (p. 46)
relative error (p. 19)
right angle (p. 30)
segment bisector (p. 24)
sides (p. 29)
space (p. 8)
supplementary angles (p. 39)
undefined terms (p. 7)
vertex (p. 29)
vertical angles (p. 37)

Exercises   Choose the letter of the term that best matches each figure.

1.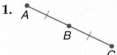

2.

3.

4.

5.

6.

a. line
b. ray
c. complementary angles
d. midpoint
e. supplementary angles
f. perpendicular
g. point
h. line segment

Lesson-by-Lesson Review

1-1 Points, Lines, and Planes

See pages 6–11.

Concept Summary
- A line is determined by two points.
- A plane is determined by three noncollinear points.

Example   Use the figure to name a plane containing point *N*.

The plane can be named as plane *P*.

You can also use any three noncollinear points to name the plane as plane *BNM*, plane *MBL*, or plane *NBL*.

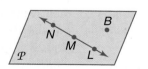

Exercises   Refer to the figure.   *See Example 1 on page 7.*

7. Name a line that contains point *I*.
8. Name a point that is not in lines *n* or *p*.
9. Name the intersection of lines *n* and *m*.
10. Name the plane containing points *E*, *J*, and *L*.

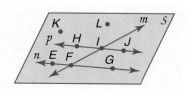

Draw and label a figure for each relationship. *See Example 3 on pages 7–8.*

11. Lines ℓ and m are coplanar and meet at point C.

12. Points S, T, and U are collinear, but points S, T, U, and V are not.

1-2 Linear Measure and Precision

See pages 13–19.

Concept Summary

- The precision of any measurement depends on the smallest unit available on the measuring device.
- The measure of a line segment is the sum of the measures of its parts.

Example Use the figure to find JK.

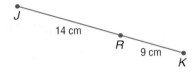

$JK = JR + RK$ Sum of parts = whole

$\quad = 14 + 9$ or 23   Substitution

So, $\overline{JK}$ is 23 centimeters long.

Exercises Find the value of the variable and PB, if P is between A and B.
See Example 4 on pages 14 and 15.

13. $AP = 7$, $PB = 3x$, $AB = 25$

14. $AP = 4c$, $PB = 2c$, $AB = 9$

15. $AP = s + 2$, $PB = 4s$, $AB = 8s - 7$

16. $AP = -2k$, $PB = k + 6$, $AB = 11$

Determine whether each pair of segments is congruent. *See Example 5 on page 16.*

17. $\overline{HI}$, $\overline{KJ}$

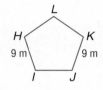

18. $\overline{AB}$, $\overline{AC}$

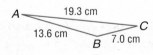

19. $\overline{VW}$, $\overline{WX}$

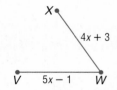

1-3 Distance and Midpoints

See pages 21–27.

Concept Summary

- Distances can be determined on a number line or the coordinate plane by using the Distance Formulas.
- The midpoint of a segment is the point halfway between the segment's endpoints.

Example Find the distance between $A(3, -4)$ and $B(-2, -10)$.

$d = \sqrt{(x_2 - x_1)^2 + (y_2 - y_1)^2}$ Distance Formula

$AB = \sqrt{(-2 - 3)^2 + [-10 - (-4)]^2}$ $(x_1, y_1) = (3, -4)$ and $(x_2, y_2) = (-2, -10)$

$\quad = \sqrt{(-5)^2 + (-6)^2}$ Simplify.

$\quad = \sqrt{61}$ or about 7.8   Simplify.

Exercises Find the distance between each pair of points.
See Example 2 on pages 21–22.

20. $A(1, 0)$, $B(-3, 2)$

21. $G(-7, 4)$, $L(3, 3)$

22. $J(0, 0)$, $K(4, -1)$

23. $M(-4, 16)$, $P(-6, 19)$

Find the coordinates of the midpoint of a segment having the given endpoints.
See Example 3 on page 23.

24. $D(0, 0)$, $E(22, -18)$

25. $U(-6, -3)$, $V(12, -7)$

26. $P(2, 5)$, $Q(-1, -1)$

27. $R(3.4, -7.3)$, $S(-2.2, -5.4)$

1-4 Angle Measure

See pages 29–36.

Concept Summary

- Angles are classified as acute, right, or obtuse according to their measure.
- An angle bisector is a ray that divides an angle into two congruent angles.

Examples a. **Name all angles that have B as a vertex.**

$\angle 6$, $\angle 4$, $\angle 7$, $\angle ABD$, $\angle EBC$

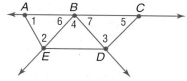

b. **Name the sides of $\angle 2$.**

$\overline{EA}$ and $\overline{EB}$

Exercises For Exercises 28–30, refer to the figure at the right. *See Example 1 on page 30.*

28. Name the vertex of $\angle 4$.

29. Name the sides of $\angle 1$.

30. Write another name for $\angle 3$.

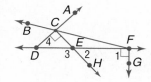

Measure each angle and classify it as *right, acute,* or *obtuse*. *See Example 2 on page 30.*

31. $\angle SQT$

32. $\angle PQT$

33. $\angle T$

34. $\angle PRT$

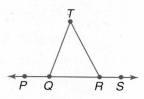

In the figure, $\overrightarrow{XW}$ bisects $\angle YXZ$ and $\overrightarrow{XV}$ bisects $\angle YXW$.
See Example 3 on page 32.

35. If $m\angle YXV = 3x$ and $m\angle VXW = 2x + 6$, find $m\angle YXW$.

36. If $m\angle YXW = 12x - 10$ and $m\angle WXZ = 8(x + 1)$, find $m\angle YXZ$.

37. If $m\angle YXZ = 9x + 17$ and $m\angle WXZ = 7x - 9$, find $m\angle YXW$.

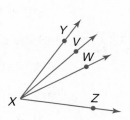

Chapter
1 For More ...

• Extra Practice, see pages 754–755.
• Mixed Problem Solving, see page 782.

1-5 Angle Relationships

See pages 37–43.

Concept Summary

• There are many special pairs of angles, such as adjacent angles, vertical angles, complementary angles, and linear pairs.

Example Find the value of x so that $\overleftrightarrow{AC}$ and $\overleftrightarrow{BD}$ are perpendicular.

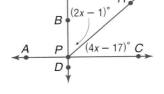

$m\angle BPC = m\angle BPR + m\angle RPC$	Sum of parts = whole
$90 = 2x - 1 + 4x - 17$	Substitution
$108 = 6x$	Simplify.
$18 = x$	Divide each side by 6.

Exercises For Exercises 38–41, use the figure at the right.
See Examples 1 and 3 on pages 38 and 40.

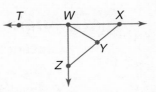

38. Name two obtuse angles.

39. Name a linear pair whose angles have vertex W.

40. If $m\angle TWZ = 2c + 36$, find c so that $\overline{TW} \perp \overline{WZ}$.

41. If $m\angle ZWY = 4k - 2$, and $m\angle YWX = 5k + 11$, find k so that $\angle ZWX$ is a right angle.

1-6 Polygons

See pages 45–50.

Concept Summary

• A polygon is a closed figure made of line segments.
• The perimeter of a polygon is the sum of the lengths of its sides.

Example Find the perimeter of the hexagon.

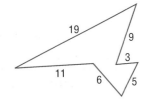

$P = s_1 + s_2 + s_3 + s_4 + s_5 + s_6$	Definition of perimeter
$= 19 + 9 + 3 + 5 + 6 + 11$ or 53	Substitution

Exercises Name each polygon by its number of sides. Then classify it as *convex* or *concave* and *regular* or *irregular*. *See Example 1 on page 46.*

42.   **43.**   **44.**

Find the perimeter of each polygon. *See Example 3 on page 47.*

45. hexagon $ABCDEF$ with vertices $A(1, 2)$, $B(5, 1)$, $C(9, 2)$, $D(9, 5)$, $E(5, 6)$, $F(1, 5)$

46. rectangle $WXYZ$ with vertices $W(-3, 5)$, $X(7, 1)$, $Y(5, -4)$, $Z(-5, 0)$

Vocabulary and Concepts

Determine whether each statement is *true* or *false*.

1. A plane contains an infinite number of lines.
2. If two angles are congruent, then their measures are equal.
3. The sum of two complementary angles is 180.
4. Two angles that form a linear pair are supplementary.

Skills and Applications

For Exercises 5–7, refer to the figure at the right.

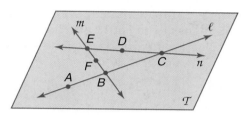

5. Name the line that contains points B and F.
6. Name a point not contained in lines ℓ or m.
7. Name the intersection of lines ℓ and n.

Find the value of the variable and VW if V is between U and W.

8. $UV = 2$, $VW = 3x$, $UW = 29$
9. $UV = r$, $VW = 6r$, $UW = 42$
10. $UV = 4p - 3$, $VW = 5p$, $UW = 15$
11. $UV = 3c + 29$, $VW = -2c - 4$, $UW = -4c$

Find the distance between each pair of points.

12. $G(0, 0)$, $H(-3, 4)$
13. $N(5, 2)$, $K(-2, 8)$
14. $A(-4, -4)$, $W(-2, 2)$

For Exercises 15–18, refer to the figure at the right.

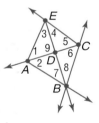

15. Name the vertex of $\angle 6$.
16. Name the sides of $\angle 4$.
17. Write another name for $\angle 7$.
18. Write another name for $\angle ADE$.

19. **ALGEBRA** The measures of two supplementary angles are $4r + 7$ and $r - 2$. Find the measures of the angles.
20. Two angles are complementary. One angle measures 26 degrees more than the other. Find the measures of the angles.

Find the perimeter of each polygon.

21. triangle PQR with vertices $P(-6, -3)$, $Q(1, -1)$, and $R(1, -5)$
22. pentagon $ABCDE$ with vertices $A(-6, 2)$, $B(-4, 7)$, $C(0, 4)$, $D(0, 0)$, and $E(-4, -3)$

DRIVING   For Exercises 23 and 24, use the following information and the diagram.
The city of Springfield is 5 miles west and 3 miles south of Capital City, while Brighton is 1 mile east and 4 miles north of Capital City. Highway 1 runs straight between Brighton and Springfield; Highway 4 connects Springfield and Capital City.

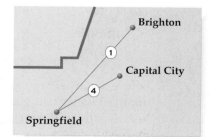

23. Find the length of Highway 1.
24. How long is Highway 4?

25. **STANDARDIZED TEST PRACTICE**   Which of the following figures is *not* a polygon?

Ⓐ   Ⓑ   Ⓒ 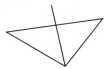   Ⓓ

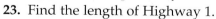

Part 1 | Multiple Choice

Record your answers on the answer sheet provided by your teacher or on a sheet of paper.

1. During a science experiment, Juanita recorded that she blinked 11 times in one minute. If this is a normal count and Juanita wakes up at 7 A.M. and goes to bed at 10 P.M., how many times will she blink during the time she is awake? (Prerequisite Skill)

   Ⓐ 165   Ⓑ 660

   Ⓒ 9900   Ⓓ 15,840

2. Find $-\sqrt{0.0225}$. (Prerequisite Skill)

   Ⓐ -0.15 Ⓑ -0.015

   Ⓒ 0.015 Ⓓ 0.15

3. Simplify $\dfrac{2x^2 + 12x + 16}{2x + 4}$. (Prerequisite Skill)

   Ⓐ 24   Ⓑ $x + 4$

   Ⓒ $4x + 12$ Ⓓ $4x^3 + x^2 + 20$

4. If two planes intersect, their intersection can be
   I. a line.
   II. three noncollinear points.
   III. two intersecting lines. (Lesson 1-1)

   Ⓐ I only   Ⓑ II only

   Ⓒ III only   Ⓓ I and II only

5. Before sonar technology, sailors determined the depth of water using a device called a sounding line. A rope with a lead weight at the end was marked in intervals called fathoms. Each fathom was equal to 6 feet. Suppose a specific ocean location has a depth of 55 fathoms. What would this distance be in yards? (Lesson 1-2)

   Ⓐ $9\frac{1}{6}$ yd   Ⓑ 110 yd

   Ⓒ 165 yd   Ⓓ 330 yd

6. An 18-foot ladder leans against the side of a house so that the bottom of the ladder is 6 feet from the house. To the nearest foot, how far up the side of the house does the top of the ladder reach? (Lesson 1-3)

   Ⓐ 12 ft

   Ⓑ 14 ft

   Ⓒ 17 ft

   Ⓓ 19 ft

   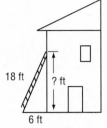

7. Ray BD is the bisector of $\angle ABC$. If $m\angle\ ABD = 2x + 14$ and $m\angle\ CBD = 5x - 10$, what is the measure of $\angle ABD$? (Lesson 1-5)

   Ⓐ 8   Ⓑ 16

   Ⓒ 30   Ⓓ 40

8. If $m\angle DEG$ is $6\frac{1}{2}$ times $m\angle FEG$, what is $m\angle DEG$? (Lesson 1-6)

   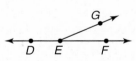

   Ⓐ 24   Ⓑ 78

   Ⓒ 130   Ⓓ 156

9. Kaitlin and Henry are participating in a treasure hunt. They are on the same straight path, walking toward each other. When Kaitlin reaches the Big Oak, she will turn 115° onto another path that leads to the treasure. At what angle will Henry turn when he reaches the Big Oak to continue on to the treasure? (Lesson 1-6)

   Ⓐ 25°   Ⓑ 35°

   Ⓒ 55°   Ⓓ 65°

Preparing for Standardized Tests
For test-taking strategies and more
practice, see pages 795–810.

Part 2 | Short Response/Grid In

**Record your answers on the answer sheet
provided by your teacher or on a sheet
of paper.**

10. Simplify $-2x + 6 + 4x^2 + x + x^2 - 5$.
    (Prerequisite Skill)

11. Solve the system of equations.
    (Prerequisite Skill)
    $$2y = 3x + 8$$
    $$y = 2x + 3$$

12. In rectangle $ABCD$, vertices A, B, and C have
    the coordinates $(-4, -1)$, $(-4, 4)$, and $(3, 4)$,
    respectively. Plot A, B, and C and find the
    coordinates of vertex D.   (Lesson 1-1)

13. The endpoints of a line segment are $(2, -1)$
    and $(-4, 3)$. What are the coordinates of its
    midpoint?   (Lesson 1-3)

14. The 200-meter race starts at point A, loops
    around the track, and finishes at point B.
    The track coach starts his stopwatch when
    the runners begin at point A and crosses the
    interior of the track so he can be at point B
    to time the runners as they cross the finish
    line. To the nearest meter, how long is $\overline{AB}$?
    (Lesson 1-3)

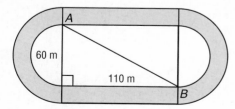

15. Mr. Lopez wants to cover the walls of his
    unfinished basement with pieces of
    plasterboard that are 8 feet high, 4 feet wide,
    and $\frac{1}{4}$ inch thick. If the basement measures
    24 feet wide, 16 feet long, and 8 feet tall,
    how many pieces of plasterboard will he
    need to cover all four walls?   (Lesson 1-4)

Test-Taking Tip   Ⓐ Ⓑ Ⓒ Ⓓ

Question 8
Most standardized tests allow you to write in the test
booklet or on scrap paper. To avoid careless errors, work
out your answers on paper rather than in your head. For
example, in Question 8, you can write an equation, such
as $x + 6\frac{1}{2}x = 180$ and solve for x. All of your solution
can be examined for accuracy when written down.

Part 3 | Extended Response

**Record your answers on a sheet of paper.
Show your work.**

16. Tami is creating a sun catcher to hang in her
    bedroom window. She makes her design on
    grid paper so that she can etch the glass
    appropriately before painting it.

    a. Graph the vertices of the glass if they are
       located at $(4, 0)$, $(-4, 0)$, $(0, -4)$, and $(0, 4)$.
       (Prerequisite Skill)

    b. Tami is putting a circle on the glass so
       that it touches the edge at the midpoint
       of each side. Find the coordinates of
       these midpoints.   (Lesson 1-3)

17. William Sparrow and his father are rebuilding
    the roof of their barn. They first build a
    system of rafters to support the roof. The
    angles formed by each beam are shown in
    the diagram.

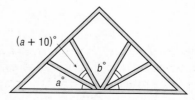

    a. If $a = 25$, what is the measure of the five
       angles formed by the beams? Justify your
       answer.   (Lesson 1-6)

    b. Classify each of the angles formed by the
       beams.   (Lesson 1-5)

Reasoning and Proof

What You'll Learn

- **Lessons 2-1 through 2-3** Make conjectures, determine whether a statement is true or false, and find counterexamples for statements.
- **Lesson 2-4** Use deductive reasoning to reach valid conclusions.
- **Lessons 2-5 and 2-6** Verify algebraic and geometric conjectures using informal and formal proof.
- **Lessons 2-7 and 2-8** Write proofs involving segment and angle theorems.

Key Vocabulary

- inductive reasoning (p. 62)
- deductive reasoning (p. 82)
- postulate (p. 89)
- theorem (p. 90)
- proof (p. 90)

Why It's Important

Logic and reasoning are used throughout geometry to solve problems and reach conclusions. There are many professions that rely on reasoning in a variety of situations. Doctors, for example, use reasoning to diagnose and treat patients.

You will investigate how doctors use reasoning in Lesson 2-4.

Getting Started

▶ **Prerequisite Skills** To be successful in this chapter, you'll need to master these skills and be able to apply them in problem-solving situations. Review these skills before beginning Chapter 2.

For Lesson 2-1
Evaluate Expressions

Evaluate each expression for the given value of *n*. *(For review, see page 736.)*

1. $3n - 2$; $n = 4$

2. $(n + 1) + n$; $n = 6$

3. $n^2 - 3n$; $n = 3$

4. $180(n - 2)$; $n = 5$

5. $n\left(\frac{n}{2}\right)$; $n = 10$

6. $\frac{n(n - 3)}{2}$; $n = 8$

For Lessons 2-6 through 2-8
Solve Equations

Solve each equation. *(For review, see pages 737 and 738.)*

7. $6x - 42 = 4x$

8. $8 - 3n = -2 + 2n$

9. $3(y + 2) = -12 + y$

10. $12 + 7x = x - 18$

11. $3x + 4 = \frac{1}{2}x - 5$

12. $2 - 2x = \frac{2}{3}x - 2$

For Lesson 2-8
Adjacent and Vertical Angles

For Exercises 13–14, refer to the figure at the right. *(For review, see Lesson 1-5.)*

13. If $m\angle AGB = 4x + 7$ and $m\angle EGD = 71$, find x.

14. If $m\angle BGC = 45$, $m\angle CGD = 8x + 4$, and $m\angle DGE = 15x - 7$, find x.

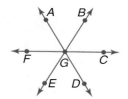

FOLDABLES™
Study Organizer

Reasoning and Proof Make this Foldable to help you organize your notes. Begin with eight sheets of $8\frac{1}{2}$" by 11" grid paper.

Step 1 **Staple**

Stack and staple the eight sheets together to form a booklet.

Step 2 **Cut Tabs**

Cut the bottom of each sheet to form a tabbed book.

Step 3 **Label**

Label each of the tabs with a lesson number. Add the chapter title to the first tab.

Reading and Writing As you read and study each lesson, use the corresponding page to write proofs and record examples of when you used logical reasoning in your daily life.

2-1 Inductive Reasoning and Conjecture

Vocabulary

- conjecture
- inductive reasoning
- counterexample

How can inductive reasoning help predict weather conditions?

Meteorologists use science and weather patterns to make predictions about future weather conditions. They are able to make accurate educated guesses based on past weather patterns.

MAKE CONJECTURES A **conjecture** is an educated guess based on known information. Examining several specific situations to arrive at a conjecture is called inductive reasoning. **Inductive reasoning** is reasoning that uses a number of specific examples to arrive at a plausible generalization or prediction.

Example 1 Patterns and Conjecture

The numbers represented below are called *triangular numbers*. Make a conjecture about the next triangular number based on the pattern.

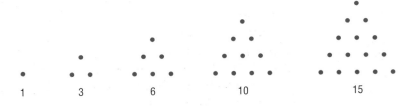

1	3	6	10	15

Observe: Each triangle is formed by adding another row of dots.

Find a Pattern:

1 ⌒ 3 ⌒ 6 ⌒ 10 ⌒ 15
+2 +3 +4 +5

The numbers increase by 2, 3, 4, and 5.

Conjecture: The next number will increase by 6. So, it will be 15 + 6 or 21.

In Chapter 1, you learned some basic geometric concepts. These concepts can be used to make conjectures in geometry.

Example 2 Geometric Conjecture

For points *P*, *Q*, and *R*, *PQ* = 9, *QR* = 15, and *PR* = 12. Make a conjecture and draw a figure to illustrate your conjecture.

Given: points *P*, *Q*, and *R*; *PQ* = 9, *QR* = 15, and *PR* = 12

Examine the measures of the segments. Since *PQ* + *PR* ≠ *QR*, the points cannot be collinear.

Conjecture: *P*, *Q*, and *R* are noncollinear.

FIND COUNTEREXAMPLES   A conjecture based on several observations may be true in most circumstances, but false in others. It takes only one false example to show that a conjecture is not true. The false example is called a **counterexample**.

Example 3 Find a Counterexample

FINANCE   Find a counterexample for the following statement based on the graph.

The rates for CDs are at least 1.5% less than the rates a year ago.

Examine the graph. The statement is true for 6-month, 1-year, and $2\frac{1}{2}$-year CDs. However, the difference in the rate for a 5-year CD is 0.74% less, which is less than 1.5%. The statement is false for a 5-year certificate of deposit. Thus, the change in the 5-year rate is a counterexample to the original statement.

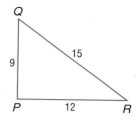

USA TODAY Snapshots®

Latest CD rates
Average certificate of deposit rates as of Wednesday:

6-month	This week	1.80%
	Last week	1.80%
	Year ago	4.55%
1-year	This week	2.12%
	Last week	2.11%
	Year ago	4.64%
2½-year	This week	2.96%
	Last week	2.96%
	Year ago	4.74%
5-year	This week	4.22%
	Last week	4.23%
	Year ago	4.96%

Source: *Bank Rate Monitor, 800-327-7717, www.bankrate.com*   USA TODAY

Check for Understanding

Concept Check   **1. Write** an example of a conjecture you have made outside of school.

2. Determine whether the following conjecture is *always, sometimes,* or *never* true based on the given information.

Given: collinear points *D*, *E*, and *F*

Conjecture: *DE* + *EF* = *DF*

3. OPEN ENDED   Write a statement. Then find a counterexample for the statement.

Guided Practice **Make a conjecture about the next item in each sequence.**

4.

5. $-8, -5, -2, 1, 4$

Make a conjecture based on the given information. Draw a figure to illustrate your conjecture.

6. $PQ = RS$ and $RS = TU$

7. $\overleftrightarrow{AB}$ and $\overleftrightarrow{CD}$ intersect at P.

Determine whether each conjecture is *true* or *false*. Give a counterexample for any false conjecture.

8. **Given:** x is an integer.
   Conjecture: $-x$ is negative.

9. **Given:** $WXYZ$ is a rectangle.
   Conjecture: $WX = YZ$ and $WZ = XY$

Application

10. **HOUSES** Most homes in the northern United States have roofs made with steep angles. In the warmer areas of the southern states, homes often have flat roofs. Make a conjecture about why the roofs are different.

Practice and Apply

Homework Help

For Exercises	See Examples
11–20	1
21–28	2
29–36	3

Extra Practice
See page 756.

Make a conjecture about the next item in each sequence.

11.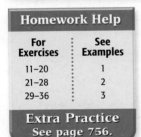

12.

13. $1, 2, 4, 8, 16$

14. $4, 6, 9, 13, 18$

15. $\frac{1}{3}, 1, \frac{5}{3}, \frac{7}{3}, 3$

16. $1, \frac{1}{2}, \frac{1}{4}, \frac{1}{8}, \frac{1}{16}$

17. $2, -6, 18, -54$

18. $-5, 25, -125, 625$

Make a conjecture about the number of blocks in the next item of each sequence.

19.

20.

Make a conjecture based on the given information. Draw a figure to illustrate your conjecture.

21. Lines ℓ and m are perpendicular.

22. $A(-2, -11), B(2, 1), C(5, 10)$

23. $\angle 3$ and $\angle 4$ are a linear pair.

24. $\overrightarrow{BD}$ is an angle bisector of $\angle ABC$.

25. $P(-1, 7), Q(6, -2), R(6, 5)$

26. $HIJK$ is a square.

27. $PQRS$ is a rectangle.

28. $\angle B$ is a right angle in $\triangle ABC$.

You can use scatter plots to make conjectures about the relationships between latitude, longitude, degree distance, and the monthly high temperature.
Visit www.geometry online.com/WebQuest to continue work on your WebQuest project.

Determine whether each conjecture is *true* **or** *false*. **Give a counterexample for any false conjecture.**

29. Given: $\angle 1$ and $\angle 2$ are complementary angles.
   Conjecture: $\angle 1$ and $\angle 2$ form a right angle.

30. Given: $m + y \geq 10$, $y \geq 4$
   Conjecture: $m \leq 6$

31. Given: points W, X, Y, and Z
   Conjecture: W, X, Y, and Z are noncollinear.

32. Given: $A(-4, 8)$, $B(3, 8)$, $C(3, 5)$
   Conjecture: $\triangle ABC$ is a right triangle.

33. Given: n is a real number.
   Conjecture: n^2 is a nonnegative number.

34. Given: $DE = EF$
   Conjecture: E is the midpoint of $\overline{DF}$.

35. Given: $JK = KL = LM = MJ$
   Conjecture: $JKLM$ forms a square.

36. Given: noncollinear points R, S, and T
   Conjecture: $\overline{RS}$, $\overline{ST}$, and $\overline{RT}$ form a triangle.

37. MUSIC Many people learn to play the piano by ear. This means that they first learned how to play without reading music. What process did they use?

CHEMISTRY **For Exercises 38–40, use the following information.**

Hydrocarbons are molecules composed of only carbon (C) and hydrogen (H) atoms. The simplest hydrocarbons are called alkanes. The first three alkanes are shown below.

Alkanes			
Compound Name	Methane	Ethane	Propane
Chemical Formula	CH_4	C_2H_6	C_3H_8
Structural Formula	H | H—C—H | H	H H | | H—C—C—H | | H H	H H H | | | H—C—C—C—H | | | H H H

38. Make a conjecture about butane, which is the next compound in the group. Write its structural formula.

39. Write the chemical formula for the 7th compound in the group.

40. Develop a rule you could use to find the chemical formula of the nth substance in the alkane group.

41. CRITICAL THINKING The expression $n^2 - n + 41$ has a prime value for $n = 1$, $n = 2$, and $n = 3$. Based on this pattern, you might conjecture that this expression always generates a prime number for any positive integral value of n. Try different values of n to test the conjecture. Answer *true* if you think the conjecture is always true. Answer *false* and give a counterexample if you think the conjecture is false.

More About. . .

Music •..................

The average medium-sized piano has about 230 strings. Each string has about 165 pounds of tension. That's a combined tension of about 18 tons.
Source: www.pianoworld.com

42. WRITING IN MATH   Answer the question that was posed at the beginning of the lesson.

How can inductive reasoning help predict weather conditions?

Include the following in your answer:

- an explanation as to how a conjecture about a weather pattern in the summer might be different from a similar weather pattern in the winter, and
- a conjecture about tomorrow's weather based on your local weather over the past several days.

43. What is the next term in the sequence 1, 1, 2, 3, 5, 8?

  (A) 11   (B) 12   (C) 13   (D) 14

44. ALGEBRA   If the average of six numbers is 18 and the average of three of the numbers is 15, then what is the sum of the remaining three numbers?

  (A) 21   (B) 45   (C) 53   (D) 63

Maintain Your Skills

Mixed Review   **Name each polygon by its number of sides and then classify it as *convex* or *concave* and *regular* or *irregular*.**   *(Lesson 1-6)*

45. 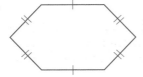   **46.**    **47.**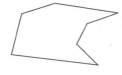

Determine whether each statement can be assumed from the figure. Explain.   *(Lesson 1-5)*

48. $\angle KJN$ is a right angle.

49. $\angle PLN \cong \angle NLM$

50. $\angle PNL$ and $\angle MNL$ are complementary.

51. $\angle KLN$ and $\angle MLN$ are supplementary.

52. $\angle KLP$ is a right angle.

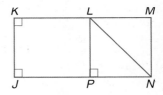

Find the coordinates of the midpoint of a segment having the given endpoints.
(Lesson 1-3)

53. $\overline{AB}$ for $A(-1, 3)$, $B(5, -5)$ **54.** $\overline{CD}$ for $C(4, 1)$, $D(-3, 7)$

55. $\overline{FG}$ for $F(4, -9)$, $G(-2, -15)$ **56.** $\overline{HJ}$ for $H(-5, -2)$, $J(7, 4)$

57. $\overline{KL}$ for $K(8, -1.8)$, $L(3, 6.2)$ **58.** $\overline{MN}$ for $M(-1.5, -6)$, $N(-4, 3)$

Find the value of the variable and *MP*, if *P* is between *M* and *N*.   *(Lesson 1-2)*

59. $MP = 7x$, $PN = 3x$, $PN = 24$ **60.** $MP = 2c$, $PN = 9c$, $PN = 63$

61. $MP = 4x$, $PN = 5x$, $MN = 36$ **62.** $MP = 6q$, $PN = 6q$, $MN = 60$

63. $MP = 4y + 3$, $PN = 2y$, $MN = 63$ **64.** $MP = 2b - 7$, $PN = 8b$, $MN = 43$

Getting Ready for the Next Lesson   **BASIC SKILL**   Determine which values in the given replacement set make the inequality true.

65. $x + 2 > 5$ **66.** $12 - x < 0$ **67.** $5x + 1 > 25$
  $\{2, 3, 4, 5\}$ $\{11, 12, 13, 14\}$ $\{4, 5, 6, 7\}$

2-2 Logic

What You'll Learn

- Determine truth values of conjunctions and disjunctions.
- Construct truth tables.

Vocabulary

- statement
- truth value
- negation
- compound statement
- conjunction
- disjunction
- truth table

How does logic apply to school?

When you answer true-false questions on a test, you are using a basic principle of logic. For example, refer to the map, and answer *true* or *false*.

Raleigh is a city in North Carolina.

You know that there is only one correct answer, either true or false.

Study Tip

Statements

A mathematical statement with one or more variables is called an *open sentence*. The truth value of an open sentence cannot be determined until values are assigned to the variables. A statement with only numeric values is a *closed sentence*.

DETERMINE TRUTH VALUES A **statement**, like the true-false example above, is any sentence that is either true or false, but not both. Unlike a conjecture, we know that a statement is either true or false. The truth or falsity of a statement is called its **truth value**.

Statements are often represented using a letter such as p or q. The statement above can be represented by p.

p: Raleigh is a city in North Carolina.   This statement is true.

The **negation** of a statement has the opposite meaning as well as an opposite truth value. For example, the negation of the statement above is *not p*.

not p: Raleigh is not a city in North Carolina.   In this case, the statement is false.

Key Concept   Negation

- **Words**   If a statement is represented by p, then *not p* is the negation of the statement.
- **Symbols**   ~p, read *not p*

Two or more statements can be joined to form a **compound statement**. Consider the following two statements.

p: Raleigh is a city in North Carolina.
q: Raleigh is the capital of North Carolina.

The two statements can be joined by the word *and*.

p and q: Raleigh is a city in North Carolina, *and* Raleigh is the capital of North Carolina.

The statement formed by joining p and q is an example of a conjunction.

> ## Key Concept   *Conjunction*
>
> - **Words** A **conjunction** is a compound statement formed by joining two or more statements with the word *and*.
>
> - **Symbols** $p \wedge q$, read *p and q*

A conjunction is true only when both statements in it are true. Since it is true that Raleigh is in North Carolina and it is the capital, the conjunction is also true.

Example 1 *Truth Values of Conjunctions*

Use the following statements to write a compound statement for each conjunction. Then find its truth value.

p: **January 1 is the first day of the year.**
q: $-5 + 11 = -6$
r: **A triangle has three sides.**

a. p and q

January is the first day of the year, and $-5 + 11 = -6$.
p and q is false, because p is true and q is false.

b. $r \wedge p$

A triangle has three sides, and January 1 is the first day of the year.
$r \wedge p$ is true, because r is true and p is true.

c. p and not r

January 1 is the first day of the year, and a triangle does not have three sides.
p and not r is false, because p is true and not r is false.

d. $\sim q \wedge r$

$-5 + 11 \neq -6$, and a triangle has three sides
$\sim q \wedge r$ is true because $\sim q$ is true and r is true.

Statements can also be joined by the word *or*. This type of statement is a disjunction. Consider the following statements.

p: Ahmed studies chemistry.

q: Ahmed studies literature.

p or q: Ahmed studies chemistry, *or* Ahmed studies literature.

> ## Key Concept   *Disjunction*
>
> - **Words** A **disjunction** is a compound statement formed by joining two or more statements with the word *or*.
>
> - **Symbols** $p \vee q$, read *p or q*

A disjunction is true if at least one of the statements is true. In the case of *p* or *q* above, the disjunction is true if Ahmed either studies chemistry or literature or both. The disjunction is false only if Ahmed studies neither chemistry nor literature.

Example 2 Truth Values of Disjunctions

Use the following statements to write a compound statement for each disjunction. Then find its truth value.

p: $100 \div 5 = 20$

q: The length of a radius of a circle is twice the length of its diameter.

r: The sum of the measures of the legs of a right triangle equals the measure of the hypotenuse.

a. *p* or *q*

$100 \div 5 = 20$, or the length of a radius of a circle is twice the length of its diameter.

p or *q* is true because *p* is true. It does not matter that *q* is false.

b. $q \vee r$

The length of a radius of a circle is twice the length of its diameter, or the sum of the measures of the legs of a right triangle equals the measure of the hypotenuse.

$q \vee r$ is false since neither statement is true.

Study Tip

Venn Diagrams
The size of the overlapping region in a Venn Diagram does not indicate how many items fall into that category.

Conjunctions can be illustrated with Venn diagrams. Refer to the statement at the beginning of the lesson. The Venn diagram at the right shows that Raleigh (R) is represented by the *intersection* of the set of cities in North Carolina and the set of state capitals. In other words, Raleigh must be in the set containing cities in North Carolina and in the set of state capitals.

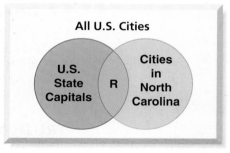

A disjunction can also be illustrated with a Venn diagram. Consider the following statements.

p: Jerrica lives in a U.S. state capital.

q: Jerrica lives in a North Carolina city.

$p \vee q$:   Jerrica lives in a U.S. state capital, or Jerrica lives in a North Carolina city.

In the Venn diagrams, the disjunction is represented by the *union* of the two sets. The union includes all U.S. capitals and all cities in North Carolina. The city in which Jerrica lives could be located in any of the three regions of the union.

The three regions represent

A U.S. state capitals excluding the capital of North Carolina,

B cities in North Carolina excluding the state capital, and

C the capital of North Carolina, which is Raleigh.

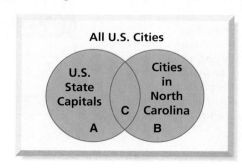

Venn diagrams can be used to solve real-world problems involving conjunctions and disjunctions.

Example 3 Use Venn Diagrams

RECYCLING   The Venn diagram shows the number of neighborhoods that have a curbside recycling program for paper or aluminum.

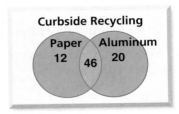

a. **How many neighborhoods recycle both paper and aluminum?**

The neighborhoods that have paper and aluminum recycling are represented by the intersection of the sets. There are 46 neighborhoods that have paper and aluminum recycling.

b. **How many neighborhoods recycle paper or aluminum?**

The neighborhoods that have paper or aluminum recycling are represented by the union of the sets. There are 12 + 46 + 20 or 78 neighborhoods that have paper or aluminum recycling.

c. **How many neighborhoods recycle paper and not aluminum?**

The neighborhoods that have paper and not aluminum recycling are represented by the nonintersecting portion of the paper region. There are 12 neighborhoods that have paper and not aluminum recycling.

TRUTH TABLES   A convenient method for organizing the truth values of statements is to use a **truth table**.

Negation	
p	$\sim p$
T	F
F	T

If p is a true statement, then $\sim p$ is a false statement.
If p is a false statement, then $\sim p$ is a true statement.

Truth tables can also be used to determine truth values of compound statements.

A conjunction is true only when both statements are true.

Conjunction		
p	q	$p \wedge q$
T	T	T
T	F	F
F	T	F
F	F	F

A disjunction is false only when both statements are false.

Disjunction		
p	q	$p \vee q$
T	T	T
T	F	T
F	T	T
F	F	F

You can use the truth values for negation, conjunction, and disjunction to construct truth tables for more complex compound statements.

Example 4 Construct Truth Tables

Construct a truth table for each compound statement.

a. $p \land \sim q$

    Step 1 Make columns with the headings p, q, $\sim q$, and $p \land \sim q$.

    Step 2 List the possible combinations of truth values for p and q.

    Step 3 Use the truth values of q to determine the truth values of $\sim q$.

    Step 4 Use the truth values for p and $\sim q$ to write the truth values for $p \land \sim q$.

Step 1 →

p	q	$\sim q$	$p \land \sim q$
T	T	F	F
T	F	T	T
F	T	F	F
F	F	T	F

Step 2   Step 3   Step 4

b. $\sim p \lor \sim q$

p	q	$\sim p$	$\sim q$	$\sim p \lor \sim q$
T	T	F	F	F
T	F	F	T	T
F	T	T	F	T
F	F	T	T	T

c. $(p \land q) \lor r$

Make columns for p, q, $p \land q$, r, and $(p \land q) \lor r$.

p	q	$p \land q$	r	$(p \land q) \lor r$
T	T	T	T	T
T	F	F	T	T
T	T	T	F	T
T	F	F	F	F
F	T	F	T	T
F	F	F	T	T
F	T	F	F	F
F	F	F	F	F

Check for Understanding

Concept Check

1. Describe how to interpret the Venn diagram for $p \land q$.

2. OPEN ENDED Write a compound statement for each condition.

    a. a true disjunction
    b. a false conjunction
    c. a true statement that includes a negation

3. Explain the difference between a conjunction and a disjunction.

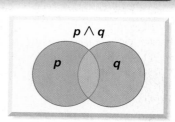

Guided Practice   Use the following statements to write a compound statement for each conjunction and disjunction. Then find its truth value.

p: $9 + 5 = 14$
q: February has 30 days.
r: A square has four sides.

4. p and q **5.** p and r **6.** $q \land r$

7. p or $\sim q$ **8.** $q \lor r$ **9.** $\sim p \lor \sim r$

10. Copy and complete the truth table.

p	q	$\sim q$	$p \land \sim q$
T	T	F	F
T	F		
F	T		
F	F		

Construct a truth table for each compound statement.

11. $p \land q$ **12.** $q \lor r$ **13.** $\sim p \land r$ **14.** $(p \lor q) \lor r$

Application   **AGRICULTURE**   For Exercises 15–17, refer to the Venn diagram that represents the states producing more than 100 million bushels of corn or wheat per year.

15. How many states produce more than 100 million bushels of corn?

16. How many states produce more than 100 million bushels of wheat?

17. How many states produce more than 100 million bushels of corn and wheat?

Grain Production

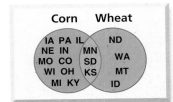

Source: U.S. Department of Agriculture

Practice and Apply

Homework Help

For Exercises	See Examples
18–29	1, 2
30–41	4
42–48	3

Extra Practice
See page 756.

Use the following statements to write a compound statement for each conjunction and disjunction. Then find its truth value.

p: $\sqrt{-64} = 8$
q: An equilateral triangle has three congruent sides.
r: $0 < 0$
s: An obtuse angle measures greater than 90° and less than 180°.

18. p and q **19.** p or q **20.** p and r

21. r and s **22.** q or r **23.** q and s

24. $p \land s$ **25.** $q \land r$ **26.** $r \lor p$

27. $s \lor q$ **28.** $(p \land q) \lor s$ **29.** $s \lor (q$ and $r)$

Copy and complete each truth table.

30.

p	q	$\sim p$	$\sim p \lor q$
T	T		
T	F		
F	T		
F	F		

31.

p	q	$\sim p$	$\sim q$	$\sim p \land \sim q$
T		F	F	
T		F	T	
F		T	F	
F		T	T	

32. Copy and complete the truth table.

p	q	r	$p \lor q$	$(p \lor q) \land r$
T	T	T		
T	T	F		
T	F	T		
T	F	F		
F	T	T		
F	T	F		
F	F	T		
F	F	F		

Construct a truth table for each compound statement.

33. q and r **34.** p or q **35.** p or r **36.** p and q

37. $q \land \sim r$ **38.** $\sim p \land \sim q$ **39.** $\sim p \lor (q \land \sim r)$ **40.** $p \land (\sim q \lor \sim r)$

MUSIC   **For Exercises 41–44, use the following information.**
A group of 400 teens were asked what type of
music they listened to. They could choose among
pop, rap, and country. The results are shown in
the Venn diagram.

Music Preference

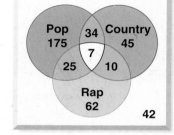

41. How many teens said that they listened to none of
these types of music?

42. How many said that they listened to all three types
of music?

43. How many said that they listened to only pop and
rap music?

44. How many teens said that they listened to pop, rap, or country music?

SCHOOL   **For Exercises 45–47, use the following information.**
In a school of 310 students, 80 participate in academic clubs, 115 participate in
sports, and 20 students participate in both.

45. Make a Venn diagram of the data.

46. How many students participate in either clubs or sports?

47. How many students do not participate in either clubs or sports?

RESEARCH   **For Exercises 48–50, use the Internet or another resource to determine
whether each statement about cities in New York is** *true* **or** *false.*

48. Albany is not located on the Hudson river.

49. Either Rochester or Syracuse is located on Lake Ontario.

50. It is false that Buffalo is located on Lake Erie.

CRITICAL THINKING   **For Exercises 51 and 52, use the following information.**
All members of Team A also belong to Team B, but only some members of
Team B also belong to Team C. Teams A and C have no members in common.

51. Draw a Venn diagram to illustrate the situation.

52. Which of the following statements is true?

   a. If a person is a member of Team C, then the person is not a member of
Team A.

   b. If a person is not a member of Team B, then the person is not a member
of Team A.

   c. No person that is a member of Team A can be a member of Team C.

53. **WRITING IN MATH**   Answer the question that was posed at the beginning of the lesson.

How does logic apply to school?

Include the following in your answer:
- an example of a conjunction using statements about your favorite subject and your favorite extracurricular activity, and
- a Venn diagram showing various characteristics of the members of your geometry class (for example, male/female, grade in school, and so on).

54. Which statement about $\triangle ABC$ has the same truth value as $AB = BC$?

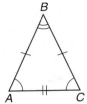

Ⓐ $m\angle A = m\angle C$ Ⓑ $m\angle A = m\angle B$

Ⓒ $AC = BC$ Ⓓ $AB = AC$

55. **ALGEBRA**   If the sum of two consecutive even integers is 78, which number is the greater of the two integers?

Ⓐ 36   Ⓑ 38

Ⓒ 40   Ⓓ 42

Maintain Your Skills

Mixed Review   **Make a conjecture about the next item in each sequence.**   *(Lesson 2-1)*

56. 3, 5, 7, 9   **57.** 1, 3, 9, 27   **58.** 6, 3, $\dfrac{3}{2}$, $\dfrac{3}{4}$

59. 17, 13, 9, 5   **60.** 64, 16, 4, 1   **61.** 5, 15, 45, 135

COORDINATE GEOMETRY   **Find the perimeter of each polygon. Round answers to the nearest tenth.**   *(Lesson 1-6)*

62. triangle ABC with vertices $A(-6, 7)$, $B(1, 3)$, and $C(-2, -7)$

63. square $DEFG$ with vertices $D(-10, -9)$, $E(-5, -2)$, $F(2, -7)$, and $G(-3, -14)$

64. quadrilateral $HIJK$ with vertices $H(5, -10)$, $I(-8, -9)$, $J(-5, -5)$, and $K(-2, -4)$

65. hexagon $LMNPQR$ with vertices $L(2, 1)$, $M(4, 5)$, $N(6, 4)$, $P(7, -4)$, $Q(5, -8)$, and $R(3, -7)$

Measure each angle and classify it as *right,* *acute,* **or** *obtuse.*   *(Lesson 1-4)*

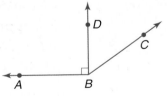

66. $\angle ABC$

67. $\angle DBC$

68. $\angle ABD$

69. **FENCING**   Michelle wanted to put a fence around her rectangular garden. The front and back measured 35 feet each, and the sides measured 75 feet each. If she wanted to make sure that she had enough feet of fencing, how much should she buy?   *(Lesson 1-2)*

Getting Ready for the Next Lesson   **PREREQUISITE SKILL**   **Evaluate each expression for the given values.**
*(To review **evaluating algebraic expressions**, see page 736.)*

70. $5a - 2b$ if $a = 4$ and $b = 3$ **71.** $4cd + 2d$ if $c = 5$ and $d = 2$

72. $4e + 3f$ if $e = -1$ and $f = -2$ **73.** $3g^2 + h$ if $g = 8$ and $h = -8$

2-3 Conditional Statements

What You'll Learn

- Analyze statements in if-then form.
- Write the converse, inverse, and contrapositive of if-then statements.

Vocabulary

- conditional statement
- if-then statement
- hypothesis
- conclusion
- related conditionals
- converse
- inverse
- contrapositive
- logically equivalent

How are conditional statements used in advertisements?

Advertisers often lure consumers into purchasing expensive items by convincing them that they are getting something for free in addition to their purchase.

Sign Up for a Six-Month Fitness Plan and Get Free Six Months

Get $1500 Cash Back When You Buy a New Car

Free Phone with Every One-Year Service Enrollment

IF-THEN STATEMENTS The statements above are examples of conditional statements. A **conditional statement** is a statement that can be written in *if-then form*. The first example above can be rewritten to illustrate this.

If you buy a car, *then* you get $1500 cash back.

Key Concept   If-Then Statement

- **Words**   An **if-then statement** is written in the form *if p, then q*. The phrase immediately following the word *if* is called the **hypothesis**, and the phrase immediately following the word *then* is called the **conclusion**.

- **Symbols** $p \rightarrow q$, read *if p then q*, or *p implies q*.

Example 1 Identify Hypothesis and Conclusion

Identify the hypothesis and conclusion of each statement.

a. **If points A, B, and C lie on line ℓ, then they are collinear.**

If points A, B, and C lie on line ℓ, then they are collinear.

$\underbrace{\text{points } A, B, \text{ and } C \text{ lie on line } \ell}_{\text{hypothesis}}$ $\underbrace{\text{they are collinear}}_{\text{conclusion}}$

Hypothesis: points A, B, and C lie on line ℓ

Conclusion: they are collinear

b. **The Tigers will play in the tournament if they win their next game.**

Hypothesis: the Tigers win their next game

Conclusion: they will play in the tournament

Study Tip

Reading Math
The word *if* is not part of the hypothesis. The word *then* is not part of the conclusion.

Identifying the hypothesis and conclusion of a statement is helpful when writing statements in if-then form.

Example 2 Write a Conditional in If-Then Form

Identify the hypothesis and conclusion of each statement. Then write each statement in if-then form.

a. **An angle with a measure greater than 90 is an obtuse angle.**

   Hypothesis: an angle has a measure greater than 90

   Conclusion: it is an obtuse angle

   If an angle has a measure greater than 90, then it is an obtuse angle.

b. **Perpendicular lines intersect.**

   Sometimes you must add information to a statement. In this case, it is necessary to know that perpendicular lines come in pairs.

   Hypothesis: two lines are perpendicular

   Conclusion: they intersect

   If two lines are perpendicular, then they intersect.

Recall that the truth value of a statement is either true or false. The hypothesis and conclusion of a conditional statement, as well as the conditional statement itself, can also be true or false.

Example 3 Truth Values of Conditionals

SCHOOL   **Determine the truth value of the following statement for each set of conditions.**

If you get 100% on your test, then your teacher will give you an A.

a. **You get 100%; your teacher gives you an A.**

   The hypothesis is true since you got 100%, and the conclusion is true because the teacher gave you an A. Since what the teacher promised is true, the conditional statement is true.

b. **You get 100%; your teacher gives you a B.**

   The hypothesis is true, but the conclusion is false. Because the result is not what was promised, the conditional statement is false.

c. **You get 98%; your teacher gives you an A.**

   The hypothesis is false, and the conclusion is true. The statement does not say what happens if you do not get 100% on the test. You could still get an A. It is also possible that you get a B. In this case, we cannot say that the statement is false. Thus, the statement is true.

d. **You get 85%; your teacher gives you a B.**

   As in part c, we cannot say that the statement is false. Therefore, the conditional statement is true.

The resulting truth values in Example 3 can be used to create a truth table for conditional statements. Notice that a conditional statement is true in all cases except where the hypothesis is true and the conclusion is false.

p	q	$p \rightarrow q$
T	T	T
T	F	F
F	T	T
F	F	T

CONVERSE, INVERSE, AND CONTRAPOSITIVE Other statements based on a given conditional statement are known as **related conditionals**.

Key Concept | | | | Related Conditionals

Statement	Formed by	Symbols	Examples
Conditional	given hypothesis and conclusion	$p \rightarrow q$	If two angles have the same measure, then they are congruent.
Converse	exchanging the hypothesis and conclusion of the conditional	$q \rightarrow p$	If two angles are congruent, then they have the same measure.
Inverse	negating both the hypothesis and conclusion of the conditional	$\sim p \rightarrow \sim q$	If two angles do not have the same measure, then they are not congruent.
Contrapositive	negating both the hypothesis and conclusion of the converse statement	$\sim q \rightarrow \sim p$	If two angles are not congruent, then they do not have the same measure.

If a given conditional is true, the converse and inverse are not necessarily true. However, the contrapositive of a true conditional is always true, and the contrapositive of a false conditional is always false. Likewise, the converse and inverse of a conditional are either both true or both false.

Statements with the same truth values are said to be **logically equivalent**. So, a conditional and its contrapositive are logically equivalent as are the converse and inverse of a conditional. These relationships are summarized below.

Study Tip

Contrapositive
The relationship of the truth values of a conditional and its contrapositive is known as the Law of Contrapositive.

p	q	Conditional $p \rightarrow q$	Converse $q \rightarrow p$	Inverse $\sim p \rightarrow \sim q$	Contrapositive $\sim q \rightarrow \sim p$
T	T	T	T	T	T
T	F	F	T	T	F
F	T	T	F	F	T
F	F	T	T	T	T

Example 4 *Related Conditionals*

Write the converse, inverse, and contrapositive of the statement *Linear pairs of angles are supplementary*. **Determine whether each statement is *true* or *false*. If a statement is false, give a counterexample.**

First, write the conditional in if-then form.

Conditional: If two angles form a linear pair, then they are supplementary. The conditional statement is true.

Write the converse by switching the hypothesis and conclusion of the conditional.

Converse: If two angles are supplementary, then they form a linear pair. The converse is false. $\angle ABC$ and $\angle PQR$ are supplementary, but are not a linear pair.

Inverse: If two angles do not form a linear pair, then they are not supplementary. The inverse is false. $\angle ABC$ and $\angle PQR$ do not form a linear pair, but they are supplementary.

The contrapositive is the negation of the hypothesis and conclusion of the converse.

Contrapositive: If two angles are not supplementary, then they do not form a linear pair. The contrapositive is true.

Concept Check

1. **Explain** why writing a conditional statement in if-then form is helpful.
2. **OPEN ENDED** Write an example of a conditional statement.
3. **Compare and contrast** the inverse and contrapositive of a conditional.

Guided Practice

Identify the hypothesis and conclusion of each statement.

4. If it rains on Monday, then I will stay home.
5. If $x - 3 = 7$, then $x = 10$.
6. If a polygon has six sides, then it is a hexagon.

Write each statement in if-then form.

7. A 32-ounce pitcher holds a quart of liquid.
8. The sum of the measures of supplementary angles is 180.
9. An angle formed by perpendicular lines is a right angle.

Determine the truth value of the following statement for each set of conditions.

If you drive faster than 65 miles per hour on the interstate, then you will receive a speeding ticket.

10. You drive 70 miles per hour, and you receive a speeding ticket.
11. You drive 62 miles per hour, and you do not receive a speeding ticket.
12. You drive 68 miles per hour, and you do not receive a speeding ticket.

Write the converse, inverse, and contrapositive of each conditional statement. Determine whether each related conditional is *true* or *false*. If a statement is false, find a counterexample.

13. If plants have water, then they will grow.
14. Flying in an airplane is safer than riding in a car.

Application

15. **FORESTRY** In different regions of the country, different variations of trees dominate the landscape. In Colorado, aspen trees cover high areas of the mountains. In Florida, cypress trees rise from swamps. In Vermont, maple trees are prevalent. Write these conditionals in if-then form.

Practice and Apply

Homework Help	
For Exercises	**See Examples**
16–21	1
22–27	2
28–39	3
40–45	4

Extra Practice
See page 756.

Identify the hypothesis and conclusion of each statement.

16. If $2x + 6 = 10$, then $x = 2$.
17. If you are a teenager, then you are at least 13 years old.
18. If you have a driver's license, then you are at least 16 years old.
19. If three points lie on a line, then they are collinear.
20. "If a man hasn't discovered something that he will die for, he isn't fit to live." (*Martin Luther King, Jr., 1963*)
21. If the measure of an angle is between 0 and 90, then the angle is acute.

Write each statement in if-then form.

22. Get a free visit with a one-year fitness plan.
23. Math teachers love to solve problems.
24. "I think, therefore I am." (*Descartes*)
25. Adjacent angles have a common side.
26. Vertical angles are congruent.
27. Equiangular triangles are equilateral.

Determine the truth value of the following statement for each set of conditions.

If you are over 18 years old, then you vote in all elections.

28. You are 19 years old and you vote.

29. You are 16 years old and you vote.

30. You are 21 years old and do not vote.

31. You are 17 years old and do not vote.

32. Your sister is 21 years old and votes.

33. Your dad is 45 years old and does not vote.

In the figure, *P*, *Q*, and *R* are collinear, *P* and *A* lie in plane *M*, and *Q* and *B* lie in plane *N*. Determine the truth value of each statement.

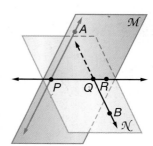

34. *P*, *Q*, and *R* lie in plane *M*.

35. $\overleftrightarrow{QB}$ lies in plane *N*.

36. *Q* lies in plane *M*.

37. *P*, *Q*, *A*, and *B* are coplanar.

38. $\overleftrightarrow{AP}$ contains *Q*.

39. Planes *M* and *N* intersect at $\overleftrightarrow{RQ}$.

Write the converse, inverse, and contrapositive of each conditional statement. Determine whether each related conditional is *true* or *false*. If a statement is false, find a counterexample.

40. If you live in Dallas, then you live in Texas.

41. If you exercise regularly, then you are in good shape.

42. The sum of two complementary angles is 90.

43. All rectangles are quadrilaterals.

44. All right angles measure 90.

45. Acute angles have measures less than 90.

•··• **SEASONS** For Exercises 46 and 47, use the following information.
Due to the movement of Earth around the sun, summer days in Alaska have more hours of daylight than darkness, and winter days have more hours of darkness than daylight.

46. Write two true conditional statements in if-then form for summer days and winter days in Alaska.

47. Write the converse of the two true conditional statements. State whether each is *true* or *false*. If a statement is false, find a counterexample.

48. CRITICAL THINKING Write a false conditional statement. Is it possible to insert the word *not* into your conditional to make it true? If so, write the true conditional.

49. **WRITING IN MATH** Answer the question that was posed at the beginning of the lesson.

How are conditional statements used in advertisements?

Include the following in your answer:
- an example of a conditional statement in if-then form, and
- an example of a conditional statement that is not in if-then form.

50. Which statement has the same truth value as the following statement?
If Ava and Willow are classmates, then they go to the same school.

Ⓐ If Ava and Willow go to the same school, then they are classmates.

Ⓑ If Ava and Willow are not classmates, then they do not go to the same school.

Ⓒ If Ava and Willow do not go to the same school, then they are not classmates.

Ⓓ If Ava and Willow go to the same school, then they are not classmates.

51. ALGEBRA In a history class with 32 students, the ratio of girls to boys is 5 to 3. How many more girls are there than boys?

Ⓐ 2   Ⓑ 8   Ⓒ 12   Ⓓ 20

Maintain Your Skills

Mixed Review Use the following statements to write a compound statement for each conjunction and disjunction. Then find its truth value. *(Lesson 2-2)*

p: George Washington was the first president of the United States.
q: A hexagon has five sides.
r: $60 \times 3 = 18$

52. $p \wedge q$ **53.** $q \vee r$ **54.** $p \vee q$

55. $\sim q \vee r$ **56.** $p \wedge \sim q$ **57.** $\sim p \wedge \sim r$

Make a conjecture based on the given information. Draw a figure to illustrate your conjecture. *(Lesson 2-1)*

58. $ABCD$ is a rectangle.   **59.** In $\triangle FGH$, $m\angle F = 45$, $m\angle G = 67$, $m\angle H = 68$.

60. $J(-3, 2)$, $K(1, 8)$, $L(5, 2)$ **61.** In $\triangle PQR$, $m\angle PQR = 90$

Use the Distance Formula to find the distance between each pair of points.
(Lesson 1-3)

62. $C(-2, -1)$, $D(0, 3)$ **63.** $J(-3, 5)$, $K(1, 0)$

64. $P(-3, -1)$, $Q(2, -3)$ **65.** $R(1, -7)$, $S(-4, 3)$

Getting Ready for the Next Lesson **PREREQUISITE SKILL** Identify the operation used to change Equation (1) to Equation (2). *(To review solving equations, see pages 737 and 738.)*

66. (1) $3x + 4 = 5x - 8$ **67.** (1) $\frac{1}{2}(a - 5) = 12$ **68.** (1) $8p = 24$
      (2) $3x = 5x - 12$ (2) $a - 5 = 24$ (2) $p = 3$

Practice Quiz 1   Lessons 2-1 through 2-3

Determine whether each conjecture is *true* or *false*. Give a counterexample for any false conjecture.
(Lesson 2-1)

1. Given: $WX = XY$
    Conjecture: W, X, and Y are collinear.

2. Given: $\angle 1$ and $\angle 2$ are complementary.
           $\angle 2$ and $\angle 3$ are complementary.
    Conjecture: $m\angle 1 = m\angle 3$

Construct a truth table for each compound statement. *(Lesson 2-2)*

3. $\sim p \wedge q$ **4.** $p \vee (q \wedge r)$

5. Write the converse, inverse, and contrapositive of the following conditional statement. Determine whether each related conditional is *true* or *false*. If a statement is false, find a counterexample. *(Lesson 2-3)*
If two angles are adjacent, then the angles have a common vertex.

Reading Mathematics

Biconditional Statements

Ashley began a new summer job, earning $10 an hour. If she works over 40 hours a week, she earns time and a half, or $15 an hour. If she earns $15 an hour, she has worked over 40 hours a week.

$p:$ Ashley earns $15 an hour
$q:$ Ashley works over 40 hours a week

$p \rightarrow q:$ If Ashley earns $15 an hour, she has worked over 40 hours a week.
$q \rightarrow p:$ If Ashley works over 40 hours a week, she earns $15 an hour.

In this case, both the conditional and its converse are true. The conjunction of the two statements is called a **biconditional**.

Key Concept — Biconditional Statement

- **Words**   A biconditional statement is the conjunction of a conditional and its converse.

- **Symbols** $(p \rightarrow q) \wedge (q \rightarrow p)$ is written $(p \leftrightarrow q)$ and read *p if and only if q*.

If and only if can be abbreviated *iff*.

So, the biconditional statement is as follows.

$p \leftrightarrow q:$ Ashley earns $15 an hour *if and only if* she works over 40 hours a week.

Examples

Write each biconditional as a conditional and its converse. Then determine whether the biconditional is *true* or *false*. If false, give a counterexample.

a. Two angle measures are complements if and only if their sum is 90.
   Conditional: If two angle measures are complements, then their sum is 90.
   Converse: If the sum of two angle measures is 90, then they are complements.
   Both the conditional and the converse are true, so the biconditional is true.

b. $x > 9$ iff $x > 0$
   Conditional: If $x > 9$, then $x > 0$.
   Converse: If $x > 0$, then $x > 9$.
   The conditional is true, but the converse is not. Let $x = 2$. Then $2 > 0$ but $2 \not> 9$.
   So, the biconditional is false.

Reading to Learn

Write each biconditional as a conditional and its converse. Then determine whether the biconditional is *true* or *false*. If false, give a counterexample.

1. A calculator will run if and only if it has batteries.

2. Two lines intersect if and only if they are not vertical.

3. Two angles are congruent if and only if they have the same measure.

4. $3x - 4 = 20$ iff $x = 7$.

5. A line is a segment bisector if and only if it intersects the segment at its midpoint.

2-4 Deductive Reasoning

What You'll Learn

- Use the Law of Detachment.
- Use the Law of Syllogism.

Vocabulary

- deductive reasoning
- Law of Detachment
- Law of Syllogism

How does deductive reasoning apply to health?

When you are ill, your doctor may prescribe an antibiotic to help you get better. Doctors may use a dose chart like the one shown to determine the correct amount of medicine you should take.

Weight (kg)	Dose (mg)
10–20	150
20–30	200
30–40	250
40–50	300
50–60	350
60–70	400

LAW OF DETACHMENT The process that doctors use to determine the amount of medicine a patient should take is called **deductive reasoning**. Unlike inductive reasoning, which uses examples to make a conjecture, deductive reasoning uses facts, rules, definitions, or properties to reach logical conclusions.

A form of deductive reasoning that is used to draw conclusions from true conditional statements is called the **Law of Detachment**.

Key Concept — Law of Detachment

- **Words** If $p \rightarrow q$ is true and p is true, then q is also true.
- **Symbols** $[(p \rightarrow q) \wedge p] \rightarrow q$

Study Tip

Validity
When you apply the Law of Detachment, make sure that the conditional is true before you test the validity of the conclusion.

Example 1 Determine Valid Conclusions

The following is a true conditional. Determine whether each conclusion is valid based on the given information. Explain your reasoning.

If a ray is an angle bisector, then it divides the angle into two congruent angles.

a. **Given:** $\overrightarrow{BD}$ bisects $\angle ABC$.

   Conclusion: $\angle ABD \cong \angle CBD$

   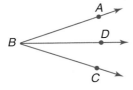

   The hypothesis states that $\overrightarrow{BD}$ is the bisector of $\angle ABC$. Since the conditional is true and the hypothesis is true, the conclusion is valid.

b. **Given:** $\angle PQT \cong \angle RQS$

   Conclusion: $\overrightarrow{QS}$ and $\overrightarrow{QT}$ are angle bisectors.

   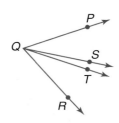

   Knowing that a conditional statement and its conclusion are true does not make the hypothesis true. An angle bisector divides an angle into two separate congruent angles. In this case, the given angles are not separated by one ray. Instead, they overlap. The conclusion is not valid.

LAW OF SYLLOGISM Another law of logic is the **Law of Syllogism**. It is similar to the Transitive Property of Equality.

> ## Key Concept
> *Law of Syllogism*
>
> - **Words** If $p \rightarrow q$ and $q \rightarrow r$ are true, then $p \rightarrow r$ is also true.
> - **Symbols** $[(p \rightarrow q) \wedge (q \rightarrow r)] \rightarrow (p \rightarrow r)$

Study Tip

Conditional Statements
Label the hypotheses and conclusions of a series of statements before applying the Law of Syllogism.

Example 2 *Determine Valid Conclusions From Two Conditionals*

CHEMISTRY Use the Law of Syllogism to determine whether a valid conclusion can be reached from each set of statements.

a. (1) If the symbol of a substance is Pb, then it is lead.
   (2) The atomic number of lead is 82.

Let p, q, and r represent the parts of the statement.

p: the symbol of a substance is Pb
q: it is lead
r: the atomic number is 82

Statement (1): $p \rightarrow q$
Statement (2): $q \rightarrow r$

Since the given statements are true, use the Law of Syllogism to conclude $p \rightarrow r$. That is, *If the symbol of a substance is Pb, then its atomic number is 82.*

b. (1) Water can be represented by H_2O.
   (2) Hydrogen (H) and oxygen (O) are in the atmosphere.

There is no valid conclusion. While both statements are true, the conclusion of each statement is not used as the hypothesis of the other.

Example 3 *Analyze Conclusions*

Determine whether statement (3) follows from statements (1) and (2) by the Law of Detachment or the Law of Syllogism. If it does, state which law was used. If it does not, write *invalid*.

a. (1) Vertical angles are congruent.
   (2) If two angles are congruent, then their measures are equal.
   (3) If two angles are vertical, then their measures are equal.

p: two angles are vertical
q: they are congruent
r: their measures are equal

Statement (3) is a valid conclusion by the Law of Syllogism.

b. (1) If a figure is a square, then it is a polygon.
   (2) Figure A is a polygon.
   (3) Figure A is a square.

Statement (1) is true, but statement (3) does not follow from statement (2). Not all polygons are squares.
Statement (3) is invalid.

Concept Check

1. **OPEN ENDED** Write an example to illustrate the correct use of the Law of Detachment.

2. **Explain** how the Transitive Property of Equality is similar to the Law of Syllogism.

3. **FIND THE ERROR** An article in a magazine states that if you get seasick, then you will get dizzy. It also says that if you get seasick, you will get an upset stomach. Suzanne says that this means that if you get dizzy, then you will get an upset stomach. Lakeisha says that she is wrong. Who is correct? Explain.

Guided Practice

Determine whether the stated conclusion is valid based on the given information. If not, write *invalid*. Explain your reasoning.

If two angles are vertical angles, then they are congruent.

4. **Given:** $\angle A$ and $\angle B$ are vertical angles.
   Conclusion: $\angle A \cong \angle B$

5. **Given:** $\angle C \cong \angle D$
   Conclusion: $\angle C$ and $\angle D$ are vertical angles.

Use the Law of Syllogism to determine whether a valid conclusion can be reached from each set of statements. If a valid conclusion is possible, write it. If not, write *no conclusion*.

6. If you are 18 years old, you are in college.
   You are in college.

7. The midpoint divides a segment into two congruent segments.
   If two segments are congruent, then their measures are equal.

Determine whether statement (3) follows from statements (1) and (2) by the Law of Detachment or the Law of Syllogism. If it does, state which law was used. If it does not, write *invalid*.

8. (1) If Molly arrives at school at 7:30 A.M., she will get help in math.
   (2) If Molly gets help in math, then she will pass her math test.
   (3) If Molly arrives at school at 7:30 A.M., then she will pass her math test.

9. (1) Right angles are congruent.
   (2) $\angle X \cong \angle Y$
   (3) $\angle X$ and $\angle Y$ are right angles.

Application

INSURANCE For Exercises 10 and 11, use the following information.
An insurance company advertised the following monthly rates for life insurance.

If you are a:	Premium for $30,000 Coverage	Premium for $50,000 Coverage
Female, age 35	$14.35	$19.00
Male, age 35	$16.50	$21.63
Female, age 45	$21.63	$25.85
Male, age 45	$23.75	$28.90

10. If Ann is 35 years old and she wants to purchase $30,000 of insurance from this company, then what is her premium?

11. If Terry paid $21.63 for life insurance, can you conclude that Terry is 35? Explain.

Practice and Apply

Homework Help

For Exercises	See Examples
12–19	1
20–23	2
24–29	3

Extra Practice
See page 757.

For Exercises 12–19, determine whether the stated conclusion is valid based on the given information. If not, write *invalid*. Explain your reasoning.

If two numbers are odd, then their sum is even.

12. Given: The sum of two numbers is 22.
  Conclusion: The two numbers are odd.

13. Given: The numbers are 5 and 7.
  Conclusion: The sum is even.

14. Given: 11 and 23 are added together.
  Conclusion: The sum of 11 and 23 is even.

15. Given: The numbers are 2 and 6.
  Conclusion: The sum is odd.

If three points are noncollinear, then they determine a plane.

16. Given: *A*, *B*, and *C* are noncollinear.
  Conclusion: *A*, *B*, and *C* determine a plane.

17. Given: *E*, *F*, and *G* lie in plane *M*.
  Conclusion: *E*, *F*, and *G* are noncollinear.

18. Given: *P* and *Q* lie on a line.
  Conclusion: *P* and *Q* determine a plane.

19. Given: $\triangle XYZ$
  Conclusion: *X*, *Y*, and *Z* determine a plane.

Use the Law of Syllogism to determine whether a valid conclusion can be reached from each set of statements. If a valid conclusion is possible, write it. If not, write *no conclusion*.

20. If you spend money on it, then it is a business.
  If you spend money on it, then it is fun.

21. If the measure of an angle is less than 90, then it is acute.
  If an angle is acute, then it is not obtuse.

22. If *X* is the midpoint of segment *YZ*, then *YX* = *XZ*.
  If the measures of two segments are equal, then they are congruent.

23. If two lines intersect to form a right angle, then they are perpendicular.
  Lines ℓ and *m* are perpendicular.

Determine whether statement (3) follows from statements (1) and (2) by the Law of Detachment or the Law of Syllogism. If it does, state which law was used. If it does not, write *invalid*.

24. (1) In-line skaters live dangerously.
  (2) If you live dangerously, then you like to dance.
  (3) If you are an in-line skater, then you like to dance.

25. (1) If the measure of an angle is greater than 90, then it is obtuse.
  (2) $m\angle ABC > 90$
  (3) $\angle ABC$ is obtuse.

26. (1) Vertical angles are congruent.
  (2) $\angle 3 \cong \angle 4$
  (3) $\angle 3$ and $\angle 4$ are vertical angles.

27. (1) If an angle is obtuse, then it cannot be acute.
  (2) $\angle A$ is obtuse.
  (3) $\angle A$ cannot be acute.

Determine whether statement (3) follows from statements (1) and (2) by the Law of Detachment or the Law of Syllogism. If it does, state which law was used. If it does not, write *invalid*.

28. (1) If you drive safely, then you can avoid accidents.
(2) Tika drives safely.
(3) Tika can avoid accidents.

29. (1) If you are a customer, then you are always right.
(2) If you are a teenager, then you are always right.
(3) If you are a teenager, then you are a customer.

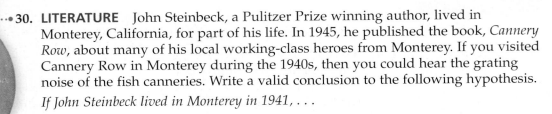
30. LITERATURE John Steinbeck, a Pulitzer Prize winning author, lived in Monterey, California, for part of his life. In 1945, he published the book, *Cannery Row*, about many of his local working-class heroes from Monterey. If you visited Cannery Row in Monterey during the 1940s, then you could hear the grating noise of the fish canneries. Write a valid conclusion to the following hypothesis.

If John Steinbeck lived in Monterey in 1941, . . .

31. SPORTS In the 2002 Winter Olympics, Canadian speed skater Catriona Le May Doan won her second Olympic title in 500-meter speed skating. Ms. Doan was in the last heat for the second round of that race. Use the two true conditional statements to reach a valid conclusion about Ms. Doan's 2002 competition.

(1) If Catriona Le May Doan skated her second 500 meters in 37.45 seconds, then she would beat the time of Germany's Monique Garbrecht-Enfeldt.

(2) If Ms. Doan beat the time of Monique Garbrecht-Enfeldt, then she would win the race.

Online Research **Data Update** Use the Internet or another resource to find the winning times for other Olympic events. Write statements using these times that can lead to a valid conclusion. Visit www.geometryonline.com/data_update to learn more.

32. CRITICAL THINKING An advertisement states that "If you like to ski, then you'll love Snow Mountain Resort." Stacey likes to ski, but when she went to Snow Mountain Resort, she did not like it very much. If you know that Stacey saw the ad, explain how her reasoning was flawed.

33. WRITING IN MATH Answer the question that was posed at the beginning of the lesson.

How does deductive reasoning apply to health?

Include the following in your answer:
- an explanation of how doctors may use deductive reasoning to prescribe medicine, and
- an example of a doctor's uses of deductive reasoning to diagnose an illness, such as strep throat or chickenpox.

Standardized Test Practice
(A) (B) (C) (D)

34. Based on the following statements, which statement must be true?
  I If Yasahiro is an athlete and he gets paid, then he is a professional athlete.
  II Yasahiro is not a professional athlete.
  III Yasahiro is an athlete.

(A) Yasahiro is an athlete and he gets paid.
(B) Yasahiro is a professional athlete or he gets paid.
(C) Yasahiro does not get paid.
(D) Yasahiro is not an athlete.

35. ALGEBRA At a restaurant, a diner uses a coupon for 15% off the cost of one meal. If the diner orders a meal regularly priced at $16 and leaves a tip of 20% of the discounted meal, how much does she pay in total?

Ⓐ $15.64 Ⓑ $16.32 Ⓒ $16.80 Ⓓ $18.72

Maintain Your Skills

Mixed Review **ADVERTISING** **For Exercises 36–38, use the following information.** *(Lesson 2-3)*

Advertising writers frequently use if-then statements to relay a message and promote their product. An ad for a type of Mexican food reads, *If you're looking for a fast, easy way to add some fun to your family's menu, try Casa Fiesta.*

36. Write the converse of the conditional.

37. What do you think the advertiser wants people to conclude about Casa Fiesta products?

38. Does the advertisement say that Casa Fiesta adds fun to your family's menu?

Construct a truth table for each compound statement. *(Lesson 2-2)*

39. $q \wedge r$ **40.** $\sim p \vee r$ **41.** $p \wedge (q \vee r)$ **42.** $p \vee (\sim q \wedge r)$

For Exercises 43–47, refer to the figure at the right. *(Lesson 1-5)*

43. Which angle is complementary to $\angle FDG$?

44. Name a pair of vertical angles.

45. Name a pair of angles that are noncongruent and supplementary.

46. Identify $\angle FDH$ and $\angle CDH$ as *congruent, adjacent, vertical, complementary, supplementary,* and/or a *linear pair.*

47. Can you assume that $\overline{DC} \cong \overline{CK}$? Explain.

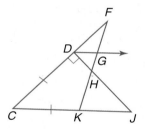

Use the Pythagorean Theorem to find the distance between each pair of points. *(Lesson 1-3)*

48. $A(1, 5)$, $B(-2, 9)$ **49.** $C(-4, -2)$, $D(2, 6)$

50. $F(7, 4)$, $G(1, 0)$ **51.** $M(-5, 0)$, $N(4, 7)$

For Exercises 52–55, draw and label a figure for each relationship. *(Lesson 1-1)*

52. $\overleftrightarrow{FG}$ lies in plane $\mathcal{M}$ and contains point H.

53. Lines r and s intersect at point W.

54. Line ℓ contains P and Q, but does not contain R.

55. Planes $\mathcal{A}$ and $\mathcal{B}$ intersect in line n.

Getting Ready for the Next Lesson **PREREQUISITE SKILL** **Write what you can assume about the segments or angles listed for each figure.** *(To review **information from figures**, see Lesson 1-5.)*

56. $\overline{AM}$, $\overline{CM}$, $\overline{CN}$, $\overline{BN}$ **57.** $\angle 1$, $\angle 2$ **58.** $\angle 4$, $\angle 5$, $\angle 6$

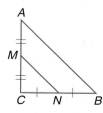   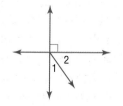   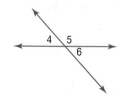

Matrix Logic

Deductive reasoning can be used in problem-solving situations. One method of solving problems uses a table. This method is called **matrix logic**.

Example

GEOLOGY On a recent test, Rashaun was given five different mineral samples to identify, along with the chart at right. Rashaun observed the following.

- Sample C is brown.
- Samples B and E are harder than glass.
- Samples D and E are red.

Identify each of the samples.

Mineral	Color	Hardness
Biotite	brown or black	softer than glass
Halite	white	softer than glass
Hematite	red	softer than glass
Feldspar	white, pink, or green	harder than glass
Jaspar	red	harder than glass

Make a table to organize the information. Mark each false condition with an X and each true condition with a √. The first observation is that Sample C is brown. Only one of the minerals, biotite, is brown, so place a check in the box that corresponds to biotite and Sample C. Then place an X in each of the other boxes in the same column and row.

Sample	A	B	C	D	E
Biotite	X	X	√	X	X
Halite		X	X	X	X
Hematite		X	X		X
Feldspar			X	X	X
Jaspar	X	X	X	X	√

The second observation is that Samples B and E are harder than glass. Place an X in each box for minerals that are softer than glass. The third observation is that Samples D and E are red. Mark the boxes accordingly. Notice that Sample E has an X in all but one box. Place a check mark in the remaining box, and an X in all other boxes in that row.

Sample	A	B	C	D	E
Biotite	X	X	√	X	X
Halite	√	X	X	X	X
Hematite	X	X	X	√	X
Feldspar	X	√	X	X	X
Jaspar	X	X	X	X	√

Then complete the table. Sample A is Halite, Sample B is Feldspar, Sample C is Biotite, Sample D is Hematite, and Sample E is Jaspar.

Exercises

1. Nate, John, and Nick just began after-school jobs. One works at a veterinarian's office, one at a computer store, and one at a restaurant. Nate buys computer games on the way to work. Nick is allergic to cat hair. John receives free meals at his job. Who works at which job?

2. Six friends live in consecutive apartments on the same side of their apartment building. Anita lives in apartment C. Kelli's apartment is just past Scott's. Anita's closest neighbors are Eric and Ava. Scott's apartment is not A through D. Eric's apartment is before Ava's. If Roberto lives in one of the apartments, who lives in which apartment?

2-5 Postulates and Paragraph Proofs

What You'll Learn

- Identify and use basic postulates about points, lines, and planes.
- Write paragraph proofs.

Vocabulary

- postulate
- axiom
- theorem
- proof
- paragraph proof
- informal proof

How were postulates used by the founding fathers of the United States?

U.S. Supreme Court Justice William Douglas stated "The First Amendment makes confidence in the common sense of our people and in the maturity of their judgment the great postulate of our democracy." The writers of the constitution assumed that citizens would act and speak with common sense and maturity. Some statements in geometry also must be assumed or accepted as true.

POINTS, LINES, AND PLANES   In geometry, a **postulate**, or **axiom**, is a statement that describes a fundamental relationship between the basic terms of geometry. Postulates are accepted as true. The basic ideas about points, lines, and planes can be stated as postulates.

Postulates

2.1 Through any two points, there is exactly one line.

2.2 Through any three points not on the same line, there is exactly one plane.

Example 1 Points and Lines

COMPUTERS   Jessica is setting up a network for her father's business. There are five computers in his office. Each computer needs to be connected to every other computer. How many connections does Jessica need to make?

Explore   There are five computers, and each is connected to four others.

Plan   Draw a diagram to illustrate the solution.

Solve   Let noncollinear points A, B, C, D, and E represent the five computers. Connect each point with every other point. Then, count the number of segments.

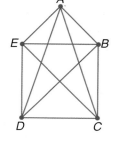

Between every two points there is exactly one segment. So, the connection between computer A and computer B is the same as the connection between computer B and computer A. For the five points, ten segments can be drawn.

Examine $\overline{AB}$, $\overline{AC}$, $\overline{AD}$, $\overline{AE}$, $\overline{BC}$, $\overline{BD}$, $\overline{BE}$, $\overline{CD}$, $\overline{CE}$, and $\overline{DE}$ each represent a connection between two computers. So there will be ten connections among the five computers.

There are other postulates that are based on relationships among points, lines, and planes.

Postulates

2.3 A line contains at least two points.

2.4 A plane contains at least three points not on the same line.

2.5 If two points lie in a plane, then the entire line containing those points lies in that plane.

2.6 If two lines intersect, then their intersection is exactly one point.

2.7 If two planes intersect, then their intersection is a line.

Example 2 Use Postulates

Determine whether each statement is *always*, *sometimes*, or *never* true. Explain.

a. **If points *A*, *B*, and *C* lie in plane *M*, then they are collinear.**

Sometimes; *A*, *B*, and *C* do not necessarily have to be collinear to lie in plane *M*.

b. **There is exactly one plane that contains noncollinear points *P*, *Q*, and *R*.**

Always; Postulate 2.2 states that through any three noncollinear points, there is exactly one plane.

c. **There are at least two lines through points *M* and *N*.**

Never; Postulate 2.1 states that through any two points, there is exactly one line.

PARAGRAPH PROOFS Undefined terms, definitions, postulates, and algebraic properties of equality are used to prove that other statements or conjectures are true. Once a statement or conjecture has been shown to be true, it is called a **theorem**, and it can be used like a definition or postulate to justify that other statements are true.

You will study and use various methods to verify or prove statements and conjectures in geometry. A **proof** is a logical argument in which each statement you make is supported by a statement that is accepted as true. One type of proof is called a **paragraph proof** or **informal proof**. In this type of proof, you write a paragraph to explain why a conjecture for a given situation is true.

Study Tip

Proofs
Before writing a proof, you should have a plan. One strategy is to *work backward*. Start with what you want to prove, and work backward step by step until you reach the given information.

Key Concept   *Proofs*

Five essential parts of a good proof:

- State the theorem or conjecture to be proven.
- List the given information.
- If possible, draw a diagram to illustrate the given information.
- State what is to be proved.
- Develop a system of deductive reasoning.

In Lesson 1-2, you learned the relationship between segments formed by the midpoint of a segment. This statement can be proven, and the result stated as a theorem.

Example 3 Write a Paragraph Proof

Given that M is the midpoint of $\overline{PQ}$, write a paragraph proof to show that $\overline{PM} \cong \overline{MQ}$.

Given: M is the midpoint of $\overline{PQ}$.

Prove: $\overline{PM} \cong \overline{MQ}$.

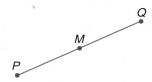

From the definition of midpoint of a segment, $PM = MQ$. This means that $\overline{PM}$ and $\overline{MQ}$ have the same measure. By the definition of congruence, if two segments have the same measure, then they are congruent. Thus, $\overline{PM} \cong \overline{MQ}$.

Once a conjecture has been proven true, it can be stated as a theorem and used in other proofs. The conjecture in Example 3 is known as the Midpoint Theorem.

Theorem 2.1

Midpoint Theorem If M is the midpoint of $\overline{AB}$, then $\overline{AM} \cong \overline{MB}$.

Check for Understanding

Concept Check
1. **Explain** how deductive reasoning is used in a proof.
2. **OPEN ENDED** Draw figures to illustrate Postulates 2.6 and 2.7.
3. **List** the types of reasons that can be used for justification in a proof.

Guided Practice **Determine the number of segments that can be drawn connecting each pair of points.**

4.

5.

6. Determine whether the following statement is *always*, *sometimes*, or *never* true. Explain.
   The intersection of three planes is two lines.

In the figure, $\overleftrightarrow{BD}$ and $\overrightarrow{BR}$ are in plane $\mathcal{P}$, and W is on $\overleftrightarrow{BD}$. State the postulate or definition that can be used to show each statement is true.

7. B, D, and W are collinear.
8. E, B, and R are coplanar.
9. R and W are collinear.

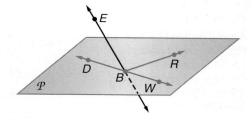

10. **PROOF** In the figure at the right, P is the midpoint of $\overline{QR}$ and $\overline{ST}$, and $\overline{QR} \cong \overline{ST}$. Write a paragraph proof to show that $PQ = PT$.

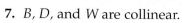

Application
11. **DANCING** Six students are participating in a dance to celebrate the opening of a new community center. The students, each connected to each of the other students with wide colored ribbons, will move in a circular motion. How many ribbons are needed?

Homework Help

For Exercises	See Examples
12–15	1
16–21	2
22–28	3

Extra Practice
See page 757.

Determine the number of segments that can be drawn connecting each pair of points.

12.

13.

14.

15.

Determine whether the following statements are *always*, *sometimes*, or *never* true. Explain.

16. Three points determine a plane.

17. Points *G* and *H* are in plane *X*. Any point collinear with *G* and *H* is in plane *X*.

18. The intersection of two planes can be a point.

19. Points *S*, *T*, and *U* determine three lines.

20. Points *A* and *B* lie in at least one plane.

21. If line ℓ lies in plane $\mathcal{P}$ and line *m* lies in plane *Q*, then lines ℓ and *m* lie in plane $\mathcal{R}$.

In the figure at the right, $\overleftrightarrow{AC}$ and $\overleftrightarrow{BD}$ lie in plane $\mathcal{J}$, and $\overleftrightarrow{BY}$ and $\overleftrightarrow{CX}$ lie in plane $\mathcal{K}$. State the postulate that can be used to show each statement is true.

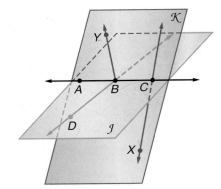

22. *C* and *D* are collinear.

23. $\overleftrightarrow{XB}$ lies in plane $\mathcal{K}$.

24. Points *A*, *C*, and *X* are coplanar.

25. $\overleftrightarrow{AD}$ lies in plane $\mathcal{J}$.

26. *X* and *Y* are collinear.

27. Points *Y*, *D*, and *C* are coplanar.

28. **PROOF**   Point *C* is the midpoint of $\overline{AB}$ and *B* is the midpoint of $\overline{CD}$. Prove that $\overline{AC} \cong \overline{BD}$.

29. **MODELS**   Faith's teacher asked her to make a figure showing the number of lines and planes formed from four points that are noncollinear and noncoplanar. Faith decided to make a mobile of straws, pipe cleaners, and colored sheets of tissue paper. She plans to glue the paper to the straws and connect the straws together to form a group of connected planes. How many planes and lines will she have?

30. **CAREERS**   Many professions use deductive reasoning and paragraph proofs. For example, a police officer uses deductive reasoning investigating a traffic accident and then writes the findings in a report. List a profession, and describe how it can use paragraph proofs.

31. CRITICAL THINKING You know that three noncollinear points lie in a single plane. In Exercise 29, you found the number of planes defined by four noncollinear points. What are the least and greatest number of planes defined by five noncollinear points?

32. WRITING IN MATH Answer the question that was posed at the beginning of the lesson.

How are postulates used in literature?

Include the following in your answer:
- an example of a postulate in historic United States' documents, and
- an example of a postulate in mathematics.

33. Which statement cannot be true?
- Ⓐ A plane can be determined using three noncollinear points.
- Ⓑ Two lines intersect at exactly one point.
- Ⓒ At least two lines can contain the same two points.
- Ⓓ A midpoint divides a segment into two congruent segments.

34. ALGEBRA For all values of x, $(8x^4 - 2x^2 + 3x - 5) - (2x^4 + x^3 + 3x + 5) =$
- Ⓐ $6x^4 - x^3 - 2x^2 - 10$.
- Ⓒ $6x^4 + x^3 - 2x^2 + 6x$.
- Ⓑ $6x^4 - 3x^2 + 6x - 10$.
- Ⓓ $6x^4 - 3x^2$.

Maintain Your Skills

Mixed Review

35. Determine whether statement (3) follows from statements (1) and (2) by the Law of Detachment or the Law of Syllogism. If it does, state which law was used. If it does not, write *invalid*. *(Lesson 2-4)*
(1) Part-time jobs require 20 hours of work per week.
(2) Jamie has a part-time job.
(3) Jamie works 20 hours per week.

Write the converse, inverse, and contrapositive of each conditional statement. Determine whether each related conditional is *true* **or** *false*. **If a statement is false, find a counterexample.** *(Lesson 2-3)*

36. If you have access to the Internet at your house, then you have a computer.

37. If $\triangle ABC$ is a right triangle, one of its angle measures is greater than 90.

38. BIOLOGY Use a Venn diagram to illustrate the following statement.
If an animal is a butterfly, then it is an arthropod. *(Lesson 2-2)*

Use the Distance Formula to find the distance between each pair of points.
(Lesson 1-3)

39. $D(3, 3)$, $F(4, -1)$ | **40.** $M(0, 2)$, $N(-5, 5)$

41. $P(-8, 2)$, $Q(1, -3)$ | **42.** $R(-5, 12)$, $S(2, 1)$

Getting Ready for the Next Lesson

PREREQUISITE SKILL Solve each equation.
*(To review **solving equations**, see pages 737 and 738.)*

43. $m - 17 = 8$ | **44.** $3y = 57$ | **45.** $\frac{y}{6} + 12 = 14$

46. $-t + 3 = 27$ | **47.** $8n - 39 = 41$ | **48.** $-6x + 33 = 0$

2-6 Algebraic Proof

What You'll Learn

- Use algebra to write two-column proofs.
- Use properties of equality in geometry proofs.

How is mathematical evidence similar to evidence in law?

Lawyers develop their cases using logical arguments based on evidence to lead a jury to a conclusion favorable to their case. At the end of a trial, a lawyer will make closing remarks summarizing the evidence and testimony that they feel proves their case. These closing arguments are similar to a proof in mathematics.

Vocabulary
- deductive argument
- two-column proof
- formal proof

Study Tip

Commutative and Associative Properties
Throughout this text, we shall assume the Commutative and Associative Properties for addition and multiplication.

ALGEBRAIC PROOF Algebra is a system with sets of numbers, operations, and properties that allow you to perform algebraic operations.

Concept Summary	Properties of Equality for Real Numbers
Reflexive Property	For every number a, $a = a$.
Symmetric Property	For all numbers a and b, if $a = b$, then $b = a$.
Transitive Property	For all numbers a, b, and c, if $a = b$ and $b = c$, then $a = c$.
Addition and Subtraction Properties	For all numbers a, b, and c, if $a = b$, then $a + c = b + c$ and $a - c = b - c$.
Multiplication and Division Properties	For all numbers a, b, and c, if $a = b$, then $a \cdot c = b \cdot c$ and if $c \neq 0$, $\frac{a}{c} = \frac{b}{c}$.
Substitution Property	For all numbers a and b, if $a = b$, then a may be replaced by b in any equation or expression
Distributive Property	For all numbers a, b, and c, $a(b + c) = ab + ac$.

The properties of equality can be used to justify each step when solving an equation. A group of algebraic steps used to solve problems form a **deductive argument**.

Example 1 Verify Algebraic Relationships

Solve $3(x - 2) = 42$.

Algebraic Steps	Properties
$3(x - 2) = 42$	Original equation
$3x - 6 = 42$	Distributive Property
$3x - 6 + 6 = 42 + 6$	Addition Property
$3x = 48$	Substitution Property
$\dfrac{3x}{3} = \dfrac{48}{3}$	Division Property
$x = 16$	Substitution Property

Example 1 is a proof of the conditional statement *If $5x + 3(x - 2) = 42$, then $x = 6$.* Notice that the column on the left is a step-by-step process that leads to a solution. The column on the right contains the reason for each statement.

In geometry, a similar format is used to prove conjectures and theorems. A **two-column proof**, or **formal proof**, contains statements and reasons organized in two columns. In a two-column proof, each step is called a *statement*, and the properties that justify each step are called *reasons*.

Example 2 Write a Two-Column Proof

Write a two-column proof.

a. If $3\left(x - \frac{5}{3}\right) = 1$, then $x = 2$

Statements	**Reasons**
1. $3\left(x - \frac{5}{3}\right) = 1$	1. Given
2. $3x - 3\left(\frac{5}{3}\right) = 1$	2. Distributive Property
3. $3x - 5 = 1$	3. Substitution
4. $3x - 5 + 5 = 1 + 5$	4. Addition Property
5. $3x = 6$	5. Substitution
6. $\frac{3x}{3} = \frac{6}{3}$	6. Division Property
7. $x = 2$	7. Substitution

b. Given: $\frac{7}{2} - n = 4 - \frac{1}{2}n$

Prove: $n = -1$

Proof:

Statements	**Reasons**
1. $\frac{7}{2} - n = 4 - \frac{1}{2}n$	1. Given
2. $2\left(\frac{7}{2} - n\right) = 2\left(4 - \frac{1}{2}n\right)$	2. Multiplication Property
3. $7 - 2n = 8 - n$	3. Distributive Property
4. $7 - 2n + n = 8 - n + n$	4. Addition Property
5. $7 - n = 8$	5. Substitution
6. $7 - n - 7 = 8 - 7$	6. Subtraction Property
7. $-n = 1$	7. Substitution
8. $\frac{-n}{-1} = \frac{1}{-1}$	8. Division Property
9. $n = -1$	9. Substitution

Study Tip

Mental Math
If your teacher permits you to do so, some steps may be eliminated by performing mental calculations. For example, in part **a** of Example 2, statements 4 and 6 could be omitted. Then the reason for statements 5 would be Addition Property and Division Property for statement 7.

GEOMETRIC PROOF Since geometry also uses variables, numbers, and operations, many of the properties of equality used in algebra are also true in geometry. For example, segment measures and angle measures are real numbers, so properties from algebra can be used to discuss their relationships. Some examples of these applications are shown below.

Property	Segments	Angles
Reflexive	$AB = AB$	$m\angle 1 = m\angle 1$
Symmetric	If $AB = CD$, then $CD = AB$.	If $m\angle 1 = m\angle 2$, then $m\angle 2 = m\angle 1$.
Transitive	If $AB = CD$ and $CD = EF$, then $AB = EF$.	If $m\angle 1 = m\angle 2$ and $m\angle 2 = m\angle 3$, then $m\angle 1 = m\angle 3$.

Example 3 *Justify Geometric Relationships*

Multiple-Choice Test Item

If $\overline{AB} \cong \overline{CD}$, and $\overline{CD} \cong \overline{EF}$, then which of the following is a valid conclusion?

   I $AB = CD$ and $CD = EF$

   II $\overline{AB} \cong \overline{EF}$

   III $AB = EF$

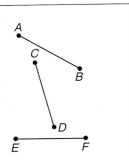

  Ⓐ I only   Ⓑ I and II

  Ⓒ I and III   Ⓓ I, II, and III

Test-Taking Tip

More than one statement may be correct. Work through each problem completely before indicating your answer.

Read the Test Item

Determine whether the statements are true based on the given information.

Solve the Test Item

Statement I:

Examine the given information, $\overline{AB} \cong \overline{CD}$ and $\overline{CD} \cong \overline{EF}$. From the definition of congruent segments, if $\overline{AB} \cong \overline{CD}$ and $\overline{CD} \cong \overline{EF}$, then $AB = CD$ and $CD = EF$. Thus, Statement I is true.

Statement II:

By the definition of congruent segments, if $AB = EF$, then $\overline{AB} \cong \overline{EF}$. Statement II is true also.

Statement III:

If $AB = CD$ and $CD = EF$, then $AB = EF$ by the Transitive Property. Thus, Statement III is true.

Because Statements I, II, and III are true, choice D is correct.

In Example 3, each conclusion was justified using a definition or property. This process is used in geometry to verify and prove statements.

Example 4 *Geometric Proof*

TIME On a clock, the angle formed by the hands at 2:00 is a 60° angle. If the angle formed at 2:00 is congruent to the angle formed at 10:00, prove that the angle at 10:00 is a 60° angle.

Given: $m\angle 2 = 60$
          $\angle 2 \cong \angle 10$

Prove: $m\angle 10 = 60$

Proof:

Statements	Reasons
1. $m\angle 2 = 60$ $\angle 2 \cong \angle 10$	1. Given
2. $m\angle 2 = m\angle 10$	2. Definition of congruent angles
3. $60 = m\angle 10$	3. Substitution
4. $m\angle 10 = 60$	4. Symmetric Property

Concept Check

1. **OPEN ENDED** Write a statement that illustrates the Substitution Property of Equality.

2. **Describe** the parts of a two-column proof.

3. **State** the part of a conditional that is related to the *Given* statement of a proof. What part is related to the *Prove* statement?

Guided Practice **State the property that justifies each statement.**

4. If $2x = 5$, then $x = \dfrac{5}{2}$.

5. If $\dfrac{x}{2} = 7$, then $x = 14$.

6. If $x = 5$ and $b = 5$, then $x = b$.

7. If $XY - AB = WZ - AB$, then $XY = WZ$.

8. Solve $\dfrac{x}{2} + 4x - 7 = 11$. List the property that justifies each step.

9. Complete the following proof.

Given: $5 - \dfrac{2}{3}x = 1$

Prove: $x = 6$

Proof:

Statements	Reasons
a. ___?___	a. Given
b. $3\left(5 - \dfrac{2}{3}x\right) = 3(1)$	b. ___?___
c. $15 - 2x = 3$	c. ___?___
d. ___?___	d. Subtraction Prop.
e. $x = 6$	e. ___?___

PROOF Write a two-column proof.

10. Prove that if $25 = -7(y - 3) + 5y$, then $-2 = y$.

11. If rectangle $ABCD$ has side lengths $AD = 3$ and $AB = 10$, then $AC = BD$.

12. The Pythagorean Theorem states that in a right triangle ABC, $c^2 = a^2 + b^2$. Prove that $a = \sqrt{c^2 - b^2}$.

Standardized Test Practice
Ⓐ Ⓑ Ⓒ Ⓓ

13. **ALGEBRA** If $8 + x = 12$, then $4 - x = $ ___?___ .

  Ⓐ 28   Ⓑ 24   Ⓒ 0   Ⓓ 4

Practice and Apply

Homework Help

For Exercises	See Examples
15, 16, 20	1
14, 17–19, 21	2
22–27	3
28, 29	4

Extra Practice
See page 757.

State the property that justifies each statement.

14. If $m\angle A = m\angle B$ and $m\angle B = m\angle C$, $m\angle A = m\angle C$.

15. If $HJ + 5 = 20$, then $HJ = 15$.

16. If $XY + 20 = YW$ and $XY + 20 = DT$, then $YW = DT$.

17. If $m\angle 1 + m\angle 2 = 90$ and $m\angle 2 = m\angle 3$, then $m\angle 1 + m\angle 3 = 90$.

18. If $\dfrac{1}{2}AB = \dfrac{1}{2}EF$, then $AB = EF$.

19. $AB = AB$

20. If $2\left(x - \frac{3}{2}\right) = 5$, which property can be used to support the statement $2x - 3 = 5$?

21. Which property allows you to state $m\angle 4 = m\angle 5$, if $m\angle 4 = 35$ and $m\angle 5 = 35$?

22. If $\frac{1}{2}AB = \frac{1}{2}CD$, which property can be used to justify the statement $AB = CD$?

23. Which property could be used to support the statement $EF = JK$, given that $EF = GH$ and $GH = JK$?

Complete each proof.

24. **Given:** $\dfrac{3x + 5}{2} = 7$

Prove: $x = 3$

Proof:

Statements	Reasons
a. $\dfrac{3x + 5}{2} = 7$	**a.** __?__
b. __?__	**b.** Mult. Prop.
c. $3x + 5 = 14$	**c.** __?__
d. $3x = 9$	**d.** __?__
e. __?__	**e.** Div. Prop.

25. **Given:** $2x - 7 = \frac{1}{3}x - 2$

Prove: $x = 3$

Proof:

Statements	Reasons
a. __?__	**a.** Given
b. __?__	**b.** Mult. Prop.
c. $6x - 21 = x - 6$	**c.** __?__
d. __?__	**d.** Subt. Prop.
e. $5x = 15$	**e.** __?__
f. __?__	**f.** Div. Prop.

PROOF **Write a two-column proof.**

26. If $4 - \frac{1}{2}a = \frac{7}{2} - a$, then $a = -1$.

27. If $-2y + \frac{3}{2} = 8$, then $y = -\frac{13}{4}$.

28. If $-\frac{1}{2}m = 9$, then $m = -18$.

29. If $5 - \frac{2}{3}z = 1$, then $z = 6$.

30. If $XZ = ZY$, $XZ = 4x + 1$, and $ZY = 6x - 13$, then $x = 7$.

31. If $m\angle ACB = m\angle ABC$, then $m\angle XCA = m\angle YBA$.

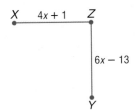

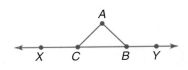

More About. . .

Physics •·············

A gymnast exhibits kinetic energy when performing on the balance beam. The movements and flips show the energy that is being displayed while the gymnast is moving.

Source: www.infoplease.com

32. **PHYSICS** Kinetic energy is the energy of motion. The formula for kinetic energy is $E_k = h \cdot f + W$, where h represents Planck's Constant, f represents the frequency of its photon, and W represents the work function of the material being used. Solve this formula for f and justify each step.

33. GARDENING Areas in the southwest and southeast have cool but mild winters. In these areas, many people plant pansies in October so that they have flowers outside year-round. In the arrangement of pansies shown, the walkway divides the two sections of pansies into four beds that are the same size. If $m\angle ACB = m\angle DCE$, what could you conclude about the relationship among $\angle ACB$, $\angle DCE$, $\angle ECF$, and $\angle ACG$?

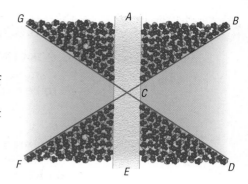

CRITICAL THINKING **For Exercises 34 and 35, use the following information.**
Below is a family tree of the Gibbs family. Clara, Carol, Cynthia, and Cheryl are all daughters of Lucy. Because they are sisters, they have a transitive and symmetric relationship. That is, Clara is a sister of Carol, Carol is a sister of Cynthia, so Clara is a sister of Cynthia.

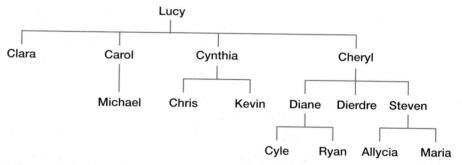

34. What other relationships in a family have reflexive, symmetric, or transitive relationships? Explain why. Remember that the child or children of each person are listed beneath that person's name. Consider relationships such as first cousin, ancestor or descendent, aunt or uncle, sibling, or any other relationship.

35. Construct your family tree on one or both sides of your family and identify the reflexive, symmetric, or transitive relationships.

36. WRITING IN MATH   Answer the question that was posed at the beginning of the lesson.
How is mathematical evidence similar to evidence in law?
Include the following in your answer:
• a description of how evidence is used to influence jurors' conclusions in court, and
• a description of the evidence used to make conclusions in mathematics.

37. In $\triangle PQR$, $m\angle P = m\angle Q$ and $m\angle R = 2(m\angle Q)$. Find $m\angle P$ if $m\angle P + m\angle Q + m\angle R = 180$.
Ⓐ 30
Ⓑ 45
Ⓒ 60
Ⓓ 90

38. ALGEBRA If 4 more than x is 5 less than y, what is x in terms of y?
Ⓐ $y - 1$
Ⓑ $y - 9$
Ⓒ $y + 9$
Ⓓ $y - 5$

Mixed Review **39. CONSTRUCTION** There are four buildings on the Medfield High School Campus, no three of which stand in a straight line. How many sidewalks need to be built so that each building is directly connected to every other building? *(Lesson 2-5)*

Determine whether the stated conclusion is valid based on the given information. If not, write *invalid.* **Explain your reasoning.** *A number is divisible by 3 if it is divisible by 6.* *(Lesson 2-4)*

40. Given: 24 is divisible by 6. **Conclusion:** 24 is divisible by 3.

41. Given: 27 is divisible by 3. **Conclusion:** 27 is divisible by 6.

42. Given: 85 is not divisible by 3. **Conclusion:** 85 is not divisible by 6.

Write each statement in if-then form. *(Lesson 2-3)*

43. "Happy people rarely correct their faults." (*La Rochefoucauld*)

44. "If you don't know where you are going, you will probably end up somewhere else." (*Laurence Peters*)

45. "A champion is afraid of losing." (*Billie Jean King*)

46. "If we would have new knowledge, we must get a whole new world of questions." (*Susanne K. Langer*)

Find the precision for each measurement. *(Lesson 1-2)*

47. 13 feet **48.** 5.9 meters **49.** 74 inches **50.** 3.1 kilometers

Getting Ready for the Next Lesson **PREREQUISITE SKILL** Find the measure of each segment. *(To review segment measures, see Lesson 1-2.)*

51. $\overline{KL}$ **52.** $\overline{QS}$ **53.** $\overline{WZ}$

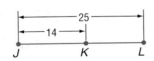

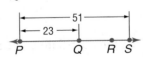

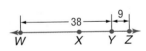

Practice Quiz 2   Lessons 2-4 through 2-6

1. Determine whether statement (3) follows from statements (1) and (2) by the Law of Detachment or the Law of Syllogism. If it does, state which law was used. If it does not, write *invalid.* *(Lesson 2-4)*

(1) If n is an integer, then n is a real number.

(2) n is a real number.

(3) n is an integer.

In the figure at the right, A, B, **and** C **are collinear. Points** A, B, C, **and** D **lie in plane** N. **State the postulate or theorem that can be used to show each statement is true.** *(Lesson 2-5)*

2. A, B, and D determine plane N.

3. $\overleftrightarrow{BE}$ intersects $\overleftrightarrow{AC}$ at B.

4. ℓ lies in plane N.

5. **PROOF** If $2(n - 3) + 5 = 3(n - 1)$, prove that $n = 2$. *(Lesson 2-6)*

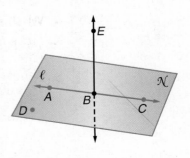

2-7 Proving Segment Relationships

What You'll Learn

- Write proofs involving segment addition.
- Write proofs involving segment congruence.

How can segment relationships be used for travel?

When leaving San Diego, the pilot said that the flight would be about 360 miles to Phoenix before continuing on to Dallas. When the plane left Phoenix, the pilot said that the flight would be flying about 1070 miles to Dallas.

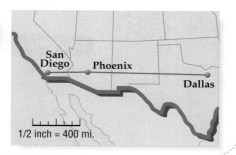

1/2 inch = 400 mi.

SEGMENT ADDITION   In Lesson 1-2, you measured segments with a ruler by placing the mark for zero on one endpoint, then finding the distance to the other endpoint. This illustrates the **Ruler Postulate.**

Postulate 2.8

Ruler Postulate   The points on any line or line segment can be paired with real numbers so that, given any two points *A* and *B* on a line, *A* corresponds to zero, and *B* corresponds to a positive real number.

The Ruler Postulate can be used to further investigate line segments.

Geometry Software Investigation

Adding Segment Measures

Construct a Figure

- Use The Geometer's Sketchpad to construct $\overline{AC}$.
- Place point *B* on $\overline{AC}$.
- Find *AB, BC,* and *AC.*

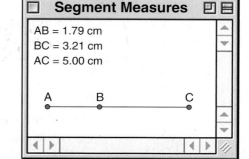

Segment Measures

AB = 1.79 cm
BC = 3.21 cm
AC = 5.00 cm

Analyze the Model

1. What is the sum *AB* + *BC*?
2. Move *B*. Find *AB, BC* and *AC*. What is the sum of *AB* + *BC*?
3. Repeat moving *B*, measuring the segments, and finding the sum *AB* + *BC* three times. Record your results.

Make a Conjecture

4. What is true about the relationship of *AB, BC,* and *AC*?
5. Is it possible to place *B* on $\overline{AC}$ so that this relationship is not true?

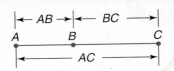
Betweenness
In general, the definition of *between* is that *B* is between *A* and *C* if *A*, *B*, and *C* are collinear and $AB + BC = AC$.

Examine the measures AB, BC, and AC in the Geometry Activity. Notice that wherever B is placed between A and C, $AB + BC = AC$. This suggests the following postulate.

Postulate 2.9

Segment Addition Postulate   If B is between A and C, then $AB + BC = AC$.

If $AB + BC = AC$, then B is between A and C.

Example 1 Proof With Segment Addition

Prove the following.

Given: $PQ = RS$

Prove: $PR = QS$

Proof:

Statements	Reasons
1. $PQ = RS$	1. Given
2. $PQ + QR = QR + RS$	2. Addition Property
3. $PQ + QR = PR$ $QR + RS = QS$	3. Segment Addition Postulate
4. $PR = QS$	4. Substitution

SEGMENT CONGRUENCE   In Lesson 2-5, you learned that once a theorem is proved, it can be used in proofs of other theorems. One theorem we can prove is similar to properties of equality from algebra.

Theorem 2.2   Segment Congruence

Congruence of segments is reflexive, symmetric, and transitive.

Reflexive Property $\overline{AB} \cong \overline{AB}$

Symmetric Property   If $\overline{AB} \cong \overline{CD}$, then $\overline{CD} \cong \overline{AB}$.

Transitive Property   If $\overline{AB} \cong \overline{CD}$, and $\overline{CD} \cong \overline{EF}$, then $\overline{AB} \cong \overline{EF}$.

You will prove the first two properties in Exercises 10 and 24.

Proof Transitive Property of Congruence

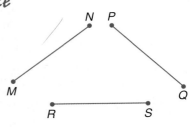

Given: $\overline{MN} \cong \overline{PQ}$
            $\overline{PQ} \cong \overline{RS}$

Prove: $\overline{MN} \cong \overline{RS}$

Proof:

Method 1   Paragraph Proof

Since $\overline{MN} \cong \overline{PQ}$ and $\overline{PQ} \cong \overline{RS}$, $MN = PQ$ and $PQ = RS$ by the definition of congruent segments. By the Transitive Property of Equality, $MN = RS$. Thus, $\overline{MN} \cong \overline{RS}$ by the definition of congruent segments.

Method 2 Two-Column Proof

Statements	Reasons
1. $\overline{MN} \cong \overline{PQ}, \overline{PQ} \cong \overline{RS}$	1. Given
2. $MN = PQ, PQ = RS$	2. Definition of congruent segments
3. $MN = RS$	3. Transitive Property
4. $\overline{MN} \cong \overline{RS}$	4. Definition of congruent segments

The theorems about segment congruence can be used to prove segment relationships.

Example **2** *Proof With Segment Congruence*

Prove the following.
Given: $\overline{JK} \cong \overline{KL}, \overline{HJ} \cong \overline{GH}, \overline{KL} \cong \overline{HJ}$
Prove: $\overline{GH} \cong \overline{JK}$
Proof:

Statements	Reasons
1. $\overline{JK} \cong \overline{KL}, \overline{KL} \cong \overline{HJ}$	1. Given
2. $\overline{JK} \cong \overline{HJ}$	2. Transitive Property
3. $\overline{HJ} \cong \overline{GH}$	3. Given
4. $\overline{JK} \cong \overline{GH}$	4. Transitive Property
5. $\overline{GH} \cong \overline{JK}$	5. Symmetric Property

Check for Understanding

Concept Check

1. Choose two cities from a United States road map. Describe the distance between the cities using the Reflexive Property.

2. **OPEN ENDED** Draw three congruent segments, and illustrate the Transitive Property using these segments.

3. **Describe** how to determine whether a point B is between points A and C.

Guided Practice

Justify each statement with a property of equality or a property of congruence.

4. $\overline{XY} \cong \overline{XY}$

5. If $\overline{GH} \cong \overline{MN}$, then $\overline{MN} \cong \overline{GH}$.

6. If $AB = AC + CB$, then $AB - AC = CB$.

7. Copy and complete the proof.
Given: $\overline{PQ} \cong \overline{RS}, \overline{QS} \cong \overline{ST}$
Prove: $\overline{PS} \cong \overline{RT}$

Proof:

Statements	Reasons
a. ___?___ , ___?___	a. Given
b. $PQ = RS, QS = ST$	b. ___?___
c. $PS = PQ + QS, RT = RS + ST$	c. ___?___
d. ___?___	d. Addition Property
e. ___?___	e. Substitution
f. $\overline{PS} \cong \overline{RT}$	f. ___?___

PROOF For Exercises 8–10, write a two-column proof.

8. **Given:** $\overline{AP} \cong \overline{CP}$
   $\overline{BP} \cong \overline{DP}$

   Prove: $\overline{AB} \cong \overline{CD}$

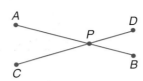

9. **Given:** $\overline{HI} \cong \overline{TU}$
   $\overline{HJ} \cong \overline{TV}$

   Prove: $\overline{IJ} \cong \overline{UV}$

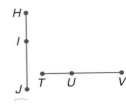

10. Symmetric Property of Congruence (Theorem 2.2)

Application 11. **GEOGRAPHY** Aberdeen in South Dakota and Helena, Miles City, and Missoula, all in Montana, are connected in a straight line by interstate highways. Missoula is 499 miles from Miles City and 972 miles from Aberdeen. Aberdeen is 473 miles from Miles City and 860 miles from Helena. Between which cities does Helena lie?

Practice and Apply

Homework Help

For Exercises	See Examples
14, 16, 17	1
12, 13, 15, 18–24	2

Extra Practice
See page 758.

Justify each statement with a property of equality or a property of congruence.

12. If $\overline{JK} \cong \overline{LM}$, then $\overline{LM} \cong \overline{JK}$.

13. If $AB = 14$ and $CD = 14$, then $AB = CD$.

14. If W, X, and Y are collinear, in that order, then $WY = WX + XY$.

15. If $\overline{MN} \cong \overline{PQ}$ and $\overline{PQ} \cong \overline{RS}$, then $\overline{MN} \cong \overline{RS}$.

16. If $EF = TU$ and $GH = VW$, then $EF + GH = TU + VW$.

17. If $JK + MN = JK + QR$, then $MN = QR$.

18. Copy and complete the proof.

   Given: $\overline{AD} \cong \overline{CE}$, $\overline{DB} \cong \overline{EB}$

   Prove: $\overline{AB} \cong \overline{CB}$

   Proof:

Statements	Reasons
a. _?_	a. Given
b. $AD = CE$, $DB = EB$	b. _?_
c. $AD + DB = CE + EB$	c. _?_
d. _?_	d. Segment Addition Postulate
e. $AB = CB$	e. _?_
f. $\overline{AB} \cong \overline{CB}$	f. _?_

PROOF Write a two-column proof.

19. If $\overline{XY} \cong \overline{WZ}$ and $\overline{WZ} \cong \overline{AB}$, then $\overline{XY} \cong \overline{AB}$.

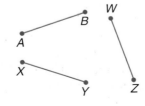

20. If $\overline{AB} \cong \overline{AC}$ and $\overline{PC} \cong \overline{QB}$, then $\overline{AP} \cong \overline{AQ}$.

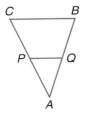

21. Copy and complete the proof.

Given: $\overline{WY} \cong \overline{ZX}$
A is the midpoint of $\overline{WY}$.
A is the midpoint of $\overline{ZX}$.

Prove: $\overline{WA} \cong \overline{ZA}$

Proof:

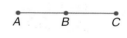

Statements	Reasons
a. $\overline{WY} \cong \overline{ZX}$ A is the midpoint of $\overline{WY}$. A is the midpoint of $\overline{ZX}$.	**a.** _?_
b. $WY = ZX$	**b.** _?_
c. _?_	**c.** Definition of midpoint
d. $WY = WA + AY$, $ZX = ZA + AX$	**d.** _?_
e. $WA + AY = ZA + AX$	**e.** _?_
f. $WA + WA = ZA + ZA$	**f.** _?_
g. $2WA = 2ZA$	**g.** _?_
h. _?_	**h.** Division Property
i. $\overline{WA} \cong \overline{ZA}$	**i.** _?_

PROOF For Exercises 22–24, write a two-column proof.

22. If $\overline{LM} \cong \overline{PN}$ and $\overline{XM} \cong \overline{XN}$,
then $\overline{LX} \cong \overline{PX}$.

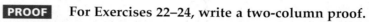

23. If $AB = BC$,
then $AC = 2BC$.

24. Reflexive Property of Congruence (Theorem 2.2)

25. **DESIGN** The front of a building has a triangular window. If $\overline{AB} \cong \overline{DE}$ and C is the midpoint of $\overline{BD}$, prove that $\overline{AC} \cong \overline{CE}$.

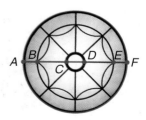

26. **LIGHTING** The light fixture in Gerrard Hall of the University of North Carolina is shown at the right. If $\overline{AB} \cong \overline{EF}$ and $\overline{BC} \cong \overline{DE}$, prove that $\overline{AC} \cong \overline{DF}$.

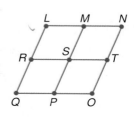

27. **CRITICAL THINKING** Given that $\overline{LN} \cong \overline{RT}$, $\overline{RT} \cong \overline{QO}$, $\overline{LQ} \cong \overline{NO}$, $\overline{MP} \cong \overline{NO}$, S is the midpoint of $\overline{RT}$, M is the midpoint of $\overline{LN}$, and P is the midpoint of $\overline{QO}$, list three statements that you could prove using the postulates, theorems, and definitions that you have learned.

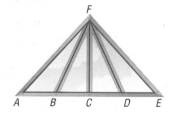

28. **WRITING IN MATH** Answer the question that was posed at the beginning of the lesson.

How can segment relationships be used for travel?

Include the following in your answer:
- an explanation of how a passenger can use the distances the pilot announced to find the total distance from San Diego to Dallas, and
- an explanation of why the Segment Addition Postulate may or may not be useful when traveling.

29. If P is the midpoint of $\overline{BC}$ and Q is the midpoint of $\overline{AD}$, what is PQ?

Ⓐ $\frac{1}{2}$

Ⓑ 1

Ⓒ 2

Ⓓ $2\frac{1}{2}$

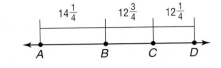

30. **GRID IN** A refreshment stand sells a large tub of popcorn for twice the price of a box of popcorn. If 60 tubs were sold for a total of $150 and the total popcorn sales were $275, how many boxes of popcorn were sold?

Maintain Your Skills

Mixed Review **State the property that justifies each statement.** *(Lesson 2-6)*

31. If $m\angle P + m\angle Q = 110$ and $m\angle R = 110$, then $m\angle P + m\angle Q = m\angle R$.

32. If $x(y + z) = a$, then $xy + xz = a$.

33. If $n - 17 = 39$, then $n = 56$.

34. If $cv = md$ and $md = 15$, then $cv = 15$.

Determine whether the following statements are *always*, *sometimes*, or *never* true. Explain. *(Lesson 2-5)*

35. A midpoint divides a segment into two noncongruent segments.

36. Three lines intersect at a single point.

37. The intersection of two planes forms a line.

38. Three single points determine three lines.

39. If the perimeter of rectangle $ABCD$ is 44 centimeters, find x and the dimensions of the rectangle.
(Lesson 1-6)

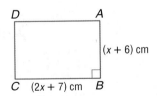

Getting Ready for the Next Lesson **PREREQUISITE SKILL** Find x.
*(To review **complementary and supplementary angles**, see Lesson 1-5.)*

40.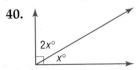

41.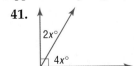

42.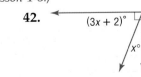

43.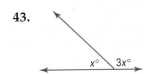

44.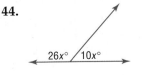

45.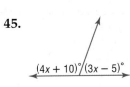

Proving Angle Relationships

What You'll Learn

- Write proofs involving supplementary and complementary angles.
- Write proofs involving congruent and right angles.

How do scissors illustrate supplementary angles?

Notice that when a pair of scissors is opened, the angle formed by the two blades, $\angle 1$, and the angle formed by a blade and a handle, $\angle 2$, are a linear pair. Likewise, the angle formed by a blade and a handle, $\angle 2$, and the angle formed by the two handles, $\angle 3$, also forms a linear pair.

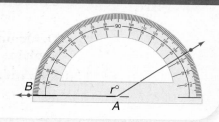

SUPPLEMENTARY AND COMPLEMENTARY ANGLES Recall that when you measure angles with a protractor, you position the protractor so that one of the rays aligns with zero degrees and then determine the position of the second ray. This illustrates the Protractor Postulate.

Postulate 2.10

Protractor Postulate Given $\overrightarrow{AB}$ and a number r between 0 and 180, there is exactly one ray with endpoint A, extending on either side of $\overrightarrow{AB}$, such that the measure of the angle formed is r.

In Lesson 2-7, you learned about the Segment Addition Postulate. A similar relationship exists between the measures of angles.

Postulate 2.11

Angle Addition Postulate If R is in the interior of $\angle PQS$, then $m\angle PQR + m\angle RQS = m\angle PQS$.

If $m\angle PQR + m\angle RQS = m\angle PQS$, then R is in the interior of $\angle PQS$.

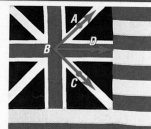
Example 1 Angle Addition

• **HISTORY** The Grand Union Flag at the left contains several angles. If $m\angle ABD = 44$ and $m\angle ABC = 88$, find $m\angle DBC$.

$$m\angle ABD + m\angle DBC = m\angle ABC \qquad \text{Angle Addition Postulate}$$
$$44 + m\angle DBC = 88 \qquad m\angle ABD = 44, m\angle ABC = 88$$
$$m\angle DBC = 44 \qquad \text{Subtraction Property}$$

The Angle Addition Postulate can be used with other angle relationships to provide additional theorems relating to angles.

Study Tip

Look Back
To review **supplementary** and **complementary angles**, see Lesson 1-5.

Theorems

2.3 **Supplement Theorem** If two angles form a linear pair, then they are supplementary angles.

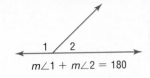

$m\angle 1 + m\angle 2 = 180$

2.4 **Complement Theorem** If the noncommon sides of two adjacent angles form a right angle, then the angles are complementary angles.

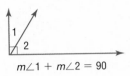

$m\angle 1 + m\angle 2 = 90$

You will prove Theorems 2.3 and 2.4 in Exercises 10 and 11.

Example 2 Supplementary Angles

If $\angle 1$ and $\angle 2$ form a linear pair and $m\angle 2 = 67$, find $m\angle 1$.

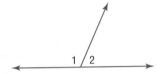

$m\angle 1 + m\angle 2 = 180$ Supplement Theorem

$m\angle 1 + 67 = 180$ $m\angle 2 = 67$

$m\angle 1 = 113$ Subtraction Property

CONGRUENT AND RIGHT ANGLES The properties of algebra that applied to the congruence of segments and the equality of their measures also hold true for the congruence of angles and the equality of their measures.

Theorem 2.5

Congruence of angles is reflexive, symmetric, and transitive.

Reflexive Property $\angle 1 \cong \angle 1$

Symmetric Property   If $\angle 1 \cong \angle 2$, then $\angle 2 \cong \angle 1$.

Transitive Property   If $\angle 1 \cong \angle 2$, and $\angle 2 \cong \angle 3$, then $\angle 1 \cong \angle 3$.

You will prove the Reflexive and Transitive Properties of Angle Congruence in Exercises 26 and 27.

Proof Symmetric Property of Congruence

Given: $\angle A \cong \angle B$

Prove: $\angle B \cong \angle A$

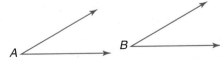

Paragraph Proof:

We are given $\angle A \cong \angle B$. By the definition of congruent angles, $m\angle A = m\angle B$. Using the Symmetric Property, $m\angle B = m\angle A$. Thus, $\angle B \cong \angle A$ by the definition of congruent angles.

Algebraic properties can be applied to prove theorems for congruence relationships involving supplementary and complementary angles.

Theorems

2.6 Angles supplementary to the same angle or to congruent angles are congruent.

Abbreviation: ⩘ suppl. to same ∠ or ≅ ⩘ are ≅.

Example: If $m\angle 1 + m\angle 2 = 180$ and $m\angle 2 + m\angle 3 = 180$, then $\angle 1 \cong \angle 3$.

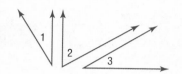

2.7 Angles complementary to the same angle or to congruent angles are congruent.

Abbreviation: ⩘ compl. to same ∠ or ≅ ⩘ are ≅.

Example: If $m\angle 1 + m\angle 2 = 90$ and $m\angle 2 + m\angle 3 = 90$, then $\angle 1 \cong \angle 3$.

You will prove Theorem 2.6 in Exercise 6.

Proof *Theorem 2.7*

Given: ∠1 and ∠3 are complementary.
∠2 and ∠3 are complementary.

Prove: $\angle 1 \cong \angle 2$

Proof:

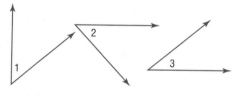

Statements	Reasons
1. ∠1 and ∠3 are complementary. ∠2 and ∠3 are complementary.	1. Given
2. $m\angle 1 + m\angle 3 = 90$ $m\angle 2 + m\angle 3 = 90$	2. Definition of complementary angles
3. $m\angle 1 + m\angle 3 = m\angle 2 + m\angle 3$	3. Substitution
4. $m\angle 3 = m\angle 3$	4. Reflexive Property
5. $m\angle 1 = m\angle 2$	5. Subtraction Property
6. $\angle 1 \cong \angle 2$	6. Definition of congruent angles

---

Example 3 *Use Supplementary Angles*

In the figure, ∠1 and ∠2 form a linear pair and ∠2 and ∠3 form a linear pair. Prove that ∠1 and ∠3 are congruent.

Given: ∠1 and ∠2 form a linear pair.
∠2 and ∠3 form a linear pair.

Prove: $\angle 1 \cong \angle 3$

Proof:

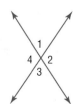

Statements	Reasons
1. ∠1 and ∠2 form a linear pair. ∠2 and ∠3 form a linear pair.	1. Given
2. ∠1 and ∠2 are supplementary. ∠2 and ∠3 are supplementary.	2. Supplement Theorem
3. $\angle 1 \cong \angle 3$	3. ⩘ suppl. to same ∠ or ≅ ⩘ are ≅.

Note that in Example 3, $\angle 1$ and $\angle 3$ are vertical angles. The conclusion in the example is a proof for the following theorem.

Study Tip

Look Back
To review **vertical angles**, see Lesson 1-5.

Theorem 2.8

Vertical Angles Theorem   If two angles are vertical angles, then they are congruent.
Abbreviation: Vert. $\angle$ are $\cong$.

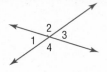

$\angle 1 \cong \angle 3$ and $\angle 2 \cong \angle 4$

Example 4 Vertical Angles

If $\angle 1$ and $\angle 2$ are vertical angles and $m\angle 1 = x$ and $m\angle 2 = 228 - 3x$, find $m\angle 1$ and $m\angle 2$.

$\angle 1 \cong \angle 2$	Vertical Angles Theorem
$m\angle 1 = m\angle 2$	Definition of congruent angles
$x = 228 - 3x$	Substitution
$4x = 228$	Add $3x$ to each side.
$x = 57$	Divide each side by 4.

$$m\angle 1 = x \qquad\qquad m\angle 2 = m\angle 1$$
$$= 57 \qquad\qquad\qquad = 57$$

The theorems you have learned can be applied to right angles. You can create right angles and investigate congruent angles by paper folding.

Geometry Activity

Right Angles

Make a Model
- Fold the paper so that one corner is folded downward.
- Fold along the crease so that the top edge meets the side edge.
- Unfold the paper and measure each of the angles formed.
- Repeat the activity three more times.

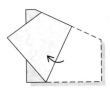

Analyze the Model
1. What do you notice about the lines formed?
2. What do you notice about each pair of adjacent angles?
3. What are the measures of the angles formed?

Make a Conjecture
4. What is true about perpendicular lines?
5. What is true about all right angles?

The following theorems support the conjectures you made in the Geometry Activity.

Theorems

2.9	Perpendicular lines intersect to form four right angles.
2.10	All right angles are congruent.
2.11	Perpendicular lines form congruent adjacent angles.
2.12	If two angles are congruent and supplementary, then each angle is a right angle.
2.13	If two congruent angles form a linear pair, then they are right angles.

Check for Understanding

Concept Check 1. **FIND THE ERROR** Tomas and Jacob wrote equations involving the angle measures shown.

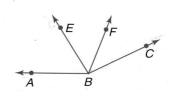

> **Tomas**
> $m\angle ABE + m\angle EBC = m\angle ABC$

> **Jacob**
> $m\angle ABE + m\angle FBC = m\angle ABC$

Who is correct? Explain your reasoning.

2. **OPEN ENDED** Draw three congruent angles. Use these angles to illustrate the Transitive Property for angle congruence.

Guided Practice **Find the measure of each numbered angle.**

3. $m\angle 1 = 65$

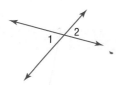

4. $\angle 6$ and $\angle 8$ are complementary. $m\angle 8 = 47$

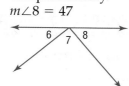

5. $m\angle 11 = x - 4$, $m\angle 12 = 2x - 5$

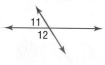

6. **PROOF** Copy and complete the proof of Theorem 2.6.

Given: $\angle 1$ and $\angle 2$ are supplementary.
$\angle 3$ and $\angle 4$ are supplementary.
$\angle 1 \cong \angle 4$

Prove: $\angle 2 \cong \angle 3$

Proof:

Statements	Reasons
a. $\angle 1$ and $\angle 2$ are supplementary. $\angle 3$ and $\angle 4$ are supplementary. $\angle 1 \cong \angle 4$	**a.** ?
b. $m\angle 1 + m\angle 2 = 180$ $m\angle 3 + m\angle 4 = 180$	**b.** ?
c. $m\angle 1 + m\angle 2 = m\angle 3 + m\angle 4$	**c.** ?
d. $m\angle 1 = m\angle 4$	**d.** ?
e. $m\angle 2 = m\angle 3$	**e.** ?
f. $\angle 2 \cong \angle 3$	**f.** ?

7. **PROOF** Write a two-column proof.

   Given: $\overrightarrow{VX}$ bisects $\angle WVY$.
   $\overrightarrow{VY}$ bisects $\angle XVZ$.

   Prove: $\angle WVX \cong \angle YVZ$

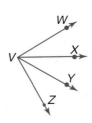

Determine whether the following statements are *always*, *sometimes*, or *never* true.

8. Two angles that are nonadjacent are __?__ vertical.

9. Two angles that are congruent are __?__ complementary to the same angle.

PROOF Write a proof for each theorem.

10. Supplement Theorem

11. Complement Theorem

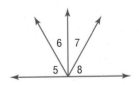

Application   **ALGEBRA**   For Exercises 12–15, use the following information.

$\angle 1$ and $\angle X$ are complementary,
$\angle 2$ and $\angle X$ are complementary,
$m\angle 1 = 2n + 2$, and $m\angle 2 = n + 32$.

12. Find n.

13. Find $m\angle 1$.

14. What is $m\angle 2$?

15. Find $m\angle X$.

Practice and Apply

Homework Help

For Exercises	See Examples
16–18	1, 2
19–24	4
25–39	3

Extra Practice
See page 758.

Find the measure of each numbered angle.

16. $m\angle 2 = 67$

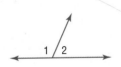

17. $m\angle 3 = 38$

18. $\angle 7$ and $\angle 8$ are complementary. $\angle 5 \cong \angle 8$ and $m\angle 6 = 29$.

19. $m\angle 9 = 2x - 4$,
    $m\angle 10 = 2x + 4$

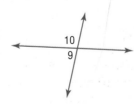

20. $m\angle 11 = 4x$,
    $m\angle 12 = 2x - 6$

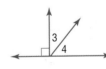

21. $m\angle 13 = 2x + 94$,
    $m\angle 14 = 7x + 49$

22. $m\angle 15 = x$,
    $m\angle 16 = 6x - 290$

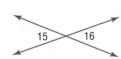

23. $m\angle 17 = 2x + 7$,
    $m\angle 18 = x + 30$

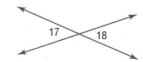

24. $m\angle 19 = 100 + 20x$,
    $m\angle 20 = 20x$

25. Prove that congruence of angles is reflexive.
26. Write a proof of the Transitive Property of Angle Congruence.

Determine whether the following statements are *always*, *sometimes*, or *never* true.
27. Two angles that are complementary __?__ form a right angle.
28. Two angles that are vertical are __?__ nonadjacent.
29. Two angles that form a right angle are __?__ complementary.
30. Two angles that form a linear pair are __?__ congruent.
31. Two angles that are supplementary are __?__ congruent.
32. Two angles that form a linear pair are __?__ supplementary.

PROOF **Use the figure to write a proof of each theorem.**
33. Theorem 2.9
34. Theorem 2.10
35. Theorem 2.11
36. Theorem 2.12
37. Theorem 2.13

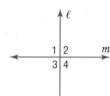

PROOF **Write a two-column proof.**

38. **Given:** $\angle ABD \cong \angle YXZ$
    Prove: $\angle CBD \cong \angle WXZ$

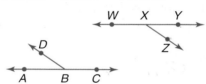

39. **Given:** $m\angle RSW = m\angle TSU$
    Prove: $m\angle RST = m\angle WSU$

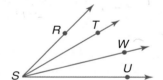

40. **RIVERS** Tributaries of rivers sometimes form a linear pair of angles when they meet the main river. The Yellowstone River forms the linear pair $\angle 1$ and $\angle 2$ with the Missouri River. If $m\angle 1$ is 28, find $m\angle 2$.

41. **HIGHWAYS** Near the city of Hopewell, Virginia, Route 10 runs perpendicular to Interstate 95 and Interstate 295. Show that the angles at the intersections of Route 10 with Interstate 95 and Interstate 295 are congruent.

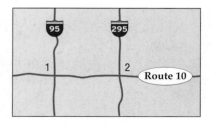

42. **CRITICAL THINKING** What conclusion can you make about the sum of $m\angle 1$ and $m\angle 4$ if $m\angle 1 = m\angle 2$ and $m\angle 3 = m\angle 4$? Explain.

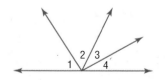

43. **WRITING IN MATH** Answer the question that was posed at the beginning of the lesson.

How do scissors illustrate supplementary angles?

Include the following in your answer:
- a description of the relationship among $\angle 1$, $\angle 2$, and $\angle 3$,
- an example of another way that you can tell the relationship between $\angle 1$ and $\angle 3$, and
- an explanation of whether this relationship is the same for two angles complementary to the same angle.

44. The measures of two complementary angles are in the ratio 4:1. What is the measure of the smaller angle?

    Ⓐ 15   Ⓑ 18   Ⓒ 24   Ⓓ 36

45. **ALGEBRA** T is the set of all positive numbers n such that $n < 50$ and $\sqrt{n}$ is an integer. What is the median of the members of set T?

    Ⓐ 4   Ⓑ 16   Ⓒ 20   Ⓓ 25

Maintain Your Skills

Mixed Review   **PROOF** **Write a two-column proof.** *(Lesson 2-7)*

46. **Given:** G is between F and H.
            H is between G and J.
     Prove: $FG + GJ = FH + HJ$

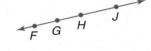

47. **Given:** X is the midpoint of $\overline{WY}$.
     Prove: $WX + YZ = XZ$

48. **PHOTOGRAPHY** Film is fed through a camera by gears that catch the perforation in the film. The distance from the left edge of the film, A, to the right edge of the image, C, is the same as the distance from the left edge of the image, B, to the right edge of the film, D. Show that the two perforated strips are the same width. *(Lesson 2-6)*

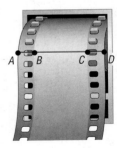

For Exercises 49–55, refer to the figure at the right.
(Lesson 1-4)

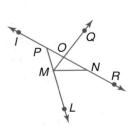

49. Name two angles that have N as a vertex.
50. If $\overrightarrow{MQ}$ bisects $\angle PMN$, name two congruent angles.
51. Name a point in the interior of $\angle LMQ$.
52. List all the angles that have O as the vertex.
53. Does $\angle QML$ appear to be acute, obtuse, right, or straight?
54. Name a pair of opposite rays.
55. List all the angles that have $\overline{MN}$ as a side.

Study Guide and Review

Vocabulary and Concept Check

axiom (p. 89)
biconditional (p. 81)
compound statement (p. 67)
conclusion (p. 75)
conditional statement (p. 75)
conjecture (p. 62)
conjunction (p. 68)
contrapositive (p. 77)

converse (p. 77)
counterexample (p. 63)
deductive argument (p. 94)
deductive reasoning (p. 82)
disjunction (p. 68)
formal proof (p. 95)
hypothesis (p. 75)
if-then statement (p. 75)

inductive reasoning (p. 62)
informal proof (p. 90)
inverse (p. 77)
Law of Detachment (p. 82)
Law of Syllogism (p. 83)
logically equivalent (p. 77)
negation (p. 67)
paragraph proof (p. 90)

postulate (p. 89)
proof (p. 90)
related conditionals (p. 77)
statement (p. 67)
theorem (p. 90)
truth table (p. 70)
truth value (p. 67)
two-column proof (p. 95)

A complete list of postulates and theorems can be found on pages R1–R8.

Exercises   Choose the correct term to complete each sentence.

1. A (*counterexample, conjecture*) is an educated guess based on known information.

2. The truth or falsity of a statement is called its (*conclusion, truth value*).

3. Two or more statements can be joined to form a (*conditional, compound*) statement.

4. A conjunction is a compound statement formed by joining two or more statements using (*or, and*).

5. The phrase immediately following the word *if* in a conditional statement is called the (*hypothesis, conclusion*).

6. The (*inverse, converse*) is formed by exchanging the hypothesis and the conclusion.

7. (*Theorems, Postulates*) are accepted as true without proof.

8. A paragraph proof is a (an) (*informal proof, formal proof*).

Lesson-by-Lesson Review

2-1 Inductive Reasoning and Conjecture

See pages
62–66.

Concept Summary

- Conjectures are based on observations and patterns.
- Counterexamples can be used to show that a conjecture is false.

Example   Given that points *P, Q,* and *R* are collinear, determine whether the conjecture that *Q* is between *P* and *R* is *true* or *false*. If the conjecture is false, give a counterexample.

In the figure, *R* is between *P* and *Q*. Since we can find a counterexample, the conjecture is false.

Exercises   Make a conjecture based on the given information. Draw a figure to illustrate your conjecture.   *See Example 2 on page 63.*

9. ∠*A* and ∠*B* are supplementary.

10. *X, Y,* and *Z* are collinear and *XY* = *YZ*.

11. In quadrilateral *LMNO*, *LM* = *LO* = *MN* = *NO*, and *m* ∠*L* = 90.

2-2 Logic

See pages 67–74.

Concept Summary

- The negation of a statement has the opposite truth value of the original statement.
- Venn diagrams and truth tables can be used to determine the truth values of statements.

Example Use the following statements to write a compound statement for each conjunction. Then find its truth value.

p: $\sqrt{15} = 5$ q: The measure of a right angle equals 90.

a. p and q
  $\sqrt{15} = 5$, and the measure of a right angle equals 90.
  p and q is false because p is false and q is true.

b. $p \vee q$
  $\sqrt{15} = 5$, or the measure of a right angle equals 90.
  $p \vee q$ is true because q is true. It does not matter that p is false.

Exercises Use the following statements to write a compound statement for each conjunction. Then find its truth value. *See Examples 1 and 2 on pages 68 and 69.*

p: $-1 > 0$ q: In a right triangle with right angle C, $a^2 + b^2 = c^2$.
r: The sum of the measures of two supplementary angles is 180.

12. p and q 13. q or r 14. $r \wedge p$
15. $p \wedge (q \vee r)$ 16. $q \vee (p \vee r)$ 17. $(q \wedge r) \wedge p$

2-3 Conditional Statements

See pages 75–80.

Concept Summary

- Conditional statements are written in if-then form.
- Form the converse, inverse, and contrapositive of an if-then statement by using negations and by exchanging the hypothesis and conclusion.

Example Identify the hypothesis and conclusion of the statement *The intersection of two planes is a line.* Then write the statement in if-then form.

Hypothesis: two planes intersect
Conclusion: their intersection is a line

If two planes intersect, then their intersection is a line.

Exercises Write the converse, inverse, and contrapositive of each conditional statement. Determine whether each related conditional is *true* or *false*. If a statement is false, find a counterexample. *See Example 4 on page 77.*

18. If an angle measure equals 120, then the angle is obtuse.
19. If the month is March, then it has 31 days.
20. If an ordered pair for a point has 0 for its x-coordinate, then the point lies on the y-axis.

Determine the truth value of the following statement for each set of conditions.

If the temperature is at most 0°C, then water freezes.   *See Example 3 on page 76.*

21. The temperature is −10°C, and water freezes.
22. The temperature is 15°C, and water freezes.
23. The temperature is −2°C, and water does not freeze.
24. The temperature is 30°C, and water does not freeze.

2-4 Deductive Reasoning

See pages 82–87.

Concept Summary

- The Law of Detachment and the Law of Syllogism can be used to determine the truth value of a compound statement.

Example Use the Law of Syllogism to determine whether a valid conclusion can be reached from the following statements.

(1) If a body in our solar system is the Sun, then it is a star.
(2) Stars are in constant motion.

p: a body in our solar system is the sun
q: it is a star
r: stars are in constant motion

Statement (1): $p \rightarrow q$ Statement (2): $q \rightarrow r$

Since the given statements are true, use the Law of Syllogism to conclude $p \rightarrow r$. That is, *If a body in our solar system is the Sun, then it is in constant motion.*

Exercises Determine whether the stated conclusion is valid based on the given information. If not, write *invalid.* Explain your reasoning.   *See Example 1 on page 82.*
If two angles are adjacent, then they have a common vertex.

25. **Given:**   ∠1 and ∠2 are adjacent angles.
    Conclusion: ∠1 and ∠2 have a common vertex.
26. **Given:**   ∠3 and ∠4 have a common vertex.
    Conclusion: ∠3 and ∠4 are adjacent angles.

**Determine whether statement (3) follows from statements (1) and (2) by the Law of Detachment or the Law of Syllogism. If it does, state which law was used. If it does not follow, write *invalid.*   *See Example 3 on page 83.*

27. (1) If a student attends North High School, then the student has an ID number.
    (2) Josh Michael attends North High School.
    (3) Josh Michael has an ID number.

28. (1) If a rectangle has four congruent sides, then it is a square.
    (2) A square has diagonals that are perpendicular.
    (3) A rectangle has diagonals that are perpendicular.

29. (1) If you like pizza with everything, then you'll like Cardo's Pizza.
    (2) If you like Cardo's Pizza, then you are a pizza connoisseur.
    (3) If you like pizza with everything, then you are a pizza connoisseur.

2-5 Postulates and Paragraph Proofs

See pages 89–93.

Concept Summary

- Use undefined terms, definitions, postulates, and theorems to prove that statements and conjectures are true.

Example   Determine whether the following statement is *always, sometimes,* or *never* true. Explain. *Two points determine a line.*

According to a postulate relating to points and lines, two points determine a line. Thus, the statement is always true.

Exercises   Determine whether the following statements are *always, sometimes,* or *never* true. Explain.   *See Example 2 on page 90.*

30. The intersection of two lines can be a line.
31. If P is the midpoint of $\overline{XY}$, then $XP = PY$.
32. If $MX = MY$, then M is the midpoint of XY.
33. Three points determine a line.
34. Points Q and R lie in at least one plane.
35. If two angles are right angles, they are adjacent.
36. An angle is congruent to itself.
37. Vertical angles are adjacent.

38. **PROOF**   Write a paragraph proof to prove that if M is the midpoint of $\overline{AB}$ and Q is the midpoint of $\overline{AM}$, then $AQ = \frac{1}{4}AB$.

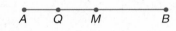

2-6 Algebraic Proof

See pages 94–100.

Concept Summary

- The properties of equality used in algebra can be applied to the measures of segments and angles to verify and prove statements.

Example   **Given:** $2x + 6 = 3 + \frac{5}{3}x$

Prove: $x = -9$

Proof:

Statements	Reasons
1. $2x + 6 = 3 + \frac{5}{3}x$	1. Given
2. $3(2x + 6) = 3\left(3 + \frac{5}{3}x\right)$	2. Multiplication Property
3. $6x + 18 = 9 + 5x$	3. Distributive Property
4. $6x + 18 - 5x = 9 + 5x - 5x$	4. Subtraction Property
5. $x + 18 = 9$	5. Substitution
6. $x + 18 - 18 = 9 - 18$	6. Subtraction Property
7. $x = -9$	7. Substitution

Exercises   State the property that justifies each statement.   *See Example 1 on page 94.*

39. If $3(x + 2) = 6$, then $3x + 6 = 6$.

40. If $10x = 20$, then $x = 2$.

41. If $AB + 20 = 45$, then $AB = 25$.

42. If $3 = CD$ and $CD = XY$, then $3 = XY$.

PROOF   Write a two-column proof.   *See Examples 2 and 4 on pages 95 and 96.*

43. If $5 = 2 - \frac{1}{2}x$, then $x = -6$.

44. If $x - 1 = \frac{x - 10}{-2}$, then $x = 4$.

45. If $AC = AB$, $AC = 4x + 1$, and $AB = 6x - 13$, then $x = 7$.

46. If $MN = PQ$ and $PQ = RS$, then $MN = RS$.

2-7 *Proving Segment Relationships*

See pages 101–106.

Concept Summary

- Use properties of equality and congruence to write proofs involving segments.

Example   **Write a two-column proof.**

Given: $QT = RT$, $TS = TP$

Prove: $QS = RP$

Proof:

Statements	Reasons
1. $QT = RT$, $TS = TP$	1. Given
2. $QT + TS = RT + TS$	2. Addition Property
3. $QT + TS = RT + TP$	3. Substitution
4. $QT + TS = QS$, $RT + TP = RP$	4. Segment Addition Postulate
5. $QS = RP$	5. Substitution

Exercises   Justify each statement with a property of equality or a property of congruence.   *See Example 1 on page 102.*

47. $PS = PS$

48. If $XY = OP$, then $OP = XY$.

49. If $AB - 8 = CD - 8$, then $AB = CD$.

50. If $EF = GH$ and $GH = LM$, then $EF = LM$.

51. If $2(XY) = AB$, then $XY = \frac{1}{2}(AB)$.

52. If $AB = CD$, then $AB + BC = CD + BC$.

Chapter
2 For More ...
• Extra Practice, see pages 756–758.
• Mixed Problem Solving, see page 783.

PROOF Write a two-column proof. *See Examples 1 and 2 on pages 102 and 103.*

53. Given: $BC = EC, CA = CD$
Prove: $BA = DE$

54. Given: $AB = CD$
Prove: $AC = BD$

2-8 Proving Angle Relationships

See pages 107–114.

Concept Summary

• The properties of equality and congruence can be applied to angle relationships.

Example **Find the measure of each numbered angle.**

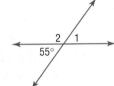

$m\angle 1 = 55$, since $\angle 1$ is a vertical angle to the $55°$ angle.
$\angle 2$ and the $55°$ angle form a linear pair.

$55 + m\angle 2 = 180$ Def. of supplementary $\angle$s
$m\angle 2 = 125$ Subtract 55 from each side.

Exercises **Find the measure of each numbered angle.** *See Example 2 on page 108.*

55. $m\angle 6$
56. $m\angle 7$
57. $m\angle 8$

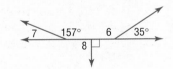

58. PROOF Copy and complete the proof.
See Example 3 on page 109.

Given: $\angle 1$ and $\angle 2$ form a linear pair.
$m\angle 2 = 2(m\angle 1)$

Prove: $m\angle 1 = 60$

Proof:

Statements	Reasons
a. $\angle 1$ and $\angle 2$ form a linear pair.	**a.** ?
b. $\angle 1$ and $\angle 2$ are supplementary.	**b.** ?
c. ?	**c.** Definition of supplementary angles
d. $m\angle 2 = 2(m\angle 1)$	**d.** ?
e. ?	**e.** Substitution
f. ?	**f.** Substitution
g. $\dfrac{3(m\angle 1)}{3} = \dfrac{180}{3}$	**g.** ?
h. ?	**h.** Substitution

Vocabulary and Concepts

1. **Explain** the difference between formal and informal proofs.
2. **Explain** how you can prove that a conjecture is false.
3. **Describe** the parts of a two-column proof.

Skills and Applications

Determine whether each conjecture is *true* or *false*. Explain your answer and give a counterexample for any false conjecture.

4. **Given:** $\angle A \cong \angle B$
   Conjecture: $\angle B \cong \angle A$

5. **Given:** y is a real number
   Conjecture: $-y > 0$

6. **Given:** $3a^2 = 48$
   Conjecture: $a = 4$

Use the following statements to write a compound statement for each conjunction or disjunction. Then find its truth value.

p: $-3 > 2$ q: $3x = 12$ when $x = 4$. r: An equilateral triangle is also equiangular.

7. p and q

8. p or q

9. $p \lor (q \land r)$

Identify the hypothesis and conclusion of each statement and write each statement in if-then form. Then write the converse, inverse, and contrapositive of each conditional.

10. An apple a day keeps the doctor away.

11. A rolling stone gathers no moss.

12. Determine whether statement (3) follows from statements (1) and (2) by the Law of Detachment or the Law of Syllogism. If it does, state which law was used. If it does not, write *invalid*.
    (1) Perpendicular lines intersect.
    (2) Lines m and n are perpendicular.
    (3) Lines m and n intersect.

Find the measure of each numbered angle.

13. $\angle 1$

14. $\angle 2$

15. $\angle 3$

16. Write a two-column proof.
    If $y = 4x + 9$ and $x = 2$, then $y = 17$.

17. Write a paragraph proof.
    Given: $AM = CN$, $MB = ND$
    Prove: $AB = CD$

18. **ADVERTISING**   Identify the hypothesis and conclusion of the following statement, then write it in if-then form. *Hard working people deserve a great vacation.*

19. **STANDARDIZED TEST PRACTICE**   If two planes intersect, their intersection can be
    I a line.   **II** three noncollinear points.   **III** two intersecting lines.
    Ⓐ I only   Ⓑ II only   Ⓒ III only   Ⓓ I and II only

Part 1 Multiple Choice

Record your answers on the answer sheet provided by your teacher or on a sheet of paper.

1. Arrange the numbers $|-7|, \frac{1}{7}, \sqrt{7}, -7^2$ in order from least to greatest. (Prerequisite Skill)

Ⓐ $|-7|, \sqrt{7}, \frac{1}{7}, -7^2$

Ⓑ $-7^2, |-7|, \frac{1}{7}, \sqrt{7}$

Ⓒ $|-7|, \frac{1}{7}, \sqrt{7}, -7^2$

Ⓓ $-7^2, \frac{1}{7}, \sqrt{7}, |-7|$

2. Points A and B lie on the line $y = 2x - 3$. Which of the following are coordinates of a point noncollinear with A and B? (Lesson 1-1)

Ⓐ $(7, 11)$ Ⓑ $(4, 5)$

Ⓒ $(-2, -10)$ Ⓓ $(-5, -13)$

3. Dana is measuring distance on a map. Which of the following tools should Dana use to make the most accurate measurement? (Lesson 1-2)

Ⓐ centimeter ruler   Ⓑ protractor

Ⓒ yardstick   Ⓓ calculator

4. Point E is the midpoint of $\overline{DF}$. If $DE = 8x - 3$ and $EF = 3x + 7$, what is x? (Lesson 1-3)

Ⓐ 1   Ⓑ 2   Ⓒ 4   Ⓓ 13

5. What is the relationship between $\angle ACF$ and $\angle DCF$? (Lesson 1-6)

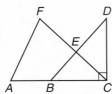

Ⓐ complementary angles

Ⓑ congruent angles

Ⓒ supplementary angles

Ⓓ vertical angles

6. Which of the following is an example of inductive reasoning? (Lesson 2-1)

Ⓐ Carlos learns that the measures of all acute angles are less than 90. He conjectures that if he sees an acute angle, its measure will be less than 90.

Ⓑ Carlos reads in his textbook that the measure of all right angles is 90. He conjectures that the measure of each right angle in a square equals 90.

Ⓒ Carlos measures the angles of several triangles and finds that their measures all add up to 180. He conjectures that the sum of the measures of the angles in any triangle is always 180.

Ⓓ Carlos knows that the sum of the measures of the angles in a square is always 360. He conjectures that if he draws a square, the sum of the measures of the angles will be 360.

7. Which of the following is the contrapositive of the statement *If Rick buys hamburgers for lunch, then Denzel buys French fries and a large soda*? (Lesson 2-2)

Ⓐ If Denzel does not buy French fries and a large soda, then Rick does not buy hamburgers for lunch.

Ⓑ If Rick does not buy hamburgers for lunch, then Denzel does not buy French fries and a large soda.

Ⓒ If Denzel buys French fries and a large soda, then Rick buys hamburgers for lunch.

Ⓓ If Rick buys hamburgers for lunch, then Denzel does not buy French fries and a large soda.

8. Which property could justify the first step in solving $3 \times \frac{14x + 6}{8} = 18$? (Lesson 2-5)

Ⓐ Division Property of Equality

Ⓑ Substitution Property of Equality

Ⓒ Addition Property of Equality

Ⓓ Transitive Property of Equality

Preparing for Standardized Tests
For test-taking strategies and more
practice, see pages 795–810.

Part 2 | Short Response/Grid In

**Record your answers on the answer sheet
provided by your teacher or on a sheet
of paper.**

9. Two cheerleaders stand at opposite corners
   of a football field. What is the shortest
   distance between them, to the nearest yard?
   (Lesson 1-3)

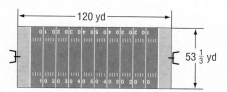

10. Consider the conditional *If I call in sick, then
    I will not get paid for the day.* Based on the
    original conditional, what is the name of the
    conditional *If I do not call in sick, then I will
    get paid for the day?*   (Lesson 2-2)

11. Examine the following statements.

    p: Martina drank a cup of soy milk.
    q: A cup is 8 ounces.
    r: Eight ounces of soy milk contains
          300 milligrams of calcium.

    Using the Law of Syllogism, how many
    milligrams of calcium did Martina get
    from drinking a cup of soy milk?
    (Lesson 2-4)

12. In the following proof, what property
    justifies statement c?   (Lesson 2-7)

    Given: $\overline{AC} \cong \overline{MN}$
    Prove: $AB + BC = MN$

    Proof:

Statements	Reasons
a. $\overline{AC} \cong \overline{MN}$	**a.** Given
b. $AC = MN$	**b.** Definition of congruent segments
c. $AC = AB + BC$	**c.** _?_
d. $AC + BC = MN$	**d.** Substitution

www.geometryonline.com/standardized_test

Test-Taking Tip Ⓐ Ⓑ Ⓒ Ⓓ

Question 6
When answering a multiple-choice question, always read
every answer choice and eliminate those you decide are
definitely wrong. This way, you may deduce the correct
answer.

Part 3 | Extended Response

**Record your answers on a sheet of paper.
Show your work.**

13. In any right triangle, the sum of the squares
    of the lengths of the legs equals the square
    of the length of the hypotenuse. From a
    single point in her yard, Marti measures and
    marks distances of 18 feet and 24 feet for
    two sides of her garden. Explain how Marti
    can ensure that the two sides of her garden
    form a right angle.   (Lesson 1-3)

14. A farmer needs to make a 100-square-foot
    rectangular enclosure for her chickens. She
    wants to save money by purchasing the
    least amount of fencing possible to enclose
    the area.   (Lesson 1-4)

    a. What whole-number dimensions, to the
    nearest yard, will require the least
    amount of fencing?

    b. Explain your procedure for finding the
    dimensions that will require the least
    amount of fencing.

    c. Explain how the amount of fencing
    required to enclose the area changes
    as the dimensions change.

15. **Given:** $\angle 1$ and $\angle 3$ are vertical angles.
    $m\angle 1 = 3x + 5, m\angle 3 = 2x + 8$

    Prove: $m\angle 1 = 14$ (Lesson 2-8)

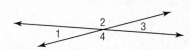

Parallel and Perpendicular Lines

What You'll Learn

- **Lessons 3-1, 3-2, and 3-5** Identify angle relationships that occur with parallel lines and a transversal, and identify and prove lines parallel from given angle relationships.
- **Lessons 3-3 and 3-4** Use slope to analyze a line and to write its equation.
- **Lesson 3-6** Find the distance between a point and a line and between two parallel lines.

Key Vocabulary

- parallel lines (p. 126)
- transversal (p. 127)
- slope (p. 139)
- equidistant (p. 160)

Why It's Important

The framework of a wooden roller coaster is composed of millions of feet of intersecting lumber that often form parallel lines and transversals. Roller coaster designers, construction managers, and carpenters must know the relationships of angles created by parallel lines and their transversals to create a safe and stable ride.

You will find how measures of angles are used in carpentry and construction in Lesson 3-2.

Getting Started

Prerequisite Skills To be successful in this chapter, you'll need to master these skills and be able to apply them in problem-solving situations. Review these skills before beginning Chapter 3.

For Lesson 3-1
Naming Segments

Name all of the lines that contain the given point.
(For review, see Lesson 1-1.)

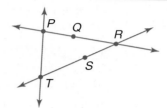

1. Q **2.** R

3. S **4.** T

For Lessons 3-2 and 3-5
Congruent Angles

Name all angles congruent to the given angle.
(For review, see Lesson 1-4.)

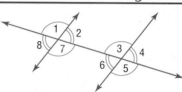

5. $\angle 2$ **6.** $\angle 5$

7. $\angle 3$ **8.** $\angle 8$

For Lessons 3-3 and 3-4
Equations of Lines

For each equation, find the value of y for the given value of x.   *(For review, see pages 736 and 738.)*

9. $y = 7x - 12$, for $x = 3$ **10.** $y = -\dfrac{2}{3}x + 4$, for $x = 8$ **11.** $2x - 4y = 18$, for $x = 6$

FOLDABLES™
Study Organizer

Parallel and Perpendicular Lines Make this Foldable to help you organize your notes. Begin with one sheet of $8\frac{1}{2}$" by 11" paper.

Step 1 Fold

Fold in half matching the short sides.

Step 2 Fold Again

Unfold and fold the long side up 2 inches to form a pocket.

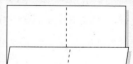

Step 3 Staple or Glue

Staple or glue the outer edges to complete the pocket.

Step 4 Label

Label each side as shown. Use index cards to record examples.

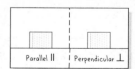

Reading and Writing As you read and study the chapter, write examples and notes about parallel and perpendicular lines on index cards. Place the cards in the appropriate pocket.

3-1 Parallel Lines and Transversals

What You'll Learn

- Identify the relationships between two lines or two planes.
- Name angles formed by a pair of lines and a transversal.

Vocabulary

- parallel lines
- parallel planes
- skew lines
- transversal
- consecutive interior angles
- alternate exterior angles
- alternate interior angles
- corresponding angles

How are parallel lines and planes used in architecture?

Architect Frank Lloyd Wright designed many buildings using basic shapes, lines, and planes. His building at the right has several examples of parallel lines, parallel planes, and skew lines.

RELATIONSHIPS BETWEEN LINES AND PLANES Lines ℓ and m are coplanar because they lie in the same plane. If the lines were extended indefinitely, they would not intersect. Coplanar lines that do not intersect are called **parallel lines**. Segments and rays contained within parallel lines are also parallel.

The symbol ‖ means *is parallel to*. Arrows are used in diagrams to indicate that lines are parallel. In the figure, the arrows indicate that $\overleftrightarrow{PQ}$ is parallel to $\overleftrightarrow{RS}$.

Similarly, two planes can intersect or be parallel. In the photograph above, the roofs of each level are contained in **parallel planes**. The walls and the floor of each level lie in intersecting planes.

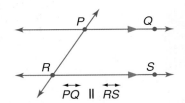

$$\overleftrightarrow{PQ} \parallel \overleftrightarrow{RS}$$

The symbol ∦ means *is not parallel to*.

Geometry Activity

Draw a Rectangular Prism

A rectangular prism can be drawn using parallel lines and parallel planes.

Step 1 Draw two parallel planes to represent the top and bottom of the prism.

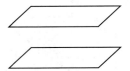

Step 2 Draw the edges. Make any hidden edges of the prism dashed.

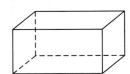

Step 3 Label the vertices.

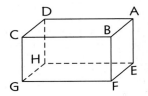

Analyze

1. Identify the parallel planes in the figure.
2. Name the planes that intersect plane *ABC* and name their intersections.
3. Identify all segments parallel to $\overline{BF}$.

Notice that in the Geometry Activity, $\overline{AE}$ and $\overline{GF}$ do not intersect. These segments are not parallel since they do not lie in the same plane. Lines that do not intersect and are not coplanar are called **skew lines**. Segments and rays contained in skew lines are also skew.

Example 1 Identify Relationships

a. **Name all planes that are parallel to plane *ABG*.**

plane *CDE*

b. **Name all segments that intersect $\overline{CH}$.**

$\overline{BC}, \overline{CD}, \overline{CE}, \overline{EH}$, and $\overline{GH}$

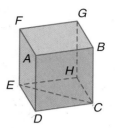

c. **Name all segments that are parallel to $\overline{EF}$.**

$\overline{AD}, \overline{BC}$, and $\overline{GH}$

d. **Name all segments that are skew to $\overline{BG}$.**

$\overline{AD}, \overline{CD}, \overline{CE}, \overline{EF}$, and $\overline{EH}$

ANGLE RELATIONSHIPS In the drawing of the railroad crossing, notice that the tracks, represented by line *t*, intersect the sides of the road, represented by lines *m* and *n*. A line that intersects two or more lines in a plane at different points is called a **transversal**.

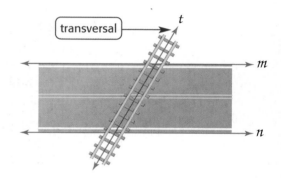

Example 2 Identify Transversals

AIRPORTS Some of the runways at O'Hare International Airport are shown below. Identify the sets of lines to which each given line is a transversal.

a. **line *q***

If the lines are extended, line *q* intersects lines *ℓ*, *n*, *p*, and *r*.

b. **line *m***

lines *ℓ*, *n*, *p*, and *r*

c. **line *n***

lines *ℓ*, *m*, *p*, and *q*

d. **line *r***

lines *ℓ*, *m*, *p*, and *q*

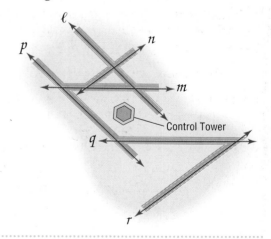

In the drawing of the railroad crossing above, notice that line *t* forms eight angles with lines *m* and *n*. These angles are given special names, as are specific pairings of these angles.

Study Tip

Same Side Interior Angles
Consecutive interior angles are also called *same side interior angles*.

Name	Angles
exterior angles	∠1, ∠2, ∠7, ∠8
interior angles	∠3, ∠4, ∠5, ∠6
consecutive interior angles	∠3 and ∠6, ∠4 and ∠5
alternate exterior angles	∠1 and ∠7, ∠2 and ∠8
alternate interior angles	∠3 and ∠5, ∠4 and ∠6
corresponding angles	∠1 and ∠5, ∠2 and ∠6, ∠3 and ∠7, ∠4 and ∠8

Transversal *p* intersects lines *q* and *r*.

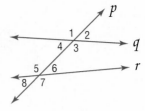

Example 3 *Identify Angle Relationships*

Refer to the figure below. Identify each pair of angles as *alternate interior, alternate exterior, corresponding,* or *consecutive interior* angles.

a. ∠1 and ∠7
    alternate exterior

b. ∠2 and ∠10
    corresponding

c. ∠8 and ∠9
    consecutive interior

d. ∠3 and ∠12
    corresponding

e. ∠4 and ∠10
    alternate interior

f. ∠6 and ∠11
    alternate exterior

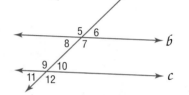

Check for Understanding

Concept Check

1. **OPEN ENDED** Draw a solid figure with parallel planes. Describe which parts of the figure are parallel.

2. **FIND THE ERROR** Juanita and Eric are naming alternate interior angles in the figure at the right. One of the angles must be ∠4.

Juanita
∠4 and ∠9
∠4 and ∠6

Eric
∠4 and ∠10
∠4 and ∠5

Who is correct? Explain your reasoning.

3. **Describe** a real-life situation in which parallel lines seem to intersect.

Guided Practice

For Exercises 4–6, refer to the figure at the right.

4. Name all planes that intersect plane *ADM*.
5. Name all segments that are parallel to $\overline{CD}$.
6. Name all segments that intersect $\overline{KL}$.

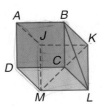

Identify the pairs of lines to which each given line is a transversal.

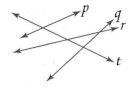

7. p
8. r
9. q
10. t

Identify each pair of angles as *alternate interior*, *alternate exterior*, *corresponding*, or *consecutive interior* angles.

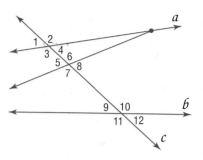

11. ∠7 and ∠10
12. ∠1 and ∠5
13. ∠4 and ∠6
14. ∠8 and ∠1

Name the transversal that forms each pair of angles. Then identify the special name for the angle pair.

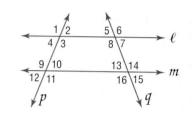

15. ∠3 and ∠10
16. ∠2 and ∠12
17. ∠8 and ∠14

Application   **MONUMENTS**   **For Exercises 18–21, refer to the photograph of the Lincoln Memorial.**

18. Describe a pair of parallel lines found on the Lincoln Memorial.
19. Find an example of parallel planes.
20. Locate a pair of skew lines.
21. Identify a transversal passing through a pair of lines.

Practice and Apply

Homework Help

For Exercises	See Examples
22–27	1
28–31	2
32–47	3

Extra Practice
See page 758.

For Exercises 22–27, refer to the figure at the right.

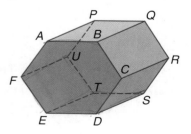

22. Name all segments parallel to $\overline{AB}$.
23. Name all planes intersecting plane BCR.
24. Name all segments parallel to $\overline{TU}$.
25. Name all segments skew to $\overline{DE}$.
26. Name all planes intersecting plane EDS.
27. Name all segments skew to $\overline{AP}$.

Identify the pairs of lines to which each given line is a transversal.

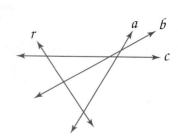

28. a
29. b
30. c
31. r

Identify each pair of angles as *alternate interior, alternate exterior, corresponding,* or *consecutive interior* angles.

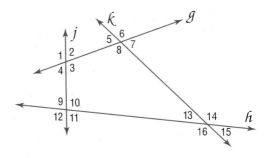

32. ∠2 and ∠10   **33.** ∠1 and ∠11

34. ∠5 and ∠3   **35.** ∠6 and ∠14

36. ∠5 and ∠15   **37.** ∠11 and ∠13

38. ∠8 and ∠3   **39.** ∠9 and ∠4

Study Tip

Make a Sketch
Use patty paper or tracing paper to copy the figure. Use highlighters or colored pencils to identify the lines that compose each pair of angles.

Name the transversal that forms each pair of angles. Then identify the special name for the angle pair.

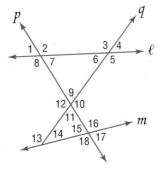

40. ∠2 and ∠9   **41.** ∠7 and ∠15

42. ∠13 and ∠17   **43.** ∠8 and ∠4

44. ∠14 and ∠16   **45.** ∠6 and ∠14

46. ∠8 and ∠6   **47.** ∠14 and ∠15

48. AVIATION Airplanes heading eastbound are assigned an altitude level that is an odd number of thousands of feet. Airplanes heading westbound are assigned an altitude level that is an even number of thousands of feet. If one airplane is flying northwest at 34,000 feet and another airplane is flying east at 25,000 feet, describe the type of lines formed by the paths of the airplanes. Explain your reasoning.

STRUCTURES For Exercises 49–51, refer to the drawing of the gazebo at the right.

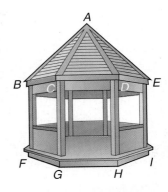

49. Name all labeled segments parallel to $\overline{BF}$.

50. Name all labeled segments skew to $\overline{AC}$.

51. Are any of the planes on the gazebo parallel to plane *ADE*? Explain.

52. COMPUTERS The word *parallel* when used with computers describes processes that occur simultaneously, or devices, such as printers, that receive more than one bit of data at a time. Find two other examples for uses of the word *parallel* in other subject areas such as history, music, or sports.

CRITICAL THINKING Suppose there is a line ℓ and a point *P* not on the line.

53. In space, how many lines can be drawn through *P* that do not intersect ℓ?

54. In space, how many lines can be drawn through *P* that are parallel to ℓ?

55. ▪WRITING IN MATH▪ Answer the question that was posed at the beginning of the lesson.

How are parallel lines and planes used in architecture?

Include the following in your answer:
- a description of where you might expect to find examples of parallel lines and parallel planes, and
- an example of skew lines and nonparallel planes.

56. ∠3 and ∠5 are __?__ angles.
- Ⓐ alternate interior
- Ⓑ alternate exterior
- Ⓒ consecutive interior
- Ⓓ corresponding

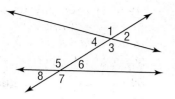

57. **GRID IN** Set M consists of all multiples of 3 between 13 and 31. Set P consists of all multiples of 4 between 13 and 31. What is one possible number in P but NOT in M?

Maintain Your Skills

Mixed Review

58. **PROOF** Write a two-column proof. *(Lesson 2-8)*
Given: $m\angle ABC = m\angle DFE$, $m\angle 1 = m\angle 4$
Prove: $m\angle 2 = m\angle 3$

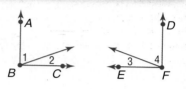

59. **PROOF** Write a paragraph proof. *(Lesson 2-7)*
Given: $\overline{PQ} \cong \overline{ZY}$, $\overline{QR} \cong \overline{XY}$
Prove: $\overline{PR} \cong \overline{XZ}$

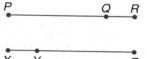

Determine whether a valid conclusion can be reached from the two true statements using the Law of Detachment or the Law of Syllogism. If a valid conclusion is possible, state it and the law that is used. If a valid conclusion does not follow, write *no conclusion*. *(Lesson 2-4)*

60. (1) If two angles are vertical, then they do not form a linear pair.
  (2) If two angles form a linear pair, then they are not congruent.

61. (1) If an angle is acute, then its measure is less than 90.
  (2) ∠EFG is acute.

Find the distance between each pair of points. *(Lesson 1-3)*

62. $A(-1, -8)$, $B(3, 4)$ **63.** $C(0, 1)$, $D(-2, 9)$ **64.** $E(-3, -12)$, $F(5, 4)$

65. $G(4, -10)$, $H(9, -25)$ **66.** $J\left(1, \frac{1}{4}\right)$, $K\left(-3, -\frac{7}{4}\right)$ **67.** $L\left(-5, \frac{8}{5}\right)$, $M\left(5, -\frac{2}{5}\right)$

Draw and label a figure for each relationship. *(Lesson 1-1)*

68. $\overleftrightarrow{AB}$ perpendicular to $\overleftrightarrow{MN}$ at point P

69. line ℓ contains R and S but not T

Getting Ready for the Next Lesson

PREREQUISITE SKILL State the measures of linear pairs of angles in each figure.
*(To review **linear pairs**, see Lesson 2-6.)*

70.

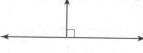

71.

72.

73.

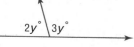

74.

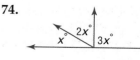

75.

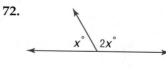
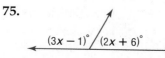

Geometry Software Investigation

Angles and Parallel Lines

You can use The Geometer's Sketchpad to investigate the measures of angles formed by two parallel lines and a transversal.

Step 1 *Draw parallel lines.*
- Place two points A and B on the screen.
- Construct a line through the points.
- Place point C so that it does not lie on $\overleftrightarrow{AB}$.
- Construct a line through C parallel to $\overleftrightarrow{AB}$.
- Place point D on this line.

Step 2 *Construct a transversal.*
- Place point E on $\overleftrightarrow{AB}$ and point F on $\overleftrightarrow{CD}$.
- Construct $\overleftrightarrow{EF}$ as a transversal through $\overleftrightarrow{AB}$ and $\overleftrightarrow{CD}$.
- Place points G and H on $\overleftrightarrow{EF}$, as shown.

Step 3 *Measure angles.*
- Measure each angle.

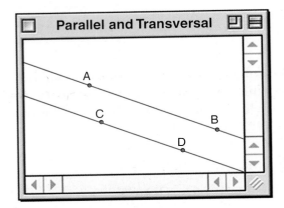

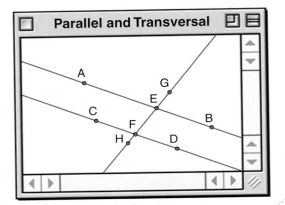

Analyze

1. List pairs of angles by the special names you learned in Lesson 3-1.
2. Which pairs of angles listed in Exercise 1 have the same measure?
3. What is the relationship between consecutive interior angles?

Make a Conjecture

4. Make a conjecture about the following pairs of angles formed by two parallel lines and a transversal. Write your conjecture in if-then form.
   a. corresponding angles
   b. alternate interior angles
   c. alternate exterior angles
   d. consecutive interior angles

5. Rotate the transversal. Are the angles with equal measures in the same relative location as the angles with equal measures in your original drawing?

6. Test your conjectures by rotating the transversal and analyzing the angles.

7. Rotate the transversal so that the measure of any of the angles is 90.
   a. What do you notice about the measures of the other angles?
   b. Make a conjecture about a transversal that is perpendicular to one of two parallel lines.

3-2 Angles and Parallel Lines

How can angles and lines be used in art?

In the painting, the artist uses lines and transversals to create patterns. The figure on the painting shows two parallel lines with a transversal passing through them. There is a special relationship between the angle pairs formed by these lines.

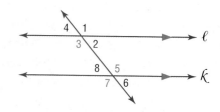

The Order of Tradition II by T.C. Stuart

PARALLEL LINES AND ANGLE PAIRS In the figure above, ∠1 and ∠2 are corresponding angles. When the two lines are parallel, there is a special relationship between these pairs of angles.

Postulate 3.1

Corresponding Angles Postulate If two parallel lines are cut by a transversal, then each pair of corresponding angles is congruent.

Examples: ∠1 ≅ ∠5, ∠2 ≅ ∠6, ∠3 ≅ ∠7, ∠4 ≅ ∠8

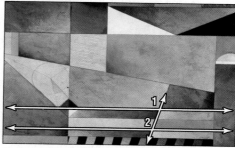

Example 1 Determine Angle Measures

In the figure, $m\angle 3 = 133$. Find $m\angle 5$.

∠3 ≅ ∠7	Corresponding Angles Postulate
∠7 ≅ ∠5	Vertical Angles Theorem
∠3 ≅ ∠5	Transitive Property
$m\angle 3 = m\angle 5$	Definition of congruent angles
$133 = m\angle 5$	Substitution

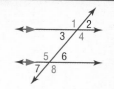

In Example 1, alternate interior angles 3 and 5 are congruent. This suggests another special relationship between angles formed by two parallel lines and a transversal. Other relationships are summarized in Theorems 3.1, 3.2, and 3.3.

Parallel Lines and Angle Pairs

Theorem	Examples
3.1 **Alternate Interior Angles Theorem** If two parallel lines are cut by a transversal, then each pair of alternate interior angles is congruent.	$\angle 4 \cong \angle 5$ $\angle 3 \cong \angle 6$
3.2 **Consecutive Interior Angles Theorem** If two parallel lines are cut by a transversal, then each pair of consecutive interior angles is supplementary.	$m\angle 4 + m\angle 6 = 180$ $m\angle 3 + m\angle 5 = 180$
3.3 **Alternate Exterior Angles Theorem** If two parallel lines are cut by a transversal, then each pair of alternate exterior angles is congruent.	$\angle 1 \cong \angle 8$ $\angle 2 \cong \angle 7$

You will prove Theorems 3.2 and 3.3 in Exercises 40 and 39, respectively.

Proof *Theorem 3.1*

Given: $a \parallel b$; p is a transversal of a and b.

Prove: $\angle 2 \cong \angle 7$, $\angle 3 \cong \angle 6$

Paragraph Proof: We are given that $a \parallel b$ with a transversal p. By the Corresponding Angles Postulate, $\angle 2 \cong \angle 4$ and $\angle 8 \cong \angle 6$. Also, $\angle 4 \cong \angle 7$ and $\angle 3 \cong \angle 8$ because vertical angles are congruent. Therefore, $\angle 2 \cong \angle 7$ and $\angle 3 \cong \angle 6$ since congruence of angles is transitive.

A special relationship occurs when the transversal is a perpendicular line.

Theorem 3.4

Perpendicular Transversal Theorem In a plane, if a line is perpendicular to one of two parallel lines, then it is perpendicular to the other.

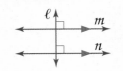

Proof *Theorem 3.4*

Given: $p \parallel q$, $t \perp p$

Prove: $t \perp q$

Proof:

Statements	Reasons
1. $p \parallel q$, $t \perp p$	1. Given
2. $\angle 1$ is a right angle.	2. Definition of $\perp$ lines
3. $m\angle 1 = 90$	3. Definition of right angle
4. $\angle 1 \cong \angle 2$	4. Corresponding Angles Postulate
5. $m\angle 1 = m\angle 2$	5. Definition of congruent angles
6. $m\angle 2 = 90$	6. Substitution Property
7. $\angle 2$ is a right angle.	7. Definition of right angles
8. $t \perp q$	8. Definition of $\perp$ lines

Example 2 Use an Auxiliary Line

Grid-In Test Item

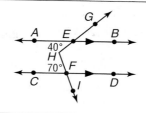

What is the measure of ∠GHI?

Read the Test Item

You need to find m∠GHI. Be sure to identify it correctly on the figure.

Solve the Test Item

Draw $\overleftrightarrow{JK}$ through H parallel to $\overleftrightarrow{AB}$ and $\overleftrightarrow{CD}$.

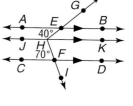

∠EHK ≅ ∠AEH	Alternate Interior Angles Theorem
m∠EHK = m∠AEH	Definition of congruent angles
m∠EHK = 40	Substitution

∠FHK ≅ ∠CFH	Alternate Interior Angles Theorem
m∠FHK = m∠CFH	Definition of congruent angles
m∠FHK = 70	Substitution

m∠GHI = m∠EHK + m∠FHK	Angle Addition Postulate
= 40 + 70 or 110	m∠EHK = 40, m∠FHK = 70

Write each digit of 110 in a column of the grid. Then shade in the corresponding bubble in each column.

Test-Taking Tip

Make a Drawing If you are allowed to write in your test booklet, sketch your drawings near the question to keep your work organized. Do not make any marks on the answer sheet except your answers.

ALGEBRA AND ANGLE MEASURES

Angles formed by two parallel lines and a transversal can be used to find unknown values.

Example 3 Find Values of Variables

ALGEBRA If m∠1 = 3x + 40, m∠2 = 2(y − 10), and m∠3 = 2x + 70, find x and y.

- Find x.

  Since $\overrightarrow{FG} \parallel \overrightarrow{EH}$, ∠1 ≅ ∠3 by the Corresponding Angles Postulate.

m∠1 = m∠3	Definition of congruent angles
3x + 40 = 2x + 70	Substitution
x = 30	Subtract 2x and 40 from each side.

- Find y.

  Since $\overrightarrow{FE} \parallel \overrightarrow{GH}$, ∠1 ≅ ∠2 by the Alternate Exterior Angles Theorem.

m∠1 = m∠2	Definition of congruent angles
3x + 40 = 2(y − 10)	Substitution
3(30) + 40 = 2(y − 10)	x = 30
130 = 2y − 20	Simplify.
150 = 2y	Add 20 to each side.
75 = y	Divide each side by 2.

Check for Understanding

Concept Check

1. **Determine** whether ∠1 is *always*, *sometimes*, or *never* congruent to ∠2. Explain.

2. **OPEN ENDED** Use a straightedge and protractor to draw a pair of parallel lines cut by a transversal so that one pair of corresponding angles measures 35°.

3. **Determine** the minimum number of angle measures you would have to know to find the measures of all of the angles in the figure for Exercise 1.

4. **State** the postulate or theorem that allows you to conclude ∠3 ≅ ∠5 in the figure at the right.

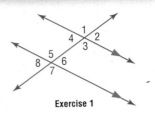

Exercise 1

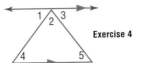

Exercise 4

Guided Practice

In the figure, $m\angle 3 = 110$ and $m\angle 12 = 55$. Find the measure of each angle.

5. ∠1
6. ∠6
7. ∠2
8. ∠10
9. ∠13
10. ∠15

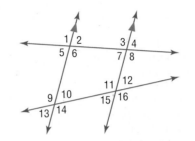

Find *x* and *y* in each figure.

11.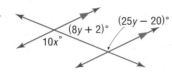

$(8y + 2)°$ $(25y - 20)°$ $10x°$

12.

$(3y + 1)°$ $(4x - 5)°$ $(3x + 11)°$

Standardized Test Practice

Ⓐ Ⓑ Ⓒ Ⓓ

13. **SHORT RESPONSE** Find $m\angle 1$.

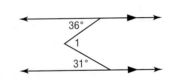

36° 1 31°

Practice and Apply

Homework Help

For Exercises	See Examples
14–31	1, 2
32–37	3

Extra Practice
See page 759.

In the figure, $m\angle 9 = 75$. Find the measure of each angle.

14. ∠3
15. ∠5
16. ∠6
17. ∠8
18. ∠11
19. ∠12

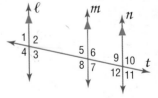

In the figure, $m\angle 3 = 43$. Find the measure of each angle.

20. ∠2
21. ∠7
22. ∠10
23. ∠11
24. ∠13
25. ∠16

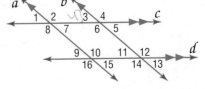

In the figure, $m\angle 1 = 50$ and $m\angle 3 = 60$. Find the measure of each angle.

26. ∠4
27. ∠5
28. ∠2
29. ∠6
30. ∠7
31. ∠8

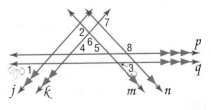

Find x and y in each figure.

32.

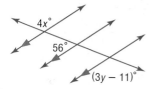

33.

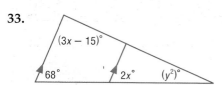
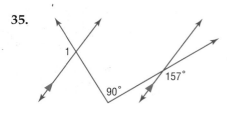
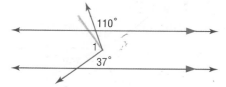

Find $m\angle 1$ in each figure.

34.

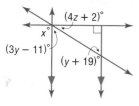

35.

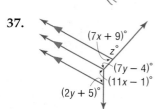
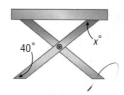

Find x, y, and z in each figure.

36.

37.

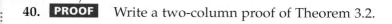

38. CARPENTRY Anthony is building a picnic table for his patio. He cut one of the legs at an angle of 40°. At what angle should he cut the other end to ensure that the top of the table is parallel to the ground?

39. PROOF Copy and complete the proof of Theorem 3.3.

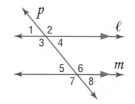

Given: $\ell \parallel m$

Prove: $\angle 1 \cong \angle 8$
$\angle 2 \cong \angle 7$

Proof:

Statements	Reasons
1. $\ell \parallel m$	1. ?
2. $\angle 1 \cong \angle 5$, $\angle 2 \cong \angle 6$	2. ?
3. $\angle 5 \cong \angle 8$, $\angle 6 \cong \angle 7$	3. ?
4. $\angle 1 \cong \angle 8$, $\angle 2 \cong \angle 7$	4. ?

40. PROOF Write a two-column proof of Theorem 3.2.

41. PROOF Write a paragraph proof of Theorem 3.4.

42. CONSTRUCTION Parallel drainage pipes are laid on each side of Polaris Street. A pipe under the street connects the two pipes. The connector pipe makes a 65° angle as shown. What is the measure of the angle it makes with the pipe on the other side of the road?

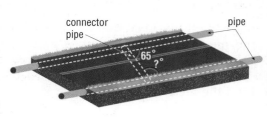

43. CRITICAL THINKING Explain why you can conclude that ∠2 and ∠6 are supplementary, but you cannot state that ∠4 and ∠6 are necessarily supplementary.

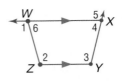

44. WRITING IN MATH Answer the question that was posed at the beginning of the lesson.

How can angles and lines be used in art?

Include the following in your answer:
- a description of how angles and lines are used to create patterns, and
- examples from two different artists that use lines and angles.

45. Line ℓ is parallel to line m. What is the value of x?

Ⓐ 30   Ⓑ 40
Ⓒ 50   Ⓓ 60

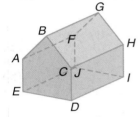

46. ALGEBRA If $ax = bx + c$, then what is the value of x in terms of a, b, and c?

Ⓐ $\dfrac{c}{a+b}$ Ⓑ $\dfrac{b}{a+c}$ Ⓒ $\dfrac{c}{a-b}$ Ⓓ $\dfrac{b+c}{a}$

Maintain Your Skills

Mixed Review

For Exercises 47–50, refer to the figure at the right. *(Lesson 3-1)*

47. Name all segments parallel to $\overline{AB}$.

48. Name all segments skew to $\overline{CH}$.

49. Name all planes parallel to AEF.

50. Name all segments intersecting $\overline{GH}$.

Find the measure of each numbered angle. *(Lesson 2-8)*

51.

52.

53° 2

Identify the hypothesis and conclusion of each statement. *(Lesson 2-3)*

53. If it rains this evening, then I will mow the lawn tomorrow.

54. A balanced diet will keep you healthy.

Getting Ready for the Next Lesson

PREREQUISITE SKILL Simplify each expression.
*(To review **simplifying expressions**, see pages 735 and 736.)*

55. $\dfrac{7-9}{8-5}$ **56.** $\dfrac{-3-6}{2-8}$ **57.** $\dfrac{14-11}{23-15}$ **58.** $\dfrac{15-23}{14-11}$ **59.** $\dfrac{2}{9} \cdot \left(-\dfrac{18}{5}\right)$

Practice Quiz 1 — Lessons 3-1 and 3-2

State the transversal that forms each pair of angles. Then identify the special name for the angle pair. *(Lesson 3-1)*

1. ∠1 and ∠8   **2.** ∠6 and ∠10   **3.** ∠11 and ∠14

Find the measure of each angle if $\ell \parallel m$ and $m\angle 1 = 105$. *(Lesson 3-2)*

4. ∠6   **5.** ∠4

3-3 Slopes of Lines

What You'll Learn

- Find slopes of lines.
- Use slope to identify parallel and perpendicular lines.

How is slope used in transportation?

Traffic signs are often used to alert drivers to road conditions. The sign at the right indicates a hill with a 6% *grade*. This means that the road will rise or fall 6 feet vertically for every 100 horizontal feet traveled.

6%

Vocabulary

- slope
- rate of change

SLOPE OF A LINE The **slope** of a line is the ratio of its vertical rise to its horizontal run.

$$\text{slope} = \frac{\text{vertical rise}}{\text{horizontal run}}$$

In a coordinate plane, the slope of a line is the ratio of the change along the y-axis to the change along the x-axis.

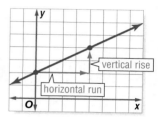

Key Concept   Slope

The slope m of a line containing two points with coordinates (x_1, y_1) and (x_2, y_2) is given by the formula

$$m = \frac{y_2 - y_1}{x_2 - x_1}, \text{ where } x_1 \neq x_2.$$

The slope of a line indicates whether the line rises to the right, falls to the right, or is horizontal. The slope of a vertical line, where $x_1 = x_2$, is undefined.

Example 1 Find the Slope of a Line

Find the slope of each line.

a.

Use the $\frac{\text{rise}}{\text{run}}$ method.

From $(-3, -2)$ to $(-1, 2)$, go up 4 units and right 2 units.

$\frac{\text{rise}}{\text{run}} = \frac{4}{2}$ or 2

b.

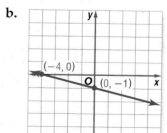

Use the slope formula.

Let $(-4, 0)$ be (x_1, y_1) and $(0, -1)$ be (x_2, y_2).

$m = \frac{y_2 - y_1}{x_2 - x_1}$

$= \frac{-1 - 0}{0 - (-4)}$ or $-\frac{1}{4}$

Study Tip

Slope

Lines with positive slope *rise* as you move from left to right, while lines with negative slope *fall* as you move from left to right.

c.

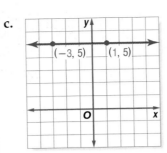

d.

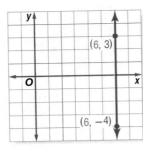

$$m = \frac{y_2 - y_1}{x_2 - x_1}$$

$$= \frac{5 - 5}{-3 - 1}$$

$$= \frac{0}{-4} \text{ or } 0$$

$$m = \frac{y_2 - y_1}{x_2 - x_1}$$

$$= \frac{3 - (-4)}{6 - 6}$$

$$= \frac{7}{0}, \text{ which is undefined}$$

The slope of a line can be used to identify the coordinates of any point on the line. It can also be used to describe a rate of change. The **rate of change** describes how a quantity is changing over time.

Example 2 *Use Rate of Change to Solve a Problem*

RECREATION   Between 1990 and 2000, annual sales of inline skating equipment increased by an average rate of $92.4 million per year. In 2000, the total sales were $1074.4 million. If sales increase at the same rate, what will the total sales be in 2008?

Let $(x_1, y_1) = (2000, 1074.4)$ and $m = 92.4$.

$m = \dfrac{y_2 - y_1}{x_2 - x_1}$	Slope formula
$92.4 = \dfrac{y_2 - 1074.4}{2008 - 2000}$	$m = 92.4$, $y_1 = 1074.4$, $x_1 = 2000$, and $x_2 = 2008$
$92.4 = \dfrac{y_2 - 1074.4}{8}$	Simplify.
$739.2 = y_2 - 1074.4$	Multiply each side by 8.
$1813.6 = y_2$	Add 1074.4 to each side.

The coordinates of the point representing the sales for 2008 are $(2008, 1813.6)$. Thus, the total sales in 2008 will be about $1813.6 million.

PARALLEL AND PERPENDICULAR LINES

Examine the graphs of lines ℓ, m, and n. Lines ℓ and m are parallel, and n is perpendicular to ℓ and m. Let's investigate the slopes of these lines.

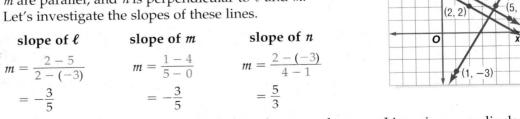

slope of ℓ	**slope of m**	**slope of n**
$m = \dfrac{2 - 5}{2 - (-3)}$	$m = \dfrac{1 - 4}{5 - 0}$	$m = \dfrac{2 - (-3)}{4 - 1}$
$= -\dfrac{3}{5}$	$= -\dfrac{3}{5}$	$= \dfrac{5}{3}$

Because lines ℓ and m are parallel, their slopes are the same. Line n is perpendicular to lines ℓ and m, and its slope is the opposite reciprocal of the slopes of ℓ and m; that is, $-\dfrac{3}{5} \cdot \dfrac{5}{3} = -1$. These results suggest two important algebraic properties of parallel and perpendicular lines.

Study Tip

Look Back
To review **if and only if statements**, see Reading Mathematics, page 81.

Postulates — *Slopes of Parallel and Perpendicular Lines*

3.2 Two nonvertical lines have the same slope if and only if they are parallel.

3.3 Two nonvertical lines are perpendicular if and only if the product of their slopes is -1.

Example 3 *Determine Line Relationships*

Determine whether $\overleftrightarrow{AB}$ and $\overleftrightarrow{CD}$ are *parallel*, *perpendicular*, or *neither*.

a. $A(-2, -5)$, $B(4, 7)$, $C(0, 2)$, $D(8, -2)$

Find the slopes of $\overleftrightarrow{AB}$ and $\overleftrightarrow{CD}$.

$$\text{slope of } \overleftrightarrow{AB} = \frac{7 - (-5)}{4 - (-2)} \qquad\qquad \text{slope of } \overleftrightarrow{CD} = \frac{-2 - 2}{8 - 0}$$

$$= \frac{12}{6} \text{ or } 2 \qquad\qquad\qquad\qquad = -\frac{4}{8} \text{ or } -\frac{1}{2}$$

The product of the slopes is $2\left(-\frac{1}{2}\right)$ or -1. So, $\overleftrightarrow{AB}$ is perpendicular to $\overleftrightarrow{CD}$.

b. $A(-8, -7)$, $B(4, -4)$, $C(-2, -5)$, $D(1, 7)$

$$\text{slope of } \overleftrightarrow{AB} = \frac{-4 - (-7)}{4 - (-8)} \qquad\qquad \text{slope of } \overleftrightarrow{CD} = \frac{7 - (-5)}{1 - (-2)}$$

$$= \frac{3}{12} \text{ or } \frac{1}{4} \qquad\qquad\qquad\qquad = \frac{12}{3} \text{ or } 4$$

The slopes are not the same, so $\overleftrightarrow{AB}$ and $\overleftrightarrow{CD}$ are not parallel. The product of the slopes is $4\left(\frac{1}{4}\right)$ or 1. So, $\overleftrightarrow{AB}$ and $\overleftrightarrow{CD}$ are neither parallel nor perpendicular.

The relationships of the slopes of lines can be used to graph a line parallel or perpendicular to a given line.

Example 4 *Use Slope to Graph a Line*

Graph the line that contains $P(-2, 1)$ and is perpendicular to $\overleftrightarrow{JK}$ with $J(-5, -4)$ and $K(0, -2)$.

First, find the slope of $\overleftrightarrow{JK}$.

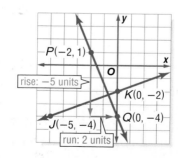

$$m = \frac{y_2 - y_1}{x_2 - x_1} \qquad \text{Slope formula}$$

$$= \frac{-2 - (-4)}{0 - (-5)} \qquad \text{Substitution}$$

$$= \frac{2}{5} \qquad \text{Simplify.}$$

Study Tip

Negative Slopes
To help determine direction with negative slopes, remember that $-\frac{5}{2} = \frac{-5}{2} = \frac{5}{-2}$.

The product of the slopes of two perpendicular lines is -1.

Since $\frac{2}{5}\left(-\frac{5}{2}\right) = -1$, the slope of the line perpendicular

to $\overleftrightarrow{JK}$ through $P(-2, 1)$ is $-\frac{5}{2}$.

Graph the line. Start at $(-2, 1)$. Move down 5 units and then move right 2 units. Label the point Q. Draw $\overleftrightarrow{PQ}$.

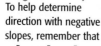 www.geometryonline.com/extra_examples

Check for Understanding

Concept Check

1. **Describe** what type of line is perpendicular to a vertical line. What type of line is parallel to a vertical line?

2. **FIND THE ERROR** Curtis and Lori calculated the slope of the line containing $A(15, 4)$ and $B(-6, -13)$. Who is correct? Explain your reasoning.

   Curtis
   $$m = \frac{4 - (-13)}{15 - (-6)}$$
   $$= \frac{17}{21}$$

   Lori
   $$m = \frac{4 - 13}{15 - 6}$$
   $$= -\frac{9}{11}$$

3. **OPEN ENDED** Give an example of a line whose slope is 0 and an example of a line whose slope is undefined.

Guided Practice

4. Determine the slope of the line that contains $A(-4, 3)$ and $B(-2, -1)$.

Find the slope of each line.

5. ℓ

6. m

7. any line perpendicular to ℓ

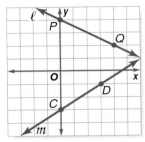

Exercises 5–7

Determine whether $\overleftrightarrow{GH}$ and $\overleftrightarrow{RS}$ are parallel, perpendicular, or neither.

8. $G(14, 13)$, $H(-11, 0)$, $R(-3, 7)$, $S(-4, -5)$

9. $G(15, -9)$, $H(9, -9)$, $R(-4, -1)$, $S(3, -1)$

Graph the line that satisfies each condition.

10. slope = 2, contains $P(1, 2)$

11. contains $A(6, 4)$, perpendicular to $\overleftrightarrow{MN}$ with $M(5, 0)$ and $N(1, 2)$

Application

MOUNTAIN BIKING For Exercises 12–14, use the following information.
A certain mountain bike trail has a section of trail with a grade of 8%.

12. What is the slope of the hill?

13. After riding on the trail, a biker is 120 meters below her original starting position. If her starting position is represented by the origin on a coordinate plane, what are the possible coordinates of her current position?

14. How far has she traveled down the hill? Round to the nearest meter.

Practice and Apply

Homework Help	
For Exercises	**See Examples**
15–18, 25–32	1
19–24	3
33–38	4
42, 43	2

Extra Practice
See page 759.

Determine the slope of the line that contains the given points.

15. $A(0, 2)$, $B(7, 3)$

16. $C(-2, -3)$, $D(-6, -5)$

17. $W(3, 2)$, $X(4, -3)$

18. $Y(1, 7)$, $Z(4, 3)$

Determine whether $\overleftrightarrow{PQ}$ and $\overleftrightarrow{UV}$ are parallel, perpendicular, or neither.

19. $P(-3, -2)$, $Q(9, 1)$, $U(3, 6)$, $V(5, -2)$

20. $P(-4, 0)$, $Q(0, 3)$, $U(-4, -3)$, $V(8, 6)$

21. $P(-10, 7)$, $Q(2, 1)$, $U(4, 0)$, $V(6, 1)$

22. $P(-9, 2)$, $Q(0, 1)$, $U(-1, 8)$, $V(-2, -1)$

23. $P(1, 1)$, $Q(9, 8)$, $U(-6, 1)$, $V(2, 8)$

24. $P(5, -4)$, $Q(10, 0)$, $U(9, -8)$, $V(5, -13)$

Find the slope of each line.

25. $\overleftrightarrow{AB}$

26. $\overleftrightarrow{PQ}$

27. $\overleftrightarrow{LM}$

28. $\overleftrightarrow{EF}$

29. a line parallel to $\overleftrightarrow{LM}$

30. a line perpendicular to $\overleftrightarrow{PQ}$

31. a line perpendicular to $\overleftrightarrow{EF}$

32. a line parallel to $\overleftrightarrow{AB}$

Graph the line that satisfies each condition.

33. slope $= -4$, passes through $P(-2, 1)$

34. contains $A(-1, -3)$, parallel to $\overleftrightarrow{CD}$ with $C(-1, 7)$ and $D(5, 1)$

35. contains $M(4, 1)$, perpendicular to $\overleftrightarrow{GH}$ with $G(0, 3)$ and $H(-3, 0)$

36. slope $= \frac{2}{5}$, contains $J(-7, -1)$

37. contains $Q(-2, -4)$, parallel to $\overleftrightarrow{KL}$ with $K(2, 7)$ and $L(2, -12)$

38. contains $W(6, 4)$, perpendicular to $\overleftrightarrow{DE}$ with $D(0, 2)$ and $E(5, 0)$.

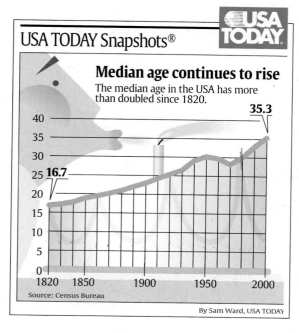

USA TODAY Snapshots®

Median age continues to rise

The median age in the USA has more than doubled since 1820.

40 — 35.3

35 —

30 —

25 16.7

20 —

15 —

10 —

5 —

0 —

1820 1850 1900 1950 2000

Source: Census Bureau

By Sam Ward, USA TODAY

POPULATION For Exercises 39–41, refer to the graph.

39. Estimate the annual rate of change of the median age from 1970 to 2000.

40. If the median age continues to increase at the same rate, what will be the median age in 2010?

41. Suppose that after 2000, the median age increases by $\frac{1}{3}$ of a year anually. In what year will the median age be 40.6?

Online Research **Data Update** Use the Internet or other resource to find the median age in the United States for years after 2000. Does the median age increase at the same rate as it did in years leading up to 2000? Visit www.geometryonline.com/data_update to learn more.

42. Determine the value of x so that a line containing $(6, 2)$ and $(x, -1)$ has a slope of $-\frac{3}{7}$. Then graph the line.

43. Find the value of x so that the line containing $(4, 8)$ and $(2, -1)$ is perpendicular to the line containing $(x, 2)$ and $(-4, 5)$. Graph the lines.

COMPUTERS For Exercises 44–46, refer to the graph at the right.

44. What is the rate of change between 1998 and 2000?

45. If the percent of classrooms with Internet access increases at the same rate as it did between 1999 and 2000, in what year will 90% of classrooms have Internet access?

46. Will the graph continue to rise indefinitely? Explain.

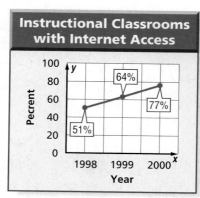

Instructional Classrooms with Internet Access

64%

77%

51%

1998 1999 2000

Year

Source: U.S. Census Bureau

47. CRITICAL THINKING The line containing the point $(5 + 2t, -3 + t)$ can be described by the equations $x = 5 + 2t$ and $y = -3 + t$. Write the slope-intercept form of the equation of this line.

48. **WRITING IN MATH** Answer the question that was posed at the beginning of the lesson.

How is slope used in transportation?

Include the following in your answer:
- an explanation of why it is important to display the grade of a road, and
- an example of slope used in transportation other than roads.

49. Find the slope of a line perpendicular to the line containing $(-5, 1)$ and $(-3, -2)$.

    Ⓐ $-\dfrac{2}{3}$ Ⓑ $-\dfrac{3}{2}$ Ⓒ $\dfrac{2}{3}$ Ⓓ $\dfrac{3}{2}$

50. ALGEBRA The winning sailboat completed a 24-mile race at an average speed of 9 miles per hour. The second-place boat finished with an average speed of 8 miles per hour. How many minutes longer than the winner did the second-place boat take to finish the race?

    Ⓐ 20 min   Ⓑ 33 min   Ⓒ 60 min   Ⓓ 120 min

Maintain Your Skills

Mixed Review   In the figure, $\overline{QR} \parallel \overline{TS}$, $\overleftrightarrow{QT} \parallel \overleftrightarrow{RS}$, and $m\angle 1 = 131$.
Find the measure of each angle. *(Lesson 3-2)*

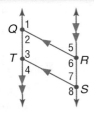

51. $\angle 6$ **52.** $\angle 7$

53. $\angle 4$ **54.** $\angle 2$

55. $\angle 5$ **56.** $\angle 8$

State the transversal that forms each pair of angles. Then identify the special name for each angle pair. *(Lesson 3-1)*

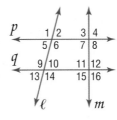

57. $\angle 1$ and $\angle 14$ **58.** $\angle 2$ and $\angle 10$

59. $\angle 3$ and $\angle 6$ **60.** $\angle 14$ and $\angle 15$

61. $\angle 7$ and $\angle 12$ **62.** $\angle 9$ and $\angle 11$

Make a conjecture based on the given information. Draw a figure to illustrate your conjecture. *(Lesson 2-1)*

63. Points H, I, and J are each located on different sides of a triangle.

64. Collinear points X, Y, and Z; Z is between X and Y.

65. $R(3, -4)$, $S(-2, -4)$, and $T(0, -4)$

Classify each angle as *right*, *acute*, or *obtuse*. *(Lesson 1-4)*

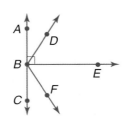

66. $\angle ABD$ **67.** $\angle DBF$

68. $\angle CBE$ **69.** $\angle ABF$

Getting Ready for the Next Lesson   **PREREQUISITE SKILL**   Solve each equation for y.
*(To review **solving equations**, see pages 737 and 738.)*

70. $2x + y = 7$ **71.** $2x + 4y = -5$ **72.** $5x - 2y + 4 = 0$

3-4 Equations of Lines

What You'll Learn

- Write an equation of a line given information about its graph.
- Solve problems by writing equations.

Vocabulary

- slope-intercept form
- point-slope form

How can the equation of a line describe the cost of cellular telephone service?

A certain cellular phone company charges a flat rate of $19.95 per month for service. All calls are charged $0.07 per minute of air time t. The total charge C for a month can be represented by the equation $C = 0.07t + 19.95$.

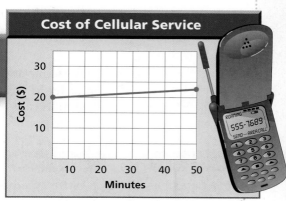

Cost of Cellular Service

WRITE EQUATIONS OF LINES You may remember from algebra that an equation of a line can be written given any of the following:

- the slope and the y-intercept,
- the slope and the coordinates of a point on the line, or
- the coordinates of two points on the line.

The graph of $C = 0.07t + 19.95$ has a slope of 0.07, and it intersects the y-axis at 19.95. These two values can be used to write an equation of the line. The **slope-intercept form** of a linear equation is $y = mx + b$, where m is the slope of the line and b is the y-intercept.

$$\overset{\text{slope}}{y = mx + b} \qquad C = 0.07t + 19.95$$
$$\underset{y\text{-intercept}}{}$$

Example 1 Slope and y-Intercept

Write an equation in slope-intercept form of the line with slope of -4 and y-intercept of 1.

$y = mx + b$ Slope-intercept form

$y = -4x + 1$ $m = -4, b = 1$

The slope-intercept form of the equation of the line is $y = -4x + 1$.

Another method used to write an equation of a line is the point-slope form of a linear equation. The **point-slope form** is $y - y_1 = m(x - x_1)$, where (x_1, y_1) are the coordinates of any point on the line and m is the slope of the line.

$$\overset{\text{given point } (x_1, y_1)}{y - y_1 = m(x - x_1)}$$
$$\underset{\text{slope}}{}$$

Choosing Forms
of Linear
Equations
If you are given a point
on a line and the slope
of the line, use point-slope
form. Otherwise, use
slope-intercept form.

Example 2 Slope and a Point

Write an equation in point-slope form of the line whose slope is $-\frac{1}{2}$ that contains $(3, -7)$.

$y - y_1 = m(x - x_1)$ Point-slope form

$y - (-7) = -\frac{1}{2}(x - 3)$ $m = -\frac{1}{2}, (x_1, y_1) = (3, -7)$

$y + 7 = -\frac{1}{2}(x - 3)$ Simplify.

The point-slope form of the equation of the line is $y + 7 = -\frac{1}{2}(x - 3)$.

Both the slope-intercept form and the point-slope form require the slope of a line in order to write an equation. There are occasions when the slope of a line is not given. In cases such as these, use two points on the line to calculate the slope. Then use the point-slope form to write an equation.

Writing Equations
Note that the point-slope
form of an equation is
different for each point
used. However, the slope-
intercept form of an
equation is unique.

Example 3 Two Points

Write an equation in slope-intercept form for line ℓ.

Find the slope of ℓ by using $A(-1, 6)$ and $B(3, 2)$.

$m = \dfrac{y_2 - y_1}{x_2 - x_1}$ Slope formula

$\quad = \dfrac{2 - 6}{3 - (-1)}$ $x_1 = -1, x_2 = 3, y_1 = 6, y_2 = 2$

$\quad = -\dfrac{4}{4}$ or -1 Simplify.

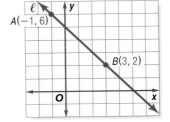

Now use the point-slope form and either point to write an equation.

Using Point A:

$y - y_1 = m(x - x_1)$ Point-slope form

$y - 6 = -1[x - (-1)]$ $m = -1, (x_1, y_1) = (-1, 6)$

$y - 6 = -1(x + 1)$ Simplify.

$y - 6 = -x - 1$ Distributive Property

$y = -x + 5$ Add 6 to each side.

Using Point B:

$y - y_1 = m(x - x_1)$ Point-slope form

$y - 2 = -1(x - 3)$ $m = -1, (x_1, y_1) = (3, 2)$

$y - 2 = -x + 3$ Distributive Property

$y = -x + 5$ Add 2 to each side.

Example 4 One Point and an Equation

Write an equation in slope-intercept form for a line containing $(2, 0)$ that is perpendicular to the line $y = -x + 5$.

Since the slope of the line $y = -x + 5$ is -1, the slope of a line perpendicular to it is 1.

$y - y_1 = m(x - x_1)$ Point-slope form

$y - 0 = 1(x - 2)$ $m = 1, (x_1, y_1) = (2, 0)$

$y = x - 2$ Distributive Property

WRITE EQUATIONS TO SOLVE PROBLEMS Many real-world situations can be modeled using linear equations. In many business applications, the slope represents a rate.

Example 5 Write Linear Equations

CELL PHONE COSTS Martina's current cellular phone plan charges $14.95 per month and $0.10 per minute of air time.

a. **Write an equation to represent the total monthly cost C for t minutes of air time.**

For each minute of air time, the cost increases $0.10. So, the rate of change, or slope, is 0.10. The y-intercept is located where 0 minutes of air time are used, or $14.95.

$C = mt + b$ Slope-intercept form
$C = 0.10t + 14.95$ $m = 0.10, b = 14.95$

The total monthly cost can be represented by the equation $C = 0.10t + 14.95$.

b. **Compare her current plan to the plan presented at the beginning of the lesson. If she uses an average of 40 minutes of air time each month, which plan offers the better rate?**

Evaluate each equation for $t = 40$.

Current plan: $C = 0.10t + 14.95$

$= 0.10(40) + 14.95$ $t = 40$

$= 18.95$ Simplify.

Alternate plan: $C = 0.07t + 19.95$

$= 0.07(40) + 19.95$ $t = 40$

$= 22.75$ Simplify.

Given her average usage, Martina's current plan offers the better rate.

Check for Understanding

Concept Check
1. **Explain** how you would write an equation of a line whose slope is $-\frac{2}{5}$ that contains $(-2, 8)$.

2. **Write** equations in slope-intercept form for two lines that contain $(-1, -5)$.

3. **OPEN ENDED** Graph a line that is not horizontal or vertical on the coordinate plane. Write the equation of the line.

Guided Practice
Write an equation in slope-intercept form of the line having the given slope and y-intercept.

4. $m = \frac{1}{2}$
     y-intercept: 4

5. $m = -\frac{3}{5}$
     intercept at $(0, -2)$

6. $m = 3$
     y-intercept: -4

Write an equation in point-slope form of the line having the given slope that contains the given point.

7. $m = \frac{3}{2}, (4, -1)$

8. $m = 3, (7, 5)$

9. $m = 1.25, (20, 137.5)$

Refer to the figure at the right. Write an equation in slope-intercept form for each line.

10. ℓ 11. k

12. the line parallel to ℓ that contains $(4, 4)$

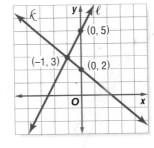

Application   **INTERNET**   For Exercises 13–14, use the following information.

Justin's current Internet service provider charges a flat rate of $39.95 per month for unlimited access. Another provider charges $4.95 per month for access and $0.95 for each hour of connection.

13. Write an equation to represent the total monthly cost for each plan.

14. If Justin is online an average of 60 hours per month, should he keep his current plan, or change to the other plan? Explain.

Practice and Apply

Homework Help

For Exercises	See Examples
15–20, 35, 36	1
21–26	2
27–30, 37–42	3
31–34, 43, 44	4
45–51	5

Extra Practice
See page 759.

Write an equation in slope-intercept form of the line having the given slope and y-intercept.

15. $m: \dfrac{1}{6}$, y-intercept: -4 16. $m: \dfrac{2}{3}$, $(0, 8)$ 17. $m: \dfrac{5}{8}$, $(0, -6)$

18. $m: \dfrac{2}{9}$, y-intercept: $\dfrac{1}{3}$ 19. $m: -1$, $b: -3$ 20. $m: -\dfrac{1}{12}$, $b: 1$

Write an equation in point-slope form of the line having the given slope that contains the given point.

21. $m = 2$, $(3, 1)$ 22. $m = -5$, $(4, 7)$ 23. $m = -\dfrac{4}{5}$, $(-12, -5)$

24. $m = \dfrac{1}{16}$, $(3, 11)$ 25. $m = 0.48$, $(5, 17.12)$ 26. $m = -1.3$, $(10, 87.5)$

Write an equation in slope-intercept form for each line.

27. k 28. ℓ

29. m 30. n

31. perpendicular to line ℓ, contains $(-1, 6)$

32. parallel to line k, contains $(7, 0)$

33. parallel to line n, contains $(0, 0)$

34. perpendicular to line m, contains $(-3, -3)$

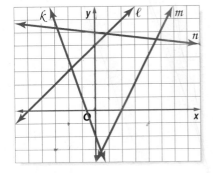

Write an equation in slope-intercept form for the line that satisfies the given conditions.

35. $m = -3$, y-intercept $= 5$ 36. $m = 0$, y-intercept $= 6$

37. x-intercept $= 5$, y-intercept $= 3$ 38. contains $(4, -1)$ and $(-2, -1)$

39. contains $(-5, -3)$ and $(10, -6)$ 40. x-intercept $= 5$, y-intercept $= -1$

41. contains $(-6, 8)$ and $(-6, -4)$ 42. contains $(-4, -1)$ and $(-8, -5)$

43. Write an equation of the line that contains $(7, -2)$ and is parallel to $2x - 5y = 8$.

44. What is an equation of the line that is perpendicular to $2y + 2 = -\dfrac{7}{4}(x - 7)$ and contains $(-2, -3)$?

45. JOBS Ann MacDonald is a salesperson at a discount appliance store. She earns $50 for each appliance that she sells plus a 5% commission on the price of the appliance. Write an equation that represents what she earned in a week in which she sold 15 appliances.

BUSINESS **For Exercises 46–49, use the following information.**
The Rainbow Paint Company sells an average of 750 gallons of paint each day.

46. How many gallons of paint will they sell in x days?

47. The store has 10,800 gallons of paint in stock. Write an equation in slope-intercept form that describes how many gallons of paint will be on hand after x days if no new stock is added.

48. Draw a graph that represents the number of gallons of paint on hand at any given time.

49. If it takes 4 days to receive a shipment of paint from the manufacturer after it is ordered, when should the store manager order more paint so that the store does not run out?

MAPS **For Exercises 50 and 51, use the following information.**
Suppose a map of Texas is placed on a coordinate plane with the western tip at the origin. Jeff Davis, Pecos, and Brewster counties meet at $(130, -70)$, and Jeff Davis, Reeves, and Pecos counties meet at $(120, -60)$.

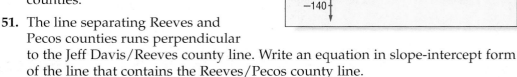

50. Write an equation in slope-intercept form that models the county line between Jeff Davis and Reeves counties.

51. The line separating Reeves and Pecos counties runs perpendicular to the Jeff Davis/Reeves county line. Write an equation in slope-intercept form of the line that contains the Reeves/Pecos county line.

52. CRITICAL THINKING The point-slope form of an equation of a line can be rewritten as $y = m(x - x_1) + y_1$. Describe how the graph of $y = m(x - x_1) + y_1$ is related to the graph of $y = mx$.

53. WRITING IN MATH   Answer the question that was posed at the beginning of the lesson.

How can the equation of a line describe cellular telephone service?

Include the following in your answer:

• an explanation of how the fee for air time affects the equation, and
• a description of how you can use equations to compare various plans.

54. What is the slope of a line perpendicular to the line represented by $2x - 8y = 16$?

   (A) -4 (B) -2 (C) $-\frac{1}{4}$ (D) $\frac{1}{4}$

55. ALGEBRA What are all of the values of y for which $y^2 < 1$?

   (A) $y < -1$ (B) $-1 < y < 1$ (C) $y > -1$ (D) $y < 1$

Mixed Review   Determine the slope of the line that contains the given points.   *(Lesson 3-3)*

56. $A(0, 6)$, $B(4, 0)$ **57.** $G(8, 1)$, $H(8, -6)$ **58.** $E(6, 3)$, $F(-6, 3)$

In the figure, $m\angle 1 = 58$, $m\angle 2 = 47$, and $m\angle 3 = 26$.
Find the measure of each angle.   *(Lesson 3-2)*

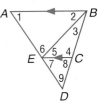

59. $\angle 7$ **60.** $\angle 5$ **61.** $\angle 6$

62. $\angle 4$ **63.** $\angle 8$ **64.** $\angle 9$

65. **PROOF**   Write a two-column proof.   *(Lesson 2-6)*

Given: $AC = DF$
$AB = DE$

Prove: $BC = EF$

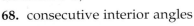

Find the perimeter of $\triangle ABC$ to the nearest hundredth, given the coordinates of its vertices.   *(Lesson 1-6)*

66. $A(10, -6)$, $B(-2, -8)$, $C(-5, -7)$ **67.** $A(-3, 2)$, $B(2, -9)$, $C(0, -10)$

Getting Ready for the Next Lesson   **PREREQUISITE SKILL**   In the figure at the right, lines s and t are intersected by the transversal m. Name the pairs of angles that meet each description.
*(To review **angles formed by two lines and a transversal**, see Lesson 3-1.)*

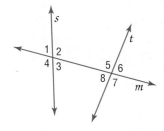

68. consecutive interior angles

69. corresponding angles

70. alternate exterior angles

71. alternate interior angles

Practice Quiz 2
Lessons 3-3 and 3-4

Determine whether $\overleftrightarrow{AB}$ and $\overleftrightarrow{CD}$ are *parallel*, *perpendicular*, or *neither*.   *(Lesson 3-3)*

1. $A(3, -1)$, $B(6, 1)$, $C(-2, -2)$, $D(2, 4)$ **2.** $A(-3, -11)$, $B(3, 13)$, $C(0, -6)$, $D(8, -8)$

For Exercises 3–8, refer to the graph at the right. Find the slope of each line.   *(Lesson 3-3)*

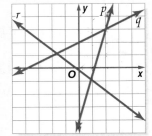

3. p

4. a line parallel to q

5. a line perpendicular to r

Write an equation in slope-intercept form for each line.   *(Lesson 3-4)*

6. q

7. parallel to r, contains $(-1, 4)$

8. perpendicular to p, contains $(0, 0)$

Write an equation in point-slope form for the line that satisfies the given condition.   *(Lesson 3-4)*

9. parallel to $y = -\frac{1}{4}x + 2$, contains $(5, -8)$

10. perpendicular to $y = -3$, contains $(-4, -4)$

3-5 Proving Lines Parallel

What You'll Learn

- Recognize angle conditions that occur with parallel lines.
- Prove that two lines are parallel based on given angle relationships.

How do you know that the sides of a parking space are parallel?

Have you ever been in a tall building and looked down at a parking lot? The parking lot is full of line segments that appear to be parallel. The workers who paint these lines must be certain that they are parallel.

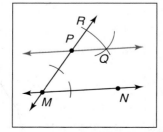

IDENTIFY PARALLEL LINES   When each stripe of a parking space intersects the center line, the angles formed are corresponding angles. If the lines are parallel, we know that the corresponding angles are congruent. Conversely, if the corresponding angles are congruent, then the lines must be parallel.

Postulate 3.4

If two lines in a plane are cut by a transversal so that corresponding angles are congruent, then the lines are parallel.

Abbreviation: *If corr. ∠s are ≅, then lines are ∥.*

Examples: If ∠1 ≅ ∠5, ∠2 ≅ ∠6, ∠3 ≅ ∠7, or ∠4 ≅ ∠8, then *m* ∥ *n*.

Postulate 3.4 justifies the construction of parallel lines.

Construction

Parallel Line Through a Point Not on Line

① Use a straightedge to draw a line. Label two points on the line as *M* and *N*. Draw a point *P* that is not on $\overleftrightarrow{MN}$. Draw $\overleftrightarrow{PM}$.

② Copy ∠PMN so that *P* is the vertex of the new angle. Label the intersection points *Q* and *R*.

③ Draw $\overleftrightarrow{PQ}$. Because ∠RPQ ≅ ∠PMN by construction and they are corresponding angles, $\overleftrightarrow{PQ}$ ∥ $\overleftrightarrow{MN}$.

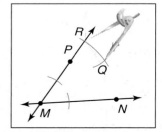

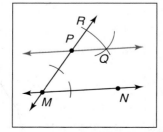

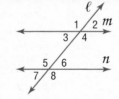

The construction establishes that there is *at least* one line through P that is parallel to $\overleftrightarrow{MN}$. In 1795, Scottish physicist and mathematician John Playfair provided the modern version of Euclid's Parallel Postulate, which states there is *exactly* one line parallel to a line through a given point not on the line.

Postulate 3.5

Parallel Postulate   If given a line and a point not on the line, then there exists exactly one line through the point that is parallel to the given line.

Parallel lines with a transversal create many pairs of congruent angles. Conversely, those pairs of congruent angles can determine whether a pair of lines is parallel.

Key Concept   *Proving Lines Parallel*

	Theorems	Examples	
3.5	If two lines in a plane are cut by a transversal so that a pair of alternate exterior angles is congruent, then the two lines are parallel. **Abbreviation:** *If alt. ext. ∠s are ≅, then lines are ∥.*	If $\angle 1 \cong \angle 8$ or if $\angle 2 \cong \angle 7$, then $m \parallel n$.	
3.6	If two lines in a plane are cut by a transversal so that a pair of consecutive interior angles is supplementary, then the lines are parallel. **Abbreviation:** *If cons. int. ∠s are suppl., then lines are ∥.*	If $m\angle 3 + m\angle 5 = 180$ or if $m\angle 4 + m\angle 6 = 180$, then $m \parallel n$.	
3.7	If two lines in a plane are cut by a transversal so that a pair of alternate interior angles is congruent, then the lines are parallel. **Abbreviation:** *If alt. int. ∠s are ≅, then lines are ∥.*	If $\angle 3 \cong \angle 6$ or if $\angle 4 \cong \angle 5$, then $m \parallel n$.	
3.8	In a plane, if two lines are perpendicular to the same line, then they are parallel. **Abbreviation:** *If 2 lines are ⊥ to the same line, then lines are ∥.*	If $\ell \perp m$ and $\ell \perp n$, then $m \parallel n$.	

Example 1   *Identify Parallel Lines*

In the figure, $\overline{BG}$ bisects $\angle ABH$. Determine which lines, if any, are parallel.

- The sum of the angle measures in a triangle must be 180, so $m\angle BDF = 180 - (45 + 65)$ or 70.
- Since $\angle BDF$ and $\angle BGH$ have the same measure, they are congruent.
- Congruent corresponding angles indicate parallel lines. So, $\overleftrightarrow{DF} \parallel \overleftrightarrow{GH}$.
- $\angle ABD \cong \angle DBF$, because $\overline{BG}$ bisects $\angle ABH$. So, $m\angle ABD = 45$.
- $\angle ABD$ and $\angle BDF$ are alternate interior angles, but they have different measures so they are not congruent.
- Thus, $\overleftrightarrow{AB}$ is not parallel to $\overleftrightarrow{DF}$ or $\overleftrightarrow{GH}$.

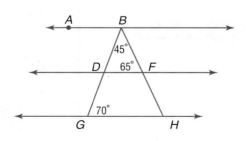

Angle relationships can be used to solve problems involving unknown values.

Example 2 Solve Problems with Parallel Lines

ALGEBRA Find x and $m\angle RSU$ so that $m \parallel n$.

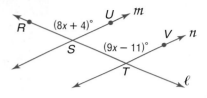

Explore From the figure, you know that $m\angle RSU = 8x + 4$ and $m\angle STV = 9x - 11$. You also know that $\angle RSU$ and $\angle STV$ are corresponding angles.

Plan For line m to be parallel to line n, the corresponding angles must be congruent. So, $m\angle RSU = m\angle STV$. Substitute the given angle measures into this equation and solve for x. Once you know the value of x, use substitution to find $m\angle RSU$.

Solve

$m\angle RSU = m\angle STV$	Corresponding angles
$8x + 4 = 9x - 11$	Substitution
$4 = x - 11$	Subtract $8x$ from each side.
$15 = x$	Add 11 to each side.

Now use the value of x to find $m\angle RSU$.

$m\angle RSU = 8x + 4$	Original equation
$= 8(15) + 4$	$x = 15$
$= 124$	Simplify.

Examine Verify the angle measure by using the value of x to find $m\angle STV$. That is, $9x - 11 = 9(15) - 11$ or 124. Since $m\angle RSU = m\angle STV$, $\angle RSU \cong \angle STV$ and $m \parallel n$.

PROVE LINES PARALLEL The angle pair relationships formed by a transversal can be used to prove that two lines are parallel.

Example 3 Prove Lines Parallel

Given: $r \parallel s$
$\angle 5 \cong \angle 6$

Prove: $\ell \parallel m$

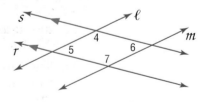

Proof:

Statements	Reasons
1. $r \parallel s$, $\angle 5 \cong \angle 6$	1. Given
2. $\angle 4$ and $\angle 5$ are supplementary.	2. Consecutive Interior Angle Theorem
3. $m\angle 4 + m\angle 5 = 180$	3. Definition of supplementary angles
4. $m\angle 5 = m\angle 6$	4. Definition of congruent angles
5. $m\angle 4 + m\angle 6 = 180$	5. Substitution Property (=)
6. $\angle 4$ and $\angle 6$ are supplementary.	6. Definition of supplementary angles
7. $\ell \parallel m$	7. If cons. int. $\angle$s are suppl., then lines are $\parallel$.

In Lesson 3-3, you learned that parallel lines have the same slope. You can use the slopes of lines to prove that lines are parallel.

Example 4 *Slope and Parallel Lines*

Determine whether $g \parallel f$.

slope of f: $m = \dfrac{4-0}{6-3}$ or $\dfrac{4}{3}$

slope of g: $m = \dfrac{4-0}{0-(-3)}$ or $\dfrac{4}{3}$

Since the slopes are the same, $g \parallel f$.

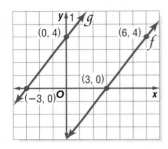

Check for Understanding

Concept Check
1. **Summarize** five different methods to prove that two lines are parallel.

2. **Find a counterexample** for the following statement.
   If lines ℓ and m are cut by transversal t so that consecutive interior angles are congruent, then lines ℓ and m are parallel and t is perpendicular to both lines.

3. **OPEN ENDED**   Describe two situations in your own life in which you encounter parallel lines. How could you verify that the lines are parallel?

Guided Practice
Given the following information, determine which lines, if any, are parallel. State the postulate or theorem that justifies your answer.

4. $\angle 16 \cong \angle 3$ 5. $\angle 4 \cong \angle 13$

6. $m\angle 14 + m\angle 10 = 180$ 7. $\angle 1 \cong \angle 7$

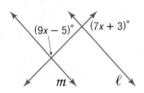

Find x so that $\ell \parallel m$.

8.
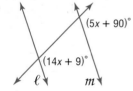

9.
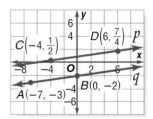

10.   **PROOF**   Write a two-column proof of Theorem 3.5.

11.   Determine whether $p \parallel q$.

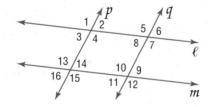

Application   12. **PHYSICS**   The Hubble Telescope gathers parallel light rays and directs them to a central focal point. Use a protractor to measure several of the angles shown in the diagram. Are the lines parallel? Explain how you know.

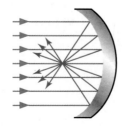

Homework Help

For Exercises	See Examples
13–24	1
26–31	2
25, 32–37	3
38–39	4

Extra Practice
See page 760.

Given the following information, determine which lines, if any, are parallel. State the postulate or theorem that justifies your answer.

13. $\angle 2 \cong \angle 8$

14. $\angle 9 \cong \angle 16$

15. $\angle 2 \cong \angle 10$

16. $\angle 6 \cong \angle 15$

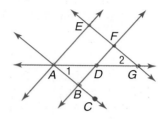

17. $\angle AEF \cong \angle BFG$

18. $\angle EAB \cong \angle DBC$

19. $\angle EFB \cong \angle CBF$

20. $m\angle GFD + m\angle CBD = 180$

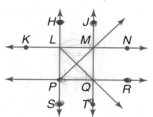

21. $\angle HLK \cong \angle JML$

22. $\angle PLQ \cong \angle MQL$

23. $m\angle MLP + \angle RPL = 180$

24. $\overleftrightarrow{HS} \perp \overrightarrow{PR}, \overleftrightarrow{JT} \perp \overrightarrow{PR}$

25. **PROOF** Copy and complete the proof of Theorem 3.8.

Given: $\ell \perp t$
$m \perp t$

Prove: $\ell \parallel m$

Proof:

Statements	Reasons
1. $\ell \perp t, m \perp t$	1. ?
2. $\angle 1$ and $\angle 2$ are right angles.	2. ?
3. $\angle 1 \cong \angle 2$	3. ?
4. $\ell \parallel m$	4. ?

WebQuest

Latitude lines are parallel, and longitude lines appear parallel in certain locations on Earth. Visit www.geometryonline.com/webquest to continue work on your WebQuest project.

Find x so that $\ell \parallel m$.

26.
$(9x - 4)°$
$140°$

27.
$(8x + 4)°$
$(9x - 11)°$

28.
$(7x - 1)°$

29.
$(4 - 5x)°$
$(7x + 100)°$

30.
$(14x + 9)°$
$(5x + 90)°$

31.
$(178 - 3x)°$
$(7x - 38)°$

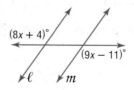
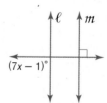
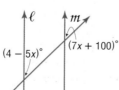
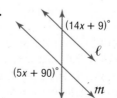
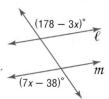

32. **PROOF** Write a two-column proof of Theorem 3.6.

33. **PROOF** Write a paragraph proof of Theorem 3.7.

PROOF Write a two-column proof for each of the following.

34. Given: $\angle 2 \cong \angle 1$
$\qquad\quad \angle 1 \cong \angle 3$
Prove: $\overline{ST} \parallel \overline{UV}$

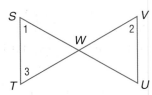

35. Given: $\overline{AD} \perp \overline{CD}$
$\qquad\quad \angle 1 \cong \angle 2$
Prove: $\overline{BC} \perp \overline{CD}$

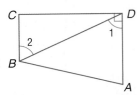

36. Given: $\overline{JM} \parallel \overline{KN}$
$\qquad\quad \angle 1 \cong \angle 2$
$\qquad\quad \angle 3 \cong \angle 4$
Prove: $\overline{KM} \parallel \overline{LN}$

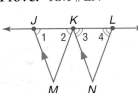

37. Given: $\angle RSP \cong \angle PQR$
$\qquad\quad \angle QRS$ and $\angle PQR$ are supplementary.
Prove: $\overline{PS} \parallel \overline{QR}$

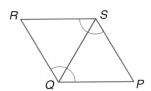

Determine whether each pair of lines is parallel. Explain why or why not.

38.

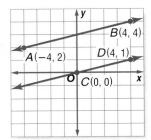

39.

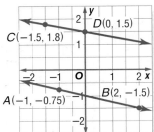

40. HOME IMPROVEMENT To build a fence, Jim positioned the fence posts and then placed a 2×4 board at an angle between the fence posts. As he placed each picket, he measured the angle that the picket made with the 2×4. Why does this ensure that the pickets will be parallel?

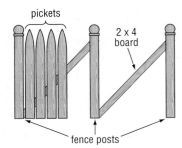

41. FOOTBALL When striping the practice football field, Mr. Hawkinson first painted the sidelines. Next he marked off 10-yard increments on one sideline. He then constructed lines perpendicular to the sidelines at each 10-yard mark. Why does this guarantee that the 10-yard lines will be parallel?

42. CRITICAL THINKING When Adeel was working on an art project, he drew a four-sided figure with two pairs of opposite parallel sides. He noticed some patterns relating to the angles in the figure. List as many patterns as you can about a 4-sided figure with two pairs of opposite parallel sides.

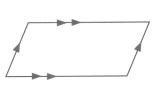

43. RESEARCH Use the Internet or other resource to find mathematicians like John Playfair who discovered new concepts and proved new theorems related to parallel lines. Briefly describe their discoveries.

44. 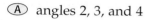 Answer the question that was posed at the beginning of the lesson.

How do you know that the sides of a parking space are parallel?

Include the following in your answer:
- a comparison of the angles at which the lines forming the edges of a parking space strike the centerline, and
- a description of the type of parking spaces that form congruent consecutive interior angles.

45. In the figure, line ℓ is parallel to line m. Line n intersects both ℓ and m. Which of the following lists includes all of the angles that are supplementary to $\angle 1$?

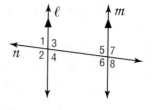

   Ⓐ angles 2, 3, and 4   Ⓑ angles 2, 3, 6, and 7

   Ⓒ angles 4, 5, and 8   Ⓓ angles 3, 4, 7, and 8

46. ALGEBRA Kendra has at least one quarter, one dime, one nickel, and one penny. If she has three times as many pennies as nickels, the same number of nickels as dimes, and twice as many dimes as quarters, then what is the least amount of money she could have?

   Ⓐ $0.41   Ⓑ $0.48   Ⓒ $0.58   Ⓓ $0.61

Maintain Your Skills

Mixed Review   **Write an equation in slope-intercept form for the line that satisfies the given conditions.** *(Lesson 3-4)*

47. $m = 0.3$, y-intercept is -6

48. $m = \frac{1}{3}$, contains $(-3, -15)$

49. contains $(5, 7)$ and $(-3, 11)$

50. perpendicular to $y = \frac{1}{2}x - 4$, contains $(4, 1)$

Find the slope of each line. *(Lesson 3-3)*

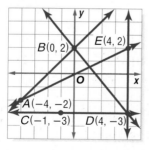

51. $\overleftrightarrow{BD}$ **52.** $\overleftrightarrow{CD}$

53. $\overrightarrow{AB}$ **54.** $\overrightarrow{EO}$

55. any line parallel to $\overleftrightarrow{DE}$

56. any line perpendicular to $\overleftrightarrow{BD}$

Construct a truth table for each compound statement. *(Lesson 2-2)*

57. p and q **58.** p or $\sim q$ **59.** $\sim p \wedge q$ **60.** $\sim p \wedge \sim q$

61. CARPENTRY A carpenter must cut two pieces of wood at angles so that they fit together to form the corner of a picture frame. What type of angles must he use to make sure that a corner results? *(Lesson 1-5)*

Getting Ready for the Next Lesson   **PREREQUISITE SKILL** Use the Distance Formula to find the distance between each pair of points. *(To review **the Distance Formula**, see Lesson 1-4.)*

62. $(2, 7)$, $(7, 19)$ **63.** $(8, 0)$, $(-1, 2)$ **64.** $(-6, -4)$, $(-8, -2)$

Graphing Calculator Investigation

Points of Intersection

You can use a TI-83 Plus graphing calculator to determine the points of intersection of a transversal and two parallel lines.

Example

Parallel lines ℓ and m are cut by a transversal t. The equations of ℓ, m, and t are $y = \frac{1}{2}x - 4$, $y = \frac{1}{2}x + 6$, and $y = -2x + 1$, respectively. Use a graphing calculator to determine the points of intersection of t with ℓ and m.

Step 1 Enter the equations in the Y= list and graph in the standard viewing window.

KEYSTROKES: [Y=] 1 [÷] 2 [X,T,θ,n] [−] 4 [ENTER] 1 [÷] 2 [X,T,θ,n] [+] 6 [ENTER] [(−)] 2 [X,T,θ,n] [+] 1 [ZOOM] 6

Step 2 Use the CALC menu to find the points of intersection.

- Find the intersection of ℓ and t.

  KEYSTROKES: [2nd] [CALC] 5 [ENTER] [▼]
  [ENTER] [ENTER]

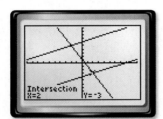

[−10, 10] scl: 1 by [−10, 10] scl: 1

Lines ℓ and t intersect at $(2, -3)$.

- Find the intersection of m and t.

  KEYSTROKES: [2nd] [CALC] 5 [▼] [ENTER]
  [ENTER] [ENTER]

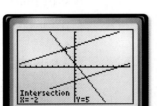

[−10, 10] scl: 1 by [−10, 10] scl: 1

Lines m and t intersect at $(-2, 5)$.

Exercises

Parallel lines a and b are cut by a transversal t. Use a graphing calculator to determine the points of intersection of t with a and b. Round to the nearest tenth.

1. a: $y = 2x - 10$
b: $y = 2x - 2$
t: $y = -\frac{1}{2}x + 4$

2. a: $y = -x - 3$
b: $y = -x + 5$
t: $y = x - 6$

3. a: $y = 6$
b: $y = 0$
t: $x = -2$

4. a: $y = -3x + 1$
b: $y = -3x - 3$
t: $y = \frac{1}{3}x + 8$

5. a: $y = \frac{4}{5}x - 2$
b: $y = \frac{4}{5}x - 7$
t: $y = -\frac{5}{4}x$

6. a: $y = -\frac{1}{6}x + \frac{2}{3}$
b: $y = -\frac{1}{6}x + \frac{5}{12}$
t: $y = 6x + 2$

www.geometryonline.com/other_calculator_keystrokes

Perpendiculars and Distance

- Find the distance between a point and a line.
- Find the distance between parallel lines.

Vocabulary

- equidistant

How does the distance between parallel lines relate to hanging new shelves?

When installing shelf brackets, it is important that the vertical bracing be parallel in order for the shelves to line up. One technique is to install the first brace and then use a carpenter's square to measure and mark two or more points the same distance from the first brace. You can then align the second brace with those marks.

DISTANCE FROM A POINT TO A LINE In Lesson 3-5, you learned that if two lines are perpendicular to the same line, then they are parallel. The carpenter's square is used to construct a line perpendicular to each pair of shelves. The space between each pair of shelves is measured along the perpendicular segment. This is to ensure that the shelves are parallel. This is an example of using lines and perpendicular segments to determine distance. The shortest segment from a point to a line is the perpendicular segment from the point to the line.

Key Concept   Distance Between a Point and a Line

- **Words** The distance from a line to a point not on the line is the length of the segment perpendicular to the line from the point.

- **Model**

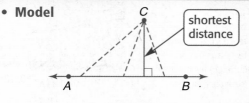

shortest distance

Study Tip

Measuring the Shortest Distance
You can use tools like the corner of a piece of paper or your book to help draw a right angle.

Example 1 Distance from a Point to a Line

Draw the segment that represents the distance from P to $\overleftrightarrow{AB}$.

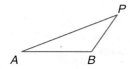

Since the distance from a line to a point not on the line is the length of the segment perpendicular to the line from the point, extend $\overline{AB}$ and draw $\overline{PQ}$ so that $\overline{PQ} \perp \overleftrightarrow{AB}$.

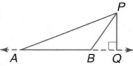

When you draw a perpendicular segment from a point to a line, you can guarantee that it is perpendicular by using the construction of a line perpendicular to a line through a point not on that line.

Example **2** *Construct a Perpendicular Segment*

COORDINATE GEOMETRY Line ℓ contains points $(-6, -9)$ and $(0, -1)$. Construct a line perpendicular to line ℓ through $P(-7, -2)$ not on ℓ. Then find the distance from P to ℓ.

① Graph line ℓ and point P. Place the compass point at point P. Make the setting wide enough so that when an arc is drawn, it intersects ℓ in two places. Label these points of intersection A and B.

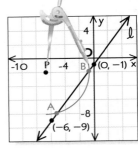

② Put the compass at point A and draw an arc below line ℓ. (*Hint:* Any compass setting greater than $\frac{1}{2}AB$ will work.)

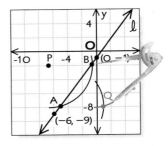

③ Using the same compass setting, put the compass at point B and draw an arc to intersect the one drawn in step 2. Label the point of intersection Q.

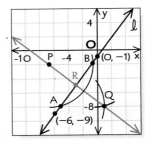

④ Draw $\overleftrightarrow{PQ}$. $\overleftrightarrow{PQ} \perp \ell$. Label point R at the intersection of $\overleftrightarrow{PQ}$ and ℓ. *Use the slopes of $\overleftrightarrow{PQ}$ and ℓ to verify that the lines are perpendicular.*

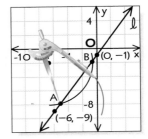

The segment constructed from point $P(-7, -2)$ perpendicular to the line ℓ, appears to intersect line ℓ at $R(-3, -5)$. Use the Distance Formula to find the distance between point P and line ℓ.

$$d = \sqrt{(x_2 - x_1)^2 + (y_2 - y_1)^2}$$
$$= \sqrt{(-7 - (-3))^2 + (-2 - (-5))^2}$$
$$= \sqrt{25} \text{ or } 5$$

The distance between P and ℓ is 5 units.

DISTANCE BETWEEN PARALLEL LINES

Two lines in a plane are parallel if they are everywhere **equidistant**. Equidistant means that the distance between two lines measured along a perpendicular line to the lines is always the same. The distance between parallel lines is the length of the perpendicular segment with endpoints that lie on each of the two lines.

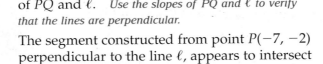

$AK = BJ = CH = DG = EF$

Distance Between Parallel Lines

The distance between two parallel lines is the distance between one of the lines and any point on the other line.

Recall that a *locus* is the set of all points that satisfy a given condition. Parallel lines can be described as the locus of points in a plane equidistant from a given line.

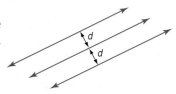

Theorem 3.9

In a plane, if two lines are equidistant from a third line, then the two lines are parallel to each other.

Example 3 *Distance Between Lines*

Find the distance between the parallel lines ℓ and m whose equations are $y = -\frac{1}{3}x - 3$ and $y = -\frac{1}{3}x + \frac{1}{3}$, respectively.

You will need to solve a system of equations to find the endpoints of a segment that is perpendicular to both ℓ and m. The slope of lines ℓ and m is $-\frac{1}{3}$.

- First, write an equation of a line p perpendicular to ℓ and m. The slope of p is the opposite reciprocal of $-\frac{1}{3}$, or 3. Use the y-intercept of line ℓ, $(0, -3)$, as one of the endpoints of the perpendicular segment.

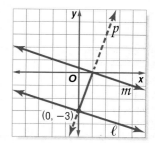

$$y - y_1 = m(x - x_1) \quad \text{Point-slope form}$$
$$y - (-3) = 3(x - 0) \quad x_1 = 0, y_1 = -3, m = 3$$
$$y + 3 = 3x \quad \text{Simplify.}$$
$$y = 3x - 3 \quad \text{Subtract 3 from each side.}$$

- Next, use a system of equations to determine the point of intersection of line m and p.

m: $y = -\frac{1}{3}x + \frac{1}{3}$ $-\frac{1}{3}x + \frac{1}{3} = 3x - 3$ Substitute $-\frac{1}{3}x + \frac{1}{3}$ for y in the second equation.

p: $y = 3x - 3$

$$-\frac{1}{3}x - 3x = -3 - \frac{1}{3} \quad \text{Group like terms on each side.}$$
$$-\frac{10}{3}x = -\frac{10}{3} \quad \text{Simplify on each side.}$$
$$x = 1 \quad \text{Divide each side by } -\frac{10}{3}.$$
$$y = 3(1) - 3 \quad \text{Substitute 1 for } x \text{ in the equation for } p.$$
$$y = 0 \quad \text{Simplify.}$$

The point of intersection is $(1, 0)$.

- Then, use the Distance Formula to determine the distance between $(0, -3)$ and $(1, 0)$.

$$d = \sqrt{(x_2 - x_1)^2 + (y_2 - y_1)^2} \quad \text{Distance Formula}$$
$$= \sqrt{(0 - 1)^2 + (-3 - 0)^2} \quad x_2 = 0, x_1 = 1, y_2 = -3, y_1 = 0$$
$$= \sqrt{10} \quad \text{Simplify.}$$

The distance between the lines is $\sqrt{10}$ or about 3.16 units.

Check for Understanding

Concept Check

1. **Explain** how to construct a segment between two parallel lines to represent the distance between them.

2. **OPEN ENDED** Make up a problem involving an everyday situation in which you need to find the distance between a point and a line or the distance between two lines. For example, find the shortest path from the patio of a house to a garden to minimize the length of a walkway and material used in its construction.

3. **Compare and contrast** three different methods that you can use to show that two lines in a plane are parallel.

Guided Practice Copy each figure. Draw the segment that represents the distance indicated.

4. L to $\overleftrightarrow{KN}$

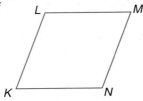

5. D to $\overleftrightarrow{AE}$

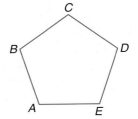

6. **COORDINATE GEOMETRY** Line ℓ contains points $(0, 0)$ and $(2, 4)$. Draw line ℓ. Construct a line perpendicular to ℓ through $A(2, -6)$. Then find the distance from A to ℓ.

Find the distance between each pair of parallel lines.

7. $y = \frac{3}{4}x - 1$

   $y = \frac{3}{4}x + \frac{1}{8}$

8. $x + 3y = 6$

   $x + 3y = -14$

9. Graph the line whose equation is $y = -\frac{3}{4}x + \frac{1}{4}$. Construct a perpendicular segment through $P(2, 5)$. Then find the distance from P to the line.

Application 10. **UTILITIES** Housing developers often locate the shortest distance from a house to the water main so that a minimum of pipe is required to connect the house to the water supply. Copy the diagram, and draw a possible location for the pipe.

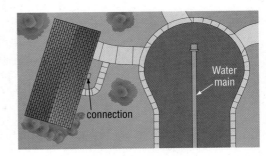

Practice and Apply

Homework Help

For Exercises	See Examples
11–16	1
17, 18	2
19–24	3
25–27	1–2

Extra Practice
See page 760.

Copy each figure. Draw the segment that represents the distance indicated.

11. C to $\overleftrightarrow{AD}$

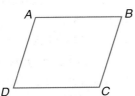

12. K to $\overleftrightarrow{JL}$

13. Q to $\overleftrightarrow{RS}$

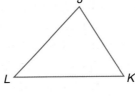

Copy each figure. Draw the segment that represents the distance indicated.

14. Y to $\overleftrightarrow{WX}$

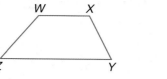

15. G to $\overleftrightarrow{HJ}$

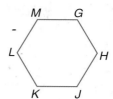

16. W to $\overleftrightarrow{UV}$

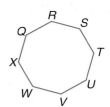

COORDINATE GEOMETRY Construct a line perpendicular to ℓ through P. Then find the distance from P to ℓ.

17. Line ℓ contains points $(-3, 0)$ and $(3, 0)$. Point P has coordinates $(4, 3)$.

18. Line ℓ contains points $(0, -2)$ and $(1, 3)$. Point P has coordinates $(-4, 4)$.

Find the distance between each pair of parallel lines.

19. $y = -3$
$y = 1$

20. $x = 4$
$x = -2$

21. $y = 2x + 2$
$y = 2x - 3$

22. $y = 4x$
$y = 4x - 17$

23. $y = 2x - 3$
$2x - y = -4$

24. $y = -\dfrac{3}{4}x - 1$
$3x + 4y = 20$

Graph each line. Construct a perpendicular segment through the given point. Then find the distance from the point to the line.

25. $y = 5, (-2, 4)$ **26.** $y = 2x + 2, (-1, -5)$ **27.** $2x - 3y = -9, (2, 0)$

28. **PROOF** Write a paragraph proof of Theorem 3.9.

29. **INTERIOR DESIGN** Theresa is installing a curtain rod on the wall above the window. In order to ensure that the rod is parallel to the ceiling, she measures and marks 9 inches below the ceiling in several places. If she installs the rod at these markings centered over the window, how does she know the curtain rod will be parallel to the ceiling?

30. **CONSTRUCTION** When framing a wall during a construction project, carpenters often use a plumb line. A *plumb line* is a string with a weight called a *plumb bob* attached on one end. The plumb line is suspended from a point and then used to ensure that wall studs are vertical. How does the plumb line help to find the distance from a point to the floor?

31. **ALGEBRA** In the coordinate plane, if a line has equation $ax + by = c$, then the distance from a point (x_1, y_1) is given by $\dfrac{|ax_1 + by_1 - c|}{\sqrt{a^2 + b^2}}$. Determine the distance from $(4, 6)$ to the line whose equation is $3x + 4y = 6$.

32. **CRITICAL THINKING** Draw a diagram that represents each description.
  a. Point P is equidistant from two parallel lines.
  b. Point P is equidistant from two intersecting lines.
  c. Point P is equidistant from two parallel planes.
  d. Point P is equidistant from two intersecting planes.
  e. A line is equidistant from two parallel planes.
  f. A plane is equidistant from two other planes that are parallel.

33.  Answer the question that was posed at the beginning of the lesson.

How does the distance between parallel lines relate to hanging new shelves?

Include the following in your answer:
- an explanation of why marking several points equidistant from the first brace will ensure that the braces are parallel, and
- a description of other types of home improvement projects that require that two or more elements are parallel.

34. GRID IN Segment *AB* is perpendicular to segment *BD*. Segment *AB* and segment *CD* bisect each other at point *X*. If *AB* = 16 and *CD* = 20, what is the length of $\overline{BD}$?

35. ALGEBRA A coin was flipped 24 times and came up heads 14 times and tails 10 times. If the first and the last flips were both heads, what is the greatest number of consecutive heads that could have occurred?

Ⓐ 7 Ⓑ 9 Ⓒ 10 Ⓓ 13

Maintain Your Skills

Mixed Review

Given the following information, determine which lines, if any, are parallel. State the postulate or theorem that justifies your answer. *(Lesson 3-5)*

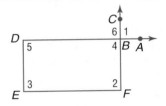

36. $\angle 5 \cong \angle 6$

37. $\angle 6 \cong \angle 2$

38. $\angle 1$ and $\angle 2$ are supplementary.

Write an equation in slope-intercept form for each line. *(Lesson 3-4)*

39. *a* **40.** *b* **41.** *c*

42. perpendicular to line *a*, contains $(-1, -4)$

43. parallel to line *c*, contains $(2, 5)$

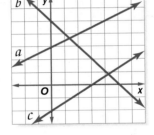

44. **PROOF** Write a two-column proof. *(Lesson 2-7)*

Given: *NL* = *NM*
  AL = *BM*

Prove: *NA* = *NB*

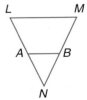

WebQuest **Internet Project** ▸ *When Is Weather Normal?*

It's time to complete your project. Use the information and data you have gathered about climate and locations on Earth to prepare a portfolio or Web page. Be sure to include graphs and/or tables in the presentation.

www.geometryonline.com/webquest

Geometry Activity

Non-Euclidean Geometry

So far in this text, we have studied **plane Euclidean geometry**, which is based on a system of points, lines, and planes. In **spherical geometry**, we study a system of points, great circles (lines), and spheres (planes). Spherical geometry is one type of **non-Euclidean geometry**.

Plane Euclidean Geometry

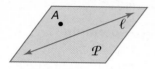

Plane $\mathcal{P}$ contains line ℓ and point A not on ℓ.

Spherical Geometry

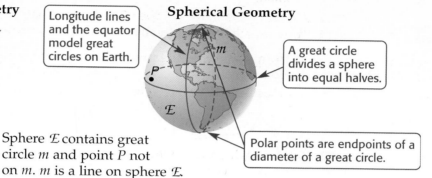

Longitude lines and the equator model great circles on Earth.

A great circle divides a sphere into equal halves.

Polar points are endpoints of a diameter of a great circle.

Sphere $\mathcal{E}$ contains great circle m and point P not on m. m is a line on sphere $\mathcal{E}$.

The table below compares and contrasts *lines* in the system of plane Euclidean geometry and *lines* (great circles) in spherical geometry.

Plane Euclidean Geometry Lines on the Plane	Spherical Geometry Great Circles (Lines) on the Sphere
1. A line segment is the shortest path between two points.	**1.** An arc of a great circle is the shortest path between two points.
2. There is a unique line passing through any two points.	**2.** There is a unique great circle passing through any pair of nonpolar points.
3. A line goes on infinitely in two directions.	**3.** A great circle is finite and returns to its original starting point.

4. If three points are collinear, exactly one is between the other two.
 	
     A   B   C
   ◄━━━●━━━━━●━━━━━●━━━►
 	
B is between *A* and *C*.
	4. If three points are collinear, any one of the three points is between the other two.
	
	A is between *B* and *C*.
	B is between *A* and *C*.
	C is between *A* and *B*.

In spherical geometry, Euclid's first four postulates and their related theorems hold true. However, theorems that depend on the parallel postulate (Postulate 5) may not be true.

In Euclidean geometry parallel lines lie in the same plane and never intersect. In spherical geometry, the sphere is the plane, and a great circle represents a line. Every great circle containing A intersects ℓ. Thus, there exists no line through point A that is parallel to ℓ.

(continued on the next page)

Every great circle of a sphere intersects all other great circles on that sphere in exactly two points. In the figure at the right, one possible line through point A intersects line ℓ at P and Q.

If two great circles divide a sphere into four congruent regions, the lines are perpendicular to each other at their intersection points. Each longitude circle on Earth intersects the equator at right angles.

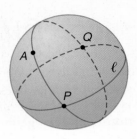

Example *Compare Plane and Spherical Geometries*

For each property listed from plane Euclidean geometry, write a corresponding statement for spherical geometry.

a. Perpendicular lines intersect at one point.

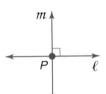

Perpendicular great circles intersect at two points.

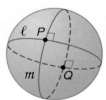

b. Perpendicular lines form four right angles.

Perpendicular great circles form eight right angles.

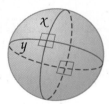

Exercises

For each property from plane Euclidean geometry, write a corresponding statement for spherical geometry.

1. A line goes on infinitely in two directions.
2. A line segment is the shortest path between two points.
3. Two distinct lines with no point of intersection are parallel.
4. Two distinct intersecting lines intersect in exactly one point.
5. A pair of perpendicular straight lines divides the plane into four infinite regions.
6. Parallel lines have infinitely many common perpendicular lines.
7. There is only one distance that can be measured between two points.

If spherical points are restricted to be nonpolar points, determine if each statement from plane Euclidean geometry is also *true* in spherical geometry. If *false*, explain your reasoning.

8. Any two distinct points determine exactly one line.
9. If three points are collinear, exactly one point is between the other two.
10. Given a line ℓ and point P not on ℓ, there exists exactly one line parallel to ℓ passing through P.

Vocabulary and Concept Check

alternate exterior angles (p. 128)
alternate interior angles (p. 128)
consecutive interior angles (p. 128)
corresponding angles (p. 128)
equidistant (p. 160)
non-Euclidean geometry (p. 165)

parallel lines (p. 126)
parallel planes (p. 126)
plane Euclidean geometry (p. 165)
point-slope form (p. 145)
rate of change (p. 140)

skew lines (p. 127)
slope (p. 139)
slope-intercept form (p. 145)
spherical geometry (p. 165)
transversal (p. 127)

A complete list of postulates and theorems can be found on pages R1–R8.

Exercises Refer to the figure and choose the term that best completes each sentence.

1. Angles 4 and 5 are *(consecutive, alternate)* interior angles.
2. The distance from point *A* to line *n* is the length of the segment *(perpendicular, parallel)* to line *n* through *A*.
3. If ∠4 and ∠6 are supplementary, lines *m* and *n* are said to be *(parallel, intersecting)* lines.
4. Line ℓ is a *(slope-intercept, transversal)* for lines *n* and *m*.
5. ∠1 and ∠8 are *(alternate interior, alternate exterior)* angles.
6. If *n* ∥ *m*, ∠6 and ∠3 are *(supplementary, congruent)*.
7. Angles 5 and 3 are *(consecutive, alternate)* interior angles.

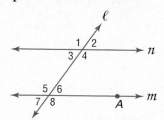

Lesson-by-Lesson Review

3-1 Parallel Lines and Transversals

See pages 126–131.

Concept Summary

- Coplanar lines that do not intersect are called *parallel*.
- When two lines are cut by a transversal, there are many angle relationships.

Example Identify each pair of angles as *alternate interior, alternate exterior, corresponding,* or *consecutive interior* angles.

a. ∠7 and ∠3
   corresponding
b. ∠4 and ∠6
   consecutive interior
c. ∠7 and ∠2
   alternate exterior
d. ∠3 and ∠6
   alternate interior

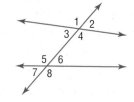

Exercises Identify each pair of angles as *alternate interior, alternate exterior, corresponding,* or *consecutive interior* angles. *See Example 3 on page 128.*

8. ∠10 and ∠6
9. ∠5 and ∠12
10. ∠8 and ∠10
11. ∠1 and ∠9
12. ∠3 and ∠6
13. ∠5 and ∠3
14. ∠2 and ∠7
15. ∠8 and ∠9

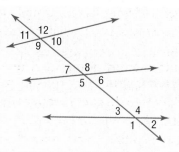

3-2   Angles and Parallel Lines

See pages 133–138.

Concept Summary

- Pairs of congruent angles formed by parallel lines and a transversal are corresponding angles, alternate interior angles, and alternate exterior angles.
- Pairs of consecutive interior angles are supplementary.

Example   In the figure, $m\angle 1 = 4p + 15$, $m\angle 3 = 3p - 10$, and $m\angle 4 = 6r + 5$. Find the values of p and r.

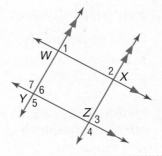

- Find p.

  Since $\overleftrightarrow{AC} \parallel \overleftrightarrow{BD}$, $\angle 1$ and $\angle 3$ are supplementary by the Consecutive Interior Angles Theorem.

  $$m\angle 1 + m\angle 3 = 180 \quad \text{Definition of supplementary angles}$$
  $$(4p + 15) + (3p - 10) = 180 \quad \text{Substitution}$$
  $$7p + 5 = 180 \quad \text{Simplify.}$$
  $$p = 25 \quad \text{Solve for } p.$$

- Find r.

  Since $\overleftrightarrow{AB} \parallel \overleftrightarrow{CD}$, $\angle 4 \cong \angle 3$ by the Corresponding Angles Postulate.

  $$m\angle 4 = m\angle 3 \quad \text{Definition of congruent angles}$$
  $$6r + 5 = 3(25) - 10 \quad \text{Substitution, } p = 25$$
  $$6r + 5 = 65 \quad \text{Simplify.}$$
  $$r = 10 \quad \text{Solve for } x.$$

Exercises   In the figure, $m\angle 1 = 53$. Find the measure of each angle.   *See Example 1 on page 133.*

16. $\angle 2$ 17. $\angle 3$
18. $\angle 4$ 19. $\angle 5$
20. $\angle 6$ 21. $\angle 7$
22. In the figure, $m\angle 1 = 3a + 40$, $m\angle 2 = 2a + 25$, and $m\angle 3 = 5b - 26$. Find a and b.
   See Example 3 on page 135.

3-3   Slopes of Lines

See pages 139–144.

Concept Summary

- The slope of a line is the ratio of its vertical rise to its horizontal run.
- Parallel lines have the same slope, while perpendicular lines have slopes whose product is -1.

Example   Determine whether $\overleftrightarrow{KM}$ and $\overleftrightarrow{LN}$ are *parallel, perpendicular,* or *neither* for $K(-3, 3)$, $M(-1, -3)$, $L(2, 5)$, and $N(5, -4)$.

slope of $\overleftrightarrow{KM}$: $m = \dfrac{-3 - 3}{-1 - (-3)}$ or -3 \qquad slope of $\overleftrightarrow{LN}$: $m = \dfrac{-4 - 5}{5 - 2}$ or -3

The slopes are the same. So $\overleftrightarrow{KM}$ and $\overleftrightarrow{LN}$ are parallel.

Exercises Determine whether $\overleftrightarrow{AB}$ and $\overleftrightarrow{CD}$ are *parallel, perpendicular, or neither.*
See Example 3 on page 141.

23. $A(-4, 1)$, $B(3, -1)$, $C(2, 2)$, $D(0, 9)$ **24.** $A(6, 2)$, $B(2, -2)$, $C(-1, -4)$, $D(5, 2)$

25. $A(1, -3)$, $B(4, 5)$, $C(1, -1)$, $D(-7, 2)$ **26.** $A(2, 0)$, $B(6, 3)$, $C(-1, -4)$, $D(3, -1)$

Graph the line that satisfies each condition. *See Example 4 on page 141.*

27. contains $(2, 3)$ and is parallel to $\overrightarrow{AB}$ with $A(-1, 2)$ and $B(1, 6)$

28. contains $(-2, -2)$ and is perpendicular to $\overrightarrow{PQ}$ with $P(5, 2)$ and $Q(3, -4)$

3-4 Equations of Lines

See pages 145–150.

Concept Summary

In general, an equation of a line can be written if you are given:

- slope and the *y*-intercept
- the slope and the coordinates of a point on the line, or
- the coordinates of two points on the line.

Example Write an equation in slope-intercept form of the line that passes through $(2, -4)$ and $(-3, 1)$.

Find the slope of the line.

$m = \dfrac{y_2 - y_1}{x_2 - x_1}$ Slope Formula

$= \dfrac{1 - (-4)}{-3 - 2}$ $(x_1, y_1) = (2, -4),$
$(x_2, y_2) = (-3, 1)$

$= \dfrac{5}{-5}$ or -1 Simplify.

Now use the point-slope form and either point to write an equation.

$y - y_1 = m(x - x_1)$ Point-slope form

$y - (-4) = -1(x - 2)$ $m = -1$, $(x_1, y_1) = (2, -4)$

$y + 4 = -x + 2$ Simplify.

$y = -x - 2$ Subtract 4 from each side.

Exercises Write an equation in slope-intercept form of the line that satisfies the given conditions. *See Examples 1–3 on pages 145 and 146.*

29. $m = 2$, contains $(1, -5)$ **30.** contains $(2, 5)$ and $(-2, -1)$

31. $m = -\dfrac{2}{7}$, *y*-intercept = 4   **32.** $m = -\dfrac{3}{2}$, contains $(2, -4)$

33. $m = 5$, *y*-intercept = -3 **34.** contains $(3, -1)$ and $(-4, 6)$

3-5 Proving Lines Parallel

See pages 151–157.

Concept Summary

When lines are cut by a transversal, certain angle relationships produce parallel lines.

- congruent corresponding angles
- congruent alternate exterior angles
- congruent alternate interior angles
- supplementary consecutive interior angles

Chapter

3 For More ...

- Extra Practice, see pages 758–760.
- Mixed Problem Solving, see page 784.

Example If ∠1 ≅ ∠8, which lines if any are parallel?

∠1 and ∠8 are alternate exterior angles for lines *r* and *s*. These lines are cut by the transversal *p*. Since the angles are congruent, lines *r* and *s* are parallel by Theorem 3.5.

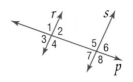

Exercises Given the following information, determine which lines, if any, are parallel. State the postulate or theorem that justifies your answer.
See Example 1 on page 152.

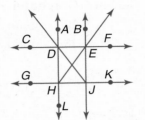

35. ∠*GHL* ≅ ∠*EJK*
36. *m*∠*ADJ* + *m*∠*DJE* = 180
37. $\overrightarrow{CF} \perp \overrightarrow{AL}, \overrightarrow{GK} \perp \overrightarrow{AL}$
38. ∠*DJE* ≅ ∠*HDJ*
39. *m*∠*EJK* + *m*∠*JEF* = 180
40. ∠*GHL* ≅ ∠*CDH*

3-6 Perpendiculars and Distance

See pages 159–164.

Concept Summary

- The distance between a point and a line is measured by the perpendicular segment from the point to the line.

Example Find the distance between the parallel lines *q* and *r* whose equations are *y* = *x* − 2 and *y* = *x* + 2, respectively.

- The slope of *q* is 1. Choose a point on line *q* such as *P*(2, 0). Let line *k* be perpendicular to *q* through *P*. The slope of line *k* is −1. Write an equation for line *k*.

$y = mx + b$	Slope-intercept form
$0 = (-1)(2) + b$	$y = 0, m = -1, x = 2$
$2 = b$	Solve for *b*. An equation for *k* is $y = -x + 2$.

- Use a system of equations to determine the point of intersection of *k* and *r*.

  $$y = x + 2$$
  $$\underline{y = -x + 2}$$
  $$2y = 4 \quad \text{Add the equations.}$$
  $$y = 2 \quad \text{Divide each side by 2.}$$

  Substitute 2 for *y* in the original equation.
  $$2 = -x + 2$$
  $$x = 0 \quad \text{Solve for } x.$$
  The point of intersection is (0, 2).

- Now use the Distance Formula to determine the distance between (2, 0) and (0, 2).

  $$d = \sqrt{(x_2 - x_1)^2 + (y_2 - y_1)^2} = \sqrt{(2 - 0)^2 + (0 - 2)^2} = \sqrt{8}$$

  The distance between the lines is $\sqrt{8}$ or about 2.83 units.

Exercises Find the distance between each pair of parallel lines.
See Example 3 on page 161.

41. *y* = 2*x* − 4, *y* = 2*x* + 1

42. $y = \frac{1}{2}x, y = \frac{1}{2}x + 5$

Vocabulary and Concepts

1. Write an equation of a line that is perpendicular to $y = 3x - \frac{2}{7}$.

2. Name a theorem that can be used to prove that two lines are parallel.

3. Find all the angles that are supplementary to $\angle 1$.

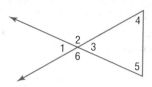

Skills and Applications

In the figure, $m\angle 12 = 64$. Find the measure of each angle.

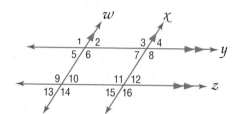

4. $\angle 8$
6. $\angle 7$
8. $\angle 3$
10. $\angle 9$

5. $\angle 13$
7. $\angle 11$
9. $\angle 4$
11. $\angle 5$

Graph the line that satisfies each condition.

12. slope $= -1$, contains $P(-2, 1)$

13. contains $Q(-1, 3)$ and is perpendicular to $\overleftrightarrow{AB}$ with $A(-2, 0)$ and $B(4, 3)$

14. contains $M(1, -1)$ and is parallel to $\overleftrightarrow{FG}$ with $F(3, 5)$ and $G(-3, -1)$

15. slope $= -\frac{4}{3}$, contains $K(3, -2)$

For Exercises 16–21, refer to the figure at the right. Find each value if $p \parallel q$.

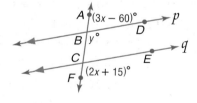

16. x
18. $m\angle FCE$
20. $m\angle BCE$

17. y
19. $m\angle ABD$
21. $m\angle CBD$

Find the distance between each pair of parallel lines.

22. $y = 2x - 1$, $y = 2x + 9$

23. $y = -x + 4$, $y = -x - 2$

24. **COORDINATE GEOMETRY** Detroit Road starts in the center of the city, and Lorain Road starts 4 miles west of the center of the city. Both roads run southeast. If these roads are put on a coordinate plane with the center of the city at $(0, 0)$, Lorain Road is represented by the equation $y = -x - 4$ and Detroit Road is represented by the equation $y = -x$. How far away is Lorain Road from Detroit Road?

25. **STANDARDIZED TEST PRACTICE** In the figure at the right, which cannot be true if $m \parallel \ell$ and $m\angle 1 = 73$?

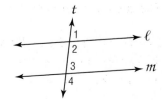

Ⓐ $m\angle 4 > 73$
Ⓑ $\angle 1 \cong \angle 4$
Ⓒ $m\angle 2 + m\angle 3 = 180$
Ⓓ $\angle 3 \cong \angle 1$

Part 1 Multiple Choice

Record your answers on the answer sheet provided by your teacher or on a sheet of paper.

1. Jahaira needed 2 meters of fabric to reupholster a chair in her bedroom. If Jahaira can only find a centimeter ruler, how much fabric should she cut?   (Prerequisite Skill)

   (A) 20 cm   (B) 200 cm

   (C) 2000 cm   (D) 20,000 cm

2. A fisherman uses a coordinate grid marked in miles to locate the nets cast at sea. How far apart are nets A and B?   (Lesson 1-3)

   (A) 3 mi

   (B) $\sqrt{28}$ mi

   (C) $\sqrt{65}$ mi

   (D) 11 mi

   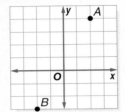

3. If $\angle ABC \cong \angle CBD$, which statement must be true?   (Lesson 1-5)

   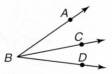

   (A) Segment BC bisects $\angle ABD$.

   (B) $\angle ABD$ is a right angle.

   (C) $\angle ABC$ and $\angle CBD$ are supplementary.

   (D) Segments AB and BD are perpendicular.

4. Valerie cut a piece of wood at a 72° angle for her project. What is the degree measure of the supplementary angle on the leftover piece of wood?   (Lesson 1-6)

   (A) 18   (B) 78   (C) 98   (D) 108

5. A pan balance scale is often used in science classes. What is the value of x to balance the scale if one side weighs $4x + 4$ units and the other weighs $6x - 8$ units?   (Lesson 2-3)

   (A) 1   (B) 2   (C) 3   (D) 6

Use the diagram below of a tandem bicycle frame for Questions 6 and 7.

6. The diagram shows the two posts on which seats are placed and several crossbars. Which term describes $\angle 6$ and $\angle 5$?   (Lesson 3-1)

   (A) alternate exterior angles

   (B) alternate interior angles

   (C) consecutive interior angles

   (D) corresponding angles

7. The quality control manager for the bicycle manufacturer wants to make sure that the two seat posts are parallel. Which angles can she measure to determine this?   (Lesson 3-5)

   (A) $\angle 2$ and $\angle 3$ (B) $\angle 1$ and $\angle 3$

   (C) $\angle 4$ and $\angle 8$ (D) $\angle 5$ and $\angle 7$

8. Which is the equation of a line that is perpendicular to the line $4y - x = 8$?   (Lesson 3-3)

   (A) $y = -\frac{1}{4}x - 2$ (B) $y = \frac{1}{4}x + 2$

   (C) $y = -4x - 15$ (D) $y = 4x + 15$

9. The graph of $y = 2x - 5$ is shown at the right. How would the graph be different if the number 2 in the equation was replaced with a 4?   (Lesson 3-4)

   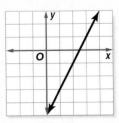

   (A) parallel to the line shown above, but shifted two units higher

   (B) parallel to the line shown above, but shifted two units lower

   (C) have a steeper slope, but intercept the y-axis at the same point

   (D) have a less steep slope, but intercept the y-axis at the same point

Preparing for Standardized Tests
For test-taking strategies and more
practice, see pages 795–810.

Part 2 Short Response/Grid In

Record your answers on the answer sheet provided by your teacher or on a sheet of paper.

10. What should statement 2 be to complete this proof? (Lesson 2-4)

Given: $\dfrac{4x - 6}{3} = 10$

Prove: $x = 9$

Statements	Reasons
1. $\dfrac{4x - 6}{3} = 10$	1. Given
2. _?_	2. Multiplication Property
3. $4x - 6 = 30$	3. Simplify.
4. $4x = 36$	4. Addition Property
5. $x = 9$	5. Division Property

The director of a high school marching band draws a diagram of a new formation as shown below. In the figure, $\overleftrightarrow{AB} \parallel \overleftrightarrow{CD}$. Use the figure for Questions 11 and 12.

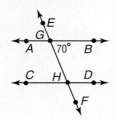

11. During the performance, a flag holder stands at point H, facing point F, and rotates right until she faces point C. What angle measure describes the flag holder's rotation? (Lesson 3-2)

12. Band members march along segment CH, turn left at point H, and continue to march along $\overline{HG}$. What is $m\angle CHG$? (Lesson 3-2)

13. What is the slope of a line containing points (3, 4) and (9, 6)? (Lesson 3-3)

www.geometryonline.com/standardized_test

Test-Taking Tip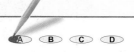

Question 13
Many standardized tests provide a reference sheet that includes formulas you may use. Quickly review the sheet before you begin so that you know what formulas are available.

Part 3 Extended Response

Record your answers on a sheet of paper. Show your work.

14. To get a player out who was running from third base to home, Kahlil threw the ball a distance of 120 feet, from second base toward home plate. Did the ball reach home plate? Show and explain your calculations to justify your answer. (Lesson 1-3)

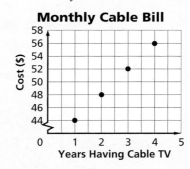

15. Brad's family has subscribed to cable television for 4 years, as shown below.

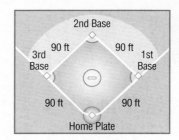

a. Find the slope of a line connecting the points on the graph. (Lesson 3-4)

b. Describe what the slope of the line represents. (Lesson 3-4)

c. If the trend continues, how much will the cable bill be in the tenth year? (Lesson 3-4)

UNIT
2

Triangles

You can use triangles and their properties to model and analyze many real-world situations. In this unit, you will learn about relationships in and among triangles, including congruence and similarity.

Chapter 4
Congruent Triangles

Chapter 5
Relationships in Triangles

Chapter 6
Proportions and Similarity

Chapter 7
Right Triangles and Trigonometry

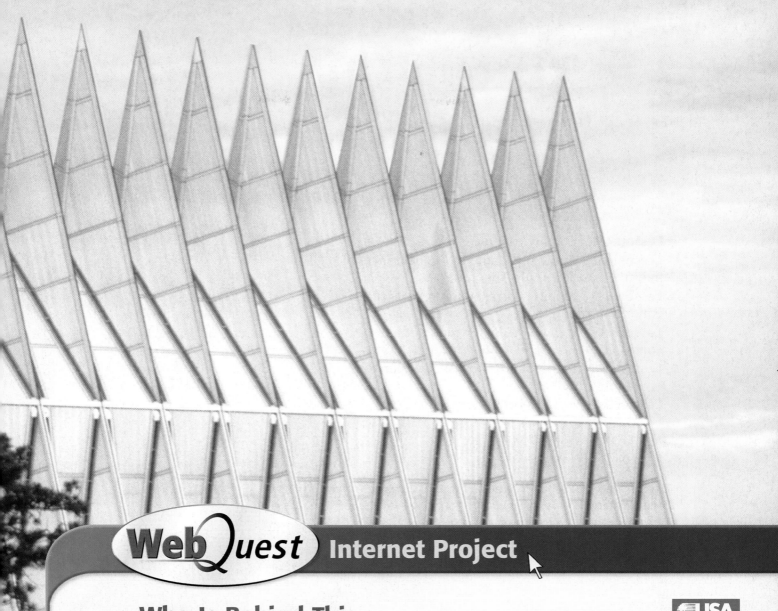

WebQuest Internet Project

Who Is Behind This Geometry Concept Anyway?

Have you ever wondered who first developed some of the ideas you are learning in your geometry class? Today, many students use the Internet for learning and research. In this project, you will be using the Internet to research a topic in geometry. You will then prepare a portfolio or poster to display your findings.

 Log on to **www.geometryonline.com/webquest**. Begin your WebQuest by reading the Task.

Continue working on your WebQuest as you study Unit 2.

Lesson	4-6	5-1	6-6	7-1
Page	218	241	325	347

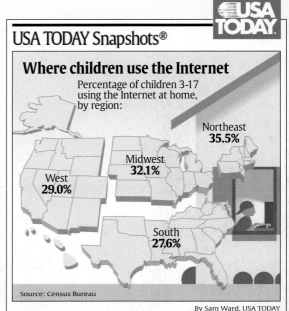

USA TODAY Snapshots®

Where children use the Internet

Percentage of children 3-17 using the Internet at home, by region:

Northeast **35.5%**

Midwest **32.1%**

West **29.0%**

South **27.6%**

Source: Census Bureau

By Sam Ward, USA TODAY

Congruent Triangles

What You'll Learn

- **Lesson 4-1** Classify triangles.
- **Lesson 4-2** Apply the Angle Sum Theorem and the Exterior Angle Theorem.
- **Lesson 4-3** Identify corresponding parts of congruent triangles.
- **Lessons 4-4 and 4-5** Test for triangle congruence using SSS, SAS, ASA, and AAS.
- **Lesson 4-6** Use properties of isosceles and equilateral triangles.
- **Lesson 4-7** Write coordinate proofs.

Key Vocabulary

- exterior angle (p. 186)
- flow proof (p. 187)
- corollary (p. 188)
- congruent triangles (p. 192)
- coordinate proof (p. 222)

Why It's Important

Triangles are found everywhere you look. Triangles with the same size and shape can even be found on the tail of a whale. *You will learn more about orca whales in Lesson 4-4.*

Getting Started

▶ **Prerequisite Skills** To be successful in this chapter, you'll need to master these skills and be able to apply them in problem-solving situations. Review these skills before beginning Chapter 4.

For Lesson 4-1   Solve Equations

Solve each equation. *(For review, see pages 737 and 738.)*

1. $2x + 18 = 5$

2. $3m - 16 = 12$

3. $4y + 12 = 16$

4. $10 = 8 - 3z$

5. $6 = 2a + \dfrac{1}{2}$

6. $\dfrac{2}{3}b + 9 = -15$

For Lessons 4-2, 4-4, and 4-5   Congruent Angles

Name the indicated angles or pairs of angles if $p \parallel q$ and $m \parallel \ell$.
(For review, see Lesson 3-1.)

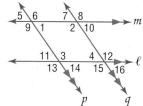

7. angles congruent to $\angle 8$

8. angles congruent to $\angle 13$

9. angles supplementary to $\angle 1$

10. angles supplementary to $\angle 12$

For Lessons 4-3 and 4-7   Distance Formula

Find the distance between each pair of points. Round to the nearest tenth.
(For review, see Lesson 1-3.)

11. $(6, 8), (-4, 3)$

12. $(-15, 12), (6, 18)$

13. $(11, -8), (-3, -4)$

14. $(-10, 4), (8, -7)$

Triangles Make this Foldable to help you organize your notes. Begin with two sheets of grid paper and one sheet of construction paper.

Step 1 **Fold and Cut**

Stack the grid paper on the construction paper. Fold diagonally as shown and cut off the excess.

Step 2 **Staple and Label**

Staple the edge to form a booklet. Then label each page with a lesson number and title.

Reading and Writing As you read and study the chapter, use your journal for sketches and examples of terms associated with triangles and sample proofs.

4-1 Classifying Triangles

What You'll Learn

- Identify and classify triangles by angles.
- Identify and classify triangles by sides.

Why are triangles important in construction?

Many structures use triangular shapes as braces for construction. The roof sections of houses are made of triangular trusses that support the roof and the house.

Vocabulary

- acute triangle
- obtuse triangle
- right triangle
- equiangular triangle
- scalene triangle
- isosceles triangle
- equilateral triangle

CLASSIFY TRIANGLES BY ANGLES

Recall that a triangle is a three-sided polygon. Triangle *ABC*, written △*ABC*, has parts that are named using the letters *A*, *B*, and *C*.

- The sides of △*ABC* are $\overline{AB}$, $\overline{BC}$, and $\overline{CA}$.
- The vertices are *A*, *B*, and *C*.
- The angles are ∠*ABC* or ∠*B*, ∠*BCA* or ∠*C*, and ∠*BAC* or ∠*A*.

There are two ways to classify triangles. One way is by their angles. All triangles have at least two acute angles, but the third angle is used to classify the triangle.

Study Tip

Common Misconceptions
These classifications are distinct groups. For example, a triangle cannot be right and acute.

Key Concept — Classifying Triangles by Angles

In an **acute triangle**, all of the angles are acute.	In an **obtuse triangle**, one angle is obtuse.	In a **right triangle**, one angle is right.
67°, 37°, 76°	13°, 142°, 25°	42°, 90°, 48°
all angle measures < 90	one angle measure > 90	one angle measure = 90

An acute triangle with all angles congruent is an **equiangular triangle**.

Example 1 — Classify Triangles by Angles

ARCHITECTURE The roof of this house is made up of three different triangles. Use a protractor to classify △*DFH*, △*DFG*, and △*HFG* as *acute*, *equiangular*, *obtuse*, or *right*.

△*DFH* has all angles with measures less than 90, so it is an acute triangle.
△*DFG* and △*HFG* both have one angle with measure equal to 90. Both of these are right triangles.

CLASSIFY TRIANGLES BY SIDES Triangles can also be classified according to the number of congruent sides they have. To indicate that sides of a triangle are congruent, an equal number of hash marks are drawn on the corresponding sides.

Key Concept — *Classifying Triangles by Sides*

No two sides of a **scalene triangle** are congruent.

At least two sides of an **isosceles triangle** are congruent.

All of the sides of an **equilateral triangle** are congruent.

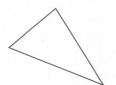

An equilateral triangle is a special kind of isosceles triangle.

Geometry Activity

Equilateral Triangles

Model

- Align three pieces of patty paper as indicated. Draw a dot at *X*.
- Fold the patty paper through *X* and *Y* and through *X* and *Z*.

Analyze

1. Is △*XYZ* equilateral? Explain.

2. Use three pieces of patty paper to make a triangle that is isosceles, but not equilateral.

3. Use three pieces of patty paper to make a scalene triangle.

Example 2 *Classify Triangles by Sides*

Identify the indicated type of triangle in the figure.

a. **isosceles triangles**

Isosceles triangles have at least two sides congruent. So, △*ABD* and △*EBD* are isosceles.

b. **scalene triangles**

Scalene triangles have no congruent sides. △*AEB*, △*AED*, △*ACB*, △*ACD*, △*BCE*, and △*DCE* are scalene.

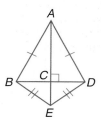

Example 3 *Find Missing Values*

ALGEBRA Find *x* and the measure of each side of equilateral triangle *RST* if $RS = x + 9$, $ST = 2x$, and $RT = 3x - 9$.

Since △*RST* is equilateral, $RS = ST$.

$x + 9 = 2x$ Substitution
$\quad 9 = x$ Subtract *x* from each side.

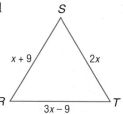

Next, substitute to find the length of each side.

$RS = x + 9$	$ST = 2x$	$RT = 3x - 9$
$= 9 + 9$ or 18	$= 2(9)$ or 18	$= 3(9) - 9$ or 18

For △*RST*, $x = 9$, and the measure of each side is 18.

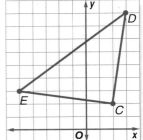

Example 4 Use the Distance Formula

COORDINATE GEOMETRY Find the measures of the sides of $\triangle DEC$. Classify the triangle by sides.

Use the Distance Formula to find the lengths of each side.

$$EC = \sqrt{(-5 - 2)^2 + (3 - 2)^2}$$
$$= \sqrt{49 + 1}$$
$$= \sqrt{50}$$

$$ED = \sqrt{(-5 - 3)^2 + (3 - 9)^2}$$
$$= \sqrt{64 + 36}$$
$$= \sqrt{100}$$

$$DC = \sqrt{(3 - 2)^2 + (9 - 2)^2}$$
$$= \sqrt{1 + 49}$$
$$= \sqrt{50}$$

Since $\overline{EC}$ and $\overline{DC}$ have the same length, $\triangle DEC$ is isosceles.

Check for Understanding

Concept Check

1. **Explain** how a triangle can be classified in two ways.

2. **OPEN ENDED** Draw a triangle that is isosceles and right.

Determine whether each of the following statements is *always*, *sometimes*, or *never* true. **Explain.**

3. Equiangular triangles are also acute.

4. Right triangles are acute.

Guided Practice

Use a protractor to classify each triangle as *acute*, *equiangular*, *obtuse*, or *right*.

5.

6.
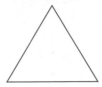

7. Identify the obtuse triangles if $\angle MJK \cong \angle KLM$, $m\angle MJK = 126$, and $m\angle JNM = 52$.

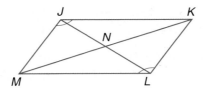

8. Identify the right triangles if $\overline{IJ} \parallel \overline{GH}$, $\overline{GH} \perp \overline{DF}$, and $\overline{GI} \perp \overline{EF}$.

9. **ALGEBRA** Find x, JM, MN, and JN if $\triangle JMN$ is an isosceles triangle with $\overline{JM} \cong \overline{MN}$.

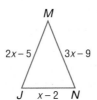

10. **ALGEBRA** Find x, QR, RS, and QS if $\triangle QRS$ is an equilateral triangle.

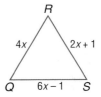

11. Find the measures of the sides of △TWZ with vertices at T(2, 6), W(4, −5), and Z(−3, 0). Classify the triangle.

Application **12. QUILTING** The star-shaped composite quilting square is made up of four different triangles. Use a ruler to classify the four triangles by sides.

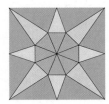

Practice and Apply

Homework Help

For Exercises	See Examples
13–18	1
19, 21–25	1, 2
26–29	3
30, 31	2
32–37, 40, 41	4

Extra Practice
See page 760.

Use a protractor to classify each triangle as *acute, equiangular, obtuse,* **or** *right.*

13.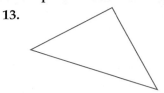

14.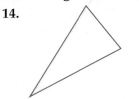

15.

16.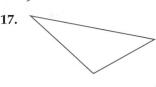

17.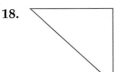

18.

19. ASTRONOMY On May 5, 2002, Venus, Saturn, and Mars were aligned in a triangular formation. Use a protractor or ruler to classify the triangle formed by sides and angles.

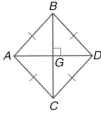

20. RESEARCH Use the Internet or other resource to find out how astronomers can predict planetary alignment.

21. ARCHITECTURE The restored and decorated Victorian houses in San Francisco are called the "Painted Ladies." Use a protractor to classify the triangles indicated in the photo by sides and angles.

Identify the indicated type of triangles in the figure if $\overline{AB} \cong \overline{BD} \cong \overline{DC} \cong \overline{CA}$ **and** $\overline{BC} \perp \overline{AD}$.

22. right

23. obtuse

24. scalene

25. isosceles

ALGEBRA Find x and the measure of each side of the triangle.

26. △GHJ is isosceles, with $\overline{HG} \cong \overline{JG}$, GH = x + 7, GJ = 3x − 5, and HJ = x − 1.

27. △MPN is equilateral with MN = 3x − 6, MP = x + 4, and NP = 2x − 1.

28. △QRS is equilateral. QR is two less than two times a number, RS is six more than the number, and QS is ten less than three times the number.

29. △JKL is isosceles with $\overline{KJ} \cong \overline{LJ}$. JL is five less than two times a number. JK is three more than the number. KL is one less than the number. Find the measure of each side.

30. CRYSTAL The top of the crystal bowl shown is circular. The diameter at the top of the bowl is MN. P is the midpoint of $\overline{MN}$, and $\overline{OP} \perp \overline{MN}$. If $MN = 24$ and $OP = 12$, determine whether $\triangle MPO$ and $\triangle NPO$ are equilateral.

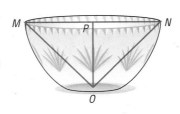

31. MAPS The total distance from Nashville, Tennessee, to Cairo, Illinois, to Lexington, Kentucky, and back to Nashville, Tennessee, is 593 miles. The distance from Cairo to Lexington is 81 more miles than the distance from Lexington to Nashville. The distance from Cairo to Nashville is 40 miles less than the distance from Nashville to Lexington. Classify the triangle formed by its sides.

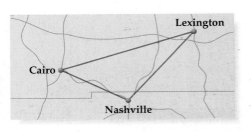

Find the measures of the sides of $\triangle ABC$ and classify each triangle by its sides.

32. $A(5, 4)$, $B(3, -1)$, $C(7, -1)$

33. $A(-4, 1)$, $B(5, 6)$, $C(-3, -7)$

34. $A(-7, 9)$, $B(-7, -1)$, $C(4, -1)$

35. $A(-3, -1)$, $B(2, 1)$, $C(2, -3)$

36. $A(0, 5)$, $B(5\sqrt{3}, 2)$, $C(0, -1)$

37. $A(-9, 0)$, $B\left(-5, 6\sqrt{3}\right)$, $C(-1, 0)$

38. **PROOF** Write a two-column proof to prove that $\triangle EQL$ is equiangular.

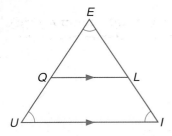

39. **PROOF** Write a paragraph proof to prove that $\triangle RPM$ is an obtuse triangle if $m\angle NPM = 33$.

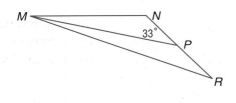

40. COORDINATE GEOMETRY Show that S is the midpoint of $\overline{RT}$ and U is the midpoint of $\overline{TV}$.

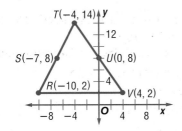

41. COORDINATE GEOMETRY Show that $\triangle ADC$ is isosceles.

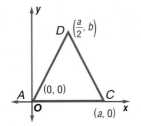

42. CRITICAL THINKING $\overline{KL}$ is a segment representing one side of isosceles right triangle KLM, with $K(2, 6)$, and $L(4, 2)$. $\angle KLM$ is a right angle, and $\overline{KL} \cong \overline{LM}$. Describe how to find the coordinates of vertex M and name these coordinates.

43. Answer the question that was posed at the beginning of the lesson.

Why are triangles important in construction?

Include the following in your answer:
- describe how to classify triangles, and
- if one type of triangle is used more often in architecture than other types.

Standardized Test Practice
Ⓐ Ⓑ Ⓒ Ⓓ

44. Classify $\triangle ABC$ with vertices $A(-1, 1)$, $B(1, 3)$, and $C(3, -1)$.

Ⓐ scalene acute Ⓑ equilateral Ⓒ isosceles acute Ⓓ isosceles right

45. ALGEBRA Find the value of y if the mean of x, y, 15, and 35 is 25 and the mean of x, 15, and 35 is 27.

Ⓐ 18 Ⓑ 19 Ⓒ 31 Ⓓ 36

Maintain Your Skills

Mixed Review **Graph each line. Construct a perpendicular segment through the given point. Then find the distance from the point to the line.** *(Lesson 3-6)*

46. $y = x + 2$, $(2, -2)$ **47.** $x + y = 2$, $(3, 3)$ **48.** $y = 7$, $(6, -2)$

Find x so that $p \parallel q$. *(Lesson 3-5)*

49.

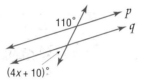

50.

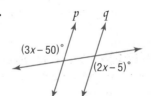

51.

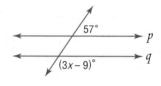

For this proof, the reasons in the right column are not in the proper order. Reorder the reasons to properly match the statements in the left column. *(Lesson 2-6)*

52. Given: $3x - 4 = x - 10$

Prove: $x = -3$

Proof:

Statements	Reasons
a. $3x - 4 = x - 10$	**1.** Subtraction Property
b. $2x - 4 = -10$	**2.** Division Property
c. $2x = -6$	**3.** Given
d. $x = -3$	**4.** Addition Property

Getting Ready for the Next Lesson **PREREQUISITE SKILL** In the figure, $\overline{AB} \parallel \overline{RQ}$, $\overline{BC} \parallel \overline{PR}$, and $\overline{AC} \parallel \overline{PQ}$. Name the indicated angles or pairs of angles.

(To review angles formed by parallel lines and a transversal, see Lessons 3-1 and 3-2.)

53. three pairs of alternate interior angles

54. six pairs of corresponding angles

55. all angles congruent to $\angle 3$

56. all angles congruent to $\angle 7$

57. all angles congruent to $\angle 11$

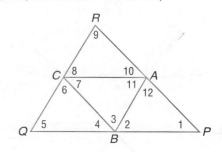

Geometry Activity

Angles of Triangles

There are special relationships among the angles of a triangle.

Activity 1   **Find the relationship among the measures of the interior angles of a triangle.**

Step 1   Draw an obtuse triangle and cut it out. Label the vertices A, B, and C.

Step 2   Find the midpoint of $\overline{AB}$ by matching A to B. Label this point D.

Step 3   Find the midpoint of $\overline{BC}$ by matching B to C. Label this point E.

Step 4   Draw $\overline{DE}$.

Step 5   Fold △ABC along $\overline{DE}$. Label the point where B touches $\overline{AC}$ as F.

Step 6   Draw $\overline{DF}$ and $\overline{FE}$. Measure each angle.

Analyze the Model

Describe the relationship between each pair.

  1. ∠A and ∠DFA   **2.** ∠B and ∠DFE   **3.** ∠C and ∠EFC

  4. What is the sum of the measures of ∠DFA, ∠DFE, and ∠EFC?

  5. What is the sum of the measures of ∠A, ∠B, and ∠C?

  6. Make a conjecture about the sum of the measures of the angles of any triangle.

In the figure at the right, ∠4 is called an *exterior angle* of the triangle. ∠1 and ∠2 are the *remote interior angles* of ∠4.

Activity 2   **Find the relationship among the interior and exterior angles of a triangle.**

Step 1   Trace △ABC from Activity 1 onto a piece of paper. Label the vertices.

Step 2   Extend $\overline{AC}$ to draw an exterior angle at C.

Step 3   Tear ∠A and ∠B off the triangle from Activity 1.

Step 4   Place ∠A and ∠B over the exterior angle.

Analyze the Model

  7. Make a conjecture about the relationship of ∠A, ∠B, and the exterior angle at C.

  8. Repeat the steps for the exterior angles of ∠A and ∠B.

  9. Is your conjecture true for all exterior angles of a triangle?

 10. Repeat Activity 2 with an acute triangle.

 11. Repeat Activity 2 with a right triangle.

 12. Make a conjecture about the measure of an exterior angle and the sum of the measures of its remote interior angles.

4-2 Angles of Triangles

What You'll Learn

- Apply the Angle Sum Theorem.
- Apply the Exterior Angle Theorem.

Vocabulary
- exterior angle
- remote interior angles
- flow proof
- corollary

How are the angles of triangles used to make kites?

The Drachen Foundation coordinates the annual Miniature Kite Contest. This kite won second place in the Most Beautiful Kite category in 2001. The overall dimensions are 10.5 centimeters by 9.5 centimeters. The wings of the beetle are triangular.

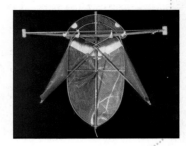

ANGLE SUM THEOREM If the measures of two of the angles of a triangle are known, how can the measure of the third angle be determined? The Angle Sum Theorem explains that the sum of the measures of the angles of any triangle is always 180.

Theorem 4.1

Angle Sum Theorem The sum of the measures of the angles of a triangle is 180.

Example: $m\angle W + m\angle X + m\angle Y = 180$

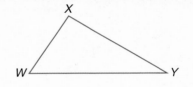

Study Tip

Look Back
Recall that sometimes extra lines have to be drawn to complete a proof. These are called *auxiliary lines.*

Proof Angle Sum Theorem

Given: $\triangle ABC$
Prove: $m\angle C + m\angle 2 + m\angle B = 180$

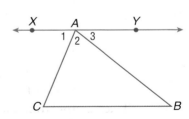

Proof:

Statements	Reasons
1. $\triangle ABC$	1. Given
2. Draw $\overleftrightarrow{XY}$ through A parallel to $\overline{CB}$.	2. Parallel Postulate
3. $\angle 1$ and $\angle CAY$ form a linear pair.	3. Def. of a linear pair
4. $\angle 1$ and $\angle CAY$ are supplementary.	4. If 2 ⫔ form a linear pair, they are supplementary.
5. $m\angle 1 + m\angle CAY = 180$	5. Def. of suppl. ⫔
6. $m\angle CAY = m\angle 2 + m\angle 3$	6. Angle Addition Postulate
7. $m\angle 1 + m\angle 2 + m\angle 3 = 180$	7. Substitution
8. $\angle 1 \cong \angle C, \angle 3 \cong \angle B$	8. Alt. Int. ⫔ Theorem
9. $m\angle 1 = m\angle C, m\angle 3 = m\angle B$	9. Def. of $\cong$ ⫔
10. $m\angle C + m\angle 2 + m\angle B = 180$	10. Substitution

If we know the measures of two angles of a triangle, we can find the measure of the third.

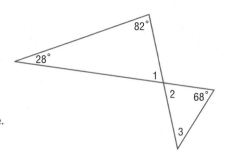

Example 1 *Interior Angles*

Find the missing angle measures.

Find $m\angle 1$ first because the measures of two angles of the triangle are known.

$m\angle 1 + 28 + 82 = 180$ Angle Sum Theorem

$\quad m\angle 1 + 110 = 180$ Simplify.

$\quad\quad\quad\quad m\angle 1 = 70$ Subtract 110 from each side.

$\angle 1$ and $\angle 2$ are congruent vertical angles. So $m\angle 2 = 70$.

$m\angle 3 + 68 + 70 = 180$ Angle Sum Theorem

$\quad m\angle 3 + 138 = 180$ Simplify.

$\quad\quad\quad\quad m\angle 3 = 42$ Subtract 138 from each side.

Therefore, $m\angle 1 = 70$, $m\angle 2 = 70$, and $m\angle 3 = 42$.

The Angle Sum Theorem leads to a useful theorem about the angles in two triangles.

Theorem 4.2

Third Angle Theorem If two angles of one triangle are congruent to two angles of a second triangle, then the third angles of the triangles are congruent.

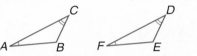 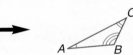

Example: If $\angle A \cong \angle F$ and $\angle C \cong \angle D$, then $\angle B \cong \angle E$.

You will prove this theorem in Exercise 44.

EXTERIOR ANGLE THEOREM

Each angle of a triangle has an exterior angle. An **exterior angle** is formed by one side of a triangle and the extension of another side. The interior angles of the triangle not adjacent to a given exterior angle are called **remote interior angles** of the exterior angle.

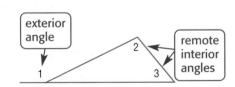

Theorem 4.3

Exterior Angle Theorem The measure of an exterior angle of a triangle is equal to the sum of the measures of the two remote interior angles.

Example: $m\angle YZP = m\angle X + m\angle Y$

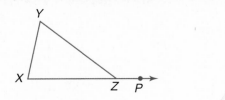

We will use a flow proof to prove this theorem. A **flow proof** organizes a series of statements in logical order, starting with the given statements. Each statement is written in a box with the reason verifying the statement written below the box. Arrows are used to indicate how the statements relate to each other.

Proof *Exterior Angle Theorem*

Write a flow proof of the Exterior Angle Theorem.

Given: $\triangle ABC$

Prove: $m\angle CBD = m\angle A + m\angle C$

Flow Proof:

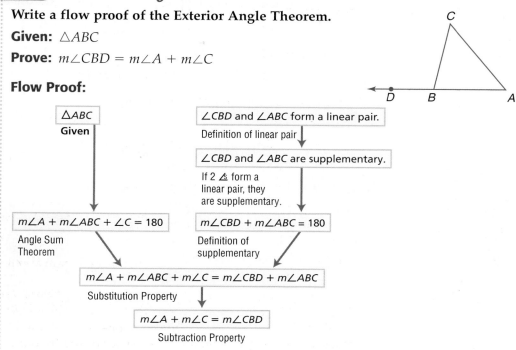

| $\triangle ABC$ |
| Given |

| $\angle CBD$ and $\angle ABC$ form a linear pair. |
| Definition of linear pair |

| $\angle CBD$ and $\angle ABC$ are supplementary. |
| If 2 $\angle$s form a linear pair, they are supplementary. |

| $m\angle A + m\angle ABC + \angle C = 180$ |
| Angle Sum Theorem |

| $m\angle CBD + m\angle ABC = 180$ |
| Definition of supplementary |

| $m\angle A + m\angle ABC + m\angle C = m\angle CBD + m\angle ABC$ |
| Substitution Property |

| $m\angle A + m\angle C = m\angle CBD$ |
| Subtraction Property |

Example 2 *Exterior Angles*

Find the measure of each numbered angle in the figure.

| $m\angle 1 = 50 + 78$ | Exterior Angle Theorem |
| $= 128$ | Simplify. |

$m\angle 1 + m\angle 2 = 180$	If 2 $\angle$s form a linear pair, they are suppl.
$128 + m\angle 2 = 180$	Substitution
$m\angle 2 = 52$	Subtract 128 from each side.

$m\angle 2 + m\angle 3 = 120$	Exterior Angle Theorem
$52 + m\angle 3 = 120$	Substitution
$m\angle 3 = 68$	Subtract 52 from each side.

| $120 + m\angle 4 = 180$ | If 2 $\angle$s form a linear pair, they are suppl. |
| $m\angle 4 = 60$ | Subtract 120 from each side. |

$m\angle 5 = m\angle 4 + 56$	Exterior Angle Theorem
$= 60 + 56$	Substitution
$= 116$	Simplify.

Therefore, $m\angle 1 = 128$, $m\angle 2 = 52$, $m\angle 3 = 68$, $m\angle 4 = 60$, and $m\angle 5 = 116$.

A statement that can be easily proved using a theorem is often called a **corollary** of that theorem. A corollary, just like a theorem, can be used as a reason in a proof.

Corollaries

4.1 The acute angles of a right triangle are complementary.

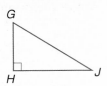

Example: $m\angle G + m\angle J = 90$

4.2 There can be at most one right or obtuse angle in a triangle.

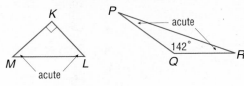

You will prove Corollaries 4.1 and 4.2 in Exercises 42 and 43.

Example 3 *Right Angles*

SKI JUMPING Ski jumper Simon Ammann of Switzerland forms a right triangle with his skis and his line of sight. Find $m\angle 2$ if $m\angle 1$ is 27.

Use Corollary 4.1 to write an equation.

$m\angle 1 + m\angle 2 = 90$

$27 + m\angle 2 = 90$ Substitution

$m\angle 2 = 63$ Subtract 27 from each side.

Check for Understanding

Concept Check

1. **OPEN ENDED** Draw a triangle. Label one exterior angle and its remote interior angles.

2. **FIND THE ERROR** Najee and Kara are discussing the Exterior Angle Theorem.

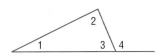

Najee

$m\angle 1 + m\angle 2 = m\angle 4$

Kara

$m\angle 1 + m\angle 2 + m\angle 4 = 180$

Who is correct? Explain your reasoning.

Guided Practice **Find the missing angle measure.**

3.

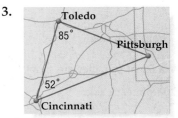

4.

Find each measure.

5. $m\angle 1$

6. $m\angle 2$

7. $m\angle 3$

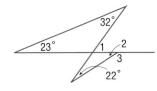

Find each measure.

8. $m\angle 1$

9. $m\angle 2$

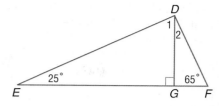

Application   10. **SKI JUMPING**   American ski jumper Eric Bergoust forms a right angle with his skis. If $m\angle 2 = 70$, find $m\angle 1$.

Practice and Apply

Homework Help	
For Exercises	**See Examples**
11–17	1
18–31	2
32–35	3
36–38	2

Extra Practice
See page 761.

Find the missing angle measures.

11.

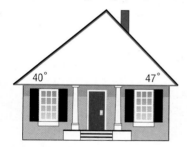

12.

13.

14.

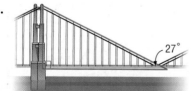

Find each measure.

15. $m\angle 1$

16. $m\angle 2$

17. $m\angle 3$

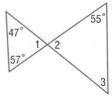

Find each measure if $m\angle 4 = m\angle 5$.

18. $m\angle 1$ 19. $m\angle 2$

20. $m\angle 3$ 21. $m\angle 4$

22. $m\angle 5$ 23. $m\angle 6$

24. $m\angle 7$

Find each measure.

25. $m\angle 1$

26. $m\angle 2$

27. $m\angle 3$

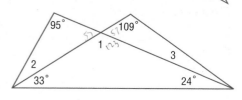

•**SPEED SKATING** For Exercises 28–31, use the following information.

Speed skater Catriona Lemay Doan of Canada forms at least two sets of triangles and exterior angles as she skates. Use the measures of given angles to find each measure.

28. $m\angle 1$

29. $m\angle 2$

30. $m\angle 3$

31. $m\angle 4$

Online Research **Data Update** Use the Internet or other resource to find the world record in speed skating. Visit www.geometryonline.com/data_update to learn more.

Find each measure if $m\angle DGF = 53$ **and** $m\angle AGC = 40$.

32. $m\angle 1$ **33.** $m\angle 2$

34. $m\angle 3$ **35.** $m\angle 4$

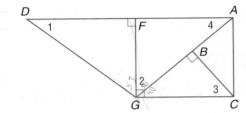

HOUSING For Exercises 36–38, use the following information.

The two braces for the roof of a house form triangles. Find each measure.

36. $m\angle 1$

37. $m\angle 2$

38. $m\angle 3$

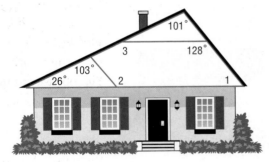

PROOF For Exercises 39–44, write the specified type of proof.

39. flow proof

    Given: $\angle FGI \cong \angle IGH$

            $\overline{GI} \perp \overline{FH}$

    Prove: $\angle F \cong \angle H$

40. two-column

    Given: $ABCD$ is a quadrilateral.

    Prove: $m\angle DAB + m\angle B +$
                $m\angle BCD + m\angle D = 360$

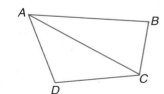

41. two-column proof of Theorem 4.3   **42.** flow proof of Corollary 4.1

43. paragraph proof of Corollary 4.2   **44.** two-column proof of Theorem 4.2

45. CRITICAL THINKING $\overrightarrow{BA}$ and $\overrightarrow{BC}$ are opposite rays. The measures of $\angle 1$, $\angle 2$, and $\angle 3$ are in a 4:5:6 ratio. Find the measure of each angle.

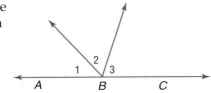

46. **WRITING IN MATH**   Answer the question that was posed at the beginning of the lesson.

How are the angles of triangles used to make kites?

Include the following in your answer:

- if two angles of two triangles are congruent, how you can find the measure of the third angle, and
- if one angle measures 90, describe the other two angles.

47. In the triangle, what is the measure of $\angle Z$?

Ⓐ 18   Ⓑ 24

Ⓒ 72   Ⓓ 90

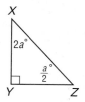

48. **ALGEBRA**   The measure of the second angle of a triangle is three times the measure of the first, and the measure of the third angle is 25 more than the measure of the first. Find the measure of each angle.

Ⓐ 25, 85, 70   Ⓑ 31, 93, 56   Ⓒ 39, 87, 54   Ⓓ 42, 54, 84

Maintain Your Skills

Mixed Review

Identify the indicated type of triangle if $\overline{BC} \cong \overline{AD}$, $\overline{EB} \cong \overline{EC}$, $\overline{AC}$ bisects $\overline{BD}$, and $m\angle AED = 125$.   *(Lesson 4-1)*

49. scalene triangles

50. obtuse triangles

51. isosceles triangles

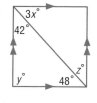

Find the distance between each pair of parallel lines.   *(Lesson 3-6)*

52. $y = x + 6$, $y = x - 10$ **53.** $y = -2x + 3$, $y = -2x - 7$

54. $4x - y = 20$, $4x - y = 3$ **55.** $2x - 3y = -9$, $2x - 3y = -6$

Find x, y, and z in each figure.   *(Lesson 3-2)*

56. 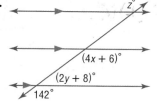   **57.** 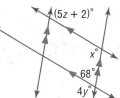   **58.**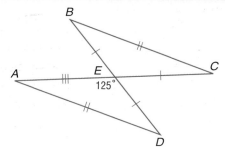

Getting Ready for the Next Lesson

PREREQUISITE SKILL   Name the property of congruence that justifies each statement.   *(To review **properties of congruence**, see Lessons 2-5 and 2-6.)*

59. $\angle 1 \cong \angle 1$ and $\overline{AB} \cong \overline{AB}$.

60. If $\overline{AB} \cong \overline{XY}$, then $\overline{XY} \cong \overline{AB}$.

61. If $\angle 1 \cong \angle 2$, then $\angle 2 \cong \angle 1$.

62. If $\angle 2 \cong \angle 3$ and $\angle 3 \cong \angle 4$, then $\angle 2 \cong \angle 4$.

63. If $\overline{PQ} \cong \overline{XY}$ and $\overline{XY} \cong \overline{HK}$, then $\overline{PQ} \cong \overline{HK}$.

64. If $\overline{AB} \cong \overline{CD}$, $\overline{CD} \cong \overline{PQ}$, and $\overline{PQ} \cong \overline{XY}$, then $\overline{AB} \cong \overline{XY}$.

Congruent Triangles

- Name and label corresponding parts of congruent triangles.
- Identify congruence transformations.

Vocabulary
- congruent triangles
- congruence transformations

Why are triangles used in bridges?

In 1930, construction started on the West End Bridge in Pittsburgh, Pennsylvania. The arch of the bridge is trussed, not solid. Steel rods are arranged in a triangular web that lends structure and stability to the bridge.

CORRESPONDING PARTS OF CONGRUENT TRIANGLES Triangles that are the same size and shape are **congruent triangles**. Each triangle has three angles and three sides. If all six of the corresponding parts of two triangles are congruent, then the triangles are congruent.

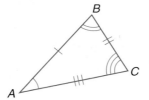 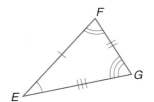

If △ABC is congruent to △EFG, the vertices of the two triangles correspond in the same order as the letters naming the triangles.

$$\triangle ABC \cong \triangle EFG$$

This correspondence of vertices can be used to name the corresponding congruent sides and angles of the two triangles.

$$\angle A \cong \angle E \qquad \angle B \cong \angle F \qquad \angle C \cong \angle G$$
$$\overline{AB} \cong \overline{EF} \qquad \overline{BC} \cong \overline{FG} \qquad \overline{AC} \cong \overline{EG}$$

The corresponding sides and angles can be determined from any congruence statement by following the order of the letters.

Study Tip

Congruent Parts
In congruent triangles, congruent sides are opposite congruent angles.

Key Concept *Definition of Congruent Triangles (CPCTC)*

Two triangles are congruent if and only if their corresponding parts are congruent.

CPCTC stands for *corresponding parts of congruent triangles are congruent.* "If and only if" is used to show that both the conditional and its converse are true.

Example 1 *Corresponding Congruent Parts*

FURNITURE DESIGN   The seat and legs of this stool form two triangles. Suppose the measures in inches are $QR = 12$, $RS = 23$, $QS = 24$, $RT = 12$, $TV = 24$, and $RV = 23$.

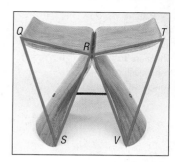

a. Name the corresponding congruent angles and sides.

$\angle Q \cong \angle T$ $\angle QRS \cong \angle TRV$ $\angle S \cong \angle V$

$\overline{QR} \cong \overline{TR}$ $\overline{RS} \cong \overline{RV}$ $\overline{QS} \cong \overline{TV}$

b. Name the congruent triangles.

$\triangle QRS \cong \triangle TRV$

Like congruence of segments and angles, congruence of triangles is reflexive, symmetric, and transitive.

Theorem 4.4	*Properties of Triangle Congruence*

Congruence of triangles is reflexive, symmetric, and transitive.

Reflexive	**Symmetric**	**Transitive**
$\triangle JKL \cong \triangle JKL$	If $\triangle JKL \cong \triangle PQR$, then $\triangle PQR \cong \triangle JKL$.	If $\triangle JKL \cong \triangle PQR$, and $\triangle PQR \cong \triangle XYZ$, then $\triangle JKL \cong \triangle XYZ$

You will prove the symmetric and reflexive parts of Theorem 4.4 in Exercises 33 and 35, respectively.

Proof *Theorem 4.4 (Transitive)*

Given: $\triangle ABC \cong \triangle DEF$

$\triangle DEF \cong \triangle GHI$

Prove: $\triangle ABC \cong \triangle GHI$

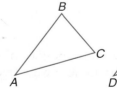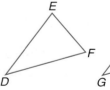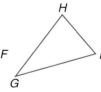

Proof:

Statements	Reasons
1. $\triangle ABC \cong \triangle DEF$	1. Given
2. $\angle A \cong \angle D$, $\angle B \cong \angle E$, $\angle C \cong \angle F$ $\overline{AB} \cong \overline{DE}$, $\overline{BC} \cong \overline{EF}$, $\overline{AC} \cong \overline{DF}$	2. CPCTC
3. $\triangle DEF \cong \triangle GHI$	3. Given
4. $\angle D \cong \angle G$, $\angle E \cong \angle H$, $\angle F \cong \angle I$ $\overline{DE} \cong \overline{GH}$, $\overline{EF} \cong \overline{HI}$, $\overline{DF} \cong \overline{GI}$	4. CPCTC
5. $\angle A \cong \angle G$, $\angle B \cong \angle H$, $\angle C \cong \angle I$	5. Congruence of angles is transitive.
6. $\overline{AB} \cong \overline{GH}$, $\overline{BC} \cong \overline{HI}$, $\overline{AC} \cong \overline{GI}$	6. Congruence of segments is transitive.
7. $\triangle ABC \cong \triangle GHI$	7. Def. of $\cong$ $\triangle$s

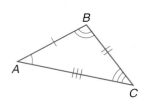
IDENTIFY CONGRUENCE TRANSFORMATIONS   In the figures below, $\triangle ABC$ is congruent to $\triangle DEF$. If you *slide* $\triangle DEF$ up and to the right, $\triangle DEF$ is still congruent to $\triangle ABC$.

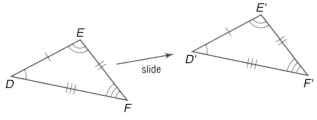

The congruency does not change whether you *turn* $\triangle DEF$ or *flip* $\triangle DEF$. $\triangle ABC$ is still congruent to $\triangle DEF$.

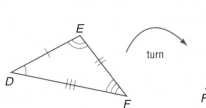

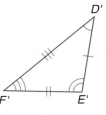
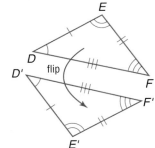

If you slide, flip, or turn a triangle, the size and shape do not change. These three transformations are called **congruence transformations**.

Example 2 *Transformations in the Coordinate Plane*

COORDINATE GEOMETRY   The vertices of $\triangle CDE$ are $C(-5, 7)$, $D(-8, 6)$, and $E(-3, 3)$. The vertices of $\triangle C'D'E'$ are $C'(5, 7)$, $D'(8, 6)$, and $E'(3, 3)$.

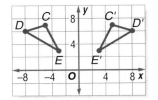

a. Verify that $\triangle CDE \cong \triangle C'D'E'$.

Use the Distance Formula to find the length of each side in the triangles.

$$DC = \sqrt{[-8 - (-5)]^2 + (6 - 7)^2} \qquad D'C' = \sqrt{(8 - 5)^2 + (6 - 7)^2}$$
$$= \sqrt{9 + 1} \text{ or } \sqrt{10} \qquad\qquad = \sqrt{9 + 1} \text{ or } \sqrt{10}$$

$$DE = \sqrt{[-8 - (-3)]^2 + (6 - 3)^2} \qquad D'E' = \sqrt{(8 - 3)^2 + (6 - 3)^2}$$
$$= \sqrt{25 + 9} \text{ or } \sqrt{34} \qquad\qquad = \sqrt{25 + 9} \text{ or } \sqrt{34}$$

$$CE = \sqrt{[-5 - (-3)]^2 + (7 - 3)^2} \qquad C'E' = \sqrt{(5 - 3)^2 + (7 - 3)^2}$$
$$= \sqrt{4 + 16} \text{ or } \sqrt{20} \qquad\qquad = \sqrt{4 + 16} \text{ or } \sqrt{20}$$

By the definition of congruence, $\overline{DC} \cong \overline{D'C'}$, $\overline{DE} \cong \overline{D'E'}$, and $\overline{CE} \cong \overline{C'E'}$.

Use a protractor to measure the angles of the triangles. You will find that the measures are the same.

In conclusion, because $\overline{DC} \cong \overline{D'C'}$, $\overline{DE} \cong \overline{D'E'}$, and $\overline{CE} \cong \overline{C'E'}$, $\angle D \cong \angle D'$, $\angle C \cong \angle C'$, and $\angle E \cong \angle E'$, $\triangle DCE \cong \triangle D'C'E'$.

b. Name the congruence transformation for $\triangle CDE$ and $\triangle C'D'E'$.

$\triangle C'D'E'$ is a flip of $\triangle CDE$.

Concept Check
1. **Explain** how slides, flips, and turns preserve congruence.

2. **OPEN ENDED** Draw a pair of congruent triangles and label the congruent sides and angles.

Guided Practice   **Identify the congruent triangles in each figure.**

3.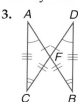

4.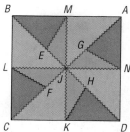

5. If $\triangle WXZ \cong \triangle STJ$, name the congruent angles and congruent sides.

6. **QUILTING** In the quilt design, assume that angles and segments that appear to be congruent are congruent. Indicate which triangles are congruent.

7. The coordinates of the vertices of $\triangle QRT$ and $\triangle Q'R'T'$ are $Q(-4, 3)$, $Q'(4, 3)$, $R(-4, -2)$, $R'(4, -2)$, $T(-1, -2)$, and $T'(1, -2)$. Verify that $\triangle QRT \cong \triangle Q'R'T'$. Then name the congruence transformation.

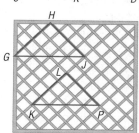

Application
8. **GARDENING** This garden lattice will be covered with morning glories in the summer. Wesley wants to save two triangular areas for artwork. If $\triangle GHJ \cong \triangle KLP$, name the corresponding congruent angles and sides.

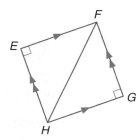

Homework Help

For Exercises	See Examples
9–22, 27–35	1
23–26	2

Extra Practice
See page 761.

Identify the congruent triangles in each figure.

9.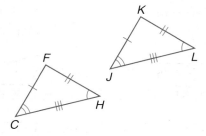

10.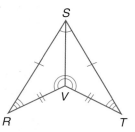

11.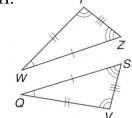

12.

Name the congruent angles and sides for each pair of congruent triangles.

13. $\triangle TUV \cong \triangle XYZ$

14. $\triangle CDG \cong \triangle RSW$

15. $\triangle BCF \cong \triangle DGH$

16. $\triangle ADG \cong \triangle HKL$

Assume that segments and angles that appear to be congruent in the numbered triangles are congruent. Indicate which triangles are congruent.

17.
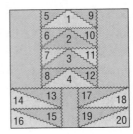

18.
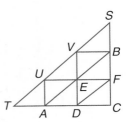

19.
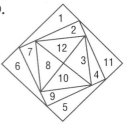

20. All of the small triangles in the figure at the right are congruent. Name three larger congruent triangles.

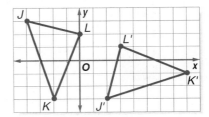

•••• **21. MOSAICS** The picture at the left is the center of a Roman mosaic. Because the four triangles connect to a square, they have at least one side congruent to a side in another triangle. What else do you need to know to conclude that the four triangles are congruent?

Verify that each of the following preserves congruence and name the congruence transformation.

22. △PQV ≅ △P'Q'V'

23. △MNP ≅ △M'N'P'

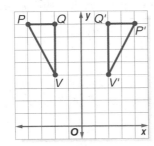

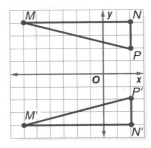

24. △GHF ≅ △G'H'F'

25. △JKL ≅ △J'K'L'

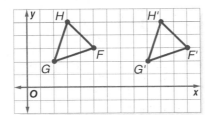

Determine whether each statement is *true* or *false*. Draw an example or counterexample for each.

26. Two triangles with corresponding congruent angles are congruent.

27. Two triangles with angles and sides congruent are congruent.

28. UMBRELLAS Umbrellas usually have eight congruent triangular sections with ribs of equal length. Are the statements △JAD ≅ △IAE and △JAD ≅ △EAI both correct? Explain.

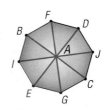

ALGEBRA **For Exercises 29 and 30, use the following information.**
$\triangle QRS \cong \triangle GHJ$, $RS = 12$, $QR = 10$, $QS = 6$, and $HJ = 2x - 4$.

29. Draw and label a figure to show the congruent triangles.

30. Find x.

ALGEBRA **For Exercises 31 and 32, use the following information.**
$\triangle JKL \cong \triangle DEF$, $m\angle J = 36$, $m\angle E = 64$, and $m\angle F = 3x + 52$.

31. Draw and label a figure to show the congruent triangles.

32. Find x.

33. **PROOF** The statements below can be used to prove that *congruence of triangles is symmetric*. Use the statements to construct a correct flow proof. Provide the reasons for each statement.

Given: $\triangle RST \cong \triangle XYZ$
Prove: $\triangle XYZ \cong \triangle RST$

 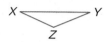

Flow Proof:

$\angle X \cong \angle R$, $\angle Y \cong$ $\angle S$, $\angle Z \cong \angle T$, $\overline{XY} \cong \overline{RS}$, $\overline{YZ} \cong$ $\overline{ST}$, $\overline{XZ} \cong \overline{RT}$?	$\angle R \cong \angle X$, $\angle S \cong$ $\angle Y$, $\angle T \cong \angle Z$, $\overline{RS} \cong \overline{XY}$, $\overline{ST} \cong$ $\overline{YZ}$, $\overline{RT} \cong \overline{XZ}$?	$\triangle RST \cong \triangle XYZ$ ___?___	$\triangle XYZ \cong \triangle RST$ ___?___

34. **PROOF** Copy the flow proof and provide the reasons for each statement.
Given: $\overline{AB} \cong \overline{CD}$, $\overline{AD} \cong \overline{CB}$, $\overline{AD} \perp \overline{DC}$,
$\overline{AB} \perp \overline{BC}$, $\overline{AD} \parallel \overline{BC}$, $\overline{AB} \parallel \overline{CD}$
Prove: $\triangle ACD \cong \triangle CAB$

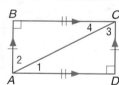

Flow Proof:

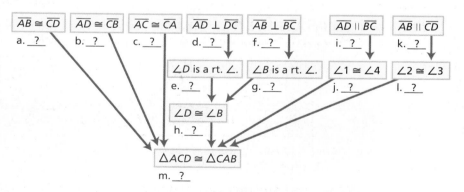

35. **PROOF** Write a flow proof to prove *Congruence of triangles is reflexive*. (Theorem 4.4)

36. **CRITICAL THINKING** $\triangle RST$ is isosceles with $RS = RT$, M, N, and P are midpoints of their sides, $\angle S \cong \angle MPS$, and $\overline{NP} \cong \overline{MP}$. What else do you need to know to prove that $\triangle SMP \cong \triangle TNP$?

 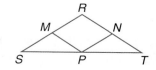

37. WRITING IN MATH Answer the question that was posed at the beginning of the lesson.

Why are triangles used in bridges?

Include the following in your answer:

- whether the shape of the triangle matters, and
- whether the triangles appear congruent.

38. Determine which statement is true given $\triangle ABC \cong \triangle XYZ$.

Ⓐ $\overline{BC} \cong \overline{ZX}$ Ⓑ $\overline{AC} \cong \overline{XZ}$ Ⓒ $\overline{AB} \cong \overline{YZ}$ Ⓓ cannot be determined

39. ALGEBRA Find the length of $\overline{DF}$ if $D(-5, 4)$ and $F(3, -7)$.

Ⓐ $\sqrt{5}$ Ⓑ $\sqrt{13}$ Ⓒ $\sqrt{57}$ Ⓓ $\sqrt{185}$

Maintain Your Skills

Mixed Review **Find x.** *(Lesson 4-2)*

40.

41.

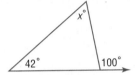

42.

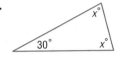

Find x and the measure of each side of the triangle. *(Lesson 4-1)*

43. $\triangle BCD$ is isosceles with $\overline{BC} \cong \overline{CD}$, $BC = 2x + 4$, $BD = x + 2$, and $CD = 10$.

44. Triangle HKT is equilateral with $HK = x + 7$ and $HT = 4x - 8$.

Write an equation in slope-intercept form for the line that satisfies the given conditions. *(Lesson 3-4)*

45. contains $(0, 3)$ and $(4, -3)$

46. $m = \dfrac{3}{4}$, y-intercept $= 8$

47. parallel to $y = -4x + 1$; contains $(-3, 1)$

48. $m = -4$, contains $(-3, 2)$

Getting Ready for the Next Lesson **PREREQUISITE SKILL** Find the distance between each pair of points.
*(To review the **Distance Formula**, see Lesson 1-4.)*

49. $(-1, 7)$, $(1, 6)$ **50.** $(8, 2)$, $(4, -2)$ **51.** $(3, 5)$, $(5, 2)$

Practice Quiz 1   Lessons 4-1 through 4-3

1. Identify the isosceles triangles in the figure, if $\overline{FH}$ and $\overline{DG}$ are congruent perpendicular bisectors. *(Lesson 4-1)*

ALGEBRA $\triangle ABC$ is equilateral with $AB = 2x$, $BC = 4x - 7$, and $AC = x + 3.5$. *(Lesson 4-1)*

2. Find x.

3. Find the measure of each side.

4. Find the measure of each numbered angle. *(Lesson 4-2)*

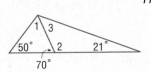

5. If $\triangle MNP \cong \triangle JKL$, name the corresponding congruent angles and sides. *(Lesson 4-3)*

Reading Mathematics

Making Concept Maps

When studying a chapter, it is wise to record the main topics and vocabulary you encounter. In this chapter, some of the new vocabulary words were *triangle*, *acute triangle*, *obtuse triangle*, *right triangle*, *equiangular triangle*, *scalene triangle*, *isosceles triangle*, and *equilateral triangle*. The triangles are all related by the size of the angles or the number of congruent sides.

A graphic organizer called a *concept map* is a convenient way to show these relationships. A concept map is shown below for the different types of triangles. The main ideas are in boxes. Any information that describes how to move from one box to the next is placed along the arrows.

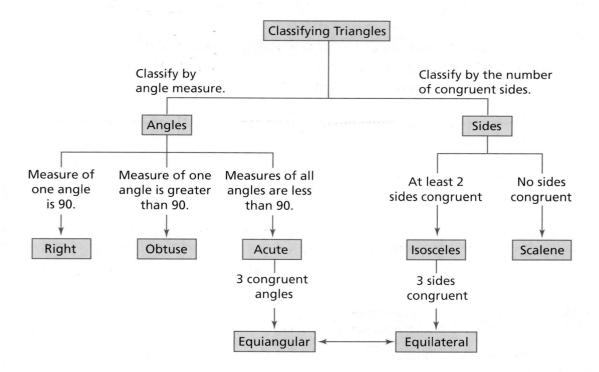

Reading to Learn

1. Describe how to use the concept map to classify triangles by their side lengths.

2. In $\triangle ABC$, $m\angle A = 48$, $m\angle B = 41$, and $m\angle C = 91$. Use the concept map to classify $\triangle ABC$.

3. Identify the type of triangle that is linked to both classifications.

4-4 Proving Congruence—SSS, SAS

What You'll Learn

- Use the SSS Postulate to test for triangle congruence.
- Use the SAS Postulate to test for triangle congruence.

Vocabulary
- included angle

How do land surveyors use congruent triangles?

Land surveyors mark and establish property boundaries. To check a measurement, they mark out a right triangle and then mark a second triangle that is congruent to the first.

SSS POSTULATE Is it always necessary to show that all of the corresponding parts of two triangles are congruent to prove that the triangles are congruent? In this lesson, we will explore two other methods to prove that triangles are congruent.

Construction

Congruent Triangles Using Sides

① Draw a triangle and label the vertices *X*, *Y*, and *Z*.

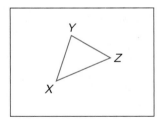

② Use a straightedge to draw any line ℓ and select a point *R*. Use a compass to construct $\overline{RS}$ on ℓ such that $\overline{RS} \cong \overline{XZ}$.

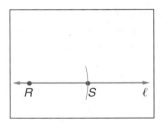

③ Using *R* as the center, draw an arc with radius equal to *XY*.

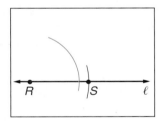

④ Using *S* as the center, draw an arc with radius equal to *YZ*.

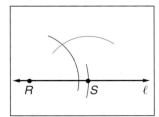

⑤ Let *T* be the point of intersection of the two arcs. Draw $\overline{RT}$ and $\overline{ST}$ to form △*RST*.

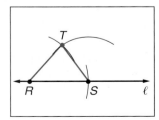

⑥ Cut out △*RST* and place it over △*XYZ*. How does △*RST* compare to △*XYZ*?

If the corresponding sides of two triangles are congruent, then the triangles are congruent. This is the Side-Side-Side Postulate, and is written as SSS.

Postulate 4.1

Side-Side-Side Congruence   If the sides of one triangle are congruent to the sides of a second triangle, then the triangles are congruent.

Abbreviation: *SSS*

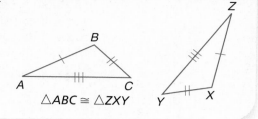

$\triangle ABC \cong \triangle ZXY$

Example 1   Use SSS in Proofs

MARINE BIOLOGY   The tail of an orca whale can be viewed as two triangles that share a common side. Write a two-column proof to prove that $\triangle BYA \cong \triangle CYA$ if $\overline{AB} \cong \overline{AC}$ and $\overline{BY} \cong \overline{CY}$.

Given: $\overline{AB} \cong \overline{AC}$; $\overline{BY} \cong \overline{CY}$

Prove: $\triangle BYA \cong \triangle CYA$

Proof:

Statements	Reasons
1. $\overline{AB} \cong \overline{AC}$; $\overline{BY} \cong \overline{CY}$	1. Given
2. $\overline{AY} \cong \overline{AY}$	2. Reflexive Property
3. $\triangle BYA \cong \triangle CYA$	3. SSS

Example 2   SSS on the Coordinate Plane

COORDINATE GEOMETRY   Determine whether $\triangle RTZ \cong \triangle JKL$ for R(2, 5), Z(1, 1), T(5, 2), L(−3, 0), K(−7, 1), and J(−4, 4). Explain.

Use the Distance Formula to show that the corresponding sides are congruent.

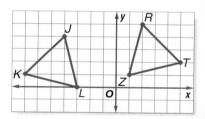

$RT = \sqrt{(2-5)^2 + (5-2)^2}$
$\quad = \sqrt{9+9}$ or $\sqrt{18}$

$TZ = \sqrt{(5-1)^2 + (2-1)^2}$
$\quad = \sqrt{16+1}$ or $\sqrt{17}$

$RZ = \sqrt{(2-1)^2 + (5-1)^2}$
$\quad = \sqrt{1+16}$ or $\sqrt{17}$

$JK = \sqrt{[-4-(-7)]^2 + (4-1)^2}$
$\quad = \sqrt{9+9}$ or $\sqrt{18}$

$KL = \sqrt{[-7-(-3)]^2 + (1-0)^2}$
$\quad = \sqrt{16+1}$ or $\sqrt{17}$

$JL = \sqrt{[-4-(-3)]^2 + (4-0)^2}$
$\quad = \sqrt{1+16}$ or $\sqrt{17}$

$RT = JK$, $TZ = KL$, and $RZ = JL$. By definition of congruent segments, all corresponding segments are congruent. Therefore, $\triangle RTZ \cong \triangle JKL$ by SSS.

SAS POSTULATE   Suppose you are given the measures of two sides and the angle they form, called the **included angle**. These conditions describe a unique triangle. Two triangles in which corresponding sides and the included pairs of angles are congruent provide another way to show that triangles are congruent.

Postulate 4.2

Side-Angle-Side Congruence If two sides and
the included angle of one triangle are congruent
to two sides and the included angle of another
triangle, then the triangles are congruent.

Abbreviation: *SAS*

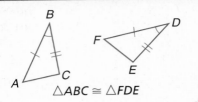

$\triangle ABC \cong \triangle FDE$

You can also construct congruent triangles given two sides and the included angle.

Construction

Congruent Triangles using Two Sides and the Included Angle

1 Draw a triangle and
label its vertices *A*,
B, and *C*.

2 Select a point *K* on
line *m*. Use a
compass to
construct $\overline{KL}$ on *m*
such that $\overline{KL} \cong \overline{BC}$.

3 Construct an angle
congruent to ∠*B*
using $\overleftrightarrow{KL}$ as a side
of the angle and
point *K* as the
vertex.

4 Construct $\overline{JK}$ such
that $\overline{JK} \cong \overline{AB}$.
Draw $\overline{JL}$ to
complete △*JKL*.

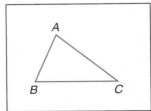
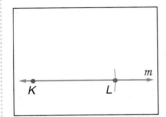
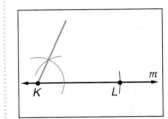
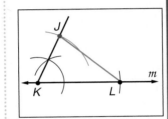

5 Cut out △*JKL* and place it over △*ABC*. How does △*JKL* compare to △*ABC*?

Example 3 Use SAS in Proofs

Write a flow proof.

Given: X is the midpoint of $\overline{BD}$.
X is the midpoint of $\overline{AC}$.

Prove: $\triangle DXC \cong \triangle BXA$

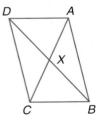

Flow Proof:

X is the midpoint of $\overline{DB}$.	→	$\overline{DX} \cong \overline{BX}$	
Given		Midpoint Theorem	

X is the midpoint of $\overline{AC}$.	→	$\overline{CX} \cong \overline{AX}$	→	△*DXC* ≅ △*BXA*
Given		Midpoint Theorem		SAS

∠*DXC* ≅ ∠*BXA*

Vertical ∠s are ≅.

Example 4 Identify Congruent Triangles

Determine which postulate can be used to prove that the triangles are congruent. If it is not possible to prove that they are congruent, write *not possible.*

a.

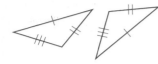

Each pair of corresponding sides are congruent. The triangles are congruent by the SSS Postulate.

b.

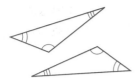

The triangles have three pairs of corresponding angles congruent. This does not match the SSS Postulate or the SAS Postulate. It is *not possible* to prove the triangles congruent.

Check for Understanding

Concept Check

1. OPEN ENDED Draw a triangle and label the vertices. Name two sides and the included angle.

2. FIND THE ERROR Carmelita and Jonathan are trying to determine whether △*ABC* is congruent to △*DEF*.

Carmelita

△*ABC* ≅ △*DEF* by SAS

Jonathan

Congruence cannot be determined.

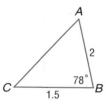

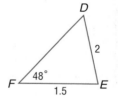

Who is correct and why?

Guided Practice

Determine whether △*EFG* ≅ △*MNP* given the coordinates of the vertices. Explain.

3. *E*(−4, −3), *F*(−2, 1), *G*(−2, −3), *M*(4, −3), *N*(2, 1), *P*(2, −3)

4. *E*(−2, −2), *F*(−4, 6), *G*(−3, 1), *M*(2, 2), *N*(4, 6), *P*(3, 1)

5. Write a flow proof.

Given: $\overline{DE}$ and $\overline{BC}$ bisect each other.

Prove: △*DGB* ≅ △*EGC*

Exercise 5

6. Write a two-column proof.

Given: $\overline{KM} \parallel \overline{JL}$, $\overline{KM} \cong \overline{JL}$

Prove: △*JKM* ≅ △*MLJ*

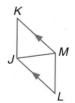

Exercise 6

Determine which postulate can be used to prove that the triangles are congruent. If it is not possible to prove that they are congruent, write *not possible.*

7.

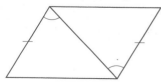

8.

Application

9. PRECISION FLIGHT The United States Navy Flight Demonstration Squadron, the Blue Angels, fly in a formation that can be viewed as two triangles with a common side. Write a two-column proof to prove that $\triangle SRT \cong \triangle QRT$ if T is the midpoint of $\overline{SQ}$ and $\overline{SR} \cong \overline{QR}$.

Practice and Apply

Homework Help

For Exercises	See Examples
10–13	2
14–19	3
20–21, 28–29	1
22–27	4

Extra Practice
See page 761.

Determine whether $\triangle JKL \cong \triangle FGH$ **given the coordinates of the vertices. Explain.**

10. $J(-3, 2)$, $K(-7, 4)$, $L(-1, 9)$, $F(2, 3)$, $G(4, 7)$, $H(9, 1)$

11. $J(-1, 1)$, $K(-2, -2)$, $L(-5, -1)$, $F(2, -1)$, $G(3, -2)$, $H(2, 5)$

12. $J(-1, -1)$, $K(0, 6)$, $L(2, 3)$, $F(3, 1)$, $G(5, 3)$, $H(8, 1)$

13. $J(3, 9)$, $K(4, 6)$, $L(1, 5)$, $F(1, 7)$, $G(2, 4)$, $H(-1, 3)$

Write a flow proof.

14. Given: $\overline{AE} \cong \overline{FC}$, $\overline{AB} \cong \overline{BC}$, $\overline{BE} \cong \overline{BF}$

Prove: $\triangle AFB \cong \triangle CEB$

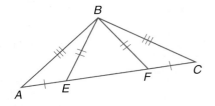

15. Given: $\overline{RQ} \cong \overline{TQ} \cong \overline{YQ} \cong \overline{WQ}$
$\angle RQY \cong \angle WQT$

Prove: $\triangle QWT \cong \triangle QYR$

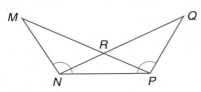

Write a two-column proof.

16. Given: $\triangle CDE$ is isosceles.
G is the midpoint of $\overline{CE}$.

Prove: $\triangle CDG \cong \triangle EDG$

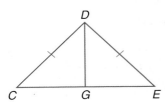

17. Given: $\triangle MRN \cong \triangle QRP$
$\angle MNP \cong \angle QPN$

Prove: $\triangle MNP \cong \triangle QPN$

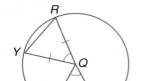

18. Given: $\overline{AC} \cong \overline{GC}$
$\overline{EC}$ bisects $\overline{AG}$.

Prove: $\triangle GEC \cong \triangle AEC$

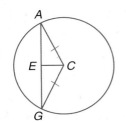

19. Given: $\triangle GHJ \cong \triangle LKJ$

Prove: $\triangle GHL \cong \triangle LKG$

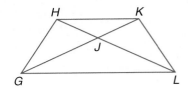

20. CATS A cat's ear is triangular in shape. Write a two-column proof to prove △RST ≅ △PNM if $\overline{RS} \cong \overline{PN}$, $\overline{RT} \cong \overline{MP}$, ∠S ≅ ∠N, and ∠T ≅ ∠M.

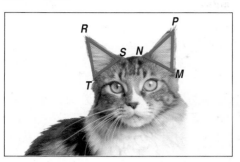

21. GEESE This photograph shows a flock of geese flying in formation. Write a two-column proof to prove that △EFG ≅ △HFG, if $\overline{EF} \cong \overline{HF}$ and G is the midpoint of $\overline{EH}$.

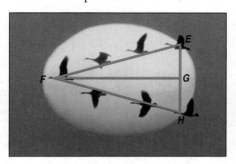

Determine which postulate can be used to prove that the triangles are congruent. If it is not possible to prove that they are congruent, write *not possible*.

22.

23.

24.

25.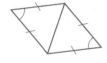

BASEBALL For Exercises 26 and 27, use the following information.
A baseball diamond is a square with four right angles and all sides congruent.

26. Write a two-column proof to prove that the distance from first base to third base is the same as the distance from home plate to second base.

27. Write a two-column proof to prove that the angle formed by second base, home plate, and third base is the same as the angle formed by second base, home plate, and first base.

28. CRITICAL THINKING Devise a plan and write a two-column proof for the following.

Given: $\overline{DE} \cong \overline{FB}$, $\overline{AE} \cong \overline{FC}$, $\overline{AE} \perp \overline{DB}$, $\overline{CF} \perp \overline{DB}$

Prove: △ABD ≅ △CDB

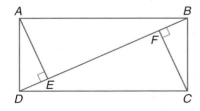

29. **WRITING IN MATH** Answer the question that was posed at the beginning of the lesson.

How do land surveyors use congruent triangles?

Include the following in your answer:
- description of three methods to prove triangles congruent, and
- another example of a career that uses properties of congruent triangles.

30. Which of the following statements about the figure is true?

Ⓐ $90 > a + b$

Ⓑ $a + b > 90$

Ⓒ $a + b = 90$

Ⓓ $a > b$

31. Classify the triangle with the measures of the angles in the ratio 3:6:7.

Ⓐ isosceles   Ⓑ acute   Ⓒ obtuse   Ⓓ right

Maintain Your Skills

Mixed Review   **Identify the congruent triangles in each figure.** *(Lesson 4-3)*

32.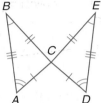

33.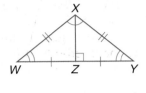

34.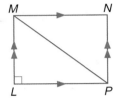

Find each measure if $\overline{PQ} \perp \overline{QR}$.
(Lesson 4-2)

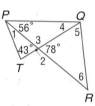

35. $m\angle 2$ **36.** $m\angle 3$

37. $m\angle 5$ **38.** $m\angle 4$

39. $m\angle 1$ **40.** $m\angle 6$

For Exercises 41–43, use the graphic at the right. *(Lesson 3-3)*

41. Find the rate of change from first quarter to the second quarter.

42. Find the rate of change from the second quarter to the third quarter.

43. Compare the rate of change from the first quarter to the second, and the second quarter to the third. Which had the greater rate of change?

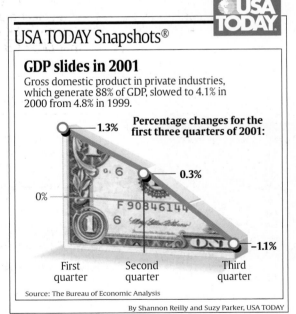

Getting Ready for the Next Lesson   **PREREQUISITE SKILL** $\overrightarrow{BD}$ and $\overrightarrow{AE}$ are angle bisectors and segment bisectors. Name the indicated segments and angles.

*(To review **bisectors of segments and angles**, see Lessons 1-5 and 1-6.)*

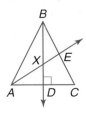

44. a segment congruent to $\overline{EC}$

45. an angle congruent to $\angle ABD$

46. an angle congruent to $\angle BDC$

47. a segment congruent to $\overline{AD}$

4-5 Proving Congruence—ASA, AAS

What You'll Learn

- Use the ASA Postulate to test for triangle congruence.
- Use the AAS Theorem to test for triangle congruence.

Vocabulary
- included side

How are congruent triangles used in construction?

The Bank of China Tower in Hong Kong has triangular trusses for structural support. These trusses form congruent triangles. In this lesson, we will explore two additional methods of proving triangles congruent.

ASA POSTULATE Suppose you were given the measures of two angles of a triangle and the side between them, the **included side**. Do these measures form a unique triangle?

Construction

Congruent Triangles Using Two Angles and Included Side

① Draw a triangle and label its vertices A, B, and C.

② Draw any line m and select a point L. Construct $\overline{LK}$ such that $\overline{LK} \cong \overline{CB}$.

③ Construct an angle congruent to ∠C at L using $\overrightarrow{LK}$ as a side of the angle.

④ Construct an angle congruent to ∠B at K using $\overrightarrow{LK}$ as a side of the angle. Label the point where the new sides of the angles meet J.

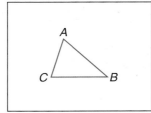 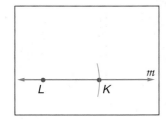 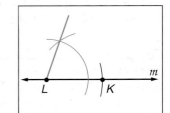 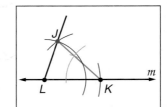

⑤ Cut out △JKL and place it over △ABC. How does △JKL compare to △ABC?

This construction leads to the Angle-Side-Angle Postulate, written as ASA.

Study Tip

Reading Math
The included side refers to the side that each of the angles share.

Postulate 4.3

Angle-Side-Angle Congruence If two angles and the included side of one triangle are congruent to two angles and the included side of another triangle, then the triangles are congruent.

Abbreviation: *ASA*

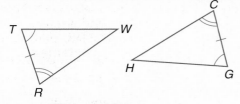

△RTW ≅ △CGH

Example 1 Use ASA in Proofs

Write a paragraph proof.

Given: $\overline{CP}$ bisects $\angle BCR$ and $\angle BPR$.

Prove: $\triangle BCP \cong \triangle RCP$

Proof:

Since $\overline{CP}$ bisects $\angle BCR$ and $\angle BPR$, $\angle BCP \cong \angle RCP$ and $\angle BPC \cong \angle RPC$. $\overline{CP} \cong \overline{CP}$ by the Reflexive Property. By ASA, $\triangle BCP \cong \triangle RCP$.

AAS THEOREM Suppose you are given the measures of two angles and a nonincluded side. Is this information sufficient to prove two triangles congruent?

Geometry Activity

Angle-Angle-Side Congruence

Model

1. Draw a triangle on a piece of patty paper. Label the vertices A, B, and C.

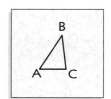

2. Copy $\overline{AB}$, $\angle B$, and $\angle C$ on another piece of patty paper and cut them out.

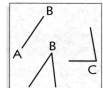

3. Assemble them to form a triangle in which the side is not the included side of the angles.

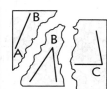

Analyze

1. Place the original $\triangle ABC$ over the assembled figure. How do the two triangles compare?

2. **Make a conjecture** about two triangles with two angles and the nonincluded side of one triangle congruent to two angles and the nonincluded side of the other triangle.

This activity leads to the Angle-Angle-Side Theorem, written as AAS.

Theorem 4.5

Angle-Angle-Side Congruence If two angles and a nonincluded side of one triangle are congruent to the corresponding two angles and side of a second triangle, then the two triangles are congruent.

Abbreviation: *AAS*

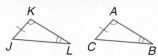

Example: $\triangle JKL \cong \triangle CAB$

Proof *Theorem 4.5*

Given: $\angle M \cong \angle S$, $\angle J \cong \angle R$, $\overline{MP} \cong \overline{ST}$

Prove: $\triangle JMP \cong \triangle RST$

Proof:

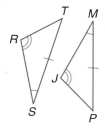

Statements	Reasons
1. $\angle M \cong \angle S$, $\angle J \cong \angle R$, $\overline{MP} \cong \overline{ST}$	1. Given
2. $\angle P \cong \angle T$	2. Third Angle Theorem
3. $\triangle JMP \cong \triangle RST$	3. ASA

Study Tip

Overlapping Triangles
When triangles overlap, it is a good idea to draw each triangle separately and label the congruent parts.

Example 2 Use AAS in Proofs

Write a flow proof.
Given: $\angle EAD \cong \angle EBC$

$\overline{AD} \cong \overline{BC}$

Prove: $\overline{AE} \cong \overline{BE}$

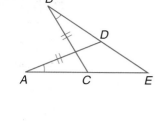

Flow Proof:

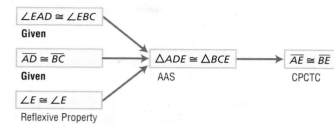

You have learned several methods for proving triangle congruence. The Concept Summary lists ways to help you determine which method to use.

Concept Summary	Methods to Prove Triangle Congruence
Definition of Congruent Triangles	All corresponding parts of one triangle are congruent to the corresponding parts of the other triangle.
SSS	The three sides of one triangle must be congruent to the three sides of the other triangle.
SAS	Two sides and the included angle of one triangle must be congruent to two sides and the included angle of the other triangle.
ASA	Two angles and the included side of one triangle must be congruent to two angles and the included side of the other triangle.
AAS	Two angles and a nonincluded side of one triangle must be congruent to two angles and side of the other triangle.

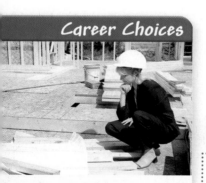
Example 3 Determine if Triangles Are Congruent

ARCHITECTURE This glass chapel was designed by Frank Lloyd Wright's son, Lloyd Wright. Suppose the redwood supports, $\overline{TU}$ and $\overline{TV}$, measure 3 feet, $TY = 1.6$ feet, and $m\angle U$ and $m\angle V$ are 31. Determine whether $\triangle TYU \cong \triangle TYV$. Justify your answer.

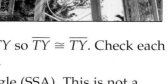

Explore We are given three measurements of each triangle. We need to determine whether the two triangles are congruent.

Plan Since $m\angle U = m\angle V$, $\angle U \cong \angle V$. Likewise, $TU = TV$ so $\overline{TU} \cong \overline{TV}$, and $TY = TY$ so $\overline{TY} \cong \overline{TY}$. Check each possibility using the five methods you know.

Solve We are given information about side-side-angle (SSA). This is not a method to prove two triangles congruent.

(continued on the next page)

Examine Use a compass, protractor, and ruler to draw a triangle with the given measurements. For simplicity of measurement, we will use centimeters instead of feet, so the measurements of the construction and those of the support beams will be proportional.

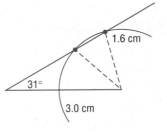

- Draw a segment 3.0 centimeters long.
- At one end, draw an angle of 31°. Extend the line longer than 3.0 centimeters.
- At the other end of the segment, draw an arc with a radius of 1.6 centimeters such that it intersects the line.

Notice that there are two possible segments that could determine the triangle. Since the given measurements do not lead to a unique triangle, we cannot show that the triangles are congruent.

Check for Understanding

Concept Check
1. **Find a counterexample** to show why AAA (Angle-Angle-Angle) cannot be used to prove triangle congruence.

2. **OPEN ENDED** Draw a triangle and label the vertices. Name two angles and the included side.

3. **Explain** why AAS is a theorem, not a postulate.

Guided Practice
Write a flow proof.

4. **Given:** $\overline{GH} \parallel \overline{KJ}, \overline{GK} \parallel \overline{HJ}$
   Prove: $\triangle GJK \cong \triangle JGH$

5. **Given:** $\overline{XW} \parallel \overline{YZ}, \angle X \cong \angle Z$
   Prove: $\triangle WXY \cong \triangle YZW$

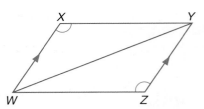

Write a paragraph proof.

6. **Given:** $\overline{QS}$ bisects $\angle RST$; $\angle R \cong \angle T$.
   Prove: $\triangle QRS \cong \triangle QTS$

7. **Given:** $\angle E \cong \angle K, \angle DGH \cong \angle DHG$
   $\overline{EG} \cong \overline{KH}$
   Prove: $\triangle EGD \cong \triangle KHD$

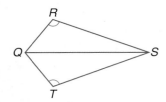

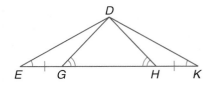

Application
8. **PARACHUTES** Suppose $\overline{ST}$ and $\overline{ML}$ each measure 7 feet, $\overline{SR}$ and $\overline{MK}$ each measure 5.5 feet, and $m\angle T = m\angle L = 49$. Determine whether $\triangle SRT \cong \triangle MKL$. Justify your answer.

Homework Help

For Exercises	See Examples
9, 11, 14, 15–18	2
10, 12, 13, 19, 20	1
21–28	3

Extra Practice
See page 762.

Write a flow proof.

9. **Given:** $\overline{EF} \parallel \overline{GH}$, $\overline{EF} \cong \overline{GH}$
   Prove: $\overline{EK} \cong \overline{KH}$

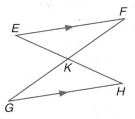

10. **Given:** $\overline{DE} \parallel \overline{JK}$, $\overline{DK}$ bisects $\overline{JE}$.
    Prove: $\triangle EGD \cong \triangle JGK$

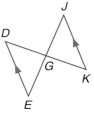

11. **Given:** $\angle V \cong \angle S$, $\overline{TV} \cong \overline{QS}$
    Prove: $\overline{VR} \cong \overline{SR}$

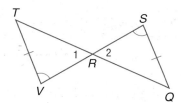

12. **Given:** $\overline{EJ} \parallel \overline{FK}$, $\overline{JG} \parallel \overline{KH}$, $\overline{EF} \cong \overline{GH}$
    Prove: $\triangle EJG \cong \triangle FKH$

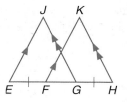

13. **Given:** $\overline{MN} \cong \overline{PQ}$, $\angle M \cong \angle Q$
    $\angle 2 \cong \angle 3$
    Prove: $\triangle MLP \cong \triangle QLN$

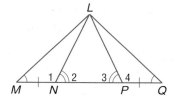

14. **Given:** Z is the midpoint of $\overline{CT}$.
    $\overline{CY} \parallel \overline{TE}$
    Prove: $\overline{YZ} \cong \overline{EZ}$

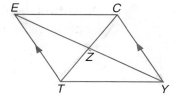

Write a paragraph proof.

15. **Given:** $\angle NOM \cong \angle POR$,
    $\overline{NM} \perp \overline{MR}$
    $\overline{PR} \perp \overline{MR}$, $\overline{NM} \cong \overline{PR}$
    Prove: $\overline{MO} \cong \overline{OR}$

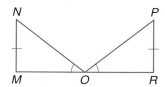

16. **Given:** $\overline{DL}$ bisects $\overline{BN}$,
    $\angle XLN \cong \angle XDB$
    Prove: $\overline{LN} \cong \overline{DB}$

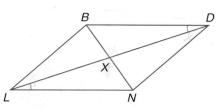

17. **Given:** $\angle F \cong \angle J$, $\angle E \cong \angle H$
    $\overline{EC} \cong \overline{GH}$
    Prove: $\overline{EF} \cong \overline{HJ}$

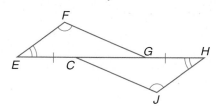

18. **Given:** $\overline{TX} \parallel \overline{SY}$
    $\angle TXY \cong \angle TSY$
    Prove: $\triangle TSY \cong \triangle YXT$

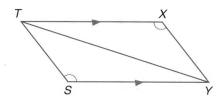

Write a two-column proof.

19. Given: $\angle MYT \cong \angle NYT$
$\angle MTY \cong \angle NTY$
Prove: $\triangle RYM \cong \triangle RYN$

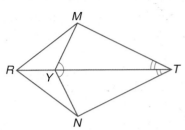

20. Given: $\triangle BMI \cong \triangle KMT$
$\overline{IP} \cong \overline{PT}$
Prove: $\triangle IPK \cong \triangle TPB$

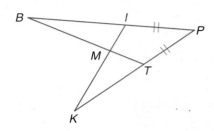

GARDENING For Exercises 21 and 22, use the following information.
Beth is planning a garden. She wants the triangular sections, $\triangle CFD$ and $\triangle HFG$, to be congruent. F is the midpoint of $\overline{DG}$, and $DG = 16$ feet.

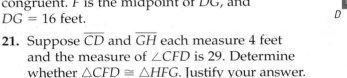

21. Suppose $\overline{CD}$ and $\overline{GH}$ each measure 4 feet and the measure of $\angle CFD$ is 29. Determine whether $\triangle CFD \cong \triangle HFG$. Justify your answer.

22. Suppose F is the midpoint of $\overline{CH}$, and $\overline{CH} \cong \overline{DG}$. Determine whether $\triangle CFD \cong \triangle HFG$. Justify your answer.

KITES For Exercises 23 and 24, use the following information.
Austin is building a kite. Suppose JL is 2 feet, JM is 2.7 feet, and the measure of $\angle NJM$ is 68.

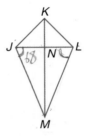

23. If N is the midpoint of $\overline{JL}$ and $\overline{KM} \perp \overline{JL}$, determine whether $\triangle JKN \cong \triangle LKN$. Justify your answer.

24. If $\overline{JM} \cong \overline{LM}$ and $\angle NJM \cong \angle NLM$, determine whether $\triangle JNM \cong \triangle LNM$. Justify your answer.

Complete each congruence statement and the postulate or theorem that applies.

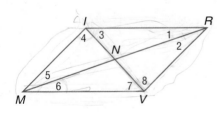

25. If $\overline{IM} \cong \overline{RV}$ and $\angle 2 \cong \angle 5$, then $\triangle INM \cong \triangle$ __?__ by __?__ .

26. If $\overline{IR} \parallel \overline{MV}$ and $\overline{IR} \cong \overline{MV}$, then $\triangle IRN \cong \triangle$ __?__ by __?__ .

27. If $\overline{IV}$ and $\overline{RM}$ bisect each other, then $\triangle RVN \cong \triangle$ __?__ by __?__ .

28. If $\angle MIR \cong \angle RVM$ and $\angle 1 \cong \angle 6$, then $\triangle MRV \cong \triangle$ __?__ by __?__ .

29. CRITICAL THINKING Aiko wants to estimate the distance between herself and a duck. She adjusts the visor of her cap so that it is in line with her line of sight to the duck. She keeps her neck stiff and turns her body to establish a line of sight to a point on the ground. Then she paces out the distance to the new point. Is the distance from the duck the same as the distance she just paced out? Explain your reasoning.

30.  Answer the question that was posed at the beginning of the lesson.

How are congruent triangles used in construction?

Include the following in your answer:
- explain how to determine whether the triangles are congruent, and
- why it is important that triangles used for structural support are congruent.

31. In $\triangle ABC$, $\overline{AD}$ and $\overline{DC}$ are angle bisectors and $m\angle B = 76$. What is the measure of $\angle ADC$?

Ⓐ 26   Ⓑ 52
Ⓒ 76   Ⓓ 128

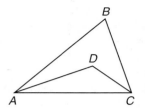

32. ALGEBRA   For a positive integer x, 1 percent of x percent of 10,000 equals

Ⓐ x.   Ⓑ $10x$.   Ⓒ $100x$.   Ⓓ $1000x$.

Maintain Your Skills

Mixed Review   **Write a flow proof.**   *(Lesson 4-4)*

33. Given: $\overline{BA} \cong \overline{DE}$, $\overline{DA} \cong \overline{BE}$
   Prove: $\triangle BEA \cong \triangle DAE$

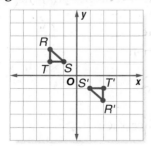

34. Given: $\overline{XZ} \perp \overline{WY}$
   $\overline{XZ}$ bisects $\overline{WY}$.
   Prove: $\triangle WZX \cong \triangle YZX$

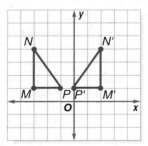

Verify that each of the following preserves congruence and name the congruence transformation.   *(Lesson 4-3)*

35.

36.

Write each statement in if-then form.   *(Lesson 2-3)*

37. Happy people rarely correct their faults.

38. A champion is afraid of losing.

Getting Ready for the Next Lesson   **PREREQUISITE SKILL   Classify each triangle according to its sides.**
(To review classification by sides, see Lesson 4-1.)

39.   **40.**   **41.**

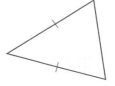

Congruence in Right Triangles

In Lessons 4-4 and 4-5, you learned theorems and postulates to prove triangles congruent. Do these theorems and postulates apply to right triangles?

Activity 1 Triangle Congruence

Model
Study each pair of right triangles.

a.
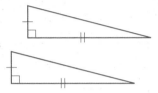

b.
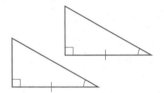

c.
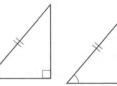

Analyze

1. Is each pair of triangles congruent? If so, which congruence theorem or postulate applies?

2. Rewrite the congruence rules from Exercise 1 using *leg*, (L), or *hypotenuse*, (H), to replace *side*. Omit the *A* for any right angle since we know that all right triangles contain a right angle and all right angles are congruent.

3. **Make a conjecture** If you know that the corresponding legs of two right triangles are congruent, what other information do you need to declare the triangles congruent? Explain.

In Lesson 4-5, you learned that SSA is not a valid test for determining triangle congruence. Can SSA be used to prove right triangles congruent?

Activity 2 SSA and Right Triangles

Make a Model
How many right triangles exist that have a hypotenuse of 10 centimeters and a leg of 7 centimeters?

Step 1 Draw $\overline{XY}$ so that $XY = 7$ centimeters.

Step 2 Use a protractor to draw a ray from Y that is perpendicular to $\overline{XY}$.

Step 3 Open your compass to a width of 10 centimeters. Place the point at X and draw a long arc to intersect the ray.

Step 4 Label the intersection Z and draw $\overline{XZ}$ to complete $\triangle XYZ$.

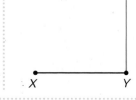

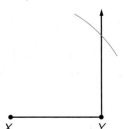

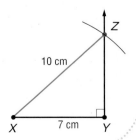

Analyze

4. Does the model yield a unique triangle?
5. Can you use the lengths of the hypotenuse and a leg to show right triangles are congruent?
6. **Make a conjecture** about the case of SSA that exists for right triangles.

The two activities provide evidence for four ways to prove right triangles congruent.

Key Concept		Right Triangle Congruence
Theorem	**Abbreviation**	**Example**
4.6 Leg-Leg Congruence If the legs of one right triangle are congruent to the corresponding legs of another right triangle, then the triangles are congruent.	LL	
4.7 Hypotenuse-Angle Congruence If the hypotenuse and acute angle of one right triangle are congruent to the hypotenuse and corresponding acute angle of another right triangle, then the two triangles are congruent.	HA	
4.8 Leg-Angle Congruence If one leg and an acute angle of one right triangle are congruent to the corresponding leg and acute angle of another right triangle, then the triangles are congruent.	LA	
Postulate		
4.4 Hypotenuse-Leg Congruence If the hypotenuse and a leg of one right triangle are congruent to the hypotenuse and corresponding leg of another right triangle, then the triangles are congruent.	HL	

PROOF Write a paragraph proof of each theorem.

7. Theorem 4.6
8. Theorem 4.7
9. Theorem 4.8 (*Hint*: There are two possible cases.)

Use the figure to write a two-column proof.

10. **Given:** $\overline{ML} \perp \overline{MK}, \overline{JK} \perp \overline{KM}$
    $\angle J \cong \angle L$
    Prove: $\overline{JM} \cong \overline{KL}$

11. **Given:** $\overline{JK} \perp \overline{KM}, \overline{JM} \cong \overline{KL}$
    $\overline{ML} \parallel \overline{JK}$
    Prove: $\overline{ML} \cong \overline{JK}$

4-6 Isosceles Triangles

What You'll Learn

- Use properties of isosceles triangles.
- Use properties of equilateral triangles.

How are triangles used in art?

The art of Lois Mailou Jones, a twentieth-century artist, includes paintings and textile design, as well as book illustration. Notice the isosceles triangles in this painting, *Damballah*.

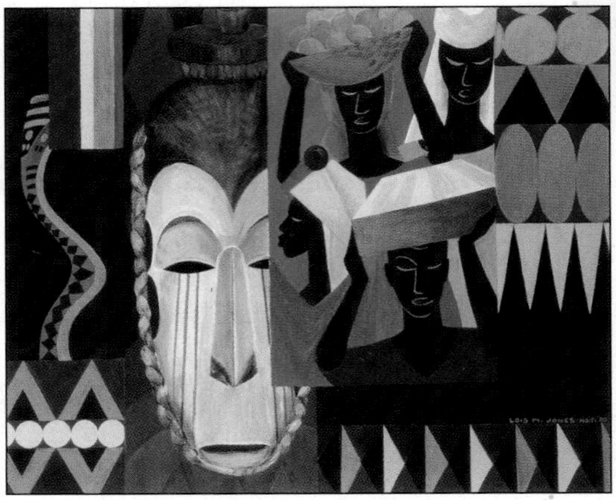

Vocabulary

- vertex angle
- base angles

PROPERTIES OF ISOSCELES TRIANGLES In Lesson 4-1, you learned that isosceles triangles have two congruent sides. Like the right triangle, the parts of an isosceles triangle have special names.

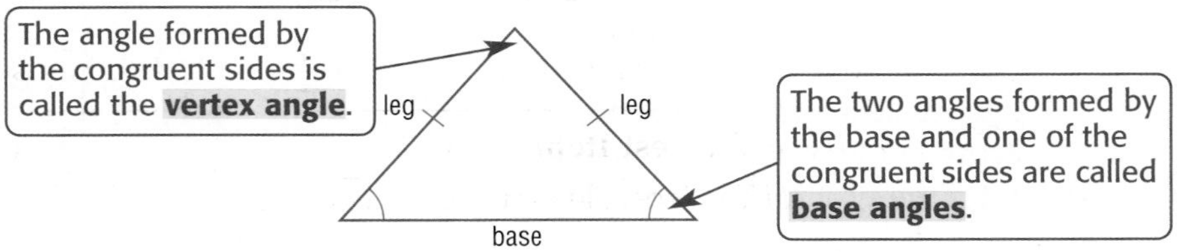
The angle formed by the congruent sides is called the **vertex angle**.
leg
leg
base
The two angles formed by the base and one of the congruent sides are called **base angles**.

In this activity, you will investigate the relationship of the base angles and legs of an isosceles triangle.

Geometry Activity

Isosceles Triangles

Model

- Draw an acute triangle on patty paper with $\overline{AC} \cong \overline{BC}$.
- Fold the triangle through C so that A and B coincide.

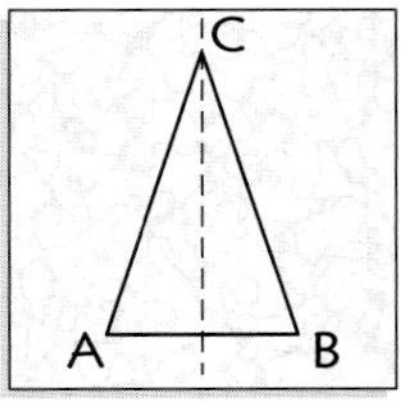
C
A
B

Analyze

1. What do you observe about $\angle A$ and $\angle B$?
2. Draw an obtuse isosceles triangle. Compare the base angles.
3. Draw a right isosceles triangle. Compare the base angles.

The results of the Geometry Activity suggest Theorem 4.9.

Theorem 4.9

Isosceles Triangle Theorem If two sides of a triangle are congruent, then the angles opposite those sides are congruent.

Example: If $\overline{AB} \cong \overline{CB}$, then $\angle A \cong \angle C$.

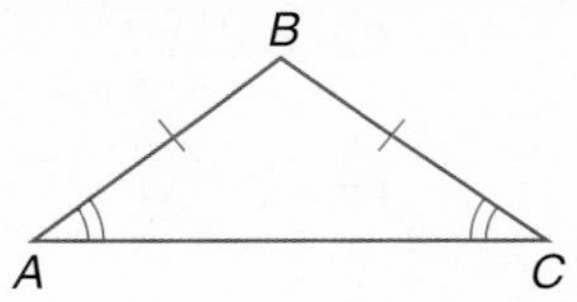
B
A
C

Example 1 *Proof of Theorem*

Write a two-column proof of the
Isosceles Triangle Theorem.

Given: $\triangle PQR$, $\overline{PQ} \cong \overline{RQ}$

Prove: $\angle P \cong \angle R$

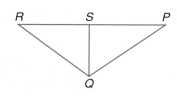

Proof:

Statements	Reasons
1. Let S be the midpoint of $\overline{PR}$.	1. Every segment has exactly one midpoint.
2. Draw an auxiliary segment $\overline{QS}$.	2. Two points determine a line.
3. $\overline{PS} \cong \overline{RS}$	3. Midpoint Theorem
4. $\overline{QS} \cong \overline{QS}$	4. Congruence of segments is reflexive.
5. $\overline{PQ} \cong \overline{RQ}$	5. Given
6. $\triangle PQS \cong \triangle RQS$	6. SSS
7. $\angle P \cong \angle R$	7. CPCTC

Example 2 *Find the Measure of a Missing Angle*

Multiple-Choice Test Item

If $\overline{GH} \cong \overline{HK}$, $\overline{HJ} \cong \overline{JK}$, and $m\angle GJK = 100$, what is the measure of $\angle HGK$?

Ⓐ 10　　Ⓑ 15　　Ⓒ 20　　Ⓓ 25

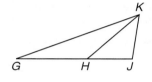

Test-Taking Tip

Diagrams Label the diagram with the given information. Use your drawing to plan the next step in solving the problem.

Read the Test Item

$\triangle GHK$ is isosceles with base $\overline{GK}$. Likewise, $\triangle HJK$ is isosceles with base $\overline{HK}$.

Solve the Test Item

Step 1 The base angles of $\triangle HJK$ are congruent. Let $x = m\angle KHJ = m\angle HKJ$.

$m\angle KHJ + m\angle HKJ + m\angle HJK = 180$　Angle Sum Theorem

$x + x + 100 = 180$　Substitution

$2x + 100 = 180$　Add.

$2x = 80$　Subtract 100 from each side.

$x = 40$　So, $m\angle KHJ = m\angle HKJ = 40$.

Step 2 $\angle GHK$ and $\angle KHJ$ form a linear pair. Solve for $m\angle GHK$.

$m\angle KHJ + m\angle GHK = 180$　Linear pairs are supplementary.

$40 + m\angle GHK = 180$　Substitution

$m\angle GHK = 140$　Subtract 40 from each side.

Step 3 The base angles of $\triangle GHK$ are congruent. Let y represent $m\angle HGK$ and $m\angle GKH$.

$m\angle GHK + m\angle HGK + m\angle GKH = 180$　Angle Sum Theorem

$140 + y + y = 180$　Substitution

$140 + 2y = 180$　Add.

$2y = 40$　Subtract 140 from each side.

$y = 20$　Divide each side by 2.

The measure of $\angle HGK$ is 20. Choice C is correct.

The converse of the Isosceles Triangle Theorem is also true.

Theorem 4.10

If two angles of a triangle are congruent, then the sides opposite those angles are congruent.

Abbreviation: *Conv. of Isos. △ Th.*

Example: If $\angle D \cong \angle F$, then $\overline{DE} \cong \overline{FE}$.

You will prove Theorem 4.10 in Exercise 33.

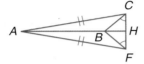

Web Quest

You can use properties of triangles to prove Thales of Miletus' important geometric ideas. Visit www.geometryonline.com/webquest to continue work on your WebQuest project.

Example 3 Congruent Segments and Angles

a. **Name two congruent angles.**

$\angle AFC$ is opposite $\overline{AC}$ and $\angle ACF$ is opposite $\overline{AF}$, so $\angle AFC \cong \angle ACF$.

b. **Name two congruent segments.**

By the converse of the Isosceles Triangle Theorem, the sides opposite congruent angles are congruent. So, $\overline{BC} \cong \overline{BF}$.

PROPERTIES OF EQUILATERAL TRIANGLES Recall that an equilateral triangle has three congruent sides. The Isosceles Triangle Theorem also applies to equilateral triangles. This leads to two corollaries about the angles of an equilateral triangle.

Corollaries

4.3 A triangle is equilateral if and only if it is equiangular.

4.4 Each angle of an equilateral triangle measures 60°.

You will prove Corollaries 4.3 and 4.4 in Exercises 31 and 32.

Example 4 Use Properties of Equilateral Triangles

$\triangle EFG$ is equilateral, and $\overline{EH}$ bisects $\angle E$.

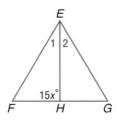

a. **Find** $m\angle 1$ **and** $m\angle 2$.

Each angle of an equilateral triangle measures 60°. So, $m\angle 1 + m\angle 2 = 60$. Since the angle was bisected, $m\angle 1 = m\angle 2$. Thus, $m\angle 1 = m\angle 2 = 30$.

b. **ALGEBRA Find** x.

$m\angle EFH + m\angle 1 + m\angle EHF = 180$	Angle Sum Theorem
$60 + 30 + 15x = 180$	$m\angle EFH = 60, m\angle 1 = 30, m\angle EHF = 15x$
$90 + 15x = 180$	Add.
$15x = 90$	Subtract 90 from each side.
$x = 6$	Divide each side by 15.

Concept Check

1. **Explain** how many angles in an isosceles triangle must be given to find the measures of the other angles.

2. **Name** the congruent sides and angles of isosceles $\triangle WXZ$ with base $\overline{WZ}$.

3. **OPEN ENDED** Describe a method to construct an equilateral triangle.

Guided Practice

Refer to the figure.

4. If $\overline{AD} \cong \overline{AH}$, name two congruent angles.

5. If $\angle BDH \cong \angle BHD$, name two congruent segments.

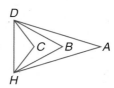

6. **ALGEBRA** Triangle GHF is equilateral with $m\angle F = 3x + 4$, $m\angle G = 6y$, and $m\angle H = 19z + 3$. Find x, y, and z.

Write a two-column proof.

7. **Given:** $\triangle CTE$ is isosceles with vertex $\angle C$.
   $m\angle T = 60$
   Prove: $\triangle CTE$ is equilateral.

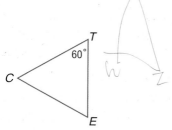

Standardized Test Practice
(A) (B) (C) (D)

8. If $\overline{PQ} \cong \overline{QS}$, $\overline{QR} \cong \overline{RS}$, and $m\angle PRS = 72$, what is the measure of $\angle QPS$?

    (A) 27   (B) 54   (C) 63   (D) 72

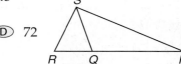

Homework Help

For Exercises	See Examples
9–14	3
15–22, 27–28, 34–37	4
23–26, 38–39	2
29–33	1

Extra Practice
See page 762.

Refer to the figure.

9. If $\overline{LT} \cong \overline{LR}$, name two congruent angles.

10. If $\overline{LX} \cong \overline{LW}$, name two congruent angles.

11. If $\overline{SL} \cong \overline{QL}$, name two congruent angles.

12. If $\angle LXY \cong \angle LYX$, name two congruent segments.

13. If $\angle LSR \cong \angle LRS$, name two congruent segments.

14. If $\angle LYW \cong \angle LWY$, name two congruent segments.

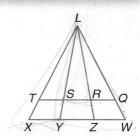

$\triangle KLN$ and $\triangle LMN$ are isosceles and $m\angle JKN = 130$. Find each measure.

15. $m\angle LNM$ 16. $m\angle M$

17. $m\angle LKN$ 18. $m\angle J$

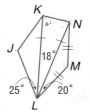

$\triangle DFG$ and $\triangle FGH$ are isosceles, $m\angle FDH = 28$ and $\overline{DG} \cong \overline{FG} \cong \overline{FH}$. Find each measure.

19. $m\angle DFG$ 20. $m\angle DGF$

21. $m\angle FGH$ 22. $m\angle GFH$

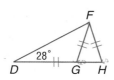

In the figure, $\overline{JM} \cong \overline{PM}$ and $\overline{ML} \cong \overline{PL}$.

23. If $m\angle PLJ = 34$, find $m\angle JPM$.
24. If $m\angle PLJ = 58$, find $m\angle PJL$.

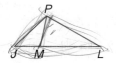

In the figure, $\overline{GK} \cong \overline{GH}$ and $\overline{HK} \cong \overline{KJ}$.

25. If $m\angle HGK = 28$, find $m\angle HJK$.
26. If $m\angle HGK = 42$, find $m\angle HJK$.

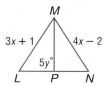

Triangle LMN is equilateral, and $\overline{MP}$ bisects $\overline{LN}$.

27. Find x and y.
28. Find the measure of each side of $\triangle LMN$.

PROOF   Write a two-column proof.

29. **Given:** $\triangle XKF$ is equilateral.
            $\overline{XJ}$ bisects $\angle X$.
   Prove: J is the midpoint of $\overline{KF}$.

30. **Given:** $\triangle MLP$ is isosceles.
            N is the midpoint of $\overline{MP}$.
   Prove: $\overline{LN} \perp \overline{MP}$

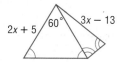

31. Corollary 4.3   32. Corollary 4.4   33. Theorem 4.10

34. **DESIGN** The basic structure covering Spaceship Earth at the Epcot Center in Orlando, Florida, is a triangle. Describe the minimum requirement to show that these triangles are equilateral.

ALGEBRA   Find x.

35.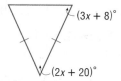

36.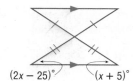

37.

ARTISANS   For Exercises 38 and 39, use the following information.

This geometric sign from the Grassfields area in Western Cameroon (Western Africa) uses approximations of isosceles triangles within and around two circles.

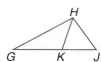

38. Trace the figure. Identify and draw one isosceles triangle from each set in the sign.

39. Describe the similarities between the different triangles.

40. **CRITICAL THINKING** In the figure, $\triangle ABC$ is isosceles, $\triangle DCE$ is equilateral, and $\triangle FCG$ is isosceles. Find the measures of the five numbered angles at vertex C.

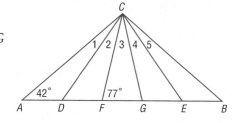

41. **WRITING IN MATH** Answer the question that was posed at the beginning of the lesson.

How are triangles used in art?

Include the following in your answer:
- at least three other geometric shapes frequently used in art, and
- a description of how isosceles triangles are used in the painting.

42. Given right triangle XYZ with hypotenuse $\overline{XY}$, YP is equal to YZ. If $m\angle PYZ = 26$, find $m\angle XZP$.

Ⓐ 13   Ⓑ 26   Ⓒ 32   Ⓓ 64

43. **ALGEBRA** A segment is drawn from (3, 5) to (9, 13). What are the coordinates of the midpoint of this segment?

Ⓐ (3, 4)   Ⓑ (12, 18)   Ⓒ (6, 8)   Ⓓ (6, 9)

Maintain Your Skills

Mixed Review   **Write a paragraph proof.** *(Lesson 4-5)*

44. Given: $\angle N \cong \angle D$, $\angle G \cong \angle I$,
$\overline{AN} \cong \overline{SD}$
Prove: $\triangle ANG \cong \triangle SDI$

45. Given: $\overline{VR} \perp \overline{RS}$, $\overline{UT} \perp \overline{SU}$
$\overline{RS} \cong \overline{US}$
Prove: $\triangle VRS \cong \triangle TUS$

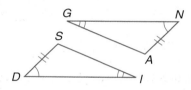
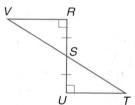

Determine whether $\triangle QRS \cong \triangle EGH$ given the coordinates of the vertices. Explain. *(Lesson 4-4)*

46. $Q(-3, 1)$, $R(1, 2)$, $S(-1, -2)$, $E(6, -2)$, $G(2, -3)$, $H(4, 1)$

47. $Q(1, -5)$, $R(5, 1)$, $S(4, 0)$, $E(-4, -3)$, $G(-1, 2)$, $H(2, 1)$

Construct a truth table for each compound statement. *(Lesson 2-2)*

48. a and b **49.** $\sim p$ or $\sim q$ **50.** k and $\sim m$ **51.** $\sim y$ or z

Getting Ready for the Next Lesson   **PREREQUISITE SKILL** Find the coordinates of the midpoint of the segment with the given endpoints. *(To review finding midpoints, see Lesson 1-5.)*

52. $A(2, 15)$, $B(7, 9)$ **53.** $C(-4, 6)$, $D(2, -12)$ **54.** $E(3, 2.5)$, $F(7.5, 4)$

Practice Quiz 2   Lessons 4-4 through 4-6

1. Determine whether $\triangle JML \cong \triangle BDG$ given that $J(-4, 5)$, $M(-2, 6)$, $L(-1, 1)$, $B(-3, -4)$, $D(-4, -2)$, and $G(1, -1)$. *(Lesson 4-4)*

2. Write a two-column proof to prove that $\overline{AJ} \cong \overline{EH}$, given $\angle A \cong \angle H$, $\angle AEJ \cong \angle HJE$. *(Lesson 4-5)*

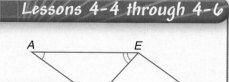

$\triangle WXY$ and $\triangle XYZ$ are isosceles and $m\angle XYZ = 128$. Find each measure. *(Lesson 4-6)*

3. $m\angle XWY$ **4.** $m\angle WXY$ **5.** $m\angle YZX$

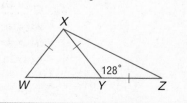

Triangles and Coordinate Proof

What You'll Learn

• Position and label triangles for use in coordinate proofs.

• Write coordinate proofs.

Vocabulary

• coordinate proof

How can the coordinate plane be useful in proofs?

In this chapter, we have used several methods of proof. You have also used the coordinate plane to identify characteristics of a triangle. We can combine what we know about triangles in the coordinate plane with algebra in a new method of proof called *coordinate proof*.

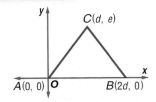

POSITION AND LABEL TRIANGLES **Coordinate proof** uses figures in the coordinate plane and algebra to prove geometric concepts. The first step in writing a coordinate proof is the placement of the figure on the coordinate plane.

Study Tip

Placement of Figures
The guidelines apply to any polygon placed on the coordinate plane.

Key Concept — Placing Figures on the Coordinate Plane

1. Use the origin as a vertex or center of the figure.

2. Place at least one side of a polygon on an axis.

3. Keep the figure within the first quadrant if possible.

4. Use coordinates that make computations as simple as possible.

Example 1 Position and Label a Triangle

Position and label isosceles triangle JKL on a coordinate plane so that base $\overline{JK}$ is a units long.

• Use the origin as vertex J of the triangle.

• Place the base of the triangle along the positive x-axis.

• Position the triangle in the first quadrant.

• Since K is on the x-axis, its y-coordinate is 0. Its x-coordinate is a because the base of the triangle is a units long.

• Since $\triangle JKL$ is isosceles, the x-coordinate of L is halfway between 0 and a or $\frac{a}{2}$. We cannot determine the y-coordinate in terms of a, so call it b.

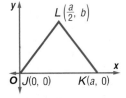

Example 2 Find the Missing Coordinates

Name the missing coordinates of isosceles right $\triangle EFG$.

Vertex F is positioned at the origin; its coordinates are $(0, 0)$. Vertex E is on the y-axis, and vertex G is on the x-axis. So $\angle EFG$ is a right angle. Since $\triangle EFG$ is isosceles, $\overline{EF} \cong \overline{GF}$. The distance from E to F is a units. The distance from F to G must be the same. So, the coordinates of G are $(a, 0)$.

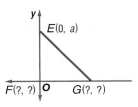

WRITE COORDINATE PROOFS After the figure has been placed on the coordinate plane and labeled, we can use coordinate proof to verify properties and to prove theorems. The Distance Formula, Slope Formula, and Midpoint Formula are often used in coordinate proof.

Example 3 Coordinate Proof

Write a coordinate proof to prove that the measure of the segment that joins the vertex of the right angle in a right triangle to the midpoint of the hypotenuse is one-half the measure of the hypotenuse.

The first step is to position and label a right triangle on the coordinate plane. Place the right angle at the origin and label it A. Use coordinates that are multiples of 2 because the Midpoint Formula takes half the sum of the coordinates.

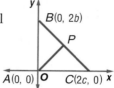

Given: right $\triangle ABC$ with right $\angle BAC$
P is the midpoint of $\overline{BC}$.

Prove: $AP = \frac{1}{2}BC$

Proof:

By the Midpoint Formula, the coordinates of P are $\left(\frac{0 + 2c}{2}, \frac{2b + 0}{2}\right)$ or (c, b). Use the Distance Formula to find AP and BC.

$$AP = \sqrt{(c - 0)^2 + (b - 0)^2} \qquad BC = \sqrt{(2c - 0)^2 + (0 - 2b)^2}$$
$$= \sqrt{c^2 + b^2} \qquad\qquad BC = \sqrt{4c^2 + 4b^2} \text{ or } 2\sqrt{c^2 + b^2}$$
$$\frac{1}{2}BC = \sqrt{c^2 + b^2}$$

Therefore, $AP = \frac{1}{2}BC$.

Example 4 Classify Triangles

ARROWHEADS Write a coordinate proof to prove that this arrowhead is shaped like an isosceles triangle. The arrowhead is 3 inches long and 1.5 inches wide.

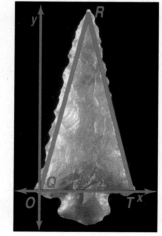

The first step is to label the coordinates of each vertex. Q is at the origin, and T is at $(1.5, 0)$. The y-coordinate of R is 3. The x-coordinate is halfway between 0 and 1.5 or 0.75. So, the coordinates of R are $(0.75, 3)$.

If the legs of the triangle are the same length, the triangle is isosceles. Use the Distance Formula to determine the lengths of QR and RT.

$$QR = \sqrt{(0.75 - 0)^2 + (3 - 0)^2}$$
$$= \sqrt{0.5625 + 9} \text{ or } \sqrt{9.5625}$$

$$RT = \sqrt{(1.5 - 0.75)^2 + (0 - 3)^2}$$
$$= \sqrt{0.5625 + 9} \text{ or } \sqrt{9.5625}$$

Since each leg is the same length, $\triangle QRT$ is isosceles. The arrowhead is shaped like an isosceles triangle.

Concept Check

1. **Explain** how to position a triangle on the coordinate plane to simplify a proof.

2. **OPEN ENDED** Draw a scalene right triangle on the coordinate plane for use in a coordinate proof. Label the coordinates of each vertex.

Guided Practice

Position and label each triangle on the coordinate plane.

3. isosceles $\triangle FGH$ with base $\overline{FH}$ that is $2b$ units long

4. equilateral $\triangle CDE$ with sides a units long

Find the missing coordinates of each triangle.

5.
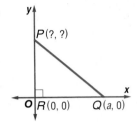

6.
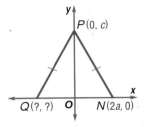

7.
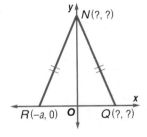

8. Write a coordinate proof for the following statement.
The midpoint of the hypotenuse of a right triangle is equidistant from each of the vertices.

Application

9. **TEPEES** Write a coordinate proof to prove that the tepee is shaped like an isosceles triangle. Suppose the tepee is 8 feet tall and 4 feet wide.

Homework Help

For Exercises	See Examples
10–15	1
16–24	2
25–29	3
30–33	4

Extra Practice
See page 762.

Position and label each triangle on the coordinate plane.

10. isosceles $\triangle QRT$ with base $\overline{QR}$ that is b units long

11. equilateral $\triangle MNP$ with sides $2a$ units long

12. isosceles right $\triangle JML$ with hypotenuse $\overline{JM}$ and legs c units long

13. equilateral $\triangle WXZ$ with sides $\frac{1}{2}b$ units long

14. isosceles $\triangle PWY$ with a base $\overline{PW}$ that is $(a + b)$ units long

15. right $\triangle XYZ$ with hypotenuse $\overline{XZ}$, $ZY = 2(XY)$, and $\overline{XY}$ b units long

Find the missing coordinates of each triangle.

16.
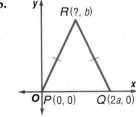

17.
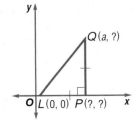

18.
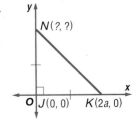

19.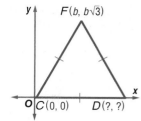
$F(b, b\sqrt{3})$
$C(0, 0)$ $D(?, ?)$

20.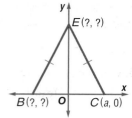
$E(?, ?)$
$B(?, ?)$ $C(a, 0)$

21.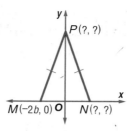
$P(?, ?)$
$M(-2b, 0)$ $N(?, ?)$

22.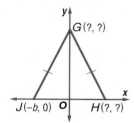
$G(?, ?)$
$J(-b, 0)$ $H(?, ?)$

23.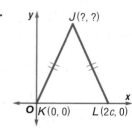
$J(?, ?)$
$K(0, 0)$ $L(2c, 0)$

24.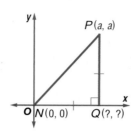
$P(a, a)$
$N(0, 0)$ $Q(?, ?)$

More About . . .

Steeplechase ●···········

The Steeplechase is a horse race two to four miles long that focuses on jumping hurdles. The rails of the fences vary in height.

Source: www.steeplechasetimes.com

Write a coordinate proof for each statement.

25. The segments joining the vertices to the midpoints of the legs of an isosceles triangle are congruent.

26. The three segments joining the midpoints of the sides of an isosceles triangle form another isosceles triangle.

27. If a line segment joins the midpoints of two sides of a triangle, then it is parallel to the third side.

28. If a line segment joins the midpoints of two sides of a triangle, then its length is equal to one-half the length of the third side.

29. STEEPLECHASE Write a coordinate proof to prove that triangles ABD and FBD are congruent. $\overline{BD}$ is perpendicular to $\overline{AF}$, and B is the midpoint of the upper bar of the hurdle.

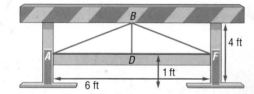

NAVIGATION **For Exercises 30 and 31, use the following information.**
A motor boat is located 800 yards east of the port. There is a ship 800 yards to the east, and another ship 800 yards to the north of the motor boat.

30. Write a coordinate proof to prove that the port, motor boat, and the ship to the north form an isosceles right triangle.

31. Write a coordinate proof to prove that the distance between the two ships is the same as the distance from the port to the northern ship.

HIKING **For Exercises 32 and 33, use the following information.**
Tami and Juan are hiking. Tami hikes 300 feet east of the camp and then hikes 500 feet north. Juan hikes 500 feet west of the camp and then 300 feet north.

32. Write a coordinate proof to prove that Juan, Tami, and the camp form a right triangle.

33. Find the distance between Tami and Juan.

Find the coordinates of point Z so △XYZ is the indicated type of triangle. Point X has coordinates (0, 0) and Y has coordinates (a, b).

34. right triangle
with right angle Z

35. isosceles triangle
with base $\overline{XZ}$

36. scalene triangle

37. **CRITICAL THINKING** Classify △ABC by its angles and its sides. Explain.

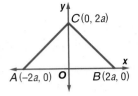

38. WRITING IN MATH   Answer the question that was posed at the beginning of the lesson.

How can the coordinate plane be useful in proofs?

Include the following in your answer:

- types of proof, and
- a theorem from this chapter that could be proved using a coordinate proof.

39. What is the length of the segment whose endpoints are at (1, −2) and (−3, 1)?

Ⓐ 3 Ⓑ 4 Ⓒ 5 Ⓓ 6

40. **ALGEBRA**   What are the coordinates of the midpoint of the line segment whose endpoints are (−5, 4) and (−2, −1)?

Ⓐ (3, 3) Ⓑ (−3.5, 1.5) Ⓒ (−1.5, 2.5) Ⓓ (3.5, −2.5)

Maintain Your Skills

Mixed Review **Write a two-column proof.** *(Lessons 4-5 and 4-6)*

41. **Given:** ∠3 ≅ ∠4
Prove: $\overline{QR} \cong \overline{QS}$

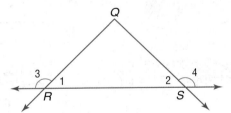

42. **Given:** isosceles triangle JKN
with vertex ∠N, $\overline{JK} \parallel \overline{LM}$
Prove: △NML is isosceles.

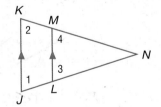

43. **Given:** $\overline{AD} \cong \overline{CE}$, $\overline{AD} \parallel \overline{CE}$
Prove: △ABD ≅ △EBC

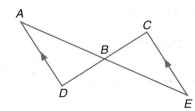

44. **Given:** $\overline{WX} \cong \overline{XY}$, ∠V ≅ ∠Z
Prove: $\overline{WV} \cong \overline{YZ}$

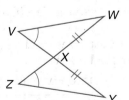

State which lines, if any, are parallel. State the postulate or theorem that justifies your answer. *(Lesson 3-5)*

45.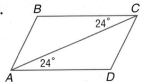

46.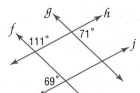

47.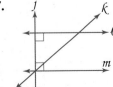

Vocabulary and Concept Check

acute triangle (p. 178)
base angles (p. 216)
congruence transformations (p. 194)
congruent triangles (p. 192)
coordinate proof (p. 222)
corollary (p. 188)

equiangular triangle (p. 178)
equilateral triangle (p. 179)
exterior angle (p. 186)
flow proof (p. 187)
included angle (p. 201)
included side (p. 207)

isosceles triangle (p. 179)
obtuse triangle (p. 178)
remote interior angles (p. 186)
right triangle (p. 178)
scalene triangle (p. 179)
vertex angle (p. 216)

A complete list of theorems and postulates can be found on pages R1–R8.

Exercises Choose the letter of the word or phrase that best matches each statement.

1. A triangle with an angle whose measure is greater than 90 is a(n) __?__ triangle.
2. A triangle with exactly two congruent sides is a(n) __?__ triangle.
3. The __?__ states that the sum of the measures of the angles of a triangle is 180.
4. If $\angle B \cong \angle E$, $\overline{AB} \cong \overline{DE}$, and $\overline{BC} \cong \overline{EF}$, then $\triangle ABC \cong \triangle DEF$ by __?__.
5. In an equiangular triangle, all angles are __?__ angles.
6. If two angles of a triangle and their included side are congruent to two angles and the included side of another triangle, this is called the __?__.
7. If $\angle A \cong \angle F$, $\angle B \cong \angle G$, and $\overline{AC} \cong \overline{FH}$, then $\triangle ABC \cong \triangle FGH$, by __?__.
8. A(n) __?__ angle of a triangle has a measure equal to the measures of the two remote interior angles of the triangle.

a. acute
b. AAS Theorem
c. ASA Theorem
d. Angle Sum Theorem
e. equilateral
f. exterior
g. isosceles
h. obtuse
i. right
j. SAS Theorem
k. SSS Theorem

Lesson-by-Lesson Review

4-1 Classifying Triangles

See pages 178–183.

Concept Summary

- Triangles can be classified by their angles as acute, obtuse, or right.
- Triangles can be classified by their sides as scalene, isosceles, or equilateral.

Example **Find the measures of the sides of $\triangle TUV$. Classify the triangle by sides.**

Use the Distance Formula to find the measure of each side.

$$TU = \sqrt{[-5 - (-2)]^2 + [4 - (-2)]^2}$$
$$= \sqrt{9 + 36} \text{ or } \sqrt{45}$$
$$UV = \sqrt{[3 - (-5)]^2 + (1 - 4)^2}$$
$$= \sqrt{64 + 9} \text{ or } \sqrt{73}$$
$$VT = \sqrt{(-2 - 3)^2 + (-2 - 1)^2}$$
$$= \sqrt{25 + 9} \text{ or } \sqrt{34}$$

Since none of the side measures are equal, $\triangle TUV$ is scalene.

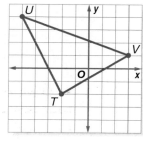

Exercises Classify each triangle by its angles and by its sides if $m\angle ABC = 100$.   *See Examples 1 and 2 on pages 178 and 179.*

9. $\triangle ABC$ 10. $\triangle BDP$ 11. $\triangle BPQ$

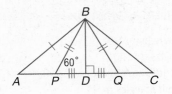

4-2 Angles of Triangles

See pages 185–191.

Concept Summary
- The sum of the measures of the angles of a triangle is 180.
- The measure of an exterior angle is equal to the sum of the measures of the two remote interior angles.

Example If $\overline{TU} \perp \overline{UV}$ and $\overline{UV} \perp \overline{VW}$, find $m\angle 1$.

$m\angle 1 + 72 + m\angle TVW = 180$ Angle Sum Theorem

$m\angle 1 + 72 + (90 - 27) = 180$ $m\angle TVW = 90 - 27$

$m\angle 1 + 135 = 180$ Simplify.

$m\angle 1 = 45$ Subtract 135 from each side.

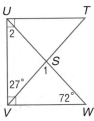

Exercises Find each measure.
See Example 1 on page 186.

12. $m\angle 1$ 13. $m\angle 2$ 14. $m\angle 3$

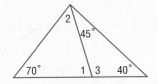

4-3 Congruent Triangles

See pages 192–198.

Concept Summary
- Two triangles are congruent when all of their corresponding parts are congruent.

Example If $\triangle EFG \cong \triangle JKL$, name the corresponding congruent angles and sides.

$\angle E \cong \angle J$, $\angle F \cong \angle K$, $\angle G \cong \angle L$, $\overline{EF} \cong \overline{JK}$, $\overline{FG} \cong \overline{KL}$, and $\overline{EG} \cong \overline{JL}$.

Exercises Name the corresponding angles and sides for each pair of congruent triangles.   *See Example 1 on page 193.*

15. $\triangle EFG \cong \triangle DCB$ 16. $\triangle LCD \cong \triangle GCF$ 17. $\triangle NCK \cong \triangle KER$

4-4 Proving Congruence—SSS, SAS

See pages 200–206.

Concept Summary
- If all of the corresponding sides of two triangles are congruent, then the triangles are congruent (SSS).
- If two corresponding sides of two triangles and the included angle are congruent, then the triangles are congruent (SAS).

Example

Determine whether $\triangle ABC \cong \triangle TUV$. Explain.

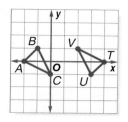

$AB = \sqrt{[-1-(-2)]^2 + (1-0)^2}$
$\quad = \sqrt{1+1}$ or $\sqrt{2}$

$TU = \sqrt{(3-4)^2 + (-1-0)^2}$
$\quad = \sqrt{1+1}$ or $\sqrt{2}$

$BC = \sqrt{[0-(-1)]^2 + (-1-1)^2}$
$\quad = \sqrt{1+4}$ or $\sqrt{5}$

$UV = \sqrt{(2-3)^2 + [1-(-1)]^2}$
$\quad = \sqrt{1+4}$ or $\sqrt{5}$

$CA = \sqrt{(-2-0)^2 + [0-(-1)]^2}$
$\quad = \sqrt{4+1}$ or $\sqrt{5}$

$VT = \sqrt{(4-2)^2 + (0-1)^2}$
$\quad = \sqrt{4+1}$ or $\sqrt{5}$

By the definition of congruent segments, all corresponding sides are congruent. Therefore, $\triangle ABC \cong \triangle TUV$ by SSS.

Exercises Determine whether $\triangle MNP \cong \triangle QRS$ given the coordinates of the vertices. Explain. *See Example 2 on page 201.*

18. $M(0, 3)$, $N(-4, 3)$, $P(-4, 6)$, $Q(5, 6)$, $R(2, 6)$, $S(2, 2)$

19. $M(3, 2)$, $N(7, 4)$, $P(6, 6)$, $Q(-2, 3)$, $R(-4, 7)$, $S(-6, 6)$

4-5 Proving Congruence—ASA, AAS

See pages 207–213.

Concept Summary

- If two pairs of corresponding angles and the included sides of two triangles are congruent, then the triangles are congruent (ASA).
- If two pairs of corresponding angles and a pair of corresponding nonincluded sides of two triangles are congruent, then the triangles are congruent (AAS).

Example

Write a proof.

Given: $\overline{JK} \parallel \overline{MN}$; L is the midpoint of $\overline{KM}$.

Prove: $\triangle JLK \cong \triangle NLM$

Flow proof:

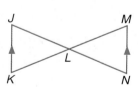

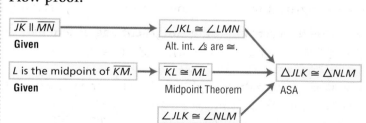

Exercises For Exercises 20 and 21, use the figure. Write a two-column proof for each of the following. *See Example 2 on page 209.*

20. Given: $\overline{DF}$ bisects $\angle CDE$.
$\overline{CE} \perp \overline{DF}$
Prove: $\triangle DGC \cong \triangle DGE$

21. Given: $\triangle DGC \cong \triangle DGE$
$\triangle GCF \cong \triangle GEF$
Prove: $\triangle DFC \cong \triangle DFE$

Chapter
4   For More ...   • Extra Practice, see pages 760–762.
                     • Mixed Problem Solving, see page 785.

4-6 Isosceles Triangles

See pages
216–221.

Concept Summary

- Two sides of a triangle are congruent if and only if the angles opposite those sides are congruent.
- A triangle is equilateral if and only if it is equiangular.

Example   If $\overline{FG} \cong \overline{GJ}$, $\overline{GJ} \cong \overline{JH}$, $\overline{FJ} \cong \overline{FH}$, and $m\angle GJH = 40$, find $m\angle H$.

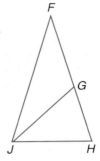

$\triangle GHJ$ is isosceles with base $\overline{GH}$, so $\angle JGH \cong \angle H$ by the Isosceles Triangle Theorem. Thus, $m\angle JGH = m\angle H$.

$$m\angle GJH + m\angle JGH + m\angle H = 180 \quad \text{Angle Sum Theorem}$$
$$40 + 2(m\angle H) = 180 \quad \text{Substitution}$$
$$2(m\angle H) = 140 \quad \text{Subtract 40 from each side.}$$
$$m\angle H = 70 \quad \text{Divide each side by 2.}$$

Exercises   For Exercises 22–25, refer to the figure at the right.
See Example 2 on page 217.

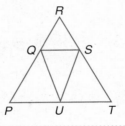

22. If $\overline{PQ} \cong \overline{UQ}$ and $m\angle P = 32$, find $m\angle PUQ$.
23. If $\overline{PQ} \cong \overline{UQ}$, $\overline{PR} \cong \overline{RT}$, and $m\angle PQU = 40$, find $m\angle R$.
24. If $\overline{RQ} \cong \overline{RS}$ and $m\angle RQS = 75$, find $m\angle R$.
25. If $\overline{RQ} \cong \overline{RS}$, $\overline{RP} \cong \overline{RT}$, and $m\angle RQS = 80$, find $m\angle P$.

4-7 Triangles and Coordinate Proof

See pages
222–226.

Concept Summary

- Coordinate proofs use algebra to prove geometric concepts.
- The Distance Formula, Slope Formula, and Midpoint Formula are often used in coordinate proof.

Example   Position and label isosceles right triangle ABC with legs of length a units on the coordinate plane.

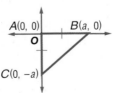

- Use the origin as the vertex of $\triangle ABC$ that has the right angle.
- Place each base along an axis.
- Since B is on the x-axis, its y-coordinate is 0. Its x-coordinate is a because the leg $\overline{AB}$ of the triangle is a units long.
- Since $\triangle ABC$ is isosceles, C should also be a distance of a units from the origin. Its coordinates are $(0, -a)$.

Exercises   Position and label each triangle on the coordinate plane.
See Example 1 on page 222.

26. isosceles $\triangle TRI$ with base $\overline{TI}$ 4a units long
27. equilateral $\triangle BCD$ with side length 6m units long
28. right $\triangle JKL$ with leg lengths of a units and b units

Vocabulary and Concepts

Choose the letter of the type of triangle that best matches each phrase.

1. triangle with no sides congruent
2. triangle with at least two sides congruent
3. triangle with all sides congruent

a. isosceles
b. scalene
c. equilateral

Skills and Applications

Identify the indicated triangles in the figure if $\overline{PB} \perp \overline{AD}$ and $\overline{PA} \cong \overline{PC}$.

4. obtuse
5. isosceles
6. right

Find the measure of each angle in the figure.

7. $m\angle 1$
8. $m\angle 2$
9. $m\angle 3$

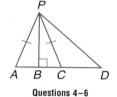

Questions 4–6

Questions 7–9

Name the corresponding angles and sides for each pair of congruent triangles.

10. $\triangle DEF \cong \triangle PQR$
11. $\triangle FMG \cong \triangle HNJ$
12. $\triangle XYZ \cong \triangle ZYX$

13. Determine whether $\triangle JKL \cong \triangle MNP$ given $J(-1, -2)$, $K(2, -3)$, $L(3, 1)$, $M(-6, -7)$, $N(-2, 1)$, and $P(5, 3)$. Explain.

14. Write a flow proof.
    Given: $\triangle JKM \cong \triangle JNM$
    Prove: $\triangle JKL \cong \triangle JNL$

In the figure, $\overline{FJ} \cong \overline{FH}$ and $\overline{GF} \cong \overline{GH}$.

15. If $m\angle JFH = 34$, find $m\angle J$.
16. If $m\angle GHJ = 152$ and $m\angle G = 32$, find $m\angle JFH$.

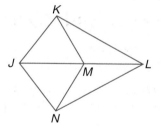

Question 14

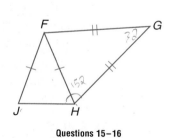

Questions 15–16

17. **LANDSCAPING** A landscaper designed a garden shaped as shown in the figure. The landscaper has decided to place point B 22 feet east of point A, point C 44 feet east of point A, point E 36 feet south of point A, and point D 36 feet south of point C. The angles at points A and C are right angles. Prove that $\triangle ABE \cong \triangle CBD$.

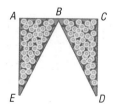

18. **STANDARDIZED TEST PRACTICE** In the figure, $\triangle FGH$ is a right triangle with hypotenuse $\overline{FH}$ and $GJ = GH$. What is the measure of $\angle JGH$?

(A) 104
(B) 62
(C) 56
(D) 28

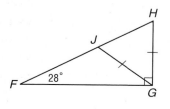

Part 1 | Multiple Choice

Record your answers on the answer sheet provided by your teacher or on a sheet of paper.

1. In 2002, Capitol City had a population of 2010, and Shelbyville had a population of 1040. If Capitol City grows at a rate of 150 people a year and Shelbyville grows at a rate of 340 people a year, when will the population of Shelbyville be greater than that of Capitol City? (Prerequisite Skill)

   Ⓐ 2004   Ⓑ 2008

   Ⓒ 2009   Ⓓ 2012

2. Which unit is most appropriate for measuring liquid in a bottle? (Lesson 1-2)

   Ⓐ grams   Ⓑ feet

   Ⓒ liters   Ⓓ meters

3. A 9-foot tree casts a shadow on the ground. The distance from the top of the tree to the end of the shadow is 12 feet. To the nearest foot, how long is the shadow? (Lesson 1-3)

   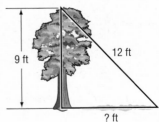

   Ⓐ 7 ft

   Ⓑ 8 ft

   Ⓒ 10 ft

   Ⓓ 13 ft

4. Which of the following is the inverse of the statement *If it is raining, then Kamika carries an umbrella*? (Lesson 2-2)

   Ⓐ If Kamika carries an umbrella, then it is raining.

   Ⓑ If Kamika does not carry an umbrella, then it is not raining.

   Ⓒ If it is not raining, then Kamika carries an umbrella.

   Ⓓ If it is not raining, then Kamika does not carry an umbrella.

5. Students in a math classroom simulated stock trading. Kris drew the graph below to model the value of his shares at closing. The graph that modeled the value of Mitzi's shares was parallel to the one Kris drew. Which equation might represent the line for Mitzi's graph? (Lesson 3-3)

   Ⓐ $-2x - y = 1$

   Ⓑ $x - 2y = 1$

   Ⓒ $x + 2y = 1$

   Ⓓ $2x - y = 1$

   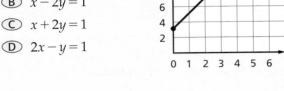

6. What is $m\angle EFG$? (Lesson 4-2)

   Ⓐ 35

   Ⓑ 70

   Ⓒ 90

   Ⓓ 110

   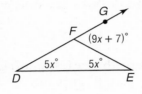

7. In the figure, $\triangle ABD \cong \triangle CBD$. If A has the coordinates $(-2, 4)$, what are the coordinates of C? (Lesson 4-3)

   Ⓐ $(-4, -2)$

   Ⓑ $(-4, 2)$

   Ⓒ $(-2, -4)$

   Ⓓ $(2, -4)$

   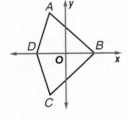

8. The wings of some butterflies can be modeled by triangles as shown. If $\overline{AC} \cong \overline{DC}$ and $\angle ACB \cong \angle ECD$, which additional statements are needed to prove that $\triangle ACB \cong \triangle ECD$? (Lesson 4-4)

   Ⓐ $\overline{BC} \cong \overline{CE}$

   Ⓑ $\overline{AB} \cong \overline{ED}$

   Ⓒ $\angle BAC \cong \angle CED$

   Ⓓ $\angle ABC \cong \angle CDE$

Preparing for Standardized Tests
For test-taking strategies and more
practice, see pages 795–810.

Part 2 | Short Response/Grid In

Record your answers on the answer sheet provided by your teacher or on a sheet of paper.

9. Find the product $3s^2(2s^3 - 7)$.
   (Prerequisite Skill)

10. After a long workout, Brian noted, "If I do not drink enough water, then I will become dehydrated." He then made another statement, "If I become dehydrated, then I did not drink enough water." How is the second statement related to the original statement? (Lesson 2-2)

11. On a coordinate map, the towns of Creston and Milford are located at $(-1, -1)$ and $(1, 3)$, respectively. A third town, Dixville, is located at $(x, -1)$ so that Creston and Dixville are endpoints of the base of the isosceles triangle formed by the three locations. What is the value of x? (Lesson 4-1)

12. A watchtower, built to help prevent forest fires, was designed as an isosceles triangle. If the side of the tower meets the ground at a 105° angle, what is the measure of the angle at the top of the tower? (Lesson 4-2)

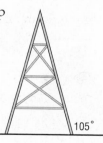

13. During a synchronized flying show, airplane A and airplane D are equidistant from the ground. They descend at the same angle to land at points B and E, respectively. Which postulate would prove that $\triangle ABC \cong \triangle DEF$? (Lesson 4-4)

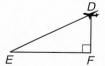

14. $\triangle ABC$ is an isosceles triangle with $\overline{AB} \cong \overline{BC}$, and the measure of vertex angle B is three times $m\angle A$. What is $m\angle C$? (Lesson 4-6)

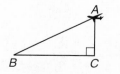
Test-Taking Tip
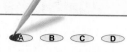
Question 8
- If you are not permitted to write in your test booklet, make a sketch of the figure on scrap paper.
- Mark the figure with all of the information you know so that you can determine the congruent triangles more easily.
- Make a list of postulates or theorems that you might use for this case.

Part 3 | Extended Response

Record your answers on a sheet of paper. Show your work.

15. Train tracks a and b are parallel lines although they appear to come together to give the illusion of distance in a drawing. All of the railroad ties are parallel to each other.

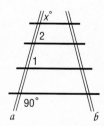

   a. What is the value of x? (Lesson 3-1)

   b. What is the relationship between the tracks and the ties that run across the tracks? (Lesson 1-5)

   c. What is the relationship between $\angle 1$ and $\angle 2$? Explain. (Lesson 3-2)

16. The measures of the angles of $\triangle ABC$ are $5x$, $4x - 1$, and $3x + 13$.
   a. Draw a figure to illustrate $\triangle ABC$. (Lesson 4-1)

   b. Find the measure of each angle of $\triangle ABC$. Explain. (Lesson 4-2)

   c. Prove that $\triangle ABC$ is an isosceles triangle. (Lesson 4-6)

What You'll Learn

- **Lesson 5-1** Identify and use perpendicular bisectors, angle bisectors, medians, and altitudes of triangles.
- **Lesson 5-2** Apply properties of inequalities relating to the measures of angles and sides of triangles.
- **Lesson 5-3** Use indirect proof with algebra and geometry.
- **Lessons 5-4 and 5-5** Apply the Triangle Inequality Theorem and SAS and SSS inequalities.

Key Vocabulary

- perpendicular bisector (p. 238)
- median (p. 240)
- altitude (p. 241)
- indirect proof (p. 255)

Why It's Important

There are several relationships among the sides and angles of triangles. These relationships can be used to compare the length of a person's stride and the rate at which that person is walking or running. *In Lesson 5-5, you will learn how to use the measure of the sides of a triangle to compare stride and rate.*

Getting Started

▶ **Prerequisite Skills** To be successful in this chapter, you'll need to master these skills and be able to apply them in problem-solving situations. Review these skills before beginning Chapter 5.

For Lesson 5-1   **Midpoint of a Segment**

Find the coordinates of the midpoint of a segment with the given endpoints.
(For review, see Lesson 1-3.)

   1. $A(-12, -5)$, $B(4, 15)$ **2.** $C(-22, -25)$, $D(10, 10)$ **3.** $E(19, -7)$, $F(-20, -3)$

For Lesson 5-2   **Exterior Angle Theorem**

Find the measure of each numbered angle if $\overline{AB} \perp \overline{BC}$.   *(For review, see Lesson 4-2.)*

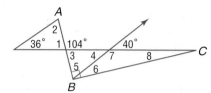

   4. $\angle 1$ **5.** $\angle 2$

   6. $\angle 3$ **7.** $\angle 4$

   8. $\angle 5$ **9.** $\angle 6$

10. $\angle 7$ **11.** $\angle 8$

For Lesson 5-3   **Deductive Reasoning**

Determine whether a valid conclusion can be reached from the two true statements using the Law of Detachment. If a valid conclusion is possible, state it. If a valid conclusion does not follow, write *no conclusion*. *(For review, see Lesson 2-4.)*

12. (1) If the three sides of one triangle are congruent to the three sides of a second triangle, then the triangles are congruent.
     (2) $\triangle ABC$ and $\triangle PQR$ are congruent.

13. (1) The sum of the measures of the angles of a triangle is 180.
     (2) Polygon JKL is a triangle.

Relationships in Triangles   Make this Foldable to help you organize your notes. Begin with one sheet of notebook paper.

Step 1   **Fold**

Fold lengthwise to the holes.

Step 2   **Cut**

Cut 5 tabs.

Step 3   **Label**

Label the edge. Then label the tabs using lesson numbers.

Reading and Writing   As you read and study each lesson, write notes and examples under the appropriate tab.

Geometry Activity

Bisectors, Medians, and Altitudes

You can use the constructions for midpoint, perpendiculars, and angle bisectors to construct special segments in triangles.

Construction 1 Construct the bisector of a side of a triangle.

1 Draw a triangle like △ABC. Adjust the compass to an opening greater than $\frac{1}{2}AC$. Place the compass at vertex A, and draw an arc above and below $\overline{AC}$.

2 Using the same compass settings, place the compass at vertex C. Draw an arc above and below $\overline{AC}$. Label the points of intersection of the arcs P and Q.

3 Use a straightedge to draw $\overrightarrow{PQ}$. Label the point where $\overrightarrow{PQ}$ bisects $\overline{AC}$ as M.

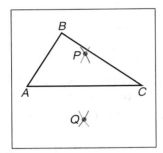

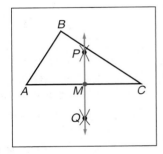

$\overline{AM} \cong \overline{MC}$ by construction and $\overline{PM} \cong \overline{PM}$ by the Reflexive Property. $\overline{AP} \cong \overline{CP}$ because the arcs were drawn with the same compass setting. Thus, △APM ≅ △CPM by SSS. By CPCTC, ∠PMA ≅ ∠PMC. A linear pair of congruent angles are right angles. So $\overrightarrow{PQ}$ is not only a bisector of $\overline{AC}$, but a perpendicular bisector.

1. Construct the perpendicular bisectors for the other two sides.
2. What do you notice about the intersection of the perpendicular bisectors?

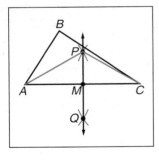

Construction 2 Construct a median of a triangle.

1 Draw intersecting arcs above and below $\overline{BC}$. Label the points of intersection R and S.

2 Use a straightedge to find the point where $\overline{RS}$ intersects $\overline{BC}$. Label the midpoint M.

3 Draw a line through A and M. $\overline{AM}$ is a median of △ABC.

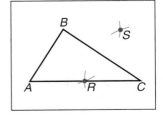

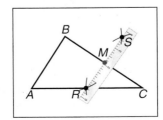

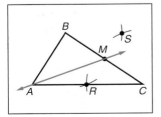

3. Construct the medians of the other two sides.
4. What do you notice about the medians of a triangle?

Construction 3   Construct an altitude of a triangle.

1 Place the compass at vertex *B* and draw two arcs intersecting $\overleftrightarrow{AC}$. Label the points where the arcs intersect the side *X* and *Y*.

2 Adjust the compass to an opening greater than $\frac{1}{2}XY$. Place the compass on *X* and draw an arc above $\overline{AC}$. Using the same setting, place the compass on *Y* and draw an arc above $\overline{AC}$. Label the intersection of the arcs *H*.

3 Use a straightedge to draw $\overleftrightarrow{BH}$. Label the point where $\overleftrightarrow{BH}$ intersects $\overline{AC}$ as *D*. $\overline{BD}$ is an altitude of $\triangle ABC$ and is perpendicular to $\overline{AC}$.

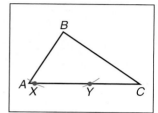
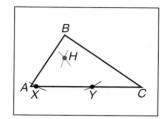
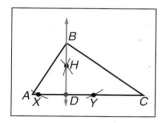

5. Construct the altitudes to the other two sides. (*Hint:* You may need to extend the lines containing the sides of your triangle.)

6. What observation can you make about the altitudes of your triangle?

Construction 4   Construct an angle bisector of a triangle.

1 Place the compass on vertex *A*, and draw arcs through $\overline{AB}$ and $\overline{AC}$. Label the points where the arcs intersect the sides as *J* and *K*.

2 Place the compass on *J*, and draw an arc. Then place the compass on *K* and draw an arc intersecting the first arc. Label the intersection *L*.

3 Use a straightedge to draw $\overrightarrow{AL}$. $\overrightarrow{AL}$ is an angle bisector of $\triangle ABC$.

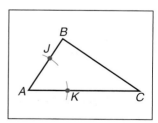
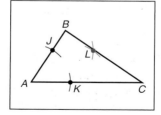
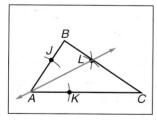

7. Construct the angle bisectors for the other two angles.

8. What do you notice about the angle bisectors?

Analyze

9. Repeat the four constructions for each type of triangle.

    a. obtuse scalene   **b.** right scalene   **c.** isosceles   **d.** equilateral

Make a Conjecture

10. Where do the lines intersect for acute, obtuse, and right triangles?

11. Under what circumstances do the special lines of triangles coincide with each other?

5-1 Bisectors, Medians, and Altitudes

What You'll Learn

- Identify and use perpendicular bisectors and angle bisectors in triangles.
- Identify and use medians and altitudes in triangles.

Vocabulary

- perpendicular bisector
- concurrent lines
- point of concurrency
- circumcenter
- incenter
- median
- centroid
- altitude
- orthocenter

How can you balance a paper triangle on a pencil point?

Acrobats and jugglers often balance objects while performing their acts. These skilled artists need to find the center of gravity for each object or body position in order to keep balanced. The center of gravity for any triangle can be found by drawing the *medians* of a triangle and locating the point where they intersect.

PERPENDICULAR BISECTORS AND ANGLE BISECTORS The first construction you made in the Geometry Activity on pages 236 and 237 was the perpendicular bisector of a side of a triangle. A **perpendicular bisector** of a side of a triangle is a line, segment, or ray that passes through the midpoint of the side and is perpendicular to that side. Perpendicular bisectors of segments have some special properties.

Theorems — Points on Perpendicular Bisectors

5.1 Any point on the perpendicular bisector of a segment is equidistant from the endpoints of the segment.

Example: If $\overline{AB} \perp \overline{CD}$ and $\overline{AB}$ bisects $\overline{CD}$, then $AC = AD$ and $BC = BD$.

5.2 Any point equidistant from the endpoints of a segment lies on the perpendicular bisector of the segment.

Example: If $AC = AD$, then A lies on the perpendicular bisector of $\overline{CD}$. If $BC = BD$, then B lies on the perpendicular bisector of $\overline{CD}$.

You will prove Theorems 5.1 and 5.2 in Exercises 10 and 31, respectively.

Study Tip

Common Misconception
Note that Theorem 5.2 states the point is on the perpendicular bisector. It does not say that any line containing that point is a perpendicular bisector.

Recall that a locus is the set of all points that satisfy a given condition. A perpendicular bisector can be described as the locus of points in a plane equidistant from the endpoints of a given segment.

Since a triangle has three sides, there are three perpendicular bisectors in a triangle. The perpendicular bisectors of a triangle intersect at a common point. When three or more lines intersect at a common point, the lines are called **concurrent lines**, and their point of intersection is called the **point of concurrency**. The point of concurrency of the perpendicular bisectors of a triangle is called the **circumcenter**.

Theorem 5.3

Circumcenter Theorem The circumcenter of a triangle is equidistant from the vertices of the triangle.

Example: If J is the circumcenter of $\triangle ABC$, then $AJ = BJ = CJ$.

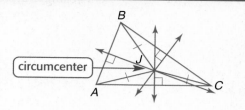

circumcenter

Proof *Theorem 5.3*

Given: ℓ, m, and n are perpendicular bisectors of $\overline{AB}$, $\overline{AC}$, and $\overline{BC}$, respectively.

Prove: $AJ = BJ = CJ$

Paragraph Proof:

Since J lies on the perpendicular bisector of $\overline{AB}$, it is equidistant from A and B. By the definition of equidistant, $AJ = BJ$. The perpendicular bisector of $\overline{BC}$ also contains J. Thus, $BJ = CJ$. By the Transitive Property of Equality, $AJ = CJ$. Thus, $AJ = BJ = CJ$.

Another special line, segment, or ray in triangles is an angle bisector.

Example 1 *Use Angle Bisectors*

Given: $\overline{PX}$ bisects $\angle QPR$, $\overline{XY} \perp \overline{PQ}$, and $\overline{XZ} \perp \overline{PR}$.

Prove: $\overline{XY} \cong \overline{XZ}$

Proof:

Statements	Reasons
1. $\overline{PX}$ bisects $\angle QPR$, $\overline{XY} \perp \overline{PQ}$, and $\overline{XZ} \perp \overline{PR}$.	1. Given
2. $\angle YPX \cong \angle ZPX$	2. Definition of angle bisector
3. $\angle PYX$ and $\angle PZX$ are right angles.	3. Definition of perpendicular
4. $\angle PYX \cong \angle PZX$	4. Right angles are congruent.
5. $\overline{PX} \cong \overline{PX}$	5. Reflexive Property
6. $\triangle PYX \cong \triangle PZX$	6. AAS
7. $\overline{XY} \cong \overline{XZ}$	7. CPCTC

In Example 1, XY and XZ are lengths representing the distance from X to each side of $\angle QPR$. This is a proof of Theorem 5.4.

Theorems *Points on Angle Bisectors*

5.4 Any point on the angle bisector is equidistant from the sides of the angle.

5.5 Any point equidistant from the sides of an angle lies on the angle bisector.

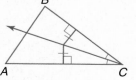

You will prove Theorem 5.5 in Exercise 32.

Study Tip

Locus

An angle bisector can be described as the locus of points in a plane equidistant from the sides of an angle. Since the sides of the angle are contained in intersecting lines, the locus of points in a plane equidistant from two intersecting lines is the angle bisector of the vertical angles formed by the lines.

 www.geometryonline.com/extra_examples

As with perpendicular bisectors, there are three angle bisectors in any triangle. The angle bisectors of a triangle are concurrent, and their point of concurrency is called the **incenter** of a triangle.

Theorem 5.6

Incenter Theorem The incenter of a triangle is equidistant from each side of the triangle.

Example: If K is the incenter of $\triangle ABC$, then $KP = KQ = KR$.

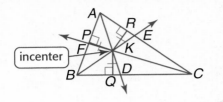

You will prove Theorem 5.6 in Exercise 33.

Study Tip

Medians as Bisectors
Because the median contains the midpoint, it is also a bisector of the side of the triangle.

MEDIANS AND ALTITUDES A **median** is a segment whose endpoints are a vertex of a triangle and the midpoint of the side opposite the vertex. Every triangle has three medians.

The medians of a triangle also intersect at a common point. The point of concurrency for the medians of a triangle is called a **centroid**. The centroid is the point of balance for any triangle.

Theorem 5.7

Centroid Theorem The centroid of a triangle is located two thirds of the distance from a vertex to the midpoint of the side opposite the vertex on a median.

Example: If L is the centroid of $\triangle ABC$, $AL = \frac{2}{3}AE$, $BL = \frac{2}{3}BF$, and $CL = \frac{2}{3}CD$.

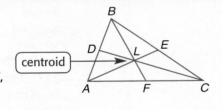

Example 2 Segment Measures

ALGEBRA Points S, T, and U are the midpoints of $\overline{DE}$, $\overline{EF}$, and $\overline{DF}$, respectively. Find x, y, and z.

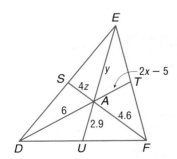

Study Tip

Eliminating Fractions
You could also multiply the equation $DA = \frac{2}{3}DT$ by 3 to eliminate the denominator.

• Find x.

$DT = DA + AT$	Segment Addition Postulate
$\quad = 6 + (2x - 5)$	Substitution
$\quad = 2x + 1$	Simplify.
$DA = \frac{2}{3}DT$	Centroid Theorem
$6 = \frac{2}{3}[2x + 1]$	$DA = 6$, $DT = 2x + 1$
$18 = 4x + 2$	Multiply each side by 3 and simplify.
$16 = 4x$	Subtract 2 from each side.
$4 = x$	Divide each side by 4.

- Find y.

$$EA = \frac{2}{3}EU \qquad \text{Centroid Theorem}$$

$$y = \frac{2}{3}(y + 2.9) \qquad EA = y, EU = y + 2.9$$

$$3y = 2y + 5.8 \qquad \text{Multiply each side by 3 and simplify.}$$

$$y = 5.8 \qquad \text{Subtract } 2y \text{ from each side.}$$

- Find z.

$$FA = \frac{2}{3}FS \qquad \text{Centroid Theorem}$$

$$4.6 = \frac{2}{3}(4.6 + 4z) \qquad FA = 4.6, FS = 4.6 + 4z$$

$$13.8 = 9.2 + 8z \qquad \text{Multiply each side by 3 and simplify.}$$

$$4.6 = 8z \qquad \text{Subtract 9.2 from each side.}$$

$$0.575 = z \qquad \text{Divide each side by 8.}$$

An **altitude** of a triangle is a segment from a vertex to the line containing the opposite side and perpendicular to the line containing that side. Every triangle has three altitudes. The intersection point of the altitudes of a triangle is called the **orthocenter**.

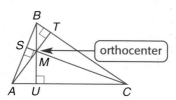

WebQuest

Finding the orthocenter can be used to help you construct your own nine-point circle. Visit www.geometry online.com/webquest to continue work on your WebQuest project.

If the vertices of a triangle are located on a coordinate plane, you can use a system of equations to find the coordinates of the orthocenter.

Example 3 *Use a System of Equations to Find a Point*

COORDINATE GEOMETRY The vertices of $\triangle JKL$ are $J(1, 3)$, $K(2, -1)$, and $L(-1, 0)$. Find the coordinates of the orthocenter of $\triangle JKL$.

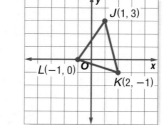

- Find an equation of the altitude from J to $\overline{KL}$.

The slope of $\overline{KL}$ is $-\frac{1}{3}$, so the slope of the altitude is 3.

$$(y - y_1) = m(x - x_1) \qquad \text{Point-slope form}$$

$$(y - 3) = 3(x - 1) \qquad x_1 = 1, y_1 = 3, m = 3$$

$$y - 3 = 3x - 3 \qquad \text{Distributive Property}$$

$$y = 3x \qquad \text{Add 3 to each side.}$$

- Next, find an equation of the altitude from K to $\overline{JL}$. The slope of $\overline{JL}$ is $\frac{3}{2}$, so the slope of the altitude to $\overline{JL}$ is $-\frac{2}{3}$.

$$(y - y_1) = m(x - x_1) \qquad \text{Point-slope form}$$

$$(y + 1) = -\frac{2}{3}(x - 2) \qquad x_1 = 2, y_1 = -1, m = -\frac{2}{3}$$

$$y + 1 = -\frac{2}{3}x + \frac{4}{3} \qquad \text{Distributive Property}$$

$$y = -\frac{2}{3}x + \frac{1}{3} \qquad \text{Subtract 1 from each side.}$$

(continued on the next page)

Study Tip

Graphing Calculator

Once you have two equations, you can graph the two lines and use the Intersect option on the Calc menu to determine where the two lines meet.

• Then, solve a system of equations to find the point of intersection of the altitudes.

Find x.

$y = -\frac{2}{3}x + \frac{1}{3}$ Equation of altitude from K

$3x = -\frac{2}{3}x + \frac{1}{3}$ Substitution, $y = 3x$

$9x = -2x + 1$ Multiply each side by 3.

$11x = 1$ Add $2x$ to each side.

$x = \frac{1}{11}$ Divide each side by 11.

Replace x with $\frac{1}{11}$ in one of the equations to find the y-coordinate.

$y = 3\left(\frac{1}{11}\right)$ $x = \frac{1}{11}$

$y = \frac{3}{11}$ Multiply.

The coordinates of the orthocenter of $\triangle JKL$ are $\left(\frac{1}{11}, \frac{3}{11}\right)$.

You can also use systems of equations to find the coordinates of the circumcenter and the centroid of a triangle graphed on a coordinate plane.

Concept Summary		Special Segments in Triangles
Name	**Type**	**Point of Concurrency**
perpendicular bisector	line, segment, or ray	circumcenter
angle bisector	line, segment, or ray	incenter
median	segment	centroid
altitude	segment	orthocenter

Check for Understanding

Concept Check

1. **Compare and contrast** a perpendicular bisector and a median of a triangle.

2. **OPEN ENDED** Draw a triangle in which the circumcenter lies outside the triangle.

3. **Find a counterexample** to the statement *An altitude and an angle bisector of a triangle are never the same segment.*

Guided Practice

4. **COORDINATE GEOMETRY** The vertices of $\triangle ABC$ are $A(-3, 3)$, $B(3, 2)$, and $C(1, -4)$. Find the coordinates of the circumcenter.

5. **PROOF** Write a two-column proof.
   Given: $\overline{XY} \cong \overline{XZ}$
          $\overline{YM}$ and $\overline{ZN}$ are medians.
   Prove: $\overline{YM} \cong \overline{ZN}$

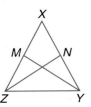

Application

6. **ALGEBRA** Lines ℓ, m, and n are perpendicular bisectors of $\triangle PQR$ and meet at T. If $TQ = 2x$, $PT = 3y - 1$, and $TR = 8$, find x, y, and z.

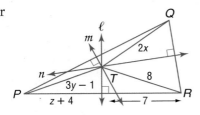

Homework Help

For Exercises	See Examples
10–12, 31–33	1
13–16, 21–26	2
7–9, 27–30	3

Extra Practice
See page 763.

COORDINATE GEOMETRY The vertices of △*DEF* are *D*(4, 0), *E*(−2, 4), and *F*(0, 6). Find the coordinates of the points of concurrency of △*DEF*.

7. centroid

8. orthocenter

9. circumcenter

10. **PROOF** Write a paragraph proof of Theorem 5.1.
   Given: $\overline{CD}$ is the perpendicular bisector of $\overline{AB}$.
   E is a point on $\overline{CD}$.
   Prove: $EB = EA$

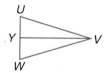

PROOF Write a two-column proof.

11. **Given:** △*UVW* is isosceles with vertex angle *UVW*.
   $\overline{YV}$ is the bisector of ∠*UVW*.
   Prove: $\overline{YV}$ is a median.

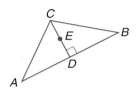

12. **Given:** $\overline{GL}$ is a median of △*EGH*.
   $\overline{JM}$ is a median of △*IJK*.
   △*EGH* ≅ △*IJK*
   Prove: $\overline{GL} ≅ \overline{JM}$

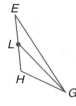
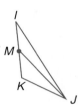

13. **ALGEBRA** Find *x* and *m*∠2 if $\overline{MS}$ is an altitude of △*MNQ*, *m*∠1 = 3*x* + 11, and *m*∠2 = 7*x* + 9.

14. **ALGEBRA** If $\overline{MS}$ is a median of △*MNQ*, *QS* = 3*a* − 14, *SN* = 2*a* + 1, and *m*∠*MSQ* = 7*a* + 1, find the value of *a*. Is $\overline{MS}$ also an altitude of △*MNQ*? Explain.

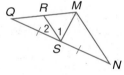

Exercises 13 and 14

15. **ALGEBRA** If $\overline{WP}$ is a median and an angle bisector, *AP* = 3*y* + 11, *PH* = 7*y* − 5, *m*∠*HWP* = *x* + 12, *m*∠*PAW* = 3*x* − 2, and *m*∠*HWA* = 4*x* − 16, find *x* and *y*. Is $\overline{WP}$ also an altitude? Explain.

16. **ALGEBRA** If $\overline{WP}$ is a perpendicular bisector, *m*∠*WHA* = 8*q* + 17, *m*∠*HWP* = 10 + *q*, *AP* = 6*r* + 4, and *PH* = 22 + 3*r*, find *r*, *q*, and *m*∠*HWP*.

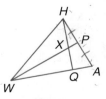

Exercises 15 and 16

State whether each sentence is *always*, *sometimes*, or *never* true.

17. The three medians of a triangle intersect at a point in the interior of the triangle.

18. The three altitudes of a triangle intersect at a vertex of the triangle.

19. The three angle bisectors of a triangle intersect at a point in the exterior of the triangle.

20. The three perpendicular bisectors of a triangle intersect at a point in the exterior of the triangle.

21. ALGEBRA Find x if $\overline{PS}$ is a median of $\triangle PQR$.

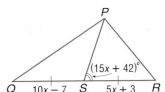

22. ALGEBRA Find x if $\overline{AD}$ is an altitude of $\triangle ABC$.

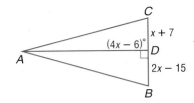

ALGEBRA For Exercises 23–26, use the following information.

In $\triangle PQR$, $ZQ = 3a - 11$, $ZP = a + 5$, $PY = 2c - 1$, $YR = 4c - 11$, $m\angle PRZ = 4b - 17$, $m\angle ZRQ = 3b - 4$, $m\angle QYR = 7b + 6$, and $m\angle PXR = 2a + 10$.

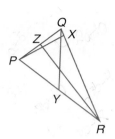

23. $\overline{PX}$ is an altitude of $\triangle PQR$. Find a.

24. If $\overline{RZ}$ is an angle bisector, find $m\angle PRZ$.

25. Find PR if $\overline{QY}$ is a median.

26. If $\overleftrightarrow{QY}$ is a perpendicular bisector of $\overline{PR}$, find b.

COORDINATE GEOMETRY For Exercises 27–30, use the following information.

$R(3, 3)$, $S(-1, 6)$, and $T(1, 8)$ are the vertices of $\triangle RST$, and $\overline{RX}$ is a median.

27. What are the coordinates of X?

28. Find RX.

29. Determine the slope of $\overleftrightarrow{RX}$.

30. Is $\overline{RX}$ an altitude of $\triangle RST$? Explain.

PROOF Write a two-column proof for each theorem.

31. Theorem 5.2

    Given: $\overline{CA} \cong \overline{CB}$
              $\overline{AD} \cong \overline{BD}$

    Prove: C and D are on the perpendicular bisector of $\overline{AB}$.

32. Theorem 5.5

33. Theorem 5.6

34. ORIENTEERING Orienteering is a competitive sport, originating in Sweden, that tests the skills of map reading and cross-country running. Competitors race through an unknown area to find various checkpoints using only a compass and topographical map. On an amateur course, clues were given to locate the first flag.

- The flag is as far from the Grand Tower as it is from the park entrance.
- If you run from Stearns Road to the flag or from Amesbury Road to the flag, you would run the same distance.

Describe how to find the first flag.

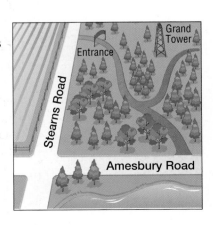

STATISTICS For Exercises 35–38, use the following information.

The *mean* of a set of data is an average value of the data. Suppose △ABC has vertices A(16, 8), B(2, 4), and C(−6, 12).

35. Find the mean of the *x*-coordinates of the vertices.

36. Find the mean of the *y*-coordinates of the vertices.

37. Graph △ABC and its medians.

38. Make a conjecture about the centroid and the means of the coordinates of the vertices.

39. CRITICAL THINKING Draw any △XYZ with median $\overline{XN}$ and altitude $\overline{XO}$. Recall that the area of a triangle is one-half the product of the measures of the base and the altitude. What conclusion can you make about the relationship between the areas of △XYN and △XZN?

40. WRITING IN MATH Answer the question that was posed at the beginning of the lesson.

How can you balance a paper triangle on a pencil point?

Include the following in your answer:

• which special point is the center of gravity, and

• a construction showing how to find this point.

41. In △FGH, which type of segment is $\overline{FJ}$?

    Ⓐ angle bisector   Ⓑ perpendicular bisector

    Ⓒ median   Ⓓ altitude

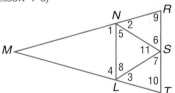

42. ALGEBRA If $xy \neq 0$ and $3x = 0.3y$, then $\dfrac{y}{x} = \underline{}$.

    Ⓐ 0.1   Ⓑ 1.0   Ⓒ 3.0   Ⓓ 10.0

Maintain Your Skills

Mixed Review **Position and label each triangle on the coordinate plane.** *(Lesson 4-7)*

43. equilateral △ABC with base $\overline{AB}$ *n* units long

44. isosceles △DEF with congruent sides 2*a* units long and base *a* units long

45. right △GHI with hypotenuse $\overline{GI}$, HI is three times GH, and GH is *x* units long

For Exercises 46–49, refer to the figure at the right. *(Lesson 4-6)*

46. If ∠9 ≅ ∠10, name two congruent segments.

47. If $\overline{NL} \cong \overline{SL}$, name two congruent angles.

48. If $\overline{LT} \cong \overline{LS}$, name two congruent angles.

49. If ∠1 ≅ ∠4, name two congruent segments.

50. INTERIOR DESIGN Stacey is installing a curtain rod on the wall above the window. To ensure that the rod is parallel to the ceiling, she measures and marks 6 inches below the ceiling in several places. If she installs the rod at these markings centered over the window, how does she know the curtain rod will be parallel to the ceiling? *(Lesson 3-6)*

Getting Ready for the Next Lesson **BASIC SKILL** Replace each ● with < or > to make each sentence true.

51. $\dfrac{3}{8}$ ● $\dfrac{5}{16}$ **52.** 2.7 ● $\dfrac{5}{3}$ **53.** −4.25 ● $-\dfrac{19}{4}$ **54.** $-\dfrac{18}{25}$ ● $-\dfrac{19}{27}$

Reading Mathematics

Math Words and Everyday Words

Several of the words and terms used in mathematics are also used in everyday language. The everyday meaning can help you to better understand the mathematical meaning and help you remember each meaning. This table shows some words used in this chapter with the everyday meanings and the mathematical meanings.

Word	Everyday Meaning	Geometric Meaning	
median	a paved or planted strip dividing a highway into lanes according to direction of travel	a segment of a triangle that connects the vertex to the midpoint of the opposite side	
altitude	the vertical elevation of an object above a surface	a segment from a vertex of a triangle that is perpendicular to the line containing the opposite side	
bisector	something that divides into two usually equal parts	a segment that divides an angle or a side into two parts of equal measure	

Source: *Merriam-Webster Collegiate Dictionary*

Notice that the geometric meaning is more specific, but related to the everyday meaning. For example, the everyday definition of *altitude* is elevation, or height. In geometry, an altitude is a segment of a triangle perpendicular to the base through the vertex. The length of an altitude is the height of the triangle.

Reading to Learn

1. How does the mathematical meaning of *median* relate to the everyday meaning?

2. **RESEARCH** Use a dictionary or other sources to find alternate definitions of *vertex*.

3. **RESEARCH** *Median* has other meanings in mathematics. Use the Internet or other sources to find alternate definitions of this term.

4. **RESEARCH** Use a dictionary or other sources to investigate definitions of *segment*.

Inequalities and Triangles

What You'll Learn

- Recognize and apply properties of inequalities to the measures of angles of a triangle.
- Recognize and apply properties of inequalities to the relationships between angles and sides of a triangle.

How can you tell which corner is bigger?

Sam is delivering two potted trees to be used on a patio. The instructions say for the trees to be placed in the two largest corners of the patio. All Sam has is a diagram of the triangular patio that shows the measurements 45 feet, 48 feet, and 51 feet. Sam can find the largest corner because the measures of the angles of a triangle are related to the measures of the sides opposite them.

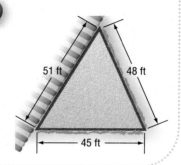

51 ft 48 ft 45 ft

ANGLE INEQUALITIES In algebra, you learned about the inequality relationship between two real numbers. This relationship is often used in proofs.

Key Concept Definition of Inequality

For any real numbers a and b, $a > b$ if and only if there is a positive number c such that $a = b + c$.

Example: If $6 = 4 + 2$, $6 > 4$ and $6 > 2$.

The properties of inequalities you studied in algebra can be applied to the measures of angles and segments.

Properties of Inequalities for Real Numbers	
For all numbers a, b, and c	
Comparison Property	$a < b$, $a = b$, or $a > b$
Transitive Property	1. If $a < b$ and $b < c$, then $a < c$.
	2. If $a > b$ and $b > c$, then $a > c$.
Addition and Subtraction Properties	1. If $a > b$, then $a + c > b + c$ and $a - c > b - c$.
	2. If $a < b$, then $a + c < b + c$ and $a - c < b - c$.
Multiplication and Division Properties	1. If $c > 0$ and $a < b$, then $ac < bc$ and $\frac{a}{c} < \frac{b}{c}$.
	2. If $c > 0$ and $a > b$, then $ac > bc$ and $\frac{a}{c} > \frac{b}{c}$.
	3. If $c < 0$ and $a < b$, then $ac > bc$ and $\frac{a}{c} > \frac{b}{c}$.
	4. If $c < 0$ and $a > b$, then $ac < bc$ and $\frac{a}{c} < \frac{b}{c}$.

Example 1 Compare Angle Measures

Determine which angle has the greatest measure.

Explore Compare the measure of $\angle 3$ to the measures of $\angle 1$ and $\angle 2$.

Plan Use properties and theorems of real numbers to compare the angle measures.

Solve Compare $m\angle 1$ to $m\angle 3$.
By the Exterior Angle Theorem, $m\angle 3 = m\angle 1 + m\angle 2$. Since angle measures are positive numbers and from the definition of inequality, $m\angle 3 > m\angle 1$.

Compare $m\angle 2$ to $m\angle 3$.
Again, by the Exterior Angle Theorem, $m\angle 3 = m\angle 1 + m\angle 2$. The definition of inequality states that if $m\angle 3 = m\angle 1 + m\angle 2$, then $m\angle 3 > m\angle 2$.

Examine $m\angle 3$ is greater than $m\angle 1$ and $m\angle 2$. Therefore, $\angle 3$ has the greatest measure.

The results from Example 1 suggest that the measure of an exterior angle is always greater than either of the measures of the remote interior angles.

Theorem 5.8

Exterior Angle Inequality Theorem If an angle is an exterior angle of a triangle, then its measure is greater than the measure of either of its corresponding remote interior angles.

Example: $m\angle 4 > m\angle 1$

$m\angle 4 > m\angle 2$

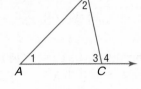

The proof of Theorem 5.8 is in Lesson 5-3.

Example 2 Exterior Angles

Use the Exterior Angle Inequality Theorem to list all of the angles that satisfy the stated condition.

a. all angles whose measures are less than $m\angle 8$

By the Exterior Angle Inequality Theorem, $m\angle 8 > m\angle 4$, $m\angle 8 > m\angle 6$, $m\angle 8 > m\angle 2$, and $m\angle 8 > m\angle 6 + m\angle 7$. Thus, the measures of $\angle 4$, $\angle 6$, $\angle 2$, and $\angle 7$ are all less than $m\angle 8$.

b. all angles whose measures are greater than $m\angle 2$

By the Exterior Angle Inequality Theorem, $m\angle 8 > m\angle 2$ and $m\angle 4 > m\angle 2$. Thus, the measures of $\angle 4$ and $\angle 8$ are greater than $m\angle 2$.

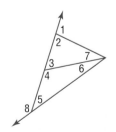

ANGLE-SIDE RELATIONSHIPS Recall that if two sides of a triangle are congruent, then the angles opposite those sides are congruent. In the following Geometry Activity, you will investigate the relationship between sides and angles when they are not congruent.

Geometry Activity

Inequalities for Sides and Angles of Triangles

Model

- Draw an acute scalene triangle, and label the vertices A, B, and C.

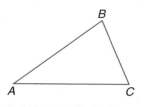

- Measure each side of the triangle. Record the measures in a table.

Side	Measure
$\overline{BC}$	
$\overline{AC}$	
$\overline{AB}$	

- Measure each angle of the triangle. Record each measure in a table.

Angle	Measure
∠A	
∠B	
∠C	

Analyze

1. Describe the measure of the angle opposite the longest side in terms of the other angles.
2. Describe the measure of the angle opposite the shortest side in terms of the other angles.
3. Repeat the activity using other triangles.

Make a Conjecture

4. What can you conclude about the relationship between the measures of sides and angles of a triangle?

The Geometry Activity suggests the following theorem.

Theorem 5.9

If one side of a triangle is longer than another side, then the angle opposite the longer side has a greater measure than the angle opposite the shorter side.

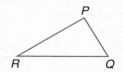

Study Tip

Theorem 5.9
The longest side in a triangle is opposite the largest angle in that triangle.

Proof *Theorem 5.9*

Given: $\triangle PQR$

$PQ < PR$

$\overline{PN} \cong \overline{PQ}$

Prove: $m\angle R < m\angle PQR$

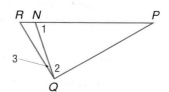

(continued on the next page)

Proof:

Statements	Reasons
1. $\triangle PQR$, $PQ < PR$, $\overline{PN} \cong \overline{PQ}$	1. Given
2. $\angle 1 \cong \angle 2$	2. Isosceles Triangle Theorem
3. $m\angle 1 = m\angle 2$	3. Definition of congruent angles
4. $m\angle R < m\angle 1$	4. Exterior Angle Inequality Theorem
5. $m\angle 2 + m\angle 3 = m\angle PQR$	5. Angle Addition Postulate
6. $m\angle 2 < m\angle PQR$	6. Definition of inequality
7. $m\angle 1 < m\angle PQR$	7. Substitution Property of Equality
8. $m\angle R < m\angle PQR$	8. Transitive Property of Inequality

Example 3 Side–Angle Relationships

Determine the relationship between the measures of the given angles.

a. $\angle ADB$, $\angle DBA$

The side opposite $\angle ADB$ is longer than the side opposite $\angle DBA$, so $m\angle ADB > m\angle DBA$.

b. $\angle CDA$, $\angle CBA$

$$m\angle DBA < m\angle ADB$$
$$m\angle CBD < m\angle CDB$$
$$m\angle DBA + m\angle CBD < m\angle ADB + m\angle CDB$$
$$m\angle CBA < m\angle CDA$$

The converse of Theorem 5.9 is also true.

Theorem 5.10

If one angle of a triangle has a greater measure than another angle, then the side opposite the greater angle is longer than the side opposite the lesser angle.

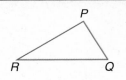

You will prove Theorem 5.10 in Lesson 5-3, Exercise 26.

Example 4 Angle–Side Relationships

TREEHOUSES Mr. Jackson is constructing the framework for part of a treehouse for his daughter. He plans to install braces at the ends of a certain floor support as shown. Which supports should he attach to A and B?

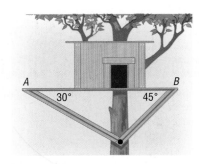

Theorem 5.9 states that if one angle of a triangle has a greater measure, then the side opposite that angle is longer than the side opposite the other angle. Therefore, Mr. Jackson should attach the longer brace at the end marked A and the shorter brace at the end marked B.

Check for Understanding

Concept Check

1. **State** whether the following statement is *always, sometimes,* or *never* true.
   In △JKL with right angle J, if m∠J is twice m∠K, then the side opposite ∠J is twice the length of the side opposite ∠K.

2. **OPEN ENDED** Draw △ABC. List the angle measures and side lengths of your triangle from greatest to least.

3. **FIND THE ERROR** Hector and Grace each labeled △QRS.

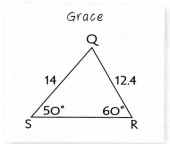

Who is correct? Explain.

Guided Practice

Determine which angle has the greatest measure.

4. ∠1, ∠2, ∠4

5. ∠2, ∠3, ∠5

6. ∠1, ∠2, ∠3, ∠4, ∠5

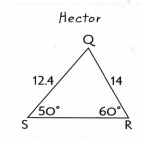

Use the Exterior Angle Inequality Theorem to list all angles that satisfy the stated condition.

7. all angles whose measures are less than m∠1

8. all angles whose measures are greater than m∠6

9. all angles whose measures are less than m∠7

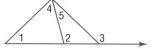

Determine the relationship between the measures of the given angles.

10. ∠WXY, ∠XYW

11. ∠XZY, ∠XYZ

12. ∠WYX, ∠XWY

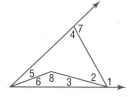

Determine the relationship between the lengths of the given sides.

13. $\overline{AE}, \overline{EB}$

14. $\overline{CE}, \overline{CD}$

15. $\overline{BC}, \overline{EC}$

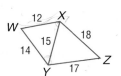

Application

16. **BASEBALL** During a baseball game, the batter hits the ball to the third baseman and begins to run toward first base. At the same time, the runner on first base runs toward second base. If the third baseman wants to throw the ball to the nearest base, to which base should he throw? Explain.

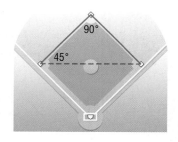

Determine which angle has the greatest measure.

17. $\angle 1, \angle 2, \angle 4$ 18. $\angle 2, \angle 4, \angle 6$

19. $\angle 3, \angle 5, \angle 7$ 20. $\angle 1, \angle 2, \angle 6$

21. $\angle 5, \angle 7, \angle 8$ 22. $\angle 2, \angle 6, \angle 8$

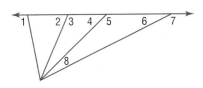

Use the Exterior Angle Inequality Theorem to list all angles that satisfy the stated condition.

23. all angles whose measures are less than $m\angle 5$

24. all angles whose measures are greater than $m\angle 6$

25. all angles whose measures are greater than $m\angle 10$

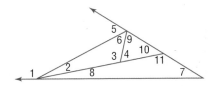

Use the Exterior Angle Inequality Theorem to list all angles that satisfy the stated condition.

26. all angles whose measures are less than $m\angle 1$

27. all angles whose measures are greater than $m\angle 9$

28. all angles whose measures are less than $m\angle 8$

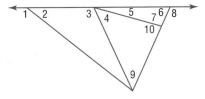

Determine the relationship between the measures of the given angles.

29. $\angle KAJ, \angle AJK$ 30. $\angle MJY, \angle JYM$

31. $\angle SMJ, \angle MJS$ 32. $\angle AKJ, \angle JAK$

33. $\angle MYJ, \angle JMY$ 34. $\angle JSY, \angle JYS$

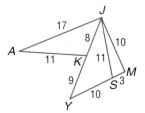

PROOF   Write a two-column proof.

35. **Given:** $\overline{JM} \cong \overline{JL}$
$\overline{JL} \cong \overline{KL}$
   Prove: $m\angle 1 > m\angle 2$

36. **Given:** $\overline{PR} \cong \overline{PQ}$
$QR > QP$
   Prove: $m\angle P > m\angle Q$

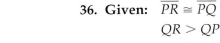

Determine the relationship between the lengths of the given sides.

37. $\overline{ZY}, \overline{YR}$ 38. $\overline{SR}, \overline{ZS}$

39. $\overline{RZ}, \overline{SR}$ 40. $\overline{ZY}, \overline{RZ}$

41. $\overline{TY}, \overline{ZY}$ 42. $\overline{TY}, \overline{ZT}$

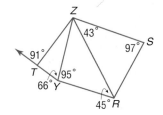

43. **COORDINATE GEOMETRY**   Triangle KLM has vertices $K(3, 2)$, $L(-1, 5)$, and $M(-3, -7)$. List the angles in order from the least to the greatest measure.

44. If $AB > AC > BC$ in $\triangle ABC$ and $\overline{AM}$, $\overline{BN}$, and $\overline{CO}$ are the medians of the triangle, list AM, BN, and CO in order from least to greatest.

45. TRAVEL A plane travels from Des Moines to Phoenix, on to Atlanta, and then completes the trip directly back to Des Moines as shown in the diagram. Write the lengths of the legs of the trip in order from greatest to least.

More About. . .

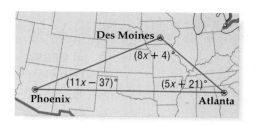

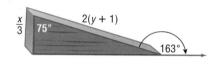

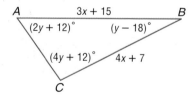

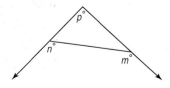

Travel

One sixth of adult Americans have never flown in a commercial aircraft.

Source: U.S. Bureau of Transportation Statistics

ALGEBRA Find the value of n. List the sides of $\triangle PQR$ in order from shortest to longest for the given angle measures.

46. $m\angle P = 9n + 29,\ m\angle Q = 93 - 5n,\ m\angle R = 10n + 2$

47. $m\angle P = 12n - 9,\ m\angle Q = 62 - 3n,\ m\angle R = 16n + 2$

48. $m\angle P = 9n - 4,\ m\angle Q = 4n - 16,\ m\angle R = 68 - 2n$

49. $m\angle P = 3n + 20,\ m\angle Q = 2n + 37,\ \angle R = 4n + 15$

50. $m\angle P = 4n + 61,\ m\angle Q = 67 - 3n,\ \angle R = n + 74$

51. DOORS The wedge at the right is used as a door stopper. The values of x and y are in inches. Write an inequality relating x and y. Then solve the inequality for y in terms of x.

52. PROOF Write a paragraph proof for the following statement.

If a triangle is not isosceles, then the measure of the median to any side of the triangle is greater than the measure of the altitude to that side.

53. CRITICAL THINKING Write and solve an inequality for x.

54. WRITING IN MATH Answer the question that was posed at the beginning of the lesson.

How can you tell which corner is bigger?

Include the following in your answer:

- the name of the theorem or postulate that lets you determine the comparison of the angle measures, and
- which angles in the diagram are the largest.

Standardized Test Practice

55. In the figure at the right, what is the value of p in terms of m and n?

Ⓐ $m + n - 180$

Ⓑ $m + n + 180$

Ⓒ $m - n + 360$

Ⓓ $360 - (m - n)$

56. ALGEBRA If $\frac{1}{2}x - 3 = 2\left(\frac{x - 1}{5}\right)$, then $x =$ ___?___.

Ⓐ 11 Ⓑ 13 Ⓒ 22 Ⓓ 26

Mixed Review   **ALGEBRA   For Exercises 57–59, use the following information.** *(Lesson 5-1)*
Two vertices of $\triangle ABC$ are $A(3, 8)$ and $B(9, 12)$. $\overline{AD}$ is a median with D at $(12, 3)$.

57. What are the coordinates of C?

58. Is $\overline{AD}$ an altitude of $\triangle ABC$? Explain.

59. The graph of point E is at $(6, 6)$. $\overline{EF}$ intersects $\overline{BD}$ at F. If F is at $\left(10\frac{1}{2}, 7\frac{1}{2}\right)$, is $\overline{EF}$ a perpendicular bisector of $\overline{BD}$? Explain.

For Exercises 60 and 61, refer to the figure. *(Lesson 4-7)*

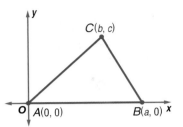

60. Find the coordinates of D if the x-coordinate of D is the mean of the x-coordinates of the vertices of $\triangle ABC$ and the y-coordinate is the mean of the y-coordinates of the vertices of $\triangle ABC$.

61. Prove that D is the intersection of the medians of $\triangle ABC$.

Name the corresponding congruent angles and sides for each pair of congruent triangles. *(Lesson 4-3)*

62. $\triangle TUV \cong \triangle XYZ$

63. $\triangle CDG \cong \triangle RSW$

64. $\triangle BCF \cong \triangle DGH$

65. Find the value of x so that the line containing points at $(x, 2)$ and $(-4, 5)$ is perpendicular to the line containing points at $(4, 8)$ and $(2, -1)$. *(Lesson 3-3)*

Getting Ready for the Next Lesson   **BASIC SKILL   Determine whether each equation is *true* or *false* if $a = 2$, $b = 5$, and $c = 6$.** *(To review **evaluating expressions**, see page 736.)*

66. $2ab = 20$

67. $c(b - a) = 15$

68. $a + c > a + b$

Practice Quiz 1   *Lessons 5-1 and 5-2*

ALGEBRA   Use $\triangle ABC$. *(Lesson 5-1)*

1. Find x if $\overline{AD}$ is a median of $\triangle ABC$.

2. Find y if $\overline{AD}$ is an altitude of $\triangle ABC$.

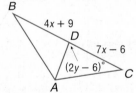

State whether each statement is *always*, *sometimes*, or *never* true. *(Lesson 5-1)*

3. The medians of a triangle intersect at one of the vertices of the triangle.

4. The angle bisectors of a triangle intersect at a point in the interior of the triangle.

5. The altitudes of a triangle intersect at a point in the exterior of the triangle.

6. The perpendicular bisectors of a triangle intersect at a point on the triangle.

7. Describe a triangle in which the angle bisectors all intersect in a point outside the triangle. If no triangle exists, write *no triangle*. *(Lesson 5-1)*

8. List the sides of $\triangle STU$ in order from longest to shortest. *(Lesson 5-2)*

Question 8

S ___ T
$24°$ $137°$
$19°$
U

ALGEBRA   In $\triangle QRS$, $m\angle Q = 3x + 20$, $m\angle R = 2x + 37$, and $m\angle S = 4x + 15$. *(Lesson 5-2)*

9. Determine the measure of each angle.

10. List the sides in order from shortest to longest.

5-3 Indirect Proof

Why

- Use indirect proof with algebra.
- Use indirect proof with geometry.

How is indirect proof used in literature?

In *The Adventure of the Blanched Soldier*, Sherlock Holmes describes his detective technique, stating, "That process starts upon the supposition that when you have eliminated all which is impossible, then whatever remains, . . . must be the truth." The method Sherlock Holmes uses is an example of *indirect reasoning*.

Vocabulary

- indirect reasoning
- indirect proof
- proof by contradiction

Study Tip

Truth Value of a Statement
Recall that a statement must be either true or false. To review **truth values**, see Lesson 2-2.

INDIRECT PROOF WITH ALGEBRA The proofs you have written so far use direct reasoning, in which you start with a true hypothesis and prove that the conclusion is true. When using **indirect reasoning**, you assume that the conclusion is false and then show that this assumption leads to a contradiction of the hypothesis, or some other accepted fact, such as a definition, postulate, theorem, or corollary. Since all other steps in the proof are logically correct, the assumption has been proven false, so the original conclusion must be true. A proof of this type is called an **indirect proof** or a **proof by contradiction**.

The following steps summarize the process of an indirect proof.

Key Concept — Steps for Writing an Indirect Proof

1. Assume that the conclusion is false.

2. Show that this assumption leads to a contradiction of the hypothesis, or some other fact, such as a definition, postulate, theorem, or corollary.

3. Point out that because the false conclusion leads to an incorrect statement, the original conclusion must be true.

Example 1 — Stating Conclusions

State the assumption you would make to start an indirect proof of each statement.

a. $AB \neq MN$
   $AB = MN$

b. **$\triangle PQR$ is an isosceles triangle.**
   $\triangle PQR$ is not an isosceles triangle.

c. **$x < 4$**
   If $x < 4$ is false, then $x = 4$ or $x > 4$. In other words, $x \geq 4$.

d. **If 9 is a factor of n, then 3 is a factor of n.**
   The conclusion of the conditional statement is *3 is a factor of n*. The negation of the conclusion is *3 is not a factor of n*.

Indirect proofs can be used to prove algebraic concepts.

Example 2 Algebraic Proof

Given: $2x - 3 > 7$

Prove: $x > 5$

Indirect Proof:

Step 1 Assume that $x \leq 5$. That is, assume that $x < 5$ or $x = 5$.

Step 2 Make a table with several possibilities for x given that $x < 5$ or $x = 5$.
This is a contradiction because when $x < 5$ or $x = 5$, $2x - 3 \leq 7$.

x	2x − 3
1	−1
2	1
3	4
4	5
5	7

Step 3 In both cases, the assumption leads to the contradiction of a known fact. Therefore, the assumption that $x \leq 5$ must be false, which means that $x > 5$ must be true.

Indirect reasoning and proof can be used in everyday situations.

Example 3 Use Indirect Proof

SHOPPING Lawanda bought two skirts for just over $60, before tax. A few weeks later, her friend Tiffany asked her how much each skirt cost. Lawanda could not remember the individual prices. Use indirect reasoning to show that at least one of the skirts cost more than $30.

Given: The two skirts cost more than $60.

Prove: At least one of the skirts cost more than $30.
That is, if $x + y > 60$, then either $x > 30$ or $y > 30$.

Indirect Proof:

Step 1 Assume that neither skirt costs more than $30. That is, $x \leq 30$ and $y \leq 30$.

Step 2 If $x \leq 30$ and $y \leq 30$, then $x + y \leq 60$. This is a contradiction because we know that the two skirts cost more than $60.

Step 3 The assumption leads to the contradiction of a known fact. Therefore, the assumption that $x \leq 30$ and $y \leq 30$ must be false. Thus, at least one of the skirts had to have cost more than $30.

INDIRECT PROOF WITH GEOMETRY Indirect reasoning can be used to prove statements in geometry.

Example 4 Geometry Proof

Given: $\ell \not\parallel m$

Prove: $\angle 1 \not\cong \angle 3$

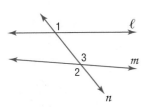

Indirect Proof:

Step 1 Assume that $\angle 1 \cong \angle 3$.

Step 2 $\angle 1$ and $\angle 3$ are corresponding angles. If two lines are cut by a transversal so that corresponding angles are congruent, the lines are parallel. This means that $\ell \parallel m$. However, this contradicts the given statement.

Step 3 Since the assumption leads to a contradiction, the assumption must be false. Therefore, $\angle 1 \not\cong \angle 3$.

Indirect proofs can also be used to prove theorems.

Proof *Exterior Angle Inequality Theorem*

Given: $\angle 1$ is an exterior angle of $\triangle ABC$.

Prove: $m\angle 1 > m\angle 3$ and $m\angle 1 > m\angle 4$

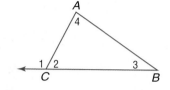

Indirect Proof:

Step 1 Make the assumption that $m\angle 1 \not> m\angle 3$ or $m\angle 1 \not> m\angle 4$. In other words, $m\angle 1 \le m\angle 3$ or $m\angle 1 \le m\angle 4$.

Step 2 You only need to show that the assumption $m\angle 1 \le m\angle 3$ leads to a contradiction as the argument for $m\angle 1 \le m\angle 4$ follows the same reasoning.

$m\angle 1 \le m\angle 3$ means that either $m\angle 1 = m\angle 3$ or $m\angle 1 < m\angle 3$.

Case 1: $m\angle 1 = m\angle 3$

$\quad m\angle 1 = m\angle 3 + m\angle 4 \qquad$ Exterior Angle Theorem

$\quad m\angle 3 = m\angle 3 + m\angle 4 \qquad$ Substitution

$\quad\quad 0 = m\angle 4 \qquad$ Subtract $m\angle 3$ from each side.

This contradicts the fact that the measure of an angle is greater than 0, so $m\angle 1 < m\angle 3$.

Case 2: $m\angle 1 < m\angle 3$

By the Exterior Angle Theorem, $m\angle 1 = m\angle 3 + m\angle 4$. Since angle measures are positive, the definition of inequality implies $m\angle 1 > m\angle 3$ and $m\angle 1 > m\angle 4$. This contradicts the assumption.

Step 3 In both cases, the assumption leads to the contradiction of a theorem or definition. Therefore, the assumption that $m\angle 1 > m\angle 3$ and $m\angle 1 > m\angle 4$ must be true.

Check for Understanding

Concept Check   **1.** **Explain** how contradicting a known fact means that an assumption is false.

2. **Compare and contrast** indirect proof and direct proof. See margin.

3. **OPEN ENDED** State a conjecture. Then write an indirect proof to prove your conjecture.

Guided Practice   **Write the assumption you would make to start an indirect proof of each statement.**

  4. If $5x < 25$, then $x < 5$.

  5. Two lines that are cut by a transversal so that alternate interior angles are congruent are parallel.

  6. If the alternate interior angles formed by two lines and a transversal are congruent, the lines are parallel.

  PROOF   Write an indirect proof.

  7. **Given:** $a > 0$
     Prove: $\dfrac{1}{a} > 0$

  8. **Given:** n is odd.
     Prove: n^2 is odd.

  9. **Given:** $\triangle ABC$
     Prove: There can be no more than one obtuse angle in $\triangle ABC$.

  10. **Given:** $m \not\parallel n$
      Prove: Lines m and n intersect at exactly one point.

  11. **PROOF**   Use an indirect proof to show that the hypotenuse of a right triangle is the longest side.

Application 12. **BICYCLING**   The Tour de France bicycle race takes place over several weeks in various stages throughout France. During two stages of the 2002 Tour de France, riders raced for just over 270 miles. Prove that at least one of the stages was longer than 135 miles.

Practice and Apply

Homework Help

For Exercises	See Examples
13–18	1
19, 20, 23	2, 3
21, 22, 24, 25	4

Extra Practice
See page 763.

Write the assumption you would make to start an indirect proof of each statement.

13. $\overline{PQ} \cong \overline{ST}$

14. If $3x > 12$, then $x > 4$.

15. If a rational number is any number that can be expressed as $\dfrac{a}{b}$, where a and b are integers, and $b \neq 0$, 6 is a rational number.

16. A median of an isosceles triangle is also an altitude.

17. Points P, Q, and R are collinear.

18. The angle bisector of the vertex angle of an isosceles triangle is also an altitude of the triangle.

  PROOF   Write an indirect proof.

19. **Given:** $\dfrac{1}{a} < 0$
    Prove: a is negative.

20. **Given:** n^2 is even.
    Prove: n^2 is divisible by 4.

21. **Given:** $\overline{PQ} \cong \overline{PR}$
    $\angle 1 \not\cong \angle 2$
    Prove: $\overline{PZ}$ is not a median of $\triangle PQR$.

22. **Given:** $m\angle 2 \neq m\angle 1$
    Prove: $\ell \not\parallel m$

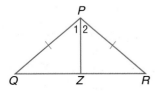

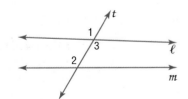

PROOF Write an indirect proof.

23. If $a > 0$, $b > 0$, and $a > b$, then $\frac{a}{b} > 1$.

24. If two sides of a triangle are not congruent, then the angles opposite those sides are not congruent.

25. **Given:** $\triangle ABC$ and $\triangle ABD$ are equilateral.
$\triangle ACD$ is not equilateral.
Prove: $\triangle BCD$ is not equilateral.

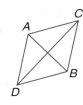

26. Theorem 5.10
Given: $m\angle A > m\angle ABC$
Prove: $BC > AC$

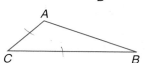

27. TRAVEL Ramon drove 175 miles from Seattle, Washington, to Portland, Oregon. It took him three hours to complete the trip. Prove that his average driving speed was less than 60 miles per hour.

EDUCATION For Exercises 28–30, refer to the graphic at the right.

28. Prove the following statement.
The majority of college-bound seniors stated that they received college information from a guidance counselor.

29. If 1500 seniors were polled for this survey, verify that 225 said they received college information from a friend.

30. Did more seniors receive college information from their parents or from teachers and friends? Explain.

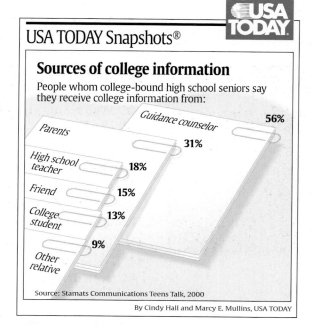

USA TODAY Snapshots®

Sources of college information

People whom college-bound high school seniors say they receive college information from:

Guidance counselor — 56%
Parents — 31%
High school teacher — 18%
Friend — 15%
College student — 13%
Other relative — 9%

Source: Stamats Communications Teens Talk, 2000

By Cindy Hall and Marcy E. Mullins, USA TODAY

31. LAW During the opening arguments of a trial, a defense attorney stated, "My client is innocent. The police report states that the crime was committed on November 6 at approximately 10:15 A.M. in San Diego. I can prove that my client was on vacation in Chicago with his family at this time. A verdict of not guilty is the only possible verdict." Explain whether this is an example of indirect reasoning.

32. RESEARCH Use the Internet or other resource to write an indirect proof for the following statement.
In the Atlantic Ocean, the percent of tropical storms that developed into hurricanes over the past five years varies from year to year.

www.geometryonline.com/self_check_quiz

33. CRITICAL THINKING Recall that a rational number is any number that can be expressed in the form $\frac{a}{b}$, where a and b are integers with no common factors and $b \neq 0$, or as a terminating or repeating decimal. Use indirect reasoning to prove that $\sqrt{2}$ is not a rational number.

34. WRITING IN MATH Answer the question that was posed at the beginning of the lesson.

How is indirect proof used in literature?

Include the following in your answer:
- an explanation of how Sherlock Holmes used indirect proof, and
- an example of indirect proof used every day.

35. Which statement about the value of x is not true?

Ⓐ $x = 60$ Ⓑ $x < 140$

Ⓒ $x + 80 = 140$ Ⓓ $x < 60$

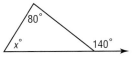

36. PROBABILITY A bag contains 6 blue marbles, 8 red marbles, and 2 white marbles. If three marbles are removed at random and no marble is returned to the bag after removal, what is the probability that all three marbles will be red?

Ⓐ $\frac{1}{10}$ Ⓑ $\frac{1}{8}$ Ⓒ $\frac{3}{8}$ Ⓓ $\frac{1}{2}$

Maintain Your Skills

Mixed Review

For Exercises 37 and 38, refer to the figure at the right.
(Lesson 5-2)

37. Which angle in $\triangle MOP$ has the greatest measure?

38. Name the angle with the least measure in $\triangle LMN$.

PROOF **Write a two-column proof.** *(Lesson 5-1)*

39. If an angle bisector of a triangle is also an altitude of the triangle, then the triangle is isosceles.

40. The median to the base of an isosceles triangle bisects the vertex angle.

41. Corresponding angle bisectors of congruent triangles are congruent.

42. ASTRONOMY The Big Dipper is a part of the larger constellation Ursa Major. Three of the brighter stars in the constellation form $\triangle RSA$. If $m\angle R = 41$ and $m\angle S = 109$, find $m\angle A$.
(Lesson 4-2)

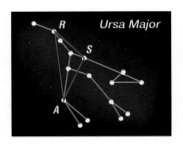

Write an equation in point-slope form of the line having the given slope that contains the given point.
(Lesson 3-4)

43. $m = 2, (4, 3)$ **44.** $m = -3, (2, -2)$ **45.** $m = 11, (-4, -9)$

Getting Ready for the Next Lesson

PREREQUISITE SKILL **Determine whether each inequality is *true* or *false*.**
(To review the meaning of inequalities, see pages 739 and 740.)

46. $19 - 10 < 11$ **47.** $31 - 17 < 12$ **48.** $38 + 76 > 109$

The Triangle Inequality

What You'll Learn

- Apply the Triangle Inequality Theorem.
- Determine the shortest distance between a point and a line.

How can you use the Triangle Inequality Theorem when traveling?

Chuck Noland travels between Chicago, Indianapolis, and Columbus as part of his job. Mr. Noland lives in Chicago and needs to get to Columbus as quickly as possible. Should he take a flight that goes from Chicago to Columbus, or a flight that goes from Chicago to Indianapolis, then to Columbus?

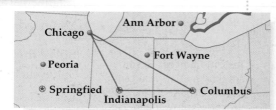

THE TRIANGLE INEQUALITY   In the example above, if you chose to fly directly from Chicago to Columbus, you probably reasoned that a straight route is shorter. This is an example of the Triangle Inequality Theorem.

Theorem 5.11

Triangle Inequality Theorem The sum of the lengths of any two sides of a triangle is greater than the length of the third side.

Examples:
$AB + BC > AC$
$BC + AC > AB$
$AC + AB > BC$

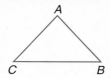

You will prove Theorem 5.11 in Exercise 40.

The Triangle Inequality Theorem can be used to determine whether three segments can form a triangle.

Example 1 Identify Sides of a Triangle

Determine whether the given measures can be the lengths of the sides of a triangle.

a. 2, 4, 5

Check each inequality.

$2 + 4 \overset{?}{>} 5$ $\qquad$ $2 + 5 \overset{?}{>} 4$ $\qquad$ $4 + 5 \overset{?}{>} 2$

$\quad 6 > 5 \; \checkmark$ $\qquad\quad$ $7 > 4 \; \checkmark$ $\qquad\quad$ $9 > 2 \; \checkmark$

All of the inequalities are true, so 2, 4, and 5 can be the lengths of the sides of a triangle.

b. 6, 8, 14

$6 + 8 \overset{?}{>} 14$ $\qquad$ Because the sum of two measures equals the measure of the

$\quad 14 \not> 14$ $\qquad$ third side, the sides cannot form a triangle.

When you know the lengths of two sides of a triangle, you can determine the range of possible lengths for the third side.

Standardized Test Practice
Ⓐ Ⓑ Ⓒ Ⓓ

Example 2 — Determine Possible Side Length

Multiple-Choice Test Item

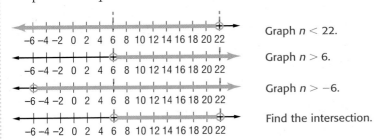

In $\triangle XYZ$, $XY = 8$, and $XZ = 14$.
Which measure cannot be YZ?
  Ⓐ 6   Ⓑ 10
  Ⓒ 14   Ⓓ 18

Read the Test Item

You need to determine which value is not valid.

Test-Taking Tip

Testing Choices If you are short on time, you can test each choice to find the correct answer and eliminate any remaining choices.

Solve the Test Item

Solve each inequality to determine the range of values for YZ.

Let $YZ = n$.

$$XY + XZ > YZ \qquad\qquad XY + YZ > XZ \qquad\qquad YZ + XZ > XY$$
$$8 + 14 > n \qquad\qquad\quad 8 + n > 14 \qquad\qquad\quad n + 14 > 8$$
$$22 > n \text{ or } n < 22 \qquad\qquad n > 6 \qquad\qquad\qquad n > -6$$

Graph the inequalities on the same number line.

Graph $n < 22$.

Graph $n > 6$.

Graph $n > -6$.

Find the intersection.

The range of values that fit all three inequalities is $6 < n < 22$.

Examine the answer choices. The only value that does not satisfy the compound inequality is 6 since $6 = 6$. Thus, the answer is choice A.

DISTANCE BETWEEN A POINT AND A LINE

Recall that the distance between point P and line ℓ is measured along a perpendicular segment from the point to the line. It was accepted without proof that $\overline{PA}$ was the shortest segment from P to ℓ. The theorems involving the relationships between the angles and sides of a triangle can now be used to prove that a perpendicular segment is the shortest distance between a point and a line.

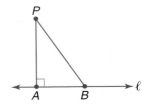

Theorem 5.12

The perpendicular segment from a point to a line is the shortest segment from the point to the line.

Example: $\overline{PQ}$ is the shortest segment from P to $\overleftrightarrow{AB}$.

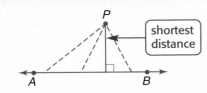

Example 3 Prove Theorem 5.12

Given: $\overline{PA} \perp \ell$

$\overline{PB}$ is any nonperpendicular segment from P to ℓ.

Prove: $PB > PA$

Proof:

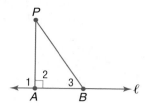

Statements	Reasons
1. $\overline{PA} \perp \ell$	1. Given
2. $\angle 1$ and $\angle 2$ are right angles.	2. $\perp$ lines form right angles.
3. $\angle 1 \cong \angle 2$	3. All right angles are congruent.
4. $m\angle 1 = m\angle 2$	4. Definition of congruent angles
5. $m\angle 1 > m\angle 3$	5. Exterior Angle Inequality Theorem
6. $m\angle 2 > m\angle 3$	6. Substitution Property
7. $PB > PA$	7. If an angle of a triangle is greater than a second angle, then the side opposite the greater angle is longer than the side opposite the lesser angle.

Corollary 5.1 follows directly from Theorem 5.12.

Corollary 5.1

The perpendicular segment from a point to a plane is the shortest segment from the point to the plane.

Example:

$\overline{QP}$ is the shortest segment from P to Plane $\mathcal{M}$.

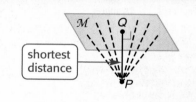

shortest distance

You will prove Corollary 5.1 in Exercise 12.

Check for Understanding

Concept Check

1. **Explain** why the distance between two nonhorizontal parallel lines on a coordinate plane cannot be found using the distance between their y-intercepts.

2. **FIND THE ERROR** Jameson and Anoki drew $\triangle EFG$ with $FG = 13$ and $EF = 5$. They each chose a possible measure for GE.

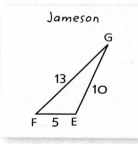

Jameson

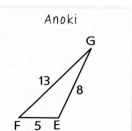

Anoki

Who is correct? Explain.

3. **OPEN ENDED** Find three numbers that can be the lengths of the sides of a triangle and three numbers that cannot be the lengths of the sides of a triangle. Justify your reasoning with a drawing.

www.geometryonline.com/extra_examples

Determine whether the given measures can be the lengths of the sides of a triangle. Write *yes* or *no*. Explain.

4. 5, 4, 3

5. 5, 15, 10

6. 30.1, 0.8, 31

7. 5.6, 10.1, 5.2

Find the range for the measure of the third side of a triangle given the measures of two sides.

8. 7 and 12

9. 14 and 23

10. 22 and 34

11. 15 and 18

12. **PROOF** Write a proof for Corollary 5.1.

    Given: $\overline{PQ} \perp$ plane $\mathcal{M}$

    Prove: $\overline{PQ}$ is the shortest segment from P to plane $\mathcal{M}$.

13. An isosceles triangle has a base 10 units long. If the congruent sides have whole number measures, what is the least possible length of the sides?

    Ⓐ 5   Ⓑ 6   Ⓒ 17   Ⓓ 21

Practice and Apply

Homework Help	
For Exercises	**See Examples**
14–25	1
26–37	2
38–40	3

Extra Practice
See page 764.

Determine whether the given measures can be the lengths of the sides of a triangle. Write *yes* or *no*. Explain.

14. 1, 2, 3

15. 2, 6, 11

16. 8, 8, 15

17. 13, 16, 29

18. 18, 32, 21

19. 9, 21, 20

20. 5, 17, 9

21. 17, 30, 30

22. 8.4, 7.2, 3.5

23. 4, 0.9, 4.1

24. 14.3, 12, 2.2

25. 0.18, 0.21, 0.52

Find the range for the measure of the third side of a triangle given the measures of two sides.

26. 5 and 11

27. 7 and 9

28. 10 and 15

29. 12 and 18

30. 21 and 47

31. 32 and 61

32. 30 and 30

33. 64 and 88

34. 57 and 55

35. 75 and 75

36. 78 and 5

37. 99 and 2

PROOF Write a two-column proof.

38. Given: $\angle B \cong \angle ACB$

    Prove: $AD + AB > CD$

39. Given: $\overline{HE} \cong \overline{EG}$

    Prove: $HE + FG > EF$

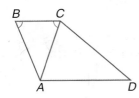

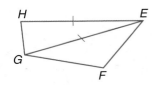

40. Given: $\angle ABC$

    Prove: $AC + BC > AB$ (Triangle Inequality Theorem)

    (*Hint:* Draw auxiliary segment $\overline{CD}$, so that C is between B and D and $\overline{CD} \cong \overline{AC}$.)

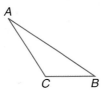

ALGEBRA   Determine whether the given coordinates are the vertices of a triangle. Explain.

41. $A(5, 8)$, $B(2, -4)$, $C(-3, -1)$ **42.** $L(-24, -19)$, $M(-22, 20)$, $N(-5, -7)$

43. $X(0, -8)$, $Y(16, -12)$, $Z(28, -15)$ **44.** $R(1, -4)$, $S(-3, -20)$, $T(5, 12)$

CRAFTS   For Exercises 45 and 46, use the following information.
Carlota has several strips of trim she wishes to use as a triangular border for a section of a decorative quilt she is going to make. The strips measure 3 centimeters, 4 centimeters, 5 centimeters, 6 centimeters, and 12 centimeters.

45. How many different triangles could Carlota make with the strips?

46. How many different triangles could Carlota make that have a perimeter that is divisible by 3?

47. HISTORY   The early Egyptians used to make triangles by using a rope with knots tied at equal intervals. Each vertex of the triangle had to occur at a knot. How many different triangles can be formed using the rope below?

PROBABILITY   For Exercises 48 and 49, use the following information.
One side of a triangle is 2 feet long. Let m represent the measure of the second side and n represent the measure of the third side. Suppose m and n are whole numbers and that $14 < m < 17$ and $13 < n < 17$.

48. List the measures of the sides of the triangles that are possible.

49. What is the probability that a randomly chosen triangle that satisfies the given conditions will be isosceles?

50. CRITICAL THINKING   State and prove a theorem that compares the measures of each side of a triangle with the differences of the measures of the other two sides.

51. WRITING IN MATH   Answer the question that was posed at the beginning of the lesson.

How can you use the Triangle Inequality when traveling?

Include the following in your answer:
- an example of a situation in which you might want to use the greater measures, and
- an explanation as to why it is not always possible to apply the Triangle Inequality when traveling.

Standardized Test Practice
ⓐ ⓑ ⓒ ⓓ

52. If two sides of a triangle measure 12 and 7, which of the following cannot be the perimeter of the triangle?

Ⓐ 29   Ⓑ 34
Ⓒ 37   Ⓓ 38

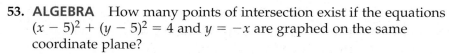

53. ALGEBRA   How many points of intersection exist if the equations $(x - 5)^2 + (y - 5)^2 = 4$ and $y = -x$ are graphed on the same coordinate plane?

Ⓐ none   Ⓑ one   Ⓒ two   Ⓓ three

Mixed Review **54.** **PROOF** Write an indirect proof. *(Lesson 5-3)*

Given: *P* is a point not on line ℓ.
Prove: $\overline{PQ}$ is the only line through *P*
perpendicular to ℓ.

ALGEBRA List the sides of △*PQR* in order from longest to shortest if the angles
of △*PQR* have the given measures. *(Lesson 5-2)*

55. $m\angle P = 7x + 8, m\angle Q = 8x - 10, m\angle R = 7x + 6$

56. $m\angle P = 3x + 44, m\angle Q = 68 - 3x, m\angle R = x + 61$

Determine whether △*JKL* ≅ △*PQR* given the coordinates of the vertices. Explain.
(Lesson 4-4)

57. $J(0, 5), K(0, 0), L(-2, 0), P(4, 8), Q(4, 3), R(6, 3)$

58. $J(6, 4), K(1, -6), L(-9, 5), P(0, 7), Q(5, -3), R(15, 8)$

59. $J(-6, -3), K(1, 5), L(2, -2), P(2, -11), Q(5, -4), R(10, -10)$

Getting Ready for
the Next Lesson **PREREQUISITE SKILL** Solve each inequality.
*(To review **solving inequalities**, see pages 739 and 740.)*

60. $3x + 54 < 90$ **61.** $8x - 14 < 3x + 19$ **62.** $4x + 7 < 180$

Practice Quiz 2   Lessons 5-3 and 5-4

Write the assumption you would make to start an indirect proof of each statement. *(Lesson 5-3)*

1. The number 117 is divisible by 13.

2. $m\angle C < m\angle D$

Write an indirect proof. *(Lesson 5-3)*

3. If $7x > 56$, then $x > 8$.

4. Given: $\overline{MO} \cong \overline{ON}, \overline{MP} \not\cong \overline{NP}$
   Prove: $\angle MOP \not\cong \angle NOP$

5. Given: $m\angle ADC \neq m\angle ADB$
   Prove: $\overline{AD}$ is not an altitude of △*ABC*.

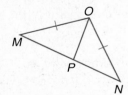

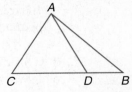

Determine whether the given measures can be the lengths of the sides of a triangle.
Write *yes* or *no*. Explain. *(Lesson 5-4)*

6. 7, 24, 25   **7.** 25, 35, 60   **8.** 20, 3, 18   **9.** 5, 10, 6

10. If the measures of two sides of a triangle are 57 and 32, what is the range of
possible measures of the third side? *(Lesson 5-4)*

5-5 Inequalities Involving Two Triangles

What You'll Learn

- Apply the SAS Inequality.
- Apply the SSS Inequality.

How does a backhoe work?

Many objects, like a backhoe, have two fixed arms connected by a joint or hinge. This allows the angle between the arms to increase and decrease. As the angle changes, the distance between the endpoints of the arms changes as well.

SAS INEQUALITY The relationship of the arms and the angle between them illustrates the following theorem.

Study Tip

SAS Inequality
The SAS Inequality Theorem is also called the *Hinge Theorem*.

Theorem 5.13

SAS Inequality/Hinge Theorem If two sides of a triangle are congruent to two sides of another triangle and the included angle in one triangle has a greater measure than the included angle in the other, then the third side of the first triangle is longer than the third side of the second triangle.

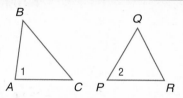

Example: Given $\overline{AB} \cong \overline{PQ}$, $\overline{AC} \cong \overline{PR}$, if $m\angle 1 > m\angle 2$, then $BC > QR$.

Proof SAS Inequality Theorem

Given: $\triangle ABC$ and $\triangle DEF$
$\overline{AC} \cong \overline{DF}$, $\overline{BC} \cong \overline{EF}$
$m\angle F > m\angle C$

Prove: $DE > AB$

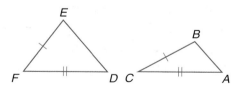

Proof:

We are given that $\overline{AC} \cong \overline{DF}$ and $\overline{BC} \cong \overline{EF}$. We also know that $m\angle F > m\angle C$. Draw auxiliary ray FZ such that $m\angle DFZ = m\angle C$ and that $\overline{ZF} \cong \overline{BC}$. This leads to two cases.

Case 1: If Z lies on $\overline{DE}$, then $\triangle FZD \cong \triangle CBA$ by SAS. Thus, $ZD = BA$ by CPCTC and the definition of congruent segments. By the Segment Addition Postulate, $DE = EZ + ZD$. Also, $DE > ZD$ by the definition of inequality. Therefore, $DE > AB$ by the Substitution Property.

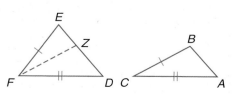

Lesson 5-5 Inequalities Involving Two Triangles **267**

Case 2: If Z does not lie on $\overline{DE}$, then let the intersection of $\overline{FZ}$ and $\overline{ED}$ be point T. Now draw another auxiliary segment $\overline{FV}$ such that V is on $\overline{DE}$ and $\angle EFV \cong \angle VFZ$.

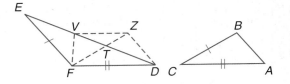

Since $\overline{FZ} \cong \overline{BC}$ and $\overline{BC} \cong \overline{EF}$, we have $\overline{FZ} \cong \overline{EF}$ by the Transitive Property. Also $\overline{VF}$ is congruent to itself by the Reflexive Property. Thus, $\triangle EFV \cong \triangle ZFV$ by SAS. By CPCTC, $\overline{EV} \cong \overline{ZV}$ or $EV = ZV$. Also, $\triangle FZD \cong \triangle CBA$ by SAS. So, $\overline{ZD} \cong \overline{BA}$ by CPCTC or $ZD = BA$. In $\triangle VZD$, $VD + ZV > ZD$ by the Triangle Inequality Theorem. By substitution, $VD + EV > ZD$. Since $ED = VD + EV$ by the Segment Addition Postulate, $ED > ZD$. Using substitution, $ED > BA$ or $DE > AB$.

Example 1 *Use SAS Inequality in a Proof*

Write a two-column proof.

Given: $\overline{YZ} \cong \overline{XZ}$
Z is the midpoint of $\overline{AC}$.
$m\angle CZY > m\angle AZX$
$\overline{BY} \cong \overline{BX}$

Prove: $BC > AB$

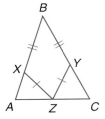

Proof:

Statements	Reasons
1. $\overline{YZ} \cong \overline{XZ}$ Z is the midpoint of $\overline{AC}$. $m\angle CZY > m\angle AZX$ $\overline{BY} \cong \overline{BX}$	1. Given
2. $CZ = AZ$	2. Definition of midpoint
3. $CY > AX$	3. SAS Inequality
4. $BY = BX$	4. Definition of congruent segments
5. $CY + BY > AX + BX$	5. Addition Property
6. $BC = CY + BY$ $AB = AX + BX$	6. Segment Addition Postulate
7. $BC > AB$	7. Substitution Property

SSS INEQUALITY The converse of the SAS Inequality Theorem is the SSS Inequality Theorem.

Theorem 5.14

SSS Inequality If two sides of a triangle are congruent to two sides of another triangle and the third side in one triangle is longer than the third side in the other, then the angle between the pair of congruent sides in the first triangle is greater than the corresponding angle in the second triangle.

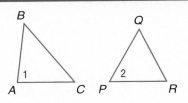

Example: Given $\overline{AB} \cong \overline{PQ}$, $\overline{AC} \cong \overline{PR}$, if $BC > QR$, then $m\angle 1 > m\angle 2$.

You will prove Theorem 5.14 in Exercise 24.

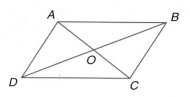

Example 2 Prove Triangle Relationships

Given: $\overline{AB} \cong \overline{CD}$
$\overline{AB} \parallel \overline{CD}$
$CD > AD$

Prove: $m\angle AOB > m\angle BOC$

Flow Proof:

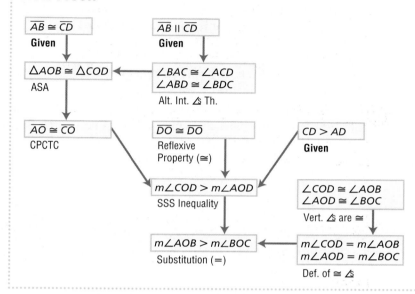

You can use algebra to relate the measures of the angles and sides of two triangles.

Example 3 Relationships Between Two Triangles

Write an inequality using the information in the figure.

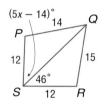

a. Compare $m\angle QSR$ and $m\angle QSP$.

In $\triangle PQS$ and $\triangle RQS$, $\overline{PS} \cong \overline{RS}$, $\overline{QS} \cong \overline{QS}$, and $QR > QP$. The SAS Inequality allows us to conclude that $m\angle QSR > m\angle QSP$.

b. Find the range of values containing x.

By the SSS Inequality, $m\angle QSR > m\angle QSP$, or $m\angle QSP < m\angle QSR$.

$m\angle QSP < m\angle QSR$	SSS Inequality
$5x - 14 < 46$	Substitution
$5x < 60$	Add 14 to each side.
$x < 12$	Divide each side by 5.

Also, recall that the measure of any angle is always greater than 0.

$5x - 14 > 0$	
$5x > 14$	Add 14 to each side.
$x > \dfrac{14}{5}$ or 2.8	Divide each side by 5.

The two inequalities can be written as the compound inequality $2.8 < x < 12$.

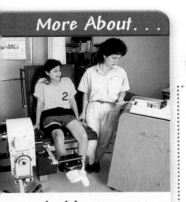
Inequalities involving triangles can be used to describe real-world situations.

Example 4 Use Triangle Inequalities

HEALTH Range of motion describes the amount that a limb can be moved from a straight position. To determine the range of motion of a person's forearm, determine the distance from his or her wrist to the shoulder when the elbow is bent as far as possible. Suppose Jessica can bend her left arm so her wrist is 5 inches from her shoulder and her right arm so her right wrist is 3 inches from her shoulder. Which of Jessica's arms has the greater range of motion? Explain.

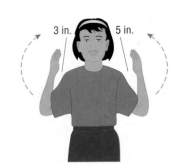

3 in.   5 in.

The distance between the wrist and shoulder is smaller on her right arm. Assuming that both her arms have the same measurements, the SSS inequality tells us that the angle formed at the elbow is smaller on the right arm. This means that the right arm has a greater range of motion.

Check for Understanding

Concept Check

1. **OPEN ENDED** Describe a real-world object that illustrates either SAS or SSS inequality.

2. **Compare and contrast** the SSS Inequality Theorem to the SSS Postulate for triangle congruence.

Guided Practice Write an inequality relating the given pair of angles or segment measures.

3. AB, CD

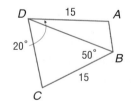

D   15   A

20°

50°   B

15

C

4. $m\angle PQS, m\angle RQS$

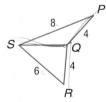

P

8   4

S   Q

6   4

R

Write an inequality to describe the possible values of x.

5.

x + 5

45°

3x − 7

6.

6   140°

12

8

7   7

6

(7x + 4)°

PROOF Write a two-column proof.

7. **Given:** $\overline{PQ} \cong \overline{SQ}$
   Prove: $PR > SR$

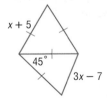

P   S

T

Q   R

8. **Given:** $\overline{TU} \cong \overline{US}$
   $\overline{US} \cong \overline{SV}$
   Prove: $ST > UV$

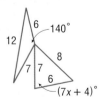

S

T   U   V

Application

9. **TOOLS** A lever is used to multiply the force applied to an object. One example of a lever is a pair of pliers. Use the SAS or SSS Inequality to explain how to use a pair of pliers.

Practice and Apply

Homework Help	
For Exercises	**See Examples**
20–24	1–2
10–19	3
25, 26	4

Extra Practice
See page 764.

Write an inequality relating the given pair of angles or segment measures.

10. AB, FD

11. $m\angle BDC$, $m\angle FDB$

12. $m\angle FBA$, $m\angle DBF$

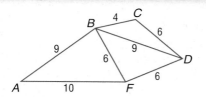

Write an inequality relating the given pair of angles or segment measures.

13. AD, DC

14. OC, OA

15. $m\angle AOD$, $m\angle AOB$

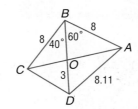

Write an inequality to describe the possible values of x.

16.

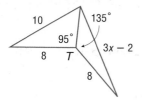

17.

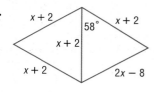

18. In the figure, $\overline{AM} \cong \overline{MB}$, $AC > BC$, $m\angle 1 = 5x + 20$ and $m\angle 2 = 8x - 100$. Write an inequality to describe the possible values of x.

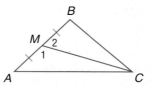

19. In the figure, $m\angle RVS = 15 + 5x$, $m\angle SVT = 10x - 20$, $RS < ST$, and $\angle RTV \cong \angle TRV$. Write an inequality to describe the possible values of x.

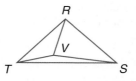

PROOF Write a two-column proof.

20. **Given:** $\triangle ABC$
    $\overline{AB} \cong \overline{CD}$
    Prove: $BC > AD$

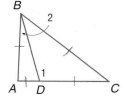

21. **Given:** $\overline{PQ} \cong \overline{RS}$
    $QR < PS$
    Prove: $m\angle 3 < m\angle 1$

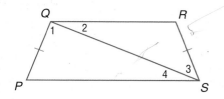

PROOF Write a two-column proof.

22. Given: $\overline{PR} \cong \overline{PQ}$
   $SQ > SR$
Prove: $m\angle 1 < m\angle 2$

23. Given: $\overline{ED} \cong \overline{DF}$
   $m\angle 1 > m\angle 2$
   D is the midpoint of $\overline{CB}$.
   $\overline{AE} \cong \overline{AF}$
Prove: $AC > AB$

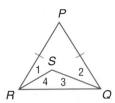

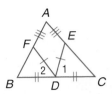

24. **PROOF** Use an indirect proof to prove the SSS Inequality Theorem (Theorem 5.14).
   Given: $\overline{RS} \cong \overline{UW}$
      $\overline{ST} \cong \overline{WV}$
      $RT > UV$
   Prove: $m\angle S > m\angle W$

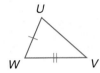

25. DOORS Open a door slightly. With the door open, measure the angle made by the door and the door frame. Measure the distance from the end of the door to the door frame. Open the door wider, and measure again. How do the measures compare?

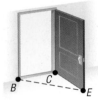

26. LANDSCAPING When landscapers plant new trees, they usually brace the tree using a stake tied to the trunk of the tree. Use the SAS or SSS Inequality to explain why this is an effective method for supporting a newly planted tree.

27. CRITICAL THINKING The SAS Inequality states that the base of an isosceles triangle gets longer as the measure of the vertex angle increases. Describe the effect of changing the measure of the vertex angle on the measure of the altitude.

BIOLOGY For Exercises 28–30, use the following information.
The velocity of a person walking or running can be estimated using the formula $v = \frac{0.78s^{1.67}}{h^{1.17}}$, where v is the velocity of the person in meters per second, s is the length of the stride in meters, and h is the height of the hip in meters.

28. Find the velocities of two people that each have a hip height of 0.85 meters and whose strides are 1.0 meter and 1.2 meters.

29. Copy and complete the table at the right for a person whose hip height is 1.1 meters.

30. Discuss how the stride length is related to either the SAS Inequality of the SSS Inequality.

Stride (m)	Velocity (m/s)
0.25	
0.50	
0.75	
1.00	
1.25	
1.50	

31. Answer the question that was posed at the beginning of the lesson.

How does a backhoe work?

Include the following in your answer:
- a description of the angle between the arms as the backhoe operator digs, and
- an explanation of how the distance between the ends of the arms is related to the angle between them.

32. If $\overline{DC}$ is a median of $\triangle ABC$ and $m\angle 1 > m\angle 2$, which of the following statements is *not* true?

Ⓐ $AD = BD$ Ⓑ $m\angle ADC = m\angle BDC$

Ⓒ $AC > BC$ Ⓓ $m\angle 1 > m\angle B$

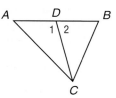

33. ALGEBRA A student bought four college textbooks that cost $99.50, $88.95, $95.90, and $102.45. She paid one half of the total amount herself and borrowed the rest from her mother. She repaid her mother in 4 equal monthly payments. How much was each of the monthly payments?

Ⓐ $24.18   Ⓑ $48.35   Ⓒ $96.70   Ⓓ $193.40

Maintain Your Skills

Mixed Review

Determine whether the given measures can be the lengths of the sides of a triangle. Write *yes* or *no*. Explain. *(Lesson 5-4)*

34. 25, 1, 21   **35.** 16, 6, 19   **36.** 8, 7, 15

Write the assumption you would make to start an indirect proof of each statement. *(Lesson 5-3)*

37. $\overline{AD}$ is a median of $\triangle ABC$.

38. If two altitudes of a triangle are congruent, then the triangle is isosceles.

Write a proof. *(Lesson 4-5)*

39. Given: $\overline{AD}$ bisects $\overline{BE}$.
           $\overline{AB} \parallel \overline{DE}$
     Prove: $\triangle ABC \cong \triangle DEC$

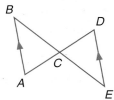

40. Given: $\overline{OM}$ bisects $\angle LMN$.
           $\overline{LM} \cong \overline{MN}$
     Prove: $\triangle MOL \cong \triangle MON$

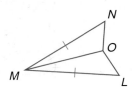

Find the measures of the sides of $\triangle EFG$ with the given vertices and classify each triangle by its sides. *(Lesson 4-1)*

41. $E(4, 6)$, $F(4, 11)$, $G(9, 6)$ **42.** $E(-7, 10)$, $F(15, 0)$, $G(-2, -1)$

43. $E(16, 14)$, $F(7, 6)$, $G(-5, -14)$ **44.** $E(9, 9)$, $F(12, 14)$, $G(14, 6)$

45. ADVERTISING An ad for Wildflowers Gift Boutique says *When it has to be special, it has to be Wildflowers*. Catalina needs a special gift. Does it follow that she should go to Wildflowers? Explain. *(Lesson 2-4)*

Vocabulary and Concept Check

altitude (p. 241)	incenter (p. 240)	orthocenter (p. 241)
centroid (p. 240)	indirect proof (p. 255)	perpendicular bisector (p. 238)
circumcenter (p. 238)	indirect reasoning (p. 255)	point of concurrency (p. 238)
concurrent lines (p. 238)	median (p. 240)	proof by contradiction (p. 255)

For a complete list of postulates and theorems, see pages R1–R8.

Exercises Choose the correct term to complete each sentence.

1. All of the angle bisectors of a triangle meet at the *(incenter, circumcenter)*.

2. In $\triangle RST$, if point P is the midpoint of $\overline{RS}$, then $\overrightarrow{PT}$ is a(n) *(angle bisector, median)*.

3. The theorem that the sum of the lengths of two sides of a triangle is greater than the length of the third side is the *(Triangle Inequality Theorem, SSS Inequality)*.

4. The three medians of a triangle intersect at the *(centroid, orthocenter)*.

5. In $\triangle JKL$, if point H is equidistant from $\overrightarrow{KJ}$ and $\overrightarrow{KL}$, then $\overrightarrow{HK}$ is an *(angle bisector, altitude)*.

6. The circumcenter of a triangle is the point where all three *(perpendicular bisectors, medians)* of the sides of the triangle intersect.

7. In $\triangle ABC$, if $\overrightarrow{AK} \perp \overrightarrow{BC}$, $\overrightarrow{BK} \perp \overrightarrow{AC}$, and $\overrightarrow{CK} \perp \overrightarrow{AB}$, then K is the *(orthocenter, incenter)* of $\triangle ABC$.

Lesson-by-Lesson Review

5-1 Bisectors, Medians, and Altitudes

See pages 238–245.

Concept Summary

- The perpendicular bisectors, angle bisectors, medians, and altitudes of a triangle are all special segments in triangles.

Example Points P, Q, and R are the midpoints of $\overline{JK}$, $\overline{KL}$, and $\overline{JL}$, respectively. Find x.

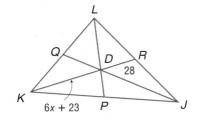

$$KD = \frac{2}{3}(KR) \qquad \text{Centroid Theorem}$$
$$6x + 23 = \frac{2}{3}(6x + 51) \qquad \text{Substitution}$$
$$6x + 23 = 4x + 34 \qquad \text{Simplify.}$$
$$2x = 11 \qquad \text{Subtract } 4x + 23 \text{ from each side.}$$
$$x = \frac{11}{2} \qquad \text{Divide each side by 2.}$$

Exercises In the figure, $\overline{CP}$ is an altitude, $\overline{CQ}$ is the angle bisector of $\angle ACB$, and R is the midpoint of $\overline{AB}$.
See Example 2 on pages 240 and 241.

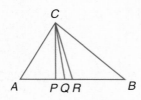

8. Find $m\angle ACQ$ if $m\angle ACB = 123 - x$ and $m\angle QCB = 42 + x$.

9. Find AB if $AR = 3x + 6$ and $RB = 5x - 14$.

10. Find x if $m\angle APC = 72 + x$.

www.geometryonline.com/vocabulary_review

5-2 Inequalities and Triangles

See pages 247–254.

Concept Summary

- The largest angle in a triangle is opposite the longest side, and the smallest angle is opposite the shortest side.

Example

Use the Exterior Angle Theorem to list all angles with measures less than $m\angle 1$.

By the Exterior Angle Theorem, $m\angle 5 < m\angle 1$, $m\angle 10 < m\angle 1$, $m\angle 7 < m\angle 1$, and $m\angle 9 + m\angle 10 < m\angle 1$. Thus, the measures of $\angle 5$, $\angle 10$, $\angle 7$, and $\angle 9$ are all less than $m\angle 1$.

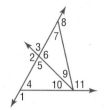

Exercises Determine the relationship between the measures of the given angles. *See Example 3 on page 250.*

11. $\angle DEF$ and $\angle DFE$

12. $\angle GDF$ and $\angle DGF$

13. $\angle DEF$ and $\angle FDE$

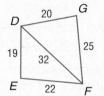

Determine the relationship between the lengths of the given sides. *See Example 4 on page 250.*

14. $\overline{SR}, \overline{SD}$ 15. $\overline{DQ}, \overline{DR}$

16. $\overline{PQ}, \overline{QR}$ 17. $\overline{SR}, \overline{SQ}$

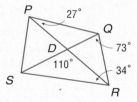

5-3 Indirect Proof

See pages 255–260.

Concept Summary

- In an indirect proof, the conclusion is assumed to be false, and a contradiction is reached.

Example

State the assumption you would make to start an indirect proof of the statement $AB < AC + BC$.

If AB is not less than $AC + BC$, then either $AB > AC + BC$ or $AB = AC + BC$. In other words, $AB \geq AC + BC$.

Exercises State the assumption you would make to start an indirect proof of each statement. *See Example 1 on page 255.*

18. $\sqrt{2}$ is an irrational number.

19. If two sides and the included angle are congruent in two triangles, then the triangles are congruent.

20. **FOOTBALL** Miguel plays quarterback for his high school team. This year, he completed 101 passes in the five games in which he played. Prove that, in at least one game, Miguel completed more than 20 passes.

Chapter

5 **For More ...**

• Extra Practice, see pages 763 and 764.
• Mixed Problem Solving, see page 786.

5-4 The Triangle Inequality

See pages
261–266.

Concept Summary

• The sum of the lengths of any two sides of a triangle is greater than the length of the third side.

Example Determine whether 7, 6, and 14 can be the measures of the sides of a triangle.

Check each inequality.

$$7 + 6 \overset{?}{>} 14 \qquad 7 + 14 \overset{?}{>} 6 \qquad 6 + 14 \overset{?}{>} 7$$

$$13 \not> 14 \qquad\qquad 21 > 6 \checkmark \qquad\qquad 20 > 7 \checkmark$$

Because the inequalities are not true in all cases, the sides cannot form a triangle.

Exercises Determine whether the given measures can be the lengths of the sides of a triangle. Write *yes* or *no*. Explain. *See Example 1 on page 261.*

21. 7, 20, 5   **22.** 16, 20, 5   **23.** 18, 20, 6

5-5 Inequalities Involving Two Triangles

See pages
267–273.

Concept Summary

• SAS Inequality: In two triangles, if two sides are congruent, then the measure of the included angle determines which triangle has the longer third side.

• SSS Inequality: In two triangles, if two sides are congruent, then the length of the third side determines which triangle has the included angle with the greater measure.

Example Write an inequality relating *LM* and *MN*.

In $\triangle LMP$ and $\triangle NMP$, $\overline{LP} \cong \overline{NP}$, $\overline{PM} \cong \overline{PM}$, and $m\angle LPM > m\angle NPM$. The SAS Inequality allows us to conclude that $LM > MN$.

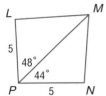

Exercises Write an inequality relating the given pair of angles or segment measures. *See Example 3 on page 269.*

24. $m\angle BAC$ and $m\angle DAC$

25. *BC* and *MD*

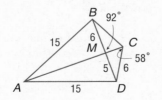

Write an inequality to describe the possible values of *x*.
See Example 3 on page 269.

26. 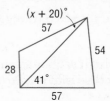   **27.**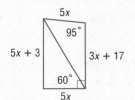

Vocabulary and Concepts

Choose the letter that best matches each description.

> a. circumcenter
> b. incenter
> c. orthocenter
> d. centroid

1. point of concurrency of the angle bisectors of a triangle
2. point of concurrency of the altitudes of a triangle
3. point of concurrency of the perpendicular bisectors of a triangle

Skills and Applications

In $\triangle GHJ$, $HP = 5x - 16$, $PJ = 3x + 8$, $m\angle GJN = 6y - 3$, $m\angle NJH = 4y + 23$, and $m\angle HMG = 4z + 14$.

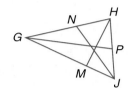

4. $\overline{GP}$ is a median of $\triangle GHJ$. Find HJ.
5. Find $m\angle GJH$ if $\overline{JN}$ is an angle bisector.
6. If $\overline{HM}$ is an altitude of $\angle GHJ$, find the value of z.

Refer to the figure at the right. Determine which angle has the greatest measure.

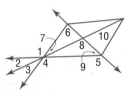

7. $\angle 8$, $\angle 5$, $\angle 7$
8. $\angle 6$, $\angle 7$, $\angle 8$
9. $\angle 1$, $\angle 6$, $\angle 9$

Write the assumption you would make to start an indirect proof of each statement.

10. If n is a natural number, then $2^n + 1$ is odd.
11. Alternate interior angles are congruent.

12. **BUSINESS** Over the course of three days, Marcus spent one and one-half hours on a teleconference for his marketing firm. Use indirect reasoning to show that, on at least one day, Marcus spent at least one half-hour on a teleconference.

Find the range for the measure of the third side of a triangle given the measures of two sides.

13. 1 and 14
14. 14 and 11
15. 13 and 19

Write an inequality for the possible values of x.

16.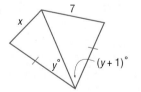

17.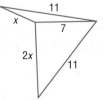

18.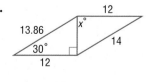

19. **GEOGRAPHY** The distance between Atlanta and Cleveland is about 554 miles. The distance between Cleveland and New York City is about 399 miles. Use the Triangle Inequality Theorem to find the possible values of the distance between New York and Atlanta.

20. **STANDARDIZED TEST PRACTICE** A given triangle has two sides with measures 8 and 11. Which of the following is *not* a possible measure for the third side?

   Ⓐ 3   Ⓑ 7   Ⓒ 12   Ⓓ 18

Part 1 Multiple Choice

Record your answers on the answer sheet provided by your teacher or on a sheet of paper.

1. Tamara works at a rug and tile store after school. The ultra-plush carpet has 80 yarn fibers per square inch. How many fibers are in a square yard? (Prerequisite Skill)

(A) 2,880

(B) 8,640

(C) 34,560

(D) 103,680

2. What is the perimeter of the figure? (Lesson 1-4)

(A) 20 units

(B) 46 units

(C) 90 units

(D) 132 units

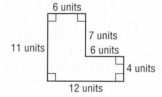

3. Which is a correct statement about the conditional and its converse below? (Lesson 2-2)

Statement: If the measure of an angle is 50°, then the angle is an acute angle.

Converse: If an angle is an acute angle, then the measure of the angle is 50°.

(A) The statement and its converse are both true.

(B) The statement is true, but its converse is false.

(C) The statement and its converse are both false.

(D) The statement is false, but its converse is true.

4. Six people attend a meeting. When the meeting is over, each person exchanges business cards with each of the other people. Use noncollinear points to determine how many exchanges are made. (Lesson 2-3)

(A) 6   (B) 15   (C) 36   (D) 60

For Questions 5 and 6, refer to the figure below.

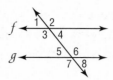

5. What is the term used to describe ∠4 and ∠5? (Lesson 3-1)

(A) alternate exterior angles

(B) alternate interior angles

(C) consecutive interior angles

(D) corresponding angles

6. Given that lines *f* and *g* are not parallel, what assumption can be made to prove that ∠3 is not congruent to ∠7? (Lesson 5-2)

(A) $f \parallel g$

(B) $\angle 3 \cong \angle 7$

(C) $\angle 3 \cong \angle 2$

(D) $m\angle 3 \cong m\angle 7$

7. $\overline{QT}$ is a median of △PST, and $\overline{RT}$ is an altitude of △PST. Which of the following line segments is shortest? (Lesson 5-4)

(A) $\overline{PT}$ (B) $\overline{QT}$ (C) $\overline{RT}$ (D) $\overline{ST}$

8. A paleontologist found the tracks of an animal that is now extinct. Which of the following lengths could be the measures of the three sides of the triangle formed by the tracks? (Lesson 5-4)

(A) 2, 9, 10

(B) 5, 8, 13

(C) 7, 11, 20

(D) 9, 13, 26

Preparing for Standardized Tests
For test-taking strategies and more
practice, see pages 795–810.

Part 2 | Short Response/Grid In

**Record your answers on the answer sheet
provided by your teacher or on a sheet of
paper.**

9. The top of an access ramp to a building is
2 feet higher than the lower end of the ramp.
If the lower end is 24 feet from the building,
what is the slope of the ramp?   (Lesson 3-3)

10. The ramps at a local skate park vary in
steepness. Find *x*.   (Lesson 4-2)

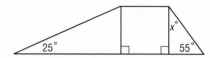

**For Questions 11 and 12, refer to the graph
below.**

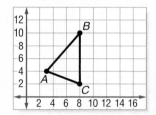

11. During a soccer game, a player stands near
the goal at point *A*. The goalposts are
located at points *B* and *C*. The goalkeeper
stands at point *P* on the goal line $\overline{BC}$ so that
$\overline{AP}$ forms a median. What is the location of
the goalkeeper?   (Lesson 5-1)

12. A defender positions herself on the goal
line $\overline{BC}$ at point *T* to assist the goalkeeper.
If $\overline{AT}$ forms an altitude of $\triangle ABC$, what is
the location of defender *T*?   (Lesson 5-1)

13. What postulate or theorem could you use to
prove that the measure of $\angle QRT$ is greater
than the measure of $\angle SRT$?   (Lesson 5-5)

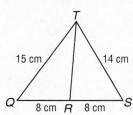

Test-Taking Tip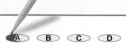
Questions 7, 10, and 11
Review any terms that you have learned before you take a
test. Remember that a median is a segment that connects
a vertex of a triangle to the midpoint of the opposite side.
An altitude is a perpendicular segment from a vertex to
the opposite side.

Part 3 | Extended Response

**Record your answers on a sheet of paper.
Show your work.**

14. Kendell is purchasing a new car stereo for
$200. He agreed to pay the same amount
each month until the $200 is paid. Kendell
made the graph below to help him figure
out when the amount will be paid.
(Lesson 3-3)

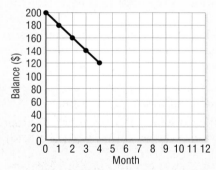

a. Use the slope of the line to write an
argument that the line intersects the
x-axis at (10, 0).

b. What does the point (10, 0) represent?

15. The vertices of $\triangle ABC$ are $A(-3, 1)$, $B(0, -2)$,
and $C(3, 4)$.

a. Graph $\triangle ABC$.   (Prerequisite Skill)

b. Use the Distance Formula to find the
length of each side to the nearest tenth.
(Lesson 1-3)

c. What type of triangle is $\triangle ABC$? How
do you know?   (Lesson 4-1)

d. Prove $\angle A \cong \angle B$.   (Lesson 4-6)

e. Prove $m\angle A > m\angle C$.   (Lesson 5-3)

Chapter 5   Standardized Test Practice   **279**

Chapter

6 Proportions and Similarity

What You'll Learn

- **Lessons 6-1, 6-2, and 6-3** Identify similar polygons, and use ratios and proportions to solve problems.
- **Lessons 6-4 and 6-5** Recognize and use proportional parts, corresponding perimeters, altitudes, angle bisectors, and medians of similar triangles to solve problems.
- **Lesson 6-6** Identify the characteristics of fractals and nongeometric iteration.

Key Vocabulary

- proportion (p. 283)
- cross products (p. 283)
- similar polygons (p. 289)
- scale factor (p. 290)
- midsegment (p. 308)

Why It's Important

Similar figures are used to represent various real-world situations involving a scale factor for the corresponding parts. For example, photography uses similar triangles to calculate distances from the lens to the object and to the image size. *You will use similar triangles to solve problems about photography in Lesson 6-5.*

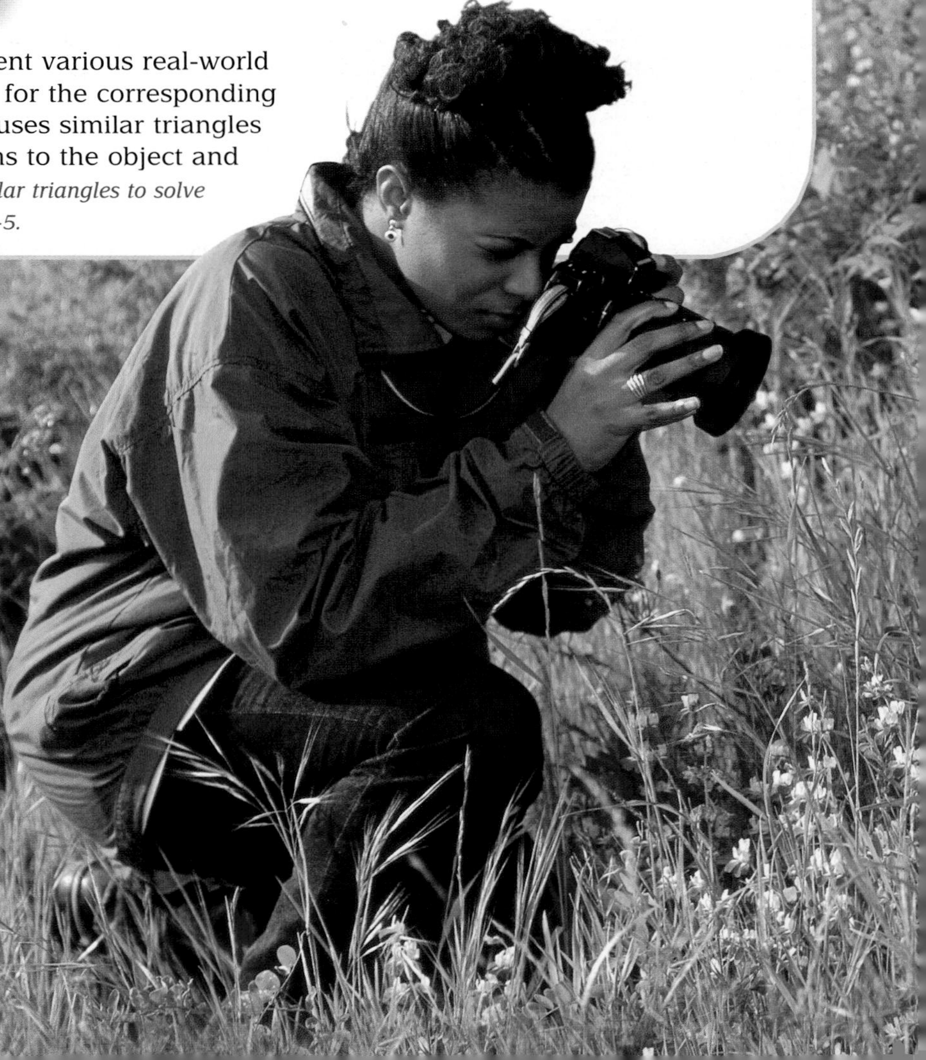

Getting Started

▶ **Prerequisite Skills** To be successful in this chapter, you'll need to master these skills and be able to apply them in problem-solving situations. Review these skills before beginning Chapter 6.

For Lesson 6-1, 6-3, and 6-4 — Solve Rational Equations

Solve each equation. *(For review, see pages 737 and 738.)*

1. $\frac{2}{3}y - 4 = 6$ **2.** $\frac{5}{6} = \frac{x-4}{12}$ **3.** $\frac{4}{3} = \frac{y+2}{y-1}$ **4.** $\frac{2y}{4} = \frac{32}{y}$

For Lesson 6-2 — Slopes of Lines

Find the slope of the line given the coordinates of two points on the line.
(For review, see Lesson 3-3.)

5. $(3, 5)$ and $(0, -1)$ **6.** $(-6, -3)$ and $(2, -3)$ **7.** $(-3, 4)$ and $(2, -2)$

For Lesson 6-5 — Show Lines Parallel

Given the following information, determine whether $a \parallel b$. State the postulate or theorem that justifies your answer.
(For review, see Lesson 3-5.)

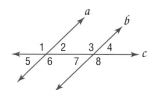

8. $\angle 1 \cong \angle 8$

9. $\angle 3 \cong \angle 6$

10. $\angle 5 \cong \angle 3$

For Lesson 6-6 — Evaluate Expressions

Evaluate each expression for $n = 1, 2, 3,$ and 4. *(For review, see page 736.)*

11. 2^n **12.** $n^2 - 2$ **13.** $3^n - 2$

FOLDABLES™
Study Organizer

Proportions and Similarity Make this Foldable to help you organize your notes. Begin with one sheet of 11" by 17" paper.

Step 1 **Punch and Fold**

Fold widthwise. Leave space to punch holes so it can be placed in your binder.

Step 2 **Divide**

Open the flap and draw lines to divide the inside into six equal parts.

Step 3 **Label**

Label each part using the lesson numbers.

Put the name of the chapter on the front flap.

Reading and Writing As you read and study the chapter, use the Foldable to write down questions you have about the concepts in each lesson. Leave room to record the answers to your questions.

6-1 Proportions

What You'll Learn

- Write ratios.
- Use properties of proportions.

Vocabulary

- ratio
- proportion
- cross products
- extremes
- means

How do artists use ratios?

Stained-glass artist Louis Comfort Tiffany used geometric shapes in his designs. In a portion of *Clematis Skylight* shown at the right, rectangular shapes are used as the background for the flowers and vines. Tiffany also used ratio and proportion in the design of this piece.

WRITE RATIOS   A **ratio** is a comparison of two quantities. The ratio of a to b can be expressed as $\frac{a}{b}$, where b is not zero. This ratio can also be written as $a{:}b$.

Example 1 Write a Ratio

SOCCER   The U.S. Census Bureau surveyed 8218 schools nationally about their girls' soccer programs. They found that 270,273 girls participated in a high school soccer program in the 1999–2000 school year. Find the ratio of girl soccer players per school to the nearest tenth.

Divide the number of girl soccer players by the number of schools.

$$\frac{\text{number of girl soccer players}}{\text{number of schools}} = \frac{270{,}273}{8{,}218} \text{ or about } 32.9$$

A ratio in which the denominator is 1 is called a *unit ratio*.

The ratio for this survey was 32.9 girl soccer players for each school.

Extended ratios can be used to compare three or more numbers. The expression $a{:}b{:}c$ means that the ratio of the first two numbers is $a{:}b$, the ratio of the last two numbers is $b{:}c$, and the ratio of the first and last numbers is $a{:}c$.

Standardized Test Practice

Example 2 Extended Ratios in Triangles

Multiple-Choice Test Item

> In a triangle, the ratio of the measures of three sides is 4:6:9, and its perimeter is 190 inches. Find the length of the longest side of the triangle.
>
> (A) 10 in.   (B) 60 in.   (C) 90 in.   (D) 100 in.

Read the Test Item

You are asked to apply the ratio to the three sides of the triangle and the perimeter to find the longest side.

Solve the Test Item

Recall that equivalent fractions can be found by multiplying the numerator and the denominator by the same number. So, $2:3 = \frac{2}{3} \cdot \frac{x}{x}$ or $\frac{2x}{3x}$. Thus, we can rewrite $4:6:9$ as $4x:6x:9x$ and use those measures for the sides of the triangle. Write an equation to represent the perimeter of the triangle as the sum of the measures of its sides.

$$4x + 6x + 9x = 190 \qquad \text{Perimeter}$$
$$19x = 190 \qquad \text{Combine like terms.}$$
$$x = 10 \qquad \text{Divide each side by 19.}$$

Use this value of x to find the measures of the sides of the triangle.

$4x = 4(10)$ or 40 inches

$6x = 6(10)$ or 60 inches

$9x = 9(10)$ or 90 inches

The longest side is 90 inches. The answer is C.

CHECK Add the lengths of the sides to make sure that the perimeter is 190.

$$40 + 60 + 90 = 190 \quad \checkmark$$

USE PROPERTIES OF PROPORTIONS An equation stating that two ratios are equal is called a **proportion**. Equivalent fractions set equal to each other form a proportion. Since $\frac{2}{3}$ and $\frac{6}{9}$ are equivalent fractions, $\frac{2}{3} = \frac{6}{9}$ is a proportion.

Every proportion has two **cross products**. The cross products in $\frac{2}{3} = \frac{6}{9}$ are 2 times 9 and 3 times 6. The **extremes** of the proportion are 2 and 9. The **means** are 3 and 6.

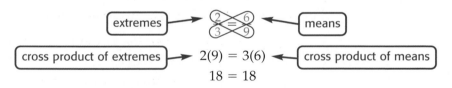

The product of the means equals the product of the extremes, so the cross products are equal. Consider the general case.

$$\frac{a}{b} = \frac{c}{d} \qquad b \neq 0, d \neq 0$$
$$(bd)\frac{a}{b} = (bd)\frac{c}{d} \qquad \text{Multiply each side by the common denominator, } bd.$$
$$da = bc \qquad \text{Simplify.}$$
$$ad = bc \qquad \text{Commutative Property}$$

Key Concept — *Property of Proportions*

- **Words** For any numbers a and c and any nonzero numbers b and d, $\frac{a}{b} = \frac{c}{d}$ if and only if $ad = bc$.

- **Examples** $\frac{4}{5} = \frac{12}{15}$ if and only if $4 \cdot 15 = 5 \cdot 12$.

To *solve a proportion* means to find the value of the variable that makes the proportion true.

 www.geometryonline.com/extra_examples

Example 3 **Solve Proportions by Using Cross Products**

Solve each proportion.

a. $\dfrac{3}{5} = \dfrac{x}{75}$

$\dfrac{3}{5} = \dfrac{x}{75}$ Original proportion

$3(75) = 5x$ Cross products

$225 = 5x$ Multiply.

$45 = x$ Divide each side by 5.

b. $\dfrac{3x - 5}{4} = \dfrac{-13}{2}$

$\dfrac{3x - 5}{4} = \dfrac{-13}{2}$ Original proportion

$(3x - 5)2 = 4(-13)$ Cross products

$6x - 10 = -52$ Simplify.

$6x = -42$ Add 10 to each side.

$x = -7$ Divide each side by 6.

Proportions can be used to solve problems involving two objects that are said to be *in proportion*. This means that if you write ratios comparing the measures of all parts of one object with the measures of comparable parts of the other object, a true proportion would always exist.

Study Tip

Common Misconception
The proportion shown in Example 4 is not the only correct proportion. There are four equivalent proportions:
$\dfrac{a}{b} = \dfrac{c}{d}, \dfrac{a}{d} = \dfrac{c}{b}$,
$\dfrac{b}{a} = \dfrac{d}{c}$, and $\dfrac{b}{c} = \dfrac{d}{a}$.
All of these have identical cross products.

Example 4 **Solve Problems Using Proportions**

AVIATION A twinjet airplane has a length of 78 meters and a wingspan of 90 meters. A toy model is made in proportion to the real airplane. If the wingspan of the toy is 36 centimeters, find the length of the toy.

Because the toy airplane and the real plane are in proportion, you can write a proportion to show the relationship between their measures. Since both ratios compare meters to centimeters, you need not convert all the lengths to the same unit of measure.

$\dfrac{\text{plane's length (m)}}{\text{model's length (cm)}} = \dfrac{\text{plane's wingspan (m)}}{\text{model's wingspan (cm)}}$

$\dfrac{78}{x} = \dfrac{90}{36}$ Substitution

$(78)(36) = x \cdot 90$ Cross products

$2808 = 90x$ Multiply.

$31.2 = x$ Divide each side by 90.

The length of the model would be 31.2 centimeters.

Check for Understanding

Concept Check

1. **Explain** how you would solve $\dfrac{28}{48} = \dfrac{21}{x}$.

2. **OPEN ENDED** Write two possible proportions having the extremes 5 and 8.

3. **FIND THE ERROR** Madeline and Suki are solving $\dfrac{15}{x} = \dfrac{3}{4}$.

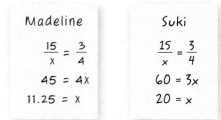

Madeline

$\dfrac{15}{x} = \dfrac{3}{4}$

$45 = 4x$

$11.25 = x$

Suki

$\dfrac{15}{x} = \dfrac{3}{4}$

$60 = 3x$

$20 = x$

Who is correct? Explain your reasoning.

Guided Practice

4. **HOCKEY**   A hockey player scored 9 goals in 12 games. Find the ratio of goals to games.

5. **SCULPTURE**   A replica of *The Thinker* is 10 inches tall. A statue of *The Thinker,* located in front of Grawemeyer Hall on the Belnap Campus of the University of Louisville in Kentucky, is 10 feet tall. What is the ratio of the replica to the statue in Louisville?

Solve each proportion.

6. $\dfrac{x}{5} = \dfrac{11}{35}$

7. $\dfrac{2.3}{4} = \dfrac{x}{3.7}$

8. $\dfrac{x-2}{2} = \dfrac{4}{5}$

9. The ratio of the measures of three sides of a triangle is $9:8:7$, and its perimeter is 144 units. Find the measure of each side of the triangle.

10. The ratio of the measures of three angles of a triangle $5:7:8$. Find the measure of each angle of the triangle.

Standardized Test Practice
(A) (B) (C) (D)

11. **GRID IN**   The scale on a map indicates that 1.5 centimeters represent 200 miles. If the distance on the map between Norfolk, Virginia, and Atlanta, Georgia, measures 2.4 centimeters, how many miles apart are the cities?

Practice and Apply

Homework Help	
For Exercises	**See Examples**
12–17, 23, 25	1
18–22	2
26, 27	4
28–35	3

Extra Practice
See page 764.

12. **BASEBALL**   A designated hitter made 8 hits in 10 games. Find the ratio of hits to games.

13. **SCHOOL**   There are 76 boys in a sophomore class of 165 students. Find the ratio of boys to girls.

14. **CURRENCY**   In a recent month, 208 South African rands were equivalent to 18 United States dollars. Find the ratio of rands to dollars.

15. **EDUCATION**   In the 2000–2001 school year, Arizona State University had 44,125 students and 1747 full-time faculty members. What was the ratio of the students to each teacher rounded to the nearest tenth?

16. Use the number line at the right to determine the ratio of *AC* to *BH*.

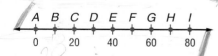

17. A cable that is 42 feet long is divided into lengths in the ratio of $3:4$. What are the two lengths into which the cable is divided?

Find the measures of the angles of each triangle.

18. The ratio of the measures of the three angles is $2:5:3$.

19. The ratio of the measures of the three angles is $6:9:10$.

Find the measures of the sides of each triangle.

20. The ratio of the measures of three sides of a triangle is $8:7:5$. Its perimeter is 240 feet.

21. The ratio of the measures of the sides of a triangle is $3:4:5$. Its perimeter is 72 inches.

22. The ratio of the measures of three sides of a triangle are $\dfrac{1}{2}:\dfrac{1}{3}:\dfrac{1}{5}$, and its perimeter is 6.2 centimeters. Find the measure of each side of the triangle.

LITERATURE For Exercises 23 and 24, use the following information.
Throughout Lewis Carroll's book, *Alice's Adventures in Wonderland,* Alice's size changes. Her normal height is about 50 inches tall. She comes across a door, about 15 inches high, that leads to a garden. Alice's height changes to 10 inches so she can visit the garden.

23. Find the ratio of the height of the door to Alice's height in Wonderland.

24. How tall would the door have been in Alice's normal world?

More About. . .

•25. ENTERTAINMENT Before actual construction of the Great Moments with Mr. Lincoln exhibit, Walt Disney and his design company built models that were in proportion to the displays they planned to build. What is the ratio of the height of the model of Mr. Lincoln compared to his actual height?

Entertainment •·······

In the model, Lincoln is 8 inches tall. In the theater, Lincoln is 6 feet 4 inches tall (his actual adult height).

Source: ©Disney Enterprises, Inc.

ICE CREAM For Exercises 26 and 27, use the following information.
There were approximately 255,082,000 people in the United States in a recent year. According to figures from the United States Census, they consumed about 4,183,344,800 pounds of ice cream that year.

26. If there were 276,000 people in the city of Raleigh, North Carolina, about how much ice cream might they have been expected to consume?

27. Find the approximate consumption of ice cream per person.

 Online Research **Data Update** Use the Internet or other resource to find the population of your community. Determine how much ice cream you could expect to be consumed each year in your community. Visit www.geometryonline.com/data_update to learn more.

ALGEBRA Solve each proportion.

28. $\dfrac{3}{8} = \dfrac{x}{5}$ **29.** $\dfrac{a}{5.18} = \dfrac{1}{4}$ **30.** $\dfrac{3x}{23} = \dfrac{48}{92}$ **31.** $\dfrac{13}{49} = \dfrac{26}{7x}$

32. $\dfrac{2x-13}{28} = \dfrac{-4}{7}$ **33.** $\dfrac{4x+3}{12} = \dfrac{5}{4}$ **34.** $\dfrac{b+1}{b-1} = \dfrac{5}{6}$ **35.** $\dfrac{3x-1}{2} = \dfrac{-2}{x+2}$

PHOTOGRAPHY For Exercises 36 and 37, use the following information.
José reduced a photograph that is 21.3 centimeters by 27.5 centimeters so that it would fit in a 10-centimeter by 10-centimeter area.

36. Find the maximum dimensions of the reduced photograph.

37. What percent of the original length is the length of the reduced photograph?

38. CRITICAL THINKING The ratios of the lengths of the sides of three polygons are given below. Make a conjecture about identifying each type of polygon.

    a. 2:2:3   **b.** 3:3:3:3   **c.** 4:5:4:5

39. WRITING IN MATH   Answer the question that was posed at the beginning of the lesson.

How do artists use ratios?

Include the following in your answer:

- four rectangles from the photo that appear to be in proportion, and
- an estimate in inches of the ratio of the width of the skylight to the length of the skylight given that the dimensions of the rectangle in the bottom left corner are approximately 3.5 inches by 5.5 inches.

40. SHORT RESPONSE In a golden rectangle, the ratio of the length of the rectangle to its width is approximately 1.618:1. Suppose a golden rectangle has a length of 12 centimeters. What is its width to the nearest tenth?

41. ALGEBRA A breakfast cereal contains wheat, rice, and oats in the ratio 3:1:2. If the manufacturer makes a mixture using 120 pounds of oats, how many pounds of wheat will be used?

    Ⓐ 60 lb   Ⓑ 80 lb   Ⓒ 120 lb   Ⓓ 180 lb

Maintain Your Skills

Mixed Review

In the figure, $\overline{SO}$ is a median of $\triangle SLN$, $\overline{OS} \cong \overline{NP}$, $m\angle 1 = 3x - 50$, and $m\angle 2 = x + 30$. Determine whether each statement is *always*, *sometimes*, or *never* true. *(Lesson 5-5)*

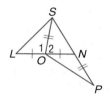

42. $LS > SN$

43. $SN < OP$

44. $x = 45$

Find the range for the measure of the third side of a triangle given the measures of two sides. *(Lesson 5-4)*

45. 16 and 31   **46.** 26 and 40   **47.** 11 and 23

48. COORDINATE GEOMETRY Given $\triangle STU$ with vertices $S(0, 5)$, $T(0, 0)$, and $U(-2, 0)$ and $\triangle XYZ$ with vertices $X(4, 8)$, $Y(4, 3)$, and $Z(6, 3)$, show that $\triangle STU \cong \triangle XYZ$. *(Lesson 4-4)*

Graph the line that satisfies each condition. *(Lesson 3-3)*

49. $m = \dfrac{3}{5}$ and contains $P(-3, -4)$

50. contains $A(5, 3)$ and $B(-1, 8)$

51. parallel to $\overleftrightarrow{JK}$ with $J(-1, 5)$ and $K(4, 3)$ and contains $E(2, 2)$

52. contains $S(8, 1)$ and is perpendicular to $\overleftrightarrow{QR}$ with $Q(6, 2)$ and $R(-4, -6)$

53. MAPS On a U.S. map, there is a scale that lists kilometers on the top and miles on the bottom.

kilometers	0	20	40	50	60	80	100
miles	0				31		62

Suppose $\overline{AB}$ and $\overline{CD}$ are segments on this map. If $AB = 100$ kilometers and $CD = 62$ miles, is $\overline{AB} \cong \overline{CD}$? Explain. *(Lesson 2-7)*

Getting Ready for the Next Lesson

PREREQUISITE SKILL Find the distance between each pair of points to the nearest tenth. *(To review the **Distance Formula**, see Lesson 1-3.)*

54. $A(12, 3)$, $B(-8, 3)$ **55.** $C(0, 0)$, $D(5, 12)$

56. $E\left(\dfrac{4}{5}, -1\right)$, $F\left(2, \dfrac{-1}{2}\right)$ **57.** $G\left(3, \dfrac{3}{7}\right)$, $H\left(4, -\dfrac{2}{7}\right)$

Spreadsheet Investigation

Fibonacci Sequence and Ratios

The Fibonacci sequence is a set of numbers that begins with 1 as its first and second terms. Each successive term is the sum of the two numbers before it. This sequence continues on indefinitely.

term	1	2	3	4	5	6	7
Fibonacci number	1	1	2	3	5	8	13

$$\uparrow \quad \uparrow \quad \uparrow \quad \uparrow \quad \uparrow$$
$$1+1 \quad 1+2 \quad 2+3 \quad 3+5 \quad 5+8$$

Example

Use a spreadsheet to create twenty terms of the Fibonacci sequence. Then compare each term with its preceding term.

Step 1 Enter the column headings in rows 1 and 2.

Step 2 Enter 1 into cell A3. Then insert the formula =A3 + 1 in cell A4. Copy this formula down the column. This will automatically calculate the number of the term.

Step 3 In column B, we will record the Fibonacci numbers. Enter 1 in cells B3 and B4 since you do not have two previous terms to add. Then insert the formula =B3 + B4 in cell B5. Copy this formula down the column.

Step 4 In column C, we will find the ratio of each term to its preceding term. Enter 1 in cell C3 since there is no preceding term. Then enter =B4/B3 in cell C4. Copy this formula down the column.

Fibonacci Table

	A	B	C
	term	Fibonacci number	ratio
1			
2	n	F(n)	F(n+1)/F(n)
3	1	1	1
4	2	1	1
5	3	2	2
6	4	3	1.5
7	5	5	1.666666667
8	6	8	1.6
9	7	13	1.625

Sheet1 / Sheet2

Exercises

1. What happens to the Fibonacci number as the number of the term increases?
2. What pattern of odd and even numbers do you notice in the Fibonacci sequence?
3. As the number of terms gets greater, what pattern do you notice in the ratio column?
4. Extend the spreadsheet to calculate fifty terms of the Fibonacci sequence. Describe any differences in the patterns you described in Exercises 1–3.

The rectangle that most humans perceive to be pleasing to the eye has a width to length ratio of about 1:1.618. This is called the *golden ratio,* and the rectangle is called the *golden rectangle.* This type of rectangle is visible in nature and architecture. The Fibonacci sequence occurs in nature in patterns that are also pleasing to the human eye, such as in sunflowers, pineapples, and tree branch structure.

5. MAKE A CONJECTURE How might the Fibonacci sequence relate to the golden ratio?

6-2 Similar Polygons

What You'll Learn

- Identify similar figures.
- Solve problems involving scale factors.

Vocabulary
- similar polygons
- scale factor

How do artists use geometric patterns?

M.C. Escher (1898–1972) was a Dutch graphic artist known for drawing impossible structures, spatial illusions, and repeating interlocking geometric patterns. The image at the right is a print of Escher's *Circle Limit IV*, which is actually a woodcutting. It includes winged images that have the same shape, but are different in size. Also note that there are not only similar dark images but also similar light images.

Circle Limit IV,
M.C. Escher (1960)

IDENTIFY SIMILAR FIGURES When polygons have the same shape but may be different in size, they are called **similar polygons**.

Study Tip

Similarity and Congruence
If two polygons are congruent, they are also similar. All of the corresponding angles are congruent, and the lengths of the corresponding sides have a ratio of 1:1.

Key Concept — Similar Polygons

- **Words** Two polygons are similar if and only if their corresponding angles are congruent and the measures of their corresponding sides are proportional.

- **Symbol** ~ is read *is similar to*

- **Example**

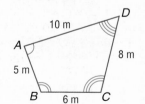
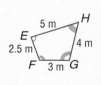

The order of the vertices in a similarity statement is important. It identifies the corresponding angles and the corresponding sides.

similarity statement	congruent angles	corresponding sides
$ABCD \sim EFGH$	$\angle A \cong \angle E$ $\angle B \cong \angle F$ $\angle C \cong \angle G$ $\angle D \cong \angle H$	$\dfrac{AB}{EF} = \dfrac{BC}{FG} = \dfrac{CD}{GH} = \dfrac{DA}{HE}$

Like congruent polygons, similar polygons may be repositioned so that corresponding parts are easy to identify.

Study Tip

Identifying
Corresponding
Parts
Using different colors to
circle letters of congruent
angles may help you
identify corresponding
parts.

Example 1 *Similar Polygons*

Determine whether each pair of figures is similar. Justify your answer.

a.
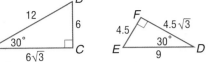

All right angles are congruent, so $\angle C \cong \angle F$.
Since $m\angle A = m\angle D$, $\angle A \cong \angle D$.
By the Third Angle Theorem, $\angle B \cong \angle E$.
Thus, all corresponding angles are congruent.

Now determine whether corresponding sides are proportional.

Sides opposite 90° angle	Sides opposite 30° angle	Sides opposite 60° angle
$\dfrac{AB}{DE} = \dfrac{12}{9}$ or $1.\overline{3}$	$\dfrac{BC}{EF} = \dfrac{6}{4.5}$ or $1.\overline{3}$	$\dfrac{AC}{DF} = \dfrac{6\sqrt{3}}{4.5\sqrt{3}}$ or $1.\overline{3}$

The ratios of the measures of the corresponding sides are equal, and the corresponding angles are congruent, so $\triangle ABC \sim \triangle DEF$.

Study Tip

Common
Misconception
When two figures have
vertices that are in
alphabetical order, this
does not mean that the
corresponding vertices in
the similarity statement will
follow alphabetical order.

b.
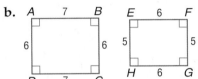

Both rectangles have all right angles and right angles are congruent.
$\dfrac{AB}{EF} = \dfrac{7}{6}$ and $\dfrac{BC}{FG} = \dfrac{6}{5}$, but $\dfrac{AB}{EF} \neq \dfrac{BC}{FG}$ because $\dfrac{7}{6} \neq \dfrac{6}{5}$. The rectangles are not similar.

SCALE FACTORS When you compare the lengths of corresponding sides of similar figures, you usually get a numerical ratio. This ratio is called the **scale factor** for the two figures. Scale factors are often given for models of real-life objects.

Example 2 *Scale Factor*

MOVIES Some special effects in movies are created using miniature models. In a recent movie, a model sports-utility vehicle (SUV) 22 inches long was created to look like a real $14\frac{2}{3}$-foot SUV. What is the scale factor of the model compared to the real SUV?

Before finding the scale factor you must make sure that both measurements use the same unit of measure.

$$14\frac{2}{3}(12) = 176 \text{ inches}$$

$$\frac{\text{length of model}}{\text{length of real SUV}} = \frac{22 \text{ inches}}{176 \text{ inches}}$$

$$= \frac{1}{8}$$

The ratio comparing the two lengths is $\frac{1}{8}$ or $1:8$. The scale factor is $\frac{1}{8}$, which means that the model is $\frac{1}{8}$ the length of the real SUV.

When finding the scale factor for two similar polygons, the scale factor will depend on the order of comparison.

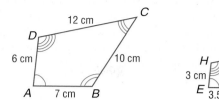

- The scale factor of quadrilateral $ABCD$ to quadrilateral $EFGH$ is 2.
- The scale factor of quadrilateral $EFGH$ to quadrilateral $ABCD$ is $\frac{1}{2}$.

Example 3 Proportional Parts and Scale Factor

The two polygons are similar.

a. Write a similarity statement. Then find x, y, and UT.

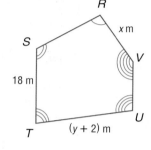

Use the congruent angles to write the corresponding vertices in order.

polygon $RSTUV \sim$ polygon $ABCDE$

Now write proportions to find x and y.

To find x:

$\dfrac{ST}{BC} = \dfrac{VR}{EA}$ Similarity proportion

$\dfrac{18}{4} = \dfrac{x}{3}$ $ST = 18$, $BC = 4$
$VR = x$, $EA = 3$

$18(3) = 4(x)$ Cross products

$54 = 4x$ Multiply.

$13.5 = x$ Divide each side by 4.

To find y:

$\dfrac{ST}{BC} = \dfrac{UT}{DC}$ Similarity proportion

$\dfrac{18}{4} = \dfrac{y+2}{5}$ $ST = 18$, $BC = 4$
$UT = y + 2$, $EA = 3$

$18(5) = 4(y + 2)$ Cross products

$90 = 4y + 8$ Multiply.

$82 = 4y$ Subtract 8 from each side.

$20.5 = y$ Divide each side by 4.

$UT = y + 2$, so $UT = 20.5 + 2$ or 22.5.

b. Find the scale factor of polygon $RSTUV$ to polygon $ABCDE$.

The scale factor is the ratio of the lengths of any two corresponding sides.

$\dfrac{ST}{BC} = \dfrac{18}{4}$ or $\dfrac{9}{2}$

You can use scale factors to produce similar figures.

Example 4 Enlargement of a Figure

Triangle ABC is similar to $\triangle XYZ$ with a scale factor of $\frac{2}{3}$. If the lengths of the sides of $\triangle ABC$ are 6, 8, and 10 inches, what are the lengths of the sides of $\triangle XYZ$?

Write proportions for finding side measures.

$\begin{array}{l} \triangle ABC \to \\ \triangle XYZ \to \end{array} \dfrac{6}{x} = \dfrac{2}{3}$ $\begin{array}{l} \triangle ABC \to \\ \triangle XYZ \to \end{array} \dfrac{8}{y} = \dfrac{2}{3}$ $\begin{array}{l} \triangle ABC \to \\ \triangle XYZ \to \end{array} \dfrac{10}{z} = \dfrac{2}{3}$

$18 = 2x$ $24 = 2y$ $30 = 2z$

$9 = x$ $12 = y$ $15 = z$

The lengths of the sides of $\triangle XYZ$ are 9, 12, and 15 inches.

Example 5 Scale Factors on Maps

MAPS The scale on the map of New Mexico is 2 centimeters = 160 miles. The distance on the map across New Mexico from east to west through Albuquerque is 4.1 centimeters. How long would it take to drive across New Mexico if you drove at an average of 60 miles per hour?

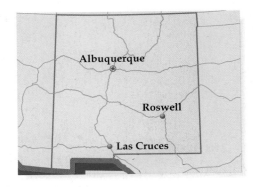

Explore   Every 2 centimeters represents 160 miles. The distance across the map is 4.1 centimeters.

Plan   Create a proportion relating the measurements to the scale to find the distance in miles. Then use the formula $d = rt$ to find the time.

Solve   centimeters → $\dfrac{2}{160} = \dfrac{4.1}{x}$ ← centimeters
             miles → ← miles

$$2x = 656 \quad \text{Cross products}$$

$$x = 328 \quad \text{Divide each side by 2.}$$

The distance across New Mexico is approximately 328 miles.

$$d = rt$$

$$328 = 60t \quad d = 328 \text{ and } r = 60$$

$$\dfrac{328}{60} = t \quad \text{Divide each side by 60.}$$

$$5\dfrac{7}{15} = t \quad \text{Simplify.}$$

It would take $5\dfrac{7}{15}$ hours or about 5 hours and 28 minutes to drive across New Mexico at an average of 60 miles per hour.

Examine   Reexamine the scale. If 2 centimeters = 160 miles, then 4 centimeters = 320 miles. The map is about 4 centimeters wide, so the distance across New Mexico is about 320 miles. The answer is about 5.5 hours and at 60 miles per hour, the trip would be 330 miles. The two distances are close estimates, so the answer is reasonable.

Study Tip

Units of Time
Remember that there are 60 minutes in an hour. When rewriting $\frac{328}{60}$ as a mixed number, you could also write $5\frac{28}{60}$, which means 5 hours 28 minutes.

Check for Understanding

Concept Check   **1. FIND THE ERROR**   Roberto and Garrett have calculated their scale factor for two similar triangles.

Roberto	Garrett
$\dfrac{AB}{PQ} = \dfrac{8}{10}$	$\dfrac{PQ}{AB} = \dfrac{10}{8}$
$= \dfrac{4}{5}$	$= \dfrac{5}{4}$

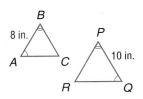

Who is correct? Explain your reasoning.

2. **Find a counterexample** for the statement *All rectangles are similar.*

3. **OPEN ENDED** Explain whether two polygons that are congruent are also similar. Then explain whether two polygons that are similar are also congruent.

Guided Practice   **Determine whether each pair of figures is similar. Justify your answer.**

4.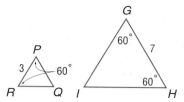

5.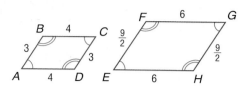

Each pair of polygons is similar. Write a similarity statement, and find *x*, the measure(s) of the indicated side(s), and the scale factor.

6. $\overline{DF}$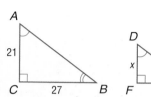

7. $\overline{FE}$, $\overline{EH}$, and $\overline{GF}$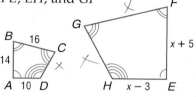

8. A rectangle with length 60 centimeters and height 40 centimeters is reduced so that the new rectangle is similar to the original and the scale factor is $\frac{1}{4}$. Find the length and height of the new rectangle.

9. A triangle has side lengths of 3 meters, 5 meters, and 4 meters. The triangle is enlarged so that the larger triangle is similar to the original and the scale factor is 5. Find the perimeter of the larger triangle.

Application   10. **MAPS** Refer to Example 5 on page 292. Draw the state of New Mexico using a scale of 2 centimeters = 100 miles. Is your drawing similar to the one in Example 4? Explain how you know.

Practice and Apply

Determine whether each pair of figures is similar. Justify your answer.

For Exercises	See Examples
11–14	1
15–20, 34–39	2, 3
21–23	4
24–26	5

Extra Practice
See page 765.

Homework Help

11.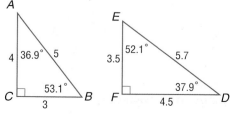

12.

13.

14.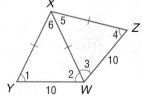

15. **ARCHITECTURE** The replica of the Eiffel Tower at an amusement park is $350\frac{2}{3}$ feet tall. The actual Eiffel Tower is 1052 feet tall. What is the scale factor comparing the amusement park tower to the actual tower?

16. **PHOTOCOPYING** Mr. Richardson walked to a copier in his office, made a copy of his proposal, and sent the original to one of his customers. When Mr. Richardson looked at his copy before filing it, he saw that the copy had been made at an 80% reduction. He needs his filing copy to be the same size as the original. What enlargement scale factor must he use on the first copy to make a second copy the same size as the original?

Each pair of polygons is similar. Write a similarity statement, and find x, the measures of the indicated sides, and the scale factor.

17. $\overline{AB}$ and $\overline{CD}$

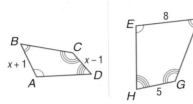

18. $\overline{AC}$ and $\overline{CE}$

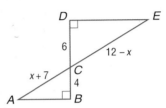

19. $\overline{BC}$ and $\overline{ED}$

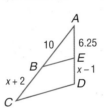

20. $\overline{GF}$ and $\overline{EG}$

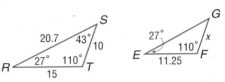

PHOTOGRAPHY For Exercises 21–23, use the following information.
A picture is enlarged by a scale factor of $\frac{5}{4}$ and then enlarged again by the same factor.

21. If the original picture was 2.5 inches by 4 inches, what were its dimensions after both enlargements?

22. Write an equation describing the enlargement process.

23. By what scale factor was the original picture enlarged?

SPORTS Make a scale drawing of each playing field using the given scale.

24. Use the information about the soccer field in Crew Stadium. Use the scale 1 millimeter = 1 meter.

25. A basketball court is 84 feet by 50 feet. Use the scale $\frac{1}{4}$ inch = 4 feet.

26. A tennis court is 36 feet by 78 feet. Use the scale $\frac{1}{8}$ inch = 1 foot.

Determine whether each statement is *always*, *sometimes*, or *never* true.

27. Two congruent triangles are similar.

28. Two squares are similar.

29. A triangle is similar to a quadrilateral.

30. Two isosceles triangles are similar.

31. Two rectangles are similar.

32. Two obtuse triangles are similar.

33. Two equilateral triangles are similar.

Each pair of polygons is similar. Find x and y. Round to the nearest hundredth if necessary.

34.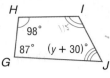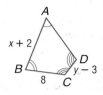

35.

36.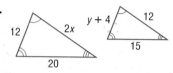

37.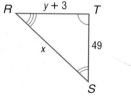

38.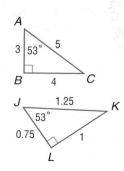

39.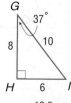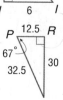

For Exercises 40–47, use the following information to find each measure.
Polygon $ABCD \sim$ polygon $AEFG$, $m\angle AGF = 108$, $GF = 14$, $AD = 12$, $DG = 4.5$, $EF = 8$, and $AB = 26$.

40. scale factor of trapezoid $ABCD$ to trapezoid $AEFG$

41. AG

42. DC

43. $m\angle ADC$

44. BC

45. perimeter of trapezoid $ABCD$

46. perimeter of trapezoid $AEFG$

47. ratio of the perimeter of polygon $ABCD$ to the perimeter of polygon $AEFG$

48. Determine which of the following right triangles are similar. Justify your answer.

COORDINATE GEOMETRY Graph the given points. Draw polygon $ABCD$ and $\overline{MN}$. Find the coordinates for vertices L and P such that $ABCD \sim NLPM$.

49. $A(2, 0)$, $B(4, 4)$, $C(0, 4)$, $D(-2, 0)$; $M(4, 0)$, $N(12, 0)$

50. $A(-7, 1)$, $B(2, 5)$, $C(7, 0)$, $D(-2, -4)$; $M(-3, 1)$, $N\left(-\dfrac{11}{2}, \dfrac{7}{2}\right)$

CONSTRUCTION For Exercises 51 and 52, use the following information.
A floor plan is given for the first floor of a new house. One inch represents 24 feet. Use the information in the plan to find the dimensions.

51. living room

52. deck

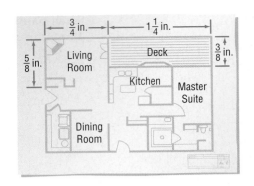

CRITICAL THINKING For Exercises 53–55, use the following information.
The area A of a rectangle is the product of its length ℓ and width w. Rectangle $ABCD$ is similar to rectangle $WXYZ$ with sides in a ratio of 4:1.

53. What is the ratio of the areas of the two rectangles?

54. Suppose the dimension of each rectangle is tripled. What is the new ratio of the sides of the rectangles?

55. What is the ratio of the areas of these larger rectangles?

STATISTICS For Exercises 56–58, refer to the graphic, which uses rectangles to represent percents.

56. Are the rectangles representing 36% and 18% similar? Explain.

57. What is the ratio of the areas of the rectangles representing 36% and 18% if area = length × width? Compare the ratio of the areas to the ratio of the percents.

58. Use the graph to make a conjecture about the overall changes in the level of professional courtesy in the workplace in the past five years.

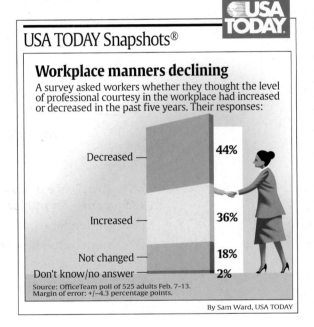

USA TODAY Snapshots®

Workplace manners declining
A survey asked workers whether they thought the level of professional courtesy in the workplace had increased or decreased in the past five years. Their responses:

Decreased — 44%
Increased — 36%
Not changed — 18%
Don't know/no answer — 2%

Source: OfficeTeam poll of 525 adults Feb. 7-13.
Margin of error: +/−4.3 percentage points.

By Sam Ward, USA TODAY

CRITICAL THINKING For Exercises 59 and 60, $\triangle ABC \sim \triangle DEF$.

59. Show that the perimeters of $\triangle ABC$ and $\triangle DEF$ have the same ratio as their corresponding sides.

60. If 6 units are added to the lengths of each side, are the new triangles similar?

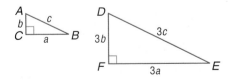

61. **WRITING IN MATH** Answer the question that was posed at the beginning of the lesson.

How do artists use geometric patterns?

Include the following in your answer:

• why Escher called the picture *Circle Limit IV,* and

• how one of the light objects and one of the dark objects compare in size.

62. In a history class with 32 students, the ratio of girls to boys is 5 to 3. How many more girls are there than boys?

　Ⓐ 2　　　　　Ⓑ 8　　　　　Ⓒ 12　　　　　Ⓓ 15

63. ALGEBRA Find x.

　Ⓐ 4.2
　Ⓑ 4.65
　Ⓒ 5.6
　Ⓓ 8.4

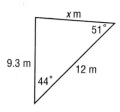

Extending the Lesson Scale factors can be used to produce similar figures. The resulting figure is an enlargement or reduction of the original figure depending on the scale factor.

Triangle ABC has vertices A(0, 0), B(8, 0), and C(2, 7). Suppose the coordinates of each vertex are multiplied by 2 to create the similar triangle A′B′C′.

64. Find the coordinates of the vertices of $\triangle A'B'C'$.

65. Graph $\triangle ABC$ and $\triangle A'B'C'$.

66. Use the Distance Formula to find the measures of the sides of each triangle.

67. Find the ratios of the sides that appear to correspond.

68. How could you use slope to determine if angles are congruent?

69. Is $\triangle ABC \sim \triangle A'B'C'$? Explain your reasoning.

Maintain Your Skills

Mixed Review　**Solve each proportion.** *(Lesson 6-1)*

70. $\dfrac{b}{7.8} = \dfrac{2}{3}$

71. $\dfrac{c-2}{c+3} = \dfrac{5}{4}$

72. $\dfrac{2}{4y+5} = \dfrac{-4}{y}$

Use the figure to write an inequality relating each pair of angle or segment measures. *(Lesson 5-5)*

73. OC, AO

74. $m\angle AOD, m\angle AOB$

75. $m\angle ABD, m\angle ADB$

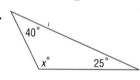

Find x. *(Lesson 4-2)*

76.

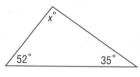

77.

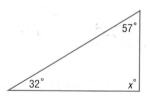

78.

79. Suppose two parallel lines are cut by a transversal and $\angle 1$ and $\angle 2$ are alternate interior angles. Find $m\angle 1$ and $m\angle 2$ if $m\angle 1 = 10x - 9$ and $m\angle 2 = 9x + 3$. *(Lesson 3-2)*

Getting Ready for the Next Lesson　**PREREQUISITE SKILL** In the figure, $\overline{AB} \parallel \overline{CD}$, $\overline{AC} \parallel \overline{BD}$, and $m\angle 4 = 118$. **Find the measure of each angle.** *(To review **angles and parallel lines**, see Lesson 3-2.)*

80. $\angle 1$　　　　**81.** $\angle 2$

82. $\angle 3$　　　　**83.** $\angle 5$

84. $\angle ABD$　　**85.** $\angle 6$

86. $\angle 7$　　　　**87.** $\angle 8$

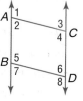

Similar Triangles

What You'll Learn

- Identify similar triangles.
- Use similar triangles to solve problems.

How do engineers use geometry?

- The Eiffel Tower was built in Paris for the 1889 world exhibition by Gustave Eiffel. Eiffel (1832–1923) was a French engineer who specialized in revolutionary steel constructions. He used thousands of triangles, some the same shape but different in size, to build the Eiffel Tower because triangular shapes result in rigid construction.

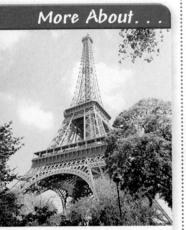

More About. . .

Eiffel Tower

The Eiffel Tower weighs 7000 tons, but the pressure per square inch it applies on the ground is only equivalent to that of a chair with a person seated in it.

Source: www.eiffel-tower.com

IDENTIFY SIMILAR TRIANGLES In Chapter 4, you learned several tests to determine whether two triangles are congruent. There are also tests to determine whether two triangles are similar.

Geometry Activity

Similar Triangles

Collect Data

- Draw △*DEF* with $m\angle D = 35$, $m\angle F = 80$, and $DF = 4$ centimeters.
- Draw △*RST* with $m\angle T = 35$, $m\angle S = 80$, and $ST = 7$ centimeters.
- Measure $\overline{EF}$, $\overline{ED}$, $\overline{RS}$, and $\overline{RT}$.
- Calculate the ratios $\dfrac{FD}{ST}$, $\dfrac{EF}{RS}$, and $\dfrac{ED}{RT}$.

Analyze the Data

1. What can you conclude about all of the ratios?
2. Repeat the activity with two more triangles with the same angle measures, but different side measures. Then repeat the activity with a third pair of triangles. Are all of the triangles similar? Explain.
3. What are the minimum requirements for two triangles to be similar?

The previous activity leads to the following postulate.

Postulate 6.1

Angle-Angle (AA) Similarity If the two angles of one triangle are congruent to two angles of another triangle, then the triangles are similar.

Example: $\angle P \cong \angle T$ and $\angle Q \cong \angle S$,
so △*PQR* ~ △*TSU*.

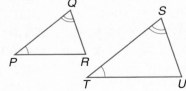

You can use the AA Similarity Postulate to prove two theorems that also verify triangle similarity.

Theorems

6.1 **Side-Side-Side (SSS) Similarity**   If the measures of the corresponding sides of two triangles are proportional, then the triangles are similar.

Example: $\frac{PQ}{ST} = \frac{QR}{SU} = \frac{RP}{UT}$, so $\triangle PQR \sim \triangle TSU$.

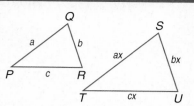

6.2 **Side-Angle-Side (SAS) Similarity**   If the measures of two sides of a triangle are proportional to the measures of two corresponding sides of another triangle and the included angles are congruent, then the triangles are similar.

Example: $\frac{PQ}{ST} = \frac{QR}{SU}$ and $\angle Q \cong \angle S$, so $\triangle PQR \sim \triangle TSU$.

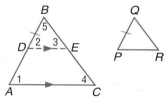

You will prove Theorem 6.2 in Exercise 34.

Proof **Theorem 6.1**

Given: $\frac{PQ}{AB} = \frac{QR}{BC} = \frac{RP}{CA}$

Prove: $\triangle BAC \sim \triangle QPR$

Locate D on $\overline{AB}$ so that $\overline{DB} \cong \overline{PQ}$ and draw $\overline{DE}$ so that $\overline{DE} \parallel \overline{AC}$.

Paragraph Proof:

Since $\overline{DB} \cong \overline{PQ}$, $DB = PQ$. $\frac{PQ}{AB} = \frac{QR}{BC} = \frac{RP}{CA}$

becomes $\frac{DB}{AB} = \frac{QR}{BC} = \frac{RP}{CA}$. Since $\overline{DE} \parallel \overline{AC}$, $\angle 2 \cong \angle 1$ and $\angle 3 \cong \angle 4$. By AA Similarity, $\triangle BDE \sim \triangle BAC$.

By the definition of similar polygons, $\frac{DB}{AB} = \frac{BE}{BC} = \frac{ED}{CA}$. By substitution, $\frac{QR}{BC} = \frac{BE}{BC}$ and $\frac{RP}{CA} = \frac{ED}{CA}$. This means that $QR = BE$ and $RP = ED$ or $\overline{QR} \cong \overline{BE}$ and $\overline{RP} \cong \overline{ED}$. With these congruences and $\overline{DB} \cong \overline{PQ}$, $\triangle BDE \cong \triangle QPR$ by SSS. By CPCTC, $\angle B \cong \angle Q$ and $\angle 2 \cong \angle P$. But $\angle 2 \cong \angle A$, so $\angle A \cong \angle P$. By AA Similarity, $\triangle BAC \sim \triangle QPR$.

Study Tip

Overlapping Triangles

When two triangles overlap, draw them separately so the corresponding parts are in the same position on the paper. Then write the corresponding angles and sides.

Example 1 *Determine Whether Triangles Are Similar*

In the figure, $\overline{FG} \cong \overline{EG}$, $BE = 15$, $CF = 20$, $AE = 9$, and $DF = 12$. Determine which triangles in the figure are similar.

Triangle FGE is an isosceles triangle. So, $\angle GFE \cong \angle GEF$. If the measures of the corresponding sides that include the angles are proportional, then the triangles are similar.

$\frac{AE}{DF} = \frac{9}{12}$ or $\frac{3}{4}$ and $\frac{BE}{CF} = \frac{15}{20}$ or $\frac{3}{4}$

By substitution, $\frac{AE}{DF} = \frac{BE}{CF}$. So, by SAS Similarity, $\triangle ABE \sim \triangle DCF$.

Like the congruence of triangles, similarity of triangles is reflexive, symmetric, and transitive.

Theorem 6.3

Similarity of triangles is reflexive, symmetric, and transitive.

Examples:

Reflexive: $\triangle ABC \sim \triangle ABC$

Symmetric: If $\triangle ABC \sim \triangle DEF$, then $\triangle DEF \sim \triangle ABC$.

Transitive: If $\triangle ABC \sim \triangle DEF$ and $\triangle DEF \sim \triangle GHI$, then $\triangle ABC \sim \triangle GHI$.

You will prove Theorem 6.3 in Exercise 38.

USE SIMILAR TRIANGLES   Similar triangles can be used to solve problems.

Example 2 Parts of Similar Triangles

ALGEBRA   Find AE and DE.

Since $\overline{AB} \parallel \overline{CD}$, $\angle BAE \cong \angle CDE$ and $\angle ABE \cong \angle DCE$ because they are the alternate interior angles. By AA Similarity, $\triangle ABE \sim \triangle DCE$. Using the definition of similar polygons, $\dfrac{AB}{DC} = \dfrac{AE}{DE}$.

$$\frac{AB}{DC} = \frac{AE}{DE}$$

$\dfrac{2}{5} = \dfrac{x-1}{x+5}$ Substitution

$2(x + 5) = 5(x - 1)$ Cross products

$2x + 10 = 5x - 5$ Distributive Property

$-3x = -15$ Subtract $5x$ and 10 from each side.

$x = 5$ Divide each side by -3.

Now find AE and ED.

$AE = x - 1$	$ED = x + 5$
$= 5 - 1$ or 4	$= 5 + 5$ or 10

Similar triangles can be used to find measurements indirectly.

Study Tip

Shadow Problems
In shadow problems, we assume that a right triangle is formed by the sun's ray from the top of the object to the end of the shadow.

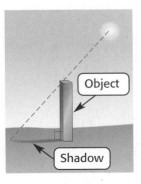

Example 3 Find a Measurement

INDIRECT MEASUREMENT   Nina was curious about the height of the Eiffel Tower. She used a 1.2 meter model of the tower and measured its shadow at 2 P.M. The length of the shadow was 0.9 meter. Then she measured the Eiffel Tower's shadow, and it was 240 meters. What is the height of the Eiffel Tower?

Assuming that the sun's rays form similar triangles, the following proportion can be written.

$$\frac{\text{height of the Eiffel Tower (m)}}{\text{height of the model tower (m)}} = \frac{\text{Eiffel Tower shadow length (m)}}{\text{model shadow length (m)}}$$

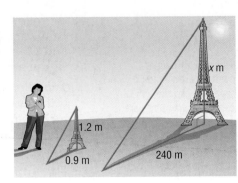

Now substitute the known values and let x be the height of the Eiffel Tower.

$$\frac{x}{1.2} = \frac{240}{0.9}$$ Substitution

$x \cdot 0.9 = 1.2(240)$ Cross products

$0.9x = 288$ Simplify.

$x = 320$ Divide each side by 0.9.

The Eiffel Tower is 320 meters tall.

Check for Understanding

Concept Check   **1. Compare and contrast** the tests to prove triangles similar with the tests to prove triangles congruent.

2. OPEN ENDED   Is it possible that $\triangle ABC$ is not similar to $\triangle RST$ and that $\triangle RST$ is not similar to $\triangle EFG$, but that $\triangle ABC$ is similar to $\triangle EFG$? Explain.

3. FIND THE ERROR   Alicia and Jason were writing proportions for the similar triangles shown at the right.

Alicia

$$\frac{r}{k} = \frac{s}{m}$$

$rm = ks$

Jason

$$\frac{r}{k} = \frac{m}{s}$$

$rs = km$

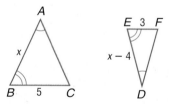

Who is correct? Explain your reasoning.

Guided Practice   **ALGEBRA**   **Identify the similar triangles. Find x and the measures of the indicated sides.**

4. DE

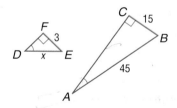

5. AB and DE

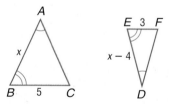

Determine whether each pair of triangles is similar. Justify your answer.

6.

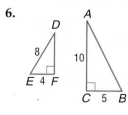

7.

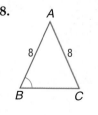

8.

Application   **9. INDIRECT MEASUREMENT**   A cell phone tower in a field casts a shadow of 100 feet. At the same time, a 4 foot 6 inch post near the tower casts a shadow of 3 feet 4 inches. Find the height of the tower in feet and inches. (*Hint*: Make a drawing.)

For Exercises	See Examples
10–17, 26–27	1
18–21, 28–31	2
38–41	3

Extra Practice
See page 765.

Determine whether each pair of triangles is similar. Justify your answer.

10.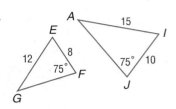

11.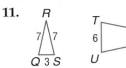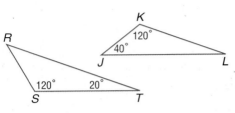

12.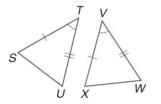

13.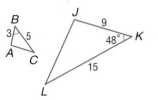

14.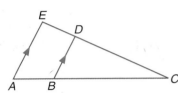

15.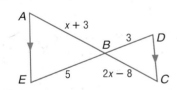

16.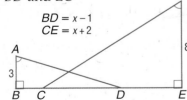

17.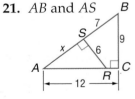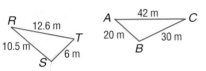

ALGEBRA Identify the similar triangles, and find x and the measures of the indicated sides.

18. *AB* and *BC*

A — x + 3
B — 3 — D
5 — 2x − 8 — C
E

19. *AB* and *AC*

A
x + 2
8 E — B
5 — 6
D — C

20. *BD* and *EC*

BD = x − 1
CE = x + 2

F
A
8
3
B C   D   E

21. *AB* and *AS*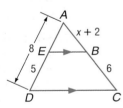

COORDINATE GEOMETRY Triangles *ABC* and *TBS* have vertices *A*(−2, −8), *B*(4, 4), *C*(−2, 7), *T*(0, −4), and *S*(0, 6).

22. Graph the triangles and prove that △*ABC* ~ △*TBS*.

23. Find the ratio of the perimeters of the two triangles.

Identify each statement as *true* or *false*. If false, state why.

24. For every pair of similar triangles, there is only one correspondence of vertices that will give you correct angle correspondence and segment proportions.

25. If △*ABC* ~ △*EFG* and △*ABC* ~ △*RST*, then △*EFG* ~ △*RST*.

302 Chapter 6 Proportions and Similarity

Identify the similar triangles in each figure. Explain your answer.

26.

27.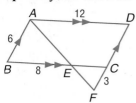

Use the given information to find each measure.

28. If $\overline{PR} \parallel \overline{WX}$, $WX = 10$, $XY = 6$, $WY = 8$, $RY = 5$, and $PS = 3$, find PY, SY, and PQ.

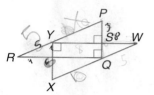

29. If $\overline{PR} \parallel \overline{KL}$, $KN = 9$, $LN = 16$, $PM = 2(KP)$, find KP, KM, MR, ML, MN, and PR.

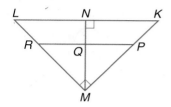

30. If $\dfrac{IJ}{XJ} = \dfrac{HJ}{YJ}$, $m\angle WXJ = 130$, and $m\angle WZG = 20$, find $m\angle YIZ$, $m\angle JHI$, $m\angle JIH$, $m\angle J$, and $m\angle JHG$.

31. If $\angle RST$ is a right angle, $\overline{SU} \perp \overline{RT}$, $\overline{UV} \perp \overline{ST}$, and $m\angle RTS = 47$, find $m\angle TUV$, $m\angle R$, $m\angle RSU$, and $m\angle SUV$.

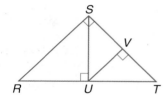

32. **HISTORY** The Greek mathematician Thales was the first to measure the height of a pyramid by using geometry. He showed that the ratio of a pyramid to a staff was equal to the ratio of one shadow to the other. If a pace is about 3 feet, approximately how tall was the pyramid at that time?

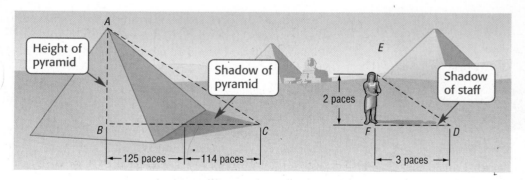

33. In the figure at the right, what relationship must be true of x and y for $\overline{BD}$ and $\overline{AE}$ to be parallel? Explain.

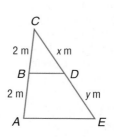

34. Write a two-column proof to show that if the measures of two sides of a triangle are proportional to the measures of two corresponding sides of another triangle and the included angles are congruent, then the triangles are similar. (Theorem 6.2)

35. a two-column proof
    Given: $\overline{LP} \parallel \overline{MN}$
    Prove: $\dfrac{LJ}{JN} = \dfrac{PJ}{JM}$

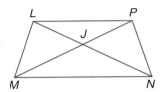

36. a paragraph proof
    Given: $\overline{EB} \perp \overline{AC}, \overline{BH} \perp \overline{AE},$
        $\overline{CJ} \perp \overline{AE}$
    Prove: a. $\triangle ABH \sim \triangle DCB$
        b. $\dfrac{BC}{BE} = \dfrac{BD}{BA}$

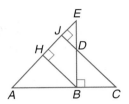

37. a two-column proof to show that if the measures of the legs of two right triangles are proportional, the triangles are similar

38. a two-column proof to prove that similarity of triangles is reflexive, symmetric, and transitive. (Theorem 6.3)

39. **SURVEYING**   Mr. Glover uses a carpenter's square, an instrument used to draw right angles, to find the distance across a stream. The carpenter's square models right angle *NOL*. He puts the square on top of a pole that is high enough to sight along $\overline{OL}$ to point *P* across the river. Then he sights along $\overline{ON}$ to point *M*. If *MK* is 1.5 feet and *OK* = 4.5 feet, find the distance *KP* across the stream.

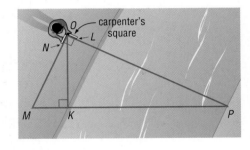

40. The lengths of three sides of triangle *ABC* are 6 centimeters, 4 centimeters, and 9 centimeters. Triangle *DEF* is similar to triangle *ABC*. The length of one of the sides of triangle *DEF* is 36 centimeters. What is the greatest perimeter possible for triangle *DEF*?

TOWERS   For Exercises 41 and 42, use the following information.
To estimate the height of the Jin Mao Tower in Shanghai, a tourist sights the top of the tower in a mirror that is 87.6 meters from the tower. The mirror is on the ground and faces upward. The tourist is 0.4 meter from the mirror, and the distance from his eyes to the ground is about 1.92 meters.

41. How tall is the tower?

42. Why is the mirror reflection a better way to indirectly measure the tower than by using shadows?

More About . . .

Towers •·············
Completed in 1999, the Jin Mao Tower is 88 stories tall and sits on a 6-story podium, making it the tallest building in China. It is structurally engineered to tolerate typhoon winds and earthquakes.
Source: www.som.com

43. FORESTRY A hypsometer as shown can be used to estimate the height of a tree. Bartolo looks through the straw to the top of the tree and obtains the readings given. Find the height of the tree.

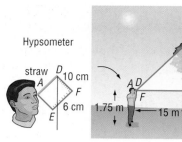

44. CRITICAL THINKING Suppose you know the height of a flagpole on the beach of the Chesapeake Bay and that it casts a shadow 4 feet long at 2:00 (EST). You also know the height of a flagpole on the shoreline of Lake Michigan whose shadow is hard to measure at 1:00 (CST). Since 2:00 (EST) = 1:00 (CST), you propose the following proportion of heights and lengths to find the length of the shadow of the Michigan flagpole. Explain whether this proportion will give an accurate measure.

$$\frac{\text{height of Chesapeake flagpole}}{\text{shadow of Chesapeake flagpole}} = \frac{\text{height of Michigan flagpole}}{\text{shadow of Michigan flagpole}}$$

COORDINATE GEOMETRY For Exercises 45 and 46, use the following information.

The coordinates of $\triangle ABC$ are $A(-10, 6)$, $B(-2, 4)$, and $C(-4, -2)$. Point $D(6, 2)$ lies on $\overrightarrow{AB}$.

45. Graph $\triangle ABC$, point D, and draw $\overline{BD}$.

46. Where should a point E be located so that $\triangle ABC \sim \triangle ADE$?

47. CRITICAL THINKING The altitude $\overline{CD}$ from the right angle C in triangle ABC forms two triangles. Triangle ABC is similar to the two triangles formed, and the two triangles formed are similar to each other. Write three similarity statements about these triangles. Why are the triangles similar to each other?

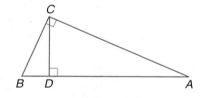

48. WRITING IN MATH   Answer the question that was posed at the beginning of the lesson.

How do engineers use geometry?

Include the following in your answer:
- why engineers use triangles in construction, and
- why you think the pressure applied to the ground from the Eiffel Tower was so small.

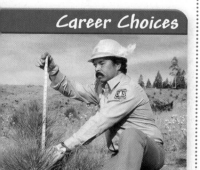
Standardized Test Practice
Ⓐ Ⓑ Ⓒ Ⓓ

49. If $\overline{EB} \parallel \overline{DC}$, find x.

Ⓐ 9.5   Ⓑ 5

Ⓒ 4   Ⓓ 2

50. ALGEBRA Solve $\frac{x+3}{6} = \frac{x}{x-2}$.

Ⓐ 6 or 1   Ⓑ 6 or −1

Ⓒ 3 or 2   Ⓓ −3 or 2

 www.geometryonline.com/self_check_quiz

Mixed Review   Each pair of polygons is similar. Write a similarity statement, find *x*, the measures of the indicated sides, and the scale factor.   *(Lesson 6-2)*

51. $\overline{BC}$, $\overline{PS}$

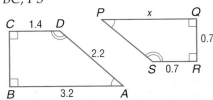

52. $\overline{EF}$, $\overline{XZ}$

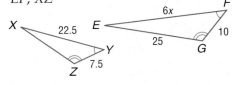

Solve each proportion.   *(Lesson 6-1)*

53. $\dfrac{1}{y} = \dfrac{3}{15}$ **54.** $\dfrac{6}{8} = \dfrac{7}{b}$ **55.** $\dfrac{20}{28} = \dfrac{m}{21}$ **56.** $\dfrac{16}{7} = \dfrac{9}{s}$

57. COORDINATE GEOMETRY $\triangle ABC$ has vertices $A(-3, -9)$, $B(5, 11)$, and $C(9, -1)$. $\overline{AT}$ is a median from A to $\overline{BC}$. Determine whether $\overline{AT}$ is an altitude. *(Lesson 5-1)*

58. ROLLER COASTERS   The sign in front of the Electric Storm roller coaster states *ALL riders must be at least 54 inches tall to ride.* If Adam is 5 feet 8 inches tall, can he ride the Electric Storm? Which law of logic leads you to this conclusion? *(Lesson 2-4)*

Getting Ready for the Next Lesson   **PREREQUISITE SKILL**   Find the coordinates of the midpoint of the segment whose endpoints are given.   *(To review **finding coordinates of midpoints**, see Lesson 1-3.)*

59. $(2, 15)$, $(9, 11)$ **60.** $(-4, 4)$, $(2, -12)$ **61.** $(0, 8)$, $(7, -13)$

Practice Quiz 1   **Lessons 6-1 through 6-3**

Determine whether each pair of figures is similar. Justify your answer.   *(Lesson 6-2)*

1.

2.

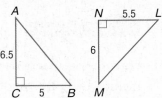

Identify the similar triangles. Find *x* and the measures of the indicated sides.   *(Lesson 6-3)*

3. $\overline{AE}$, $\overline{DE}$

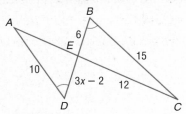

4. $\overline{PT}$, $\overline{ST}$

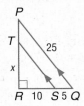

5. MAPS   The scale on a map shows that 1.5 centimeters represents 100 miles. If the distance on the map from Atlanta, Georgia, to Los Angeles, California, is 29.2 centimeters, approximately how many miles apart are the two cities?   *(Lesson 6-1)*

6-4 Parallel Lines and Proportional Parts

What You'll Learn

- Use proportional parts of triangles.
- Divide a segment into parts.

Vocabulary

- midsegment

How do city planners use geometry?

Street maps frequently have parallel and perpendicular lines. In Chicago, because of Lake Michigan, Lake Shore Drive runs at an angle between Oak Street and Ontario Street. City planners need to take this angle into account when determining dimensions of available land along Lake Shore Drive.

PROPORTIONAL PARTS OF TRIANGLES Nonparallel transversals that intersect parallel lines can be extended to form similar triangles. So the sides of the triangles are proportional.

Theorem 6.4

Triangle Proportionality Theorem If a line is parallel to one side of a triangle and intersects the other two sides in two distinct points, then it separates these sides into segments of proportional lengths.

Example: If $\overline{BD} \parallel \overline{AE}$, $\dfrac{BA}{CB} = \dfrac{DE}{CD}$.

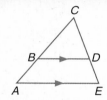

Study Tip

Overlapping Triangles
Trace two copies of $\triangle ACE$. Cut along $\overline{BD}$ to form $\triangle BCD$. Now $\triangle ACE$ and $\triangle BCD$ are no longer overlapping. Place the triangles side-by-side to compare corresponding angles and sides.

Proof Theorem 6.4

Given: $\overline{BD} \parallel \overline{AE}$

Prove: $\dfrac{BA}{CB} = \dfrac{DE}{CD}$

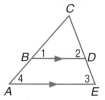

Paragraph Proof:

Since $\overline{BD} \parallel \overline{AE}$, $\angle 4 \cong \angle 1$ and $\angle 3 \cong \angle 2$ because they are corresponding angles. Then, by AA Similarity, $\triangle ACE \sim \triangle BCD$. From the definition of similar polygons, $\dfrac{CA}{CB} = \dfrac{CE}{CD}$. By the Segment Addition Postulate, $CA = BA + CB$ and $CE = DE + CD$. Substituting for CA and CE in the ratio, we get the following proportion.

$$\frac{BA + CB}{CB} = \frac{DE + CD}{CD}$$

$$\frac{BA}{CB} + \frac{CB}{CB} = \frac{DE}{CD} + \frac{CD}{CD} \qquad \text{Rewrite as a sum.}$$

$$\frac{BA}{CB} + 1 = \frac{DE}{CD} + 1 \qquad \frac{CB}{CB} = 1 \text{ and } \frac{CD}{CD} = 1$$

$$\frac{BA}{CB} = \frac{DE}{CD} \qquad \text{Subtract 1 from each side.}$$

Example 1 **Find the Length of a Side**

In $\triangle EFG$, $\overline{HL} \parallel \overline{EF}$, $EH = 9$, $HG = 21$, and $FL = 6$. Find LG.

From the Triangle Proportionality Theorem, $\dfrac{EH}{HG} = \dfrac{FL}{LG}$.

Substitute the known measures.

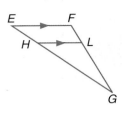

$$\dfrac{9}{21} = \dfrac{6}{LG}$$

$9(LG) = (21)6$ Cross products

$9(LG) = 126$ Multiply.

$LG = 14$ Divide each side by 9.

Study Tip

Using Fractions
You can also rewrite $\dfrac{9}{21}$ as $\dfrac{3}{7}$. Then use your knowledge of fractions to find the missing denominator.

$$\dfrac{3}{7} = \dfrac{6}{?}$$
$\times 2$

The correct denominator is 14.

Proportional parts of a triangle can also be used to prove the converse of Theorem 6.4.

Theorem 6.5

Converse of the Triangle Proportionality Theorem

If a line intersects two sides of a triangle and separates the sides into corresponding segments of proportional lengths, then the line is parallel to the third side.

Example: If $\dfrac{BA}{CB} = \dfrac{DE}{CD}$, then $\overline{BD} \parallel \overline{AE}$.

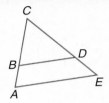

You will prove Theorem 6.5 in Exercise 38.

Example 2 **Determine Parallel Lines**

In $\triangle HKM$, $HM = 15$, $HN = 10$, and $\overline{HJ}$ is twice the length of $\overline{JK}$. Determine whether $\overline{NJ} \parallel \overline{MK}$. Explain.

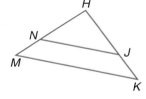

$HM = HN + NM$ Segment Addition Postulate

$15 = 10 + NM$ $HM = 15$, $HN = 10$

$5 = NM$ Subtract 10 from each side.

In order to show $\overline{NJ} \parallel \overline{MK}$, we must show that $\dfrac{HN}{NM} = \dfrac{HJ}{JK}$. $HN = 10$ and $NM = HM - HN$ or 5. So $\dfrac{HN}{NM} = \dfrac{10}{5}$ or 2. Let $JK = x$. Then $HJ = 2x$. So, $\dfrac{HJ}{JK} = \dfrac{2x}{x}$ or 2. Thus, $\dfrac{HN}{NM} = \dfrac{HJ}{JK} = 2$. Since the sides have proportional lengths, $\overline{NJ} \parallel \overline{MK}$.

A **midsegment** of a triangle is a segment whose endpoints are the midpoints of two sides of the triangle.

Theorem 6.6

Triangle Midsegment Theorem A midsegment of a triangle is parallel to one side of the triangle, and its length is one-half the length of that side.

Example: If B and D are midpoints of $\overline{AC}$ and $\overline{EC}$ respectively, $\overline{BD} \parallel \overline{AE}$ and $BD = \dfrac{1}{2}AE$.

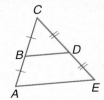

You will prove Theorem 6.6 in Exercise 39.

Example **3** *Midsegment of a Triangle*

Triangle *ABC* has vertices $A(-4, 1)$, $B(8, -1)$, and $C(-2, 9)$. $\overline{DE}$ is a midsegment of $\triangle ABC$.

a. Find the coordinates of *D* and *E*.

Use the Midpoint Formula to find the midpoints of $\overline{AB}$ and $\overline{CB}$.

$$D\left(\frac{-4 + 8}{2}, \frac{1 + (-1)}{2}\right) = D(2, 0)$$

$$E\left(\frac{-2 + 8}{2}, \frac{9 + (-1)}{2}\right) = E(3, 4)$$

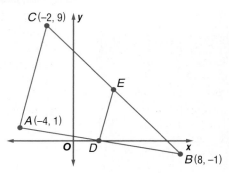

Study Tip

Look Back
To review the **Distance and Midpoint Formulas**, see Lesson 1-3.

b. Verify that $\overline{AC}$ is parallel to $\overline{DE}$.

If the slopes of $\overline{AC}$ and $\overline{DE}$ are equal, $\overline{AC} \parallel \overline{DE}$.

slope of $\overline{AC} = \dfrac{9 - 1}{-2 - (-4)}$ or 4

slope of $\overline{DE} = \dfrac{4 - 0}{3 - 2}$ or 4

Because the slopes of $\overline{AC}$ and $\overline{DE}$ are equal, $\overline{AC} \parallel \overline{DE}$.

c. Verify that $DE = \frac{1}{2}AC$.

First, use the Distance Formula to find *AC* and *DE*.

$$AC = \sqrt{[-2 - (-4)]^2 + (9 - 1)^2} \qquad DE = \sqrt{(3 - 2)^2 + (4 - 0)^2}$$
$$= \sqrt{4 + 64} \qquad\qquad\qquad\quad = \sqrt{1 + 16}$$
$$= \sqrt{68} \qquad\qquad\qquad\qquad\; = \sqrt{17}$$

$$\frac{DE}{AC} = \frac{\sqrt{17}}{\sqrt{68}}$$
$$= \sqrt{\frac{1}{4}} \text{ or } \frac{1}{2}$$

If $\dfrac{DE}{AC} = \dfrac{1}{2}$, then $DE = \dfrac{1}{2}AC$.

DIVIDE SEGMENTS PROPORTIONALLY We have seen that parallel lines cut the sides of a triangle into proportional parts. Three or more parallel lines also separate transversals into proportional parts. If the ratio is 1, they separate the transversals into congruent parts.

Study Tip

Three Parallel Lines
Corollary 6.1 is a special case of Theorem 6.4. In some drawings, the transversals are not shown to intersect. But, if extended, they will intersect and therefore, form triangles with each parallel line and the transversals.

Corollaries

6.1 If three or more parallel lines intersect two transversals, then they cut off the transversals proportionally.

Example: If $\overleftrightarrow{DA} \parallel \overleftrightarrow{EB} \parallel \overleftrightarrow{FC}$, then $\dfrac{AB}{BC} = \dfrac{DE}{EF}$,

$\dfrac{AC}{DF} = \dfrac{BC}{EF}$, and $\dfrac{AC}{BC} = \dfrac{DF}{EF}$.

6.2 If three or more parallel lines cut off congruent segments on one transversal, then they cut off congruent segments on every transversal.
Example: If $\overline{AB} \cong \overline{BC}$, then $\overline{DE} \cong \overline{EF}$.

More About. . .

Example 4 Proportional Segments

• **MAPS** Refer to the map at the beginning of the lesson. The streets from Oak Street to Ontario Street are all parallel to each other. The distance from Oak Street to Ontario along Michigan Avenue is about 3800 feet. The distance between the same two streets along Lake Shore Drive is about 4430 feet. If the distance from Delaware Place to Walton Street along Michigan Avenue is about 411 feet, what is the distance between those streets along Lake Shore Drive?

Make a sketch of the streets in the problem. Notice that the streets form the bottom portion of a triangle that is cut by parallel lines. So you can use the Triangle Proportionality Theorem.

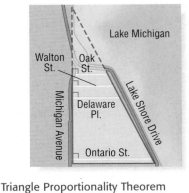

$$\frac{\text{Michigan Ave.}}{\underset{\text{Oak to Ontario}}{\text{Delaware to Walton}}} = \frac{\text{Lake Shore Drive}}{\underset{\text{Oak to Ontario}}{\text{Delaware to Walton}}} \qquad \text{Triangle Proportionality Theorem}$$

$$\frac{411}{3800} = \frac{x}{4430} \qquad \text{Substitution}$$

$$3800 \cdot x = 411(4430) \qquad \text{Cross products}$$

$$3800x = 1{,}820{,}730 \qquad \text{Multiply.}$$

$$x = 479 \qquad \text{Divide each side by 3800.}$$

The distance from Delaware Place to Oak Street along Lake Shore Drive is about 479 feet.

Example 5 Congruent Segments

Find x and y.

To find x:

$$AB = BC \qquad \text{Given}$$

$$3x - 4 = 6 - 2x \qquad \text{Substitution}$$

$$5x - 4 = 6 \qquad \text{Add } 2x \text{ to each side.}$$

$$5x = 10 \qquad \text{Add 4 to each side.}$$

$$x = 2 \qquad \text{Divide each side by 5.}$$

To find y:

$$\overline{DE} \cong \overline{EF} \qquad \text{Parallel lines that cut off congruent segments on one transversal cut off congruent segments on every transversal.}$$

$$DE \cong EF \qquad \text{Definition of congruent segments}$$

$$3y = \frac{5}{3}y + 1 \qquad \text{Substitution}$$

$$9y = 5y + 3 \qquad \text{Multiply each side by 3 to eliminate the denominator.}$$

$$4y = 3 \qquad \text{Subtract } 5y \text{ from each side.}$$

$$y = \frac{3}{4} \qquad \text{Divide each side by 4.}$$

It is possible to separate a segment into two congruent parts by constructing the perpendicular bisector of a segment. However, a segment cannot be separated into three congruent parts by constructing perpendicular bisectors. To do this, you must use parallel lines and the similarity theorems from this lesson.

Construction

Trisect a Segment

1 Draw $\overline{AB}$ to be trisected. Then draw $\overrightarrow{AM}$.

2 With the compass at A, mark off an arc that intersects $\overrightarrow{AM}$ at X. Use the same compass setting to construct $\overline{XY}$ and $\overline{YZ}$ congruent to $\overline{AX}$.

3 Draw $\overline{ZB}$. Then construct lines through Y and X that are parallel to $\overline{ZB}$. Label the intersection points on $\overline{AB}$ as P and Q.

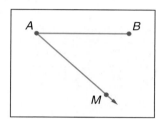

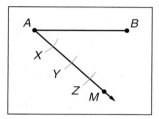

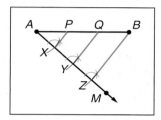

Conclusion: Because parallel lines cut off congruent segments on transversals, $\overline{AP} \cong \overline{PQ} \cong \overline{QB}$.

Check for Understanding

Concept Check
1. **Explain** how you would know if a line that intersects two sides of a triangle is parallel to the third side.

2. **OPEN ENDED** Draw two segments that are intersected by three lines so that the parts are proportional. Then draw a counterexample.

3. **Compare and contrast** Corollary 6.1 and Corollary 6.2.

Guided Practice
For Exercises 4 and 5, refer to $\triangle RST$.

4. If $RL = 5$, $RT = 9$, and $WS = 6$, find RW.

5. If $TR = 8$, $LR = 3$, and $RW = 6$, find WS.

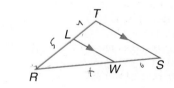

COORDINATE GEOMETRY **For Exercises 6–8, use the following information.**
Triangle ABC has vertices $A(-2, 6)$, $B(-4, 0)$, and $C(10, 0)$. $\overline{DE}$ is a midsegment.

6. Find the coordinates of D and E.

7. Verify that $\overline{DE}$ is parallel to $\overline{BC}$.

8. Verify that $DE = \frac{1}{2}BC$.

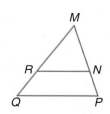

9. In $\triangle MQP$, $MP = 25$, $MN = 9$, $MR = 4.5$, and $MQ = 12.5$. Determine whether $\overline{RN} \parallel \overline{QP}$. Justify your answer.

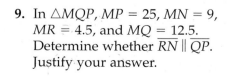

10. In $\triangle ACE$, $ED = 8$, $DC = 20$, $BC = 25$, and $AB = 12$. Determine whether $\overline{DB} \parallel \overline{AE}$.

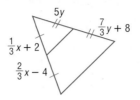

11. Find x and y.

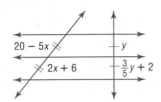

12. Find x and y.

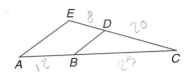

Application **13. MAPS** The distance along Talbot Road from the Triangle Park entrance to the Walkthrough is 880 yards. The distance along Talbot Road from the Walkthrough to Clay Road is 1408 yards. The distance along Woodbury Avenue from the Walkthrough to Clay Road is 1760 yards. If the Walkthrough is parallel to Clay Road, find the distance from the entrance to the Walkthrough along Woodbury.

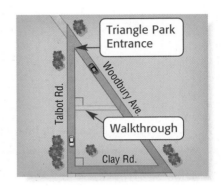

Practice and Apply

Homework Help

For Exercises	See Examples
14–19	1
20–26	2
27, 28	3
35–37, 43	4
33, 34	5

Extra Practice
See page 765.

For Exercises 14 and 15, refer to $\triangle XYZ$.

14. If $XM = 4$, $XN = 6$, and $NZ = 9$, find XY.

15. If $XN = t - 2$, $NZ = t + 1$, $XM = 2$, and $XY = 10$, solve for t.

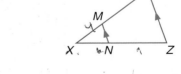

16. If $DB = 24$, $AE = 3$, and $EC = 18$, find AD.

17. Find x and ED if $AE = 3$, $AB = 2$, $BC = 6$, and $ED = 2x - 3$.

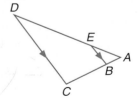

18. Find x, AC, and CD if $AC = x - 3$, $BE = 20$, $AB = 16$, and $CD = x + 5$.

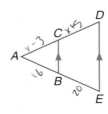

19. Find BC, FE, CD, and DE if $AB = 6$, $AF = 8$, $BC = x$, $CD = y$, $DE = 2y - 3$, and $FE = x + \dfrac{10}{3}$.

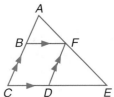

Find x so that $\overline{GJ} \parallel \overline{FK}$.

20. $GF = 12$, $HG = 6$, $HJ = 8$, $JK = x - 4$

21. $HJ = x - 5$, $JK = 15$, $FG = 18$, $HG = x - 4$

22. $GH = x + 3.5$, $HJ = x - 8.5$, $FH = 21$, $HK = 7$

Determine whether $\overline{QT} \parallel \overline{RS}$. Justify your answer.

23. $PR = 30$, $PQ = 9$, $PT = 12$, and $PS = 18$

24. $QR = 22$, $RP = 65$, and SP is 3 times TS.

25. $TS = 8.6$, $PS = 12.9$, and PQ is half RQ.

26. $PQ = 34.88$, $RQ = 18.32$, $PS = 33.25$, and $TS = 11.45$

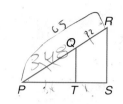

27. Find the length of $\overline{BC}$ if $\overline{BC} \parallel \overline{DE}$ and $\overline{DE}$ is a midsegment of $\triangle ABC$.

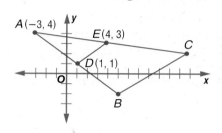

28. Show that $\overline{WM} \parallel \overline{TS}$ and determine whether $\overline{WM}$ is a midsegment.

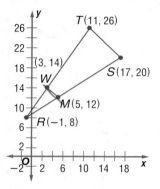

COORDINATE GEOMETRY For Exercises 29 and 30, use the following information.

Triangle ABC has vertices $A(-1, 6)$, $B(-4, -3)$, and $C(7, -5)$. $\overline{DE}$ is a midsegment.

29. Verify that $\overline{DE}$ is parallel to $\overline{AB}$.

30. Verify that $DE = \frac{1}{2}AB$.

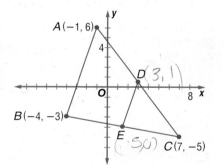

31. **COORDINATE GEOMETRY** Given $A(2, 12)$ and $B(5, 0)$, find the coordinates of P such that P separates $\overrightarrow{AB}$ into two parts with a ratio of 2 to 1.

32. **COORDINATE GEOMETRY** In $\triangle LMN$, $\overline{PR}$ divides $\overline{NL}$ and $\overline{MN}$ proportionally. If the vertices are $N(8, 20)$, $P(11, 16)$, and $R(3, 8)$ and $\frac{LP}{PN} = \frac{2}{1}$, find the coordinates of L and M.

ALGEBRA Find x and y.

33.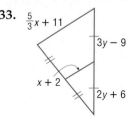

34.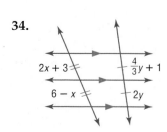

CONSTRUCTION For Exercises 35–37, use the following information and drawing.

Two poles, 30 feet and 50 feet tall, are 40 feet apart and perpendicular to the ground. The poles are supported by wires attached from the top of each pole to the bottom of the other, as in the figure. A coupling is placed at *C* where the two wires cross.

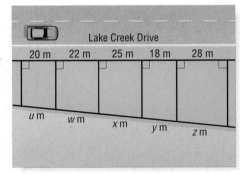

35. Find *x*, the distance from *C* to the taller pole.

36. How high above the ground is the coupling?

37. How far down the wire from the smaller pole is the coupling?

PROOF Write a two-column proof of each theorem.

38. Theorem 6.5

39. Theorem 6.6

CONSTRUCTION Construct each segment as directed.

40. a segment 8 centimeters long, separated into three congruent segments

41. a segment separated into four congruent segments

42. a segment separated into two segments in which their lengths have a ratio of 1 to 4

43. REAL ESTATE In Lake Creek, the lots on which houses are to be built are laid out as shown. What is the lake frontage for each of the five lots if the total frontage is 135.6 meters?

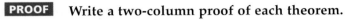

44. CRITICAL THINKING Copy the figure that accompanies Corollary 6.1 on page 309. Draw $\overline{DC}$. Let *G* be the intersection point of $\overline{DC}$ and $\overline{BE}$. Using that segment, explain how you could prove $\dfrac{AB}{BC} = \dfrac{DE}{EF}$.

45. **WRITING IN MATH** Answer the question that was posed at the beginning of the lesson.

How do city planners use geometry?

Include the following in your answer:
- why maps are important to city planners, and
- what geometry facts a city planner needs to know to explain why the block between Chestnut and Pearson is longer on Lake Shore Drive than on Michigan Avenue.

46. Find *x*.

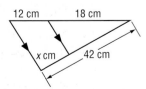

Ⓐ 16

Ⓑ 16.8

Ⓒ 24

Ⓓ 28.4

47. GRID IN The average of *a* and *b* is 18, and the ratio of *a* to *b* is 5 to 4. What is the value of $a - b$?

48. MIDPOINTS IN POLYGONS Draw any quadrilateral *ABCD* on a coordinate plane. Points *E, F, G,* and *H* are midpoints of $\overline{AB}$, $\overline{BC}$, $\overline{CD}$, and $\overline{DA}$, respectively.

   a. Connect the midpoints to form quadrilateral *EFGH*. Describe what you know about the sides of quadrilateral *EFGH*.

   b. Will the same reasoning work with five-sided polygons? Explain why or why not.

Maintain Your Skills

Mixed Review **Determine whether each pair of triangles is similar. Justify your answer.** *(Lesson 6-3)*

49.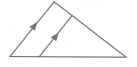

50.

51.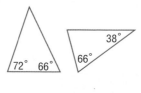

Each pair of polygons is similar. Find *x* and *y*. *(Lesson 6-2)*

52.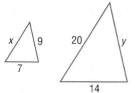

53.

Determine the relationship between the measures of the given angles. *(Lesson 5-2)*

54. ∠*ADB*, ∠*ABD*

55. ∠*ABD*, ∠*BAD*

56. ∠*BCD*, ∠*CDB*

57. ∠*CBD*, ∠*BCD*

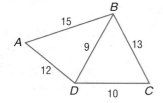

ARCHITECTURE **For Exercises 58 and 59, use the following information.**
The geodesic dome was developed by Buckminster Fuller in the 1940s as an energy-efficient building. The figure at the right shows the basic structure of one geodesic dome. *(Lesson 4-1)*

58. How many equilateral triangles are in the figure?

59. How many obtuse triangles are in the figure?

Determine the truth value of the following statement for each set of conditions.
If you have a fever, then you are sick. *(Lesson 2-3)*

60. You do not have a fever, and you are sick.

61. You have a fever, and you are not sick.

62. You do not have a fever, and you are not sick.

63. You have a fever, and you are sick.

Getting Ready for the Next Lesson **PREREQUISITE SKILL** **Write all the pairs of corresponding parts for each pair of congruent triangles.** *(To review **corresponding congruent parts**, see Lesson 4-3.)*

64. △*ABC* ≅ △*DEF*   **65.** △*RST* ≅ △*XYZ*   **66.** △*PQR* ≅ △*KLM*

6-5 Parts of Similar Triangles

What You'll Learn

- Recognize and use proportional relationships of corresponding perimeters of similar triangles.
- Recognize and use proportional relationships of corresponding angle bisectors, altitudes, and medians of similar triangles.

How is geometry related to photography?

- The camera lens was 6.16 meters from this Dale Chihuly glass sculpture when this photograph was taken. The image on the film is 35 millimeters tall. Similar triangles enable us to find the height of the actual sculpture.

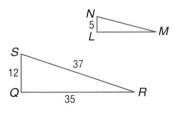
PERIMETERS Triangle ABC is similar to $\triangle DEF$ with a scale factor of $1:3$. You can use variables and the scale factor to compare their perimeters. Let the measures of the sides of $\triangle ABC$ be a, b, and c. The measures of the corresponding sides of $\triangle DEF$ would be $3a$, $3b$, and $3c$.

$$\frac{\text{perimeter of } \triangle ABC}{\text{perimeter of } \triangle DEF} = \frac{a + b + c}{3a + 3b + 3c}$$

$$= \frac{1(a + b + c)}{3(a + b + c)} \text{ or } \frac{1}{3}$$

The perimeters are in the same proportion as the side measures of the two similar figures. This suggests Theorem 6.7, the Proportional Perimeters Theorem.

Theorem 6.7

Proportional Perimeters Theorem If two triangles are similar, then the perimeters are proportional to the measures of corresponding sides.

You will prove Theorem 6.7 in Exercise 8.

Example 1 Perimeters of Similar Triangles

If $\triangle LMN \sim \triangle QRS$, $QR = 35$, $RS = 37$, $SQ = 12$, and $NL = 5$, find the perimeter of $\triangle LMN$.

Let x represent the perimeter of $\triangle LMN$. The perimeter of $\triangle QRS = 35 + 37 + 12$ or 84.

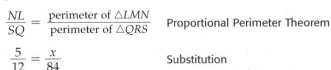

$\dfrac{NL}{SQ} = \dfrac{\text{perimeter of } \triangle LMN}{\text{perimeter of } \triangle QRS}$ Proportional Perimeter Theorem

$\dfrac{5}{12} = \dfrac{x}{84}$ Substitution

$12x = 420$ Cross products

$x = 35$ Divide each side by 12.

The perimeter of $\triangle LMN$ is 35 units.

SPECIAL SEGMENTS OF SIMILAR TRIANGLES Think about a triangle drawn on a piece of paper being placed on a copy machine and either enlarged or reduced. The copy is similar to the original triangle. Now suppose you drew in special segments of a triangle, such as the altitudes, medians, or angle bisectors, on the original. When you enlarge or reduce that original triangle, all of those segments are enlarged or reduced at the same rate. This conjecture is formally stated in Theorems 6.8, 6.9, and 6.10.

Theorems	*Special Segments of Similar Triangles*

6.8 If two triangles are similar, then the measures of the corresponding altitudes are proportional to the measures of the corresponding sides.

Abbreviation: *~ △s have corr. altitudes proportional to the corr. sides.*

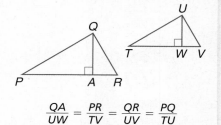

$$\frac{QA}{UW} = \frac{PR}{TV} = \frac{QR}{UV} = \frac{PQ}{TU}$$

6.9 If two triangles are similar, then the measures of the corresponding angle bisectors of the triangles are proportional to the measures of the corresponding sides.

Abbreviation: *~ △s have corr. ∠ bisectors proportional to the corr. sides.*

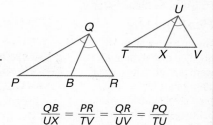

$$\frac{QB}{UX} = \frac{PR}{TV} = \frac{QR}{UV} = \frac{PQ}{TU}$$

6.10 If two triangles are similar, then the measures of the corresponding medians are proportional to the measures of the corresponding sides.

Abbreviation: *~ △s have corr. medians proportional to the corr. sides.*

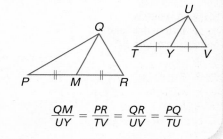

$$\frac{QM}{UY} = \frac{PR}{TV} = \frac{QR}{UV} = \frac{PQ}{TU}$$

You will prove Theorems 6.8 and 6.10 in Exercises 30 and 31, respectively.

Example 2 Write a Proof

Write a paragraph proof of Theorem 6.9.

Since the corresponding angles to be bisected are chosen at random, we need not prove this for every pair of bisectors.

Given: $\triangle RTS \sim \triangle EGF$
$\overline{TA}$ and $\overline{GB}$ are angle bisectors.

Prove: $\dfrac{TA}{GB} = \dfrac{RT}{EG}$

Paragraph Proof: Because corresponding angles of similar triangles are congruent, $\angle R \cong \angle E$ and $\angle RTS \cong \angle EGF$. Since $\angle RTS$ and $\angle EGF$ are bisected, we know that $\frac{1}{2}m\angle RTS = \frac{1}{2}m\angle EGF$ or $m\angle RTA = m\angle EGB$. This makes $\angle RTF \cong \angle EGB$ and $\triangle RTF \sim \triangle EGB$ by AA Similarity. Thus, $\frac{TA}{GB} = \frac{RT}{EG}$.

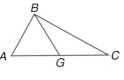

Example 3 *Medians of Similar Triangles*

In the figure, $\triangle ABC \sim \triangle DEF$. $\overline{BG}$ is a median of $\triangle ABC$, and $\overline{EH}$ is a median of $\triangle DEF$. Find *EH* if $BC = 30$, $BG = 15$, and $EF = 15$.

Let x represent *EH*.

$\dfrac{BG}{EH} = \dfrac{BC}{EF}$ Write a proportion.

$\dfrac{15}{x} = \dfrac{30}{15}$ $BG = 15$, $EH = x$, $BC = 30$, and $EF = 15$

$30x = 225$ Cross products

$x = 7.5$ Divide each side by 30.

Thus, $EH = 7.5$.

The theorems about the relationships of special segments in similar triangles can be used to solve real-life problems.

Example 4 *Solve Problems with Similar Triangles*

•· **PHOTOGRAPHY** Refer to the application at the beginning of the lesson. The drawing below illustrates the position of the camera and the distance from the lens of the camera to the film. Find the height of the sculpture.

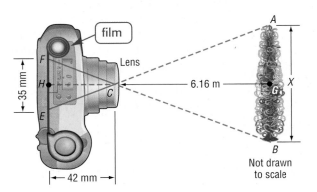

Not drawn to scale

$\triangle ABC$ and $\triangle EFC$ are similar. The distance from the lens to the film in the camera is $CH = 42$ mm. $\overline{CG}$ and $\overline{CH}$ are altitudes of $\triangle ABC$ and $\triangle EFC$, respectively. If two triangles are similar, then the measures of the corresponding altitudes are proportional to the measures of the corresponding sides. This leads to the proportion $\dfrac{AB}{EF} = \dfrac{GC}{HC}$.

$\dfrac{AB}{EF} = \dfrac{GC}{HC}$ Write the proportion.

$\dfrac{x \text{ m}}{35 \text{ mm}} = \dfrac{6.16 \text{ m}}{42 \text{ mm}}$ $AB = x$ m, $EF = 35$ m, $GC = 6.16$ m, $HC = 42$ mm

$x \cdot 42 = 35(6.16)$ Cross products

$42x = 215.6$ Simplify.

$x \approx 5.13$ Divide each side by 42.

The sculpture is about 5.13 meters tall.

An angle bisector also divides the side of the triangle opposite the angle proportionally.

> ### Theorem 6.11
>
> **Angle Bisector Theorem** An angle bisector in a triangle separates the opposite side into segments that have the same ratio as the other two sides.
>
> **Example:** $\dfrac{AD}{DB} = \dfrac{AC}{BC}$ ← segments with vertex A
> ← segments with vertex B
>
>

You will prove this theorem in Exercise 32.

Check for Understanding

Concept Check **1. Explain** what must be true about $\triangle ABC$ and $\triangle MNQ$ before you can conclude that $\dfrac{AD}{MR} = \dfrac{BA}{NM}$.

2. OPEN ENDED The perimeter of one triangle is 24 centimeters, and the perimeter of a second triangle is 36 centimeters. If the length of one side of the smaller triangle is 6, find possible lengths of the other sides of the triangles so that they are similar.

Guided Practice **Find the perimeter of the given triangle.**

3. $\triangle DEF$, if $\triangle ABC \sim \triangle DEF$, $AB = 5$, $BC = 6$, $AC = 7$, and $DE = 3$

4. $\triangle WZX$, if $\triangle WZX \sim \triangle SRT$, $ST = 6$, $WX = 5$, and the perimeter of $\triangle SRT = 15$

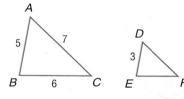

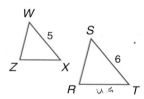

Find x.

5.

6.

7.

8. PROOF Write a paragraph proof of Theorem 6.7.

Given: $\triangle ABC \sim \triangle DEF$

$\dfrac{AB}{DE} = \dfrac{m}{n}$

Prove: $\dfrac{\text{perimeter of } \triangle ABC}{\text{perimeter of } \triangle DEF} = \dfrac{m}{n}$

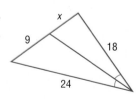

Application **9. PHOTOGRAPHY** The distance from the film to the lens in a camera is 10 centimeters. The film image is 5 centimeters high. Tamika is 165 centimeters tall. How far should she be from the camera in order for the photographer to take a full-length picture?

Homework Help

For Exercises	See Examples
10–15	1
16, 17, 28	4
18–27	3
30–37	2

Extra Practice
See page 766.

Find the perimeter of the given triangle.

10. △BCD, if △BCD ~ △FDE, CD = 12, FD = 5, FE = 4, and DE = 8

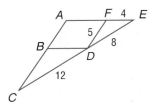

11. △ADF, if △ADF ~ △BCE, BC = 24, EB = 12, CE = 18, and DF = 21

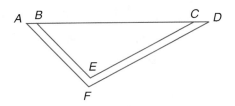

12. △CBH, if △CBH ~ △FEH, ADEG is a parallelogram, CH = 7, FH = 10, FE = 11, and EH = 6

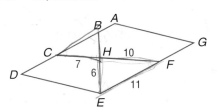

13. △DEF, if △DEF ~ △CBF, perimeter of △CBF = 27, DF = 6, and FC = 8

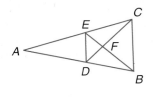

14. △ABC, if △ABC ~ △CBD, CD = 4, DB = 3, and CB = 5

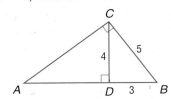

15. △ABC, if △ABC ~ △CBD, AD = 5, CD = 12, and BC = 31.2

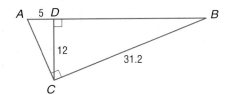

16. DESIGN Rosario wants to enlarge the dimensions of an 18-centimeter by 24-centimeter picture by 30%. She plans to line the inside edge of the frame with blue cord. The store only had 110 centimeters of blue cord in stock. Will this be enough to fit on the inside edge of the frame? Explain.

17. PHYSICAL FITNESS A park has two similar triangular jogging paths as shown. The dimensions of the inner path are 300 meters, 350 meters, and 550 meters. The shortest side of the outer path is 600 meters. Will a jogger on the inner path run half as far as one on the outer path? Explain.

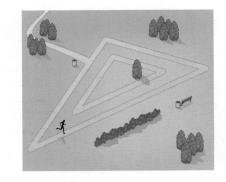

18. Find EG if △ACB ~ △EGF, $\overline{AD}$ is an altitude of △ACB, $\overline{EH}$ is an altitude of △EGF, AC = 17, AD = 15, and EH = 7.5.

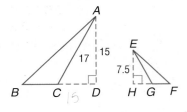

19. Find EH if $\triangle ABC \sim \triangle DEF$, $\overline{BG}$ is an altitude of $\triangle ABC$, $\overline{EH}$ is an altitude of $\triangle DEF$, $BG = 3$, $BF = 4$, $FC = 2$, and $CE = 1$.

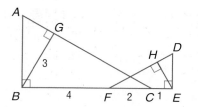

20. Find FB if $\overline{SA}$ and $\overline{FB}$ are altitudes and $\triangle RST \sim \triangle EFG$.

21. Find DC if $\overline{DG}$ and $\overline{JM}$ are altitudes and $\triangle KJL \sim \triangle EDC$.

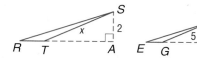
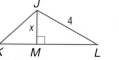

Find x.

22.

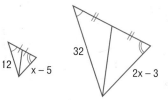

23.

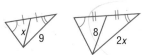

24.

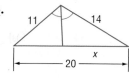

25.

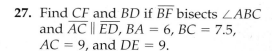

26. Find UB if $\triangle RST \sim \triangle UVW$, $\overline{TA}$ and $\overline{WB}$ are medians, $TA = 8$, $RA = 3$, $WB = 3x - 6$, and $UB = x + 2$.

27. Find CF and BD if $\overline{BF}$ bisects $\angle ABC$ and $\overline{AC} \parallel \overline{ED}$, $BA = 6$, $BC = 7.5$, $AC = 9$, and $DE = 9$.

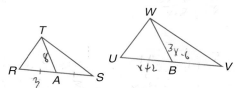
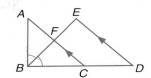

28. PHOTOGRAPHY One of the first cameras invented was called a *camera obscura*. Light entered an opening in the front, and an image was reflected in the back of the camera, upside down, forming similar triangles. If the image of the person on the back of the camera is 12 inches, the distance from the opening to the person is 7 feet, and the camera itself is 15 inches long, how tall is the person being photographed?

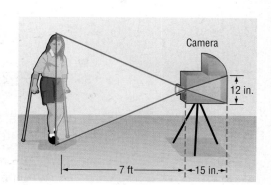

29. CRITICAL THINKING $\overline{CD}$ is an altitude to the hypotenuse $\overline{AB}$. Make a conjecture about x, y, and z. Justify your reasoning.

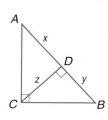

PROOF   Write the indicated type of proof.

30. a paragraph proof of Theorem 6.8

31. a two-column proof of Theorem 6.10

32. a two-column proof of the Angle Bisector Theorem (Theorem 6.11)

Given: $\overline{CD}$ bisects $\angle ACB$

By construction, $\overline{AE} \parallel \overline{CD}$.

Prove: $\dfrac{AD}{AC} = \dfrac{BD}{BC}$

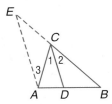

33. a paragraph proof

Given: $\triangle ABC \sim \triangle PQR$

$\overline{BD}$ is an altitude of $\triangle ABC$.

$\overline{QS}$ is an altitude of $\triangle PQR$.

Prove: $\dfrac{QP}{BA} = \dfrac{QS}{BD}$

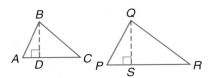

34. a flow proof

Given: $\angle C \cong \angle BDA$

Prove: $\dfrac{AC}{DA} = \dfrac{AD}{BA}$

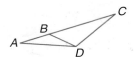

35. a two-column proof

Given: $\overline{JF}$ bisects $\angle EFG$.

$\overline{EH} \parallel \overline{FG}$, $\overline{EF} \parallel \overline{HG}$

Prove: $\dfrac{EK}{KF} = \dfrac{GJ}{JF}$

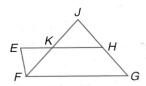

36. a two-column proof

Given: $\overline{RU}$ bisects $\angle SRT$;

$\overline{VU} \parallel \overline{RT}$.

Prove: $\dfrac{SV}{VR} = \dfrac{SR}{RT}$

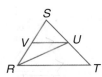

37. a flow proof

Given: $\triangle RST \sim \triangle ABC$; W and D are midpoints of $\overline{TS}$ and $\overline{CB}$.

Prove: $\triangle RWS \sim \triangle ADB$

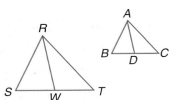

38. **WRITING IN MATH**   Answer the question that was posed at the beginning of the lesson.

How is geometry related to photography?

Include the following in your answer:

- a sketch of how a camera works showing the image and the film, and
- why the two isosceles triangles are similar.

39. GRID IN   Triangle ABC is similar to $\triangle DEF$. If $AC = 10.5$, $AB = 6.5$, and $DE = 8$, find DF.

40. ALGEBRA   The sum of three numbers is 180. Two of the numbers are the same, and each of them is one-third of the greatest number. What is the least number?

   Ⓐ 30   Ⓑ 36   Ⓒ 45   Ⓓ 72

Mixed Review

Determine whether $\overline{MN} \parallel \overline{OP}$. Justify your answer. *(Lesson 6-4)*

41. $LM = 7, LN = 9, LO = 14, LP = 16$

42. $LM = 6, MN = 4, LO = 9, OP = 6$

43. $LN = 12, NP = 4, LM = 15, MO = 5$

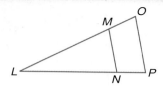

Identify the similar triangles. Find x and the measure(s) of the indicated side(s). *(Lesson 6-3)*

44. VW and WX

45. PQ

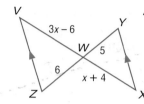

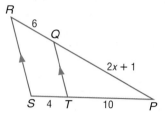

Write an equation in slope-intercept form for the line that satisfies the given conditions. *(Lesson 3-4)*

46. x-intercept is 3, y-intercept is -3 **47.** $m = 2$, contains $(-1, -1)$

Getting Ready for the Next Lesson

PREREQUISITE SKILL Name the next two numbers in each pattern.
*(To review **patterns**, see Lesson 2-1.)*

48. 5, 12, 19, 26, 33, …   **49.** 10, 20, 40, 80, 160, …   **50.** 0, 5, 4, 9, 8, 13, …

Practice Quiz 2

Refer to $\triangle ABC$. *(Lesson 6-4)*

1. If $AD = 8$, $AE = 12$, and $EC = 18$, find AB.

2. If $AE = m - 2$, $EC = m + 4$, $AD = 4$, and $AB = 20$, find m.

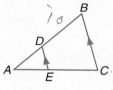

Determine whether $\overline{YZ} \parallel \overline{VW}$. Justify your answer. *(Lesson 6-4)*

3. $XY = 30$, $XV = 9$, $XW = 12$, and $XZ = 18$

4. $XV = 34.88$, $VY = 18.32$, $XZ = 33.25$, and $WZ = 11.45$

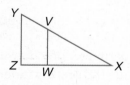

Find the perimeter of the given triangle. *(Lesson 6-5)*

5. $\triangle DEF$ if $\triangle DEF \sim \triangle GFH$

6. $\triangle RUW$ if $\triangle RUW \sim \triangle STV$, $ST = 24$, $VS = 12$, $VT = 18$, and $UW = 21$

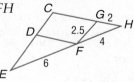

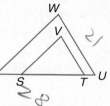

Find x. *(Lesson 6-5)*

7.

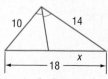

8.

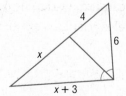

9.

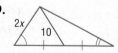

10. LANDSCAPING Paulo is designing two gardens shaped like similar triangles. One garden has a perimeter of 53.5 feet, and the longest side is 25 feet. He wants the second garden to have a perimeter of 32.1 feet. Find the length of the longest side of this garden. *(Lesson 6-5)*

Sierpinski Triangle

Collect Data

Stage 0 On isometric dot paper, draw an equilateral triangle in which each side is 16 units long.

Stage 1 Connect the midpoints of each side to form another triangle. Shade the center triangle.

Stage 2 Repeat the process using the three nonshaded triangles. Connect the midpoints of each side to form other triangles.

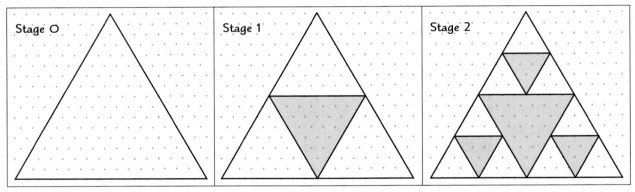

If you repeat this process indefinitely, the pattern that results is called the **Sierpinski Triangle.** Since this figure is created by repeating the same procedure over and over again, it is an example of a geometric shape called a *fractal*.

Analyze the Data

1. Continue the process through Stage 4. How many nonshaded triangles do you have at Stage 4?
2. What is the perimeter of a nonshaded triangle in Stage 0 through Stage 4?
3. If you continue the process indefinitely, describe what will happen to the perimeter of each nonshaded triangle.
4. Study △DFM in Stage 2 of the Sierpenski Triangle shown at the right. Is this an equilateral triangle? Are △BCE, △GHL, or △IJN equilateral?
5. Is △BCE ~ △DFM? Explain your answer.
6. How many Stage 1 Sierpinski triangles are there in Stage 2?

Make a Conjecture

7. How can three copies of a Stage 2 triangle be combined to form a Stage 3 triangle?

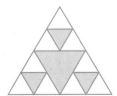 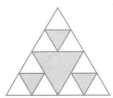 

8. Combine three copies of the Stage 4 Sierpinski triangle. Which stage of the Sierpinski Triangle is this?
9. How many copies of the Stage 4 triangle would you need to make a Stage 6 triangle?

6-6 Fractals and Self-Similarity

What You'll Learn

- Recognize and describe characteristics of fractals.
- Identify nongeometric iteration.

How is mathematics found in nature?

Patterns can be found in many objects in nature, including broccoli. If you take a piece of broccoli off the stalk, the small piece resembles the whole. This pattern of repeated shapes at different scales is part of fractal geometry.

CHARACTERISTICS OF FRACTALS Benoit Mandelbrot, a mathematician, coined the term *fractal* to describe things in nature that are irregular in shape, such as clouds, coastlines, or the growth of a tree. The patterns found in nature are analyzed and then recreated on a computer, where they can be studied more closely. These patterns are created using a process called **iteration**. Iteration is a process of repeating the same procedure over and over again. A **fractal** is a geometric figure that is created using iteration. The pattern's structure appears to go on infinitely.

WebQuest

By creating a Sierpinski Triangle, you can find a pattern in the area and perimeter of this well-known fractal. Visit www.geometry online.com/webquest to continue work on your WebQuest project.

Compare the pictures of a human circulatory system and the mouth of the Ganges in Bangladesh. Notice how the branches of the tributaries have the same pattern as the branching of the blood vessels.

One characteristic of fractals is that they are **self similar**. That is, the smaller and smaller details of a shape have the same geometric characteristics as the original form.

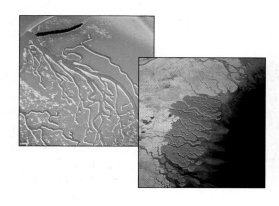

The Sierpinski Triangle is a fractal that is self-similar. Stage 1 is formed by drawing the midsegments of an equilateral triangle and shading in the triangle formed by them. Stage 2 repeats the process in the unshaded triangles. This process can continue indefinitely with each part still being similar to the original.

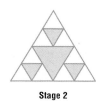

| Stage 0 | Stage 1 | Stage 2 | Stage 4 |

The Sierpinski Triangle is said to be *strictly self-similar*, which means that any of its parts, no matter where they are located or what size is selected, contain a figure that is similar to the whole.

Example 1 *Self-Similarity*

Prove that a triangle formed in Stage 2 of a Sierpinski triangle is similar to the triangle in Stage 0.

The argument will be the same for any triangle in Stage 2, so we will use only △CGJ from Stage 2.

Given: △ABC is equilateral.
D, E, F, G, J, and H are midpoints of $\overline{AB}$, $\overline{BC}$, $\overline{CA}$, $\overline{FC}$, $\overline{CE}$, and $\overline{FE}$, respectively.

Prove: △CGJ ~ △CAB

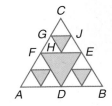

Study Tip

Look Back
To review **midsegment**, see Lesson 6-4.

Statements	Reasons
1. △ABC is equilateral; D, E, F are midpoints of $\overline{AB}$, $\overline{BC}$, $\overline{CA}$; G, J, and H are midpoints of $\overline{FC}$, $\overline{CE}$, $\overline{FE}$.	1. Given
2. $\overline{FE}$ is a midsegment of △CAB; $\overline{GJ}$ is a midsegment of △CFE.	2. Definition of a Triangle Midsegment
3. $\overline{FE} \parallel \overline{AB}$; $\overline{GJ} \parallel \overline{FE}$	3. Triangle Midsegment Theorem
4. $\overline{GJ} \parallel \overline{AB}$	4. Two segments parallel to the same segment are parallel.
5. ∠CGJ ≅ ∠CAB	5. Corresponding ∠ Postulate
6. ∠C ≅ ∠C	6. Reflexive Property
7. △CGJ ~ △CAB	7. AA Similarity

Thus, using the same reasoning, every triangle in Stage 2 is similar to the original triangle in Stage 0.

You can generate many other fractal images using an iterative process.

Example 2 *Create a Fractal*

Draw a segment and trisect it. Create a fractal by replacing the middle third of the segment with two segments of the same length as the removed segment.

After the first geometric iteration, repeat the process on each of the four segments in Stage 1. Continue to repeat the process.

This fractal image is called a *Koch curve*.

Study Tip

Common Misconceptions
Not all repeated patterns are self-similar. The Koch curve is one such example when applied to a triangle.

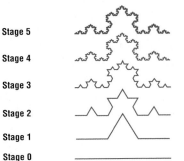

Stage 5
Stage 4
Stage 3
Stage 2
Stage 1
Stage 0

If the first stage is an equilateral triangle, instead of a segment, this iteration will produce a fractal called *Koch's snowflake*.

NONGEOMETRIC ITERATION An iterative process does not always include manipulation of geometric shapes. Iterative processes can be translated into formulas or algebraic equations. These are called *recursive formulas*.

Example 3 *Evaluate a Recursive Formula*

Find the value of x^2, where x initially equals 2. Then use that value as the next x in the expression. Repeat the process four times and describe your observations.

The iterative process is to square the value repeatedly. Begin with $x = 2$. The value of x^2 becomes the next value for x.

x	2	4	16	256	65,536
x^2	4	16	256	65,536	4,294,967,296

The values grow greater with each iteration, approaching infinity.

Example 4 *Find a Recursive Formula*

PASCAL'S TRIANGLE *Pascal's Triangle* **is a numerical pattern in which each row begins and ends with 1 and all other terms in the row are the sum of the two numbers above it.**

a. **Find a formula in terms of the row number for the sum of the values in any row in the Pascal's triangle.**

   To find the sum of the values in the tenth row, we can investigate a simpler problem. What is the sum of values in the first four rows of the triangle?

Row	Pascal's Triangle	Sum	Pattern
1	1	1	$2^0 = 2^{1-1}$
2	1 1	2	$2^1 = 2^{2-1}$
3	1 2 1	4	$2^2 = 2^{3-1}$
4	1 3 3 1	8	$2^3 = 2^{4-1}$
5	1 4 6 4 1	16	$2^4 = 2^{5-1}$

   It appears that the sum of any row is a power of 2. The formula is 2 to a power that is one less than the row number: $A_n = 2^{n-1}$.

b. **What is the sum of the values in the tenth row of Pascal's triangle?**

   The sum of the values in the tenth row will be 2^{10-1} or 512.

Example 5 *Solve a Problem Using Iteration*

BANKING Felisa has $2500 in a money market account that earns 3.2% interest. If the interest is compounded annually, find the balance of her account after 3 years.

First, write an equation to find the balance after one year.

current balance + (current balance × interest rate) = new balance

$$2500 + (2500 \cdot 0.032) = 2580$$
$$2580 + (2580 \cdot 0.032) = 2662.56$$
$$2662.56 + (2662.56 \cdot 0.032) = 2747.76$$

After 3 years, Felisa will have $2747.76 in her account.

Concept Check

1. **Describe** a *fractal* in your own words. Include characteristics of fractals in your answer.

2. **Explain** why computers provide an efficient way to generate fractals.

3. **OPEN ENDED** Find an example of fractal geometry in nature, excluding those mentioned in the lesson.

Guided Practice

For Exercises 4–6, use the following information.
A *fractal tree* can be drawn by making two new branches from the endpoint of each original branch, each one-third as long as the previous branch.

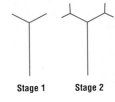

Stage 1   Stage 2

4. Draw Stages 3 and 4 of a fractal tree. How many total branches do you have in Stages 1 through 4? (Do not count the stems.)

5. Find a pattern to predict the number of branches at each stage.

6. Is a fractal tree strictly self-similar? Explain.

For Exercises 7–9, use a calculator.

7. Find the square root of 2. Then find the square root of the result.

8. Find the square root of the result in Exercise 7. What would be the result after 100 repeats of taking the square root?

9. Determine whether this is an iterative process. Explain.

Application

10. **BANKING** Jamir has $4000 in a savings account. The annual percent interest rate is 1.1%. Find the amount of money Jamir will have after the interest is compounded four times.

Practice and Apply

Homework Help

For Exercises	See Examples
11–13, 21–23	1, 2
14–20, 25, 28	4
24–29	2
30–37	3
38–40	5

Extra Practice
See page 766.

For Exercises 11–13, Stage 1 of a fractal is drawn on grid paper so that each side of the large square is 27 units long. The trisection points of the sides are connected to make 9 smaller squares with the middle square shaded. The shaded square is known as a *hole*.

11. Copy Stage 1 on your paper. Then draw Stage 2 by repeating the Stage 1 process in each of the outer eight squares. How many holes are in this stage?

12. Draw Stage 3 by repeating the Stage 1 process in each unshaded square of Stage 2. How many holes are in Stage 3?

13. If you continue the process indefinitely, will the figure you obtain be strictly self-similar? Explain.

14. Count the number of dots in each arrangement. These numbers are called *triangular numbers*. The second triangular number is 3 because there are three dots in the array. How many dots will be in the seventh triangular number?

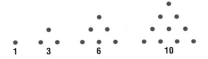

1   3   6   10

For Exercises 15–20, refer to Pascal's triangle on page 327. Look at the third diagonal from either side, starting at the top of the triangle.

15. Describe the pattern.

16. Explain how Pascal's triangle relates to the triangular numbers.

17. Generate eight rows of Pascal's triangle. Replace each of the even numbers with a 0 and each of the odd numbers with a 1. Color each 1 and leave the 0s uncolored. Describe the picture.

18. Generate eight rows of Pascal's triangle. Divide each entry by 3. If the remainder is 1 or 2, shade the number cell black. If the remainder is 0, leave the cell unshaded. Describe the pattern that emerges.

19. Find the sum of the first 25 numbers in the outside diagonal of Pascal's triangle.

20. Find the sum of the first 50 numbers in the second diagonal.

The three shaded interior triangles shown were made by trisecting the three sides of an equilateral triangle and connecting the points.

21. Prove that one of the nonshaded triangles is similar to the original triangle.

22. Repeat the iteration once more.

23. Is the new figure strictly self-similar?

24. How many nonshaded triangles are in Stages 1 and 2?

Refer to the Koch Curve on page 326.

25. What is a formula for the number of segments in terms of the stage number? Use your formula to predict the number of segments in Stage 8 of a Koch curve.

26. If the length of the original segment is 1 unit, how long will the segments be in each of the first four stages? What will happen to the length of each segment as the number of stages continues to increase?

Refer to the Koch Snowflake on page 326. At Stage 1, the length of each side is 1 unit.

27. What is the perimeter at each of the first four stages of a Koch snowflake?

28. What is a formula for the perimeter in terms of the stage number? Describe the perimeter as the number of stages continues to increase.

29. Write a paragraph proof to show that the triangles generated on the sides of a Koch Snowflake in Stage 1, are similar to the original triangle.

Find the value of each expression. Then use that value as the next x in the expression. Repeat the process four times, and describe your observations.

30. $\sqrt{x}$, where x initially equals 12

31. $\frac{1}{x}$, where x initially equals 5

32. $x^{\frac{1}{3}}$, where x initially equals 0.3

33. 2^x, where x initially equals 0

Find the first three iterates of each expression.

34. $2x + 1$, x initially equals 1

35. $x - 5$, where x initially equals 5

36. $x^2 - 1$, x initially equals 2

37. $3(2 - x)$, where x initially equals 4

38. **BANKING** Raini has a credit card balance of $1250 and a monthly interest rate of 1.5%. If he makes payments of $100 each month, what will the balance be after 3 months?

WEATHER For Exercises 39 and 40, use the following information.
There are so many factors that affect the weather that it is difficult for meteorologists to make accurate long term predictions. Edward N. Lorenz called this dependence *the Butterfly Effect* and posed the question "Can the flap of a butterfly's wings in Brazil cause a tornado in Texas?"

39. Use a calculator to find the first ten iterates of $4x(1 - x)$ when x initially equals 0.200 and when the initial value is 0.201. Did the change in initial value affect the tenth value?

40. Why do you think this is called the Butterfly Effect?

41. ART Describe how artist Jean-Paul Agosti used iteration and self-similarity in his painting *Jardin du Creuset*.

42. NATURE Some of these pictures are of real objects and others are fractal images of objects.
   a. Compare the pictures and identify those you think are of real objects.
   b. Describe the characteristics of fractals shown in the images.

| flower | mountain | feathers | moss |

43. RESEARCH Use the Internet or other sources to find the names and pictures of the other fractals Waclaw Sierpinski developed.

44. CRITICAL THINKING Draw a right triangle on grid paper with 6 and 8 units for the lengths of the perpendicular sides. Shade the triangle formed by the three midsegments. Repeat the process for each unshaded triangle. Find the perimeter of the shaded triangle in Stage 1. What is the total perimeter of all the shaded triangles in Stage 2?

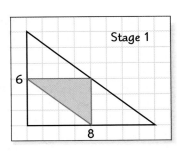

Stage 1

6

8

45. WRITING IN MATH   Answer the question that was posed at the beginning of the lesson.

How is mathematics related to nature?

Include the following in your answer:
- explain why broccoli is an example of fractal geometry, and
- how scientists can use fractal geometry to better understand nature.

46. GRID IN A triangle has side lengths of 3 inches, 6 inches, and 8 inches. A similar triangle is 24 inches on one side. Find the maximum perimeter, in inches, of the second triangle.

47. ALGEBRA A repair technician charges $80 for the first thirty minutes of each house call plus $2 for each additional minute. The repair technician charged a total of $170 for a job. How many minutes did the repair technician work?

Ⓐ 45 min　　　Ⓑ 55 min　　　Ⓒ 75 min　　　Ⓓ 85 min

Maintain Your Skills

Mixed Review **Find x.** *(Lesson 6-5)*

48.

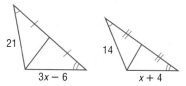

49.

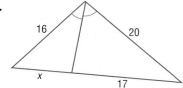

50.

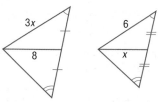

51.

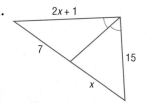

For Exercises 52–54, refer to $\triangle JKL$. *(Lesson 6-4)*

52. If $JL = 27$, $BL = 9$, and $JK = 18$, find JA.

53. If $AB = 8$, $KL = 10$, and $JB = 13$, find JL.

54. If $JA = 25$, $AK = 10$, and $BL = 14$, find JB.

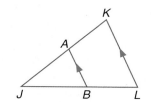

55. FOLKLORE The Bermuda Triangle is an imaginary region located off the southeastern Atlantic coast of the United States. It is the subject of many stories about unexplained losses of ships, small boats, and aircraft. Use the vertex locations to name the angles in order from least measure to greatest measure. *(Lesson 5-4)*

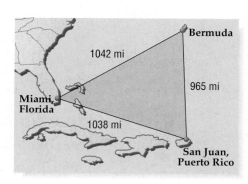

Find the length of each side of the polygon for the given perimeter. *(Lesson 1-6)*

56. $P = 60$ centimeters　　　**57.** $P = 54$ feet　　　**58.** $P = 57$ units

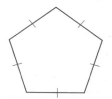

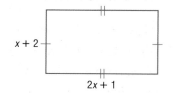

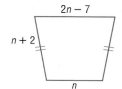

Chapter 6 Study Guide and Review

Vocabulary and Concept Check

cross products (p. 283)   means (p. 283)   scale factor (p. 290)
extremes (p. 283)   midsegment (p. 308)   self-similar (p. 325)
fractal (p. 325)   proportion (p. 283)   similar polygons (p. 289)
iteration (p. 325)   ratio (p. 282)

A complete list of postulates and theorems can be found on pages R1–R8.

Exercises   State whether each sentence is *true* or *false*. If false, replace the underlined expression to make a true sentence.

1. A midsegment of a triangle is a segment whose endpoints are the midpoints of two sides of the triangle.
2. Two polygons are similar if and only if their corresponding angles are congruent and the measures of the corresponding sides are congruent.
3. If two angles of one triangle are congruent to two angles of another triangle, then the triangles are similar.
4. If two triangles are similar, then the perimeters are proportional to the measures of the corresponding angles.
5. A fractal is a geometric figure that is created using *recursive formulas*.
6. A midsegment of a triangle is parallel to one side of the triangle, and its length is twice the length of that side.
7. For any numbers a and c and any nonzero numbers b and d, $\frac{a}{b} = \frac{c}{d}$ if and only if $ad = bc$.
8. If two triangles are similar, then the measures of the corresponding angle bisectors of the triangle are proportional to the measures of the corresponding sides.
9. If a line intersects two sides of a triangle and separates the sides into corresponding segments of proportional lengths, then the line is equal to one-half the length of the third side.

Lesson-by-Lesson Review

6-1 Proportions

See pages 282–287.

Concept Summary
- A ratio is a comparison of two quantities.
- A proportion is an equation stating that two ratios are equal.

Example Solve $\frac{z}{40} = \frac{5}{8}$.

$\frac{z}{40} = \frac{5}{8}$ Original proportion

$z \cdot 8 = 40(5)$ Cross products

$8z = 200$ Multiply.

$z = 25$ Divide each side by 8.

www.geometryonline.com/vocabulary_review

Exercises   Solve each proportion.   *See Example 3 on page 284.*

10. $\dfrac{3}{4} = \dfrac{x}{12}$

11. $\dfrac{7}{3} = \dfrac{28}{z}$

12. $\dfrac{x+2}{5} = \dfrac{14}{10}$

13. $\dfrac{3}{7} = \dfrac{7}{y-3}$

14. $\dfrac{4-x}{3+x} = \dfrac{16}{25}$

15. $\dfrac{x-12}{6} = \dfrac{x+7}{-4}$

16. BASEBALL   A player's slugging percentage is the ratio of the number of total bases from hits to the number of total at-bats. The ratio is converted to a decimal (rounded to three places) by dividing. If Alex Rodriguez of the Texas Rangers has 263 total bases in 416 at-bats, what is his slugging percentage?

17. A 108-inch-long board is cut into two pieces that have lengths in the ratio 2:7. How long is each new piece?

6-2 Similar Polygons

See pages 289–297.

Concept Summary

- In similar polygons, corresponding angles are congruent, and corresponding sides are in proportion.
- The ratio of two corresponding sides in two similar polygons is the scale factor.

Example

Determine whether the pair of triangles is similar. Justify your answer.

$\angle A \cong \angle D$ and $\angle C \cong \angle F$, so by the Third Angle Theorem, $\angle B \cong \angle E$. All of the corresponding angles are congruent.

Now, check to see if corresponding sides are in proportion.

$\dfrac{AB}{DE} = \dfrac{10}{8}$

$\dfrac{BC}{EF} = \dfrac{11}{8.8}$

$\dfrac{CA}{FD} = \dfrac{16}{12.8}$

$= \dfrac{5}{4}$ or 1.25

$= \dfrac{5}{4}$ or 1.25

$= \dfrac{5}{4}$ or 1.25

The corresponding angles are congruent, and the ratios of the measures of the corresponding sides are equal, so $\triangle ABC \sim \triangle DEF$.

Exercises   Determine whether each pair of figures is similar. Justify your answer.
See Example 1 on page 290.

18.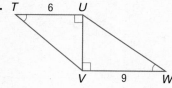

19.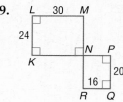

Each pair of polygons is similar. Write a similarity statement, and find *x*, the measures of the indicated sides, and the scale factor.   *See Example 3 on page 291.*

20. $\overline{AB}$ and $\overline{AG}$

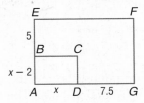

21. $\overline{PQ}$ and $\overline{QS}$

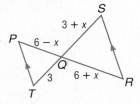

6-3 Similar Triangles

See pages 298–306.

Concept Summary

- AA, SSS, and SAS Similarity can all be used to prove triangles similar.
- Similarity of triangles is reflexive, symmetric, and transitive.

Example **INDIRECT MEASUREMENT** Alonso wanted to determine the height of a tree on the corner of his block. He knew that a certain fence by the tree was 4 feet tall. At 3 P.M., he measured the shadow of the fence to be 2.5 feet tall. Then he measured the tree's shadow to be 11.3 feet. What is the height of the tree?

Since the triangles formed are similar, a proportion can be written. Let *x* be the height of the tree.

$$\frac{\text{height of the tree}}{\text{height of the fence}} = \frac{\text{tree shadow length}}{\text{fence shadow length}}$$

$$\frac{x}{4} = \frac{11.3}{2.5} \qquad \text{Substitution}$$

$$x \cdot 2.5 = 4(11.3) \qquad \text{Cross products}$$

$$2.5x = 45.2 \qquad \text{Simplify.}$$

$$x = 18.08 \qquad \text{Divide each side by 2.5.}$$

The height of the tree is 18.08 feet.

Exercises Determine whether each pair of triangles is similar. Justify your answer. *See Example 1 on page 299.*

22.

23.

24.

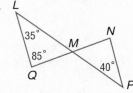

Identify the similar triangles. Find *x*.   *See Example 2 on page 300.*

25.

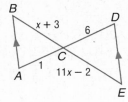

26.

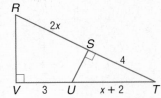

6-4 Parallel Lines and Proportional Parts

See pages 307–315.

Concept Summary

- A segment that intersects two sides of a triangle and is parallel to the third side divides the two intersected sides in proportion.
- If two lines divide two segments in proportion, then the lines are parallel.

Example In $\triangle TRS$, $TS = 12$. Determine whether $\overline{MN} \parallel \overline{SR}$.

If $TS = 12$, then $MS = 12 - 9$ or 3. Compare the segment lengths to determine if the lines are parallel.

$$\frac{TM}{MS} = \frac{9}{3} = 3 \qquad \frac{TN}{NR} = \frac{10}{5} = 2$$

Because $\frac{TM}{MS} \neq \frac{TN}{NR}$, $\overline{MN} \not\parallel \overline{SR}$.

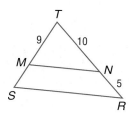

Exercises Determine whether $\overline{GL} \parallel \overline{HK}$. Justify your answer.
See Example 2 on page 308.

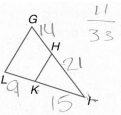

27. $IH = 21$, $HG = 14$, $LK = 9$, $KI = 15$
28. $GH = 10$, $HI = 35$, $IK = 28$, $IL = 36$
29. $GH = 11$, $HI = 22$, and IL is three times the length of $\overline{KL}$.
30. $LK = 6$, $KI = 18$, and IG is three times the length of $\overline{HI}$.

Refer to the figure at the right. *See Example 1 on page 308.*

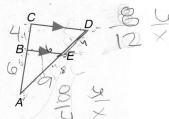

31. Find ED if $AB = 6$, $BC = 4$, and $AE = 9$.
32. Find AE if $AB = 12$, $AC = 16$, and $ED = 5$.
33. Find CD if $AE = 8$, $ED = 4$, and $BE = 6$.
34. Find BC if $BE = 24$, $CD = 32$, and $AB = 33$.

6-5 Parts of Similar Triangles

See pages 316–323.

Concept Summary

- Similar triangles have perimeters proportional to the corresponding sides.
- Corresponding angle bisectors, medians, and altitudes of similar triangles have lengths in the same ratio as corresponding sides.

Example If $\overline{FB} \parallel \overline{EC}$, $\overline{AD}$ is an angle bisector of $\angle A$, $BF = 6$, $CE = 10$, and $AD = 5$, find AM.

By AA Similarity using $\angle AFE \cong \angle ABF$ and $\angle A \cong \angle A$, $\triangle ABF \sim \triangle ACE$.

$\dfrac{AM}{AD} = \dfrac{BF}{CE}$ ~$\triangle$s have angle bisectors in the same proportion as the corresponding sides.

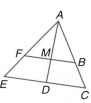

$\dfrac{x}{5} = \dfrac{6}{10}$ $AD = 5$, $AF = 6$, $FE = 4$, $AM = x$

$10x = 30$ Cross products

$x = 3$ Divide each side by 10.

Thus, $AM = 3$.

Chapter
6 **For More ...** • Extra Practice, see pages 764–766.
• Mixed Problem Solving, see page 787.

Exercises Find the perimeter of the given triangle. *See Example 1 on page 316.*

35. △DEF if △DEF ~ △ABC

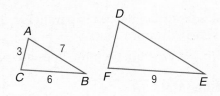

36. △QRS if △QRS ~ △QTP

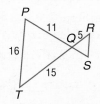

37. △CPD if the perimeter of △BPA is 12, $BM = \sqrt{13}$, and $CN = 3\sqrt{13}$

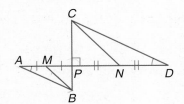

38. △PQR, if △PQM ~ △PRQ

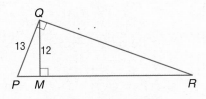

6-6 ## Fractals and Self-Similarity

See pages 325–331.

Concept Summary

• Iteration is the creation of a sequence by repetition of the same operation.
• A fractal is a geometric figure created by iteration.
• An iterative process involving algebraic equations is a recursive formula.

Example Find the value of $\frac{x}{2} + 4$, where x initially equals -8. Use that value as the next x in the expression. Repeat the process five times and describe your observations.

Make a table to organize each iteration.

Iteration	1	2	3	4	5	6
x	-8	0	4	6	7	7.5
$\frac{x}{2}+4$	0	4	6	7	7.5	7.75

The x values appear to get closer to the number 8 with each iteration.

Exercises Draw Stage 2 of the fractal shown below. Determine whether Stage 2 is similar to Stage 1. *See Example 2 on page 326.*

39.

_____   ⌐‾⌐
                    __| |__

Stage 0   Stage 1

Find the first three iterates of each expression. *See Example 3 on page 327.*

40. $x^3 - 4$, x initially equals 2

41. $3x + 4$, x initially equals -4

42. $\frac{1}{x}$, x initially equals 10

43. $\frac{x}{10} - 9$, x initially equals 30

Vocabulary and Concepts

Choose the answer that best matches each phrase.

1. an equation stating that two ratios are equal
2. the ratio between corresponding sides of two similar figures
3. the means multiplied together and the extremes multiplied together

a. scale factor
b. proportion
c. cross products

Skills and Applications

Solve each proportion.

4. $\dfrac{x}{14} = \dfrac{1}{2}$

5. $\dfrac{4x}{3} = \dfrac{108}{x}$

6. $\dfrac{k+2}{7} = \dfrac{k-2}{3}$

Each pair of polygons is similar. Write a similarity statement and find the scale factor.

7.
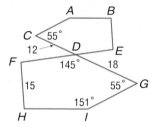

8.
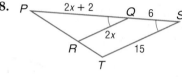

9.
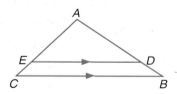

Determine whether each pair of triangles is similar. Justify your answer.

10.

11.

12.
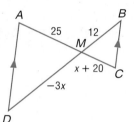

Refer to the figure at the right.

13. Find KJ if $GJ = 8$, $GH = 12$, and $HI = 4$.
14. Find GK if $GI = 14$, $GH = 7$, and $KJ = 6$.
15. Find GI if $GH = 9$, $GK = 6$, and $KJ = 4$.

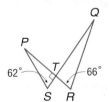

Find the perimeter of the given triangle.

16. $\triangle DEF$, if $\triangle DEF \sim \triangle ACB$

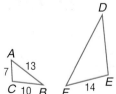

17. $\triangle ABC$

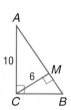

18. Find the first three iterates of $5x + 27$ when x initially equals -3.

19. **BASKETBALL** Terry wants to measure the height of the top of the backboard of his basketball hoop. At 4:00, the shadow of a 4-foot fence is 20 inches, and the shadow of the backboard is 65 inches. What is the height of the top of the backboard?

20. **STANDARDIZED TEST PRACTICE** If a person's weekly salary is $\$X$ and $\$Y$ is saved, what part of the weekly salary is spent?

Ⓐ $\dfrac{X}{Y}$ Ⓑ $\dfrac{X-Y}{X}$ Ⓒ $\dfrac{X-Y}{Y}$ Ⓓ $\dfrac{Y-X}{Y}$

Part 1 Multiple Choice

Record your answers on the answer sheet provided by your teacher or on a sheet of paper.

1. Which of the following is equivalent to $|-8 + 2|$? (Prerequisite Skill)

Ⓐ 10   Ⓑ 6   Ⓒ −6   Ⓓ −10

2. Kip's family moved to a new house. He used a coordinate plane with units in miles to locate his new house and school in relation to his old house. What is the distance between his new house and school? (Lesson 1-3)

Ⓐ 12 miles

Ⓑ $\sqrt{229}$ miles

Ⓒ 17 miles

Ⓓ $\sqrt{425}$ miles

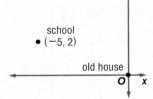

3. The diagonals of rectangle $ABCD$ are $\overline{AC}$ and $\overline{BD}$. Hallie found that the distances from the point where the diagonals intersect to each vertex were the same. Which of the following conjectures could Hallie make? (Lesson 2-1)

Ⓐ Diagonals of a rectangle are congruent.

Ⓑ Diagonals of a rectangle create equilateral triangles.

Ⓒ Diagonals of a rectangle intersect at more than one point.

Ⓓ Diagonals of a rectangle are congruent to the width.

4. If two sides of a triangular sail are congruent, which of the following terms *cannot* be used to describe the shape of the sail? (Lesson 4-1)

Ⓐ acute   Ⓑ equilateral

Ⓒ obtuse   Ⓓ scalene

5. Miguel is using centimeter grid paper to make a scale drawing of his favorite car. Miguel's drawing is 11.25 centimeters wide. How many feet long is the actual car? (Lesson 6-1)

Ⓐ 15.0 ft

Ⓑ 18.75 ft

Ⓒ 22.5 ft

Ⓓ 33.0 ft

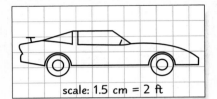

scale: 1.5 cm = 2 ft

6. Joely builds a large corkboard for her room that is 45 inches tall and 63 inches wide. She wants to build a smaller corkboard with a similar shape for the kitchen. Which of the following could be the dimensions of that corkboard? (Lesson 6-2)

Ⓐ 4 in. by 3 in.

Ⓑ 7 in. by 5 in.

Ⓒ 12 in. by 5 in.

Ⓓ 21 in. by 14 in.

7. If $\triangle PQR$ and $\triangle STU$ are similar, which of the following is a correct proportion? (Lesson 6-3)

Ⓐ $\dfrac{s}{u} = \dfrac{t}{q}$

Ⓑ $\dfrac{s}{u} = \dfrac{p}{q}$

Ⓒ $\dfrac{s}{u} = \dfrac{p}{r}$

Ⓓ $\dfrac{s}{u} = \dfrac{r}{p}$

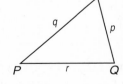

8. In $\triangle ABC$, D is the midpoint of $\overline{AB}$, and E is the midpoint of $\overline{AC}$. Which of the following is *not* true? (Lesson 6-4)

Ⓐ $\dfrac{AD}{DB} = \dfrac{AE}{EC}$

Ⓑ $\overline{DE} \parallel \overline{BC}$

Ⓒ $\triangle ABC \sim \triangle ADE$

Ⓓ $\angle 1 \cong \angle 4$

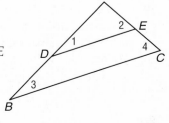

Part 2 Short Response/Grid In

Record your answers on the answer sheet provided by your teacher or on a sheet of paper.

9. During his presentation, Dante showed a picture of several types of balls used in sports. From this picture, he conjectured that all balls used in sports are spheres. Brianna then showed another ball. What is this type of example called? (Lesson 2-1)

Dante   Brianna

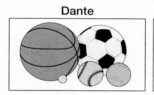

10. What is an equation of the line with slope 3 that contains $A(2, 2)$? (Lesson 3-4)

11. In $\triangle DEF$, P is the midpoint of $\overline{DE}$, and Q is the midpoint of side $\overline{DF}$. If $EF = 3x + 4$ and $PQ = 20$, what is the value of x? (Lesson 6-4)

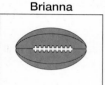

12. A city planner designs a triangular traffic median on Main Street to provide more green space in the downtown area. The planner builds a model so that the section of the median facing Main Street East measures 20 centimeters. What is the perimeter, in centimeters, of the model of the traffic median? (Lesson 6-5)

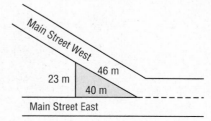

www.geometryonline.com/standardized_test

Preparing for Standardized Tests
For test-taking strategies and more practice, see pages 795–810.

Test-Taking Tip   A   B   C   D

Question 7
In similar triangles, corresponding angles are congruent and corresponding sides are proportional. When you set up a proportion, be sure that it compares corresponding sides. In this question, p corresponds to s, q corresponds to t, and r corresponds to u.

Part 3 Extended Response

Record your answers on a sheet of paper. Show your work.

13. A cable company charges a one-time connection fee plus a monthly flat rate as shown in the graph.

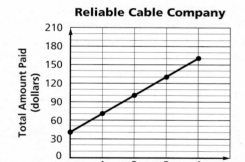

a. What is the slope of the line that joins the points on the graph? (Lesson 3-3)

b. Discuss what the value of the slope represents. (Lesson 3-3)

c. Write an equation of the line. (Lesson 3-4)

d. If the company presents a special offer that lowers the monthly rate by $5, how will the equation and graph change? (Lesson 3-4)

14. Given $\triangle ADE$ and $\overline{BC}$ is equidistant from $\overline{DE}$.

a. Prove that $\triangle ABC \sim \triangle ADE$. (Lessons 6-3 and 6-4)

b. Suppose $AB = 3500$ feet, $BD = 1500$ feet, and $BC = 1400$ feet. Find DE. (Lesson 6-3)

Right Triangles and Trigonometry

What You'll Learn

- **Lessons 7-1, 7-2, and 7-3** Solve problems using the geometric mean, the Pythagorean Theorem, and its converse.
- **Lessons 7-4 and 7-5** Use trigonometric ratios to solve right triangle problems.
- **Lessons 7-6 and 7-7** Solve triangles using the Law of Sines and the Law of Cosines.

Key Vocabulary

- geometric mean (p. 342)
- Pythagorean triple (p. 352)
- trigonometric ratio (p. 364)
- Law of Sines (p. 377)
- Law of Cosines (p. 385)

Why It's Important

Trigonometry is used to find the measures of the sides and angles of triangles. These ratios are frequently used in real-world applications such as architecture, aviation, and surveying. *You will learn how surveyors use trigonometry in Lesson 7-6.*

Getting Started

▶ **Prerequisite Skills** To be successful in this chapter, you'll need to master these skills and be able to apply them in problem-solving situations. Review these skills before beginning Chapter 7.

For Lesson 7-1　　　　　　　　　　　　　　　　　　　　　　　**Proportions**

Solve each proportion. Round to the nearest hundredth, if necessary. *(For review, see Lesson 6-1.)*

1. $\dfrac{3}{4} = \dfrac{12}{a}$　　　　**2.** $\dfrac{c}{5} = \dfrac{8}{3}$　　　　**3.** $\dfrac{e}{20} = \dfrac{6}{5} = \dfrac{f}{10}$　　　　**4.** $\dfrac{4}{3} = \dfrac{6}{y} = \dfrac{1}{z}$

For Lesson 7-2　　　　　　　　　　　　　　　　　　　**Pythagorean Theorem**

Find the measure of the hypotenuse of each right triangle having legs with the given measures. Round to the nearest hundredth, if necessary. *(For review, see Lesson 1-3.)*

5. 5 and 12　　　　**6.** 6 and 8　　　　**7.** 15 and 15　　　　**8.** 14 and 27

For Lessons 7-3 and 7-4　　　　　　　　　　　　　　　**Radical Expressions**

Simplify each expression. *(For review, see pages 744 and 745.)*

9. $\sqrt{8}$　　　　**10.** $\sqrt{10^2 - 5^2}$　　　　**11.** $\sqrt{39^2 - 36^2}$　　　　**12.** $\dfrac{7}{\sqrt{2}}$

For Lessons 7-5 through 7-7　　　　　　　　　　　　　　**Angle Sum Theorem**

Find *x*. *(For review, see Lesson 4-2.)*

13. 　　　**14.** 　　　**15.**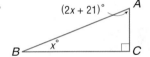

 **FOLDABLES™ Study Organizer**

Right Triangles and Trigonometry Make this Foldable to help you organize your notes. Begin with seven sheets of grid paper.

Step 1　Fold

Fold each sheet along the diagonal from the corner of one end to 2.5 inches away from the corner of the other end.

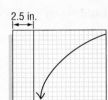

2.5 in.

Step 2　Stack

Stack the sheets, and fold the rectangular part in half.

Step 3　Staple

Staple the sheets in three places.

Step 4　Label

Label each sheet with a lesson number, and the rectangular part with the chapter title.

Reading and Writing As you read and study the chapter, write notes, define terms, and solve problems in your Foldable.

Geometric Mean

- Find the geometric mean between two numbers.
- Solve problems involving relationships between parts of a right triangle and the altitude to its hypotenuse.

Vocabulary
- geometric mean

How can the geometric mean be used to view paintings?

When you look at a painting, you should stand at a distance that allows you to see all of the details in the painting. The distance that creates the best view is the geometric mean of the distance from the top of the painting to eye level and the distance from the bottom of the painting to eye level.

GEOMETRIC MEAN   The **geometric mean** between two numbers is the positive square root of their product.

Key Concept   *Geometric Mean*

For two positive numbers a and b, the geometric mean is the positive number x where the proportion $a : x = x : b$ is true. This proportion can be written using fractions as $\frac{a}{x} = \frac{x}{b}$ or with cross products as $x^2 = ab$ or $x = \sqrt{ab}$.

Example 1 *Geometric Mean*

Find the geometric mean between each pair of numbers.

a. **4 and 9**

Let x represent the geometric mean.

$\frac{4}{x} = \frac{x}{9}$ Definition of geometric mean

$x^2 = 36$ Cross products

$x = \sqrt{36}$ Take the positive square root of each side.

$x = 6$ Simplify.

b. **6 and 15**

$\frac{6}{x} = \frac{x}{15}$ Definition of geometric mean

$x^2 = 90$ Cross products

$x = \sqrt{90}$ Take the positive square root of each side.

$x = 3\sqrt{10}$ Simplify.

$x \approx 9.5$ Use a calculator.

ALTITUDE OF A TRIANGLE Consider right triangle *XYZ* with altitude $\overline{WZ}$ drawn from the right angle *Z* to the hypotenuse $\overline{XY}$. A special relationship exists for the three right triangles, △*XYZ*, △*XZW*, and △*ZYW*.

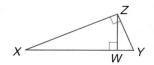

Geometry Software Investigation

Right Triangles Formed by the Altitude

Use The Geometer's Sketchpad to draw a right triangle *XYZ* with right angle *Z*. Draw the altitude $\overline{ZW}$ from the right angle to the hypotenuse. Explore the relationships among the three right triangles formed.

Think and Discuss

1. Find the measures of ∠*X*, ∠*XZY*, ∠*Y*, ∠*XWZ*, ∠*XZW*, ∠*YWZ*, and ∠*YZW*.

2. What is the relationship between the measures of ∠*X* and ∠*YZW*? What is the relationship between the measures of ∠*Y* and ∠*XZW*?

3. Drag point *Z* to another position. Describe the relationship between the measures of ∠*X* and ∠*YZW* and between the measures of ∠*Y* and ∠*XZW*.

4. Make a conjecture about △*XYZ*, △*XZW*, and △*ZYW*.

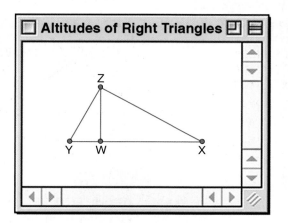
Altitudes of Right Triangles

Study Tip

Altitudes of a Right Triangle
The altitude drawn to the hypotenuse originates from the right angle. The other two altitudes of a right triangle are the legs.

The results of the Geometry Software Investigation suggest the following theorem.

Theorem 7.1

If the altitude is drawn from the vertex of the right angle of a right triangle to its hypotenuse, then the two triangles formed are similar to the given triangle and to each other.

Example: △*XYZ* ~ △*XWY* ~ △*YWZ*

You will prove this theorem in Exercise 45.

By Theorem 7.1, since △*XWY* ~ △*YWZ*, the corresponding sides are proportional. Thus, $\frac{XW}{YW} = \frac{YW}{ZW}$. Notice that $\overline{XW}$ and $\overline{ZW}$ are segments of the hypotenuse of the largest triangle.

Theorem 7.2

The measure of an altitude drawn from the vertex of the right angle of a right triangle to its hypotenuse is the geometric mean between the measures of the two segments of the hypotenuse.

Example: *YW* is the geometric mean of *XW* and *ZW*.

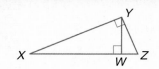

You will prove this theorem in Exercise 46.

Example 2 **Altitude and Segments of the Hypotenuse**

In $\triangle PQR$, $RS = 3$ and $QS = 14$. Find PS.

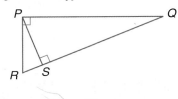

Let $x = PS$.

$$\frac{RS}{PS} = \frac{PS}{QS}$$

$$\frac{3}{x} = \frac{x}{14} \qquad RS = 3,\ QS = 14,\ \text{and } PS = x$$

$$x^2 = 42 \qquad \text{Cross products}$$

$$x = \sqrt{42} \qquad \text{Take the positive square root of each side.}$$

$$x \approx 6.5 \qquad \text{Use a calculator.}$$

PS is about 6.5.

Ratios in right triangles can be used to solve problems.

Example 3 **Altitude and Length of the Hypotenuse**

ARCHITECTURE   Mr. Martinez is designing a walkway that must pass over an elevated train. To find the height of the elevated train, he holds a carpenter's square at eye level and sights along the edges from the street to the top of the train. If Mr. Martinez's eye level is 5.5 feet above the street and he is 8.75 feet from the train, find the distance from the street to the top of the train. Round to the nearest tenth.

Draw a diagram. Let $\overline{YX}$ be the altitude drawn from the right angle of $\triangle WYZ$.

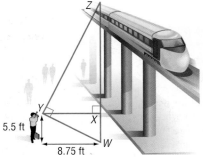

$$\frac{WX}{YX} = \frac{YX}{ZX}$$

$$\frac{5.5}{8.75} = \frac{8.75}{ZX} \qquad WX = 5.5 \text{ and } YX = 8.75$$

$$5.5ZX = 76.5625 \qquad \text{Cross products}$$

$$ZX \approx 13.9 \qquad \text{Divide each side by 5.5.}$$

Mr. Martinez estimates that the elevated train is $5.5 + 13.9$ or about 19.4 feet high.

The altitude to the hypotenuse of a right triangle determines another relationship between the segments.

Theorem 7.3

If the altitude is drawn from the vertex of the right angle of a right triangle to its hypotenuse, then the measure of a leg of the triangle is the geometric mean between the measures of the hypotenuse and the segment of the hypotenuse adjacent to that leg.

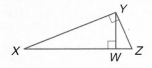

Example: $\dfrac{XZ}{XY} = \dfrac{XY}{XW}$ and $\dfrac{XZ}{YZ} = \dfrac{ZY}{WZ}$

You will prove Theorem 7.3 in Exercise 47.

Example 4 Hypotenuse and Segment of Hypotenuse

Find x and y in $\triangle PQR$.

$\overline{PQ}$ and $\overline{RQ}$ are legs of right triangle PQR.
Use Theorem 7.3 to write a proportion
for each leg and then solve.

<table>
<tr><td>

$\dfrac{PR}{PQ} = \dfrac{PQ}{PS}$

$\dfrac{6}{y} = \dfrac{y}{2}$ $PS = 2, PQ = y, PR = 6$

$y^2 = 12$ Cross products

$y = \sqrt{12}$ Take the square root.

$y = 2\sqrt{3}$ Simplify.

$y \approx 3.5$ Use a calculator.
</td><td>

$\dfrac{PR}{RQ} = \dfrac{RQ}{SR}$

$\dfrac{6}{x} = \dfrac{x}{4}$ $RS = 4, RQ = x, PR = 6$

$x^2 = 24$ Cross products

$x = \sqrt{24}$ Take the square root.

$x = 2\sqrt{6}$ Simplify.

$x \approx 4.9$ Use a calculator.
</td></tr>
</table>

Study Tip

Simplifying Radicals

Remember that $\sqrt{12} = \sqrt{4} \cdot \sqrt{3}$. Since $\sqrt{4} = 2$, $\sqrt{12} = 2\sqrt{3}$. For more practice simplifying radicals, see pages 744 and 745.

Check for Understanding

Concept Check

1. **OPEN ENDED** Find two numbers whose geometric mean is 12.

2. **Draw and label** a right triangle with an altitude drawn from the right angle. From your drawing, explain the meaning of *the hypotenuse and the segment of the hypotenuse adjacent to that leg* in Theorem 7.3.

3. **FIND THE ERROR** $\triangle RST$ is a right isosceles triangle. Holly and Ian are finding the measure of altitude $\overline{SU}$.

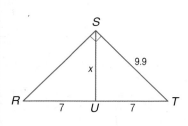

Holly

$\dfrac{RS}{SU} = \dfrac{SU}{RT}$

$\dfrac{9.9}{x} = \dfrac{x}{14}$

$x^2 = 138.6$

$x = \sqrt{138.6}$

$x \approx 11.8$

Ian

$\dfrac{RU}{SU} = \dfrac{SU}{UT}$

$\dfrac{7}{x} = \dfrac{x}{7}$

$x^2 = 49$

$x = 7$

Who is correct? Explain your reasoning.

Guided Practice

Find the geometric mean between each pair of numbers.

4. 9 and 4   5. 36 and 49   6. 6 and 8   7. $2\sqrt{2}$ and $3\sqrt{2}$

Find the measure of the altitude drawn to the hypotenuse.

8.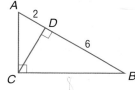
9.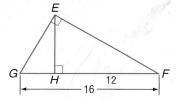

Find x and y.

10.

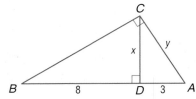

11.

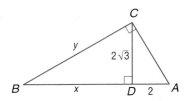

Application **12. DANCES** Khaliah is making a banner for the dance committee. The banner is to be as high as the wall of the gymnasium. To find the height of the wall, Khaliah held a book up to her eyes so that the top and bottom of the wall were in line with the top edge and binding of the cover. If Khaliah's eye level is 5 feet off the ground and she is standing 12 feet from the wall, how high is the wall?

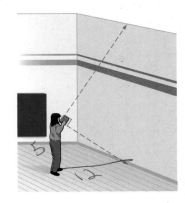

Practice and Apply

Homework Help

For Exercises	See Examples
13–20	1
21–26	2
27–32	3, 4

Extra Practice
See page 766.

Find the geometric mean between each pair of numbers.

13. 5 and 6

14. 24 and 25

15. $\sqrt{45}$ and $\sqrt{80}$

16. $\sqrt{28}$ and $\sqrt{1372}$

17. $\frac{3}{5}$ and 1

18. $\frac{8\sqrt{3}}{5}$ and $\frac{6\sqrt{3}}{5}$

19. $\frac{2\sqrt{2}}{6}$ and $\frac{5\sqrt{2}}{6}$

20. $\frac{13}{7}$ and $\frac{5}{7}$

Find the measure of the altitude drawn to the hypotenuse.

21.

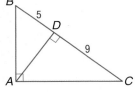

22.

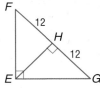

23.

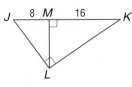

24.

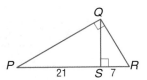

25.

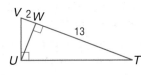

26.

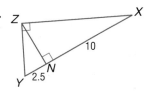

Find x, y, and z.

27.

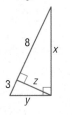

28.

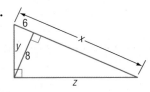

29.

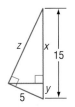

30.

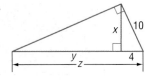

31.

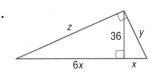

32.

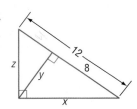

The geometric mean and one extreme are given. Find the other extreme.

33. $\sqrt{17}$ is the geometric mean between a and b. Find b if $a = 7$.

34. $\sqrt{12}$ is the geometric mean between x and y. Find x if $y = \sqrt{3}$.

Determine whether each statement is *always, sometimes,* or *never* true.

35. The geometric mean for consecutive positive integers is the average of the two numbers.

36. The geometric mean for two perfect squares is a positive integer.

37. The geometric mean for two positive integers is another integer.

38. The measure of the altitude of a triangle is the geometric mean between the measures of the segments of the side it intersects.

39. **BIOLOGY** The shape of the shell of a chambered nautilus can be modeled by a geometric mean. Consider the sequence of segments $\overline{OA}, \overline{OB}, \overline{OC}, \overline{OD}, \overline{OE}, \overline{OF}, \overline{OG}, \overline{OH}, \overline{OI}$, and $\overline{OJ}$. The length of each of these segments is the geometric mean between the lengths of the preceding segment and the succeeding segment. Explain this relationship. (*Hint:* Consider $\triangle FGH$.)

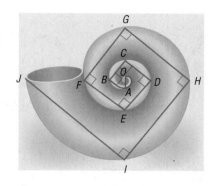

WebQuest

You can use geometric mean and the Quadratic Formula to discover the golden mean. Visit www.geometry online.com/webquest to continue work on your WebQuest project.

40. **RESEARCH** Refer to the information at the left. Use the Internet or other resource to write a brief description of the golden ratio.

41. **CONSTRUCTION** In the United States, most building codes limit the steepness of the slope of a roof to $\frac{4}{3}$, as shown at the right. A builder wants to put a support brace from point C perpendicular to $\overline{AP}$. Find the length of the brace.

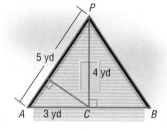

SOCCER For Exercises 42 and 43, refer to the graphic.

42. Find the geometric mean between the number of players from Indiana and North Carolina.

43. Are there two schools whose geometric mean is the same as the geometric mean between UCLA and Clemson? If so, which schools?

44. **CRITICAL THINKING** Find the exact value of DE, given $AD = 12$ and $BD = 4$.

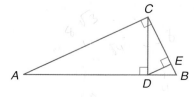

USA TODAY Snapshots®

Bruins bring skills to MLS
Universities producing the most players in Major League Soccer this season:

UCLA 15
Indiana 10
Virginia 9
North Carolina 7
Washington 7
Clemson 6

Source: MLS

By Ellen J. Horrow and Adrienne Lewis, USA TODAY

PROOF Write the specified type proof for each theorem.

45. two-column proof of Theorem 7.1

46. paragraph proof of Theorem 7.2

47. two-column proof of Theorem 7.3

48. **WRITING IN MATH** Answer the question that was posed at the beginning of the lesson.

How can the geometric mean be used to view paintings?

Include the following in your answer:

- an explanation of what happens when you are too far or too close to a painting, and
- an explanation of how the curator of a museum would determine where to place roping in front of paintings on display.

Standardized Test Practice

49. Find x and y.

Ⓐ 4 and 6   Ⓑ 2.5 and 7.5

Ⓒ 3.6 and 6.4   Ⓓ 3 and 7

50. ALGEBRA Solve $5x^2 + 405 = 1125$.

Ⓐ ±15   Ⓑ ±12

Ⓒ ±4$\sqrt{3}$ Ⓓ ±4

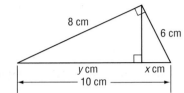

Maintain Your Skills

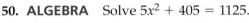

Mixed Review Find the first three iterations of each expression. *(Lesson 6-6)*

51. $x + 3$, where x initially equals 12

52. $3x + 2$, where x initially equals 4

53. $x^2 - 2$, where x initially equals 3

54. $2(x - 3)$, where x initially equals 1

55. The measures of the sides of a triangle are 20, 24, and 30. Find the measures of the segments formed where the bisector of the smallest angle meets the opposite side. *(Lesson 6-5)*

Use the Exterior Angle Inequality Theorem to list all angles that satisfy the stated condition. *(Lesson 5-2)*

56. all angles with a measure less than $m\angle 8$

57. all angles with a measure greater than $m\angle 1$

58. all angles with a measure less than $m\angle 7$

59. all angles with a measure greater than $m\angle 6$

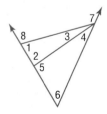

Write an equation in slope-intercept form for the line that satisfies the given conditions. *(Lesson 3-4)*

60. $m = 2$, y-intercept $= 4$

61. x-intercept is 2, y-intercept $= -8$

62. passes through $(2, 6)$ and $(-1, 0)$

63. $m = -4$, passes through $(-2, -3)$

Getting Ready for the Next Lesson **PREREQUISITE SKILL** Use the Pythagorean Theorem to find the length of the hypotenuse of each right triangle.

*(To review **using the Pythagorean Theorem**, see Lesson 1-4.)*

64.
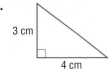

65.
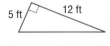

66.

The Pythagorean Theorem

In Chapter 1, you learned that the Pythagorean Theorem relates the measures of the legs and the hypotenuse of a right triangle. Ancient cultures used the Pythagorean Theorem before it was officially named in 1909.

Use square pieces of patty paper and algebra. Then you too can discover this relationship among the measures of the sides of a right triangle.

Activity

Use paper folding to develop the Pythagorean Theorem.

Step 1 On a piece of patty paper, make a mark along one side so that the two resulting segments are not congruent. Label one as a and the other as b.

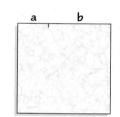

Step 2 Copy these measures on the other sides in the order shown at the right. Fold the paper to divide the square into four sections. Label the area of each section.

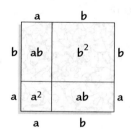

Step 3 On another sheet of patty paper, mark the same lengths a and b on the sides in the different pattern shown at the right.

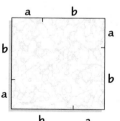

Step 4 Use your straightedge and pencil to connect the marks as shown at the right. Let c represent the length of each hypotenuse.

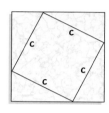

Step 5 Label the area of each section, which is $\frac{1}{2} ab$ for each triangle and c^2 for the square.

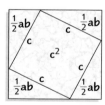

Step 6 Place the squares side by side and color the corresponding regions that have the same area. For example, $ab = \frac{1}{2}ab + \frac{1}{2}ab$.

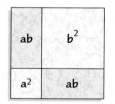 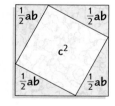

The parts that are not shaded tell us that $a^2 + b^2 = c^2$.

Model

1. Use a ruler to find actual measures for a, b, and c. Do these measures confirm that $a^2 + b^2 = c^2$?

2. Repeat the activity with different a and b values. What do you notice?

Analyze the model

3. **Explain** why the drawing at the right is an illustration of the Pythagorean Theorem.

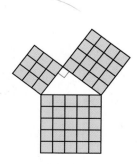

The Pythagorean Theorem and Its Converse

7-2

What You'll Learn

- Use the Pythagorean Theorem.
- Use the converse of the Pythagorean Theorem.

Vocabulary
- Pythagorean triple

How are right triangles used to build suspension bridges?

The Talmadge Memorial Bridge over the Savannah River has two soaring towers of suspension cables. Note the right triangles being formed by the roadway, the perpendicular tower, and the suspension cables. The Pythagorean Theorem can be used to find measures in any right triangle.

Study Tip

Look Back
To review **finding the hypotenuse of a right triangle,** see Lesson 1-3.

THE PYTHAGOREAN THEOREM In Lesson 1-3, you used the Pythagorean Theorem to find the distance between two points by finding the length of the hypotenuse when given the lengths of the two legs of a right triangle. You can also find the measure of any side of a right triangle given the other two measures.

Theorem 7.4

Pythagorean Theorem In a right triangle, the sum of the squares of the measures of the legs equals the square of the measure of the hypotenuse.

Symbols: $a^2 + b^2 = c^2$

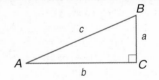

The geometric mean can be used to prove the Pythagorean Theorem.

Proof Pythagorean Theorem

Given: $\triangle ABC$ with right angle at C
Prove: $a^2 + b^2 = c^2$
Proof:

Draw right triangle ABC so C is the right angle. Then draw the altitude from C to $\overline{AB}$. Let $AB = c$, $AC = b$, $BC = a$, $AD = x$, $DB = y$, and $CD = h$.

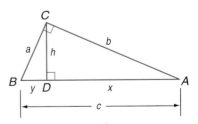

Two geometric means now exist.

$$\frac{c}{a} = \frac{a}{y} \qquad \text{and} \qquad \frac{c}{b} = \frac{b}{x}$$

$$a^2 = cy \qquad \text{and} \qquad b^2 = cx \quad \text{Cross products}$$

Add the equations.

$$a^2 + b^2 = cy + cx$$
$$a^2 + b^2 = c(y + x) \quad \text{Factor.}$$
$$a^2 + b^2 = c^2 \quad \text{Since } c = y + x, \text{ substitute } c \text{ for } (y + x).$$

Example 1 · Find the Length of the Hypotenuse

LONGITUDE AND LATITUDE   NASA Dryden is located at about 117 degrees longitude and 34 degrees latitude. NASA Ames is located at about 122 degrees longitude and 37 degrees latitude. Use the lines of longitude and latitude to find the degree distance to the nearest tenth between NASA Dryden and NASA Ames.

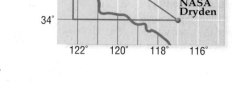

The change in longitude between the two locations is $|117-122|$ or 5 degrees. Let this distance be a.

The change in latitude is $|37-34|$ or 3 degrees latitude. Let this distance be b.

Use the Pythagorean Theorem to find the distance in degrees from NASA Dryden to NASA Ames, represented by c.

$$a^2 + b^2 = c^2 \quad \text{Pythagorean Theorem}$$
$$5^2 + 3^2 = c^2 \quad a = 5, b = 3$$
$$25 + 9 = c^2 \quad \text{Simplify.}$$
$$34 = c^2 \quad \text{Add.}$$
$$\sqrt{34} = c \quad \text{Take the square root of each side.}$$
$$5.8 \approx c \quad \text{Use a calculator.}$$

The degree distance between NASA Dryden and NASA Ames is about 5.8 degrees.

Example 2 · Find the Length of a Leg

Find x.

$$(XY)^2 + (YZ)^2 = (XZ)^2 \quad \text{Pythagorean Theorem}$$
$$7^2 + x^2 = 14^2 \quad XY = 7, XZ = 14$$
$$49 + x^2 = 196 \quad \text{Simplify.}$$
$$x^2 = 147 \quad \text{Subtract 49 from each side.}$$
$$x = \sqrt{147} \quad \text{Take the square root of each side.}$$
$$x = 7\sqrt{3} \quad \text{Simplify.}$$
$$x \approx 12.1 \quad \text{Use a calculator.}$$

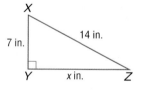

CONVERSE OF THE PYTHAGOREAN THEOREM   The converse of the Pythagorean Theorem can help you determine whether three measures of the sides of a triangle are those of a right triangle.

Theorem 7.5

Converse of the Pythagorean Theorem   If the sum of the squares of the measures of two sides of a triangle equals the square of the measure of the longest side, then the triangle is a right triangle.

Symbols: If $a^2 + b^2 = c^2$, then $\triangle ABC$ is a right triangle.

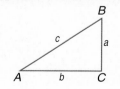

You will prove this theorem in Exercise 38.

Example 3 *Verify a Triangle is a Right Triangle*

COORDINATE GEOMETRY Verify that $\triangle PQR$ is a right triangle.

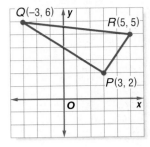

Use the Distance Formula to determine the lengths of the sides.

$$PQ = \sqrt{(-3 - 3)^2 + (6 - 2)^2} \quad x_1 = 3, y_1 = 2, x_2 = -3, y_2 = 6$$
$$= \sqrt{(-6)^2 + 4^2} \qquad \text{Subtract.}$$
$$= \sqrt{52} \qquad \text{Simplify.}$$

$$QR = \sqrt{[5 - (-3)]^2 + (5 - 6)^2} \quad x_1 = -3, y_1 = 6, x_2 = 5, y_2 = 5$$
$$= \sqrt{8^2 + (-1)^2} \qquad \text{Subtract.}$$
$$= \sqrt{65} \qquad \text{Simplify.}$$

$$PR = \sqrt{(5 - 3)^2 + (5 - 2)^2} \quad x_1 = 3, y_1 = 2, x_2 = 5, y_2 = 5$$
$$= \sqrt{2^2 + 3^2} \qquad \text{Subtract.}$$
$$= \sqrt{13} \qquad \text{Simplify.}$$

By the converse of the Pythagorean Theorem, if the sum of the squares of the measures of two sides of a triangle equals the square of the measure of the longest side, then the triangle is a right triangle.

$$PQ^2 + PR^2 = QR^2 \qquad \text{Converse of the Pythagorean Theorem}$$
$$\left(\sqrt{52}\right)^2 + \left(\sqrt{13}\right)^2 \stackrel{?}{=} \left(\sqrt{65}\right)^2 \quad PQ = \sqrt{52}, PR = \sqrt{13}, QR = \sqrt{65}$$
$$52 + 13 \stackrel{?}{=} 65 \qquad \text{Simplify.}$$
$$65 = 65 \qquad \text{Add.}$$

Since the sum of the squares of two sides equals the square of the longest side, $\triangle PQR$ is a right triangle.

A **Pythagorean triple** is three whole numbers that satisfy the equation $a^2 + b^2 = c^2$, where c is the greatest number. One common Pythagorean triple is 3-4-5, in which the sides of a right triangle are in the ratio 3:4:5. If the measures of the sides of any right triangle are whole numbers, the measures form a Pythagorean triple.

Example 4 *Pythagorean Triples*

Determine whether each set of measures can be the sides of a right triangle. Then state whether they form a Pythagorean triple.

a. 8, 15, 16

Since the measure of the longest side is 16, 16 must be c, and a or b are 15 and 8.

$$a^2 + b^2 = c^2 \qquad \text{Pythagorean Theorem}$$
$$8^2 + 15^2 \stackrel{?}{=} 16^2 \qquad a = 8, b = 15, c = 16$$
$$64 + 225 \stackrel{?}{=} 256 \qquad \text{Simplify.}$$
$$289 \neq 256 \qquad \text{Add.}$$

Since $289 \neq 256$, segments with these measures cannot form a right triangle. Therefore, they do not form a Pythagorean triple.

Study Tip

Distance Formula
When using the Distance Formula, be sure to follow the order of operations carefully. Perform the operation inside the parentheses first, square each term, and then add.

b. 20, 48, and 52

$$a^2 + b^2 = c^2 \qquad \text{Pythagorean Theorem}$$
$$20^2 + 48^2 \overset{?}{=} 52^2 \qquad a = 20, b = 48, c = 52$$
$$400 + 2304 \overset{?}{=} 2704 \qquad \text{Simplify.}$$
$$2704 = 2704 \qquad \text{Add.}$$

These segments form the sides of a right triangle since they satisfy the Pythagorean Theorem. The measures are whole numbers and form a Pythagorean triple.

c. $\dfrac{\sqrt{3}}{5}, \dfrac{\sqrt{6}}{5}$, and $\dfrac{3}{5}$

$$a^2 + b^2 = c^2 \qquad \text{Pythagorean Theorem}$$
$$\left(\frac{\sqrt{3}}{5}\right)^2 + \left(\frac{\sqrt{6}}{5}\right)^2 \overset{?}{=} \left(\frac{3}{5}\right)^2 \qquad a = \frac{\sqrt{3}}{5}, b = \frac{\sqrt{6}}{5}, c = \frac{3}{5}$$
$$\frac{3}{25} + \frac{6}{25} \overset{?}{=} \frac{9}{25} \qquad \text{Simplify.}$$
$$\frac{9}{25} = \frac{9}{25} \qquad \text{Add.}$$

Since $\dfrac{9}{25} = \dfrac{9}{25}$, segments with these measures form a right triangle. However, the three numbers are not whole numbers. Therefore, they do not form a Pythagorean triple.

Check for Understanding

Concept Check

1. FIND THE ERROR Maria and Colin are determining whether 5-12-13 is a Pythagorean triple.

Colin

$a^2 + b^2 = c^2$
$13^2 + 5^2 \overset{?}{=} 12^2$
$169 + 25 \overset{?}{=} 144$
$193 \neq 144$
no

Maria

$a^2 + b^2 = c^2$
$5^2 + 12^2 \overset{?}{=} 13^2$
$25 + 144 \overset{?}{=} 169$
$169 = 169$
yes

Who is correct? Explain your reasoning.

2. Explain why a Pythagorean triple can represent the measures of the sides of a right triangle.

3. OPEN ENDED Draw a pair of similar right triangles. List the corresponding sides, the corresponding angles, and the scale factor. Are the measures of the sides of each triangle a Pythagorean triple?

Guided Practice Find x.

4.

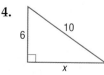

5.

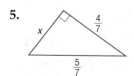

6.

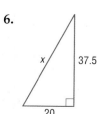

7. **COORDINATE GEOMETRY** Determine whether △*JKL* with vertices *J*(−2, 2), *K*(−1, 6), and *L*(3, 5) is a right triangle. Explain.

Determine whether each set of numbers can be the measures of the sides of a right triangle. Then state whether they form a Pythagorean triple.

8. 15, 36, 39

9. $\sqrt{40}$, 20, 21

10. $\sqrt{44}$, 8, $\sqrt{108}$

Application 11. **COMPUTERS** Computer monitors are usually measured along the diagonal of the screen. A 19-inch monitor has a diagonal that measures 19 inches. If the height of the screen is 11.5 inches, how wide is the screen?

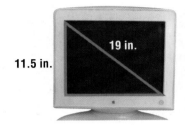
19 in.
11.5 in.

Practice and Apply

Homework Help

For Exercises	See Examples
14, 15	1
12, 13, 16, 17	2
18–21	3
22–29	4

Extra Practice
See page 767.

Find *x*.

12.

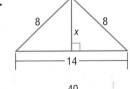
8
8
x
14

13.

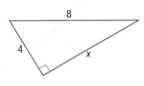
8
4
x

14.

x
28
20

15.

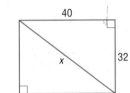
40
x
32

16.

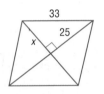
33
25
x

17.

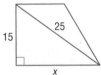
25
15
x

COORDINATE GEOMETRY Determine whether △*QRS* is a right triangle for the given vertices. Explain.

18. *Q*(1, 0), *R*(1, 6), *S*(9, 0)

19. *Q*(3, 2), *R*(0, 6), *S*(6, 6)

20. *Q*(−4, 6), *R*(2, 11), *S*(4, −1)

21. *Q*(−9, −2), *R*(−4, −4), *S*(−6, −9)

Determine whether each set of numbers can be the measures of the sides of a right triangle. Then state whether they form a Pythagorean triple.

22. 8, 15, 17

23. 7, 24, 25

24. 20, 21, 31

25. 37, 12, 34

26. $\frac{1}{5}, \frac{1}{7}, \frac{\sqrt{74}}{35}$

27. $\frac{\sqrt{3}}{2}, \frac{\sqrt{2}}{3}, \frac{35}{36}$

28. $\frac{3}{5}, \frac{4}{5}, 1$

29. $\frac{6}{7}, \frac{8}{7}, \frac{10}{7}$

For Exercises 30–35, use the table of Pythagorean triples.

30. Copy and complete the table.

31. A *primitive* Pythagorean triple is a Pythagorean triple with no common factors except 1. Name any primitive Pythagorean triples contained in the table.

32. Describe the pattern that relates these sets of Pythagorean triples.

a	b	c
5	12	13
10	24	?
15	?	39
?	48	52

33. These Pythagorean triples are called a *family*. Why do you think this is?

34. Are the triangles described by a family of Pythagorean triples similar? Explain.

35. For each Pythagorean triple, find two triples in the same family.
  a. 8, 15, 17
  b. 9, 40, 41
  c. 7, 24, 25

GEOGRAPHY For Exercises 36 and 37, use the following information.

Denver is located at about 105 degrees longitude and 40 degrees latitude. San Francisco is located at about 122 degrees longitude and 38 degrees latitude. Las Vegas is located at about 115 degrees longitude and 36 degrees latitude. Using the lines of longitude and latitude, find each degree distance.

36. San Francisco to Denver

37. Las Vegas to Denver

38. **PROOF** Write a paragraph proof of Theorem 7.5.

39. **PROOF** Use the Pythagorean Theorem and the figure at the right to prove the Distance Formula.

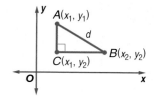

40. **PAINTING** A painter sets a ladder up to reach the bottom of a second-story window 16 feet above the ground. The base of the ladder is 12 feet from the house. While the painter mixes the paint, a neighbor's dog bumps the ladder, which moves the base 2 feet farther away from the house. How far up the side of the house does the ladder reach?

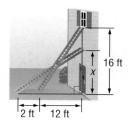

41. **SAILING** The mast of a sailboat is supported by wires called *shrouds*. What is the total length of wire needed to form these shrouds?

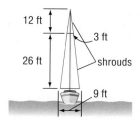

42. **LANDSCAPING** Six congruent square stones are arranged in an L-shaped walkway through a garden. If $x = 15$ inches, then find the area of the L-shaped walkway.

43. **NAVIGATION** A fishing trawler off the coast of Alaska was ordered by the U.S. Coast Guard to change course. They were to travel 6 miles west and then sail 12 miles south to miss a large iceberg before continuing on the original course. How many miles out of the way did the trawler travel?

44. **CRITICAL THINKING** The figure at the right is a rectangular prism with $AB = 8$, $BC = 6$, and $BF = 8$, and M is the midpoint of BD. Find BD and HM. How are EM, FM, and GM related to HM?

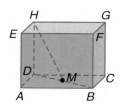

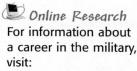
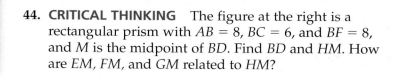

45. WRITING IN MATH Answer the question that was posed at the beginning of the lesson.

How are right triangles used to build suspension bridges?

Include the following in your answer:
- the locations of the right triangles, and
- an explanation of which parts of the right triangle are formed by the cables.

46. In the figure, if $AE = 10$, what is the value of h?

  Ⓐ 6   Ⓑ 8

  Ⓒ 10   Ⓓ 12

47. ALGEBRA If $x^2 + 36 = (9 - x)^2$, then find x.

  Ⓐ 6   Ⓑ no solution

  Ⓒ 2.5   Ⓓ 10

Graphing Calculator

PROGRAMMING For Exercises 48 and 49, use the following information.

The TI-83 Plus program uses a procedure for finding *Pythagorean triples* that was developed by Euclid around 320 B.C. Run the program to generate a list of Pythagorean triples.

48. List all the members of the 3-4-5 family that are generated by the program.

49. A geometry student made the conjecture that if three whole numbers are a Pythagorean triple, then their product is divisible by 60. Does this conjecture hold true for each triple that is produced by the program?

PROGRAM: PYTHTRIP	
:For (X, 2, 6)	:Disp B,A,C
:For (Y, 1, 5)	:Else
:If X > Y	:Disp A,B,C
:Then	:End
:int (X² − Y² + 0.5)→A	:End
:2XY→B	:Pause
:int (X²+Y²+0.5)→C	:Disp " "
:If A > B	:End
:Then	:End
	:Stop

Maintain Your Skills

Mixed Review **Find the geometric mean between each pair of numbers.** *(Lesson 7-1)*

50. 3 and 12   **51.** 9 and 12   **52.** 11 and 7

53. 6 and 9   **54.** 2 and 7   **55.** 2 and 5

Find the value of each expression. Then use that value as the next x in the expression. Repeat the process and describe your observations. *(Lesson 6-6)*

56. $\sqrt{2x}$, where x initially equals 5   **57.** 3^x, where x initially equals 1

58. $x^{\frac{1}{2}}$, where x initially equals 4   **59.** $\frac{1}{x}$, where x initially equals 4

60. Determine whether the sides of a triangle could have the lengths 12, 13, and 25. Explain. *(Lesson 5-4)*

Getting Ready for the Next Lesson **PREREQUISITE SKILL Simplify each expression by rationalizing the denominator.**
*(To review **simplifying radical expressions**, see pages 744 and 745.)*

61. $\dfrac{7}{\sqrt{3}}$ **62.** $\dfrac{18}{\sqrt{2}}$ **63.** $\dfrac{\sqrt{14}}{\sqrt{2}}$ **64.** $\dfrac{3\sqrt{11}}{\sqrt{3}}$ **65.** $\dfrac{24}{\sqrt{2}}$

66. $\dfrac{12}{\sqrt{3}}$ **67.** $\dfrac{2\sqrt{6}}{\sqrt{3}}$ **68.** $\dfrac{15}{\sqrt{3}}$ **69.** $\dfrac{2}{\sqrt{8}}$ **70.** $\dfrac{25}{\sqrt{10}}$

7-3 Special Right Triangles

What You'll Learn

- Use properties of 45°-45°-90° triangles.
- Use properties of 30°-60°-90° triangles.

How is triangle tiling used in wallpaper design?

Triangle tiling is the process of laying copies of a single triangle next to each other to fill an area. One type of triangle tiling is *wallpaper tiling*. There are exactly 17 types of triangle tiles that can be used for wallpaper tiling.

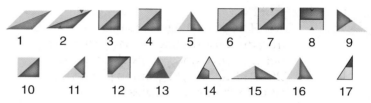

Tile 4 is made up of two 45°-45°-90° triangles that form a square. This tile is rotated to make the wallpaper design shown at the right.

PROPERTIES OF 45°-45°-90° TRIANGLES

Facts about 45°-45°-90° triangles are used to solve many geometry problems. The Pythagorean Theorem allows us to discover special relationships that exist among the sides of a 45°-45°-90° triangle.

Draw a diagonal of a square. The two triangles formed are isosceles right triangles. Let x represent the measure of each side and let d represent the measure of the hypotenuse.

$$d^2 = x^2 + x^2 \qquad \text{Pythagorean Theorem}$$
$$d^2 = 2x^2 \qquad \text{Add.}$$
$$d = \sqrt{2x^2} \qquad \text{Take the positive square root of each side.}$$
$$d = \sqrt{2} \cdot \sqrt{x^2} \qquad \text{Factor.}$$
$$d = x\sqrt{2} \qquad \text{Simplify.}$$

This algebraic proof verifies that the length of the hypotenuse of any 45°-45°-90° triangle is $\sqrt{2}$ times the length of its leg. The ratio of the sides is $1 : 1 : \sqrt{2}$.

Theorem 7.6

In a 45°-45°-90° triangle, the length of the hypotenuse is $\sqrt{2}$ times the length of a leg.

Example 1 · Find the Measure of the Hypotenuse

WALLPAPER TILING Assume that the length of one of the legs of the 45°-45°-90° triangles in the wallpaper in the figure is 4 inches. What is the length of the diagonal of the entire wallpaper square?

The length of each leg of the 45°-45°-90° triangle is 4 inches. The length of the hypotenuse is $\sqrt{2}$ times as long as a leg. The length of the hypotenuse of one of the triangles is $4\sqrt{2}$. There are four 45°-45°-90° triangles along the diagonal of the square. So, the length of the diagonal of the square is $4(4\sqrt{2})$ or $16\sqrt{2}$ inches.

Example 2 · Find the Measure of the Legs

Find x.

The length of the hypotenuse of a 45°-45°-90° triangle is $\sqrt{2}$ times the length of a leg of the triangle.

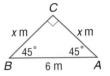

$$AB = (AC)\sqrt{2}$$

$6 = x\sqrt{2}$ $AB = 6, AC = x$

$\dfrac{6}{\sqrt{2}} = x$ Divide each side by $\sqrt{2}$.

$\dfrac{6}{\sqrt{2}} \cdot \dfrac{\sqrt{2}}{\sqrt{2}} = x$ Rationalize the denominator.

$\dfrac{6\sqrt{2}}{2} = x$ Multiply.

$3\sqrt{2} = x$ Divide.

> **Study Tip**
>
> **Rationalizing Denominators**
> To rationalize a denominator, multiply the fraction by 1 in the form of a radical over itself so that the product in the denominator is a rational number.

PROPERTIES OF 30°-60°-90° TRIANGLES

There is also a special relationship among the measures of the sides of a 30°-60°-90° triangle.

When an altitude is drawn from any vertex of an equilateral triangle, two congruent 30°-60°-90° triangles are formed. $\overline{LM}$ and $\overline{KM}$ are congruent segments, so let $LM = x$ and $KM = x$. By the Segment Addition Postulate, $LM + KM = KL$. Thus, $KL = 2x$. Since $\triangle JKL$ is an equilateral triangle, $KL = JL = JK$. Therefore, $JL = 2x$ and $JK = 2x$.

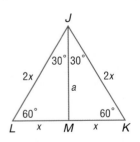

Let a represent the measure of the altitude. Use the Pythagorean Theorem to find a.

$(JM)^2 + (LM)^2 = (JL)^2$ Pythagorean Theorem

$a^2 + x^2 = (2x)^2$ $JM = a, LM = x, JL = 2x$

$a^2 + x^2 = 4x^2$ Simplify.

$a^2 = 3x^2$ Subtract x^2 from each side.

$a = \sqrt{3x^2}$ Take the positive square root of each side.

$a = \sqrt{3} \cdot \sqrt{x^2}$ Factor.

$a = x\sqrt{3}$ Simplify.

So, in a 30°-60°-90° triangle, the measures of the sides are x, $x\sqrt{3}$, and $2x$. The ratio of the sides is $1:\sqrt{3}:2$.

The relationship of the side measures leads to Theorem 7.7.

Theorem 7.7

In a 30°-60°-90° triangle, the length of the hypotenuse is twice the length of the shorter leg, and the length of the longer leg is $\sqrt{3}$ times the length of the shorter leg.

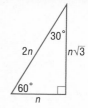

Example 3 30°-60°-90° Triangles

Find AC.

$\overline{AC}$ is the longer leg, $\overline{AB}$ is the shorter leg, and $\overline{BC}$ is the hypotenuse.

$AB = \frac{1}{2}(BC)$

$\quad = \frac{1}{2}(14)$ or 7 $\qquad BC = 14$

$AC = \sqrt{3}(AB)$

$\quad = \sqrt{3}(7)$ or $7\sqrt{3}$ $\quad AB = 7$

Example 4 Special Triangles in a Coordinate Plane

COORDINATE GEOMETRY Triangle *PCD* is a 30°-60°-90° triangle with right angle *C*. $\overline{CD}$ is the longer leg with endpoints *C*(3, 2) and *D*(9, 2). Locate point *P* in Quadrant I.

Graph *C* and *D*. $\overline{CD}$ lies on a horizontal gridline of the coordinate plane. Since $\overline{PC}$ will be perpendicular to $\overline{CD}$, it lies on a vertical gridline. Find the length of $\overline{CD}$.

$CD = |9 - 3| = 6$

$\overline{CD}$ is the longer leg. $\overline{PC}$ is the shorter leg.
So, $CD = \sqrt{3}(PC)$. Use *CD* to find *PC*.

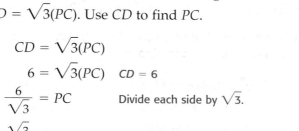

$\qquad CD = \sqrt{3}(PC)$

$\qquad 6 = \sqrt{3}(PC) \quad CD = 6$

$\qquad \frac{6}{\sqrt{3}} = PC \qquad$ Divide each side by $\sqrt{3}$.

$\qquad \frac{6}{\sqrt{3}} \cdot \frac{\sqrt{3}}{\sqrt{3}} = PC \qquad$ Rationalize the denominator.

$\qquad \frac{6\sqrt{3}}{3} = PC \qquad$ Multiply.

$\qquad 2\sqrt{3} = PC \qquad$ Simplify.

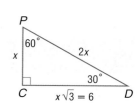

Point *P* has the same *x*-coordinate as *C*. *P* is located $2\sqrt{3}$ units above *C*. So, the coordinates of *P* are $(3, 2 + 2\sqrt{3})$ or about (3, 5.46).

Check for Understanding

Concept Check

1. **OPEN ENDED** Draw a 45°-45°-90° triangle. Be sure to label the angles and the sides and to explain how you made the drawing.

2. **Explain** how to draw a 30°-60°-90° triangle with the shorter leg 2 centimeters long.

3. **Write** an equation to find the length of a rectangle that has a diagonal twice as long as its width.

Guided Practice **Find x and y.**

4.
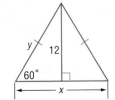

5.

6.

Find the missing measures.

7. If $c = 8$, find a and b.

8. If $b = 18$, find a and c.

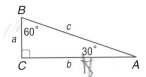

Triangle *ABD* is a 30°-60°-90° triangle with right angle *B* and with $\overline{AB}$ as the shorter leg. Graph *A* and *B*, and locate point *D* in Quadrant I.

9. $A(8, 0)$, $B(8, 3)$

10. $A(6, 6)$, $B(2, 6)$

Application 11. **SOFTBALL** Find the distance from home plate to second base if the bases are 90 feet apart.

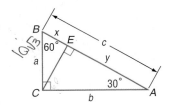

Practice and Apply

Find x and y.

Homework Help	
For Exercises	**See Examples**
12, 13, 17, 22, 25	1, 2
14–16, 18–21, 23, 24	3
27–31	4

Extra Practice
See page 767.

12.
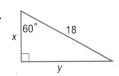

13.
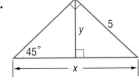

14.
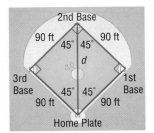

15.
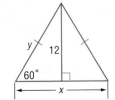

16.
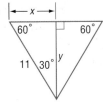

17.

For Exercises 18 and 19, use the figure at the right.

18. If $a = 10\sqrt{3}$, find CE and y.

19. If $x = 7\sqrt{3}$, find a, CE, y, and b.

20. The length of an altitude of an equilateral triangle is 12 feet. Find the length of a side of the triangle.

21. The perimeter of an equilateral triangle is 45 centimeters. Find the length of an altitude of the triangle.

22. The length of a diagonal of a square is $22\sqrt{2}$ millimeters. Find the perimeter of the square.

23. The altitude of an equilateral triangle is 7.4 meters long. Find the perimeter of the triangle.

24. The diagonals of a rectangle are 12 inches long and intersect at an angle of 60°. Find the perimeter of the rectangle.

25. The sum of the squares of the measures of the sides of a square is 256. Find the measure of a diagonal of the square.

26. Find x, y, z, and the perimeter of $ABCD$.

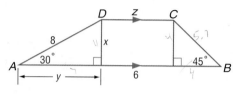

27. $\triangle PAB$ is a 45°-45°-90° triangle with right angle B. Find the coordinates of P in Quadrant I for $A(-3, 1)$ and $B(4, 1)$.

28. $\triangle PGH$ is a 45°-45°-90° triangle with $m\angle P = 90$. Find the coordinates of P in Quadrant I for $G(4, -1)$ and $H(4, 5)$.

29. $\triangle PCD$ is a 30°-60°-90° triangle with right angle C and $\overline{CD}$ the longer leg. Find the coordinates of P in Quadrant III for $C(-3, -6)$ and $D(-3, 7)$.

30. $\triangle PCD$ is a 30°-60°-90° triangle with $m\angle C = 30$ and hypotenuse $\overline{CD}$. Find the coordinates of P for $C(2, -5)$ and $D(10, -5)$ if P lies above $\overline{CD}$.

31. If $\overline{PQ} \parallel \overline{SR}$, use the figure to find a, b, c, and d.

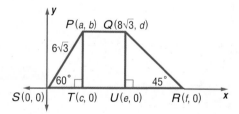

TRIANGLE TILING For Exercises 32–35, use the following information.
Triangle tiling refers to the process of taking many copies of a single triangle and laying them next to each other to fill an area. For example, the pattern shown is composed of tiles like the one outlined.

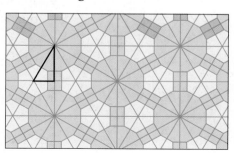

32. How many 30°-60°-90° triangles are used to create the basic circular pattern?

33. Which angle of the 30°-60°-90° triangle is being rotated to make the basic shape?

34. Explain why there are no gaps in the basic pattern.

35. Use grid paper to cut out 30°-60°-90° triangles. Color the same pattern on each triangle. Create one basic figure that would be part of a wallpaper tiling.

36. Find x, y, and z.

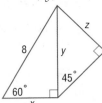

37. If $BD = 8\sqrt{3}$ and $m\angle DHB = 60$, find BH.

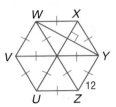

38. Each triangle in the figure is a 30°-60°-90° triangle. Find x.

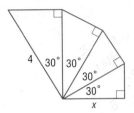

39. In regular hexagon $UVWXYZ$, each side is 12 centimeters long. Find WY.

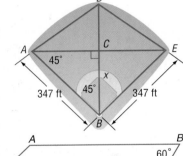

40. BASEBALL The diagram at the right shows some dimensions of Comiskey Park in Chicago, Illinois. $\overline{BD}$ is a segment from home plate to dead center field, and $\overline{AE}$ is a segment from the left field foul-ball pole to the right field foul-ball pole. If the center fielder is standing at C, how far is he from home plate?

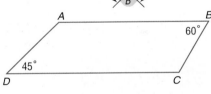

41. CRITICAL THINKING Given figure $ABCD$, with $\overline{AB} \parallel \overline{DC}$, $m\angle B = 60$, $m\angle D = 45$, $BC = 8$, and $AB = 24$, find the perimeter.

42. **WRITING IN MATH** Answer the question that was posed at the beginning of the lesson.

How is triangle tiling used in wallpaper design?

Include the following in your answer:
- which of the numbered designs contain 30°-60°-90° triangles and which contain 45°-45°-90° triangles, and
- a reason why rotations of the basic design left no holes in the completed design.

43. In the right triangle, what is AB if $BC = 6$?

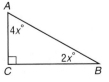

Ⓐ 12 units

Ⓑ $6\sqrt{2}$ units

Ⓒ $4\sqrt{3}$ units

Ⓓ $2\sqrt{3}$ units

44. SHORT RESPONSE For real numbers a and b, where $b \neq 0$, if $a \star b = \dfrac{a^2}{b^2}$, then $(3 \star 4)(5 \star 3) = \underline{}$.

Mixed Review

Determine whether each set of measures can be the sides of a right triangle. Then state whether they form a Pythagorean triple. *(Lesson 7-2)*

45. 3, 4, 5

46. 9, 40, 41

47. 20, 21, 31

48. 20, 48, 52

49. 7, 24, 25

50. 12, 34, 37

Find x, y, and z. *(Lesson 7-1)*

51.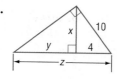

52.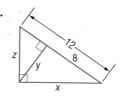

53.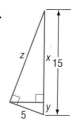

Write an inequality or equation relating each pair of angles. *(Lesson 5-5)*

54. $m\angle ALK$, $m\angle ALN$

55. $m\angle ALK$, $m\angle NLO$

56. $m\angle OLK$, $m\angle NLO$

57. $m\angle KLO$, $m\angle ALN$

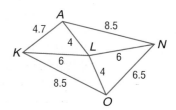

58. Determine whether $\triangle JKL$ with vertices $J(-3, 2)$, $K(-1, 5)$, and $L(4, 4)$ is congruent to $\triangle RST$ with vertices $R(-6, 6)$, $S(-4, 3)$, and $T(1, 4)$. Explain. *(Lesson 4-4)*

Getting Ready for the Next Lesson

PREREQUISITE SKILL Solve each equation.
(To review solving equations, see pages 737 and 738.)

59. $5 = \dfrac{x}{3}$

60. $\dfrac{x}{9} = 0.14$

61. $0.5 = \dfrac{10}{k}$

62. $0.2 = \dfrac{13}{g}$

63. $\dfrac{7}{n} = 0.25$

64. $9 = \dfrac{m}{0.8}$

65. $\dfrac{24}{x} = 0.4$

66. $\dfrac{35}{y} = 0.07$

Practice Quiz 1

Lessons 7-1 through 7-3

Find the measure of the altitude drawn to the hypotenuse. *(Lesson 7-1)*

1.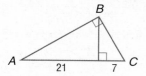

2.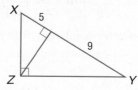

3. Determine whether $\triangle ABC$ with vertices $A(2, 1)$, $B(4, 0)$, and $C(5, 7)$ is a right triangle. Explain. *(Lesson 7-2)*

Find x and y. *(Lesson 7-3)*

4.

5.

 www.geometryonline.com/self_check_quiz

7-4 Trigonometry

- Find trigonometric ratios using right triangles.
- Solve problems using trigonometric ratios.

Vocabulary

- trigonometry
- trigonometric ratio
- sine
- cosine
- tangent

How can surveyors determine angle measures?

The old surveyor's telescope shown at right is called a theodolite (thee AH duh lite). It is an optical instrument used to measure angles in surveying, navigation, and meteorology. It consists of a telescope fitted with a level and mounted on a tripod so that it is free to rotate about its vertical and horizontal axes. After measuring angles, surveyors apply trigonometry to calculate distance or height.

TRIGONOMETRIC RATIOS The word **trigonometry** comes from two Greek terms, *trigon*, meaning triangle, and *metron*, meaning measure. The study of trigonometry involves triangle measurement. A ratio of the lengths of sides of a right triangle is called a **trigonometric ratio**. The three most common trigonometric ratios are **sine**, **cosine**, and **tangent**.

Key Concept — Trigonometric Ratios

Words	Symbols	Models
sine of $\angle A = \dfrac{\text{measure of leg opposite } \angle A}{\text{measure of hypotenuse}}$ sine of $\angle B = \dfrac{\text{measure of leg opposite } \angle B}{\text{measure of hypotenuse}}$	$\sin A = \dfrac{BC}{AB}$ $\sin B = \dfrac{AC}{AB}$	A, hypotenuse, leg opposite $\angle A$, leg opposite $\angle B$, B, C
cosine of $\angle A = \dfrac{\text{measure of leg adjacent to } \angle A}{\text{measure of hypotenuse}}$ cosine of $\angle B = \dfrac{\text{measure of leg adjacent to } \angle B}{\text{measure of hypotenuse}}$	$\cos A = \dfrac{AC}{AB}$ $\cos B = \dfrac{BC}{AB}$	A, hypotenuse, leg adjacent to $\angle B$, leg adjacent to $\angle A$, B, C
tangent of $\angle A = \dfrac{\text{measure of leg opposite } \angle A}{\text{measure of leg adjacent to } \angle A}$ tangent of $\angle B = \dfrac{\text{measure of leg opposite } \angle B}{\text{measure of leg adjacent to } \angle B}$	$\tan A = \dfrac{BC}{AC}$ $\tan B = \dfrac{AC}{BC}$	A, hypotenuse, leg opposite $\angle A$ and adjacent to $\angle B$, leg adjacent to $\angle A$ and opposite $\angle B$, B, C

Trigonometric ratios are related to the acute angles of a right triangle, *not* the right angle.

Geometry Activity

Trigonometric Ratios

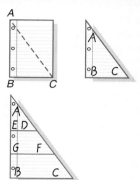

- Fold a rectangular piece of paper along a diagonal from *A* to *C*. Then cut along the fold to form right triangle *ABC*. Write the name of each angle on the inside of the triangle.
- Fold the triangle so that there are two segments perpendicular to $\overline{BA}$. Label points *D*, *E*, *F*, and *G* as shown. Use a ruler to measure $\overline{AC}$, $\overline{AB}$, $\overline{BC}$, $\overline{AF}$, $\overline{AG}$, $\overline{FG}$, $\overline{AD}$, $\overline{AE}$, and $\overline{DE}$ to the nearest millimeter.

Analyze

1. What is true of △AED, △AGF, and △ABC?
2. Copy the table. Write the ratio of the side lengths for each trigonometric ratio. Then calculate a value for each ratio to the nearest ten-thousandth.

	In △AED	in △AGF	In △ABC
sin *A*			
cos *A*			
tan *A*			

3. Study the table. Write a sentence about the patterns you observe with the trigonometric ratios.
4. What is true about $m\angle A$ in each triangle?

As the Geometry Activity suggests, the value of a trigonometric ratio depends *only* on the measure of the angle. It does not depend on the size of the triangle.

Example 1 *Find Sine, Cosine, and Tangent Ratios*

Find sin *R*, cos *R*, tan *R*, sin *S*, cos *S*, and tan *S*.
Express each ratio as a fraction and as a decimal.

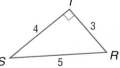

$$\sin R = \frac{\text{opposite leg}}{\text{hypotenuse}}$$
$$= \frac{ST}{RS}$$
$$= \frac{4}{5} \text{ or } 0.8$$

$$\cos R = \frac{\text{adjacent leg}}{\text{hypotenuse}}$$
$$= \frac{RT}{RS}$$
$$= \frac{3}{5} \text{ or } 0.6$$

$$\tan R = \frac{\text{opposite leg}}{\text{adjacent leg}}$$
$$= \frac{ST}{RT}$$
$$= \frac{4}{3} \text{ or } 1.\overline{3}$$

$$\sin S = \frac{\text{opposite leg}}{\text{hypotenuse}}$$
$$= \frac{RT}{RS}$$
$$= \frac{3}{5} \text{ or } 0.6$$

$$\cos S = \frac{\text{adjacent leg}}{\text{hypotenuse}}$$
$$= \frac{ST}{RS}$$
$$= \frac{4}{5} \text{ or } 0.8$$

$$\tan S = \frac{\text{opposite leg}}{\text{adjacent leg}}$$
$$= \frac{RT}{ST}$$
$$= \frac{3}{4} \text{ or } 0.75$$

Example 2 *Use a Calculator to Evaluate Expressions*

Use a calculator to find each value to the nearest ten thousandth.

a. cos 39°

  KEYSTROKES: [COS] 39 [ENTER]

  cos 39° ≈ 0.7771

b. sin 67°

  KEYSTROKES: [SIN] 67 [ENTER]

  sin 67° ≈ 0.9205

USE TRIGONOMETRIC RATIOS   You can use trigonometric ratios to find the missing measures of a right triangle if you know the measures of two sides of a triangle or the measure of one side and one acute angle.

Example 3 *Use Trigonometric Ratios to Find a Length*

SURVEYING   Dakota is standing on the ground 97 yards from the base of a cliff. Using a theodolite, he noted that the angle formed by the ground and the line of sight to the top of the cliff was 56°. Find the height of the cliff to the nearest yard.

Let x be the height of the cliff in yards.

$$\tan 56° = \frac{x}{97} \qquad \tan = \frac{\text{leg opposite}}{\text{leg adjacent}}$$

$$97 \tan 56° = x \qquad \text{Multiply each side by 97.}$$

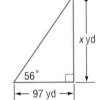

Use a calculator to find x.

KEYSTROKES: 97 [TAN] 56 [ENTER]   *143.8084139*

The cliff is about 144 yards high.

Study Tip

Tangent
The tangent of angle A is the same as the slope of the line through the origin that forms an angle of measure A with the positive x-axis.

When solving equations like $3x = -27$, you use the inverse of multiplication to find x. In trigonometry, you can find the measure of the angle by using the inverse of sine, cosine, or tangent.

Given equation	To find the angle	Read as
sin $A = x$	$A = \sin^{-1}(x)$	A equals *the inverse sine of x.*
cos $A = y$	$A = \cos^{-1}(y)$	A equals *the inverse cosine of y.*
tan $A = z$	$A = \tan^{-1}(z)$	A equals *the inverse tangent of z.*

Example 4 *Use Trigonometric Ratios to Find an Angle Measure*

COORDINATE GEOMETRY   Find $m\angle A$ in right triangle ABC for $A(1, 2)$, $B(6, 2)$, and $C(5, 4)$.

Explore   You know the coordinates of the vertices of a right triangle and that $\angle C$ is the right angle. You need to find the measure of one of the angles.

Plan   Use the Distance Formula to find the measure of each side. Then use one of the trigonometric ratios to write an equation. Use the inverse to find $m\angle A$.

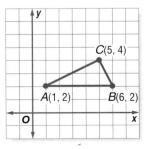

Solve

$$AB = \sqrt{(6-1)^2 + (2-2)^2} \qquad BC = \sqrt{(5-6)^2 + (4-2)^2}$$
$$= \sqrt{25+0} \text{ or } 5 \qquad\qquad = \sqrt{1+4} \text{ or } \sqrt{5}$$

$$AC = \sqrt{(5-1)^2 + (4-2)^2}$$
$$= \sqrt{16+4}$$
$$= \sqrt{20} \text{ or } 2\sqrt{5}$$

Use the cosine ratio.

$$\cos A = \frac{AC}{AB} \qquad\qquad \cos = \frac{\text{leg adjacent}}{\text{hypotenuse}}$$

$$\cos A = \frac{2\sqrt{5}}{5} \qquad\qquad AC = 2\sqrt{5} \text{ and } AB = 5$$

$$A = \cos^{-1}\left(\frac{2\sqrt{5}}{5}\right) \qquad \text{Solve for } A.$$

Use a calculator to find $m\angle A$.

KEYSTROKES: 2nd [COS⁻¹] 2 2nd [√] 5) ÷ 5 ENTER

$m\angle A \approx 26.56505118$

The measure of $\angle A$ is about 26.6.

Examine   Use the sine ratio to check the answer.

$$\sin A = \frac{BC}{AB} \qquad \sin = \frac{\text{leg opposite}}{\text{hypotenuse}}$$

$$\sin A = \frac{\sqrt{5}}{5} \qquad BC = \sqrt{5} \text{ and } AB = 5$$

KEYSTROKES: 2nd [SIN⁻¹] 2nd [√] 5) ÷ 5 ENTER

$m\angle A \approx 26.56505118$

The answer is correct.

Check for Understanding

Concept Check

1. **Explain** why trigonometric ratios do not depend on the size of the right triangle.

2. **OPEN ENDED**   Draw a right triangle and label the measures of one acute angle and the measure of the side opposite that angle. Then solve for the remaining measures.

3. **Compare and contrast** the sine, cosine, and tangent ratios.

4. **Explain** the difference between $\tan A = \frac{x}{y}$ and $\tan^{-1}\left(\frac{x}{y}\right) = A$.

Guided Practice   Use $\triangle ABC$ to find $\sin A$, $\cos A$, $\tan A$, $\sin B$, $\cos B$, and $\tan B$. Express each ratio as a fraction and as a decimal to the nearest hundredth.

5. $a = 14$, $b = 48$, and $c = 50$

6. $a = 8$, $b = 15$, and $c = 17$

Use a calculator to find each value. Round to the nearest ten-thousandth.

7. $\sin 57°$

8. $\cos 60°$

9. $\cos 33°$

10. $\tan 30°$

11. $\tan 45°$

12. $\sin 85°$

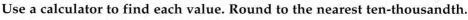

Lesson 7-4 Trigonometry   **367**

Find the measure of each angle to the nearest tenth of a degree.

13. $\tan A = 1.4176$

14. $\sin B = 0.6307$

COORDINATE GEOMETRY Find the measure of the angle to the nearest tenth in each right triangle ABC.

15. $\angle A$ in $\triangle ABC$, for $A(6, 0)$, $B(-4, 2)$, and $C(0, 6)$

16. $\angle B$ in $\triangle ABC$, for $A(3, -3)$, $B(7, 5)$, and $C(7, -3)$

Application **17. SURVEYING** Maureen is standing on horizontal ground level with the base of the CN Tower in Toronto, Ontario. The angle formed by the ground and the line segment from her position to the top of the tower is 31.2°. She knows that the height of the tower to the top of the antennae is about 1815 feet. Find her distance from the CN Tower to the nearest foot.

Practice and Apply

Homework Help

For Exercises	See Examples
18–21, 28–36	1
22–27	2
43–48	3
37–42, 52–54	4

Extra Practice
See page 767.

Use $\triangle PQR$ with right angle R to find $\sin P$, $\cos P$, $\tan P$, $\sin Q$, $\cos Q$, and $\tan Q$. Express each ratio as a fraction, and as a decimal to the nearest hundredth.

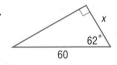

18. $p = 12$, $q = 35$, and $r = 37$

19. $p = \sqrt{6}$, $q = 2\sqrt{3}$, and $r = 3\sqrt{2}$

20. $p = \dfrac{3}{2}$, $q = \dfrac{3\sqrt{3}}{2}$, and $r = 3$

21. $p = 2\sqrt{3}$, $q = \sqrt{15}$, and $r = 3\sqrt{3}$

Use a calculator to find each value. Round to the nearest ten-thousandth.

22. $\sin 6°$

23. $\tan 42.8°$

24. $\cos 77°$

25. $\sin 85.9°$

26. $\tan 12.7°$

27. $\cos 22.5°$

Use the figure to find each trigonometric ratio. Express answers as a fraction and as a decimal rounded to the nearest ten-thousandth.

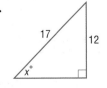

28. $\sin A$

29. $\tan B$

30. $\cos A$

31. $\sin x°$

32. $\cos x°$

33. $\tan A$

34. $\cos B$

35. $\sin y°$

36. $\tan x°$

Find the measure of each angle to the nearest tenth of a degree.

37. $\sin B = 0.7245$

38. $\cos C = 0.2493$

39. $\tan E = 9.4618$

40. $\sin A = 0.4567$

41. $\cos D = 0.1212$

42. $\tan F = 0.4279$

Find x. Round to the nearest tenth.

43.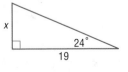

44.

45.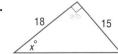

46.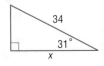

47.

48.

49. AVIATION A plane is one mile above sea level when it begins to climb at a constant angle of 3° for the next 60 ground miles. About how far above sea level is the plane after its climb?

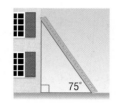

SAFETY For Exercises 50 and 51, use the following information.
To guard against a fall, a ladder should make an angle of 75° or less with the ground.

50. What is the maximum height that a 20-foot ladder can reach safely?

51. How far from the building is the base of the ladder at the maximum height?

COORDINATE GEOMETRY Find the measure of each angle to the nearest tenth in each right triangle.

52. $\angle J$ in $\triangle JCL$ for $J(2, 2)$, $C(2, -2)$, and $L(7, -2)$

53. $\angle C$ in $\triangle BCD$ for $B(-1, -5)$, $C(-6, -5)$, and $D(-1, 2)$

54. $\angle X$ in $\triangle XYZ$ for $X(-5, 0)$, $Y(7, 0)$, and $Z(0, \sqrt{35})$

55. Find the perimeter of $\triangle ABC$ if $m\angle A = 35$, $m\angle C = 90$, and $AB = 20$ inches.

Find x and y. Round to the nearest tenth.

56.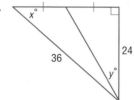

57.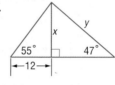

58.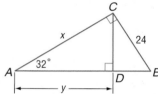

ASTRONOMY For Exercises 59 and 60, use the following information.
One way to find the distance between the sun and a relatively close star is to determine the angles of sight for the star exactly six months apart. Half the measure formed by these two angles of sight is called the *stellar parallax*. Distances in space are sometimes measured in *astronomical units*. An astronomical unit is equal to the average distance between Earth and the sun.

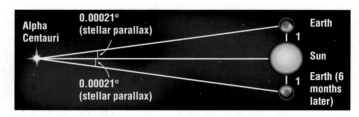

59. Find the distance between Alpha Centauri and the sun.

60. Make a conjecture as to why this method is used only for close stars.

61. CRITICAL THINKING Use the figure at the right to find sin $x°$.

62. **WRITING IN MATH** Answer the question that was posed at the beginning of the lesson.

How do surveyors determine angle measures?

Include the following in your answer:
- where theodolites are used, and
- the kind of information one obtains from a theodolite.

Exercise 61

63. Find cos C.

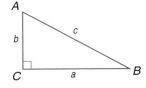

  Ⓐ $\dfrac{3}{5}$　　　　　Ⓑ $\dfrac{3}{4}$

  Ⓒ $\dfrac{4}{5}$　　　　　Ⓓ $\dfrac{5}{4}$

64. ALGEBRA　If $x^2 = 15^2 + 24^2 - 15(24)$, find x.

  Ⓐ 20.8　　　　Ⓑ 21　　　　Ⓒ 12　　　　Ⓓ 9

*Extending
the Lesson* Each of the basic trigonometric ratios has a
reciprocal ratio. The reciprocals of the sine,
cosine, and tangent are called the *cosecant*,
secant, and the *cotangent*, respectively.

Reciprocal	Trigonometric Ratio	Abbreviation	Definition	
$\dfrac{1}{\sin A}$	cosecant of $\angle A$	csc A	$\dfrac{\text{measure of the hypotenuse}}{\text{measure of the leg opposite } \angle A}$	$= \dfrac{c}{a}$
$\dfrac{1}{\cos A}$	secant of $\angle A$	sec A	$\dfrac{\text{measure of the hypotenuse}}{\text{measure of the leg adjacent } \angle A}$	$= \dfrac{c}{b}$
$\dfrac{1}{\tan A}$	cotangent of $\angle A$	cot A	$\dfrac{\text{measure of the leg adjacent } \angle A}{\text{measure of the leg opposite } \angle A}$	$= \dfrac{b}{a}$

Use $\triangle ABC$ to find csc A, sec A, cot A, csc B, sec B, and cot B. Express each ratio as
a fraction or as a radical in simplest form.

65. $a = 3$, $b = 4$, and $c = 5$　　　　　**66.** $a = 12$, $b = 5$, and $c = 13$

67. $a = 4$, $b = 4\sqrt{3}$, and $c = 8$　　　　**68.** $a = 2\sqrt{2}$, $b = 2\sqrt{2}$, and $c = 4$

Maintain Your Skills

Mixed Review **Find each measure.** *(Lesson 7-3)*

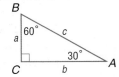

69. If $a = 4$, find b and c.

70. If $b = 3$, find a and c.

71. If $c = 5$, find a and b.

**Determine whether each set of measures can be the sides of a right triangle.
Then state whether they form a Pythagorean triple.** *(Lesson 7-2)*

72. 4, 5, 6　　　**73.** 5, 12, 13　　　**74.** 9, 12, 15　　　**75.** 8, 12, 16

76. TELEVISION　During a 30-minute television program, the ratio of minutes of
commercials to minutes of the actual show is 4 : 11. How many minutes are
spent on commercials? *(Lesson 6-1)*

*Getting Ready for
the Next Lesson* **PREREQUISITE SKILL**　**Find each angle measure if $h \parallel k$.** *(To review
angles formed by parallel lines and a transversal, see Lesson 3-2.)*

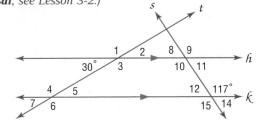

77. $m\angle 15$　　　　**78.** $m\angle 7$

79. $m\angle 3$　　　　**80.** $m\angle 12$

81. $m\angle 11$　　　　**82.** $m\angle 4$

7-5 Angles of Elevation and Depression

What You'll Learn

- Solve problems involving angles of elevation.
- Solve problems involving angles of depression.

Vocabulary

- angle of elevation
- angle of depression

How do airline pilots use angles of elevation and depression?

A pilot is getting ready to take off from Mountain Valley airport. She looks up at the peak of a mountain immediately in front of her. The pilot must estimate the speed needed and the angle formed by a line along the runway and a line from the plane to the peak of the mountain to clear the mountain.

ANGLES OF ELEVATION

An **angle of elevation** is the angle between the line of sight and the horizontal when an observer looks upward.

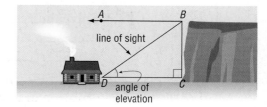

Example 1 Angle of Elevation

AVIATION The peak of Goose Bay Mountain is 400 meters higher than the end of a local airstrip. The peak rises above a point 2025 meters from the end of the airstrip. A plane takes off from the end of the runway in the direction of the mountain at an angle that is kept constant until the peak has been cleared. If the pilot wants to clear the mountain by 50 meters, what should the angle of elevation be for the takeoff to the nearest tenth of a degree?

Make a drawing.

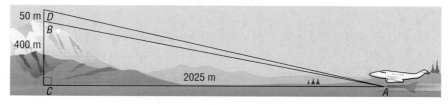

Since CB is 400 meters and BD is 50 meters, CD is 450 meters. Let x represent $m\angle DAC$.

$\tan x° = \dfrac{CD}{AC}$ $\tan = \dfrac{\text{opposite}}{\text{adjacent}}$

$\tan x° = \dfrac{450}{2025}$ $CD = 450, AC = 2025$

$x = \tan^{-1}\left(\dfrac{450}{2025}\right)$ Solve for x.

$x \approx 12.5$ Use a calculator.

The angle of elevation for the takeoff should be more than 12.5°.

ANGLES OF DEPRESSION An **angle of depression** is the angle between the line of sight when an observer looks downward, and the horizontal.

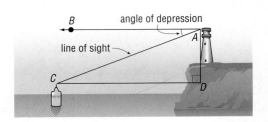

Example 2 *Angle of Depression*

Short-Response Test Item

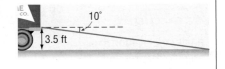

The tailgate of a moving van is 3.5 feet above the ground. A loading ramp is attached to the rear of the van at an incline of 10°. Find the length of the ramp to the nearest tenth foot.

Read the Test Item

The angle of depression between the ramp and the horizontal is 10°. Use trigonometry to find the length of the ramp.

Solve the Test Item

Method 1

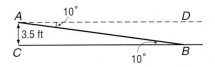

The ground and the horizontal level with the back of the van are parallel. Therefore, $m\angle DAB = m\angle ABC$ since they are alternate interior angles.

$$\sin 10° = \frac{3.5}{AB}$$

$$AB \sin 10° = 3.5$$

$$AB = \frac{3.5}{\sin 10°}$$

$$AB \approx 20.2$$

The ramp is about 20.2 feet long.

Method 2

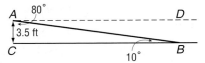

The horizontal line from the back of the van and the segment from the ground to the back of the van are perpendicular. So, $\angle DAB$ and $\angle BAC$ are complementary angles. Therefore, $m\angle BAC = 90 - 10$ or 80.

$$\cos 80° = \frac{3.5}{AB}$$

$$AB \cos 80° = 3.5$$

$$AB = \frac{3.5}{\cos 80°}$$

$$AB \approx 20.2$$

Angles of elevation or depression to two different objects can be used to find the distance between those objects.

Example 3 *Indirect Measurement*

Olivia is in a lighthouse on a cliff. She observes two sailboats due east of the lighthouse. The angles of depression to the two boats are 33° and 57°. Find the distance between the two sailboats to the nearest foot.

△*CDA* and △*CDB* are right triangles, and and *CD* = 110 + 85 or 195. The distance between the boats is *AB* or *BD* − *AD*. Use the right triangles to find these two lengths.

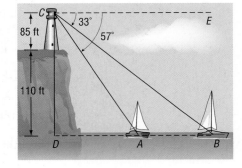

Because $\overline{CE}$ and $\overline{DB}$ are horizontal lines, they are parallel. Thus, $\angle ECB \cong \angle CBD$ and $\angle ECA \cong \angle CAD$ because they are alternate interior angles. This means that $m\angle CBD = 33$ and $m\angle CAD = 57$.

$\tan 33° = \dfrac{195}{DB}$ $\tan = \dfrac{\text{opposite}}{\text{adjacent}}; m\angle CBD = 33$

$DB \tan 33° = 195$ Multiply each side by DB.

$DB = \dfrac{195}{\tan 33°}$ Divide each side by tan 33°.

$DB \approx 300.27$ Use a calculator.

$\tan 57° = \dfrac{195}{DA}$ $\tan = \dfrac{\text{opposite}}{\text{adjacent}}; m\angle CAD = 57$

$DA \tan 57° = 195$ Multiply each side by DA.

$DA = \dfrac{195}{\tan 57°}$ Divide each side by tan 57°.

$DA \approx 126.63$ Use a calculator.

The distance between the boats is $DB - DA$.

$DB - DA \approx 300.27 - 126.63$ or about 174 feet.

Check for Understanding

Concept Check

1. **OPEN ENDED** Find a real-life example of an angle of depression. Draw a diagram and identify the angle of depression.

2. **Explain** why an angle of elevation is given that name.

3. **Name** the angles of depression and elevation in the figure.

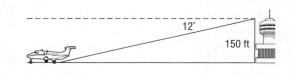

Guided Practice

4. **AVIATION** A pilot is flying at 10,000 feet and wants to take the plane up to 20,000 feet over the next 50 miles. What should be his angle of elevation to the nearest tenth? (*Hint:* There are 5280 feet in a mile.)

5. **SHADOWS** Find the angle of elevation of the sun when a 7.6-meter flagpole casts a 18.2-meter shadow. Round to the nearest tenth of a degree.

6. **SALVAGE** A salvage ship uses sonar to determine that the angle of depression to a wreck on the ocean floor is 13.25°. The depth chart shows that the ocean floor is 40 meters below the surface. How far must a diver lowered from the salvage ship walk along the ocean floor to reach the wreck?

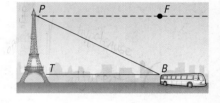

Standardized Test Practice
Ⓐ Ⓑ Ⓒ Ⓓ

7. **SHORT RESPONSE** From the top of a 150-foot high tower, an air traffic controller observes an airplane on the runway. To the nearest foot, how far from the base of the tower is the airplane?

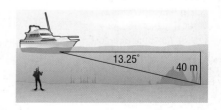

www.geometryonline.com/extra_examples

Homework Help

For Exercises	See Examples
12, 14–18	1
9–11	2
8, 13, 19, 20	3

Extra Practice
See page 768.

8. **BOATING**   Two boats are observed by a parasailer 75 meters above a lake. The angles of depression are 12.5° and 7°. How far apart are the boats?

9. **GOLF**   A golfer is standing at the tee, looking up to the green on a hill. If the tee is 36 yards lower than the green and the angle of elevation from the tee to the hole is 12°, find the distance from the tee to the hole.

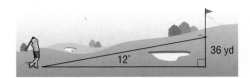

10. **AVIATION**   After flying at an altitude of 500 meters, a helicopter starts to descend when its ground distance from the landing pad is 11 kilometers. What is the angle of depression for this part of the flight?

11. **SLEDDING**   A sledding run is 300 yards long with a vertical drop of 27.6 yards. Find the angle of depression of the run.

12. **RAILROADS**   The Monongahela Incline overlooks the city of Pittsburgh, Pennsylvania. Refer to the information at the left to determine the incline of the railway.

13. **AMUSEMENT PARKS**   From the top of a roller coaster, 60 yards above the ground, a rider looks down and sees the merry-go-round and the Ferris wheel. If the angles of depression are 11° and 8° respectively, how far apart are the merry-go-round and the Ferris wheel?

CIVIL ENGINEERING   For Exercises 14 and 15, use the following information.
The percent grade of a highway is the ratio of the vertical rise or fall over a given horizontal distance. The ratio is expressed as a percent to the nearest whole number. Suppose a highway has a vertical rise of 140 feet for every 2000 feet of horizontal distance.

14. Calculate the percent grade of the highway.

15. Find the angle of elevation that the highway makes with the horizontal.

16. **SKIING**   A ski run has an angle of elevation of 24.4° and a vertical drop of 1100 feet. To the nearest foot, how long is the ski run?

GEYSERS   For Exercises 17 and 18, use the following information.
Kirk visits Yellowstone Park and Old Faithful on a perfect day. His eyes are 6 feet from the ground, and the geyser can reach heights ranging from 90 feet to 184 feet.

17. If Kirk stands 200 feet from the geyser and the eruption rises 175 feet in the air, what is the angle of elevation to the top of the spray to the nearest tenth?

18. In the afternoon, Kirk returns and observes the geyser's spray reach a height of 123 feet when the angle of elevation is 37°. How far from the geyser is Kirk standing to the nearest tenth of a foot?

19. **BIRDWATCHING**   Two observers are 200 feet apart, in line with a tree containing a bird's nest. The angles of elevation to the bird's nest are 30° and 60°. How far is each observer from the base of the tree?

More About. . .

Railroads •·············

The Monongahela Incline is 635 feet long with a vertical rise of 369.39 feet. It was built at a cost of $50,000 and opened on May 28, 1870. It is still used by commuters to and from Pittsburgh.

Source: www.portauthority.org

20. METEOROLOGY The altitude of the base of a cloud formation is called the *ceiling*. To find the ceiling one night, a meteorologist directed a spotlight vertically at the clouds. Using a theodolite placed 83 meters from the spotlight and 1.5 meters above the ground, he found the angle of elevation to be 62.7°. How high was the ceiling?

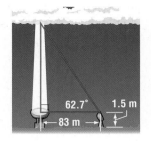

MEDICINE   For Exercises 21–23, use the following information.
A doctor is using a treadmill to assess the strength of a patient's heart. At the beginning of the exam, the 48-inch long treadmill is set at an incline of 10°.

21. How far off the horizontal is the raised end of the treadmill at the beginning of the exam?

22. During one stage of the exam, the end of the treadmill is 10 inches above the horizontal. What is the incline of the treadmill to the nearest degree?

23. Suppose the exam is divided into five stages and the incline of the treadmill is increased 2° for each stage. Does the end of the treadmill rise the same distance between each stage?

24. TRAVEL Kwan-Yong uses a theodolite to measure the angle of elevation from the ground to the top of Ayers Rock to be 15.85°. He walks half a kilometer closer and measures the angle of elevation to be 25.6°. How high is Ayers Rock to the nearest meter?

25. AEROSPACE On July 20, 1969, Neil Armstrong became the first human to walk on the moon. During this mission, *Apollo 11* orbited the moon three miles above the surface. At one point in the orbit, the onboard guidance system measured the angles of depression to the far and near edges of a large crater. The angles measured 16° and 29°, respectively. Find the distance across the crater.

Online Research **Data Update** Use the Internet to determine the angle of depression formed by someone aboard the international space station looking down to your community. Visit www.geometryonline.com/data_update to learn more.

26. CRITICAL THINKING Two weather observation stations are 7 miles apart. A weather balloon is located between the stations. From Station 1, the angle of elevation to the weather balloon is 33°. From Station 2, the angle of elevation to the balloon is 52°. Find the altitude of the balloon to the nearest tenth of a mile.

27. **WRITING IN MATH** Answer the question that was posed at the beginning of the lesson.

How do airline pilots use angles of elevation and depression?

Include the following in your answer:
- when pilots use angles of elevation or depression, and
- the difference between angles of elevation and depression.

28. The top of a signal tower is 120 meters above sea level. The angle of depression from the top of tower to a passing ship is 25°. How many meters from the foot of the tower is the ship?

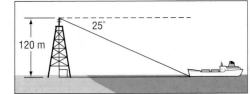

Ⓐ 283.9 m   Ⓑ 257.3 m
Ⓒ 132.4 m   Ⓓ 56 m

29. ALGEBRA If $\frac{y}{28} = \frac{x}{16}$, then find x when $y = \frac{1}{2}$.

Ⓐ $\frac{2}{7}$ Ⓑ $\frac{4}{7}$ Ⓒ $\frac{7}{4}$ Ⓓ $3\frac{1}{2}$

Maintain Your Skills

Mixed Review **Find the measure of each angle to the nearest tenth of a degree.** *(Lesson 7-4)*

30. $\cos A = 0.6717$ **31.** $\sin B = 0.5127$ **32.** $\tan C = 2.1758$
33. $\cos D = 0.3421$ **34.** $\sin E = 0.1455$ **35.** $\tan F = 0.3541$

Find x and y. *(Lesson 7-3)*

36.   **37.** 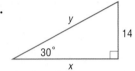   **38.**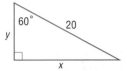

39. HOBBIES A twin-engine airplane used for medium-range flights has a length of 78 meters and a wingspan of 90 meters. If a scale model is made with a wingspan of 36 centimeters, find its length. *(Lesson 6-2)*

40. Copy and complete the flow proof. *(Lesson 4-6)*

Given: $\angle 5 \cong \angle 6$
  $\overline{FR} \cong \overline{GS}$

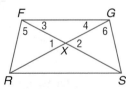

Prove: $\angle 4 \cong \angle 3$

Proof:

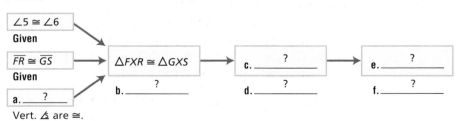

Getting Ready for the Next Lesson
PREREQUISITE SKILL Solve each proportion.
*(To review **solving proportions**, see Lesson 6-1.)*

41. $\frac{x}{6} = \frac{35}{42}$ **42.** $\frac{3}{x} = \frac{5}{45}$ **43.** $\frac{12}{17} = \frac{24}{x}$ **44.** $\frac{24}{36} = \frac{x}{15}$

45. $\frac{12}{13} = \frac{48}{x}$ **46.** $\frac{x}{18} = \frac{5}{8}$ **47.** $\frac{28}{15} = \frac{7}{x}$ **48.** $\frac{x}{40} = \frac{3}{26}$

7-6 The Law of Sines

- Use the Law of Sines to solve triangles.
- Solve problems by using the Law of Sines.

Vocabulary

- Law of Sines
- solving a triangle

How are triangles used in radio astronomy?

The Very Large Array (VLA), one of the world's premier astronomical radio observatories, consists of 27 radio antennas in a Y-shaped configuration on the Plains of San Agustin in New Mexico. Astronomers use the VLA to make pictures from the radio waves emitted by astronomical objects. Construction of the antennas is supported by a variety of triangles, many of which are not right triangles.

Study Tip

Obtuse Angles
There are also values for sin A, cos A, and tan A, when $A \geq 90°$. Values of the ratios for these angles will be found using the trigonometric functions on your calculator.

THE LAW OF SINES In trigonometry, the **Law of Sines** can be used to find missing parts of triangles that are not right triangles.

Key Concept — Law of Sines

Let $\triangle ABC$ be any triangle with a, b, and c representing the measures of the sides opposite the angles with measures A, B, and C, respectively. Then
$$\frac{\sin A}{a} = \frac{\sin B}{b} = \frac{\sin C}{c}.$$

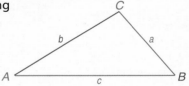

Proof — Law of Sines

$\triangle ABC$ is a triangle with an altitude from C that intersects $\overline{AB}$ at D. Let h represent the measure of $\overline{CD}$. Since $\triangle ADC$ and $\triangle BDC$ are right triangles, we can find sin A and sin B.

$\sin A = \dfrac{h}{b}$	$\sin B = \dfrac{h}{a}$	Definition of sine
$b \sin A = h$	$a \sin B = h$	Cross products
$b \sin A = a \sin B$		Substitution
$\dfrac{\sin A}{a} = \dfrac{\sin B}{b}$		Divide each side by ab.

The proof can be completed by using a similar technique with the other altitudes to show that $\dfrac{\sin A}{a} = \dfrac{\sin C}{c}$ and $\dfrac{\sin B}{b} = \dfrac{\sin C}{c}$.

Example 1 Use the Law of Sines

a. Find b. Round to the nearest tenth.

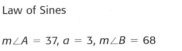

Use the Law of Sines to write a proportion.

$$\frac{\sin A}{a} = \frac{\sin B}{b} \qquad \text{Law of Sines}$$

$$\frac{\sin 37°}{3} = \frac{\sin 68°}{b} \qquad m\angle A = 37,\ a = 3,\ m\angle B = 68$$

$$b \sin 37° = 3 \sin 68° \qquad \text{Cross products}$$

$$b = \frac{3 \sin 68°}{\sin 37°} \qquad \text{Divide each side by } \sin 37°.$$

$$b \approx 4.6 \qquad \text{Use a calculator.}$$

Study Tip

Rounding
If you round before the final answer, your results may differ from results in which rounding was not done until the final answer.

b. Find $m\angle Z$ to the nearest degree in $\triangle XYZ$ if $y = 17$, $z = 14$, and $m\angle Y = 92$.

Write a proportion relating $\angle Y$, $\angle Z$, y, and z.

$$\frac{\sin Y}{y} = \frac{\sin Z}{z} \qquad \text{Law of Sines}$$

$$\frac{\sin 92°}{17} = \frac{\sin Z}{14} \qquad m\angle Y = 92,\ y = 17,\ z = 14$$

$$14 \sin 92° = 17 \sin Z \qquad \text{Cross products}$$

$$\frac{14 \sin 92°}{17} = \sin Z \qquad \text{Divide each side by 17.}$$

$$\sin^{-1}\left(\frac{14 \sin 92°}{17}\right) = Z \qquad \text{Solve for } Z.$$

$$55° \approx Z \qquad \text{Use a calculator.}$$

So, $m\angle Z \approx 55$.

The Law of Sines can be used to solve a triangle. **Solving a triangle** means finding the measures of all of the angles and sides of a triangle.

Example 2 Solve Triangles

a. Solve $\triangle ABC$ if $m\angle A = 33$, $m\angle B = 47$, and $b = 14$. Round angle measures to the nearest degree and side measures to the nearest tenth.

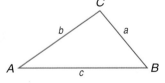

We know the measures of two angles of the triangle. Use the Angle Sum Theorem to find $m\angle C$.

$$m\angle A + m\angle B + m\angle C = 180 \qquad \text{Angle Sum Theorem}$$

$$33 + 47 + m\angle C = 180 \qquad m\angle A = 33,\ m\angle B = 47$$

$$80 + m\angle C = 180 \qquad \text{Add.}$$

$$m\angle C = 100 \qquad \text{Subtract 80 from each side.}$$

Since we know $m\angle B$ and b, use proportions involving $\dfrac{\sin B}{b}$.

Study Tip

Look Back
To review the **Angle Sum Theorem**, see Lesson 4-2.

To find a:

$$\frac{\sin B}{b} = \frac{\sin A}{a} \qquad \text{Law of Sines}$$

$$\frac{\sin 47°}{14} = \frac{\sin 33°}{a} \qquad \text{Substitute.}$$

$$a \sin 47° = 14 \sin 33° \qquad \text{Cross products}$$

$$a = \frac{14 \sin 33°}{\sin 47°} \qquad \text{Divide each side by } \sin 47°.$$

$$a \approx 10.4 \qquad \text{Use a calculator.}$$

To find c:

$$\frac{\sin B}{b} = \frac{\sin C}{c}$$

$$\frac{\sin 47°}{14} = \frac{\sin 100°}{c}$$

$$c \sin 47° = 14 \sin 100°$$

$$c = \frac{14 \sin 100°}{\sin 47°}$$

$$c \approx 18.9$$

Therefore, $m\angle C = 100$, $a \approx 10.4$, and $c \approx 18.9$.

b. Solve $\triangle ABC$ if $m\angle C = 98$, $b = 14$, and $c = 20$. Round angle measures to the nearest degree and side measures to the nearest tenth.

We know the measures of two sides and an angle opposite one of the sides.

$\dfrac{\sin B}{b} = \dfrac{\sin C}{c}$	Law of Sines
$\dfrac{\sin B}{14} = \dfrac{\sin 98°}{20}$	$m\angle C = 98$, $b = 14$, and $c = 20$
$20 \sin B = 14 \sin 98°$	Cross products
$\sin B = \dfrac{14 \sin 98°}{20}$	Divide each side by 20.
$B = \sin^{-1}\left(\dfrac{14 \sin 98°}{20}\right)$	Solve for B.
$B \approx 44°$	Use a calculator.

$m\angle A + m\angle B + m\angle C = 180$	Angle Sum Theorem
$m\angle A + 44 + 98 = 180$	$m\angle B = 44$ and $m\angle C = 98$
$m\angle A + 142 = 180$	Add.
$m\angle A = 38$	Subtract 142 from each side.

$\dfrac{\sin A}{a} = \dfrac{\sin C}{c}$	Law of Sines
$\dfrac{\sin 38°}{a} = \dfrac{\sin 98°}{20}$	$m\angle A = 38$, $m\angle C = 98$, and $c = 20$
$20 \sin 38° = a \sin 98°$	Cross products
$\dfrac{20 \sin 38°}{\sin 98°} = a$	Divide each side by $\sin 98°$.
$12.4 \approx a$	Use a calculator.

Therefore, $A \approx 38°$, $B \approx 44°$, and $a \approx 12.4$.

USE THE LAW OF SINES TO SOLVE PROBLEMS   The Law of Sines is very useful in solving direct and indirect measurement applications.

Example 3 *Indirect Measurement*

When the angle of elevation to the sun is 62°, a telephone pole tilted at an angle of 7° from the vertical casts a shadow of 30 feet long on the ground. Find the length of the telephone pole to the nearest tenth of a foot.

Draw a diagram.

Draw $\overline{SD} \perp \overline{GD}$. Then find the $m\angle GDP$ and $m\angle GPD$.

$m\angle GDP = 90 - 7$ or 83

$m\angle GPD + 62 + 83 = 180$ or $m\angle GPD = 35$

Since you know the measures of two angles of the triangle, $m\angle GDP$ and $m\angle GPD$, and the length of a side opposite one of the angles ($\overline{GD}$ is opposite $\angle GPD$) you can use the Law of Sines to find the length of the pole.

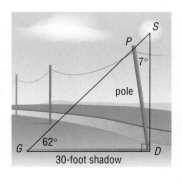

(continued on the next page)

 www.geometryonline.com/extra_examples

$$\frac{PD}{\sin \angle DGP} = \frac{GD}{\sin \angle GPD} \quad \text{Law of Sines}$$

$$\frac{PD}{\sin 62°} = \frac{30}{\sin 35°} \quad m\angle DGP = 62, m\angle GPD = 35, \text{ and } GD = 30$$

$$PD \sin 35° = 30 \sin 62° \quad \text{Cross products}$$

$$PD = \frac{30 \sin 62°}{\sin 35°} \quad \text{Divide each side by } \sin 35°.$$

$$PD \approx 46.2 \quad \text{Use a calculator.}$$

The telephone pole is about 46.2 feet long.

Concept Summary   Law of Sines

The Law of Sines can be used to solve a triangle in the following cases.

Case 1 You know the measures of two angles and any side of a triangle. (AAS or ASA)

Case 2 You know the measures of two sides and an angle opposite one of these sides of the triangle. (SSA)

Check for Understanding

Concept Check

1. **FIND THE ERROR** Makayla and Felipe are trying to find d in $\triangle DEF$.

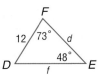

Makayla
$$\sin 59° = \frac{d}{12}$$

Felipe
$$\frac{\sin 59°}{d} = \frac{\sin 48°}{12}$$

Who is correct? Explain your reasoning.

2. **OPEN ENDED** Draw an acute triangle and label the measures of two angles and the length of one side. Explain how to solve the triangle.

3. **Compare** the two cases for the Law of Sines.

Guided Practice

Find each measure using the given measures of $\triangle XYZ$. Round angle measures to the nearest degree and side measures to the nearest tenth.

4. Find y if $x = 3$, $m\angle X = 37$, and $m\angle Y = 68$.

5. Find x if $y = 12.1$, $m\angle X = 57$, and $m\angle Z = 72$.

6. Find $m\angle Y$ if $y = 7$, $z = 11$, and $m\angle Z = 37$.

7. Find $m\angle Z$ if $y = 17$, $z = 14$, and $m\angle Y = 92$.

Solve each $\triangle PQR$ described below. Round angle measures to the nearest degree and side measures to the nearest tenth.

8. $m\angle R = 66$, $m\angle Q = 59$, $p = 72$

9. $p = 32$, $r = 11$, $m\angle P = 105$

10. $m\angle P = 33$, $m\angle R = 58$, $q = 22$

11. $p = 28$, $q = 22$, $m\angle P = 120$

12. $m\angle P = 50$, $m\angle Q = 65$, $p = 12$

13. $q = 17.2$, $r = 9.8$, $m\angle Q = 110.7$

14. Find the perimeter of parallelogram $ABCD$ to the nearest tenth.

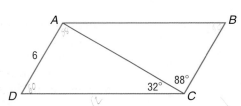

15. SURVEYING To find the distance between
two points *A* and *B* that are on opposite sides
of a river, a surveyor measures the distance to
point *C* on the same side of the river as point *A*.
The distance from *A* to *C* is 240 feet. He then
measures the angle from *A* to *B* as 62° and
measures the angle from *C* to *B* as 55°. Find the
distance from *A* to *B*.

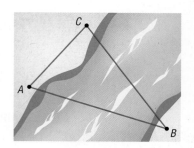

Practice and Apply

Homework Help

For Exercises	See Examples
16–21	1
22–29	2
30–38	3

Extra Practice
See page 768.

**Find each measure using the given measures of △*KLM*. Round angle measures
to the nearest degree and side measures to the nearest tenth.**

16. If $m\angle L = 45$, $m\angle K = 63$, and $\ell = 22$, find k.

17. If $k = 3.2$, $m\angle L = 52$, and $m\angle K = 70$, find ℓ.

18. If $m = 10.5$, $k = 18.2$, and $m\angle K = 73$, find $m\angle M$.

19. If $k = 10$, $m = 4.8$, and $m\angle K = 96$, find $m\angle M$.

20. If $m\angle L = 88$, $m\angle K = 31$, and $m = 5.4$, find ℓ.

21. If $m\angle M = 59$, $\ell = 8.3$, and $m = 14.8$, find $m\angle L$.

Solve each △*WXY* described below. Round measures to the nearest tenth.

22. $m\angle Y = 71$, $y = 7.4$, $m\angle X = 41$

23. $x = 10.3$, $y = 23.7$, $m\angle Y = 96$

24. $m\angle X = 25$, $m\angle W = 52$, $y = 15.6$

25. $m\angle Y = 112$, $x = 20$, $y = 56$

26. $m\angle W = 38$, $m\angle Y = 115$, $w = 8.5$

27. $m\angle W = 36$, $m\angle Y = 62$, $w = 3.1$

28. $w = 30$, $y = 9.5$, $m\angle W = 107$

29. $x = 16$, $w = 21$, $m\angle W = 88$

30. An isosceles triangle has a base of 46 centimeters and a vertex angle of 44°.
Find the perimeter.

31. Find the perimeter of quadrilateral *ABCD* to the
nearest tenth.

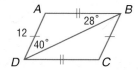

32. GARDENING Elena is planning a triangular garden. She wants to build a fence
around the garden to keep out the deer. The length of one side of the garden is
26 feet. If the angles at the end of this side are 78° and 44°, find the length of
fence needed to enclose the garden.

33. AVIATION Two radar stations that are 20 miles apart located a plane at the
same time. The first station indicated that the position of the plane made an
angle of 43° with the line between the stations. The second station indicated
that it made an angle of 48° with the same line. How far is each station from
the plane?

34. SURVEYING Maria Lopez is a surveyor who must determine the distance
across a section of the Rio Grande Gorge in New Mexico. Standing on one side
of the ridge, she measures the angle formed by the edge of the ridge and the line
of sight to a tree on the other side of the ridge. She then walks along the ridge
315 feet, passing the tree and measures the angle formed by the edge of the
ridge and the new line of sight to the same tree. If the first angle is 80° and the
second angle is 85°, find the distance across the gorge.

35. REAL ESTATE A house is built on a triangular plot of land. Two sides of the plot are 160 feet long, and they meet at an angle of 85°. If a fence is to be placed along the perimeter of the property, how much fencing material is needed?

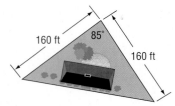

MIRRORS For Exercises 36 and 37, use the following information.
Kayla, Jenna, and Paige live in a town nicknamed "Perpendicular City" because the planners and builders took great care to have all the streets oriented north-south or east-west. The three of them play a game where they signal each other using mirrors. Kayla and Jenna signal each other from a distance of 1433 meters. Jenna turns 27° to signal Paige. Kayla rotates 40° to signal Paige.

36. To the nearest tenth of a meter, how far apart are Kayla and Paige?

37. To the nearest tenth of a meter, how far apart are Jenna and Paige?

AVIATION For Exercises 38 and 39, use the following information.
Keisha Jefferson is flying a small plane due west. To avoid the jet stream, she must change her course. She turns the plane 27° to the south and flies 60 miles. Then she makes a turn of 124° heads back to her original course.

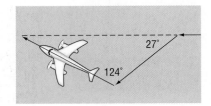

38. How far must she fly after the second turn to return to the original course?

39. How many miles did she add to the flight by changing course?

40. CRITICAL THINKING Does the Law of Sines apply to the acute angles of a right triangle? Explain your answer.

41. WRITING IN MATH   Answer the question that was posed at the beginning of the lesson.

How are triangles used in radio astronomy?

Include the following in your answer:
- a description of what the VLA is and the purpose of the VLA, and
- the purpose of the triangles in the VLA.

42. SHORT RESPONSE In $\triangle XYZ$, if $x = 12$, $m\angle X = 48$, and $m\angle Y = 112$, solve the triangle to the nearest tenth.

43. ALGEBRA The table below shows the customer ratings of five restaurants in the *Metro City Guide to Restaurants*. The rating scale is from 1, the worst, to 10, the best. Which of the five restaurants has the best average rating?

Restaurant	Food	Decor	Service	Value
Del Blanco's	7	9	4	7
Aquavent	8	9	4	6
Le Circus	10	8	3	5
Sushi Mambo	7	7	5	6
Metropolis Grill	9	8	7	7

Ⓐ Metropolis Grill   Ⓑ Le Circus
Ⓒ Aquavent   Ⓓ Del Blanco's

Mixed Review   **ARCHITECTURE**   For Exercises 44 and 45, use the following information.

Mr. Martinez is an architect who designs houses so that the windows receive minimum sun in the summer and maximum sun in the winter. For Columbus, Ohio, the angle of elevation of the sun at noon on the longest day is 73.5° and on the shortest day is 26.5°. Suppose a house is designed with a south-facing window that is 6 feet tall. The top of the window is to be installed 1 foot below the overhang.   *(Lesson 7-5)*

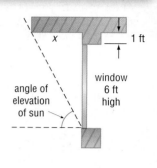

44. How long should the architect make the overhang so that the window gets no direct sunlight at noon on the longest day?

45. Using the overhang from Exercise 44, how much of the window will get direct sunlight at noon on the shortest day?

Use △JKL to find sin J, cos J, tan J, sin L, cos L, and tan L. Express each ratio as a fraction and as a decimal to the nearest hundredth.   *(Lesson 7-4)*

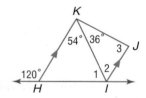

46. $j = 8, k = 17, \ell = 15$ **47.** $j = 20, k = 29, \ell = 21$

48. $j = 12, k = 24, \ell = 12\sqrt{3}$ **49.** $j = 7\sqrt{2}, k = 14, \ell = 7\sqrt{2}$

If $\overline{KH}$ is parallel to $\overline{JI}$, find the measure of each angle.   *(Lesson 4-2)*

50. ∠1

51. ∠2

52. ∠3

Getting Ready for the Next Lesson   **PREREQUISITE SKILL**   Evaluate $\dfrac{c^2 - a^2 - b^2}{-2ab}$ for the given values of a, b, and c.
*(To review **evaluating expressions**, see page 736.)*

53. $a = 7, b = 8, c = 10$ **54.** $a = 4, b = 9, c = 6$ **55.** $a = 5, b = 8, c = 10$

56. $a = 16, b = 4, c = 13$ **57.** $a = 3, b = 10, c = 9$ **58.** $a = 5, b = 7, c = 11$

𝒫ractice Quiz 2   *Lessons 7-4 through 7-6*

Find x to the nearest tenth.   *(Lesson 7-4)*

1.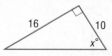

2.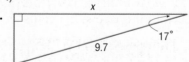

3.

4. COMMUNICATIONS   To secure a 500-foot radio tower against high winds, guy wires are attached to the tower 5 feet from the top. The wires form a 15° angle with the tower. Find the distance from the centerline of the tower to the anchor point of the wires.
(Lesson 7-5)

5. Solve △DEF.   *(Lesson 7-6)*

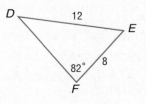

Geometry Software Investigation

The Ambiguous Case of the Law of Sines

In Lesson 7-6, you learned that you could solve a triangle using the Law of Sines if you know the measures of two angles and any side of the triangle (AAS or ASA). You can also solve a triangle by the Law of Sines if you know the measures of two sides and an angle opposite one of the sides (SSA). When you use SSA to solve a triangle, and the given angle is acute, sometimes it is possible to find two different triangles. You can use The Geometer's Sketchpad to explore this case, called the **ambiguous case**, of the Law of Sines.

Model

Step 1 Construct $\overrightarrow{AB}$ and $\overrightarrow{AC}$. Construct a circle whose center is B so that it intersects $\overrightarrow{AC}$ at two points. Then, construct any radius $\overline{BD}$.

Step 2 Find the measures of $\overline{BD}$, $\overline{AB}$, and $\angle A$.

Step 3 Use the rotate tool to move D so that it lies on one of the intersection points of circle B and $\overrightarrow{AC}$. In $\triangle ABD$, find the measures of $\angle ABD$, $\angle BDA$, and $\overline{AD}$.

Step 4 Using the rotate tool, move D to the other intersection point of circle B and $\overrightarrow{AC}$.

Step 5 Note the measures of $\angle ABD$, $\angle BDA$, and $\overline{AD}$ in $\triangle ABD$.

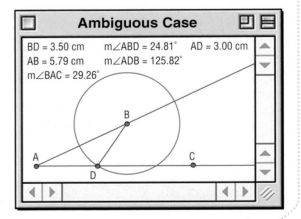

Analyze

1. Which measures are the same in both triangles?

2. Repeat the activity using different measures for $\angle A$, $\overline{BD}$, and $\overline{AB}$. How do the results compare to the earlier results?

Make a Conjecture

3. Compare your results with those of your classmates. How do the results compare?

4. What would have to be true about circle B in order for there to be one unique solution? Test your conjecture by repeating the activity.

5. Is it possible, given the measures of $\overline{BD}$, $\overline{AB}$, and $\angle A$, to have no solution? Test your conjecture and explain.

7-7 The Law of Cosines

What You'll Learn

- Use the Law of Cosines to solve triangles.
- Solve problems by using the Law of Cosines.

Vocabulary

- Law of Cosines

How are triangles used in building design?

The Chicago Metropolitan Correctional Center is a 27-story triangular federal detention center. The cells are arranged around a lounge-like common area. The architect found that a triangular floor plan allowed for the maximum number of cells to be most efficiently centered around the lounge.

THE LAW OF COSINES Suppose you know the lengths of the sides of the triangular building and want to solve the triangle. The **Law of Cosines** allows us to solve a triangle when the Law of Sines cannot be used.

Key Concept — Law of Cosines

Let $\triangle ABC$ be any triangle with a, b, and c representing the measures of sides opposite angles A, B, and C, respectively. Then the following equations are true.

$$a^2 = b^2 + c^2 - 2bc \cos A$$

$$b^2 = a^2 + c^2 - 2ac \cos B$$

$$c^2 = a^2 + b^2 - 2ab \cos C$$

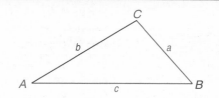

The Law of Cosines can be used to find missing measures in a triangle if you know the measures of two sides and their included angle.

Example 1 Two Sides and the Included Angle

Find a if $c = 8$, $b = 10$, and $m\angle A = 60$.

Use the Law of Cosines since the measures of two sides and the included are known.

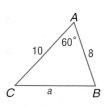

$a^2 = b^2 + c^2 - 2bc \cos A$	Law of Cosines
$a^2 = 10^2 + 8^2 - 2(10)(8) \cos 60°$	$b = 10$, $c = 8$, and $m\angle A = 60$
$a^2 = 164 - 160 \cos 60°$	Simplify.
$a = \sqrt{164 - 160 \cos 60°}$	Take the square root of each side.
$a \approx 9.2$	Use a calculator.

You can also use the Law of Cosines to find the measures of angles of a triangle when you know the measures of the three sides.

Example 2 *Three Sides*

Find $m\angle R$.

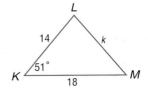

$r^2 = q^2 + s^2 - 2qs \cos R$	Law of Cosines
$23^2 = 37^2 + 18^2 - 2(37)(18) \cos R$	$r = 23, q = 37, s = 18$
$529 = 1693 - 1332 \cos R$	Simplify.
$-1164 = -1332 \cos R$	Subtract 1693 from each side.
$\dfrac{-1164}{-1332} = \cos R$	Divide each side by –1332.
$R = \cos^{-1}\left(\dfrac{1164}{1332}\right)$	Solve for R.
$R \approx 29.1°$	Use a calculator.

USE THE LAW OF COSINES TO SOLVE PROBLEMS

Most problems can be solved using more than one method. Choosing the most efficient way to solve a problem is sometimes not obvious.

When solving right triangles, you can use sine, cosine, or tangent ratios. When solving other triangles, you can use the Law of Sines or the Law of Cosines. You must decide how to solve each problem depending on the given information.

Example 3 *Select a Strategy*

Solve $\triangle KLM$. Round angle measure to the nearest degree and side measure to the nearest tenth.

We do not know whether $\triangle KLM$ is a right triangle, so we must use the Law of Cosines or the Law of Sines. We know the measures of two sides and the included angle. This is SAS, so use the Law of Cosines.

$k^2 = \ell^2 + m^2 - 2\ell m \cos K$	Law of Cosines
$k^2 = 18^2 + 14^2 - 2(18)(14) \cos 51°$	$\ell = 18, m = 14,$ and $m\angle K = 51$
$k = \sqrt{18^2 + 14^2 - 2(18)(14) \cos 51°}$	Take the square root of each side.
$k \approx 14.2$	Use a calculator.

Next, we can find $m\angle L$ or $m\angle M$. If we decide to find $m\angle L$, we can use either the Law of Sines or the Law of Cosines to find this value. In this case, we will use the Law of Sines.

$\dfrac{\sin L}{\ell} = \dfrac{\sin K}{k}$	Law of Sines
$\dfrac{\sin L}{18} \approx \dfrac{\sin 51°}{14.2}$	$\ell = 18, k \approx 14.2,$ and $m\angle K = 51$
$14.2 \sin L \approx 18 \sin 51°$	Cross products
$\sin L \approx \dfrac{18 \sin 51°}{14.2}$	Divide each side by 14.2.
$L \approx \sin^{-1}\left(\dfrac{18 \sin 51°}{14.2}\right)$	Take the inverse sine of each side.
$L \approx 80°$	Use a calculator.

Use the Angle Sum Theorem to find $m\angle M$.

$$m\angle K + m\angle L + m\angle M = 180 \quad \text{Angle Sum Theorem}$$
$$51 + 80 + m\angle M \approx 180 \quad m\angle K = 51 \text{ and } m\angle L \approx 80$$
$$m\angle M \approx 49 \quad \text{Subtract 131 from each side.}$$

Therefore, $k \approx 14.2$, $m\angle K \approx 80$, and $m\angle M \approx 49$.

Example 4 · Use Law of Cosines to Solve Problems

REAL ESTATE Ms. Jenkins is buying some property that is shaped like quadrilateral $ABCD$. Find the perimeter of the property.

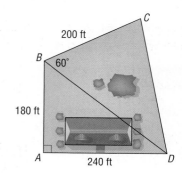

Use the Pythagorean Theorem to find BD in $\triangle ABD$.

$$(AB)^2 + (AD)^2 = (BD)^2 \quad \text{Pythagorean Theorem}$$
$$180^2 + 240^2 = (BD)^2 \quad AB = 180, AD = 240$$
$$90{,}000 = (BD)^2 \quad \text{Simplify.}$$
$$300 = BD \quad \text{Take the square root of each side.}$$

Next, use the Law of Cosines to find CD in $\triangle BCD$.

$$(CD)^2 = (BC)^2 + (BD)^2 - 2(BC)(BD)\cos\angle CBD \quad \text{Law of Cosines}$$
$$(CD)^2 = 200^2 + 300^2 - 2(200)(300)\cos 60° \quad BC = 200, BD = 300, m\angle CBD = 60$$
$$(CD)^2 = 130{,}000 - 120{,}000\cos 60° \quad \text{Simplify.}$$
$$CD = \sqrt{130{,}000 - 120{,}000\cos 60°} \quad \text{Take the square root of each side.}$$
$$CD \approx 264.6 \quad \text{Use a calculator.}$$

The perimeter of the property is $180 + 200 + 264.6 + 240$ or about 884.6 feet.

Check for Understanding

Concept Check
1. **OPEN ENDED** Draw and label one acute and one obtuse triangle, illustrating when you can use the Law of Cosines to find the missing measures.

2. **Explain** when you should use the Law of Sines or the Law of Cosines to solve a triangle.

3. **Find a counterexample** for the following statement.
   The Law of Cosines can be used to find the length of a missing side in any triangle.

Guided Practice
In $\triangle BCD$, given the following measures, find the measure of the missing side.
4. $c = \sqrt{2}$, $d = 5$, $m\angle B = 45$
5. $b = 107$, $c = 94$, $m\angle D = 105$

In $\triangle RST$, given the lengths of the sides, find the measure of the stated angle to the nearest degree.
6. $r = 33$, $s = 65$, $t = 56$; $m\angle S$
7. $r = 2.2$, $s = 1.3$, $t = 1.6$; $m\angle R$

Solve each triangle using the given information. Round angle measure to the nearest degree and side measure to the nearest tenth.
8. $\triangle XYZ$: $x = 5$, $y = 10$, $z = 13$
9. $\triangle KLM$: $k = 20$, $m = 24$, $m\angle L = 47$

Application **10. CRAFTS** Jamie, age 25, is creating a logo for herself and two cousins, ages 10 and 5. She is using a quarter (25 cents), a dime (10 cents), and a nickel (5 cents) to represent their ages. She will hold the coins together by soldering a triangular piece of silver wire so that the three vertices of the triangle lie at the centers of the three circular coins. The diameter of the quarter is 24 millimeters, the diameter of the nickel is 22 millimeters, and the diameter of a dime is 10 millimeters. Find the measures of the three angles in the triangle.

Practice and Apply

Homework Help

For Exercises	See Examples
11–14	1
15–18, 38	2
22–37, 39–41	3

Extra Practice
See page 768.

In △*TUV*, given the following measures, find the measure of the missing side.

11. $t = 9.1, v = 8.3, m\angle U = 32$ **12.** $t = 11, u = 17, m\angle V = 78$

13. $u = 11, v = 17, m\angle T = 105$ **14.** $v = 11, u = 17, m\angle T = 59$

In △*EFG*, given the lengths of the sides, find the measure of the stated angle to the nearest degree.

15. $e = 9.1, f = 8.3, g = 16.7; m\angle F$ **16.** $e = 14, f = 19, g = 32; m\angle E$

17. $e = 325, f = 198, g = 208; m\angle F$ **18.** $e = 21.9, f = 18.9, g = 10; m\angle G$

Solve each triangle using the given information. Round angle measures to the nearest degree and side measures to the nearest tenth.

19.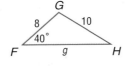

20.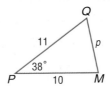

21.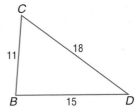

22. △*ABC*: $m\angle A = 42, m\angle C = 77, c = 6$ **23.** △*ABC*: $a = 10.3, b = 9.5, m\angle C = 37$

24. △*ABC*: $a = 15, b = 19, c = 28$ **25.** △*ABC*: $m\angle A = 53, m\angle C = 28, c = 14.9$

26. KITES Beth is building a kite like the one at the right. If $\overline{AB}$ is 5 feet long, $\overline{BC}$ is 8 feet long, and $\overline{BD}$ is $7\frac{2}{3}$ feet long, find the measure of the angle between the short sides and the angle between the long sides to the nearest degree.

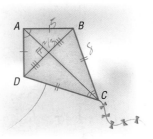

Solve each △*LMN* described below. Round measures to the nearest tenth.

27. $m = 44, \ell = 54, m\angle L = 23$ **28.** $m = 18, \ell = 24, n = 30$

29. $m = 19, n = 28, m\angle L = 49$ **30.** $m\angle M = 46, m\angle L = 55, n = 16$

31. $m = 256, \ell = 423, n = 288$ **32.** $m\angle M = 55, \ell = 6.3, n = 6.7$

33. $m\angle M = 27, \ell = 5, n = 10$ **34.** $n = 17, m = 20, \ell = 14$

35. $\ell = 14, n = 21, m\angle M = 60$ **36.** $\ell = 14, m = 15, n = 16$

37. $m\angle L = 51, \ell = 40, n = 35$ **38.** $\ell = 10, m = 11, n = 12$

39. In quadrilateral $ABCD$, $AC = 188$, $BD = 214$, $m\angle BPC = 70$, and P is the midpoint of $\overline{AC}$ and $\overline{BD}$. Find the perimeter of quadrilateral $ABCD$.

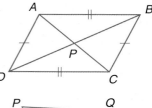

40. In quadrilateral $PQRS$, $PQ = 721$, $QR = 547$, $RS = 593$, $PS = 756$, and $m\angle P = 58$. Find QS, $m\angle PQS$, and $m\angle R$.

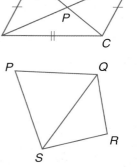

More About. . .

41. BUILDINGS Refer to the information at the left. Find the measures of the angles of the triangular building to the nearest tenth.

42. SOCCER Carlos and Adam are playing soccer. Carlos is standing 40 feet from one post of the goal and 50 feet from the other post. Adam is standing 30 feet from one post of the goal and 22 feet from the other post. If the goal is 24 feet wide, which player has a greater angle to make a shot on goal?

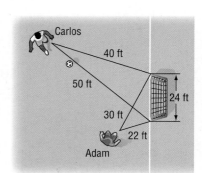

Buildings

The Swissôtel in Chicago, Illinois, is built in the shape of a triangular prism. The lengths of the sides of the triangle are 180 feet, 186 feet, and 174 feet.

Source: Swissôtel

43. **PROOF** Justify each statement for the derivation of the Law of Cosines.

Given: $\overline{AD}$ is an altitude of $\triangle ABC$.

Prove: $c^2 = a^2 + b^2 - 2ab \cos C$

Proof:

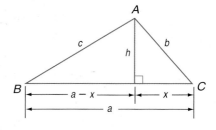

Statement	Reasons
a. $c^2 = (a - x)^2 + h^2$	**a.** ?
b. $c^2 = a^2 - 2ax + x^2 + h^2$	**b.** ?
c. $x^2 + h^2 = b^2$	**c.** ?
d. $c^2 = a^2 - 2ax + b^2$	**d.** ?
e. $\cos C = \frac{x}{b}$	**e.** ?
f. $b \cos C = x$	**f.** ?
g. $c^2 = a^2 - 2a(b \cos C) + b^2$	**g.** ?
h. $c^2 = a^2 + b^2 - 2ab \cos C$	**h.** ?

44. CRITICAL THINKING Graph $A(-6, -8)$, $B(10, -4)$, $C(6, 8)$, and $D(5, 11)$ on a coordinate plane. Find the measure of interior angle ABC and the measure of exterior angle DCA.

45. **WRITING IN MATH** Answer the question that was posed at the beginning of the lesson.

How are triangles used in building construction?

Include the following in your answer:
- why the building was triangular instead of rectangular, and
- why the Law of Sines could not be used to solve the triangle.

Standardized Test Practice
A B C D

46. For $\triangle DEF$, find d to the nearest tenth if $e = 12$, $f = 15$, and $m\angle D = 75$.
Ⓐ 18.9   Ⓑ 16.6   Ⓒ 15.4   Ⓓ 9.8

47. ALGEBRA Ms. LaHue earns a monthly base salary of $1280 plus a commission of 12.5% of her total monthly sales. At the end of one month, Ms. LaHue earned $4455. What were her total sales for the month?
Ⓐ $3175   Ⓑ $10,240   Ⓒ $25,400   Ⓓ $35,640

Maintain Your Skills
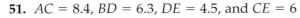

Mixed Review

Find each measure using the given measures from $\triangle XYZ$. Round angle measure to the nearest degree and side measure to the nearest tenth. *(Lesson 7-6)*

48. If $y = 4.7$, $m\angle X = 22$, and $m\angle Y = 49$, find x.

49. If $y = 10$, $x = 14$, and $m\angle X = 50$, find $m\angle Y$.

50. SURVEYING A surveyor is 100 meters from a building and finds that the angle of elevation to the top of the building is 23°. If the surveyor's eye level is 1.55 meters above the ground, find the height of the building. *(Lesson 7-5)*

For Exercises 51–54, determine whether $\overline{AB} \parallel \overline{CD}$. *(Lesson 6-4)*

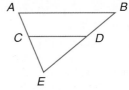

51. $AC = 8.4$, $BD = 6.3$, $DE = 4.5$, and $CE = 6$

52. $AC = 7$, $BD = 10.5$, $BE = 22.5$, and $AE = 15$

53. $AB = 8$, $AE = 9$, $CD = 4$, and $CE = 4$

54. $AB = 5.4$, $BE = 18$, $CD = 3$, and $DE = 10$

Use the figure at the right to write a paragraph proof. *(Lesson 6-3)*

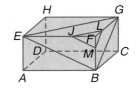

55. Given: $\triangle JFM \sim \triangle EFB$
           $\triangle LFM \sim \triangle GFB$
    Prove: $\triangle JFL \sim \triangle EFG$

56. Given: $\overline{JM} \parallel \overline{EB}$
           $\overline{LM} \parallel \overline{GB}$
    Prove: $\overline{JL} \parallel \overline{EG}$

COORDINATE GEOMETRY The vertices of $\triangle XYZ$ are $X(8, 0)$, $Y(-4, 8)$, and $Z(0, 12)$. Find the coordinates of the points of concurrency of $\triangle XYZ$ to the nearest tenth. *(Lesson 5-1)*

57. orthocenter   **58.** centroid   **59.** circumcenter

Web Quest **Internet Project**

Who is Behind This Geometry Idea Anyway?

It's time to complete your project. Use the information and data you have gathered about your research topic, two mathematicians, and a geometry problem to prepare a portfolio or Web page. Be sure to include illustrations and/or tables in the presentation.

www.geometryonline.com/webquest

Trigonometric Identities

In algebra, the equation $2(x + 2) = 2x + 4$ is called an *identity* because the equation is true for all values of x. There are equations involving trigonometric ratios that are true for all values of the angle measure. These are called **trigonometric identities**.

In the figure, $P(x, y)$ is in Quadrant I. The Greek letter theta (pronounced THAY tuh) θ, represents the measure of the angle formed by the x-axis and $\overline{OP}$. Triangle POR is a right triangle. Let r represent the length of the hypotenuse. Then the following are true.

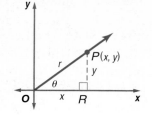

$$\sin \theta = \frac{y}{r} \qquad \cos \theta = \frac{x}{r} \qquad \tan \theta = \frac{y}{x}$$

$$\csc \theta = \frac{r}{y} \qquad \sec \theta = \frac{r}{x} \qquad \cot \theta = \frac{x}{y}$$

Notice that $\dfrac{1}{\sin \theta} = \dfrac{1}{\frac{y}{r}}$ ▷ $\dfrac{1}{\frac{y}{r}} = 1 \div \dfrac{y}{r}$ ▷ $1 \div \dfrac{y}{r} = 1 \cdot \dfrac{r}{y}$ or $\dfrac{r}{y}$ ▷ $\dfrac{r}{y} = \csc \theta$.

So, $\dfrac{1}{\sin \theta} = \csc \theta$. This is known as one of the **reciprocal identities**.

Activity

Verify that $\cos^2 \theta + \sin^2 \theta = 1$.

The expression $\cos^2 \theta$ means $(\cos \theta)^2$. To verify an identity, work on only one side of the equation and use what you know to show how that side is equivalent to the other side.

$\cos^2 \theta + \sin^2 \theta \overset{?}{=} 1$	Original equation
$\left(\dfrac{x}{r}\right)^2 + \left(\dfrac{y}{r}\right)^2 \overset{?}{=} 1$	Substitute.
$\dfrac{x^2}{r^2} + \dfrac{y^2}{r^2} \overset{?}{=} 1$	Simplify.
$\dfrac{x^2 + y^2}{r^2} \overset{?}{=} 1$	Combine fractions with like denominators.
$\dfrac{r^2}{r^2} \overset{?}{=} 1$	Pythagorean Theorem: $x^2 + y^2 = r^2$
$1 = 1 \checkmark$	Simplify.

Since $1 = 1$, $\cos^2 \theta + \sin^2 \theta = 1$.

Analyze

1. The identity $\cos^2 \theta + \sin^2 \theta = 1$ is known as a **Pythagorean identity**. Why do you think the word *Pythagorean* is used to name this?

2. Find two more reciprocal identities involving $\dfrac{1}{\cos \theta}$ and $\dfrac{1}{\tan \theta}$.

Verify each identity.

3. $\dfrac{\sin \theta}{\cos \theta} = \tan \theta$ 4. $\cot \theta = \dfrac{\cos \theta}{\sin \theta}$ 5. $\tan^2 \theta + 1 = \sec^2 \theta$ 6. $\cot^2 \theta + 1 = \csc^2 \theta$

Vocabulary and Concept Check

ambiguous case (p. 384)   geometric mean (p. 342)   Pythagorean triple (p. 352)   tangent (p. 364)
angle of depression (p. 372)   Law of Cosines (p. 385)   reciprocal identities (p. 391)   trigonometric identity (p. 391)
angle of elevation (p. 371)   Law of Sines (p. 377)   sine (p. 364)   trigonometric ratio (p. 364)
cosine (p. 364)   Pythagorean identity (p. 391)   solving a triangle (p. 378)   trigonometry (p. 364)

A complete list of postulates and theorems can be found on pages R1–R8.

Exercises State whether each statement is *true* or *false*. If false, replace the underlined word or words to make a true sentence.

1. The Law of Sines can be applied if you know the measures of two sides and an angle <u>opposite</u> one of these sides of the triangle.
2. The tangent of $\angle A$ is the measure of the leg <u>adjacent</u> to $\angle A$ divided by the measure of the leg <u>opposite</u> $\angle A$.
3. In <u>any</u> triangle, the sum of the squares of the measures of the legs equals the square of the measure of the hypotenuse.
4. An angle of <u>elevation</u> is the angle between the line of sight and the horizontal when an observer looks upward.
5. The geometric mean between two numbers is the positive square root of their <u>product</u>.
6. In a <u>30°-60°-90°</u> triangle, two of the sides will have the same length.
7. Looking at a city while flying in a plane is an example that uses angle of <u>elevation</u>.

Lesson-by-Lesson Review

7-1 Geometric Mean

See pages 342–348.

Concept Summary

- The geometric mean of two numbers is the square root of their product.
- You can use the geometric mean to find the altitude of a right triangle.

Examples **1** Find the geometric mean between 10 and 30.

$$\frac{10}{x} = \frac{x}{30}$$ Definition of geometric mean

$$x^2 = 300$$ Cross products

$$x = \sqrt{300} \text{ or } 10\sqrt{3}$$ Take the square root of each side.

2 Find NG in $\triangle TGR$.

The measure of the altitude is the geometric mean between the measures of the two hypotenuse segments.

$$\frac{TN}{GN} = \frac{GN}{RN}$$ Definition of geometric mean

$$\frac{2}{GN} = \frac{GN}{4}$$ $TN = 2$, $RN = 4$

$$8 = (GN)^2$$ Cross products

$$\sqrt{8} \text{ or } 2\sqrt{2} = GN$$ Take the square root of each side.

 www.geometryonline.com/vocabulary_review

Exercises Find the geometric mean between each pair of numbers.
See Example 1 on page 342.

8. 4 and 16   9. 4 and 81   10. 20 and 35   11. 18 and 44

12. In $\triangle PQR$, $PS = 8$, and $QS = 14$.
Find RS. *See Example 2 on page 344.*

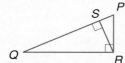

7-2 The Pythagorean Theorem and Its Converse

See pages 350–356.

Concept Summary

- The Pythagorean Theorem can be used to find the measures of the sides of a right triangle.
- If the measures of the sides of a triangle form a Pythagorean triple, then the triangle is a right triangle.

Example Find k.

$$a^2 + (LK)^2 = (JL)^2 \quad \text{Pythagorean Theorem}$$
$$a^2 + 8^2 = 13^2 \quad LK = 8 \text{ and } JL = 13$$
$$a^2 + 64 = 169 \quad \text{Simplify.}$$
$$a^2 = 105 \quad \text{Subtract 64 from each side.}$$
$$a = \sqrt{105} \quad \text{Take the square root of each side.}$$
$$a \approx 10.2 \quad \text{Use a calculator.}$$

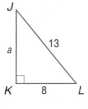

Exercises Find x. *See Example 2 on page 351.*

13.

14.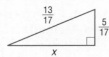

15.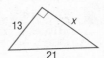

7-3 Special Right Triangles

See pages 357–363.

Concept Summary

- The measure of the hypotenuse of a 45°-45°-90° triangle is $\sqrt{2}$ times the length of the legs of the triangle. The measures of the sides are x, x, and $x\sqrt{2}$.
- In a 30°-60°-90° triangle, the measures of the sides are x, $x\sqrt{3}$, and $2x$.

Examples 1 Find x.

The measure of the shorter leg $\overline{XZ}$ of $\triangle XYZ$ is half the measure of the hypotenuse $\overline{XY}$. Therefore, $XZ = \frac{1}{2}(26)$ or 13. The measure of the longer leg is $\sqrt{3}$ times the measure of the shorter leg. So, $x = 13\sqrt{3}$.

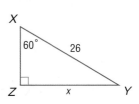

2 Find x.

The measure of the hypotenuse of a 45°-45°-90° triangle is $\sqrt{2}$ times the length of a leg of the triangle.

$$x\sqrt{2} = 4$$

$$x = \frac{4}{\sqrt{2}}$$

$$x = \frac{4}{\sqrt{2}} \cdot \frac{\sqrt{2}}{\sqrt{2}} \text{ or } 2\sqrt{2}$$

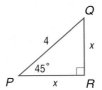

Exercises Find x and y.   *See Examples 1 and 3 on pages 358 and 359.*

16.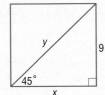

17.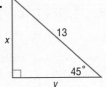

18.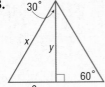

For Exercises 19 and 20, use the figure at the right.
See Example 3 on page 359.

19. If $y = 18$, find z and a.

20. If $x = 14$, find a, z, b, and y.

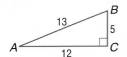

7-4 Trigonometry

See pages 364–370.

Concept Summary

- Trigonometric ratios can be used to find measures in right triangles.

Example Find $\sin A$, $\cos A$, and $\tan A$. Express each ratio as a fraction and as a decimal.

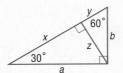

$$\sin A = \frac{\text{opposite leg}}{\text{hypotenuse}}$$
$$= \frac{BC}{AB}$$
$$= \frac{5}{13} \text{ or about } 0.38$$

$$\cos A = \frac{\text{adjacent leg}}{\text{hypotenuse}}$$
$$= \frac{AC}{AB}$$
$$= \frac{12}{13} \text{ or about } 0.92$$

$$\tan A = \frac{\text{opposite leg}}{\text{adjacent leg}}$$
$$= \frac{BC}{AC}$$
$$= \frac{5}{12} \text{ or about } 0.42$$

Exercises Use $\triangle FGH$ to find $\sin F$, $\cos F$, $\tan F$, $\sin G$, $\cos G$, and $\tan G$. Express each ratio as a fraction and as a decimal to the nearest hundredth.   *See Example 1 on page 365.*

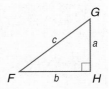

21. $a = 9$, $b = 12$, $c = 15$ **22.** $a = 7$, $b = 24$, $c = 25$

Find the measure of each angle to the nearest tenth of a degree.
See Example 4 on pages 366 and 367.

23. $\sin P = 0.4522$ **24.** $\cos Q = 0.1673$ **25.** $\tan R = 0.9324$

7-5 Angles of Elevation and Depression

See pages
371–376.

Concept Summary

- Trigonometry can be used to solve problems related to angles of elevation and depression.

Example   A store has a ramp near its front entrance. The ramp measures 12 feet, and has a height of 3 feet. What is the angle of elevation?

Make a drawing.

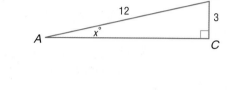

Let x represent $m\angle BAC$.

$\sin x° = \dfrac{BC}{AB}$ $\sin x = \dfrac{\text{opposite}}{\text{hypotenuse}}$

$\sin x° = \dfrac{3}{12}$ $BC = 3$ and $AB = 12$

$x = \sin^{-1}\left(\dfrac{3}{12}\right)$ Find the inverse.

$x \approx 14.5$ Use a calculator.

The angle of elevation for the ramp is about 14.5°.

Exercises   Determine the angles of elevation or depression in each situation.
See Examples 1 and 2 on pages 371 and 372.

26. An airplane must clear a 60-foot pole at the end of a runway 500 yards long.

27. An escalator descends 100 feet for each horizontal distance of 240 feet.

28. A hot-air balloon descends 50 feet for every 1000 feet traveled horizontally.

29. **DAYLIGHT**   At a certain time of the day, the angle of elevation of the sun is 44°. Find the length of a shadow cast by a building that is 30 yards high.

30. **RAILROADS**   A railroad track rises 30 feet for every 400 feet of track. What is the measure of the angle of elevation of the track?

7-6 The Law of Sines

See pages
377–383.

Concept Summary

- To find the measures of a triangle by using the Law of Sines, you must either know the measures of two angles and any side (AAS or ASA), or two sides and an angle opposite one of these sides (SSA) of the triangle.
- To solve a triangle means to find the measures of all sides and angles.

Example   Solve $\triangle XYZ$ if $m\angle X = 32$, $m\angle Y = 61$, and $y = 15$. Round angle measures to the nearest degree and side measures to the nearest tenth.

Find the measure of $\angle Z$.

$m\angle X + m\angle Y + m\angle Z = 180$ Angle Sum Theorem

$32 + 61 + m\angle Z = 180$ $m\angle X = 32$ and $m\angle Y = 61$

$93 + m\angle Z = 180$ Add.

$m\angle Z = 87$ Subtract 93 from each side.

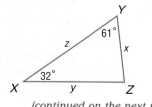

(continued on the next page)

Chapter
7 For More ...

• Extra Practice, see pages 766-768.
• Mixed Problem Solving, see page 788.

Since we know $m\angle Y$ and y, use proportions involving $\sin Y$ and y.

To find x:

$$\frac{\sin Y}{y} = \frac{\sin X}{x} \qquad \text{Law of Sines}$$

$$\frac{\sin 61°}{15} = \frac{\sin 32°}{x} \qquad \text{Substitute.}$$

$$x \sin 61° = 15 \sin 32° \qquad \text{Cross products}$$

$$x = \frac{15 \sin 32°}{\sin 61°} \qquad \text{Divide.}$$

$$x \approx 9.1 \qquad \text{Use a calculator.}$$

To find z:

$$\frac{\sin Y}{y} = \frac{\sin Z}{z}$$

$$\frac{\sin 61°}{15} = \frac{\sin 87°}{z}$$

$$z \sin 61° = 15 \sin 87°$$

$$z = \frac{15 \sin 87°}{\sin 61°}$$

$$z \approx 17.1$$

Exercises Find each measure using the given measures of $\triangle FGH$. Round angle measures to the nearest degree and side measures to the nearest tenth. *See Example 1 on page 378.*

31. Find f if $g = 16$, $m\angle G = 48$, and $m\angle F = 82$.

32. Find $m\angle H$ if $h = 10.5$, $g = 13$, and $m\angle G = 65$.

Solve each $\triangle ABC$ described below. Round angle measures to the nearest degree and side measures to the nearest tenth. *See Example 2 on pages 378 and 379.*

33. $a = 15$, $b = 11$, $m\angle A = 64$

34. $c = 12$, $m\angle C = 67$, $m\angle A = 55$

35. $m\angle A = 29$, $a = 4.8$, $b = 8.7$

36. $m\angle A = 29$, $m\angle B = 64$, $b = 18.5$

7-7 The Law of Cosines

See pages 385-390.

Concept Summary

• The Law of Cosines can be used to solve triangles when you know the measures of two sides and the included angle (SAS) or the measures of the three sides (SSS).

Example Find a if $b = 23$, $c = 19$, and $m\angle A = 54$.

Since the measures of two sides and the included angle are known, use the Law of Cosines.

$$a^2 = b + c^2 - 2bc \cos A \qquad \text{Law of Cosines}$$

$$a^2 = 23^2 + 19^2 - 2(23)(19) \cos 54° \qquad b = 23,\ c = 19,\ \text{and}\ m\angle A = 54$$

$$a = \sqrt{23^2 + 19^2 - 2(23)(19) \cos 54°} \qquad \text{Take the square root of each side.}$$

$$a \approx 19.4 \qquad \text{Use a calculator.}$$

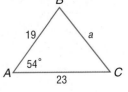

Exercises In $\triangle XYZ$, given the following measures, find the measure of the missing side. *See Example 1 on page 385.*

37. $x = 7.6$, $y = 5.4$, $m\angle Z = 51$

38. $x = 21$, $m\angle Y = 73$, $z = 16$

Solve each triangle using the given information. Round angle measure to the nearest degree and side measure to the nearest tenth. *See Example 3 on pages 386 and 387.*

39. $c = 18$, $b = 13$, $m\angle A = 64$

40. $b = 5.2$, $m\angle C = 53$, $c = 6.7$

Practice Test

Vocabulary and Concepts

1. **State** the Law of Cosines for $\triangle ABC$ used to find $m\angle C$.

2. **Determine** whether the geometric mean of two perfect squares will always be rational. Explain.

3. **Give** an example of side measures of a 30°-60°-90° triangle.

Skills and Applications

Find the geometric mean between each pair of numbers.

4. 7 and 63

5. 6 and 24

6. 10 and 50

Find the missing measures.

7.

8.

9.

10.

11.
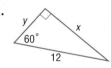

12.

Use the figure to find each trigonometric ratio. Express answers as a fraction.

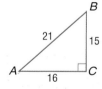

13. $\cos B$

14. $\tan A$

15. $\sin A$

Find each measure using the given measures from $\triangle FGH$. Round to the nearest tenth.

16. Find g if $m\angle F = 59$, $f = 13$, and $m\angle G = 71$.

17. Find $m\angle H$ if $m\angle F = 52$, $f = 10$, and $h = 12.5$.

18. Find f if $g = 15$, $h = 13$, and $m\angle F = 48$.

19. Find h if $f = 13.7$, $g = 16.8$, and $m\angle H = 71$.

Solve each triangle. Round each angle measure to the nearest degree and each side measure to the nearest tenth.

20. $a = 15$, $b = 17$, $m\angle C = 45$

21. $a = 12.2$, $b = 10.9$, $m\angle B = 48$

22. $a = 19$, $b = 23.2$, $c = 21$

23. **TRAVEL** From an airplane, Janara looked down to see a city. If she looked down at an angle of 9° and the airplane was half a mile above the ground, what was the horizontal distance to the city?

24. **CIVIL ENGINEERING** A section of freeway has a steady incline of 10°. If the horizontal distance from the beginning of the incline to the end is 5 miles, how high does the incline reach?

25. **STANDARDIZED TEST PRACTICE** Find tan X.

(A) $\dfrac{5}{12}$ (B) $\dfrac{12}{13}$ (C) $\dfrac{17}{12}$ (D) $\dfrac{12}{5}$

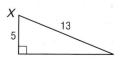

Part 1 | Multiple Choice

Record your answers on the answer sheet provided by your teacher or on a sheet of paper.

1. If ∠4 and ∠3 are supplementary, which reason could you use as the first step in proving that ∠1 and ∠2 are supplementary? (Lesson 2-7)

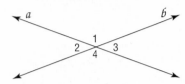

  (A) Definition of similar angles

  (B) Definition of perpendicular lines

  (C) Definition of a vertical angle

  (D) Division Property

2. In △*ABC*, $\overline{BD}$ is a median. If $AD = 3x + 5$ and $CD = 5x - 1$, find *AC*. (Lesson 5-1)

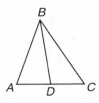

  (A) 3   (B) 11

  (C) 14   (D) 28

3. If pentagons *ABCDE* and *PQRST* are similar, find *SR*. (Lesson 6-2)

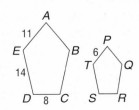

  (A) $14\frac{2}{3}$ (B) $4\frac{4}{11}$

  (C) 3   (D) $1\frac{5}{6}$

4. In △*ABC*, $\overline{CD}$ is an altitude and $m\angle ACB = 90°$. If $AD = 12$ and $BD = 3$, find *AC* to the nearest tenth. (Lesson 7-1)

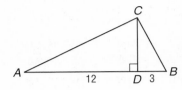

  (A) 6.5   (B) 9.0   (C) 13.4   (D) 15.0

5. What is the length of $\overline{RT}$? (Lesson 7-3)

  (A) 5 cm   (B) $5\sqrt{2}$ cm

  (C) $5\sqrt{3}$ cm   (D) 10 cm

6. An earthquake damaged a communication tower. As a result, the top of the tower broke off at a point 60 feet above the base. If the fallen portion of the tower made a 36° angle with the ground, what was the approximate height of the original tower? (Lesson 7-4)

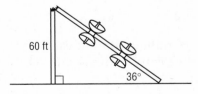

  (A) 35 ft   (B) 95 ft

  (C) 102 ft   (D) 162 ft

7. Miraku drew a map showing Regina's house, Steve's apartment, and Trina's workplace. The three locations formed △*RST*, where $m\angle R = 34$, $r = 14$, and $s = 21$. Which could be $m\angle S$? (Lesson 7-6)

  (A) 15   (B) 22   (C) 57   (D) 84

Preparing for Standardized Tests
For test-taking strategies and more
practice, see pages 795–810.

Part 2 | Short Response/Grid In

Record your answers on the answer sheet
provided by your teacher or on a sheet of
paper.

8. Find $m\angle ABC$
   if $m\angle CDA = 61$.
   (Lesson 1-6)

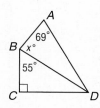

For Questions 9 and 10, refer to the graph.

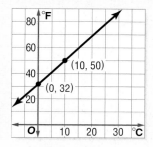

9. At the International Science Fair, a Canadian
   student recorded temperatures in degrees
   Celsius. A student from the United States
   recorded the same temperatures in degrees
   Fahrenheit. They used their data to plot a
   graph of Celsius versus Fahrenheit. What is
   the slope of their graph?   (Lesson 3-3)

10. Students used the equation of the line for
    the temperature graph of Celsius versus
    Fahrenheit to convert degrees Celsius to
    degrees Fahrenheit. If the line goes through
    points (0, 32) and (10, 50), what equation can
    the students use to convert degrees Celsius
    to degrees Fahrenheit?   (Lesson 3-4)

11. $\triangle TUV$ and $\triangle XYZ$ are similar. Calculate the
    ratio $\dfrac{YZ}{UV}$.   (Lesson 6-3)

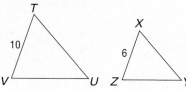

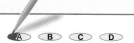

Test-Taking Tip Ⓐ Ⓑ Ⓒ Ⓓ

Questions 6, 7, and 12
If a standardized test question involves trigonometric
functions, draw a diagram that represents the problem
and use a calculator (if allowed) or the table of
trigonometric relationships provided with the test to
help you find the answer.

12. Dee is parasailing at the ocean. The angle of
    depression from her line of sight to the boat
    is 41°. If the cable attaching Dee to the boat
    is 500 feet long, how many feet is Dee above
    the water?   (Lesson 7-5)

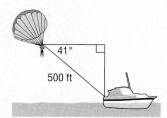

Part 3 | Extended Response

Record your answers on a sheet of paper.
Show your work.

13. Toby, Rani, and Sasha are practicing for a
    double Dutch rope-jumping tournament.
    Toby and Rani are standing at points T
    and R and are turning the ropes. Sasha is
    standing at S, equidistant from both Toby
    and Rani. Sasha will jump into the middle
    of the turning rope to point X. Prove that
    when Sasha jumps into the rope, she will
    be at the midpoint between Toby and Rani.
    (Lessons 4-5 and 4-6)

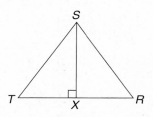

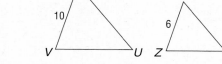 www.geometryonline.com/standardized_test

UNIT
3

Quadrilaterals and Circles

Two-dimensional shapes such as quadrilaterals and circles can be used to describe and model the world around us. In this unit, you will learn about the properties of quadrilaterals and circles and how these two-dimensional figures can be transformed.

Chapter 8
Quadrilaterals

Chapter 9
Transformations

Chapter 10
Circles

WebQuest Internet Project

"Geocaching" Sends Folks on a Scavenger Hunt

Source: *USA TODAY,* July 26, 2001

"N42 DEGREES 02.054 W88 DEGREES 12.329 – Forget the poison ivy and needle-sharp brambles.

Dave April is a man on a mission. Clutching a palm-size Global Positioning System (GPS) receiver in one hand and a computer printout with latitude and longitude coordinates in the other, the 37-year-old software developer trudges doggedly through a suburban Chicago forest preserve, intent on finding a geek's version of buried treasure." Geocaching is one of the many new ways that people are spending their leisure time. In this project, you will use quadrilaterals, circles, and geometric transformations to give clues for a treasure hunt.

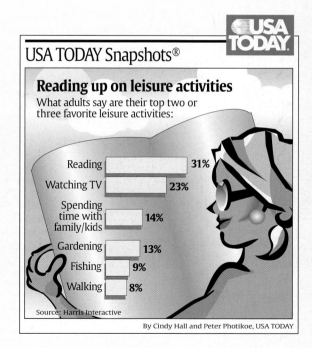

USA TODAY Snapshots®

Reading up on leisure activities
What adults say are their top two or three favorite leisure activities:

Reading — 31%
Watching TV — 23%
Spending time with family/kids — 14%
Gardening — 13%
Fishing — 9%
Walking — 8%

Source: Harris Interactive

By Cindy Hall and Peter Photikoe, USA TODAY

Log on to **www.geometryonline.com/webquest**. Begin your WebQuest by reading the Task.

Then continue working on your WebQuest as you study Unit 3.

Lesson	8-6	9-1	10-1
Page	444	469	527

Chapter 8

Quadrilaterals

What You'll Learn

- **Lesson 8-1** Investigate interior and exterior angles of polygons.
- **Lessons 8-2 and 8-3** Recognize and apply the properties of parallelograms.
- **Lessons 8-4 through 8-6** Recognize and apply the properties of rectangles, rhombi, squares, and trapezoids.
- **Lesson 8-7** Position quadrilaterals for use in coordinate proof.

Key Vocabulary

- parallelogram (p. 411)
- rectangle (p. 424)
- rhombus (p. 431)
- square (p. 432)
- trapezoid (p. 439)

Why It's Important

Several different geometric shapes are examples of quadrilaterals. These shapes each have individual characteristics. A rectangle is a type of quadrilateral. Tennis courts are rectangles, and the properties of the rectangular court are used in the game. *You will learn more about tennis courts in Lesson 8-4.*

Getting Started

▶ **Prerequisite Skills** To be successful in this chapter, you'll need to master these skills and be able to apply them in problem-solving situations. Review these skills before beginning Chapter 8.

For Lesson 8-1
Exterior Angles of Triangles

Find *x* for each figure. *(For review, see Lesson 4-2.)*

1.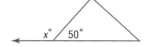

2.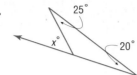

3.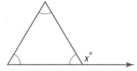

For Lessons 8-4 and 8-5
Perpendicular Lines

Find the slopes of $\overline{RS}$ and $\overline{TS}$ for the given points, R, T, and S. Determine whether $\overline{RS}$ and $\overline{TS}$ are *perpendicular* or *not perpendicular*. *(For review, see Lesson 3-6.)*

4. $R(4, 3)$, $S(-1, 10)$, $T(13, 20)$
5. $R(-9, 6)$, $S(3, 8)$, $T(1, 20)$
6. $R(-6, -1)$, $S(5, 3)$, $T(2, 5)$
7. $R(-6, 4)$, $S(-3, 8)$, $T(5, 2)$

For Lesson 8-7
Slope

Write an expression for the slope of a segment given the coordinates of the endpoints. *(For review, see Lesson 3-3.)*

8. $\left(\dfrac{c}{2}, \dfrac{d}{2}\right)$, $(-c, d)$
9. $(0, a)$, $(b, 0)$
10. $(-a, c)$, $(-c, a)$

 **Quadrilaterals** Make this Foldable to help you organize your notes. Begin with a sheet of notebook paper.

Step 1 Fold

Fold lengthwise to the left margin.

Step 2 Cut

Cut 4 tabs.

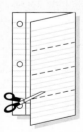

Step 3 Label

Label the tabs using the lesson concepts.

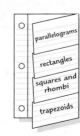

Reading and Writing As you read and study the chapter, use your Foldable to take notes, define terms, and record concepts about quadrilaterals.

8-1 Angles of Polygons

B

A

What You'll Learn

- Find the sum of the measures of the interior angles of a polygon.
- Find the sum of the measures of the exterior angles of a polygon.

Vocabulary
- diagonal

How does a scallop shell illustrate the angles of polygons?

This scallop shell resembles a 12-sided polygon with diagonals drawn from one of the vertices. A **diagonal** of a polygon is a segment that connects any two nonconsecutive vertices. For example, $\overline{AB}$ is one of the diagonals of this polygon.

Study Tip

Look Back
To review the **sum of the measures of the angles of a triangle**, see Lesson 4-2.

SUM OF MEASURES OF INTERIOR ANGLES Polygons with more than three sides have diagonals. The polygons below show all of the possible diagonals drawn from one vertex.

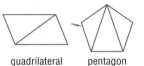

quadrilateral   pentagon   hexagon   heptagon   octagon

In each case, the polygon is separated into triangles. Each angle of the polygon is made up of one or more angles of triangles. The sum of the measures of the angles of each polygon can be found by adding the measures of the angles of the triangles. Since the sum of the measures of the angles in a triangle is 180, we can easily find this sum. Make a table to find the sum of the angle measures for several convex polygons.

Convex Polygon	Number of Sides	Number of Triangle	Sum of Angle Measures
triangle	3	1	(1 · 180) or 180
quadrilateral	4	2	(2 · 180) or 360
pentagon	5	3	(3 · 180) or 540
hexagon	6	4	(4 · 180) or 720
heptagon	7	5	(5 · 180) or 900
octagon	8	6	(6 · 180) or 1080

Look for a pattern in the sum of the angle measures. In each case, the sum of the angle measures is 2 less than the number of sides in the polygon times 180. So in an n-gon, the sum of the angle measures will be $(n - 2)180$ or $180(n - 2)$.

Theorem 8.1

Interior Angle Sum Theorem If a convex polygon has n sides and S is the sum of the measures of its interior angles, then $S = 180(n - 2)$.

Example:

$n = 5$
$S = 180(n - 2)$
$\quad = 180(5 - 2)$ or 540

Example 1   Interior Angles of Regular Polygons

CHEMISTRY   The benzene molecule, C_6H_6, consists of six carbon atoms in a regular hexagonal pattern with a hydrogen atom attached to each carbon atom. Find the sum of the measures of the interior angles of the hexagon.

Since the molecule is a convex polygon, we can use the Interior Angle Sum Theorem.

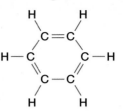

$S = 180(n - 2)$ Interior Angle Sum Theorem

$\quad = 180(6 - 2)$ $n = 6$

$\quad = 180(4)$ or 720 Simplify.

The sum of the measures of the interior angles is 720.

The Interior Angle Sum Theorem can also be used to find the number of sides in a regular polygon if you are given the measure of one interior angle.

Example 2   Sides of a Polygon

The measure of an interior angle of a regular polygon is 108. Find the number of sides in the polygon.

Use the Interior Angle Sum Theorem to write an equation to solve for n, the number of sides.

$S = 180(n - 2)$ Interior Angle Sum Theorem

$(108)n = 180(n - 2)$ $S = 108n$

$108n = 180n - 360$ Distributive Property

$0 = 72n - 360$ Subtract 108n from each side.

$360 = 72n$ Add 360 to each side.

$5 = n$ Divide each side by 72.

The polygon has 5 sides.

In Example 2, the Interior Angle Sum Theorem was applied to a regular polygon. In Example 3, we will apply this theorem to a quadrilateral that is not a regular polygon.

Example 3   Interior Angles

ALGEBRA   Find the measure of each interior angle.

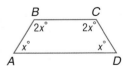

Since $n = 4$, the sum of the measures of the interior angles is $180(4 - 2)$ or 360. Write an equation to express the sum of the measures of the interior angles of the polygon.

$360 = m\angle A + m\angle B + m\angle C + m\angle D$ Sum of measures of angles

$360 = x + 2x + 2x + x$ Substitution

$360 = 6x$ Combine like terms.

$60 = x$ Divide each side by 6.

Use the value of x to find the measure of each angle.

$m\angle A = 60$, $m\angle B = 2 \cdot 60$ or 120, $m\angle C = 2 \cdot 60$ or 120, and $m\angle D = 60$.

SUM OF MEASURES OF EXTERIOR ANGLES   The Interior Angle Sum Theorem relates the interior angles of a convex polygon to the number of sides. Is there a relationship among the exterior angles of a convex polygon?

Geometry Activity

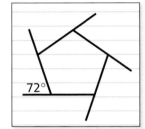

Sum of the Exterior Angles of a Polygon

Collect Data

• Draw a triangle, a convex quadrilateral, a convex pentagon, a convex hexagon, and a convex heptagon.

• Extend the sides of each polygon to form exactly one exterior angle at each vertex.

• Use a protractor to measure each exterior angle of each polygon and record it on your drawing.

72°

Analyze the Data

1. Copy and complete the table.

Polygon	triangle	quadrilateral	pentagon	hexagon	heptagon
number of exterior angles					
sum of measure of exterior angles					

2. What conjecture can you make?

The Geometry Activity suggests Theorem 8.2.

Theorem 8.2

Exterior Angle Sum Theorem   If a polygon is convex, then the sum of the measures of the exterior angles, one at each vertex, is 360.

Example:

$$m\angle 1 + m\angle 2 + m\angle 3 + m\angle 4 + m\angle 5 = 360$$

You will prove Theorem 8.2 in Exercise 42.

Example 4   Exterior Angles

Find the measures of an exterior angle and an interior angle of convex regular octagon *ABCDEFGH*.

At each vertex, extend a side to form one exterior angle. The sum of the measures of the exterior angles is 360. A convex regular octagon has 8 congruent exterior angles.

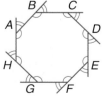

$8n = 360$ n = measure of each exterior angle

$n = 45$ Divide each side by 8.

The measure of each exterior angle is 45. Since each exterior angle and its corresponding interior angle form a linear pair, the measure of the interior angle is $180 - 45$ or 135.

Check for Understanding

Concept Check

1. **Explain** why the Interior Angle Sum Theorem and the Exterior Angle Sum Theorem only apply to convex polygons.

2. **Determine** whether the Interior Angle Sum Theorem and the Exterior Angle Sum Theorem apply to polygons that are not regular. Explain.

3. **OPEN ENDED** Draw a regular convex polygon and a convex polygon that is not regular with the same number of sides. Find the sum of the interior angles for each.

Guided Practice

Find the sum of the measures of the interior angles of each convex polygon.

4. pentagon

5. dodecagon

The measure of an interior angle of a regular polygon is given. Find the number of sides in each polygon.

6. 60

7. 90

ALGEBRA Find the measure of each interior angle.

8.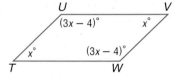

9.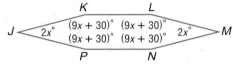

Find the measures of an exterior angle and an interior angle given the number of sides of each regular polygon.

10. 6

11. 18

Application

12. **AQUARIUMS** The regular polygon at the right is the base of a fish tank. Find the sum of the measures of the interior angles of the pentagon.

Practice and Apply

Homework Help

For Exercises	See Examples
13–20	1
21–26	2
27–34	3
35–44	4

Extra Practice
See page 769.

Find the sum of the measures of the interior angles of each convex polygon.

13. 32-gon

14. 18-gon

15. 19-gon

16. 27-gon

17. 4y-gon

18. 2x-gon

19. **GARDENING** Carlotta is designing a garden for her backyard. She wants a flower bed shaped like a regular octagon. Find the sum of the measures of the interior angles of the octagon.

20. **GAZEBOS** A company is building regular hexagonal gazebos. Find the sum of the measures of the interior angles of the hexagon.

The measure of an interior angle of a regular polygon is given. Find the number of sides in each polygon.

21. 140

22. 170

23. 160

24. 165

25. $157\frac{1}{2}$

26. $176\frac{2}{5}$

ALGEBRA Find the measure of each interior angle using the given information.

27.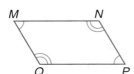

28.

29. parallelogram $MNPQ$ with $m\angle M = 10x$ and $m\angle N = 20x$

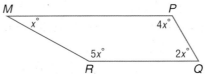

30. isosceles trapezoid $TWYZ$ with $\angle Z \cong \angle Y$, $m\angle Z = 30x$, $\angle T \cong \angle W$, and $m\angle T = 20x$

31. decagon in which the measures of the interior angles are $x + 5$, $x + 10$, $x + 20$, $x + 30$, $x + 35$, $x + 40$, $x + 60$, $x + 70$, $x + 80$, and $x + 90$

32. polygon $ABCDE$ with $m\angle A = 6x$, $m\angle B = 4x + 13$, $m\angle C = x + 9$, $m\angle D = 2x - 8$, and $m\angle E = 4x - 1$

33. quadrilateral in which the measures of the angles are consecutive multiples of x

34. quadrilateral in which the measure of each consecutive angle increases by 10

Find the measures of each exterior angle and each interior angle for each regular polygon.

35. decagon

36. hexagon

37. nonagon

38. octagon

Find the measures of an interior angle and an exterior angle given the number of sides of each regular polygon. Round to the nearest tenth if necessary.

39. 11

40. 7

41. 12

42. **PROOF** Use algebra to prove the Exterior Angle Sum Theorem.

43. **ARCHITECTURE** The Pentagon building in Washington, D.C., was designed to resemble a regular pentagon. Find the measure of an interior angle and an exterior angle of the courtyard.

44. **ARCHITECTURE** Compare the dome to the architectural elements on each side of the dome. Are the interior and exterior angles the same? Find the measures of the interior and exterior angles.

45. **CRITICAL THINKING** Two formulas can be used to find the measure of an interior angle of a regular polygon: $s = \dfrac{180(n - 2)}{n}$ and $s = 180 - \dfrac{360}{n}$. Show that these are equivalent.

46.  Answer the question that was posed at the beginning of the lesson.

How does a scallop shell illustrate the angles of polygons?

Include the following in your answer:
- explain how triangles are related to the Interior Angle Sum Theorem, and
- describe how to find the measure of an exterior angle of a polygon.

47. A regular pentagon and a square share a mutual vertex X. The sides $\overline{XY}$ and $\overline{XZ}$ are sides of a third regular polygon with a vertex at X. How many sides does this polygon have?

   (A) 19   (B) 20

   (C) 28   (D) 32

48. GRID IN If $6x + 3y = 48$ and $\dfrac{9y}{2x} = 9$, then $x = ?$

Maintain Your Skills

Mixed Review

In $\triangle ABC$, given the lengths of the sides, find the measure of the given angle to the nearest tenth. *(Lesson 7-7)*

49. $a = 6, b = 9, c = 11; m\angle C$ **50.** $a = 15.5, b = 23.6, c = 25.1; m\angle B$

51. $a = 47, b = 53, c = 56; m\angle A$ **52.** $a = 12, b = 14, c = 16; m\angle C$

Solve each $\triangle FGH$ described below. Round angle measures to the nearest degree and side measures to the nearest tenth. *(Lesson 7-6)*

53. $f = 15, g = 17, m\angle F = 54$ **54.** $m\angle F = 47, m\angle H = 78, g = 31$

55. $m\angle G = 56, m\angle H = 67, g = 63$ **56.** $g = 30.7, h = 32.4, m\angle G = 65$

57. **PROOF** Write a two-column proof. *(Lesson 4-5)*

   Given: $\overline{JL} \parallel \overline{KM}$
   $\overline{JK} \parallel \overline{LM}$

   Prove: $\triangle JKL \cong \triangle MLK$

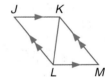

State the transversal that forms each pair of angles. Then identify the special name for the angle pair.
(Lesson 3-1)

58. $\angle 3$ and $\angle 11$

59. $\angle 6$ and $\angle 7$

60. $\angle 8$ and $\angle 10$

61. $\angle 12$ and $\angle 16$

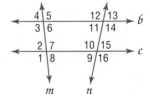

Getting Ready for the Next Lesson

PREREQUISITE SKILL In the figure, $\overline{AB} \parallel \overline{DC}$ and $\overline{AD} \parallel \overline{BC}$. Name all pairs of angles for each type indicated. *(To review angles formed by parallel lines and a transversal, see Lesson 3-1.)*

62. consecutive interior angles

63. alternate interior angles

64. corresponding angles

65. alternate exterior angles

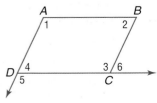

Spreadsheet Investigation

Angles of Polygons

It is possible to find the interior and exterior measurements along with the sum of the interior angles of any regular polygon with n number of sides using a spreadsheet.

Example

Design a spreadsheet using the following steps.

- Label the columns as shown in the spreadsheet below.

- Enter the digits 3–10 in the first column.

- The number of triangles formed by diagonals from the same vertex in a polygon is 2 less than the number of sides. Write a formula for Cell B2 to subtract 2 from each number in Cell A2.

- Enter a formula for Cell C2 so the spreadsheet will find the sum of the measures of the interior angles. Remember that the formula is $S = (n - 2)180$.

- Continue to enter formulas so that the indicated computation is performed. Then, copy each formula through Row 9. The final spreadsheet will appear as below.

Polygons and Angles

	A	B	C	D	E	F	G
1	Number of Sides	Number of Triangles	Sum of Measures of Interior Angles	Measure of Each Interior Angle	Measure of Each Exterior Angle	Sum of Measures of Exterior Angles	
2	3	1	180	60	120	360	
3	4	2	360	90	90	360	
4	5	3	540	108	72	360	
5	6	4	720	120	60	360	
6	7	5	900	128.57	51.43	360	
7	8	6	1080	135	45	360	
8	9	7	1260	140	40	360	
9	10	8	1440	144	36	360	
10							

Sheet1 / Sheet2 / Sheet3

Exercises

1. Write the formula to find the measure of each interior angle in the polygon.
2. Write the formula to find the sum of the measures of the exterior angles.
3. What is the measure of each interior angle if the number of sides is 1? 2?
4. Is it possible to have values of 1 and 2 for the number of sides? Explain.

For Exercises 5–8, use the spreadsheet.

5. How many triangles are in a polygon with 15 sides?
6. Find the measure of the exterior angle of a polygon with 15 sides.
7. Find the measure of the interior angle of a polygon with 110 sides.
8. If the measure of the exterior angles is 0, find the measure of the interior angles. Is this possible? Explain.

Parallelograms

Vocabulary
• parallelogram

What You'll Learn

• Recognize and apply properties of the sides and angles of parallelograms.

• Recognize and apply properties of the diagonals of parallelograms.

How are parallelograms used to represent data?

The graphic shows the percent of Global 500 companies that use the Internet to find potential employees. The top surfaces of the wedges of cheese are all polygons with a similar shape. However, the size of the polygon changes to reflect the data. What polygon is this?

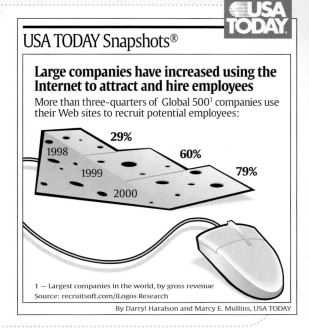

USA TODAY Snapshots®

Large companies have increased using the Internet to attract and hire employees

More than three-quarters of Global 500[1] companies use their Web sites to recruit potential employees:

1998 — 29%
1999 — 60%
2000 — 79%

1 — Largest companies in the world, by gross revenue
Source: recruitsoft.com/iLogos Research

By Darryl Haralson and Marcy E. Mullins, USA TODAY

SIDES AND ANGLES OF PARALLELOGRAMS A quadrilateral with parallel opposite sides is called a **parallelogram**.

Study Tip

Reading Math
Recall that the matching arrow marks on the segments mean that the sides are parallel.

Key Concept — Parallelogram

• **Words** A parallelogram is a quadrilateral with both pairs of opposite sides parallel.

• **Symbols** □ABCD

• **Example**

There are two pairs of parallel sides.
$\overline{AB}$ and $\overline{DC}$
$\overline{AD}$ and $\overline{BC}$

This activity will help you make conjectures about the sides and angles of a parallelogram.

Geometry Activity

Properties of Parallelograms

Make a model

Step 1 Draw two sets of intersecting parallel lines on patty paper. Label the vertices *FGHJ*.

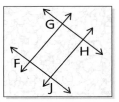

(continued on the next page)

Step 2   Trace *FGHJ*. Label the second parallelogram *PQRS* so ∠F and ∠P are congruent.

Step 3   Rotate ▱*PQRS* on ▱*FGHJ* to compare sides and angles.

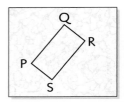

Analyze
1. List all of the segments that are congruent.
2. List all of the angles that are congruent.
3. Describe the angle relationships you observed.

The Geometry Activity leads to four properties of parallelograms.

Key Concept		*Properties of Parallelograms*
Theorem	**Example**	
8.3 Opposite sides of a parallelogram are congruent. **Abbreviation:** *Opp. sides of ▱ are ≅.*	$\overline{AB} \cong \overline{DC}$ $\overline{AD} \cong \overline{BC}$	
8.4 Opposite angles in a parallelogram are congruent. **Abbreviation:** *Opp. ∠ of ▱ are ≅.*	∠A ≅ ∠C ∠B ≅ ∠D	
8.5 Consecutive angles in a parallelogram are supplementary. **Abbreviation:** *Cons. ∠ in ▱ are suppl.*	$m\angle A + m\angle B = 180$ $m\angle B + m\angle C = 180$ $m\angle C + m\angle D = 180$ $m\angle D + m\angle A = 180$	
8.6 If a parallelogram has one right angle, it has four right angles. **Abbreviation:** *If ▱ has 1 rt. ∠, it has 4 rt. ∠.*	$m\angle G = 90$ $m\angle H = 90$ $m\angle J = 90$ $m\angle K = 90$	

You will prove Theorems 8.3, 8.5, and 8.6 in Exercises 41, 42, and 43, respectively.

Study Tip

Including a Figure
Theorems are presented in general terms. In a proof, you must include a drawing so that you can refer to segments and angles specifically.

Example 1   Proof of Theorem 8.4

Write a two-column proof of Theorem 8.4.

Given: ▱*ABCD*

Prove: ∠A ≅ ∠C
∠D ≅ ∠B

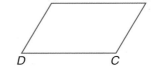

Proof:

Statements	**Reasons**
1. ▱*ABCD*	1. Given
2. $\overline{AB} \parallel \overline{DC}$, $\overline{AD} \parallel \overline{BC}$	2. Definition of parallelogram
3. ∠A and ∠D are supplementary. ∠D and ∠C are supplementary. ∠C and ∠B are supplementary.	3. If parallel lines are cut by a transversal, consecutive interior angles are supplementary.
4. ∠A ≅ ∠C ∠D ≅ ∠B	4. Supplements of the same angles are congruent.

Example 2 Properties of Parallelograms

ALGEBRA Quadrilateral *LMNP* is a parallelogram. Find $m\angle PLM$, $m\angle LMN$, and *d*.

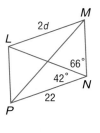

$m\angle MNP = 66 + 42$ or 108   Angle Addition Theorem

$\angle PLM \cong \angle MNP$	Opp. $\angle$s of $\square$ are $\cong$.
$m\angle PLM = m\angle MNP$	Definition of congruent angles
$m\angle PLM = 108$	Substitution

$m\angle PLM + m\angle LMN = 180$	Cons. $\angle$s of $\square$ are suppl.
$108 + m\angle LMN = 180$	Substitution
$m\angle LMN = 72$	Subtract 108 from each side.

$\overline{LM} \cong \overline{PN}$	Opp. sides of $\square$ are $\cong$.
$LM = PN$	Definition of congruent segments
$2d = 22$	Substitution
$d = 11$	Substitution

DIAGONALS OF PARALLELOGRAMS

In parallelogram *JKLM*, $\overline{JL}$ and $\overline{KM}$ are diagonals. Theorem 8.7 states the relationship between diagonals of a parallelogram.

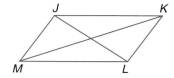

Theorem 8.7

The diagonals of a parallelogram bisect each other.

Abbreviation: *Diag. of $\square$ bisect each other.*

Example: $\overline{RQ} \cong \overline{QT}$ and $\overline{SQ} \cong \overline{QU}$

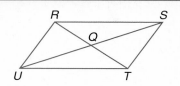

You will prove Theorem 8.7 in Exercise 44.

Standardized Test Practice
Ⓐ Ⓑ Ⓒ Ⓓ

Example 3 Diagonals of a Parallelogram

Multiple-Choice Test Item

> What are the coordinates of the intersection of the diagonals of parallelogram *ABCD* with vertices $A(2, 5)$, $B(6, 6)$, $C(4, 0)$, and $D(0, -1)$?
>
> Ⓐ (4, 2)   Ⓑ (4.5, 2)   Ⓒ $\left(\frac{7}{6}, \frac{-5}{2}\right)$ Ⓓ (3, 2.5)

Read the Test Item

Since the diagonals of a parallelogram bisect each other, the intersection point is the midpoint of $\overline{AC}$ and $\overline{BD}$.

Test-Taking Tip

Check Answers Always check your answer. To check the answer to this problem, find the coordinates of the midpoint of $\overline{BD}$.

Solve the Test Item

Find the midpoint of $\overline{AC}$.

$$\left(\frac{x_1 + x_2}{2}, \frac{y_1 + y_2}{2}\right) = \left(\frac{2 + 4}{2}, \frac{5 + 0}{2}\right) \quad \text{Midpoint Formula}$$

$$= (3, 2.5)$$

The coordinates of the intersection of the diagonals of parallelogram *ABCD* are (3, 2.5). The answer is D.

Theorem 8.8 describes another characteristic of the diagonals of a parallelogram.

Theorem 8.8

Each diagonal of a parallelogram separates the parallelogram into two congruent triangles.

Abbreviation: *Diag. separates □ into 2 ≅ △s.*

Example: △ACD ≅ △CAB

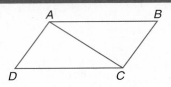

You will prove Theorem 8.8 in Exercise 45.

Check for Understanding

Concept Check

1. **Describe** the characteristics of the sides and angles of a parallelogram.

2. **Describe** the properties of the diagonals of a parallelogram.

3. **OPEN ENDED**   Draw a parallelogram with one side twice as long as another side.

Guided Practice

Complete each statement about □QRST. Justify your answer.

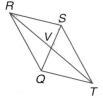

4. $\overline{SV} \cong$ __?__

5. △VRS ≅ __?__

6. ∠TSR is supplementary to __?__ .

Use □JKLM to find each measure or value if JK = 2b + 3 and JM = 3a.

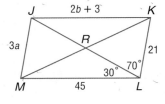

7. $m\angle MJK$ 8. $m\angle JML$

9. $m\angle JKL$ 10. $m\angle KJL$

11. a 12. b

PROOF   **Write the indicated type of proof.**

13. two-column

   Given: □VZRQ and □WQST
   Prove: ∠Z ≅ ∠T

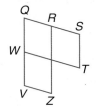

14. paragraph

   Given: □XYRZ, $\overline{WZ} \cong \overline{WS}$
   Prove: ∠XYR ≅ ∠S

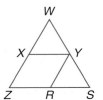

Standardized Test Practice
Ⓐ Ⓑ Ⓒ Ⓓ

15. **MULTIPLE CHOICE**   Find the coordinates of the intersection of the diagonals of parallelogram GHJK with vertices G(−3, 4), H(1, 1), J(3, −5), and K(−1, −2).

   Ⓐ (0, 0.5)   Ⓑ (6, −1)   Ⓒ (0, −0.5)   Ⓓ (5, 0)

Homework Help

For Exercises	See Examples
16–33	2
34–40	3
41–47	1

Extra Practice
See page 769.

Complete each statement about ▱ABCD. Justify your answer.

16. $\angle DAB \cong$?

17. $\angle ABD \cong$?

18. $\overline{AB} \parallel$?

19. $\overline{BG} \cong$?

20. $\triangle ABD \cong$?

21. $\angle ACD \cong$?

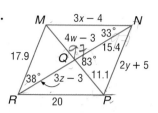

ALGEBRA Use ▱MNPR to find each measure or value.

22. $m\angle MNP$

23. $m\angle NRP$

24. $m\angle RNP$

25. $m\angle RMN$

26. $m\angle MQN$

27. $m\angle MQR$

28. x

29. y

30. w

31. z

DRAWING For Exercises 32 and 33, use the following information.
The frame of a pantograph is a parallelogram.

32. Find x and EG if $EJ = 2x + 1$ and $JG = 3x$.

33. Find y and FH if $HJ = \frac{1}{2}y + 2$ and $JF = y - \frac{1}{2}$.

34. **DESIGN** The chest of drawers shown at the right is called *Side 2*. It was designed by Shiro Kuramata. Describe the properties of parallelograms the artist used to place each drawer pull.

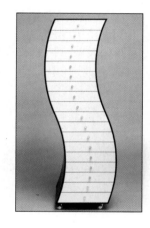

35. **ALGEBRA** Parallelogram $ABCD$ has diagonals $\overline{AC}$ and $\overline{DB}$ that intersect at point P. If $AP = 3a + 18$, $AC = 12a$, $PB = a + 2b$, and $PD = 3b + 1$, find a, b, and DB.

36. **ALGEBRA** In parallelogram $ABCD$, $AB = 2x + 5$, $m\angle BAC = 2y$, $m\angle B = 120$, $m\angle CAD = 21$, and $CD = 21$. Find x and y.

COORDINATE GEOMETRY For Exercises 37–39, refer to ▱EFGH.

37. Use the Distance Formula to verify that the diagonals bisect each other.

38. Determine whether the diagonals of this parallelogram are congruent.

39. Find the slopes of $\overline{EH}$ and $\overline{EF}$. Are the consecutive sides perpendicular? Explain.

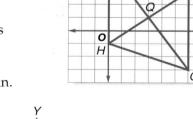

40. Determine the relationship among $ACBX$, $ABYC$, and $ABCZ$ if $\triangle XYZ$ is equilateral and A, B, and C are midpoints of $\overline{XZ}$, $\overline{XY}$, and $\overline{ZY}$, respectively.

PROOF Write the indicated type of proof.

41. two-column proof of Theorem 8.3

42. two-column proof of Theorem 8.5

43. paragraph proof of Theorem 8.6

44. paragraph proof of Theorem 8.7

45. two-column proof of Theorem 8.8

PROOF Write a two-column proof.

46. Given: $\square DGHK$, $\overline{FH} \perp \overline{GD}$, $\overline{DJ} \perp \overline{HK}$
Prove: $\triangle DJK \cong \triangle HFG$

47. Given: $\square BCGH$, $\overline{HD} \cong \overline{FD}$
Prove: $\angle F \cong \angle GCB$

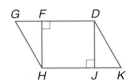

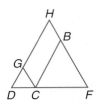

48. CRITICAL THINKING Find the ratio of MS to SP, given that $MNPQ$ is a parallelogram with $MR = \frac{1}{4} MN$.

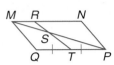

49. **WRITING IN MATH** Answer the question that was posed at the beginning of the lesson.

How are parallelograms used to represent data?

Include the following in your answer:

• properties of parallelograms, and
• a display of the data in the graphic with a different parallelogram.

Standardized Test Practice
Ⓐ Ⓑ Ⓒ Ⓓ

50. SHORT RESPONSE Two consecutive angles of a parallelogram measure $(3x + 42)°$ and $(9x - 18)°$. Find the measures of the angles.

51. ALGEBRA The perimeter of the rectangle $ABCD$ is equal to p and $x = \frac{y}{5}$. What is the value of y in terms of p?

Ⓐ $\frac{p}{3}$ Ⓑ $\frac{5p}{12}$ Ⓒ $\frac{5p}{8}$ Ⓓ $\frac{5p}{6}$

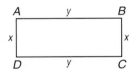

Maintain Your Skills

Mixed Review **Find the sum of the measures of the interior angles of each convex polygon.**
(Lesson 8-1)

52. 14-gon   **53.** 22-gon   **54.** 17-gon   **55.** 36-gon

Determine whether the *Law of Sines* or the *Law of Cosines* should be used to solve each triangle. Then solve each triangle. Round to the nearest tenth.
(Lesson 7-7)

56.
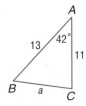

57.
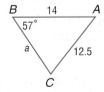

58.

Use Pascal's Triangle for Exercises 59 and 60.   *(Lesson 6-6)*

59. Find the sum of the first 30 numbers in the outside diagonal of Pascal's triangle.

60. Find the sum of the first 70 numbers in the second diagonal.

Getting Ready for the Next Lesson **PREREQUISITE SKILL** The vertices of a quadrilateral are $A(-5, -2)$, $B(-2, 5)$, $C(2, -2)$, and $D(-1, -9)$. **Determine whether each segment is a side or a diagonal of the quadrilateral, and find the slope of each segment.**
*(To review **slope**, see Lesson 3-3.)*

61. $\overline{AB}$ **62.** $\overline{BD}$ **63.** $\overline{CD}$

8-3 Tests for Parallelograms

What You'll Learn

- Recognize the conditions that ensure a quadrilateral is a parallelogram.
- Prove that a set of points forms a parallelogram in the coordinate plane.

How are parallelograms used in architecture?

The roof of the covered bridge appears to be a parallelogram. Each pair of opposite sides looks like they are the same length. How can we know for sure if this shape is really a parallelogram?

CONDITIONS FOR A PARALLELOGRAM   By definition, the opposite sides of a parallelogram are parallel. So, if a quadrilateral has each pair of opposite sides parallel it is a parallelogram. Other tests can be used to determine if a quadrilateral is a parallelogram.

Geometry Activity

Testing for a Parallelogram

Model

- Cut two straws to one length and two other straws to a different length.
- Connect the straws by inserting a pipe cleaner in one end of each size of straw to form a quadrilateral like the one shown at the right.
- Shift the sides to form quadrilaterals of different shapes.

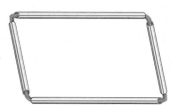

Analyze

1. Measure the distance between the opposite sides of the quadrilateral in at least three places. Repeat this process for several figures. What can you conclude about opposite sides?
2. Classify the quadrilaterals that you formed.
3. Compare the measures of pairs of opposite sides.
4. Measure the four angles in several of the quadrilaterals. What relationships do you find?

Make a Conjecture

5. What conditions are necessary to verify that a quadrilateral is a parallelogram?

Theorem	Example
8.9 If both pairs of opposite sides of a quadrilateral are congruent, then the quadrilateral is a parallelogram. **Abbreviation:** *If both pairs of opp. sides are ≅, then quad. is ▱.*	
8.10 If both pairs of opposite angles of a quadrilateral are congruent, then the quadrilateral is a parallelogram. **Abbreviation:** *If both pairs of opp. ⦤ are ≅, then quad. is ▱.*	
8.11 If the diagonals of a quadrilateral bisect each other, then the quadrilateral is a parallelogram. **Abbreviation:** *If diag. bisect each other, then quad. is ▱.*	
8.12 If one pair of opposite sides of a quadrilateral is both parallel and congruent, then the quadrilateral is a parallelogram. **Abbreviation:** *If one pair of opp. sides is ∥ and ≅, then the quad. is a ▱.*	

You will prove Theorems 8.9, 8.11, and 8.12 in Exercises 39, 40, and 41, respectively.

Example 1 **Write a Proof**

PROOF Write a paragraph proof for Theorem 8.10

Given: ∠A ≅ ∠C, ∠B ≅ ∠D

Prove: *ABCD* is a parallelogram.

Paragraph Proof:

Because two points determine a line, we can draw $\overline{AC}$. We now have two triangles. We know the sum of the angle measures of a triangle is 180, so the sum of the angle measures of two triangles is 360. Therefore, $m\angle A + m\angle B + m\angle C + m\angle D = 360$.

Since ∠A ≅ ∠C and ∠B ≅ ∠D, $m\angle A = m\angle C$ and $m\angle B = m\angle D$. Substitute to find that $m\angle A + m\angle A + m\angle B + m\angle B = 360$, or $2(m\angle A) + 2(m\angle B) = 360$. Dividing each side of the equation by 2 yields $m\angle A + m\angle B = 180$. This means that consecutive angles are supplementary and $\overline{AD} \parallel \overline{BC}$.

Likewise, $2m\angle A + 2m\angle D = 360$, or $m\angle A + m\angle D = 180$. These consecutive supplementary angles verify that $\overline{AB} \parallel \overline{DC}$. Opposite sides are parallel, so *ABCD* is a parallelogram.

More About. . .

Art •·········

Ellsworth Kelly created *Sculpture for a Large Wall* in 1957. The sculpture is made of 104 aluminum panels. The piece is over 65 feet long, 11 feet high, and 2 feet deep.

Source: www.moma.org

Example 2 **Properties of Parallelograms**

•·**ART** Some panels in the sculpture appear to be parallelograms. Describe the information needed to determine whether these panels are parallelograms.

A panel is a parallelogram if both pairs of opposite sides are congruent, or if one pair of opposite sides is congruent and parallel. If the diagonals bisect each other, or if both pairs of opposite angles are congruent, then the panel is a parallelogram.

Example 3 Properties of Parallelograms

Determine whether the quadrilateral is a parallelogram. Justify your answer.

Each pair of opposite angles have the same measure. Therefore, they are congruent. If both pairs of opposite angles are congruent, the quadrilateral is a parallelogram.

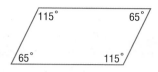

A quadrilateral is a parallelogram if any one of the following is true.

Concept Summary — Tests for a Parallelogram

1. Both pairs of opposite sides are parallel. (Definition)

2. Both pairs of opposite sides are congruent. (Theorem 8.9)

3. Both pairs of opposite angles are congruent. (Theorem 8.10)

4. Diagonals bisect each other. (Theorem 8.11)

5. A pair of opposite sides is both parallel and congruent. (Theorem 8.12)

Study Tip

Common Misconceptions
If a quadrilateral meets one of the five tests, it is a parallelogram. All of the properties of parallelograms need not be shown.

Example 4 Find Measures

ALGEBRA Find x and y so that each quadrilateral is a parallelogram.

a.

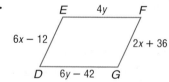

Opposite sides of a parallelogram are congruent.

$\overline{EF} \cong \overline{DG}$	Opp. sides of ▱ are ≅.	$\overline{DE} \cong \overline{FG}$	Opp. sides of ▱ are ≅.
$EF = DG$	Def. of ≅ segments	$DE = FG$	Def. of ≅ segments
$4y = 6y - 42$	Substitution	$6x - 12 = 2x + 36$	Substitution
$-2y = -42$	Subtract 6y.	$4x = 48$	Subtract 2x and add 12.
$y = 21$	Divide by −2.	$x = 12$	Divide by 4.

So, when x is 12 and y is 21, $DEFG$ is a parallelogram.

b.

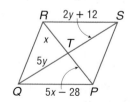

Diagonals in a parallelogram bisect each other.

$\overline{QT} \cong \overline{TS}$	Opp. sides of ▱ are ≅.	$\overline{RT} \cong \overline{TP}$	Opp. sides of ▱ are ≅.
$QT = TS$	Def. of ≅ segments	$RT = TP$	Def. of ≅ segments
$5y = 2y + 12$	Substitution	$x = 5x - 28$	Substitution
$3y = 12$	Subtract 2y.	$-4x = -28$	Subtract 5x.
$y = 4$	Divide by 3.	$x = 7$	Divide by −4.

$PQRS$ is a parallelogram when $x = 7$ and $y = 4$.

PARALLELOGRAMS ON THE COORDINATE PLANE We can use the Distance Formula and the Slope Formula to determine if a quadrilateral is a parallelogram in the coordinate plane.

Study Tip

Coordinate Geometry
The Midpoint Formula can also be used to show that a quadrilateral is a parallelogram by Theorem 8.11.

Example 5 *Use Slope and Distance*

COORDINATE GEOMETRY Determine whether the figure with the given vertices is a parallelogram. Use the method indicated.

a. $A(3, 3)$, $B(8, 2)$, $C(6, -1)$, $D(1, 0)$; **Slope Formula**

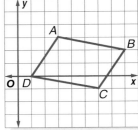

If the opposite sides of a quadrilateral are parallel, then it is a parallelogram.

slope of $\overline{AB} = \dfrac{2-3}{8-3}$ or $\dfrac{-1}{5}$ slope of $\overline{DC} = \dfrac{-1-0}{6-1}$ or $\dfrac{-1}{5}$

slope of $\overline{AD} = \dfrac{3-0}{3-1}$ or $\dfrac{3}{2}$ slope of $\overline{BC} = \dfrac{-1-2}{6-8}$ or $\dfrac{3}{2}$

Since opposite sides have the same slope, $\overline{AB} \parallel \overline{DC}$ and $\overline{AD} \parallel \overline{BC}$. Therefore, $ABCD$ is a parallelogram by definition.

b. $P(5, 3)$, $Q(1, -5)$, $R(-6, -1)$, $S(-2, 7)$; **Distance and Slope Formulas**

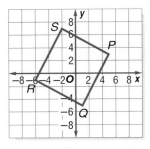

First use the Distance Formula to determine whether the opposite sides are congruent.

$PS = \sqrt{[5 - (-2)]^2 + (3 - 7)^2}$

$\quad = \sqrt{7^2 + (-4)^2}$ or $\sqrt{65}$

$QR = \sqrt{[1 - (-6)]^2 + [-5 - (-1)]^2}$

$\quad = \sqrt{7^2 + (-4)^2}$ or $\sqrt{65}$

Since $PS = QR$, $\overline{PS} \cong \overline{QR}$.

Next, use the Slope Formula to determine whether $\overline{PS} \parallel \overline{QR}$.

slope of $\overline{PS} = \dfrac{3-7}{5-(-2)}$ or $-\dfrac{4}{7}$ slope of $\overline{QR} = \dfrac{-5-(-1)}{1-(-6)}$ or $-\dfrac{4}{7}$

$\overline{PS}$ and $\overline{QR}$ have the same slope, so they are parallel. Since one pair of opposite sides is congruent and parallel, $PQRS$ is a parallelogram.

Check for Understanding

Concept Check

1. **List** and describe four tests for parallelograms.

2. **OPEN ENDED** Draw a parallelogram. Label the congruent angles.

3. **FIND THE ERROR** Carter and Shaniqua are describing ways to show that a quadrilateral is a parallelogram.

> **Carter**
>
> A quadrilateral is a parallelogram if one pair of opposite sides is congruent and one pair of opposite sides is parallel.

> **Shaniqua**
>
> A quadrilateral is a parallelogram if one pair of opposite sides is congruent and parallel.

Who is correct? Explain your reasoning.

Guided Practice **Determine whether each quadrilateral is a parallelogram. Justify your answer.**

4.

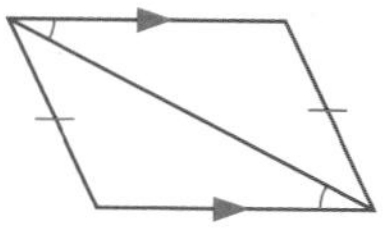

5.

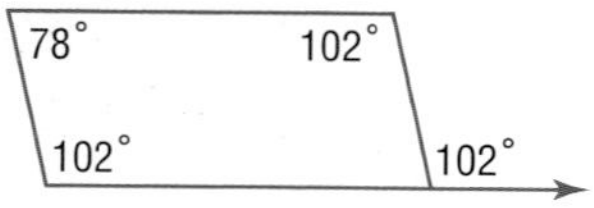
78°
102°
102°
102°

ALGEBRA **Find x and y so that each quadrilateral is a parallelogram.**

6.

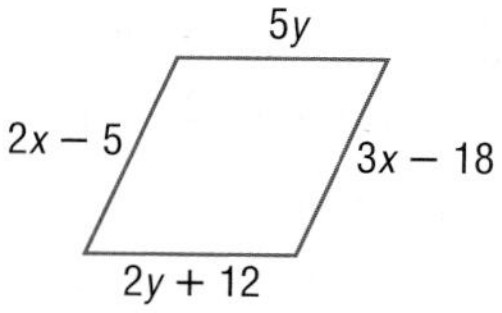
5y
2x − 5
3x − 18
2y + 12

7.

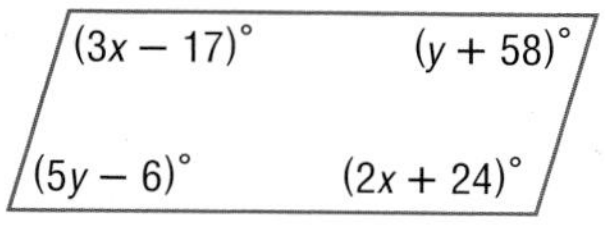
(3x − 17)°
(y + 58)°
(5y − 6)°
(2x + 24)°

COORDINATE GEOMETRY **Determine whether the figure with the given vertices is a parallelogram. Use the method indicated.**

8. $B(0, 0)$, $C(4, 1)$, $D(6, 5)$, $E(2, 4)$; Slope Formula

9. $A(-4, 0)$, $B(3, 1)$, $C(1, 4)$, $D(-6, 3)$; Distance and Slope Formulas

10. $E(-4, -3)$, $F(4, -1)$, $G(2, 3)$, $H(-6, 2)$; Midpoint Formula

11. **PROOF** Write a two-column proof to prove that $PQRS$ is a parallelogram given that $\overline{PT} \cong \overline{TR}$ and $\angle TSP \cong \angle TQR$.

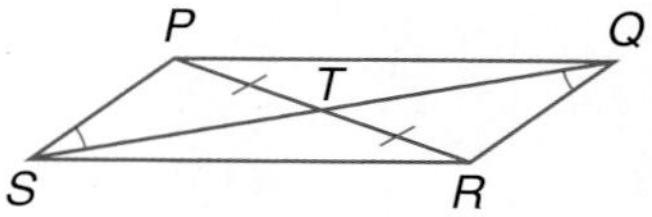
P
Q
T
S
R

Application **12. TANGRAMS** A tangram set consists of seven pieces: a small square, two small congruent right triangles, two large congruent right triangles, a medium-sized right triangle, and a quadrilateral. How can you determine the shape of the quadrilateral? Explain.

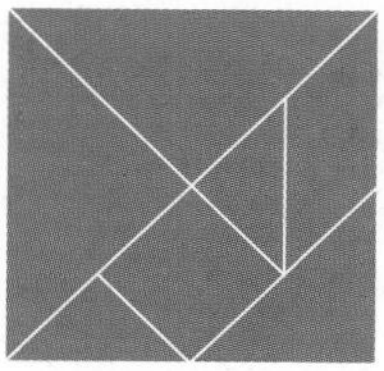

Practice and Apply

Homework Help

For Exercises	See Examples
13–18	3
19–24	4
25–36	5
37–38	2
39–42	1

Extra Practice
See page 769.

Determine whether each quadrilateral is a parallelogram. Justify your answer.

13.

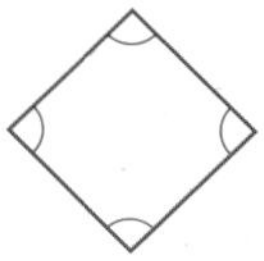

14.

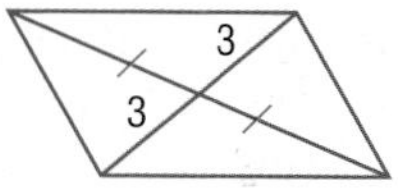
3
3

15.

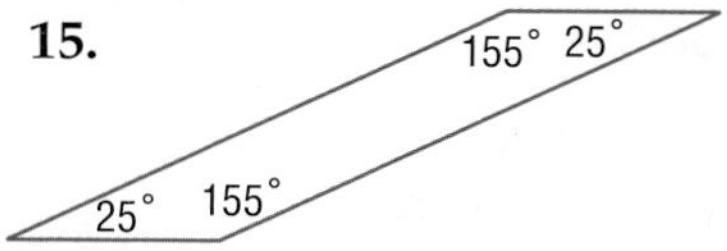
155°
25°
25°
155°

16.

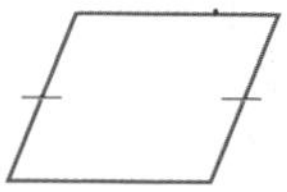

17.

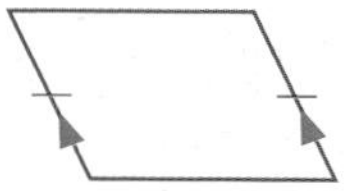

18.

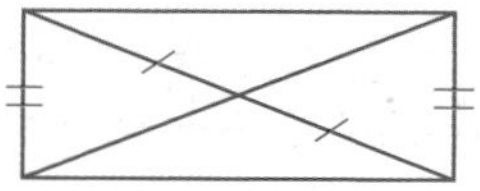

ALGEBRA **Find x and y so that each quadrilateral is a parallelogram.**

19.

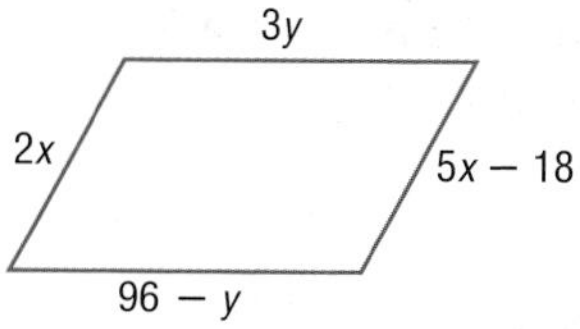
3y
2x
5x − 18
96 − y

20.

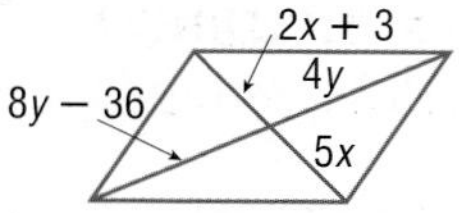
2x + 3
4y
8y − 36
5x

21.

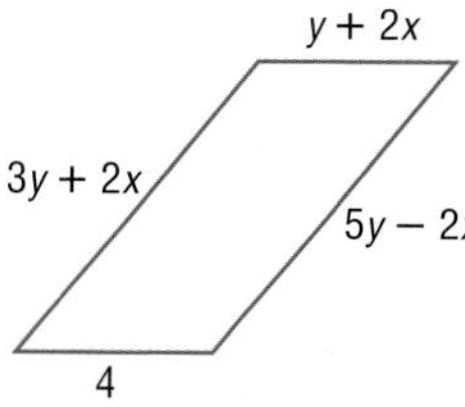
y + 2x
3y + 2x
5y − 2x
4

22.

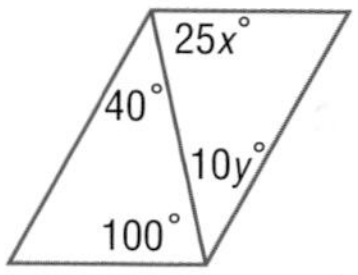
25x°
40°
10y°
100°

23.

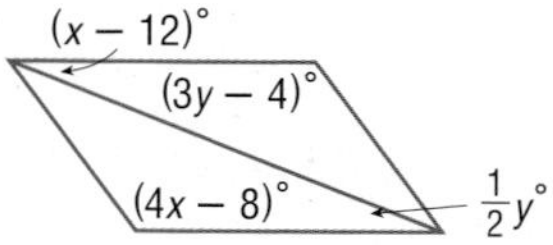
(x − 12)°
(3y − 4)°
(4x − 8)°
1/2 y°

24.

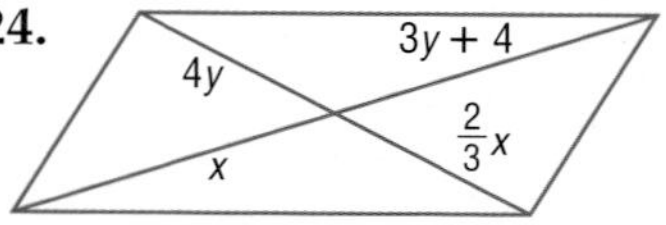
4y
3y + 4
2/3 x
x

COORDINATE GEOMETRY Determine whether a figure with the given vertices is a parallelogram. Use the method indicated.

25. $B(-6, -3)$, $C(2, -3)$, $E(4, 4)$, $G(-4, 4)$; Slope Formula
26. $Q(-3, -6)$, $R(2, 2)$, $S(-1, 6)$, $T(-5, 2)$; Slope Formula
27. $A(-5, -4)$, $B(3, -2)$, $C(4, 4)$, $D(-4, 2)$; Distance Formula
28. $W(-6, -5)$, $X(-1, -4)$, $Y(0, -1)$, $Z(-5, -2)$; Midpoint Formula
29. $G(-2, 8)$, $H(4, 4)$, $J(6, -3)$, $K(-1, -7)$; Distance and Slope Formulas
30. $H(5, 6)$, $J(9, 0)$, $K(8, -5)$, $L(3, -2)$; Distance Formula
31. $S(-1, 9)$, $T(3, 8)$, $V(6, 2)$, $W(2, 3)$; Midpoint Formula
32. $C(-7, 3)$, $D(-3, 2)$, $F(0, -4)$, $G(-4, -3)$; Distance and Slope Formulas

33. Quadrilateral $MNPR$ has vertices $M(-6, 6)$, $N(-1, -1)$, $P(-2, -4)$, and $R(-5, -2)$. Determine how to move one vertex to make $MNPR$ a parallelogram.

34. Quadrilateral $QSTW$ has vertices $Q(-3, 3)$, $S(4, 1)$, $T(-1, -2)$, and $W(-5, -1)$. Determine how to move one vertex to make $QSTW$ a parallelogram.

COORDINATE GEOMETRY The coordinates of three of the vertices of a parallelogram are given. Find the possible coordinates for the fourth vertex.

35. $A(1, 4)$, $B(7, 5)$, and $C(4, -1)$.
36. $Q(-2, 2)$, $R(1, 1)$, and $S(-1, -1)$.

37. **STORAGE** Songan purchased an expandable hat rack that has 11 pegs. In the figure, H is the midpoint of $\overline{KM}$ and $\overline{JL}$. What type of figure is $JKLM$? Explain.

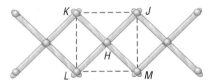

38. **METEOROLOGY** To show the center of a storm, television stations superimpose a "watchbox" over the weather map. Describe how you know that the watchbox is a parallelogram.

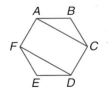

Online Research **Data Update** Each hurricane is assigned a name as the storm develops. What is the name of the most recent hurricane or tropical storm in the Atlantic or Pacific Oceans? Visit www.geometryonline.com/data_update to learn more.

PROOF Write a two-column proof of each theorem.
39. Theorem 8.9   40. Theorem 8.11   41. Theorem 8.12

42. Li-Cheng claims she invented a new geometry theorem. *A diagonal of a parallelogram bisects its angles.* Determine whether this theorem is true. Find an example or counterexample.

43. **CRITICAL THINKING** Write a proof to prove that $FDCA$ is a parallelogram if $ABCDEF$ is a regular hexagon.

44. **WRITING IN MATH** Answer the question that was posed at the beginning of the lesson.

How are parallelograms used in architecture?

Include the following in your answer:
- the information needed to prove that the roof of the covered bridge is a parallelogram, and
- another example of parallelograms used in architecture.

Career Choices

Atmospheric Scientist

Atmospheric scientists, or meteorologists, study weather patterns. They can work for private companies, the Federal Government or television stations.

 Online Research For information about a career as an atmospheric scientist, visit: www.geometryonline.com/careers

45. A parallelogram has vertices at $(-2, 2)$, $(1, -6)$, and $(8, 2)$. Which ordered pair could represent the fourth vertex?

Ⓐ $(5, 6)$ Ⓑ $(11, -6)$ Ⓒ $(14, 3)$ Ⓓ $(8, -8)$

46. ALGEBRA Find the distance between $X(5, 7)$ and $Y(-3, -4)$.

Ⓐ $\sqrt{19}$ Ⓑ $3\sqrt{15}$ Ⓒ $\sqrt{185}$ Ⓓ $5\sqrt{29}$

Maintain Your Skills

Mixed Review

Use ▱$NQRM$ to find each measure or value. *(Lesson 8-2)*

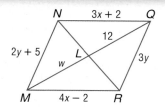

47. w

48. x

49. NQ

50. QR

The measure of an interior angle of a regular polygon is given. Find the number of sides in each polygon. *(Lesson 8-1)*

51. 135   **52.** 144   **53.** 168

54. 162   **55.** 175   **56.** 175.5

Find x and y. *(Lesson 7-3)*

57.   **58.**   **59.**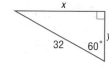

Getting Ready for the Next Lesson

PREREQUISITE SKILL Use slope to determine whether $\overline{AB}$ and $\overline{BC}$ are *perpendicular* or *not perpendicular*. *(To review slope and perpendicularity, see Lesson 3-3.)*

60. $A(2, 5)$, $B(6, 3)$, $C(8, 7)$ **61.** $A(-1, 2)$, $B(0, 7)$, $C(4, 1)$

62. $A(0, 4)$, $B(5, 7)$, $C(8, 3)$ **63.** $A(-2, -5)$, $B(1, -3)$, $C(-1, 0)$

Practice Quiz 1   *Lessons 8-1 through 8-3*

1. The measure of an interior angle of a regular polygon is $147\frac{3}{11}$. Find the number of sides in the polygon. *(Lesson 8-1)*

Use ▱$WXYZ$ to find each measure. *(Lesson 8-2)*

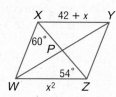

2. $WZ = \underline{\quad ? \quad}$.

3. $m\angle XYZ = \underline{\quad ? \quad}$.

ALGEBRA Find x and y so that each quadrilateral is a parallelogram. *(Lesson 8-3)*

4.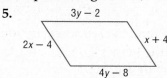

5.

8-4 Rectangles

What You'll Learn

- Recognize and apply properties of rectangles.
- Determine whether parallelograms are rectangles.

Vocabulary

- rectangle

How are rectangles used in tennis?

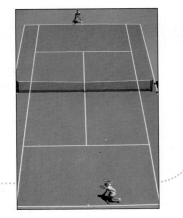

Many sports are played on fields marked by parallel lines. A tennis court has parallel lines at half-court for each player. Parallel lines divide the court for singles and doubles play. The service box is marked by perpendicular lines.

PROPERTIES OF RECTANGLES A **rectangle** is a quadrilateral with four right angles. Since both pairs of opposite angles are congruent, it follows that it is a special type of parallelogram. Thus, a rectangle has all the properties of a parallelogram. Because the right angles make a rectangle a rigid figure, the diagonals are also congruent.

Theorem 8.13

If a parallelogram is a rectangle, then the diagonals are congruent.

Abbreviation: *If ▱ is rectangle, diag. are ≅.*

$\overline{AC} \cong \overline{BD}$

You will prove Theorem 8.13 in Exercise 40.

If a quadrilateral is a rectangle, then the following properties are true.

Key Concept · Rectangle

Words A rectangle is a quadrilateral with four right angles.

Properties	Examples
1. Opposite sides are congruent and parallel.	$\overline{AB} \cong \overline{DC}$ $\overline{AB} \parallel \overline{DC}$ $\overline{BC} \cong \overline{AD}$ $\overline{BC} \parallel \overline{AD}$
2. Opposite angles are congruent.	$\angle A \cong \angle C$ $\angle B \cong \angle D$
3. Consecutive angles are supplementary.	$m\angle A + m\angle B = 180$ $m\angle B + m\angle C = 180$ $m\angle C + m\angle D = 180$ $m\angle D + m\angle A = 180$
4. Diagonals are congruent and bisect each other.	$\overline{AC}$ and $\overline{BD}$ bisect each other. $\overline{AC} \cong \overline{BD}$
5. All four angles are right angles.	$m\angle DAB = m\angle BCD = $ $m\angle ABC = m\angle ADC = 90$

Example 1 Diagonals of a Rectangle

ALGEBRA Quadrilateral *MNOP* is a rectangle. If $MO = 6x + 14$ and $PN = 9x + 5$, find x.

The diagonals of a rectangle are congruent, so $\overline{MO} \cong \overline{PN}$.

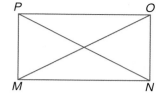

$\overline{MO} \cong \overline{PN}$	Diagonals of a rectangle are $\cong$.
$MO = PN$	Definition of congruent segments
$6x + 14 = 9x + 5$	Substitution
$14 = 3x + 5$	Subtract 6x from each side.
$9 = 3x$	Subtract 5 from each side.
$3 = x$	Divide each side by 3.

Rectangles can be constructed using perpendicular lines.

Study Tip

Look Back

To review **constructing perpendicular lines through a point**, see Lesson 3-6.

Construction

Rectangle

① Use a straightedge to draw line ℓ. Label a point P on ℓ. Place the point at P and locate point Q on ℓ. Now construct lines perpendicular to ℓ through P and through Q. Label them m and n.

② Place the compass point at P and mark off a segment on m. Using the same compass setting, place the compass at Q and mark a segment on n. Label these points R and S. Draw $\overline{RS}$.

③ Locate the compass setting that represents PR and compare to the setting for QS. The measures should be the same.

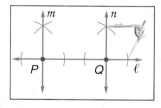
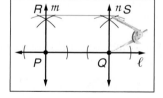
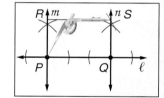

Example 2 Angles of a Rectangle

ALGEBRA Quadrilateral *ABCD* is a rectangle.

a. Find x.

$\angle DAB$ is a right angle, so $m\angle DAB = 90$.

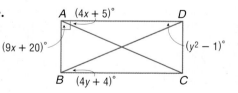

$m\angle DAC + m\angle BAC = m\angle DAB$	Angle Addition Theorem
$4x + 5 + 9x + 20 = 90$	Substitution
$13x + 25 = 90$	Simplify.
$13x = 65$	Subtract 25 from each side.
$x = 5$	Divide each side by 13.

 www.geometryonline.com/extra_examples

b. Find y.

Since a rectangle is a parallelogram, opposite sides are parallel. So, alternate interior angles are congruent.

$\angle ADB \cong \angle CBD$	Alternate Interior Angles Theorem
$m\angle ADB = m\angle CBD$	Definition of $\cong$ angles
$y^2 - 1 = 4y + 4$	Substitution
$y^2 - 4y - 5 = 0$	Subtract $4y$ and 4 from each side.
$(y - 5)(y + 1) = 0$	Factor.

$$y - 5 = 0 \qquad y + 1 = 0$$
$$y = 5 \qquad\quad y = -1 \quad \text{Disregard } y = -1 \text{ because it yields angle measures of 0.}$$

PROVE THAT PARALLELOGRAMS ARE RECTANGLES The converse of Theorem 8.13 is also true.

Theorem 8.14

If the diagonals of a parallelogram are congruent, then the parallelogram is a rectangle.

Abbreviation: *If diagonals of $\square$ are $\cong$, $\square$ is a rectangle.*

$\overline{AC} \cong \overline{BD}$

You will prove Theorem 8.14 in Exercise 41.

Example 3 *Diagonals of a Parallelogram*

WINDOWS   Trent is building a tree house for his younger brother. He has measured the window opening to be sure that the opposite sides are congruent. He measures the diagonals to make sure that they are congruent. This is called *squaring* the frame. How does he know that the corners are 90° angles?

First draw a diagram and label the vertices. We know that $\overline{WX} \cong \overline{ZY}$, $\overline{XY} \cong \overline{WZ}$, and $\overline{WY} \cong \overline{XZ}$.

Because $\overline{WX} \cong \overline{ZY}$ and $\overline{XY} \cong \overline{WZ}$, WXYZ is a parallelogram.

$\overline{XZ}$ and $\overline{WY}$ are diagonals and they are congruent. A parallelogram with congruent diagonals is a rectangle. So, the corners are 90° angles.

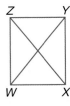

Example 4 *Rectangle on a Coordinate Plane*

COORDINATE GEOMETRY   Quadrilateral *FGHJ* has vertices $F(-4, -1)$, $G(-2, -5)$, $H(4, -2)$, and $J(2, 2)$. Determine whether *FGHJ* is a rectangle.

Method 1:   Use the Slope Formula, $m = \dfrac{y_2 - y_1}{x_2 - x_1}$, to see if consecutive sides are perpendicular.

slope of $\overline{FJ} = \dfrac{2 - (-1)}{2 - (-4)}$ or $\dfrac{1}{2}$

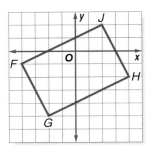

slope of $\overline{GH} = \dfrac{-2 - (-5)}{4 - (-2)}$ or $\dfrac{1}{2}$

slope of $\overline{FG} = \dfrac{-5 - (-1)}{-2 - (-4)}$ or -2

slope of $\overline{JH} = \dfrac{2 - (-2)}{2 - 4}$ or -2

Because $\overline{FJ} \parallel \overline{GH}$ and $\overline{FG} \parallel \overline{JH}$, quadrilateral $FGHJ$ is a parallelogram.

The product of the slopes of consecutive sides is -1. This means that $\overline{FJ} \perp \overline{FG}$, $\overline{FJ} \perp \overline{JH}$, $\overline{JH} \perp \overline{GH}$, and $\overline{FG} \perp \overline{GH}$. The perpendicular segments create four right angles. Therefore, by definition $FGHJ$ is a rectangle.

Method 2:   Use the Distance Formula, $d = \sqrt{(x_2 - x_1)^2 + (y_2 - y_1)^2}$, to determine whether opposite sides are congruent.

First, we must show that quadrilateral $FGHJ$ is a parallelogram.

$$FJ = \sqrt{(-4 - 2)^2 + (-1 - 2)^2}$$
$$= \sqrt{36 + 9}$$
$$= \sqrt{45}$$

$$GH = \sqrt{(-2 - 4)^2 + [-5 - (-2)]^2}$$
$$= \sqrt{36 + 9}$$
$$= \sqrt{45}$$

$$FG = \sqrt{[-4 - (-2)]^2 + [-1 - (-5)]^2}$$
$$= \sqrt{4 + 16}$$
$$= \sqrt{20}$$

$$JH = \sqrt{(2 - 4)^2 + [2 - (-2)]^2}$$
$$= \sqrt{4 + 16}$$
$$= \sqrt{20}$$

Since each pair of opposite sides of the quadrilateral have the same measure, they are congruent. Quadrilateral $FGHJ$ is a parallelogram.

$$FH = \sqrt{(-4 - 4)^2 + [-1 - (-2)]^2}$$
$$= \sqrt{64 + 1}$$
$$= \sqrt{65}$$

$$GJ = \sqrt{(-2 - 2)^2 + (-5 - 2)^2}$$
$$= \sqrt{16 + 49}$$
$$= \sqrt{65}$$

The length of each diagonal is $\sqrt{65}$. Since the diagonals are congruent, $FGHJ$ is a rectangle by Theorem 8.14.

Check for Understanding

Concept Check   1. How can you determine whether a parallelogram is a rectangle?

2. **OPEN ENDED**   Draw two congruent right triangles with a common hypotenuse. Do the legs form a rectangle?

3. **FIND THE ERROR**   McKenna and Consuelo are defining a rectangle for an assignment.

McKenna

A rectangle is a parallelogram with one right angle.

Consuelo

A rectangle has a pair of parallel opposite sides and a right angle.

Who is correct? Explain.

Guided Practice

4. ALGEBRA *ABCD* is a rectangle. If $AC = 30 - x$ and $BD = 4x - 60$, find x.

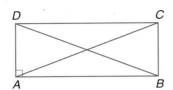

5. ALGEBRA *MNQR* is a rectangle. If $NR = 2x + 10$ and $NP = 2x - 30$, find *MP*.

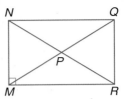

ALGEBRA Quadrilateral *QRST* is a rectangle. Find each measure or value.

6. x

7. $m\angle RPS$

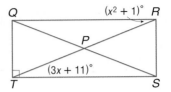

8. COORDINATE GEOMETRY Quadrilateral *EFGH* has vertices $E(-4, -3)$, $F(3, -1)$, $G(2, 3)$, and $H(-5, 1)$. Determine whether *EFGH* is a rectangle.

Application

9. FRAMING Mrs. Walker has a rectangular picture that is 12 inches by 48 inches. Because this is not a standard size, a special frame must be built. What can the framer do to guarantee that the frame is a rectangle? Justify your reasoning.

Practice and Apply

Homework Help

For Exercises	See Examples
10–15, 36–37	1
16–24	2
25–26, 35, 38–46	3
27–34	4

Extra Practice
See page 770.

ALGEBRA Quadrilateral *JKMN* is a rectangle.

10. If $NQ = 5x - 3$ and $QM = 4x + 6$, find *NK*.
11. If $NQ = 2x + 3$ and $QK = 5x - 9$, find *JQ*.
12. If $NM = 8x - 14$ and $JK = x^2 + 1$, find *JK*.
13. If $m\angle NJM = 2x - 3$ and $m\angle KJM = x + 5$, find x.
14. If $m\angle NKM = x^2 + 4$ and $m\angle KNM = x + 30$, find $m\angle JKN$.
15. If $m\angle JKN = 2x^2 + 2$ and $m\angle NKM = 14x$, find x.

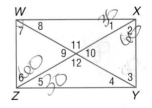

WXYZ is a rectangle. Find each measure if $m\angle 1 = 30$.

16. $m\angle 1$
17. $m\angle 2$
18. $m\angle 3$
19. $m\angle 4$
20. $m\angle 5$
21. $m\angle 6$
22. $m\angle 7$
23. $m\angle 8$
24. $m\angle 9$

25. PATIOS A contractor has been hired to pour a rectangular concrete patio. How can he be sure that the frame in which to pour the concrete is rectangular?

26. TELEVISION Television screens are measured on the diagonal. What is the measure of the diagonal of this screen?

21 in.

36 in.

COORDINATE GEOMETRY Determine whether *DFGH* is a rectangle given each set of vertices. Justify your answer.

27. *D*(9, −1), *F*(9, 5), *G*(−6, 5), *H*(−6, 1)

28. *D*(6, 2), *F*(8, −1), *G*(10, 6), *H*(12, 3)

29. *D*(−4, −3), *F*(−5, 8), *G*(6, 9), *H*(7, −2)

COORDINATE GEOMETRY The vertices of *WXYZ* are *W*(2, 4), *X*(−2, 0), *Y*(−1, −7), and *Z*(9, 3).

30. Find *WY* and *XZ*.

31. Find the coordinates of the midpoints of $\overline{WY}$ and $\overline{XZ}$.

32. Is *WXYZ* a rectangle? Explain.

COORDINATE GEOMETRY The vertices of parallelogram *ABCD* are *A*(−4, −4), *B*(2, −1), *C*(0, 3), and *D*(−6, 0).

33. Determine whether *ABCD* is a rectangle.

34. If *ABCD* is a rectangle and *E, F, G,* and *H* are midpoints of its sides, what can you conclude about *EFGH*?

35. **MINIATURE GOLF** The windmill section of a miniature golf course will be a rectangle 10 feet long and 6 feet wide. Suppose the contractor placed stakes and strings to mark the boundaries with the corners at *A, B, C* and *D*. The contractor measured $\overline{BD}$ and $\overline{AC}$ and found that $\overline{AC} > \overline{BD}$. Describe where to move the stakes *L* and *K* to make *ABCD* a rectangle. Explain.

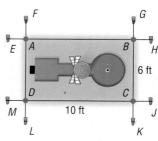

GOLDEN RECTANGLES **For Exercises 36 and 37, use the following information.**
Many artists have used *golden rectangles* in their work. In a golden rectangle, the ratio of the length to the width is about 1.618. This ratio is known as the *golden ratio*.

36. A rectangle has dimensions of 19.42 feet and 12.01 feet. Determine if the rectangle is a golden rectangle. Then find the length of the diagonal.

37. **RESEARCH** Use the Internet or other sources to find examples of golden rectangles.

38. What are the minimal requirements to justify that a parallelogram is a rectangle?

39. Draw a counterexample to the statement *If the diagonals are congruent, the quadrilateral is a rectangle.*

PROOF Write a two-column proof.

40. Theorem 8.13

41. Theorem 8.14

42. Given: *PQST* is a rectangle.
$\overline{QR} \cong \overline{VT}$

Prove: $\overline{PR} \cong \overline{VS}$

43. Given: *DEAC* and *FEAB* are rectangles.
$\angle GKH \cong \angle JHK$
$\overline{GJ}$ and $\overline{HK}$ intersect at *L*.

Prove: *GHJK* is a parallelogram.

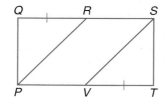

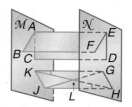

44. CRITICAL THINKING Using four of the twelve points as corners, how many rectangles can be drawn?

SPHERICAL GEOMETRY The figure shows a *Saccheri quadrilateral* on a sphere. Note that it has four sides with $\overline{CT} \perp \overline{TR}$, $\overline{AR} \perp \overline{TR}$, and $\overline{CT} \cong \overline{AR}$.

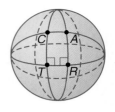

45. Is $\overline{CT}$ parallel to $\overline{AR}$? Explain.

46. How does AC compare to TR?

47. Can a rectangle exist in spherical geometry? Explain.

48. WRITING IN MATH   Answer the question that was posed at the beginning of the lesson.

How are rectangles used in tennis?

Include the following in your answer:

- the number of rectangles on one side of a tennis court, and
- a method to ensure the lines on the court are parallel

Standardized Test Practice
Ⓐ Ⓑ Ⓒ Ⓓ

49. In the figure, $\overline{AB} \parallel \overline{CE}$. If $DA = 6$, what is DB?

　Ⓐ 6　　　　　　Ⓑ 7

　Ⓒ 8　　　　　　Ⓓ 9

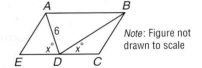

Note: Figure not drawn to scale

50. ALGEBRA   A rectangular playground is surrounded by an 80-foot long fence. One side of the playground is 10 feet longer than the other. Which of the following equations could be used to find s, the shorter side of the playground?

　Ⓐ $10s + s = 80$ 　　　　　　　Ⓑ $4s + 10 = 80$

　Ⓒ $s(s + 10) = 80$ 　　　　　　Ⓓ $2(s + 10) + 2s = 80$

Maintain Your Skills

Mixed Review　**51. TEXTILE ARTS**   The Navajo people are well known for their skill in weaving. The design at the right, known as the Eye-Dazzler, became popular with Navajo weavers in the 1880s. How many parallelograms, not including rectangles, are in the pattern?　*(Lesson 8-3)*

For Exercises 52–57, use $\square ABCD$. Find each measure or value.　*(Lesson 8-2)*

52. $m\angle AFD$ 　　　　　　**53.** $m\angle CDF$

54. $m\angle FBC$ 　　　　　　**55.** $m\angle BCF$

56. y 　　　　　　　　　　**57.** x

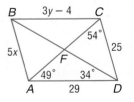

Find the measure of the altitude drawn to the hypotenuse.　*(Lesson 7-1)*

58.　　　　　　　　**59.**　　　　　　　　**60.**

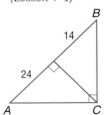

Getting Ready for the Next Lesson　**PREREQUISITE SKILL**   Find the distance between each pair of points.

*(To review the **Distance Formula**, see Lesson 1-4.)*

61. $(1, -2), (-3, 1)$ 　　　　**62.** $(-5, 9), (5, 12)$ 　　　　**63.** $(1, 4), (22, 24)$

Rhombi and Squares

What You'll Learn

- Recognize and apply the properties of rhombi.
- Recognize and apply the properties of squares.

Vocabulary
- rhombus
- square

How can you ride a bicycle with square wheels?

Professor Stan Wagon at Macalester College in St. Paul, Minnesota, developed a bicycle with square wheels. There are two front wheels so the rider can balance without turning the handlebars. Riding over a specially curved road ensures a smooth ride.

PROPERTIES OF RHOMBI A square is a special type of parallelogram called a rhombus. A **rhombus** is a quadrilateral with all four sides congruent. All of the properties of parallelograms can be applied to rhombi. There are three other characteristics of rhombi described in the following theorems.

Key Concept — Rhombus

	Theorem	Example
8.15	The diagonals of a rhombus are perpendicular.	$\overline{AC} \perp \overline{BD}$
8.16	If the diagonals of a parallelogram are perpendicular, then the parallelogram is a rhombus. (Converse of Theorem 8.15)	If $\overline{BD} \perp \overline{AC}$, then $\square ABCD$ is a rhombus.
8.17	Each diagonal of a rhombus bisects a pair of opposite angles.	$\angle DAC \cong \angle BAC \cong \angle DCA \cong \angle BCA$ $\angle ABD \cong \angle CBD \cong \angle ADB \cong \angle CDB$

You will prove Theorems 8.16 and 8.17 in Exercises 35 and 36, respectively.

Study Tip

Proof
Since a rhombus has four congruent sides, one diagonal separates the rhombus into two congruent isosceles triangles. Drawing two diagonals separates the rhombus into four congruent right triangles.

Example 1 — Proof of Theorem 8.15

Given: $PQRS$ is a rhombus.

Prove: $\overline{PR} \perp \overline{SQ}$

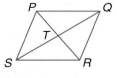

Proof:

By the definition of a rhombus, $\overline{PQ} \cong \overline{QR} \cong \overline{RS} \cong \overline{PS}$. A rhombus is a parallelogram and the diagonals of a parallelogram bisect each other, so $\overline{QS}$ bisects $\overline{PR}$ at T. Thus, $\overline{PT} \cong \overline{RT}$. $\overline{QT} \cong \overline{QT}$ because congruence of segments is reflexive. Thus, $\triangle PQT \cong \triangle RQT$ by SSS. $\angle QTP \cong \angle QTR$ by CPCTC. $\angle QTP$ and $\angle QTR$ also form a linear pair. Two congruent angles that form a linear pair are right angles. $\angle QTP$ is a right angle, so $\overline{PR} \perp \overline{SQ}$ by the definition of perpendicular lines.

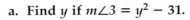

Example 2 *Measures of a Rhombus*

ALGEBRA Use rhombus *QRST* and the given
information to find the value of each variable.

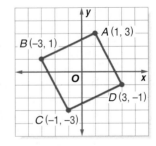

a. Find *y* if $m\angle 3 = y^2 - 31$.

$m\angle 3 = 90$ The diagonals of a rhombus are perpendicular.

$y^2 - 31 = 90$ Substitution

$y^2 = 121$ Add 31 to each side.

$y = \pm 11$ Take the square root of each side.

The value of *y* can be 11 or −11.

b. Find $m\angle TQS$ if $m\angle RST = 56$.

$m\angle TQR = m\angle RST$ Opposite angles are congruent.

$m\angle TQR = 56$ Substitution

The diagonals of a rhombus bisect the angles. So, $m\angle TQS$ is $\frac{1}{2}(56)$ or 28.

PROPERTIES OF SQUARES If a quadrilateral is both a rhombus and a
rectangle, then it is a **square**. All of the properties of parallelograms and rectangles
can be applied to squares.

Example 3 *Squares*

COORDINATE GEOMETRY Determine whether
parallelogram *ABCD* is a *rhombus*, a *rectangle*,
or a *square*. List all that apply. Explain.

Explore   Plot the vertices on a coordinate plane.

Plan   If the diagonals are perpendicular, then
            ABCD is either a rhombus or a square.
            The diagonals of a rectangle are congruent.
            If the diagonals are congruent and
            perpendicular, then *ABCD* is a square.

Solve   Use the Distance Formula to compare the lengths of the diagonals.

$DB = \sqrt{[3 - (-3)]^2 + (-1 - 1)^2}$ $AC = \sqrt{(1 + 1)^2 + (3 + 3)^2}$

$ = \sqrt{36 + 4}$ $ = \sqrt{4 + 36}$

$ = \sqrt{40}$ $ = \sqrt{40}$

Use slope to determine whether the diagonals are perpendicular.

slope of $\overline{DB} = \dfrac{1 - (-1)}{-3 - 3}$ or $-\dfrac{1}{3}$ slope of $\overline{AC} = \dfrac{-3 - 3}{-1 - 1}$ or 3

Since the slope of $\overline{AC}$ is the negative reciprocal of the slope of $\overline{DB}$, the
diagonals are perpendicular. The lengths of $\overline{DB}$ and $\overline{AC}$ are the same
so the diagonals are congruent. *ABCD* is a rhombus, a rectangle, and
a square.

Examine   You can verify that *ABCD* is a square by finding the measure and slope
            of each side. All four sides are congruent and consecutive sides are
            perpendicular.

Construction

Rhombus

1 Draw any segment $\overline{AD}$. Place the compass point at A, open to the width of AD, and draw an arc above $\overline{AD}$.

2 Label any point on the arc as B. Using the same setting, place the compass at B, and draw an arc to the right of B.

3 Place the compass at D, and draw an arc to intersect the arc drawn from B. Label the point of intersection C.

4 Use a straightedge to draw $\overline{AB}$, $\overline{BC}$, and $\overline{CD}$.

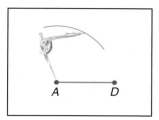
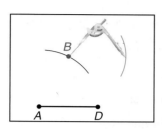
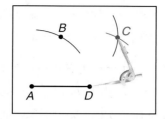
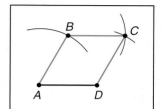

Conclusion: Since all of the sides are congruent, quadrilateral $ABCD$ is a rhombus.

Example 4 · Diagonals of a Square

BASEBALL The infield of a baseball diamond is a square, as shown at the right. Is the pitcher's mound located in the center of the infield? Explain.

Since a square is a parallelogram, the diagonals bisect each other. Since a square is a rhombus, the diagonals are congruent. Therefore, the distance from first base to third base is equal to the distance between home plate and second base.

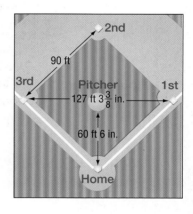

Thus, the distance from home plate to the center of the infield is 127 feet $3\frac{3}{8}$ inches divided by 2 or 63 feet $7\frac{11}{16}$ inches. This distance is longer than the distance from home plate to the pitcher's mound so the pitcher's mound is not located in the center of the field. It is about 3 feet closer to home.

If a quadrilateral is a rhombus or a square, then the following properties are true.

Study Tip

Square and Rhombus
A square is a rhombus, but a rhombus is not necessarily a square.

Concept Summary · Properties of Rhombi and Squares

Rhombi	Squares
1. A rhombus has all the properties of a parallelogram.	**1.** A square has all the properties of a parallelogram.
2. All sides are congruent.	**2.** A square has all the properties of a rectangle.
3. Diagonals are perpendicular.	**3.** A square has all the properties of a rhombus.
4. Diagonals bisect the angles of the rhombus.	

Concept Check

1. **Draw a diagram** to demonstrate the relationship among parallelograms, rectangles, rhombi, and squares.

2. **OPEN ENDED** Draw a quadrilateral that has the characteristics of a rectangle, a rhombus, and a square.

3. **Explain** the difference between a square and a rectangle.

Guided Practice

ALGEBRA In rhombus $ABCD$, $AB = 2x + 3$ and $BC = 5x$.

4. Find x.
5. Find AD.
6. Find $m\angle AEB$.
7. Find $m\angle BCD$ if $m\angle ABC = 83.2$.

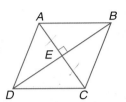

COORDINATE GEOMETRY Given each set of vertices, determine whether $\square MNPQ$ is a *rhombus*, a *rectangle*, or a *square*. List all that apply. Explain your reasoning.

8. $M(0, 3)$, $N(-3, 0)$, $P(0, -3)$, $Q(3, 0)$
9. $M(-4, 0)$, $N(-3, 3)$, $P(2, 2)$, $Q(1, -1)$

10. **PROOF** Write a two-column proof.
    Given: $\triangle KGH$, $\triangle HJK$, $\triangle GHJ$, and $\triangle JKG$ are isosceles.
    Prove: $GHJK$ is a rhombus.

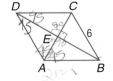

Application

11. **REMODELING** The Steiner family is remodeling their kitchen. Each side of the floor measures 10 feet. What other measurements should be made to determine whether the floor is a square?

Homework Help

For Exercises	See Examples
12–19	2
20–23	3
24–36	4
37–42	1

Extra Practice
See page 770.

In rhombus $ABCD$, $m\angle DAB = 2m\angle ADC$ and $CB = 6$.

12. Find $m\angle ACD$.
13. Find $m\angle DAB$.
14. Find DA.
15. Find $m\angle ADB$.

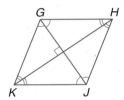

ALGEBRA Use rhombus $XYZW$ with $m\angle WYZ = 53$, $VW = 3$, $XV = 2a - 2$, and $ZV = \dfrac{5a + 1}{4}$.

16. Find $m\angle YZV$.
17. Find $m\angle XYW$.
18. Find XZ.
19. Find XW.

COORDINATE GEOMETRY Given each set of vertices, determine whether $\square EFGH$ is a *rhombus*, a *rectangle*, or a *square*. List all that apply. Explain your reasoning.

20. $E(1, 10)$, $F(-4, 0)$, $G(7, 2)$, $H(12, 12)$
21. $E(-7, 3)$, $F(-2, 3)$, $G(1, 7)$, $H(-4, 7)$
22. $E(1, 5)$, $F(6, 5)$, $G(6, 10)$, $H(1, 10)$
23. $E(-2, -1)$, $F(-4, 3)$, $G(1, 5)$, $H(3, 1)$

Construct each figure using a compass and ruler.

24. a square with one side 3 centimeters long

25. a square with a diagonal 5 centimeters long

Use the Venn diagram to determine whether each statement is *always*, *sometimes*, or *never* true.

26. A parallelogram is a square.

27. A square is a rhombus.

28. A rectangle is a parallelogram.

29. A rhombus is a rectangle.

30. A rhombus is a square.

31. A square is a rectangle.

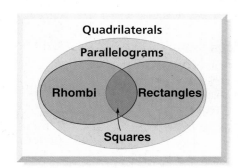

More About. . .

Design •·····

The plant stand is constructed from painted wood and metal. The overall dimensions are $36\frac{1}{2}$ inches tall by $15\frac{3}{4}$ inches wide.

Source: www.metmuseum.org

32. DESIGN Otto Prutscher designed the plant stand at the left in 1903. The base is a square, and the base of each of the five boxes is also a square. Suppose each smaller box is one half as wide as the base. Use the information at the left to find the dimensions of the base of one of the smaller boxes.

33. PERIMETER The diagonals of a rhombus are 12 centimeters and 16 centimeters long. Find the perimeter of the rhombus.

34. ART This piece of art is Dorthea Rockburne's *Egyptian Painting: Scribe.* The diagram shows three of the shapes shown in the piece. Use a ruler or a protractor to determine which type of quadrilateral is represented by each figure.

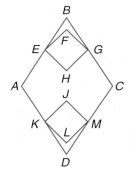

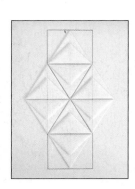

Write a paragraph proof for each theorem.

35. Theorem 8.16

36. Theorem 8.17

SQUASH For Exercises 37 and 38, use the diagram of the court for squash, a game similar to racquetball and tennis.

37. The diagram labels the diagonal as 11,665 millimeters. Is this correct? Explain.

38. The service boxes are squares. Find the length of the diagonal.

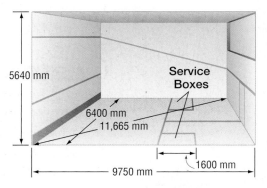

• **39. FLAGS** Study the flags shown below. Use a ruler and protractor to determine if any of the flags contain parallelograms, rectangles, rhombi, or squares.

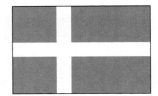

Denmark

St. Vincent and The Grenadines

Trinidad and Tobago

PROOF Write a two-column proof.

40. Given: $\triangle WZY \cong \triangle WXY$, $\triangle WZY$ and $\triangle XYZ$ are isosceles.

Prove: WXYZ is a rhombus.

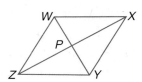

41. Given: $\triangle TPX \cong \triangle QPX \cong \triangle QRX \cong \triangle TRX$

Prove: TPQR is a rhombus.

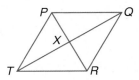

42. Given: $\triangle LGK \cong \triangle MJK$
GHJK is a parallelogram.

Prove: GHJK is a rhombus.

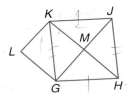

43. Given: QRST and QRTV are rhombi.

Prove: $\triangle QRT$ is equilateral.

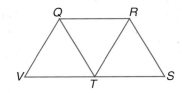

44. CRITICAL THINKING
The pattern at the right is a series of rhombi that continue to form a hexagon that increases in size. Copy and complete the table.

Hexagon	Number of rhombi
1	3
2	12
3	27
4	48
5	
6	
x	

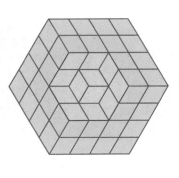

45. WRITING IN MATH Answer the question that was posed at the beginning of the lesson.

How can you ride a bicycle with square wheels?

Include the following in your answer:
• difference between squares and rhombi, and
• how nonsquare rhombus-shaped wheels would work with the curved road.

46. Points *A*, *B*, *C*, and *D* are on a square. The area of the square is 36 square units. Which of the following statements is true?

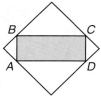

Ⓐ The perimeter of rectangle *ABCD* is greater than 24 units.

Ⓑ The perimeter of rectangle *ABCD* is less than 24 units.

Ⓒ The perimeter of rectangle *ABCD* is equal to 24 units.

Ⓓ The perimeter of rectangle *ABCD* cannot be determined from the information given.

47. ALGEBRA For all integers $x \neq 2$, let $<x> = \dfrac{1+x}{x-2}$. Which of the following has the greatest value?

Ⓐ $<0>$ Ⓑ $<1>$ Ⓒ $<3>$ Ⓓ $<4>$

Maintain Your Skills

Mixed Review

ALGEBRA Use rectangle *LMNP*, parallelogram *LKMJ*, and the given information to solve each problem. *(Lesson 8-4)*

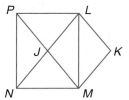

48. If $LN = 10$, $LJ = 2x + 1$, and $PJ = 3x - 1$, find x.

49. If $m\angle PLK = 110$, find $m\angle LKM$.

50. If $m\angle MJN = 35$, find $m\angle MPN$.

51. If $MK = 6x$, $KL = 3x + 2y$, and $JN = 14 - x$, find x and y.

52. If $m\angle LMP = m\angle PMN$, find $m\angle PJL$.

COORDINATE GEOMETRY Determine whether the points are the vertices of a parallelogram. Use the method indicated. *(Lesson 8-3)*

53. $P(0, 2)$, $Q(6, 4)$, $R(4, 0)$, $S(-2, -2)$; Distance Formula

54. $F(1, -1)$, $G(-4, 1)$, $H(-3, 4)$, $J(2, 1)$; Distance Formula

55. $K(-3, -7)$, $L(3, 2)$, $M(1, 7)$, $N(-3, 1)$; Slope Formula

56. $A(-4, -1)$, $B(-2, -5)$, $C(1, 7)$, $D(3, 3)$; Slope Formula

Refer to △*PQS*. *(Lesson 6-4)*

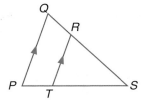

57. If $RT = 16$, $QP = 24$, and $ST = 9$, find PS.

58. If $PT = y - 3$, $PS = y + 2$, $RS = 12$, and $QS = 16$, solve for y.

59. If $RT = 15$, $QP = 21$, and $PT = 8$, find TS.

Refer to the figure. *(Lesson 4-6)*

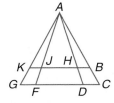

60. If $\overline{AG} \cong \overline{AC}$, name two congruent angles.

61. If $\overline{AJ} \cong \overline{AH}$, name two congruent angles.

62. If $\angle AFD \cong \angle ADF$, name two congruent segments.

63. If $\angle AKB \cong \angle ABK$, name two congruent segments.

Getting Ready for the Next Lesson

PREREQUISITE SKILL Solve each equation.
*(To review **solving equations**, see pages 737 and 738.)*

64. $\frac{1}{2}(8x - 6x - 7) = 5$

65. $\frac{1}{2}(7x + 3x + 1) = 12.5$

66. $\frac{1}{2}(4x + 6 + 2x + 13) = 15.5$

67. $\frac{1}{2}(7x - 2 + 3x + 3) = 25.5$

Kites

A **kite** is a quadrilateral with exactly two distinct pairs of adjacent congruent sides. In kite $ABCD$, diagonal $\overline{BD}$ separates the kite into two congruent triangles. Diagonal $\overline{AC}$ separates the kite into two noncongruent isosceles triangles.

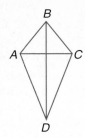

Activity
Construct a kite $QRST$.

① Draw $\overline{RT}$.

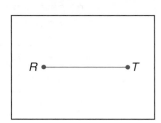

② Choose a compass setting greater than $\frac{1}{2}\overline{RT}$. Place the compass at point R and draw an arc above $\overline{RT}$. Then without changing the compass setting, move the compass to point T and draw an arc that intersects the first one. Label the intersection point Q. Increase the compass setting. Place the compass at R and draw an arc below $\overline{RT}$. Then, without changing the compass setting, draw an arc from point T to intersect the other arc. Label the intersection point S.

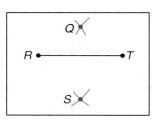

③ Draw $QRST$.

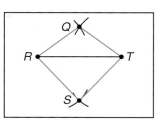

Model
1. Draw $\overline{QS}$ in kite $QRST$. Use a protractor to measure the angles formed by the intersection of $\overline{QS}$ and $\overline{RT}$.
2. Measure the interior angles of kite $QRST$. Are any congruent?
3. Label the intersection of $\overline{QS}$ and $\overline{RT}$ as point N. Find the lengths of $\overline{QN}$, $\overline{NS}$, $\overline{TN}$, and $\overline{NR}$. How are they related?
4. How many pairs of congruent triangles can be found in kite $QRST$?
5. Construct another kite $JKLM$. Repeat Exercises 1–4.

Analyze
6. Use your observations and measurements of kites $QRST$ and $JKLM$ to make conjectures about the angles, sides, and diagonals of kites.

8-6 Trapezoids

What You'll Learn

- Recognize and apply the properties of trapezoids.
- Solve problems involving the medians of trapezoids.

Vocabulary

- trapezoid
- isosceles trapezoid
- median

How are trapezoids used in architecture?

The Washington Monument in Washington, D.C., is an obelisk made of white marble. The width of the base is longer than the width at the top. Each face of the monument is an example of a trapezoid.

PROPERTIES OF TRAPEZOIDS A **trapezoid** is a quadrilateral with exactly one pair of parallel sides. The parallel sides are called *bases*. The *base angles* are formed by a base and one of the legs. The nonparallel sides are called *legs*.

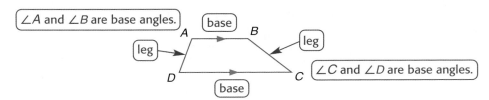

$\angle A$ and $\angle B$ are base angles.

$\angle C$ and $\angle D$ are base angles.

If the legs are congruent, then the trapezoid is an **isosceles trapezoid**. Theorems 8.18 and 8.19 describe two characteristics of isosceles trapezoids.

Study Tip

Isosceles Trapezoid
If you extend the legs of an isosceles trapezoid until they meet, you will have an isosceles triangle. Recall that the base angles of an isosceles triangle are congruent.

Theorems

8.18 Both pairs of base angles of an isosceles trapezoid are congruent.

8.19 The diagonals of an isosceles trapezoid are congruent.

Example:
$\angle DAB \cong \angle CBA$
$\angle ADC \cong \angle BCD$

$\overline{AC} \cong \overline{BD}$

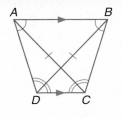

You will prove Theorem 8.18 in Exercise 36.

Example 1 Proof of Theorem 8.19

Write a flow proof of Theorem 8.19.

Given: $MNOP$ is an isosceles trapezoid.

Prove: $\overline{MO} \cong \overline{NP}$

Proof:

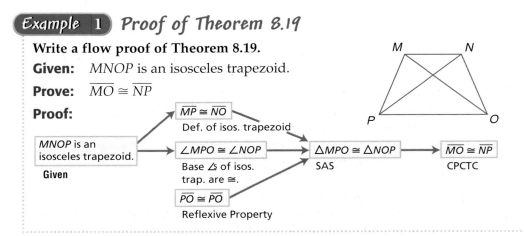

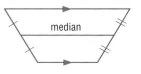

Example 2 *Identify Isoceles Trapezoids*

ART The sculpture pictured is *Zim Zum I* by Barnett Newman. The walls are connected at right angles. In perspective, the rectangular panels appear to be trapezoids. Use a ruler and protractor to determine if the images of the front panels are isosceles trapezoids. Explain.

The panel on the left is an isosceles trapezoid. The bases are parallel and are different lengths. The legs are not parallel and they are the same length.

The panel on the right is not an isosceles trapezoid. Each side is a different length.

More About . . .

Art •·········

Barnett Newman designed this piece to be 50% larger. This piece was built for an exhibition in Japan but it could not be built as large as the artist wanted because of size limitations on cargo from New York to Japan.

Source: www.sfmoma.org

Example 3 *Identify Trapezoids*

COORDINATE GEOMETRY *JKLM* is a quadrilateral with vertices *J*(−18, −1), *K*(−6, 8), *L*(18, 1), and *M*(−18, −26).

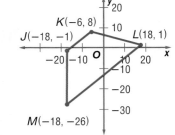

a. Verify that *JKLM* is a trapezoid.

A quadrilateral is a trapezoid if exactly one pair of opposite sides are parallel. Use the Slope Formula.

$$\text{slope of } \overline{JK} = \frac{-1 - 8}{-18 - (-6)} \qquad \text{slope of } \overline{ML} = \frac{1 - (-26)}{18 - (-18)}$$

$$= \frac{-9}{-12} \text{ or } \frac{3}{4} \qquad\qquad = \frac{27}{36} \text{ or } \frac{3}{4}$$

$$\text{slope of } \overline{JM} = \frac{-1 - (-26)}{-18 - (-18)} \qquad \text{slope of } \overline{KL} = \frac{1 - 8}{18 - (-6)}$$

$$= \frac{25}{0} \text{ or undefined} \qquad\qquad = \frac{-7}{24}$$

Exactly one pair of opposite sides are parallel, $\overline{JK}$ and $\overline{ML}$. So, *JKLM* is a trapezoid.

b. Determine whether *JKLM* is an isosceles trapezoid. Explain.

First use the Distance Formula to show that the legs are congruent.

$$JM = \sqrt{[-18 - (-18)]^2 + [-1 - (-26)]^2} \qquad KL = \sqrt{(-6 - 18)^2 + (8 - 1)^2}$$

$$= \sqrt{0 + 625} \qquad\qquad\qquad = \sqrt{576 + 49}$$

$$= \sqrt{625} \text{ or } 25 \qquad\qquad\qquad = \sqrt{625} \text{ or } 25$$

Since the legs are congruent, *JKLM* is an isosceles trapezoid.

MEDIANS OF TRAPEZOIDS The segment that joins midpoints of the legs of a trapezoid is the **median**. The median of a trapezoid can also be called a *midsegment*. Recall from Lesson 6-4 that the midsegment of a triangle is the segment joining the midpoints of two sides. The median of a trapezoid has the same properties as the midsegment of a triangle. You can construct the median of a trapezoid using a compass and a straightedge.

median

Geometry Activity

Median of a Trapezoid

Model

1. Draw trapezoid $\overline{WXYZ}$ with legs $\overline{XY}$ and $\overline{WZ}$.

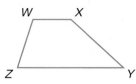

2. Construct the perpendicular bisectors of $\overline{XY}$ and $\overline{WZ}$. Label the midpoints M and N.

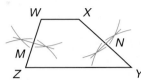

3. Draw $\overline{MN}$.

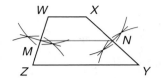

Analyze

1. Measure $\overline{WX}$, $\overline{ZY}$, and $\overline{MN}$ to the nearest millimeter.
2. Make a conjecture based on your observations.

The results of the Geometry Activity suggest Theorem 8.20.

Theorem 8.20

The median of a trapezoid is parallel to the bases, and its measure is one-half the sum of the measures of the bases.

Example: $EF = \frac{1}{2}(AB + DC)$

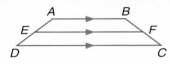

Example 4 Median of a Trapezoid

ALGEBRA $QRST$ is an isosceles trapezoid with median $\overline{XY}$.

a. Find TS if $QR = 22$ and $XY = 15$.

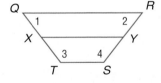

$XY = \frac{1}{2}(QR + TS)$ Theorem 8.20

$15 = \frac{1}{2}(22 + TS)$ Substitution

$30 = 22 + TS$ Multiply each side by 2.

$8 = TS$ Subtract 22 from each side.

b. Find $m\angle 1$, $m\angle 2$, $m\angle 3$, and $m\angle 4$ if $m\angle 1 = 4a - 10$ and $m\angle 3 = 3a + 32.5$.

Since $\overline{QR} \parallel \overline{TS}$, $\angle 1$ and $\angle 3$ are supplementary. Because this is an isosceles trapezoid, $\angle 1 \cong \angle 2$ and $\angle 3 \cong \angle 4$.

$m\angle 1 + m\angle 3 = 180$ Consecutive Interior Angles Theorem

$4a - 10 + 3a + 32.5 = 180$ Substitution

$7a + 22.5 = 180$ Combine like terms.

$7a = 157.5$ Subtract 22.5 from each side.

$a = 22.5$ Divide each side by 7.

If $a = 22.5$, then $m\angle 1 = 80$ and $m\angle 3 = 100$.

Because $\angle 1 \cong \angle 2$ and $\angle 3 \cong \angle 4$, $m\angle 2 = 80$ and $m\angle 4 = 100$.

Concept Check   **1.** **List** the minimum requirements to show that a quadrilateral is a trapezoid.

2. **Make a chart** comparing the characteristics of the diagonals of a trapezoid, a rectangle, a square, and a rhombus. (*Hint:* Use the types of quadrilaterals as column headings and the properties of diagonals as row headings.)

3. **OPEN ENDED**   Draw an isosceles trapezoid and a trapezoid that is not isosceles. Draw the median for each. Is the median parallel to the bases in both trapezoids?

Guided Practice   **COORDINATE GEOMETRY**   *QRST* is a quadrilateral with vertices *Q*(−3, 2), *R*(−1, 6), *S*(4, 6), *T*(6, 2).

4. Verify that *QRST* is a trapezoid.

5. Determine whether *QRST* is an isosceles trapezoid. Explain.

6. **PROOF**   *CDFG* is an isosceles trapezoid with bases $\overline{CD}$ and $\overline{FG}$. Write a flow proof to prove ∠*DGF* ≅ ∠*CFG*.

7. **ALGEBRA**   *EFGH* is an isosceles trapezoid with bases $\overline{EF}$ and $\overline{GH}$ and median $\overline{YZ}$. If *EF* = 3*x* + 8, *HG* = 4*x* − 10, and *YZ* = 13, find *x*.

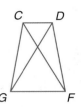

Application   **8.** **PHOTOGRAPHY**   Photographs can show a building in a perspective that makes it appear to be a different shape. Identify the types of quadrilaterals in the photograph.

Homework Help	
For Exercises	**See Examples**
9–12, 23–32	3
13–19, 39	4
20–22, 38	2
33–37	1

Extra Practice
See page 770.

COORDINATE GEOMETRY   For each quadrilateral whose vertices are given,
a. verify that the quadrilateral is a trapezoid, and
b. determine whether the figure is an isosceles trapezoid.

9. *A*(−3, 3), *B*(−4, −1), *C*(5, −1), *D*(2, 3)

10. *G*(−5, −4), *H*(5, 4), *J*(0, 5), *K*(−5, 1)

11. *C*(−1, 1), *D*(−5, −3), *E*(−4, −10), *F*(6, 0)

12. *Q*(−12, 1), *R*(−9, 4), *S*(−4, 3), *T*(−11, −4)

ALGEBRA   Find the missing measure(s) for the given trapezoid.

13. For trapezoid *DEGH*, *X* and *Y* are midpoints of the legs. Find *DE*.

14. For trapezoid *RSTV*, *A* and *B* are midpoints of the legs. Find *VT*.

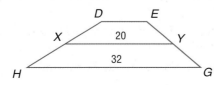

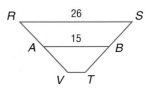

15. For isosceles trapezoid $XYZW$, find the length of the median, $m\angle W$, and $m\angle Z$.

W 8 Z

70°

X 20 Y

16. For trapezoid $QRST$, A and B are midpoints of the legs. Find AB, $m\angle Q$, and $m\angle S$.

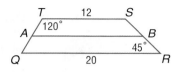

T 12 S

120°

A

B

45°

Q 20 R

For Exercises 17 and 18, use trapezoid $QRST$.

17. Let $\overline{GH}$ be the median of $RSBA$. Find GH.

18. Let $\overline{JK}$ be the median of $ABTQ$. Find JK.

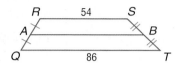

R 54 S

A B

Q 86 T

19. ALGEBRA $JKLM$ is a trapezoid with $\overline{JK} \parallel \overline{LM}$ and median $\overline{RP}$. Find RP if $JK = 2(x + 3)$, $RP = 5 + x$, and $ML = \frac{1}{2}x - 1$.

20. DESIGN The bagua is a tool used in Feng Shui design. This bagua consists of two regular octagons centered around a yin-yang symbol. How can you determine the type of quadrilaterals in the bagua?

21. SEWING Madison is making a valance for a window treatment. She is using striped fabric cut on the bias, or diagonal, to create a chevron pattern. Identify the polygons formed in the fabric.

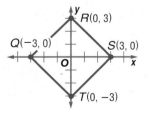

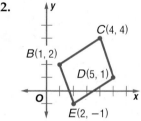
COORDINATE GEOMETRY Determine whether each figure is a *trapezoid*, a *parallelogram*, a *square*, a *rhombus*, or a *quadrilateral*. Choose the most specific term. Explain.

22.

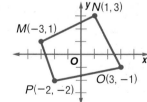

C(4, 4)

B(1, 2)

D(5, 1)

O x

E(2, −1)

23.

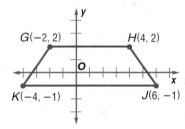

G(−2, 2) H(4, 2)

O x

K(−4, −1) J(6, −1)

24.

N(1, 3)

M(−3, 1)

O x

O(3, −1)

P(−2, −2)

25.

R(0, 3)

Q(−3, 0) S(3, 0)

O x

T(0, −3)

COORDINATE GEOMETRY For Exercises 26–28, refer to quadrilateral $PQRS$.

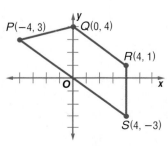

P(−4, 3) Q(0, 4)

R(4, 1)

O x

S(4, −3)

26. Determine whether the figure is a trapezoid. If so, is it isosceles? Explain.

27. Find the coordinates of the midpoints of $\overline{PQ}$ and $\overline{RS}$, and label them A and B.

28. Find AB without using the Distance Formula.

COORDINATE GEOMETRY   For Exercises 29–31, refer to quadrilateral *DEFG*.

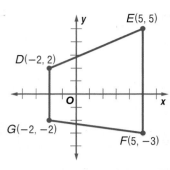

29. Determine whether the figure is a trapezoid. If so, is it isosceles? Explain.

30. Find the coordinates of the midpoints of $\overline{DE}$ and $\overline{GF}$, and label them *W* and *V*.

31. Find *WV* without using the Distance Formula.

WebQuest

You can use a map of Seattle to locate and draw a quadrilateral that will help you begin to find the hidden treasure. Visit www.geometry online.com/webquest to continue work on your WebQuest project.

PROOF   Write a flow proof.

32. **Given:** $\overline{HJ} \parallel \overline{GK}$, $\triangle HGK \cong \triangle JKG$, $\overline{HG} \not\parallel \overline{JK}$
    Prove: *GHJK* is an isosceles trapezoid.

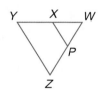

33. **Given:** $\triangle TZX \cong \triangle YXZ$, $\overline{WX} \not\parallel \overline{ZY}$
    Prove: *XYZW* is a trapezoid.

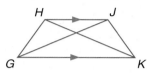

34. **Given:** *ZYXP* is an isosceles trapezoid.
    Prove: $\triangle PWX$ is isosceles.

35. **Given:** *E* and *C* are midpoints of $\overline{AD}$ and $\overline{DB}$. $\overline{AD} \cong \overline{DB}$
    Prove: *ABCE* is an isosceles trapezoid.

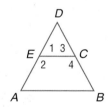

36. Write a paragraph proof of Theorem 8.18.

CONSTRUCTION   Use a compass and ruler to construct each figure.

37. an isosceles trapezoid

38. trapezoid with a median 2 centimeters long

39. **CRITICAL THINKING**   In *RSTV*, *RS* = 6, *VT* = 3, and *RX* is twice the length of *XV*. Find *XY*.

40. **CRITICAL THINKING**   Is it possible for an isosceles trapezoid to have two right base angles? Explain.

41. **WRITING IN MATH**   Answer the question that was posed at the beginning of the lesson.

    How are trapezoids used in architecture?

    Include the following in your answer:
    • the characteristics of a trapezoid, and
    • other examples of trapezoids in architecture.

Standardized Test Practice
(A) (B) (C) (D)

42. **SHORT RESPONSE** What type of quadrilateral is *WXYZ*? Justify your answer.

43. ALGEBRA In the figure, which point lies within the shaded region?

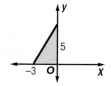

- Ⓐ $(-2, 4)$
- Ⓑ $(-1, 3)$
- Ⓒ $(1, -3)$
- Ⓓ $(2, -4)$

Maintain Your Skills

Mixed Review **ALGEBRA** In rhombus $LMPQ$, $m\angle QLM = 2x^2 - 10$, $m\angle QPM = 8x$, and $MP = 10$. *(Lesson 8-5)*

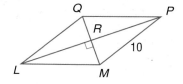

44. Find $m\angle LPQ$.   **45.** Find QL.
46. Find $m\angle LQP$.   **47.** Find $m\angle LQM$.
48. Find the perimeter of $LMPQ$.

COORDINATE GEOMETRY For Exercises 49–51, refer to quadrilateral $RSTV$. *(Lesson 8-4)*

49. Find RS and TV.

50. Find the coordinates of the midpoints of $\overline{RT}$ and $\overline{SV}$.

51. Is $RSTV$ a rectangle? Explain.

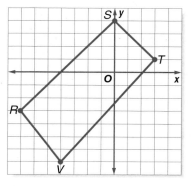

Solve each proportion. *(Lesson 6-1)*

52. $\dfrac{16}{38} = \dfrac{24}{y}$ **53.** $\dfrac{y}{6} = \dfrac{17}{30}$ **54.** $\dfrac{5}{y + 4} = \dfrac{20}{28}$ **55.** $\dfrac{2y}{9} = \dfrac{52}{36}$

Getting Ready for the Next Lesson **PREREQUISITE SKILL** Write an expression for the slope of the segment given the coordinates of the endpoints. *(To review **slope**, see Lesson 3-3.)*

56. $(0, a)$, $(-a, 2a)$ **57.** $(-a, b)$, (a, b) **58.** (c, c), (c, d)
59. $(a, -b)$, $(2a, b)$ **60.** $(3a, 2b)$, $(b, -a)$ **61.** (b, c), $(-b, -c)$

Practice Quiz 2   *Lessons 8-4 through 8-6*

Quadrilateral $ABCD$ is a rectangle. *(Lesson 8-4)*
1. Find x.
2. Find y.

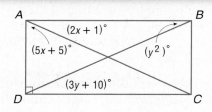

3. COORDINATE GEOMETRY Determine whether $MNPQ$ is a *rhombus*, a *rectangle*, or a *square* for $M(-5, -3)$, $N(-2, 3)$, $P(1, -3)$, and $Q(-2, -9)$. List all that apply. Explain. *(Lesson 8-5)*

For trapezoid $TRSV$, M and N are midpoints of the legs. *(Lesson 8-6)*
4. If $VS = 21$ and $TR = 44$, find MN.
5. If $TR = 32$ and $MN = 25$, find VS.

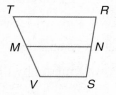

Reading Mathematics

Hierarchy of Polygons

A *hierarchy* is a ranking of classes or sets of things. Examples of some classes of polygons are rectangles, rhombi, trapezoids, parallelograms, squares, and quadrilaterals. These classes are arranged in the hierarchy below.

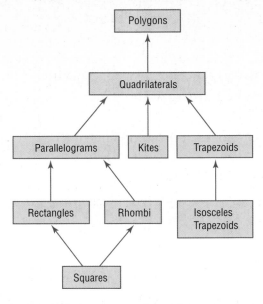

Use the following information to help read the hierarchy diagram.

• The class that is the broadest is listed first, followed by the other classes in order. For example, *polygons* is the broadest class in the hierarchy diagram above, and *squares* is a very specific class.

• Each class is contained within any class linked above it in the hierarchy. For example, *all* squares are also rhombi, rectangles, parallelograms, quadrilaterals, and polygons. However, an isosceles trapezoid is not a square or a kite.

• Some, but not all, elements of each class are contained within lower classes in the hierarchy. For example, some trapezoids are isosceles trapezoids, and some rectangles are squares.

Reading to Learn

Refer to the hierarchy diagram at the right. Write *true*, *false*, or *not enough information* for each statement.

1. All mogs are jums.

2. Some jebs are jums.

3. All lems are jums.

4. Some wibs are jums.

5. All mogs are bips.

6. Draw a hierarchy diagram to show these classes: equilateral triangles, polygons, isosceles triangles, triangles, and scalene triangles.

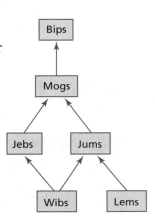

8-7 Coordinate Proof With Quadrilaterals

What You'll Learn

- Position and label quadrilaterals for use in coordinate proofs.
- Prove theorems using coordinate proofs.

How can you use a coordinate plane to prove theorems about quadrilaterals?

In Chapter 4, you learned that variable coordinates can be assigned to the vertices of triangles. Then the Distance and Midpoint Formulas and coordinate proofs were used to prove theorems. The same is true for quadrilaterals.

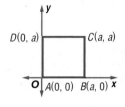

POSITION FIGURES   The first step to using a coordinate proof is to place the figure on the coordinate plane. The placement of the figure can simplify the steps of the proof.

Study Tip

Look Back

To review **placing a figure on a coordinate plane**, see Lesson 4-7.

Example 1   Positioning a Square

Position and label a square with sides *a* units long on the coordinate plane.

- Let *A*, *B*, *C*, and *D* be vertices of a square with sides *a* units long.

- Place the square with vertex *A* at the origin, $\overline{AB}$ along the positive *x*-axis, and $\overline{AD}$ along the *y*-axis. Label the vertices *A*, *B*, *C*, and *D*.

- The *y*-coordinate of *B* is 0 because the vertex is on the *x*-axis. Since the side length is *a*, the *x*-coordinate is *a*.

- *D* is on the *y*-axis so the *x*-coordinate is 0. The *y*-coordinate is $0 + a$ or *a*.

- The *x*-coordinate of *C* is also *a*. The *y*-coordinate is $0 + a$ or *a* because the side $\overline{BC}$ is *a* units long.

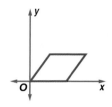

Some examples of quadrilaterals placed on the coordinate plane are given below. Notice how the figures have been placed so the coordinates of the vertices are as simple as possible.

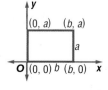

rectangle

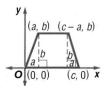

parallelogram

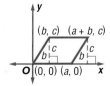

isosceles trapezoid

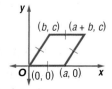

rhombus

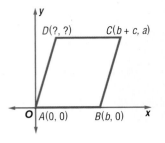

Example 2 Find Missing Coordinates

Name the missing coordinates for the parallelogram.

Opposite sides of a parallelogram are congruent and parallel. So, the y-coordinate of D is a.

The length of $\overline{AB}$ is b, and the length of $\overline{DC}$ is b. So, the x-coordinate of D is $(b + c) - b$ or c.

The coordinates of D are (c, a).

PROVE THEOREMS Once a figure has been placed on the coordinate plane, we can prove theorems using the Slope, Midpoint, and Distance Formulas.

Geometry Software Investigation

Quadrilaterals

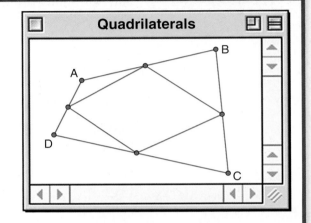

Model

- Use The Geometer's Sketchpad to draw a quadrilateral *ABCD* with no two sides parallel or congruent.
- Construct the midpoints of each side.
- Draw the quadrilateral formed by the midpoints of the segments.

Analyze

1. Measure each side of the quadrilateral determined by the midpoints of *ABCD*.

2. What type of quadrilateral is formed by the midpoints? Justify your answer.

In this activity, you discover that the quadrilateral formed from the midpoints of any quadrilateral is a parallelogram.   You will prove this in Exercise 22.

Study Tip

Problem Solving
To prove that a quadrilateral is a square, you can also show that all sides are congruent and that the diagonals bisect each other.

Example 3 Coordinate Proof

Place a square on a coordinate plane. Label the midpoints of the sides, *M*, *N*, *P*, and *Q*. Write a coordinate proof to prove that *MNPQ* is a square.

The first step is to position a square on the coordinate plane. Label the vertices to make computations as simple as possible.

Given: *ABCD* is a square.
  M, *N*, *P*, and *Q* are midpoints.

Prove: *MNPQ* is a square.

Proof:

By the Midpoint Formula, the coordinates of *M*, *N*, *P*, and *Q* are as follows.

$$M\left(\frac{2a + 0}{2}, \frac{0 + 0}{2}\right) = (a, 0)$$

$$N\left(\frac{2a + 2a}{2}, \frac{2a + 0}{2}\right) = (2a, a)$$

$$P\left(\frac{0 + 2a}{2}, \frac{2a + 2a}{2}\right) = (a, 2a)$$

$$Q\left(\frac{0 + 0}{2}, \frac{0 + 2a}{2}\right) = (0, a)$$

Find the slopes of $\overline{QP}$, $\overline{MN}$, $\overline{QM}$, and $\overline{PN}$.

slope of $\overline{QP} = \dfrac{2a - a}{a - 0}$ or 1 slope of $\overline{MN} = \dfrac{a - 0}{2a - a}$ or 1

slope of $\overline{QM} = \dfrac{a - 0}{0 - a}$ or -1 slope of $\overline{PN} = \dfrac{2a - a}{a - 2a}$ or -1

Each pair of opposite sides is parallel, so they have the same slope. Consecutive sides form right angles because their slopes are negative reciprocals.

Use the Distance Formula to find the length of $\overline{QP}$ and $\overline{QM}$.

$$QP = \sqrt{(a - 0)^2 + (2a - a)^2}$$
$$= \sqrt{a^2 + a^2}$$
$$= \sqrt{2a^2} \text{ or } a\sqrt{2}$$

$$QM = \sqrt{(a - 0)^2 + (0 - a)^2}$$
$$= \sqrt{a^2 + a^2}$$
$$= \sqrt{2a^2} \text{ or } a\sqrt{2}$$

$MNPQ$ is a square because each pair of opposite sides is parallel, and consecutive sides form right angles and are congruent.

Example 4 *Properties of Quadrilaterals*

PARKING   Write a coordinate proof to prove that the sides of the parking space are parallel.

Given: $14x - 6y = 0$; $7x - 3y = 56$

Prove: $\overline{AD} \parallel \overline{BC}$

Proof: Rewrite both equations in slope-intercept form.

$14x - 6y = 0$ $7x - 3y = 56$

$\dfrac{-6y}{-6} = \dfrac{-14x}{-6}$ $\dfrac{-3y}{-3} = \dfrac{-7x + 56}{-3}$

$y = \dfrac{7}{3}x$ $y = \dfrac{7}{3}x - \dfrac{56}{3}$

Since $\overline{AD}$ and $\overline{BC}$ have the same slope, they are parallel.

Check for Understanding

Concept Check   1. **Explain** how to position a quadrilateral to simplify the steps of the proof.

2. **OPEN ENDED**   Position and label a trapezoid with two vertices on the y-axis.

Guided Practice   **Position and label the quadrilateral on the coordinate plane.**

3. rectangle with length a units and height $a + b$ units

Name the missing coordinates for each quadrilateral.

4.
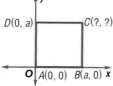

5.
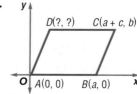

Write a coordinate proof for each statement.

6. The diagonals of a parallelogram bisect each other.

7. The diagonals of a square are perpendicular.

www.geometryonline.com/extra_examples

Application

8. **STATES** The state of Tennessee can be separated into two shapes that resemble quadrilaterals. Write a coordinate proof to prove that *DEFG* is a trapezoid. All measures are approximate and given in kilometers.

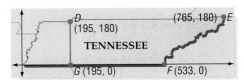

Practice and Apply

Homework Help

For Exercises	See Examples
9–10	1
11–16, 23	2
17–22	3
24–26	4

Extra Practice
See page 771.

Position and label each quadrilateral on the coordinate plane.

9. isosceles trapezoid with height c units, bases a units and $a + 2b$ units

10. parallelogram with side length c units and height b units

Name the missing coordinates for each parallelogram or trapezoid.

11.

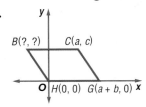

12.

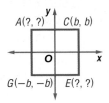

13.

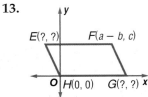

14.

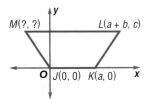

15.

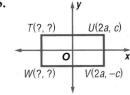

16.

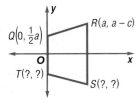

Position and label each figure on the coordinate plane. Then write a coordinate proof for each of the following.

17. The diagonals of a rectangle are congruent.

18. If the diagonals of a parallelogram are congruent, then it is a rectangle.

19. The diagonals of an isosceles trapezoid are congruent.

20. The median of an isosceles trapezoid is parallel to the bases.

21. The segments joining the midpoints of the sides of a rectangle form a rhombus.

22. The segments joining the midpoints of the sides of a quadrilateral form a parallelogram.

23. **CRITICAL THINKING** *A* has coordinates $(0, 0)$, and *B* has coordinates (a, b). Find the coordinates of *C* and *D* so *ABCD* is an isosceles trapezoid.

ARCHITECTURE For Exercises 24–26, use the following information.

The Leaning Tower of Pisa is approximately 60 meters tall, from base to belfry. The tower leans about 5.5° so the top level is 4.5 meters over the first level.

24. Position and label the tower on a coordinate plane.

25. Is it possible to write a coordinate proof to prove that the sides of the tower are parallel? Explain.

26. From the given information, what conclusion can be drawn?

27. **WRITING IN MATH** Answer the question that was posed at the beginning of the lesson.

How is the coordinate plane used in proofs?

Include the following in your answer:
- guidelines for placing a figure on a coordinate grid, and
- an example of a theorem from this chapter that could be proved using the coordinate plane.

Standardized Test Practice
Ⓐ Ⓑ Ⓒ Ⓓ

28. In the figure, *ABCD* is a parallelogram. What are the coordinates of point *D*?

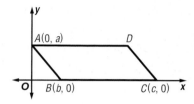

Ⓐ $(a, c + b)$ Ⓑ $(c + b, a)$

Ⓒ $(b - c, a)$ Ⓓ $(c - b, a)$

29. **ALGEBRA** If $p = -5$, then $5 - p^2 - p = $ ___?___ .

Ⓐ -15 Ⓑ -5

Ⓒ 10 Ⓓ 30

Maintain Your Skills

Mixed Review

30. **PROOF** Write a two-column proof. *(Lesson 8-6)*

Given: *MNOP* is a trapezoid with bases $\overline{MN}$ and $\overline{OP}$. $\overline{MN} \cong \overline{QO}$

Prove: *MNOQ* is a parallelogram.

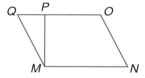

JKLM is a rectangle. *MLPR* is a rhombus. $\angle JMK \cong \angle RMP$, $m\angle JMK = 55$, and $m\angle MRP = 70$. *(Lesson 8-5)*

31. Find $m\angle MPR$.

32. Find $m\angle KML$.

33. Find $m\angle KLP$.

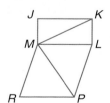

Find the geometric mean between each pair of numbers. *(Lesson 7-1)*

34. 7 and 14

35. $2\sqrt{5}$ and $6\sqrt{5}$

Write an expression relating the given pair of angle measures. *(Lesson 5-5)*

36. $m\angle WVX, m\angle VXY$

37. $m\angle XVZ, m\angle VXZ$

38. $m\angle XYV, m\angle VXY$

39. $m\angle XZY, m\angle ZXY$

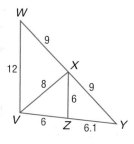

Vocabulary and Concept Check

diagonal (p. 404)
isosceles trapezoid (p. 439)
kite (p. 438)

median (p. 440)
parallelogram (p. 411)
rectangle (p. 424)

rhombus (p. 431)
square (p. 432)
trapezoid (p. 439)

A complete list of postulates and theorems can be found on pages R1–R8.

Exercises State whether each sentence is *true* or *false*. If false, replace the underlined term to make a true sentence.

1. The diagonals of a <u>rhombus</u> are perpendicular.
2. All <u>squares</u> are rectangles.
3. If a parallelogram is a <u>rhombus</u>, then the diagonals are congruent.
4. Every <u>parallelogram</u> is a quadrilateral.
5. A(n) <u>rhombus</u> is a quadrilateral with exactly one pair of parallel sides.
6. Each diagonal of a <u>rectangle</u> bisects a pair of opposite angles.
7. If a quadrilateral is both a rhombus and a rectangle, then it is a <u>square</u>.
8. Both pairs of base angles in a(n) <u>isosceles trapezoid</u> are congruent.

Lesson-by-Lesson Review

8-1 Angles of Polygons

See pages 404–409.

Concept Summary

- If a convex polygon has n sides and the sum of the measures of its interior angles is S, then $S = 180(n - 2)$.
- The sum of the measures of the exterior angles of a convex polygon is 360.

Example Find the measure of an interior angle of a regular decagon.

$S = 180(n - 2)$ Interior Angle Sum Theorem

$= 180(10 - 2)$ $n = 10$

$= 180(8)$ or 1440 Simplify.

The measure of each interior angle is $1440 \div 10$, or 144.

Exercises Find the measure of each interior angle of a regular polygon given the number of sides. *See Example 1 on page 405.*

9. 6   10. 15   11. 4   12. 20

ALGEBRA Find the measure of each interior angle. *See Example 3 on page 405.*

13.

14.

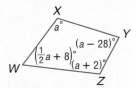

www.geometryonline.com/vocabulary_review

8-2 Parallelograms

See pages 411–416.

Concept Summary

- In a parallelogram, opposite sides are parallel and congruent, opposite angles are congruent, and consecutive angles are supplementary.
- The diagonals of a parallelogram bisect each other.

Example

WXYZ is a parallelogram.
Find $m\angle YZW$ and $m\angle XWZ$.

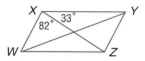

$m\angle YZW = m\angle WXY$ Opp. $\angle$s of $\square$ are $\cong$.

$m\angle YZW = 82 + 33$ or 115 $m\angle WXY = m\angle WXZ + m\angle YXZ$

$m\angle XWZ + m\angle WXY = 180$ Cons. $\angle$s in $\square$ are suppl.

$m\angle XWZ + (82 + 33) = 180$ $m\angle WXY = m\angle WXZ + m\angle YXZ$

$m\angle XWZ + 115 = 180$ Simplify.

$m\angle XWZ = 65$ Subtract 115 from each side.

Exercises   Use $\square ABCD$ to find each measure.
See Example 2 on page 413.

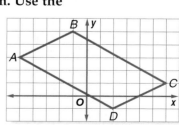

15. $m\angle BCD$
16. AF
17. $m\angle BDC$
18. BC
19. CD
20. $m\angle ADC$

8-3 Tests for Parallelograms

See pages 417–423.

Concept Summary

A quadrilateral is a parallelogram if any one of the following is true.

- Both pairs of opposite sides are parallel and congruent.
- Both pairs of opposite angles are congruent.
- Diagonals bisect each other.
- A pair of opposite sides is both parallel and congruent.

Example

COORDINATE GEOMETRY   Determine whether the figure with vertices $A(-5, 3)$, $B(-1, 5)$, $C(6, 1)$, and $D(2, -1)$ is a parallelogram. Use the Distance and Slope Formulas.

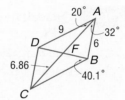

$AB = \sqrt{[-5 - (-1)]^2 + (3 - 5)^2}$

$\quad = \sqrt{(-4)^2 + (-2)^2}$ or $\sqrt{20}$

$CD = \sqrt{(6 - 2)^2 + [1 - (-1)]^2}$

$\quad = \sqrt{4^2 + 2^2}$ or $\sqrt{20}$

Since $AB = CD$, $\overline{AB} \cong \overline{CD}$.

slope of $\overline{AB} = \dfrac{5 - 3}{-1 - (-5)}$ or $\dfrac{1}{2}$ slope of $\overline{CD} = \dfrac{-1 - 1}{2 - 6}$ or $\dfrac{1}{2}$

$\overline{AB}$ and $\overline{CD}$ have the same slope, so they are parallel. Since one pair of opposite sides is congruent and parallel, $ABCD$ is a parallelogram.

Exercises   Determine whether the figure with the given vertices is a parallelogram. Use the method indicated.   *See Example 5 on page 420.*

21. $A(-2, 5)$, $B(4, 4)$, $C(6, -3)$, $D(-1, -2)$; Distance Formula

22. $H(0, 4)$, $J(-4, 6)$, $K(5, 6)$, $L(9, 4)$; Midpoint Formula

23. $S(-2, -1)$, $T(2, 5)$, $V(-10, 13)$, $W(-14, 7)$; Slope Formula

---

8-4 Rectangles

See pages 424–430.

Concept Summary

- A rectangle is a quadrilateral with four right angles and congruent diagonals.
- If the diagonals of a parallelogram are congruent, then the parallelogram is a rectangle.

Example   **Quadrilateral $KLMN$ is a rectangle. If $PL = x^2 - 1$ and $PM = 4x + 11$, find x.**

The diagonals of a rectangle are congruent and bisect each other, so $\overline{PL} \cong \overline{PM}$.

$\overline{PL} \cong \overline{PM}$	Diag. are $\cong$ and bisect each other.
$PL = PM$	Def. of $\cong$ angles
$x^2 - 1 = 4x + 11$	Substitution
$x^2 - 1 - 4x = 11$	Subtract $4x$ from each side.
$x^2 - 4x - 12 = 0$	Subtract 11 from each side.
$(x + 2)(x - 6) = 0$	Factor.

$$x + 2 = 0 \qquad x - 6 = 0$$
$$x = -2 \qquad x = 6$$

The value of x is -2 or 6.

Exercises   *ABCD* is a rectangle.
See Examples 1 and 2 on pages 425 and 426.

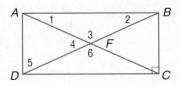

24. If $AC = 9x - 1$ and $AF = 2x + 7$, find AF.

25. If $m\angle 1 = 12x + 4$ and $m\angle 2 = 16x - 12$, find $m\angle 2$.

26. If $CF = 4x + 1$ and $DF = x + 13$, find x.

27. If $m\angle 2 = 70 - 4x$ and $m\angle 5 = 18x - 8$, find $m\angle 5$.

COORDINATE GEOMETRY   Determine whether *RSTV* is a rectangle given each set of vertices. Justify your answer.   *See Example 4 on pages 426 and 427.*

28. $R(-3, -5)$, $S(0, -5)$, $T(3, 4)$, $V(0, 4)$

29. $R(0, 0)$, $S(6, 3)$, $T(4, 7)$, $V(-2, 4)$

8-5 Rhombi and Squares

See pages 431–437.

Concept Summary

- A rhombus is a quadrilateral with each side congruent, diagonals that are perpendicular, and each diagonal bisecting a pair of opposite angles.
- A quadrilateral that is both a rhombus and a rectangle is a square.

Example Use rhombus $JKLM$ to find $m\angle JMK$ and $m\angle KJM$.

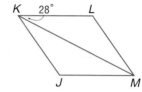

The opposite sides of a rhombus are parallel, so $\overline{KL} \parallel \overline{JM}$. $\angle JMK \cong \angle LKM$ because alternate interior angles are congruent.

$$m\angle JMK = m\angle LKM \quad \text{Definition of congruence}$$
$$= 28 \quad \text{Substitution}$$

The diagonals of a rhombus bisect the angles, so $\angle JKM \cong \angle LKM$.

$$m\angle KJM + m\angle JKL = 180 \quad \text{Cons. } \angle \text{s in } \square \text{ are suppl.}$$
$$m\angle KJM + (m\angle JKM + m\angle LKM) = 180 \quad m\angle JKL = m\angle JKM + m\angle LKM$$
$$m\angle KJM + (28 + 28) = 180 \quad \text{Substitution}$$
$$m\angle KJM + 56 = 180 \quad \text{Add.}$$
$$m\angle KJM = 124 \quad \text{Subtract 56 from each side.}$$

Exercises Use rhombus $ABCD$ with $m\angle 1 = 2x + 20$, $m\angle 2 = 5x - 4$, $AC = 15$, and $m\angle 3 = y^2 + 26$. *See Example 2 on page 432.*

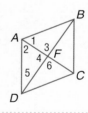

30. Find x.
31. Find AF.
32. Find y.

8-6 Trapezoids

See pages 439–445.

Concept Summary

- In an isosceles trapezoid, both pairs of base angles are congruent and the diagonals are congruent.
- The median of a trapezoid is parallel to the bases, and its measure is one-half the sum of the measures of the bases.

Example $RSTV$ is a trapezoid with bases $\overline{RV}$ and $\overline{ST}$ and median $\overline{MN}$. Find x if $MN = 60$, $ST = 4x - 1$, and $RV = 6x + 11$.

$$MN = \tfrac{1}{2}(ST + RV)$$
$$60 = \tfrac{1}{2}[(4x - 1) + (6x + 11)] \quad \text{Substitution}$$
$$120 = 4x - 1 + 6x + 11 \quad \text{Multiply each side by 2.}$$
$$120 = 10x + 10 \quad \text{Simplify.}$$
$$110 = 10x \quad \text{Subtract 10 from each side.}$$
$$11 = x \quad \text{Divide each side by 10.}$$

Exercises Find the missing value for the given trapezoid.

See Example 4 on page 441.

33. For isosceles trapezoid *ABCD*, *X* and *Y* are midpoints of the legs. Find $m\angle XBC$ if $m\angle ADY = 78$.

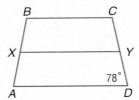

34. For trapezoid *JKLM*, *A* and *B* are midpoints of the legs. If $AB = 57$ and $KL = 21$, find *JM*.

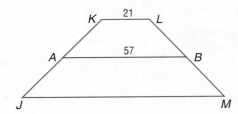

8-7 *Coordinate Proof with Quadrilaterals*

See pages 447–451.

Concept Summary

- Position a quadrilateral so that a vertex is at the origin and at least one side lies along an axis.

Example Position and label rhombus *RSTV* on the coordinate plane. Then write a coordinate proof to prove that each pair of opposite sides is parallel.

First, draw rhombus *RSTV* on the coordinate plane. Label the coordinates of the vertices.

Given: *RSTV* is a rhombus.

Prove: $\overline{RV} \parallel \overline{ST}, \overline{RS} \parallel \overline{VT}$

Proof:

slope of $\overline{RV} = \dfrac{c - 0}{b - 0}$ or $\dfrac{c}{b}$ slope of $\overline{ST} = \dfrac{c - 0}{(a + b) - a}$ or $\dfrac{c}{b}$

slope of $\overline{RS} = \dfrac{0 - 0}{a - 0}$ or 0 slope of $\overline{VT} = \dfrac{c - c}{(a + b) - b}$ or 0

$\overline{RV}$ and $\overline{ST}$ have the same slope. So $\overline{RV} \parallel \overline{ST}$. $\overline{RS}$ and $\overline{VT}$ have the same slope, and $\overline{RS} \parallel \overline{VT}$.

Exercises Position and label each figure on the coordinate plane. Then write a coordinate proof for each of the following. *See Example 3 on pages 448 and 449.*

35. The diagonals of a square are perpendicular.

36. A diagonal separates a parallelogram into two congruent triangles.

Name the missing coordinates for each quadrilateral. *See Example 2 on page 448.*

37.
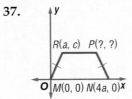

38.
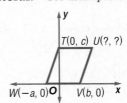

Vocabulary and Concepts

Determine whether each conditional is *true* or *false*. If false, draw a counterexample.

1. If a quadrilateral has four right angles, then it is a rectangle.
2. If a quadrilateral has all four sides congruent, then it is a square.
3. If the diagonals of a quadrilateral are perpendicular, then it is a rhombus.

Skills and Applications

Complete each statement about $\square FGHK$. Justify your answer.

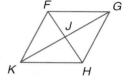

4. $\overline{HK} \cong$ __?__.

5. $\angle FKJ \cong$ __?__.

6. $\angle FKH \cong$ __?__.

7. $\overline{GH} \parallel$ __?__.

Determine whether the figure with the given vertices is a parallelogram. Justify your answer.

8. $A(4, 3), B(6, 0), C(4, -8), D(2, -5)$

9. $S(-2, 6), T(2, 11), V(3, 8), W(-1, 3)$

10. $F(7, -3), G(4, -2), H(6, 4), J(12, 2)$

11. $W(-4, 2), X(-3, 6), Y(2, 7), Z(1, 3)$

ALGEBRA $QRST$ is a rectangle.

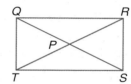

12. If $QP = 3x + 11$ and $PS = 4x + 8$, find QS.

13. If $m\angle QTR = 2x^2 - 7$ and $m\angle SRT = x^2 + 18$, find $m\angle QTR$.

COORDINATE GEOMETRY Determine whether $\square ABCD$ is a *rhombus*, a *rectangle*, or a *square*. List all that apply. Explain your reasoning.

14. $A(12, 0), B(6, -6), C(0, 0), D(6, 6)$

15. $A(-2, 4), B(5, 6), C(12, 4), D(5, 2)$

Name the missing coordinates for each quadrilateral.

16.

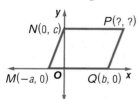

17.

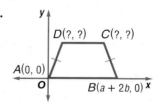

18. Position and label an isosceles trapezoid on the coordinate plane. Write a coordinate proof to prove that the median is parallel to each base.

19. **SAILING** Many large sailboats have a *keel* to keep the boat stable in high winds. A keel is shaped like a trapezoid with its top and bottom parallel. If the root chord is 9.8 feet and the tip chord is 7.4 feet, find the length of the mid-chord.

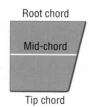

Root chord

Mid-chord

Tip chord

20. **STANDARDIZED TEST PRACTICE** The measure of an interior angle of a regular polygon is 108. Find the number of sides.

   (A) 8   (B) 6   (C) 5   (D) 3

Part 1 Multiple Choice

Record your answers on the answer sheet provided by your teacher or on a sheet of paper.

1. A trucking company wants to purchase a ramp to use when loading heavy objects onto a truck. The closest that the truck can get to the loading area is 5 meters. The height from the ground to the bed of the truck is 3 meters. To the nearest meter, what should the length of the ramp be? **(Lesson 1-3)**

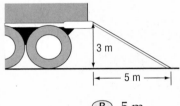

   (A) 4 m   (B) 5 m

   (C) 6 m   (D) 7 m

2. Which of the following is the contrapositive of the statement below? **(Lesson 2-3)**

   If an astronaut is in orbit, then he or she is weightless.

   (A) If an astronaut is weightless, then he or she is in orbit.

   (B) If an astronaut is not in orbit, then he or she is not weightless.

   (C) If an astronaut is on Earth, then he or she is weightless.

   (D) If an astronaut is not weightless, then he or she is not in orbit.

3. Rectangle *QRST* measures 7 centimeters long and 4 centimeters wide. Which of the following could be the dimensions of a rectangle similar to rectangle *QRST*?
   (Lesson 6-2)

   (A) 28 cm by 14 cm

   (B) 21 cm by 12 cm

   (C) 14 cm by 4 cm

   (D) 7 cm by 8 cm

4. A 24 foot ladder, leaning against a house, forms a 60° angle with the ground. How far up the side of the house does the ladder reach? **(Lesson 7-3)**

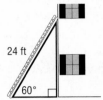

   (A) 12 ft

   (B) $12\sqrt{2}$ ft

   (C) $12\sqrt{3}$ ft

   (D) 20 ft

5. In rectangle *JKLM* shown below, $\overline{JL}$ and $\overline{MK}$ are diagonals. If $JL = 2x + 5$ and $KM = 4x - 11$, what is x? **(Lesson 8-4)**

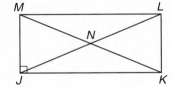

   (A) 10

   (B) 8

   (C) 6

   (D) 5

6. Joaquin bought a set of stencils for his younger sister. One of the stencils is a quadrilateral with perpendicular diagonals that bisect each other, but are **not** congruent. What kind of quadrilateral is this piece? **(Lesson 8-5)**

   (A) square   (B) rectangle

   (C) rhombus   (D) trapezoid

7. In the diagram below, *ABCD* is a trapezoid with diagonals $\overline{AC}$ and $\overline{BD}$ intersecting at point *E*.

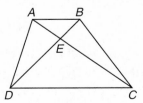

   Which statement is true? **(Lesson 8-6)**

   (A) $\overline{AB}$ is parallel to $\overline{CD}$.

   (B) $\angle ADC$ is congruent to $\angle BCD$.

   (C) $\overline{CE}$ is congruent to $\overline{DE}$.

   (D) $\overline{AC}$ and $\overline{BD}$ bisect each other.

Part 2 Short Response/Grid In

Record your answers on the answer sheet provided by your teacher or on a sheet of paper.

8. At what point does the graph of $y = -4x + 5$ cross the x-axis on a coordinate plane?   (Prerequisite Skill)

9. Candace and Julio are planning to see a movie together. They decide to meet at the house that is closer to the theater. From the locations shown on the diagram, whose house is closer to the theater?   (Lesson 5-3)

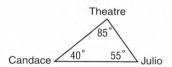

10. In the diagram, $\overline{CE}$ is the mast of a sailboat with sail $\triangle ABC$.

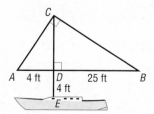

   Marcia wants to calculate the length, in feet, of the mast. Write an equation in which the geometric mean is represented by x. (Lesson 7-1)

11. $\overline{AC}$ is a diagonal of rhombus $ABCD$. If $m\angle CDE$ is 116, what is $m\angle ACD$? (Lesson 8-4)

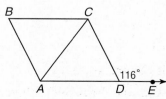

Test-Taking Tip

Question 10
Read the question carefully to check that you answered the question that was asked. In question 10, you are asked to write an equation, not to find the length of the mast.

Part 3 Extended Response

Record your answers on a sheet of paper. Show your work.

12. On the tenth hole of a golf course, a sand trap is located right before the green at point M. Matt is standing 126 yards away from the green at point N. Quintashia is standing 120 yards away from the beginning of the sand trap at point Q.

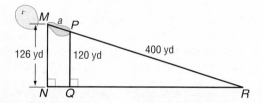

   a. Explain why $\triangle MNR$ is similar to $\triangle PQR$. (Lesson 6-3)

   b. Write and solve a proportion to find the distance across the sand trap, a. (Lesson 6-3)

13. Quadrilateral $ABCD$ has vertices with coordinates: $A(0, 0)$, $B(a, 0)$, $C(a + b, c)$, and $D(b, c)$.

   a. Position and label $ABCD$ on the coordinate plane. Prove that $ABCD$ is a parallelogram.   (Lesson 8-2 and 8-7)

   b. If $a^2 = b^2 + c^2$, what can you determine about the slopes of the diagonals $\overline{AC}$ and $\overline{BD}$?   (Lesson 8-7)

   c. What kind of parallelogram is $ABCD$? (Lesson 8-7)

9 Transformations

What You'll Learn

- **Lesson 9-1, 9-2, 9-3, and 9-5** Name, draw, and recognize figures that have been reflected, translated, rotated, or dilated.
- **Lesson 9-4** Identify and create different types of tessellations.
- **Lesson 9-6** Find the magnitude and direction of vectors and perform operations on vectors.
- **Lesson 9-7** Use matrices to perform transformations on the coordinate plane.

Key Vocabulary

- **reflection** (p. 463)
- **translation** (p. 470)
- **rotation** (p. 476)
- **tessellation** (p. 483)
- **dilation** (p. 490)
- **vector** (p. 498)

Why It's Important

Transformations, lines of symmetry, and tessellations can be seen in artwork, nature, interior design, quilts, amusement parks, and marching band performances. These geometric procedures and characteristics make objects more visually pleasing.

You will learn how quilts are created by using transformations in Lesson 9-3.

Getting Started

Prerequisite Skills To be successful in this chapter, you'll need to master these skills and be able to apply them in problem-solving situations. Review these skills before beginning Chapter 9.

For Lessons 9-1 through 9-5   **Graph Points**

Graph each pair of points. *(For review, see pages 728 and 729.)*

1. $A(1, 3)$, $B(-1, 3)$ **2.** $C(-3, 2)$, $D(-3, -2)$ **3.** $E(-2, 1)$, $F(-1, -2)$

4. $G(2, 5)$, $H(5, -2)$ **5.** $J(-7, 10)$, $K(-6, 7)$ **6.** $L(3, -2)$, $M(6, -4)$

For Lesson 9-6   **Distance and Slope**

Find $m\angle A$. Round to the nearest tenth. *(For review, see Lesson 7-4.)*

7. $\tan A = \dfrac{3}{4}$ **8.** $\tan A = \dfrac{5}{8}$ **9.** $\sin A = \dfrac{2}{3}$

10. $\sin A = \dfrac{4}{5}$ **11.** $\cos A = \dfrac{9}{12}$ **12.** $\cos A = \dfrac{15}{17}$

For Lesson 9-7   **Multiply Matrices**

Find each product. *(For review, see pages 752 and 753.)*

13. $\begin{bmatrix} 0 & 1 \\ 1 & -1 \end{bmatrix} \cdot \begin{bmatrix} 5 & 4 \\ -5 & -1 \end{bmatrix}$ **14.** $\begin{bmatrix} -1 & 0 \\ 1 & 1 \end{bmatrix} \cdot \begin{bmatrix} 0 & -2 \\ -2 & 3 \end{bmatrix}$

15. $\begin{bmatrix} 0 & 1 \\ -1 & 0 \end{bmatrix} \cdot \begin{bmatrix} -3 & 4 & 5 \\ -2 & -5 & 1 \end{bmatrix}$ **16.** $\begin{bmatrix} -1 & 0 \\ 0 & -1 \end{bmatrix} \cdot \begin{bmatrix} -1 & -3 & -3 & 2 \\ 3 & -1 & -2 & 1 \end{bmatrix}$

Transformations Make this Foldable to help you organize your notes. Begin with one sheet of notebook paper.

Step 1 Fold

Fold a sheet of notebook paper in half lengthwise.

Step 2 Cut

Cut on every third line to create 8 tabs.

Step 3 Label

Label each tab with a vocabulary word from this chapter.

Reading and Writing As you read and study the chapter, use each tab to write notes and examples of transformations, tessellations, and vectors on the coordinate plane.

Geometry Activity

Transformations

In a plane, you can slide, flip, turn, enlarge, or reduce figures to create new figures. These corresponding figures are frequently designed into wallpaper borders, mosaics, and artwork. Each figure that you see will correspond to another figure. These corresponding figures are formed using transformations.

A **transformation** maps an initial figure, called a preimage, onto a final figure, called an image. Below are some of the types of transformations. The red lines show some corresponding points.

translation
A figure can be slid in any direction.

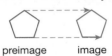

preimage   image

reflection
A figure can be flipped over a line.

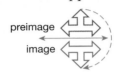

preimage
image

rotation
A figure can be turned around a point.

preimage
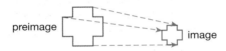
image

dilation
A figure can be enlarged or reduced.

preimage

image

Exercises **Identify the following transformations. The blue figure is the preimage.**

1.

2.

3.

4.

5.

6.

7.

8.

9.

10.

Make a Conjecture

11. An *isometry* is a transformation in which the resulting image is congruent to the preimage. Which transformations are isometries?

9-1 Reflections

What You'll Learn

- Draw reflected images.
- Recognize and draw lines of symmetry and points of symmetry.

Vocabulary

- reflection
- line of reflection
- isometry
- line of symmetry
- point of symmetry

Where are reflections found in nature?

On a clear, bright day glacial-fed lakes can provide vivid reflections of the surrounding vistas. Note that each point above the water line has a corresponding point in the image in the lake. The distance that a point lies above the water line appears the same as the distance its image lies below the water.

DRAW REFLECTIONS A **reflection** is a transformation representing a flip of a figure. Figures may be reflected in a point, a line, or a plane.

The figure shows a reflection of $ABCDE$ in line m. Note that the segment connecting a point and its image is perpendicular to line m and is bisected by line m. Line m is called the **line of reflection** for $ABCDE$ and its image $A'B'C'D'E'$. Because E lies on the line of reflection, its preimage and image are the same point.

A', A'', A''', and so on, name corresponding points for one or more transformations.

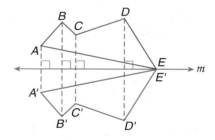

It is possible to reflect a preimage in a point. In the figure below, polygon $UVWXYZ$ is reflected in point P.

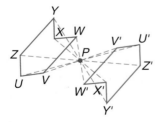

Note that P is the midpoint of each segment connecting a point with its image.

$\overline{UP} \cong \overline{PU'}, \overline{VP} \cong \overline{PV'},$

$\overline{WP} \cong \overline{PW'}, \overline{XP} \cong \overline{PX'},$

$\overline{YP} \cong \overline{PY'}, \overline{ZP} \cong \overline{PZ'}$

Study Tip

Look Back
To review **congruence transformations**, see Lesson 4-3.

When reflecting a figure in a line or in a point, the image is congruent to the preimage. Thus, a reflection is a *congruence transformation*, or an **isometry**. That is, reflections preserve distance, angle measure, betweenness of points, and collinearity. In the figure above, polygon $UVWXYZ \cong$ polygon $U'V'W'X'Y'Z'$.

Corresponding Sides	Corresponding Angles
$\overline{UV} \cong \overline{U'V'}$	$\angle UVW \cong \angle U'V'W'$
$\overline{VW} \cong \overline{V'W'}$	$\angle VWX \cong \angle V'W'X'$
$\overline{WX} \cong \overline{W'X'}$	$\angle WXY \cong \angle W'X'Y'$
$\overline{XY} \cong \overline{X'Y'}$	$\angle XYZ \cong \angle X'Y'Z'$
$\overline{YZ} \cong \overline{Y'Z'}$	$\angle YZU \cong \angle Y'Z'U'$
$\overline{UZ} \cong \overline{U'Z'}$	$\angle ZUV \cong \angle Z'U'V'$

Example **1** *Reflecting a Figure in a Line*

Draw the reflected image of quadrilateral *DEFG* in line *m*.

Step 1 Since *D* is on line *m*, *D* is its own reflection. Draw segments perpendicular to line *m* from *E*, *F*, and *G*.

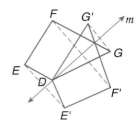

Step 2 Locate *E′*, *F′*, and *G′* so that line *m* is the perpendicular bisector of $\overline{EE'}$, $\overline{FF'}$, and $\overline{GG'}$. Points *E′*, *F′*, and *G′* are the respective images of *E*, *F*, and *G*.

Step 3 Connect vertices *D*, *E′*, *F′*, and *G′*.

Since points *D*, *E′*, *F′*, and *G′* are the images of points *D*, *E*, *F*, and *G* under reflection in line *m*, then quadrilateral *DE′F′G′* is the reflection of quadrilateral *DEFG* in line *m*.

Reflections can also occur in the coordinate plane.

Example **2** *Reflection in the x-axis*

COORDINATE GEOMETRY Quadrilateral *KLMN* has vertices *K*(2, −4), *L*(−1, 3), *M*(−4, 2), and *N*(−3, −4). Graph *KLMN* and its image under reflection in the *x*-axis. Compare the coordinates of each vertex with the coordinates of its image.

Use the vertical grid lines to find a corresponding point for each vertex so that the *x*-axis is equidistant from each vertex and its image.

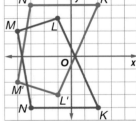

$K(2, -4) \rightarrow K'(2, 4)$ $L(-1, 3) \rightarrow L'(-1, -3)$

$M(-4, 2) \rightarrow M'(-4, -2)$ $N(-3, -4) \rightarrow N'(-3, 4)$

Plot the reflected vertices and connect to form the image *K′L′M′N′*. The *x*-coordinates stay the same, but the *y*-coordinates are opposites. That is, $(a, b) \rightarrow (a, -b)$.

Example **3** *Reflection in the y-axis*

COORDINATE GEOMETRY Suppose quadrilateral *KLMN* from Example 2 is reflected in the *y*-axis. Graph *KLMN* and its image under reflection in the *y*-axis. Compare the coordinates of each vertex with the coordinates of its image.

Use the horizontal grid lines to find a corresponding point for each vertex so that the *y*-axis is equidistant from each vertex and its image.

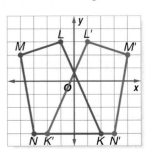

$K(2, -4) \rightarrow K'(-2, -4)$ $L(-1, 3) \rightarrow L'(1, 3)$

$M(-4, 2) \rightarrow M'(4, 2)$ $N(-3, -4) \rightarrow N'(3, -4)$

Plot the reflected vertices and connect to form the image *K′L′M′N′*. The *x*-coordinates are opposites and the *y*-coordinates are the same. That is, $(a, b) \rightarrow (-a, b)$.

Example 4 **Reflection in the Origin**

COORDINATE GEOMETRY Suppose quadrilateral *KLMN* from Example 2 is reflected in the origin. Graph *KLMN* and its image under reflection in the origin. Compare the coordinates of each vertex with the coordinates of its image.

Since $\overline{KK'}$ passes through the origin, use the horizontal and vertical distances from *K* to the origin to find the coordinates of *K'*. From *K* to the origin is 4 units up and 2 units left. *K'* is located by repeating that pattern from the origin. Four units up and 2 units left yields *K'*(−2, 4).

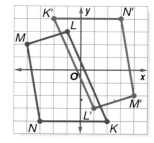

$K(2, -4) \rightarrow K'(-2, 4)$　　　$L(-1, 3) \rightarrow L'(1, -3)$

$M(-4, 2) \rightarrow M'(4, -2)$　　　$N(-3, -4) \rightarrow N'(3, 4)$

Plot the reflected vertices and connect to form the image *K'L'M'N'*. Comparing coordinates shows that $(a, b) \rightarrow (-a, -b)$.

Example 5 **Reflection in the Line** $y = x$

COORDINATE GEOMETRY Suppose quadrilateral *KLMN* from Example 2 is reflected in the line $y = x$. Graph *KLMN* and its image under reflection in the line $y = x$. Compare the coordinates of each vertex with the coordinates of its image.

The slope of $y = x$ is 1. $\overline{KK'}$ is perpendicular to $y = x$, so its slope is −1. From *K* to the line $y = x$, move up three units and left three units. From the line $y = x$ move up three units and left three units to *K'*(−4, 2).

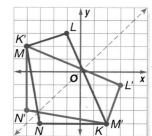

$K(2, -4) \rightarrow K'(-4, 2)$　　　$L(-1, 3) \rightarrow L'(3, -1)$

$M(-4, 2) \rightarrow M'(2, -4)$　　　$N(-3, -4) \rightarrow N'(-4, -3)$

Plot the reflected vertices and connect to form the image *K'L'M'N'*. Comparing coordinates shows that $(a, b) \rightarrow (b, a)$.

Concept Summary　　　Reflections in the Coordinate Plane

Reflection	*x*-axis	*y*-axis	origin	$y = x$
Preimage to Image	$(a, b) \rightarrow (a, -b)$	$(a, b) \rightarrow (-a, b)$	$(a, b) \rightarrow (-a, -b)$	$(a, b) \rightarrow (b, a)$
How to find coordinates	Multiply the *y*-coordinate by −1.	Multiply the *x*-coordinate by −1.	Multiply both coordinates by −1.	Interchange the *x*- and *y*-coordinates.
Example	$B(-3, 1)$ $A(2, 3)$ $B'(-3, -1)$ $A'(2, -3)$	$A'(-3, 2)$ $A(3, 2)$ $B'(-1, -2)$ $B(1, -2)$	$A(3, 2)$ $B'(-3, 1)$ $B(3, -1)$ $A'(-3, -2)$	$B(-3, 2)$ $A(1, 3)$ $A'(3, 1)$ $B'(2, -3)$

Example 6 *Use Reflections*

GOLF Adeel and Natalie are playing miniature golf. Adeel says that he read how to use reflections to help make a hole-in-one on most miniature golf holes. Describe how he should putt the ball to make a hole-in-one.

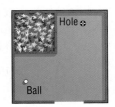

If Adeel tries to putt the ball directly to the hole, he will strike the border as indicated by the blue line. So, he can mentally reflect the hole in the line that contains the right border. If he putts the ball at the reflected image of the hole, the ball will strike the border, and it will rebound on a path toward the hole.

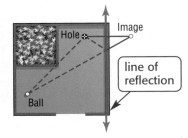

LINES AND POINTS OF SYMMETRY   Some figures can be folded so that the two halves match exactly. The fold is a line of reflection called a **line of symmetry**. For some figures, a point can be found that is a common point of reflection for all points on a figure. This common point of reflection is called a **point of symmetry**.

Lines of Symmetry

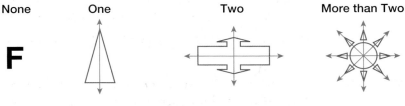

Points of Symmetry

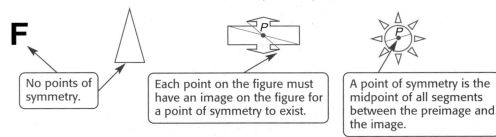

Example 7 *Draw Lines of Symmetry*

Determine how many lines of symmetry a square has. Then determine whether a square has point symmetry.

A square has four lines of symmetry.

A square has point symmetry. P is the point of symmetry such that $AP = PA'$, $BP = PB'$, $CP = PC'$, and so on.

Check for Understanding

Concept Check

1. Find a counterexample to disprove the statement *Two or more lines of symmetry for a plane figure intersect in a point of symmetry.*

2. OPEN ENDED Draw a figure on the coordinate plane and then reflect it in the line $y = x$. Label the coordinates of the preimage and the image.

3. Identify four properties that are preserved in reflections.

Guided Practice

4. Copy the figure at the right. Draw its reflected image in line m.

Determine how many lines of symmetry each figure has. Then determine whether the figure has point symmetry.

5.   **6.**   **7.**

COORDINATE GEOMETRY Graph each figure and its image under the given reflection.

8. $\overline{AB}$ with endpoints $A(2, 4)$ and $B(-3, -3)$ reflected in the x-axis

9. $\triangle ABC$ with vertices $A(-1, 4)$, $B(4, -2)$, and $C(0, -3)$ reflected in the y-axis

10. $\triangle DEF$ with vertices $D(-1, -3)$, $E(3, -2)$, and $F(1, 1)$ reflected in the origin

11. $\square GHIJ$ with vertices $G(-1, 2)$, $H(2, 3)$, $I(6, 1)$, and $J(3, 0)$ reflected in the line $y = x$

Application

NATURE Determine how many lines of symmetry each object has. Then determine whether each object has point symmetry.

12.    **13.**   **14.**

Practice and Apply

Homework Help	
For Exercises	**See Examples**
15–26, 38, 39	1
29, 34, 40, 41	2
30, 33, 41	3
27, 28	4
31, 32, 40	5
42	6
35–37, 44–47	7

Extra Practice
See page 771.

Refer to the figure at the right. Name the image of each figure under a reflection in:

line ℓ	line m	point Z
15. $\overline{WX}$	**18.** T	**21.** U
16. $\overline{WZ}$	**19.** $\overline{UY}$	**22.** $\angle TXZ$
17. $\angle XZY$	**20.** $\triangle YVW$	**23.** $\triangle YUZ$

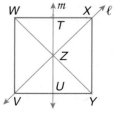

Copy each figure. Draw the image of each figure under a reflection in line ℓ.

24. 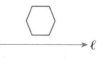   **25.**   **26.**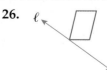

COORDINATE GEOMETRY Graph each figure and its image under the given reflection.

27. rectangle $MNPQ$ with vertices $M(2, 3)$, $N(2, -3)$, $P(-2, -3)$, and $Q(-2, 3)$ in the origin

28. quadrilateral $GHIJ$ with vertices $G(-2, -2)$, $H(2, 0)$, $I(3, 3)$, and $J(-2, 4)$ in the origin

29. square $QRST$ with vertices $Q(-1, 4)$, $R(2, 5)$, $S(3, 2)$, and $T(0, 1)$ in the x-axis

30. trapezoid with vertices $D(4, 0)$, $E(-2, 4)$, $F(-2, -1)$, and $G(4, -3)$ in the y-axis

31. $\triangle BCD$ with vertices $B(5, 0)$, $C(-2, 4)$, and $D(-2, -1)$ in the line $y = x$

32. $\triangle KLM$ with vertices $K(4, 0)$, $L(-2, 4)$, and $M(-2, 1)$ in the line $y = 2$

33. The reflected image of $\triangle FGH$ has vertices $F'(1, 4)$, $G'(4, 2)$, and $H'(3, -2)$. Describe the reflection in the y-axis.

34. The reflected image of $\triangle XYZ$ has vertices $X'(1, 4)$, $Y'(2, 2)$, and $Z'(-2, -3)$. Describe the reflection in the line $x = -1$.

Determine how many lines of symmetry each figure has. Then determine whether the figure has point symmetry.

35.

36.

37.

Copy each figure and then reflect the figure in line m first and then reflect that image in line n. Compare the preimage with the final image.

38.

39.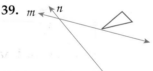

40. **COORDINATE GEOMETRY** Square $DEFG$ with vertices $D(-1, 4)$, $E(2, 8)$, $F(6, 5)$, and $G(3, 1)$ is reflected first in the x-axis, then in the line $y = x$. Find the coordinates of $D''E''F''G''$.

41. **COORDINATE GEOMETRY** Triangle ABC has been reflected in the x-axis, then the y-axis, then the origin. The result has coordinates $A'''(4, 7)$, $B'''(10, -3)$, and $C'''(-6, -8)$. Find the coordinates of A, B, and C.

42. **BILLIARDS** Tonya is playing billiards. She wants to pocket the eight ball in the lower right pocket using the white cue ball. Copy the diagram and sketch the path the eight ball must travel after being struck by the cue ball.

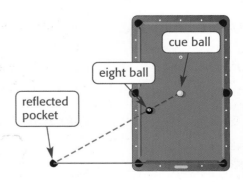

43. **CRITICAL THINKING** Show that the image of a point upon reflection in the origin is the same image obtained when reflecting a point in the x-axis and then the y-axis.

A reflection of the quadrilateral on the map will help you come closer to locating the hidden treasure. Visit www.geometry online.com/webquest to continue work on your WebQuest project.

DIAMONDS For Exercises 44–47, use the following information.
Diamond jewelers offer a variety of cuts. For each top view, identify any lines or points of symmetry.

44. round cut

45. pear cut

46. heart cut

47. emerald cut

48. **WRITING IN MATH** Answer the question that was posed at the beginning of the lesson.

Where are reflections found in nature?

Include the following in your answer:
- three examples in nature having line symmetry, and
- an explanation of how the distance from each point above the water line relates to the image in the water.

Standardized Test Practice

49. The image of $A(-2, 5)$ under a reflection is $A'(2, -5)$. Which reflection or group of reflections was used?
  I. reflected in the *x*-axis   **II.** reflected in the *y*-axis   **III.** reflected in the origin
  Ⓐ I or III   Ⓑ II and III   Ⓒ I and II   Ⓓ I and II, or III

50. ALGEBRA If $a \star c = 2a + b + 2c$, find $a \star c$ when $a = 25$, $b = 18$, and $c = 45$.
  Ⓐ 176   Ⓑ 158   Ⓒ 133   Ⓓ 88

Maintain Your Skills

Mixed Review **Write a coordinate proof for each of the following.** *(Lesson 8-7)*

51. The segments joining the midpoints of the opposite sides of a quadrilateral bisect each other.

52. The segments joining the midpoints of the sides of an isosceles trapezoid form a rhombus.

Refer to trapezoid $ACDF$. *(Lesson 8-6)*

53. Find BE.

54. Let $\overline{XY}$ be the median of $BCDE$. Find XY.

55. Let $\overline{WZ}$ be the median of $ABEF$. Find WZ.

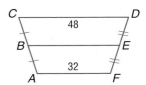

Solve each $\triangle FGH$ **described below. Round angle measures to the nearest degree and side measures to the nearest tenth.** *(Lesson 7-6)*

56. $m\angle G = 53$, $m\angle H = 71$, $f = 48$

57. $g = 21$, $m\angle G = 45$, $m\angle F = 59$

58. $h = 13.2$, $m\angle F = 106$, $f = 14.5$

Getting Ready for the Next Lesson **PREREQUISITE SKILL** Find the exact length of each side of quadrilateral $EFGH$.
(To review the Distance Formula, see Lesson 3-3.)

59. $\overline{EF}$ **60.** $\overline{FG}$

61. $\overline{GH}$ **62.** $\overline{HE}$

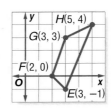

9-2 Translations

What You'll Learn

- Draw translated images using coordinates.
- Draw translated images by using repeated reflections.

Vocabulary

- translation
- composition
- glide reflection

How are translations used in a marching band show?

The sights and pageantry of a marching band performance can add to the excitement of a sporting event.

The movements of each band member as they progress through the show are examples of *translations*.

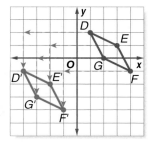

TRANSLATIONS USING COORDINATES A **translation** is a transformation that moves all points of a figure the same distance in the same direction. Translations on the coordinate plane can be drawn if you know the direction and how far the figure is moving horizontally and/or vertically. For the fixed values of a and b, a translation moves every point $P(x, y)$ of a plane figure to an image $P'(x + a, y + b)$. One way to symbolize a transformation is to write $(x, y) \rightarrow (x + a, y + b)$.

In the figure, quadrilateral *DEFG* has been translated 5 units to the left and three units down. This can be written as $(x, y) \rightarrow (x - 5, y - 3)$.

$D(1, 2) \rightarrow D'(1 - 5, 2 - 3)$ or $D'(-4, -1)$
$E(3, 1) \rightarrow E'(3 - 5, 1 - 3)$ or $E'(-2, -2)$
$F(4, -1) \rightarrow F'(4 - 5, -1 - 3)$ or $F'(-1, -4)$
$G(2, 0) \rightarrow G'(2 - 5, 0 - 3)$ or $G'(-3, -3)$

Example 1 Translations in the Coordinate Plane

Rectangle *PQRS* has vertices $P(-3, 5)$, $Q(-4, 2)$, $R(3, 0)$, and $S(4, 3)$. Graph *PQRS* and its image for the translation $(x, y) \rightarrow (x + 8, y - 5)$.

This translation moved every point of the preimage 8 units right and 5 units down.

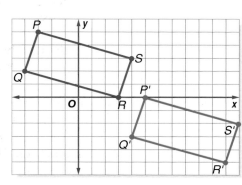

$P(-3, 5) \rightarrow P'(-3 + 8, 5 - 5)$ or $P'(5, 0)$
$Q(-4, 2) \rightarrow Q'(-4 + 8, 2 - 5)$ or $Q'(4, -3)$
$R(3, 0) \rightarrow R'(3 + 8, 0 - 5)$ or $R'(11, -5)$
$S(4, 3) \rightarrow S'(4 + 8, 3 - 5)$ or $S'(12, -2)$

Plot the translated vertices and connect to form rectangle $P'Q'R'S'$.

Example 2 *Repeated Translations*

ANIMATION Computers are often used to create animation. The graph shows repeated translations that result in animation of the star. Find the translation that moves star 1 to star 2 and the translation that moves star 4 to star 5.

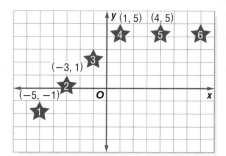

To find the translation from star 1 to star 2, use the coordinates at the top of each star. Use the coordinates $(-5, -1)$ and $(-3, 1)$ in the formula.

$$(x, y) \rightarrow (x + a, y + b)$$
$$(-5, -1) \rightarrow (-3, 1)$$

$x + a = -3$	$y + b = 1$
$-5 + a = -3$ $x = -5$	$-1 + b = 1$ $y = -1$
$a = 2$ Add 5 to each side.	$b = 2$ Add 1 to each side.

The translation is $(x, y) \rightarrow (x + 2, y + 2)$.

Use the coordinates $(1, 5)$ and $(4, 5)$ to find the translation from star 4 to star 5.

$$(x, y) \rightarrow (x + a, y + b)$$
$$(1, 5) \rightarrow (4, 5)$$

$x + a = 4$	$y + b = 5$
$1 + a = 4$ $x = 1$	$5 + b = 5$ $y = 5$
$a = 3$ Subtract 1 from each side.	$b = 0$ Subtract 5 from each side.

The translation is $(x, y) \rightarrow (x + 3, y)$ from star 4 to star 5 and from star 5 to star 6.

TRANSLATIONS BY REPEATED REFLECTIONS Another way to find a translation is to perform a reflection in the first of two parallel lines and then reflect the image in the other parallel line. A transformation made up of successive transformations is called a **composition**.

Example 3 *Find a Translation Using Reflections*

In the figure, lines m and n are parallel. Determine whether the red figure is a translation image of the blue preimage, quadrilateral $ABCD$.

Reflect quadrilateral $ABCD$ in line m. The result is the green image, quadrilateral $A'B'C'D'$. Then reflect the green image, quadrilateral $A'B'C'D'$ in line n. The red image, quadrilateral $A''B''C''D''$, has the same orientation as quadrilateral $ABCD$.

Quadrilateral $A''B''C''D''$ is the translation image of quadrilateral $ABCD$.

Since translations are compositions of two reflections, all translations are isometries. Thus, all properties preserved by reflections are preserved by translations. These properties include betweenness of points, collinearity, and angle and distance measure.

Check for Understanding

Concept Check
1. **OPEN ENDED** Choose integer coordinates for any two points A and B. Then describe how you could count to find the translation of point A to point B.

2. **Explain** which properties are preserved in a translation and why they are preserved.

3. **FIND THE ERROR** Allie and Tyrone are describing the transformation in the drawing.

Allie	Tyrone
This is a translation right 3 units and down 2 units.	This is a reflection in the y-axis and then the x-axis.

Who is correct? Explain your reasoning.

Guided Practice
In each figure, $m \parallel n$. Determine whether the red figure is a translation image of the blue figure. Write *yes* or *no*. Explain your answer.

4.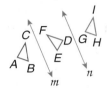

5.

Application
COORDINATE GEOMETRY Graph each figure and its image under the given translation.

6. $\overline{DE}$ with endpoints $D(-3, -4)$ and $E(4, 2)$ under the translation $(x, y) \rightarrow (x + 1, y + 3)$

7. $\triangle KLM$ with vertices $K(5, -2)$, $L(-3, -1)$, and $M(0, 5)$ under the translation $(x, y) \rightarrow (x - 3, y - 4)$

8. **ANIMATION** Find the translations that move the hexagon on the coordinate plane in the order given.

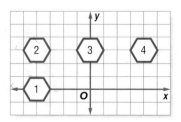

Practice and Apply

In each figure, $a \parallel b$. Determine whether the red figure is a translation image of the blue figure. Write *yes* or *no*. Explain your answer.

9.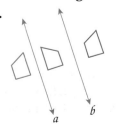

10.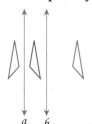

11.

Homework Help

For Exercises	See Examples
9–14, 28, 29	3
15–20, 24–27	1
19, 21–23	2

Extra Practice
See page 771.

12.

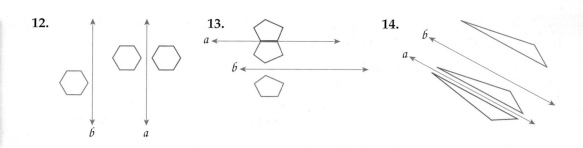

13.

14.

COORDINATE GEOMETRY Graph each figure and its image under the given translation.

15. $\overline{PQ}$ with endpoints $P(2, -4)$ and $Q(4, 2)$ under the translation left 3 units and up 4 units

16. $\overline{AB}$ with endpoints $A(-3, 7)$ and $B(-6, -6)$ under the translation 4 units to the right and down 2 units

17. $\triangle MJP$ with vertices $M(-2, -2)$, $J(-5, 2)$, and $P(0, 4)$ under the translation $(x, y) \rightarrow (x + 1, y + 4)$

18. $\triangle EFG$ with vertices $E(0, -4)$, $F(-4, -4)$, and $G(0, 2)$ under the translation $(x, y) \rightarrow (x + 2, y - 1)$

19. quadrilateral $PQRS$ with vertices $P(1, 4)$, $Q(-1, 4)$, $R(-2, -4)$, and $S(2, -4)$ under the translation $(x, y) \rightarrow (x - 5, y + 3)$

20. pentagon $VWXYZ$ with vertices $V(-3, 0)$, $W(-3, 2)$, $X(-2, 3)$, $Y(0, 2)$, and $Z(-1, 0)$ under the translation $(x, y) \rightarrow (x + 4, y - 3)$

21. CHESS The bishop shown in square f8 can only move diagonally along dark squares. If the bishop is in c1 after two moves, describe the translation.

22. RESEARCH Use the Internet or other resource to write a possible translation for each chess piece for a single move.

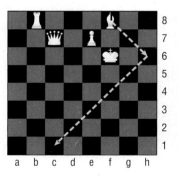

MOSAICS For Exercises 23–25, use the following information.
The mosaic tiling shown on the right is a thirteenth-century Roman inlaid marble tiling. Suppose this pattern is a floor design where the length of a side of the small white equilateral triangle is 12 inches. All triangles and hexagons are regular. Describe the translations in inches represented by each line.

23. green line

24. blue line

25. red line

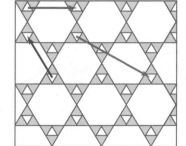

26. CRITICAL THINKING Triangle TWY has vertices $T(3, -7)$, $W(7, -4)$, and $Y(9, -8)$. Triangle BDG has vertices $B(3, 3)$, $D(7, 6)$, and $G(9, 2)$. If $\triangle BDG$ is the translation image of $\triangle TWY$ with respect to two parallel lines, find the equations that represent two possible parallel lines.

COORDINATE GEOMETRY Graph each figure and the image under the given translation.

27. $\triangle PQR$ with vertices $P(-3, -2)$, $Q(-1, 4)$, and $R(2, -2)$ under the translation $(x, y) \rightarrow (x + 2, y - 4)$

28. $\triangle RST$ with vertices $R(-4, -1)$, $S(-1, 3)$, and $T(-1, 1)$ reflected in $y = 2$ and then reflected in $y = -2$

29. Under $(x, y) \rightarrow (x - 4, y + 5)$, $\triangle ABC$ has translated vertices $A'(-8, 5)$, $B'(2, 7)$, and $C'(3, 1)$. Find the coordinates of A, B, and C.

30. Triangle FGH is translated to $\triangle MNP$. Given $F(3, 9)$, $G(-1, 4)$, $M(4, 2)$, and $P(6, -3)$, find the coordinates of H and N. Then write the coordinate form of the translation.

STUDENTS For Exercises 31–33, refer to the graphic at the right. Each bar of the graph is made up of a boy-girl-boy unit.

31. Which categories show a boy-girl-boy unit that is translated within the bar?

32. Which categories show a boy-girl-boy unit that is reflected within the bar?

33. Does each person shown represent the same percent? Explain.

Online Research
Data Update How much allowance do teens receive? Visit www.geometryonline.com/data_update to learn more.

34. **WRITING IN MATH** Answer the question that was posed at the beginning of the lesson.

How are translations used in a marching band show?

Include the following in your answer:
• the types of movements by band members that are translations, and
• a description of a simple pattern for a band member.

GLIDE REFLECTION For Exercises 35–37, use the following information.
A **glide reflection** is a composition of a translation and a reflection in a line parallel to the direction of the translation.

35. Is a glide reflection an *isometry*? Explain.

36. Triangle DEF has vertices $D(4, 3)$, $E(2, -2)$ and $F(0, 1)$. Sketch the image of a glide reflection composed of the translation $(x, y) \rightarrow (x, y - 2)$ and a reflection in the y-axis.

37. Triangle ABC has vertices $A(-3, -2)$, $B(-1, -3)$ and $C(2, -1)$. Sketch the image of a glide reflection composed of the translation $(x, y) \rightarrow (x + 3, y)$ and a reflection in $y = 1$.

38. Triangle *XYZ* with vertices *X*(5, 4), *Y*(3, −1), and *Z*(0, 2) is translated so that *X'* is at (3, 1). State the coordinates of *Y'* and *Z'*.

Ⓐ *Y'*(5, 2) and *Z'*(2, 5)   Ⓑ *Y'*(0, −3) and *Z'*(−3, 0)

Ⓒ *Y'*(1, −4) and *Z'*(−2, −1)   Ⓓ *Y'*(11, 4) and *Z'*(8, 6)

39. ALGEBRA Find the slope of a line through *P*(−2, 5) and *T*(2, −1).

Ⓐ $-\frac{3}{2}$ Ⓑ −1   Ⓒ 0   Ⓓ $\frac{2}{3}$

Maintain Your Skills

Mixed Review

Copy each figure. Draw the reflected image of each figure in line *m*. *(Lesson 9-1)*

40.
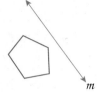

41.

42.

Name the missing coordinates for each quadrilateral. *(Lesson 8-7)*

43. *QRST* is an isosceles trapezoid.

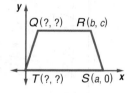

44. *ABCD* is a parallelogram.

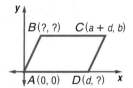

45. LANDSCAPING Juanna is a landscaper. She wishes to determine the height of a tree. Holding a drafter's 45° triangle so that one leg is horizontal, she sights the top of the tree along the hypotenuse as shown at the right. If she is 6 yards from the tree and her eyes are 5 feet from the ground, find the height of the tree. *(Lesson 7-3)*

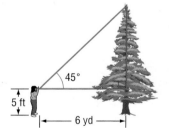

State the assumption you would make to start an indirect proof of each statement. *(Lesson 5-3)*

46. Every shopper that walks through the door is greeted by a salesperson.

47. If you get a job, you have filled out an application.

48. If $4y + 17 \leq 41$, $y \leq 6$.

49. If two lines are cut by a transversal and a pair of alternate interior angles are congruent, then the two lines are parallel.

Find the distance between each pair of parallel lines. *(Lesson 3-6)*

50. $x = -2$
$x = 5$

51. $y = -6$
$y = -1$

52. $y = 2x + 3$
$y = 2x - 7$

53. $y = x + 2$
$y = x - 4$

Getting Ready for the Next Lesson

PREREQUISITE SKILL Use a protractor and draw an angle for each of the following degree measures. *(To review drawing angles, see Lesson 1-4.)*

54. 30   **55.** 45   **56.** 52   **57.** 60   **58.** 105   **59.** 150

9-3 Rotations

- Draw rotated images using the angle of rotation.
- Identify figures with rotational symmetry.

Vocabulary

- rotation
- center of rotation
- angle of rotation
- rotational symmetry
- invariant points
- direct isometry
- indirect isometry

How do some amusement rides illustrate rotations?

In 1926, Herbert Sellner invented the Tilt-A-Whirl. Today, no carnival is complete without these cars that send riders tipping and spinning as they make their way around a circular track.

The Tilt-A-Whirl provides an example of rotation.

Study Tip

Turns
A rotation, sometimes called a turn, is generally measured as a counter-clockwise turn. A half-turn is 180° and a full turn is 360°.

DRAW ROTATIONS A **rotation** is a transformation that turns every point of a preimage through a specified angle and direction about a fixed point. The fixed point is called the **center of rotation**.

In the figure, R is the center of rotation for the preimage $ABCD$. The measures of angles ARA', BRB', CRC', and DRD' are equal. Any point P on the preimage $ABCD$ has an image P' on $A'B'C'D'$ such that the measure of $\angle PRP'$ is a constant measure. This is called the **angle of rotation**.

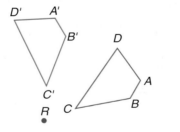
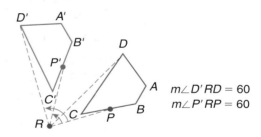

$m\angle D'RD = 60$
$m\angle P'RP = 60$

A rotation exhibits all of the properties of isometries, including preservation of distance and angle measure. Therefore, it is an isometry.

Example 1 Draw a Rotation

Triangle ABC has vertices $A(2, 3)$, $B(6, 3)$, and $C(5, 5)$. Draw the image of $\triangle ABC$ under a rotation of 60° counterclockwise about the origin.

- First graph $\triangle ABC$.
- Draw a segment from the origin O, to point A.
- Use a protractor to measure a 60° angle counterclockwise with $\overline{OA}$ as one side.
- Draw $\overrightarrow{OR}$.
- Use a compass to copy $\overline{OA}$ onto $\overrightarrow{OR}$. Name the segment $\overline{OA'}$

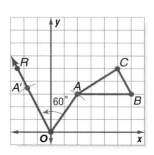

- Repeat with points *B* and *C*.
  △*A'B'C'* is the image of △*ABC* under a
  60° counterclockwise rotation about the origin.

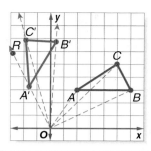

Another way to perform rotations is by reflecting a figure successively in two intersecting lines. Reflecting a figure once and then reflecting the image in a second line is another example of a composition of reflections.

Geometry Software Investigation

Reflections in Intersecting Lines

Construct a Figure

- Use The Geometer's Sketchpad to construct scalene triangle *ABC*.
- Construct lines *m* and *n* so that they intersect outside △*ABC*.
- Label the point of intersection *P*.

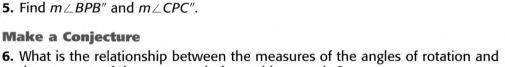

Analyze

1. Reflect △*ABC* in line *m*. Then, reflect △*A'B'C'* in line *n*.

2. Describe the transformation of △*ABC* to △*A"B"C"*.

3. Measure the acute angle formed by *m* and *n*.

4. Construct a segment from *A* to *P* and from *A"* to *P*. Find the measure of the angle of rotation, ∠*APA"*.

5. Find *m*∠*BPB"* and *m*∠*CPC"*.

Make a Conjecture

6. What is the relationship between the measures of the angles of rotation and the measure of the acute angle formed by *m* and *n*?

When rotating a figure by reflecting it in two intersecting lines, there is a relationship between the angle of rotation and the angle formed by the intersecting lines.

Key Concept

Postulate 9.1   In a given rotation, if *A* is the preimage, *A"* is the image, and *P* is the center of rotation, then the measure of the angle of rotation ∠*APA"* is twice the measure of the acute or right angle formed by the intersecting lines of reflection.

Corollary 9.1   Reflecting an image successively in two perpendicular lines results in a 180° rotation.

www.geometryonline.com/extra_examples

Common
Misconception
The order in which
you reflect a figure in
two nonperpendicular
intersecting lines produces
rotations of the same
degree measure, but one
is clockwise and the other
is counterclockwise.

Example 2 Reflections in Intersecting Lines

Find the image of rectangle *DEFG* under reflections in line *m* and then line *n*.

First reflect rectangle *DEFG* in line *m*.
Then label the image *D'E'F'G'*.

Next, reflect the image in line *n*.

Rectangle *D"E"F"G"* is the image of rectangle *DEFG* under reflections in lines *m* and *n*. How can you transform *DEFG* directly to *D"E"F"G"* by using a rotation?

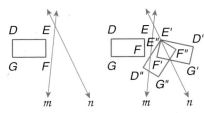

ROTATIONAL SYMMETRY

Some objects have rotational symmetry. If a figure can be rotated less than 360 degrees about a point so that the image and the preimage are indistinguishable, then the figure has **rotational symmetry**.

In the figure, the pentagon has rotational symmetry of *order* 5 because there are 5 rotations of less than 360° (including 0 degrees) that produce an image indistinguishable from the original. The rotational symmetry has a *magnitude* of 72° because 360 degrees divided by the order, in this case 5, produces the magnitude of the symmetry.

Example 3 Identifying Rotational Symmetry

QUILTS One example of rotational symmetry artwork is quilt patterns. A quilt made by Judy Mathieson of Sebastopol, California, won the Masters Award for Contemporary Craftsmanship at the International Quilt Festival in 1999. Identify the order and magnitude of the symmetry in each part of the quilt.

a. large star in center of quilt

The large star in the center of the quilt has rotational symmetry of order 20 and magnitude of 18°.

b. entire quilt

The entire quilt has rotational symmetry of order 4 and magnitude of 90°.

Check for Understanding

Concept Check

1. **OPEN ENDED** Draw a figure on the coordinate plane in Quadrant I. Rotate the figure clockwise 90 degrees about the origin. Then rotate the figure 90 degrees counterclockwise. Describe the results using the coordinates.

2. **Explain** two techniques that can be used to rotate a figure.

3. **Compare and contrast** translations and rotations.

Guided Practice

4. Copy △*BCD* and rotate the triangle 60° counterclockwise about point *G*.

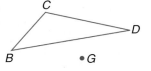

Copy each figure. Use a composition of reflections to find the rotation image with respect to lines ℓ and m.

5.

6.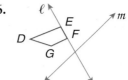

7. $\overline{XY}$ has endpoints $X(-5, 8)$ and $Y(0, 3)$. Draw the image of $\overline{XY}$ under a rotation of 45° clockwise about the origin.

8. $\triangle PQR$ has vertices $P(-1, 8)$, $Q(4, -2)$, and $R(-7, -4)$. Draw the image of $\triangle PQR$ under a rotation of 90° counterclockwise about the origin.

9. Identify the order and magnitude of the rotational symmetry in a regular hexagon.

10. Identify the order and magnitude of the rotational symmetry in a regular octagon.

Application 11. **FANS** The blades of a fan exhibit rotational symmetry. Identify the order and magnitude of the symmetry of the blades of each fan in the pictures.

Practice and Apply

Homework Help

For Exercises	See Examples
12–15, 27	1
19–26	2
16–28, 28	3

Extra Practice
See page 772.

12. Copy pentagon *BCDEF*. Then rotate the pentagon 110° counterclockwise about point *R*.

13. Copy $\triangle MNP$. Then rotate the triangle 180° counterclockwise around point *Q*.

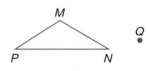

COORDINATE GEOMETRY Draw the rotation image of each figure 90° in the given direction about the center point and label the vertices.

14. $\triangle XYZ$ with vertices $X(0, -1)$, $Y(3,1)$, and $Z(1, 5)$ counterclockwise about the point $P(-1, 1)$

15. $\triangle RST$ with vertices $R(0, 1)$, $S(5,1)$, and $T(2, 5)$ clockwise about the point $P(-2, 5)$

RECREATION For Exercises 16–18, use the following information.
A Ferris wheel's motion is an example of a rotation. The Ferris wheel shown has 20 cars.

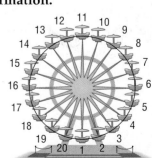

16. Identify the order and magnitude of the symmetry of a 20-seat Ferris wheel.

17. What is the measure of the angle of rotation if seat 1 of a 20-seat Ferris wheel is moved to the seat 5 position?

18. If seat 1 of a 20-seat Ferris wheel is rotated 144°, find the original seat number of the position it now occupies.

Copy each figure. Use a composition of reflections to find the rotation image with respect to lines *m* and *t*.

19.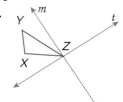

20.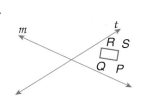

21.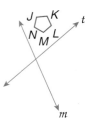

COORDINATE GEOMETRY Draw the rotation image of each triangle by reflecting the triangles in the given lines. State the coordinates of the rotation image and the angle of rotation.

22. △*TUV* with vertices *T*(4, 0), *U*(2, 3), and *V*(1, 2), reflected in the *y*-axis and then the *x*-axis

23. △*KLM* with vertices *K*(5, 0), *L*(2, 4), and *M*(−2, 4), reflected in the line *y* = *x* and then the *x*-axis

24. △*XYZ* with vertices *X*(5, 0), *Y*(3, 4), and *Z*(−3, 4), reflected in the line *y* = −*x* and then the line *y* = *x*

25. COORDINATE GEOMETRY The point at (2, 0) is rotated 30° counterclockwise about the origin. Find the exact coordinates of its image.

26. MUSIC A five-disc CD changer rotates as each CD is played. Identify the magnitude of the rotational symmetry as the changer moves from one CD to another.

More About. . .

Music •

Some manufacturers now make CD changers that are referred to as CD jukeboxes. These can hold hundreds of CDs, which may be more than a person's entire CD collection.

Source: www.usatoday.com

Determine whether the indicated composition of reflections is a rotation. Explain.

27.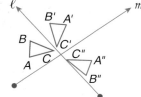

28.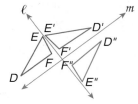

AMUSEMENT RIDES For each ride, determine whether the rider undergoes a rotation. Write *yes* or *no*.

29. spinning teacups

30. scrambler

31. roller coaster loop

32. ALPHABET Which capital letters of the alphabet produce the same letter after being rotated 180°?

33. CRITICAL THINKING In △*ABC*, *m*∠*BAC* = 40. Triangle *AB'C* is the image of △*ABC* under reflection and △*AB'C'* is the image of △*AB'C* under reflection. How many such reflections would be necessary to map △*ABC* onto itself?

34. CRITICAL THINKING If a rotation is performed on a coordinate plane, what angles of rotation would make the rotations easier? Explain.

35. COORDINATE GEOMETRY Quadrilateral *QRST* is rotated 90° clockwise about the origin. Describe the transformation using coordinate notation.

36. Triangle *FGH* is rotated 80° clockwise and then rotated 150° counterclockwise about the origin. To what counterclockwise rotation about the origin is this equivalent?

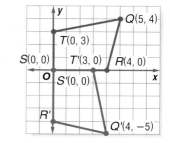

CRITICAL THINKING For Exercises 37–39, use the following information.
Points that do not change position under a transformation are called **invariant points**. For each of the following transformations, identify any invariant points.

37. reflection in a line

38. a rotation of $x°(0 < x < 360)$ about point *P*

39. $(x, y) \rightarrow (x + a, y + b)$, where *a* and *b* are not 0

40. WRITING IN MATH   Answer the question that was posed at the beginning of the lesson.

How do amusement rides exemplify rotations?

Include the following in your answer:
• a description of how the Tilt-A-Whirl actually rotates two ways, and
• other amusement rides that use rotation.

Standardized Test Practice

41. In the figure, describe the rotation that moves triangle 1 to triangle 2.

Ⓐ 180° clockwise

Ⓑ 135° clockwise

Ⓒ 135° counterclockwise

Ⓓ 90° counterclockwise

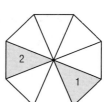

42. ALGEBRA Suppose *x* is $\frac{2}{5}$ of *y* and *y* is $\frac{1}{3}$ of *z*. If *x* = 6, then *z* = ?

Ⓐ $\frac{4}{5}$ Ⓑ $\frac{18}{5}$ Ⓒ 5 Ⓓ 45

Extending the Lesson   A **direct isometry** is one in which the image of a figure is found by moving the figure intact within the plane. An **indirect isometry** cannot be performed by maintaining the clockwise orientation of the points as in a direct isometry.

43. Copy and complete the table below. Determine whether each transformation preserves the given property. Write *yes* or *no*.

Transformation	angle measure	betweenness of points	orientation	collinearity	distance measure
reflection					
translation					
rotation					

Identify each type of transformation as a direct isometry or an indirect isometry.

44. reflection **45.** translation **46.** rotation

Mixed Review In each figure, $a \parallel b$. Determine whether the blue figure is a translation image of the red figure. Write *yes* or *no*. Explain your answer. *(Lesson 9-2)*

47.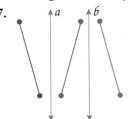

48.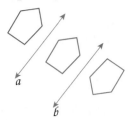

49.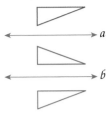

For the figure to the right, name the reflected image of each image. *(Lesson 9-1)*

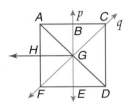

50. $\overline{AG}$ in line p

51. F in point G

52. $\overline{GE}$ in line q

53. $\angle CGD$ in line p

Complete each statement about $\square PQRS$. Justify your answer. *(Lesson 8-2)*

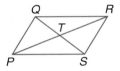

54. $\overline{QR} \parallel$ _____?_____

55. $\overline{PT} \cong$ _____?_____

56. $\angle SQR \cong$ _____?_____

57. $\angle QPS \cong$ _____?_____

58. SURVEYING A surveyor is 100 meters from a building and finds that the angle of elevation to the top of the building is 23°. If the surveyor's eye level is 1.55 meters above the ground, find the height of the building. *(Lesson 7-5)*

Determine whether the given measures can be the lengths of the sides of a triangle. Write *yes* or *no*. *(Lesson 5-4)*

59. 6, 8, 16

60. 12, 17, 20

61. 22, 23, 37

Getting Ready for the Next Lesson **PREREQUISITE SKILL** Find whole number values for each variable so each equation is true. *(To review solving problems by making a table, see pages 737–738.)*

62. $180a = 360$

63. $180a + 90b = 360$

64. $135a + 45b = 360$

65. $120a + 30b = 360$

66. $180a + 60b = 360$

67. $180a + 30b = 360$

Practice Quiz 1 *Lesson 9-1 through 9-3*

Graph each figure and the image in the given reflection. *(Lesson 9-1)*

1. $\triangle DEF$ with vertices $D(-1, 1)$, $E(1, 4)$ and $F(3, 2)$ in the origin

2. quadrilateral $ABCD$ with vertices $A(0, 2)$, $B(2, 2)$, $C(3, 0)$, and $D(-1, 1)$ in the line $y = x$

Graph each figure and the image under the given translation. *(Lesson 9-2)*

3. $\overline{PQ}$ with endpoints $P(1, -4)$ and $Q(4, -1)$ under the translation left 3 units and up 4 units

4. $\triangle KLM$ with vertices $K(-2, 0)$, $L(-4, 2)$, and $M(0, 4)$ under the translation $(x, y) \rightarrow (x + 1, y - 4)$

5. Identify the order and magnitude of the symmetry of a 36-horse carousel. *(Lesson 9-3)*

9-4 Tessellations

What You'll Learn

- Identify regular tessellations.
- Create tessellations with specific attributes.

Vocabulary

- tessellation
- regular tessellation
- uniform
- semi-regular tessellation

How are tessellations used in art?

M.C. Escher (1898–1972) was a Dutch graphic artist famous for repeating geometric patterns. He was also well known for his spatial illusions, impossible buildings, and techniques in wood-cutting and lithography. In the picture, figures can be reduced to basic regular polygons. Equilateral triangles and regular hexagons are prominent in the repeating patterns.

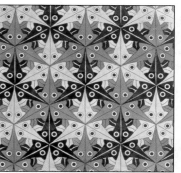

Symmetry Drawing E103,
M.C. Escher

REGULAR TESSELLATIONS Reflections, translations, and rotations can be used to create patterns using polygons. A pattern that covers a plane by transforming the same figure or set of figures so that there are no overlapping or empty spaces is called a **tessellation**.

In a tessellation, the sum of the measures of the angles of the polygons surrounding a point (at a vertex) is 360.

You can use what you know about angle measures in regular polygons to help determine which polygons tessellate.

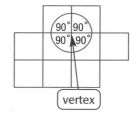

Geometry Activity

Tessellations of Regular Polygons

Model and Analyze

- Study a set of pattern blocks to determine which shapes are regular.
- Make a tessellation with each type of regular polygon.

1. Which shapes in the set are regular?
2. Write an expression showing the sum of the angles at each vertex of the tessellation.
3. Copy and complete the table below.

Regular Polygon	triangle	square	pentagon	hexagon	heptagon	octagon
Measure of One Interior Angle						
Does it tessellate?						

Make a Conjecture

4. What must be true of the angle measure of a regular polygon for a regular tessellation to occur?

The tessellations you formed in the Geometry Activity are regular tessellations. A **regular tessellation** is a tessellation formed by only one type of regular polygon. In the activity, you found that if a regular polygon has an interior angle with a measure that is a factor of 360, then the polygon will tessellate the plane.

Study Tip

Look Back
To review finding the measure of an interior angle of a regular polygon, see Lesson 8-1.

Example 1 Regular Polygons

Determine whether a regular 24-gon tessellates the plane. Explain.

Let $\angle 1$ represent one interior angle of a regular 24-gon.

$$m\angle 1 = \frac{180(n-2)}{n} \qquad \text{Interior Angle Formula}$$
$$= \frac{180(24-2)}{24} \qquad \text{Substitution}$$
$$= 165 \qquad \text{Simplify.}$$

Since 165 is not a factor of 360, a 24-gon will not tessellate the plane.

TESSELLATIONS WITH SPECIFIC ATTRIBUTES   A tessellation pattern can contain any type of polygon. Tessellations containing the same arrangement of shapes and angles at each vertex are called **uniform**.

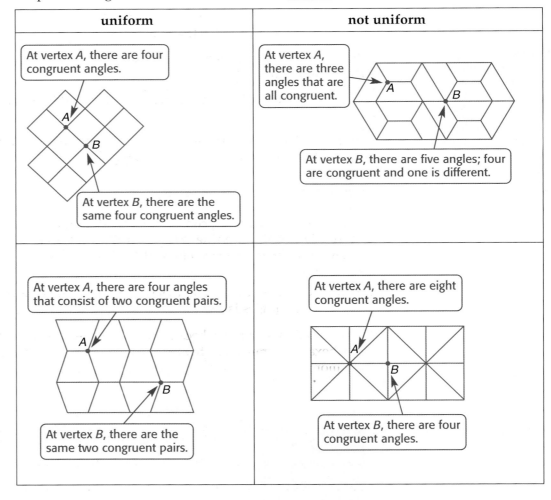

uniform	not uniform
At vertex *A*, there are four congruent angles.	At vertex *A*, there are three angles that are all congruent.
At vertex *B*, there are the same four congruent angles.	At vertex *B*, there are five angles; four are congruent and one is different.
At vertex *A*, there are four angles that consist of two congruent pairs.	At vertex *A*, there are eight congruent angles.
At vertex *B*, there are the same two congruent pairs.	At vertex *B*, there are four congruent angles.

Tessellations can be formed using more than one type of polygon. A uniform tessellation formed using two or more regular polygons is called a **semi-regular tessellation**.

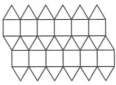

Example 2 Semi-Regular Tessellation

Determine whether a semi-regular tessellation can be created from regular hexagons and equilateral triangles, all having sides 1 unit long.

Method 1 Make a model.
Two semi-regular models are shown. You will notice that the spaces at each vertex can be filled in with equilateral triangles. Model 1 has two hexagons and two equilateral triangles arranged in an alternating pattern around each vertex. Model 2 has one hexagon and four equilateral triangles at each vertex.

Model 1

Model 2

Method 2 Solve algebraically.
Each interior angle of a regular hexagon measures $\frac{180(6-2)}{6}$ or $120°$.

Each angle of an equilateral triangle measures $60°$. Find whole-number values for h and t so that $120h + 60t = 360$.

Let $h = 1$.		Let $h = 2$.
$120(1) + 60t = 360$	Substitution	$120(2) + 60t = 360$
$120 + 60t = 360$	Simplify.	$240 + 60t = 360$
$60t = 240$	Subtract from each side.	$60t = 120$
$t = 4$	Divide each side by 60.	$t = 2$

When $h = 1$ and $t = 4$, there is one hexagon with four equilateral triangles at each vertex. (Model 2)

When $h = 2$ and $t = 2$, there are two hexagons and two equilateral triangles. (Model 1)

Note if $h = 0$ and $t = 6$ or $h = 3$ and $t = 0$, then the tessellations are regular because there would be only one regular polygon.

Example 3 **Classify Tessellations**

FLOORING Tile flooring comes in many shapes and patterns. Determine whether the pattern is a tessellation. If so, describe it as *uniform, not uniform, regular,* or *semi-regular.*

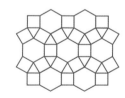

The pattern is a tessellation because at the different vertices the sum of the angles is $360°$.

The tessellation is uniform because at each vertex there are two squares, a triangle, and a hexagon arranged in the same order. The tessellation is also semi-regular since more than one regular polygon is used.

Check for Understanding

Concept Check

1. **Compare and contrast** a semi-regular tessellation and a uniform tessellation.

2. **OPEN ENDED** Use these pattern blocks that are 1 unit long on each side to create a tessellation.

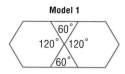

3. **Explain** why the tessellation is *not* a regular tessellation.

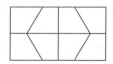

 www.geometryonline.com/extra_examples

Guided Practice **Determine whether each regular polygon tessellates the plane. Explain.**

    4. decagon   **5.** 30-gon

Determine whether a semi-regular tessellation can be created from each set of figures. Assume that each figure has a side length of 1 unit.

    6. a square and a triangle   **7.** an octagon and a square

Determine whether each pattern is a tessellation. If so, describe it as *uniform, not uniform, regular,* or *semi-regular.*

    8.    **9.**

Application   **10. QUILTING** The "Postage Stamp" pattern can be used in quilting. Explain why this is a tessellation and what kind it is.

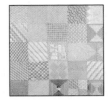

Practice and Apply

Homework Help

For Exercises	See Examples
11–16	1
17–20	2
21–28	3

Extra Practice
See page 772.

Determine whether each regular polygon tessellates the plane. Explain.

  11. nonagon   **12.** hexagon   **13.** equilateral triangle

  14. dodecagon   **15.** 23-gon   **16.** 36-gon

Determine whether a semi-regular tessellation can be created from each set of figures. Assume that each figure has a side length of 1 unit.

  17. regular octagons and non-square rhombi

  18. regular dodecagons and equilateral triangles

  19. regular dodecagons, squares, and equilateral triangles

  20. regular heptagons, squares, and equilateral triangles

Determine whether each figure tessellates the plane. If so, describe the tessellation as *uniform, not uniform, regular,* or *semi-regular.*

  21. parallelogram   **22.** kite

  23. quadrilateral   **24.** pentagon and square

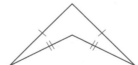   

Determine whether each pattern is a tessellation. If so, describe it as *uniform, not uniform, regular,* or *semi-regular.*

  25.    **26.**

27.

28.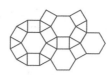

29. BRICKWORK   In the picture, suppose the side of the octagon is the same length as the side of the square. What kind of tessellation is this?

Determine whether each statement is *always, sometimes,* **or** *never* **true. Justify your answers.**

30. Any triangle will tessellate the plane.

31. Semi-regular tessellations are not uniform.

32. Uniform tessellations are semi-regular.

33. Every quadrilateral will tessellate the plane.

34. Regular 16-gons will tessellate the plane.

INTERIOR DESIGN   **For Exercises 35 and 36, use the following information.**
Kele's family is tiling the entry floor with the tiles in the pattern shown.

35. Determine whether the pattern is a tessellation.

36. Is the tessellation *uniform, regular,* or *semi-regular*?

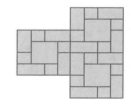

37. BEES   A honeycomb is composed of hexagonal cells made of wax in which bees store honey. Determine whether this pattern is a tessellation. If so, describe it as *uniform, not uniform, regular,* or *semi-regular.*

38. CRITICAL THINKING   What could be the measures of the interior angles in a pentagon that tessellates a plane? Is this tessellation regular? Is it uniform?

39. **WRITING IN MATH**   Answer the question that was posed at the beginning of the lesson.

How are tessellations used in art?
Include the following in your answer:
- how equilateral triangles and regular hexagons form a tessellation, and
- other geometric figures that can be found in the picture.

40. Find the measure of an interior angle of a regular nonagon.
  Ⓐ 150   Ⓑ 147   Ⓒ 140   Ⓓ 115

41. ALGEBRA   Evaluate $\dfrac{360(x-2)}{2x} - \dfrac{180}{x}$ if $x = 12$.
  Ⓐ 135   Ⓑ 150   Ⓒ 160   Ⓓ 225

Mixed Review **COORDINATE GEOMETRY** Draw the rotation image of each figure 90° in the given direction about the center point and label the coordinates. *(Lesson 9-3)*

42. △ABC with A(8, 1), B(2, −6), and C(−4, −2) counterclockwise about P(−2, 2)

43. △DEF with D(6, 2), E(6, −3), and F(2, 3) clockwise about P(3, −2)

44. ▱GHIJ with G(−1, 2), H(−3, −3), I(−5, −6), and J(−3, −1) counterclockwise about P(−2, −3)

45. rectangle KLMN with K(−3, −5), L(3, 3), M(7, 0), and N(1, −8) counterclockwise about P(−2, 0)

46. REMODELING The diagram at the right shows the floor plan of Justin's kitchen. Each square on the diagram represents a 3 foot by 3 foot area. While remodeling his kitchen, Justin moved his refrigerator from square A to square B. Describe the move. *(Lesson 9-2)*

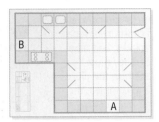

ALGEBRA Find x and y so that each quadrilateral is a parallelogram. *(Lesson 8-3)*

47.
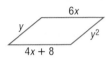

48.
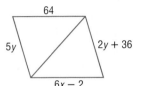

49.
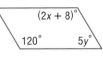

Determine whether each set of numbers can be the measures of the sides of a right triangle. Then state whether they form a Pythagorean triple. *(Lesson 7-2)*

50. 12, 16, 20   **51.** 9, 10, 15   **52.** 2.5, 6, 6.5

53. 14, $14\sqrt{3}$, 28   **54.** 14, 48, 50   **55.** $\dfrac{1}{2}, \dfrac{1}{3}, \dfrac{1}{4}$

Points A, B, and C are the midpoints of $\overline{DF}, \overline{DE},$ and $\overline{EF}$, respectively. *(Lesson 6-4)*

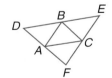

56. If BC = 11, AC = 13, and AB = 15, find the perimeter of △DEF.

57. If DE = 18, DA = 10, and FC = 7, find AB, BC, and AC.

COORDINATE GEOMETRY For Exercises 58–61, use the following information.
The vertices of quadrilateral PQRS are P(5, 2), Q(1, 6), R(−3, 2), and S(1, −2). *(Lesson 3-3)*

58. Show that the opposite sides of quadrilateral PQRS are parallel.

59. Show that the adjacent sides of quadrilateral PQRS are perpendicular

60. Determine the length of each side of quadrilateral PQRS.

61. What type of figure is quadrilateral PQRS?

Getting Ready for the Next Lesson **PREREQUISITE SKILL** If quadrilateral ABCD ~ quadrilateral WXYZ, find each of the following. *(To review similar polygons, see Lesson 6-2.)*

62. scale factor of ABCD to WXYZ

63. XY

64. YZ

65. WZ

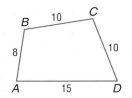
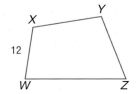

Geometry Activity
A Follow-up of Lesson 9-4

Tessellations and Transformations

Activity 1   Make a tessellation using a translation.

Step 1   Start by drawing a square. Then copy the figure shown below.

Step 2   Translate the shape on the top side to the bottom side.

Step 3   Translate the figure on the left side and the dot to the right side to complete the pattern.

Step 4   Repeat this pattern on a tessellation of squares.

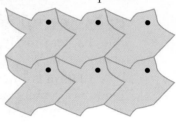

Activity 2   Make a tessellation using a rotation.

Step 1   Start by drawing an equilateral triangle. Then draw a trapezoid inside the right side of the triangle.

Step 2   Rotate the trapezoid so you can copy the change on the side indicated.

Step 3   Repeat this pattern on a tessellation of equilateral triangles. Alternating colors can be used to best show the tessellation.

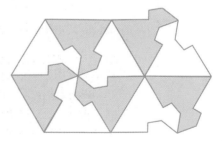

Model and Analyze

1. Is the area of the square in Step 1 of Activity 1 the same as the area of the new shape in Step 2? Explain.

2. Describe how you would create the unit for the pattern shown at the right.

Make a tessellation for each pattern described. Use a tessellation of two rows of three squares as your base.

3.

4.

5.

9-5 Dilations

What You'll Learn

- Determine whether a dilation is an enlargement, a reduction, or a congruence transformation.
- Determine the scale factor for a given dilation.

Vocabulary
- dilation
- similarity transformation

How do you use dilations when you use a computer?

Have you ever tried to paste an image into a word processing document and the picture was too large? Many word processing programs allow you to scale the size of the picture so that you can fit it in your document. Scaling a picture is an example of a dilation.

Study Tip

Scale Factor
When discussing dilations, scale factor has the same meaning as with proportions. The letter r usually represents the scale factor.

CLASSIFY DILATIONS   All of the transformations you have studied so far in this chapter produce images that are congruent to the original figure. A dilation is another type of transformation.

A **dilation** is a transformation that may change the size of a figure. A dilation requires a center point and a scale factor. The figures below show how dilations can result in a larger figure and a smaller figure than the original.

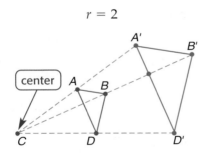

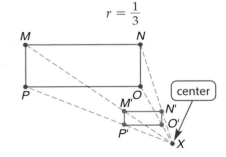

Triangle $A'B'D'$ is a dilation of $\triangle ABD$.

$CA' = 2(CA)$

$CB' = 2(CB)$

$CD' = 2(CD)$

$\triangle A'B'D'$ is larger than $\triangle ABD$.

Rectangle $M'N'O'P'$ is a dilation of rectangle $MNOP$.

$XM' = \frac{1}{3}(XM) \quad XN' = \frac{1}{3}(XN)$

$XO' = \frac{1}{3}(XO) \quad XP' = \frac{1}{3}(XP)$

Rectangle $M'N'O'P'$ is smaller than rectangle $MNOP$.

The value of r determines whether the dilation is an enlargement or a reduction.

Key Concept   Dilation

If $|r| > 1$, the dilation is an enlargement.

If $0 < |r| < 1$, the dilation is a reduction.

If $|r| = 1$, the dilation is a congruence transformation.

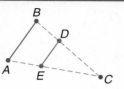

Study Tip

Isometry Dilation
A dilation with a scale factor of 1 produces an image that coincides with the preimage. The two are congruent.

As you can see in the figures on the previous page, dilation preserves angle measure, betweenness of points, and collinearity, but does not preserve distance. That is, dilations produce similar figures. Therefore, a dilation is a **similarity transformation**.

This means that $\triangle ABD \sim \triangle A'B'D'$ and $\square MNOP \sim \square M'N'O'P'$. This implies that $\dfrac{A'B'}{AB} = \dfrac{B'D'}{BD} = \dfrac{A'D'}{AD}$ and $\dfrac{M'N'}{MN} = \dfrac{N'O'}{NO} = \dfrac{O'P'}{OP} = \dfrac{M'P'}{MP}$. The ratios of measures of the corresponding parts is equal to the absolute value scale factor of the dilation, $|r|$. So, $|r|$ determines the size of the image as compared to the size of the preimage.

Theorem 9.1

If a dilation with center C and a scale factor of r transforms A to E and B to D, then $ED = |r|(AB)$.

You will prove Theorem 9.1 in Exercise 41.

Example 1 Determine Measures Under Dilations

Find the measure of the dilation image $\overline{A'B'}$ or the preimage $\overline{AB}$ using the given scale factor.

a. $AB = 12$, $r = -2$

$A'B' = |r|(AB)$
$A'B' = 2(12)$ $|r| = 2, AB = 12$
$A'B' = 24$ Multiply.

b. $A'B' = 36$, $r = \dfrac{1}{4}$

$A'B' = |r|(AB)$
$36 = \dfrac{1}{4}(AB)$ $A'B' = 36, |r| = \dfrac{1}{4}$
$144 = AB$ Multiply each side by 4.

When the scale factor is negative, the image falls on the opposite side of the center than the preimage.

Key Concept   **Dilations**

If $r > 0$, P' lies on $\overrightarrow{CP}$, and $CP' = r \cdot CP$.

If $r < 0$, P' lies on $\overrightarrow{CP'}$ the ray opposite $\overrightarrow{CP}$, and $CP' = |r| \cdot CP$.

The center of a dilation is always its own image.

Example 2 Draw a Dilation

Draw the dilation image of $\triangle JKL$ with center C and $r = -\dfrac{1}{2}$.

Since $0 < |r| < 1$, the dilation is a reduction of $\triangle JKL$.

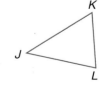

Draw $\overline{CJ}$, $\overline{CK}$, and $\overline{CL}$. Since r is negative, J', K', and L' will lie on $\overrightarrow{CJ'}$, $\overrightarrow{CK'}$, and $\overrightarrow{CL'}$, respectively. Locate J', K', and L' so that $CJ' = \dfrac{1}{2}(CJ)$, $CK' = \dfrac{1}{2}(CK)$, and $CL' = \dfrac{1}{2}(CL)$. Draw $\triangle J'K'L'$.

In the coordinate plane, you can use the scale factor to determine the coordinates of the image of dilations centered at the origin.

Theorem 9.2

If $P(x, y)$ is the preimage of a dilation centered at the origin with a scale factor r, then the image is $P'(rx, ry)$.

Example 3 · Dilations in the Coordinate Plane

COORDINATE GEOMETRY Triangle ABC has vertices $A(7, 10)$, $B(4, -6)$, and $C(-2, 3)$. Find the image of $\triangle ABC$ after a dilation centered at the origin with a scale factor of 2. Sketch the preimage and the image.

Preimage (x, y)	Image (2x, 2y)
$A(7, 10)$	$A'(14, 20)$
$B(4, -6)$	$B'(8, -12)$
$C(-2, 3)$	$C'(-4, 6)$

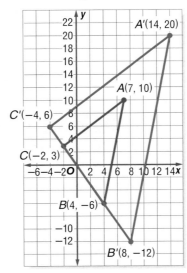

IDENTIFY THE SCALE FACTOR In Chapter 6, you found scale factors of similar figures. If you know the measurement of a figure and its dilated image, you can determine the scale factor.

Example 4 · Identify Scale Factor

Determine the scale factor for each dilation with center C. Then determine whether the dilation is an *enlargement*, *reduction*, or *congruence transformation*.

Study Tip

Look Back
To review **scale factor**, see Lesson 6-2.

a.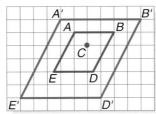

b.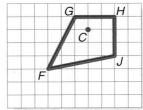

a.
$$\text{scale factor} = \frac{\text{image length}}{\text{preimage length}}$$

$$= \frac{6 \text{ units}}{3 \text{ units}} \quad \begin{array}{l} \leftarrow \text{image length} \\ \leftarrow \text{preimage length} \end{array}$$

$$= 2 \qquad \text{Simplify.}$$

Since the scale factor is greater than 1, the dilation is an enlargement.

b.
$$\text{scale factor} = \frac{\text{image length}}{\text{preimage length}}$$

$$= \frac{4 \text{ units}}{4 \text{ units}} \quad \begin{array}{l} \leftarrow \text{image length} \\ \leftarrow \text{preimage length} \end{array}$$

$$= 1 \qquad \text{Simplify.}$$

Since the scale factor is 1, the dilation is a congruence transformation.

Example 5 Scale Drawing

Multiple-Choice Test Item

Jacob wants to make a scale drawing of a painting in an art museum. The painting is 4 feet wide and 8 feet long. Jacob decides on a dilation reduction factor of $\frac{1}{6}$. What size paper will he need to make a complete sketch?

  Ⓐ $8\frac{1}{2}$ in. by 11 in.   Ⓑ 9 in. by 12 in.   Ⓒ 11 in. by 14 in.   Ⓓ 11 in. by 17 in.

Test-Taking Tip

Compare Measurements
Compare the measurements given in the problem to those in the answers. These answers are in inches, so convert feet to inches before using the scale factor. It may make calculations easier.

Read the Test Item

The painting's dimensions are given in feet, and the paper choices are in inches. You need to convert from feet to inches in the problem.

Solve the Test Item

Step 1 Convert feet to inches.

  4 feet = 4(12) or 48 inches

  8 feet = 8(12) or 96 inches

Step 2 Use the scale factor to find the image dimensions.

  $w = \frac{1}{6}(48)$ or 8 $\ell = \frac{1}{6}(96)$ or 16

Step 3 The dimensions of the image are 8 inches by 16 inches. Choice D is the only size paper on which the scale drawing will fit.

Check for Understanding

Concept Check   **1.** Find a counterexample to disprove the statement *All dilations are isometries.*

2. OPEN ENDED   Draw a figure on the coordinate plane. Then show a dilation of the figure that is a reduction and a dilation of the figure that is an enlargement.

3. FIND THE ERROR   Desiree and Trey are trying to describe the effect of a negative *r* value for a dilation of quadrilateral *WXYZ*.

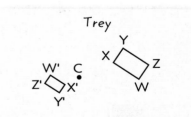

Who is correct? Explain your reasoning.

Guided Practice   **Draw the dilation image of each figure with center *C* and the given scale factor.**

  4. $r = 4$ **5.** $r = \frac{1}{5}$ **6.** $r = -2$

Find the measure of the dilation image $\overline{A'B'}$ or the preimage $\overline{AB}$ using the given scale factor.

7. $AB = 3, r = 4$

8. $A'B' = 8, r = -\dfrac{2}{5}$

9. $\overline{PQ}$ has endpoints $P(9, 0)$ and $Q(0, 6)$. Find the image of $\overline{PQ}$ after a dilation centered at the origin with a scale factor $r = \dfrac{1}{3}$. Sketch the preimage and the image.

10. Triangle KLM has vertices $K(5, 8)$, $L(-3, 4)$, and $M(-1, -6)$. Find the image of $\triangle KLM$ after a dilation centered at the origin with scale factor of 3. Sketch the preimage and the image.

Determine the scale factor for each dilation with center C. Determine whether the dilation is an *enlargement*, *reduction*, or *congruence transformation*.

11.

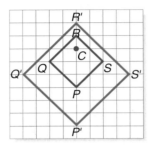

12.

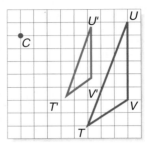

Standardized Test Practice

Ⓐ Ⓑ Ⓒ Ⓓ

13. Alexis made a scale drawing of the plan for her spring garden. It will be a rectangle measuring 18 feet by 12 feet. On the scaled version, it measures 8 inches on the longer sides. What is the measure of each of the shorter sides?

Ⓐ $6\dfrac{1}{2}$ in.　　　Ⓑ $5\dfrac{1}{2}$ in.　　　Ⓒ $5\dfrac{1}{3}$ in.　　　Ⓓ $4\dfrac{3}{5}$ in.

Practice and Apply

Homework Help

For Exercises	See Examples
14–19	1
20–25	2
26–29	3
32–37	4
30–31, 38–40	5

Extra Practice
See page 772.

Draw the dilation image of each figure with center C and the given scale factor.

14. $r = 3$

15. $r = 2$

16. $r = \dfrac{1}{2}$

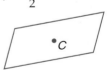

17. $r = \dfrac{2}{5}$

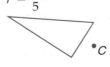

18. $r = -1$

19. $r = -\dfrac{1}{4}$

Find the measure of the dilation image $\overline{S'T'}$ or the preimage $\overline{ST}$ using the given scale factor.

20. $ST = 6, r = -1$

21. $ST = \dfrac{4}{5}, r = \dfrac{3}{4}$

22. $S'T' = 12, r = \dfrac{2}{3}$

23. $S'T' = \dfrac{12}{5}, r = -\dfrac{3}{5}$

24. $ST = 32, r = -\dfrac{5}{4}$

25. $ST = 2.25, r = 0.4$

COORDINATE GEOMETRY Find the image of each polygon, given the vertices, after a dilation centered at the origin with a scale factor of 2. Then graph a dilation centered at the origin with a scale factor of $\frac{1}{2}$.

26. $F(3, 4)$, $G(6, 10)$, $H(-3, 5)$

27. $X(1, -2)$, $Y(4, -3)$, $Z(6, -1)$

28. $P(1, 2)$, $Q(3, 3)$, $R(3, 5)$, $S(1, 4)$

29. $K(4, 2)$, $L(-4, 6)$, $M(-6, -8)$, $N(6, -10)$

Determine the scale factor for each dilation with center C. Determine whether the dilation is an *enlargement*, *reduction*, or *congruence transformation*.

30.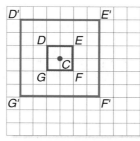

31.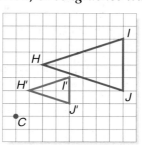

32.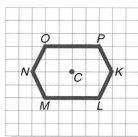

33.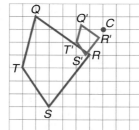

34.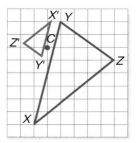

35.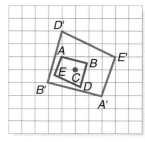

36. AIRPLANES Etay is building a model of the SR-71 Blackbird. If the wingspan of his model is 14 inches, what is the approximate scale factor of the model?

PHOTOCOPY For Exercises 37 and 38, refer to the following information.
A 10-inch by 14-inch rectangular design is being reduced on a photocopier by a factor of 75%.

37. What are the new dimensions of the design?

38. How has the area of the preimage changed?

For Exercises 39 and 40, use the following information.
A dilation on a rectangle has a scale factor of 4.

39. What is the effect of the dilation on the perimeter of the rectangle?

40. What is the effect of the dilation on the area of the rectangle?

41. **PROOF** Write a paragraph proof of Theorem 9.1.

42. Triangle ABC has vertices $A(12, 4)$, $B(4, 8)$, and $C(8, -8)$. After two successive dilations centered at the origin with the same scale factor, the final image has vertices $A''(3, 1)$, $B''(1, 2)$, and $C''(2, -2)$. Determine the scale factor r of each dilation from $\triangle ABC$ to $\triangle A''B''C''$.

43. Segment XY has endpoints $X(4, 2)$ and $Y(0, 5)$. After a dilation, the image has endpoints of $X'(7, 17)$ and $Y'(15, 11)$. Find the absolute value of the scale factor.

More About. . .

Airplanes

The SR-71 Blackbird is 107 feet 5 inches long with a wingspan of 55 feet 7 inches and can fly at speeds over 2200 miles per hour. It can fly nonstop from Los Angeles to Washington, D.C., in just over an hour, while a standard commercial jet takes about five hours to complete the trip.
Source: NASA

DIGITAL PHOTOGRAPHY For Exercises 44–46, use the following information.
Dinah is editing a digital photograph that is 640 pixels wide and 480 pixels high on her monitor.

44. If Dinah zooms the image on her monitor 150%, what are the dimensions of the image?

45. Suppose that Dinah wishes to use the photograph on a web page and wants the image to be 32 pixels wide. What scale factor should she use to reduce the image?

46. Dinah resizes the photograph so that it is 600 pixels high. What scale factor did she use?

47. **DESKTOP PUBLISHING** Grace is creating a template for her class newsletter. She has a photograph that is 10 centimeters by 12 centimeters, but the maximum space available for the photograph is 6 centimeters by 8 centimeters. She wants the photograph to be as large as possible on the page. When she uses a scanner to save the photograph, at what percent of the original photograph's size should she save the image file?

For Exercises 48–50, use quadrilateral *ABCD*.

48. Find the perimeter of quadrilateral *ABCD*.

49. Graph the image of quadrilateral *ABCD* after a dilation centered at the origin with scale factor −2.

50. Find the perimeter of quadrilateral *A'B'C'D'* and compare it to the perimeter of quadrilateral *ABCD*.

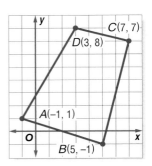

51. Triangle *TUV* has vertices *T*(6, −5), *U*(3, −8), and *V*(−1, −2). Find the coordinates of the final image of triangle *TUV* after a reflection in the *x*-axis, a translation with $(x, y) \rightarrow (x + 4, y − 1)$, and a dilation centered at the origin with a scale factor of $\frac{1}{3}$. Sketch the preimage and the image.

52. **CRITICAL THINKING** In order to perform a dilation not centered at the origin, you must first translate all of the points so the center is the origin, dilate the figure, and translate the points back. Consider a triangle with vertices *G*(3, 5), *H*(7, −4), and *I*(−1, 0). State the coordinates of the vertices of the image after a dilation centered at (3, 5) with a scale factor of 2.

53. WRITING IN MATH   Answer the question that was posed at the beginning of the lesson.

How do you use dilations when you use a computer?

Include the following in your answer:
- how a "cut and paste" in word processing may be an example of a dilation, and
- other examples of dilations when using a computer.

54. The figure shows two regular pentagons. Find the perimeter of the larger pentagon.

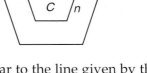

Ⓐ 5*n*   Ⓑ 10*n*

Ⓒ 15*n*   Ⓓ 60*n*

55. **ALGEBRA**   What is the slope of a line perpendicular to the line given by the equation $3x + 5y = 12$?

Ⓐ $\frac{5}{3}$ Ⓑ $\frac{3}{5}$ Ⓒ $-\frac{3}{5}$ Ⓓ $-\frac{5}{3}$

Maintain Your Skills

Mixed Review Determine whether a semi-regular tessellation can be created from each figure. Assume that each figure is regular and has a side length of 1 unit. *(Lesson 9-4)*

56. a triangle and a pentagon **57.** an octagon and a hexagon

58. a square and a triangle **59.** a hexagon and a dodecagon

COORDINATE GEOMETRY Draw the rotation image of each figure 90° in the given direction about the center point and label the vertices. *(Lesson 9-3)*

60. △*ABC* with *A*(7, −1), *B*(5, 0), and *C*(1, 6) counterclockwise about *P*(−1, 4)

61. ▱*DEFG* with *D*(−4, −2), *E*(−3, 3), *F*(3, 1), and *G*(2, −4) clockwise about *P*(−4, −6)

62. CONSTRUCTION The Vanamans are building an addition to their house. Ms. Vanaman is cutting an opening for a new window. If she measures to see that the opposite sides are the same length and that the diagonal measures are the same, can Ms. Vanaman be sure that the window opening is rectangular? Explain. *(Lesson 8-4)*

63. Given: ∠*J* ≅ ∠*L*

 B is the midpoint of $\overline{JL}$.

 Prove: △*JHB* ≅ △*LCB*

(Lesson 4-4)

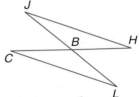

Getting Ready for the Next Lesson **PREREQUISITE SKILL** Find *m*∠*A* to the nearest tenth.

*(To review **finding angles using inverses of trigonometric ratios**, see Lesson 7-3.)*

64. **65.** **66.**

    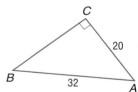

Practice Quiz 2 *Lessons 9-4 and 9-5*

Determine whether each pattern is a tessellation. If so, describe it as *uniform, regular, semi-regular,* or *not uniform.* *(Lesson 9-4)*

1. **2.**

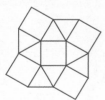 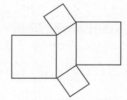

Draw the dilation image of each figure with center *C* and given scale factor. *(Lesson 9-5)*

3. $r = \dfrac{3}{4}$ **4.** $r = -2$

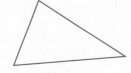

5. Triangle *ABC* has vertices *A*(10, 2), *B*(1, 6), and *C*(−4, 4). Find the image of △*ABC* after a dilation centered at the origin with scale factor of $-\dfrac{1}{2}$. Sketch the preimage and the image. *(Lesson 9-5)*

---

 www.geometryonline.com/self_check_quiz

9-6 Vectors

What You'll Learn

- Find magnitudes and directions of vectors.
- Perform translations with vectors.

How do vectors help a pilot plan a flight?

Commercial pilots must submit flight plans prior to departure. These flight plans take into account the speed and direction of the plane as well as the speed and direction of the wind.

Vocabulary

- vector
- magnitude
- direction
- standard position
- component form
- equal vectors
- parallel vectors
- resultant
- scalar
- scalar multiplication

MAGNITUDE AND DIRECTION   The speed and direction of a plane and the wind can be represented by vectors.

Key Concept   Vectors

- **Words**   A **vector** is a quantity that has both **magnitude**, or length, and **direction**, and is represented by a directed segment.

- **Symbols** $\vec{v}$

    $\overrightarrow{AB}$, where A is the initial point and B is the endpoint

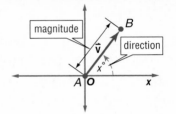

Study Tip

Common Misconception
The notation for a vector from C to D, $\overrightarrow{CD}$, is similar to the notation for a ray from C to D, $\overrightarrow{CD}$. Be sure to use the correct arrow above the letters when writing each.

A vector in **standard position** has its initial point at the origin. In the diagram, $\overrightarrow{CD}$ is in standard position and can be represented by the ordered pair $\langle 4, 2 \rangle$.

A vector can also be drawn anywhere in the coordinate plane. To write such a vector as an ordered pair, find the change in the x values and the change in y values, $\langle$change in x, change in $y\rangle$, from the tip to the tail of the directed segment. The ordered pair representation of a vector is called the **component form** of the vector.

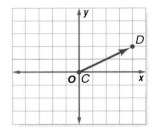

Example 1   Write Vectors in Component Form

Write the component form of $\overrightarrow{EF}$.

Find the change in x-values and the corresponding change in y-values.

$\overrightarrow{EF} = \langle x_2 - x_1, y_2 - y_1 \rangle$ Component form of vector

$\quad\;\; = \langle 7 - 1, 4 - 5 \rangle$ $x_1 = 1, y_1 = 5, x_2 = 7, y_2 = 4$

$\quad\;\; = \langle 6, -1 \rangle$ Simplify.

Because the magnitude and direction of a vector are not changed by translation, the vector $\langle 6, -1 \rangle$ represents the same vector as $\overrightarrow{EF}$.

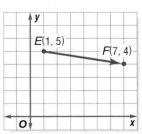

The Distance Formula can be used to find the magnitude of a vector. The symbol for the magnitude of $\overrightarrow{AB}$ is $|\overrightarrow{AB}|$. The *direction* of a vector is the measure of the angle that the vector forms with the positive x-axis or any other horizontal line. You can use the trigonometric ratios to find the direction of a vector.

Example 2 *Magnitude and Direction of a Vector*

Find the magnitude and direction of $\overrightarrow{PQ}$ for $P(3, 8)$ and $Q(-4, 2)$.

Find the magnitude.

$$|\overrightarrow{PQ}| = \sqrt{(x_2 - x_1)^2 + (y_2 - y_1)^2} \qquad \text{Distance Formula}$$

$$= \sqrt{(-4 - 3)^2 + (2 - 8)^2} \qquad x_1 = 3, y_1 = 8, x_2 = -4, y_2 = 2$$

$$= \sqrt{85} \qquad \text{Simplify.}$$

$$\approx 9.2 \qquad \text{Use a calculator.}$$

Graph $\overrightarrow{PQ}$ to determine how to find the direction. Draw a right triangle that has $\overrightarrow{PQ}$ as its hypotenuse and an acute angle at P.

$\tan P = \dfrac{y_2 - y_1}{x_2 - x_1} \qquad \tan = \dfrac{\text{length of opposite side}}{\text{length of adjacent side}}$

$= \dfrac{2 - 8}{-4 - 3} \qquad \text{Substitution}$

$= \dfrac{6}{7} \qquad \text{Simplify.}$

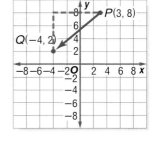

$m\angle P = \tan^{-1}\dfrac{6}{7}$

$\approx 40.6 \qquad \text{Use a calculator.}$

A vector in standard position that is equal to $\overrightarrow{PQ}$ lies in the third quadrant and forms an angle with the negative x-axis that has a measure equal to $m\angle P$. The x-axis is a straight angle with a measure that is 180. So, the direction of $\overrightarrow{PQ}$ is $m\angle P + 180$ or about $220.6°$.

Thus, $\overrightarrow{PQ}$ has a magnitude of about 9.2 units and a direction of about $220.6°$.

In Example 1, we stated that a vector in standard position with magnitude of 9.2 units and a direction of $220.6°$ was equal to $\overrightarrow{PQ}$. This leads to a definition of equal vectors.

Key Concept

Equal Vectors	Two vectors are equal if and only if they have the same magnitude and direction.	
Example	$\vec{v} = \vec{z}$	
Nonexample	$\vec{v} \neq \vec{u}, \vec{w} \neq \vec{y}$	
Parallel Vectors	Two vectors are parallel if and only if they have the same or opposite direction.	
Example	$\vec{v} \parallel \vec{w} \parallel \vec{y} \parallel \vec{z}$	
Nonexample	$\vec{v} \not\parallel \vec{x}$	

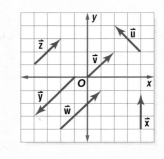

TRANSLATIONS WITH VECTORS   Vectors can be used to describe translations.

Example 3 *Translations with Vectors*

Graph the image of △*ABC* with vertices *A*(−3, −1), *B*(−1, −2), and *C*(−3, −3) under the translation $\vec{v}$ = ⟨4, 3⟩.

First, graph △*ABC*. Next, translate each vertex by $\vec{v}$, 4 units right and 3 units up. Connect the vertices to form △*A'B'C'*.

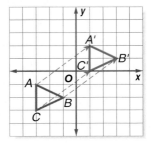

Vectors can be combined to perform a composition of translations by adding the vectors. To add vectors, add their corresponding components. The sum of two vectors is called the **resultant**.

Key Concept   *Vector Addition*

- **Words**   To add two vectors, add the corresponding components.

- **Symbols**   If $\vec{a}$ = ⟨a_1, a_2⟩ and = $\vec{b}$ ⟨b_1, b_2⟩, then $\vec{a}$ + $\vec{b}$ = ⟨a_1 + b_1, a_2 + b_2⟩, and $\vec{b}$ + $\vec{a}$ = ⟨b_1 + a_1, b_2 + a_2⟩.

- **Model**

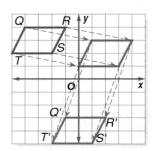

Example 4 *Add Vectors*

Graph the image of ▱*QRST* with vertices *Q*(−4, 4), *R*(−1, 4), *S*(−2, 2), and *T*(−5, 2) under the translation $\vec{m}$ = ⟨5, −1⟩ and $\vec{n}$ = ⟨−2, −6⟩.

Graph ▱*QRST*.

Method 1   Translate two times.
Translate ▱*QRST* by $\vec{m}$. Then translate the image of ▱*QRST* by $\vec{n}$.

Translate each vertex 5 units right and 1 unit down.

Then translate each vertex 2 units left and 6 units down.

Label the image ▱*Q'R'S'T'*.

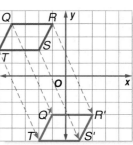

Method 2   Find the resultant, and then translate.
Add $\vec{m}$ and $\vec{n}$.

$\vec{m}$ + $\vec{n}$ = ⟨5 − 2, −1 − 6⟩

         = ⟨3, −7⟩

Translate each vertex 3 units right and 7 units down.

Notice that the vertices for the image are the same for either method.

Comparing Magnitude and Components of Vectors

Model and Analyze

• Draw $\vec{a}$ in standard position.
• Draw $\vec{b}$ in standard position with the same direction as $\vec{a}$, but with a magnitude twice the magnitude of $\vec{a}$.

1. Write $\vec{a}$ and $\vec{b}$ in component form.
2. What do you notice about the components of $\vec{a}$ and $\vec{b}$?
3. Draw $\vec{b}$ so that its magnitude is three times that of $\vec{a}$. How do the components of $\vec{a}$ and $\vec{b}$ compare?

Make a Conjecture

4. Describe the vector magnitude and direction of a vector $\langle x, y \rangle$ after the components are multiplied by n.

In the Geometry Activity, you found that a vector can be multiplied by a positive constant, called a **scalar**, that will change the magnitude of the vector, but not affect its direction. Multiplying a vector by a positive scalar is called **scalar multiplication**.

Key Concept
Scalar Multiplication

• **Words**   To multiply a vector by a scalar multiply each component by the scalar.

• **Symbols**   If $\vec{a} = \langle a_1, a_2 \rangle$ has a magnitude $|\vec{a}|$ and direction d, then $n\vec{a} = n\langle a_1, a_2 \rangle = \langle na_1, na_2 \rangle$, where n is a positive real number, the magnitude is $|n\vec{a}|$, and its direction is d.

• **Model**

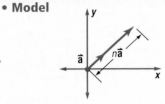

Example 5   Solve Problems Using Vectors

AVIATION   Refer to the application at the beginning of the lesson.

a. **Suppose a pilot begins a flight along a path due north flying at 250 miles per hour. If the wind is blowing due east at 20 miles per hour, what is the resultant velocity and direction of the plane?**

The initial path of the plane is due north, so a vector representing the path lies on the positive y-axis 250 units long. The wind is blowing due east, so a vector representing the wind will be parallel to the positive x-axis 20 units long. The resultant path can be represented by a vector from the initial point of the vector representing the plane to the terminal point of the vector representing the wind.

Use the Pythagorean Theorem.

$c^2 = a^2 + b^2$ Pythagorean Theorem

$c^2 = 250^2 + 20^2$ $a = 250, b = 20$

$c^2 = 62{,}900$ Simplify.

$c = \sqrt{62{,}900}$ Take the square root of each side.

$c \approx 250.8$

The resultant speed of the plane is about 250.8 miles per hour.

(continued on the next page)

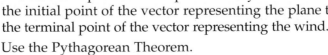

Use the tangent ratio to find the direction of the plane.

$$\tan \theta = \frac{20}{250} \qquad \text{side opposite} = 20, \text{ side adjacent} = 250$$

$$\theta = \tan^{-1} \frac{20}{250} \qquad \text{Solve for } \theta.$$

$$\theta \approx 4.6 \qquad \text{Use a calculator.}$$

The resultant direction of the plane is about 4.6° east of due north. Therefore, the resultant vector is 250.8 miles per hour at 4.6° east of due north.

b. If the wind velocity doubles, what is the resultant path and velocity of the plane?

Use scalar multiplication to find the magnitude of the vector for wind velocity.

$$n|\mathbf{a}| = 2|20| \qquad \text{Magnitude of } n\mathbf{a}; n = 2, |\mathbf{a}| = 20$$

$$= 2(20) \text{ or } 40 \quad \text{Simplify.}$$

Next, use the Pythagorean Theorem to find the magnitude of the resultant vector.

$$c^2 = a^2 + b^2 \qquad \text{Pythagorean Theorem}$$

$$c^2 = 250^2 + 40^2 \quad a = 250, b = 40$$

$$c^2 = 64{,}100 \qquad \text{Simplify.}$$

$$c = \sqrt{64{,}100} \qquad \text{Take the square root of each side.}$$

$$c \approx 253.2$$

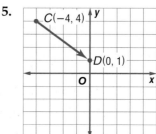

Then, use the tangent ratio to find the direction of the plane.

$$\tan \theta = \frac{40}{250} \qquad \text{side opposite} = 40, \text{ side adjacent} = 250$$

$$\theta = \tan^{-1} \frac{40}{250} \qquad \text{Solve for } \theta.$$

$$\theta \approx 9.1 \qquad \text{Use a calculator.}$$

If the wind velocity doubles, the plane flies along a path approximately 9.1° east of due north at about 253.2 miles per hour.

Check for Understanding

Concept Check

1. **OPEN ENDED** Draw a pair of vectors on a coordinate plane. Label each vector in component form and then find their sum.

2. **Compare and contrast** equal vectors and parallel vectors.

3. **Discuss** the similarity of using vectors to translate a figure and using an ordered pair.

Guided Practice **Write the component form of each vector.**

4.
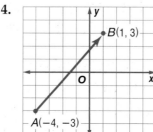

5.
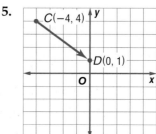

Find the magnitude and direction of $\overrightarrow{AB}$ for the given coordinates.

6. $A(2, 7)$, $B(-3, 3)$

7. $A(-6, 0)$, $B(-12, -4)$

8. What are the magnitude and direction of $\vec{v} = \langle 8, -15 \rangle$?

Graph the image of each figure under a translation by the given vector.

9. $\triangle JKL$ with vertices $J(2, -1)$, $K(-7, -2)$, $L(-2, 8)$; $\vec{t} = \langle -1, 9 \rangle$

10. trapezoid $PQRS$ with vertices $P(1, 2)$, $Q(7, 3)$, $R(15, 1)$, $S(3, -1)$; $\vec{u} = \langle 3, -3 \rangle$

11. Graph the image of $\square WXYZ$ with vertices $W(6, -6)$, $X(3, -8)$, $Y(-4, -4)$, and $Z(-1, -2)$ under the translation by $\vec{e} = \langle -1, 6 \rangle$ and $\vec{f} = \langle 8, -5 \rangle$.

Application **Find the magnitude and direction of each resultant for the given vectors.**

12. $\vec{g} = \langle 4, 0 \rangle$, $\vec{h} = \langle 0, 6 \rangle$

13. $\vec{t} = \langle 0, -9 \rangle$, $\vec{u} = \langle 12, -9 \rangle$

14. BOATING Raphael sails his boat due east at a rate of 10 knots. If there is a current of 3 knots moving 30° south of east, what is the resultant speed and direction of the boat?

Practice and Apply

Homework Help

For Exercises	See Examples
15–20	1
21–36	2
37–42	3
43–46	4
47-57	5

Extra Practice
See page 773.

Write the component form of each vector.

15.

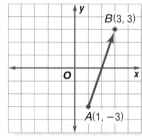

16.
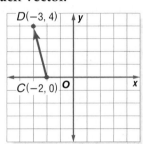

17.
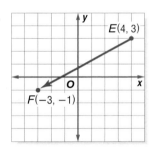

18.
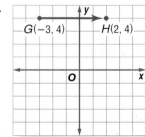

19.
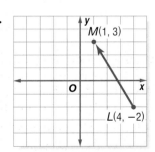

20.
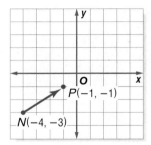

Find the magnitude and direction of $\overrightarrow{CD}$ for the given coordinates. Round to the nearest tenth.

21. $C(4, 2)$, $D(9, 2)$

22. $C(-2, 1)$, $D(2, 5)$

23. $C(-5, 10)$, $D(-3, 6)$

24. $C(0, -7)$, $D(-2, -4)$

25. $C(-8, -7)$, $D(6, 0)$

26. $C(10, -3)$, $D(-2, -2)$

27. What are the magnitude and direction of $\vec{t} = \langle 7, 24 \rangle$?

28. What are the magnitude and direction of $\vec{u} = \langle -12, 15 \rangle$?

29. What are the magnitude and direction of $\vec{v} = \langle -25, -20 \rangle$?

30. What are the magnitude and direction of $\vec{w} = \langle 36, -15 \rangle$?

Find the magnitude and direction of $\overrightarrow{MN}$ for the given coordinates. Round to the nearest tenth.

31. $M(-3, 3)$, $N(-9, 9)$
32. $M(8, 1)$, $N(2, 5)$
33. $M(0, 2)$, $N(-12, -2)$
34. $M(-1, 7)$, $N(6, -8)$
35. $M(-1, 10)$, $N(1, -12)$
36. $M(-4, 0)$, $N(-6, -4)$

Graph the image of each figure under a translation by the given vector.

37. $\triangle ABC$ with vertices $A(3, 6)$, $B(3, -7)$, $C(-6, 1)$; $\vec{a} = \langle 0, -6 \rangle$

38. $\triangle DEF$ with vertices $D(-12, 6)$, $E(7, 6)$, $F(7, -3)$; $\vec{b} = \langle -3, -9 \rangle$

39. square $GHIJ$ with vertices $G(-1, 0)$, $H(-6, -3)$, $I(-9, 2)$, $J(-4, 5)$; $\vec{c} = \langle 3, -8 \rangle$

40. quadrilateral $KLMN$ with vertices $K(0, 8)$, $L(4, 6)$, $M(3, -3)$, $N(-4, 8)$; $\vec{x} = \langle -10, 2 \rangle$

41. pentagon $OPQRS$ with vertices $O(5, 3)$, $P(5, -3)$, $Q(0, -4)$, $R(-5, 0)$, $S(0, 4)$; $\vec{y} = \langle -5, 11 \rangle$

42. hexagon $TUVWXY$ with vertices $T(4, -2)$, $U(3, 3)$, $V(6, 4)$, $W(9, 3)$, $X(8, -2)$, $Y(6, -5)$; $\vec{z} = \langle -18, 12 \rangle$

Graph the image of each figure under a translation by the given vectors.

43. $\square ABCD$ with vertices $A(-1, -6)$, $B(4, -8)$, $C(-3, -11)$, $D(-8, -9)$; $\vec{p} = \langle 11, 6 \rangle$, $\vec{q} = \langle -9, -3 \rangle$

44. $\triangle XYZ$ with vertices $X(3, -5)$, $Y(9, 4)$, $Z(12, -2)$; $\vec{p} = \langle 2, 2 \rangle$, $\vec{q} = \langle -4, -7 \rangle$

45. quadrilateral $EFGH$ with vertices $E(-7, -2)$, $F(-3, 8)$, $G(4, 15)$, $H(5, -1)$; $\vec{p} = \langle -6, 10 \rangle$, $\vec{q} = \langle 1, -8 \rangle$

46. pentagon $STUVW$ with vertices $S(1, 4)$, $T(3, 8)$, $U(6, 8)$, $V(6, 6)$, $W(4, 4)$; $\vec{p} = \langle -4, 5 \rangle$, $\vec{q} = \langle 12, 11 \rangle$

Find the magnitude and direction of each resultant for the given vectors.

47. $\vec{a} = \langle 5, 0 \rangle$, $\vec{b} = \langle 0, 12 \rangle$
48. $\vec{c} = \langle 0, -8 \rangle$, $\vec{d} = \langle -8, 0 \rangle$
49. $\vec{e} = \langle -4, 0 \rangle$, $\vec{f} = \langle 7, -4 \rangle$
50. $\vec{u} = \langle 12, 6 \rangle$, $\vec{v} = \langle 0, 6 \rangle$
51. $\vec{w} = \langle 5, 6 \rangle$, $\vec{x} = \langle -1, -4 \rangle$
52. $\vec{y} = \langle 9, -10 \rangle$, $\vec{z} = \langle -10, -2 \rangle$

More About . . .

Rivers •·················

The Congo River is one of the fastest rivers in the world. It has no dry season, because it has tributaries both north and south of the Equator. The river flows so quickly that it doesn't form a delta where it ends in the Atlantic like most rivers do when they enter an ocean.

Source: Compton's Encyclopedia

53. **SHIPPING** A freighter has to go around an oil spill in the Pacific Ocean. The captain sails due east for 35 miles. Then he turns the ship and heads due south for 28 miles. What is the distance and direction of the ship from its original point of course correction?

54. **RIVERS** Suppose a section of the New River in West Virginia has a current of 2 miles per hour. If a swimmer can swim at a rate of 4.5 miles per hour, how does the current in the New River affect the speed and direction of the swimmer as she tries to swim directly across the river?

AVIATION For Exercises 55–57, use the following information.
A jet is flying northwest, and its velocity is represented by $\langle -450, 450 \rangle$ in miles per hour. The wind is from the west, and its velocity is represented by $\langle 100, 0 \rangle$ in miles per hour.

55. Find the resultant vector for the jet in component form.

56. Find the magnitude of the resultant.

57. Find the direction of the resultant.

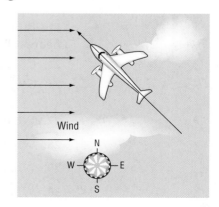

58. CRITICAL THINKING If two vectors have opposite directions but the same magnitude, the resultant is $\langle 0, 0 \rangle$ when they are added. Find three vectors of equal magnitude, each with its tail at the origin, the sum of which is $\langle 0, 0 \rangle$.

59.  Answer the question that was posed at the beginning of the lesson.

How do vectors help a pilot plan a flight?

Include the following in your answer:
- an explanation of how a wind from the west affects the overall velocity of a plane traveling east, and
- an explanation as to why planes traveling from Hawaii to the continental U.S. take less time than planes traveling from the continental U.S. to Hawaii.

60. If $\vec{q} = \langle 5, 10 \rangle$ and $\vec{r} = \langle 3, 5 \rangle$, find the magnitude for the sum of these two vectors.

   Ⓐ 23   Ⓑ 17   Ⓒ 7   Ⓓ $\sqrt{29}$

61. ALGEBRA If $5^b = 125$, then find $4^b \times 3$.

   Ⓐ 48   Ⓑ 64   Ⓒ 144   Ⓓ 192

Maintain Your Skills

Mixed Review

Find the measure of the dilation image $\overline{A'B'}$ or the preimage of $\overline{AB}$ using the given scale factor. *(Lesson 9-5)*

62. $AB = 8, r = 2$ **63.** $AB = 12, r = \frac{1}{2}$

64. $A'B' = 15, r = 3$ **65.** $A'B' = 12, r = \frac{1}{4}$

Determine whether each pattern is a tessellation. If so, describe it as *uniform*, *not uniform*, *regular*, or *semi-regular*. *(Lesson 9-4)*

66.

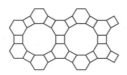

67.

ALGEBRA Use rhombus $WXYZ$ with $m\angle XYZ = 5m\angle WZY$ and $YZ = 12$. *(Lesson 8-5)*

68. Find $m\angle XYZ$.   **69.** Find WX.

70. Find $m\angle XZY$.   **71.** Find $m\angle WXY$.

72. Each side of a rhombus is 30 centimeters long. One diagonal makes a 25° angle with a side. What is the length of each diagonal to the nearest tenth? *(Lesson 7-4)*

Getting Ready for the Next Lesson

PREREQUISITE SKILL Perform the indicated operation.
(To review operations with matrices, see pages 752–753.)

73. $\begin{bmatrix} -5 & 5 \\ -3 & -2 \end{bmatrix} + \begin{bmatrix} 1 & -8 \\ -7 & 6 \end{bmatrix}$ **74.** $\begin{bmatrix} -2 & 2 & -2 \\ -7 & -2 & -5 \end{bmatrix} + \begin{bmatrix} -8 & -8 & -8 \\ 1 & 1 & 1 \end{bmatrix}$

75. $3\begin{bmatrix} -9 & -5 & -1 \\ 9 & 1 & 5 \end{bmatrix}$ **76.** $\frac{1}{2}\begin{bmatrix} -4 & -5 & 0 & 2 \\ 4 & 4 & 6 & 0 \end{bmatrix}$

77. $\begin{bmatrix} -4 & -4 \\ 2 & 2 \end{bmatrix} + 2\begin{bmatrix} 8 & 4 \\ -3 & -7 \end{bmatrix}$ **78.** $\begin{bmatrix} 1 & -1 \\ -1 & 1 \end{bmatrix} \cdot \begin{bmatrix} 2 & -3 \\ -2 & -4 \end{bmatrix}$

Transformations with Matrices

What You'll Learn

- Use matrices to determine the coordinates of translations and dilations.
- Use matrices to determine the coordinates of reflections and rotations.

Vocabulary

- column matrix
- vertex matrix
- translation matrix
- reflection matrix
- rotation matrix

How can matrices be used to make movies?

Many movie directors use computers to create special effects that cannot be easily created in real life. A special effect is often a simple image that is enhanced using transformations. Complex images can be broken down into simple polygons, which are moved and resized using matrices to define new vertices for the polygons.

TRANSLATIONS AND DILATIONS In Lesson 9-6, you learned that a vector can be represented by the ordered pair $\langle x, y \rangle$. A vector can also be represented by a **column matrix** $\begin{bmatrix} x \\ y \end{bmatrix}$. Likewise, polygons can be represented by placing all of the column matrices of the coordinates of the vertices into one matrix, called a **vertex matrix**.

Triangle PQR with vertices $P(3, 5)$, $Q(1, -2)$, and $R(-4, 4)$ can be represented by the vertex matrix at the right.

$$\triangle PQR = \begin{matrix} P & Q & R \\ \begin{bmatrix} 3 & 1 & -4 \\ 5 & -2 & 4 \end{bmatrix} \end{matrix} \begin{matrix} \leftarrow \text{x-coordinates} \\ \leftarrow \text{y-coordinates} \end{matrix}$$

Like vectors, matrices can be used to perform translations. You can use matrix addition and a **translation matrix** to find the coordinates of a translated figure.

Study Tip

Translation Matrix
A translation matrix contains the same number of rows and columns as the vertex matrix of a figure.

Example 1 Translate a Figure

Use a matrix to find the coordinates of the vertices of the image of $\square ABCD$ with $A(3, 2)$, $B(1, -3)$, $C(-3, -1)$, and $D(-1, 4)$ under the translation $(x, y) \rightarrow (x + 5, y - 3)$.

Write the vertex matrix for $\square ABCD$. $\begin{bmatrix} 3 & 1 & -3 & -1 \\ 2 & -3 & -1 & 4 \end{bmatrix}$

To translate the figure 5 units to the right, add 5 to each x-coordinate. To translate the figure 3 units down, add -3 to each y-coordinate. This can be done by adding the translation matrix $\begin{bmatrix} 5 & 5 & 5 & 5 \\ -3 & -3 & -3 & -3 \end{bmatrix}$ to the vertex matrix of $\square ABCD$.

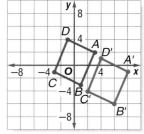

$$\begin{matrix} \text{Vertex Matrix} \\ \text{of } \square ABCD \end{matrix} \qquad \begin{matrix} \text{Translation} \\ \text{Matrix} \end{matrix} \qquad \begin{matrix} \text{Vertex Matrix} \\ \text{of } \square A'B'C'D' \end{matrix}$$

$$\begin{bmatrix} 3 & 1 & -3 & -1 \\ 2 & -3 & -1 & 4 \end{bmatrix} + \begin{bmatrix} 5 & 5 & 5 & 5 \\ -3 & -3 & -3 & -3 \end{bmatrix} = \begin{bmatrix} 8 & 6 & 2 & 4 \\ -1 & -6 & -4 & 1 \end{bmatrix}$$

The coordinates of $\square A'B'C'D'$ are $A'(8, -1)$, $B'(6, -6)$, $C'(2, -4)$, and $D'(4, 1)$.

Scalars can be used with matrices to perform dilations.

Example 2 Dilate a Figure

Triangle *FGH* has vertices *F*(−3, 1), *G*(−1, 2), and *H*(1, −1). Use scalar multiplication to dilate △*FGH* centered at the origin so that its perimeter is 3 times the original perimeter.

If the perimeter of a figure is 3 times the original perimeter, then the lengths of the sides of the figure will be 3 times the measures of the original lengths. Multiply the vertex matrix by a scale factor of 3.

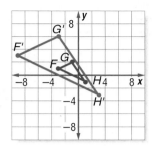

$$3 \begin{bmatrix} -3 & -1 & 1 \\ 1 & 2 & -1 \end{bmatrix} = \begin{bmatrix} -9 & -3 & 3 \\ 3 & 6 & -3 \end{bmatrix}$$

The coordinates of the vertices of △*F'G'H'* are *F'*(−9, 3), *G'*(−3, 6), and *H'*(3, −3).

REFLECTIONS AND ROTATIONS

A **reflection matrix** can be used to multiply the vertex matrix of a figure to find the coordinates of the image. The matrices used for four common reflections are shown below.

Study Tip

Reflection Matrices

The matrices used for reflections and rotations always have two rows and two columns no matter what the number of columns in the vertex matrix.

Concept Summary				**Reflection Matrices**
For a reflection in the:	*x*-axis	*y*-axis	origin	line *y* = *x*
Multiply the vertex matrix on the left by:	$\begin{bmatrix} 1 & 0 \\ 0 & -1 \end{bmatrix}$	$\begin{bmatrix} -1 & 0 \\ 0 & 1 \end{bmatrix}$	$\begin{bmatrix} -1 & 0 \\ 0 & -1 \end{bmatrix}$	$\begin{bmatrix} 0 & 1 \\ 1 & 0 \end{bmatrix}$
The product of the reflection matrix and the vertex matrix $\begin{bmatrix} a & b \\ c & d \end{bmatrix}$ would be:	$\begin{bmatrix} a & b \\ -c & -d \end{bmatrix}$	$\begin{bmatrix} -a & -b \\ c & d \end{bmatrix}$	$\begin{bmatrix} -a & -b \\ -c & -d \end{bmatrix}$	$\begin{bmatrix} c & d \\ a & b \end{bmatrix}$

Example 3 Reflections

Use a matrix to find the coordinates of the vertices of the image of $\overline{TU}$ with *T*(−4, −4) and *U*(3, 2) after a reflection in the *x*-axis.

Write the ordered pairs as a vertex matrix. Then multiply the vertex matrix by the reflection matrix for the *x*-axis.

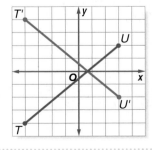

$$\begin{bmatrix} 1 & 0 \\ 0 & -1 \end{bmatrix} \cdot \begin{bmatrix} -4 & 3 \\ -4 & 2 \end{bmatrix} = \begin{bmatrix} -4 & 3 \\ 4 & -2 \end{bmatrix}$$

The coordinates of the vertices of $\overline{T'U'}$ are *T'*(−4, 4) and *U'*(3, −2). Graph $\overline{TU}$ and $\overline{T'U'}$.

Matrices can also be used to determine the vertices of a figure's image by rotation using a **rotation matrix**. Commonly used rotation matrices are summarized on the next page.

Study Tip

Graphing
Calculator
It may be helpful to store
the reflection and rotation
matrices in your calculator
to avoid reentering them.

Concept Summary — Rotation Matrices

For a counterclockwise rotation about the origin of:	90°	180°	270°
Multiply the vertex matrix on the left by:	$\begin{bmatrix} 0 & -1 \\ 1 & 0 \end{bmatrix}$	$\begin{bmatrix} -1 & 0 \\ 0 & -1 \end{bmatrix}$	$\begin{bmatrix} 0 & 1 \\ -1 & 0 \end{bmatrix}$
The product of the rotation matrix and the vertext matrix $\begin{bmatrix} a & b \\ c & d \end{bmatrix}$ would be:	$\begin{bmatrix} -c & -d \\ a & b \end{bmatrix}$	$\begin{bmatrix} -a & -b \\ -c & -d \end{bmatrix}$	$\begin{bmatrix} c & d \\ -a & -b \end{bmatrix}$

Example 4 Use Rotations

COMPUTERS A software designer is creating a screen saver by transforming the figure at the right.

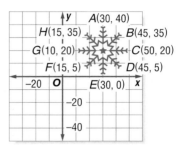

a. **Write a vertex matrix for the figure.**

One possible vertex matrix is
$$\begin{bmatrix} 30 & 45 & 50 & 45 & 30 & 15 & 10 & 15 \\ 40 & 35 & 20 & 5 & 0 & 5 & 20 & 35 \end{bmatrix}.$$

b. **Use a matrix to find the coordinates of the figure under a 90° counterclockwise rotation about the origin.**

Enter the rotation matrix in your calculator as matrix A and enter the vertex matrix as matrix B. Then multiply.

KEYSTROKES: 2nd [MATRX] 1 2nd [MATRX] 2 ENTER

$$AB = \begin{bmatrix} -40 & -35 & -20 & -5 & 0 & -5 & -20 & -35 \\ 30 & 45 & 50 & 45 & 30 & 15 & 10 & 15 \end{bmatrix}$$

The vertices of the figure are $A'(-40, 30)$, $B'(-35, 45)$, $C'(-20, 50)$, $D'(-5, 45)$, $E'(0, 30)$, $F'(-5, 15)$, $G'(-20, 10)$, and $H'(-35, 15)$.

c. **What are the coordinates of the image if the pattern is enlarged to twice its size before a 180° rotation about the origin?**

Enter the rotation matrix as matrix C. A dilation is performed before the rotation. Multiply the matrices by 2.

KEYSTROKES: 2 2nd [MATRX] 3 2nd [MATRX] 2 ENTER

$$2AC = \begin{bmatrix} -60 & -90 & -100 & -90 & -60 & -30 & -20 & -30 \\ -80 & -70 & -40 & -10 & 0 & -10 & -40 & -70 \end{bmatrix}$$

The vertices of the figure are $A'(-60, -80)$, $B'(-90, -70)$, $C'(-100, -40)$, $D'(-90, -10)$, $E'(-60, 0)$, $F'(-30, -10)$, $G'(-20, -40)$, and $H'(-30, -70)$.

Check for Understanding

Concept Check

1. **Write** the reflection matrix for $\triangle ABC$ and its image $\triangle A'B'C'$ at the right.

2. **Discuss** the similarities of using coordinates, vectors, and matrices to translate a figure.

3. **OPEN ENDED** Graph any $\square PQRS$ on a coordinate grid. Then write a translation matrix that moves $\square PQRS$ down and left on the grid.

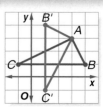

Guided Practice **Use a matrix to find the coordinates of the vertices of the image of each figure under the given translation.**

4. $\triangle ABC$ with $A(5, 4)$, $B(3, -1)$, and $C(0, 2)$; $(x, y) \rightarrow (x - 2, y - 1)$

5. $\square DEFG$ with $D(-1, 3)$, $E(5, 3)$, $F(3, 0)$, and $G(-3, 0)$; $(x, y) \rightarrow (x, y + 6)$

Use scalar multiplication to find the coordinates of the vertices of each figure for a dilation centered at the origin with the given scale factor.

6. $\triangle XYZ$ with $X(3, 4)$, $Y(6, 10)$, and $Z(-3, 5)$; $r = 2$

7. $\square ABCD$ with $A(1, 2)$, $B(3, 3)$, $C(3, 5)$, and $D(1, 4)$; $r = -\dfrac{1}{4}$

Use a matrix to find the coordinates of the endpoints or vertices of the image of each figure under the given reflection.

8. $\overline{EF}$ with $E(-2, 4)$ and $F(5, 1)$; x-axis

9. quadrilateral $HIJK$ with $H(-5, 4)$, $I(-1, -1)$, $J(-3, -6)$, and $K(-7, -3)$; y-axis

Use a matrix to find the coordinates of the endpoints or vertices of the image of each figure under the given rotation.

10. $\overline{LM}$ with $L(-2, 1)$ and $M(3, 5)$; 90° counterclockwise

11. $\triangle PQR$ with $P(6, 3)$, $Q(6, 7)$, and $R(2, 7)$; 270° counterclockwise

12. Use matrices to find the coordinates of the image of quadrilateral $STUV$ with $S(-4, 1)$, $T(-2, 2)$, $U(0, 1)$, and $V(-2, -2)$ after a dilation by a scale factor of 2 and a rotation 90° counterclockwise about the origin.

Application **LANDSCAPING   For Exercises 13 and 14, use the following information.**
A garden design is drawn on a coordinate grid. The original plan shows a rose bed with vertices at $(3, -1)$, $(7, -3)$, $(5, -7)$, and $(1, -5)$. Changes to the plan require that the rose bed's perimeter be half the original perimeter with respect to the origin, while the shape remains the same.

13. What are the new coordinates for the vertices of the rose bed?

14. If the center of the rose bed was originally located at $(4, -4)$, what will be the coordinates of the center after the changes have been made?

Practice and Apply

Homework Help

For Exercises	See Examples
15–18, 28, 31, 40, 41	1
19–22, 27, 32, 39, 42	2
23–26, 30, 33, 39, 41 29, 34,	3
35–38, 40, 42	4

Extra Practice
See page 773.

Use a matrix to find the coordinates of the endpoints or vertices of the image of each figure under the given translation.

15. $\overline{EF}$ with $E(-4, 1)$, and $F(-1, 3)$; $(x, y) \rightarrow (x - 2, y + 5)$

16. $\triangle JKL$ with $J(-3, 5)$, $K(4, 8)$, and $L(7, 5)$; $(x, y) \rightarrow (x - 3, y - 4)$

17. $\square MNOP$ with $M(-2, 7)$, $N(2, 9)$, $O(2, 7)$, and $P(-2, 5)$; $(x, y) \rightarrow (x + 3, y - 6)$

18. trapezoid $RSTU$ with $R(2, 3)$, $S(6, 2)$, $T(6, -1)$, and $U(-2, 1)$; $(x, y) \rightarrow (x - 6, y - 2)$

Use scalar multiplication to find the coordinates of the vertices of each figure for a dilation centered at the origin with the given scale factor.

19. $\triangle ABC$ with $A(6, 5)$, $B(4, 5)$, and $C(3, 7)$; $r = 2$

20. $\triangle DEF$ with $D(-1, 4)$, $E(0, 1)$, and $F(2, 3)$; $r = -\dfrac{1}{3}$

21. quadrilateral $GHIJ$ with $G(4, 2)$, $H(-4, 6)$, $I(-6, -8)$, and $J(6, -10)$; $r = -\dfrac{1}{2}$

22. pentagon $KLMNO$ with $K(1, -2)$, $L(3, -1)$, $M(6, -1)$, $N(4, -3)$, and $O(3, -3)$; $r = 4$

Use a matrix to find the coordinates of the endpoints or vertices of the image of each figure under the given reflection.

23. $\overline{XY}$ with $X(2, 2)$, and $Y(4, -1)$; y-axis

24. $\triangle ABC$ with $A(5, -3)$, $B(0, -5)$, and $C(-1, -3)$; $y = x$

25. quadrilateral $DEFG$ with $D(-4, 5)$, $E(2, 6)$, $F(3, 1)$, and $G(-3, -4)$; x-axis

26. quadrilateral $HIJK$ with $H(9, -1)$, $I(2, -6)$, $J(-4, -3)$, and $K(-2, 4)$; y-axis

Find the coordinates of the vertices of the image of $\triangle VWX$ under the given transformation.

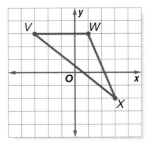

27. dilation by scale factor $\dfrac{2}{3}$

28. translation $(x, y) \rightarrow (x - 4, y - 1)$

29. rotation 90° counterclockwise about the origin

30. reflection in the line $y = x$

Find the coordinates of the vertices of the image of polygon $PQRST$ under the given transformation.

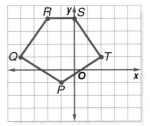

31. translation $(x, y) \rightarrow (x + 3, y - 2)$

32. dilation by scale factor -3

33. reflection in the y-axis

34. rotation 180° counterclockwise about the origin

Use a matrix to find the coordinates of the endpoints or vertices of the image of each figure under the given rotation.

35. $\overline{MN}$ with $M(12, 1)$ and $N(-3, 10)$; 90° counterclockwise

36. $\triangle PQR$ with $P(5, 1)$, $Q(1, 2)$, and $R(1, -4)$; 180° counterclockwise

37. $\square STUV$ with $S(2, 1)$, $T(6, 1)$, $U(5, -3)$, and $V(1, -3)$; 90° counterclockwise

38. pentagon $ABCDE$ with $A(-1, 1)$, $B(6, 0)$, $C(4, -8)$, $D(-4, -10)$, and $E(-5, -3)$; 270° counterclockwise

Find the coordinates of the image under the stated transformations.

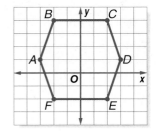

39. dilation by scale factor $\dfrac{1}{3}$, then a reflection in the x-axis

40. translation $(x, y) \rightarrow (x - 5, y + 2)$ then a rotation 90° counterclockwise about the origin

41. reflection in the line $y = x$ then the translation $(x, y) \rightarrow (x + 1, y + 4)$

42. rotation 180° counterclockwise about the origin, then a dilation by scale factor -2

PALEONTOLOGY For Exercises 43 and 44, use the following information.
Paleontologists sometimes discover sets of fossilized dinosaur footprints like those shown at the right.

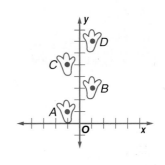

43. Describe the transformation combination shown.

44. Write two matrix operations that could be used to find the coordinates of point C.

CONSTRUCTION   For Exercises 45 and 46, use the following information.
House builders often use one set of blueprints for many
projects. By changing the orientation of a floor plan,
a builder can make several different looking houses.

45. Write a transformation matrix that could be used
    to create a floor plan with the garage on the left.

46. If the current plan is of a house that faces south,
    write a transformation matrix that could be used
    to create a floor plan for a house that faces east.

47. **CRITICAL THINKING**   Write a matrix to represent a
    reflection in the line $y = -x$.

48. WRITING IN MATH   Answer the question that was posed at the beginning of
    the lesson.

    How can matrices be used to make movies?

    Include the following in your answer:
    - an explanation of how transformations are used in movie production, and
    - an everyday example of transformation that can be modeled using matrices.

49. **SHORT RESPONSE**   Quadrilateral $ABCD$ is
    rotated 90° clockwise about the origin. Write
    the transformation matrix.

50. **ALGEBRA**   A video store stocks 2500 different
    movie titles. If 26% of the titles are action movies
    and 14% are comedies, how many are neither action
    movies nor comedies?
    Ⓐ 1000
    Ⓒ 1850
    Ⓑ 1500
    Ⓓ 2150

Maintain Your Skills

Mixed Review   **Graph the image of each figure under a translation by the given vector.**   *(Lesson 9-6)*

51. $\triangle ABC$ with $A(-6, 1)$, $B(4, 8)$, and $C(1, -4)$; $\vec{v} = \langle -1, -5 \rangle$

52. quadrilateral $DEFG$ with $D(3, -3)$, $E(1, 2)$, $F(8, -1)$, and $G(4, -6)$; $\vec{w} = \langle -7, 8 \rangle$

53. Determine the scale factor used for the dilation
    at the right, centered at C. Determine whether the
    dilation is an *enlargement, reduction,* or *congruence
    transformation.*   *(Lesson 9-5)*

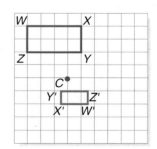

**Find the measures of an exterior angle and an interior
angle given the number of sides of a regular polygon.**
(Lesson 8-1)

54. 5
55. 6
56. 8
57. 10

58. **FORESTRY**   To estimate the height of a tree,
    Lara sights the top of the tree in a mirror
    that is 34.5 meters from the tree. The mirror
    is on the ground and faces upward. Lara is
    standing 0.75 meter from the mirror, and
    the distance from her eyes to the ground is
    1.75 meters. How tall is the tree?   *(Lesson 6-3)*

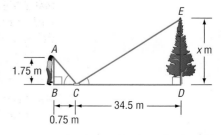

Vocabulary and Concept Check

angle of rotation (p. 476)	line of reflection (p. 463)	scalar (p. 501)
center of rotation (p. 476)	line of symmetry (p. 466)	scalar multiplication (p. 501)
column matrix (p. 506)	magnitude (p. 498)	semi-regular tessellation (p. 484)
component form (p. 498)	parallel vectors (p. 499)	similarity transformation (p. 491)
composition (p. 471)	point of symmetry (p. 466)	standard position (p. 498)
dilation (p. 490)	reflection (p. 463)	tessellation (p. 483)
direct isometry (p. 481)	reflection matrix (p. 507)	transformation (p. 462)
direction (p. 498)	regular tessellation (p. 484)	translation (p. 470)
equal vectors (p. 499)	resultant (p. 500)	translation matrix (p. 506)
glide reflection (p. 474)	rotation (p. 476)	uniform (p. 484)
indirect isometry (p. 481)	rotation matrix (p. 507)	vector (p. 498)
invariant points (p. 481)	rotational symmetry (p. 478)	vertex matrix (p. 506)
isometry (p. 463)		

A complete list of postulates and theorems can be found on pages R1–R8.

Exercises State whether each sentence is *true* or *false*. If false, replace the underlined word or phrase to make a true sentence.

1. A dilation can change the distance between each point on the figure and the given *line of symmetry*.

2. A tessellation is *uniform* if the same combination of shapes and angles is present at every vertex.

3. Two vectors can be added easily if you know their *magnitude*.

4. Scalar multiplication affects only the *direction* of a vector.

5. In a rotation, the figure is turned about the *point of symmetry*.

6. A *reflection* is a transformation determined by a figure and a line.

7. A *congruence transformation* is the amount by which a figure is enlarged or reduced in a dilation.

8. A *scalar multiple* is the sum of two other vectors.

Lesson-by-Lesson Review

Reflections

See pages
463–469.

Concept Summary

- The line of symmetry in a figure is a line where the figure could be folded in half so that the two halves match exactly.

Example **Copy the figure. Draw the image of the figure under a reflection in line ℓ.**

The green triangle is the reflected image of the blue triangle.

 www.geometryonline.com/vocabulary_review

Exercises Graph each figure and its image under the given reflection.
See Example 2 on page 464.

9. triangle *ABC* with *A*(2, 1), *B*(5, 1), and *C*(2, 3) in the *x*-axis

10. parallelogram *WXYZ* with *W*(−4, 5), *X*(−1, 5), *Y*(−3, 3), and *Z*(−6, 3) in the line *y* = *x*

11. rectangle *EFGH* with *E*(−4, −2), *F*(0, −2), *G*(0, −4), and *H*(−4, −4) in the line *x* = 1

9-2 Translations

See pages 470–475.

Concept Summary

- A translation moves all points of a figure the same distance in the same direction.
- A translation can be represented as a composition of reflections.

Example **COORDINATE GEOMETRY** Triangle *ABC* has vertices *A*(2, 1), *B*(4, −2), and *C*(1, −4). Graph △*ABC* and its image for the translation $(x, y) \to (x - 5, y + 3)$.

(x, y)	$(x - 5, y + 3)$
(2, 1)	(−3, 4)
(4, −2)	(−1, 1)
(1, −4)	(−4, −1)

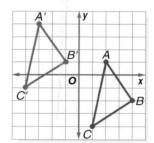

This translation moved every point of the preimage 5 units left and 3 units up.

Exercises Graph each figure and the image under the given translation.
See Example 1 on page 470.

12. quadrilateral *EFGH* with *E*(2, 2), *F*(6, 2), *G*(4, −2), *H*(1, −1) under the translation $(x, y) \to (x - 4, y - 4)$

13. $\overline{ST}$ with endpoints *S*(−3, −5), *T*(−1, −1) under the translation $(x, y) \to (x + 2, y + 4)$

14. △*XYZ* with *X*(2, 5), *Y*(1, 1), *Z*(5, 1) under the translation $(x, y) \to (x + 1, y - 3)$

9-3 Rotations

See pages 476–482.

Concept Summary

- A rotation turns each point in a figure through the same angle about a fixed point.
- An object has rotational symmetry when you can rotate it less than 360° and the preimage and image are indistinguishable.

Example **Identify the order and magnitude of the rotational symmetry in the figure.**

The figure has rotational symmetry of order 12 because there are 12 rotations of less than 360° (including 0°) that produce an image indistinguishable from the original.

The magnitude is 360° ÷ 12 or 30°.

Exercises Draw the rotation image of each triangle by reflecting the triangles in the given lines. State the coordinates of the rotation image and the angle of rotation. *See Example 2 on page 478.*

15. $\triangle BCD$ with vertices $B(-3, 5)$, $C(-3, 3)$, and $D(-5, 3)$ reflected in the x-axis and then the y-axis

16. $\triangle FGH$ with vertices $F(0, 3)$, $G(-1, 0)$, $H(-4, 1)$ reflected in the line $y = x$ and then the line $y = -x$

17. $\triangle LMN$ with vertices $L(2, 2)$, $M(5, 3)$, $N(3, 6)$ reflected in the line $y = -x$ and then the x-axis

The figure at the right is a regular nonagon.
See Exercise 3 on page 478.

18. Identify the order and magnitude of the symmetry.

19. What is the measure of the angle of rotation if vertex 2 is moved counterclockwise to the current position of vertex 6?

20. If vertex 5 is rotated 280° counterclockwise, find its new position.

9-4 Tessellations

See pages 483–488.

Concept Summary

- A tessellation is a repetitious pattern that covers a plane without overlap.
- A regular tessellation contains the same combination of shapes and angles at every vertex.

Example **Classify the tessellation at the right.**

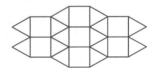

The tessellation is uniform, because at each vertex there are two squares and three equilateral triangles. Both the square and equilateral triangle are regular polygons.

Since there is more than one regular polygon in the tessellation, it is a semi-regular tessellation.

Exercises Determine whether each pattern is a tessellation. If so, describe it as *uniform, not uniform, regular,* or *semi-regular.* *See Example 3 on page 485.*

21.

22.

23.

Determine whether each regular polygon will tessellate the plane. Explain.
See Example 1 on page 484.

24. pentagon   25. triangle   26. decagon

9-5 Dilations

See pages 490–497.

Concept Summary

- Dilations can be enlargements, reductions, or congruence transformations.

Example Triangle *EFG* has vertices *E*(−4, −2), *F*(−3, 2), and *G*(1, 1). Find the image of △*EFG* after a dilation centered at the origin with a scale factor of $\frac{3}{2}$.

Preimage (*x*, *y*)	Image $\left(\frac{3}{2}x, \frac{3}{2}y\right)$
E(−4, −2)	*E*′(−6, −3)
F(−3, 2)	$F'\left(-\frac{9}{2}, 3\right)$
G(1, 1)	$G'\left(\frac{3}{2}, \frac{3}{2}\right)$

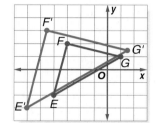

Exercises Find the measure of the dilation image $\overline{C'D'}$ or preimage of $\overline{CD}$ using the given scale factor. *See Example 1 on page 491.*

27. *CD* = 8, *r* = 3
28. $CD = \frac{2}{3}, r = -6$
29. *C*′*D*′ = 24, *r* = 6

30. $C'D' = 60, r = \frac{10}{3}$
31. $CD = 12, r = -\frac{5}{6}$
32. $C'D' = \frac{55}{2}, r = \frac{5}{4}$

Find the image of each polygon, given the vertices, after a dilation centered at the origin with a scale factor of −2. *See Example 3 on page 492.*

33. *P*(−1, 3), *Q*(2, 2), *R*(1, −1)
34. *E*(−3, 2), *F*(1, 2), *G*(1, −2), *H*(−3, −2)

9-6 Vectors

See pages 498–505.

Concept Summary

- A vector is a quantity with both magnitude and direction.
- Vectors can be used to translate figures on the coordinate plane.

Example Find the magnitude and direction of $\overrightarrow{PQ}$ for *P*(−8, 4) and *Q*(6, 10).

Find the magnitude.

$$|\overrightarrow{PQ}| = \sqrt{(x_2 - x_1)^2 + (y_2 - y_1)^2} \quad \text{Distance Formula}$$

$$= \sqrt{(6 + 8)^2 + (10 - 4)^2} \quad x_1 = -8, y_1 = 4, x_2 = 6, y_2 = 10$$

$$= \sqrt{232} \quad \text{Simplify.}$$

$$\approx 15.2 \quad \text{Use a calculator.}$$

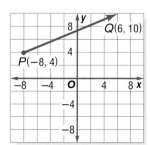

Find the direction.

$$\tan P = \frac{y_2 - y_1}{x_2 - x_1} \quad \frac{\text{length of opposite side}}{\text{length of adjacent side}}$$

$$= \frac{10 - 4}{6 + 8} \quad \text{Substitution}$$

$$= \frac{6}{14} \text{ or } \frac{3}{7} \quad \text{Simplify.}$$

$$m\angle P = \tan^{-1} \frac{3}{7}$$

$$\approx 23.2 \quad \text{Use a calculator.}$$

Chapter
9 For More ...
• Extra Practice, see pages 771–773.
• Mixed Problem Solving, see page 790.

Exercises Write the component form of each vector. *See Example 1 on page 498.*

35.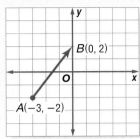

36.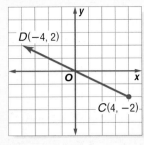

37.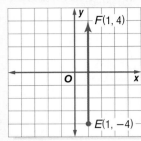

Find the magnitude and direction of $\overrightarrow{AB}$ for the given coordinates.
See Example 2 on page 499.

38. $A(-6, 4)$, $B(-9, -3)$

39. $A(8, 5)$, $B(-5, -2)$

40. $A(-14, 2)$, $B(15, -5)$

41. $A(16, 40)$, $B(-45, 0)$

9-7 ## Transformations with Matrices

See pages
506–511.

Concept Summary

• The vertices of a polygon can be represented by a vertex matrix.
• Matrix operations can be used to perform transformations.

Example Use a matrix to find the coordinates of the vertices of the image of $\triangle ABC$ with $A(1, -1)$, $B(2, -4)$, $C(7, -1)$ after a reflection in the y-axis.

Write the ordered pairs in a vertex matrix. Then use a calculator to multiply the vertex matrix by the reflection matrix.

$$\begin{bmatrix} -1 & 0 \\ 0 & 1 \end{bmatrix} \cdot \begin{bmatrix} 1 & 2 & 7 \\ -1 & -4 & -1 \end{bmatrix} = \begin{bmatrix} -1 & -2 & -7 \\ -1 & -4 & -1 \end{bmatrix}$$

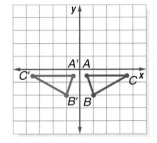

The coordinates of $\triangle A'B'C'$ are $A'(-1, -1)$, $B'(-2, -4)$, and $C'(-7, -1)$.

Exercises Use a matrix to find the coordinates of the vertices of the image after the stated transformation.
See Example 1 on page 506.

42. translation $(x, y) \rightarrow (x - 3, y - 6)$

43. dilation by scale factor $\dfrac{4}{5}$

44. reflection in the line $y = x$

45. rotation 270° counterclockwise about the origin

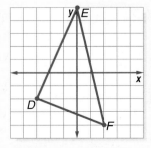

Find the coordinates of the image after the stated transformations.
See Examples 2–4 on pages 507 and 508.

46. $\triangle PQR$ with $P(9, 2)$, $Q(1, -1)$, and $R(4, 5)$; $(x, y) \rightarrow (x + 2, y - 5)$, then a reflection in the x-axis

47. quadrilateral $WXYZ$ with $W(-8, 1)$, $X(-2, 3)$, $Y(-1, 0)$, and $Z(-6, -3)$; a rotation 180° counterclockwise, then a dilation by scale factor -2

Vocabulary and Concepts

Choose the correct term to complete each sentence.

1. If a dilation does not change the size of the object, then it is a(n) (*isometry*, *reflection*).
2. Tessellations with the same shapes and angles at each vertex are called (*uniform*, *regular*).
3. A vector multiplied by a (*vector*, *scalar*) results in another vector.

Skills and Applications

Name the reflected image of each figure under a reflection in line *m*.

4. *A*
5. $\overline{BC}$
6. $\triangle DCE$

COORDINATE GEOMETRY Graph each figure and its image under the given translation.

7. $\triangle PQR$ with $P(-3, 5)$, $Q(-2, 1)$, and $R(-4, 2)$ under the translation right 3 units and up 1 unit
8. Parallelogram *WXYZ* with $W(-2, -5)$, $X(1, -5)$, $Y(2, -2)$, and $Z(-1, -2)$ under the translation up 5 units and left 3 units
9. $\overline{FG}$ with $F(3, 5)$ and $G(6, -1)$ under the translation $(x, y) \rightarrow (x - 4, y - 1)$

Draw the rotation image of each triangle by reflecting the triangles in the given lines. State the coordinates of the rotation image and the angle of rotation.

10. $\triangle JKL$ with $J(-1, -2)$, $K(-3, -4)$, $L(1, -4)$ reflected in the *y*-axis and then the *x*-axis
11. $\triangle ABC$ with $A(-3, -2)$, $B(-1, 1)$, $C(3, -1)$ reflected in the line $y = x$ and then the line $y = -x$
12. $\triangle RST$ with $R(1, 6)$, $S(1, 1)$, $T(3, -2)$ reflected in the *y*-axis and then the line $y = x$

Determine whether each pattern is a tessellation. If so, describe it as *uniform*, *not uniform*, *regular*, or *semi-regular*.

13.
14.
15.

Find the measure of the dilation image $\overline{M'N'}$ or preimage of $\overline{MN}$ using the given scale factor.

16. $MN = 5$, $r = 4$
17. $MN = 8$, $r = \frac{1}{4}$
18. $M'N' = 36$, $r = 3$
19. $MN = 9$, $r = -\frac{1}{5}$
20. $M'N' = 20$, $r = \frac{2}{3}$
21. $M'N' = \frac{29}{5}$, $r = -\frac{3}{5}$

Find the magnitude and direction of each vector.

22. $\vec{v} = \langle -3, 2 \rangle$
23. $\vec{w} = \langle -6, -8 \rangle$

24. **TRAVEL** In trying to calculate how far she must travel for an appointment, Gunja measured the distance between Richmond, Virginia, and Charlotte, North Carolina, on a map. The distance on the map was 2.25 inches, and the scale factor was 1 inch equals 150 miles. How far must she travel?

25. **STANDARDIZED TEST PRACTICE** What reflections could be used to create the image $(3, 4)$ from $(3, -4)$?
    I. reflection in the *x*-axis   II. reflection in the *y*-axis   III. reflection in the origin

    (A) I only   (B) III only   (C) I and III   (D) I and II

Part 1 | Multiple Choice

Record your answers on the answer sheet provided by your teacher or on a sheet of paper.

1. Ms. Lee told her students, "If you do not get enough rest, you will be tired. If you are tired, you will not be able to concentrate." Which of the following is a logical conclusion that could follow Ms. Lee's statements? (Lesson 2-4)

   (A) If you get enough rest, you will be tired.

   (B) If you are tired, you will be able to concentrate.

   (C) If you do not get enough rest, you will be able to concentrate.

   (D) If you do not get enough rest, you will not be able to concentrate.

2. Which of the following statements is true? (Lesson 3-5)

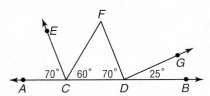

   (A) $\overline{CE} \parallel \overline{DF}$

   (B) $\overline{CF} \parallel \overline{DG}$

   (C) $\overline{CF} \cong \overline{DF}$

   (D) $\overline{CE} \cong \overline{DF}$

3. Which of the following would *not* prove that quadrilateral *QRST* is a parallelogram? (Lesson 8-2)

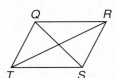

   (A) Both pairs of opposite angles are congruent.

   (B) Both pairs of opposite sides are parallel.

   (C) Diagonals bisect each other.

   (D) A pair of opposite sides is congruent.

4. If $Q(4, 2)$ is reflected in the *y*-axis, what will be the coordinates of Q'? (Lesson 9-1)

   (A) $(-4, -2)$ (B) $(-4, 2)$

   (C) $(2, -4)$ (D) $(2, 4)$

5. Which of the following statements about the figures below is true? (Lesson 9-2)

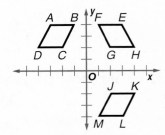

   (A) Parallelogram *JKLM* is a reflection image of ▱*ABCD*.

   (B) Parallelogram *EFGH* is a translation image of ▱*ABCD*.

   (C) Parallelogram *JKLM* is a translation image of ▱*EFGH*.

   (D) Parallelogram *JKLM* is a translation image of ▱*ABCD*.

6. Which of the following is **not** necessarily preserved in a congruence transformation? (Lesson 9-2)

   (A) angle and distance measure

   (B) orientation

   (C) collinearity

   (D) betweenness of points

7. Which transformation is used to map △*ABC* to △*A′B′C′*? (Lesson 9-3)

   (A) rotation

   (B) reflection

   (C) dilation

   (D) translation

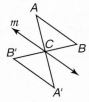

Preparing for Standardized Tests
For test-taking strategies and more
practice, see pages 795-810.

Part 2 | Short Response/Grid In

Record your answers on the answer sheet provided by your teacher or on a sheet of paper.

8. A new logo was designed for GEO Company. The logo is shaped like a symmetrical hexagon. What are the coordinates of the missing vertex of the logo?   (Lesson 1-1)

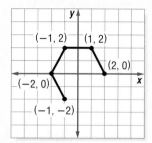

9. A soccer coach is having her players practice penalty kicks. She places two cones equidistant from the goal and asks the players to line up behind each cone. What is the value of x?   (Lesson 4-6)

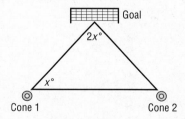

Cone 1   Cone 2

10. A steel cable, which supports a tram, needs to be replaced. To determine the length x of the cable currently in use, the engineer makes several measurements and draws the diagram below of two right triangles, $\triangle ABC$ and $\triangle EDC$. If $m\angle ACB = m\angle ECD$, what is the length x of the cable currently in use? Round the result to the nearest meter.
(Lesson 6-3)

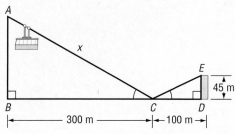

Test-Taking Tip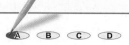
Question 4
To check your answer, remember the following rule. In a reflection over the x-axis, the x-coordinate remains the same, and the y-coordinate changes its sign. In a reflection over the y-axis, the y-coordinate remains the same, and the x-coordinate changes its sign.

Part 3 | Extended Response

Record your answers on a sheet of paper. Show your work.

11. Kelli drew the diagram below to show the front view of a circus tent. Prove that $\triangle ABD$ is congruent to $\triangle ACE$.   (Lessons 4-5 and 4-6)

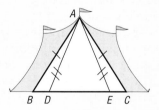

12. Paul is studying to become a landscape architect. He drew a map view of a park with the following vertices: $Q(2, 2)$, $R(-2, 4)$, $S(-3, -2)$, and $T(3, -4)$

   a. On a coordinate plane, graph quadrilateral $QRST$.   (Prerequisite Skill)
   b. Paul's original drawing appears small on his paper. His instructor says that he should dilate the image with the origin as center and a scale factor of 2. Graph and label the coordinates of the dilation image $Q'R'S'T'$.   (Lesson 9-5)
   c. Explain how Paul can determine the coordinates of the vertices of $Q'R'S'T'$ without using a coordinate plane. Use one of the vertices for a demonstration of your method.   (Lesson 9-5)
   d. Dilations are similarity transformations. What properties are preserved during an enlargement? reduction? congruence transformation?   (Lesson 9-5)

What You'll Learn

- **Lessons 10-1** Identify parts of a circle and solve problems involving circumference.
- **Lessons 10-2, 10-3, 10-4, and 10-6** Find arc and angle measures in a circle.
- **Lessons 10-5 and 10-7** Find measures of segments in a circle.
- **Lesson 10-8** Write the equation of a circle.

Key Vocabulary

- chord (p. 522)
- circumference (p. 523)
- arc (p. 530)
- tangent (p. 552)
- secant (p. 561)

Why It's Important

A circle is a unique geometric shape in which the angles, arcs, and segments intersecting that circle have special relationships. You can use a circle to describe a safety zone for fireworks, a location on Earth seen from space, and even a rainbow. *You will learn about angles of a circle when satellites send signals to Earth in Lesson 10-6.*

Getting Started

▶ **Prerequisite Skills** To be successful in this chapter, you'll need to master these skills and be able to apply them in problem-solving situations. Review these skills before beginning Chapter 10.

For Lesson 10-1   Solve Equations

Solve each equation for the given variable.   *(For review, see pages 737 and 738.)*

1. $\frac{4}{9}p = 72$ for p **2.** $6.3p = 15.75$

3. $3x + 12 = 8x$ for x **4.** $7(x + 2) = 3(x - 6)$

5. $C = 2pr$ for r **6.** $r = \frac{C}{6.28}$ for C

For Lesson 10-5   Pythagorean Theorem

Find x. Round to the nearest tenth if necessary.   *(For review, see Lesson 7-2.)*

7.

8.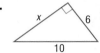

9.

For Lesson 10-7   Quadratic Formula

Solve each equation by using the Quadratic Formula. Round to the nearest tenth.

10. $x^2 - 4x = 10$ **11.** $3x^2 - 2x - 4 = 0$

12. $x^2 = x + 15$ **13.** $2x^2 + x = 15$

Circles   Make this Foldable to help you organize your notes. Begin with five sheets of plain $8\frac{1}{2}$" by 11" paper, and cut out five large circles that are the same size.

Step 1 **Fold and Cut**

Fold two of the circles in half and cut one-inch slits at each end of the folds.

Step 2 **Fold and Cut**

Fold the remaining three circles in half and cut a slit in the middle of the fold.

Step 3 **Slide**

Slide the two circles with slits on the ends through the large slit of the other circles.

Step 4 **Label**

Fold to make a booklet. Label the cover with the title of the chapter and each sheet with a lesson number.

Reading and Writing   As you read and study each lesson, take notes and record concepts on the appropriate page of your Foldable.

10-1 Circles and Circumference

What You'll Learn

- Identify and use parts of circles.
- Solve problems involving the circumference of a circle.

Vocabulary

- circle
- center
- chord
- radius
- diameter
- circumference
- pi (π)

How far does a carousel animal travel in one rotation?

The largest carousel in the world still in operation is located in Spring Green, Wisconsin. It weighs 35 tons and contains 260 animals, none of which is a horse! The rim of the carousel base is a circle. The width, or diameter, of the circle is 80 feet. The distance that one of the animals on the outer edge travels can be determined by special segments in a circle.

PARTS OF CIRCLES   A **circle** is the locus of all points in a plane equidistant from a given point called the **center** of the circle. A circle is usually named by its center point. The figure below shows circle C, which can be written as ⊙C. Several special segments in circle C are also shown.

Any segment with endpoints that are on the circle is a **chord** of the circle. $\overline{AF}$ and $\overline{BE}$ are chords.

A chord that passes through the center is a **diameter** of the circle. $\overline{BE}$ is a diameter.

Any segment with endpoints that are the center and a point on the circle is a **radius**. $\overline{CD}$, $\overline{CB}$, and $\overline{CE}$ are radii of the circle.

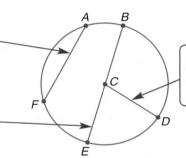

Note that diameter $\overline{BE}$ is made up of collinear radii $\overline{CB}$ and $\overline{CE}$.

Example 1 Identify Parts of a Circle

a. Name the circle.

The circle has its center at K, so it is named circle K, or ⊙K.

In this textbook, the center of a circle will always be shown in the figure with a dot.

b. Name a radius of the circle.

Five radii are shown: $\overline{KN}$, $\overline{KO}$, $\overline{KP}$, $\overline{KQ}$, and $\overline{KR}$.

c. Name a chord of the circle.

Two chords are shown: $\overline{NO}$ and $\overline{RP}$.

d. Name a diameter of the circle.

$\overline{RP}$ is the only chord that goes through the center, so $\overline{RP}$ is a diameter.

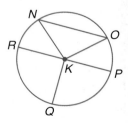

By the definition of a circle, the distance from the center to any point on the circle is always the same. Therefore, all radii are congruent. A diameter is composed of two radii, so all diameters are congruent. The letters d and r are usually used to represent diameter and radius in formulas. So, $d = 2r$ and $r = \frac{d}{2}$ or $\frac{1}{2}d$.

Example 2 *Find Radius and Diameter*

Circle A has diameters $\overline{DF}$ and $\overline{PG}$.

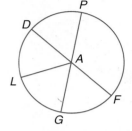

a. **If $DF = 10$, find DA.**

$r = \frac{1}{2}d$ Formula for radius

$r = \frac{1}{2}(10)$ or 5   Substitute and simplify.

b. **If $PA = 7$, find PG.**

$d = 2r$ Formula for diameter

$d = 2(7)$ or 14   Substitute and simplify.

c. **If $AG = 12$, find LA.**

Since all radii are congruent, $LA = AG$. So, $LA = 12$.

Circles can intersect. The segment connecting the centers of the two intersecting circles contains a radius of each circle.

Example 3 *Find Measures in Intersecting Circles*

The diameters of $\odot A$, $\odot B$, and $\odot C$ are 10 inches, 20 inches, and 14 inches, respectively.

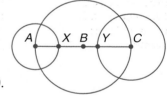

a. **Find XB.**

Since the diameter of $\odot A$ is 10, $AX = 5$.
Since the diameter of $\odot B$ is 20, $AB = 10$ and $BC = 10$.
$\overline{XB}$ is part of radius $\overline{AB}$.

$AX + XB = AB$ Segment Addition Postulate

$5 + XB = 10$ Substitution

$XB = 5$ Subtract 5 from each side.

b. **Find BY.**

$\overline{BY}$ is part of $\overline{BC}$.
Since the diameter of $\odot C$ is 14, $YC = 7$.

$BY + YC = BC$ Segment Addition Postulate

$BY + 7 = 10$ Substitution

$BY = 3$ Subtract 7 from each side.

CIRCUMFERENCE   The **circumference** of a circle is the distance around the circle. Circumference is most often represented by the letter C.

Geometry Activity

Circumference Ratio

A special relationship exists between the circumference of a circle and its diameter.

Gather Data and Analyze

Collect ten round objects.

1. Measure the circumference and diameter of each object using a millimeter measuring tape. Record the measures in a table like the one at the right.

2. Compute the value of $\frac{C}{d}$ to the nearest hundredth for each object. Record the result in the fourth column of the table.

Object	C	d	$\frac{C}{d}$
1			
2			
3			
⋮			
10			

Make a Conjecture

3. What seems to be the relationship between the circumference and the diameter of the circle?

The Geometry Activity suggests that the circumference of any circle can be found by multiplying the diameter by a number slightly larger than 3. By definition, the ratio $\frac{C}{d}$ is an irrational number called **pi**, symbolized by the Greek letter **π**. Two formulas for the circumference can be derived using this definition.

$\frac{C}{d} = \pi$ Definition of pi

$C = \pi d$ Multiply each side by d.

$C = \pi d$

$C = \pi(2r)$ $d = 2r$

$C = 2\pi r$ Simplify.

Key Concept   *Circumference*

For a circumference of C units and a diameter of d units or a radius of r units,
$$C = \pi d \text{ or } C = 2\pi r.$$

If you know the diameter or radius, you can find the circumference. Likewise, if you know the circumference, you can find the diameter or radius.

Example 4 *Find Circumference, Diameter, and Radius*

a. Find C if r = 7 centimeters.

$C = 2\pi r$ Circumference formula

$= 2\pi(7)$ Substitution

$= 14\pi$ or about 43.98 cm

b. Find C if d = 12.5 inches.

$C = \pi d$ Circumference formula

$= \pi(12.5)$ Substitution

$= 12.5\pi$ or 39.27 in.

c. Find d and r to the nearest hundredth if C = 136.9 meters.

$C = \pi d$ Circumference formula

$136.9 = \pi d$ Substitution

$\frac{136.9}{\pi} = d$ Divide each side by π.

$43.58 \approx d$ Use a calculator.

$d \approx 43.58$ m

$r = \frac{1}{2}d$ Radius formula

$\approx \frac{1}{2}(43.58)$ $d \approx 43.58$

≈ 21.79 m   Use a calculator.

You can also use other geometric figures to help you find the circumference of a circle.

Standardized Test Practice
Ⓐ Ⓑ Ⓒ Ⓓ

Example 5 *Use Other Figures to Find Circumference*

Multiple-Choice Test Item

Find the exact circumference of $\odot P$.

Ⓐ 13 cm
Ⓑ 12π cm
Ⓒ 40.84 cm
Ⓓ 13π cm

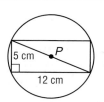

Read the Test Item

You are given a figure that involves a right triangle and a circle. You are asked to find the exact circumference of the circle.

Test-Taking Tip

Notice that the problem asks for an exact answer. Since you know that an exact circumference contains π, you can eliminate choices A and C.

Solve the Test Item

The diameter of the circle is the same as the hypotenuse of the right triangle.

$a^2 + b^2 = c^2$ Pythagorean Theorem

$5^2 + 12^2 = c^2$ Substitution

$169 = c^2$ Simplify.

$13 = c$ Take the square root of each side.

So the diameter of the circle is 13 centimeters.

$C = \pi d$ Circumference formula

$C = \pi(13)$ or 13π Substitution

Because we want the exact circumference, the answer is D.

Check for Understanding

Concept Check

1. **Describe** how the value of π can be calculated.

2. **Write** two equations that show how the diameter of a circle is related to the radius of a circle.

3. **OPEN ENDED** Explain why a diameter is the longest chord of a circle.

Guided Practice

For Exercises 4–9, refer to the circle at the right.

4. Name the circle.
5. Name a radius.
6. Name a chord.
7. Name a diameter.
8. Suppose $BD = 12$ millimeters. Find the radius of the circle.
9. Suppose $CE = 5.2$ inches. Find the diameter of the circle.

Circle W has a radius of 4 units, $\odot Z$ has a radius of 7 units, and $XY = 2$. Find each measure.

10. YZ
11. IX
12. IC

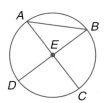

The radius, diameter, or circumference of a circle is given. Find the missing measures. Round to the nearest hundredth if necessary.

13. $r = 5$ m, $d = $? , $C = $?

14. $C = 2368$ ft, $d = $? , $r = $?

Standardized Test Practice
Ⓐ Ⓑ Ⓒ Ⓓ

15. Find the exact circumference of the circle.
   Ⓐ 4.5π mm
   Ⓑ 9π mm
   Ⓒ 18π mm
   Ⓓ 81π mm

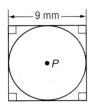
9 mm

Practice and Apply

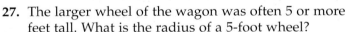

Homework Help

For Exercises	See Examples
16–25	1
26–31	2
32–43	3
48–51	4
52	5

Extra Practice
See page 773.

For Exercises 16–20, refer to the circle at the right.
16. Name the circle.
17. Name a radius.
18. Name a chord.
19. Name a diameter.
20. Name a radius not contained in a diameter.

HISTORY   For Exercises 21–31, refer to the model of a Conestoga wagon wheel.
21. Name the circle.
22. Name a radius of the circle.
23. Name a chord of the circle.
24. Name a diameter of the circle.
25. Name a radius not contained in a diameter.
26. Suppose the radius of the circle is 2 feet. Find the diameter.
27. The larger wheel of the wagon was often 5 or more feet tall. What is the radius of a 5-foot wheel?
28. If $TX = 120$ centimeters, find TR.
29. If $RZ = 32$ inches, find ZW.
30. If $UR = 18$ inches, find RV.
31. If $XT = 1.2$ meters, find UR.

The diameters of $\odot A$, $\odot B$, and $\odot C$ are 10, 30, and 10 units, respectively. Find each measure if $\overline{AZ} \cong \overline{CW}$ and $CW = 2$.

32. AZ 33. ZX
34. BX 35. BY
36. YW 37. AC

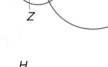

Circles G, J, and K all intersect at L. If $GH = 10$, find each measure.
38. FG 39. FH
40. GL 41. GJ
42. JL 43. JK

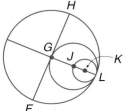

The radius, diameter, or circumference of a circle is given. Find the missing measures. Round to the nearest hundredth if necessary.

44. $r = 7$ mm, $d = $ __?__ , $C = $ __?__

45. $d = 26.8$ cm, $r = $ __?__ , $C = $ __?__

46. $C = 26\pi$ mi, $d = $ __?__ , $r = $ __?__

47. $C = 76.4$ m, $d = $ __?__ , $r = $ __?__

48. $d = 12\frac{1}{2}$yd, $r = $ __?__ , $C = $ __?__

49. $r = 6\frac{3}{4}$in., $d = $ __?__ , $C = $ __?__

50. $d = 2a$, $r = $ __?__ , $C = $ __?__

51. $r = \frac{a}{6}$, $d = $ __?__ , $C = $ __?__

Find the exact circumference of each circle.

52.
30 m
16 m

53.
3 ft
4 ft

54.
10 in.

55.
4√2 cm

56. **PROBABILITY** Find the probability that a segment with endpoints that are the center of the circle and a point on the circle is a radius. Explain.

57. **PROBABILITY** Find the probability that a chord that does not contain the center of a circle is the longest chord of the circle.

FIREWORKS For Exercises 58–60, use the following information.
Every July 4th Boston puts on a gala with the Boston Pops Orchestra, followed by a huge fireworks display. The fireworks are shot from a barge in the river. There is an explosion circle inside which all of the fireworks will explode. Spectators sit outside a safety circle that is 800 feet from the center of the fireworks display.

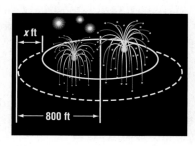
x ft
800 ft

58. Find the approximate circumference of the safety circle.

59. If the safety circle is 200 to 300 feet farther from the center than the explosion circle, find the range of values for the radius of the explosion circle.

60. Find the least and maximum circumference of the explosion circle to the nearest foot.

Online Research **Data Update** Find the largest firework ever made. How does its dimension compare to the Boston display? Visit www.geometryonline.com/data_update to learn more.

61. **CRITICAL THINKING** In the figure, O is the center of the circle, and $x^2 + y^2 + p^2 + t^2 = 288$. What is the exact circumference of $\odot O$?

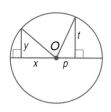
y
x
O
t
p

62. **WRITING IN MATH** Answer the question that was posed at the beginning of the lesson.

How far does a carousel animal travel in one rotation?

Include the following in your answer:
- a description of how the circumference of a circle relates to the distance traveled by the animal, and
- whether an animal located one foot from the outside edge of the carousel travels a mile when it makes 22 rotations for each ride.

63. GRID IN In the figure, the radius of circle A is twice the radius of circle B and four times the radius of circle C. If the sum of the circumferences of the three circles is 42π, find the measure of $\overline{AC}$.

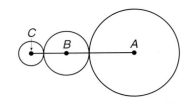

64. ALGEBRA There are k gallons of gasoline available to fill a tank. After d gallons have been pumped, what percent of gasoline, in terms of k and d, has been pumped?

Ⓐ $\dfrac{100d}{k}\%$ Ⓑ $\dfrac{k}{100d}\%$ Ⓒ $\dfrac{100k}{d}\%$ Ⓓ $\dfrac{100k-d}{k}\%$

*Extending
the Lesson*

65. CONCENTRIC CIRCLES Circles that have the same center, but different radii, are called *concentric circles*. Use the figure at the right to find the exact circumference of each circle. List the circumferences in order from least to greatest.

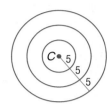

Maintain Your Skills

Mixed Review

Find the magnitude to the nearest tenth and direction to the nearest degree of each vector. *(Lesson 9-6)*

66. $\overrightarrow{AB} = \langle 1, 4 \rangle$ **67.** $\vec{v} = \langle 4, 9 \rangle$

68. $\overrightarrow{AB}$ if $A(4, 2)$ and $B(7, 22)$ **69.** $\overrightarrow{CD}$ if $C(0, -20)$ and $D(40, 0)$

Find the measure of the dilation image of $\overline{AB}$ for each scale factor k. *(Lesson 9-5)*

70. $AB = 5, k = 6$ **71.** $AB = 16, k = 1.5$ **72.** $AB = \dfrac{2}{3}, k = -\dfrac{1}{2}$

73. **PROOF** Write a two-column proof. *(Lesson 5-3)*

Given: $\overline{RQ}$ bisects $\angle SRT$.

Prove: $m\angle SQR > m\angle SRQ$

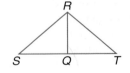

74. COORDINATE GEOMETRY Name the missing coordinates if $\triangle DEF$ is isosceles with vertex angle E. *(Lesson 4-3)*

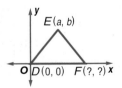

*Getting Ready for
the Next Lesson*

PREREQUISITE SKILL Find x. *(To review **angle addition**, see Lesson 1-4.)*

75.
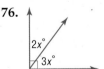

76.

77.
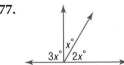

78.
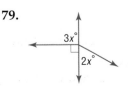

79.

80.
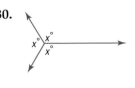

10-2 Angles and Arcs

What You'll Learn

- Recognize major arcs, minor arcs, semicircles, and central angles and their measures.
- Find arc length.

Vocabulary

- central angle
- arc
- minor arc
- major arc
- semicircle

What kinds of angles do the hands on a clock form?

Most clocks on electronic devices are digital, showing the time as numerals. Analog clocks are often used in decorative furnishings and wrist watches. An analog clock has moving hands that indicate the hour, minute, and sometimes the second. This clock face is a circle. The three hands form three central angles of the circle.

ANGLES AND ARCS In Chapter 1, you learned that a degree is $\frac{1}{360}$ of the circular rotation about a point. This means that the sum of the measures of the angles about the center of the clock above is 360. Each of the angles formed by the clock hands is called a central angle. A **central angle** has the center of the circle as its vertex, and its sides contain two radii of the circle.

Key Concept — Sum of Central Angles

- **Words** The sum of the measures of the central angles of a circle with no interior points in common is 360.

- **Example** $m\angle 1 + m\angle 2 + m\angle 3 = 360$

Example 1 Measures of Central Angles

ALGEBRA Refer to $\odot O$.

a. Find $m\angle AOD$.

$\angle AOD$ and $\angle DOB$ are a linear pair, and the angles of a linear pair are supplementary.

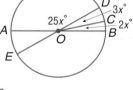

$$m\angle AOD + m\angle DOB = 180$$
$$m\angle AOD + m\angle DOC + m\angle COB = 180 \quad \text{Angle Sum Theorem}$$
$$25x + 3x + 2x = 180 \quad \text{Substitution}$$
$$30x = 180 \quad \text{Simplify.}$$
$$x = 6 \quad \text{Divide each side by 60.}$$

Use the value of x to find $m\angle AOD$.

$$m\angle AOD = 25x \quad \text{Given}$$
$$= 25(6) \text{ or } 150 \quad \text{Substitution}$$

b. Find m∠AOE.

∠AOE and ∠AOD form a linear pair.

m∠AOE + m∠AOD = 180   Linear pairs are supplementary.

m∠AOE + 150 = 180   Substitution

m∠AOE = 30   Subtract 150 from each side.

A central angle separates the circle into two parts, each of which is an **arc**. The measure of each arc is related to the measure of its central angle.

Key Concept			**Arcs of a Circle**
Type of Arc:	minor arc	major arc	semicircle
Example:	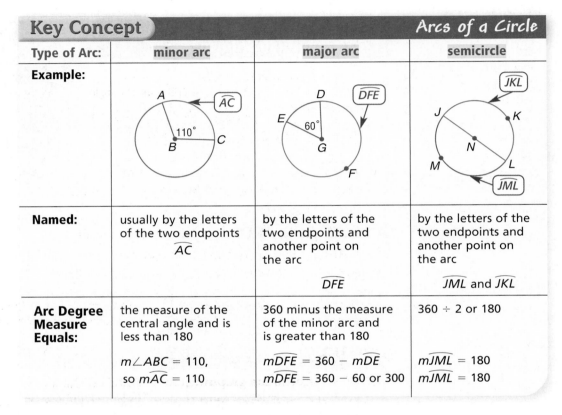		
Named:	usually by the letters of the two endpoints $\overarc{AC}$	by the letters of the two endpoints and another point on the arc $\overarc{DFE}$	by the letters of the two endpoints and another point on the arc $\overarc{JML}$ and $\overarc{JKL}$
Arc Degree Measure Equals:	the measure of the central angle and is less than 180 $m\angle ABC = 110$, so $m\overarc{AC} = 110$	360 minus the measure of the minor arc and is greater than 180 $m\overarc{DFE} = 360 - m\overarc{DE}$ $m\overarc{DFE} = 360 - 60$ or 300	360 ÷ 2 or 180 $m\overarc{JML} = 180$ $m\overarc{JML} = 180$

Arcs with the same measure in the same circle or in congruent circles are congruent.

Theorem 10.1

In the same or in congruent circles, two arcs are congruent if and only if their corresponding central angles are congruent.

You will prove Theorem 10.1 in Exercise 54.

Arcs of a circle that have exactly one point in common are *adjacent arcs*. Like adjacent angles, the measures of adjacent arcs can be added.

Postulate 10.1

Arc Addition Postulate   The measure of an arc formed by two adjacent arcs is the sum of the measures of the two arcs.

Example: In ⊙S, $m\overarc{PQ} + m\overarc{QR} = m\overarc{PQR}$.

Example 2 Measures of Arcs

In $\odot F$, $m\angle DFA = 50$ and $\overline{CF} \perp \overline{FB}$. Find each measure.

a. $m\widehat{BE}$

$\widehat{BE}$ is a minor arc, so $m\widehat{BE} = m\angle BFE$.

$\angle BFE \cong \angle DFA$	Vertical angles are congruent.
$m\angle BFE = m\angle DFA$	Definition of congruent angles
$m\widehat{BE} = m\angle DFA$	Transitive Property
$m\widehat{BE} = 50$	Substitution

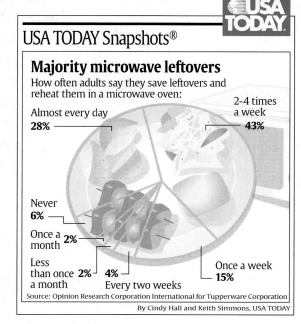

b. $m\widehat{CBE}$

$\widehat{CBE}$ is composed of adjacent arcs, $\widehat{CB}$ and $\widehat{BE}$.

$m\widehat{CB} = m\angle CFB$
$\qquad\quad = 90 \qquad\qquad$ $\angle CFB$ is a right angle.

$m\widehat{CBE} = m\widehat{CB} + m\widehat{BE} \qquad$ Arc Addition Postulate
$m\widehat{CBE} = 90 + 50$ or $140 \qquad$ Substitution

c. $m\widehat{ACE}$

One way to find $m\widehat{ACE}$ is by using $\widehat{ACB}$ and $\widehat{BE}$.
$\widehat{ACB}$ is a semicircle.

$m\widehat{ACE} = m\widehat{ACB} + \widehat{BE} \qquad$ Arc Addition Postulate
$m\widehat{ACE} = 180 + 50$ or $230 \qquad$ Substitution

In a circle graph, the central angles divide a circle into wedges to represent data, often expressed as a percent. The size of the angle is proportional to the percent.

Example 3 Circle Graphs

FOOD   Refer to the graphic.

a. Find the measurement of the central angle for each category.

The sum of the percents is 100% and represents the whole. Use the percents to determine what part of the whole circle (360°) each central angle contains.

$2\%(360°) = 7.2°$
$6\%(360°) = 21.6°$
$28\%(360°) = 100.8°$
$43\%(360°) = 154.8°$
$15\%(360°) = 54°$
$4\%(360°) = 14.4°$

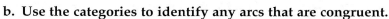

b. Use the categories to identify any arcs that are congruent.

The arcs for the wedges named *Once a month* and *Less than once a month* are congruent because they both represent 2% or 7.2° of the circle.

ARC LENGTH   Another way to measure an arc is by its length. An arc is part of the circle, so the length of an arc is a part of the circumference.

Study Tip

Look Back
To review proportions, see Lesson 6-1.

Example 4   Arc Length

In ⊙P, $PR = 15$ and $m\angle QPR = 120$. Find the length of $\overset{\frown}{QR}$.

In ⊙P, $r = 15$, so $C = 2\pi(15)$ or 30π and $m\overset{\frown}{QR} = m\angle QPR$ or 120. Write a proportion to compare each part to its whole.

$$\begin{array}{l}\text{degree measure of arc} \to \\ \text{degree measure of whole circle} \to\end{array} \frac{120}{360} = \frac{\ell}{30\pi} \begin{array}{l}\leftarrow \text{arc length} \\ \leftarrow \text{circumference}\end{array}$$

Now solve the proportion for ℓ.

$$\frac{120}{360} = \frac{\ell}{30\pi}$$

$$\frac{120}{360}(30\pi) = \ell \qquad \text{Multiply each side by } 30\pi.$$

$$10\pi = \ell \qquad \text{Simplify.}$$

The length of $\overset{\frown}{QR}$ is 10π units or about 31.42 units.

The proportion used to find the arc length in Example 4 can be adapted to find the arc length in any circle.

Key Concept   Arc Length

$$\begin{array}{l}\text{degree measure of arc} \to \\ \text{degree measure of whole circle} \to\end{array} \frac{A}{360} = \frac{\ell}{2\pi r} \begin{array}{l}\leftarrow \text{arc length} \\ \leftarrow \text{circumference}\end{array}$$

This can also be expressed as $\frac{A}{360} \cdot C = \ell$.

Check for Understanding

Concept Check   1. **OPEN-ENDED**   Draw a circle and locate three points on the circle. Name all of the arcs determined by the three points and use a protractor to find the measure of each arc.

2. **Explain** why it is necessary to use three letters to name a semicircle.

3. **Describe** the difference between *concentric* circles and *congruent* circles.

Guided Practice   **ALGEBRA**   Find each measure.

4. $m\angle NCL$ 5. $m\angle RCL$

6. $m\angle RCM$ 7. $m\angle RCN$

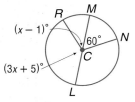

In ⊙A, $m\angle EAD = 42$. Find each measure.

8. $m\overset{\frown}{BC}$ 9. $m\overset{\frown}{CBE}$

10. $m\overset{\frown}{EDB}$ 11. $m\overset{\frown}{CD}$

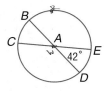

12. Points T and R lie on ⊙W so that $WR = 12$ and $m\angle TWR = 60$. Find the length of $\overset{\frown}{TR}$.

Application **13. SURVEYS** The graph shows the results of a survey of 1400 chief financial officers who were asked how many hours they spend working on the weekend. Determine the measurement of each angle of the graph. Round to the nearest degree.

Executives Working on the Weekend

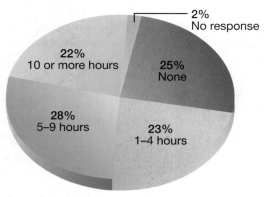

Source: Accountemps

Practice and Apply

Homework Help

For Exercises	See Examples
14–23	1
24–39	2
40–43	3
44–45	4

Extra Practice
See page 774.

Find each measure.

14. $m\angle CGB$ **15.** $m\angle BGE$
16. $m\angle AGD$ **17.** $m\angle DGE$
18. $m\angle CGD$ **19.** $m\angle AGE$

ALGEBRA **Find each measure.**

20. $m\angle ZXV$ **21.** $m\angle YXW$
22. $m\angle ZXY$ **23.** $m\angle VXW$

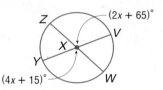

In $\odot O$, $\overline{EC}$ **and** $\overline{AB}$ **are diameters, and** $\angle BOD \cong \angle DOE \cong \angle EOF \cong \angle FOA$. **Find each measure.**

24. $m\widehat{BC}$ **25.** $m\widehat{AC}$
26. $m\widehat{AE}$ **27.** $m\widehat{EB}$
28. $m\widehat{ACB}$ **29.** $m\widehat{AD}$
30. $m\widehat{CBF}$ **31.** $m\widehat{ADC}$

ALGEBRA **In** $\odot Z$, $\angle WZX \cong \angle XZY$, $m\angle VZU = 4x$, $m\angle UZY = 2x + 24$, **and** $\overline{VY}$ **and** $\overline{WU}$ **are diameters. Find each measure.**

32. $m\widehat{UY}$ **33.** $m\widehat{WV}$
34. $m\widehat{WX}$ **35.** $m\widehat{XY}$
36. $m\widehat{WUY}$ **37.** $m\widehat{YVW}$
38. $m\widehat{XVY}$ **39.** $m\widehat{WUX}$

The diameter of $\odot C$ **is 32 units long. Find the length of each arc for the given angle measure.**

40. $\widehat{DE}$ if $m\angle DCE = 100$ **41.** $\widehat{DHE}$ if $m\angle DCE = 90$
42. $\widehat{HDF}$ if $m\angle HCF = 125$ **43.** $\widehat{HD}$ if $m\angle DCH = 45$

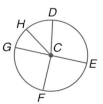

ONLINE MUSIC For Exercises 44–46, refer to the table and use the following information.
A recent survey asked online users how many legally free music files they have collected. The results are shown in the table.

44. If you were to construct a circle graph of this information, how many degrees would be needed for each category?

45. Describe the kind of arc associated with each category.

46. Construct a circle graph for these data.

Free Music Downloads	
How many free music files have you collected?	
100 files or less	76%
101 to 500 files	16%
501 to 1000 files	5%
More than 1000 files	3%

Source: QuickTake.com

Determine whether each statement is *sometimes*, *always*, or *never* true.

47. The measure of a major arc is greater than 180.

48. The central angle of a minor arc is an acute angle.

49. The sum of the measures of the central angles of a circle depends on the measure of the radius.

50. The semicircles of two congruent circles are congruent.

51. **CRITICAL THINKING** Central angles 1, 2, and 3 have measures in the ratio 2 : 3 : 4. Find the measure of each angle.

52. **CLOCKS** The hands of a clock form the same angle at various times of the day. For example, the angle formed at 2:00 is congruent to the angle formed at 10:00. If a clock has a diameter of 1 foot, what is the distance along the edge of the clock from the minute hand to the hour hand at 2:00?

53. **IRRIGATION** Some irrigation systems spray water in a circular pattern. You can adjust the nozzle to spray in certain directions. The nozzle in the diagram is set so it does not spray on the house. If the spray has a radius of 12 feet, what is the approximate length of the arc that the spray creates?

54. **PROOF** Write a proof of Theorem 10.1.

55. **CRITICAL THINKING** The circles at the right are concentric circles that both have point E as their center. If $m\angle 1 = 42$, determine whether $\overarc{AB} \cong \overarc{CD}$. Explain.

56. **WRITING IN MATH** Answer the question that was posed at the beginning of the lesson.

What kind of angles do the hands of a clock form?

Include the following in your answer:
• the kind of angle formed by the hands of a clock, and
• several times of day when these angles are congruent.

57. Compare the circumference of circle *E* with the perimeter of rectangle *ABCD*. Which statement is true?

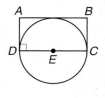

Ⓐ The perimeter of *ABCD* is greater than the circumference of circle *E*.

Ⓑ The circumference of circle *E* is greater than the perimeter of *ABCD*.

Ⓒ The perimeter of *ABCD* equals the circumference of circle *E*.

Ⓓ There is not enough information to determine this comparison.

58. SHORT RESPONSE A circle is divided into three central angles that have measures in the ratio 3 : 5 : 10. Find the measure of each angle.

Maintain Your Skills

Mixed Review

The radius, diameter, or circumference of a circle is given. Find the missing measures. Round to the nearest hundredth if necessary. *(Lesson 10-1)*

59. $r = 10, d =$ _?_, $C =$ _?_

60. $d = 13, r =$ _?_, $C =$ _?_

61. $C = 28\pi, d =$ _?_, $r =$ _?_

62. $C = 75.4, d =$ _?_, $r =$ _?_

63. SOCCER Two soccer players kick the ball at the same time. One exerts a force of 72 newtons east. The other exerts a force of 45 newtons north. What are the magnitude to the nearest tenth and direction to the nearest degree of the resultant force on the soccer ball? *(Lesson 9-6)*

ALGEBRA Find *x*. *(Lesson 6-5)*

64.

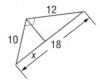

65.

Find the exact distance between each point and line or pair of lines. *(Lesson 3-6)*

66. point $Q(6, -2)$ and the line with the equation $y - 7 = 0$

67. parallel lines with the equations $y = x + 3$ and $y = x - 4$

68. Angle *A* has a measure of 57.5. Find the measures of the complement and supplement of $\angle A$. *(Lesson 2-8)*

Use the following statement for Exercises 69 and 70.
If ABC is a triangle, then ABC has three sides. *(Lesson 2-3)*

69. Write the converse of the statement.

70. Determine the truth value of the statement and its converse.

Getting Ready for the Next Lesson

PREREQUISITE SKILL Find *x*. *(To review isosceles triangles, see Lesson 4-6.)*

71.

72.

73.

74.

75.

76.

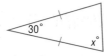

10-3 Arcs and Chords

What You'll Learn

- Recognize and use relationships between arcs and chords.
- Recognize and use relationships between chords and diameters.

Vocabulary
- inscribed
- circumscribed

How do the grooves in a Belgian waffle iron model segments in a circle?

Waffle irons have grooves in each heated plate that result in the waffle pattern when the batter is cooked. One model of a Belgian waffle iron is round, and each groove is a chord of the circle.

ARCS AND CHORDS The endpoints of a chord are also endpoints of an arc. If you trace the waffle pattern on patty paper and fold along the diameter, $\overline{AB}$ and $\overline{CD}$ match exactly, as well as $\overset{\frown}{AB}$ and $\overset{\frown}{CD}$. This suggests the following theorem.

Theorem 10.2

In a circle or in congruent circles, two minor arcs are congruent if and only if their corresponding chords are congruent.

Abbreviations:

In ⊙, 2 minor arcs are ≅, corr. chords are ≅.
In ⊙, 2 chords are ≅, corr. minor arcs are ≅

Examples

If $\overline{AB} \cong \overline{CD}$,
$\overset{\frown}{AB} \cong \overset{\frown}{CD}$.

If $\overset{\frown}{AB} \cong \overset{\frown}{CD}$,
$\overline{AB} \cong \overline{CD}$.

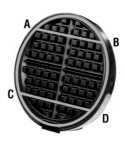

> **Study Tip**
>
> *Reading Mathematics*
> Remember that the phrase *if and only if* means that the conclusion and the hypothesis can be switched and the statement is still true.

You will prove part 2 of Theorem 10.2 in Exercise 4.

Example 1 Prove Theorems

PROOF Theorem 10.2 (part 1)

Given: ⊙X, $\overset{\frown}{UV} \cong \overset{\frown}{YW}$

Prove: $\overline{UV} \cong \overline{YW}$

Proof:

Statements	Reasons
1. ⊙X, $\overset{\frown}{UV} \cong \overset{\frown}{YW}$	1. Given
2. $\angle UXV \cong \angle WXY$	2. If arcs are ≅, their corresponding central ∡ are ≅.
3. $\overline{UX} \cong \overline{XV} \cong \overline{XW} \cong \overline{XY}$	3. All radii of a circle are congruent.
4. $\triangle UXV \cong \triangle WXY$	4. SAS
5. $\overline{UV} \cong \overline{YW}$	5. CPCTC

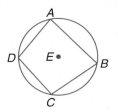

Inscribed and Circumscribed
A circle can also be inscribed in a polygon, so that the polygon is circumscribed about the circle. You will learn about this in Lesson 10-5.

The chords of adjacent arcs can form a polygon. Quadrilateral *ABCD* is an **inscribed** polygon because all of its vertices lie on the circle. Circle E is **circumscribed** about the polygon because it contains all the vertices of the polygon.

Example 2 *Inscribed Polygons*

SNOWFLAKES The main veins of a snowflake create six congruent central angles. Determine whether the hexagon containing the flake is regular.

$\angle 1 \cong \angle 2 \cong \angle 3 \cong \angle 4 \cong \angle 5 \cong \angle 6$ Given

$\overarc{KL} \cong \overarc{LM} \cong \overarc{MN} \cong \overarc{NO} \cong \overarc{OJ} \cong \overarc{JK}$ If central $\angle$s are $\cong$, corresponding arcs are $\cong$.

$\overline{KL} \cong \overline{LM} \cong \overline{MN} \cong \overline{NO} \cong \overline{OJ} \cong \overline{JK}$ In $\odot$, 2 minor arcs $\cong$, corr. chords are $\cong$.

Because all the central angles are congruent, the measure of each angle is $360 \div 6$ or 60.

Let *x* be the measure of each base angle in the triangle containing $\overline{KL}$.

$m\angle 1 + x + x = 180$ Angle Sum Theorem

$60 + 2x = 180$ Substitution

$2x = 120$ Subtract 60 from each side.

$x = 60$ Divide each side by 2.

This applies to each triangle in the figure, so each angle of the hexagon is 2(60) or 120. Thus the hexagon has all sides congruent and all vertex angles congruent.

DIAMETERS AND CHORDS   Diameters that are perpendicular to chords create special segment and arc relationships. Suppose you draw circle C and one of its chords $\overline{WX}$ on a piece of patty paper and fold the paper to construct the perpendicular bisector. You will find that the bisector also cuts $\overarc{WX}$ in half and passes through the center of the circle, making it contain a diameter.

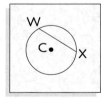 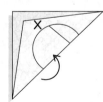 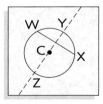

This is formally stated in the next theorem.

Theorem 10.3

In a circle, if a diameter (or radius) is perpendicular to a chord, then it bisects the chord and its arc.

Example: If $\overline{BA} \perp \overline{TV}$, then $\overline{UT} \cong \overline{UV}$ and $\overarc{AT} \cong \overarc{AV}$.

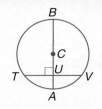

You will prove Theorem 10.3 in Exercise 36.

Example 3 Radius Perpendicular to a Chord

Circle O has a radius of 13 inches. Radius $\overline{OB}$ is perpendicular to chord $\overline{CD}$, which is 24 inches long.

a. If $m\overarc{CD} = 134$, find $m\overarc{CB}$.

$\overline{OB}$ bisects $\overarc{CD}$, so $m\overarc{CB} = \frac{1}{2}m\overarc{CD}$.

$m\overarc{CB} = \frac{1}{2}m\overarc{CD}$ Definition of arc bisector

$m\overarc{CB} = \frac{1}{2}(134)$ or 67 $m\overarc{CD} = 134$

b. Find OX.

Draw radius $\overline{OC}$. $\triangle CXO$ is a right triangle.

$CO = 13$ $r = 13$

$\overline{OB}$ bisects $\overline{CD}$.   A radius perpendicular to a chord bisects it.

$CX = \frac{1}{2}(CD)$ Definition of segment bisector

     $= \frac{1}{2}(24)$ or 12 $CD = 24$

Use the Pythagorean Theorem to find XO.

$(CX)^2 + (OX)^2 = (CO)^2$ Pythagorean Theorem

$12^2 + (OX)^2 = 13^2$ $CX = 12, CO = 13$

$144 + (OX)^2 = 169$ Simplify.

$(OX)^2 = 25$ Subtract 144 from each side.

$OX = 5$ Take the square root of each side.

In the next activity, you will discover another property of congruent chords.

Geometry Activity

Congruent Chords and Distance

Model

Step 1 Use a compass to draw a large circle on patty paper. Cut out the circle.

Step 2 Fold the circle in half.

Step 3 Without opening the circle, fold the edge of the circle so it does not intersect the first fold.

Step 4 Unfold the circle and label as shown.

Step 5 Fold the circle, laying point V onto T to bisect the chord. Open the circle and fold again to bisect $\overline{WY}$. Label as shown.

Analyze

1. What is the relationship between $\overline{SU}$ and $\overline{VT}$? $\overline{SX}$ and $\overline{WY}$?
2. Use a centimeter ruler to measure $\overline{VT}$, $\overline{WY}$, $\overline{SU}$, and $\overline{SX}$. What do you find?
3. **Make a conjecture** about the distance that two chords are from the center when they are congruent.

The Geometry Activity suggests the following theorem.

Theorem 10.4

In a circle or in congruent circles, two chords are congruent if and only if they are equidistant from the center.

You will prove Theorem 10.4 in Exercises 37 and 38.

Example 4 Chords Equidistant from Center

Chords $\overline{AC}$ and $\overline{DF}$ are equidistant from the center. If the radius of $\odot G$ is 26, find AC and DE.

$\overline{AC}$ and $\overline{DF}$ are equidistant from G, so $\overline{AC} \cong \overline{DF}$.

Draw $\overline{AG}$ and $\overline{GF}$ to form two right triangles. Use the Pythagorean Theorem.

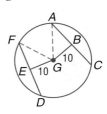

$(AB)^2 + (BG)^2 = (AG)^2$ Pythagorean Theorem

$(AB)^2 + 10^2 = 26^2$ $BG = 10, AG = 26$

$(AB)^2 + 100 = 676$ Simplify.

$(AB)^2 = 576$ Subtract 100 from each side.

$AB = 24$ Take the square root of each side.

$AB = \frac{1}{2}(AC)$, so $AC = 2(24)$ or 48.

$\overline{AC} \cong \overline{DF}$, so DF also equals 48. $DE = \frac{1}{2}DF$, so $DE = \frac{1}{2}(48)$ or 24.

Check for Understanding

Concept Check

1. Explain the difference between an inscribed polygon and a circumscribed circle.

2. **OPEN ENDED** Construct a circle and inscribe any polygon. Draw the radii to the vertices of the polygon and use a protractor to determine whether any sides of the polygon are congruent.

3. **FIND THE ERROR** Lucinda and Tokei are writing conclusions about the chords in $\odot F$. Who is correct? Explain your reasoning.

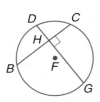

Lucinda

Because $\overline{DG} \perp \overline{BC}$, $\angle DHB \cong \angle DHC \cong \angle CHG \cong \angle BHG$, and $\overline{DG}$ bisects $\overline{BC}$.

Tokei

$\overline{DG} \perp \overline{BC}$, but $\overline{DG}$ does not bisect $\overline{BC}$ because it is not a diameter.

Guided Practice

4. **PROOF** Prove part 2 of Theorem 10.2.

Given: $\odot X$, $\overline{UV} \cong \overline{WY}$

Prove: $\overarc{UV} \cong \overarc{WY}$

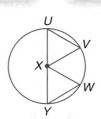

Circle O has a radius of 10, $AB = 10$, and $m\overarc{AB} = 60$. Find each measure.

5. $m\overarc{AY}$ 6. AX 7. OX

In $\odot P$, $PD = 10$, $PQ = 10$, and $QE = 20$. Find each measure.

8. AB 9. PE

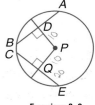

Exercises 5–7

Exercises 8–9

Application 10. **TRAFFIC SIGNS** A yield sign is an equilateral triangle. Find the measure of each arc of the circle circumscribed about the yield sign.

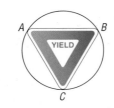

Practice and Apply

Homework Help

For Exercises	See Examples
11–22	3
23–25	2
26–33	4
36–38	1

Extra Practice
See page 774.

In ⊙X, AB = 30, CD = 30, and $m\widehat{CZ}$ = 40. Find each measure.

11. *AM*
12. *MB*
13. *CN*
14. *ND*
15. $m\widehat{DZ}$
16. $m\widehat{CD}$
17. $m\widehat{AB}$
18. $m\widehat{YB}$

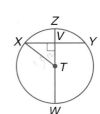

The radius of ⊙P is 5 and PR = 3. Find each measure.

19. *QR*
20. *QS*

In ⊙T, ZV = 1, and TW = 13. Find each measure.

21. *XV*
22. *XY*

Exercises 19–20

Exercises 21–22

TRAFFIC SIGNS Determine the measure of each arc of the circle circumscribed about the traffic sign.

23. regular octagon
24. square
25. rectangle

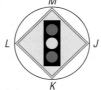

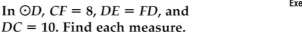

In ⊙F, $\overline{FH} \cong \overline{FL}$ and FK = 17. Find each measure.

26. *LK*
27. *KM*
28. *JG*
29. *JH*

Exercises 26–29

In ⊙D, CF = 8, DE = FD, and DC = 10. Find each measure.

30. *FB*
31. *BC*
32. *AB*
33. *ED*

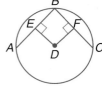

Exercises 30–33

34. **ALGEBRA** In ⊙Z, PZ = ZQ, XY = 4a − 5, and ST = −5a + 13. Find SQ.

35. **ALGEBRA** In ⊙B, the diameter is 20 units long, and m∠ACE = 45. Find x.

Exercise 34

Exercise 35

36. **PROOF** Copy and complete the flow proof of Theorem 10.3.

Given: $\odot P$, $\overline{AB} \perp \overline{TK}$
Prove: $\overline{AR} \cong \overline{BR}$, $\overarc{AK} \cong \overarc{BK}$

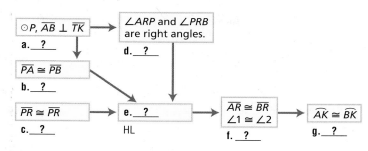

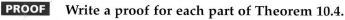

PROOF Write a proof for each part of Theorem 10.4.

37. In a circle, if two chords are equidistant from the center, then they are congruent.

38. In a circle, if two chords are congruent, then they are equidistant from the center.

39. **SAYINGS** An old adage states that "You can't fit a square peg in a round hole." Actually, you can, it just won't fill the hole. If a hole is 4 inches in diameter, what is the approximate width of the largest square peg that fits in the round hole?

For Exercises 40–43, draw and label a figure. Then solve.

40. The radius of a circle is 34 meters long, and a chord of the circle is 60 meters long. How far is the chord from the center of the circle?

41. The diameter of a circle is 60 inches, and a chord of the circle is 48 inches long. How far is the chord from the center of the circle?

42. A chord of a circle is 48 centimeters long and is 10 centimeters from the center of the circle. Find the radius.

43. A diameter of a circle is 32 yards long. A chord is 11 yards from the center. How long is the chord?

44. **CARPENTRY** Mr. Ortega wants to drill a hole in the center of a round picnic table for an umbrella pole. To locate the center of the circle, he draws two chords of the circle and uses a ruler to find the midpoint for each chord. Then he uses a framing square to draw a line perpendicular to each chord at its midpoint. Explain how this process locates the center of the tabletop.

45. **CRITICAL THINKING** A diameter of $\odot P$ has endpoints A and B. Radius $\overline{PQ}$ is perpendicular to $\overline{AB}$. Chord $\overline{DE}$ bisects $\overline{PQ}$ and is parallel to $\overline{AB}$. Does $DE = \frac{1}{2}(AB)$? Explain.

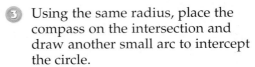

CONSTRUCTION Use the following steps for each construction in Exercises 46 and 47.

① Construct a circle, and place a point on the circle.

③ Using the same radius, place the compass on the intersection and draw another small arc to intercept the circle.

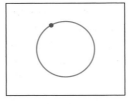

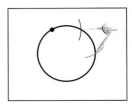

② Using the same radius, place the compass on the point and draw a small arc to intercept the circle.

④ Continue the process in Step 3 until you return to the original point.

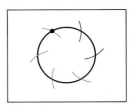

46. Connect the intersections with chords of the circle. What type of figure is formed? Verify you conjecture.

47. Repeat the construction. Connect every other intersection with chords of the circle. What type of figure is formed? Verify your conjecture.

COMPUTERS For Exercises 48 and 49, use the following information.
The hard drive of a computer contains platters divided into tracks, which are defined by concentric circles, and sectors, which are defined by radii of the circles.

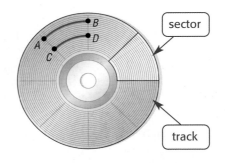

48. In the diagram of a hard drive platter at the right, what is the relationship between $m\widehat{AB}$ and $m\widehat{CD}$?

49. Are $\widehat{AB}$ and $\widehat{CD}$ congruent? Explain.

50. CRITICAL THINKING The figure shows two concentric circles with $\overline{OX} \perp \overline{AB}$ and $\overline{OY} \perp \overline{CD}$. Write a statement relating $\overline{AB}$ and $\overline{CD}$. Verify your reasoning.

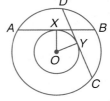

51. **WRITING IN MATH** Answer the question that was posed at the beginning of the lesson.

How do the grooves in a Belgian waffle iron model segments in a circle?

Include the following in your answer:
- a description of how you might find the length of a groove without directly measuring it, and
- a sketch with measurements for a waffle iron that is 8 inches wide.

52. Refer to the figure. Which of the following statements is true?
 I. $\overline{DB}$ bisects $\overline{AC}$. **II.** $\overline{AC}$ bisects $\overline{DB}$. **III.** $OA = OC$

   Ⓐ I and II   Ⓑ II and III
   Ⓒ I and III   Ⓓ I, II, and III

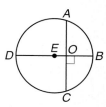

53. SHORT RESPONSE According to the 2000 census, the population of Bridgeworth was 204 thousand, and the population of Sutterly was 216 thousand. If the population of each city increased by exactly 20% ten years later, how many more people will live in Sutterly than in Bridgeworth in 2010?

Maintain Your Skills

Mixed Review In $\odot S$, $m\angle TSR = 42$. Find each measure.
(Lesson 10-2)

54. $m\widehat{KT}$ **55.** $m\widehat{ERT}$ **56.** $m\widehat{KRT}$

Refer to $\odot M$. *(Lesson 10-1)*

57. Name a chord that is not a diameter.
58. If $MD = 7$, find RI.
59. Name congruent segments in $\odot M$.

Exercises 54–56

Exercises 57–59

Getting Ready for the Next Lesson
PREREQUISITE SKILL Solve each equation.
(To review solving equations, see pages 742 and 743.)

60. $\frac{1}{2}x = 120$ **61.** $\frac{1}{2}x = 25$ **62.** $2x = \frac{1}{2}(45 + 35)$

63. $3x = \frac{1}{2}(120 - 60)$ **64.** $45 = \frac{1}{2}(4x + 30)$ **65.** $90 = \frac{1}{2}(6x + 3x)$

Practice Quiz 1   *Lessons 10-1 through 10-3*

PETS For Exercises 1–6, refer to the front circular edge of the hamster wheel shown at the right. *(Lessons 10-1 and 10-2)*

1. Name three radii of the wheel.
2. If $BD = 3x$ and $CB = 7x - 3$, find AC.
3. If $m\angle CBD = 85$, find $m\widehat{AD}$.
4. If $r = 3$ inches, find the circumference of circle B to the nearest tenth of an inch.
5. There are 40 equally-spaced rungs on the wheel. What is the degree measure of an arc connecting two consecutive rungs?
6. What is the length of $\widehat{CAD}$ to the nearest tenth if $m\angle ABD = 150$ and $r = 3$?

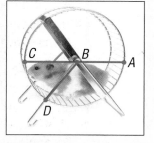

Find each measure. *(Lesson 10-3)*

7. $m\angle CAM$ **8.** $m\widehat{ES}$ **9.** SC **10.** x

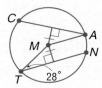
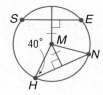
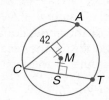
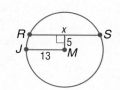

10-4 Inscribed Angles

What You'll Learn

- Find measures of inscribed angles.
- Find measures of angles of inscribed polygons.

Vocabulary
- intercepted

How is a socket like an inscribed polygon?

A socket is a tool that comes in varying diameters. It is used to tighten or unscrew nuts or bolts. The "hole" in the socket is a hexagon cast in a metal cylinder.

INSCRIBED ANGLES   In Lesson 10-3, you learned that a polygon that has its vertices on a circle is called an inscribed polygon. Likewise, an *inscribed angle* is an angle that has its vertex on the circle and its sides contained in chords of the circle.

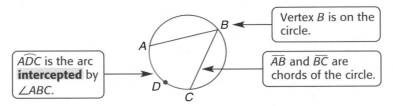

$\overset{\frown}{ADC}$ is the arc **intercepted** by $\angle ABC$.

Vertex *B* is on the circle.

$\overline{AB}$ and $\overline{BC}$ are chords of the circle.

Geometry Activity

Measure of Inscribed Angles

Model
- Use a compass to draw a circle and label the center *W*.
- Draw an inscribed angle and label it *XYZ*.
- Draw $\overline{WX}$ and $\overline{WZ}$.

Analyze
1. Measure $\angle XYZ$ and $\angle XWZ$.
2. Find $m\overset{\frown}{XZ}$ and compare it with $m\angle XYZ$.
3. **Make a conjecture** about the relationship of the measure of an inscribed angle and the measure of its intercepted arc.

This activity suggests the following theorem.

Theorem 10.5

Inscribed Angle Theorem   If an angle is inscribed in a circle, then the measure of the angle equals one-half the measure of its intercepted arc (or the measure of the intercepted arc is twice the measure of the inscribed angle).

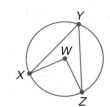

Example: $m\angle ABC = \frac{1}{2}(m\overset{\frown}{ADC})$ or $2(m\angle ABC) = m\overset{\frown}{ADC}$

To prove Theorem 10.5, you must consider three cases.

	Case 1	Case 2	Case 3
Model of Angle Inscribed in ⊙O			
Location of center of circle	on a side of the angle	in the interior of the angle	in the exterior of the angle

Proof *Theorem 10.5 (Case 1)*

Given: ∠ABC inscribed in ⊙D and $\overline{AB}$ is a diameter.

Prove: $m\angle ABC = \frac{1}{2}m\widehat{AC}$

Draw $\overline{DC}$ and let $m\angle B = x$.

Proof:

Since $\overline{DB}$ and $\overline{DC}$ are congruent radii, $\triangle BDC$ is isosceles and $\angle B \cong \angle C$. Thus, $m\angle B = m\angle C = x$. By the Exterior Angle Theorem, $m\angle ADC = m\angle B + m\angle C$. So $m\angle ADC = 2x$. From the definition of arc measure, we know that $m\widehat{AC} = m\angle ADC$ or $2x$. Comparing $m\widehat{AC}$ and $m\angle ABC$, we see that $m\widehat{AC} = 2(m\angle ABC)$ or that $m\angle ABC = \frac{1}{2}m\widehat{AC}$.

You will prove Cases 2 and 3 of Theorem 10.5 in Exercises 35 and 36.

Study Tip

Using Variables
You can also assign a variable to an unknown measure. So, if you let $m\widehat{AD} = x$, the second equation becomes $140 + 100 + x + x = 360$, or $240 + 2x = 360$. This last equation may seem simpler to solve.

Example **1** *Measures of Inscribed Angles*

In ⊙O, $m\widehat{AB} = 140$, $m\widehat{BC} = 100$, and $m\widehat{AD} = m\widehat{DC}$.

Find the measures of the numbered angles.

First determine $m\widehat{DC}$ and $m\widehat{AD}$.

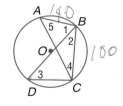

$$m\widehat{AB} + m\widehat{BC} + m\widehat{DC} + m\widehat{AD} = 360 \quad \text{Arc Addition Theorem}$$

$$140 + 100 + m\widehat{DC} + m\widehat{DC} = 360 \quad \begin{array}{l}m\widehat{AB} = 140, m\widehat{BC} = 100, \\ m\widehat{DC} = m\widehat{AD}\end{array}$$

$$240 + 2(m\widehat{DC}) = 360 \quad \text{Simplify.}$$

$$2(m\widehat{DC}) = 120 \quad \text{Subtract 240 from each side.}$$

$$m\widehat{DC} = 60 \quad \text{Divide each side by 2.}$$

So, $m\widehat{DC} = 60$ and $m\widehat{AD} = 60$.

$m\angle 1 = \frac{1}{2}m\widehat{AD}$ $\qquad$ $m\angle 2 = \frac{1}{2}m\widehat{DC}$

$\quad = \frac{1}{2}(60)$ or 30 $\qquad$ $\quad = \frac{1}{2}(60)$ or 30

$m\angle 3 = \frac{1}{2}m\widehat{BC}$ $\qquad$ $m\angle 4 = \frac{1}{2}m\widehat{AB}$

$\quad = \frac{1}{2}(100)$ or 50 $\qquad$ $\quad = \frac{1}{2}(140)$ or 70

$m\angle 5 = \frac{1}{2}m\widehat{BC}$

$\quad = \frac{1}{2}(100)$ or 50

In Example 1, note that ∠3 and ∠5 intercept the same arc and are congruent.

Theorem 10.6

If two inscribed angles of a circle (or congruent circles) intercept congruent arcs or the same arc, then the angles are congruent.

Abbreviations:
Inscribed ∠s of ≅ arcs are ≅.
Inscribed ∠s of same arc are ≅.

Examples:

∠DAC ≅ ∠DBC   ∠FAE ≅ ∠CBD

You will prove Theorem 10.6 in Exercise 37.

Example 2 Proofs with Inscribed Angles

Given: ⊙P with $\overline{CD} \cong \overline{AB}$
Prove: △AXB ≅ △CXD

Proof:

Statements	Reasons
1. ∠DAB intercepts $\overset{\frown}{DB}$. ∠DCB intercepts $\overset{\frown}{DB}$.	1. Definition of intercepted arc
2. ∠DAB ≅ ∠DCB	2. Inscribed ∠s of same arc are ≅.
3. ∠1 ≅ ∠2	3. Vertical ∠s are ≅.
4. $\overline{CD} \cong \overline{AB}$	4. Given
5. △AXB ≅ △CXD	5. AAS

You can also use the measure of an inscribed angle to determine probability of a point lying on an arc.

Example 3 Inscribed Arcs and Probability

PROBABILITY   Points A and B are on a circle so that $m\overset{\frown}{AB} = 60$. Suppose point D is randomly located on the same circle so that it does not coincide with A or B. What is the probability that $m\angle ADB = 30$?

Since the angle measure is half the arc measure, inscribed ∠ADB must intercept $\overset{\frown}{AB}$, so D must lie on major arc AB. Draw a figure and label any information you know.

$$m\overset{\frown}{BDA} = 360 - m\overset{\frown}{AB}$$
$$= 360 - 60 \text{ or } 300$$

Since ∠ADB must intercept $\overset{\frown}{AB}$, the probability that $m\angle ADB = 30$ is the same as the probability of D being contained in $\overset{\frown}{BDA}$.

The probability that D is located on $\overset{\frown}{ADB}$ is $\frac{5}{6}$. So, the probability that $m\angle ADB = 30$ is also $\frac{5}{6}$.

ANGLES OF INSCRIBED POLYGONS   An inscribed triangle with a side that is a diameter is a special type of triangle.

Theorem 10.7

If an inscribed angle intercepts a semicircle, the angle is a right angle.

Example: $\overset{\frown}{ADC}$ is a semicircle, so $m\angle ABC = 90$.

You will prove Theorem 10.7 in Exercise 38.

Example 4 Angles of an Inscribed Triangle

ALGEBRA   **Triangles *ABD* and *ADE* are inscribed in $\odot F$ with $\overset{\frown}{AB} \cong \overset{\frown}{BD}$. Find the measure of each numbered angle if $m\angle 1 = 12x - 8$ and $m\angle 2 = 3x + 8$.**

AED is a right angle because $\overset{\frown}{AED}$ is a semicircle.

$m\angle 1 + m\angle 2 + m\angle AED = 180$	Angle Sum Theorem
$(12x - 8) + (3x + 8) + 90 = 180$	$m\angle 1 = 12x - 8, m\angle 2 = 3x + 8, m\angle AED = 90$
$15x + 90 = 180$	Simplify.
$15x = 90$	Subtract 90 from each side.
$x = 6$	Divide each side by 15.

Use the value of *x* to find the measures of $\angle 1$ and $\angle 2$.

$m\angle 1 = 12x - 8$	Given		$m\angle 2 = 3x + 8$	Given	
$= 12(6) - 8$	$x = 6$		$= 3(6) + 8$	$x = 6$	
$= 64$	Simplify.		$= 26$	Simplify.	

Angle *ABD* is a right angle because it intercepts a semicircle. Because $\overset{\frown}{AB} \cong \overset{\frown}{BD}$, $\overline{AB} \cong \overline{BD}$, which leads to $\angle 3 \cong \angle 4$. Thus, $m\angle 3 = m\angle 4$.

$m\angle 3 + m\angle 4 + m\angle ABD = 180$	Angle Sum Theorem
$m\angle 3 + m\angle 3 + 90 = 180$	$m\angle 3 = m\angle 4, m\angle ABD = 90$
$2(m\angle 3) + 90 = 180$	Simplify.
$2(m\angle 3) = 90$	Subtract 90 from each side.
$m\angle 3 = 45$	Divide each side by 2.

Since $m\angle 3 = m\angle 4$, $m\angle 4 = 45$.

Example 5 Angles of an Inscribed Quadrilateral

Quadrilateral *ABCD* is inscribed in $\odot P$. If $m\angle B = 80$ and $m\angle C = 40$, find $m\angle A$ and $m\angle D$.

Draw a sketch of this situation.

To find $m\angle A$, we need to know $m\overset{\frown}{BCD}$.

To find $m\overset{\frown}{BCD}$, first find $m\overset{\frown}{DAB}$.

$m\overset{\frown}{DAB} = 2(m\angle C)$	Inscribed Angle Theorem
$= 2(40)$ or 80	$m\angle C = 40$

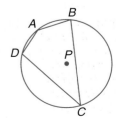

(continued on the next page)

$$m\widehat{BCD} + m\widehat{DAB} = 360 \qquad \text{Sum of angles in circle} = 360$$
$$m\widehat{BCD} + 80 = 360 \qquad m\widehat{DAB} = 80$$
$$m\widehat{BCD} = 280 \qquad \text{Subtract 80 from each side.}$$
$$m\widehat{BCD} = 2(m\angle A) \qquad \text{Inscribed Angle Theorem}$$
$$280 = 2(m\angle A) \qquad \text{Substitution}$$
$$140 = m\angle A \qquad \text{Divide each side by 2.}$$

To find $m\angle D$, we need to know $m\widehat{ABC}$, but first we must find $m\widehat{ADC}$.

$$m\widehat{ADC} = 2(m\angle B) \qquad \text{Inscribed Angle Theorem}$$
$$m\widehat{ADC} = 2(80) \text{ or } 160 \qquad m\angle B = 80$$
$$m\widehat{ABC} + m\widehat{ADC} = 360 \qquad \text{Sum of angles in circle} = 360$$
$$m\widehat{ABC} + 160 = 360 \qquad m\widehat{ADC} = 160$$
$$m\widehat{ABC} = 200 \qquad \text{Subtract 160 from each side.}$$
$$m\widehat{ABC} = 2(m\angle D) \qquad \text{Inscribed Angle Theorem}$$
$$200 = 2(m\angle D) \qquad \text{Substitution}$$
$$100 = m\angle D \qquad \text{Divide each side by 2.}$$

In Example 5, note that the opposite angles of the quadrilateral are supplementary. This is stated in Theorem 10.8 and can be verified by considering that the arcs intercepted by opposite angles of an inscribed quadrilateral form a circle.

Theorem 10.8

If a quadrilateral is inscribed in a circle, then its opposite angles are supplementary.

Example:
Quadrilateral $ABCD$ is inscribed in $\odot P$.
$\angle A$ and $\angle C$ are supplementary.
$\angle B$ and $\angle D$ are supplementary.

You will prove this theorem in Exercise 39.

Check for Understanding

Concept Check
1. **OPEN ENDED** Draw a counterexample of an inscribed trapezoid. If possible, include at least one angle that is an inscribed angle.

2. **Compare and contrast** an inscribed angle and a central angle that intercepts the same arc.

Guided Practice
3. In $\odot R$, $m\widehat{MN} = 120$ and $m\widehat{MQ} = 60$. Find the measure of each numbered angle.

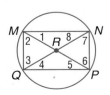

4. **PROOF** Write a paragraph proof.
   Given: Quadrilateral $ABCD$ is inscribed in $\odot P$.
   $$m\angle C = \frac{1}{2}m\angle B$$

   Prove: $m\widehat{CDA} = 2(m\widehat{DAB})$

5. **ALGEBRA** In $\odot A$ at the right, $\overset{\frown}{PQ} \cong \overset{\frown}{RS}$. Find the measure of each numbered angle if $m\angle 1 = 6x + 11$, $m\angle 2 = 9x + 19$, $m\angle 3 = 4y - 25$, and $m\angle 4 = 3y - 9$.

6. Suppose quadrilateral $VWXY$ is inscribed in $\odot C$. If $m\angle X = 28$ and $m\angle W = 110$, find $m\angle V$ and $m\angle Y$.

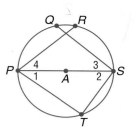

Application 7. **PROBABILITY** Points X and Y are endpoints of a diameter of $\odot W$. Point Z is another point on the circle. Find the probability that $\angle XZY$ is a right angle.

Practice and Apply

<table>
<tr><td colspan="2">**Homework Help**</td></tr>
<tr><td>For
Exercises</td><td>See
Examples</td></tr>
<tr><td>8–10</td><td>1</td></tr>
<tr><td>11–12,
35–39</td><td>2</td></tr>
<tr><td>13–17</td><td>4</td></tr>
<tr><td>18–21,
26–29</td><td>5</td></tr>
<tr><td>31–34</td><td>3</td></tr>
</table>

Extra Practice
See page 774.

Find the measure of each numbered angle for each figure.

8. $\overset{\frown}{PQ} \cong \overset{\frown}{RQ}$, $m\overset{\frown}{PS} = 45$, and $m\overset{\frown}{SR} = 75$

9. $m\angle BDC = 25$, $m\overset{\frown}{AB} = 120$, and $m\overset{\frown}{CD} = 130$

10. $m\overset{\frown}{XZ} = 100$, $\overline{XY} \perp \overline{ST}$, and $\overline{ZW} \perp \overline{ST}$

PROOF Write a two-column proof.

11. **Given:** $\overset{\frown}{AB} \cong \overset{\frown}{DE}$, $\overset{\frown}{AC} \cong \overset{\frown}{CE}$
    Prove: $\triangle ABC \cong \triangle EDC$

12. **Given:** $\odot P$
    Prove: $\triangle AXB \sim \triangle CXD$

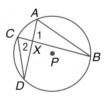

ALGEBRA Find the measure of each numbered angle for each figure.

13. $m\angle 1 = x$, $m\angle 2 = 2x - 13$

14. $m\overset{\frown}{AB} = 120$

15. $m\angle R = \frac{1}{3}x + 5$, $m\angle K = \frac{1}{2}x$

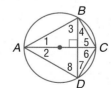

16. $PQRS$ is a rhombus inscribed in a circle. Find $m\angle QRP$ and $m\overset{\frown}{SP}$.

17. In $\odot D$, $\overline{DE} \cong \overline{EC}$, $m\overset{\frown}{CF} = 60$, and $\overline{DE} \perp \overline{EC}$. Find $m\angle 4$, $m\angle 5$, and $m\overset{\frown}{AF}$.

Exercise 16

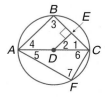

Exercise 17

18. Quadrilateral *WRTZ* is inscribed in a circle. If $m\angle W = 45$ and $m\angle R = 100$, find $m\angle T$ and $m\angle Z$.

19. Trapezoid *ABCD* is inscribed in a circle. If $m\angle A = 60$, find $m\angle B$, $m\angle C$, and $m\angle D$.

20. Rectangle *PDQT* is inscribed in a circle. What can you conclude about $\overline{PQ}$?

21. Square *EDFG* is inscribed in a circle. What can you conclude about $\overline{EF}$?

Equilateral pentagon *PQRST* is inscribed in ⊙*U*.
Find each measure.

22. $m\widehat{QR}$ **23.** $m\angle PSR$

24. $m\angle PQR$ **25.** $m\widehat{PTS}$

Quadrilateral *ABCD* is inscribed in ⊙*Z* such that
$m\angle BZA = 104$, $m\widehat{CB} = 94$, and $\overline{AB} \parallel \overline{DC}$. **Find each**
measure.

26. $m\widehat{BA}$ **27.** $m\widehat{ADC}$

28. $m\angle BDA$ **29.** $m\angle ZAC$

30. SCHOOL RINGS Some designs of class rings involve adding gold or silver to the surface of the round stone. The design at the right includes two inscribed angles. If $m\angle ABC = 50$ and $m\widehat{DBF} = 128$, find $m\widehat{AC}$ and $m\angle DEF$.

PROBABILITY For Exercises 31–34, use the following information.
Point *T* is randomly selected on ⊙*C* so that it does not coincide with points *P*, *Q*, *R*, or *S*. $\overline{SQ}$ is a diameter of ⊙*C*.

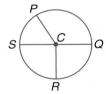

31. Find the probability that $m\angle PTS = 20$ if $m\widehat{PS} = 40$.

32. Find the probability that $m\angle PTR = 55$ if $m\widehat{PSR} = 110$.

33. Find the probability that $m\angle STQ = 90$.

34. Find the probability that $m\angle PTQ = 180$.

PROOF Write the indicated proof for each theorem.

35. two-column proof:
Case 2 of Theorem 10.5
Given: *T* lies inside $\angle PRQ$.
$\overline{RK}$ is a diameter of ⊙*T*.
Prove: $m\angle PRQ = \frac{1}{2}m\widehat{PKQ}$

36. two-column proof:
Case 3 of Theorem 10.5
Given: *T* lies outside $\angle PRQ$.
$\overline{RK}$ is a diameter of ⊙*T*.
Prove: $m\angle PRQ = \frac{1}{2}m\widehat{PQ}$

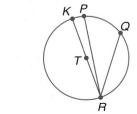

37. two-column proof:
Theorem 10.6

38. paragraph proof:
Theorem 10.7

39. paragraph proof:
Theorem 10.8

STAINED GLASS   In the stained glass window design, all of the small arcs around the circle are congruent. Suppose the center of the circle is point O.

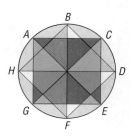

40. What is the measure of each of the small arcs?

41. What kind of figure is $\triangle AOC$? Explain.

42. What kind of figure is quadrilateral $BDFH$? Explain.

43. What kind of figure is quadrilateral $ACEG$? Explain.

44. CRITICAL THINKING   A trapezoid $ABCD$ is inscribed in $\odot O$. Explain how you can verify that $ABCD$ must be an isosceles trapezoid.

45. [**WRITING IN MATH**]   Answer the question that was posed at the beginning of the lesson.

How is a socket like an inscribed polygon?

Include the following in your answer:
- a definition of an inscribed polygon, and
- the side length of a regular hexagon inscribed in a circle $\frac{3}{4}$ inch wide.

46. What is the ratio of the measure of $\angle ACB$ to the measure of $\angle AOB$?

  Ⓐ $1 : 1$ Ⓑ $2 : 1$

  Ⓒ $1 : 2$ Ⓓ not enough information

47. GRID IN   The daily newspaper always follows a particular format. Each even-numbered page contains six articles, and each odd-numbered page contains seven articles. If today's paper has 36 pages, how many articles does it contain?

Maintain Your Skills

Mixed Review   **Find each measure.**   *(Lesson 10-3)*

48. If $AB = 60$ and $DE = 48$, find CF.

49. If $AB = 32$ and $FC = 11$, find FE.

50. If $DE = 60$ and $FC = 16$, find AB.

Points Q and R lie on $\odot P$. Find the length of $\overparen{QR}$ for the given radius and angle measure.   *(Lesson 10-2)*

51. $PR = 12$, and $m\angle QPR = 60$ **52.** $m\angle QPR = 90$, $PR = 16$

Complete each sentence with *sometimes*, *always*, or *never*.   *(Lesson 4-1)*

53. Equilateral triangles are ___?___ isosceles.

54. Acute triangles are ___?___ equilateral.

55. Obtuse triangles are ___?___ scalene.

Getting Ready for the Next Lesson   **PREREQUISITE SKILL   Determine whether each figure is a right triangle.**
*(To review the **Pythagorean Theorem**, see Lesson 7-2.)*

56.

57.

58.
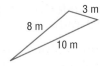

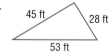

10-5 Tangents

What You'll Learn

- Use properties of tangents.
- Solve problems involving circumscribed polygons.

Vocabulary

- tangent
- point of tangency

How are tangents related to track and field events?

In July 2001, Yipsi Moreno of Cuba won her first major title in the hammer throw at the World Athletic Championships in Edmonton, Alberta, Canada, with a throw of 70.65 meters. The hammer is a metal ball, usually weighing 16 pounds, attached to a steel wire at the end of which is a grip. The ball is spun around by the thrower and then released, with the greatest distance thrown winning the event.

TANGENTS The figure models the hammer throw event. Circle *A* represents the circular area containing the spinning thrower. Ray *BC* represents the path the hammer takes when released. $\overrightarrow{BC}$ is **tangent** to $\odot A$, because the line containing $\overrightarrow{BC}$ intersects the circle in exactly one point. This point is called the **point of tangency**.

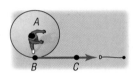

Geometry Software Investigation

Tangents and Radii

Study Tip

Tangent Lines
All of the theorems applying to tangent lines also apply to parts of the line that are tangent to the circle.

Model

- Use The Geometer's Sketchpad to draw a circle with center *W*. Then draw a segment tangent to $\odot W$. Label the point of tangency as *X*.
- Choose another point on the tangent and name it *Y*. Draw $\overline{WY}$.

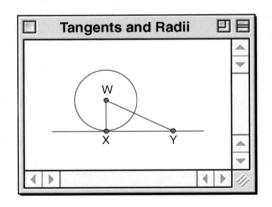

Think and Discuss

1. What is $\overline{WX}$ in relation to the circle?

2. Measure $\overline{WY}$ and $\overline{WX}$. Write a statement to relate *WX* and *WY*.

3. Move point *Y* along the tangent. How does the location of *Y* affect the statement you wrote in Exercise 2?

4. Measure $\angle WXY$. What conclusion can you make?

5. **Make a conjecture** about the shortest distance from the center of the circle to a tangent of the circle.

This investigation suggests an indirect proof of Theorem 10.9.

Theorem 10.9

If a line is tangent to a circle, then it is perpendicular to the radius drawn to the point of tangency.

Example: If $\overleftrightarrow{RT}$ is a tangent, $\overline{OR} \perp \overleftrightarrow{RT}$.

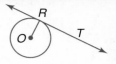

Example 1 Find Lengths

ALGEBRA $\overline{ED}$ is tangent to $\odot F$ at point E. Find x.

Because the radius is perpendicular to the tangent at the point of tangency, $\overline{EF} \perp \overline{DE}$. This makes $\angle DEF$ a right angle and $\triangle DEF$ a right triangle. Use the Pythagorean Theorem to find x.

$$(EF)^2 + (DE)^2 = (DF)^2 \quad \text{Pythagorean Theorem}$$
$$3^2 + 4^2 = x^2 \quad\quad EF = 3, DE = 4, DF = x$$
$$25 = x^2 \quad\quad\quad \text{Simplify.}$$
$$\pm 5 = x \quad\quad\quad \text{Take the square root of each side.}$$

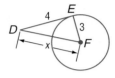

Because x is the length of $\overline{DF}$, ignore the negative result. Thus, $x = 5$.

The converse of Theorem 10.9 is also true.

Theorem 10.10

If a line is perpendicular to a radius of a circle at its endpoint on the circle, then the line is tangent to the circle.

Example: If $\overline{OR} \perp \overleftrightarrow{RT}$, $\overleftrightarrow{RT}$ is a tangent.

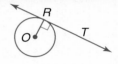

You will prove this theorem in Exercise 22.

Study Tip

Identifying Tangents
Never assume that a segment is tangent to a circle by appearance unless told otherwise. The figure must either have a right angle symbol or include the measurements that confirm a right angle.

Example 2 Identify Tangents

a. **Determine whether $\overline{MN}$ is tangent to $\odot L$.**

First determine whether $\triangle LMN$ is a right triangle by using the converse of the Pythagorean Theorem.

$$(LM)^2 + (MN)^2 \stackrel{?}{=} (LN)^2 \quad \text{Converse of Pythagorean Theorem}$$
$$3^2 + 4^2 \stackrel{?}{=} 5^2 \quad\quad LM = 3, MN = 4, LN = 3 + 2 \text{ or } 5$$
$$25 = 25 \checkmark \quad\quad \text{Simplify.}$$

Because the converse of the Pythagorean Theorem is true, $\triangle LMN$ is a right triangle and $\angle LMN$ is a right angle. Thus, $\overline{LM} \perp \overline{MN}$, making $\overline{MN}$ a tangent to $\odot L$.

b. **Determine whether $\overline{PQ}$ is tangent to $\odot R$.**

Since $RQ = RS$, $RP = 4 + 4$ or 8 units.

$$(RQ)^2 + (PQ)^2 \stackrel{?}{=} (RP)^2 \quad \text{Converse of Pythagorean Theorem}$$
$$4^2 + 5^2 \stackrel{?}{=} 8^2 \quad\quad RQ = 4, PQ = 5, RP = 8$$
$$41 \neq 64 \quad\quad \text{Simplify.}$$

Because the converse of the Pythagorean Theorem did not prove true in this case, $\triangle RQP$ is not a right triangle.

So, $\overline{PQ}$ is not tangent to $\odot R$.

 www.geometryonline.com/extra_examples

More than one line can be tangent to the same circle. In the figure, $\overline{AB}$ and $\overline{BC}$ are tangent to $\odot D$. So, $(AB)^2 + (AD)^2 = (DB)^2$ and $(BC)^2 + (CD)^2 = (DB)^2$.

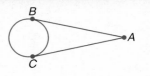

$(AB)^2 + (AD)^2 = (BC)^2 + (CD)^2$ Substitution

$(AB)^2 + (AD)^2 = (BC)^2 + (AD)^2$ $AD = CD$

$\quad\quad(AB)^2 = (BC)^2$ Subtract $(AD)^2$ from each side.

$\quad\quad\quad AB = BC$ Take the square root of each side.

The last statement implies that $\overline{AB} \cong \overline{BC}$. This is a proof of Theorem 10.10.

Theorem 10.11

If two segments from the same exterior point are tangent to a circle, then they are congruent.

Example: $\overline{AB} \cong \overline{AC}$

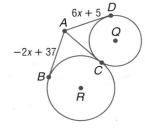

You will prove this theorem in Exercise 27.

Example 3   Solve a Problem Involving Tangents

ALGEBRA   Find x. Assume that segments that appear tangent to circles are tangent.

$\overline{AD}$ and $\overline{AC}$ are drawn from the same exterior point and are tangent to $\odot Q$, so $\overline{AD} \cong \overline{AC}$. $\overline{AC}$ and $\overline{AB}$ are drawn from the same exterior point and are tangent to $\odot R$, so $\overline{AC} \cong \overline{AB}$. By the Transitive Property, $\overline{AD} \cong \overline{AB}$.

$\quad\quad AD = AB$ Definition of congruent segments

$6x + 5 = -2x + 37$ Substitution

$8x + 5 = 37$ Add $2x$ to each side.

$\quad\quad 8x = 32$ Subtract 5 from each side.

$\quad\quad\quad x = 4$ Divide each side by 8.

Construction

Line Tangent to a Circle Through a Point Exterior to the Circle

1 Construct a circle. Label the center C. Draw a point outside $\odot C$. Then draw $\overline{CA}$.

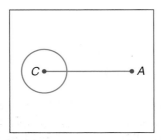

2 Construct the perpendicular bisector of $\overline{CA}$ and label it line ℓ. Label the intersection of ℓ and $\overline{CA}$ as point X.

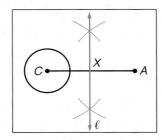

3 Construct circle X with radius XC. Label the points where the circles intersect as D and E.

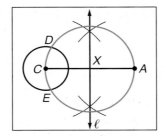

4 Draw $\overline{AD}$. $\triangle ADC$ is inscribed in a semicircle. So $\angle ADC$ is a right angle, and $\overline{AD}$ is a tangent.

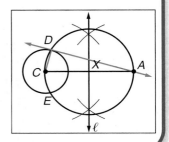

You will construct a line tangent to a circle through a point on the circle in Exercise 21.

CIRCUMSCRIBED POLYGONS   In Lesson 10-3, you learned that circles can be circumscribed about a polygon. Likewise, polygons can be circumscribed about a circle, or the circle is inscribed in the polygon. Notice that the vertices of the polygon *do not* lie on the circle, but every side of the polygon is tangent to the circle.

Polygons are circumscribed.   Polygons are *not* circumscribed.

Example 4 Triangles Circumscribed About a Circle

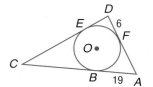

Triangle *ADC* is circumscribed about ⊙*O*. Find the perimeter of △*ADC* if *EC = DE + AF*.

Use Theorem 10.10 to determine the equal measures.
AB = AF = 19, *FD = DE* = 6, and *EC = CB*.
We are given that *EC = DE + AF*, so *EC* = 6 + 19 or 25.

$P = AB + BC + EC + DE + FD + AF$ Definition of perimeter

$\quad = 19 + 25 + 25 + 6 + 6 + 19$ or 100   Substitution

The perimeter of △*ADC* is 100 units.

Check for Understanding

Concept Check   1. **Determine** the number of tangents that can be drawn to a circle for each point. Explain your reasoning.

    a. containing a point outside the circle

    b. containing a point inside the circle

    c. containing a point on the circle

2. Write an argument to support or provide a counterexample to the statement *If two lines are tangent to the same circle, they intersect.*

3. **OPEN ENDED**   Draw an example of a circumscribed polygon and an example of an inscribed polygon.

Guided Practice   **For Exercises 4 and 5, use the figure at the right.**

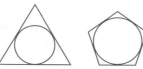

4. Tangent $\overline{MP}$ is drawn to ⊙*O*. Find *x* if *MO* = 20.

5. If *RO* = 13, determine whether $\overline{PR}$ is tangent to ⊙*O*.

6. Rhombus *ABCD* is circumscribed about ⊙*P* and has a perimeter of 32. Find *x*.

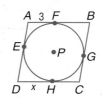

Application   7. **AGRICULTURE**   A pivot-circle irrigation system waters part of a fenced square field. If the spray extends to a distance of 72 feet, what is the total length of the fence around the field?

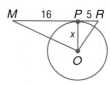

Determine whether each segment is tangent to the given circle.

8. $\overline{BC}$

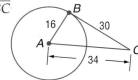

9. $\overline{DE}$

10. $\overline{GH}$

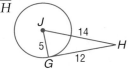

11. $\overline{KL}$

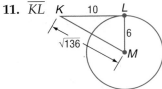

Find x. Assume that segments that appear to be tangent are tangent.

12.

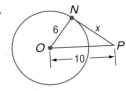

13.

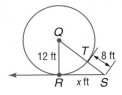

14.

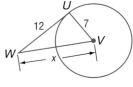

15.

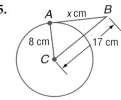

16.

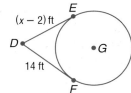

17.

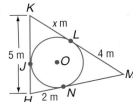

18.

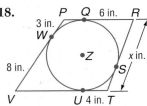

19.

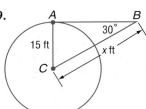

20.

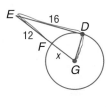

21. **CONSTRUCTION** Construct a line tangent to a circle through a point on the circle following these steps.
   - Construct a circle with center T.
   - Locate a point P on $\odot T$ and draw $\overrightarrow{TP}$.
   - Construct a perpendicular to $\overrightarrow{TP}$ through point P.

22. **PROOF** Write an indirect proof of Theorem 10.10 by assuming that ℓ is not tangent to $\odot A$.

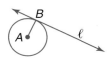

   Given: $\ell \perp \overline{AB}$, $\overline{AB}$ is a radius of $\odot A$.

   Prove: Line ℓ is tangent to $\odot A$.

Find the perimeter of each polygon for the given information.

23.

24. $ST = 18$, radius of $\odot P = 5$

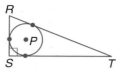

25. $BY = CZ = AX = 2.5$
    diameter of $\odot G = 5$

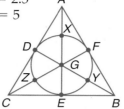

26. $CF = 6(3 - x)$, $DB = 12y - 4$

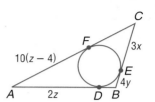

27. **PROOF** Write a two-column proof to show that if two segments from the same exterior point are tangent to a circle, then they are congruent. (Theorem 10.11)

28. **PHOTOGRAPHY** The film in a 35-mm camera unrolls from a cylinder, travels across an opening for exposure, and then is forwarded into another circular chamber as each photograph is taken. The roll of film has a diameter of 25 millimeters, and the distance from the center of the roll to the intake of the chamber is 100 millimeters. To the nearest millimeter, how much of the film would be exposed if the camera were opened before the roll had been totally used?

ASTRONOMY For Exercises 29 and 30, use the following information.
A solar eclipse occurs when the moon blocks the sun's rays from hitting Earth. Some areas of the world will experience a total eclipse, others a partial eclipse, and some no eclipse at all, as shown in the diagram below.

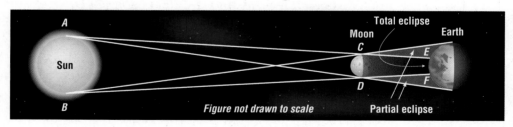

29. The blue section denotes a total eclipse on that part of Earth. Which tangents define the blue area?

30. The pink areas denote the portion of Earth that will have a partial eclipse. Which tangents define the northern and southern boundaries of the partial eclipse?

31. **CRITICAL THINKING** Find the measure of tangent $\overline{GN}$. Explain your reasoning.

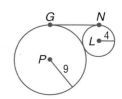

32. WRITING IN MATH Answer the question that was posed at the beginning of the lesson.

How are tangents related to track and field events?

Include the following in your answer:

- how the hammer throw models a tangent, and
- the distance the hammer landed from the athlete if the wire and handle are 1.2 meters long and the athlete's arm is 0.8 meter long.

Standardized Test Practice
Ⓐ Ⓑ Ⓒ Ⓓ

33. GRID IN $\overline{AB}$, $\overline{BC}$, $\overline{CD}$, and $\overline{AD}$ are tangent to a circle. If $AB = 19$, $BC = 6$, and $CD = 14$, find AD.

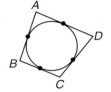

34. ALGEBRA Find the mean of all of the numbers from 1 to 1000 that end in 2.

Ⓐ 496　　　Ⓑ 497
Ⓒ 498　　　Ⓓ 500

Extending the Lesson

A line that is tangent to two circles in the same plane is called a *common tangent*.

Common internal tangents intersect the segment connecting the centers.	*Common external tangents* do not intersect the segment connecting the centers.
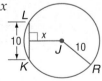 Lines k and j are common internal tangents.	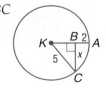 Lines ℓ and m are common external tangents.

Refer to the diagram of the eclipse on page 557.

35. Name two common internal tangents. **36.** Name two common external tangents.

Maintain Your Skills

Mixed Review

37. LOGOS Circles are often used in logos for commercial products. The logo at the right shows two inscribed angles and two central angles. If $\overarc{AC} \cong \overarc{BD}$, $m\overarc{AF} = 90$, $m\overarc{FE} = 45$, and $m\overarc{ED} = 90$, find $m\angle AFC$ and $m\angle BED$. *(Lesson 10-4)*

Find each measure. *(Lesson 10-3)*

38. x

39. BC

40. AP

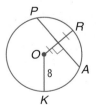

41. PROOF Write a coordinate proof to show that if E is the midpoint of $\overline{AB}$ in rectangle $ABCD$, then $\triangle CED$ is isosceles. *(Lesson 8-7)*

Getting Ready for the Next Lesson

PREREQUISITE SKILL Solve each equation.
(To review solving equations, see pages 737 and 738.)

42. $x + 3 = \frac{1}{2}[(4x + 6) - 10]$

43. $2x - 5 = \frac{1}{2}[(3x + 16) - 20]$

44. $2x + 4 = \frac{1}{2}[(x + 20) - 10]$

45. $x + 3 = \frac{1}{2}[(4x + 10) - 45]$

Inscribed and Circumscribed Triangles

In Lesson 5-1, you learned that there are special points of concurrency in a triangle. Two of these will be used in these activities.

- The *incenter* is the point at which the angle bisectors meet. It is equidistant from the sides of the triangle.
- The *circumcenter* is the point at which the perpendicular bisectors of the sides intersect. It is equidistant from the vertices of the triangle.

Activity 1

Construct a circle inscribed in a triangle. *The triangle is circumscribed about the circle.*

1 Draw a triangle and label its vertices A, B, and C. Construct two angle bisectors of the triangle to locate the incenter. Label it D.

2 Construct a segment perpendicular to a side of $\triangle ABC$ through the incenter. Label the intersection E.

3 Use the compass to measure DE. Then put the point of the compass on D, and draw a circle with that radius.

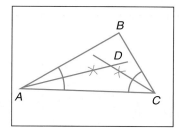
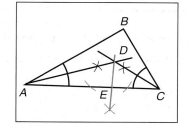
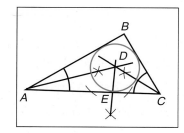

Activity 2

Construct a circle through any three noncollinear points.
This construction may be referred to as circumscribing a circle about a triangle.

1 Draw a triangle and label its vertices A, B, and C. Construct perpendicular bisectors of two sides of the triangle to locate the circumcenter. Label it D.

2 Use the compass to measure the distance from the circumcenter D to any of the three vertices.

3 Using that setting, place the compass point at D, and draw a circle about the triangle.

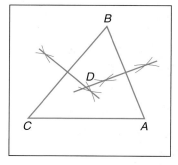
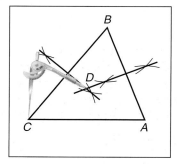
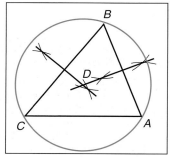

(continued on the next page)

For the next activity, refer to the construction of an inscribed regular hexagon on page 542.

Activity 3

Construct an equilateral triangle circumscribed about a circle.

① Construct a circle and divide it into six congruent arcs.

② Place a point at every other arc. Draw rays from the center through these points.

③ Construct a line perpendicular to each of the rays through the points.

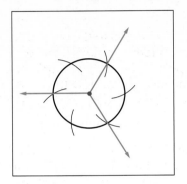

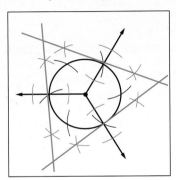

Model

1. Draw an obtuse triangle and inscribe a circle in it.

2. Draw a right triangle and circumscribe a circle about it.

3. Draw a circle of any size and circumscribe an equilateral triangle about it.

Analyze

Refer to Activity 1.

4. Why do you only have to construct the perpendicular to one side of the triangle?

5. How can you use the Incenter Theorem to explain why this construction is valid?

Refer to Activity 2.

6. Why do you only have to measure the distance from the circumcenter to any one vertex?

7. How can you use the Circumcenter Theorem to explain why this construction is valid?

Refer to Activity 3.

8. What is the measure of each of the six congruent arcs?

9. Write a convincing argument as to why the lines constructed in Step 3 form an equilateral triangle.

10. Why do you think the terms *incenter* and *circumcenter* are good choices for the points they define?

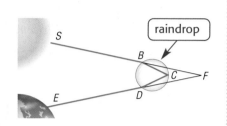

Secants, Tangents, and Angle Measures

10-6

What You'll Learn

- Find measures of angles formed by lines intersecting on or inside a circle.
- Find measures of angles formed by lines intersecting outside the circle.

Vocabulary
- secant

How is a rainbow formed by segments of a circle?

Droplets of water in the air refract or bend sunlight as it passes through them, creating a rainbow. The various angles of refraction result in an arch of colors. In the figure, the sunlight from point *S* enters the raindrop at *B* and is bent. The light proceeds to the back of the raindrop, and is reflected at *C* to leave the raindrop at point *D* heading to Earth. Angle *F* represents the measure of how the resulting ray of light deviates from its original path.

INTERSECTIONS ON OR INSIDE A CIRCLE A line that intersects a circle in exactly two points is called a **secant**. In the figure above, $\overleftrightarrow{SF}$ and $\overleftrightarrow{EF}$ are secants of the circle. When two secants intersect inside a circle, the angles formed are related to the arcs they intercept.

Theorem 10.12

If two secants intersect in the interior of a circle, then the measure of an angle formed is one-half the sum of the measure of the arcs intercepted by the angle and its vertical angle.

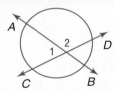

Examples: $m\angle 1 = \frac{1}{2}(m\widehat{AC} + m\widehat{BD})$

$m\angle 2 = \frac{1}{2}(m\widehat{AD} + m\widehat{BC})$

Proof Theorem 10.12

Given: secants $\overleftrightarrow{RT}$ and $\overleftrightarrow{SU}$

Prove: $m\angle 1 = \frac{1}{2}(m\widehat{ST} + m\widehat{RU})$

Draw $\overline{RS}$. Label $\angle TRS$ as $\angle 2$ and $\angle USR$ as $\angle 3$.

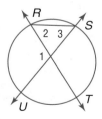

Proof:

Statements	Reasons
1. $m\angle 1 = m\angle 2 + m\angle 3$	1. Exterior Angle Theorem
2. $m\angle 2 = \frac{1}{2}m\widehat{ST}, m\angle 3 = \frac{1}{2}m\widehat{RU}$	2. The measure of inscribed $\angle$ = half the measure of the intercepted arc.
3. $m\angle 1 = \frac{1}{2}m\widehat{ST} + \frac{1}{2}m\widehat{RU}$	3. Substitution
4. $m\angle 1 = \frac{1}{2}(m\widehat{ST} + m\widehat{RU})$	4. Distributive Property

Example 1 *Secant-Secant Angle*

Find $m\angle 2$ if $m\widehat{BC} = 30$ and $m\widehat{AD} = 20$.

Method 1

$m\angle 1 = \frac{1}{2}(m\widehat{BC} + m\widehat{AD})$

$\quad\quad = \frac{1}{2}(30 + 20)$ or 25   Substitution

$m\angle 2 = 180 - m\angle 1$

$\quad\quad = 180 - 25$ or 155

Method 2

$m\angle 2 = \frac{1}{2}(m\widehat{AB} + m\widehat{DEC})$

Find $m\widehat{AB} + m\widehat{DEC}$.

$m\widehat{AB} + m\widehat{DEC} = 360 - (m\widehat{BC} + m\widehat{AD})$

$\quad\quad\quad\quad\quad = 360 - (30 + 20)$

$\quad\quad\quad\quad\quad = 360 - 50$ or 310

$m\angle 2 = \frac{1}{2}(m\widehat{AB} + m\widehat{DEC})$

$\quad\quad = \frac{1}{2}(310)$ or 155

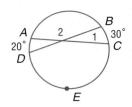

A secant can also intersect a tangent at the point of tangency. Angle *ABC* intercepts $\widehat{BC}$, and $\angle DBC$ intercepts $\widehat{BEC}$. Each angle formed has a measure half that of the arc it intercepts.

$$m\angle ABC = \frac{1}{2}m\widehat{BC} \quad\quad m\angle DBC = \frac{1}{2}m\widehat{BEC}$$

This is stated formally in Theorem 10.13.

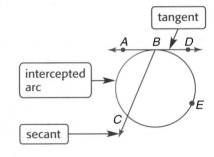

Theorem 10.13

If a secant and a tangent intersect at the point of tangency, then the measure of each angle formed is one-half the measure of its intercepted arc.

You will prove this theorem in Exercise 43.

Example 2 *Secant-Tangent Angle*

Find $m\angle ABC$ if $m\widehat{AB} = 102$.

$m\widehat{ADB} = 360 - m\widehat{AB}$

$\quad\quad\quad = 360 - 102$ or 258

$m\angle ABC = \frac{1}{2}m\widehat{ADC}$

$\quad\quad\quad = \frac{1}{2}(258)$ or 129

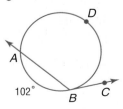

INTERSECTIONS OUTSIDE A CIRCLE Secants and tangents can also meet outside a circle. The measure of the angle formed also involves half of the measures of the arcs they intercept.

Theorem 10.14

If two secants, a secant and a tangent, or two tangents intersect in the exterior of a circle, then the measure of the angle formed is one-half the positive difference of the measures of the intercepted arcs.

Two Secants	Secant-Tangent	Two Tangents
	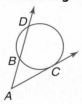	
$m\angle A = \frac{1}{2}(m\widehat{DE} - m\widehat{BC})$	$m\angle A = \frac{1}{2}(m\widehat{DC} - m\widehat{BC})$	$m\angle A = \frac{1}{2}(m\widehat{BDC} - m\widehat{BC})$

You will prove this theorem in Exercise 40.

Example 3 *Secant-Secant Angle*

Find x.

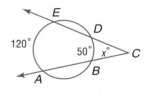

$m\angle C = \frac{1}{2}(m\widehat{EA} - m\widehat{DB})$

$x = \frac{1}{2}(120 - 50)$ Substitution

$x = \frac{1}{2}(70)$ or 35   Simplify.

Example 4 *Tangent-Tangent Angle*

SATELLITES Suppose a geostationary satellite S orbits about 35,000 kilometers above Earth rotating so that it appears to hover directly over the equator. Use the figure to determine the arc measure on the equator visible to this geostationary satellite.

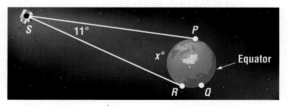

$\widehat{PR}$ represents the arc along the equator visible to the satellite S. If $x = m\widehat{PR}$, then $m\widehat{PQR} = 360 - x$. Use the measure of the given angle to find $m\widehat{PR}$.

$m\angle S = \frac{1}{2}(m\widehat{PQR} - m\widehat{PR})$

$11 = \frac{1}{2}[(360 - x) - x]$ Substitution

$22 = 360 - 2x$ Multiply each side by 2 and simplify.

$-338 = -2x$ Subtract 360 from each side.

$169 = x$ Divide each side by -2.

The measure of the arc on Earth visible to the satellite is 169.

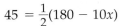 **Example 5** *Secant-Tangent Angle*

Find x.

$\widehat{WRV}$ is a semicircle because $\overline{WV}$ is a diameter.
So, $m\widehat{WRV} = 180$.

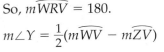

$m\angle Y = \frac{1}{2}(m\widehat{WV} - m\widehat{ZV})$

$\quad 45 = \frac{1}{2}(180 - 10x)$ Substitution

$\quad 90 = 180 - 10x$ Multiply each side by 2.

$\quad -90 = -10x$ Subtract 180 from each side.

$\quad 9 = x$ Divide each side by −10.

Check for Understanding

Concept Check

1. **Describe** the difference between a secant and a tangent.

2. **OPEN ENDED** Draw a circle and one of its diameters. Call the diameter $\overline{AC}$. Draw a line tangent to the circle at A. What type of angle is formed by the tangent and the diameter? Explain.

Guided Practice Find each measure.

3. $m\angle 1$

4. $m\angle 2$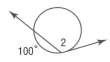

Find x.

5.

6.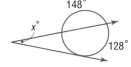

7.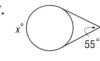

Application **CIRCUS** For Exercises 8–11, refer to the figure and the information below.

One of the acrobatic acts in the circus requires the artist to balance on a board that is placed on a round drum as shown at the right. Find each measure if $\overline{SA} \parallel \overline{LK}$, $m\angle SLK = 78$, and $m\widehat{SA} = 46$.

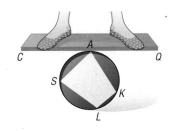

8. $m\angle CAS$ 9. $m\angle QAK$
10. $m\widehat{KL}$ 11. $m\widehat{SL}$

Practice and Apply

Find each measure.

12. $m\angle 3$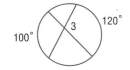

13. $m\angle 4$

14. $m\angle 5$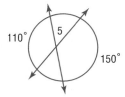

Homework Help

For Exercises	See Examples
12–15	1
16–19	2
21–24, 31	3
25–28, 32	5
29–30	4

Extra Practice
See page 775.

15. $m\angle 6$

16. $m\angle 7$

17. $m\angle 8$

18. $m\angle 9$

19. $m\angle 10$

20. $m\widehat{AC}$

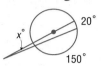

Find x. Assume that any segment that appears to be tangent is tangent.

21.

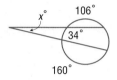

22.

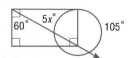

23.

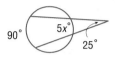

24.

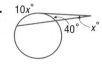

25.

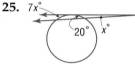

26.

27.

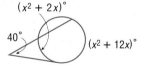

28.

29.

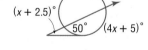

30.

31.

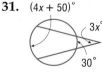

32.

33. WEAVING Once yarn is woven from wool fibers, it is often dyed and then threaded along a path of pulleys to dry. One set of pulleys is shown below. Note that the yarn appears to intersect itself at C, but in reality it does not. Use the information from the diagram to find $m\widehat{BH}$.

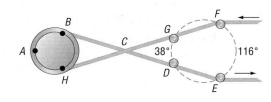

Find each measure if $m\widehat{FE} = 118$, $m\widehat{AB} = 108$, $m\angle EGB = 52$, and $m\angle EFB = 30$.

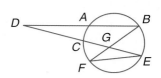

34. $m\widehat{AC}$

35. $m\widehat{CF}$

36. $m\angle EDB$

·•**LANDMARKS** **For Exercises 37–39, use the following information.**
Stonehenge is a British landmark made of huge stones arranged in a circular pattern that reflects the movements of Earth and the moon. The diagram shows that the angle formed by the north/south axis and the line aligned from the station stone to the northmost moonrise position measures 23.5°.

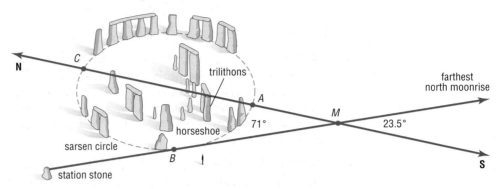

37. Find $m\widehat{BC}$.

38. Is $\widehat{ABC}$ a semicircle? Explain.

39. If the circle measures about 100 feet across, approximately how far would you walk around the circle from point B to point C?

40. **PROOF** Write a two-column proof of Theorem 10.14. Consider each case.

a. Case 1: Two Secants
Given: $\overleftrightarrow{AC}$ and $\overleftrightarrow{AT}$ are secants to the circle.
Prove: $m\angle CAT = \frac{1}{2}(m\widehat{CT} - m\widehat{BR})$

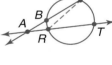

b. Case 2: Secant and a Tangent
Given: $\overleftrightarrow{DG}$ is a tangent to the circle.
$\overleftrightarrow{DF}$ is a secant to the circle.
Prove: $m\angle FDG = \frac{1}{2}(m\widehat{FG} - m\widehat{GE})$

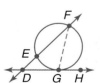

c. Case 3: Two Tangents
Given: $\overleftrightarrow{HI}$ and $\overleftrightarrow{HJ}$ are tangents to the circle.
Prove: $m\angle IHJ = \frac{1}{2}(m\widehat{IXJ} - m\widehat{IJ})$

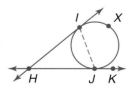

41. **CRITICAL THINKING** Circle E is inscribed in rhombus $ABCD$. The diagonals of the rhombus are 10 centimeters and 24 centimeters long. To the nearest tenth centimeter, how long is the radius of circle E? (*Hint:* Draw an altitude from E.)

42. TELECOMMUNICATION The signal from a telecommunication tower follows a ray that has its endpoint on the tower and is tangent to Earth. Suppose a tower is located at sea level as shown in the figure. Determine the measure of the arc intercepted by the two tangents.

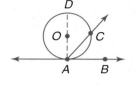

Note: Art not drawn to scale

43. **PROOF** Write a paragraph proof of Theorem 10.13
   a. Given: $\overleftrightarrow{AB}$ is a tangent of $\odot O$.
           $\overleftrightarrow{AC}$ is a secant of $\odot O$.
           $\angle CAB$ is acute.
    Prove: $m\angle CAB = \frac{1}{2}m\widehat{CA}$

   b. Prove Theorem 10.13 if the angle in part **a** is obtuse.

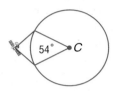

44. SATELLITES A satellite is orbiting so that it maintains a constant altitude above the equator. The camera on the satellite can detect an arc of 6000 kilometers on Earth's surface. This arc measures 54°. What is the measure of the angle of view of the camera located on the satellite?

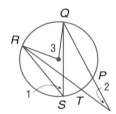

45. CRITICAL THINKING In the figure, $\angle 3$ is a central angle. List the numbered angles in order from greatest measure to least measure. Explain your reasoning.

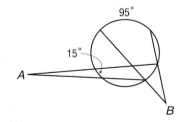

46. **WRITING IN MATH** Answer the question that was posed at the beginning of the lesson.

How is a rainbow formed by segments of a circle?

Include the following in your answer:
- the types of segments represented in the figure on page 561, and
- how you would calculate the angle representing how the light deviates from its original path.

Standardized Test Practice
Ⓐ Ⓑ Ⓒ Ⓓ

47. What is the measure of $\angle B$ if $m\angle A = 10$?
   Ⓐ 30   Ⓑ 35
   Ⓒ 47.5   Ⓓ 90

48. ALGEBRA Which of the following sets of data can be represented by a linear equation?

Ⓐ
x	y
1	2
2	4
3	8
4	16

Ⓑ
x	y
1	4
2	2
3	2
4	4

Ⓒ
x	y
2	2
4	3
6	4
8	5

Ⓓ
x	y
1	1
3	9
5	25
7	49

Mixed Review **Find x. Assume that segments that appear to be tangent are tangent.** *(Lesson 10-5)*

49.

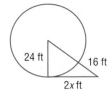

24 ft   16 ft

2x ft

50.

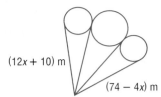

(12x + 10) m

(74 − 4x) m

In ⊙P, m$\widehat{EN}$ = 66 and m∠GPM = 89.
Find each measure. *(Lesson 10-4)*

51. m∠EGN

52. m∠GME

53. m∠GNM

RAMPS **Use the following information for Exercises 54 and 55.**
The *Americans with Disabilities Act* (ADA), which went into effect in 1990, requires
that wheelchair ramps have at least a 12-inch run for each rise of 1 inch. *(Lesson 3-3)*

54. Determine the slope represented by this requirement.

55. The maximum length the law allows for a ramp is 30 feet. How many inches tall
is the highest point of this ramp?

56. **PROOF** Write a paragraph proof to show that
AB = CF if $\overline{AC} \cong \overline{BF}$. *(Lesson 2-5)*

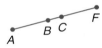

Getting Ready for **PREREQUISITE SKILL** **Solve each equation by factoring.**
the Next Lesson *(To review **solving equations by factoring**, see pages 750 and 751.)*

57. $x^2 + 6x - 40 = 0$ **58.** $2x^2 + 7x - 30 = 0$ **59.** $3x^2 - 24x + 45 = 0$

Practice Quiz 2   *Lessons 10-4 through 10-6*

1. AMUSEMENT RIDES A Ferris wheel is shown at the right. If the distances
between the seat axles are the same, what is the measure of an angle formed
by the braces attaching consecutive seats? *(Lesson 10-4)*

2. Find the measure of each numbered angle.
(Lesson 10-4)

68°

Find x. Assume that any segment that appears to be tangent is tangent. *(Lessons 10-5 and 10-6)*

3.

6 m

x m

4.

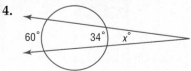

60°   34°   x°

5.

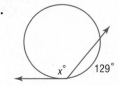

x°   129°

10-7 Special Segments in a Circle

What You'll Learn

- Find measures of segments that intersect in the interior of a circle.
- Find measures of segments that intersect in the exterior of a circle.

How are lengths of intersecting chords related?

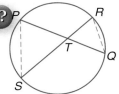

The star is inscribed in a circle. It was formed by intersecting chords. Segments *AD* and *EB* are two of those chords. When two chords intersect, four smaller segments are defined.

SEGMENTS INTERSECTING INSIDE A CIRCLE In Lesson 10-2, you learned how to find lengths of parts of a chord that is intersected by the perpendicular diameter. But how do you find lengths for other intersecting chords?

Geometry Activity

Intersecting Chords

Make A Model

- Draw a circle and two intersecting chords.
- Name the chords $\overline{PQ}$ and $\overline{RS}$ intersecting at *T*.
- Draw $\overline{PS}$ and $\overline{RQ}$.

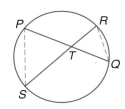

Analyze

1. Name pairs of congruent angles. Explain your reasoning.
2. How are $\triangle PTS$ and $\triangle RTQ$ related? Why?
3. **Make a conjecture** about the relationship of $\overline{PT}$, $\overline{TQ}$, $\overline{RT}$, and $\overline{ST}$.

The results of the activity suggest a proof for Theorem 10.15.

Theorem 10.15

If two chords intersect in a circle, then the products of the measures of the segments of the chords are equal.

Example: $AE \cdot EC = BE \cdot ED$

You will prove Theorem 10.15 in Exercise 21.

Example 1 Intersection of Two Chords

Find *x*.

$AE \cdot EB = CE \cdot ED$

$$x \cdot 6 = 3 \cdot 4 \qquad \text{Substitution}$$
$$6x = 12 \qquad \text{Multiply.}$$
$$x = 2 \qquad \text{Divide each side by 6.}$$

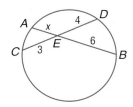

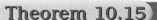

Intersecting chords can also be used to measure arcs.

Example 2 Solve Problems

TUNNELS Tunnels are constructed to allow roadways to pass through mountains. What is the radius of the circle containing the arc if the opening is not a semicircle?

Draw a model using a circle. Let x represent the unknown measure of the segment of diameter $\overline{AB}$. Use the products of the lengths of the intersecting chords to find the length of the diameter.

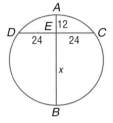

$AE \cdot EB = DE \cdot EC$	Segment products
$12x = 24 \cdot 24$	Substitution
$x = 48$	Divide each side by 12.
$AB = AE + EB$	Segment Addition Postulate
$AB = 12 + 48$ or 60	Substitution and addition

Since the diameter is 60, $r = 30$.

SEGMENTS INTERSECTING OUTSIDE A CIRCLE
Nonparallel chords of a circle can be extended to form secants that intersect in the exterior of a circle. The special relationship among secant segments excludes the chord.

Theorem 10.16

If two secant segments are drawn to a circle from an exterior point, then the product of the measures of one secant segment and its external secant segment is equal to the product of the measures of the other secant segment and its external secant segment.

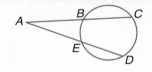

Example: $AB \cdot AC = AE \cdot AD$

You will prove this theorem in Exercise 30.

Example 3 Intersection of Two Secants

Find RS if $PQ = 12$, $QR = 2$, and $TS = 3$.
Let $RS = x$.

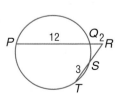

$QR \cdot PR = RS \cdot RT$	Secant Segment Products
$2 \cdot (12 + 2) = x \cdot (x + 3)$	Substitution
$28 = x^2 + 3x$	Distributive Property
$0 = x^2 + 3x - 28$	Subtract 28 from each side.
$0 = (x + 7)(x - 4)$	Factor.

$x + 7 = 0 \qquad x - 4 = 0$

$x = -7 \qquad x = 4$ Disregard negative value.

The same secant segment product can be used with a secant segment and a tangent. In this case, the tangent is both the exterior part and the whole segment. This is stated in Theorem 10.17.

Theorem 10.17

If a tangent segment and a secant segment are drawn to a circle from an exterior point, then the square of the measure of the tangent segment is equal to the product of the measures of the secant segment and its external secant segment.

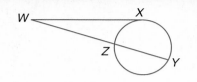

Example: $WX \cdot WX = WZ \cdot WY$

You will prove this theorem in Exercise 31.

Example 4 *Intersection of a Secant and a Tangent.*

Find x. Assume that segments that appear to be tangent are tangent.

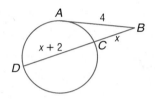

$(AB)^2 = BC \cdot BD$

$4^2 = x(x + x + 2)$

$16 = x(2x + 2)$

$16 = 2x^2 + 2x$

$0 = 2x^2 + 2x - 16$

$0 = x^2 + x - 8$

This expression is not factorable. Use the Quadratic Formula.

$x = \dfrac{-b \pm \sqrt{b^2 - 4ac}}{2a}$ Quadratic Formula

$= \dfrac{-1 \pm \sqrt{1^2 - 4(1)(-8)}}{2(1)}$ $a = 1, b = 1, c = -8$

$= \dfrac{-1 + \sqrt{33}}{2}$ or $x = \dfrac{-1 - \sqrt{33}}{2}$ Disregard the negative solution.

≈ 2.37 Use a calculator.

Check for Understanding

Concept Check

1. **Show** how the products for secant segments are similar to the products for a tangent and a secant segment.

2. **FIND THE ERROR** Becky and Latisha are writing products to find x. Who is correct? Explain your reasoning.

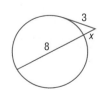

Becky	Latisha
$3^2 = x \cdot 8$	$3^2 = x(x + 8)$
$9 = 8x$	$9 = x^2 + 8x$
$\dfrac{9}{8} = x$	$0 = x^2 + 8x - 9$
	$0 = (x + 9)(x - 1)$
	$x = 1$

3. **OPEN ENDED** Draw a circle with two secant segments and one tangent segment that intersect at the same point.

Guided Practice **Find x. Round to the nearest tenth if necessary. Assume that segments that appear to be tangent are tangent.**

4.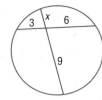

5.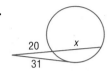

6.

Application 7. **HISTORY** The Roman Coliseum has many "entrances" in the shape of a door with an arched top. The ratio of the arch width to the arch height is 7:3. Find the ratio of the arch width to the radius of the circle that contains the arch.

Practice and Apply

Homework Help

For Exercises	See Examples
8–11	1
12–15	4
16–19	3
20, 27	2

Extra Practice
See page 775.

Find x. Round to the nearest tenth if necessary. Assume that segments that appear to be tangent are tangent.

8.

9.

10.

11.

12.

13.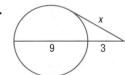

14.

15.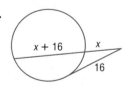

16.

17.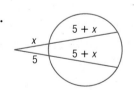

18.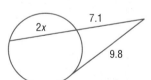

19.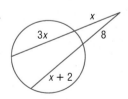

20. **KNOBS** If you remove a knob from a kitchen appliance, you may notice that the hole is not completely round. Suppose the flat edge is 4 millimeters long and the distance from the curved edge to the flat edge is about 4.25 millimeters. Find the radius of the circle containing the hole.

21. PROOF Copy and complete the proof of Theorem 10.15.

Given: $\overline{WY}$ and $\overline{ZX}$ intersect at T.

Prove: $WT \cdot TY = ZT \cdot TX$

Statements	Reasons
a. $\angle W \cong \angle Z, \angle X \cong \angle Y$	**a.** ?
b. ?	**b.** AA Similarity
c. $\dfrac{WT}{ZT} = \dfrac{TX}{TY}$	**c.** ?
d. ?	**d.** Cross products

Find each variable. Round to the nearest tenth, if necessary.

22.

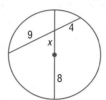

23.

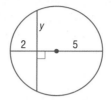

24.

25.

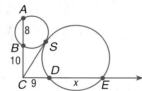

26.

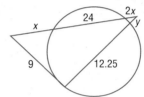

27.

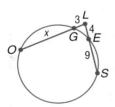

28.

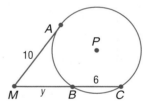

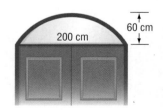

29. CONSTRUCTION An arch over a courtroom door is 60 centimeters high and 200 centimeters wide. Find the radius of the circle containing the arc of the arch.

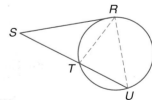

30. PROOF Write a two-column proof of Theorem 10.16.

Given: secants $\overline{EC}$ and $\overline{EB}$

Prove: $EA \cdot EC = ED \cdot EB$

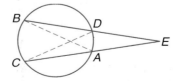

31. PROOF Write a two-column proof of Theorem 10.17.

Given: tangent $\overline{RS}$, secant $\overline{SU}$

Prove: $(RS)^2 = ST \cdot SU$

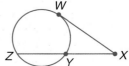

32. CRITICAL THINKING In the figure, Y is the midpoint of $\overline{XZ}$. Find WX in terms of XY. Explain your reasoning.

33. WRITING IN MATH   Answer the question that was posed at the beginning of the lesson.

How are the lengths of intersecting chords related?

Include the following in your answer:
- the segments formed by intersecting segments, $\overline{AD}$ and $\overline{EB}$, and
- the relationship among these segments.

34. Find two possible values for x from the information in the figure.

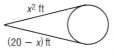

    Ⓐ $-4, -5$ Ⓑ $-4, 5$

    Ⓒ $4, 5$ Ⓓ $4, -5$

35. **ALGEBRA** Mr. Rodriguez can wash his car in 15 minutes, while his son Marcus takes twice as long to do the same job. If they work together, how long will it take them to wash the car?

    Ⓐ 5 min   Ⓑ 7.5 min   Ⓒ 10 min   Ⓓ 22.5 min

Maintain Your Skills

Mixed Review **Find the measure of each numbered angle. Assume that segments that appear tangent are tangent.** *(Lesson 10-6)*

36.

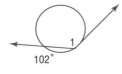

37.

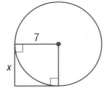

38.

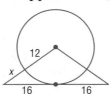

Find x. Assume that segments that appear to be tangent are tangent. *(Lesson 10-5)*

39.

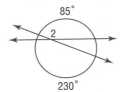

40.

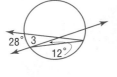

41.

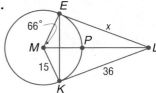

42. **INDIRECT MEASUREMENT** Joseph Blackarrow is measuring the width of a stream on his land to build a bridge over it. He picks out a rock across the stream as landmark A and places a stone on his side as point B. Then he measures 5 feet at a right angle from $\overline{AB}$ and marks this C. From C, he sights a line to point A on the other side of the stream and measures the angle to be about $67°$. How far is it across the stream rounded to the nearest whole foot? *(Lesson 7-5)*

Classify each triangle by its sides and by its angles. *(Lesson 4-1)*

43.

44.

45.

Getting Ready for
the Next Lesson

PREREQUISITE SKILL   Find the distance between each pair of points.

*(To review the **Distance Formula**, see Lesson 1-3.)*

46. $C(-2, 7)$, $D(10, 12)$ **47.** $E(1, 7)$, $F(3, 4)$ **48.** $G(9, -4)$, $H(15, -2)$

10-8 Equations of Circles

What You'll Learn

- Write the equation of a circle.
- Graph a circle on the coordinate plane.

What kind of equations describes the ripples of a splash?

When a rock enters the water, ripples move out from the center forming concentric circles. If the rock is assigned coordinates, each ripple can be modeled by an equation of a circle.

EQUATION OF A CIRCLE

The fact that a circle is the *locus* of points in a plane equidistant from a given point creates an equation for any circle.

Suppose the center is at (3, 2) and the radius is 4. The radius is the distance from the center. Let $P(x, y)$ be the endpoint of any radius.

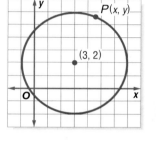

$d = \sqrt{(x_2 - x_1)^2 + (y_2 - y_1)^2}$ Distance Formula

$4 = \sqrt{(x - 3)^2 + (y - 2)^2}$ $d = 4, (x_1, y_1) = (3, 2)$

$16 = (x - 3)^2 + (y - 2)^2$ Square each side.

Applying this same procedure to an unknown center (h, k) and radius r yields a general equation for any circle.

Key Concept   Standard Equation of a Circle

An equation for a circle with center at (h, k) and radius of r units is $(x - h)^2 + (y - k)^2 = r^2$.

Study Tip

Equation of Circles
Note that the equation of a circle is kept in the form shown above. The terms being squared are not expanded.

Example 1 Equation of a Circle

Write an equation for each circle.

a. center at (−2, 4), d = 4

If $d = 4, r = 2$.

$(x - h)^2 + (y - k)^2 = r^2$ Equation of a circle

$[x - (-2)]^2 + [y - 4]^2 = 2^2$ $(h, k) = (-2, 4), r = 2$

$(x + 2)^2 + (y - 4)^2 = 4$ Simplify.

b. center at origin, r = 3

$(x - h)^2 + (y - k)^2 = r^2$ Equation of a circle

$(x - 0)^2 + (y - 0)^2 = 3^2$ $(h, k) = (0, 0), r = 3$

$x^2 + y^2 = 9$ Simplify.

Other information about a circle can be used to find the equation of the circle.

Example 2 Use Characteristics of Circles

A circle with a diameter of 14 has its center in the third quadrant. The lines $y = -1$ and $x = 4$ are tangent to the circle. Write an equation of the circle.

Sketch a drawing of the two tangent lines.

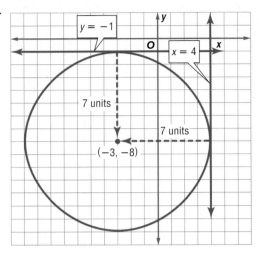

Since $d = 14$, $r = 7$. The line $x = 4$ is perpendicular to a radius. Since $x = 4$ is is a vertical line, the radius lies on a horizontal line. Count 7 units to the left from $x = 4$. Find the value of h.

$$h = 4 - 7 \text{ or } -3$$

Likewise, the radius perpendicular to the line $y = -1$ lies on a vertical line. The value of k is 7 units down from -1.

$$k = -1 - 7 \text{ or } -8$$

The center is at $(-3, -8)$, and the radius is 7. An equation for the circle is $(x + 3)^2 + (y + 8)^2 = 49$.

GRAPH CIRCLES You can analyze the equation of a circle to find information that will help you graph the circle on a coordinate plane.

Study Tip

Graphing Calculator
To use the center and radius to graph a circle, select a suitable window that contains the center of the circle. For a TI-83 Plus, press ZOOM 5. Then use **9: Circle (** on the **Draw** menu. Put in the coordinates of the center and then the radius so that the screen shows "Circle (−2, 3, 4)". Then press ENTER.

Example 3 Graph a Circle

a. Graph $(x + 2)^2 + (y - 3)^2 = 16$.

Compare each expression in the equation to the standard form.

$$(x - h)^2 = (x + 2)^2 \qquad (y - k)^2 = (y - 3)^2$$
$$x - h = x + 2 \qquad\qquad y - k = y - 3$$
$$-h = 2 \qquad\qquad\qquad -k = -3$$
$$h = -2 \qquad\qquad\qquad k = 3$$

$r^2 = 16$, so $r = 4$.

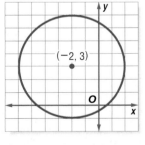

The center is at $(-2, 3)$, and the radius is 4. Graph the center. Use a compass set at a width of 4 grid squares to draw the circle.

b. Graph $x^2 + y^2 = 9$.

Write the equation in standard form.
$$(x - 0)^2 + (y - 0)^2 = 3^2$$

The center is at $(0, 0)$, and the radius is 3. Draw a circle with radius 3, centered at the origin.

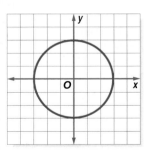

If you know three points on the circle, you can find the center and radius of the circle and write its equation.

Locus
The center of the circle is the locus of points equidistant from the three given points. This is a **compound locus** because the point satisfies more than one condition.

Example 4 A Circle Through Three Points

CELL PHONES Cell phones work by the transfer of phone signals from one tower to another via satellite. Cell phone companies try to locate towers so that they service multiple communities. Suppose three large metropolitan areas are modeled by the points $A(4, 4)$, $B(0, -12)$, and $C(-4, 6)$, and each unit equals 100 miles. Determine the location of a tower equidistant from all three cities, and write an equation for the circle.

Explore You are given three points that lie on a circle.

Plan Graph $\triangle ABC$. Construct the perpendicular bisectors of two sides to locate the center, which is the location of the tower. Find the length of a radius. Use the center and radius to write an equation.

Solve Graph $\triangle ABC$ and construct the perpendicular bisectors of two sides. The center appears to be at $(-2, -3)$. This is the location of the tower.

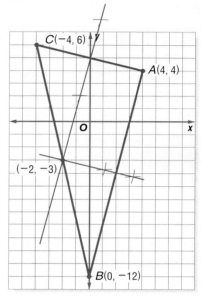

Find r by using the Distance Formula with the center and any of the three points.

$$r = \sqrt{[-2 - 4]^2 + [-3 - 4]^2}$$
$$= \sqrt{85}$$

Write an equation.

$$[x - (-2)]^2 + [y - (-3)]^2 = \left(\sqrt{85}\right)^2$$
$$(x + 2)^2 + (y + 3)^2 = 85$$

Examine You can verify the location of the center by finding the equations of the two bisectors and solving a system of equations. You can verify the radius by finding the distance between the center and another of the three points on the circle.

Check for Understanding

Concept Check **1. OPEN ENDED** Draw an obtuse triangle on a coordinate plane and construct the circle that circumscribes it.

2. Explain how the definition of a circle leads to its equation.

Guided Practice **Write an equation for each circle.**

3. center at $(-3, 5)$, $r = 10$

4. center at origin, $r = \sqrt{7}$

5. diameter with endpoints at $(2, 7)$ and $(-6, 15)$

Graph each equation.

6. $(x + 5)^2 + (y - 2)^2 = 9$

7. $(x - 3)^2 + y^2 = 16$

8. Write an equation of a circle that contains $M(-2, -2)$, $N(2, -2)$, and $Q(2, 2)$. Then graph the circle.

Application 9. **WEATHER** Meteorologists track severe storms using Doppler radar. A polar grid is used to measure distances as the storms progress. If the center of the radar screen is the origin and each ring is 10 miles farther from the center, what is the equation of the fourth ring?

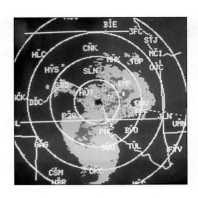

Practice and Apply

Homework Help

For Exercises	See Examples
10–17	1
18–23	2
24–29	3
30–31	4

Extra Practice
See page 776.

Write an equation for each circle.

10. center at origin, $r = 3$

11. center at $(-2, -8)$, $r = 5$

12. center at $(1, -4)$, $r = \sqrt{17}$

13. center at $(0, 0)$, $d = 12$

14. center at $(5, 10)$, $r = 7$

15. center at $(0, 5)$, $d = 20$

16. center at $(-8, 8)$, $d = 16$

17. center at $(-3, -10)$, $d = 24$

18. a circle with center at $(-3, 6)$ and a radius with endpoint at $(0, 6)$

19. a circle with a diameter that has endpoints at $(2, -2)$ and $(-2, 2)$

20. a circle with a diameter that has endpoints at $(-7, -2)$ and $(-15, 6)$

21. a circle with center at $(-2, 1)$ and a radius with endpoint at $(1, 0)$

22. a circle with $d = 12$ and a center translated 18 units left and 7 units down from the origin

23. a circle with its center in quadrant I, radius of 5 units, and tangents $x = 2$ and $y = 3$

Graph each equation.

24. $x^2 + y^2 = 25$

25. $x^2 + y^2 = 36$

26. $x^2 + y^2 - 1 = 0$

27. $x^2 + y^2 - 49 = 0$

28. $(x - 2)^2 + (y - 1)^2 = 4$

29. $(x + 1)^2 + (y + 2)^2 = 9$

Write an equation of the circle containing each set of points. Copy and complete the graph of the circle.

30.

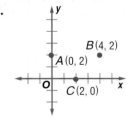

31.

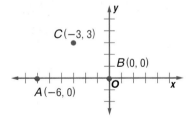

32. Find the radius of a circle with equation $(x - 2)^2 + (y - 2)^2 = r^2$ that contains the point at $(2, 5)$.

33. Find the radius of a circle with equation $(x - 5)^2 + (y - 3)^2 = r^2$ that contains the point at $(5, 1)$.

34. **COORDINATE GEOMETRY** Refer to the Examine part of Example 4. Verify the coordinates of the center by solving a system of equations that represent the perpendicular bisectors.

AERODYNAMICS For Exercises 35–37, use the following information.
The graph shows cross sections of spherical sound waves produced by a supersonic airplane. When the radius of the wave is 1 unit, the plane is 2 units from the origin. A wave of radius 3 occurs when the plane is 6 units from the center.

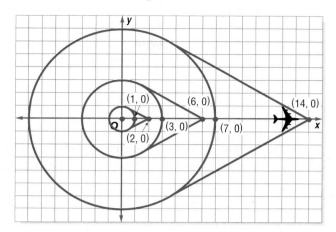

35. Write the equation of the circle when the plane is 14 units from the center.

36. What type of circles are modeled by the cross sections?

37. What is the radius of the circle for a plane 26 units from the center?

38. The equation of a circle is $(x - 6)^2 + (y + 2)^2 = 36$. Determine whether the line $y = 2x - 2$ is a secant, a tangent, or neither of the circle. Explain.

39. The equation of a circle is $x^2 - 4x + y^2 + 8y = 16$. Find the center and radius of the circle.

40. **WEATHER** The geographic center of Tennessee is near Murfreesboro. The closest Doppler weather radar is in Nashville. If Murfreesboro is designated as the origin, then Nashville has coordinates $(-58, 55)$, where each unit is one mile. If the radar has a radius of 80 miles, write an equation for the circle that represents the radar coverage from Nashville.

41. **RESEARCH** Use the Internet or other materials to find the closest Doppler radar to your home. Write an equation of the circle for the radar coverage if your home is the center.

42. **SPACE TRAVEL** Apollo 8 was the first manned spacecraft to orbit the moon at an average altitude of 185 kilometers above the moon's surface. Determine an equation to model a single circular orbit of the Apollo 8 command module if the radius of the moon is 1740 kilometers. Let the center of the moon be at the origin.

43. **CRITICAL THINKING** Determine the coordinates of any intersection point of the graphs of each pair of equations.
a. $x^2 + y^2 = 9$, $y = x + 3$
b. $x^2 + y^2 = 25$, $x^2 + y^2 = 9$
c. $(x + 3)^2 + y^2 = 9$, $(x - 3)^2 + y^2 = 9$

44. WRITING IN MATH Answer the question that was posed at the beginning of the lesson.

What kind of equations describe the ripples of a splash?

Include the following in your answer:
- the general form of the equation of a circle, and
- the equations of five ripples if each ripple is 3 inches farther from the center.

45. Which of the following is an equation of a circle with center at $(-2, 7)$ and a diameter of 18?

    Ⓐ $x^2 + y^2 - 4x + 14y + 53 = 324$ Ⓑ $x^2 + y^2 + 4x - 14y + 53 = 81$

    Ⓒ $x^2 + y^2 - 4x + 14y + 53 = 18$ Ⓓ $x^2 + y^2 + 4x - 14y + 53 = 3$

46. ALGEBRA Jordan opened a one-gallon container of milk and poured one pint of milk into his glass. What is the fractional part of one gallon left in the container?

    Ⓐ $\frac{1}{8}$ Ⓑ $\frac{1}{2}$ Ⓒ $\frac{3}{4}$ Ⓓ $\frac{7}{8}$

Maintain Your Skills

Mixed Review **Find each measure if $EX = 24$ and $DE = 7$.** *(Lesson 10-7)*

47. AX **48.** DX

49. QX **50.** TX

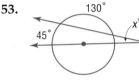

Find x. *(Lesson 10-6)*

51.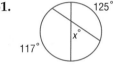
    52.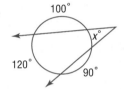
    53.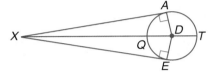

For Exercises 54 and 55, use the following information.
Triangle ABC has vertices $A(-3, 2)$, $B(4, -1)$, and $C(0, -4)$.

54. What are the coordinates of the image after moving $\triangle ABC$ 3 units left and 4 units up? *(Lesson 9-2)*

55. What are the coordinates of the image of $\triangle ABC$ after a reflection in the y-axis? *(Lesson 9-1)*

56. CRAFTS For a Father's Day present, a kindergarten class is making foam plaques. The edge of each plaque is covered with felt ribbon all the way around with 1 inch overlap. There are 25 children in the class. How much ribbon does the teacher need to buy for all 25 children to complete this craft? *(Lesson 1-6)*

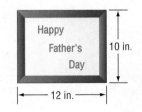

WebQuest **Internet Project**

"Geocaching" Sends Folks on a Scavenger Hunt

It's time to complete your project. Use the information and data you have gathered about designing a treasure hunt to prepare a portfolio or Web page. Be sure to include illustrations and/or tables in the presentation.

www.geometryonline.com/webquest

Vocabulary and Concept Check

arc (p. 530)	circumference (p. 523)	major arc (p. 530)	radius (p. 522)
center (p. 522)	circumscribed (p. 537)	minor arc (p. 530)	secant (p. 561)
central angle (p. 529)	diameter (p. 522)	pi (π) (p. 524)	semicircle (p. 530)
chord (p. 522)	inscribed (p. 537)	point of tangency (p. 552)	tangent (p. 552)
circle (p. 522)	intercepted (p. 544)		

A complete list of postulates and theorems can be found on pages R1–R8.

Exercises Choose the letter of the term that best matches each phrase.

1. arcs of a circle that have exactly one point in common
2. a line that intersects a circle in exactly one point
3. an angle with a vertex that is on the circle and with sides containing chords of the circle
4. a line that intersects a circle in exactly two points
5. an angle with a vertex that is at the center of the circle
6. arcs that have the same measure
7. the distance around a circle
8. circles that have the same radius
9. a segment that has its endpoints on the circle
10. circles that have different radii, but the same center

a. adjacent arcs
b. central angle
c. chord
d. circumference
e. concentric circles
f. congruent arcs
g. congruent circles
h. inscribed angle
i. secant
j. tangent

Lesson-by-Lesson Review

10-1 Circles and Circumference

See pages 522–528.

Concept Summary

- The diameter of a circle is twice the radius.
- The circumference C of a circle with diameter d or a radius of r can be written in the form $C = \pi d$ or $C = 2\pi r$.

Example Find r to the nearest hundredth if $C = 76.2$ feet.

$C = 2\pi r$ Circumference formula

$76.2 = 2\pi r$ Substitution

$\dfrac{76.2}{2\pi} = r$ Divide each side by 2π.

$12.13 \approx r$ Use a calculator.

Exercises The radius, diameter, or circumference of a circle is given. Find the missing measures. Round to the nearest hundredth if necessary. *See Example 4 on page 524.*

11. $d = 15$ in., $r = $ __?__ , $C = $ __?__
12. $r = 6.4$ m, $d = $ __?__ , $C = $ __?__
13. $C = 68$ yd, $r = $ __?__ , $d = $ __?__
14. $d = 52$ cm, $r = $ __?__ , $C = $ __?__
15. $C = 138$ ft, $r = $ __?__ , $d = $ __?__
16. $r = 11$ mm, $d = $ __?__ , $C = $ __?__

10-2 Angles and Arcs

See pages 529–535.

Concept Summary

- The sum of the measures of the central angles of a circle with no interior points in common is 360.
- The measure of each arc is related to the measure of its central angle.
- The length of an arc is proportional to the length of the circumference.

Examples In $\odot P$, $m\angle MPL = 65$ and $\overline{NP} \perp \overline{PL}$.

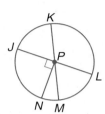

1 Find $m\widehat{NM}$.

$\widehat{NM}$ is a minor arc, so $m\widehat{NM} = m\angle NPM$.

$\angle JPN$ is a right angle and $m\angle MPL = 65$, so $m\angle NPM = 25$.

$m\widehat{NM} = 25$

2 Find $m\widehat{NJK}$.

$\widehat{NJK}$ is composed of adjacent arcs, $\widehat{NJ}$ and $\widehat{JK}$. $\angle MPL \cong \angle JPK$, so $m\angle JPK = 65$.

$m\widehat{NJ} = m\angle NPJ$ or 90 $\angle NPJ$ is a right angle

$m\widehat{NJK} = m\widehat{NJ} + m\widehat{JK}$ Arc Addition Postulate

$m\widehat{NJK} = 90 + 65$ or 155   Substitution

Exercises Find each measure.
See Example 1 on page 529.

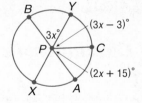

17. $m\widehat{YC}$
18. $m\widehat{BC}$
19. $m\widehat{BX}$
20. $m\widehat{BCA}$

In $\odot G$, $m\angle AGB = 30$ and $\overline{CG} \perp \overline{GD}$. **Find each measure.** *See Example 2 on page 531.*

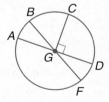

21. $m\widehat{AB}$ 22. $m\widehat{BC}$

23. $m\widehat{FD}$ 24. $m\widehat{CDF}$

25. $m\widehat{BCD}$ 26. $m\widehat{FAB}$

Find the length of the indicated arc in each $\odot I$. *See Example 4 on page 532.*

27. $\widehat{DG}$ if $m\angle DGI = 24$ and $r = 6$

28. $\widehat{WN}$ if $\triangle IWN$ is equilateral and $WN = 5$

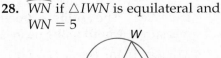

10-3 Arcs and Chords

See pages 536–543.

Concept Summary

- The endpoints of a chord are also the endpoints of an arc.
- Diameters perpendicular to chords bisect chords and intercepted arcs.

Examples

Circle L has a radius of 32 centimeters. $\overline{LH} \perp \overline{GJ}$, and $GJ = 40$ centimeters. Find LK.

Draw radius $\overline{LJ}$. $LJ = 32$ and $\triangle LKJ$ is a right triangle.

$\overline{LH}$ bisects $\overline{GJ}$, since they are perpendicular.

$KJ = \frac{1}{2}(GJ)$ Definition of segment bisector

$\quad = \frac{1}{2}(40)$ or 20 $GJ = 40$, and simplify.

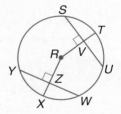

Use the Pythagorean Theorem to find LK.

$(LK)^2 + (KJ)^2 = (LJ)^2$ Pythagorean Theorem

$(LK)^2 + 20^2 = 32^2$ $KJ = 20$, $LJ = 32$

$(LK)^2 + 400 = 1024$ Simplify.

$(LK)^2 = 624$ Subtract 400 from each side.

$LK = \sqrt{624}$ Take the square root of each side.

$LK \approx 24.98$ Use a calculator.

Exercises In $\odot R$, $SU = 20$, $YW = 20$, and $m\widehat{YX} = 45$. **Find each measure.**   *See Example 3 on page 538.*

29. SV 30. WZ

31. UV 32. $m\widehat{YW}$

33. $m\widehat{ST}$ 34. $m\widehat{SU}$

10-4 Inscribed Angles

See pages 544–551.

Concept Summary

- The measure of the inscribed angle is half the measure of its intercepted arc.
- The angles of inscribed polygons can be found by using arc measures.

Example

ALGEBRA Triangles FGH and FHJ are inscribed in $\odot K$ with $\widehat{FG} \cong \widehat{FJ}$. Find x if $m\angle 1 = 6x - 5$, and $m\angle 2 = 7x + 4$.

FJH is a right angle because $\widehat{FJH}$ is a semicircle.

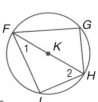

$m\angle 1 + m\angle 2 + m\angle FJH = 180$ Angle Sum Theorem

$(6x - 5) + (7x + 4) + 90 = 180$ $m\angle 1 = 6x - 5$, $m\angle 2 = 7x + 4$, $m\angle FJH = 90$

$13x + 89 = 180$ Simplify.

$x = 7$ Solve for x.

Exercises Find the measure of each numbered angle.
See Example 1 on page 545.

35.

36.

37.

Find the measure of each numbered angle for each situation given.
See Example 4 on page 547.

38. $m\widehat{GH} = 78$

39. $m\angle 2 = 2x, m\angle 3 = x$

40. $m\widehat{JH} = 114$

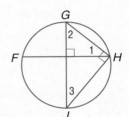

10-5 Tangents

See pages 552–558.

Concept Summary

- A line that is tangent to a circle intersects the circle in exactly one point.
- A tangent is perpendicular to a radius of a circle.
- Two segments tangent to a circle from the same exterior point are congruent.

Example **ALGEBRA** Given that the perimeter of $\triangle ABC = 25$, find x. Assume that segments that appear tangent to circles are tangent.

In the figure, $\overline{AB}$ and $\overline{AC}$ are drawn from the same exterior point and are tangent to $\odot Q$. So $\overline{AB} \cong \overline{AC}$.

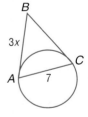

The perimeter of the triangle, $AB + BC + AC$, is 25.

$AB + BC + AC = 25$ Definition of perimeter

$3x + 3x + 7 = 25$ $AB = BC = 3x, AC = 7$

$6x + 7 = 25$ Simplify.

$6x = 18$ Subtract 7 from each side.

$x = 3$ Divide each side by 6.

Exercises Find x. Assume that segments that appear to be tangent are tangent.
See Example 3 on page 554.

41.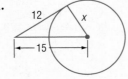

42.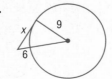

43.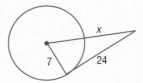

10-6 Secants, Tangents, and Angle Measures

See pages 561–568.

Concept Summary

- The measure of an angle formed by two secant lines is half the positive difference of its intercepted arcs.
- The measure of an angle formed by a secant and tangent line is half its intercepted arc.

Example Find x.

$$m\angle V = \frac{1}{2}\left(m\widehat{XT} - m\widehat{WU}\right)$$

$34 = \frac{1}{2}(128 - x)$ Substitution

$-30 = -\frac{1}{2}x$ Simplify.

$x = 60$ Multiply each side by -2.

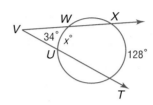

Exercises Find x. *See Example 3 on page 563.*

44.

45.

46.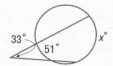

10-7 Special Segments in a Circle

See pages 569–574.

Concept Summary

- The lengths of intersecting chords in a circle can be found by using the products of the measures of the segments.
- The secant segment product also applies to segments that intersect outside the circle, and to a secant segment and a tangent.

Example Find a, if $FG = 18$, $GH = 42$, and $FK = 15$.

Let $KJ = a$.

$FK \cdot FJ = FG \cdot FH$ Secant Segment Products

$15(a + 15) = 18(18 + 42)$ Substitution

$15a + 225 = 1080$ Distributive Property

$15a = 855$ Subtract 225 from each side.

$a = 57$ Divide each side by 15.

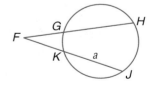

Exercises Find x to the nearest tenth. Assume that segments that appear to be tangent are tangent. *See Examples 3 and 4 on pages 570 and 571.*

47.

48.

49.

Chapter
10 For More ... • Extra Practice, see pages 773–776.
• Mixed Problem Solving, see page 791.

10-8 Equations of Circles

See pages
575–580.

Concept Summary

- The coordinates of the center of a circle (h, k) and its radius r can be used to write an equation for the circle in the form $(x - h)^2 + (y - k)^2 = r^2$.

- A circle can be graphed on a coordinate plane by using the equation written in standard form.

- A circle can be graphed through any three noncollinear points on the coordinate plane.

Examples 1 **Write an equation of a circle with center $(-1, 4)$ and radius 3.**

Since the center is at $(-1, 4)$ and the radius is 3, $h = -1$, $k = 4$, and $r = 3$.

$(x - h)^2 + (y - k)^2 = r^2$ Equation of a circle

$[x - (-1)]^2 + (y - 4)^2 = 3^2$ $h = -1, k = 4$, and $r = 3$

$(x + 1)^2 + (y - 4)^2 = 9$ Simplify.

2 **Graph $(x - 2)^2 + (y + 3)^2 = 6.25$.**

Identify the values of h, k, and r by writing the equation in standard form.

$(x - 2)^2 + (y + 3)^2 = 6.25$

$(x - 2)^2 + [y - (-3)]^2 = 2.5^2$

$h = 2, k = -3$, and $r = 2.5$

Graph the center $(2, -3)$ and use a compass to construct a circle with radius 2.5 units.

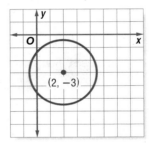

Exercises **Write an equation for each circle.** *See Examples 1 and 2 on pages 575 and 576.*

50. center at $(0, 0)$, $r = \sqrt{5}$

51. center at $(-4, 8)$, $d = 6$

52. diameter with endpoints at $(0, -4)$ and $(8, -4)$

53. center at $(-1, 4)$ and is tangent to $x = 1$

Graph each equation. *See Example 3 on page 576.*

54. $x^2 + y^2 = 2.25$ **55.** $(x - 4)^2 + (y + 1)^2 = 9$

For Exercises 56 and 57, use the following information.

A circle graphed on a coordinate plane contains $A(0, 6)$, $B(6, 0)$, and $C(6, 6)$.
See Example 4 on page 577.

56. Write an equation of the circle.

57. Graph the circle.

Vocabulary and Concepts

1. **Describe** the differences among a tangent, a secant, and a chord of a circle.
2. **Explain** how to find the center of a circle given the coordinates of the endpoints of a diameter.

Skills and Applications

3. Determine the radius of a circle with circumference 25π units. Round to the nearest tenth.

For Questions 4–11, refer to ⊙N.

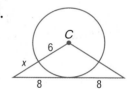

4. Name the radii of ⊙N.
5. If $AD = 24$, find CN.
6. Is $ED > AD$? Explain.
7. If AN is 5 meters long, find the exact circumference of ⊙N.
8. If $m\angle BNC = 20$, find $m\widehat{BC}$.
9. If $m\widehat{BC} = 30$ and $\widehat{AB} \cong \widehat{CD}$, find $m\widehat{AB}$.
10. If $\overline{BE} \cong \overline{ED}$ and $m\widehat{ED} = 120$, find $m\widehat{BE}$.
11. If $m\widehat{AE} = 75$, find $m\angle ADE$.

Find x. Assume that segments that appear to be tangent are tangent.

12.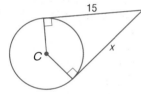

13.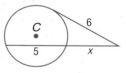

14.

15.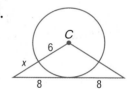

16.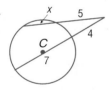

17.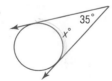

18.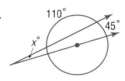

19.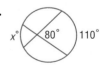

20. **AMUSEMENT RIDES** Suppose a Ferris wheel is 50 feet wide. Approximately how far does a rider travel in one rotation of the wheel?

21. Write an equation of a circle with center at $(-2, 5)$ and a diameter of 50.

22. Graph $(x - 1)^2 + (y + 2)^2 = 4$.

23. **PROOF** Write a two-column proof.
    Given: ⊙X with diameters $\overline{RS}$ and $\overline{TV}$
    Prove: $\widehat{RT} \cong \widehat{VS}$

24. **CRAFTS** Takita is making bookends out of circular wood pieces as shown at the right. What is the height of the cut piece of wood?

25. **STANDARDIZED TEST PRACTICE** Circle C has radius r and $ABCD$ is a rectangle. Find DB.

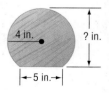

Ⓐ r Ⓑ $r\dfrac{\sqrt{2}}{2}$ Ⓒ $r\sqrt{3}$ Ⓓ $r\dfrac{\sqrt{3}}{2}$

Part 1 Multiple Choice

Record your answer on the answer sheet provided by your teacher or on a sheet of paper.

1. Which of the following shows the graph of $3y = 6x - 9$? (Prerequisite Skill)

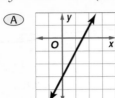 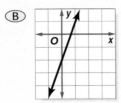

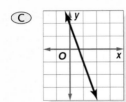 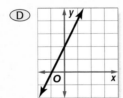

2. In Hyde Park, Main Street and Third Avenue do not meet at right angles. Use the figure below to determine the measure of $\angle 1$ if $m\angle 1 = 6x - 5$ and $m\angle 2 = 3x + 13$. (Lesson 1-5)

   (A) 6
   (B) 18
   (C) 31
   (D) 36

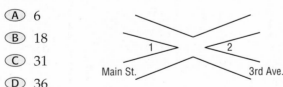

3. Part of a proof is shown below. What is the reason to justify Step b? (Lesson 2-5)

   Given: $4x + \frac{4}{3} = 12$ **Prove:** $x = \frac{8}{3}$

Statements	Reasons
a. $4x + \frac{4}{3} = 12$	a. Given
b. $3\left(4x + \frac{4}{3}\right) = 3(12)$	b. ___?___

   (A) Multiplication Property
   (B) Distributive Property
   (C) Cross products
   (D) none of the above

4. If an equilateral triangle has a perimeter of $(2x + 9)$ miles and one side of the triangle measures $(x + 2)$ miles, how long (in miles) is the side of the triangle? (Lesson 4-1)

   (A) 3   (B) 5   (C) 9   (D) 15

5. A pep team is holding up cards to spell out the school name. What symmetry does the card shown below have? (Lesson 9-1)

   (A) only line symmetry
   (B) only point symmetry
   (C) both line and point symmetry
   (D) neither line nor point symmetry

   $\boxed{\text{A}}$

Use the figure below for Questions 6 and 7.

6. In circle F, which are chords? (Lesson 10-1)

   (A) $\overline{AD}$ and $\overline{EF}$
   (B) $\overline{AF}$ and $\overline{BC}$
   (C) $\overline{EF}$, $\overline{DF}$, and $\overline{AF}$
   (D) $\overline{AD}$ and $\overline{BC}$

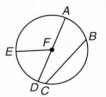

7. In circle F, what is the measure of $\overarc{EA}$ if $m\angle DFE$ is 36? (Lesson 10-2)

   (A) 54   (B) 104   (C) 144   (D) 324

8. Which statement is false? (Lesson 10-3)

   (A) Two chords that are equidistant from the center of a circle are congruent.
   (B) A diameter of a circle that is perpendicular to a chord bisects the chord and its arc.
   (C) The measure of a major arc is the measure of its central angle.
   (D) Minor arcs in the same circle are congruent if their corresponding chords are congruent.

9. Which of the segments described could be a secant of a circle? (Lesson 10-6)

   (A) intersects exactly one point on a circle
   (B) has its endpoints on a circle
   (C) one endpoint at the center of the circle
   (D) intersects exactly two points on a circle

Preparing for Standardized Tests
For test-taking strategies and more
practice, see pages 795–810.

Part 2 | Short Response/Grid In

Record your answers on the answer sheet provided by your teacher or on a sheet of paper.

10. What is the shortest side of quadrilateral *DEFG*?   (Lesson 5-3)

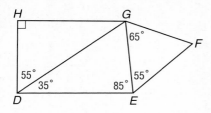

11. An architect designed a house and a garage that are similar in shape. How many feet long is $\overline{ST}$?   (Lesson 6-2)

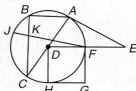

12. Two triangles are drawn on a coordinate grid. One has vertices at (0, 1), (0, 7), and (6, 4). The other has vertices at (7, 7), (10, 7), and (8.5, 10). What scale factor can be used to compare the smaller triangle to the larger?   (Lesson 9-5)

Use the figure below for Questions 13–15.

13. Point *D* is the center of the circle. What is $m\angle ABC$? (Lesson 10-4)

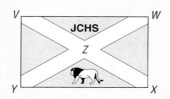

14. $\overline{AE}$ is a tangent. If $AD = 12$ and $FE = 18$, how long is $\overline{AE}$ to the nearest tenth unit? (Lesson 10-5)

15. Chords $\overline{JF}$ and $\overline{BC}$ intersect at *K*. If $BK = 8$, $KC = 12$, and $KF = 16$, find *JK*.   (Lesson 10-7)

 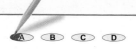
Part 3 | Extended Response

Record your answers on a sheet of paper. Show your work.

16. The Johnson County High School flag is shown below. Points have been added for reference.

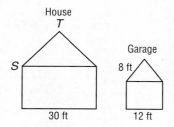

a. Which diagonal segments would have to be congruent for *VWXY* to be a rectangle? (Lesson 8-3)

b. Suppose the length of rectangle *VWXY* is 2 more than 3 times the width and the perimeter is 164 inches. What are the dimensions of the flag?   (Lesson 1-6)

17. The segment with endpoints $A(1, -2)$ and $B(1, 6)$ is the diameter of a circle.

a. Graph the points and draw the circle. (Lesson 10-1)

b. What is the center of the circle? (Lesson 10-1)

c. What is the length of the radius? (Lesson 10-8)

d. What is the circumference of the circle? (Lesson 10-8)

e. What is the equation of the circle? (Lesson 10-8)

Area and Volume

Area and volume can be used to analyze real-world situations. In this unit, you will learn about formulas used to find the areas of two-dimensional figures and the surface areas and volumes of three-dimensional figures.

Chapter 11
Areas of Polygons and Circles

Chapter 12
Surface Area

Chapter 13
Volume

WebQuest Internet Project

Town With Major D-Day Losses Gets Memorial

Source: *USA TODAY*, May 27, 2001

"BEDFORD, Va. For years, World War II was a sore subject that many families in this small farming community avoided. 'We lost so many men,' said Boyd Wilson, 79, who joined Virginia's 116th National Guard before it was sent to war. 'It was just painful.' The war hit Bedford harder than perhaps any other small town in America, taking 19 of its sons, fathers and brothers in the opening moments of the Allied invasion of Normandy. Within a week, 23 of Bedford's 35 soldiers were dead. It was the highest per capita loss for any U.S. community." In this project, you will use scale drawings, surface area, and volume to design a memorial to honor war veterans.

 Log on to **www.geometryonline.com/webquest**. Begin your WebQuest by reading the Task.

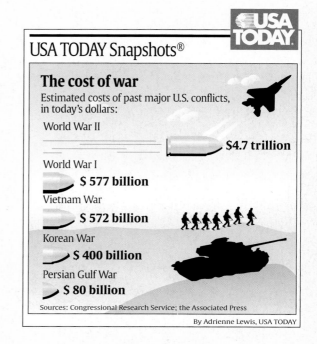

USA TODAY Snapshots®

The cost of war
Estimated costs of past major U.S. conflicts, in today's dollars:

World War II — **$4.7 trillion**

World War I — **$ 577 billion**

Vietnam War — **$ 572 billion**

Korean War — **$ 400 billion**

Persian Gulf War — **$ 80 billion**

Sources: Congressional Research Service; the Associated Press

By Adrienne Lewis, USA TODAY

Continue working on your WebQuest as you study Unit 4.

Lesson	11-4	12-5	13-3
Page	618	662	703

Areas of Polygons and Circles

What You'll Learn

- **Lessons 11-1, 11-2, and 11-3** Find areas of parallelograms, triangles, rhombi, trapezoids, regular polygons, and circles.
- **Lesson 11-4** Find areas of irregular figures.
- **Lesson 11-5** Find geometric probability and areas of sectors and segments of circles.

Why It's Important

Skydivers use geometric probability when they attempt to land on a target marked on the ground. They can determine the chances of landing in the center of the target. *You will learn about skydiving in Lesson 11-5.*

Key Vocabulary

- **apothem** (p. 610)
- **irregular figure** (p. 617)
- **geometric probability** (p. 622)
- **sector** (p. 623)
- **segment** (p. 624)

Getting Started

▶ **Prerequisite Skills** To be successful in this chapter, you'll need to master these skills and be able to apply them in problem-solving situations. Review these skills before beginning Chapter 11.

For Lesson 11-1   Area of a Rectangle

The area and width of a rectangle are given. Find the length of the rectangle.
(For review, see pages 732–733.)

1. $A = 150, w = 15$ **2.** $A = 38, w = 19$ **3.** $A = 21.16, w = 4.6$

4. $A = 2000, w = 32$ **5.** $A = 450, w = 25$ **6.** $A = 256, w = 20$

For Lessons 11-2 and 11-4   Evaluate a Given Expression

Evaluate each expression if $a = 6$, $b = 8$, $c = 10$, and $d = 11$.   *(For review, see page 736.)*

7. $\frac{1}{2}a(b + c)$ **8.** $\frac{1}{2}ab$ **9.** $\frac{1}{2}(2b + c)$

10. $\frac{1}{2}d(a + c)$ **11.** $\frac{1}{2}(b + c)$ **12.** $\frac{1}{2}cd$

For Lesson 11-3   Height of a Triangle

Find h in each triangle.   *(For review, see Lesson 7-3.)*

13.   **14.**    **15.**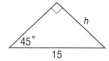

 **Study Organizer**

Areas of Polygons and Circles Make this Foldable to help you organize your notes. Begin with five sheets of notebook paper.

Step 1 **Stack**

Stack 4 of the 5 sheets of notebook paper as illustrated.

Step 2 **Cut**

Cut in about 1 inch along the heading line on the top sheet of paper.

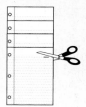

Step 3 **Cut**

Cut the margins off along the right edge.

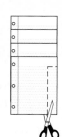

Step 4 **Stack**

Stack in order of cuts, placing the uncut fifth sheet at the back. Label the tabs as shown.

Reading and Writing As you read and study the chapter, take notes and record examples of areas of polygons and circles.

Reading Mathematics

Prefixes

Many of the words used in mathematics use the same prefixes as other everyday words. Understanding the meaning of the prefixes can help you understand the terminology better.

Prefix	Meaning	Everyday Words	Meaning
bi-	2	bicycle	a 2-wheeled vehicle
		bipartisan	involving members of 2 political parties
tri-	3	triangle	closed figure with 3 sides
		tricycle	a 3-wheeled vehicle
		triplet	one of 3 children born at the same time
quad-	4	quadrilateral	closed figure with 4 sides
		quadriceps	muscles with 4 parts
		quadruple	four times as many
penta-	5	pentagon	closed figure with 5 sides
		pentathlon	athletic contest with 5 events
hexa-	6	hexagon	closed figure with 6 sides
hept-	7	heptagon	closed figure with 7 sides
oct-	8	octagon	closed figure with 8 sides
		octopus	animal with 8 legs
dec-	10	decagon	closed figure with 10 sides
		decade	a period of 10 years
		decathlon	athletic contest with 10 events

Several pairs of words in the chart have different prefixes, but the same root word. *Pentathlon* and *decathlon* are both athletic contests. *Heptagon* and *octagon* are both closed figures. Knowing the meaning of the root of the term as well as the prefix can help you learn vocabulary.

Reading to Learn

Use a dictionary to find the meanings of the prefix and root for each term. Then write a definition of the term.

1. bisector   **2.** polygon   **3.** equilateral

4. concentric   **5.** circumscribe   **6.** collinear

7. RESEARCH Use a dictionary to find the meanings of the prefix and root of *circumference*.

8. RESEARCH Use a dictionary or the Internet to find as many words as you can with the prefix *poly-* and the definition of each.

11-1 Areas of Parallelograms

What You'll Learn

- Find perimeters and areas of parallelograms.
- Determine whether points on a coordinate plane define a parallelogram.

How is area related to garden design?

This composition of square-cut granite and moss was designed by Shigemori Mirei in Kyoto, Japan. How could you determine how much granite was used in this garden?

AREAS OF PARALLELOGRAMS

Recall that a *parallelogram* is a quadrilateral with both pairs of opposite sides parallel. Any side of a parallelogram can be called a base. For each base, there is a corresponding altitude that is perpendicular to the base.

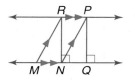

In □*MNPR*, if $\overline{MN}$ is the base, $\overline{RN}$ and $\overline{PQ}$ are altitudes. The length of an altitude is called the *height* of the parallelogram. If $\overline{MR}$ is the base, then the altitudes are $\overline{PT}$ and $\overline{NS}$.

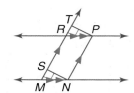

Geometry Activity

Area of a Parallelogram

Model

- Draw parallelogram *ABCD* on grid paper. Label the vertices on the interior of the angles with letters *A*, *B*, *C*, and *D*.

- Fold □*ABCD* so that *A* lies on *B* and *C* lies on *D*, forming a rectangle.

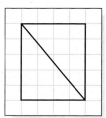

Analyze

1. What is the area of the rectangle?

2. How many rectangles form the parallelogram?

3. What is the area of the parallelogram?

4. How do the base and altitude of the parallelogram relate to the length and width of the rectangle?

5. **Make a conjecture** Use what you observed to write a formula for the area of a parallelogram.

The Geometry Activity leads to the formula for the area of a parallelogram.

Study Tip

Units
Length is measured in linear units, and area is measured in square units.

Key Concept   *Area of a Parallelogram*

If a parallelogram has an area of A square units, a base of b units, and a height of h units, then $A = bh$.

Study Tip

Look Back
To review **perimeter of polygons**, see Lesson 1-6.

Example 1   *Perimeter and Area of a Parallelogram*

Find the perimeter and area of $\square TRVW$.

Base and Side: Each pair of opposite sides of a parallelogram has the same measure. Each base is 18 inches long, and each side is 12 inches long.

Perimeter: The perimeter of a polygon is the sum of the measures of its sides. So, the perimeter of $\square TRVW$ is $2(18) + 2(12)$ or 60 inches.

Height: Use a 30°-60°-90° triangle to find the height. Recall that if the measure of the leg opposite the 30° angle is x, then the length of the hypotenuse is $2x$, and the length of the leg opposite the 60° angle is $x\sqrt{3}$.

$12 = 2x$ Substitute 12 for the hypotenuse.

$6 = x$ Divide each side by 2.

So, the height of the parallelogram is $x\sqrt{3}$ or $6\sqrt{3}$ inches.

Area: $A = bh$ Area of a parallelogram

$ = 18\left(6\sqrt{3}\right)$ $b = 18, h = 6\sqrt{3}$

$ = 108\sqrt{3}$ or about 187.1

The perimeter of $\square TRVW$ is 60 inches, and the area is about 187.1 square inches.

Example 2   *Use Area to Solve a Real-World Problem*

INTERIOR DESIGN   **The Waroners are planning to recarpet part of the first floor of their house. Find the amount of carpeting needed to cover the living room, den, and hall.**

To estimate how much they can spend on carpeting, they need to find the square yardage of each room.

Living Room: $w = 13$ ft, $\ell = 15$ ft

Den: $w = 9$ ft, $\ell = 15$ ft

Hall: It is the same width as the living room, so $w = 13$. The total length of the house is 35 feet. So, $\ell = 35 - 15 - 15$ or 5 feet.

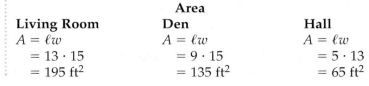

Area

Living Room	Den	Hall
$A = \ell w$	$A = \ell w$	$A = \ell w$
$= 13 \cdot 15$	$= 9 \cdot 15$	$= 5 \cdot 13$
$= 195$ ft^2	$= 135$ ft^2	$= 65$ ft^2

The total area is $195 + 135 + 65$ or 395 square feet. There are 9 square feet in one square yard, so divide by 9 to convert from square feet to square yards.

$$395 \text{ ft}^2 \div \frac{9 \text{ ft}^2}{1 \text{ yd}^2} = 395 \text{ ft}^2 \times \frac{1 \text{ yd}^2}{9 \text{ ft}^2}$$

$$\approx 43.9 \text{ yd}^2$$

Therefore, 44 square yards of carpeting are needed to cover these areas.

PARALLELOGRAMS ON THE COORDINATE PLANE
Recall the properties of quadrilaterals that you studied in Chapter 8. Using these properties as well as the formula for slope and the Distance Formula, you can find the areas of quadrilaterals on the coordinate plane.

Study Tip

Look Back
To review **properties of parallelograms**, **rectangles**, and **squares**, see Lessons 8-3, 8-4, and 8-5.

Example 3 *Area on the Coordinate Plane*

COORDINATE GEOMETRY   The vertices of a quadrilateral are $A(-4, -3)$, $B(2, -3)$, $C(4, -6)$, and $D(-2, -6)$.

a. **Determine whether the quadrilateral is a *square*, a *rectangle*, or a *parallelogram*.**

First graph each point and draw the quadrilateral. Then determine the slope of each side.

slope of $\overline{AB} = \dfrac{-3 - (-3)}{-4 - 2}$

$= \dfrac{0}{-6}$ or 0

slope of $\overline{CD} = \dfrac{-6 - (-6)}{4 - (-2)}$

$= \dfrac{0}{6}$ or 0

slope of $\overline{BC} = \dfrac{-3 - (-6)}{2 - 4}$

$= \dfrac{3}{-2}$

slope of $\overline{AD} = \dfrac{-3 - (-6)}{-4 - (-2)}$

$= \dfrac{3}{-2}$

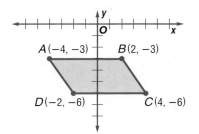

Opposite sides have the same slope, so they are parallel. $ABCD$ is a parallelogram. The slopes of the consecutive sides are *not* negative reciprocals of each other, so the sides are not perpendicular. Thus, the parallelogram is neither a square nor a rectangle.

b. **Find the area of quadrilateral $ABCD$.**

Base: $\overline{CD}$ is parallel to the x-axis, so subtract the x-coordinates of the endpoints to find the length: $CD = \left| 4 - (-2) \right|$ or 6.

Height: Since $\overline{AB}$ and $\overline{CD}$ are horizontal segments, the distance between them, or the height, can be measured on any vertical segment. Reading from the graph, the height is 3.

$A = bh$ Area formula

$= 6(3)$ $b = 6$, $h = 3$

$= 18$ Simplify.

The area of $\square ABCD$ is 18 square units.

Concept Check

1. **Compare and contrast** finding the area of a rectangle and the area of a parallelogram.

2. **OPEN ENDED** Make and label a scale drawing of your bedroom. Then find its area in square yards.

Guided Practice

Find the perimeter and area of each parallelogram. Round to the nearest tenth if necessary.

3.

4.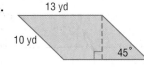

5.

Given the coordinates of the vertices of quadrilateral *TVXY*, determine whether it is a *square*, a *rectangle*, or a *parallelogram*. Then find the area of *TVXY*.

6. $T(0, 0)$, $V(2, 6)$, $X(6, 6)$, $Y(4, 0)$ 7. $T(10, 16)$, $V(2, 18)$, $X(-3, -2)$, $Y(5, -4)$

Application

8. **DESIGN** Mr. Kang is planning to stain his deck. To know how much stain to buy, he needs to find the area of the deck. What is the area?

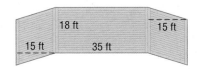

Homework Help

For Exercises	See Examples
9–14	1
15–17, 27, 28, 31	2
20–25	3

Extra Practice
See page 776.

Find the perimeter and area of each parallelogram. Round to the nearest tenth if necessary.

9.
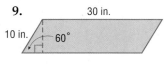

10.
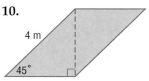

11.
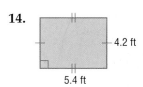

12.
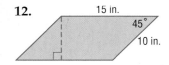

13.
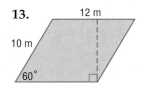

14.

Find the area of each shaded region. Round to the nearest tenth if necessary.

15.
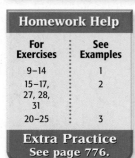

16.
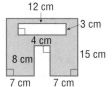

17.
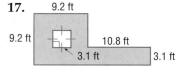

Find the height and base of each parallelogram given its area.

18. 100 square units

19. 2000 square units

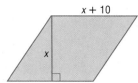

COORDINATE GEOMETRY   Given the coordinates of the vertices of a quadrilateral, determine whether it is a *square*, a *rectangle*, or a *parallelogram*. Then find the area of the quadrilateral.

20. $A(0, 0)$, $B(4, 0)$, $C(5, 5)$, $D(1, 5)$

21. $E(-5, -3)$, $F(3, -3)$, $G(5, 4)$, $H(-3, 4)$

22. $J(-1, -4)$, $K(4, -4)$, $L(6, 6)$, $M(1, 6)$

23. $N(-6, 2)$, $O(2, 2)$, $P(4, -6)$, $Q(-4, -6)$

24. $R(-2, 4)$, $S(8, 4)$, $T(8, -3)$, $U(-2, -3)$

25. $V(1, 10)$, $W(4, 8)$, $X(2, 5)$, $Y(-1, 7)$

26. INTERIOR DESIGN   The Bessos are planning to have new carpet installed in their guest bedroom, family room, and hallway. Find the number of square yards of carpet they should order.

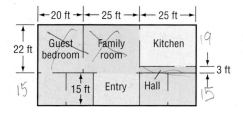

Find the area of each figure.

27.

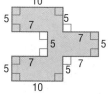

28.

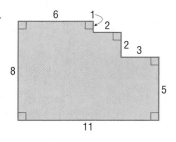

••• **ART**   For Exercises 29 and 30, use the following information.
A *triptych* painting is a series of three pieces with a similar theme displayed together. Suppose the center panel is a 12-inch square and the panels on either side are 12 inches by 5 inches. The panels are 2 inches apart with a 3 inch wide border around the edges.

29. Determine whether the triptych will fit a 45-inch by 20-inch frame. Explain.

30. Find the area of the artwork.

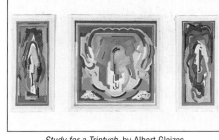

Study for a Triptych, by Albert Gleizes

31. CROSSWALKS   A crosswalk with two stripes each 52 feet long is at a 60° angle to the curb. The width of the crosswalk at the curb is 16 feet. Find the perpendicular distance between the stripes of the crosswalk.

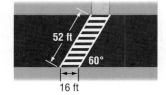

VARYING DIMENSIONS   For Exercises 32–34, use the following information.
A parallelogram has a base of 8 meters, sides of 11 meters, and a height of 10 meters.

32. Find the perimeter and area of the parallelogram.

33. Suppose the dimensions of the parallelogram were divided in half. Find the perimeter and the area.

34. Compare the perimeter and area of the parallelogram in Exercise 33 with the original.

35. CRITICAL THINKING   A piece of twine 48 inches long is cut into two lengths. Each length is then used to form a square. The sum of the areas of the two squares is 74 square inches. Find the length of each side of the smaller square and the larger square.

36. WRITING IN MATH Answer the question that was posed at the beginning of the lesson.

How is area related to garden design?

Include the following in your answer:
- how to determine the total area of granite squares, and
- other uses for area.

37. What is the area of $\square ABCD$?

Ⓐ 24 m^2 Ⓑ 30 m^2 Ⓒ 48 m^2 Ⓓ 60 m^2

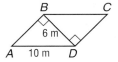

38. ALGEBRA Which statement is correct?

Ⓐ $x^2 > (x-1)^2$

Ⓑ $x^2 = (x-1)^2$

Ⓒ $x^2 < (x-1)^2$

Ⓓ The relationship cannot be determined.

Maintain Your Skills

Mixed Review **Determine the coordinates of the center and the measure of the radius for each circle with the given equation.** *(Lesson 10-8)*

39. $(x-5)^2 + (y-2)^2 = 49$

40. $(x+3)^2 + (y+9)^2 - 81 = 0$

41. $\left(x + \frac{2}{3}\right)^2 + \left(y - \frac{1}{9}\right)^2 - \frac{4}{9} = 0$

42. $(x-2.8)^2 + (y+7.6)^2 = 34.81$

Find x. Assume that segments that appear to be tangent are tangent. *(Lesson 10-7)*

43.

44.

45.

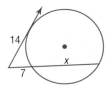

COORDINATE GEOMETRY **Draw the rotation image of each triangle by reflecting the triangles in the given lines. State the coordinates of the rotation image and the angle of rotation.** *(Lesson 9-3)*

46. $\triangle ABC$ with vertices $A(-1, 3)$, $B(-4, 6)$, and $C(-5, 1)$, reflected in the y-axis and then the x-axis

47. $\triangle FGH$ with vertices $F(0, 4)$, $G(-2, 2)$, and $H(2, 2)$, reflected in $y = x$ and then the y-axis

48. $\triangle LMN$ with vertices $L(2, 0)$, $M(3, -3)$, and $N(1, -4)$, reflected in the y-axis and then the line $y = -x$

49. BIKES Nate is making a ramp for bike jumps. The ramp support forms a right angle. The base is 12 feet long, and the height is 5 feet. What length of plywood does Nate need for the ramp? *(Lesson 7-2)*

Getting Ready for the Next Lesson **PREREQUISITE SKILL** **Evaluate each expression if $w = 8$, $x = 4$, $y = 2$, and $z = 5$.**
*(To review **evaluating expressions**, see page 736.)*

50. $\frac{1}{2}(7y)$

51. $\frac{1}{2}wx$

52. $\frac{1}{2}z(x + y)$

53. $\frac{1}{2}x(y + w)$

Areas of Triangles, Trapezoids, and Rhombi

What You'll Learn

- Find areas of triangles.
- Find areas of trapezoids and rhombi.

How is the area of a triangle related to beach umbrellas?

Umbrellas can protect you from rain, wind, and sun. The umbrella shown at the right is made of triangular panels. To cover the umbrella frame with canvas panels, you need to know the area of each panel.

AREAS OF TRIANGLES You have learned how to find the areas of squares, rectangles, and parallelograms. The formula for the area of a triangle is related to these formulas.

Geometry Activity

Area of a Triangle

•• Model

You can determine the area of a triangle by using the area of a rectangle.

- Draw a triangle on grid paper so that one edge is along a horizontal line. Label the vertices on the interior of the angles of the triangle as *A*, *B*, and *C*.
- Draw a line perpendicular to $\overline{AC}$ through *A*.
- Draw a line perpendicular to $\overline{AC}$ through *C*.
- Draw a line parallel to $\overline{AC}$ through *B*.
- Label the points of intersection of the lines drawn as *D* and *E* as shown.
- Find the area of rectangle *ACDE* in square units.
- Cut out rectangle *ACDE*. Then cut out △*ABC*. Place the two smaller pieces over △*ABC* to completely cover the triangle.

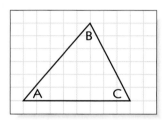

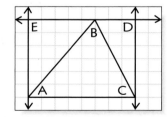

Analyze

1. What do you observe about the two smaller triangles and △*ABC*?
2. What fraction of rectangle *ACDE* is △*ABC*?
3. Derive a formula that could be used to find the area of △*ABC*.

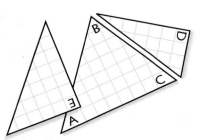

The Geometry Activity suggests the formula for finding the area of a triangle.

Key Concept   Area of a Triangle

If a triangle has an area of A square units, a base of b units, and a corresponding height of h units, then $A = \frac{1}{2}bh$.

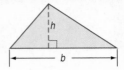

Example 1 **Areas of Triangles**

Find the area of quadrilateral $XYZW$ if $XZ = 39$, $HW = 20$, and $YG = 21$.
The area of the quadrilateral is equal to the sum of the areas of $\triangle XWZ$ and $\triangle XYZ$.

area of $XYZW$ = area of $\triangle XYZ$ + area of $\triangle XWZ$

$$= \frac{1}{2} bh_1 + \frac{1}{2} bh_2$$

$$= \frac{1}{2}(39)(21) + \frac{1}{2}(39)(20) \quad \text{Substitution}$$

$$= 409.5 + 390 \quad \text{Simplify.}$$

$$= 799.5$$

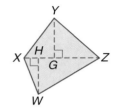

The area of quadrilateral $XYZW$ is 799.5 square units.

AREAS OF TRAPEZOIDS AND RHOMBI   The formulas for the areas of trapezoids and rhombi are related to the formula for the area of a triangle.

Trapezoid $MNPQ$ has diagonal $\overline{QN}$ with parallel bases $\overline{MN}$ and $\overline{PQ}$. Therefore, the altitude h from vertex Q to the extension of base $\overline{MN}$ is the same length as the altitude from vertex N to the base $\overline{QP}$. Since the area of the trapezoid is the area of two nonoverlapping parts, we can write the following equation.

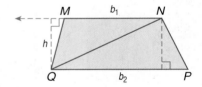

area of   area of   area of
trapezoid $MNPQ$ = $\triangle MNQ$ + $\triangle NPQ$

$$A = \frac{1}{2}(b_1)h + \frac{1}{2}(b_2)h \quad \text{Let the area be } A, MN \text{ be } b_1, \text{ and } QP \text{ be } b_2.$$

$$A = \frac{1}{2}(b_1 + b_2)h \quad \text{Factor.}$$

$$A = \frac{1}{2}h(b_1 + b_2) \quad \text{Commutative Property}$$

This is the formula for the area of any trapezoid.

Key Concept   Area of a Trapezoid

If a trapezoid has an area of A square units, bases of b_1 units and b_2 units, and a height of h units, then $A = \frac{1}{2}h(b_1 + b_2)$.

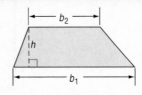

Example 2 Area of a Trapezoid on the Coordinate Plane

COORDINATE GEOMETRY Find the area of trapezoid $TVWZ$ with vertices $T(-3, 4)$, $V(3, 4)$, $W(6, -1)$, and $Z(-5, -1)$.

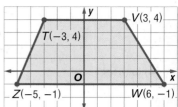

Bases: Since $\overline{TV}$ and $\overline{ZW}$ are horizontal, find their length by subtracting the x-coordinates of their endpoints.

$$TV = |-3 - 3| \qquad ZW = |-5 - 6|$$
$$= |-6| \text{ or } 6 \qquad = |-11| \text{ or } 11$$

Height: Because the bases are horizontal segments, the distance between them can be measured on a vertical line. That is, subtract the y-coordinates.

$$h = |4 - (-1)| \text{ or } 5$$

Area: $A = \frac{1}{2}h(b_1 + b_2)$ Area of a trapezoid

$\qquad\quad = \frac{1}{2}(5)(6 + 11)$ $h = 5, b_1 = 6, b_2 = 11$

$\qquad\quad = 42.5$ Simplify.

The area of trapezoid $TVWZ$ is 42.5 square units.

The formula for the area of a triangle can also be used to derive the formula for the area of a rhombus.

Key Concept   *Area of a Rhombus*

If a rhombus has an area of A square units and diagonals of d_1 and d_2 units, then $A = \frac{1}{2}d_1d_2$.
Example: $A = \frac{1}{2}(AC)(BD)$

You will derive this formula in Exercise 46.

Example 3 Area of a Rhombus on the Coordinate Plane

COORDINATE GEOMETRY Find the area of rhombus $EFGH$ with vertices at $E(-1, 3)$, $F(2, 7)$, $G(5, 3)$, and $H(2, -1)$.

Explore To find the area of the rhombus, we need to know the lengths of each diagonal.

Plan Use coordinate geometry to find the length of each diagonal. Use the formula to find the area of rhombus $EFGH$.

Solve Let $\overline{EG}$ be d_1 and $\overline{FH}$ be d_2.

Subtract the x-coordinates of E and G to find that d_1 is 6.
Subtract the y-coordinates of F and H to find that d_2 is 8.

$A = \frac{1}{2}d_1d_2$ Area of a rhombus

$\quad = \frac{1}{2}(6)(8)$ or 24 $d_1 = 6, d_2 = 8$

Examine The area of rhombus $EFGH$ is 24 square units.

If you know all but one measure in a quadrilateral, you can solve for the missing measure using the appropriate area formula.

Example 4 Algebra: Find Missing Measures

a. Rhombus *WXYZ* has an area of 100 square meters. Find *WY* if *XZ* = 10 meters.

Use the formula for the area of a rhombus and solve for d_2.

$$A = \frac{1}{2}d_1d_2$$

$$100 = \frac{1}{2}(10)(d_2)$$

$$100 = 5d_2$$

$$20 = d_2$$

WY is 20 meters long.

b. Trapezoid *PQRS* has an area of 250 square inches. Find the height of *PQRS*.

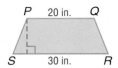

Use the formula for the area of a trapezoid and solve for *h*.

$$A = \frac{1}{2}h(b_1 + b_2)$$

$$250 = \frac{1}{2}h(20 + 30)$$

$$250 = \frac{1}{2}(50)h$$

$$250 = 25h$$

$$10 = h$$

The height of trapezoid *PQRS* is 10 inches.

Since the dimensions of congruent figures are equal, the areas of congruent figures are also equal.

Postulate 11.1

Congruent figures have equal areas.

Study Tip

Look Back
To review **the properties of rhombi and trapezoids**, see Lessons 8-5 and 8-6.

Example 5 Area of Congruent Figures

QUILTING This quilt block is composed of twelve congruent rhombi arranged in a regular hexagon. The height of the hexagon is 8 inches. If the total area of the rhombi is 48 square inches, find the lengths of each diagonal and the area of one rhombus.

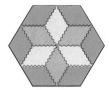

First, find the area of one rhombus. From Postulate 11.1, the area of each rhombus is the same. So, the area of each rhombus is 48 ÷ 12 or 4 square inches.

Next, find the length of one diagonal. The height of the hexagon is equal to the sum of the long diagonals of two rhombi. Since the rhombi are congruent, the long diagonals must be congruent. So, the long diagonal is equal to 8 ÷ 2, or 4 inches.

Use the area formula to find the length of the other diagonal.

$$A = \frac{1}{2}d_1d_2 \quad \text{Area of a rhombus}$$

$$4 = \frac{1}{2}(4)\,d_2 \quad A = 4, d_1 = 4$$

$$2 = d_2 \quad \text{Solve for } d_2.$$

Each rhombus in the pattern has an area of 4 square inches and diagonals 4 inches and 2 inches long.

Concept Check

1. **OPEN ENDED** Draw an isosceles trapezoid that contains at least one isosceles triangle.

2. **FIND THE ERROR** Robert and Kiku are finding the area of trapezoid *JKLM*.

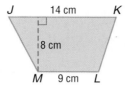

Robert
$A = \frac{1}{2}(8)(14 + 9)$
$= \frac{1}{2}(8)(14) + 9$
$= 56 + 9$
$= 65 \text{ cm}^2$

Kiku
$A = \frac{1}{2}(8)(14 + 9)$
$= \frac{1}{2}(8)(23)$
$= 4(23)$
$= 92 \text{ cm}^2$

Who is correct? Explain your reasoning.

3. **Determine** whether it is *always, sometimes,* or *never* true that rhombi with the same area have the same diagonal lengths. Explain your reasoning.

Guided Practice   Find the area of each quadrilateral.

4.
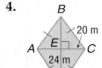

5.
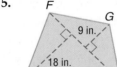

6.
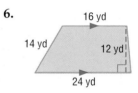

COORDINATE GEOMETRY   Find the area of each figure given the coordinates of the vertices.

7. $\triangle ABC$ with $A(2, -3)$, $B(-5, -3)$, and $C(-1, 3)$

8. trapezoid *FGHJ* with $F(-1, 8)$, $G(5, 8)$, $H(3, 4)$, and $J(1, 4)$

9. rhombus *LMPQ* with $L(-4, 3)$, $M(-2, 4)$, $P(0, 3)$, and $Q(-2, 2)$

ALGEBRA   Find the missing measure for each quadrilateral.

10. Trapezoid *NOPQ* has an area of 250 square inches. Find the height of *NOPQ*.

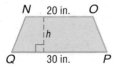

11. Rhombus *RSTU* has an area of 675 square meters. Find *SU*.

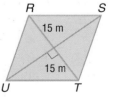

Application   12. **INTERIOR DESIGN** Jacques is designing a window hanging composed of 13 congruent rhombi. The total width of the window hanging is 15 inches, and the total area is $82\frac{7}{8}$ square inches. Find the length of each diagonal and the area of one rhombus.

Homework Help

For Exercises	See Examples
13, 14, 19–21	1
15, 16, 22–25	2
17, 18, 26–29	3
30–35, 40–44	4
36–39	5

Extra Practice
See page 776.

Find the area of each figure. Round to the nearest tenth if necessary.

13.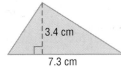
3.4 cm
7.3 cm

14.
7 ft
10.2 ft

15.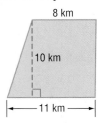
8 km
10 km
11 km

16.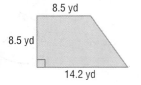
8.5 yd
8.5 yd
14.2 yd

17.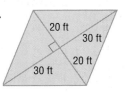
20 ft
30 ft
30 ft
20 ft

18.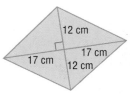
12 cm
17 cm
17 cm
12 cm

19.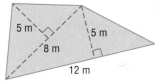
5 m
5 m
8 m
12 m

20.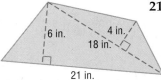
6 in.
4 in.
18 in.
21 in.

21.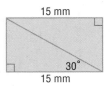
15 mm
30°
15 mm

COORDINATE GEOMETRY Find the area of trapezoid *PQRT* given the coordinates of the vertices.

22. $P(0, 3), Q(3, 7), R(5, 7), T(6, 3)$
23. $P(-4, -5), Q(-2, -5), R(4, 6), T(-4, 6)$
24. $P(-3, 8), Q(6, 8), R(6, 2), T(1, 2)$
25. $P(-6, 3), Q(1, 3), R(-2, -2), T(-4, -2)$

COORDINATE GEOMETRY Find the area of rhombus *JKLM* given the coordinates of the vertices.

26. $J(2, 1), K(7, 4), L(12, 1), M(7, -2)$
27. $J(-1, 2), K(1, 7), L(3, 2), M(1, -3)$
28. $J(-1, -4), K(2, 2), L(5, -4), M(2, -10)$
29. $J(2, 4), K(6, 6), L(10, 4), M(6, 2)$

ALGEBRA Find the missing measure for each figure.

30. Trapezoid *ABCD* has an area of 750 square meters. Find the height of *ABCD*.

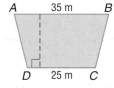
A   35 m   B
D   25 m   C

31. Trapezoid *GHJK* has an area of 188.35 square feet. If *HJ* is 16.5 feet, find *GK*.

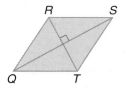
H   16.5 ft   J
8.7 ft
G   K

32. Rhombus *MNPQ* has an area of 375 square inches. If *MP* is 25 inches, find *NQ*.

N
M   P
Q

33. Rhombus *QRST* has an area of 137.9 square meters. If *RT* is 12.2 meters, find *QS*.

R   S
Q   T

34. Triangle *WXY* has an area of 248 square inches. Find the length of the base.

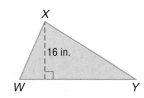
X
16 in.
W   Y

35. Triangle *PQS* has an area of 300 square centimeters. Find the height.

Q
P   30 cm   S

GARDENS For Exercises 36 and 37, use the following information.
Keisha designed a garden that is shaped like two congruent rhombi. She wants the long diagonals lined with a stone walkway. The total area of the garden is 150 square feet, and the shorter diagonals are each 12 feet long.

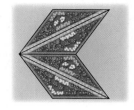

36. Find the length of each stone walkway.

37. Find the length of each side of the garden.

• **REAL ESTATE** For Exercises 38 and 39, use the following information.
The map shows the layout and dimensions of several lot parcels in Linworth Village. Suppose Lots 35 and 12 are trapezoids.

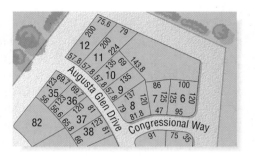

38. If the height of Lot 35 is 122.81 feet, find the area of this lot.

39. If the height of Lot 12 is 199.8 feet, find the area of this lot.

Online Research **Data Update** Use the Internet or other resource to find the median price of homes in the United States. How does this compare to the median price of homes in your community? Visit www.geometryonline.com/data_update to learn more.

Find the area of each figure.

40. rhombus with a perimeter of 20 meters and a diagonal of 8 meters

41. rhombus with a perimeter of 52 inches and a diagonal of 24 inches

42. isosceles trapezoid with a perimeter of 52 yards; the measure of one base is 10 yards greater than the other base, the measure of each leg is 3 yards less than twice the length of the shorter base

43. equilateral triangle with a perimeter of 15 inches

44. scalene triangle with sides that measure 34.0 meters, 81.6 meters, and 88.4 meters.

45. Find the area of $\triangle JKM$.

46. Derive the formula for the area of a rhombus using the formula for the area of a triangle.

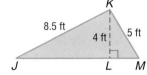

47. Determine whether the statement *Two triangles that have the same area also have the same perimeter* is true or false. Give an example or counterexample.

Each pair of figures is similar. Find the area and perimeter of each figure. Describe how changing the dimensions affects the perimeter and area.

48.

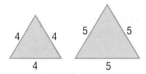

49.

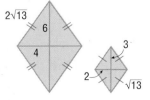

50. **RECREATION** Becky wants to cover a kite frame with decorative paper. If the length of one diagonal is 20 inches and the other diagonal measures 25 inches, find the area of the surface of the kite.

SIMILAR FIGURES   For Exercises 51–56, use the
following information.
Triangle *ABC* is similar to triangle *DEF*.

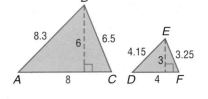

51. Find the scale factor.

52. Find the perimeter of each triangle.

53. Compare the ratio of the perimeters of the
triangles to the scale factor.

54. Find the area of each triangle.

55. Compare the ratio of the areas of the triangles to the scale factor.

56. Compare the ratio of the areas of the triangles to the ratio of the perimeters of
the triangles.

57. **CRITICAL THINKING**   In the figure, the vertices of
quadrilateral *ABCD* intersect square *EFGH* and divide its
sides into segments with measures that have a ratio of 1:2.
Find the area of *ABCD*. Describe the relationship between
the areas of *ABCD* and *EFGH*.

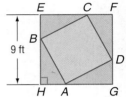

58. WRITING IN MATH   Answer the question that was posed at
the beginning of the lesson.

How is the area of a triangle related to beach umbrellas?

Include the following in your answer:
- how to find the area of a triangle, and
- how the area of a triangle can help you find the areas of rhombi and trapezoids.

59. In the figure, if point *B* lies on the perpendicular
bisector of $\overline{AC}$, what is the area of △*ABC*?

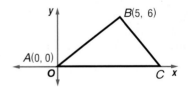

Ⓐ 15 units^2 Ⓑ 30 units^2

Ⓒ 50 units^2 Ⓓ 1602 units^2

60. **ALGEBRA**   What are the solutions of the equation $(2x − 7)(x + 10) = 0$?

Ⓐ −3.5 and 10   Ⓑ 7 and −10   Ⓒ $\frac{2}{7}$ and −10   Ⓓ 3.5 and −10

*Extending
the Lesson*
Trigonometric Ratios and the Areas of Triangles
The area of any triangle can be found given the measures of two
sides of the triangle and the measure of the included angle.
Suppose we are given *AC* = 15, *BC* = 8, and *m*∠*C* = 60. To
find the height of the triangle, use the sine ratio, $\sin C = \frac{h}{BC}$.
Then use the value of *h* in the formula for the area of a
triangle. So, the area is $\frac{1}{2}(15)(8 \sin 60°)$ or 52.0 square meters.

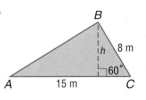

61. Derive a formula to find the area of any triangle, given the measures of two
sides of the triangle and their included angle.

Find the area of each triangle. Round to the nearest hundredth.

62.   63.   64.

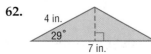   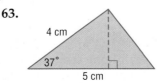   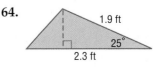

Mixed Review **Find the area of each figure. Round to the nearest tenth.** *(Lesson 11-1)*

65.
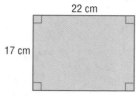
22 cm
17 cm

66.
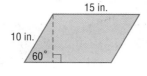
15 in.
10 in.
60°

67.
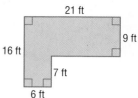
21 ft
9 ft
16 ft
7 ft
6 ft

Write an equation of circle R based on the given information. *(Lesson 10-8)*

68. center: $R(1, 2)$
radius: 7

69. center: $R\left(-4, \frac{1}{2}\right)$
radius: $\frac{11}{2}$

70. center: $R(-1.3, 5.6)$
radius: 3.5

71. CRAFTS Andria created a pattern to appliqué flowers onto a quilt by first drawing a regular pentagon that was 3.5 inches long on each side. Then she added a semicircle onto each side of the pentagon to create the appearance of five petals. How many inches of gold trim does she need to edge 10 flowers? *(Lesson 10-1)*

Given the magnitude and direction of a vector, find the component form with values rounded to the nearest tenth. *(Lesson 9-6)*

72. magnitude of 136 at a direction of 25 degrees with the positive *x*-axis

73. magnitude of 280 at a direction of 52 degrees with the positive *x*-axis

Getting Ready for the Next Lesson **PREREQUISITE SKILL** Find *x*. Round to the nearest tenth.
*(To review **trigonometric ratios in right triangles**, see Lesson 7-4.)*

74.

46
x
73°

75.
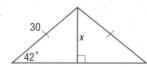
30
x
42°

76.
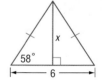
x
58°
6

The coordinates of the vertices of quadrilateral *JKLM* are *J*(−8, 4), *K*(−4, 0), *L*(0, 4), and *M*(−4, 8). *(Lesson 11-1)*

1. Determine whether *JKLM* is a *square*, a *rectangle*, or a *parallelogram*.

2. Find the area of *JKLM*.

Find the area of each trapezoid. *(Lesson 11-2)*

3.
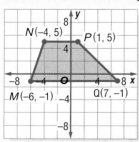
N(−4, 5)
P(1, 5)
M(−6, −1)
Q(7, −1)

4.
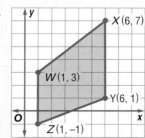
X(6, 7)
W(1, 3)
Y(6, 1)
Z(1, −1)

5. The area of a rhombus is 546 square yards. If d_1 is 26 yards long, find the length of d_2. *(Lesson 11-2)*

Areas of Regular Polygons and Circles

What You'll Learn

- Find areas of regular polygons.
- Find areas of circles.

Vocabulary

- apothem

How can you find the area of a polygon?

The foundations of most gazebos are shaped like regular hexagons. Suppose the owners of this gazebo would like to install tile on the floor. If tiles are sold in square feet, how can they find out the actual area of tiles needed to cover the floor?

AREAS OF REGULAR POLYGONS   In regular hexagon $ABCDEF$ inscribed in circle G, $\overline{GA}$ and $\overline{GF}$ are radii from the center of the circle G to two vertices of the hexagon. $\overline{GH}$ is drawn from the center of the regular polygon perpendicular to a side of the polygon. This segment is called an **apothem**.

Study Tip

Look Back
To review **apothem,**
see Lesson 1-4.

Triangle GFA is an isosceles triangle, since the radii are congruent. If all of the radii were drawn, they would separate the hexagon into 6 nonoverlapping congruent isosceles triangles.

The area of the hexagon can be determined by adding the areas of the triangles. Since $\overline{GH}$ is perpendicular to $\overline{AF}$, it is an altitude of $\triangle AGF$. Let a represent the length of $\overline{GH}$ and let s represent the length of a side of the hexagon.

$$\text{Area of } \triangle AGF = \frac{1}{2}bh$$

$$= \frac{1}{2}sa$$

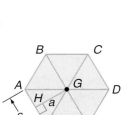

The area of one triangle is $\frac{1}{2}sa$ square units. So the area of the hexagon is $6\left(\frac{1}{2}sa\right)$ square units. Notice that the perimeter P of the hexagon is $6s$ units. We can substitute P for $6s$ in the area formula. So, $A = 6\left(\frac{1}{2}sa\right)$ becomes $A = \frac{1}{2}Pa$. This formula can be used for the area of any regular polygon.

Key Concept
Area of a Regular Polygon

If a regular polygon has an area of A square units, a perimeter of P units, and an apothem of a units, then $A = \frac{1}{2}Pa$.

Example 1 Area of a Regular Polygon

Find the area of a regular pentagon with a perimeter of 40 centimeters.

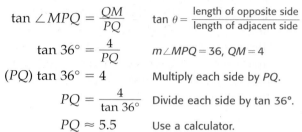

Apothem: The central angles of a regular pentagon are all congruent. Therefore, the measure of each angle is $\frac{360}{5}$ or 72. $\overline{PQ}$ is an apothem of pentagon *JKLMN*. It bisects $\angle NPM$ and is a perpendicular bisector of $\overline{NM}$. So, $m\angle MPQ = \frac{1}{2}(72)$ or 36. Since the perimeter is 40 centimeters, each side is 8 centimeters and $QM = 4$ centimeters. Write a trigonometric ratio to find the length of $\overline{PQ}$.

$$\tan \angle MPQ = \frac{QM}{PQ} \qquad \tan \theta = \frac{\text{length of opposite side}}{\text{length of adjacent side}}$$

$$\tan 36° = \frac{4}{PQ} \qquad m\angle MPQ = 36, \; QM = 4$$

$$(PQ)\tan 36° = 4 \qquad \text{Multiply each side by } PQ.$$

$$PQ = \frac{4}{\tan 36°} \qquad \text{Divide each side by } \tan 36°.$$

$$PQ \approx 5.5 \qquad \text{Use a calculator.}$$

Area:

$$A = \frac{1}{2}Pa \qquad \text{Area of a regular polygon}$$

$$\approx \frac{1}{2}(40)(5.5) \qquad P = 40, \; a \approx 5.5$$

$$\approx 110 \qquad \text{Simplify.}$$

So, the area of the pentagon is about 110 square centimeters.

AREAS OF CIRCLES

You can use a calculator to help derive the formula for the area of a circle from the areas of regular polygons.

Geometry Activity

Area of a Circle

Collect Data

Suppose each regular polygon is inscribed in a circle of radius *r*.

1. Copy and complete the following table. Round to the nearest hundredth.

Inscribed Polygon						
Number of Sides	3	5	8	10	20	50
Measure of a Side	1.73*r*	1.18*r*	0.77*r*	0.62*r*	0.31*r*	0.126*r*
Measure of Apothem	0.5*r*	0.81*r*	0.92*r*	0.95*r*	0.99*r*	0.998*r*
Area						

Analyze the Data

2. What happens to the appearance of the polygon as the number of sides increases?

3. What happens to the areas as the number of sides increases?

4. **Make a conjecture** about the formula for the area of a circle.

You can see from the Geometry Activity that the more sides a regular polygon has, the more closely it resembles a circle.

Key Concept — Area of a Circle

If a circle has an area of A square units and a radius of r units, then $A = \pi r^2$.

Example 2 — Use Area of a Circle to Solve a Real-World Problem

SEWING A caterer has a 48-inch diameter table that is 34 inches tall. She wants a tablecloth that will touch the floor. Find the area of the tablecloth in square yards.

Study Tip

Square Yards
A square yard measures 36 inches by 36 inches or 1296 square inches.

The diameter of the table is 48 inches, and the tablecloth must extend 34 inches in each direction. So the diameter of the tablecloth is $34 + 48 + 34$ or 116 inches. Divide by 2 to find that the radius is 58 inches.

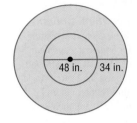

$$A = \pi r^2 \qquad \text{Area of a circle}$$
$$= \pi (58)^2 \qquad \text{Substitution}$$
$$\approx 10{,}568.3 \qquad \text{Use a calculator.}$$

The area of the tablecloth is 10,568.3 square inches. To convert to square yards, divide by 1296. The area of the tablecloth is 8.2 square yards to the nearest tenth.

You can use the properties of circles and regular polygons to find the areas of inscribed and circumscribed polygons.

Study Tip

Look Back
To review **inscribed and circumscribed polygons,** see Lesson 10-4.

Example 3 — Area of an Inscribed Polygon

Find the area of the shaded region. Assume that the triangle is equilateral.

The area of the shaded region is the difference between the area of the circle and the area of the triangle. First, find the area of the circle.

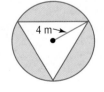

$$A = \pi r^2 \qquad \text{Area of a circle}$$
$$= \pi (4)^2 \qquad \text{Substitution}$$
$$\approx 50.3 \qquad \text{Use a calculator.}$$

To find the area of the triangle, use properties of 30°-60°-90° triangles. First, find the length of the base. The hypotenuse of $\triangle ABC$ is 4, so BC is $2\sqrt{3}$. Since $EC = 2(BC)$, $EC = 4\sqrt{3}$.

Next, find the height of the triangle, DB. Since $m\angle DCB$ is 60, $DB = 2\sqrt{3}(\sqrt{3})$ or 6.

Use the formula to find the area of the triangle.

$$A = \frac{1}{2}bh \qquad \text{Area of a triangle}$$
$$= \frac{1}{2}(4\sqrt{3})(6) \qquad b = 4\sqrt{3},\ h = 6$$
$$\approx 20.8 \qquad \text{Use a calculator.}$$

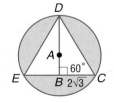

The area of the shaded region is $50.3 - 20.8$ or 29.5 square meters to the nearest tenth.

Check for Understanding

Concept Check

1. **Explain** how to derive the formula for the area of a regular polygon.

2. **OPEN ENDED** Describe a method for finding the base or height of a right triangle given one acute angle and the length of one side.

Guided Practice

Find the area of each polygon. Round to the nearest tenth.

3. a regular hexagon with a perimeter of 42 yards

4. a regular nonagon with a perimeter of 108 meters

Find the area of each shaded region. Assume that all polygons that appear to be regular are regular. Round to the nearest tenth.

5.

6.

Application

7. **UPHOLSTERY** Tyra wants to cover the cushions of her papasan chair with a different fabric. If there are seven circular cushions that are the same size with a diameter of 12 inches, around a center cushion with a diameter of 20 inches, find the area of fabric in square yards that she will need to cover both sides of the cushions. Allow an extra 3 inches of fabric around each cushion.

Practice and Apply

Homework Help

For Exercises	See Examples
8–13, 26, 27	1
14–23, 37–42	3
24, 25, 28–31	2

Extra Practice
See page 777.

Find the area of each polygon. Round to the nearest tenth.

8. a regular octagon with a perimeter of 72 inches

9. a square with a perimeter of $84\sqrt{2}$ meters

10. a square with apothem length of 12 centimeters

11. a regular hexagon with apothem length of 24 inches

12. a regular triangle with side length of 15.5 inches

13. a regular octagon with side length of 10 kilometers

Find the area of each shaded region. Assume that all polygons that appear to be regular are regular. Round to the nearest tenth.

14.

15.

16.

17.

18.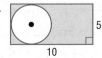

19.

20.

21.

22.

23. ALGEBRA A circle is inscribed in a square, which is circumscribed by another circle. If the diagonal of the square is $2x$, find the ratio of the area of the large circle to the area of the small circle.

24. CAKE A bakery sells single-layer mini-cakes that are 3 inches in diameter for $4 each. They also have a 9-inch cake for $15. If both cakes are the same thickness, which option gives you more cake for the money, nine mini-cakes or one 9-inch cake? Explain.

25. PIZZA A pizza shop sells 8-inch pizzas for $5 and 16-inch pizzas for $10. Which would give you more pizza, two 8-inch pizzas or one 16-inch pizza? Explain.

COORDINATE GEOMETRY **The coordinates of the vertices of a regular polygon are given. Find the area of each polygon to the nearest tenth.**

26. $T(0, 0), U(-7, -7), V(0, -14), W(7, -7)$

27. $G(-12, 0), H(0, 4\sqrt{3}), J(0, -4\sqrt{3})$

28. $J(5, 0), K(2.5\sqrt{2}, -2.5\sqrt{2}), L(0, -5), M(-2.5\sqrt{2}, -2.5\sqrt{2}), N(-5, 0),$
$P(-2.5\sqrt{2}, 2.5\sqrt{2}), Q(0, 5), R(2.5\sqrt{2}, 2.5\sqrt{2})$

29. $A(-2\sqrt{2}, 2\sqrt{2}), B(0, 4), C(2\sqrt{2}, 2\sqrt{2}), D(4, 0), E(2\sqrt{2}, -2\sqrt{2}), F(0, -4),$
$G(-2\sqrt{2}, -2\sqrt{2}), H(-4, 0)$

Find the area of each circle. Round to the nearest tenth.

30. $C = 34\pi$ **31.** $C = 17\pi$ **32.** $C = 54.8$ **33.** $C = 91.4$

SWIMMING POOL **For Exercises 34 and 35, use the following information.**
The area of a circular pool is approximately 7850 square feet. The owner wants to replace the tiling at the edge of the pool.

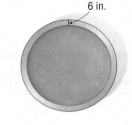
6 in.

34. The edging is 6 inches wide, so she plans to use 6-inch square tiles to form a continuous inner edge. How many tiles will she need to purchase?

35. Once the square tiles are in place around the pool, there will be extra space between the tiles. What shape of tile will best fill this space? How many tiles of this shape should she purchase?

AVIATION **For Exercises 36–38, refer to the circle graph.**

36. Suppose the radius of the circle on the graph is 1.3 centimeters. Find the area of the circle on the graph.

37. Francesca wants to use this circle graph for a presentation. She wants the circle to use as much space on a 22″ by 28″ sheet of poster board as possible. Find the area of the circle.

38. CRITICAL THINKING Make a conjecture about how you could determine the area of the region representing the pilots who are certified to fly private airplanes.

Study Tip

Look Back
To review **circle graphs**, see Lesson 10-2.

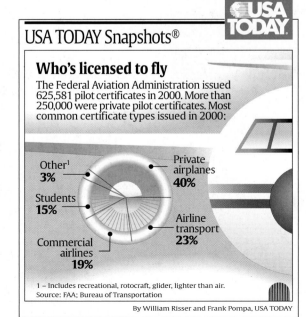

USA TODAY Snapshots®

Who's licensed to fly
The Federal Aviation Administration issued 625,581 pilot certificates in 2000. More than 250,000 were private pilot certificates. Most common certificate types issued in 2000:

Other[1] 3%
Private airplanes 40%
Students 15%
Airline transport 23%
Commercial airlines 19%

1 – Includes recreational, rotocraft, glider, lighter than air.
Source: FAA; Bureau of Transportation

By William Risser and Frank Pompa, USA TODAY

Find the area of each shaded region. Round to the nearest tenth.

39.

7, 3

40.

12, 9

41.

20

42.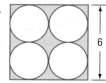

6

43.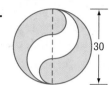

30

44.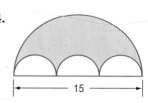

15

GARDENS   For Exercises 45–47, use the following information.

The Elizabeth Park Rose Garden in Hartford, Connecticut, was designed with a gazebo surrounded by two concentric rose garden plots. Wide paths emanate from the center, dividing the garden into square and circular sections.

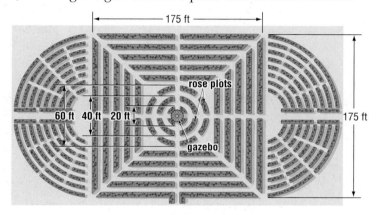
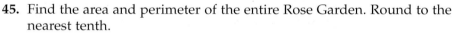

175 ft

rose plots

60 ft   40 ft   20 ft

gazebo

175 ft

45. Find the area and perimeter of the entire Rose Garden. Round to the nearest tenth.

46. What is the total of the circumferences of the three concentric circles formed by the gazebo and the two circular rose garden plots? (Ignore the width of the rose plots and the width of the paths.)

47. Each rose plot has a width of 5 feet. What is the area of the path between the outer two complete circles of rose garden plots?

····•**48. ARCHITECTURE**   The Anraku-ji Temple in Japan is composed of four octagonal floors of different sizes that are separated by four octagonal roofs of different sizes. Refer to the information at the left. Determine whether the areas of each of the four floors are in the same ratio as their sizes. Explain.

SIMILAR FIGURES   For Exercises 49–54, use the following information.

Polygons *FGHJK* and *VWXUZ* are similar regular pentagons.

49. Find the scale factor.

50. Find the perimeter of each pentagon.

51. Compare the ratio of the perimeters of the pentagons to the scale factor.

52. Find the area of each pentagon.

53. Compare the ratio of the areas of the pentagons to the scale factor.

54. Compare the ratio of the areas of the pentagons to the ratio of the perimeters of the pentagons.

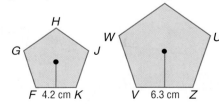

H, G, J, W, X, U, F 4.2 cm K, V 6.3 cm Z

55. CRITICAL THINKING A circle inscribes one regular hexagon and circumscribes another. If the radius of the circle is 10 units long, find the ratio of the area of the smaller hexagon to the area of the larger hexagon.

56. **WRITING IN MATH** Answer the question that was posed at the beginning of the lesson.

How can you find the area of a polygon?

Include the following in your answer:
- information needed about the gazebo floor to find the area, and
- how to find the number of tiles needed to cover the floor.

Standardized Test Practice
Ⓐ Ⓑ Ⓒ Ⓓ

57. A square is inscribed in a circle of area 18π square units. Find the length of a side of the square.

Ⓐ 3 units   Ⓑ 6 units

Ⓒ $3\sqrt{2}$ units   Ⓓ $6\sqrt{2}$ units

58. ALGEBRA The average of x numbers is 15. If the sum of the x numbers is 90, what is the value of x?

Ⓐ 5   Ⓑ 6   Ⓒ 8   Ⓓ 15

Maintain Your Skills

Mixed Review   **Find the area of each quadrilateral.** *(Lesson 11-2)*

59.
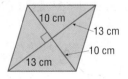

60.
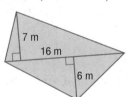

61.
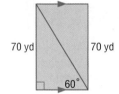

COORDINATE GEOMETRY **Given the coordinates of the vertices of a quadrilateral, determine whether it is a *square*, a *rectangle*, or a *parallelogram*. Then find the area of the quadrilateral.** *(Lesson 11-1)*

62. $A(-3, 2)$, $B(4, 2)$, $C(2, -1)$, $D(-5, -1)$

63. $F(4, 1)$, $G(4, -5)$, $H(-2, -5)$, $J(-2, 1)$

64. $K(-1, -3)$, $L(-2, 5)$, $M(1, 5)$, $N(2, -3)$

65. $P(5, -7)$, $Q(-1, -7)$, $R(-1, -2)$, $S(5, -2)$

Refer to trapezoid *CDFG* with median $\overline{HE}$.
(Lesson 8-6)

66. Find *GF*.

67. Let $\overline{WX}$ be the median of *CDEH*. Find *WX*.

68. Let $\overline{YZ}$ be the median of *HEFG*. Find *YZ*.

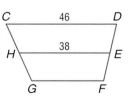

Getting Ready for the Next Lesson   **PREREQUISITE SKILL** **Find h.** *(To review **special right triangles**, see Lesson 7-3.)*

69.

70.

71.

72.

Areas of Irregular Figures

What You'll Learn

- Find areas of irregular figures.
- Find areas of irregular figures on the coordinate plane.

Vocabulary

- irregular figure
- irregular polygon

How do windsurfers use area?

The sail for a windsurf board cannot be classified as a triangle or a parallelogram. However, it can be separated into figures that can be identified, such as trapezoids and a triangle.

IRREGULAR FIGURES An **irregular figure** is a figure that cannot be classified into the specific shapes that we have studied. To find areas of irregular figures, separate the figure into shapes of which we can find the area. The sum of the areas of each is the area of the figure.

Auxiliary lines are drawn in quadrilateral $ABCD$. $\overline{DE}$ and $\overline{DF}$ separate the figure into $\triangle ADE$, $\triangle CDF$, and rectangle $BEDF$.

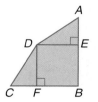

Postulate 11.2

The area of a region is the sum of all of its nonoverlapping parts.

Example 1 Area of an Irregular Figure

Find the area of the figure.

The figure can be separated into a rectangle with dimensions 6 units by 19 units, an equilateral triangle with sides each measuring 6 units, and a semicircle with a radius of 3 units.

Use 30°-60°-90° relationships to find that the height of the triangle is $3\sqrt{3}$.

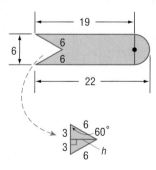

area of irregular figure = area of rectangle − area of triangle + area of semicircle

$$= \ell w - \frac{1}{2}bh + \frac{1}{2}\pi r^2 \qquad \text{Area formulas}$$

$$= 19 \cdot 6 - \frac{1}{2}(6)(3\sqrt{3}) + \frac{1}{2}\pi(3^2) \qquad \text{Substitution}$$

$$= 114 - 9\sqrt{3} + \frac{9}{2}\pi \qquad \text{Simplify.}$$

$$\approx 112.5 \qquad \text{Use a calculator.}$$

The area of the irregular figure is 112.5 square units to the nearest tenth.

www.geometryonline.com/extra_examples

Example 2 Find the Area of an Irregular Figure to Solve a Problem

WebQuest

Identifying the polygons forming a region such as a tessellation will help you determine the type of tessellation. Visit www.geometryonline.com/webquest to continue work on your WebQuest project.

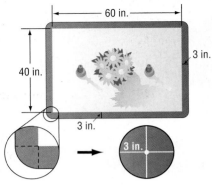

FURNITURE Melissa's dining room table has hardwood around the outside. Find the area of wood around the edge of the table.

First, draw auxiliary lines to separate the figure into regions. The table can be separated into four rectangles and four corners.

The four corners of the table form a circle with radius 3 inches.

area of wood edge = area of rectangles + area of circle

$$= 2\ell w + 2\ell w + \pi r^2 \qquad \text{Area formulas}$$

$$= 2(3)(60) + 2(3)(40) + \pi(3^2) \qquad \text{Substitution}$$

$$= 360 + 240 + 9\pi \qquad \text{Simplify.}$$

$$\approx 628.3 \qquad \text{Use a calculator.}$$

The area of the wood edge of the table is 628.3 square inches to the nearest tenth.

IRREGULAR FIGURES ON THE COORDINATE PLANE The formula for the area of a regular polygon does not apply to an **irregular polygon**, a polygon that is not regular. To find the area of an irregular polygon on the coordinate plane, separate the polygon into known figures.

Study Tip

Estimation
Estimate the area of the simple closed curves by counting the unit squares. Use the estimate to determine if your answer is reasonable.

Example 3 Coordinate Plane

COORDINATE GEOMETRY Find the area of polygon *RSTUV*.

First, separate the figure into regions. Draw an auxiliary line from *S* to *U*. This divides the figure into triangle *STU* and trapezoid *RSUV*.

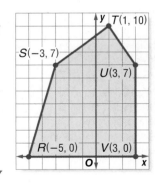

Find the difference between *x*-coordinates to find the length of the base of the triangle and the lengths of the bases of the trapezoid. Find the difference between *y*-coordinates to find the heights of the triangle and trapezoid.

area of *RSTUV* = area of △*STU* + area of trapezoid *RSUV*

$$= \frac{1}{2}bh + \frac{1}{2}h(b_1 + b_2) \qquad \text{Area formulas}$$

$$= \frac{1}{2}(6)(3) + \frac{1}{2}(7)(8 + 6) \qquad \text{Substitution}$$

$$= 58 \qquad \text{Simplify.}$$

The area of *RSTUV* is 58 square units.

Concept Check

1. **OPEN ENDED** Sketch an irregular figure on a coordinate plane and find its area.

2. **Describe** the difference between an irregular figure and an irregular polygon.

Guided Practice   Find the area of each figure. Round to the nearest tenth if necessary.

3.
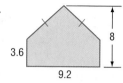

4.
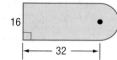

COORDINATE GEOMETRY   Find the area of each figure.

5.
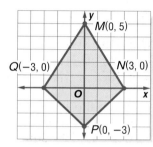

6.
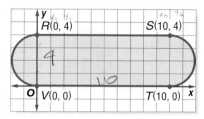

Application

7. **GATES**   The Roths have a series of interlocking gates to form a play area for their baby. Find the area enclosed by the wall and gates.

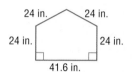

Homework Help	
For Exercises	**See Examples**
8–13	1
14, 15, 23–27	2
16–22	3

Extra Practice
See page 777.

Find the area of each figure. Round to the nearest tenth if necessary.

8.
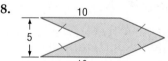

9.

10.
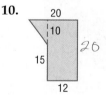

11.
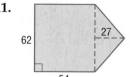

12.
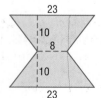

13.
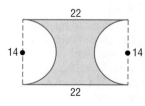

WINDOWS   For Exercises 14 and 15, use the following information.

Mr. Cortez needs to replace this window in his house. The window panes are rectangles and sectors.

14. Find the perimeter of the window.

15. Find the area of the window.

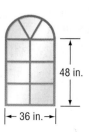

48 in.

36 in.

COORDINATE GEOMETRY Find the area of each figure. Round to the nearest tenth if necessary.

16.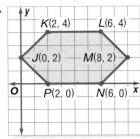

17.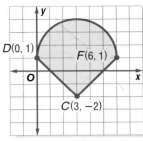

18.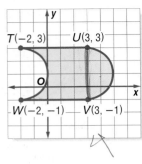

COORDINATE GEOMETRY The vertices of an irregular figure are given. Find the area of each figure.

19. $M(-4, 0)$, $N(0, 3)$, $P(5, 3)$, $Q(5, 0)$

20. $T(-4, -2)$, $U(-2, 2)$, $V(3, 4)$, $W(3, -2)$

21. $G(-3, -1)$, $H(-3, 1)$, $I(2, 4)$, $J(5, -1)$, $K(1, -3)$

22. $P(-8, 7)$, $Q(3, 7)$, $R(3, -2)$, $S(-1, 3)$, $T(-11, 1)$

23. **GEOGRAPHY** Estimate the area of the state of Alabama. Each square on the grid represents 2500 square miles.

24. **RESEARCH** Find a map of your state or a state of your choice. Estimate the area. Then use the Internet or other source to check the accuracy of your estimate.

CALCULUS For Exercises 25–27, use the following information.
The irregular region under the curve has been approximated by rectangles of equal width.

25. Use the rectangles to approximate the area of the region.

26. Analyze the estimate. Do you think the actual area is larger or smaller than your estimate? Explain.

27. How could the irregular region be separated to give an estimate of the area that is more accurate?

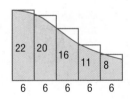

28. **CRITICAL THINKING** Find the ratio of the area of △ABC to the area of square $BCDE$.

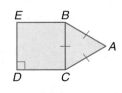

29. **WRITING IN MATH** Answer the question that was posed at the beginning of the lesson.

How do windsurfers use area?

Include the following in your answer:
- describe how to find the area of the sail, and
- another example of an irregular figure.

30. In the figure consisting of squares A, B, and C, $JK = 2KL$ and $KL = 2LM$. If the perimeter of the figure is 66 units, what is the area?

(A) 117 units²   (B) 189 units²

(C) 224 units²   (D) 258 units²

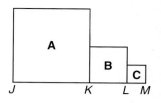

31. ALGEBRA For all integers n, $\boxed{n} = n^2$ if n is odd and $\boxed{n} = \sqrt{n}$ if n is even. What is the value of $\boxed{16} + \boxed{9}$?

(A) 7   (B) 25   (C) 85   (D) 97

Maintain Your Skills

Mixed Review Find the area of each shaded region. Assume that all polygons are regular unless otherwise stated. Round to the nearest tenth. *(Lesson 11-3)*

32.

33.

34.

Find the area of each figure. Round to the nearest tenth if necessary. *(Lesson 11-2)*

35. equilateral triangle with perimeter of 57 feet

36. rhombus with a perimeter of 40 yards and a diagonal of 12 yards

37. isosceles trapezoid with a perimeter of 90 meters if the longer base is 5 meters less than twice as long as the other base, each leg is 3 meters less than the shorter base, and the height is 15.43 meters

38. COORDINATE GEOMETRY The point (6, 0) is rotated 45° clockwise about the origin. Find the exact coordinates of its image. *(Lesson 9-3)*

Getting Ready for the Next Lesson **BASIC SKILL** Express each fraction as a decimal to the nearest hundredth.

39. $\frac{5}{8}$ **40.** $\frac{13}{16}$ **41.** $\frac{9}{47}$ **42.** $\frac{10}{21}$

Practice Quiz 2   Lessons 11–3 and 11–4

Find the area of each polygon. Round to the nearest tenth. *(Lesson 11-3)*

1. regular hexagon with apothem length of 14 millimeters

2. regular octagon with a perimeter of 72 inches

Find the area of each shaded region. Assume that all polygons are regular. Round to the nearest tenth. *(Lesson 11-3)*

3.

4.

5. COORDINATE GEOMETRY Find the area of $CDGHJ$ with vertices $C(-3, -2)$, $D(1, 3)$, $G(5, 5)$, $H(8, 3)$, and $J(5, -2)$. *(Lesson 11-4)*

11-5 Geometric Probability

What You'll Learn

- Solve problems involving geometric probability.
- Solve problems involving sectors and segments of circles.

Vocabulary

- geometric probability
- sector
- segment

How can geometric probability help you win a game of darts?

To win at darts, you have to throw a dart at either the center or the part of the dartboard that earns the most points. In games, probability can sometimes be used to determine chances of winning. Probability that involves a geometric measure such as length or area is called **geometric probability**.

GEOMETRIC PROBABILITY In Chapter 1, you learned that the probability that a point lies on a part of a segment can be found by comparing the length of the part to the length of the whole segment. Similarly, you can find the probability that a point lies in a part of a two-dimensional figure by comparing the area of the part to the area of the whole figure.

Study Tip

Look Back
To review **probability with line segments,** see page 20.

> ### Key Concept — Probability and Area
>
> If a point in region A is chosen at random, then the probability $P(B)$ that the point is in region B, which is in the interior of region A, is
>
> $$P(B) = \frac{\text{area of region } B}{\text{area of region } A}.$$
>
>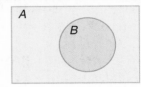

When determining geometric probability with targets, we assume
- that the object lands within the target area, and
- it is equally likely that the object will land anywhere in the region.

Standardized Test Practice
Ⓐ Ⓑ Ⓒ Ⓓ

Example 1 Probability with Area

Grid-In Test Item

> A square game board has black and white stripes of equal width as shown. What is the chance that a dart thrown at the board will land on a white stripe?
>
>

Read the Test Item

You want to find the probability of landing on a white stripe, not a black stripe.

Solve the Test Item

We need to divide the area of the white stripes by the total area of the game board. Extend the sides of each stripe. This separates the square into 36 small unit squares.

The white stripes have an area of 15 square units. The total area is 36 square units.

The probability of tossing a chip onto the white stripes is $\frac{15}{36}$ or $\frac{5}{12}$.

Fill in the Grid

Write $\frac{5}{12}$ as 5/12 in the top row of the grid-in.

Then shade in the appropriate bubble under each entry.

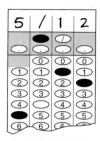

SECTORS AND SEGMENTS OF CIRCLES

Sometimes you need to know the area of a sector of a circle in order to find a geometric probability. A **sector** of a circle is a region of a circle bounded by a central angle and its intercepted arc.

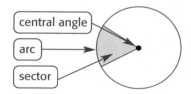

central angle

arc

sector

Key Concept — *Area of a Sector*

If a sector of a circle has an area of A square units, a central angle measuring $N°$, and a radius of r units, then $A = \frac{N}{360}\pi r^2$.

Study Tip

Common Misconceptions
The probability of an event can be expressed as a decimal or a fraction. These numbers are also sometimes represented by a percent.

Example 2 *Probability with Sectors*

a. **Find the area of the blue sector.**

Use the formula to find the area of the sector.

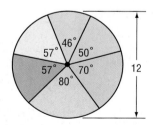

$A = \dfrac{N}{360}\pi r^2$ Area of a sector

$= \dfrac{46}{360}\pi(6^2)$ $N = 46, r = 6$

$= 4.6\pi$ Simplify.

b. **Find the probability that a point chosen at random lies in the blue region.**

To find the probability, divide the area of the sector by the area of the circle. The area of the circle is πr^2 with a radius of 6.

$P(\text{blue}) = \dfrac{\text{area of sector}}{\text{area of circle}}$ Geometric probability formula

$= \dfrac{4.6\pi}{\pi \cdot 6^2}$ Area of sector = 4.6π, area of circle = $\pi \cdot 6^2$

≈ 0.13 Use a calculator.

The probability that a random point is in the blue sector is about 0.13 or 13%.

The region of a circle bounded by an arc and a chord is called a **segment** of a circle. To find the area of a segment, subtract the area of the triangle formed by the radii and the chord from the area of the sector containing the segment.

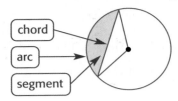

Example 3 Probability with Segments

A regular hexagon is inscribed in a circle with a diameter of 14.

a. **Find the area of the red segment.**

Area of the sector:

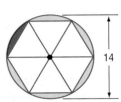

$A = \dfrac{N}{360}\pi r^2$ Area of a sector

$\quad = \dfrac{60}{360}\pi(7^2)$ $N = 60, r = 7$

$\quad = \dfrac{49}{6}\pi$ Simplify.

$\quad \approx 25.66$ Use a calculator.

Area of the triangle:

Since the hexagon was inscribed in the circle, the triangle is equilateral, with each side 7 units long. Use properties of 30°-60°-90° triangles to find the apothem. The value of x is 3.5, the apothem is $x\sqrt{3}$ or $3.5\sqrt{3}$ which is approximately 6.06.

Next, use the formula for the area of a triangle.

$A = \dfrac{1}{2}bh$ Area of a triangle

$\quad = \dfrac{1}{2}(7)(6.06)$ $b = 7, h = 6.06$

$\quad \approx 21.22$ Simplify.

Area of the segment:

area of segment = area of sector − area of triangle

$\quad\quad\quad \approx 25.66 - 21.22$ Substitution

$\quad\quad\quad \approx 4.44$ Simplify.

b. **Find the probability that a point chosen at random lies in the red region.**

Divide the area of the sector by the area of the circle to find the probability. First, find the area of the circle. The radius is 7, so the area is $\pi(7^2)$ or about 153.94 square units.

$P(\text{blue}) = \dfrac{\text{area of segment}}{\text{area of circle}}$

$\quad\quad\quad \approx \dfrac{4.44}{153.94}$

$\quad\quad\quad \approx 0.03$

The probability that a random point is on the red segment is about 0.03 or 3%.

Concept Check

1. **Explain** how to find the area of a sector of a circle.

2. **OPEN ENDED**   List three games that involve geometric probability.

3. **FIND THE ERROR**   Rachel and Taimi are finding the probability that a point chosen at random lies in the green region.

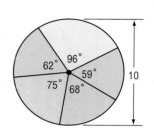

Rachel

$$A = \frac{N}{360}\pi r^2$$

$$= \frac{59+62}{360}\pi(5^2)$$

$$\approx 26.4$$

$$P(green) \approx \frac{26.4}{25\pi} \approx 0.34$$

Taimi

$$A = \frac{N}{360}\pi r^2$$

$$= \frac{59}{360}\pi(5^2) + \frac{62}{360}$$

$$\approx 13.0$$

$$P(green) \approx \frac{13.0}{25\pi} \approx 0.17$$

Who is correct? Explain your answer.

Guided Practice

Find the area of the blue region. Then find the probability that a point chosen at random will be in the blue region.

4.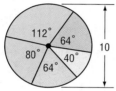

5.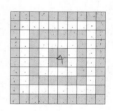

Standardized Test Practice
Ⓐ Ⓑ Ⓒ Ⓓ

6. What is the chance that a point chosen at random lies in the shaded region?

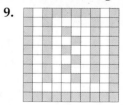

Homework Help

For Exercises	See Examples
7–9, 16, 24–30	1
10–15, 20–23	2
17–19	3

Extra Practice
See page 777.

Find the probability that a point chosen at random lies in the shaded region.

7.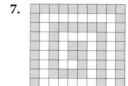

8.

9.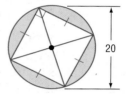

Find the area of the indicated sector. Then find the probability of spinning the color indicated if the diameter of each spinner is 15 centimeters.

10. blue

11. pink

12. purple

Find the area of the indicated sector. Then find the probability of choosing the color indicated if the diameter of each spinner is 15 centimeters.

13. red

14. green

15. yellow

16. PARACHUTES A skydiver must land on a target of three concentric circles. The diameter of the center circle is 2 yards, and the circles are spaced 1 yard apart. Find the probability that she will land on the shaded area.

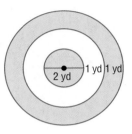

Find the area of the shaded region. Then find the probability that a point chosen at random is in the shaded region. Assume all inscribed polygons are regular.

17.

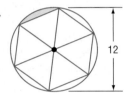

18.

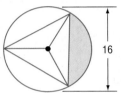

19.

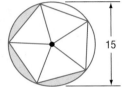

SURVEYS For Exercises 20–23, use the following information.

A survey was taken at a high school, and the results were put in a circle graph. The students were asked to list their favorite colors. The measurement of each central angle is shown. If a person is chosen at random from the school, find the probability of each response.

What's Your Favorite Color?

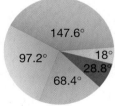

20. Favorite color is red.

21. Favorite color is blue or green.

22. Favorite color is *not* red or blue.

23. Favorite color is *not* orange or green.

TENNIS For Exercises 24 and 25, use the following information.

A tennis court has stripes dividing it into rectangular regions. For singles play, the inbound region is defined by segments $\overline{AB}$ and $\overline{CD}$. The doubles court is bound by the segments $\overline{EF}$ and $\overline{GH}$.

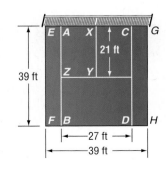

24. Find the probability that a ball in a singles game will land inside the court, but out of bounds.

25. When serving, the ball must land within *AXYZ*, the service box. Find the probability that a ball will land in the service box, relative to the court.

DARTS For Exercises 26–30, use the following information.

Each sector of the dartboard has congruent central angles. Find the probability that the dart will land on the indicated color. The diameter of the center circle is 2 units.

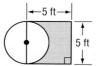

26. black　　　　27. white　　　　28. red

29. Point values are assigned to each color. Should any of the colors have the same point value? Explain.

30. Which color should have the lowest point value? Explain.

31. **CRITICAL THINKING** Study each spinner in Exercises 13–15.
   a. Are the chances of landing on each color equal? Explain.
   b. Would this be considered a fair spinner to use in a game? Explain.

32. WRITING IN MATH　Answer the question that was posed at the beginning of the lesson.

   How can geometric probability help you win a game of darts?

   Include the following in your answer:

   • an explanation of how to find the geometric probability of landing on a red sector, and

   • an explanation of how to find the geometric probability of landing in the center circle.

33. One side of a square is a diameter of a circle. The length of one side of the square is 5 feet. To the nearest hundredth, what is the probability that a point chosen at random is in the shaded region?

   Ⓐ 0.08　　　Ⓑ 0.22　　　Ⓒ 0.44　　　Ⓓ 0.77

34. **ALGEBRA** If $4y = 16$, then $12 \div y =$

   Ⓐ 1.　　　Ⓑ 2.　　　Ⓒ 3.　　　Ⓓ 4.

Maintain Your Skills

Mixed Review　**Find the area of each figure. Round to the nearest tenth, if necessary.** *(Lesson 11-4)*

35.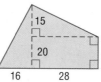

36.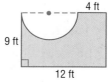

Find the area of each polygon. Round to the nearest tenth, if necessary. *(Lesson 11-3)*

37. a regular triangle with a perimeter of 48 feet

38. a square with a side length of 21 centimeters

39. a regular hexagon with an apothem length of 8 inches

ALGEBRA Find the measure of each angle on ⊙F with diameter $\overline{AC}$. *(Lesson 10-2)*

40. $\angle AFB$　　41. $\angle CFD$　　42. $\angle AFD$　　43. $\angle DFB$

Find the length of the third side of a triangle given the measures of two sides and the included angle of the triangle. Round to the nearest tenth. *(Lesson 7-7)*

44. $m = 6.8$, $n = 11.1$, $m\angle P = 57$

45. $f = 32$, $h = 29$, $m\angle G = 41$

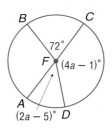

Vocabulary and Concept Check

apothem (p. 610)
geometric probability (p. 622)

irregular figure (p. 617)
irregular polygon (p. 618)

sector (p. 623)
segment (p. 624)

A complete list of postulates and theorems can be found on pages R1–R8.

Exercises Choose the formula to find the area of each shaded figure.

1.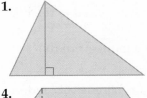

2.

3.

4.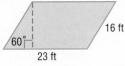

5.

6.

a. $A = \pi r^2$

b. $A = \dfrac{N}{360}\pi r^2$

c. $A = \dfrac{1}{2}bh$

d. $A = \dfrac{1}{2}Pa$

e. $A = bh$

f. $A = \dfrac{1}{2}h(b_1 + b_2)$

Lesson-by-Lesson Review

11-1 Area of Parallelograms

See pages 595–600.

Concept Summary

• The area of a parallelogram is the product of the base and the height.

Example Find the area of $\square GHJK$.

The area of a parallelogram is given by the formula $A = bh$.

$A = bh$ Area of a parallelogram

   $= 14(9)$ or 126 $b = 14, h = 9$

The area of the parallelogram is 126 square units.

Exercises Find the perimeter and area of each parallelogram.   *See Example 1 on page 596.*

7.

8.

COORDINATE GEOMETRY Given the coordinates of the vertices of a quadrilateral, determine whether it is a *square*, a *rectangle*, or a *parallelogram*. Then find the area of the quadrilateral.   *See Example 3 on page 597.*

9. $A(-6, 1), B(1, 1), C(1, -6), D(-6, -6)$

10. $E(7, -2), F(1, -2), G(2, 2), H(8, 2)$

11. $J(-1, -4), K(-5, 0), L(-5, 5), M(-1, 1)$

12. $P(-7, -1), Q(-3, 3), R(-1, 1), S(-5, -3)$

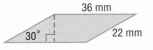
www.geometryonline.com/vocabulary_review

11-2 Areas of Triangles, Rhombi, and Trapezoids

See pages 601–609.

Concept Summary

- The formula for the area of a triangle can be used to find the areas of many different figures.
- Congruent figures have equal areas.

Example Trapezoid $MNPQ$ has an area of 360 square feet. Find the length of $\overline{MN}$.

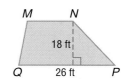

$A = \frac{1}{2}h(b_1 + b_2)$ Area of a trapezoid

$360 = \frac{1}{2}(18)(b_1 + 26)$ $A = 360, h = 18, b_2 = 26$

$360 = 9b_1 + 234$ Multiply.

$14 = b_1$ Solve for b_1.

The length of $\overline{MN}$ is 14 feet.

Exercises Find the missing measure for each quadrilateral. *See Example 4 on page 604.*

13. Triangle CDE has an area of 336 square inches. Find CE.

14. Trapezoid $GHJK$ has an area of 75 square meters. Find the height.

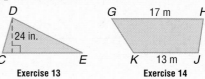

Exercise 13   Exercise 14

11-3 Areas of Regular Polygons and Circles

See pages 610–616.

Concept Summary

- A regular n-gon is made up of n congruent isosceles triangles.
- The area of a circle of radius r units is πr^2 square units.

Example Find the area of a regular hexagon with a perimeter of 72 feet.

Since the perimeter is 72 feet, the measure of each side is 12 feet. The central angle of a hexagon is 60°. Use the properties of 30°-60°-90° triangles to find that the apothem is $6\sqrt{3}$ feet.

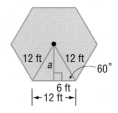

$A = \frac{1}{2}Pa$ Area of a regular polygon

$= \frac{1}{2}(72)(6\sqrt{3})$ $P = 72, a = 6\sqrt{3}$

$= 216\sqrt{3}$ Simplify.

≈ 374.1

The area of the regular hexagon is 374.1 square feet to the nearest tenth.

Exercises Find the area of each polygon. Round to the nearest tenth.
See Example 1 on page 611.

15. a regular pentagon with perimeter of 100 inches

16. a regular decagon with side length of 12 millimeters

Chapter
11 For More ...
• Extra Practice, see pages 776–777.
• Mixed Problem Solving, see page 792.

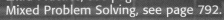

11-4 Areas of Irregular Figures

See pages 617–621.

Concept Summary

• The area of an irregular figure is the sum of the areas of its nonoverlapping parts.

Example **Find the area of the figure.**

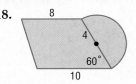

Separate the figure into a rectangle and a triangle.

$$\begin{array}{l} \text{area of} \\ \text{irregular figure} \end{array} = \begin{array}{l} \text{area of} \\ \text{rectangle} \end{array} - \begin{array}{l} \text{area of} \\ \text{semicircle} \end{array} + \begin{array}{l} \text{area of} \\ \text{triangle} \end{array}$$

$$= \ell w - \frac{1}{2}\pi r^2 + \frac{1}{2}bh \qquad \text{Area formulas}$$

$$= (6)(8) - \frac{1}{2}\pi(4^2) + \frac{1}{2}(8)(8) \qquad \text{Substitution}$$

$$= 48 - 8\pi + 32 \text{ or about } 54.9 \qquad \text{Simplify.}$$

The area of the irregular figure is 54.9 square units to the nearest tenth.

Exercises **Find the area of each figure to the nearest tenth.** *See Example 1 on page 617.*

17.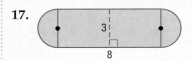

18.

11-5 Geometric Probability

See pages 622–627.

Concept Summary

• To find a geometric probability, divide the area of a part of a figure by the total area.

Example **Find the probability that a point chosen at random will be in the blue sector.**

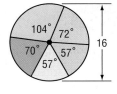

First find the area of the blue sector.

$$A = \frac{N}{360}\pi r^2 \qquad \text{Area of a sector}$$

$$= \frac{104}{360}\pi(8^2) \text{ or about } 58.08 \qquad \text{Substitute and simplify.}$$

To find the probability, divide the area of the sector by the area of the circle.

$$P(\text{blue}) = \frac{\text{area of sector}}{\text{area of circle}} \qquad \text{Geometric probability formula}$$

$$= \frac{58.08}{\pi 8^2} \text{ or about } 0.29 \qquad \text{The probability is about 0.29 or 29\%.}$$

Exercises **Find the probability that a point chosen at random will be in the sector of the given color.** *See Example 2 on page 623.*

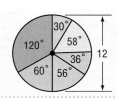

19. red

20. purple or green

Practice Test

Vocabulary and Concepts

Choose the letter of the correct area formula for each figure.

1. regular polygon
2. trapezoid
3. triangle

> **a.** $A = \frac{1}{2}Pa$
>
> **b.** $A = \frac{1}{2}bh$
>
> **c.** $A = \frac{1}{2}h(b_1 + b_2)$

Skills and Applications

COORDINATE GEOMETRY Given the coordinates of the vertices of a quadrilateral, determine whether it is a *square*, a *rectangle*, or a *parallelogram*. Then find the area of the quadrilateral.

4. $R(-6, 8)$, $S(-1, 5)$, $T(-1, 1)$, $U(-6, 4)$
5. $R(7, -1)$, $S(9, 3)$, $T(5, 5)$, $U(3, 1)$
6. $R(2, 0)$, $S(4, 5)$, $T(7, 5)$, $U(5, 0)$
7. $R(3, -6)$, $S(9, 3)$, $T(12, 1)$, $U(6, -8)$

Find the area of each figure. Round to the nearest tenth if necessary.

8.

9.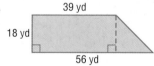

10.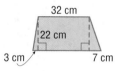

11. a regular octagon with apothem length of 3 ft
12. a regular pentagon with a perimeter of 115 cm

Each spinner has a diameter of 12 inches. Find the probability of spinning the indicated color.

13. red

14. orange

15. green

Find the area of each figure. Round to the nearest tenth.

16.

17.

18.

19. **SOCCER BALLS** The surface of a soccer ball is made of a pattern of regular pentagons and hexagons. If each hexagon on a soccer ball has a perimeter of 9 inches, what is the area of a hexagon?

20. **STANDARDIZED TEST PRACTICE** What is the area of a quadrilateral with vertices at $(-3, -1)$, $(-1, 4)$, $(7, 4)$, and $(5, -1)$?

   Ⓐ 50 units^2 Ⓑ 45 units^2 Ⓒ $8\sqrt{29}$ units^2 Ⓓ 40 units^2

Part 1 Multiple Choice

Record your answers on the answer sheet provided by your teacher or on a sheet of paper.

1. Solve $3\left(\dfrac{2x - 4}{-6}\right) = 18$.   (Prerequisite Skill)

   Ⓐ −19   Ⓑ −16   Ⓒ 4   Ⓓ 12

2. Sam rode his bike along the path from the library to baseball practice. What type of angle did he form during the ride?   (Lesson 1-5)

   Ⓐ straight

   Ⓑ obtuse

   Ⓒ acute

   Ⓓ right

   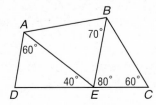

3. What is the logical conclusion of these statements?

   If you exercise, you will maintain better health.
   If you maintain better health, you will live longer.
   (Lesson 2-4)

   Ⓐ If you exercise, you will live longer.

   Ⓑ If you do not exercise, you will not live longer.

   Ⓒ If you do not exercise, you will not maintain better health.

   Ⓓ If you maintain better health, you will not live longer.

4. Which segments are parallel?   (Lesson 3-5)

   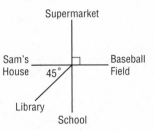

   Ⓐ $\overline{AB}$ and $\overline{CD}$ Ⓑ $\overline{AD}$ and $\overline{BC}$

   Ⓒ $\overline{AD}$ and $\overline{BE}$ Ⓓ $\overline{AE}$ and $\overline{BC}$

5. The front view of a pup tent resembles an isosceles triangle. The entrance to the tent is an angle bisector. The tent is secured by stakes. What is the distance between the two stakes?
   (Lesson 5-1)

   Ⓐ 3 ft

   Ⓑ 4 ft

   Ⓒ 5 ft

   Ⓓ 6 ft

   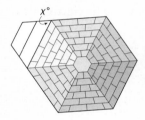

6. A carpenter is building steps leading to a hexagonal gazebo. The outside edges of the steps need to be cut at an angle. Find x.
   (Lesson 8-1)

   Ⓐ 180   Ⓑ 120   Ⓒ 72   Ⓓ 60

7. Which statement is *always* true?   (Lesson 10-4)

   Ⓐ When an angle is inscribed in a circle, the angle's measure equals one-half of the measure of the intercepted arc.

   Ⓑ In a circle, an inscribed quadrilateral will have consecutive angles that are supplementary.

   Ⓒ In a circle, an inscribed angle that intercepts a semicircle is obtuse.

   Ⓓ If two inscribed angles of a circle intercept congruent arcs, then the angles are complementary.

8. The apothem of a regular hexagon is 7.8 centimeters. If the length of each side is 9 centimeters, what is the area of the hexagon?   (Lesson 11-3)

   Ⓐ 35.1 cm²   Ⓑ 70.2 cm²

   Ⓒ 210.6 cm²   Ⓓ 421.2 cm²

Preparing for Standardized Tests
For test-taking strategies and more
practice, see pages 795–810.

Part 2 | Short Response/Grid In

**Record your answers on the answer sheet
provided by your teacher or on a sheet
of paper.**

9. The post office is located halfway between
the fire station and the library. What are the
coordinates of the post office? **(Lesson 1-3)**

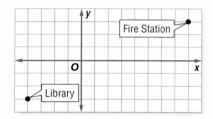

10. What is the slope of a line perpendicular
to the line represented by the equation
$3x - 6y = 12$? **(Lesson 3-3)**

11. $\triangle RST$ is a right triangle. Find $m\angle R$.
(Lesson 4-2)

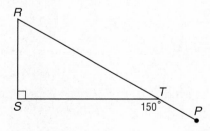

12. If $\angle A$ and $\angle E$ are congruent, find AB, the
distance in feet across the pond.
(Lesson 6-3)

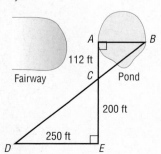

13. If point $J(6, -3)$ is translated 5 units up and
then reflected over the y-axis, what will the
new coordinates of J' be? **(Lesson 9-2)**

 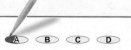
Part 3 | Extended Response

**Record your answers on a sheet of paper.
Show your work.**

14. Lori and her family are camping near a
mountain. Their campground is in a clearing
next to a stretch of forest.

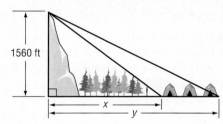

a. The angle of elevation from Lori's line of
sight at the edge of the forest to the top
of the mountain, is 38°. Find the distance
x from the base of the mountain to the
edge of the forest. Round to the nearest
foot. **(Lesson 7-5)**

b. The angle of elevation from the far edge
of the campground to the top of the
mountain is 35°. Find the distance y
from the base of the mountain to the far
edge of the campground. Round to the
nearest foot. **(Lesson 7-5)**

c. What is the width of the campground?
Round to the nearest foot. **(Lesson 7-5)**

15. Parallelogram $ABCD$ has vertices
$A(0, 0)$, $B(3, 4)$, and $C(8, 4)$.

a. Find the possible coordinates for D.
(Lesson 8-2)

b. Find the area of $ABCD$. **(Lesson 11-1)**

What You'll Learn

- **Lesson 12-1** Identify three-dimensional figures.
- **Lesson 12-2** Draw two-dimensional models for solids.
- **Lessons 12-3 through 12-6** Find the lateral areas and surface areas of prisms, cylinders, pyramids, and cones.
- **Lesson 12-7** Find the surface areas of spheres and hemispheres.

Key Vocabulary

- polyhedron (p. 637)
- net (p. 644)
- surface area (p. 644)
- lateral area (p. 649)

Why It's Important

Diamonds and other gems are cut to enhance the beauty of the stones. The stones are cut into regular geometric shapes. Each cut has a special name.

You will learn more about gemology in Lesson 12-1.

Getting Started

▶ **Prerequisite Skills**   To be successful in this chapter, you'll need to master these skills and be able to apply them in problem-solving situations. Review these skills before beginning Chapter 12.

For Lesson 12-1   Parallel Lines and Planes

In the figure, $\overline{AC} \parallel \ell$. Determine whether each statement is *true, false,* or *cannot be determined.*   *(For review, see Lesson 3-1.)*

1. $\triangle ADC$ lies in plane $\mathcal{N}$.

2. $\triangle ABC$ lies in plane $\mathcal{K}$.

3. The line containing $\overline{AB}$ is parallel to plane $\mathcal{K}$.

4. The line containing $\overline{AC}$ lies in plane $\mathcal{K}$.

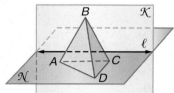

For Lessons 12-3 and 12-5   Areas of Triangles and Trapezoids

Find the area of each figure. Round to the nearest tenth if necessary.
(For review, see Lesson 11-2.)

5.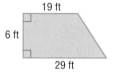

6.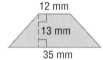

7.

For Lessons 12-4, 12-6, and 12-7   Area of Circles

Find the area of each circle with the given radius or diameter. Round to the nearest tenth.
(For review, see Lesson 11-3.)

8. $d = 19.0$ cm

9. $r = 1.5$ yd

10. $d = 10.4$ m

Surface Area   Make this Foldable to help you organize your notes. Begin with a sheet of 11" by 17" paper.

Step 1   **Fold Lengthwise**

Fold lengthwise leaving a two-inch tab.

Step 2   **Fold**

Fold the paper into five sections.

Step 3   **Cut**

Open. Cut along each fold to make five tabs.

Step 4   **Label**

Label as shown.

Reading and Writing   As you read and study the chapter, define terms and write notes about surface area for each three-dimensional figure.

Three-Dimensional Figures

What You'll Learn

- Use orthogonal drawings of three-dimensional figures to make models.
- Identify and use three-dimensional figures.

Why are drawings of three-dimensional structures valuable to archaeologists?

Archaeologists and Egyptologists continue to study the Great Pyramids of Egypt. Drawings of these three-dimensional structures are helpful in their study.

Vocabulary

- orthogonal drawing
- corner view
- perspective view
- polyhedron
- face
- edges
- prism
- bases
- regular prism
- pyramid
- regular polyhedron
- Platonic solids
- cylinder
- cone
- sphere
- cross section
- reflection symmetry

DRAWINGS OF THREE-DIMENSIONAL FIGURES

If you see a three-dimensional object from only one viewpoint, you may not know its true shape. Here are four views of the pyramid of Menkaure in Giza, Egypt. The two-dimensional views of the top, left, front, and right sides of an object are called an **orthogonal drawing**.

top view   left view   front view   right view

This sculpture is *Stacked Pyramid* by Jackie Ferrara. How can we show the stacks of blocks on each side of the piece in a two-dimensional drawing? Let the edge of each block represent a unit of length and use a dark segment to indicate a break in the surface.

The view of a figure from a corner is called the **corner view** or **perspective view**. You can use isometric dot paper to draw the corner view of a solid figure. One corner view of a cube is shown at the right.

Example 1 Use Orthogonal Drawings

a. **Draw the back view of a figure given its orthogonal drawing.**

Use blocks to make a model. Then use your model to draw the back view.

- The top view indicates two rows and two columns of different heights.
- The front view indicates that the left side is 5 blocks high and the right side is 3 blocks high. The dark segments indicate breaks in the surface.

top view   left view   front view   right view

- The right view indicates that the right front column is only one block high. The left front column is 4 blocks high. The right back column is 3 blocks high.
- Check the left side of your model. All of the blocks should be flush.
- Check to see that all views correspond to the model.

Now that your model is accurate, turn it around to the back and draw what you see. The blocks are flush, so no heavy segments are needed.

b. Draw the corner view of the figure.

Turn your model so you are looking at the corners of the blocks. The lowest columns should be in front so the differences in height between the columns is visible.

Connect the dots on the isometric dot paper to represent the edges of the solid. Shade the tops of each column.

IDENTIFY THREE-DIMENSIONAL FIGURES

A solid with all flat surfaces that enclose a single region of space is called a **polyhedron**. Each flat surface, or **face**, is a polygon. The line segments where the faces intersect are called **edges**. Edges intersect at a point called a *vertex*.

A **prism** is a polyhedron with two parallel congruent faces called **bases**. The other faces are parallelograms. The intersection of three edges is a vertex. Prisms are named by the shape of their bases. A **regular prism** is a prism with bases that are regular polygons. A cube is an example of a regular prism. Some common prisms are shown below.

Name	Triangular Prism	Rectangular Prism	Pentagonal Prism
Model			
Shape of Base	triangle	rectangle	pentagon

A polyhedron with all faces (except for one) intersecting at one vertex is a **pyramid**. Pyramids are named for their bases, which can be any polygon.

A polyhedron is a **regular polyhedron** if all of its faces are regular congruent polygons and all of the edges are congruent.

square pyramid

There are exactly five types of regular polyhedra. These are called the **Platonic solids** because Plato described them extensively in his writings.

Platonic Solids					
Name	tetrahedron	hexahedron	octahedron	dodecahedron	icosahedron
Model					
Faces	4	6	8	12	20
Shape of Face	equilateral triangle	square	equilateral triangle	regular pentagon	equilateral triangle

Plato was a teacher of mathematics and philosophy. Around 387 B.C., he founded a school in Athens, Greece, called the "Academy."

Source: www.infoplease.com

There are solids that are *not* polyhedrons. All of the faces in each solid are not polygons. A **cylinder** is a solid with congruent circular bases in a pair of parallel planes. A **cone** has a circular base and a vertex. A **sphere** is a set of points in space that are a given distance from a given point.

Name	cylinder	cone	sphere
Model	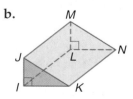		

Example 2 Identify Solids

Identify each solid. Name the bases, faces, edges, and vertices.

a.

The base is a rectangle, and the other four faces meet in a point. So this solid is a rectangular pyramid.

Base: □$ABCD$

Faces: □$ABCD$, △AED, △DEC, △CEB, △AEB

Edges: $\overline{AB}$, $\overline{BC}$, $\overline{CD}$, $\overline{DA}$, $\overline{AE}$, $\overline{DE}$, $\overline{CE}$, $\overline{BE}$

Vertices: A, B, C, D, E

Study Tip

Common Misconception
Prisms can be oriented so the bases are not the top and bottom of the solid.

b.

The bases are right triangles. So this is a triangular prism.

Bases: △IJK, △LMN

Faces: △IJK, △LMN, □$ILNK$, □$KJMN$, □$IJML$

Edges: $\overline{IL}$, $\overline{LN}$, $\overline{NK}$, $\overline{IK}$, $\overline{IJ}$, $\overline{LM}$, $\overline{JM}$, $\overline{MN}$, $\overline{JK}$

Vertices: I, J, K, L, M, N

c.

This solid has a circle for a base and a vertex. So it is a cone.

Base: ⊙Q

Vertex: P

Interesting shapes occur when a plane intersects, or slices, a solid figure. If the plane is parallel to the base or bases of the solid, then the intersection of the plane and solid is called a **cross section** of the solid.

Example 3 *Slicing Three-Dimensional Figures*

CARPENTRY A carpenter purchased a section of a tree trunk. He wants to cut the trunk into a circle, an oval, and a rectangle. How could he cut the tree trunk to get each shape?

The tree trunk has a cylindrical shape. If the blade of the saw was placed parallel to the bases, the cross section would be a circle.

If the blade was placed at an angle to the bases of the tree trunk, the slice would be an oval shape, or an ellipse.

To cut a rectangle from the cylinder, place the blade perpendicular to the bases. The slice is a rectangle.

Check for Understanding

Concept Check
1. **Explain** how the Platonic Solids are different from other polyhedra.

2. **Explain** the difference between a square pyramid and a square prism.

3. **OPEN ENDED** Draw a rectangular prism.

Guided Practice
4. **Draw** the back view and corner view of a figure given its orthogonal drawing.

top view   left view   front view   right view

Identify each solid. Name the bases, faces, edges, and vertices.

5.

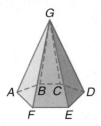

6.

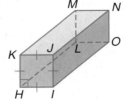

7.

Application
8. **DELICATESSEN** A slicer is used to cut whole pieces of meat and cheese for sandwiches. Suppose a customer wants slices of cheese that are round and slices that are rectangular. How can the cheese be placed on the slicer to get each shape?

Homework Help

For Exercises	See Examples
9–15, 23, 24	1
16–22, 36–41	2
25–35	3

Extra Practice
See page 778.

Draw the back view and corner view of a figure given each orthogonal drawing.

9.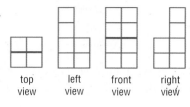

top view   left view   front view   right view

10.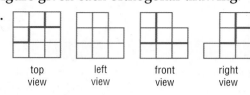

top view   left view   front view   right view

11.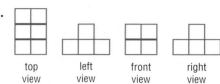

top view   left view   front view   right view

12.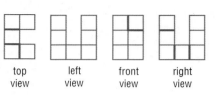

top view   left view   front view   right view

Given the corner view of a figure, sketch the orthogonal drawing.

13.   14.   15.

Identify each solid. Name the bases, faces, edges, and vertices.

16. 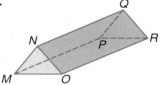   17. 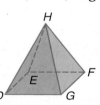   18.

19.   20.   21.

22. **EULER'S FORMULA** The number of faces F, vertices V, and edges E of a polyhedron are related by Euler's (OY luhrz) formula: $F + V = E + 2$. Determine whether Euler's formula is true for each of the figures in Exercises 16–21.

SPEAKERS For Exercises 23 and 24, use the following information.
The top and front views of a speaker for a stereo system are shown.

23. Is it possible to determine the shape of the speaker? Explain.

24. Describe possible shapes for the speaker. Draw the left and right views of one of the possible shapes.

Top View   Front View

Determine the shape resulting from each slice of the cone.

25.   26.   27.

Determine the shape resulting from each slice of the rectangular prism.

28.

29.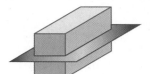

30.

Draw a diagram and describe how a plane can slice a tetrahedron to form the following shapes.

31. equilateral triangle   32. isosceles triangle   33. quadrilateral

Describe the solid that results if the number of sides of each base increases infinitely. The bases of each solid are regular polygons inscribed in a circle.

34. pyramid   35. prism

GEMOLOGY For Exercises 36–38, use the following information.
A well-cut diamond enhances the natural beauty of the stone. These cuts are called *facets*.

uncut

emerald cut

round cut

36. Describe the shapes seen in an uncut diamond.

37. What shapes are seen in the emerald-cut diamond?

38. List the shapes seen in the round-cut diamond.

For Exercises 39–41, use the following table.

Number of Faces	Prism	Pyramid
4	none	tetrahedron
5	a. _____?_____	square or rectangular
6	b. _____?_____	c. _____?_____
7	pentagonal	d. _____?_____
8	e. _____?_____	heptagonal

39. Name the type of prism or pyramid that has the given number of faces.

40. Analyze the information in the table. Is there a pattern between the number of faces and the bases of the corresponding prisms and pyramids?

41. Is it possible to classify a polyhedron given only the number of faces? Explain.

42. **CRITICAL THINKING** Construct a Venn diagram that shows the relationship among polyhedra, Platonic solids, prisms, and pyramids.

43. **WRITING IN MATH** Answer the question that was posed at the beginning of the lesson.

Why are drawings of three-dimensional structures valuable to archaeologists?

Include the following in your answer:
- types of two-dimensional models and drawings, and
- the views of a structure used to show three dimensions.

44. All of the following can be formed by the intersection of a cube and a plane *except*

(A) a triangle.   (B) a rectangle.   (C) a point.   (D) a circle.

45. ALGEBRA For which of the following values of x is $\dfrac{x^3}{x^4}$ the least?

(A) -4 (B) -3 (C) -2 (D) -1

Extending the Lesson

SYMMETRY AND SOLIDS In a two-dimensional plane, figures are symmetric with respect to a line or a point. In three-dimensional space, solids are symmetric with respect to a plane. This is called **reflection symmetry**. A square pyramid has four planes of symmetry. Two pass through the altitude and one pair of opposite vertices of the base. Two pass through the altitude and the midpoint of one pair of opposite edges of the base.

For each solid, determine the number of planes of symmetry.

46. tetrahedron   **47.** cylinder   **48.** sphere

Maintain Your Skills

Mixed Review

SURVEYS For Exercises 49–52, use the following information.
The results of a restaurant survey are shown in the circle graph with the measurement of each central angle. Each customer was asked to choose a favorite entrée. If a customer is chosen at random, find the probability of each response. *(Lesson 11-5)*

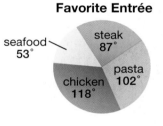

Favorite Entrée

49. steak   **50.** not seafood

51. either pasta or chicken   **52.** neither pasta nor steak

COORDINATE GEOMETRY The coordinates of the vertices of an irregular figure are given. Find the area of each figure. *(Lesson 11-4)*

53. $A(1, 4)$, $B(4, 1)$, $C(1, -2)$, $D(-3, 1)$

54. $F(-2, -4)$, $G(-2, -1)$, $H(1, 1)$, $J(4, 1)$, $K(6, -4)$

55. $L(-2, 2)$, $M(0, 1)$, $N(0, -2)$, $P(-4, -2)$

Find the perimeter and area of each parallelogram. Round to the nearest tenth if necessary. *(Lesson 11-1)*

56.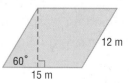

57.

58.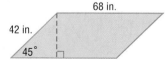

Getting Ready for the Next Lesson

PREREQUISITE SKILL Find the area of each rectangle. Round to the nearest tenth if necessary. *(To review **finding the area of a rectangle**, see pages 732–733.)*

59. 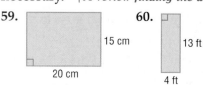   **60.**   **61.**   **62.**

12-2 Nets and Surface Area

What You'll Learn

- Draw two-dimensional models for three-dimensional figures.
- Find surface area.

Vocabulary

- net
- surface area

Why is surface area important to car manufacturers?

Have you wondered why cars have evolved from boxy shapes to sleeker shapes with rounded edges? Car manufacturers use aerodynamics, or the study of wind resistance, and the shapes of surfaces to design cars that are faster and more efficient.

Study Tip

Isometric Dot Paper
Note that right angles of the prism are 60° and 120° angles on isometric dot paper. This is to show perspective.

MODELS FOR THREE-DIMENSIONAL FIGURES You have used isometric dot paper to draw corner views of solids given the orthogonal view. In this lesson, isometric dot paper will be used to draw two-dimensional models of geometric solids.

Example 1 Draw a Solid

Sketch a rectangular prism 2 units high, 5 units long, and 3 units wide using isometric dot paper.

Step 1 Draw the corner of the solid; 2 units down, 5 units to the left, and 3 units to the right.

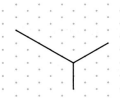

Step 2 Draw a parallelogram for the top of the solid.

Step 3 Draw segments 2 units down from each vertex for the vertical edges.

Step 4 Connect the corresponding vertices. Use dashed lines for the hidden edges. Shade the top of the solid.

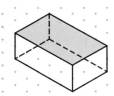

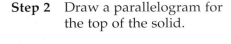

If you cut a cardboard box at the edges and lay it flat, you will have a pattern, or **net**, for the three-dimensional solid. Nets can be made for most solid figures. This net is a pattern for the cube. It can be folded into the shape of the cube without any overlap.

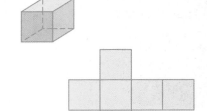

Standardized Test Practice
Ⓐ Ⓑ Ⓒ Ⓓ

Example 2 Nets for a Solid

Multiple-Choice Test Item

Which net could be folded into a pyramid if folds are made only along the dotted lines?

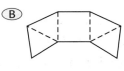
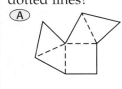
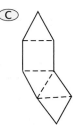
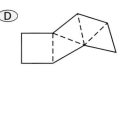

Test-Taking Tip

Nets One figure may have several different nets that represent the shape.

Read the Test Item

You are given four nets, only one of which can be folded into a pyramid.

Solve the Test Item

Each of the answer choices has one square and four triangles. So the square is the base of the pyramid. Each triangle in the sketch represents a face of the pyramid. The faces must meet at a point and cannot overlap. Analyze each answer choice carefully.

 This net has overlapping triangles.

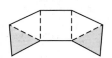 This net also has two triangles that overlap.

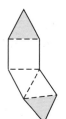 This also has overlapping triangles.

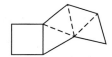 None of the triangles overlap. Each face of the pyramid is represented. This choice is correct.

The answer is D.

SURFACE AREA Nets are very useful in visualizing the polygons that make up the surface of the solid. A net for tetrahedron *QRST* is shown at the right. The **surface area** is the sum of the areas of each face of the solid. Add the areas of △*QRT*, △*QTS*, △*QRS*, and △*RST* to find the surface area of tetrahedron *QRST*.

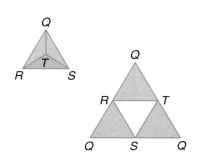

Drawing Nets
It is helpful to let each unit of the dot paper represent one unit of measure. When the numbers are large, let each unit of dot paper represent two units of measure.

Example 3 Nets and Surface Area

a. Draw a net for the right triangular prism shown.

Use the Pythagorean Theorem to find the height of the triangular base.

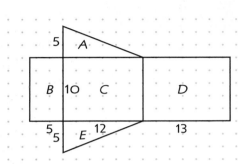

$$13^2 = 12^2 + h^2 \quad \text{Pythagorean Theorem}$$
$$169 = 144 + h^2 \quad \text{Simplify.}$$
$$25 = h^2 \quad \text{Subtract 144 from each side.}$$
$$5 = h \quad \text{Take the square root of each side.}$$

Use rectangular dot paper to draw a net. Let one unit on the dot paper represent 2 centimeters.

b. Use the net to find the surface area of the triangular prism.

To find the surface area of the prism, add the areas of the three rectangles and the two triangles.

Write an equation for the surface area.

$$\text{Surface area} = B + C + D + A + E$$
$$= 10 \cdot 5 + 10 \cdot 12 + 10 \cdot 13 + \frac{5 \cdot 12}{2} + \frac{5 \cdot 12}{2}$$
$$= 50 + 120 + 130 + 30 + 30 \text{ or } 360$$

The surface area of the right triangular prism is 360 square centimeters.

Check for Understanding

Concept Check

1. **OPEN ENDED** Draw a net for a cube different from the one on page 644.

2. **Compare and contrast** isometric dot paper and rectangular dot paper. When is each type of paper useful?

Guided Practice

Sketch each solid using isometric dot paper.

3. rectangular prism 4 units high, 2 units long, and 3 units wide

4. cube 2 units on each edge

For each solid, draw a net and find the surface area.

5.
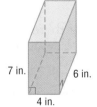
7 in. 6 in. 4 in.

6.
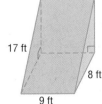
17 ft 8 ft 9 ft

7.
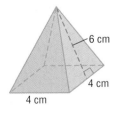
6 cm 4 cm 4 cm

Standardized Test Practice
Ⓐ Ⓑ Ⓒ Ⓓ

8. Which shape *cannot* be folded to make a pyramid?

Ⓐ   Ⓑ   Ⓒ   Ⓓ

Homework Help

For Exercises	See Examples
9–14, 25–27	1
15–24, 35–38	2
28–34	3

Extra Practice
See page 778.

Sketch each solid using isometric dot paper.

9. rectangular prism 3 units high, 4 units long, and 5 units wide

10. cube 5 units on each edge

11. cube 4 units on each edge

12. rectangular prism 6 units high, 6 units long, and 3 units wide

13. triangular prism 4 units high, with bases that are right triangles with legs 5 units and 4 units long

14. triangular prism 2 units high, with bases that are right triangles with legs 3 units and 7 units long

For each solid, draw a net and find the surface area. Round to the nearest tenth if necessary.

15.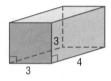

16.

17.

18.

19.

20.

21.

22.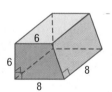

23.

24. **FOOD** In 1999, Marks & Spencer, a British grocery store, created the biggest sandwich ever made. The tuna and cucumber sandwich was in the form of a triangular prism. Suppose each slice of bread was 8 inches thick. Draw a net of the sandwich, and find the surface area in square feet to the nearest tenth.

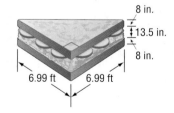

8 in.
13.5 in.
8 in.
6.99 ft   6.99 ft

Online Research **Data Update** Are there records for other types of sandwiches? Visit www.geometryonline.com/data_update to learn more.

Given the net of a solid, use isometric dot paper to draw the solid.

25.

26.

27.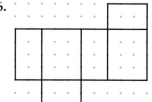

Given each polyhedron, copy its net and label the remaining vertices.

28.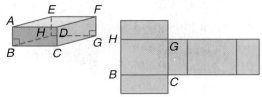

29.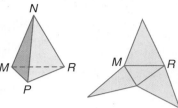

30.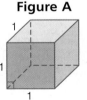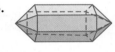

31.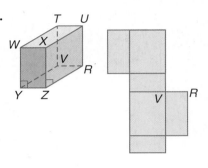

GEOLOGY For Exercises 32–34, use the following information.
Many minerals have a crystalline structure. The forms of three minerals are shown below. Draw a net of each crystal.

32.

tourmaline

33.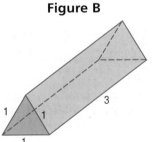

quartz

34.

calcite

VARYING DIMENSIONS For Exercises 35–38, use Figures A, B, and C.

35. Draw a net for each solid and find its surface area.

Figure A   **Figure B**   **Figure C**

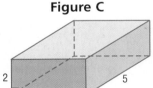

36. Double the dimensions of each figure. Find the surface areas.

37. How does the surface area change when the dimensions are doubled? Explain.

38. Make a conjecture about the surface area of a solid whose dimensions have been tripled. Check your conjecture by finding the surface area.

39. CRITICAL THINKING Many board games use a standard die like the one shown. The sum of the number of dots on each pair of opposite faces is 7. Determine whether the net represents a standard die. Explain.

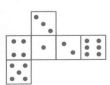

40. **WRITING IN MATH** Answer the question that was posed at the beginning of the lesson.

Why is surface area important to car manufacturers?

Include the following in your answer:
- compare the surface area of a subcompact car and a large truck, and
- explain which two-dimensional models of cars would be helpful to designers.

Standardized Test Practice
Ⓐ Ⓑ Ⓒ Ⓓ

41. Which shape could be folded into a rectangular prism if folds are made only along the dotted lines?

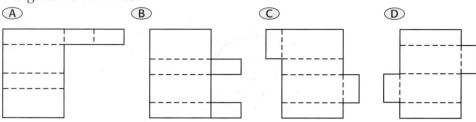

Ⓐ   Ⓑ   Ⓒ   Ⓓ

42. **ALGEBRA** What is the complete factorization of $16a^3 - 54b^3$?
- Ⓐ $(2a - 3b)(4a^2 + 6ab + 9b^2)$
- Ⓑ $2(2a - 3b)(4a^2 + 6ab + 9b^2)$
- Ⓒ $2(2a - 3b)(4a^2 - 6ab + 9b^2)$
- Ⓓ $2(2a + 3b)(4a^2 + 6ab + 9b^2)$

Maintain Your Skills

Mixed Review

Determine the shape resulting from each slice of the triangular prism.
(Lesson 12-1)

43.   **44.**   **45.**

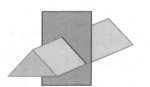

46. **PROBABILITY** A rectangular garden is 100 feet long and 200 feet wide and includes a square flower bed that is 20 feet on each side. Find the probability that a butterfly in the garden is somewhere in the flower bed. *(Lesson 11-5)*

Equilateral hexagon *FGHJKL* is inscribed in ⊙*M*. Find each measure. *(Lesson 10-4)*

47. $m\angle FHJ$

48. $m\widehat{LK}$

49. $m\angle LFG$

Getting Ready for the Next Lesson

PREREQUISITE SKILL **Find the area of each figure.**
*(To review finding **areas of parallelograms, triangles, and trapezoids**, see Lessons 11-1 and 11-2.)*

50.   **51.**   **52.**   **53.**

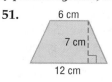 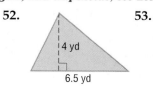 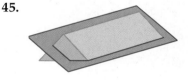 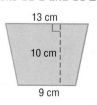

12-3 Surface Areas of Prisms

What You'll Learn

- Find lateral areas of prisms.
- Find surface areas of prisms.

How do brick masons know how many bricks to order for a project?

The owner of a house wants to build a new unattached brick garage. The sides of the garage will be brick. The brick mason has to estimate the number of bricks needed to complete the project.

Vocabulary

- lateral faces
- lateral edges
- right prism
- oblique prism
- lateral area

LATERAL AREAS OF PRISMS

Most buildings are prisms or combinations of prisms. The garage shown above could be separated into a rectangular prism and a rectangular pyramid. Prisms have the following characteristics.

- The bases are congruent faces in parallel planes.

- The rectangular faces that are not bases are called **lateral faces**.

- The lateral faces intersect at the **lateral edges**. Lateral edges are parallel segments.

- A segment perpendicular to the bases, with an endpoint in each plane, is called an *altitude* of the prism. The height of a prism is the length of the altitude.

- A prism with lateral edges that are also altitudes is called a **right prism**. If the lateral edges are not perpendicular to the bases, it is an **oblique prism**.

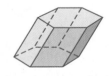

right hexagonal prism

oblique hexagonal prism

Study Tip

Reading Math
From this point in the text, you can assume that solids are right solids. If a solid is oblique, it will be clearly stated.

The **lateral area** L is the sum of the areas of the lateral faces.

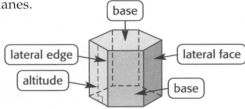

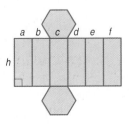

$$L = ah + bh + ch + dh + eh + fh$$
$$= h(a + b + c + d + e + f) \quad \text{Distributive Property}$$
$$= Ph \quad\quad\quad\quad\quad\quad\quad P = a + b + c + d + e + f$$

> ## Key Concept
> **Lateral Area of a Prism**
>
> If a right prism has a lateral area of L square units, a height of h units, and each base has a perimeter of P units, then $L = Ph$.

Example 1   Lateral Area of a Pentagonal Prism

Find the lateral area of the regular pentagonal prism.

The bases are regular pentagons. So the perimeter of one base is 5(14) or 70 centimeters.

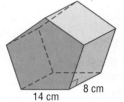

$$L = Ph \qquad \text{Lateral area of a prism}$$
$$= (70)(8) \qquad P = 70,\ h = 8$$
$$= 560 \qquad \text{Multiply.}$$

The lateral area is 560 square centimeters.

SURFACE AREAS OF PRISMS   The surface area of a prism is the lateral area plus the areas of the bases. The bases are congruent, so the areas are equal.

Study Tip

Right Prisms
The bases of a right prism are congruent, but the faces are not always congruent.

> ## Key Concept
> **Surface Area of a Prism**
>
> If the surface area of a right prism is T square units, its height is h units, and each base has an area of B square units and a perimeter of P units, then $T = L + 2B$.

Example 2   Surface Area of a Triangular Prism

Find the surface area of the triangular prism.

First, find the measure of the third side of the triangular base.

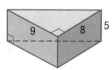

$$c^2 = a^2 + b^2 \qquad \text{Pythagorean Theorem}$$
$$c^2 = 8^2 + 9^2 \qquad \text{Substitution}$$
$$c^2 = 145 \qquad \text{Simplify.}$$
$$c = \sqrt{145} \qquad \text{Take the square root of each side.}$$

$$T = L + 2B \qquad\qquad\qquad \text{Surface area of a prism}$$
$$= Ph + 2B \qquad\qquad\qquad L = Ph$$
$$= \left(8 + 9 + \sqrt{145}\right)5 + 2\left[\tfrac{1}{2}(8 \cdot 9)\right] \quad \text{Substitution}$$
$$\approx 217.2 \qquad\qquad\qquad \text{Use a calculator.}$$

The surface area is approximately 217.2 square units.

Example 3 *Use Surface Area to Solve a Problem*

FURNITURE Rick wants to have an ottoman reupholstered. Find the surface area that will be reupholstered.

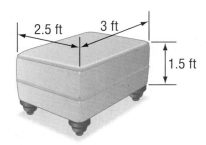

The ottoman is shaped like a rectangular prism. Since the bottom of the ottoman is not covered with fabric, find the lateral area and then add the area of one base. The perimeter of a base is $2(3) + 2(2.5)$ or 11 feet. The area of a base is $3(2.5)$ or 7.5 square feet.

$T = L + B$ Formula for surface area

$\ \ \ = (11)(1.5) + 7.5$ $P = 11$, $h = 1.5$, and $B = 7.5$

$\ \ \ = 24$ Simplify.

The total area that will be reupholstered is 24 square feet.

Check for Understanding

Concept Check   1. **Explain** the difference between a right prism and an oblique prism.

2. **OPEN ENDED**   Draw a prism and label the bases, lateral faces, and lateral edges.

Guided Practice   **Find the lateral area and surface area of each prism.**

3.
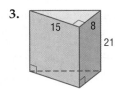

4.
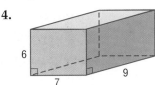

Application   5. **PAINTING**   Eva and Casey are planning to paint the walls and ceiling of their living room. The room is 20 feet long, 15 feet wide, and 12 feet high. Find the surface area to be painted.

Practice and Apply

Homework Help

For Exercises	See Examples
6–11, 14, 15	1
12, 13, 16–21	2
22–36	3

Extra Practice
See page 778.

Find the lateral area of each prism or solid. Round to the nearest tenth if necessary.

6.

7.
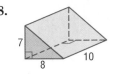

8.

9.

10.
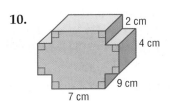

11.
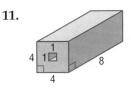

www.geometryonline.com/extra_examples

12. The surface area of a cube is 864 square inches. Find the length of the lateral edge of the cube.

13. The surface area of a triangular prism is 540 square centimeters. The bases are right triangles with legs measuring 12 centimeters and 5 centimeters. Find the height.

14. The lateral area of a rectangular prism is 156 square inches. What are the possible whole-number dimensions of the prism if the height is 13 inches?

15. The lateral area of a rectangular prism is 96 square meters. What are the possible whole-number dimensions of the prism if the height is 4 meters?

Find the surface area of each prism. Round to the nearest tenth if necessary.

16.
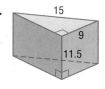

17.

18.
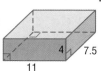

19.

20.

21.
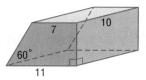

PAINTING For Exercises 22–24, use the following information.
A gallon of paint costs $16 and covers 400 square feet. Two coats of paint are recommended for even coverage. The room to be painted is 10 feet high, 15 feet long, and 15 feet wide. Only $1\frac{1}{2}$ gallons of paint are left to paint the room.

22. Is this enough paint for the walls of the room? Explain.

23. How many gallons of paint are needed to paint the walls?

24. How much would it cost to paint the walls and ceiling?

••• **TOURISM** For Exercises 25–27, use the following information.
The World's Only Corn Palace is located in Mitchell, South Dakota. The sides of the building are covered with huge murals made from corn and other grains.

25. Estimate the area of the Corn Palace to be covered if its base is 310 by 185 feet and it is 45 feet tall, not including the turrets.

26. Suppose a bushel of grain can cover 15 square feet. How many bushels of grain does it take to cover the Corn Palace?

27. Will the actual amount of grain needed be higher or lower than the estimate? Explain.

28. GARDENING This greenhouse is designed for a home gardener. The frame on the back of the greenhouse attaches to one wall of the house. The outside of the greenhouse is covered with tempered safety glass. Find the surface area of the glass covering the greenhouse.

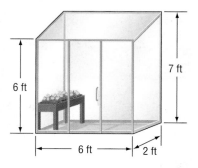

For Exercises 29–33, use prisms A, B, and C.

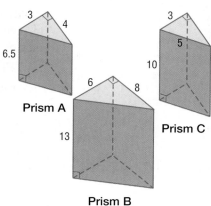

29. Compare the bases of each prism.

30. Write three ratios to compare the perimeters of the bases of the prisms.

31. Write three ratios to compare the areas of the bases of the prisms.

32. Write three ratios to compare the surface areas of the prisms.

33. Which pairs of prisms have the same ratio of base areas as ratio of surface areas? Why do you think this is so?

STATISTICS For Exercises 34–36, use the graphic at the right.
Malik plans to build a three-dimensional model of the data from the graph.

- A rectangular prism will represent each category.

- Each prism will be 30 centimeters wide and 20 centimeters deep.

- The length of the prism for TV will be 84 centimeters.

34. Find the surface area of each prism that Malik builds.

35. Will the surface area of the finished product be the sum of the surface areas of each prism? Explain.

36. Find the total surface area of the finished model.

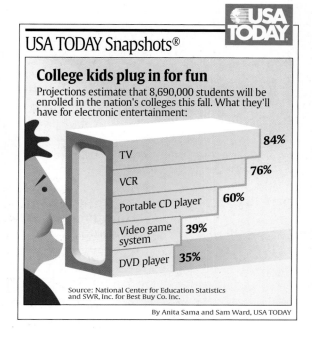

37. **CRITICAL THINKING** Suppose the lateral area of a right rectangular prism is 144 square centimeters. If the length is three times the width and the height is twice the width, find the surface area.

38. WRITING IN MATH Answer the question that was posed at the beginning of the lesson.

How do brick masons know how many bricks to order for a project?

Include the following in your answer:
- how lateral area is used, and
- why overestimation is important in the process.

39. The surface area of a cube is 121.5 square meters. What is the length of each edge?

  (A) 4.05 m   (B) 4.5 m   (C) 4.95 m   (D) 5 m

40. **ALGEBRA** For all $a \neq 4$, $\dfrac{a^2 - 16}{4a - 16} = \underline{\quad ? \quad}$.

  (A) $a + 16$ (B) $a + 1$ (C) $\dfrac{a - 4}{4}$ (D) $\dfrac{a + 4}{4}$

OBLIQUE PRISMS   The altitude of an oblique prism is not the length of a lateral edge. For an oblique rectangular prism, the bases are rectangles, two faces are rectangles and two faces are parallelograms. To find the lateral area and the surface area, apply the definitions of each.

Find the lateral area and surface area of each oblique prism.

41.

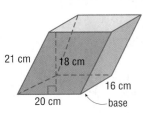

21 cm 18 cm
20 cm 16 cm
base

42.

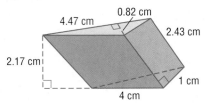

4.47 cm 0.82 cm 2.43 cm
2.17 cm 1 cm
4 cm

43. RESEARCH   Use a dictionary to find the meaning of the term *oblique*. How is the everyday meaning related to the mathematical meaning?

Maintain Your Skills

For each solid, draw a net and find the surface area.   *(Lesson 12-2)*

44.

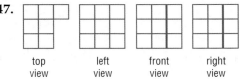

8
12 6

45.

6
3 4

46.

3
4
5

Draw the back view and corner view of the figure given the orthogonal drawing. *(Lesson 12-1)*

47.

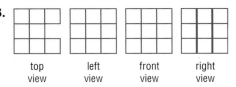

top view   left view   front view   right view

48.

top view   left view   front view   right view

Circle Q has a radius of 24 units, ⊙R has a radius of 16 units, and BC = 5. Find each measure.   *(Lesson 10-1)*

49. *AB*   **50.** *AD*   **51.** *QR*

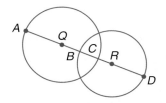

52. NAVIGATION   An airplane is three miles above sea level when it begins to climb at a 3.5° angle. If this angle is constant, how far above sea level is the airplane after flying 50 miles?   *(Lesson 7-4)*

53. ART   Kiernan drew a sketch of a house. If the height of the house in her drawing was 5.5 inches and the actual height of the house was 33 feet, find the scale factor of the drawing.   *(Lesson 6-1)*

PREREQUISITE SKILL   Find the area of each circle. Round to the nearest hundredth.   *(To review **finding the area of a circle**, see Lesson 11-3.)*

54.

40 cm

55.

50 in.

56.

3.5 ft

57.

82 mm

12-4 Surface Areas of Cylinders

What You'll Learn

- Find lateral areas of cylinders.
- Find surface areas of cylinders.

Vocabulary

- axis
- right cylinder
- oblique cylinder

How are cylinders used in extreme sports?

Extreme sports, such as in-line skating, biking, skateboarding, and snowboarding use a cylindrical-shaped ramp called a half-pipe. The half-pipe looks like half of a cylinder. Usually there is a flat section in the middle with sides almost 8 feet high. Near the top, the sides are almost vertical.

LATERAL AREAS OF CYLINDERS   The **axis** of the cylinder is the segment with endpoints that are centers of the circular bases. If the axis is also the altitude, then the cylinder is called a **right cylinder**. Otherwise, the cylinder is an **oblique cylinder**.

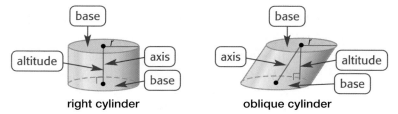

right cylinder   oblique cylinder

The net of a cylinder is composed of two congruent circles and a rectangle. The area of this rectangle is the lateral area. The length of the rectangle is the same as the circumference of the base, $2\pi r$. So, the lateral area of a right cylinder is $2\pi rh$.

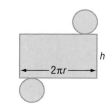

Study Tip

Formulas
An alternate formula for the lateral area of a cylinder is $L = \pi dh$, with πd as the circumference of a circle.

Key Concept — Lateral Area of a Cylinder

If a right cylinder has a lateral area of L square units, a height of h units, and the bases have radii of r units, then $L = 2\pi rh$.

Example 1 Lateral Area of a Cylinder

MANUFACTURING   An office has recycling barrels for cans and paper. The barrels are cylindrical with cardboard sides and plastic lids and bases. Each barrel is 3 feet tall, and the diameter is 30 inches. How many square feet of cardboard are used to make each barrel?

The cardboard section of the barrel represents the lateral area of the cylinder. If the diameter of the lid is 30 inches, then the radius is 15 inches. The height is 3 feet or 36 inches. Use the formula to find the lateral area.   *(continued on the next page)*

 www.geometryonline.com/extra_examples

$$L = 2\pi rh \qquad \text{Lateral area of a cylinder}$$

$$= 2\pi(15)(36) \quad r = 15, h = 36$$

$$\approx 3392.9 \qquad \text{Use a calculator.}$$

Each barrel uses approximately 3393 square inches of cardboard. Because 144 square inches equal one square foot, there are 3393 ÷ 144 or about 23.6 square feet of cardboard per barrel.

SURFACE AREAS OF CYLINDERS To find the surface area of a cylinder, first find the lateral area and then add the areas of the bases. This leads to the formula for the surface area of a right cylinder.

Key Concept Surface Area of a Cylinder

If a right cylinder has a surface area of T square units, a height of h units, and the bases have radii of r units, then $T = 2\pi rh + 2\pi r^2$.

Example 2 Surface Area of a Cylinder

Find the surface area of the cylinder.

The radius of the base and the height of the cylinder are given. Substitute these values in the formula to find the surface area.

$$T = 2\pi rh + 2\pi r^2 \qquad \text{Surface area of a cylinder}$$

$$= 2\pi(8.3)(6.6) + 2\pi(8.3)^2 \quad r = 8.3, h = 6.6$$

$$\approx 777.0 \qquad \text{Use a calculator.}$$

The surface area is approximately 777.0 square feet.

Example 3 Find Missing Dimensions

Find the radius of the base of a right cylinder if the surface area is 128π square centimeters and the height is 12 centimeters.

Use the formula for surface area to write and solve an equation for the radius.

$$T = 2\pi rh + 2\pi r^2 \qquad \text{Surface area of a cylinder}$$

$$128\pi = 2\pi(12)r + 2\pi r^2 \quad \text{Substitution}$$

$$128\pi = 24\pi r + 2\pi r^2 \qquad \text{Simplify.}$$

$$64 = 12r + r^2 \qquad \text{Divide each side by } 2\pi.$$

$$0 = r^2 + 12r - 64 \qquad \text{Subtract 64 from each side.}$$

$$0 = (r - 4)(r + 16) \qquad \text{Factor.}$$

$$r = 4 \text{ or } -16$$

Since the radius of a circle cannot have a negative value, -16 is eliminated. So, the radius of the base is 4 centimeters.

Check for Understanding

Concept Check

1. **Explain** how to find the surface area of a cylinder.

2. **OPEN ENDED** Draw a net for a cylinder.

3. **FIND THE ERROR** Jamie and Dwayne are finding the surface area of a cylinder with one base.

9 in.

4 in.

Jamie

$T = 2\pi(4)(9) + \pi(4^2)$

$T = 72\pi + 16\pi$

$T = 88\pi$ in^2

Dwayne

$T = 2\pi(4)(9) + 2\pi(4^2)$

$T = 72\pi + 32\pi$

$T = 104\pi$ in^2

Who is correct? Explain.

Guided Practice

4. Find the surface area of a cylinder with a radius of 4 feet and height of 6 feet. Round to the nearest tenth.

5. Find the surface area of the cylinder. Round to the nearest tenth.

11 m

22 m

Find the radius of the base of each cylinder.

6. The surface area is 96π square centimeters, and the height is 8 centimeters.

7. The surface area is 140π square feet, and the height is 9 feet.

Application

8. **CONTESTS** Mrs. Fairway's class is collecting labels from soup cans to raise money for the school. The students collected labels from 3258 cans. If the cans are 4 inches high with a diameter of 2.5 inches, find the area of the labels that were collected.

Practice and Apply

Homework Help	
For Exercises	**See Examples**
9–16	2
17–20	3
21, 22, 14–25	1

Extra Practice
See page 779.

Find the surface area of a cylinder with the given dimensions. Round to the nearest tenth.

9. $r = 13$ m, $h = 15.8$ m

10. $d = 13.6$ ft, $h = 1.9$ ft

11. $d = 14.2$ in., $h = 4.5$ in.

12. $r = 14$ mm, $h = 14$ mm

Find the surface area of each cylinder. Round to the nearest tenth.

13.
4 ft

6 ft

14.
8.2 yd

7.2 yd

15.
4.4 cm

0.9 cm

16.
9.6 m

3.4 m

Find the radius of the base of each cylinder.

17. The surface area is 48π square centimeters, and the height is 5 centimeters.

18. The surface area is 340π square inches, and the height is 7 inches.

19. The surface area is 320π square meters, and the height is 12 meters.

20. The surface area is 425.1 square feet, and the height is 6.8 feet.

21. **KITCHENS** Raul purchased a set of canisters with diameters of 5 inches and heights of 9 inches, 6 inches, and 3 inches. Make a conjecture about the relationship between the heights of the canisters and their lateral areas. Check your conjecture.

22. **CAMPING** Campers can use a solar cooker to cook food. You can make a solar cooker from supplies you have on hand. The reflector in the cooker shown at the right is half of a cardboard cylinder covered with aluminum foil. The reflector is 18 inches long and has a diameter of $5\frac{1}{2}$ inches. How much aluminum foil was needed to cover the inside of the reflector?

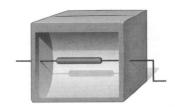

23. Suppose the height of a right cylinder is tripled. Is the surface area or lateral area tripled? Explain.

AGRICULTURE **For Exercises 24 and 25, use the following information.**
The acid from the contents of a silo can weaken its concrete walls and seriously damage the silo's structure. So the inside of the silo must occasionally be resurfaced. The cost of the resurfacing is a function of the lateral area of the inside of the silo.

24. Find the lateral area of a silo 13 meters tall with an interior diameter of 5 meters.

25. A second grain silo is 26 meters tall. If both silos have the same lateral area, find the radius of the second silo.

26. **CRITICAL THINKING** Some pencils are cylindrical, and others are hexagonal prisms. If the diameter of the cylinder is the same length as the longest diagonal of the hexagon, which has the greater surface area? Explain. Assume that each pencil is 11 inches long and unsharpened.

27. **WRITING IN MATH** Answer the question that was posed at the beginning of the lesson.

How are cylinders used in extreme sports?

Include the following in your answer:
- how to find the lateral area of a semicylinder, and
- how to determine if the half-pipe ramp is a semicylinder.

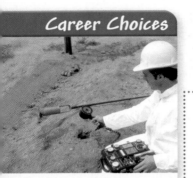

28. A cylinder has a height of 13.4 centimeters and a diameter of 8.2 centimeters. To the nearest tenth, what is the surface area of the cylinder?
Ⓐ 51.5 cm²   Ⓑ 450.8 cm²   Ⓒ 741.9 cm²   Ⓓ 1112.9 cm²

29. **ALGEBRA** For the band concert, student tickets cost $2 and adult tickets cost $5. A total of 200 tickets were sold. If the total sales were more than $500, what was the minimum number of adult tickets sold?
Ⓐ 30   Ⓑ 33   Ⓒ 34   Ⓓ 40

Standardized Test Practice
Ⓐ Ⓑ Ⓒ Ⓓ

Extending the Lesson

LOCUS A cylinder can be defined in terms of locus. The locus of points in space a given distance from a line is the lateral surface of a cylinder.

Draw a figure and describe the locus of all points in space that satisfy each set of conditions.

30. 5 units from a given line

31. equidistant from two opposite vertices of a face of a cube

Mixed Review   Find the lateral area of each prism. *(Lesson 12-3)*

32.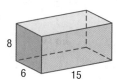

33.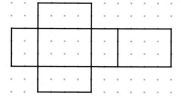

34.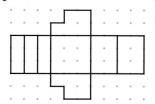

Given the net of a solid, use isometric dot paper to draw the solid. *(Lesson 12-2)*

35.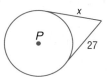

36.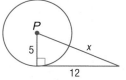

Find *x*. Assume that segments that appear to be tangent are tangent. *(Lesson 10-5)*

37.

38.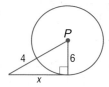

39.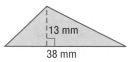

Solve each △*ABC* described below. Round to the nearest tenth if necessary. *(Lesson 7-7)*

40. $m\angle A = 54$, $b = 6.3$, $c = 7.1$ **41.** $m\angle B = 47$, $m\angle C = 69$, $a = 15$

Getting Ready for the Next Lesson   **PREREQUISITE SKILL**   Find the area of each figure.
(To review finding areas of triangles and trapezoids, see Lesson 11-2.)

42.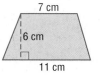

17 in.

20 in.

43.

7 cm

6 cm

11 cm

44.

13 mm

38 mm

Practice Quiz 1   *Lessons 12-1 through 12-4*

1. Draw a corner view of the figure given the orthogonal drawing. *(Lesson 12-1)*

2. Sketch a rectangular prism 2 units wide, 3 units long, and 2 units high using isometric dot paper. *(Lesson 12-2)*

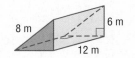

top view   left view   front view   right view

3. Find the lateral area of the prism. Round to the nearest tenth. *(Lesson 12-3)*

4. Find the surface area of the prism. Round to the nearest tenth. *(Lesson 12-4)*

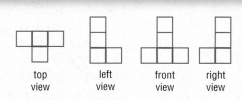

8 m   12 m   6 m

5. Find the radius of the base of a right cylinder if the surface area is 560 square feet and the height is 11 feet. Round to the nearest tenth. *(Lesson 12-4)*

Surface Areas of Pyramids

What You'll Learn

- Find lateral areas of regular pyramids.
- Find surface areas of regular pyramids.

Vocabulary

- regular pyramid
- slant height

How are pyramids used in architecture?

In 1989, a new entrance was completed in the courtyard of the Louvre museum in Paris, France. Visitors can enter the museum through a glass pyramid that stands 71 feet tall. The pyramid is glass with a structural system of steel rods and cables.

LATERAL AREAS OF REGULAR PYRAMIDS

Pyramids have the following characteristics.

- All of the faces, except the base, intersect at one point called the *vertex*.
- The base is always a polygon.
- The faces that intersect at the vertex are called *lateral faces* and form triangles. The edges of the lateral faces that have the vertex as an endpoint are called *lateral edges*.
- The *altitude* is the segment from the vertex perpendicular to the base.

Study Tip

Right Pyramid
In a *right pyramid*, the altitude is the segment with endpoints that are the center of the base and the vertex. But the base is not always a regular polygon.

If the base of a pyramid is a regular polygon and the segment with endpoints that are the center of the base and the vertex is perpendicular to the base, then the pyramid is called a **regular pyramid**. They have specific characteristics. The altitude is the segment with endpoints that are the center of the base and the vertex. All of the lateral faces are congruent isosceles triangles. The height of each lateral face is called the **slant height** ℓ of the pyramid.

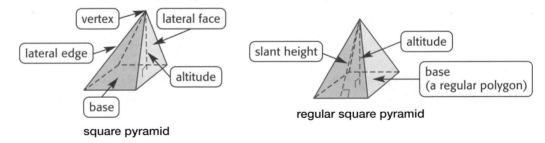

square pyramid

regular square pyramid

The figure below is a regular hexagonal pyramid. Its lateral area L can be found by adding the areas of all its congruent triangular faces as shown in its net.

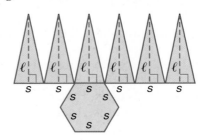

Area of the net

$$L = \frac{1}{2}s\ell + \frac{1}{2}s\ell + \frac{1}{2}s\ell + \frac{1}{2}s\ell + \frac{1}{2}s\ell + \frac{1}{2}s\ell \quad \text{Sum of the areas of the lateral faces}$$

$$= \frac{1}{2}\ell(s + s + s + s + s + s) \qquad \text{Distributive Property}$$

$$= \frac{1}{2}P\ell \qquad\qquad\qquad P = s + s + s + s + s + s$$

Key Concept — Lateral Area of a Regular Pyramid

If a regular pyramid has a lateral area of L square units, a slant height of ℓ units, and its base has a perimeter of P units, then $L = \frac{1}{2}P\ell$.

Example 1 Use Lateral Area to Solve a Problem

BIRDHOUSES   The roof of a birdhouse is a regular hexagonal pyramid. The base of the pyramid has sides of 4 inches, and the slant height of the roof is 12 inches. If the roof is made of copper, find the amount of copper used for the roof.

We need to find the lateral area of the hexagonal pyramid. The sides of the base measure 4 inches, so the perimeter is 6(4) or 24 inches.

$$L = \frac{1}{2}P\ell \qquad \text{Lateral area of a regular pyramid}$$

$$= \frac{1}{2}(24)(12) \quad P = 24, \ell = 12$$

$$= 144 \qquad \text{Multiply.}$$

So, 144 square inches of copper are used to cover the roof of the birdhouse.

Study Tip

Making Connections
The total surface area for a pyramid is $L + B$, because there is only one base to consider.

SURFACE AREAS OF REGULAR PYRAMIDS   The surface area of a regular pyramid is the sum of the lateral area and the area of the base.

Key Concept — Surface Area of a Regular Pyramid

If a regular pyramid has a surface area of T square units, a slant height of ℓ units, and its base has a perimeter of P units and an area of B square units, then $T = \frac{1}{2}P\ell + B$.

Example 2 Surface Area of a Square Pyramid

Find the surface area of the square pyramid.

To find the surface area, first find the slant height of the pyramid. The slant height is the hypotenuse of a right triangle with legs that are the altitude and a segment with a length that is one-half the side measure of the base.

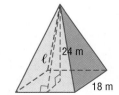

$$c^2 = a^2 + b^2 \qquad \text{Pythagorean Theorem}$$

$$\ell^2 = 9^2 + 24^2 \qquad a = 9, b = 24, c = \ell$$

$$\ell = \sqrt{657} \qquad \text{Simplify.}$$

(continued on the next page)

Now find the surface area of a regular pyramid. The perimeter of the base is 4(18) or 72 meters, and the area of the base is 18^2 or 324 square meters.

$$T = \frac{1}{2}P\ell + B \qquad \text{Surface area of a regular pyramid}$$

$$T = \frac{1}{2}(72)\sqrt{657} + 324 \quad P = 72, \ell = \sqrt{657}, B = 324$$

$$T \approx 1246.8 \qquad \text{Use a calculator.}$$

The surface area is 1246.8 square meters to the nearest tenth.

Study Tip

Look Back
To review **finding the areas of regular polygons,** see Lesson 11-3. To review **trigonometric ratios,** see Lesson 7-4.

Example 3 *Surface Area of Pentagonal Pyramid*

Find the surface area of the regular pyramid.

The altitude, slant height, and apothem form a right triangle. Use the Pythagorean Theorem to find the apothem. Let a represent the length of the apothem.

$$c^2 = a^2 + b^2 \qquad \text{Pythagorean Theorem}$$

$$(17)^2 = a^2 + 15^2 \quad b = 15, c = 17$$

$$8 = a \qquad \text{Simplify.}$$

Now find the length of the sides of the base. The central angle of the pentagon measures $\frac{360°}{5}$ or 72°. Let x represent the measure of the angle formed by a radius and the apothem. Then, $x = \frac{72}{2}$ or 36.

Use trigonometry to find the length of the sides.

$$\tan 36° = \frac{\frac{1}{2}s}{8} \quad \tan x° = \frac{\text{opposite}}{\text{adjacent}}$$

$$8(\tan 36°) = \frac{1}{2}s \quad \text{Multiply each side by 8.}$$

$$16(\tan 36°) = s \quad \text{Multiply each side by 2.}$$

$$11.6 \approx s \quad \text{Use a calculator.}$$

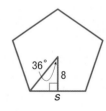

Next, find the perimeter and area of the base.

$$P = 5s$$

$$\approx 5(11.6) \text{ or } 58$$

$$B = \frac{1}{2}Pa$$

$$\approx \frac{1}{2}(58)(8) \text{ or } 232$$

Finally, find the surface area.

$$T = \frac{1}{2}P\ell + B \qquad \text{Surface area of a regular pyramid}$$

$$\approx \frac{1}{2}(58)(17) + 232 \quad P \approx 58, \ell = 17, B \approx 232$$

$$\approx 726.5 \qquad \text{Simplify.}$$

The surface area is approximately 726.5 square inches.

Web Quest

Making a sketch of a pyramid can help you find its slant height, lateral area, and base area. Visit www.geometryonline. com/webquest to continue work on your WebQuest project.

Concept Check

1. **OPEN ENDED** Draw a regular pyramid and a pyramid that is not regular.

2. **Explain** whether a regular pyramid can also be a regular polyhedron.

Guided Practice

Find the surface area of each regular pyramid. Round to the nearest tenth if necessary.

3.
7 ft
4 ft

4.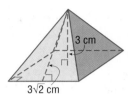
3 cm
3√2 cm

5.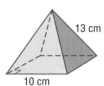
13 cm
10 cm

Application

6. **DECORATIONS** Minowa purchased 3 decorative three-dimensional stars. Each star is composed of 6 congruent square pyramids with faces of paper and a base of cardboard. If the base is 2 inches on each side and the slant height is 4 inches, find the amount of paper used for one star.

Practice and Apply

Homework Help

For Exercises	See Examples
7, 9, 10, 14, 15	2
8, 11–13	3
16, 18–24	1

Extra Practice
See page 779.

Find the surface area of each regular pyramid. Round to the nearest tenth if necessary.

7.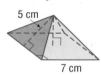
5 cm
7 cm

8.
6 in.
4.5 in.

9.
10 ft
8 ft

10.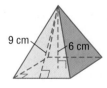
9 cm
6 cm

11.
8 yd
6 yd

12.
6.4 m
3.2 m

13.
13 in.
12 in.

14.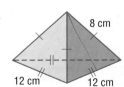
8 cm
12 cm   12 cm

15.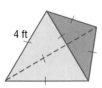
4 ft

16. **CONSTRUCTION** The roof on a building is a square pyramid with no base. If the altitude of the pyramid measures 5 feet and the slant height measures 20 feet, find the area of the roof.

17. **PERFUME BOTTLES** Some perfumes are packaged in square pyramidal containers. The base of one bottle is 3 inches square, and the slant height is 4 inches. A second bottle has the same surface area, but the slant height is 6 inches long. Find the dimensions of the base of the second bottle.

18. STADIUMS The Pyramid Arena in Memphis, Tennessee, is the third largest pyramid in the world. The base is 360,000 square feet, and the pyramid is 321 feet tall. Find the lateral area of the pyramid. (Assume that the base is a square).

19. HOTELS The Luxor Hotel in Las Vegas is a black glass pyramid. The base is a square with edges 646 feet long. The hotel is 350 feet tall. Find the area of the glass.

20. HISTORY Each side of the square base of Khafre's Pyramid is 214.5 meters. The sides rise at an angle of about 53°. Find the lateral area of the pyramid.

For Exercises 21–23, use the following information.
This solid is a composite of a cube and square pyramid. The base of the solid is the base of the cube. Find the indicated measurements for the solid.

21. Find the height.

22. Find the lateral area.

23. Find the surface area.

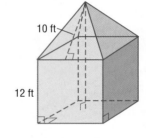

24. A *frustum* is the part of a solid that remains after the top portion has been cut by a plane parallel to the base. Find the lateral area of the frustum of a regular pyramid.

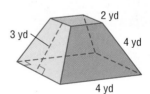

25. CRITICAL THINKING This square prism measures 1 inch on each side. The corner of the cube is cut off, or *truncated*. Does this change the surface area of the cube? Include the surface area of the original cube and that of the truncated cube in your answer.

26. WRITING IN MATH   Answer the question that was posed at the beginning of the lesson.

How are pyramids used in architecture?

Include the following in your answer:

• information needed to find the lateral area and surface area, and

• other examples of pyramidal shapes used in architecture.

27. The base of a square pyramid has a perimeter of 20 centimeters, and the slant height is 10 centimeters. What is the surface area of the pyramid?

   (A) 96.8 cm²   (B) 116 cm²   (C) 121.8 cm²   (D) 125 cm²

28. ALGEBRA If $x \otimes y = \dfrac{1}{x - y}$, what is the value of $\dfrac{1}{2} \otimes \dfrac{3}{4}$?

Ⓐ -4　　　Ⓑ $-\dfrac{1}{4}$　　　Ⓒ $\dfrac{4}{5}$　　　Ⓓ $\dfrac{5}{4}$

Maintain Your Skills

Mixed Review　**Find the surface area of each cylinder. Round to the nearest tenth.** *(Lesson 12-4)*

29.
15 m　7 m

30.
14 cm　22 cm

31.
9 yd　23 yd

32. FOOD Most cereals are packaged in cardboard boxes. If a box of cereal is 14 inches high, 6 inches wide, and 2.5 inches deep, find the surface area of the box. *(Lesson 12-3)*

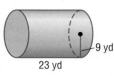
2.5 in.　14 in.　6 in.　Cereal

Find the perimeter and area of each figure. Round to the nearest tenth if necessary. *(Lesson 11-1)*

33.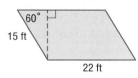
60°　15 ft　22 ft

34.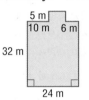
5 m　10 m　6 m　32 m　24 m

35.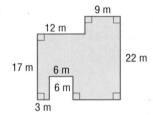
9 m　12 m　17 m　6 m　6 m　22 m　3 m

For Exercises 36–39, refer to the figure at the right. Name the reflected image of each figure. *(Lesson 9-1)*

36. $\overline{FM}$ in line b

37. $\overline{JK}$ in line a

38. L in point M

39. $\overline{GM}$ in line a

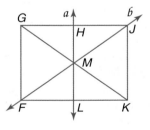
G　a　b　H　J　M　F　L　K

Determine whether each statement is *true* or *false*. Explain. *(Lesson 8-3)*

40. If two pairs of consecutive sides of a quadrilateral are congruent, then the quadrilateral must be a parallelogram.

41. If all four sides of a quadrilateral are congruent, then the quadrilateral is a parallelogram.

Getting Ready for the Next Lesson　**PREREQUISITE SKILL** Use the Pythagorean Theorem to solve for the missing length in each triangle. Round to the nearest tenth.
*(To review the **Pythagorean Theorem**, see Lesson 7-2.)*

42.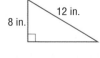
12 in.　8 in.

43.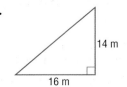
14 m　16 m

44.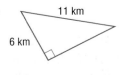
11 km　6 km

Surface Areas of Cones

What You'll Learn

- Find lateral areas of cones.
- Find surface areas of cones.

Vocabulary
- circular cone
- right cone
- oblique cone

How is the lateral area of a cone used to cover tepees?

Native American tribes on the Great Plains typically lived in tepees, or tipis. Tent poles were arranged in a conical shape, and animal skins were stretched over the frame for shelter. The top of a tepee was left open for smoke to escape.

LATERAL AREAS OF CONES   The shape of a tepee suggests a **circular cone**. Cones have the following characteristics.

- The base is a circle and the vertex is the point V.
- The *axis* is the segment with endpoints that are the vertex and the center of the base.
- The segment that has the vertex as one endpoint and is perpendicular to the base is called the *altitude* of the cone.

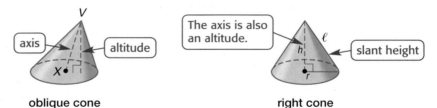

oblique cone   right cone

Study Tip

Reading Math
From this point in the text, you can assume that cones are right circular cones. If the cone is oblique, it will be clearly stated.

A cone with an axis that is also an altitude is called a **right cone**. Otherwise, it is called an **oblique cone**. The measure of any segment joining the vertex of a right cone to the edge of the circular base is called the *slant height*, ℓ. The measure of the altitude is the height h of the cone.

We can use the net for a cone to derive the formula for the lateral area of a cone. The lateral region of the cone is a sector of a circle with radius ℓ, the slant height of the cone. The arc length of the sector is the same as the circumference of the base, or $2\pi r$. The circumference of the circle containing the sector is $2\pi\ell$. The area of the sector is proportional to the area of the circle.

$$\frac{\text{area of sector}}{\text{area of circle}} = \frac{\text{measure of arc}}{\text{circumference of circle}}$$ Write a proportion.

$$\frac{\text{area of sector}}{\pi\ell^2} = \frac{2\pi r}{2\pi\ell}$$ Substitution

$$\text{area of sector} = \frac{(\pi\ell^2)(2\pi r)}{2\pi\ell}$$ Multiply each side by $\pi\ell^2$.

$$\text{area of sector} = \pi r\ell$$ Simplify.

This derivation leads to the formula for the lateral area of a right circular cone.

Key Concept — Lateral Area of a Cone

If a right circular cone has a lateral area of L square units, a slant height of ℓ units, and the radius of the base is r units, then $L = \pi r \ell$.

Example 1 Lateral Area of a Cone

LAMPS Diego has a conical lamp shade with an altitude of 6 inches and a diameter of 12 inches. Find the lateral area of the lampshade.

Explore We are given the altitude and the diameter of the base. We need to find the slant height of the cone.

Plan The radius of the base, height, and slant height form a right triangle. Use the Pythagorean Theorem to solve for the slant height. Then use the formula for the lateral area of a right circular cone.

Solve Write an equation and solve for ℓ.

$\ell^2 = 6^2 + 6^2$ Pythagorean Theorem

$\ell^2 = 72$ Simplify.

$\ell = \sqrt{72}$ or $6\sqrt{2}$ Take the square root of each side.

Next, use the formula for the lateral area of a right circular cone.

$L = \pi r \ell$ Lateral area of a cone

$\approx \pi(6)(6\sqrt{2})$ $r = 6$, $\ell = 6\sqrt{2}$

≈ 159.9 Use a calculator.

The lateral area is approximately 159.9 square inches.

Examine Use estimation to check the reasonableness of this result. The lateral area is approximately $3 \cdot 6 \cdot 9$ or 162 square inches. Compared to the estimate, the answer is reasonable.

SURFACE AREAS OF CONES
To find the surface area of a cone, add the area of the base to the lateral area.

Key Concept — Surface Area of a Cone

If a right circular cone has a surface area of T square units, a slant height of ℓ units, and the radius of the base is r units, then $T = \pi r \ell + \pi r^2$.

Example 2 Surface Area of a Cone

Find the surface area of the cone.

$T = \pi r \ell + \pi r^2$ Surface area of a cone

$= \pi(4.7)(13.6) + \pi(4.7)^2$ $r = 4.7$, $\ell = 13.6$

≈ 270.2 Use a calculator.

The surface area is approximately 270.2 square centimeters.

Study Tip

Storing Values in Calculator Memory

You can store the calculated value of ℓ by $\sqrt{}$ 72 STO▸ ALPHA [L]. To find the lateral area, use 2nd [π] × 6 × ALPHA [L] ENTER.

Study Tip

Making Connections

The surface area of a cone is like the surface area of a pyramid, $T = L + B$.

Check for Understanding

Concept Check

1. **OPEN ENDED** Draw an oblique cone. Mark the vertex and the center of the base.

2. **Explain** why the formula for the lateral area of a right circular cone does not apply to oblique cones.

Guided Practice **Find the surface area of each cone. Round to the nearest tenth.**

3.
17 cm
10 cm

4.
12 ft
10 ft

5.
8 in.
8 in.

Application

6. **TOWERS** In 1921, Italian immigrant Simon Rodia bought a home in Los Angeles, California, and began building conical towers in his backyard. The structures are made of steel mesh and cement mortar. Suppose the height of one tower is 55 feet and the diameter of the base is 8.5 feet, find the lateral area of the tower.

Practice and Apply

Homework Help

For Exercises	See Examples
7–21, 31	2
22–30	1

Extra Practice
See page 779.

Find the surface area of each cone. Round to the nearest tenth.

7.
12 cm
5 cm

8.
8 ft
10 ft

9.
9 in.
9 in.

10.
15 ft   17 ft

11.
7.5 m
12 m

12.
6.4 yd
2.6 yd

For Exercises 13–16, round to the nearest tenth.

13. Find the surface area of the cone if the height is 16 inches and the slant height is 18 inches.

14. Find the surface area of the cone if the height is 8.7 meters and the slant height is 19.1 meters.

15. The surface area of a cone is 1020 square meters and the radius is 14.5 meters. Find the slant height.

16. The surface area of a cone is 293.2 square feet and the radius is 6.1 feet. Find the slant height.

Find the radius of a cone given the surface area and slant height. Round to the nearest tenth.

17. $T = 359$ ft^2, $\ell = 15$ ft

18. $T = 523$ m^2, $\ell = 12.1$ m

Find the surface area of each solid. Round to the nearest tenth.

19.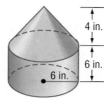
4 in.

6 in.

6 in.

20. 5 ft

3 ft

5 ft

21.
28 m

6.2 m

14 m

More About. . .

Tepees

The Saamis Tepee in Medicine Hat, Alberta, Canada, is the world's largest tepee. The frame is made of steel, instead of wood. It weighs 200 tons.

Source:
ww.medicinehatchamber.com

22. TEPEES Find the area of canvas used to cover a tepee if the diameter of the base is 42 feet and the slant height is 47.9 feet.

23. PARTY HATS Shelley plans to make eight conical party hats for her niece's birthday party. She wants each hat to be 18 inches tall and the bases of each to be 22 inches in circumference. How much material will she use to make the hats?

24. WINTER STORMS Many states use a cone structure to store salt used to melt snow on highways and roads. Find the lateral area of one of these cone structures if the building measures 24 feet tall and the diameter is 45 feet.

25. SPOTLIGHTS A yellow-pink spotlight was positioned directly above a performer. If the surface area of the cone of light was approximately 500 square feet and the slant height was 20 feet, find the diameter of light on stage.

The height of a cone is 7 inches, and the radius is 4 inches. Round final answers to the nearest ten-thousandth.

26. Find the lateral area of the cone using the store feature of a calculator.

27. Round the slant height to the nearest tenth and then calculate the lateral area of the cone.

28. Round the slant height to the nearest hundredth and then calculate the lateral area of the cone.

29. Compare the lateral areas for Exercises 26–28. Which is most accurate? Explain.

Determine whether each statement is *sometimes*, *always*, or *never* true. Explain.

30. If the diagonal of the base of a square pyramid is equal to the diameter of the base of a cone and the heights of both solids are equal, then the pyramid and cone have equal lateral areas.

31. The ratio of the radii of the bases of two cones is equal to the ratio of the surface areas of the cones.

32. CRITICAL THINKING If you were to move the vertex of a right cone down the axis toward the center of the base, explain what would happen to the lateral area and surface area of the cone.

33. WRITING IN MATH   Answer the question that was posed at the beginning of the lesson.

How is the lateral area of a cone used to cover tepees?

Include the following in your answer:

• information needed to find the lateral area of the canvas covering, and

• how the open top of a tepee affects the lateral area of the canvas covering it.

Lesson 12-6 Surface Areas of Cones 669

34. The lateral area of the cone is 91.5π square feet. What is the radius of the base?

Ⓐ 5.9 ft   Ⓑ 6.1 ft   Ⓒ 7.5 ft   Ⓓ 10 ft

15 ft

35. ALGEBRA Three times the first of three consecutive odd integers is 3 more than twice the third. Find the third integer.

Ⓐ 9   Ⓑ 11   Ⓒ 13   Ⓓ 15

Maintain Your Skills

Mixed Review

36. ARCHITECTURE The Transamerica Tower in San Francisco is a regular pyramid with a square base that is 149 feet on each side and a height of 853 feet. Find its lateral area. *(Lesson 12-5)*

Find the radius of the base of the right cylinder. Round to the nearest tenth. *(Lesson 12-4)*

37. The surface area is 563 square feet, and the height is 9.5 feet.

38. The surface area is 185 square meters, and the height is 11 meters.

39. The surface area is 470 square yards, and the height is 6.5 yards.

40. The surface area is 951 square centimeters, and the height is 14 centimeters.

In $\odot M$, $FL = 24$, $HJ = 48$, and $m\widehat{HP} = 45$.
Find each measure. *(Lesson 10-3)*

41. FG

42. NJ

43. HN

44. LG

45. $m\widehat{PJ}$

46. $m\widehat{HJ}$

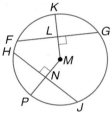

Find the geometric mean between each pair of numbers. *(Lesson 7-1)*

47. 7 and 63   **48.** 8 and 18   **49.** 16 and 44

Getting Ready for the Next Lesson

PREREQUISITE SKILL Find the circumference of each circle given the radius or the diameter. Round to the nearest tenth.
(To review finding the circumference of a circle, see Lesson 10-1.)

50. $r = 6$ **51.** $d = 8$ **52.** $d = 18$

53. $r = 8.2$ **54.** $d = 19.8$ **55.** $r = 4.1$

Practice Quiz 2   Lessons 12-5 and 12-6

Find the surface area of each solid. Round to the nearest tenth. *(Lessons 12-5 and 12-6)*

1.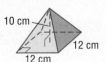
10 cm
12 cm
12 cm

2.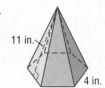
11 in.
4 in.

3.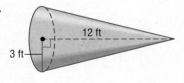
12 ft
3 ft

4. Find the surface area of a cone if the radius is 6 meters and the height is 2 meters. Round to the nearest tenth. *(Lesson 12-6)*

5. Find the slant height of a cone if the lateral area is 123 square inches and the radius is 10 inches. Round to the nearest tenth. *(Lesson 12-6)*

Surface Areas of Spheres

What You'll Learn

- Recognize and define basic properties of spheres.
- Find surface areas of spheres.

How do manufacturers of sports equipment use the surface areas of spheres?

The sports equipment industry has grown significantly in recent years because people are engaging in more physical activities. Balls are used in many sports, such as golf, basketball, baseball, and soccer. Some are hollow, while others have solid inner cores. Each ball is identifiable by its design, texture, color, and size.

Vocabulary
- great circle
- hemisphere

PROPERTIES OF SPHERES To visualize a sphere, consider infinitely many congruent circles in space, all with the same point for their center. Considered together, these circles form a sphere. In space, a sphere is the locus of all points that are a given distance from a given point called its *center*.

Study Tip

Circles and Spheres
The shortest distance between any two points on a sphere is the length of the arc of a great circle passing through those two points.

There are several special segments and lines related to spheres.

- A segment with endpoints that are the center of the sphere and a point on the sphere is a *radius* of the sphere. In the figure, $\overline{DC}$, $\overline{DA}$, and $\overline{DB}$ are radii.

- A *chord* of a sphere is a segment with endpoints that are points on the sphere. In the figure, $\overline{GF}$ and $\overline{AB}$ are chords.

- A chord that contains the center of the sphere is a *diameter* of the sphere. In the figure, $\overline{AB}$ is a diameter.

- A *tangent* to a sphere is a line that intersects the sphere in exactly one point. In the figure, $\overleftrightarrow{JH}$ is tangent to the sphere at E.

The intersection of a plane and a sphere can be a point or a circle. When a plane intersects a sphere so that it contains the center of the sphere, the intersection is called a **great circle**. A great circle has the same center as the sphere, and its radii are also radii of the sphere.

a point   a circle   a great circle

Study Tip

Great Circles
A sphere has an infinite
number of great circles.

Each great circle separates a sphere into two congruent halves, each called
a **hemisphere**. Note that a hemisphere has a circular base.

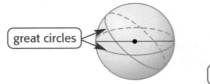

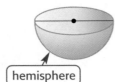

Example 1 Spheres and Circles

**In the figure, *O* is the center of the sphere, and plane $\mathcal{R}$
intersects the sphere in circle *A*. If *AO* = 3 centimeters and
OB = 10 centimeters, find *AB*.**

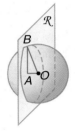

The radius of circle *A* is the segment $\overline{AB}$, *B* is a point on circle *A* and
on sphere *O*. Use the Pythagorean Theorem for right triangle *ABO*
to solve for *AB*.

$OB^2 = AB^2 + AO^2$ Pythagorean Theorem

$10^2 = AB^2 + 3^2$ *OB* = 10, *AO* = 3

$100 = AB^2 + 9$ Simplify.

$91 = AB^2$ Subtract 9 from each side.

$9.5 \approx AB$ Use a calculator.

AB is approximately 9.5 centimeters.

SURFACE AREAS OF SPHERES You will investigate the surface area of a
sphere in the geometry activity.

Geometry Activity

Surface Area of a Sphere

Model

- Cut a polystyrene ball along a great circle. Trace the great
  circle onto a piece of paper. Then cut out the circle.

- Fold the circle into eight sectors. Then unfold and
  cut the pieces apart. Tape the pieces back together
  in the pattern shown at the right.

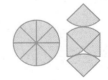

- Use tape or glue to put the two pieces of the ball
  together. Tape the paper pattern to the sphere.

Analyze

1. Approximately what fraction of the surface of the sphere is covered by the
pattern?

2. What is the area of the pattern in terms of *r*, the radius of the sphere?

Make a Conjecture

3. Make a conjecture about the formula for the surface area of a sphere.

The activity leads us to the formula for the surface area of a sphere.

> **Key Concept** **Surface Area of a Sphere**
>
> If a sphere has a surface area of *T* square units and a radius of *r* units, then $T = 4\pi r^2$.
>
>

Example 2 Surface Area

a. **Find the surface area of the sphere given the area of the great circle.**

From the activity, we find that the surface area of a sphere is four times the area of the great circle.

$T = 4\pi r^2$ Surface area of a sphere

$\approx 4(201.1)$ $\pi r^2 \approx 201.1$

≈ 804.4 Multiply.

$A \approx 201.1 \text{ in}^2$

The surface area is approximately 804.4 square inches.

b. **Find the surface area of the hemisphere.**

4.2 cm

A hemisphere is half of a sphere. To find the surface area, find half of the surface area of the sphere and add the area of the great circle.

$\text{surface area} = \frac{1}{2}(4\pi r^2) + \pi r^2$ Surface area of a hemisphere

$= \frac{1}{2}[4\pi(4.2)^2] + \pi(4.2)^2$ Substitution

≈ 166.3 Use a calculator.

The surface area is approximately 166.3 square centimeters.

More About. . .

Baseball •·············

A great circle of a standard baseball has a circumference between 9 and $9\frac{1}{4}$ inches.

Source: www.mlb.com

Example 3 Surface Area

• BASEBALL Find the surface area of a baseball with a circumference of 9 inches to determine how much leather is needed to cover the ball.

First, find the radius of the sphere.

$C = 2\pi r$ Circumference of a circle

$9 = 2\pi r$ $C = 9$

$\frac{9}{2\pi} = r$ Divide each side by 2π.

$1.4 \approx r$ Use a calculator.

Next, find the surface area of a sphere.

$T = 4\pi r^2$ Surface area of a sphere

$\approx 4\pi(1.4)^2$ $r \approx 1.4$

≈ 25.8 Use a calculator.

The surface area is approximately 25.8 square inches.

Check for Understanding

Concept Check **1. OPEN ENDED** Draw a sphere and a great circle.

2. FIND THE ERROR Loesha and Tim are finding the surface area of a hemisphere with a radius of 6 centimeters.

6 cm

Loesha
$$T = \frac{1}{2}(4\pi r^2)$$
$$T = 2\pi(6^2)$$
$$T = 72\pi$$

Tim
$$T = \frac{1}{2}(4\pi r^2) + \pi r^2$$
$$T = 2\pi(6^2) + \pi(6^2)$$
$$T = 72\pi + 36\pi$$
$$T = 108\pi$$

Who is correct? Explain.

Guided Practice **In the figure, A is the center of the sphere, and plane $\mathcal{M}$ intersects the sphere in circle C.**

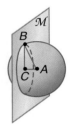

3. If $AC = 9$ and $BC = 12$, find AB.

4. If the radius of the sphere is 15 units and the radius of the circle is 10 units, find AC. $\approx$

5. If Q is a point on $\odot C$ and $AB = 18$, find AQ.

Find the surface area of each sphere or hemisphere. Round to the nearest tenth.

6. a sphere with radius 6.8 inches

7. a hemisphere with the circumference of a great circle 8π centimeters

8. a sphere with the area of a great circle approximately 18.1 square meters

Application **9. BASKETBALL** An NCAA (National Collegiate Athletic Association) basketball has a radius of $4\frac{3}{4}$ inches. Find the surface area.

Practice and Apply

For Exercises	See Examples
10–15, 25–29	1
16, 30–38	3
17–24, 39, 40	2

Extra Practice
See page 780.

In the figure, P is the center of the sphere, and plane $\mathcal{K}$ intersects the sphere in circle T.

10. If $PT = 4$ and $RT = 3$, find PR.

11. If $PT = 3$ and $RT = 8$, find PR.

12. If the radius of the sphere is 13 units and the radius of $\odot T$ is 12 units, find PT.

13. If the radius of the sphere is 17 units and the radius of $\odot T$ is 15 units, find PT.

14. If X is a point on $\odot T$ and $PR = 9.4$, find PX.

15. If Y is a point on $\odot T$ and $PR = 12.8$, find PY.

16. GRILLS A hemispherical barbecue grill has two racks, one for the food and one for the charcoal. The food rack is a great circle of the grill and has a radius of 11 inches. The charcoal rack is 5 inches below the food rack. Find the difference in the areas of the two racks.

5 in.

Find the surface area of each sphere or hemisphere. Round to the nearest tenth.

17.
25 in.

18.
14.5 cm

19.
450 m

20.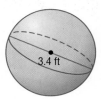
3.4 ft

21. Hemisphere: The circumference of a great circle is 40.8 inches.
22. Sphere: The circumference of a great circle is 30.2 feet.
23. Sphere: The area of a great circle is 814.3 square meters.
24. Hemisphere: The area of a great circle is 227.0 square kilometers.

Determine whether each statement is *true* or *false*. If false, give a counterexample.

25. The radii of a sphere are congruent to the radius of its great circle.
26. In a sphere, two different great circles intersect in only one point.
27. Two spheres with congruent radii can intersect in a circle.
28. A sphere's longest chord will pass through the center of the circle.
29. Two spheres can intersect in one point.

EARTH   For Exercises 30–32, use the following information.
The diameter of Earth is 7899.83 miles from the North Pole to the South Pole and 7926.41 miles from opposite points at the equator.

30. Approximate the surface area of Earth using each measure.
31. If the atmosphere of Earth extends to about 100 miles above the surface, find the surface area of the atmosphere surrounding Earth. Use the mean of the two diameters.
32. About 75% of Earth's surface is covered by water. Find the surface area of water on Earth, using the mean of the two diameters.

33. **IGLOOS**   Use the information at the left to find the surface area of the living area if the diameter is 13 feet.

34. Find the ratio of the surface area of two spheres if the radius of one is twice the radius of the second sphere.

35. Find the ratio of the radii of two spheres if the surface area of one is one half the surface area of the other.

36. Find the ratio of the surface areas of two spheres if the radius of one is three times the radius of the other.

ASTRONOMY   For Exercises 37 and 38, use the following information.
In 2002, NASA's Chandra X-Ray Observatory found two unusual neutron stars. These two stars are smaller than previously found neutron stars, but they have the mass of a larger neutron star, causing astronomers to think this star may not be made of neutrons, but a different form of matter.

37. Neutron stars have diameters from 12 to 20 miles in size. Find the range of the surface area.
38. One of the new stars has a diameter of 7 miles. Find the surface area of this star.

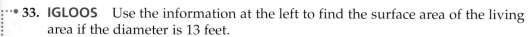

39. A sphere is inscribed in a cube. Describe how the radius of the sphere is related to the dimensions of the cube.

40. A sphere is circumscribed about a cube. Find the length of the radius of the sphere in terms of the dimensions of the cube.

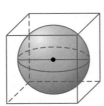

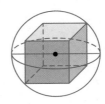

41. CRITICAL THINKING In spherical geometry, a plane is the surface of a sphere and a line is a great circle. How many lines exist that contain point X and do not intersect line g?

42. WRITING IN MATH   Answer the question that was posed at the beginning of the lesson.

How do manufacturers of sports equipment use the surface area of spheres?

Include the following in your answer:
- how to find the surface area of a sphere, and
- other examples of sports that use spheres.

43. A rectangular solid that is 4 inches long, 5 inches high, and 7 inches wide is inscribed in a sphere. What is the radius of this sphere?

　Ⓐ 4.74 in.　　Ⓑ 5.66 in.　　Ⓒ 7.29 in.　　Ⓓ 9.49 in.

44. ALGEBRA Solve $\sqrt{x^2 + 7} - 2 = x - 1$.

　Ⓐ -3　　Ⓑ $\frac{1}{3}$　　Ⓒ 3　　Ⓓ no solution

Maintain Your Skills

Mixed Review   **Find the surface area of each cone. Round to the nearest tenth.**   *(Lesson 12-6)*

45. $h = 13$ inches, $\ell = 19$ inches　　**46.** $r = 7$ meters, $h = 10$ meters

47. $r = 4.2$ cm, $\ell = 15.1$ cm　　**48.** $d = 11.2$ ft, $h = 7.4$ ft

Find the surface area of each regular pyramid. Round to the nearest tenth if necessary.   *(Lesson 12-5)*

49.

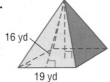

16 yd

19 yd

50.

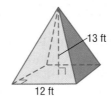

13 ft

12 ft

51.

24 cm

11 cm

52. RECREATION   Find the area of fabric needed to cover one side of a frisbee with a diameter of 9 inches. Allow an additional 3 inches around the frisbee.   *(Lesson 11-3)*

Write an equation for each circle.   *(Lesson 10-8)*

53. a circle with center at $(-2, 7)$ and a radius with endpoint at $(3, 2)$

54. a diameter with endpoints at $(6, -8)$ and $(2, 5)$

Geometry Activity

A Follow-Up of Lesson 12-7

Locus and Spheres

Spheres are defined in terms of a locus of points in space. The definition of a sphere is the set of all points that are a given distance from a given point.

Activity 1

Find the locus of points a given distance from the endpoints of a segment.

Collect the Data

- Draw a given line segment with endpoints S and T.
- Create a set of points that are equidistant from S and a set of points that are equidistant from T.

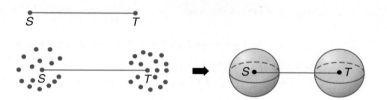

Analyze

1. Draw a figure and describe the locus of points in space that are 5 units from each endpoint of a given segment that is 25 units long.
2. Are the two spheres congruent?
3. What are the radii and diameters of each sphere?
4. Find the distance between the two spheres.

Activity 2

Investigate spheres that intersect.

Find the locus of all points that are equidistant from the centers of two intersecting spheres with the same radius.

Collect the Data

- Draw a line segment.
- Draw congruent overlapping spheres, with the centers at the endpoints of the given line segment.

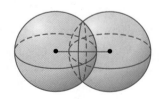

Analyze

5. What is the shape of the intersection of the spheres?
6. Can this be described as a locus of points in space or on a plane? Explain.
7. Describe the intersection as a locus.
8. **MINING** What is the locus of points that describes how particles will disperse in an explosion at ground level if the expected distance a particle could travel is 300 feet?

Vocabulary and Concept Check

axis (p. 655)
bases (p. 637)
circular cone (p. 666)
cone (p. 638)
corner view (p. 636)
cross section (p. 639)
cylinder (p. 638)
edges (p. 637)
face (p. 637)

great circle (p. 671)
hemisphere (p. 672)
lateral area (p. 649)
lateral edges (p. 649)
lateral faces (p. 649)
net (p. 644)
oblique cone (p. 666)
oblique cylinder (p. 655)

oblique prism (p. 649)
orthogonal drawing (p. 636)
perspective view (p. 636)
Platonic solids (p. 638)
polyhedron (p. 637)
prism (p. 637)
pyramid (p. 637)
reflection symmetry (p. 642)
regular polyhedron (p. 637)

regular prism (p. 637)
regular pyramid (p. 660)
right cone (p. 666)
right cylinder (p. 655)
right prism (p. 649)
slant height (p. 660)
sphere (p. 638)
surface area (p. 644)

A complete list of postulates and theorems can be found on pages R1–R8.

Exercises   Match each expression with the correct formula.

1. lateral area of a prism
2. surface area of a prism
3. lateral area of a cylinder
4. surface area of a cylinder
5. lateral area of a regular pyramid
6. surface area of a regular pyramid
7. lateral area of a cone
8. surface area of a cone
9. surface area of a sphere
10. surface area of a cube

a. $L = \frac{1}{2}P\ell$

b. $L = 2\pi rh$

c. $T = 4\pi r^2$

d. $L = Ph$

e. $L = \pi r\ell$

f. $T = 6s^2$

g. $T = \pi r\ell + \pi r^2$

h. $T = 2\pi rh + 2\pi r^2$

i. $T = Ph + 2B$

j. $T = \frac{1}{2}P\ell + B$

Lesson-by-Lesson Review

12-1 Three-Dimensional Figures

See pages 636–642.

Concept Summary

- A solid can be determined from its orthogonal drawing.
- Solids can be classified by bases, faces, edges, and vertices.

Examples   Identify each solid. Name the bases, faces, edges, and vertices.

a.

The base is a rectangle, and all of the lateral faces intersect at point T, so this solid is a rectangular pyramid.
Base: $\square PQRS$
Faces: $\triangle TPQ$, $\triangle TQR$, $\triangle TRS$, $\triangle TSP$
Edges: $\overline{PQ}$, $\overline{QR}$, $\overline{RS}$, $\overline{PS}$, $\overline{PT}$, $\overline{QT}$, $\overline{RT}$, $\overline{ST}$
Vertices: P, Q, R, S, T

b.

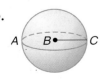

This solid has no bases, faces, or edges. It is a sphere.

www.geometryonline.com/vocabulary_review

Exercises Identify each solid. Name the bases, faces, edges, and vertices.
See Example 2 on page 638.

11.

12.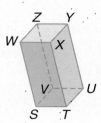

13.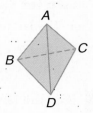

12-2 Nets and Surface Area

See pages 643–648.

Concept Summary

- Every three-dimensional solid can be represented by one or more two-dimensional nets.
- The area of the net of a solid is the same as the surface area of the solid.

Examples Draw a net and find the surface area for the right rectangular prism shown.

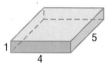

Use rectangular dot paper to draw a net. Since each face is a rectangle, opposite sides have the same measure.

To find the surface area of the prism, add the areas of the six rectangles.

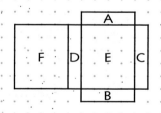

$$\text{Surface area} = A + B + C + D + E + F$$
$$= 4 \cdot 1 + 4 \cdot 1 + 5 \cdot 1 + 5 \cdot 1 + 4 \cdot 5 + 4 \cdot 5$$
$$= 4 + 4 + 5 + 5 + 20 + 20$$
$$= 58$$

The surface area is 58 square units.

Exercises For each solid, draw a net and find the surface area.
See Example 3 on page 645.

14.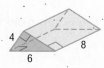

15.

16.

17.

18.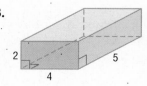

19.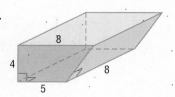

12-3 Surface Areas of Prisms

See pages 649–654.

Concept Summary

- The lateral faces of a prism are the faces that are not bases of the prism.

- The lateral surface area of a right prism is the perimeter of a base of the prism times the height of the prism. .

Example **Find the lateral area of the regular hexagonal prism.**

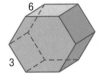

The bases are regular hexagons. So the perimeter of one base is 6(3) or 18. Substitute this value into the formula.

$L = Ph$ Lateral area of a prism

$= (18)(6)$ $P = 18, h = 6$

$= 108$ Multiply.

The lateral area is 108 square units.

Exercises   **Find the lateral area of each prism.**   *See Example 1 on page 650.*

20.

21.
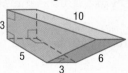

22.

12-4 Surface Areas of Cylinders

See pages 655–659.

Concept Summary

- The lateral surface area of a cylinder is 2π multiplied by the product of the radius of a base of the cylinder and the height of the cylinder.

- The surface area of a cylinder is the lateral surface area plus the area of both circular bases.

Example **Find the surface area of a cylinder with a radius of 38 centimeters and a height of 123 centimeters.**

$T = 2\pi rh + 2\pi r^2$ Surface area of a cylinder

$= 2\pi(38)(123) + 2\pi(38)^2$ $r = 38, h = 123$

$\approx 38{,}440.5$ Use a calculator.

The surface area of the cylinder is approximately 38,440.5 square centimeters.

Exercises   **Find the surface area of a cylinder with the given dimensions. Round to the nearest tenth.**   *See Example 2 on page 656.*

23. $d = 4$ in., $h = 12$ in.

24. $r = 6$ ft, $h = 8$ ft

25. $r = 4$ mm, $h = 58$ mm

26. $d = 4$ km, $h = 8$ km

12-5 Surface Areas of Pyramids

See pages 660–665.

Concept Summary

- The slant height ℓ of a regular pyramid is the length of an altitude of a lateral face.

- The lateral area of a pyramid is $\frac{1}{2}P\ell$, where ℓ is the slant height of the pyramid and P is the perimeter of the base of the pyramid.

Example **Find the surface area of the regular pyramid.**

The perimeter of the base is 4(5) or 20 units, and the area of the base is 5^2 or 25 square units. Substitute these values into the formula for the surface area of a pyramid.

$$T = \frac{1}{2}P\ell + B \qquad \text{Surface area of a regular pyramid}$$

$$= \frac{1}{2}(20)(12) + 25 \quad P = 20, \ell = 12, B = 25$$

$$= 145 \qquad \text{Simplify.}$$

The surface area is 145 square units.

Exercises   **Find the surface area of each regular pyramid. Round to the nearest tenth if necessary.**   *See Example 2 on pages 661 and 662.*

27.   28.   29.

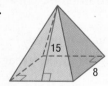   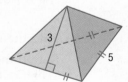

12-6 Surface Areas of Cones

See pages 666–670.

Concept Summary

- A cone is a solid with a circular base and a single vertex.

- The lateral area of a right cone is $\pi r\ell$, where ℓ is the slant height of the cone and r is the radius of the circular base.

Example **Find the surface area of the cone.**

Substitute the known values into the formula for the surface area of a right cylinder.

$$T = \pi r\ell + \pi r^2 \qquad \text{Surface area of a cone}$$

$$T = \pi(3)(12) + \pi(3)^2 \quad r = 3, \ell = 12$$

$$T \approx 141.4 \qquad \text{Use a calculator.}$$

The surface area is approximately 141.4 square meters.

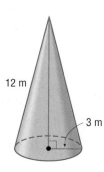

Chapter
12 For More ... • Extra Practice, see pages 778–780.
 • Mixed Problem Solving, see page 793.

Exercises Find the surface area of each cone. Round to the nearest tenth.
See Example 2 on page 667.

30.

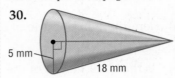

5 mm

18 mm

31.

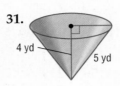

4 yd

5 yd

32.

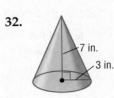

7 in.

3 in.

12-7 *Surface Areas of Spheres*

See pages
671–676.

Concept Summary

• The set of all points in space a given distance from one point is a sphere.
• The surface area of a sphere is $4\pi r^2$, where r is the radius of the sphere.

Examples **a.** **Find the surface area of a sphere with a diameter of 10 centimeters.**

10 cm

$T = 4\pi r^2$ Surface area of a sphere

$= 4\pi(5)^2$ $r = 5$

≈ 314.2 Use a calculator.

The surface area is approximately 314.2 square centimeters.

b. **Find the surface area of a hemisphere with radius 6.3 inches.**

To find the surface area of a hemisphere, add the area of the great circle to half of the surface area of the sphere.

surface area $= \frac{1}{2}(4\pi r^2) + \pi r^2$ Surface area of a hemisphere

$= \frac{1}{2}[4\pi(6.3)^2] + \pi(6.3)^2$ $r = 6.3$

≈ 374.1 Use a calculator.

The surface area is approximately 374.1 square inches.

Exercises Find the surface area of each sphere or hemisphere. Round to the nearest tenth if necessary. *See Example 2 on page 673.*

33.

18.2 ft

34.

3.9 cm

35. Area of great
circle = 121 mm²

36. Area of great
circle = 218 in²

37. a hemisphere with radius 16 ft

38. a sphere with diameter 5 m

39. a sphere that has a great circle with an area of 220 ft²

40. a hemisphere that has a great circle with an area of 30 cm²

Vocabulary and Concepts

Match each expression to the correct formula.
1. surface area of a prism
2. surface area of a cylinder
3. surface area of a regular pyramid

a. $T = 2\pi rh + 2\pi r^2$
b. $T = \frac{1}{2}P\ell + B$
c. $T = Ph + 2B$

Skills and Applications

Identify each solid. Name the bases, faces, edges, and vertices.

4.

5.

6.

For each solid, draw a net and find the surface area.

7.

8.

Find the lateral area of each prism.

9.

10.

11.

Find the surface area of a cylinder with the given dimensions. Round to the nearest tenth.

12. $r = 8$ ft, $h = 22$ ft

13. $r = 3$ mm, $h = 2$ mm

14. $r = 78$ m, $h = 100$ m

The figure at the right is a composite solid of a tetrahedron and a triangular prism. Find each measure in the solid. Round to the nearest tenth if necessary.

15. height

16. lateral area

17. surface area

Find the surface area of each cone. Round to the nearest tenth.

18. $h = 24$, $r = 7$

19. $h = 3$ m, $\ell = 4$ m

20. $r = 7$, $\ell = 12$

Find the surface area of each sphere. Round to the nearest tenth if necessary.

21. $r = 15$ in.

22. $d = 14$ m

23. The area of a great circle of the sphere is 116 square feet.

24. **GARDENING** The surface of a greenhouse is covered with plastic or glass. Find the amount of plastic needed to cover the greenhouse shown.

25. **STANDARDIZED TEST PRACTICE** A cube has a surface area of 150 square centimeters. What is the length of each edge?
Ⓐ 25 cm   Ⓑ 15 cm   Ⓒ 12.5 cm   Ⓓ 5 cm

 www.geometryonline.com/chapter_test

Part 1 Multiple Choice

Record your answers on the answer sheet provided by your teacher or on a sheet of paper.

1. A decorative strip of wood is used to bisect the center window pane sector. What is the measure of each angle formed when the pane is bisected? (Lesson 1-5)

    Ⓐ 12°
    Ⓑ 25°
    Ⓒ 50°
    Ⓓ 59°

    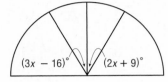

2. The coordinates of the endpoints of a segment are (0, 1) and (6, 9). A congruent segment has one endpoint at (10, 6). Which could be the coordinates of the other endpoint? (Lesson 4-3)

    Ⓐ (10, 15)   Ⓑ (14, 14)
    Ⓒ (16, 15)   Ⓓ (16, 14)

3. Which piece of additional information is enough to prove that △ABC is similar to △DEF? (Lesson 6-3)

    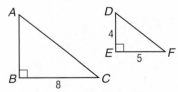

    Ⓐ △ABC and △DEF are right triangles.
    Ⓑ The length of $\overline{AB}$ is proportional to the length of $\overline{DE}$.
    Ⓒ ∠A is congruent to ∠D.
    Ⓓ The length of $\overline{BC}$ is twice the length of $\overline{DE}$.

4. Which of the following lists the sides of △ABC in order from longest to shortest? (Lesson 5-3)

    Ⓐ $\overline{BC}, \overline{AB}, \overline{AC}$
    Ⓑ $\overline{AC}, \overline{AB}, \overline{BC}$
    Ⓒ $\overline{AC}, \overline{BC}, \overline{AB}$
    Ⓓ $\overline{BC}, \overline{AC}, \overline{AB}$

5. What is the approximate length of $\overline{AB}$? (Lesson 7-2)

    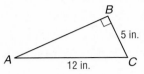

    Ⓐ 8.9 in.   Ⓑ 10.9 in.
    Ⓒ 12 in.   Ⓓ 13 in.

6. The diameter of circle *P* is 18 inches. What is the area of the shaded region? (Lesson 11-5)

    Ⓐ 27 π
    Ⓑ 54 π
    Ⓒ 81 π
    Ⓓ 216 π

7. Which statement is false? (Lesson 12-1)

    Ⓐ A pyramid is a polyhedron.
    Ⓑ The bases of a cylinder are in parallel planes.
    Ⓒ All of the Platonic Solids are regular prisms.
    Ⓓ A cone has a vertex.

8. Shelly bought a triangular prism at the science museum. The bases of the prism are equilateral triangles with side lengths of 2 centimeters. The height of the prism is 4 centimeters. What is the surface area of Shelly's prism to the nearest square centimeter? (Lesson 12-3)

    Ⓐ 16 cm²   Ⓑ 27 cm²
    Ⓒ 28 cm²   Ⓓ 31 cm²

9. A spherical weather balloon has a diameter of 4 feet. What is the surface area of the balloon to the nearest square foot? (Lesson 12-7)

    Ⓐ 50 ft²   Ⓑ 25 ft²
    Ⓒ 16 ft²   Ⓓ 13 ft²

Part 2 | Short Response/Grid In

Record your answers on the answer sheet provided by your teacher or on a sheet of paper.

10. Samantha wants to estimate the height of the tree.

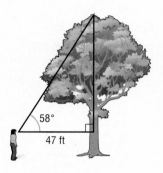

58°
47 ft

If Samantha is 5 feet tall, what is the height of the tree to the nearest foot?   (Lesson 7-5)

11. If the base and height of the triangle are decreased by 2x units, what is the area of the resulting triangle?
(Lesson 11-2)

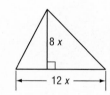

8 x

12 x

12. Mr. Jiliana built a wooden deck around half of his circular swimming pool. He needs to know the area of the deck so he can buy cans of stain. What is the area, to the nearest square foot, of the deck?   (Lesson 11-4)

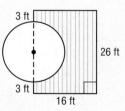

3 ft
26 ft
3 ft
16 ft

13. What is the surface area of this regular pentagonal pyramid to the nearest tenth of a square centimeter?   (Lesson 12-5)

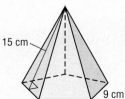

15 cm
9 cm

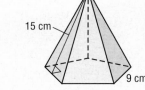
Test-Taking Tip

Question 2
Sometimes backsolving is the best technique to use to solve a multiple-choice test question. Check each of the given choices until you find the correct coordinates.

Part 3 | Extended Response

Record your answers on a sheet of paper. Show your work.

14. A cylindrical pole is being used to support a large tent. The diameter of the base is 18 inches, and the height is 15 feet.   (Lessons 12-2 and 12-4)

  a. Draw a net of the cylinder and label the dimensions.

  b. What is the lateral area of the pole to the nearest square foot?

  c. What is the surface area of the pole to the nearest square foot?

15. Aliya is constructing a model of a rocket. She uses a right cylinder for the base and a right cone for the top.   (Lessons 12-4 and 12-6)

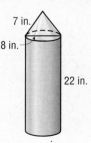

7 in.
8 in.
22 in.

  a. What is the surface area of the cone to the nearest square inch?

  b. What is the surface area of the cylinder to the nearest square inch?

  c. What is the surface area of the rocket, once it is assembled? Round to the nearest square inch.

Chapter 13 Volume

What You'll Learn

- **Lessons 13-1, 13-2, and 13-3** Find volumes of prisms, cylinders, pyramids, cones, and spheres.
- **Lesson 13-4** Identify congruent and similar solids, and state the properties of similar solids.
- **Lesson 13-5** Graph solids in space, and use the Distance and Midpoint Formulas in space.

Key Vocabulary

- volume (p. 688)
- similar solids (p. 707)
- congruent solids (p. 707)
- ordered triple (p. 714)

Why It's Important

Volcanoes like those found in Lassen Volcanic National Park, California, are often shaped like cones. By applying the formula for volume of a cone, you can find the amount of material in a volcano. *You will learn more about volcanoes in Lesson 13-2.*

Getting Started

▶ **Prerequisite Skills** To be successful in this chapter, you'll need to master these skills and be able to apply them in problem-solving situations. Review these skills before beginning Chapter 13.

For Lesson 13-1   **Pythagorean Theorem**

Find the value of the variable in each equation.   *(For review, see Lesson 7-2.)*

1. $a^2 + 12^2 = 13^2$ **2.** $\left(4\sqrt{3}\right)^2 + b^2 = 8^2$ **3.** $a^2 + a^2 = \left(3\sqrt{2}\right)^2$

4. $b^2 + 3b^2 = 192$ **5.** $256 + 7^2 = c^2$ **6.** $144 + 12^2 = c^2$

For Lesson 13-2   **Area of Polygons**

Find the area of each regular polygon. Round to the nearest tenth.   *(For review, see Lesson 11-3.)*

7. hexagon with side length 7.2 cm   **8.** hexagon with side length 7 ft

9. octagon with side length 13.4 mm   **10.** octagon with side length 10 in.

For Lesson 13-4   **Exponential Expressions**

Simplify.   *(For review, see page 746.)*

11. $(5b)^2$ **12.** $\left(\dfrac{n}{4}\right)^2$ **13.** $\left(\dfrac{3x}{4y}\right)^2$ **14.** $\left(\dfrac{4y}{7}\right)^2$

For Lesson 13-5   **Midpoint Formula**

W is the midpoint of $\overline{AB}$. For each pair of points, find the coordinates of the third point.
(For review, see Lesson 1-3.)

15. $A(0, -1), B(-5, 4)$ **16.** $A(5, 0), B(-3, 6)$

17. $A(1, -1), W(10, 10)$ **18.** $W(0, 0), B(-2, 2)$

Volume   Make this Foldable to help you organize your notes. Begin with one sheet of $8\frac{1}{2}$" by 11" paper.

Step 1   **Fold**

Fold in thirds.

Step 2   **Fold and Label**

Fold in half lengthwise. Label as shown.

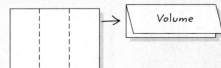

Volume

Step 3   **Label**

Unfold book. Draw lines along the folds and label as shown.

Prisms	Cylinders	Pyramids
Cones	Spheres	Similar

Reading and Writing   As you read and study the chapter, write examples and notes about the volume of each solid and about similar solids.

Volumes of Prisms and Cylinders

What You'll Learn

- Find volumes of prisms.
- Find volumes of cylinders.

Vocabulary

- volume

How is mathematics used in comics?

Creators of comics occasionally use mathematics.

SHOE

In the comic above, the teacher is getting ready to teach a geometry lesson on volume. Shoe seems to be confused about the meaning of volume.

VOLUMES OF PRISMS The **volume** of a figure is the measure of the amount of space that a figure encloses. Volume is measured in cubic units. You can create a rectangular prism from different views of the figure to investigate its volume.

Geometry Activity

Volume of a Rectangular Prism

• Model

Use cubes to make a model of the solid with the given orthogonal drawing.

top view   left view   front view   right view

Analyze

1. How many cubes make up the prism?
2. Find the product of the length, width, and height of the prism.
3. Compare the number of cubes to the product of the length, width, and height.
4. Repeat the activity with a prism of different dimensions.
5. **Make a conjecture** about the formula for the volume of a right rectangular prism.

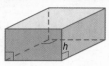

Study Tip

Volume and Area
Area is two-dimensional, so it is measured in square units. Volume is three-dimensional, so it is measured in cubic units.

The Geometry activity leads to the formula for the volume of a prism.

Key Concept   **Volume of a Prism**

If a prism has a volume of V cubic units, a height of h units, and each base has an area of B square units, then $V = Bh$.

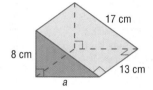

Area of base = B

Example 1 *Volume of a Triangular Prism*

Find the volume of the triangular prism.

Use the Pythagorean Theorem to find the leg of the base of the prism.

$a^2 + b^2 = c^2$ Pythagorean Theorem

$a^2 + 8^2 = 17^2$ $b = 8$, $c = 17$

$a^2 + 64 = 289$ Multiply.

$a^2 = 225$ Subtract 64 from each side.

$a = 15$ Take the square root of each side.

Next, find the volume of the prism.

$V = Bh$ Volume of a prism

$= \frac{1}{2}(8)(15)(13)$ $B = \frac{1}{2}(8)(15)$, $h = 13$

$= 780$ Simplify.

The volume of the prism is 780 cubic centimeters.

The volume formula can be used to solve real-world problems.

Example 2 *Volume of a Rectangular Prism*

SNOW   The weight of wet snow is 0.575 times the volume of snow in cubic inches divided by 144. How many pounds of wet snow would a person shovel in a rectangular driveway 25 feet by 10 feet after 12 inches of snow have fallen?

First, make a drawing.
Then convert feet to inches.
25 feet = 25 × 12 or 300 inches
10 feet = 10 × 12 or 120 inches

To find the pounds of wet snow shoveled, first find the volume of snow on the driveway.

$V = Bh$ Volume of a prism

$= 300(120)(12)$ $B = 300(120)$, $h = 12$

$= 432,000$ The volume is 432,000 cubic inches.

Now multiply the volume by 0.575 and divide by 144.

$\dfrac{0.575(432,000)}{144} = 1725$ Simplify.

A person shoveling 12 inches of snow on a rectangular driveway 25 feet by 10 feet would shovel 1725 pounds of snow.

VOLUMES OF CYLINDERS Like the volume of a prism, the volume of a cylinder is the product of the area of the base and the height.

> ### Key Concept | Volume of a Cylinder
>
> If a cylinder has a volume of V cubic units, a height of h units, and the bases have radii of r units, then $V = Bh$ or $V = \pi r^2 h$.
>
>
>
> Area of base $= \pi r^2$

Study Tip

Square Roots
When you take the square root of each side of $h^2 = 144$, the result is actually ± 12. Since this is a measure, the negative value is discarded.

Example 3 · Volume of a Cylinder

Find the volume of each cylinder.

a.

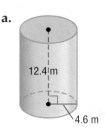

12.4 m

4.6 m

The height h is 12.4 meters, and the radius r is 4.6 meters.

$V = \pi r^2 h$ Volume of a cylinder

$ = \pi(4.6^2)(12.4)$ $r = 4.6, h = 12.4$

$ \approx 824.3$ Use a calculator.

The volume is approximately 824.3 cubic meters.

b.

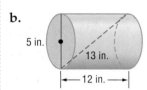

5 in.

13 in.

12 in.

The diameter of the base, the diagonal, and the lateral edge of the cylinder form a right triangle. Use the Pythagorean Theorem to find the height.

$a^2 + b^2 = c^2$ Pythagorean Theorem

$h^2 + 5^2 = 13^2$ $a = h, b = 5$, and $c = 13$

$h^2 + 25 = 169$ Multiply.

$h^2 = 144$ Subtract 25 from each side.

$h = 12$ Take the square root of each side.

Now find the volume.

$V = \pi r^2 h$ Volume of a cylinder

$ = \pi(2.5^2)(12)$ $r = 2.5$ and $h = 12$

$ \approx 235.6$ Use a calculator.

The volume is approximately 235.6 cubic inches.

Study Tip

Look Back
To review **oblique solids**, see Lesson 12-3.

    Thus far, we have only studied the volumes of right solids. Do the formulas for volume apply to oblique solids as well as right solids?

    Study the two stacks of quarters. The stack on the left represents a right cylinder, and the stack on the right represents an oblique cylinder. Since each stack has the same number of coins, with each coin the same size and shape, the two cylinders must have the same volume. Cavalieri, an Italian mathematician of the seventeenth century, was credited with making this observation first.

Study Tip

Cavalieri's Principle
This principle applies to the volumes of all solids.

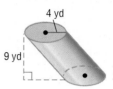

Key Concept   *Cavalieri's Principle*

If two solids have the same height and the same cross-sectional area at every level, then they have the same volume.

If a cylinder has a base with an area of B square units and a height of h units, then its volume is Bh cubic units, whether it is right or oblique.

Example 4   *Volume of an Oblique Cylinder*

Find the volume of the oblique cylinder.

To find the volume, use the formula for a right cylinder.

$$V = \pi r^2 h \qquad \text{Volume of a cylinder}$$
$$= \pi(4^2)(9) \qquad r = 4, h = 9$$
$$\approx 452.4 \qquad \text{Use a calculator.}$$

The volume is approximately 452.4 cubic yards.

Check for Understanding

Concept Check   **1. OPEN ENDED**   List three objects that are cylinders and three that are prisms.

                  2. FIND THE ERROR   Che and Julia are trying to find the number of cubic feet in a cubic yard.

Che	Julia
$V = Bh$	$V = Bh$
$= 3 \times 3 \times 3$	$= 3 \times 3 \times 3$
$= 9$	$= 27$
There are 9 cubic feet in one cubic yard.	There are 27 cubic feet in one cubic yard.

Who is correct? Explain your reasoning.

Guided Practice   **Find the volume of each prism or cylinder. Round to the nearest tenth if necessary.**

3.

12 cm
8 cm
6 cm

4.   8 in.
17 in.

5.
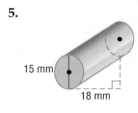
15 mm
18 mm

Application   **6. DIGITAL CAMERA**   The world's most powerful digital camera is located in New Mexico at the Apache Point Observatory. It is surrounded by a rectangular prism made of aluminum that protects the camera from wind and unwanted light. If the prism is 12 feet long, 12 feet wide, and 14 feet high, find its volume to the nearest cubic foot.

Homework Help

For Exercises	See Examples
7, 10, 17	3
8, 11, 18	2
9, 12	1
13–16	4

Extra Practice
See page 780.

Find the volume of each prism or cylinder. Round to the nearest tenth if necessary.

7.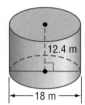
9 cm
15 cm

8.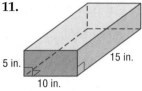
18.7 in.
3.6 in.
12.2 in.

9.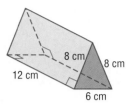
8 cm
8 cm
12 cm
6 cm

10.
12.4 m
18 m

11.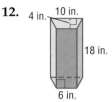
15 in.
5 in.
10 in.

12.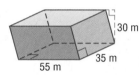
4 in.   10 in.
18 in.
6 in.

Find the volume of each oblique prism or cylinder. Round to the nearest tenth if necessary.

13.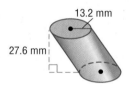
3.2 ft
3.5 ft
2.5 ft

14.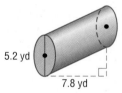
30 m
55 m
35 m

15.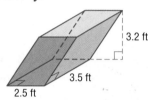
13.2 mm
27.6 mm

16.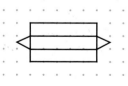
5.2 yd
7.8 yd

17. The volume of a cylinder is 615.8 cubic meters, and its height is 4 meters. Find the diameter of the cylinder.

18. The volume of a right rectangular prism is 1152 cubic inches, and the area of each base is 64 square inches. Find the length of the lateral edge of the prism.

Find the volume of the solid formed by each net. Round to the nearest tenth if necessary.

19.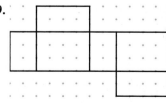

20.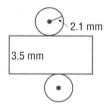

21.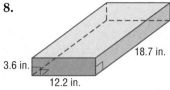
2.1 mm
3.5 mm

Find the volume of each solid. Round to the nearest tenth if necessary.

22.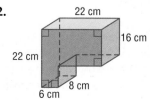
22 cm
16 cm
22 cm
8 cm
6 cm

23.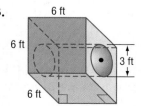
6 ft
6 ft
3 ft
6 ft

24.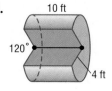
10 ft
120°
4 ft

25. **MANUFACTURING** A can is 12 centimeters tall and has a diameter of 6.5 centimeters. It fits into a rubberized cylindrical holder that is 11.5 centimeters tall, including 1 centimeter, which is the thickness of the base of the holder. The thickness of the rim of the holder is 1 centimeter. What is the volume of the rubberized material that makes up the holder?

26. **ARCHITECTURE** The Marina Towers in Chicago are cylindrical shaped buildings that are 586 feet tall. There is a 35-foot-diameter cylindrical core in the center of each tower. If the core extends 40 feet above the roof of the tower, find the volume of the core.

27. **AQUARIUM** The New England Aquarium in Boston, Massachusetts, has one of the world's largest cylindrical tanks. The Giant Ocean tank holds approximately 200,000 gallons and is 23 feet deep. If it takes about $7\frac{1}{2}$ gallons of water to fill a cubic foot, what is the radius of the Giant Ocean Tank?

28. **SWIMMING** A swimming pool is 50 meters long and 25 meters wide. The adjustable bottom of the pool can be up to 3 meters deep for competition and as shallow as 0.3 meter deep for recreation. The pool was filled to the recreational level, and then the floor was lowered to the competition level. If the volume of a liter of water is 0.001 cubic meter, how much water had to be added to fill the pool?

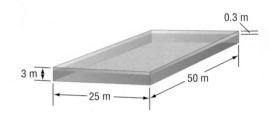

•**ENGINEERING** For Exercises 29 and 30, use the following information.
Machinists make parts for intricate pieces of equipment. Suppose a part has a regular hexagonal hole drilled in a brass block.

29. Find the volume of the resulting part.

30. The *density* of a substance is its mass per unit volume. At room temperature, the density of brass is 8.0 grams per cubic centimeter. What is the mass of this block of brass?

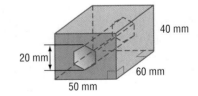

31. **CRITICAL THINKING** Find the volume of a regular pentagonal prism with a height of 5 feet and a perimeter of 20 feet.

32. **WRITING IN MATH** Answer the question that was posed at the beginning of the lesson.

How is mathematics used in comics?

Include the following in your answer:
• the meaning of volume that Shoe has in the comic, and
• the mathematical meaning of volume.

33. A rectangular swimming pool has a volume of 16,320 cubic feet, a depth of 8 feet, and a length of 85 feet. What is the width of the swimming pool?

Ⓐ 24 ft   Ⓑ 48 ft   Ⓒ 192 ft   Ⓓ 2040 ft

34. ALGEBRA   Factor $\pi r^2 h - 2\pi rh$ completely.

Ⓐ $\pi(r^2h - 2rh)$ Ⓑ $\pi rh(r - 2)$

Ⓒ $\pi rh(rh - 2rh)$ Ⓓ $2\pi r(rh - h)$

Maintain Your Skills

Mixed Review   **Find the surface area of each sphere. Round to the nearest tenth.**   *(Lesson 12-7)*

35.
12 ft

36.
41 cm

37.
18 m

38.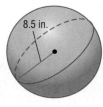
8.5 in.

Find the surface area of each cone. Round to the nearest tenth.   *(Lesson 12-6)*

39. slant height = 11 m, radius = 6 m

40. diameter = 16 cm, slant height = 13.5 cm

41. radius = 5 in., height = 12 in.

42. diameter = 14 in., height = 24 in.

43. HOUSING   Martin lost a file at his home, and he only has time to search three of the rooms before he has to leave for work. If the shaded parts of his home will be searched, what is the probability that he finds his file?   *(Lesson 11-5)*

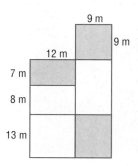
9 m
9 m
12 m
7 m
8 m
13 m

Find the area of each polygon. Round to the nearest tenth.   *(Lesson 11-3)*

44. a regular hexagon with a perimeter of 156 inches

45. a regular octagon with an apothem 7.5 meters long and a side 6.2 meters long

Find x to the nearest tenth. Assume that any segment that appears to be tangent is tangent.   *(Lesson 10-7)*

46.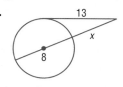
13
x
8

47.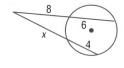
8
6
x
4

48.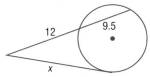
9.5
12
x

Getting Ready for the Next Lesson

PREREQUISITE SKILL   **Find the area of each polygon with given side length, s. Round to the nearest hundredth.**

*(To review **finding areas of regular polygons**, see Lesson 11-3)*

49. equilateral triangle, $s = 7$ in.   **50.** regular hexagon, $s = 12$ cm

51. regular pentagon, $s = 6$ m   **52.** regular octagon, $s = 50$ ft

Spreadsheet Investigation

Prisms

Changing the dimensions of a prism affects the surface area and the volume of the prism. You can investigate the changes by using a spreadsheet.

- Create a spreadsheet by entering the length of the rectangular prism in column B, the width in column C, and the height in column D.

- In cell E2, enter the formula for the total surface area.

- Copy the formula in cell E2 to the other cells in column E.

- Write a formula to find the volume of the prism. Enter the formula in cell F2.

- Copy the formula in cell F2 into the other cells in column F.

	A	B	C	D	E	F
					Prisms.xls	
1	Prism	*l*	*w*	*h*	Surface Area	Volume
2	1	1	2	3	22	6
3	2					
4	3					
5	4					
6	5					
7						

Sheet1 / Sheet2

Use your spreadsheet to find the surface areas and volumes of prisms with the dimensions given in the table below.

Prism	Length	Width	Height	Surface Area	Volume
1	1	2	3		
2	2	4	6		
3	3	6	9		
4	4	8	12		
5	8	16	24		

Exercises

1. Compare the dimensions of prisms 1 and 2, prisms 2 and 4, and prisms 4 and 5.

2. Compare the surface areas of prisms 1 and 2, prisms 2 and 4, and prisms 4 and 5.

3. Compare the volumes of prisms 1 and 2, prisms 2 and 4, and prisms 4 and 5.

4. Write a statement about the change in the surface area and volume of a prism when the dimensions are doubled.

Volumes of Pyramids and Cones

What You'll Learn

- Find volumes of pyramids.
- Find volumes of cones.

Why do architects use geometry?

The Transamerica Pyramid is the tallest skyscraper in San Francisco. The 48-story building is a square pyramid. The building was designed to allow more light to reach the street.

VOLUMES OF PYRAMIDS The pyramid and the prism at the right share a base and have the same height. As you can see, the volume of the pyramid is less than the volume of the prism.

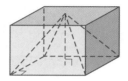

Geometry Activity

Investigating the Volume of a Pyramid

Activity

- Draw each net on card stock.
- Cut out the nets. Fold on the dashed lines.
- Tape the edges together to form models of the solids with one face removed.
- Estimate how much greater the volume of the prism is than the volume of the pyramid.
- Fill the pyramid with rice. Then pour this rice into the prism. Repeat until the prism is filled.

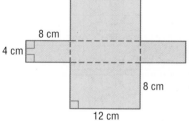

Analyze

1. How many pyramids of rice did it take to fill the prism?
2. Compare the areas of the bases of the prism and pyramid.
3. Compare the heights of the prism and the pyramid.
4. Make a conjecture about the formula for the volume of a pyramid.

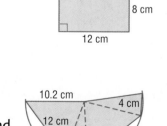

The Geometry Activity leads to the formula for the volume of a pyramid.

Key Concept — Volume of a Pyramid

If a pyramid has a volume of V cubic units, a height of h units, and a base with an area of B square units, then $V = \frac{1}{3}Bh$.

Area of base $= B$

Example 1 Volume of a Pyramid

NUTRITION Travis is making a plaster model of the Food Guide Pyramid for a class presentation. The model is a square pyramid with a base edge of 12 inches and a height of 15 inches. Find the volume of plaster needed to make the model.

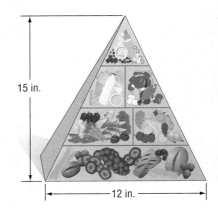

15 in.

12 in.

$$V = \frac{1}{3}Bh \qquad \text{Volume of a pyramid}$$

$$= \frac{1}{3}s^2h \qquad B = s^2$$

$$= \frac{1}{3}(12^2)(15) \qquad s = 12, h = 15$$

$$= 720 \qquad \text{Multiply.}$$

Travis needs 720 cubic inches of plaster to make the model.

VOLUMES OF CONES The derivation of the formula for the volume of a cone is similar to that of a pyramid. If the areas of the bases of a cone and a cylinder are the same and if the heights are equal, then the volume of the cylinder is three times as much as the volume of the cone.

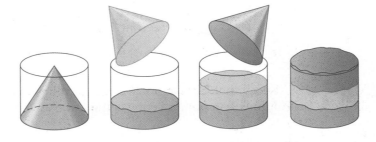

Key Concept | Volume of a Cone

If a right circular cone has a volume of V cubic units, a height of h units, and the base has a radius of r units, then $V = \frac{1}{3}Bh$ or $V = \frac{1}{3}\pi r^2h$.

Area of base $= \pi r^2$

Example 2 Volumes of Cones

Find the volume of each cone.

a.

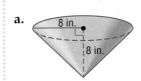

8 in.

8 in.

$$V = \frac{1}{3}\pi r^2h \qquad \text{Volume of a cone}$$

$$= \frac{1}{3}\pi(8^2)(8) \qquad r = 8, h = 8$$

$$\approx 536.165 \qquad \text{Use a calculator.}$$

The volume of the cone is approximately 536.2 cubic inches.

b.

Use trigonometry to find the radius of the base.

$\tan A = \dfrac{\text{opposite}}{\text{adjacent}}$ Definition of tangent

$\tan 48° = \dfrac{10}{r}$ $A = 48°$, opposite $= 10$, and adjacent $= r$

$r = \dfrac{10}{\tan 48°}$ Solve for r.

$r \approx 9.0$ Use a calculator.

Now find the volume.

$V = \dfrac{1}{3}Bh$ Volume of a cone

$= \dfrac{1}{3}\pi r^2 h$ $B = \pi r^2$

$\approx \dfrac{1}{3}\pi (9^2)(10)$ $r \approx 9$, $h = 10$

≈ 848.992 Use a calculator.

The volume of the cone is approximately 849.0 cubic inches.

Recall that Cavalieri's Principle applies to all solids. So, the formula for the volume of an oblique cone is the same as that of a right cone.

Study Tip

Common Misconceptions
The formula for the surface area of a cone only applies to right cones. However, a right cone and an oblique cone with the same radius and height have the same volume but different surface areas.

Example 3   *Volume of an Oblique Cone*

Find the volume of the oblique cone.

$V = \dfrac{1}{3}Bh$ Volume of a cone

$= \dfrac{1}{3}\pi r^2 h$ $B = \pi r^2$

$= \dfrac{1}{3}\pi (8.6^2)(12)$ $r = 8.6$, $h = 12$

≈ 929.4 Use a calculator.

The volume of the oblique cone is approximately 929.4 cubic inches.

Check for Understanding

Concept Check

1. **Describe** the effect on the volumes of a cone and a pyramid if the dimensions are doubled.

2. **Explain** how the volume of a pyramid is related to that of a prism with the same height and a base congruent to that of the pyramid.

3. **OPEN ENDED** Draw and label two cones with different dimensions, but with the same volume.

Guided Practice **Find the volume of each pyramid or cone. Round to the nearest tenth if necessary.**

4.
12 in.
10 in.
16 in.

5.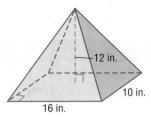
12 mm
60°

6.
8 in.
20 in.

Application **7. MECHANICAL ENGINEERING**
The American Heritage Center at the University of Wyoming is a conical building. If the height is 77 feet, and the area of the base is about 38,000 square feet, find the volume of air that the heating and cooling systems would have to accommodate. Round to the nearest tenth.

Practice and Apply

Homework Help	
For Exercises	**See Examples**
8–10, 17–18, 24–27, 30	1
11–13, 19–23, 28–29	2
14–16	3

Extra Practice
See page 780.

Find the volume of each pyramid or cone. Round to the nearest tenth if necessary.

8.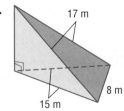
10 cm
6 cm

9.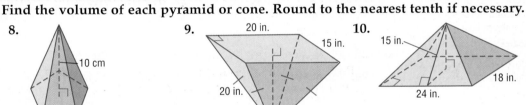
20 in.
15 in.
20 in.

10.
15 in.
18 in.
24 in.

11.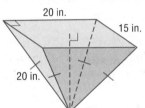
30 mm
36 mm

12.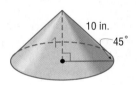
10 in.
45°

13.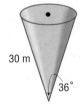
30 m
36°

14.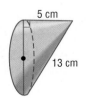
17 m
8 m
15 m

15.
5 cm
13 cm

16.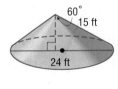
60°
15 ft
24 ft

Find the volume of each solid. Round to the nearest tenth.

17.
10 mm
10 mm
12 mm
12 mm
12 mm

18.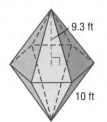
9.3 ft
10 ft

19.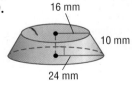
16 mm
10 mm
24 mm

• VOLCANOES   For Exercises 20–23, use the following information.
The slope of a volcano is the angle made by the side of the cone and a horizontal line. Find the volume of material in each volcano, assuming that it is a solid cone.

Volcano	Location	Type	Characteristics
Mauna Loa	Hawaii, United States	shield dome	4170 m tall, 103 km across at base
Mount Fuji	Honshu, Japan	composite	3776 m tall, slope of 9°
Paricutín	Michoacán, Mexico	cinder cone	410 m tall, 33° slope
Vesuvius	Campania, Italy	composite	22.3 km across at base, 1220 m tall

20. Mauna Loa   **21.** Mount Fuji   **22.** Paricutin   **23.** Vesuvius

24. The shared base of the pyramids that make up the solid on the left is congruent to the base of the solid on the right. Write a ratio comparing the volumes of the solids. Explain your answer.

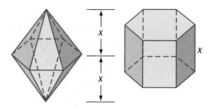

HISTORY   For Exercises 25–27, use the following information.
The Great Pyramid of Khufu is a square pyramid. The lengths of the sides of the base are 755 feet. The original height was 481 feet. The current height is 449 feet.

25. Find the original volume of the pyramid.

26. Find the present day volume of the pyramid.

27. Compare the volumes of the pyramid. What volume of material has been lost?

28. PROBABILITY   What is the probability of randomly choosing a point inside the cylinder, but not inside the cone that has the same base and height as the cylinder.

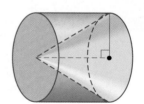

29. A pyramid with a square base is next to a circular cone as shown at the right. The circular base is inscribed in the square. Isosceles $\triangle ABC$ is perpendicular to the base of the pyramid. DE is the slant height of the cone. Find the volume of the figure.

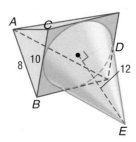

30. ARCHITECTURE   In an attempt to rid Florida's Lower Sugarloaf Key of mosquitoes, Richter Perky built a tower to attract bats. The Perky Bat Tower is a frustum of a pyramid with a square base. Each side of the base of the tower is 15 feet long, the top is a square with sides 8 feet long, and the tower is 35 feet tall. How many cubic feet of space does the tower supply for bats? (*Hint*: Draw the pyramid that contains the frustrum.)

31. CRITICAL THINKING   Find the volume of a regular tetrahedron with one side measuring 12 inches.

32. WRITING IN MATH Answer the question that was posed at the beginning of the lesson.

How do architects use geometry?

Include the following in your answer:

- compare the available office space on the first floor and on the top floor, and
- explain how a pyramidal building allows more light to reach the street than a rectangular prism building.

33. Which of the following is the volume of the square pyramid if $b = 2h$?

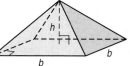

Ⓐ $\frac{h^3}{3}$ Ⓑ $\frac{4h^3}{3}$ Ⓒ $4h^3$ Ⓓ $\frac{8h^3 - 4h}{3}$

34. ALGEBRA If the volume of a right prism is represented by $x^3 \pm 9x$, find the factors that could represent the length, width, and height.

Ⓐ $x, x - 3, x + 3$ Ⓑ $x, x - 9, x + 1$
Ⓒ $x, x + 9, x \pm 1$ Ⓓ $x, x - 3, x \pm 3$

Maintain Your Skills

Mixed Review   Find the volume of each prism or cylinder. Round to the nearest tenth if necessary. *(Lesson 13-1)*

35.
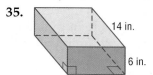

36.

37.
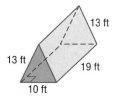

Find the surface area of each sphere. Round to the nearest tenth if necessary. *(Lesson 12-7)*

38. The circumference of a great circle is 86 centimeters.

39. The area of a great circle is 64.5 square yards.

40. BASEBALL A baseball field has the shape of a rectangle with a corner cut out as shown at the right. What is the total area of the baseball field? *(Lesson 11-4)*

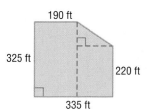

Getting Ready for the Next Lesson   **PREREQUISITE SKILL** Evaluate each expression. Round to the nearest hundredth.
*(To review **evaluating expressions**, see page 736.)*

41. $4\pi r^2, r = 3.4$ **42.** $\frac{4}{3}\pi r^3, r = 7$ **43.** $4\pi r^2, r = 12$

Practice Quiz 1
Lessons 13-1 and 13-2

1. FOOD A canister of oatmeal is 10 inches tall with a diameter of 4 inches. Find the maximum volume of oatmeal that the canister can hold to the nearest tenth. *(Lesson 13-1)*

Find the volume of each solid. Round to the nearest tenth. *(Lessons 13-1 and 13-2)*

2.

3.

4.

5.
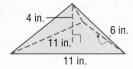

13-3 Volumes of Spheres

What You'll Learn

- Find volumes of spheres.
- Solve problems involving volumes of spheres.

How can you find the volume of Earth?

Eratosthenes was an ancient Greek mathematician who estimated the circumference of Earth. He assumed that Earth was a sphere and estimated that the circumference was about 40,000 kilometers. From the circumference, the radius of Earth can be calculated. Then the volume of Earth can be determined.

VOLUMES OF SPHERES You can relate finding a formula for the volume of a sphere to finding the volume of a right pyramid and the surface area of a sphere.

Suppose the space inside a sphere is separated into infinitely many near-pyramids, all with vertices located at the center of the sphere. Observe that the height of these pyramids is equal to the radius r of the sphere. The sum of the areas of all the pyramid bases equals the surface area of the sphere.

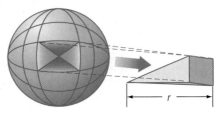

Each pyramid has a volume of $\frac{1}{3}Bh$, where B is the area of its base and h is its height. The volume of the sphere is equal to the sum of the volumes of all of the small pyramids.

$$V = \frac{1}{3}B_1h_1 + \frac{1}{3}B_2h_2 + \frac{1}{3}B_3h_3 + \dots + \frac{1}{3}B_nh_n \quad \text{Sum of the volumes of all of the pyramids}$$

$$= \frac{1}{3}B_1r + \frac{1}{3}B_2r + \frac{1}{3}B_3r + \dots + \frac{1}{3}B_nr \quad \text{Replace } h \text{ with } r.$$

$$= \frac{1}{3}r(B_1 + B_2 + B_3 + \dots + B_n) \quad \text{Distributive Property}$$

$$= \frac{1}{3}r(4\pi r^2) \quad \text{Replace } B_1 + B_2 + B_3 + \dots + B_n \text{ with } 4\pi r^2.$$

$$= \frac{4}{3}\pi r^3 \quad \text{Simplify.}$$

Study Tip

Look Back
Recall that the surface area of a sphere, $4\pi r^2$, is equal to $B_1 + B_2 + B_3 + \dots + B_n$. To review **surface area of a sphere**, see Lesson 12-7.

Key Concept   *Volume of a Sphere*

If a sphere has a volume of V cubic units and a radius of r units, then $V = \frac{4}{3}\pi r^3$.

Example 1 Volumes of Spheres

Find the volume of each sphere.

a.

$$V = \frac{4}{3}\pi r^3 \qquad \text{Volume of a sphere}$$

$$= \frac{4}{3}\pi(24^3) \qquad r = 24$$

$$\approx 57{,}905.8 \text{ in}^3 \quad \text{Use a calculator.}$$

b. $C = 36$ cm

WebQuest

The formulas for the surface area and volume of a sphere can help you find the surface area and volume of the hemisphere.
Visit www.geometry online.com/webquest to continue work on your WebQuest project.

First find the radius of the sphere.

$$C = 2\pi r \quad \text{Circumference of a circle}$$

$$36 = 2\pi r \quad C = 36$$

$$\frac{18}{\pi} = r \qquad \text{Solve for } r.$$

Now find the volume.

$$V = \frac{4}{3}\pi r^3 \qquad \text{Volume of a sphere}$$

$$= \frac{4}{3}\pi\left(\frac{18}{\pi}\right)^3 \qquad r = \frac{18}{\pi}$$

$$\approx 787.9 \text{ cm}^3 \quad \text{Use a calculator.}$$

Example 2 Volume of a Hemisphere

Find the volume of the hemisphere.

The volume of a hemisphere is one-half the volume of the sphere.

$$V = \frac{1}{2}\left(\frac{4}{3}\pi r^3\right) \quad \text{Volume of a hemisphere}$$

$$= \frac{2}{3}\pi(2^3) \qquad r = 2$$

$$\approx 16.8 \text{ ft}^3 \qquad \text{Use a calculator.}$$

SOLVE PROBLEMS INVOLVING VOLUMES OF SPHERES Often spherical objects are contained in other solids. A comparison of the volumes is necessary to know if one object can be contained in the other.

Standardized Test Practice
A B C D

Example 3 Volume Comparison

Short-Response Test Item

> Compare the volumes of the sphere and the cylinder. Determine which quantity is greater.

Read the Test Item

You are asked to compare the volumes of the sphere and the cylinder.

(continued on the next page)

Solve the Test Item

Volume of the sphere: $\frac{4}{3}\pi r^3$

Volume of the cylinder: $\pi r^2 h = \pi r^2 (2r)$ $h = 2r$

$= 2\pi r^3$ Simplify.

Compare $2\pi r^3$ to $\frac{4}{3}\pi r^3$. Since 2 is greater than $\frac{4}{3}$, the volume of the cylinder is greater than the volume of the sphere.

Check for Understanding

Concept Check

1. **Explain** how to find the formula for the volume of a sphere.

2. **FIND THE ERROR**   Winona and Kenji found the volume of a sphere with a radius of 12 centimeters.

Winona	Kenji
$V = \frac{4}{3}\pi(12)^3$	$V = \frac{4}{3}\pi(12)^3$
$= 4\pi(4)^3$	$= \frac{4}{3}\pi(1728)$
$= 256\pi$ cm^3	$= 2304\pi$ cm^3

Who is correct? Explain your reasoning.

Guided Practice

Find the volume of each sphere or hemisphere. Round to the nearest tenth.

3. The radius is 13 inches long.

4. The diameter of the sphere is 12.5 centimeters.

5.
4 in.

6. C = 18 cm

7.
8.4 m

Standardized Test Practice
Ⓐ Ⓑ Ⓒ Ⓓ

8. **SHORT RESPONSE**   Compare the volumes of a sphere with a radius of 5 inches and a cone with a height of 20 inches and a base with a diameter of 10 inches.

Practice and Apply

Homework Help

For Exercises	See Examples
9–19	1–2
20–22, 28–29, 32	3

Extra Practice
See page 781.

Find the volume of each sphere or hemisphere. Round to the nearest tenth.

9. The radius of the sphere is 7.62 meters.

10. The diameter of the sphere is 33 inches.

11. The diameter of the sphere is 18.4 feet.

12. The radius of the sphere is $\frac{\sqrt{3}}{2}$ centimeters.

13. C = 24 in.

14.
35.8 mm

15.
3.2 m

16.
28 ft

17.
12 in.

18. C = 48 cm

19. ASTRONOMY   The diameter of the moon is 3476 kilometers. Find the volume of the moon.

20. SPORTS   If a golf ball has a diameter of 4.3 centimeters and a tennis ball has a diameter of 6.9 centimeters, find the difference between the volumes of the two balls.

FOOD   For Exercises 21 and 22, use the following information.
Suppose a sugar cone is 10 centimeters deep and has a diameter of 4 centimeters. A spherical scoop of ice cream with a diameter of 4 centimeters rests on the top of the cone.

21. If all the ice cream melts into the cone, will the cone overflow? Explain.

22. If the cone does not overflow, what percent of the cone will be filled?

FAMILY   For Exercises 23–26, use the following information.
Suppose the bubble in the graphic is a sphere with a radius of 17 millimeters.

23. What is the volume of the bubble?

24. What is the volume of the portion of the bubble in which the kids had just the right amount of time with their mother?

25. What is the surface area of that portion of the bubble in which the kids wish they could spend more time with their mother?

26. What is the area of the two-dimensional sector of the circle in which the kids wish they could spend less time with their mother?

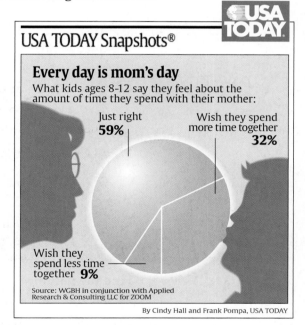
27. PROBABILITY   Find the probability of choosing a point at random inside a sphere that has a radius of 6 centimeters and is inscribed in a cylinder.

28. TENNIS   Find the volume of the empty space in a cylindrical tube of three tennis balls. The diameter of each ball is about 2.5 inches. The cylinder is 2.5 inches in diameter and is 7.5 inches tall.

29. Find the volume of a sphere that is circumscribed about a cube with a volume of 216 cubic inches.

Find the volume of each sphere or hemisphere. Round to the nearest tenth.

30. The surface area of a sphere is 784π square inches.

31. A hemisphere has a surface area of 18.75π square meters.

32. ARCHITECTURE The Pantheon in Rome is able to contain a perfect sphere. The building is a cylinder 142 feet in diameter with a hemispherical domed roof. The total height is 142 feet. Find the volume of the interior of the Pantheon.

33. CRITICAL THINKING A vitamin capsule consists of a right cylinder with a hemisphere on each end. The capsule is 16 millimeters long and 4 millimeters thick. What is the volume of the capsule?

34. WRITING IN MATH   Answer the question that was posed at the beginning of the lesson.

How can you find the volume of Earth?

Include the following in your answer:
- the important dimension you must have to find the volume of Earth, and
- the radius and volume of Earth from this estimate.

35. RESEARCH Use the Internet or other source to find the most current calculations for the volume of Earth.

Standardized Test Practice
Ⓐ Ⓑ Ⓒ Ⓓ

36. If the radius of a sphere is increased from 3 units to 5 units, what percent would the volume of the smaller sphere be of the volume of the larger sphere?

Ⓐ 21.6%   Ⓑ 40%   Ⓒ 60%   Ⓓ 463%

37. ALGEBRA Simplify $\frac{1}{2}(4\pi r^2) + \pi r^2 h + \frac{1}{2}(4\pi r^2)$.

Ⓐ $\pi r^2(4 + h)$ Ⓑ $4\pi r^2 h$ Ⓒ $\pi r^2(9 + h)$ Ⓓ $2\pi r^2(2 + h)$

Maintain Your Skills

Mixed Review   **Find the volume of each cone. Round to the nearest tenth.**   *(Lesson 13-2)*

38. height = 9.5 meters, radius = 6 meters

39. height = 7 meters, diameter = 15 meters

40. REFRIGERATORS A refrigerator has a volume of 25.9 cubic feet. If the interior height is 5.0 feet and the width is 2.4 feet, find the depth.   *(Lesson 13-1)*

Write an equation for each circle.   *(Lesson 10-8)*

41. center at $(2, -1)$, $r = 8$

42. center at $(-4, -3)$, $r = \sqrt{19}$

43. diameter with endpoints at $(5, -4)$ and $(-1, 6)$

Getting Ready for the Next Lesson   **PREREQUISITE SKILL   Simplify.**

*(To review **simplifying expressions involving exponents**, see pages 746 and 747.)*

44. $(2a)^2$ **45.** $(3x)^3$ **46.** $\left(\frac{5a}{b}\right)^2$ **47.** $\left(\frac{2k}{5}\right)^3$

Congruent and Similar Solids

What You'll Learn

- Identify congruent or similar solids.
- State the properties of similar solids.

Vocabulary

- similar solids
- congruent solids

How are similar solids applied to miniature collectibles?

People collect miniatures of race cars, farm equipment, and monuments such as the Statue of Liberty. The scale factors commonly used for miniatures include 1:16, 1:24, 1:32, and 1:64. One of the smallest miniatures has a scale factor of 1:1000.

If a car is 108 inches long, then a 1:24 scale model would be 108 ÷ 24 or 4.5 inches long.

CONGRUENT OR SIMILAR SOLIDS **Similar solids** are solids that have exactly the same shape but not necessarily the same size. You can determine if two solids are similar by comparing the ratios of corresponding linear measurements. In two similar polyhedra, all of the corresponding faces are similar, and all of the corresponding edges are proportional. *All spheres are similar just like all circles are similar.*

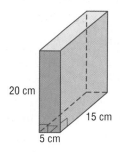
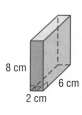

Similar Solids

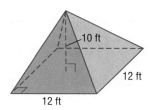
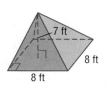

Nonsimilar Solids

Study Tip

Look Back
To review **scale factor**, see Lessons 6-2 and 9-5.

In the similar solids above, $\frac{8}{20} = \frac{2}{5} = \frac{6}{15}$. Recall that the ratio of the measures is called the *scale factor*.

If the ratio of corresponding measurements of two solids is 1:1, then the solids are congruent. For two solids to be congruent, all of the following conditions must be met.

Key Concept   *Congruent Solids*

Two solids are congruent if:

- the corresponding angles are congruent,
- the corresponding edges are congruent,
- the corresponding faces are congruent, and
- the volumes are equal.

Congruent solids are exactly the same shape and exactly the same size. They are a special case of similar solids. They have a scale factor of 1.

Example **1** *Similar and Congruent Solids*

Determine whether each pair of solids are *similar, congruent,* or *neither.*

a. Find the ratios between the corresponding parts of the regular hexagonal pyramids.

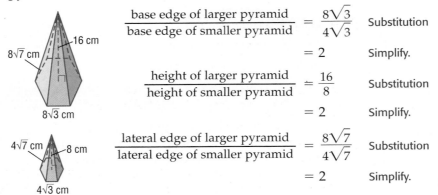

$$\frac{\text{base edge of larger pyramid}}{\text{base edge of smaller pyramid}} = \frac{8\sqrt{3}}{4\sqrt{3}} \quad \text{Substitution}$$

$$= 2 \quad \text{Simplify.}$$

$$\frac{\text{height of larger pyramid}}{\text{height of smaller pyramid}} = \frac{16}{8} \quad \text{Substitution}$$

$$= 2 \quad \text{Simplify.}$$

$$\frac{\text{lateral edge of larger pyramid}}{\text{lateral edge of smaller pyramid}} = \frac{8\sqrt{7}}{4\sqrt{7}} \quad \text{Substitution}$$

$$= 2 \quad \text{Simplify.}$$

The ratios of the measures are equal, so we can conclude that the pyramids are similar. Since the scale factor is not 1, the solids are not congruent.

b. Compare the ratios between the corresponding parts of the cones.

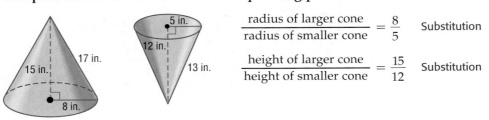

$$\frac{\text{radius of larger cone}}{\text{radius of smaller cone}} = \frac{8}{5} \quad \text{Substitution}$$

$$\frac{\text{height of larger cone}}{\text{height of smaller cone}} = \frac{15}{12} \quad \text{Substitution}$$

Since the ratios are not the same, there is no need to find the ratio of the slant heights. The cones are not similar.

PROPERTIES OF SIMILAR SOLIDS You can investigate the relationships between similar solids using spreadsheets.

Spreadsheet Investigation

Explore Similar Solids

Collect the Data

Step 1 In Column A, enter the labels *length, width, height, surface area, volume, scale factor, ratios of surface area,* and the *ratios of volume.* Columns B, C, D, and E will be used for four similar prisms.

Step 2 Enter the formula for the surface area of the prism in cell B4. Copy the formula into the other cells in row 4.

Step 3 Write a similar formula to find the volume of the prism. Copy the formula in the cells in row 5.

Step 4 Enter the formula =C1/B1 in cell C6, enter =D1/B1 in cell D6, and so on. These formulas find the scale factor of prism B and each other solid.

Step 5 Type the formula =C4/B4 in cell C7, type =D4/B4 in cell D7, and so on. This formula will find the ratio of the surface area of prism B to the surface areas of each of the other prisms.

Step 6 Write a formula for the ratio of the volume of prism C to the volume of prism B. Enter the formula in cell C8. Enter similar formulas in the cells in row 8.

Step 7 Use the spreadsheet to find the surface areas, volumes, and ratios for prisms with the dimensions given.

Similar Solids.xls

	A	B	C	D	E	F
1	length	1	2	3	4	
2	width	4	8	12	16	
3	height	6	12	18	24	
4	surface area	68	272	612	1088	
5	volume	24	192	648	1536	
6	scale factor		2	3	4	
7	ratios of surface area		4	9	16	
8	ratios of volume		8	27	64	
9						
10						

Sheet1 / Sheet2

Analyze

1. Compare the ratios in cells 6, 7, and 8 of columns C, D, and E. What do you observe?

2. Write a statement about the ratio of the surface areas of two solids if the scale factor is $a:b$.

3. Write a statement about the ratio of the volumes of two solids if the scale factor is $a:b$.

The Spreadsheet Investigation suggests the following theorem.

Theorem 13.1

If two solids are similar with a scale factor of $a:b$, then the surface areas have a ratio of $a^2:b^2$, and the volumes have a ratio of $a^3:b^3$.

Example:

Scale factor 3:2
Ratio of surface areas $3^2:2^2$ or 9:4
Ratio of volumes $3^3:2^3$ or 27:8

Example 2 *Mirror Balls*

ENTERTAINMENT Mirror balls are spheres that are covered with reflective tiles. One ball has a diameter of 4 inches, and another has a diameter of 20 inches.

a. Find the scale factor of the two spheres.

Write the ratio of the corresponding measures of the spheres.

$$\frac{\text{diameter of the smaller sphere}}{\text{diameter of the larger sphere}} = \frac{4}{20} \qquad \text{Substitution}$$

$$= \frac{1}{5} \qquad \text{Simplify.}$$

The scale factor is 1:5.

b. Find the ratio of the surface areas of the two spheres.

If the scale factor is $a:b$, then the ratio of the surface areas is $a^2:b^2$.

$$\frac{\text{surface area of the smaller sphere}}{\text{surface area of the larger sphere}} = \frac{a^2}{b^2} \quad \text{Theorem 13.1}$$

$$= \frac{1^2}{5^2} \quad a = 1 \text{ and } b = 5$$

$$= \frac{1}{25} \quad \text{Simplify.}$$

The ratio of the surface areas is $1:25$.

c. Find the ratio of the volumes of the two spheres.

If the scale factor is $a:b$, then the ratio of the volumes is $a^3:b^3$.

$$\frac{\text{volume of the smaller sphere}}{\text{volume of the larger sphere}} = \frac{a^3}{b^3} \quad \text{Theorem 13.1}$$

$$= \frac{1^3}{5^3} \quad a = 1 \text{ and } b = 5$$

$$= \frac{1}{125} \quad \text{Simplify.}$$

The ratio of the volumes of the two spheres is $1:125$.

Check for Understanding

Concept Check
1. **OPEN ENDED** Draw and label the dimensions of a pair of cones that are similar and a pair of cones that are neither similar nor congruent.

2. **Explain** the relationship between the surface areas of similar solids and volumes of similar solids.

Guided Practice **Determine whether each pair of solids are *similar*, *congruent*, or *neither*.**

3.

4.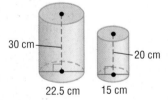

For Exercises 5–7, refer to the pyramids on the right.

5. Find the scale factor of the two pyramids.

6. Find the ratio of the surface areas of the two pyramids.

7. Find the ratio of the volumes of the two pyramids.

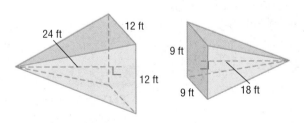

Application **LAWN ORNAMENTS** **For Exercises 8–10, use the following information.**
There are two gazing balls in a garden. One has a diameter of 2 inches, and the other has a diameter of 16 inches.

8. Find the scale factor of the two gazing balls.

9. Determine the ratio of the surface areas of the two spheres.

10. What is the ratio of the volumes of the gazing balls?

Practice and Apply

Homework Help

For Exercises	See Examples
11–16, 38, 39	1
17–37	2

Extra Practice
See page 781.

Determine whether each pair of solids are *similar, congruent,* or *neither.*

11.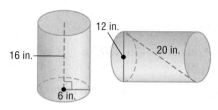

12.

13.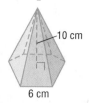

14.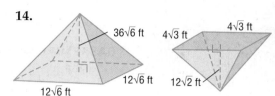

15.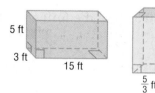

16.

17. ARCHITECTURE To encourage recycling, the people of Rome, Italy, built a model of Basilica di San Pietro from empty beverage cans. The model was built to a 1:5 scale. The model measured 26 meters high, 49 meters wide, and 93 meters long. Find the dimensions of the actual Basilica di San Pietro.

Determine whether each statement is *sometimes, always,* or *never* true. Justify your answer.

18. Two spheres are similar.

19. Congruent solids have equal surface areas.

20. Similar solids have equal volumes.

21. A pyramid is similar to a cone.

22. Cones and cylinders with the same height and base are similar.

23. Nonsimilar solids have different surface areas.

MINIATURES For Exercises 24–26, use the information at the left.

24. If the door handle of the full-sized car is 15 centimeters long, how long is the door handle on the Micro-Car?

25. If the surface area of the Micro-Car is x square centimeters, what is the surface area of the full-sized car?

26. If the scale factor was 1:18 instead of 1:1000, find the length of the miniature door handle.

For Exercises 27–30, refer to the two similar right prisms.

27. Find the ratio of the perimeters of the bases.

28. What is the ratio of the surface areas?

29. What is the ratio of the volumes?

30. Suppose the volume of the smaller prism is 48 cubic inches. Find the volume of the larger prism.

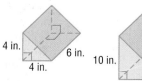 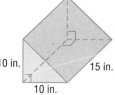

31. The diameters of two similar cones are in the ratio 5 to 6. If the volume of the smaller cone is 125π cubic centimeters and the diameter of the larger cone is 12 centimeters, what is the height of the larger cone?

32. FESTIVALS The world's largest circular pumpkin pie was made for the Circleville Pumpkin Show in Circleville, Ohio. The diameter was 5 feet. Most pies are 8 inches in diameter. If the pies are similar, what is the ratio of the volumes?

 Online Research **Data Update** How many pies do Americans purchase in a year? Visit www.geometryonline.com/data_update to learn more.

BASKETBALL For Exercises 33–35, use the information at the left. Find the indicated ratio of the smaller ball to the larger ball.

33. scale factor   **34.** ratio of surface areas **35.** ratio of the volumes

TOURISM For Exercises 36 and 37, use the following information.
Dale Ungerer, a farmer in Hawkeye, Iowa, constructed a gigantic ear of corn to attract tourists to his farm. The ear of corn is 32 feet long and has a circumference of 12 feet. Each "kernel" is a one-gallon milk jug with a volume of 231 cubic inches.

36. If a real ear of corn is 10 inches long, what is the scale factor between the gigantic ear of corn and the similar real ear of corn?

37. Estimate the volume of a kernel of the real ear of corn.

For Exercises 38 and 39, use the following information.
When a cone is cut by a plane parallel to its base, a cone similar to the original is formed.

38. What is the ratio of the volume of the frustum to that of the original cone? to the smaller cone?

39. What is the ratio of the lateral area of the frustum to that of the original cone? to the smaller cone?

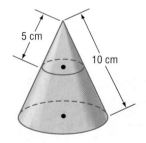

CRITICAL THINKING For Exercises 40 and 41, refer to the figure.

40. Is it possible for the two cones inside the cylinder to be congruent? Explain.

41. Is the volume of the cone on the right *equal to, greater than,* or *less than* the sum of the volume of the cones inside the cylinder? Explain.

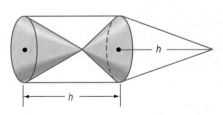

42. WRITING IN MATH Answer the question that was posed at the beginning of the lesson.

How is the geometry of similar solids applied to miniature collectibles?

Include the following in your answer:
- the scale factors that are commonly used, and
- the answer to this question: If a miniature is 4.5 inches with a scale factor of 1:24, then how long is the actual object?

More About. . .

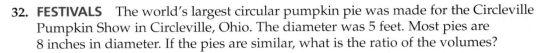

Basketball

The National Collegiate Athletic Association (NCAA) states that the maximum circumference of a basketball for men is 30 inches. The maximum circumference of a women's basketball is 29 inches.

Source: www.ncaa.org

43. If the ratio of the surface areas of two similar solids is 4:9, find the ratio of the volumes.

Ⓐ 64:729   Ⓑ 2:3   Ⓒ 8:27   Ⓓ 16:81

44. ALGEBRA If $xyz = 4$ and $y^2z = 5$, then what is the value of $\frac{x}{y}$?

Ⓐ 20.0   Ⓑ 2.0   Ⓒ 1.5   Ⓓ 0.8

Maintain Your Skills

Mixed Review   **Find the volume of each sphere. Round to the nearest tenth.** *(Lesson 13-3)*

45. diameter = 8 feet   **46.** radius = 9.5 meters

47. radius = 15.1 centimeters   **48.** diameter = 23 inches

Find the volume of each pyramid or cone. Round to the nearest tenth. *(Lesson 13-2)*

49.   **50.** 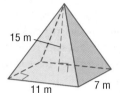   **51.**

Find the radius of the base of each cylinder. Round to the nearest tenth.
(Lesson 12-4)

52. The surface area is 430 square centimeters, and the height is 7.4 centimeters.

53. The surface area is 224.7 square yards, and the height is 10 yards.

NAVIGATION   **For Exercises 54–56, use the following information.**
As part of a scuba diving exercise, a 12-foot by 3-foot rectangular-shaped rowboat was sunk in a quarry. A boat takes a scuba diver to a random spot in the enclosed section of the quarry and anchors there so that the diver can search for the rowboat. *(Lesson 11-5)*

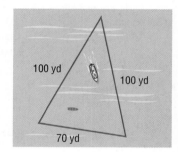

54. What is the approximate area of the enclosed section of the quarry?

55. What is the area of the rowboat?

56. What is the probability that the boat will anchor over the sunken rowboat?

Getting Ready for the Next Lesson   **PREREQUISITE SKILL**   **Determine whether the ordered pair is on the graph of the given equation. Write *yes* or *no*.** *(To review **graphs in the coordinate plane**, see Lesson 1-1.)*

57. $y = 3x + 5$, (4, 17)   **58.** $y = -4x + 1$, (-2, 9)   **59.** $y = 7x - 4$, (-1, 3)

Practice Quiz 2
Lessons 13-3 and 13-4

Find the volume of each sphere. Round to the nearest tenth. *(Lesson 13-3)*

1. radius = 25.3 ft   **2.** diameter = 36.8 cm

The two square pyramids are similar. *(Lesson 13-4)*

3. Find the scale factor of the pyramids.

4. What is the ratio of the surface areas?

5. What is the ratio of the volumes?

www.geometryonline.com/self_check_quiz

Coordinates in Space

What You'll Learn

- Graph solids in space.
- Use the Distance and Midpoint Formulas for points in space.

Vocabulary

- ordered triple

How is three-dimensional graphing used in computer animation?

The initial step in computer animation is creating a three-dimensional image. A *mesh* is created first. This is an outline that shows the size and shape of the image. Then the image is *rendered*, adding color and texture. The image is animated using software. There is a way to describe the location of each point in the image.

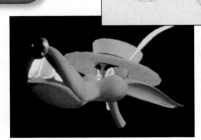

Study Tip

Reading Math
When the three planes intersect to form the three dimensional coordinate system, eight regions are formed. These regions are called *octants*.

GRAPH SOLIDS IN SPACE To describe the location of a point on the coordinate plane, we use an ordered pair of two coordinates. In space, each point requires three numbers, or coordinates, to describe its location because space has three dimensions. In space, the x-, y-, and z-axes are perpendicular to each other.

A point in space is represented by an **ordered triple** of real numbers (x, y, z). In the figure at the right, the ordered triple $(2, 3, 6)$ locates point P. Notice that a rectangular prism is used to show perspective.

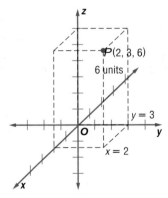

Study Tip

Drawing in Three Dimensions
Use the properties of a rectangular prism to correctly locate the z-coordinate. A is the vertex farthest from the origin.

Example 1 *Graph a Rectangular Solid*

Graph a rectangular solid that has $A(-4, 2, 4)$ and the origin as vertices. Label the coordinates of each vertex.

- Plot the x-coordinate first. Draw a segment from the origin 4 units in the negative direction.

- To plot the y-coordinate, draw a segment 2 units in the positive direction.

- Next, to plot the z-coordinate, draw a segment 4 units long in the positive direction.

- Label the coordinate A.

- Draw the rectangular prism and label each vertex.

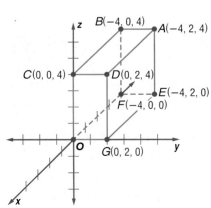

DISTANCE AND MIDPOINT FORMULA
Recall that the Distance Formula is derived from the Pythagorean Theorem. The Pythagorean Theorem can also be used to find the formula for the distance between two points in space.

Study Tip

Look Back
To review **Distance and Midpoint Formulas**, see Lesson 1-3.

Key Concept   *Distance Formula in Space*

Given two points $A(x_1, y_1, z_1)$ and $B(x_2, y_2, z_2)$ in space, the distance between A and B is given by the following equation.

$$d = \sqrt{(x_2 - x_1)^2 + (y_2 - y_1)^2 + (z_2 - z_1)^2}$$

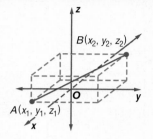

This formula is an extension of the Distance Formula in two dimensions. The Midpoint Formula can also be extended to the three-dimensions.

Key Concept   *Midpoint Formula in Space*

Given two points $A(x_1, y_1, z_1)$ and $B(x_2, y_2, z_2)$ in space, the midpoint of $\overline{AB}$ is at
$$\left(\frac{x_1 + x_2}{2}, \frac{y_1 + y_2}{2}, \frac{z_1 + z_2}{2}\right).$$

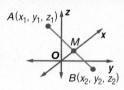

Example 2   *Distance and Midpoint Formulas in Space*

Study Tip

Look Back
To review **translations**, see Lesson 9-2.

a. Determine the distance between $T(6, 0, 0)$ and $Q(-2, 4, 2)$.

$TQ = \sqrt{(x_2 - x_1)^2 + (y_2 - y_1)^2 + (z_2 - z_1)^2}$ Distance Formula in Space

$ = \sqrt{[6 - (-2)]^2 + (0 - 4)^2 + (0 - 2)^2}$ Substitution

$ = \sqrt{84}$ or $2\sqrt{21}$ Simplify.

b. Determine the coordinates of the midpoint M of $\overline{TQ}$.

$M = \left(\frac{x_1 + x_2}{2}, \frac{y_1 + y_2}{2}, \frac{z_1 + z_2}{2}\right)$ Midpoint Formula in Space

$ = \left(\frac{6 - 2}{2}, \frac{0 + 4}{2}, \frac{0 + 2}{2}\right)$ Substitution

$ = (2, 2, 1)$ Simplify.

Example 3   *Translating a Solid*

ELEVATORS   Suppose an elevator is 5 feet wide, 6 feet deep, and 8 feet tall. Position the elevator on the ground floor at the origin of a three dimensional space. If the distance between the floors of a warehouse is 10 feet, write the coordinates of the vertices of the elevator after going up to the third floor.

Explore   Since the elevator is a rectangular prism, use positive values for $x, y,$ and z. Write the coordinates of each corner. The points on the elevator will rise 10 feet for each floor. When the elevator ascends to the third floor, it will have traveled 20 feet.

(continued on the next page)

Plan Use the translation $(x, y, z) \rightarrow (x, y, z + 20)$ to find the coordinates of each vertex of the rectangular prism that represents the elevator.

Solve

Coordinates of the vertices, (x, y, z) Preimage	Translated coordinates, $(x, y, z + 20)$ Image
$J(0, 5, 8)$	$J'(0, 5, 28)$
$K(6, 5, 8)$	$K'(6, 5, 28)$
$L(6, 0, 8)$	$L'(6, 0, 28)$
$M(0, 0, 8)$	$M'(0, 0, 28)$
$N(6, 0, 0)$	$N'(6, 0, 20)$
$O(0, 0, 0)$	$O'(0, 0, 20)$
$P(0, 5, 0)$	$P'(0, 5, 20)$
$Q(6, 5, 0)$	$Q'(6, 5, 20)$

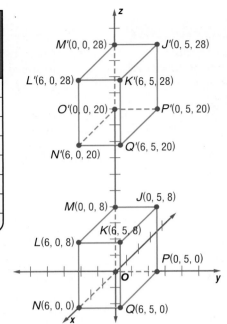

Examine Check that the distance between corresponding vertices is 20 feet.

Matrices can be used for transformations in space such as dilations.

Study Tip

Look Back
To review **matrices and transformations**, see Lesson 9-7.

Example 4 *Dilation with Matrices*

Dilate the prism by a scale factor of 2. Graph the image under the dilation.

First, write a vertex matrix for the rectangular prism.

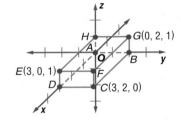

$$\begin{array}{c} \\ x \\ y \\ z \end{array} \begin{array}{cccccccc} A & B & C & D & E & F & G & H \\ \left[\begin{array}{cccccccc} 0 & 0 & 3 & 3 & 3 & 3 & 0 & 0 \\ 0 & 2 & 2 & 0 & 0 & 2 & 2 & 0 \\ 0 & 0 & 0 & 0 & 1 & 1 & 1 & 1 \end{array}\right] \end{array}$$

Next, multiply each element of the vertex matrix by the scale factor, 2.

$$2\begin{array}{cccccccc} A & B & C & D & E & F & G & H \\ \left[\begin{array}{cccccccc} 0 & 0 & 3 & 3 & 3 & 3 & 0 & 0 \\ 0 & 2 & 2 & 0 & 0 & 2 & 2 & 0 \\ 0 & 0 & 0 & 0 & 1 & 1 & 1 & 1 \end{array}\right] = \begin{array}{cccccccc} A' & B' & C' & D' & E' & F' & G' & H' \\ \left[\begin{array}{cccccccc} 0 & 0 & 6 & 6 & 6 & 6 & 0 & 0 \\ 0 & 4 & 4 & 0 & 0 & 4 & 4 & 0 \\ 0 & 0 & 0 & 0 & 2 & 2 & 2 & 2 \end{array}\right] \end{array}$$

The coordinates of the vertices of the dilated image are $A'(0, 0, 0)$, $B'(0, 4, 0)$, $C'(6, 4, 0)$, $D'(6, 0, 0)$, $E'(6, 0, 2)$, $F'(6, 4, 2)$, $G'(0, 4, 2)$, and $H'(0, 0, 2)$.

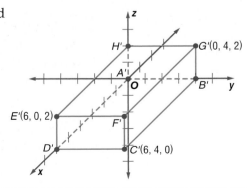

Check for Understanding

1. **Compare and contrast** the number of regions on the coordinate plane and in three-dimensional coordinate space.

2. **OPEN ENDED** Draw and label the vertices of a rectangular prism that has a volume of 24 cubic units.

3. **Find a counterexample** for the following statement.
   Every rectangular prism will be congruent to its image from any type of transformation.

Guided Practice

Graph a rectangular solid that contains the given point and the origin as vertices. Label the coordinates of each vertex.

4. $A(2, 1, 5)$

5. $P(-1, 4, 2)$

Determine the distance between each pair of points. Then determine the coordinates of the midpoint M of the segment joining the pair of points.

6. $D(0, 0, 0)$ and $E(1, 5, 7)$

7. $G(-3, -4, 6)$ and $H(5, -3, -5)$

8. The vertices of a rectangular prism are $M(0, 0, 0)$, $N(-3, 0, 0)$, $P(-3, 4, 0)$, $Q(0, 4, 0)$, $R(0, 0, 2)$, $S(0, 4, 2)$, $T(-3, 4, 2)$, and $V(-3, 0, 2)$. Dilate the prism by a scale factor of 2. Graph the image under the dilation.

Application

9. **STORAGE** A storage container is 12 feet deep, 8 feet wide, and 8 feet high. To allow the storage company to locate and identify the container, they assign ordered triples to the corners using positive x, y, and z values. If the container is stored 16 feet up and 48 feet back in the warehouse, find the ordered triples of the vertices describing the new location. Use the translation $(x, y, z) \rightarrow (x - 48, y, z + 16)$.

Practice and Apply

For Exercises	See Examples
10–15	1
16–21, 32–34	2
22, 25–28, 31, 35	3
23–24, 29–30	4

Extra Practice
See page 781.

Graph a rectangular solid that contains the given point and the origin as vertices. Label the coordinates of each vertex.

10. $C(-2, 2, 2)$

11. $R(3, -4, 1)$

12. $P(4, 6, -3)$

13. $G(4, 1, -3)$

14. $K(-2, -4, -4)$

15. $W(-1, -3, -6)$

Determine the distance between each pair of points. Then determine the coordinates of the midpoint M of the segment joining the pair of points.

16. $K(2, 2, 0)$ and $L(-2, -2, 0)$

17. $P(-2, -5, 8)$ and $Q(3, -2, -1)$

18. $F\left(\frac{3}{5}, 0, \frac{4}{5}\right)$ and $G(0, 3, 0)$

19. $G(1, -1, 6)$ and $H\left(\frac{1}{5}, -\frac{2}{5}, 2\right)$

20. $S\left(6\sqrt{3}, 4, 4\sqrt{2}\right)$ and $T\left(4\sqrt{3}, 5, \sqrt{2}\right)$

21. $B\left(\sqrt{3}, 2, 2\sqrt{2}\right)$ and $C\left(-2\sqrt{3}, 4, 4\sqrt{2}\right)$

22. **AVIATION** An airplane at an elevation of 2 miles is 50 miles east and 100 miles north of an airport. This location can be written as (50, 100, 2). A second airplane is at an elevation of 2.5 miles and is located 240 miles west and 140 miles north of the airport. This location can be written as (−240, 140, 2.5). Find the distance between the airplanes to the nearest tenth of a mile.

Lesson 13-5 Coordinates in Space 717

Dilate each prism by the given scale factor. Graph the image under the dilation.

23. scale factor of 3

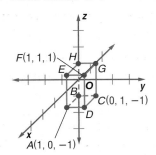

F(1, 1, 1) H
E
G
O
B
C(0, 1, −1)
D
A(1, 0, −1)

24. scale factor of 2

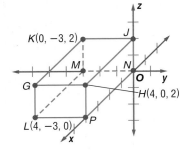

K(0, −3, 2) J
M N
O y
G
H(4, 0, 2)
L(4, −3, 0) P

Consider a rectangular prism with the given coordinates. Find the coordinates of the vertices of the prism after the translation.

25. $P(−2, −3, 3), Q(−2, 0, 3), R(0, 0, 3), S(0, −3, 3) T(−2, 0, 0), U(−2, −3, 0),$ $V(0, −3, 0),$ and $W(0, 0, 0); (x, y, z) \rightarrow (x + 2, y + 5, z − 5)$

26. $A(2, 0, 1), B(2, 0, 0), C(2, 1, 0), D(2, 1, 1), E(0, 0, 1), F(0, 1, 1), G(0, 1, 0),$ and $H(0, 0, 0); (x, y, z) \rightarrow (x − 2, y + 1, z − 1).$

Consider a cube with coordinates $A(3, 3, 3), B(3, 0, 3), C(0, 0, 3), D(0, 3, 3), E(3, 3, 0),$ $F(3, 0, 0), G(0, 0, 0),$ and $H(0, 3, 0).$ Find the coordinates of the image under each transformation. Graph the preimage and the image.

27. Use the translation $(x, y, z) \rightarrow (x + 1, y + 2, z − 2).$

28. Use the translation $(x, y, z) \rightarrow (x − 2, y − 3, z + 2).$

29. Dilate the cube by a factor of 2. What is the volume of the image?

30. Dilate the cube by a factor of $\frac{1}{3}$. What is the ratio of the volumes for these two cubes?

31. RECREATION Two hot-air balloons take off from the same site. One hot-air balloon is 12 miles west and 12 miles south of the takeoff point and 0.4 mile above the ground. The other balloon is 4 miles west and 10 miles south of the takeoff site and 0.3 mile above the ground. Find the distance between the two balloons to the nearest tenth of a mile.

32. If $M(5, 1, 2)$ is the midpoint of segment $\overline{AB}$ and point A has coordinates $(2, 4, 7),$ then what are the coordinates of point B?

33. The center of a sphere is at $(4, −2, 6),$ and the endpoint of a diameter is at $(8, 10, −2).$ What are the coordinates of the other endpoint of the diameter?

34. Find the center and the radius of a sphere if the diameter has endpoints at $(−12, 10, 12)$ and $(14, −8, 2).$

35. GAMES The object of a video game is to move a rectangular prism around to fit with other solids. The prism has moved to combine with the red L-shaped solid. Write the translation that moved the prism to the new location.

36. CRITICAL THINKING A sphere with its center at $(2, 4, 6)$ and a radius of 4 units is inscribed in a cube. Graph the cube and determine the coordinates of the vertices.

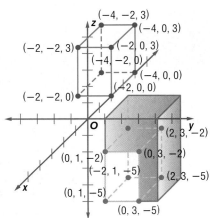

(−4, −2, 3)
(−4, 0, 3)
(−2, −2, 3)
(−2, 0, 3)
(−4, −2, 0)
(−4, 0, 0)
(−2, 0, 0)
(−2, −2, 0)
(2, 3, −2)
(0, 1, −2)
(0, 3, −2)
(−2, 1, −5)
(2, 3, −5)
(0, 1, −5)
(0, 3, −5)

37. WRITING IN MATH Answer the question that was posed at the beginning of the lesson.

How is three-dimensional graphing used in computer animation?

Include the following in your answer:
- the purpose of using an ordered triple, and
- why three-dimensional graphing is used instead of two-dimensional graphing.

Standardized Test Practice
Ⓐ Ⓑ Ⓒ Ⓓ

38. The center of a sphere is at $(4, -5, 3)$, and the endpoint of a diameter is at $(5, -4, -2)$. What are the coordinates of the other endpoint of the diameter?

Ⓐ $(-1, -1, 5)$　　Ⓑ $\left(-\frac{1}{2}, -\frac{1}{2}, \frac{5}{2}\right)$　　Ⓒ $(3, -6, 8)$　　Ⓓ $(13, -14, 4)$

39. ALGEBRA Solve $\sqrt{x+1} = x - 1$.

Ⓐ 0 and 3　　Ⓑ 3　　Ⓒ -2 and 1　　Ⓓ -3 and 0

Extending the Lesson

LOCUS The locus of points in space with coordinates that satisfy the equation $y = 2x - 6$ is a plane perpendicular to the xy-plane whose intersection with the xy-plane is the graph of $y = 2x - 6$ in the xy-plane.

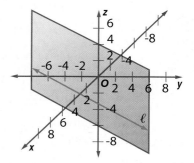

40. Describe the locus of points in space that satisfy the equation $x + y = -5$.

41. Describe the locus of points in space that satisfy the equation $x + z = 4$.

Maintain Your Skills

Mixed Review **Determine whether each pair of solids are** *similar, congruent,* **or** *neither.*
(Lesson 13-4)

42.
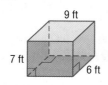

43.
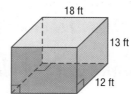

Find the volume of a sphere having the given radius or diameter. Round to the nearest tenth. *(Lesson 13-3)*

44. $r = 10$ cm　　**45.** $d = 13$ yd　　**46.** $r = 17.2$ m　　**47.** $d = 29$ ft

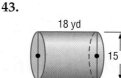 **Internet Project**

Town With Major D-Day Losses Gets Memorial

It's time to complete your project. Use the information and data you have gathered about designing your memorial. Add some additional data or pictures to your portfolio or Web page. Be sure to include your scale drawings and calculations in the presentation.

www.geometryonline.com/webquest

Vocabulary and Concept Check

congruent solids (p. 707) ordered triple (p. 714) similar solids (p. 707) volume (p. 688)

A complete list of postulates and theorems can be found on pages R1–R8.

Exercises Complete each sentence with the correct italicized term.

1. You can use $V = \frac{1}{3}Bh$ to find the volume of a (*prism, pyramid*).

2. (*Similar, Congruent*) solids always have the same volume.

3. Every point in space can be represented by (*an ordered triple, an ordered pair*).

4. $V = \pi r^2 h$ is the formula for the volume of a (*sphere, cylinder*).

5. In (*similar, congruent*) solids, if $a \neq b$ and $a : b$ is the ratio of the lengths of corresponding edges, then $a^3 : b^3$ is the ratio of the volumes.

6. The formula $V = Bh$ is used to find the volume of a (*prism, pyramid*).

7. To find the length of an edge of a pyramid, you can use (*the Distance Formula in Space, Cavalieri's Principle*).

8. You can use $V = \frac{4}{3}\pi r^3$ to find the volume of a (*cylinder, sphere*).

9. To find the volume of an oblique pyramid, you can use (*Cavalieri's Principle, the Distance Formula in Space*).

10. The formula $V = \frac{1}{3}Bh$ is used to find the volume of a (*cylinder, cone*).

Lesson-by-Lesson Review

13-1 Volumes of Prisms and Cylinders

See pages 688–694.

Concept Summary

• The volumes of prisms and cylinders are given by the formula $V = Bh$.

Example Find the volume of the cylinder.

$V = \pi r^2 h$ Volume of a cylinder

$= \pi(12^2)(5)$ $r = 12$ and $h = 5$

≈ 2261.9 Use a calculator.

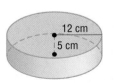

The volume is approximately 2261.9 cubic centimeters.

Exercises Find the volume of each prism or cylinder. Round to the nearest tenth if necessary. *See Examples 1 and 3 on pages 689 and 690.*

11.
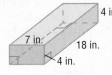

12.
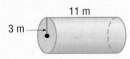

13.

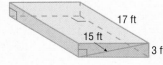
www.geometryonline.com/vocabulary_review

13-2 Volumes of Pyramids and Cones

See pages
696–701.

Concept Summary

- The volume of a pyramid is given by the formula $V = \frac{1}{3}Bh$.

- The volume of a cone is given by the formula $V = \frac{1}{3}\pi r^2 h$.

Example **Find the volume of the square pyramid.**

$$V = \frac{1}{3}Bh \qquad \text{Volume of a pyramid}$$

$$= \frac{1}{3}(21^2)(19) \quad B = 21^2 \text{ and } h = 19$$

$$= 2793 \qquad \text{Simplify.}$$

The volume of the pyramid is 2793 cubic inches.

Exercises   **Find the volume of each pyramid or cone. Round to the nearest tenth.**
See Examples 1 and 2 on pages 697 and 698.

14.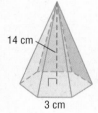

15.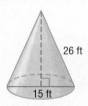

16.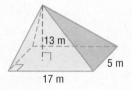

13-3 Volume of Spheres

See pages
702–706.

Concept Summary

- The volume of a sphere is given by the formula $V = \frac{4}{3}\pi r^3$.

Example **Find the volume of the sphere.**

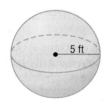

$$V = \frac{4}{3}\pi r^3 \qquad \text{Volume of a sphere}$$

$$= \frac{4}{3}\pi(5^3) \quad r = 5$$

$$\approx 523.6 \qquad \text{Use a calculator.}$$

The volume of the sphere is about 523.6 cubic feet.

Exercises   **Find the volume of each sphere. Round to the nearest tenth.**
See Example 1 on page 703.

17. The radius of the sphere is 2 feet.

18. The diameter of the sphere is 4 feet.

19. The circumference of the sphere is 65 millimeters.

20. The surface area of the sphere is 126 square centimeters.

21. The area of a great circle of the sphere is 25π square units.

Chapter
13 **For More ...**
• Extra Practice, see pages 780 and 781.
• Mixed Problem Solving, see page 794.

13-4 Congruent and Similar Solids

See pages 707–713.

Concept Summary

• Similar solids have the same shape, but not necessarily the same size.
• Congruent solids are similar solids with a scale factor of 1.

Example Determine whether the two cylinders are *congruent, similar,* or *neither.*

$$\frac{\text{diameter of larger cylinder}}{\text{diameter of smaller cylinder}} = \frac{6}{3} \quad \text{Substitution}$$

$$= 2 \quad \text{Simplify.}$$

$$\frac{\text{height of larger cylinder}}{\text{height of smaller cylinder}} = \frac{15}{5} \quad \text{Substitution}$$

$$= 3 \quad \text{Simplify.}$$

The ratios of the measures are not equal, so the cylinders are not similar.

Exercises Determine whether the two solids are *congruent, similar,* or *neither.*
See Example 1 on page 708.

22. $T = 232$ cm² $T = 232$ cm²

23.

13-5 Coordinates in Space

See pages 714–719.

Concept Summary

• The Distance Formula in Space is $d = \sqrt{(x_2 - x_1)^2 + (y_2 - y_1)^2 + (z_2 - z_1)^2}$.
• Given $A(x_1, y_1, z_1)$ and $B(x_2, y_2, z_2)$, the midpoint of $\overline{AB}$ is at $\left(\frac{x_1 + x_2}{2}, \frac{y_1 + y_2}{2}, \frac{z_1 + z_2}{2}\right)$.

Example Consider $\triangle ABC$ with vertices $A(13, 7, 10)$, $B(17, 18, 6)$, and $C(15, 10, 10)$. Find the length of the median from A to $\overline{BC}$ of ABC.

$$M = \left(\frac{17 + 15}{2}, \frac{18 + 10}{2}, \frac{6 + 10}{2}\right) \quad \text{Formula for the midpoint of } \overline{BC}$$

$$= (16, 14, 8) \quad \text{Simplify.}$$

$\overline{AM}$ is the desired median, so AM is the length of the median.

$$AM = \sqrt{(16 - 13)^2 + (14 - 7)^2 + (8 - 10)^2} \text{ or } \sqrt{62} \quad \text{Distance Formula in Space}$$

Exercises Determine the distance between each pair of points. Then determine the coordinates of the midpoint M of the segment joining the pair of points.
See Example 2 on page 715.

24. $A(-5, -8, -2)$ and $B(3, -8, 4)$
25. $C(-9, 2, 4)$ and $D(-9, 9, 7)$
26. $E(-4, 5, 5)$ and the origin
27. $F(5\sqrt{2}, 3\sqrt{7}, 6)$ and $G(-2\sqrt{2}, 3\sqrt{7}, -12)$

Vocabulary and Concepts

Write the letter of the formula used to find the volume of each of the following figures.

1. right cylinder
2. right pyramid
3. sphere

a. $V = \frac{4}{3}\pi r^3$
b. $V = \pi r^2 h$
c. $V = \frac{1}{3}Bh$

Skills and Applications

Find the volume of each solid. Round to the nearest tenth if necessary.

4.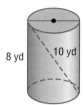
8 yd   10 yd

5.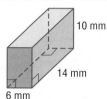
10 mm
14 mm
6 mm

6.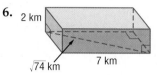
2 km
$\sqrt{74}$ km
7 km

7.
3 ft
5 ft
5 ft

8.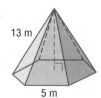
13 m
5 m

9.
8.2 cm
6.8 cm

10. $C = 22\pi$

9 in.

11. **SPORTS** The diving pool at the Georgia Tech Aquatic Center was used for the springboard and platform diving competitions of the 1996 Olympic Games. The pool is 78 feet long and 17 feet deep, and it is 110.3 feet from one corner on the surface of the pool to the opposite corner on the surface. If it takes about 7.5 gallons of water to fill one cubic foot of space, approximately how many gallons of water are needed to fill the diving pool?

Find the volume of each sphere. Round to the nearest tenth.

12. The radius has a length of 3 cm.
13. The circumference of the sphere is 34 ft.
14. The surface area of the sphere is 184 in².
15. The area of a great circle is 157 mm².

The two cylinders at the right are similar.

16. Find the ratio of the radii of the bases of the cylinders.
17. What is the ratio of the surface areas?
18. What is the ratio of the volumes?

15   10

Determine the distance between each pair of points in space. Then determine the coordinates of the midpoint M of the segment joining the pair of points.

19. the origin and $(0, -3, 5)$
20. the origin and $(-1, 10, -5)$
21. the origin and $(9, 5, -7)$
22. $(-2, 2, 2)$ and $(-3, -5, -4)$
23. $(9, 3, 4)$ and $(-9, -7, 6)$
24. $(8, -6, 1)$ and $(-3, 5, 10)$

25. **STANDARDIZED TEST PRACTICE** A rectangular prism has a volume of 360 cubic feet. If the prism has a length of 15 feet and a height of 2 feet, what is the width?

(A) 30 ft   (B) 24 ft   (C) 12 ft   (D) 7.5 ft

Part 1 | Multiple Choice

Record your answers on the answer sheet provided by your teacher or on a sheet of paper.

1. *ABCD* is a rectangle. What is the relationship between $\angle ACD$ and $\angle ACB$? **(Lesson 1-6)**

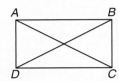

  (A) They are complementary angles.

  (B) They are perpendicular angles.

  (C) They are supplementary angles.

  (D) They are corresponding angles.

2. What is the measure of $\angle DEF$? **(Lesson 4-2)**

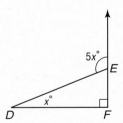

  (A) 22.5 (B) 67.5 (C) 112.5 (D) 157.5

3. Two sides of a triangle measure 13 and 21 units. Which could be the measure of the third side? **(Lesson 5-4)**

  (A) 5 (B) 8 (C) 21 (D) 34

4. $\triangle QRS$ is similar to $\triangle TUV$. Which statement is true? **(Lesson 6-3)**

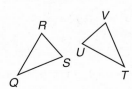

  (A) $m\angle Q = m\angle V$ (B) $m\angle Q = m\angle S$

  (C) $m\angle Q = m\angle T$ (D) $m\angle Q = m\angle U$

5. The wooden block shown must be able to slide onto a cylindrical rod. What is the volume of the block after the hole is drilled? Round to the nearest tenth. **(Lesson 13-1)**

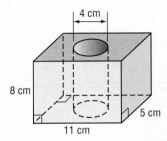

  (A) 100.5 cm³ (B) 339.5 cm³

  (C) 402.0 cm³ (D) 440.0 cm³

6. The circumference of a regulation soccer ball is 25 inches. What is the volume of the soccer ball to the nearest cubic inch? **(Lesson 13-3)**

  (A) 94 in³ (B) 264 in³

  (C) 333 in³ (D) 8177 in³

7. If the two cylinders are similar, what is the volume of the larger cylinder to the nearest tenth of a cubic centimeter? **(Lesson 13-4)**

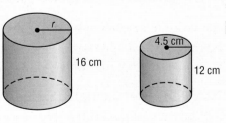

  (A) 730.0 cm³ (B) 1017.4 cm³

  (C) 1809.6 cm³ (D) 2122.6 cm³

8. The center of a sphere has coordinates (3, 1, 4). A point on the surface of the sphere has coordinates (9, −2, −2). What is the measure of the radius of the sphere? **(Lesson 13-5)**

  (A) 7 (B) $\sqrt{61}$

  (C) $\sqrt{73}$ (D) 9

Preparing for Standardized Tests
For test-taking strategies and more
practice, see pages 795–810.

Part 2 | Short Response/Grid In

**Record your answers on the answer sheet
provided by your teacher or on a sheet of
paper.**

9. Find $\dfrac{12z^5 + 27z^2 - 6z}{3z}$. (Prerequisite Skill)

10. Sierra said, "If math is my favorite subject,
then I like math." Carlos then said, "If I do
not like math, then it is not my favorite
subject." Carlos formed the __?__ of Sierra's
statement. (Lesson 2-3)

11. Describe the information needed about two
triangles to prove that they are congruent by
the SSS Postulate. (Lesson 4-1)

12. The figure is a regular octagon. Find x.
(Lesson 8-1)

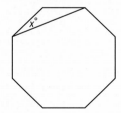

13. *ABCD* is an isosceles trapezoid. What are
the coordinates of *A*? (Lesson 8-7)

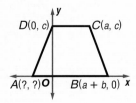

14. What is the volume of the cone?
(Lesson 13-2)

Test-Taking Tip

Question 6
Sometimes more than one step is required to find the
answer. In this question, you need to use the circumference
formula, $C = 2\pi r$, to find the length of the radius. Then
you can use the surface area formula, $T = 4\pi r^2$.

Part 3 | Extended Response

**Record your answers on a sheet of paper.
Show your work.**

15. A manufacturing company packages their
product in the small cylindrical can shown
in the diagram. During a promotion for the
product, they doubled the height of the cans
and sold them for the same price.

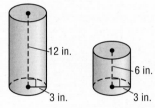

a. Find the surface area of each can. Explain
the effect of doubling the height on the
amount of material used to produce
the can. (Lesson 12-4)

b. Find the volume of each can. Explain the
effect doubling the height on the amount
of product that can fit inside. (Lesson 13-1)

16. Engineering students designed an enlarged
external fuel tank for a space shuttle as part
of an assignment.

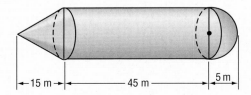

What is the volume of the entire fuel tank to
the nearest cubic meter? Show your work.
(Lesson 13-1)

Student Handbook

Prerequisite Skills

① Graphing Ordered Pairs

- Points in the coordinate plane are named by **ordered pairs** of the form (x, y). The first number, or **x-coordinate**, corresponds to a number on the x-axis. The second number, or **y-coordinate**, corresponds to a number on the y-axis.

Example 1 Write the ordered pair for each point.

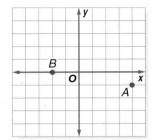

a. **A**

The x-coordinate is 4.
The y-coordinate is -1.
The ordered pair is $(4, -1)$.

b. **B**

The x-coordinate is -2.
The point lies on the x-axis,
so its y-coordinate is 0.
The ordered pair is $(-2, 0)$.

- The x-axis and y-axis separate the coordinate plane into four regions, called **quadrants**. The point at which the axes intersect is called the **origin**. The axes and points on the axes are not located in any of the quadrants.

Example 2 Graph and label each point on a coordinate plane. Name the quadrant in which each point is located.

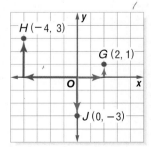

a. **G(2, 1)**

Start at the origin. Move 2 units right, since the x-coordinate is 2. Then move 1 unit up, since the y-coordinate is 1. Draw a dot, and label it G. Point $G(2, 1)$ is in Quadrant I.

b. **H(−4, 3)**

Start at the origin. Move 4 units left, since the x-coordinate is -4. Then move 3 units up, since the y-coordinate is 3. Draw a dot, and label it H. Point $H(-4, 3)$ is in Quadrant II.

c. **J(0, −3)**

Start at the origin. Since the x-coordinate is 0, the point lies on the y-axis. Move 3 units down, since the y-coordinate is -3. Draw a dot, and label it J. Because it is on one of the axes, point $J(0, -3)$ is not in any quadrant.

Example 3 Graph a polygon with vertices $A(-3, 3)$, $B(1, 3)$, $C(0, 1)$, and $D(-4, 1)$.

Graph the ordered pairs on a coordinate plane.
Connect each pair of consecutive points.
The polygon is a parallelogram.

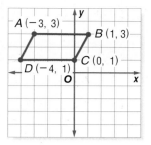

Example 4 Graph four points that satisfy the equation $y = 4 - x$.

Make a table.
Choose four values for x.
Evaluate each value of x for $4 - x$.

Plot the points.

x	$4 - x$	y	(x, y)
0	$4 - 0$	4	$(0, 4)$
1	$4 - 1$	3	$(1, 3)$
2	$4 - 2$	2	$(2, 2)$
3	$4 - 3$	1	$(3, 1)$

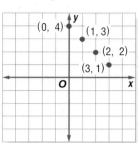

Exercises Write the ordered pair for each point shown at the right.

1. B
2. C
3. D
4. E
5. F
6. G
7. H
8. I
9. J
10. K
11. W
12. M
13. N
14. P
15. Q

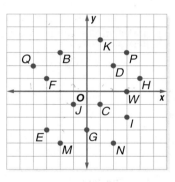

Graph and label each point on a coordinate plane. Name the quadrant in which each point is located.

16. $M(-1, 3)$
17. $S(2, 0)$
18. $R(-3, -2)$
19. $P(1, -4)$
20. $B(5, -1)$
21. $D(3, 4)$
22. $T(2, 5)$
23. $L(-4, -3)$
24. $A(-2, 2)$
25. $N(4, 1)$
26. $H(-3, -1)$
27. $F(0, -2)$
28. $C(-3, 1)$
29. $E(1, 3)$
30. $G(3, 2)$
31. $I(3, -2)$

Graph the following geometric figures.

32. a square with vertices $W(-3, 3)$, $X(-3, -1)$, $Y(1, 3)$, and $Z(1, -1)$

33. a polygon with vertices $J(4, 2)$, $K(1, -1)$, $L(-2, 2)$, and $M(1, 5)$

34. a triangle with vertices $F(2, 4)$, $G(-3, 2)$, and $H(-1, -3)$

35. a rectangle with vertices $P(-2, -1)$, $Q(4, -1)$, $R(-2, 1)$, and $S(4, 1)$

Graph four points that satisfy each equation.

36. $y = 2x$
37. $y = 1 + x$
38. $y = 3x - 1$
39. $y = 2 - x$

2 Changing Units of Measure within Systems

Metric Units of Length
1 kilometer (km) = 1000 meters (m)
1 m = 100 centimeters (cm)
1 cm = 10 millimeters (mm)

Customary Units of Length
1 foot (ft) = 12 inches (in.)
1 yard (yd) = 3 ft
1 mile (mi) = 5280 ft

- To convert from larger units to smaller units, multiply.
- To convert from smaller units to larger units, divide.

Example 1 **State which metric unit you would use to measure the length of your pen.**

Since a pen has a small length, the *centimeter* is the appropriate unit of measure.

Example 2 **Complete each sentence.**

a. 4.2 km = ? m

There are 1000 meters in a kilometer.
4.2 km × 1000 = 4200 m

b. 125 mm = ? cm

There are 10 millimeters in a centimeter.
125 mm ÷ 10 = 12.5 cm

c. 16 ft = ? in.

There are 12 inches in a foot.
16 ft × 12 = 192 in.

d. 39 ft = ? yd

There are 3 feet in a yard.
39 ft ÷ 3 = 13 yd

Example 3 **Complete each sentence.**

a. 17 mm = ? m

There are 100 centimeters in a meter. First change *millimeters* to *centimeters*.

17 mm = ? cm	smaller unit → larger unit
17 mm ÷ 10 = 1.7 cm	Since 10 mm = 1 cm, divide by 10.

Then change *centimeters* to *meters*.

1.7 cm = ? m	smaller unit → larger unit
1.7 cm ÷ 100 = 0.017 m	Since 100 cm = 1 m, divide by 100.

b. 6600 yd = ? mi

There are 5280 feet in one mile. First change *yards* to *feet*.

6600 yd = ? ft	larger unit → smaller unit
6600 yd × 3 = 19,800 ft	Since 3 ft = 1 yd, multiply by 3.

Then change *feet* to *miles*.

19,800 ft = ? mi	smaller unit → larger unit
19,800 ft ÷ 5280 = $3\frac{3}{4}$ or 3.75 mi	Since 5280 ft = 1 mi, divide by 5280.

Metric Units of Capacity
1 liter (L) = 1000 millimeters (mL)

Customary Units of Capacity			
1 cup (c) = 8 fluid ounces (fl oz)		1 quart (qt) = 2 pt	
1 pint (pt) = 2 c		1 gallon (gal) = 4 qt	

Example 4 **Complete each sentence.**

a. 3.7 L = ? mL

There are 1000 milliliters in a liter.
3.7 L × 1000 = 3700 mL

b. 16 qt = ? gal

There are 4 quarts in a gallon.
16 qt ÷ 4 = 4 gal

- Examples c and d involve two-step conversions.

   c. 7 pt = __?__ fl oz

   There are 8 fluid ounces in a cup.
   First change *pints* to *cups*.

   7 pt = __?__ c

   7 pt × 2 = 14 c

   Then change *cups* to *fluid ounces*.

   14 c = __?__ fl oz

   14 c × 8 = 112 fl oz

   d. 4 gal = __?__ pt

   There are 4 quarts in a gallon.
   First change *gallons* to *quarts*.

   4 gal = __?__ qt

   4 gal × 4 = 16 qt

   Then change *quarts* to *pints*.

   16 qt = __?__ pt

   16 qt × 2 = 32 pt

- The mass of an object is the amount of matter that it contains.

Metric Units of Mass
1 kilogram (kg) = 1000 grams (g)
1 g = 1000 milligrams (mg)

Customary Units of Weight
1 pound (lb) = 16 ounces (oz)
1 ton (T) = 2000 lb

Example 5   Complete each sentence.

   a. 2300 mg = __?__ g

   There are 1000 milligrams in a gram.

   2300 mg ÷ 1000 = 2.3 g

   b. 120 oz = __?__ lb

   There are 16 ounces in a pound.

   120 oz ÷ 16 = 7.5 lb

- Examples c and d involve two-step conversions.

   c. 5.47 kg = __?__ mg

   There are 1000 milligrams in a gram.
   Change *kilograms* to *grams*.

   5.47 kg = __?__ g

   5.47 kg × 1000 = 5470 g

   Then change *grams* to *milligrams*.

   5470 g = __?__ mg

   5470 g × 1000 = 5,470,000 mg

   d. 5 T = __?__ oz

   There are 16 ounces in a pound.
   Change *tons* to *pounds*.

   5 T = __?__ lb

   5 T × 2000 = 10,000 lb

   Then change *pounds* to *ounces*.

   10,000 lb = __?__ oz

   10,000 lb × 16 = 160,000 oz

Exercises   State which metric unit you would probably use to measure each item.

1. radius of a tennis ball
2. length of a notebook
3. mass of a textbook
4. mass of a beach ball
5. width of a football field
6. thickness of a penny
7. amount of liquid in a cup
8. amount of water in a bath tub

Complete each sentence.

9. 120 in. = __?__ ft
10. 18 ft = __?__ yd
11. 10 km = __?__ m
12. 210 mm = __?__ cm
13. 180 mm = __?__ m
14. 3100 m = __?__ km
15. 90 in. = __?__ yd
16. 5280 yd = __?__ mi
17. 8 yd = __?__ ft
18. 0.62 km = __?__ m
19. 370 mL = __?__ L
20. 12 L = __?__ mL
21. 32 fl oz = __?__ c
22. 5 qt = __?__ c
23. 10 pt = __?__ qt
24. 48 c = __?__ gal
25. 4 gal = __?__ qt
26. 36 mg = __?__ g
27. 13 lb = __?__ oz
28. 130 g = __?__ kg
29. 9.05 kg = __?__ g

3 Perimeter and Area of Rectangles and Squares

Perimeter is the distance around a figure whose sides are segments. Perimeter is measured in linear units.

Perimeter of a Rectangle		Perimeter of a Square	
Words	Multiply two times the sum of the length and width.	**Words**	Multiply 4 times the length of a side.
Formula	$P = 2(\ell + w)$	**Formula**	$P = 4s$

Area is the number of square units needed to cover a surface. Area is measured in square units.

Area of a Rectangle		Area of a Square	
Words	Multiply the length and width.	**Words**	Square the length of a side.
Formula	$A = \ell w$	**Formula**	$A = s^2$

Example 1 Find the perimeter and area of each rectangle.

a.

$$P = 2(\ell + w) \quad \text{Perimeter formula}$$
$$= 2(4 + 9) \quad \text{Replace } \ell \text{ with 4 and } w \text{ with 9.}$$
$$= 26 \quad \text{Simplify.}$$

$$A = \ell w \quad \text{Area formula}$$
$$= 4 \cdot 9 \quad \text{Replace } \ell \text{ with 4 and } w \text{ with 9.}$$
$$= 36 \quad \text{Multiply.}$$

The perimeter is 26 units, and the area is 36 square units.

b. a rectangle with length 8 units and width 3 units.

$P = 2(\ell + w)$ Perimeter formula

$\quad = 2(8 + 3)$ Replace ℓ with 8 and w with 3.

$\quad = 22$ Simplify.

$A = \ell \cdot w$ Area formula

$\quad = 8 \cdot 3$ Replace ℓ with 8 and w with 3.

$\quad = 24$ Multiply

The perimeter is 22 units, and the area is 24 square units.

Example 2 **Find the perimeter and area of a square that has a side of length 14 feet.**

$P = 4s$ Perimeter formula

$\quad = 4(14)$ $s = 14$

$\quad = 56$ Multiply.

$A = s^2$ Area formula

$\quad = 14^2$ $s = 14$

$\quad = 196$ Multiply.

The perimeter is 56 feet, and the area is 196 square feet.

Exercises **Find the perimeter and area of each figure.**

1.
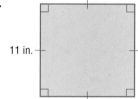
11 in.

2.
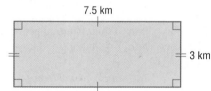
7.5 km
3 km

3.

3.5 yd

4.

4 ft
2.5 ft

5.
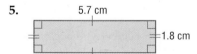
5.7 cm
1.8 cm

6.
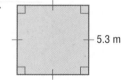
5.3 m

7. a rectangle with length 7 meters and width 11 meters

8. a square with length 4.5 inches

9. a rectangular sandbox with length 2.4 meters and width 1.6 meters

10. a square with length 6.5 yards

11. a square office with length 12 feet

12. a rectangle with length 4.2 inches and width 15.7 inches

13. a square with length 18 centimeters

14. a rectangle with length 5.3 feet and width 7 feet

15. FENCING Jansen purchased a lot that was 121 feet in width and 360 feet in length. If he wants to build a fence around the entire lot, how many feet of fence does he need?

16. CARPETING Leonardo's bedroom is 10 feet wide and 11 feet long. If the carpet store has a remnant whose area is 105 square feet, could it be used to cover his bedroom floor? Explain.

④ Operations with Integers

- The absolute value of any number n is its distance from zero on a number line and is written as $|n|$. Since distance cannot be less than zero, the absolute value of a number is always greater than or equal to zero.

Example 1 Evaluate each expression.

a. $|3|$

$|3| = 3$ Definition of absolute value

b. $|-7|$

$|-7| = 7$ Definition of absolute value

c. $|-4 + 2|$

$|-4 + 2| = |-2|$ $-4 + 2 = -2$

$ = 2$ Simplify.

- To add integers with the same sign, add their absolute values. Give the result the same sign as the integers. To add integers with different signs, subtract their absolute values. Give the result the same sign as the integer with the greater absolute value.

Example 2 Find each sum.

a. $-3 + (-5)$ Both numbers are negative, so the sum is negative.

$-3 + (-5) = -8$ Add $|-3|$ and $|-5|$.

b. $-4 + 2$ The sum is negative because $|-4| > |2|$.

$-4 + 2 = -2$ Subtract $|2|$ from $|-4|$.

c. $6 + (-3)$ The sum is positive because $|6| > |-3|$.

$6 + (-3) = 3$ Subtract $|-3|$ from $|6|$.

d. $1 + 8$ Both numbers are positive, so the sum is positive

$1 + 8 = 9$ Add $|1|$ and $|8|$.

- To subtract an integer, add its additive inverse.

Example 3 Find each difference.

a. $4 - 7$

$4 - 7 = 4 + (-7)$ To subtract 7, add -7.

$ = -3$

b. $2 - (-4)$

$2 - (-4) = 2 + 4$ To subtract -4, add 4.

$ = 6$

- The product of two integers with different signs is negative. The product of two integers with the same sign is positive. Similarly, the quotient of two integers with different signs is negative, and the quotient of two integers with the same sign is positive.

Example 4 Find each product or quotient.

  a. **4(−7)**   The factors have different signs.
     $4(-7) = -28$ The product is negative.

  b. **−64 ÷ (−8)**   The dividend and divisor have the same sign.
     $-64 \div (-8) = 8$ The quotient is positive.

  c. **−9(−6)**   The factors have the same sign.
     $-9(-6) = 54$ The product is positive.

  d. **−55 ÷ 5**   The dividend and divisor have different signs.
     $-55 \div 5 = -11$ The quotient is negative.

  e. $\dfrac{24}{-3}$ The dividend and divisor have different signs.

     $\dfrac{24}{-3} = -8$ The quotient is negative.

- To evaluate expressions with absolute value, evaluate the absolute values first and then perform the operation.

Example 5 Evaluate each expression.

  a. $\left| -3 \right| - \left| 5 \right|$

     $\left| -3 \right| - \left| 5 \right| = 3 - 5$ $\left| -3 \right| = 3,\ \left| 5 \right| = 5$

     $\qquad\qquad\quad = -2$ Simplify.

  b. $\left| -5 \right| + \left| -2 \right|$

     $\left| -5 \right| + \left| -2 \right| = 5 + 2$ $\left| -5 \right| = 5,\ \left| -2 \right| = 2$

     $\qquad\qquad\quad = 7$ Simplify.

Exercises **Evaluate each absolute value.**

1. $\left| -3 \right|$ 2. $\left| 4 \right|$ 3. $\left| 0 \right|$ 4. $\left| -5 \right|$

Find each sum or difference.

5. $-4 - 5$ 6. $3 + 4$ 7. $9 - 5$ 8. $-2 - 5$

9. $3 - 5$ 10. $-6 + 11$ 11. $-4 + (-4)$ 12. $5 - 9$

13. $-3 + 1$ 14. $-4 + (-2)$ 15. $2 - (-8)$ 16. $7 + (-3)$

17. $-4 - (-2)$ 18. $3 - (-3)$ 19. $3 + (-4)$ 20. $-3 - (-9)$

Evaluate each expression.

21. $\left| -4 \right| - \left| 6 \right|$ 22. $\left| -7 \right| + \left| -1 \right|$ 23. $\left| 1 \right| + \left| -2 \right|$ 24. $\left| 2 \right| - \left| -5 \right|.$

25. $\left| -5 + 2 \right|$ 26. $\left| 6 + 4 \right|$ 27. $\left| 3 - 7 \right|$ 28. $\left| -3 - 3 \right|$

Find each product or quotient.

29. $-36 \div 9$ 30. $-3(-7)$ 31. $6(-4)$ 32. $-25 \div 5$

33. $-6(-3)$ 34. $7(-8)$ 35. $-40 \div (-5)$ 36. $11(3)$

37. $44 \div (-4)$ 38. $-63 \div (-7)$ 39. $6(5)$ 40. $-7(12)$

41. $-10(4)$ 42. $80 \div (-16)$ 43. $72 \div 9$ 44. $39 \div 3$

5 Evaluating Algebraic Expressions

An expression is an algebraic expression if it contains sums and/or products of variables and numbers. To evaluate an algebraic expression, replace the variable or variables with known values, and then use the order of operations.

Order of Operations
Step 1 Evaluate expressions inside grouping symbols.
Step 2 Evaluate all powers.
Step 3 Do all multiplications and/or divisions from left to right.
Step 4 Do all additions and/or subtractions from left to right.

Example 1 Evaluate each expression.

a. $x - 5 + y$ if $x = 15$ and $y = -7$

$$
\begin{aligned}
x - 5 + y &= 15 - 5 + (-7) &&x = 15, y = -7 \\
&= 10 + (-7) &&\text{Subtract 5 from 15.} \\
&= 3 &&\text{Add.}
\end{aligned}
$$

b. $6ab^2$ if $a = -3$ and $b = 3$

$$
\begin{aligned}
6ab^2 &= 6(-3)(3)^2 &&a = -3, b = 3 \\
&= 6(-3)(9) &&3^2 = 9 \\
&= (-18)(9) &&\text{Multiply.} \\
&= -162 &&\text{Multiply.}
\end{aligned}
$$

Example 2 Evaluate each expression if $m = -2$, $n = -4$, and $p = 5$.

a. $\dfrac{2m + n}{p - 3}$

The division bar is a grouping symbol. Evaluate the numerator and denominator before dividing.

$$
\begin{aligned}
\frac{2m + n}{p - 3} &= \frac{2(-2) + (-4)}{5 - 3} &&\text{Replace } m \text{ with } -2, n \text{ with } -4, \text{ and } p \text{ with 5.} \\
&= \frac{-4 - 4}{5 - 3} &&\text{Multiply.} \\
&= \frac{-8}{2} &&\text{Subtract.} \\
&= -4 &&\text{Simplify.}
\end{aligned}
$$

b. $-3(m^2 + 2n)$

$$
\begin{aligned}
-3(m^2 + 2n) &= -3[(-2)^2 + 2(-4)] &&\text{Replace } m \text{ with } -2 \text{ and } n \text{ with } -4. \\
&= -3[4 + (-8)] &&\text{Multiply.} \\
&= -3(-4) &&\text{Add.} \\
&= 12 &&\text{Multiply.}
\end{aligned}
$$

Example 3 Evaluate $3|a - b| + 2|c - 5|$ if $a = -2$, $b = -4$, and $c = 3$.

$$
\begin{aligned}
3|a - b| + 2|c - 5| &= 3|-2 - (-4)| + 2|3 - 5| &&\text{Substitute for } a, b, \text{ and } c. \\
&= 3|2| + 2|-2| &&\text{Simplify.} \\
&= 3(2) + 2(2) &&\text{Find absolute values.} \\
&= 10 &&\text{Simplify.}
\end{aligned}
$$

Exercises
Evaluate each expression if $a = 2$, $b = -3$, $c = -1$, and $d = 4$.

1. $2a + c$
2. $\dfrac{bd}{2c}$
3. $\dfrac{2d - a}{b}$
4. $3d - c$
5. $\dfrac{3b}{5a + c}$
6. $5bc$
7. $2cd + 3ab$
8. $\dfrac{c - 2d}{a}$

Evaluate each expression if $x = 2$, $y = -3$, and $z = 1$.

9. $24 + |x - 4|$
10. $13 + |8 + y|$
11. $|5 - z| + 11$
12. $|2y - 15| + 7$
13. $|y| - 7$
14. $11 - 7 + |-x|$
15. $|x| - |2z|$
16. $|z - y| + 6$

6 Solving Linear Equations

- If the same number is added to or subtracted from each side of an equation, the resulting equation is true.

Example 1 Solve each equation.

a. $x - 7 = 16$

$$x - 7 = 16 \quad \text{Original equation}$$
$$x - 7 + 7 = 16 + 7 \quad \text{Add 7 to each side.}$$
$$x = 23 \quad \text{Simplify.}$$

b. $m + 12 = -5$

$$m + 12 = -5 \quad \text{Original equation}$$
$$m + 12 + (-12) = -5 + (-12) \quad \text{Add } -12 \text{ to each side.}$$
$$m = -17 \quad \text{Simplify.}$$

c. $k + 31 = 10$

$$k + 31 = 10 \quad \text{Original equation}$$
$$k + 31 - 31 = 10 - 31 \quad \text{Subtract 31 from each side.}$$
$$k = -21 \quad \text{Simplify.}$$

- If each side of an equation is multiplied or divided by the same number, the resulting equation is true.

Example 2 Solve each equation.

a. $4d = 36$

$$4d = 36 \quad \text{Original equation}$$
$$\frac{4d}{4} = \frac{36}{4} \quad \text{Divide each side by 4.}$$
$$x = 9 \quad \text{Simplify.}$$

b. $-\dfrac{t}{8} = -7$

$$-\frac{t}{8} = -7 \quad \text{Original equation.}$$
$$-8\left(-\frac{t}{8}\right) = -8(-7) \quad \text{Multiply each side by } -8.$$
$$t = 56 \quad \text{Simplify.}$$

c. $\dfrac{3}{5}x = -8$

$$\frac{3}{5}x = -8 \quad \text{Original equation.}$$
$$\frac{5}{3}\left(\frac{3}{5}x\right) = \frac{5}{3}(-8) \quad \text{Multiply each side by } \frac{5}{3}.$$
$$x = -\frac{40}{3} \quad \text{Simplify.}$$

- To solve equations with more than one operation, often called *multi-step equations*, undo operations by working backward.

Example 3 Solve each equation.

a. $12 - m = 20$

$$12 - m = 20 \quad \text{Original equation}$$
$$12 - m - 12 = 20 - 12 \quad \text{Subtract 12 from each side.}$$
$$-m = 8 \quad \text{Simplify.}$$
$$m = -8 \quad \text{Divide each side by } -1.$$

b. $8q - 15 = 49$

$$8q - 15 = 49 \qquad \text{Original equation}$$
$$8q - 15 + 15 = 49 + 15 \qquad \text{Add 15 to each side.}$$
$$8q = 64 \qquad \text{Simplify.}$$
$$\frac{8q}{8} = \frac{64}{8} \qquad \text{Divide each side by 8.}$$
$$q = 8 \qquad \text{Simplify.}$$

c. $12y + 8 = 6y - 5$

$$12y + 8 = 6y - 5 \qquad \text{Original equation}$$
$$12y + 8 - 8 = 6y - 5 - 8 \qquad \text{Subtract 8 from each side.}$$
$$12y = 6y - 13 \qquad \text{Simplify.}$$
$$12y - 6y = 6y - 13 - 6y \qquad \text{Subtract 6y from each side.}$$
$$6y = -13 \qquad \text{Simplify.}$$
$$\frac{6y}{6} = \frac{-13}{6} \qquad \text{Divide each side by 6.}$$
$$y = -\frac{13}{6} \qquad \text{Simplify.}$$

- When solving equations that contain grouping symbols, first use the Distributive Property to remove the grouping symbols.

Example 4 Solve $3(x - 5) = 13$.

$$3(x - 5) = 13 \qquad \text{Original equation}$$
$$3x - 15 = 13 \qquad \text{Distributive Property}$$
$$3x - 15 + 15 = 13 + 15 \qquad \text{Add 15 to each side.}$$
$$3x = 28 \qquad \text{Simplify.}$$
$$x = \frac{28}{3} \qquad \text{Divide each side by 3.}$$

Exercises **Solve each equation.**

1. $r + 11 = 3$
2. $n + 7 = 13$
3. $d - 7 = 8$
4. $\frac{8}{5}a = -6$
5. $-\frac{p}{12} = 6$
6. $\frac{x}{4} = 8$
7. $\frac{12}{5}f = -18$
8. $\frac{y}{7} = -11$
9. $\frac{6}{7}y = 3$
10. $c - 14 = -11$
11. $t - 14 = -29$
12. $p - 21 = 52$
13. $b + 2 = -5$
14. $q + 10 = 22$
15. $-12q = 84$
16. $5s = 30$
17. $5c - 7 = 8c - 4$
18. $2\ell + 6 = 6\ell - 10$
19. $\frac{m}{10} + 15 = 21$
20. $-\frac{m}{8} + 7 = 5$
21. $8t + 1 = 3t - 19$
22. $9n + 4 = 5n + 18$
23. $5c - 24 = -4$
24. $3n + 7 = 28$
25. $-2y + 17 = -13$
26. $-\frac{t}{13} - 2 = 3$
27. $\frac{2}{9}x - 4 = \frac{2}{3}$
28. $9 - 4g = -15$
29. $-4 - p = -2$
30. $21 - b = 11$
31. $-2(n + 7) = 15$
32. $5(m - 1) = -25$
33. $-8a - 11 = 37$
34. $\frac{7}{4}q - 2 = -5$
35. $2(5 - n) = 8$
36. $-3(d - 7) = 6$

7 Solving Inequalities in One Variable

Statements with **greater than** ($>$), **less than** ($<$), **greater than or equal to** ($\geq$), or **less than or equal to** ($\leq$) are **inequalities**.

- If any number is added or subtracted to each side of an inequality, the resulting inequality is true.

Example 1 Solve each inequality.

a. $x - 17 > 12$

$$x - 17 > 12 \qquad \text{Original inequality}$$
$$x - 17 + 17 > 12 + 17 \qquad \text{Add 17 to each side.}$$
$$x > 29 \qquad \text{Simplify.}$$

The solution set is $\{x \mid x > 29\}$.

b. $y + 11 \leq 5$

$$y + 11 \leq 5 \qquad \text{Original inequality}$$
$$y + 11 - 11 \leq 5 - 11 \qquad \text{Subtract 11 from each side.}$$
$$y \leq -6 \qquad \text{Simplify.}$$

The solution set is $\{y \mid y \leq -6\}$.

- If each side of an inequality is multiplied or divided by a positive number, the resulting inequality is true.

Example 2 Solve each inequality.

a. $\dfrac{t}{6} \geq 11$

$$\frac{t}{6} \geq 11 \qquad \text{Original inequality}$$
$$(6)\frac{t}{6} \geq (6)11 \qquad \text{Multiply each side by 6.}$$
$$t \geq 66 \qquad \text{Simplify.}$$

The solution set is $\{t \mid t > 66\}$.

b. $8p < 72$

$$8p < 72 \qquad \text{Original inequality}$$
$$\frac{8p}{8} < \frac{72}{8} \qquad \text{Divide each side by 8.}$$
$$p < 9 \qquad \text{Simplify.}$$

The solution set is $\{p \mid p < 9\}$.

- If each side of an inequality is multiplied or divided by the same negative number, the direction of the inequality symbol must be *reversed* so that the resulting inequality is true.

Example 3 Solve each inequality.

a. $-5c > 30$

$$-5c > 30 \qquad \text{Original inequality}$$
$$\frac{-5c}{-5} < \frac{30}{-5} \qquad \text{Divide each side by } -5. \text{ Change } > \text{ to } <.$$
$$c < -6 \qquad \text{Simplify.}$$

The solution set is $\{c \mid c < -6\}$.

b. $-\dfrac{d}{13} \le -4$

$$-\dfrac{d}{13} \le -4 \qquad \text{Original inequality}$$

$$(-13)\left(-\dfrac{d}{13}\right) \ge (-13)(-4) \qquad \text{Multiply each side by } -13. \text{ Change } \le \text{ to } \ge.$$

$$d \ge 52 \qquad \text{Simplify.}$$

The solution set is $\{d \,|\, d \ge 52\}$.

• Inequalities involving more than one operation can be solved by undoing the operations in the same way you would solve an equation with more than one operation.

Example 4 **Solve each inequality.**

a. $-6a + 13 < -7$

$$-6a + 13 < -7 \qquad \text{Original inequality}$$

$$-6a + 13 - 13 < -7 - 13 \qquad \text{Subtract 13 from each side.}$$

$$-6a < -20 \qquad \text{Simplify.}$$

$$\dfrac{-6a}{-6} > \dfrac{-20}{-6} \qquad \text{Divide each side by } -6. \text{ Change } < \text{ to } >.$$

$$a > \dfrac{10}{3} \qquad \text{Simplify.}$$

The solution set is $\left\{a \,\middle|\, a > \dfrac{10}{3}\right\}$.

b. $4z + 7 \ge 8z - 1$

$$4z + 7 \ge 8z - 1 \qquad \text{Original inequality.}$$

$$4z + 7 - 7 \ge 8z - 1 - 7 \qquad \text{Subtract 7 from each side.}$$

$$4z \ge 8z - 8 \qquad \text{Simplify.}$$

$$4z - 8z \ge 8z - 8 - 8z \qquad \text{Subtract } 8z \text{ from each side.}$$

$$-4z \ge -8 \qquad \text{Simplify.}$$

$$\dfrac{-4z}{-4} \le \dfrac{-8}{-4} \qquad \text{Divide each side by } -4. \text{ Change } \ge \text{ to } \le.$$

$$z \le 2 \qquad \text{Simplify.}$$

The solution set is $\left\{z \,\middle|\, z \le 2\right\}$.

Exercises **Solve each inequality.**

1. $x - 7 < 6$

2. $4c + 23 \le -13$

3. $-\dfrac{p}{5} \ge 14$

4. $-\dfrac{a}{8} < 5$

5. $\dfrac{t}{6} > -7$

6. $\dfrac{a}{11} \le 8$

7. $d + 8 \le 12$

8. $m + 14 > 10$

9. $2z - 9 < 7z + 1$

10. $6t - 10 \ge 4t$

11. $3z + 8 < 2$

12. $a + 7 \ge -5$

13. $m - 21 < 8$

14. $x - 6 \ge 3$

15. $-3b \le 48$

16. $4y < 20$

17. $12k \ge -36$

18. $-4h > 36$

19. $\dfrac{2}{5}b - 6 \le -2$

20. $\dfrac{8}{3}t + 1 > -5$

21. $7q + 3 \ge -4q + 25$

22. $-3n - 8 > 2n + 7$

23. $-3w + 1 \le 8$

24. $-\dfrac{4}{5}k - 17 > 11$

8 Graphing Using Intercepts and Slope

- The x-coordinate of the point at which a line crosses the x-axis is called the **x-intercept**. The y-coordinate of the point at which a line crosses the y-axis is called the **y-intercept**. Since two points determine a line, one method of graphing a linear equation is to find these intercepts.

Example 1 Determine the x-intercept and y-intercept of $4x - 3y = 12$. Then graph the equation.

To find the x-intercept, let $y = 0$.

$4x - 3y = 12$	Original equation
$4x - 3(0) = 12$	Replace y with 0.
$4x = 12$	Simplify.
$x = 3$	Divide each side by 4.

To find the y-intercept, let $x = 0$.

$4x - 3y = 12$	Original equation
$4(0) - 3y = 12$	Replace x with 0.
$-3y = 12$	Divide each side by -3.
$y = -4$	Simplify.

Put a point on the x-axis at 3 and a point on the y-axis at -4. Draw the line through the two points.

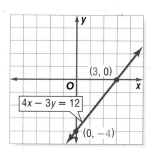

- A linear equation of the form $y = mx + b$ is in *slope-intercept* form, where m is the slope and b is the y-intercept. When an equation is written in this form, you can graph the equation quickly.

Example 2 Graph $y = \frac{3}{4}x - 2$.

Step 1 The y-intercept is -2. So, plot a point at $(0, -2)$.

Step 2 The slope is $\frac{3}{4}$. $\frac{\text{rise}}{\text{run}}$
From $(0, -2)$, move up 3 units and right 4 units. Plot a point.

Step 3 Draw a line connecting the points.

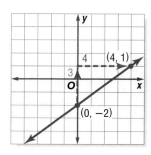

Exercises Graph each equation using both intercepts.

1. $-2x + 3y = 6$
2. $2x + 5y = 10$
3. $3x - y = 3$
4. $-x + 2y = 2$
5. $3x + 4y = 12$
6. $4y + x = 4$

Graph each equation using the slope and y-intercept.

7. $y = -x + 2$
8. $y = x - 2$
9. $y = x + 1$
10. $y = 3x - 1$
11. $y = -2x + 3$
12. $y = -3x - 1$

Graph each equation using either method.

13. $y = \frac{2}{3}x - 3$
14. $y = \frac{1}{2}x - 1$
15. $y = 2x - 2$
16. $-6x + y = 2$
17. $2y - x = -2$
18. $3x + 4y = -12$
19. $4x - 3y = 6$
20. $4x + y = 4$
21. $y = 2x - \frac{3}{2}$

9 Solving Systems of Linear Equations

- Two or more equations that have common variables are called a **system of equations**. The solution of a system of equations in two variables is an ordered pair of numbers that satisfies both equations. A system of two linear equations can have zero, one, or an infinite number of solutions. There are three methods by which systems of equations can be solved: graphing, elimination, and substitution.

Example 1 Solve each system of equations by graphing. Then determine whether each system has *no* solution, *one* solution, or *infinitely many* solutions.

a. $y = -x + 3$
$y = 2x - 3$

The graphs appear to intersect at $(2, 1)$.
Check this estimate by replacing x with 2 and y with 1 in each equation.

Check: $y = -x + 3 \qquad y = 2x - 3$
$1 \stackrel{?}{=} -2 + 3 \qquad 1 \stackrel{?}{=} 2(2) - 3$
$1 = 1 \checkmark \qquad 1 = 1 \checkmark$

The system has one solution at $(2, 1)$.

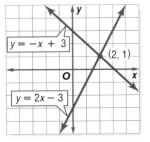

b. $y - 2x = 6$
$3y - 6x = 9$

The graphs of the equations are parallel lines. Since they do not intersect, there are no solutions of this system of equations. Notice that the lines have the same slope but different y-intercepts. Equations with the same slope *and* the same y-intercepts have an infinite number of solutions.

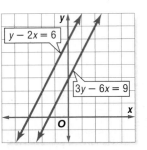

- It is difficult to determine the solution of a system when the two graphs intersect at noninteger values. There are algebraic methods by which an exact solution can be found. One such method is **substitution**.

Example 2 Use substitution to solve the system of equations.

$y = -4x$

$2y + 3x = 8$

Since $y = -4x$, substitute $-4x$ for y in the second equation.

$2y + 3x = 8 \qquad$ Second equation
$2(-4x) + 3x = 8 \qquad y = -4x$
$-8x + 3x = 8 \qquad$ Simplify.
$-5x = 8 \qquad$ Combine like terms.
$\dfrac{-5x}{-5} = \dfrac{8}{-5} \qquad$ Divide each side by -5.
$x = -\dfrac{8}{5} \qquad$ Simplify.

Use $y = -4x$ to find the value of y.

$y = -4x \qquad$ First equation
$y = -4\left(-\dfrac{8}{5}\right) \qquad x = -\dfrac{8}{5}$
$y = \dfrac{32}{5} \qquad$ Simplify.

The solution is $\left(-\dfrac{8}{5}, \dfrac{32}{5}\right)$.

- Sometimes adding or subtracting two equations together will eliminate one variable. Using this step to solve a system of equations is called **elimination**.

Example 3 Use elimination to solve the system of equations.

$3x + 5y = 7$
$4x + 2y = 0$

Either x or y can be eliminated. In this example, we will eliminate x.

$3x + 5y = 7$ Multiply by 4. $12x + 20y = 28$

$4x + 2y = 0$ Multiply by -3. $+ \underline{-12x - 6y = 0}$

$14y = 28$ Add the equations.

$\dfrac{14y}{14} = \dfrac{28}{14}$ Divide each side by 14.

$y = 2$ Simplify.

Now substitute 2 for y in either equation to find the value of x.

$4x + 2y = 0$ Second equation
$4x + 2(2) = 0$ $y = 2$
$4x + 4 = 0$ Simplify.
$4x + 4 - 4 = 0 - 4$ Subtract 4 from each side.
$4x = -4$ Simplify.
$\dfrac{4x}{4} = \dfrac{-4}{4}$ Divide each side by 4.
$x = -1$ Simplify.

The solution is $(-1, 2)$.

Exercises Solve by graphing.

1. $y = -x + 2$
$y = -\dfrac{1}{2}x + 1$

2. $y = 3x - 3$
$y = x + 1$

3. $y - 2x = 1$
$2y - 4x = 1$

4. $2x - 4y = -2$
$-6x + 12y = 6$

5. $4x + 3y = 12$
$3x - y = 9$

6. $3y + x = -3$
$y - 3x = -1$

Solve by substitution.

7. $-5x + 3y = 12$
$x + 2y = 8$

8. $x - 4y = 22$
$2x + 5y = -21$

9. $y + 5x = -3$
$3y - 2x = 8$

10. $y - 2x = 2$
$7y + 4x = 23$

11. $2x - 3y = -8$
$-x + 2y = 5$

12. $4x + 2y = 5$
$3x - y = 10$

Solve by elimination.

13. $-3x + y = 7$
$3x + 2y = 2$

14. $3x + 4y = -1$
$-9x - 4y = 13$

15. $-4x + 5y = -11$
$2x + 3y = 11$

16. $6x - 5y = 1$
$-2x + 9y = 7$

17. $3x - 2y = 8$
$5x - 3y = 16$

18. $4x + 7y = -17$
$3x + 2y = -3$

Name an appropriate method to solve each system of equations. Then solve the system.

19. $4x - y = 11$
$2x - 3y = 3$

20. $4x + 6y = 3$
$-10x - 15y = -4$

21. $3x - 2y = 6$
$5x - 5y = 5$

22. $3y + x = 3$
$-2y + 5x = 15$

23. $4x - 7y = 8$
$-2x + 5y = -1$

24. $x + 3y = 6$
$4x - 2y = -32$

10 Square Roots and Simplifying Radicals

- A radical expression is an expression that contains a square root. The expression is in simplest form when the following three conditions have been met.

- No radicands have perfect square factors other than 1.

- No radicands contain fractions.

- No radicals appear in the denominator of a fraction.

- The **Product Property** states that for two numbers a and $b \geq 0$, $\sqrt{ab} = \sqrt{a} \cdot \sqrt{b}$.

Example 1 Simplify.

   a. $\sqrt{45}$

$$\sqrt{45} = \sqrt{3 \cdot 3 \cdot 5} \quad \text{Prime factorization of 45}$$
$$= \sqrt{3^2} \cdot \sqrt{5} \quad \text{Product Property of Square Roots}$$
$$= 3\sqrt{5} \quad \text{Simplify.}$$

   b. $\sqrt{3} \cdot \sqrt{3}$

$$\sqrt{3} \cdot \sqrt{3} = \sqrt{3 \cdot 3} \quad \text{Product Property}$$
$$= \sqrt{9} \text{ or } 3 \quad \text{Simplify.}$$

   c. $\sqrt{6} \cdot \sqrt{15}$

$$\sqrt{6} \cdot \sqrt{15} = \sqrt{6 \cdot 15} \quad \text{Product Property}$$
$$= \sqrt{3 \cdot 2 \cdot 3 \cdot 5} \quad \text{Prime factorization}$$
$$= \sqrt{3^2} \cdot \sqrt{10} \quad \text{Product Property}$$
$$= 3\sqrt{10} \quad \text{Simplify.}$$

- For radical expressions in which the exponent of the variable inside the radical is *even* and the resulting simplified exponent is *odd*, you must use absolute value to ensure nonnegative results.

Example 2 $\sqrt{20x^3y^5z^6}$

$$\sqrt{20x^3y^5z^6} = \sqrt{2^2 \cdot 5 \cdot x^3 \cdot y^5 \cdot z^6} \quad \text{Prime factorization}$$
$$= \sqrt{2^2} \cdot \sqrt{5} \cdot \sqrt{x^3} \cdot \sqrt{y^5} \cdot \sqrt{z^6} \quad \text{Product Property}$$
$$= 2 \cdot \sqrt{5} \cdot x \cdot \sqrt{x} \cdot y^2 \cdot \sqrt{y} \cdot |z^3| \quad \text{Simplify.}$$
$$= 2xy^2|z^3|\sqrt{5xy} \quad \text{Simplify.}$$

- The **Quotient Property** states that for any numbers a and b, where $a \geq 0$ and $b \geq 0$,

$$\sqrt{\frac{a}{b}} = \frac{\sqrt{a}}{\sqrt{b}}.$$

Example 3 Simplify $\sqrt{\frac{25}{16}}$.

$$\sqrt{\frac{25}{16}} = \frac{\sqrt{25}}{\sqrt{16}} \quad \text{Quotient Property}$$
$$= \frac{5}{4} \quad \text{Simplify.}$$

- Rationalizing the denominator of a radical expression is a method used to eliminate radicals from the denominator of a fraction. To rationalize the denominator, multiply the expression by a fraction equivalent to 1 such that the resulting denominator is a perfect square.

Example 4 Simplify.

a. $\dfrac{2}{\sqrt{3}}$

$$\dfrac{2}{\sqrt{3}} = \dfrac{2}{\sqrt{3}} \cdot \dfrac{\sqrt{3}}{\sqrt{3}} \qquad \text{Multiply by } \dfrac{\sqrt{3}}{\sqrt{3}}.$$

$$= \dfrac{2\sqrt{3}}{3} \qquad \text{Simplify.}$$

b. $\dfrac{\sqrt{13y}}{\sqrt{18}}$

$$\dfrac{\sqrt{13y}}{\sqrt{18}} = \dfrac{\sqrt{13y}}{\sqrt{2 \cdot 3 \cdot 3}} \qquad \text{Prime factorization}$$

$$= \dfrac{\sqrt{13y}}{3\sqrt{2}} \qquad \text{Product Property}$$

$$= \dfrac{\sqrt{13y}}{3\sqrt{2}} \cdot \dfrac{\sqrt{2}}{\sqrt{2}} \qquad \text{Multiply by } \dfrac{\sqrt{2}}{\sqrt{2}}.$$

$$= \dfrac{\sqrt{26y}}{6} \qquad \text{Product Property}$$

- Sometimes, conjugates are used to simplify radical expressions. Conjugates are binomials of the form $p\sqrt{q} + r\sqrt{s}$ and $p\sqrt{q} - r\sqrt{s}$.

Example 5 Simplify $\dfrac{3}{5 - \sqrt{2}}$.

$$\dfrac{3}{5 - \sqrt{2}} = \dfrac{3}{5 - \sqrt{2}} \cdot \dfrac{5 + \sqrt{2}}{5 + \sqrt{2}} \qquad \dfrac{5 + \sqrt{2}}{5 + \sqrt{2}} = 1$$

$$= \dfrac{3(5 + \sqrt{2})}{5^2 - (\sqrt{2})^2} \qquad (a - b)(a + b) = a^2 - b^2$$

$$= \dfrac{15 + 3\sqrt{2}}{25 - 2} \qquad \text{Multiply. } (\sqrt{2})^2 = 2$$

$$= \dfrac{15 + 3\sqrt{2}}{23} \qquad \text{Simplify.}$$

Exercises Simplify.

1. $\sqrt{32}$

2. $\sqrt{75}$

3. $\sqrt{50} \cdot \sqrt{10}$

4. $\sqrt{12} \cdot \sqrt{20}$

5. $\sqrt{6} \cdot \sqrt{6}$

6. $\sqrt{16} \cdot \sqrt{25}$

7. $\sqrt{98x^3y^6}$

8. $\sqrt{56a^2b^4c^5}$

9. $\sqrt{\dfrac{81}{49}}$

10. $\sqrt{\dfrac{121}{16}}$

11. $\sqrt{\dfrac{63}{8}}$

12. $\sqrt{\dfrac{288}{147}}$

13. $\dfrac{\sqrt{10p^3}}{\sqrt{27}}$

14. $\dfrac{\sqrt{108}}{\sqrt{2q^6}}$

15. $\dfrac{4}{5 - 2\sqrt{3}}$

16. $\dfrac{7\sqrt{3}}{5 - 2\sqrt{6}}$

17. $\dfrac{3}{\sqrt{48}}$

18. $\dfrac{\sqrt{24}}{\sqrt{125}}$

19. $\dfrac{3\sqrt{5}}{2 - \sqrt{2}}$

20. $\dfrac{3}{-2 + \sqrt{13}}$

11 Multiplying Polynomials

- The **Product of Powers** rule states that for any number a and all integers m and n, $a^m \cdot a^n = a^{m+n}$.

Example 1 Simplify each expression.

a. $(4p^5)(p^4)$

$$
\begin{aligned}
(4p^5)(p^4) &= (4)(1)(p^5 \cdot p^4) && \text{Commutative and Associative Properties} \\
&= (4)(1)(p^{5+4}) && \text{Product of powers} \\
&= 4p^9 && \text{Simplify.}
\end{aligned}
$$

b. $(3yz^5)(-9y^2z^2)$

$$
\begin{aligned}
(3yz^5)(-9y^2z^2) &= (3)(-9)(y \cdot y^2)(z^5 \cdot z^2) && \text{Commutative and Associative Properties} \\
&= -27(y^{1+2})(z^{5+2}) && \text{Product of powers} \\
&= -27y^3z^7 && \text{Simplify.}
\end{aligned}
$$

- The Distributive Property can be used to multiply a monomial by a polynomial.

Example 2 Simplify $3x^3(-4x^2 + x - 5)$.

$$
\begin{aligned}
3x^3(-4x^2 + x - 5) &= 3x^3(-4x^2) + 3x^3(x) - 3x^3(5) && \text{Distributive Property} \\
&= -12x^5 + 3x^4 - 15x^3 && \text{Multiply.}
\end{aligned}
$$

- To find the power of a power, multiply the exponents. This is called the **Power of a Power** rule.

Example 3 Simplify each expression.

a. $(-3x^2y^4)^3$

$$
\begin{aligned}
(-3x^2y^4)^3 &= (-3)^3(x^2)^3(y^4)^3 && \text{Power of a product} \\
&= -27x^6y^{12} && \text{Power of a power}
\end{aligned}
$$

b. $(xy)^3(-2x^4)^2$

$$
\begin{aligned}
(xy)^3(-2x^4)^2 &= x^3y^3(-2)^2(x^4)^2 && \text{Power of a product} \\
&= x^3y^3(4)x^8 && \text{Power of a power} \\
&= 4x^3 \cdot x^8 \cdot y^3 && \text{Commutative Property} \\
&= 4x^{11}y^3 && \text{Product of powers}
\end{aligned}
$$

- To multiply two binomials, find the sum of the products of

    F   the *First* terms,
    O   the *Outer* terms,
    I   the *Inner* terms, and
    L   the *Last* terms.

Example 4 Find each product.

a. $(2x - 3)(x + 1)$

$$
\begin{aligned}
(2x - 3)(x + 1) &= \overset{F}{(2x)(x)} + \overset{O}{(2x)(1)} + \overset{I}{(-3)(x)} + \overset{L}{(-3)(1)} && \text{FOIL method} \\
&= 2x^2 + 2x - 3x - 3 && \text{Multiply.} \\
&= 2x^2 - x - 3 && \text{Combine like terms.}
\end{aligned}
$$

b. $(x + 6)(x + 5)$

$$
\begin{aligned}
(x + 6)(x + 5) &= \overset{F}{(x)(x)} + \overset{O}{(x)(5)} + \overset{I}{(6)(x)} + \overset{L}{(6)(5)} && \text{FOIL method} \\
&= x^2 + 5x + 6x + 30 && \text{Multiply.} \\
&= x^2 + 11x + 30 && \text{Combine like terms.}
\end{aligned}
$$

- The Distributive Property can be used to multiply any two polynomials.

Example 5 Find $(3x - 2)(2x^2 + 7x - 4)$.

$$\begin{aligned}(3x - 2)(2x^2 + 7x - 4) &= 3x(2x^2 + 7x - 4) - 2(2x^2 + 7x - 4) \quad \text{Distributive Property} \\ &= 6x^3 + 21x^2 - 12x - 4x^2 - 14x + 8 \quad \text{Distributive Property} \\ &= 6x^3 + 17x^2 - 26x + 8 \quad \text{Combine like terms.}\end{aligned}$$

- Three special products are:
$\quad (a + b)^2 = a^2 + 2ab + b^2,$
$\quad (a - b)^2 = a^2 - 2ab + b^2,$ and
$\quad (a + b)(a - b) = a^2 - b^2.$

Example 6 Find each product.

a. $(2x - z)^2$

$$\begin{aligned}(a - b)^2 &= a^2 - 2ab + b^2 \quad &\text{Square of a difference} \\ (2x - z)^2 &= (2x)^2 - 2(2x)(z) + (z)^2 \quad &a = 2x \text{ and } b = z \\ &= 4x^2 - 4xz + z^2 \quad &\text{Simplify.}\end{aligned}$$

b. $(3x + 7)(3x - 7)$

$$\begin{aligned}(a + b)(a - b) &= a^2 - b^2 \quad &\text{Product of sum and difference} \\ (3x + 7)(3x - 7) &= (3x)^2 - (7)^2 \quad &a = 3x \text{ and } b = 7 \\ &= 9x^2 - 49 \quad &\text{Simplify.}\end{aligned}$$

Exercises Find each product.

1. $(3q^2)(q^5)$

2. $(5m)(4m^3)$

3. $\left(\frac{9}{2}c\right)(8c^5)$

4. $(n^6)(10n^2)$

5. $(fg^8)(15f^2g)$

6. $(6j^4k^4)(j^2k)$

7. $(2ab^3)(4a^2b^2)$

8. $\left(\frac{8}{5}x^3y\right)(4x^3y^2)$

9. $-2q^2(q^2 + 3)$

10. $5p(p - 18)$

11. $15c(-3c^2 + 2c + 5)$

12. $8x(-4x^2 - x + 11)$

13. $4m^2(-2m^2 + 7m - 5)$

14. $8y^2(5y^3 - 2y + 1)$

15. $\left(\frac{3}{2}m^3n^2\right)^2$

16. $(-2c^3d^2)^2$

17. $(-5wx^5)^3$

18. $(6a^5b)^3$

19. $(k^2\ell)^3(13k^2)^2$

20. $(-5w^3x^2)^2(2w^5)^2$

21. $(-7y^3z^2)(4y^2)^4$

22. $\left(\frac{1}{2}p^2q^2\right)^2(4pq^3)^3$

23. $(m - 1)(m - 4)$

24. $(s - 7)(s - 2)$

25. $(x - 3)(x + 4)$

26. $(a + 3)(a - 6)$

27. $(5d + 3)(d - 4)$

28. $(q + 2)(3q + 5)$

29. $(2q + 3)(5q + 2)$

30. $(2a - 3)(2a - 5)$

31. $(d + 1)(d - 1)$

32. $(4a - 3)(4a + 3)$

33. $(s - 5)^2$

34. $(3f - g)^2$

35. $(2r - 5)^2$

36. $\left(t + \frac{8}{3}\right)^2$

37. $(x + 4)(x^2 - 5x - 2)$

38. $(x - 2)(x^2 + 3x - 7)$

39. $(3b - 2)(3b^2 + b + 1)$

40. $(2j + 7)(j^2 - 2j + 4)$

12 Dividing Polynomials

- The **Quotient of Powers** rule states that for any nonzero number a and all integers m and n, $\dfrac{a^m}{a^n} = a^{m-n}$.

- To find the power of a quotient, find the power of the numerator and the power of the denominator.

Example 1 Simplify.

a. $\dfrac{x^5 y^8}{-xy^3}$

$$\dfrac{x^5 y^8}{-xy^3} = \left(\dfrac{x^5}{-x}\right)\left(\dfrac{y^8}{y^3}\right) \qquad \text{Group powers that have the same base.}$$

$$= -(x^{5-1})(y^{8-3}) \qquad \text{Quotient of powers}$$

$$= -x^4 y^5 \qquad \text{Simplify.}$$

b. $\left(\dfrac{4z^3}{3}\right)^3$

$$\left(\dfrac{4z^3}{3}\right)^3 = \dfrac{(4z^3)^3}{3^3} \qquad \text{Power of a quotient}$$

$$= \dfrac{4^3(z^3)^3}{3^3} \qquad \text{Power of a product}$$

$$= \dfrac{64z^9}{27} \qquad \text{Power of a product}$$

c. $\dfrac{w^{-2}x^4}{2w^{-5}}$

$$\dfrac{w^{-2}x^4}{2w^{-5}} = \dfrac{1}{2}\left(\dfrac{w^{-2}}{w^{-5}}\right)x^4 \qquad \text{Group powers that have the same base.}$$

$$= \dfrac{1}{2}(w^{-2-(-5)})x^4 \qquad \text{Quotient of powers}$$

$$= \dfrac{1}{2}w^3x^4 \qquad \text{Simplify.}$$

- You can divide a polynomial by a monomial by separating the terms of the numerator.

Example 2 Simplify $\dfrac{15x^3 - 3x^2 + 12x}{3x}$.

$$\dfrac{15x^3 - 3x^2 + 12x}{3x} = \dfrac{15x^3}{3x} - \dfrac{3x^2}{3x} + \dfrac{12x}{3x} \qquad \text{Divide each term by } 3x.$$

$$= 5x^2 - x + 4 \qquad \text{Simplify.}$$

- Division can sometimes be performed using factoring.

Example 3 Find $(n^2 - 8n - 9) \div (n - 9)$.

$$(n^2 - 8n - 9) \div (n - 9) = \dfrac{n^2 - 8n - 9}{(n - 9)} \qquad \text{Write as a rational expression.}$$

$$= \dfrac{(n - 9)(n + 1)}{(n - 9)} \qquad \text{Factor the numerator.}$$

$$= \dfrac{(n - 9)(n + 1)}{(n - 9)} \qquad \text{Divide by the GCF.}$$

$$= n + 1 \qquad \text{Simplify.}$$

- When you cannot factor, you can use a long division process similar to the one you use in arithmetic.

Example 4 Find $(n^3 - 4n^2 - 9) \div (n - 3)$.

In this case, there is no n term, so you must rename the dividend using 0 as the coefficient of the missing term.

$(n^3 - 4n^2 + 9) \div (n - 3) = (n^3 - 4n^2 + 0n + 9) \div (n - 3)$

Divide the first term of the dividend, n^3, by the first term of the divisor, n.

$$
\begin{array}{r}
n^2 - n - 3 \\
n - 3 \overline{)n^3 - 4n^2 + 0n + 12} \\
\underline{(-)\ n^3 - 3n^2} \\
-n^2 + 0n \\
\underline{(-)-n^2 + 3n} \\
-3n + 12 \\
\underline{(-)-3n + 9} \\
3
\end{array}
$$

Multiply n^2 and $n - 3$.

Subtract and bring down $0n$.

Multiply $-n$ and $n - 3$.

Subtract and bring down 12.

Multiply -3 and $n - 3$.

Subtract.

Therefore, $(n^3 - 4n^2 + 9) \div (n - 3) = n^2 - n - 3 + \dfrac{3}{n - 3}$. Since the quotient has a nonzero remainder, $n - 3$ is not a factor of $n^3 - 4n^2 + 9$.

Exercises **Find each quotient.**

1. $\dfrac{a^2c^2}{2a}$

2. $\dfrac{5q^5r^3}{q^2r^2}$

3. $\dfrac{b^2d^5}{8b^{-2}d^3}$

4. $\dfrac{5p^{-3}x}{2p^{-7}}$

5. $\dfrac{3r^{-3}s^2t^4}{2r^2st^{-3}}$

6. $\dfrac{3x^3y^{-1}z^5}{xyz^2}$

7. $\left(\dfrac{w^4}{6}\right)^3$

8. $\left(\dfrac{-3q^2}{5}\right)^3$

9. $\left(\dfrac{-2y^2}{7}\right)^2$

10. $\left(\dfrac{5m^2}{3}\right)^4$

11. $\dfrac{4z^2 - 16z - 36}{4z}$

12. $(5d^2 + 8d - 20) \div 10d$

13. $(p^3 - 12p^2 + 3p + 8) \div 4p$

14. $(b^3 + 4b^2 + 10) \div 2b$

15. $\dfrac{a^3 - 6a^2 + 4a - 3}{a^2}$

16. $\dfrac{8x^2y - 10xy^2 + 6x^3}{2x^2}$

17. $\dfrac{s^2 - 2s - 8}{s - 4}$

18. $(r^2 + 9r + 20) \div (r + 5)$

19. $(t^2 - 7t + 12) \div (t - 3)$

20. $(c^2 + 3c - 54) \div (c + 9)$

21. $(2q^2 - 9q - 5) \div (q - 5)$

22. $\dfrac{3z^2 - 2z - 5}{z + 1}$

23. $\dfrac{(m^3 + 3m^2 - 5m + 1)}{m - 1}$

24. $(d^3 - 2d^2 + 4d + 24) \div (d + 2)$

25. $(2j^3 + 5j + 26) \div (j + 2)$

26. $\dfrac{2x^3 + 3x^2 - 176}{x - 4}$

27. $(x^2 + 6x - 3) \div (x + 4)$

28. $\dfrac{h^3 + 2h^2 - 6h + 1}{h - 2}$

Factoring to Solve Equations

- Some polynomials can be factored using the Distributive Property.

 Example 1 Factor $5t^2 + 15t$.

Find the greatest common factor (GCF) of $5t^2$ and $15t$.

$5t^2 = 5 \cdot t \cdot t$, $15t = 3 \cdot 5 \cdot t$ GCF: $5 \cdot t$ or $5t$

$5t^2 + 15t = 5t(t) + 5t(3)$ Rewrite each term using the GCF.

$\quad\quad = 5t(t + 3)$ Distributive Property

- To factor polynomials of the form $x^2 + bx + c$, find two integers m and n so that $mn = c$ and $m + n = b$. Then write $x^2 + bx + c$ using the pattern $(x + m)(x + n)$.

Example 2 Factor each polynomial.

a. $x^2 + 7x + 10$

In this equation, b is 7 and c is 10. Find two numbers with a product of 10 and with a sum of 7.

Both b and c are positive.

Factors of 10	Sum of Factors
1, 10	11
2, 5	7

$x^2 + 7x + 10 = (x + m)(x + n)$ The correct factors are 2 and 5.

$\quad\quad = (x + 2)(x + 5)$ Write the pattern; $m = 2$ and $n = 5$.

b. $x^2 - 8x + 15$

In this equation, b is -8 and c is 15. This means that $m + n$ is negative and mn is positive. So m and n must both be negative.

b is negative and c is positive.

Factors of 15	Sum of Factors
$-1, -15$	-16
$-3, -5$	-8

$x^2 - 8x + 15 = (x + m)(x + n)$ The correct factors are -3 and -5.

$\quad\quad = (x - 3)(x - 5)$ Write the pattern; $m = -3$ and $n = -5$.

- To factor polynomials of the form $ax^2 + bx + c$, find two integers m and n with a product equal to ac and with a sum equal to b. Write $ax^2 + bx + c$ using the pattern $ax^2 + mx + nx + c$. Then factor by grouping.

 c. $5x^2 - 19x - 4$

b is negative and c is negative.

In this equation, a is 5, b is -19, and c is -4. Find two numbers with a product of -20 and with a sum of -19.

Factors of -20	Sum of Factors
$-2, 10$	8
$2, -10$	-8
$-1, 20$	19
$1, -20$	-19

The correct factors are 1 and -20.

$5x^2 - 19x - 4 = 5x^2 + mx + nx - 4$ Write the pattern.

$\quad\quad = 5x^2 + x + (-20)x - 4$ $m = 1$ and $n = -20$

$\quad\quad = (5x^2 + x) - (20x + 4)$ Group terms with common factors.

$\quad\quad = x(5x + 1) - 4(5x + 1)$ Factor the GCF from each group.

$\quad\quad = (x - 4)(5x + 1)$ Distributive Property

Prerequisite Skills (side tab)

- Here are some special products.

Perfect Square Trinomials

$a^2 + 2ab + b^2 = (a + b)(a + b)$
$\qquad\qquad\quad = (a + b)^2$

$a^2 - 2ab + b^2 = (a - b)(a - b)$
$\qquad\qquad\quad = (a - b)^2$

Difference of Squares

$a^2 - b^2 = (a + b)(a - b)$

Example 3 Factor each polynomial.

a. $9x^2 + 6x + 1$

> The first and last terms are perfect squares, and the middle term is equal to $2(3x)(1)$.

$9x^2 + 6x + 1 = (3x)^2 + 2(3x)(1) + 1^2$ Write as $a^2 + 2ab + b^2$.
$\qquad\qquad\quad = (3x + 1)^2$ Factor using the pattern.

b. $x^2 - 9 = 0$

> This is a difference of squares.

$x^2 - 9 = x^2 - (3)^2$ Write in the form $a^2 - b^2$.
$\qquad\quad = (x - 3)(x + 3)$ Factor the difference of squares.

- The binomial $x - a$ is a factor of the polynomial $f(x)$ if and only if $f(a) = 0$. Since 0 times any number is equal to zero, this implies that we can use factoring to solve equations.

Example 4 Solve $x^2 - 5x + 4 = 0$ by factoring.

Factor the polynomial. This expression is of the form $x^2 + bx + c$.

$x^2 - 5x + 4 = 0$ Original equation
$(x - 1)(x - 4) = 0$ Factor the polynomial.

If $ab = 0$, then $a = 0$, $b = 0$, or both equal 0. Let each factor equal 0.

$x - 1 = 0$ or $x - 4 = 0$
$\quad x = 1$ $x = 4$

Exercises Factor each polynomial.

1. $u^2 - 12u$
2. $w^2 + 4w$
3. $7j^2 - 28j$
4. $2g^2 + 24g$
5. $6x^2 + 2x$
6. $5t^2 - 30t$
7. $z^2 + 10z + 21$
8. $n^2 + 8n + 15$
9. $h^2 + 8h + 12$
10. $x^2 + 14x + 48$
11. $m^2 + 6m - 7$
12. $b^2 + 2b - 24$
13. $q^2 - 9q + 18$
14. $p^2 - 5p + 6$
15. $a^2 - 3a - 4$
16. $k^2 - 4k - 32$
17. $n^2 - 7n - 44$
18. $y^2 - 3y - 88$
19. $3z^2 + 4z - 4$
20. $2y^2 + 9y - 5$
21. $5x^2 + 7x + 2$
22. $3s^2 + 11s - 4$
23. $6r^2 - 5r + 1$
24. $8a^2 + 15a - 2$
25. $w^2 - \dfrac{9}{4}$
26. $c^2 - 64$
27. $r^2 + 14r + 49$
28. $b^2 + 18b + 81$
29. $j^2 - 12j + 36$
30. $4t^2 - 25$

Solve each equation by factoring.

31. $10r^2 - 35r = 0$
32. $3x^2 + 15x = 0$
33. $k^2 + 13k + 36 = 0$
34. $w^2 - 8w + 12 = 0$
35. $c^2 - 5c - 14 = 0$
36. $z^2 - z - 42 = 0$
37. $2y^2 - 5y - 12 = 0$
38. $3b^2 - 4b - 15 = 0$
39. $t^2 + 12t + 36 = 0$
40. $u^2 + 5u + \dfrac{25}{4} = 0$
41. $q^2 - 8q + 16 = 0$
42. $a^2 - 6a + 9 = 0$

14 Operations with Matrices

- A **matrix** is a rectangular arrangement of numbers in rows and columns. Each entry in a matrix is called an **element**. A matrix is usually described by its **dimensions**, or the number of **rows** and **columns**, with the number of rows stated first.

- For example, matrix A has dimensions 3×2 and matrix B has dimensions 2×4.

$$\text{matrix } A = \begin{bmatrix} 6 & -2 \\ 0 & 5 \\ -4 & 10 \end{bmatrix} \qquad \text{matrix } B = \begin{bmatrix} 7 & -1 & -2 & 0 \\ 3 & 6 & -5 & 2 \end{bmatrix}$$

- If two matrices have the same dimensions, you can add or subtract them. To do this, add or subtract corresponding elements of the two matrices.

Example 1 If $A = \begin{bmatrix} 12 & 7 & -3 \\ 0 & -1 & -6 \end{bmatrix}$, $B = \begin{bmatrix} -3 & 0 & 5 \\ 2 & 7 & -7 \end{bmatrix}$, and $C = \begin{bmatrix} 9 & 1 & -5 \\ 0 & -1 & 15 \end{bmatrix}$, find the sum and difference.

a. $A + B$

$$A + B = \begin{bmatrix} 12 & 7 & -3 \\ 0 & -1 & -6 \end{bmatrix} + \begin{bmatrix} -3 & 0 & 5 \\ 2 & 7 & -7 \end{bmatrix} \qquad \text{Substitution}$$

$$= \begin{bmatrix} 12 + (-3) & 7 + 0 & -3 + 5 \\ 0 + 2 & -1 + 7 & -6 + (-7) \end{bmatrix} \qquad \text{Definition of matrix addition}$$

$$= \begin{bmatrix} 9 & 7 & 2 \\ 2 & 6 & -13 \end{bmatrix} \qquad \text{Simplify.}$$

b. $B - C$

$$B - C = \begin{bmatrix} -3 & 0 & 5 \\ 2 & 7 & -7 \end{bmatrix} - \begin{bmatrix} 9 & 1 & -5 \\ 0 & -1 & 15 \end{bmatrix} \qquad \text{Substitution}$$

$$= \begin{bmatrix} -3 - 9 & 0 - 1 & 5 - (-5) \\ 2 - 0 & 7 - (-1) & -7 - 15 \end{bmatrix} \qquad \text{Definition of matrix subtraction}$$

$$= \begin{bmatrix} -12 & -1 & 10 \\ 2 & 8 & -22 \end{bmatrix} \qquad \text{Simplify.}$$

- You can multiply any matrix by a constant called a *scalar*. This is called **scalar multiplication**. To perform scalar multiplication, multiply each element by the scalar.

Example 2 If $D = \begin{bmatrix} -4 & 6 & -1 \\ 0 & 7 & 2 \\ -3 & -8 & -4 \end{bmatrix}$, find $2D$.

$$2D = 2\begin{bmatrix} -4 & 6 & -1 \\ 0 & 7 & 2 \\ -3 & -8 & -4 \end{bmatrix} \qquad \text{Substitution}$$

$$= \begin{bmatrix} 2(-4) & 2(6) & 2(-1) \\ 2(0) & 2(7) & 2(2) \\ 2(-3) & 2(-8) & 2(-4) \end{bmatrix} \qquad \text{Definition of scalar multiplication}$$

$$= \begin{bmatrix} -8 & 12 & -2 \\ 0 & 14 & 4 \\ -6 & -16 & -8 \end{bmatrix} \qquad \text{Simplify.}$$

- You can multiply two matrices if and only if the number of columns in the first matrix is equal to the number of rows in the second matrix. The product of two matrices is found by multiplying columns and rows. The entry in the first row and first column of AB, the resulting product, is found by multiplying corresponding elements in the first row of A and the first column of B and then adding.

Example 3 Find EF if $E = \begin{bmatrix} 3 & -2 \\ 0 & 6 \end{bmatrix}$ and $F = \begin{bmatrix} -1 & 5 \\ 6 & -3 \end{bmatrix}$.

$$EF = \begin{bmatrix} 3 & -2 \\ 0 & 6 \end{bmatrix} \cdot \begin{bmatrix} -1 & 5 \\ 6 & -3 \end{bmatrix}$$

Multiply the numbers in the first row of E by the numbers in the first column of F and add the products.

$$EF = \begin{bmatrix} 3 & -2 \\ 0 & 6 \end{bmatrix} \cdot \begin{bmatrix} -1 & 5 \\ 6 & -3 \end{bmatrix} = \begin{bmatrix} 3(-1) + (-2)(6) \end{bmatrix}$$

Multiply the numbers in the first row of E by the numbers in the second column of F and add the products.

$$EF = \begin{bmatrix} 3 & -2 \\ 0 & 6 \end{bmatrix} \cdot \begin{bmatrix} -1 & 5 \\ 6 & -3 \end{bmatrix} = \begin{bmatrix} 3(-1) + (-2)(6) & 3(5) + (-2)(-3) \end{bmatrix}$$

Multiply the numbers in the second row of E by the numbers in the first column of F and add the products.

$$EF = \begin{bmatrix} 3 & -2 \\ 0 & 6 \end{bmatrix} \cdot \begin{bmatrix} -1 & 5 \\ 6 & -3 \end{bmatrix} = \begin{bmatrix} 3(-1) + (-2)(6) & 3(5) + (-2)(-3) \\ 0(-1) + 6(6) \end{bmatrix}$$

Multiply the numbers in the second row of E by the numbers in the second column of F and add the products.

$$EF = \begin{bmatrix} 3 & -2 \\ 0 & 6 \end{bmatrix} \cdot \begin{bmatrix} -1 & 5 \\ 6 & -3 \end{bmatrix} = \begin{bmatrix} 3(-1) + (-2)(6) & 3(5) + (-2)(-3) \\ 0(-1) + 6(6) & 0(5) + 6(-3) \end{bmatrix}$$

Simplify the matrix.

$$\begin{bmatrix} 3(-1) + (-2)(6) & 3(5) + (-2)(-3) \\ 0(-1) + 6(6) & 0(5) + 6(-3) \end{bmatrix} = \begin{bmatrix} -15 & 21 \\ 36 & -18 \end{bmatrix}$$

Exercises If $A = \begin{bmatrix} 10 & -9 \\ 4 & -3 \\ -1 & 11 \end{bmatrix}$, $B = \begin{bmatrix} -1 & -3 \\ 2 & 8 \\ 7 & 6 \end{bmatrix}$, and $C = \begin{bmatrix} 8 & 0 \\ -2 & 2 \\ -10 & 6 \end{bmatrix}$, find each sum,

difference, or product.

1. $A + B$	**2.** $B + C$	**3.** $A - C$	**4.** $C - B$
5. $3A$	**6.** $5B$	**7.** $-4C$	**8.** $\frac{1}{2}C$
9. $2A + C$	**10.** $A - 5C$	**11.** $\frac{1}{2}C + B$	**12.** $3A - 3B$

If $X = \begin{bmatrix} 2 & -8 \\ 10 & 4 \end{bmatrix}$, $Y = \begin{bmatrix} -1 & 0 \\ 6 & -5 \end{bmatrix}$, and $Z = \begin{bmatrix} 4 & -8 \\ -7 & 0 \end{bmatrix}$, find each sum, difference,

or product.

13. $X + Z$	**14.** $Y + Z$	**15.** $X - Y$	**16.** $3Y$
17. $-6X$	**18.** $\frac{1}{2}X + Z$	**19.** $5Z - 2Y$	**20.** XY
21. YZ	**22.** XZ	**23.** $\frac{1}{2}(XZ)$	**24.** $XY + 2Z$

Extra Practice

Lesson 1-1

(pages 6–12)

For Exercises 1–7, refer to the figure.

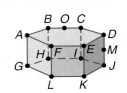

1. How many planes are shown in the figure?
2. Name three collinear points.
3. Name all planes that contain point G.
4. Name the intersection of plane ABD and plane DJK.
5. Name two planes that do not intersect.
6. Name a plane that contains $\overleftrightarrow{FK}$ and $\overleftrightarrow{EL}$.
7. Is the intersection of plane ACD and plane EDJ a point or a line? Explain.

Draw and label a figure for each relationship.

8. Line a intersects planes $\mathcal{A}$, $\mathcal{B}$, and C at three distinct points.
9. Planes X and Z intersect in line m. Line b intersects the two planes in two distinct points.

Lesson 1-2

(pages 13–19)

Find the precision for each measurement. Explain its meaning.

1. 42 in.
2. 86 mm
3. 251 cm
4. 33.5 in.
5. $5\frac{1}{4}$ ft
6. 89 m

Find the value of the variable and BC if B is between A and C.

7. $AB = 4x$, $BC = 5x$; $AB = 16$
8. $AB = 17$, $BC = 3m$, $AC = 32$
9. $AB = 9a$, $BC = 12a$, $AC = 42$
10. $AB = 25$, $BC = 3b$, $AC = 7b + 13$
11. $AB = 5n + 5$, $BC = 2n$; $AC = 54$
12. $AB = 6c - 8$, $BC = 3c + 1$, $AC = 65$

Lesson 1-3

(pages 21–27)

Use the Pythagorean Theorem to find the distance between each pair of points.

1. $A(0, 0)$, $B(-3, 4)$
2. $C(-1, 2)$, $N(5, 10)$
3. $X(-6, -2)$, $Z(6, 3)$
4. $M(-5, -8)$, $O(3, 7)$
5. $T(-10, 2)$, $R(6, -10)$
6. $F(5, -6)$, $N(-5, 6)$

Use the Distance Formula to find the distance between each pair of points.

7. $D(0, 0)$, $M(8, -7)$
8. $X(-1, 1)$, $Y(1, -1)$
9. $Z(-4, 0)$, $A(-3, 7)$
10. $K(6, 6)$, $D(-3, -3)$
11. $T(-1, 3)$, $N(0, 2)$
12. $S(7, 2)$, $E(-6, 7)$

Find the coordinates of the midpoint of a segment having the given endpoints.

13. $A(0, 0)$, $D(-2, -8)$
14. $D(-4, -3)$, $E(2, 2)$
15. $K(-4, -5)$, $M(5, 4)$
16. $R(-10, 5)$, $S(8, 4)$
17. $B(2.8, -3.4)$, $Z(1.2, 5.6)$
18. $D(-6.2, 7)$, $K(3.4, -4.8)$

Find the coordinates of the missing endpoint given that B is the midpoint of $\overline{AC}$.

19. $C(0, 0)$, $B(5, -6)$
20. $C(-7, -4)$, $B(3, 5)$
21. $C(8, -4)$, $B(-10, 2)$
22. $C(6, 8)$, $B(-3, 5)$
23. $C(6, -8)$, $B(3, -4)$
24. $C(-2, -4)$, $B(0, 5)$

Lesson 1-4

(pages 29–36)

For Exercises 1–14, use the figure at the right.
Name the vertex of each angle.

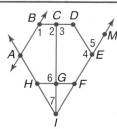

1. ∠1
2. ∠4
3. ∠6
4. ∠7

Name the sides of each angle.

5. ∠AIE
6. ∠4
7. ∠6
8. ∠AHF

Write another name for each angle.

9. ∠3
10. ∠DEF
11. ∠2

Measure each angle and classify it as *right,* *acute,* **or** *obtuse.*

12. ∠ABC
13. ∠CGF
14. ∠HIF

Lesson 1-5

(pages 37–43)

For Exercises 1–7, refer to the figure.

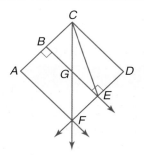

1. Name two acute vertical angles.
2. Name two obtuse vertical angles.
3. Name a pair of complementary adjacent angles.
4. Name a pair of supplementary adjacent angles.
5. Name a pair of congruent supplementary adjacent angles.
6. If $m\angle BGC = 4x + 5$ and $m\angle FGE = 6x - 15$, find $m\angle BGF$.
7. If $m\angle BCG = 5a + 5$, $m\angle GCE = 3a - 4$, and $m\angle ECD = 4a - 7$, find the value of a so that $\overline{AC} \perp \overline{CD}$.

8. The measure of ∠A is nine less than the measure of ∠B. If ∠A and ∠B form a linear pair, what are their measures?

9. The measure of an angle's complement is 17 more than the measure of the angle. Find the measure of the angle and its complement.

Lesson 1-6

(pages 45–50)

Name each polygon by its number of sides. Classify it as *convex* **or** *concave* **and**
regular **or** *irregular.* **Then find the perimeter.**

1.
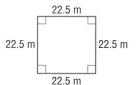

2.
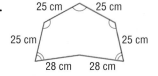

3.
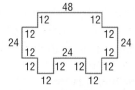

All measurements in inches.

Find the perimeter of each polygon.

4. triangle with vertices at $X(3, 3)$, $Y(-2, 1)$, and $Z(1, -3)$
5. pentagon with vertices at $P(-2, 3)$, $E(-5, 0)$, $N(-2, -4)$, $T(2, -1)$, and $A(2, 2)$
6. hexagon with vertices at $H(0, 4)$, $E(-3, 2)$, $X(-3, -2)$, $G(0, -5)$, $O(5, -2)$, and $N(5, 2)$

Lesson 2-1

(pages 62–66)

Make a conjecture based on the given information. Draw a figure to illustrate your conjecture.

1. Lines j and k are parallel.
3. $\overline{AB}$ bisects $\overline{CD}$ at K.

2. $A(-1, -7), B(4, -7), C(4, -3), D(-1, -3)$
4. $\overrightarrow{SR}$ is an angle bisector of $\angle TSU$.

Determine whether each conjecture is *true* or *false*. Give a counterexample for any false conjecture.

5. **Given:** *EFG* is an equilateral triangle.
   Conjecture: $EF = FG$

6. **Given:** r is a rational number.
   Conjecture: r is a whole number.

7. **Given:** n is a whole number.
   Conjecture: n is a rational number.

8. **Given:** $\angle 1$ and $\angle 2$ are supplementary angles.
   Conjecture: $\angle 1$ and $\angle 2$ form a linear pair.

Lesson 2-2

(pages 67–74)

Use the following statements to write a compound statement for each conjunction and disjunction. Then find its truth value.

p: $(-3)^2 = 9$ q: A robin is a fish. r: An acute angle measures less than 90°.

1. p and q
4. p or r
7. $q \wedge r$

2. p or q
5. $\sim p$ or q
8. $(p \wedge q) \vee r$

3. p and r
6. p or $\sim r$
9. $\sim p \vee \sim r$

Copy and complete each truth table.

10.

p	q	$\sim q$	$p \vee \sim q$
T			
T			
F			
F			

11.

p	q	$\sim p$	$\sim q$	$\sim p \vee \sim q$
T	T			
T	F			
F	T			
F	F			

Lesson 2-3

(pages 75–80)

Identify the hypothesis and conclusion of each statement.

1. If no sides of a triangle are equal, then it is a scalene triangle.
2. If it rains today, you will be wearing your raincoat.
3. If $6 - x = 11$, then $x = -5$.
4. If you are in college, you are at least 18 years old.

Write each statement in if-then form.

5. The sum of the measures of two supplementary angles is 180.
6. A triangle with two congruent sides is an isosceles triangle.
7. Two lines that do not intersect are parallel lines.
8. A Saint Bernard is a dog.

Write the converse, inverse, and contrapositive of each conditional statement. Determine whether each related conditional is *true* or *false*. If a statement is false, find a counterexample.

9. All triangles are polygons.
10. If two angles are congruent angles, then they have the same measure.
11. If three points lie on the same line, then they are collinear.
12. If $\overrightarrow{PQ}$ is a perpendicular bisector of $\overline{LM}$, then a right angle is formed.

Lesson 2-4

(pages 82–87)

Use the Law of Syllogism to determine whether a valid conclusion can be reached from each set of statements. If a valid conclusion is possible, write it. If not, write *no conclusion.*

1. (1) If it rains, then the field will be muddy.

(2) If the field is muddy, then the game will be cancelled.

2. (1) If you read a book, then you enjoy reading.

(2) If you are in the 10th grade, then you passed the 9th grade.

Determine if statement (3) follows from statements (1) and (2) by the Law of Detachment or the Law of Syllogism. If it does, state which law was used. If it does not, write *invalid.*

3. (1) If it snows outside, you will wear your winter coat.

(2) It is snowing outside.

(3) You will wear your winter coat.

4. (1) Two complementary angles are both acute angles.

(2) $\angle 1$ and $\angle 2$ are acute angles.

(3) $\angle 1$ and $\angle 2$ are complementary angles.

Lesson 2-5

(pages 89–93)

Determine whether the following statements are *always, sometimes,* or *never* true. Explain.

1. $\overleftrightarrow{RS}$ is perpendicular to $\overleftrightarrow{PS}$.

2. Three points will lie on one line.

3. Points B and C are in plane $\mathcal{K}$. A line perpendicular to line BC is in plane $\mathcal{K}$.

For Exercises 4–7, use the figure at the right. In the figure, $\overleftrightarrow{EC}$ and $\overrightarrow{CD}$ are in plane $\mathcal{R}$, and F is on $\overleftrightarrow{CD}$. State the postulate that can be used to show each statement is true.

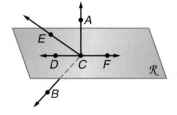

4. $\overleftrightarrow{DF}$ lies in plane $\mathcal{R}$.

5. E and C are collinear.

6. D, F, and E are coplanar.

7. E and F are collinear.

Lesson 2-6

(pages 94–100)

State the property that justifies each statement.

1. If $x - 5 = 6$, then $x = 11$.

2. If $AB = CD$ and $CD = EF$, then $AB = EF$.

3. If $a - b = r$, then $r = a - b$.

4. Copy and complete the following proof.

Given: $\dfrac{5x - 1}{8} = 3$

Prove: $x = 5$

Proof:

Statements	Reasons
a. ___?___	**a.** Given
b. ___?___	**b.** Multiplication Prop.
c. $5x - 1 = 24$	**c.** ___?___
d. $5x = 25$	**d.** ___?___
e. ___?___	**e.** Division Property

Lesson 2-7

(pages 101–106)

Justify each statement with a property of equality or a property of congruence.

1. If $CD = OP$, then $CD + GH = OP + GH$.
2. If $\overline{MN} \cong \overline{PQ}$, then $\overline{PQ} \cong \overline{MN}$.
3. If $\overline{TU} \cong \overline{JK}$ and $\overline{JK} \cong \overline{DF}$, then $\overline{TU} \cong \overline{DF}$.
4. If $AB = 10$ and $CD = 10$, then $AB = CD$.
5. $\overline{XB} \cong \overline{XB}$
6. If $GH = RS$, then $GH - VW = RS - VW$.
7. If $EF = XY$, then $EF + KL = XY + KL$.
8. If $\overline{JK} \cong \overline{XY}$ and $\overline{XY} \cong \overline{LM}$, then $\overline{JK} \cong \overline{LM}$.

Write a two-column proof.

9. **Given:** $\overline{AB} \cong \overline{AF}, \overline{AF} \cong \overline{ED}, \overline{ED} \cong \overline{CD}$
   Prove: $\overline{AB} \cong \overline{CD}$

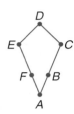

10. **Given:** $AC = DF, AB = DE$
    Prove: $BC = EF$

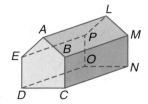

Lesson 2-8

(pages 107–114)

Find the measure of each numbered angle.

1. $m\angle 9 = 141 + x$
   $m\angle 10 = 25 + x$

2. $m\angle 11 = x + 40$
   $m\angle 12 = x + 10$
   $m\angle 13 = 3x + 30$

3. $m\angle 14 = x + 25$
   $m\angle 15 = 4x + 50$
   $m\angle 16 = x + 45$

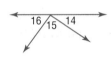

Determine whether the following statements are *always*, *sometimes*, or *never* true.

4. Two angles that are complementary are congruent.
5. Two angles that form a linear pair are complementary.
6. Two congruent angles are supplementary.
7. Perpendicular lines form four right angles.
8. Two right angles are supplementary.
9. Two lines intersect to form four right angles.

Lesson 3-1

(pages 126–131)

For Exercises 1–3, refer to the figure at the right.

1. Name all segments parallel to $\overline{AE}$.
2. Name all planes intersecting plane BCN.
3. Name all segments skew to $\overline{DC}$.

Identify each pair of angles as *alternate interior*, *alternate exterior*, *corresponding*, or *consecutive interior* angles.

4. $\angle 2$ and $\angle 5$
5. $\angle 9$ and $\angle 13$
6. $\angle 12$ and $\angle 13$
7. $\angle 3$ and $\angle 6$

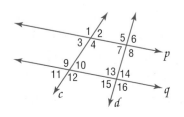

Lesson 3-2

(pages 133–138)

In the figure, $m\angle 5 = 72$ and $m\angle 9 = 102$.
Find the measure of each angle.

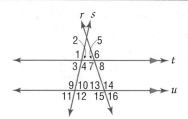

1. $m\angle 1$ **2.** $m\angle 13$

3. $m\angle 4$ **4.** $m\angle 10$

5. $m\angle 7$ **6.** $m\angle 16$

Find x and y in each figure.

7.

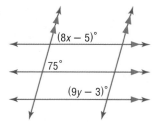

8.

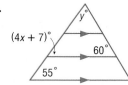

Lesson 3-3

(pages 139–144)

Find the slope of each line.

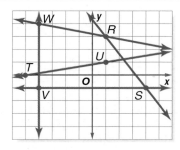

1. $\overleftrightarrow{RS}$ **2.** $\overleftrightarrow{TU}$

3. $\overleftrightarrow{WV}$ **4.** $\overleftrightarrow{WR}$

5. a line parallel to $\overleftrightarrow{TU}$

6. a line perpendicular to $\overleftrightarrow{WR}$

7. a line perpendicular to $\overleftrightarrow{WV}$

Determine whether $\overleftrightarrow{RS}$ and $\overleftrightarrow{TU}$ are *parallel*, *perpendicular*, or *neither*.

8. $R(3, 5)$, $S(5, 6)$, $T(-2, 0)$, $U(4, 3)$ **9.** $R(5, 11)$, $S(2, 2)$, $T(-1, 0)$, $U(2, 1)$

10. $R(-1, 4)$, $S(-3, 7)$, $T(5, -1)$, $U(8, 1)$ **11.** $R(-2, 5)$, $S(-4, 1)$, $T(3, 3)$, $U(1, 5)$

Lesson 3-4

(pages 145–150)

Write an equation in slope-intercept form of the line having the given slope and
y-intercept.

1. $m = 1$, y-intercept: -5 **2.** $m = -\frac{1}{2}$, y-intercept: $\frac{1}{2}$ **3.** $m = 3$, $b = -\frac{1}{4}$

Write an equation in point-slope form of the line having the given slope that
contains the given point.

4. $m = 3$, $(-2, 4)$ **5.** $m = -4$, $(0, 3)$ **6.** $m = \frac{2}{3}$, $(5, -7)$

For Exercises 7–14, use the graph at the right.
Write an equation in slope-intercept form for each line.

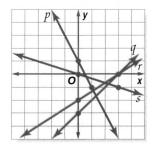

7. p **8.** q

9. r **10.** s

11. parallel to line q, contains $(2, -5)$

12. perpendicular to line r, contains $(0, 1)$

13. parallel to line s, contains $(-2, -2)$

14. perpendicular to line p, contains $(0, 0)$

Lesson 3-5

(pages 151–157)

Given the following information, determine which lines, if any, are parallel. State the postulate or theorem that justifies your answer.

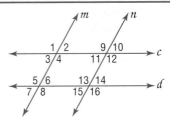

1. $\angle 9 \cong \angle 16$
2. $\angle 10 \cong \angle 16$
3. $\angle 12 \cong \angle 13$
4. $m\angle 12 + m\angle 14 = 180$

Find x so that $r \parallel s$.

5.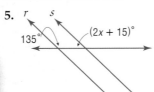

6.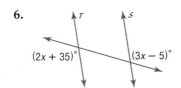

7.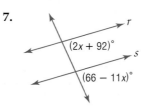

Lesson 3-6

(pages 159–164)

Copy each figure. Draw the segment that represents the distance indicated.

1. P to $\overleftrightarrow{RS}$

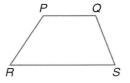

2. J to $\overleftrightarrow{KL}$

3. B to $\overrightarrow{FE}$

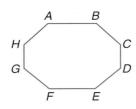

Find the distance between each pair of parallel lines.

4. $y = \frac{2}{3}x - 2$

$y = \frac{2}{3}x + \frac{1}{2}$

5. $y = 2x + 4$

$y - 2x = -5$

6. $x + 4y = -6$

$x + 4y = 4$

COORDINATE GEOMETRY Construct a line perpendicular to ℓ through P. Then find the distance from P to ℓ.

7. Line ℓ contains points $(0, 4)$ and $(-4, 0)$. Point P has coordinates $(2, -1)$.
8. Line ℓ contains points $(3, -2)$ and $(0, 2)$. Point P has coordinates $(-2.5, 3)$.

Lesson 4-1

(pages 178–183)

Use a protractor to classify each triangle as *acute, equiangular, obtuse,* or *right.*

1.

2.

3.

Identify the indicated type of triangles in the figure if $\overline{AB} \cong \overline{CD}, \overline{AD} \cong \overline{BC}, \overline{AE} \cong \overline{BE} \cong \overline{EC} \cong \overline{ED}$, and $m\angle BAD = m\angle ABC = m\angle BCD = m\angle ADC = 90$.

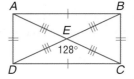

4. right
5. obtuse
6. acute
7. isosceles

8. Find a and the measure of each side of equilateral triangle MNO if $MN = 5a$, $NO = 4a + 6$, and $MO = 7a - 12$.

9. Triangle TAC is an isosceles triangle with $\overline{TA} \cong \overline{AC}$. Find b, TA, AC, and TC if $TA = 3b + 1$, $AC = 4b - 11$, and $TC = 6b - 2$.

Lesson 4-2

(pages 185–191)

Find the measure of each angle.

1. ∠1
2. ∠2
3. ∠3
4. ∠4
5. ∠5
6. ∠6
7. ∠7
8. ∠8
9. ∠9
10. ∠10

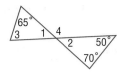

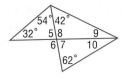

Lesson 4-3

(pages 192–198)

Identify the congruent triangles in each figure.

1.

2.

3.

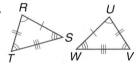

4.

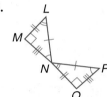

5. Write a two-column proof.

Given: △ANG ≅ △NGA

△NGA ≅ △GAN

Prove: △AGN is equilateral and equiangular.

Lesson 4-4

(pages 200–206)

Determine whether △RST ≅ △JKL given the coordinates of the vertices. Explain.

1. $R(-6, 2)$, $S(-4, 4)$, $T(-2, 2)$, $J(6, -2)$, $K(4, -4)$, $L(2, -2)$

2. $R(-6, 3)$, $S(-4, 7)$, $T(-2, 3)$, $J(2, 3)$, $K(5, 7)$, $L(6, 3)$

Write a two-column proof.

3. **Given:** △GWN is equilateral.

$\overline{WS} \cong \overline{WI}$

∠SWG ≅ ∠IWN

Prove: △SWG ≅ △IWN

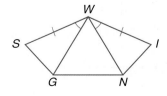

4. **Given:** △ANM ≅ △ANI

$\overline{DI} \cong \overline{OM}$

$\overline{ND} \cong \overline{NO}$

Prove: △DIN ≅ △OMN

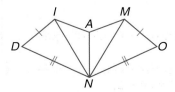

Extra Practice

Write a paragraph proof.

1. Given: $\triangle TEN$ is isosceles with base $\overline{TN}$.
   $\angle 1 \cong \angle 4$, $\angle T \cong \angle N$

Prove: $\triangle TEC \cong \triangle NEA$

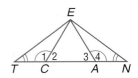

2. Given: $\angle S \cong \angle W$
   $\overline{SY} \cong \overline{YW}$

Prove: $\overline{ST} \cong \overline{WV}$

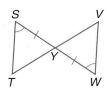

Write a flow proof.

3. Given: $\angle 1 \cong \angle 2$, $\angle 3 \cong \angle 4$

Prove: $\overline{PT} \cong \overline{LX}$

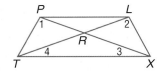

4. Given: $\overline{FP} \parallel \overline{ML}$, $\overline{FL} \parallel \overline{MP}$

Prove: $\overline{MP} \cong \overline{FL}$

Refer to the figure for Exercises 1–6.

1. If $\overline{AD} \cong \overline{BD}$, name two congruent angles.

2. If $\overline{BF} \cong \overline{FG}$, name two congruent angles.

3. If $\overline{BE} \cong \overline{BG}$, name two congruent angles.

4. If $\angle FBE \cong \angle FEB$, name two congruent segments.

5. If $\angle BCA \cong \angle BAC$, name two congruent segments.

6. If $\angle DBC \cong \angle BCD$, name two congruent segments.

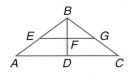

Position and label each triangle on the coordinate plane.

1. isosceles $\triangle ABC$ with base $\overline{BC}$ that is r units long

2. equilateral $\triangle XYZ$ with sides $4b$ units long

3. isosceles right $\triangle RST$ with hypotenuse $\overline{ST}$ and legs $(3 + a)$ units long

4. equilateral $\triangle CDE$ with base $\overline{DE}$ $\frac{1}{4}b$ units long.

Name the missing coordinates of each triangle.

5.

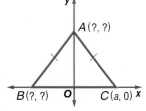

6.

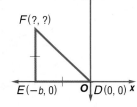

7.

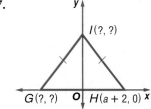

Lesson 5-1

(pages 238–246)

For Exercises 1–4, refer to the figures at the right.

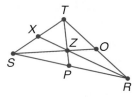

1. Suppose $CP = 7x - 1$ and $PB = 6x + 3$. If S is the circumcenter of $\triangle ABC$, find x and CP.
2. Suppose $m\angle ACT = 15a - 8$ and $m\angle ACB = 74$. If S is the incenter of $\triangle ABC$, find a and $m\angle ACT$.
3. Suppose $TO = 7b + 5$, $OR = 13b - 10$, and $TR = 18b$. If Z is the centroid of $\triangle TRS$, find b and TR.
4. Suppose $XR = 19n - 14$ and $ZR = 10n + 4$. If Z is the centroid of $\triangle TRS$, find n and ZR.

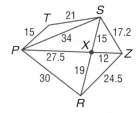

State whether each sentence is *always*, *sometimes*, or *never* true.

5. The circumcenter and incenter of a triangle are the same point.
6. The three altitudes of a triangle intersect at a point inside the triangle.
7. In an equilateral triangle, the circumcenter, incenter, and centroid are the same point.
8. The incenter is inside of a triangle.

Lesson 5-2

(pages 247–254)

Determine the relationship between the measures of the given angles.

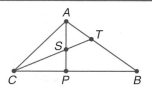

1. $\angle TPS$, $\angle TSP$
2. $\angle PRZ$, $\angle ZPR$
3. $\angle SPZ$, $\angle SZP$
4. $\angle SPR$, $\angle SRP$

5. **Given:** $FH > FG$
   Prove: $m\angle 1 > m\angle 2$

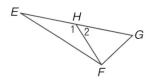

6. **Given:** $\overline{RQ}$ bisects $\angle SRT$.
   Prove: $m\angle SQR > m\angle SRQ$

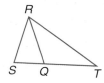

Lesson 5-3

(pages 255–260)

State the assumption you would make to start an indirect proof of each statement.

1. $\angle ABC \cong \angle XYZ$
2. An angle bisector of an equilateral triangle is also a median.
3. $\overrightarrow{RS}$ bisects $\angle ARC$

Write an indirect proof.

4. **Given:** $\angle AOY \cong \angle AOX$
   $\overline{XO} \not\cong \overline{YO}$
   Prove: $\overrightarrow{AO}$ is not the angle bisector of $\angle XAY$.

5. **Given:** $\triangle RUN$
   Prove: There can be no more than one right angle in $\triangle RUN$.

Lesson 5-4

(pages 261–266)

Determine whether the given measures can be the lengths of the sides of a triangle. Write *yes* or *no*.

1. 2, 2, 6   **2.** 2, 3, 4   **3.** 6, 8, 10   **4.** 1, 1, 2
5. 15, 20, 30   **6.** 1, 3, 5   **7.** 2.5, 3.5, 6.5   **8.** 0.3, 0.4, 0.5

Find the range for the measure of the third side of a triangle given the measures of two sides.

9. 6 and 10   **10.** 2 and 5   **11.** 20 and 12   **12.** 8 and 8
13. 18 and 36   **14.** 32 and 34   **15.** 2 and 29   **16.** 80 and 25

Write a two-column proof.

17. Given: $RS = RT$
  Prove: $UV + VS > UT$

18. Given: quadrilateral $ABCD$
  Prove: $AD + CD + AB > BC$

Lesson 5-5

(pages 267–273)

Write an inequality relating the given pair of angle or segment measures.

1. XZ, OZ

2. $m\angle ZIO$, $m\angle ZUX$

3. $m\angle AEZ$, $m\angle AZE$

4. IO, AE

5. $m\angle AZE$, $m\angle IZO$

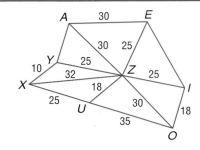

Write an inequality to describe the possible values of x.

6.

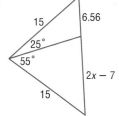

7.

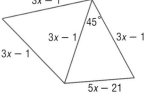

Lesson 6-1

(pages 282–287)

1. ARCHITECTURE   The ratio of the height of a model of a house to the actual house is 1:63. If the width of the model is 16 inches, find the width of the actual house in feet.

2. CONSTRUCTION   A 64-inch long board is divided into lengths in the ratio 2:3. What are the two lengths into which the board is divided?

ALGEBRA   Solve each proportion.

3. $\dfrac{x+4}{26} = -\dfrac{1}{3}$ **4.** $\dfrac{3x+1}{14} = \dfrac{5}{7}$ **5.** $\dfrac{x-3}{4} = \dfrac{x+1}{5}$ **6.** $\dfrac{2x+2}{2x-1} = \dfrac{1}{3}$

7. Find the measures of the sides of a triangle if the ratio of the measures of three sides of a triangle is 9:6:5, and its perimeter is 100 inches.

8. Find the measures of the angles in a triangle if the ratio of the measures of the three angles is 13:16:21.

Lesson 6-2

(pages 289–297)

Determine whether each pair of figures is similar. Justify your answer.

1.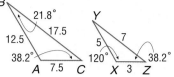

2.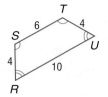

For Exercises 3 and 4, use $\triangle RST$ with vertices $R(3, 6)$, $S(1, 2)$, and $T(3, -1)$. Explain.

3. If the coordinates of each vertex are decreased by 3, describe the new figure. Is it similar to $\triangle RST$?

4. If the coordinates of each vertex are multiplied by 0.5, describe the new figure. Is it similar to $\triangle RST$?

Lesson 6-3

(pages 298–306)

Determine whether each pair of triangles is similar. Justify your answer.

1.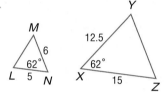

2.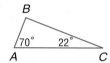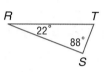

ALGEBRA Identify the similar triangles. Find x and the measures of the indicated sides.

3. RT and SV

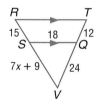

4. PN and MN

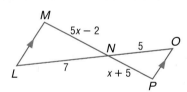

Lesson 6-4

(pages 307–315)

1. If $HI = 28$, $LH = 21$, and $LK = 8$, find IJ.

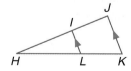

2. Find x, AD, DR, and QR if $AU = 15$, $QU = 25$, $AD = 3x + 6$, $DR = 8x - 2$, and $UD = 15$.

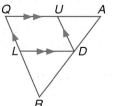

Find x so that $\overline{XY} \parallel \overline{LM}$.

3. $XL = 3$, $YM = 5$, $LD = 9$, $MD = x + 3$

4. $YM = 3$, $LD = 3x + 1$, $XL = 4$, $MD = x + 7$

5. $MD = 5x - 6$, $YM = 3$, $LD = 5x + 1$, $XL = 5$

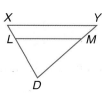

Lesson 6-5

(pages 316–323)

Find the perimeter of each triangle.

1. $\triangle ABC$ if $\triangle ABC \sim \triangle DBE$, $AB = 17.5$, $BC = 15$, $BE = 6$, and $DE = 5$

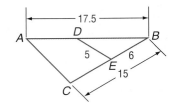

2. $\triangle RST$ if $\triangle RST \sim \triangle XYZ$, $RT = 12$, $XZ = 8$, and the perimeter of $\triangle XYZ = 22$

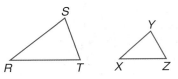

3. $\triangle LMN$ if $\triangle LMN \sim \triangle NXY$, $NX = 14$, $YX = 11$, $YN = 9$, and $LN = 27$

4. $\triangle GHI$ if $\triangle ABC \sim \triangle GHI$, $AB = 6$, $GH = 10$, and the perimeter of $\triangle ABC = 25$

Lesson 6-6

(pages 325–331)

Stage 1 of a fractal is shown drawn on grid paper. Stage 1 is made by dividing a square into 4 congruent squares and shading the top left-hand square.

1. Draw Stage 2 by repeating the Stage 1 process in each of the 3 remaining unshaded squares. How many shaded squares are at this stage?

2. Draw Stage 3 by repeating the Stage 1 process in each of the unshaded squares in Stage 2. How many shaded squares are at this stage?

Find the value of each expression. Then, use that value as the next x in the expression. Repeat the process and describe your observations.

3. $x^{\frac{1}{4}}$, where x initially equals 6

4. 4^x, where x initially equals 0.4

5. x^3, where x initially equals 0.5

6. 3^x, where x initially equals 10

Lesson 7-1

(pages 342–348)

Find the geometric mean between each pair of numbers. State exact answers and answers to the nearest tenth.

1. 8 and 12

2. 15 and 20

3. 1 and 2

4. 4 and 16

5. $3\sqrt{2}$ and $6\sqrt{2}$

6. $\frac{1}{2}$ and 10

7. $\frac{3}{8}$ and $\frac{1}{2}$

8. $\frac{\sqrt{2}}{2}$ and $\frac{3\sqrt{2}}{2}$

9. $\frac{1}{10}$ and $\frac{7}{10}$

Find the altitude of each triangle.

10.

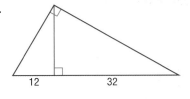

11.

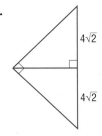

12.

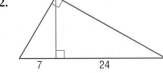

Lesson 7-2

(pages 350–356)

Determine whether △DEF is a right triangle for the given vertices. Explain.

1. D(0, 1), E(3, 2), F(2, 3)

2. D(−2, 2), E(3, −1), F(−4, −3)

3. D(2, −1), E(−2, −4), F(−4, −1)

4. D(1, 2), E(5, −2), F(−2, −1)

Determine whether each set of measures are the sides of a right triangle. Then state whether they form a Pythagorean triple.

5. 1, 1, 2

6. 21, 28, 35

7. 3, 5, 7

8. 2, 5, 7

9. 24, 45, 51

10. $\dfrac{1}{3}, \dfrac{5}{3}, \dfrac{\sqrt{26}}{3}$

11. $\dfrac{6}{11}, \dfrac{8}{11}, \dfrac{10}{11}$

12. $\dfrac{1}{2}, \dfrac{1}{2}, 1$

13. $\dfrac{\sqrt{6}}{3}, \dfrac{\sqrt{10}}{5}, \dfrac{\sqrt{240}}{15}$

Lesson 7-3

(pages 357–363)

Find the measures of x and y.

1.

2.

3.
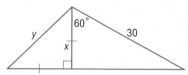

4.
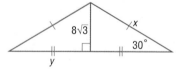

5.
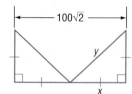

6.
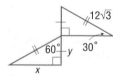

Lesson 7-4

(pages 364–370)

Use △MAN with right angle N to find sin M, cos M, tan M, sin A, cos A, and tan A. Express each ratio as a fraction, and as a decimal to the nearest hundredth.

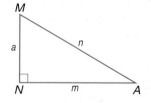

1. m = 21, a = 28, n = 35

2. $m = \sqrt{2}, a = \sqrt{3}, n = \sqrt{5}$

3. $m = \dfrac{\sqrt{2}}{2}, a = \dfrac{\sqrt{2}}{2}, n = 1$

4. $m = 3\sqrt{5}, a = 5\sqrt{3}, n = 2\sqrt{30}$

Find the measure of each angle to the nearest tenth of a degree.

5. cos A = 0.6293

6. sin B = 0.5664

7. tan C = 0.2665

8. sin D = 0.9352

9. tan M = 0.0808

10. cos R = 0.1097

Find x. Round to the nearest tenth.

11.
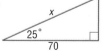

12.
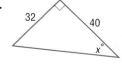

13.

Lesson 7-5

(pages 371–376)

1. **COMMUNICATIONS** A house is located below a hill that has a satellite dish. If $MN = 450$ feet and $RN = 120$ feet, what is the measure of the angle of elevation to the top of the hill?

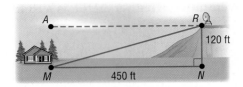

2. **AMUSEMENT PARKS** Mandy is at the top of the Mighty Screamer roller coaster. Her friend Bryn is at the bottom of the coaster waiting for the next ride. If the angle of depression from Mandy to Bryn is $26°$ and OL is 75 feet, what is the distance from L to C?

3. **SKIING** Mitchell is at the top of the Bridger Peak ski run. His brother Scott is looking up from the ski lodge at I. If the angle of elevation from Scott to Mitchell is $13°$ and the distance from K to I is 2000 ft, what is the length of the ski run SI?

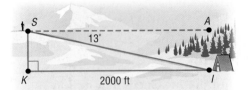

Lesson 7-6

(pages 377–383)

Find each measure using the given measures from $\triangle ANG$. Round angle measures to the nearest degree and side measures to the nearest tenth.

1. If $m\angle N = 32$, $m\angle A = 47$, and $n = 15$, find a.
2. If $a = 10.5$, $m\angle N = 26$, $m\angle A = 75$, find n.
3. If $n = 18.6$, $a = 20.5$, $m\angle A = 65$, find $m\angle N$.
4. If $a = 57.8$, $n = 43.2$, $m\angle A = 33$, find $m\angle N$.

Solve each $\triangle AKX$ described below. Round angle measures to the nearest degree and side measures to the nearest tenth.

5. $m\angle X = 62$, $a = 28.5$, $m\angle K = 33$
6. $k = 3.6$, $x = 3.7$, $m\angle X = 55$
7. $m\angle K = 35$, $m\angle A = 65$, $x = 50$
8. $m\angle A = 122$, $m\angle X = 15$, $a = 33.2$

Lesson 7-7

(pages 385–390)

In $\triangle CDE$, given the lengths of the sides, find the measure of the stated angle to the nearest tenth.

1. $c = 100$, $d = 125$, $e = 150$; $m\angle E$
2. $c = 5$, $d = 6$, $e = 9$; $m\angle C$
3. $c = 1.2$, $d = 3.5$, $e = 4$; $m\angle D$
4. $c = 42.5$, $d = 50$, $e = 81.3$; $m\angle E$

Solve each triangle using the given information. Round angle measures to the nearest degree and side measures to the nearest tenth.

5.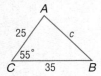

6.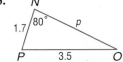

7.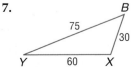

Lesson 8-1

(pages 404–409)

Find the sum of the measures of the interior angles of each convex polygon.

1. 25-gon
2. 30-gon
3. 22-gon
4. 17-gon
5. 5*a*-gon
6. *b*-gon

The measure of an interior angle of a regular polygon is given. Find the number of sides in each polygon.

7. 156
8. 168
9. 162

Find the measures of an interior angle and an exterior angle given the number of sides of a regular polygon. Round to the nearest tenth.

10. 15
11. 13
12. 42

Lesson 8-2

(pages 411–416)

Complete each statement about ▱*RSTU*. Justify your answer.

1. $\angle SRU \cong$ __?__
2. $\angle UTS$ is supplementary to __?__
3. $\overline{RU} \parallel$ __?__
4. $\overline{RU} \cong$ __?__
5. $\triangle RST \cong$ __?__
6. $\overline{SV} \cong$ __?__

ALGEBRA Use ▱*ABCD* to find each measure or value.

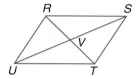

7. $m\angle BAE =$ __?__
8. $m\angle BCE =$ __?__
9. $m\angle BEC =$ __?__
10. $m\angle CED =$ __?__
11. $m\angle ABE =$ __?__
12. $m\angle EBC =$ __?__
13. $a =$ __?__
14. $b =$ __?__
15. $c =$ __?__
16. $d =$ __?__

Lesson 8-3

(pages 417–423)

Determine whether each quadrilateral is a parallelogram. Justify your answer.

1.
2.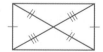
3.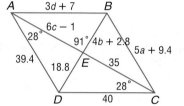

ALGEBRA Find *x* and *y* so that each quadrilateral is a parallelogram.

4.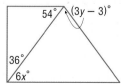
5.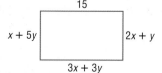
6.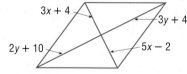

Determine whether a figure with the given vertices is a parallelogram. Use the method indicated.

7. $L(-3, 2)$, $M(5, 2)$, $N(3, -6)$, $O(-5, -6)$; Slope Formula
8. $W(-5, 6)$, $X(2, 5)$, $Y(-3, -4)$, $Z(-8, -2)$; Distance Formula
9. $Q(-5, 4)$, $R(0, 6)$, $S(3, -1)$, $T(-2, -3)$; Midpoint Formula
10. $G(-5, 0)$, $H(-13, 5)$, $I(-10, 9)$, $J(-2, 4)$; Distance and Slope Formulas

Lesson 8-4

(pages 424–430)

ALGEBRA   Refer to rectangle QRST.

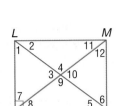

1. If $QU = 2x + 3$ and $UT = 4x - 9$, find SU.
2. If $RU = 3x - 6$ and $UT = x + 9$, find RS.
3. If $QS = 3x + 40$ and $RT = 16 - 3x$, find QS.
4. If $m\angle STQ = 5x + 3$ and $m\angle RTQ = 3 - x$, find x.
5. If $m\angle SRQ = x^2 + 6$ and $m\angle RST = 36 - x$, find $m\angle SRT$.
6. If $m\angle TQR = x^2 + 16$ and $m\angle QTR = x + 32$, find $m\angle TQS$.

Find each measure in rectangle LMNO if $m\angle 5 = 38$.

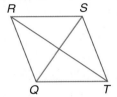

7. $m\angle 1$
8. $m\angle 2$
9. $m\angle 3$
10. $m\angle 4$
11. $m\angle 6$
12. $m\angle 7$
13. $m\angle 8$
14. $m\angle 9$
15. $m\angle 10$
16. $m\angle 11$
17. $m\angle 12$
18. $m\angle OLM$

Lesson 8-5

(pages 431–437)

In rhombus QRST, $m\angle QRS = m\angle TSR - 40$ and $TS = 15$.

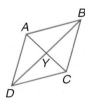

1. Find $m\angle TSQ$.
2. Find $m\angle QRS$.
3. Find $m\angle SRT$.
4. Find QR.

ALGEBRA   Use rhombus ABCD with $AY = 6$, $DY = 3r + 3$, and $BY = \dfrac{10r - 4}{2}$.

5. Find $m\angle ACB$.
6. Find $m\angle ABD$.
7. Find BY.
8. Find AC.

Lesson 8-6

(pages 439–445)

COORDINATE GEOMETRY   For each quadrilateral with the given vertices,
a. verify that the quadrilateral is a trapezoid, and
b. determine whether the figure is an isosceles trapezoid.

1. $A(0, 9)$, $B(3, 4)$, $C(-5, 4)$, $D(-2, 9)$
2. $Q(1, 4)$, $R(4, 6)$, $S(10, 7)$, $T(1, 1)$
3. $L(1, 2)$, $M(4, -1)$, $N(3, -5)$, $O(-3, 1)$
4. $W(1, -2)$, $X(3, -1)$, $Y(7, -2)$, $Z(1, -5)$

5. For trapezoid $ABDC$, E and F are midpoints of the legs. Find CD.

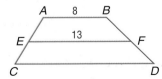

6. For trapezoid $LMNO$, P and Q are midpoints of the legs. Find PQ, $m\angle M$, and $m\angle O$.

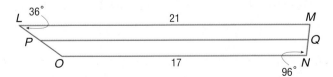

7. For isosceles trapezoid $QRST$, find the length of the median, $m\angle S$, and $m\angle R$.

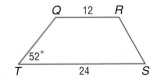

8. For trapezoid $XYZW$, A and B are midpoints of the legs. For trapezoid $XYBA$, C and D are midpoints of the legs. Find CD.

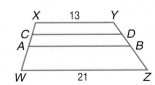

770   Extra Practice

Lesson 8-7

Name the missing coordinates for each quadrilateral.

1. isosceles trapezoid *ABCD*

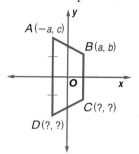

2. rectangle *QRST*

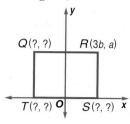

Position and label each figure on the coordinate plane. Then write a coordinate proof for each of the following.

3. The diagonals of a square are congruent.

4. Quadrilateral *EFGH* with vertices $E(0, 0)$, $F(a\sqrt{2}, a\sqrt{2})$, $G(2a + a\sqrt{2}, a\sqrt{2})$, and $H(2a, 0)$ is a rhombus.

Lesson 9-1

(pages 463–469)

COORDINATE GEOMETRY   Graph each figure and its image under the given reflection.

1. $\triangle ABN$ with vertices $A(2, 2)$, $B(3, -2)$, and $N(-3, -1)$ in the *x*-axis

2. rectangle *BARN* with vertices $B(3, 3)$, $A(3, -4)$, $R(-1, -4)$, and $N(-1, 3)$ in the line $y = x$

3. trapezoid *ZOID* with vertices $Z(2, 3)$, $O(2, -4)$, $I(-3, -3)$, and $D(-3, 1)$ in the origin

4. $\triangle PQR$ with vertices $P(-2, 1)$, $Q(2, -2)$, and $R(-3, -4)$ in the *y*-axis

5. square *BDFH* with vertices $B(-4, 4)$, $D(-1, 4)$, $F(-1, 1)$, and $H(-4, 1)$ in the origin

6. quadrilateral *QUAD* with vertices $Q(1, 3)$, $U(3, 1)$, $A(-1, 0)$, and $D(-3, 4)$ in the line $y = -1$

7. $\triangle CAB$ with vertices $C(0, 4)$, $A(1, -3)$, and $B(-4, 0)$ in the line $x = -2$

Lesson 9-2

(pages 470–475)

In each figure, $c \parallel d$. Determine whether the red figure is a translation image of the blue figure. Write *yes* or *no*. Explain your answer.

1.

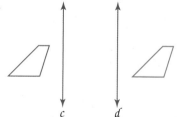

2.

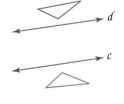

3.

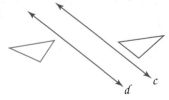

COORDINATE GEOMETRY   Graph each figure and its image under the given translation.

4. $\overline{LM}$ with endpoints $L(2, 3)$ and $M(-4, 1)$ under the translation $(x, y) \rightarrow (x + 2, y + 1)$

5. $\triangle DEF$ with vertices $D(1, 2)$, $E(-2, 1)$, and $F(-3, -1)$ under the translation $(x, y) \rightarrow (x - 1, y - 3)$

6. quadrilateral *WXYZ* with vertices $W(1, 1)$, $X(-2, 3)$, $Y(-3, -2)$, and $Z(2, -2)$ under the translation $(x, y) \rightarrow (x + 1, y - 1)$

7. pentagon *ABCDE* with vertices $A(1, 3)$, $B(-1, 1)$, $C(-1, -2)$, $D(3, -2)$, and $E(3, 1)$ under the translation $(x, y) \rightarrow (x - 2, y + 3)$

8. $\triangle RST$ with vertices $R(-4, 3)$, $S(-2, -3)$, and $T(2, -1)$ under the translation $(x, y) \rightarrow (x + 3, y - 2)$

GEOMETRY Draw the rotation image of each figure 90° in the given ...bout the center point and label the vertices with coordinates.

1. ..LM with vertices $K(4, 2)$, $L(1, 3)$, and $M(2, 1)$ counterclockwise about the point $P(1, -1)$

2. $\triangle FGH$ with vertices $F(-3, -3)$, $G(2, -4)$, and $H(-1, -1)$ clockwise about the point $P(0, 0)$

COORDINATE GEOMETRY Draw the rotation image of each triangle by reflecting the triangle in the given lines. State the coordinates of the rotation image and the angle of rotation.

3. $\triangle HIJ$ with vertices $H(2, 2)$, $I(-2, 1)$, and $J(-1, -2)$, reflected in the x-axis and then in the y-axis

4. $\triangle NOP$ with vertices $N(3, 1)$, $O(5, -3)$, and $P(2, -3)$, reflected in the y-axis and then in the line $y = x$

5. $\triangle QUA$ with vertices $Q(0, 4)$, $U(-3, 2)$, and $A(1, 1)$, reflected in the x-axis and then in the line $y = x$

6. $\triangle AEO$ with vertices $A(-5, 3)$, $E(-4, 1)$, and $O(-1, 2)$, reflected in the line $y = -x$ and then in the y-axis

Lesson 9-4

Determine whether a semi-regular tessellation can be created from each set of figures. Assume each figure has a side length of 1 unit.

1. regular hexagons and squares

2. squares and regular pentagons

3. regular hexagons and regular octagons

Determine whether each statement is *always, sometimes,* or *never* true.

4. Any right isosceles triangle forms a uniform tessellation.

5. A semi-regular tessellation is uniform.

6. A polygon that is not regular can tessellate the plane.

7. If the measure of one interior angle of a regular polygon is greater than 120, it cannot tessellate the plane.

Lesson 9-5

Find the measure of the dilation image or the preimage of $\overline{OM}$ with the given scale factor.

1. $OM = 1$, $r = -2$

2. $OM = 3$, $r = \dfrac{1}{3}$

3. $O'M' = \dfrac{3}{4}$, $r = 3$

4. $OM = \dfrac{7}{8}$, $r = -\dfrac{5}{7}$

5. $O'M' = 4$, $r = -\dfrac{2}{3}$

6. $O'M' = 4.5$, $r = -1.5$

COORDINATE GEOMETRY Find the image of each polygon, given the vertices, after a dilation centered at the origin with scale factor $r = 3$. Then graph a dilation with $r = \dfrac{1}{3}$.

7. $T(1, 1)$, $R(-1, 2)$, $I(-2, 0)$

8. $E(2, 1)$, $I(3, -3)$, $O(-1, -2)$

9. $A(0, -1)$, $B(-1, 1)$, $C(0, 2)$, $D(1, 1)$

10. $B(1, 0)$, $D(2, 0)$, $F(3, -2)$, $H(0, -2)$

Lesson 9-6

(pages 498–505)

Find the magnitude and direction of $\overrightarrow{XY}$ for the given coordinates.

1. $X(1, 1), Y(-2, 3)$
2. $X(-1, -1), Y(2, 2)$
3. $X(-5, 4), Y(-2, -3)$
4. $X(2, 1), Y(-4, -4)$
5. $X(-2, -1), Y(2, -2)$
6. $X(3, -1), Y(-3, 1)$

Graph the image of each figure under a translation by the given vector.

7. $\triangle HIJ$ with vertices $H(2, 3), I(-4, 2), J(-1, 1); \vec{a} = \langle 1, 3 \rangle$
8. quadrilateral $RSTW$ with vertices $R(4, 0), S(0, 1), T(-2, -2), W(3, -1); \vec{x} = \langle -3, 4 \rangle$
9. pentagon $AEIOU$ with vertices $A(-1, 3), E(2, 3), I(2, 0), O(-1, -2), U(-3, 0); \vec{b} = \langle -2, -1 \rangle$

Find the magnitude and direction of each resultant for the given vectors.

10. $\vec{c} = \langle 2, 3 \rangle, \vec{d} = \langle 3, 4 \rangle$
11. $\vec{a} = \langle 1, 3 \rangle, \vec{b} = \langle -4, 3 \rangle$
12. $\vec{x} = \langle 1, 2 \rangle, \vec{y} = \langle 4, -6 \rangle$
13. $\vec{s} = \langle 2, 5 \rangle, \vec{t} = \langle -6, -8 \rangle$
14. $\vec{m} = \langle 2, -3 \rangle, \vec{n} = \langle -2, 3 \rangle$
15. $\vec{u} = \langle -7, 2 \rangle, \vec{v} = \langle 4, 1 \rangle$

Lesson 9-7

(pages 506–511)

Find the coordinates of the image under the stated transformation.

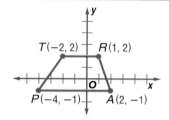

1. reflection in the x-axis

2. rotation 90° clockwise about the origin

3. translation $(x, y) \rightarrow (x - 4, y + 3)$

4. dilation by scale factor -4

Use a matrix to find the coordinates of the vertices of the image of each figure after the stated transformation.

5. $\triangle DEF$ with $D(2, 4), E(-2, -4)$, and $F(4, -6)$; dilation by a scale factor of 2.5

6. $\triangle RST$ with $R(3, 4), S(-6, -2)$, and $T(5, -3)$; reflection in the x-axis

7. quadrilateral $CDEF$ with $C(1, 1), D(-2, 5), E(-2, 0)$, and $F(-1, -2)$; rotation of 90° counterclockwise

8. quadrilateral $WXYZ$ with $W(0, 4), X(-5, 0), Y(0, -3)$, and $Z(5, -2)$; translation $(x, y) \rightarrow (x + 1, y - 4)$

9. quadrilateral $JKLM$ with $J(-6, -2), K(-2, -8), L(4, -4)$, and $M(6, 6)$; dilation by a scale factor of $-\frac{1}{2}$

10. pentagon $ABCDE$ with $A(2, 2), B(0, 4), C(-3, 2), D(-3, -4)$, and $E(2, -4)$; reflection in the line $y = x$

Lesson 10-1

(pages 522–528)

The radius, diameter, or circumference of a circle is given. Find the missing measures to the nearest hundredth.

1. $r = 18$ in., $d =$ __?__ , $C =$ __?__
2. $d = 34.2$ ft, $r =$ __?__ , $C =$ __?__
3. $C = 12\pi$ m, $r =$ __?__ , $d =$ __?__
4. $C = 84.8$ mi, $r =$ __?__ , $d =$ __?__
5. $d = 8.7$ cm, $r =$ __?__ , $C =$ __?__
6. $r = 3b$ in., $d =$ __?__ , $C =$ __?__

Find the exact circumference of each circle.

7.
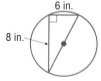
6 in.
8 in.

8.
6 cm

9.
12 yd

10.
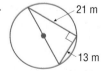
21 m
13 m

...**2**...sure.

2. $m\angle LKJ$

4. $m\angle LKG$

6. $m\angle HKJ$

...$\angle LKI$

5. $m\angle HKI$

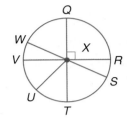

In ⊙X, $\overline{WS}$, $\overline{VR}$, and $\overline{QT}$ are diameters, $m\angle WXV = 25$ and $m\angle VXU = 45$. Find each measure.

7. $m\widehat{QR}$ **8.** $m\widehat{QW}$

9. $m\widehat{TU}$ **10.** $m\widehat{WRV}$

11. $m\widehat{SV}$ **12.** $m\widehat{TRW}$

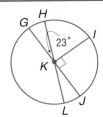

Lesson 10-3

In ⊙S, $HJ = 22$, $LG = 18$, $m\widehat{IJ} = 35$, and $m\widehat{LM} = 30$. Find each measure.

1. HR **2.** RJ

3. LT **4.** TG

5. $m\widehat{HJ}$ **6.** $m\widehat{LG}$

7. $m\widehat{MG}$ **8.** $m\widehat{HI}$

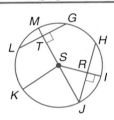

In ⊙R, $CR = RF$, and $ED = 30$. Find each measure.

9. AB **10.** EF

11. DF **12.** BC

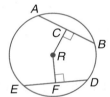

Lesson 10-4

Find the measure of each numbered angle for each figure.

1. $m\widehat{AB} = 176$, and $m\widehat{BC} = 42$ **2.** $\overline{WX} \cong \overline{ZY}$, and $m\widehat{ZW} = 120$ **3.** $m\widehat{QR} = 40$, and $m\widehat{TS} = 110$

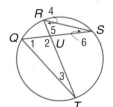

4. ▱$ABCD$ is a rectangle, and $m\widehat{BC} = 70$.

5. $m\widehat{TR} = 100$, and $\overline{SR} \perp \overline{QT}$

6. $m\widehat{UIY} = m\widehat{XZ} = 56$ and $m\widehat{UIV} = m\widehat{XW} = 56$

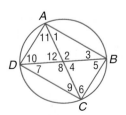

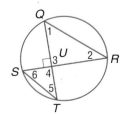

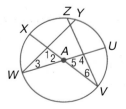

7. Rhombus $ABCD$ is inscribed in a circle. What can you conclude about $\overline{BD}$?

8. Triangle RST is inscribed in a circle. If the measure of $\widehat{RS}$ is 170, what is the measure of $\angle T$?

Extra Practice

Lesson 10-5

(pages 552–558)

Determine whether each segment is tangent to the given circle.

1.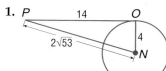

2.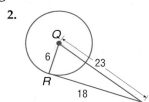

Find *x*. Assume that segments that appear to be tangent are tangent.

3.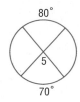

4.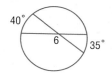

5.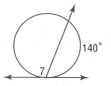

Lesson 10-6

(pages 561–568)

Find each measure.

1. $m\angle 5$

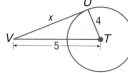

2. $m\angle 6$

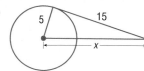

3. $m\angle 7$

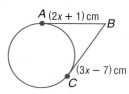

Find *x*. Assume that any segment that appears to be tangent is tangent.

4.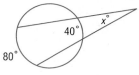

5.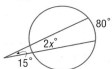

6.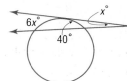

Lesson 10-7

(pages 569–574)

Find *x*. Assume that segments that appear to be tangent are tangent.

1.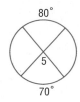

2.

3.

Find each variable to the nearest tenth.

4.

5.

6.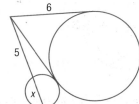

Extra Practice

Lesson 10-8

(pages 575–580)

Write an equation for each circle.

1. center at $(1, -2)$, $r = 2$
2. center at origin, $r = 4$
3. center at $(-3, -4)$, $r = \sqrt{11}$
4. center at $(3, -1)$, $d = 6$
5. center at $(6, 12)$, $r = 7$
6. center at $(4, 0)$, $d = 8$
7. center at $(6, -6)$, $d = 22$
8. center at $(-5, 1)$, $d = 2$

Graph each equation.

9. $x^2 + y^2 = 25$
10. $x^2 + y^2 - 3 = 1$
11. $(x - 3)^2 + (y + 1)^2 = 9$
12. $(x - 1)^2 + (y - 4)^2 = 1$

13. Find the radius of a circle whose equation is $(x + 3)^2 + (y - 1)^2 = r^2$ and contains $(-2, 1)$.

14. Find the radius of a circle whose equation is $(x - 4)^2 + (y - 3)^2 = r^2$ and contains $(8, 3)$.

Lesson 11-1

(pages 595–600)

Find the area and perimeter of each parallelogram. Round to the nearest tenth if necessary.

1.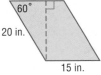
2.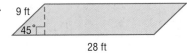
3.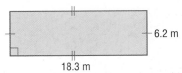

COORDINATE GEOMETRY   Given the coordinates of the vertices of a quadrilateral, determine whether it is a *square*, a *rectangle*, or a *parallelogram*. Then find the area of the quadrilateral.

4. $Q(-3, 3)$, $R(-1, 3)$, $S(-1, 1)$, $T(-3, 1)$
5. $A(-7, -6)$, $B(-2, -6)$, $C(-2, -3)$, $D(-7, -3)$
6. $L(5, 3)$, $M(8, 3)$, $N(9, 7)$, $O(6, 7)$
7. $W(-1, -2)$, $X(-1, 1)$, $Y(2, 1)$, $Z(2, -2)$

Lesson 11-2

(pages 601–609)

Find the area of each quadrilateral.

1.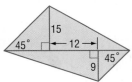
2.
3.

COORDINATE GEOMETRY   Find the area of trapezoid *ABCD* given the coordinates of the vertices.

4. $A(1, 1)$, $B(2, 3)$, $C(4, 3)$, $D(7, 1)$
5. $A(-2, 2)$, $B(2, 2)$, $C(7, -3)$, $D(-4, -3)$
6. $A(1, -1)$, $B(4, -1)$, $C(8, 5)$, $D(1, 5)$
7. $A(-2, 2)$, $B(4, 2)$, $C(3, -2)$, $D(1, -2)$

COORDINATE GEOMETRY   Find the area of rhombus *LMNO* given the coordinates of the vertices.

8. $L(-3, 0)$, $M(1, -2)$, $N(-3, -4)$, $O(-7, -2)$
9. $L(-3, -2)$, $M(-4, 2)$, $N(-3, 6)$, $O(-2, 2)$
10. $L(-1, -4)$, $M(3, 4)$, $N(-1, 12)$, $O(-5, 4)$
11. $L(-2, -2)$, $M(4, 4)$, $N(10, -2)$, $O(4, -8)$

Lesson 11-3

(pages 610–616)

Find the area of each regular polygon. Round to the nearest tenth.

1. a square with perimeter 54 feet

2. a triangle with side length 9 inches

3. an octagon with side length 6 feet

4. a decagon with apothem length of 22 centimeters

Find the area of each shaded region. Assume that all polygons that appear to be regular are regular. Round to the nearest tenth.

5.

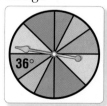

6 cm

6.

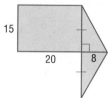

6 ft

15 ft

7.

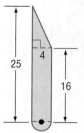

7 in.

Lesson 11-4

(pages 617–621)

Find the area of each figure. Round to the nearest tenth if necessary.

1.

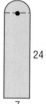

24

7

2.

15

20   8

3.

25   4

16

COORDINATE GEOMETRY The vertices of an irregular figure are given. Find the area of each figure.

4. $R(0, 5), S(3, 3), T(3, 0)$

5. $A(-5, -3), B(-3, 0), C(2, -1), D(2, -3)$

6. $L(-1, 4), M(3, 2), N(3, -1), O(-1, -2), P(-3, 1)$

Lesson 11-5

(pages 622–627)

Find the total area of the sectors of the indicated color. Then find the probability of spinning the color indicated if the diameter of each spinner is 20 inches.

1. orange

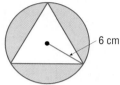

36°

2. blue

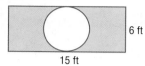

75° 40°
52°
60° 68°
65°

3. green

89° 110°
80° 81°

Find the area of the shaded region. Then find the probability that a point chosen at random is in the shaded region.

4.

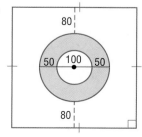

80

50 100 50

80

5.

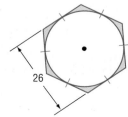

26

6.

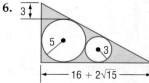

3

5

3

16 + 2√15

Extra Practice

Lesson 12-1

(pages 636–642)

Draw the back view and corner view of a figure given its orthogonal drawing.

1.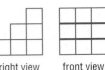

top view   right view   front view   left view

2.

top view   right view   front view   left view

Identify each solid. Name the bases, faces, edges, and vertices.

3.

4.

5.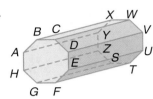

Lesson 12-2

(pages 643–648)

Sketch each solid using isometric dot paper.

1. rectangular prism 2 units high, 3 units long, and 2 units wide
2. rectangular prism 1 unit high, 2 units long, and 3 units wide
3. triangular prism 3 units high with bases that are right triangles with legs 3 units and 4 units long
4. triangular prism 5 units high with bases that are right triangles with legs 4 units and 6 units long

For each solid, draw a net and find the surface area. Round to the nearest tenth if necessary.

5.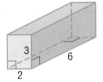

6.

7.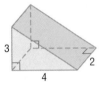

Lesson 12-3

(pages 649–654)

Find the lateral area and the surface area of each prism. Round to the nearest tenth if necessary.

1.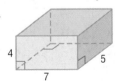

2.

3.

4.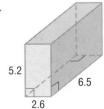

5.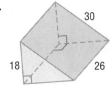

6.

7. The surface area of a right triangular prism is 228 square inches. The base is a right triangle with legs measuring 6 inches and 8 inches. Find the height of the prism.

8. The surface area of a right triangular prism with height 18 inches is 1380 square inches. The base is a right triangle with a leg measuring 15 inches and a hypotenuse of length 25 inches. Find the length of the other leg of the base.

Lesson 12-4

(pages 655–659)

Find the surface area of a cylinder with the given dimensions. Round to the nearest tenth.

1. $r = 2$ ft, $h = 3.5$ ft

2. $d = 15$ in., $h = 20$ in.

3. $r = 3.7$ m, $h = 6.2$ m

4. $d = 19$ mm, $h = 32$ mm

Find the surface area of each cylinder. Round to the nearest tenth.

5.
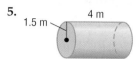
1.5 m · 4 m

6.
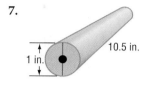
14 ft · 32.5 ft

7.
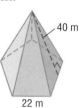
10.5 in. · 1 in.

8.
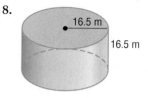
16.5 m · 16.5 m

Lesson 12-5

(pages 660–665)

Find the surface area of each regular pyramid. Round to the nearest tenth.

1.

9 cm · 7 cm · 7 cm

2.
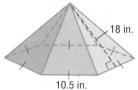
18 in. · 10.5 in.

3.

40 m · 22 m

4.

15 cm · 10 cm · 10 cm

5.

17 ft · 15 ft

6.
3 cm

Lesson 12-6

(pages 666–670)

Find the surface area of each cone. Round to the nearest tenth.

1.

10 in. · 6 in.

2.

30 ft · 34 ft

3.

17 cm · 17 cm

4.
10.5 in. · 2.2 in.

5.
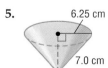
6.25 cm · 7.0 cm

6.
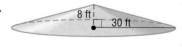
8 ft · 30 ft

7. Find the surface area of a cone if the height is 28 inches and the slant height is 40 inches.

8. Find the surface area of a cone if the height is 7.5 centimeters and the radius is 2.5 centimeters.

Lesson 12-7

(pages 671–676)

Find the surface area of each sphere or hemisphere. Round to the nearest tenth.

1.

120 ft

2.

42.5 m

3.

2520 mi

4.

33 cm

5. a hemisphere with the circumference of a great circle 14.1 cm

6. a sphere with the circumference of a great circle 50.3 in.

7. a sphere with the area of a great circle 98.5 m²

8. a hemisphere with the circumference of a great circle 3.1 in.

9. a hemisphere with the area of a great circle 31,415.9 ft²

Lesson 13-1

(pages 688–694)

Find the volume of each prism or cylinder. Round to the nearest tenth if necessary.

1.
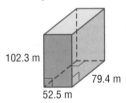
102.3 m
79.4 m
52.5 m

2.

8 ft
30 ft

3.
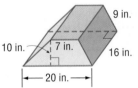
9 in.
10 in.
7 in.
16 in.
20 in.

Find the volume of each solid to the nearest tenth.

4.
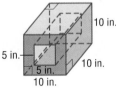
10 in.
5 in.
5 in.
10 in.
10 in.

5.

21 cm
9√2 cm

6.
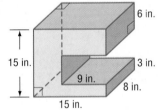
6 in.
15 in.
3 in.
9 in.
8 in.
15 in.

Lesson 13-2

(pages 696–701)

Find the volume of each cone or pyramid. Round to the nearest tenth if necessary.

1.

7.5 ft
5 ft

2.

40 mm
20 mm

3.
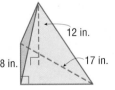
12 in.
8 in.
17 in.

4.

13 m
5 m

5.
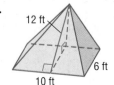
12 ft
6 ft
10 ft

6.
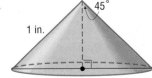
45°
1 in.

Lesson 13-3
(pages 702–706)

Find the volume of each sphere or hemisphere. Round to the nearest tenth.

1.

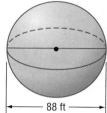

88 ft

2. $C = 4$ m

3.

17 mm

4. The diameter of the sphere is 3 cm.

5. The radius of the hemisphere is $7\sqrt{2}$ m.

6. The diameter of the hemisphere is 90 ft.

7. The radius of the sphere is 0.5 in.

Lesson 13-4
(pages 707–713)

Determine whether each pair of solids are *similar*, *congruent*, or *neither*.

1.

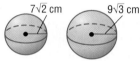

$7\sqrt{2}$ cm $9\sqrt{3}$ cm

2.

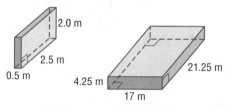

2.0 m
2.5 m
0.5 m
4.25 m
17 m
21.25 m

3.

16 ft   18 ft   16 ft   43 ft   31 ft   31 ft

4.

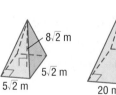

6 mm
8 mm
6 mm
10 mm

5.

16 in.   15 in.   30 in.   34 in.

6.

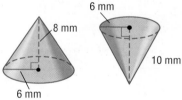

$8\sqrt{2}$ m $5\sqrt{2}$ m $5\sqrt{2}$ m   32 m   20 m   20 m

Lesson 13-5
(pages 714–719)

Graph the rectangular solid that contains the given point and the origin. Label the coordinates of each vertex.

1. $A(3, -3, -3)$

2. $E(-1, 2, -3)$

3. $I(3, -1, 2)$

4. $Z(2, -1, 3)$

5. $Q(-4, -2, -4)$

6. $Y(-3, 1, -4)$

Determine the distance between each pair of points. Then determine the coordinates of the midpoint, *M*, of the segment joining the pair of points.

7. $A(-3, 3, 1)$ and $B(3, -3, -1)$

8. $O(2, -1, -3)$ and $P(-2, 4, -4)$

9. $D(0, -5, -3)$ and $E(0, 5, 3)$

10. $J(-1, 3, 5)$ and $K(3, -5, -3)$

11. $A(2, 1, 6)$ and $Z(-4, -5, -3)$

12. $S(-8, 3, -5)$ and $T(6, -1, 2)$

Mixed Problem Solving and Proof

Chapter 1 Points, Lines, Planes, and Angles

(pages 4–59)

ARCHITECTURE For Exercises 1–4, use the following information.
The Burj Al Arab in Dubai, United Arab Emirates, is one of the world's tallest hotels. *(Lesson 1-1)*

1. Trace the outline of the building on your paper.

2. Label three different planes suggested by the outline.

3. Highlight three lines in your drawing that, when extended, do not intersect.

4. Label three points on your sketch. Determine if they are coplanar and collinear.

SKYSCRAPERS For Exercises 5–7, use the following information. *(Lesson 1-2)*

Tallest Buildings in San Antonio, TX	
Name	**Height (ft)**
Tower of the Americas	622
Marriot Rivercenter	546
Weston Centre	444
Tower Life	404

Source: www.skyscrapers.com

5. What is the precision for the measures of the heights of the buildings?

6. What does the precision mean for the measure of the Tower of the Americas?

7. What is the difference in height between Weston Centre and Tower Life?

PERIMETER For Exercises 8–11, use the following information. *(Lesson 1-3)*
The coordinates of the vertices of $\triangle ABC$ are $A(0, 6)$, $B(-6, -2)$, and $C(8, -4)$. Round to the nearest tenth.

8. Find the perimeter of $\triangle ABC$.

9. Find the coordinates of the midpoints of each side of $\triangle ABC$.

10. Suppose the midpoints are connected to form a triangle. Find the perimeter of this triangle.

11. Compare the perimeters of the two triangles.

12. **TRANSPORTATION** Mile markers are used to name the exits on Interstate 70 in Kansas. The exit for Hays is 3 miles farther than halfway between Exits 128 and 184. What is the exit number for the Hays exit? *(Lesson 1-3)*

13. **ENTERTAINMENT** The Ferris wheel at the Navy Pier in Chicago has forty gondolas. What is the measure of an angle with a vertex that is the center of the wheel and with sides that are two consecutive spokes on the wheel? Assume that the gondolas are equally spaced. *(Lesson 1-4)*

CONSTRUCTION For Exercises 14–15, use the following information.
A framer is installing a cathedral ceiling in a newly built home. A protractor and a plumb bob are used to check the angle at the joint between the ceiling and wall. The wall is vertical, so the angle between the vertical plumb line and the ceiling is the same as the angle between the wall and the ceiling. *(Lesson 1-5)*

14. How are $\angle ABC$ and $\angle CBD$ related?

15. If $m\angle ABC = 110$, what is $m\angle CBD$?

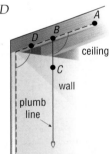

STRUCTURES For Exercises 16–17, use the following information. *(Lesson 1-6)*
The picture shows the Hongkong and Shanghai Bank located in Hong Kong, China.

16. Name five different polygons suggested by the picture.

17. Classify each polygon you identified as *convex* or *concave* and *regular* or *irregular*.

POPULATION For Exercises 1–2, use the table showing the population density for various states in 1960, 1980, and 2000. The figures represent the number of people per square mile. *(Lesson 2-1)*

State	1960	1980	2000
CA	100.4	151.4	217.2
CT	520.6	637.8	702.9
DE	225.2	307.6	401.0
HI	98.5	150.1	188.6
MI	137.7	162.6	175.0

Source: U.S. Census Bureau

1. Find a counterexample for the following statement. The population density for each state in the table increased by at least 30 during each 20-year period.

2. Write two conjectures for the year 2010.

STATES For Exercises 3–5, refer to the Venn diagram. *(Lesson 2-2)*

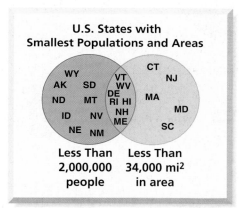

U.S. States with Smallest Populations and Areas

Source: *World Almanac*

3. How many states have less than 2,000,000 people?

4. How many states have less than 34,000 square miles in area?

5. How many states have less than 2,000,000 people and are less than 34,000 square miles in area?

LITERATURE For Exercises 6–7, use the following quote from Lewis Carroll's *Alice's Adventures in Wonderland*. *(Lesson 2-3)*

"Then you should say what you mean," the March Hare went on.

"I do," Alice hastily replied; "at least—at least I mean what I say—that's the same thing, you know."

"Not the same thing a bit!" said the Hatter.

6. Who is correct? Explain.

7. How are the phrases *say what you mean* and *mean what you say* related?

8. **AIRLINE SAFETY** Airports in the United States post a sign stating *If any unknown person attempts to give you any items including luggage to transport on your flight, do not accept it and notify airline personnel immediately.* Write a valid conclusion to the hypothesis, *If a person Candace does not know attempts to give her an item to take on her flight, . . .* *(Lesson 2-4)*

9. **PROOF** Write a paragraph proof to show that $\overline{AB} \cong \overline{CD}$ if B is the midpoint of $\overline{AC}$ and C is the midpoint of $\overline{BD}$. *(Lesson 2-5)*

10. **CONSTRUCTION** Engineers consider the expansion and contraction of materials used in construction. The coefficient of linear expansion, k, is dependent on the change in length and the change in temperature and is found by the formula, $k = \dfrac{\Delta \ell}{\ell(T - t)}$. Solve this formula for T and justify each step. *(Lesson 2-6)*

11. **PROOF** Write a two-column proof. *(Lesson 2-7)*

Given: $ABCD$ has 4 congruent sides.
$DH = BF = AE; EH = FE$

Prove: $AB + BE + AE = AD + AH + DH$

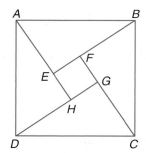

ILLUSIONS This drawing was created by German psychologist Wilhelm Wundt. *(Lesson 2-8)*

12. Describe the relationship between each pair of vertical lines.

13. A close-up of the angular lines is shown below. If $\angle 4 \cong \angle 2$, write a two-column proof to show that $\angle 3 \cong \angle 1$.

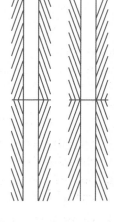

1. **OPTICAL ILLUSIONS** Lines ℓ and m are parallel, but appear to be bowed due to the transversals drawn through ℓ and m. Make a conjecture about the relationship between $\angle 1$ and $\angle 2$. *(Lesson 3-1)*

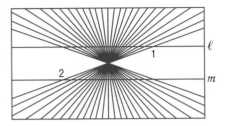

ARCHITECTURE For Exercises 2–10, use the following information.
The picture shows one of two towers of the Puerta de Europa in Madrid, Spain. Lines a, b, c, and d are parallel. The lines are cut by transversals e and f. If $m\angle 1 = m\angle 2 = 75$, find the measure of each angle. *(Lesson 3-2)*

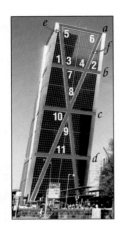

2. $\angle 3$	**3.** $\angle 4$
4. $\angle 5$	**5.** $\angle 6$
6. $\angle 7$	**7.** $\angle 8$
8. $\angle 9$	**9.** $\angle 10$
10. $\angle 11$	

11. **PROOF** Write a two-column proof. *(Lesson 3-2)*
    Given: $\overline{MQ} \parallel \overline{NP}$
    $\qquad\quad \angle 4 \cong \angle 3$
    Prove: $\angle 1 \cong \angle 5$

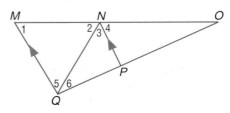

12. **EDUCATION** Between 1995 and 2000, the average cost for tuition and fees for American universities increased by an average rate of $84.20 per year. In 2000, the average cost was $2600. If costs increase at the same rate, what will the total average cost be in 2010? *(Lesson 3-3)*

RECREATION For Exercises 13 and 14, use the following information. *(Lesson 3-4)*
The Three Forks community swimming pool holds 74,800 gallons of water. At the end of the summer, the pool is drained and winterized.

13. If the pool drains at the rate of 1200 gallons per hour, write an equation to describe the number of gallons left after x hours.

14. How many hours will it take to drain the pool?

15. **CONSTRUCTION** An *engineer and carpenter square* is used to draw parallel line segments. Martin makes two cuts at an angle of 120° with the edge of the wood through points D and P. Explain why these cuts will be parallel. *(Lesson 3-5)*

16. **PROOF** Write a two-column proof. *(Lesson 3-5)*
    Given: $\angle 1 \cong \angle 3$
    $\qquad\quad \overline{AB} \parallel \overline{DC}$
    Prove: $\overline{BC} \parallel \overline{AD}$

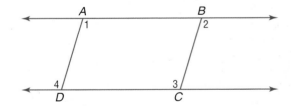

17. **CITIES** The map shows a portion of Seattle, Washington. Describe a segment that represents the shortest distance from the Bus Station to Denny Way. Can you walk the route indicated by your segment? Explain. *(Lesson 3-6)*

QUILTING   For Exercises 1 and 2, trace the quilt pattern square below. *(Lesson 4-1)*

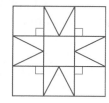

1. Shade all right triangles red. Do these triangles appear to be scalene or isosceles? Explain.

2. Shade all acute triangles blue. Do these triangles appear to be scalene, isoscles, or equilateral? Explain.

3. **ASTRONOMY**   Leo is a constellation that represents a lion. Three of the brighter stars in the constellation form △*LEO*. If the angles have measures as shown in the figure, find *m*∠*OLE*. *(Lesson 4-2)*

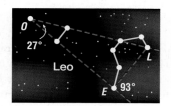

4. **ARCHITECTURE**   The diagram shows an A-frame house with various points labeled. Assume that segments and angles that appear to be congruent in the diagram are congruent. Indicate which triangles are congruent. *(Lesson 4-3)*

RECREATION   For Exercises 5–7, use the following information.

Tapatan is a game played in the Philippines on a square board, like the one shown at the top right. Players take turns placing each of their three pieces on a different point of intersection. After all the pieces have been played, the players take turns moving a piece along a line to another intersection. A piece cannot jump over another piece. A player who gets all their pieces in a straight line wins. Point *E* bisects all four line segments that pass through it. All sides are congruent, and the diagonals are congruent. Suppose a letter is assigned to each intersection. *(Lesson 4-4)*

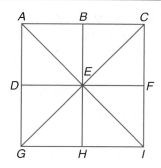

5. Is △*GHE* ≅ △*CBE*? Explain.

6. Is △*AEG* ≅ △*IEG*? Explain.

7. Write a flow proof to show that △*ACI* ≅ △*CAG*.

8. **HISTORY**   It is said that Thales determined the distance from the shore to the Greek ships by sighting the angle to the ship from a point *P* on the shore, walking to point *Q*, and then sighting the angle to the ship from *Q*. He then reproduced the angles on the other side of $\overline{PQ}$ and continued these lines until they intersected. Is this method valid? Explain. *(Lesson 4-5)*

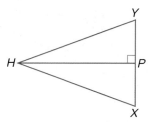

9. **PROOF**   Write a two-column proof. *(Lesson 4-6)*

   Given: $\overline{PH}$ bisects ∠*YHX*.
   $\overline{PH} \perp \overline{YX}$

   Prove: △*YHX* is an isosceles triangle.

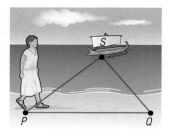

10. **PROOF**   △*ABC* is a right isosceles triangle with hypotenuse $\overline{AB}$. *M* is the midpoint of $\overline{AB}$. Write a coordinate proof to show that $\overline{CM}$ is perpendicular to $\overline{AB}$. *(Lesson 4-7)*

CONSTRUCTION For Exercises 1–4, draw a large, acute scalene triangle. Use a compass and straightedge to make the required constructions. *(Lesson 5-1)*

1. Find the circumcenter. Label it *C*.

2. Find the centroid of the triangle. Label it *D*.

3. Find the orthocenter. Label it *O*.

4. Find the incenter of the triangle. Label it *I*.

RECREATION For Exercises 5–7, use the following information. *(Lesson 5-2)*
Kailey plans to fly over the route marked on the map of Oahu in Hawaii.

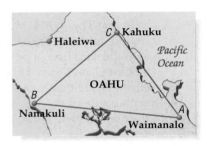

5. The measure of angle *A* is two degrees more than the measure of angle *B*. The measure of angle *C* is fourteen degrees less than twice the measure of angle *B*. What are the measures of the three angles?

6. Write the lengths of the legs of Kailey's trip in order from least to greatest.

7. The length of the entire trip is about 68 miles. The middle leg is 11 miles greater than one-half the length of the shortest leg. The longest leg is 12 miles greater than three-fourths of the shortest leg. What are the lengths of the legs of the trip?

8. **LAW** A man is accused of comitting a crime. If the man is telling the truth when he says, "I work every Tuesday from 3:00 P.M. to 11:00 P.M.," what fact about the crime could be used to prove by indirect reasoning that the man was innocent? *(Lesson 5-3)*

TRAVEL For Exercises 9 and 10, use the following information.
The total air distance to fly from Bozeman, Montana, to Salt Lake City, Utah, to Boise, Idaho is just over 634 miles.

9. Write an indirect proof to show that at least one of the legs of the trip is longer than 317 miles. *(Lesson 5-3)*

10. The air distance from Bozeman to Salt Lake City is 341 miles and the distance from Salt Lake to Boise is 294 miles. Find the range for the distance from Bozeman to Boise. *(Lesson 5-4)*

11. **PROOF** Write a two-column proof.
    Given: $\angle ZST \cong \angle ZTS$
    $\angle XRA \cong \angle XAR$
    $TA = 2AX$
    Prove: $2XR + AZ > SZ$
    (Lesson 5-4)

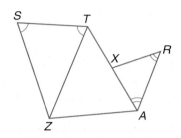

12. **GEOGRAPHY** The map shows a portion of Nevada. The distance from Tonopah to Round Mountain is the same as the distance from Tonopah to Warm Springs. The distance from Tonopah to Hawthorne is the same as the distance from Tonopah to Beatty. Use the angle measures to determine which distance is greater, Round Mountain to Hawthorne or Warm Springs to Beatty. *(Lesson 5-5)*

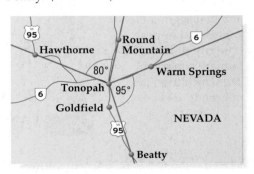

13. **PROOF** Write a two-column proof. *(Lesson 5-5)*
    Given: $\overline{DB}$ is a median of $\triangle ABC$.
    $m\angle 1 > m\angle 2$
    Prove: $m\angle C > m\angle A$

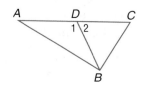

1. **TOYS** In 2000, $34,554,900,000 was spent on toys in the U.S. The U.S. population in 2000 was 281,421,906, with 21.4% of the population 14 years and under. If all of the toys purchased in 2000 were for children 14 years and under, what was the average amount spent per child? *(Lesson 6-1)*

QUILTING **For Exercises 2–4, use the following information.** *(Lesson 6-2)*
Felicia found a pattern for a quilt square. The pattern measures three-quarters of an inch on a side. Felicia wants to make a quilt that is 77 inches by 110 inches when finished.

2. If Felicia wants to use only whole quilt squares, what is the greatest side length she can use for each square?

3. How many quilt squares will she need for the quilt?

4. By what scale factor will she need to increase the pattern for the quilt square?

PROOF **For Exercises 5 and 6, write a paragraph proof.** *(Lesson 6-3)*

5. **Given:** $\triangle WYX \sim \triangle QYR$,
   $\triangle ZYX \sim \triangle SYR$
   Prove: $\triangle WYZ \sim \triangle QYS$

6. **Given:** $\overline{WX} \parallel \overline{QR}$,
   $\overline{ZX} \parallel \overline{SR}$
   Prove: $\overline{WZ} \parallel \overline{QS}$

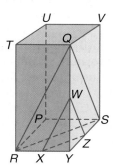

HISTORY **For Exercises 7 and 8, use the following information.** *(Lesson 6-4)*
In the fifteenth century, mathematicians and artists tried to construct the perfect letter. Damiano da Moile used a square as a frame to design the letter "A" as shown in the diagram. The thickness of the major stroke of the letter was to be $\frac{1}{12}$ of the height of the letter.

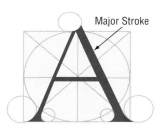
Major Stroke

7. Explain why the bar through the middle of the A is half the length between the outside bottom corners of the sides of the letter.

8. If the letter were 3 centimeters tall, how wide would the major stroke of the A be?

9. **PROOF** Write a two-column proof. *(Lesson 6-5)*
   Given: $\overline{WS}$ bisects $\angle RWT$. $\angle 1 \cong \angle 2$
   Prove: $\dfrac{VW}{WT} = \dfrac{RS}{ST}$

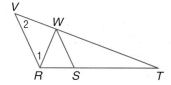

ART **For Exercises 10 and 11, use the diagram of a square mosaic tile.** $AB = BC = CD = \frac{1}{3}AD$ and $DE = EF = FG = \frac{1}{3}DG$. *(Lesson 6-5)*

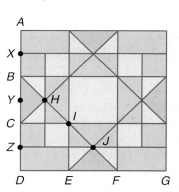

10. What is the ratio of the perimeter of $\triangle BDF$ to the perimeter of $\triangle BCI$? Explain.

11. Find two triangles such that the ratio of their perimeters is $2:3$. Explain.

12. **TRACK** A triangular track is laid out as shown. $\triangle RST \sim \triangle WVU$. If $UV = 500$ feet, $VW = 400$ feet, $UW = 300$ feet, and $ST = 1000$ feet, find the perimeter of $\triangle RST$. *(Lesson 6-5)*

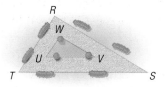

13. **BANKING** Ashante has $5000 in a savings account with a yearly interest rate of 2.5%. The interest is compounded twice per year. What will be the amount in the savings account after 5 years? *(Lesson 6-6)*

1. **PROOF**   Write a two-column proof. *(Lesson 7-1)*

   Given: *D* is the midpoint of $\overline{BE}$, $\overline{BD}$ is an altitude of right triangle △*ABC*

   Prove: $\dfrac{AD}{DE} = \dfrac{DE}{DC}$

   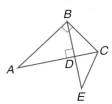

2. **AMUSEMENT PARKS**   The map shows the locations of four rides at an amusement park. Find the length of the path from the roller coaster to the bumper boats. Round to the nearest tenth. *(Lesson 7-1)*

   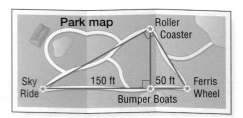

3. **CONSTRUCTION**   Carlotta drew a diagram of a right triangular brace with side measures of 2.7 centimeters, 3.0 centimeters, and 5.3 centimeters. Is the diagram correct? Explain. *(Lesson 7-2)*

   DESIGN   For Exercises 4–5, use the following information. *(Lesson 7-3)*
   Kwan designed the pinwheel. The blue triangles are congruent equilateral triangles each with an altitude of 4 inches. The red triangles are congruent isosceles right triangles. The hypotenuse of a red triangle is congruent to a side of a blue triangle.

   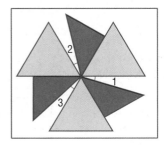

4. If angles 1, 2, and 3 are congruent, find the measure of each angle.

5. Find the perimeter of the pinwheel. Round to the nearest inch.

COMMUNICATION   For Exercises 6–9, use the following information. *(Lesson 7-4)*
The diagram shows a radio tower secured by four pairs of guy wires that are equally spaced apart with *DX* = 60 feet. Round to the nearest tenth if necessary.

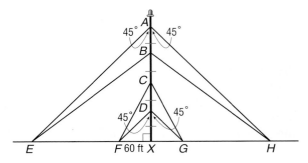

6. Name the isosceles triangles in the diagram.

7. Find $m\angle BEX$ and $m\angle CFX$.

8. Find *AE*, *EB*, *CF*, and *DF*.

9. Find the total amount of wire used to support the tower.

10. **METEOROLOGY**   A searchlight is 6500 feet from a weather station. If the angle of elevation to the spot of light on the clouds above the station is 47°, how high is the cloud ceiling? *(Lesson 7-5)*

GARDENING   For Exercises 11 and 12, use the information below. *(Lesson 7-6)*
A flower bed at Magic City Rose Garden is in the shape of an obtuse scalene triangle with the shortest side measuring 7.5 feet. Another side measures 14 feet and the measure of the opposite angle is 103°.

11. Find the measures of the other angles of the triangle. Round to the nearest degree.

12. Find the perimeter of the garden. Round to the nearest tenth.

13. **HOUSING**   Mr. and Mrs. Abbott bought a lot at the end of a cul-de-sac. They want to build a fence on three sides of the lot, excluding $\overline{HE}$. To the nearest foot, how much fencing will they need to buy? *(Lesson 7-7)*

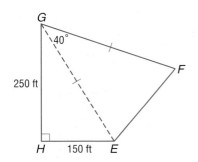

ENGINEERING   For Exercises 1–2, use the following information.

The London Eye in London, England, is the world's largest observation wheel. The ride has 32 enclosed capsules for riders. *(Lesson 8-1)*

1. Suppose each capsule is connected with a straight piece of metal forming a 32-gon. Find the sum of the measures of the interior angles.

2. What is the measure of one interior angle of the 32-gon?

3. **QUILTING**   The quilt square shown is called the Lone Star pattern. Describe two ways that the quilter could ensure that the pieces will fit properly. *(Lesson 8-2)*

4. **PROOF**   Write a two-column proof. *(Lesson 8-3)*
   Given: $\square ABCD$, $\overline{AE} \cong \overline{CF}$
   Prove: Quadrilateral $EBFD$ is a parallelogram.

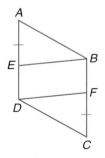

5. **MUSIC**   Why will the keyboard stand shown always remain parallel to the floor? *(Lesson 8-3)*

6. **PROOF**   Write a two-column proof. *(Lesson 8-4)*
   Given: $\square WXZY$, $\angle 1$ and $\angle 2$ are complementary.
   Prove: $WXZY$ is a rectangle.

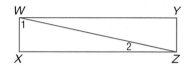

7. **PROOF**   Write a paragraph proof. *(Lesson 8-4)*
   Given: $\square KLMN$
   Prove: $PQRS$ is a rectangle.

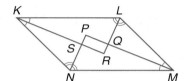

8. **CONSTRUCTION**   Mr. Redwing is building a sandbox. He placed stakes at what he believes will be the four vertices of a square with a distance of 5 feet between each stake. How can he be sure that the sandbox will be a square? *(Lesson 8-5)*

DESIGN   For Exercises 9 and 10, use the square floor tile design shown below. *(Lesson 8-6)*

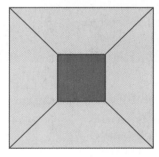

9. Explain how you know that the trapezoids in the design are isosceles.

10. The perimeter of the floor tile is 48 inches, and the perimeter of the interior red square is 16 inches. Find the perimeter of one trapezoid.

11. **PROOF**   Position a quadrilateral on the coordinate plane with vertices $Q(-a, 0)$, $R(a, 0)$, $S(b, c)$, and $T(-b, c)$. Prove that the quadrilateral is an isosceles trapezoid. *(Lesson 8-7)*

QUILTING **For Exercises 1 and 2, use the diagram of a quilt square.** *(Lesson 9-1)*

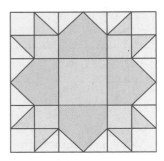

1. How many lines of symmetry are there for the entire quilt square?

2. Consider different sections of the quilt square. Describe at least three different lines of reflection and the figures reflected in those lines.

3. **ENVIRONMENT** A cloud of dense gas and dust pours out of Surtsey, a volcanic island off the south coast of Iceland. If the cloud blows 40 miles north and then 30 miles east, make a sketch to show the translation of the smoke particles. Then find the distance of the shortest path that would take the particles to the same position. *(Lesson 9-2)*

ART **For Exercises 4–7, use the mosaic tile.**

4. Identify the order and magnitude of rotation that takes a yellow triangle to a blue triangle. *(Lesson 9-3)*

5. Identify the order and magnitude of rotation that takes a blue triangle to a yellow triangle. *(Lesson 9-3)*

6. Identify the magnitude of rotation that takes a trapezoid to a consecutive trapezoid. *(Lesson 9-3)*

7. Can the mosaic tile tessellate the plane? Explain. *(Lesson 9-4)*

8. **CRAFTS** Eduardo found a pattern for cross-stitch on the Internet. The pattern measures 2 inches by 3 inches. He would like to enlarge the piece to 4 inches by 6 inches. The copy machine available to him enlarges 150% or less by increments of whole number percents. Find two whole number percents by which he can consecutively enlarge the piece and get as close to the desired dimensions as possible without exceeding them. *(Lesson 9-5)*

AVIATION **For Exercises 9 and 10, use the following information.** *(Lesson 9-6)*
A small aircraft flies due south at an average speed of 190 miles per hour. The wind is blowing due west at 30 miles per hour.

9. Draw a diagram using vectors to represent this situation.

10. Find the resultant velocity and direction of the plane.

GRAPHICS **For Exercises 11–14, use the graphic shown on the computer screen.** *(Lesson 9-7)*

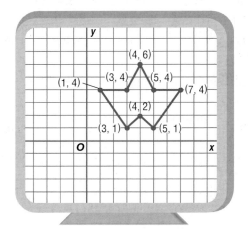

11. Suppose you want the figure to move to Quadrant III but be upside down. Write two matrices that make this transformation, if they are applied consecutively.

12. Write one matrix that can be used to do the same transformation as in Exercise 11. What type of transformation is this?

13. Compare the two matrices in Exercise 11 to the matrix in Exercise 12. What do you notice?

14. Write the vertex matrix for the figure in Quadrant III and graph it on the coordinate plane.

1. **CYCLING**   A bicycle tire travels about 50.27 inches during one rotation of the wheel. What is the diameter of the tire? *(Lesson 10-1)*

SPACE   **For Exercises 2–4, use the following information.** *(Lesson 10-2)*
School children were recently surveyed about what they believe to be the most important reason to explore Mars. They were given five choices and the table below shows the results.

Reason to Visit Mars	Number of Students
Learn about life beyond Earth	910
Learn more about Earth	234
Seek potential for human inhabitance	624
Use as a base for further exploration	364
Increase human knowledge	468

Source: *USA TODAY*

2. If you were to construct a circle graph of this data, how many degrees would be allotted to each category?

3. Describe the type of arc associated with each category.

4. Construct a circle graph for these data.

5. **CRAFTS**   Yvonne uses wooden spheres to make paperweights to sell at craft shows. She cuts off a flat surface for each base. If the original sphere has a radius of 4 centimeters and the diameter of the flat surface is 6 centimeters, what is the height of the paperweight? *(Lesson 10-3)*

6. **PROOF**   Write a two-column proof. *(Lesson 10-4)*

   Given: $\overarc{MHT}$ is a semicircle.

   $\overline{RH} \perp \overline{TM}$

   Prove: $\dfrac{TR}{RH} = \dfrac{TH}{HM}$

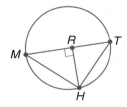

7. **PROOF**   Write a paragraph proof. *(Lesson 10-5)*

   Given: $\overline{GR}$ is tangent to $\odot D$ at G.

   $\overline{AG} \cong \overline{DG}$

   Prove: $\overline{AG}$ bisects $\overline{RD}$.

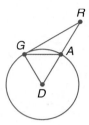

8. **METEOROLOGY**   A rainbow is really a full circle with a center at a point in the sky directly opposite the Sun. The position of a rainbow varies according to the viewer's position, but its angular size, $\angle ABC$, is always 42°. If $m\overarc{CD} = 160$, find the measure of the visible part of the rainbow, $m\overarc{AC}$. *(Lesson 10-6)*

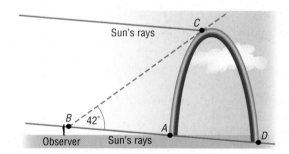

9. **CONSTRUCTION**   An arch over an entrance is 100 centimeters wide and 30 centimeters high. Find the radius of the circle that contains the arch. *(Lesson 10-7)*

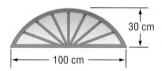

10. **SPACE**   Objects that have been left behind in Earth's orbit from space missions are called "space junk." These objects are a hazard to current space missions and satellites. Eighty percent of space junk orbits Earth at a distance of 1,200 miles from the surface of Earth, which has a diameter of 7,926 miles. Write an equation to model the orbit of 80% of space junk with Earth's center at the origin. *(Lesson 10-8)*

Mixed Problem Solving and Proof

REMODELING For Exercises 1–3, use the following information.

The diagram shows the floor plan of the home that the Summers are buying. They want to replace the patio with a larger sunroom to increase their living space by one-third. *(Lesson 11-1)*

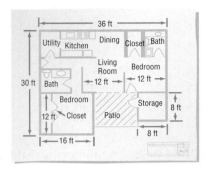

1. Excluding the patio and storage area, how many square feet of living area are in the current house?

2. What area should be added to the house to increase the living area by one-third?

3. The Summers want to connect the bedroom and storage area with the sunroom. What will be the dimensions of the sunroom?

HOME REPAIR For Exercises 4 and 5, use the following information.

Scott needs to replace the shingles on the roof of his house. The roof is composed of two large isosceles trapezoids, two smaller isosceles trapezoids, and a rectangle. Each trapezoid has the same height. *(Lesson 11-2)*

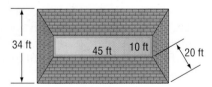

4. Find the height of the trapezoids.

5. Find the area of the roof covered by shingles.

6. **SPORTS** The Moore High School basketball team wants to paint their basketball court as shown. They want the center circle and the free throw areas painted blue. What is the area of the court that they will paint blue? *(Lesson 11-3)*

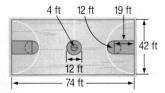

MUSEUMS For Exercises 7–9, use the following information.

The Hyalite Hills Museum plans to install the square mosaic pattern shown below in the entry hall. It is 10 feet on each side with each small black or red square tile measuring 2 feet on each side. *(Lesson 11-4)*

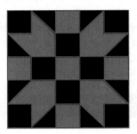

7. Find the area of black tiles.

8. Find the area of red tiles.

9. Which is greater, the total perimeter of the red tiles or the total perimeter of the black tiles? Explain.

10. **GAMES** If the dart lands on the target, find the probability that it lands in the blue region. *(Lesson 11-5)*

11. **ACCOMMODATIONS** The convention center in Washington, D.C., lies in the northwest sector of the city between New York and Massachusetts Avenues, which intersect at a 130° angle. If the amount of hotel space is evenly distributed over an area with that intersection as the center and a radius of 1.5 miles, what is the probability that a vistor, randomly assigned to a hotel, will be housed in the sector containing the convention center? *(Lesson 11-5)*

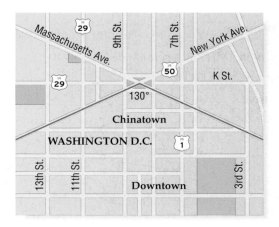

1. **ARCHITECTURE**
   Sketch an orthogonal drawing of the Eiffel Tower. *(Lesson 12-1)*

2. **CONSTRUCTION**   The roof shown below is a hip-and-valley style. Use the dimensions given to find the area of the roof that would need to be shingled. *(Lesson 12-2)*

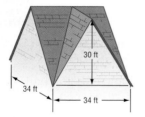

3. **AERONAUTICAL ENGINEERING**   The surface area of the wing on an aircraft is used to determine a design factor known as wing loading. If the total weight of the aircraft and its load is w and the total surface area of its wings is s, then the formula for the wing loading factor, ℓ, is $\ell = \dfrac{w}{s}$. If the wing loading factor is exceeded, the pilot must either reduce the fuel load or remove passengers or cargo. Find the wing loading factor for a plane if it had a take-off weight of 750 pounds and the surface area of the wings was 532 square feet. *(Lesson 12-2)*

4. **MANUFACTURING**   Many baking pans are given a special nonstick coating. A rectangular cake pan is 9 inches by 13 inches by 2 inches deep. What is the area of the inside of the pan that needs to be coated? *(Lesson 12-3)*

5. **COMMUNICATIONS**   Coaxial cable is used to transmit long-distance telephone calls, cable television programming, and other communications. A typical coaxial cable contains 22 copper tubes and has a diameter of 3 inches. What is the lateral area of a coaxial cable that is 500 feet long? *(Lesson 12-4)*

COLLECTIONS   For Exercises 6 and 7, use the following information.
Soledad collects unique salt-and-pepper shakers. She inherited a pair of tetrahedral shakers from her mother. *(Lesson 12-5)*

6. Each edge of a shaker measures 3 centimeters. Make a sketch of one shaker.

7. Find the total surface area of one shaker.

8. **FARMING**   The picture below shows a combination hopper cone and bin used by farmers to store grain after harvest. The cone at the bottom of the bin allows the grain to be emptied more easily. Use the dimensions shown in the diagram to find the entire surface area of the bin with a conical top and bottom. Write the exact answer and the answer rounded to the nearest square foot. *(Lesson 12-6)*

GEOGRAPHY   For Exercises 9–11, use the following information.
Joaquin is buying Dennis a globe for his birthday. The globe has a diameter of 16 inches. *(Lesson 12-7)*

9. What is the surface area of the globe?

10. If the diameter of Earth is 7926 miles, find the surface area of Earth.

11. The continent of Africa occupies about 11,700,000 square miles. How many square inches will be used to represent Africa on the globe?

Mixed Problem Solving and Proof

1. **METEOROLOGY**   The TIROS weather satellites were a series of weather satellites, the first being launched on April 1, 1960. These satellites carried television and infrared cameras and were covered by solar cells. If the cylinder-shaped body of a TIROS had a diameter of 42 inches and a height of 19 inches, what was the volume available for carrying instruments and cameras? Round to the nearest tenth. *(Lesson 13-1)*

2. **SPACECRAFT**   The smallest manned spacecraft, used by astronauts for jobs outside the Space Shuttle, is the Manned Maneuvering Unit. It is 4 feet tall, 2 feet 8 inches wide, and 3 feet 8 inches deep. Find the volume of this spacecraft in cubic feet. Round to the nearest tenth. *(Lesson 13-1)*

3. **MUSIC**   To play a concertina, you push and pull the end plates and press the keys. The air pressure causes vibrations of the metal reeds that make the notes. When fully expanded, the concertina is 36 inches from end to end. If the concertina is compressed, it is 7 inches from end to end. Find the volume of air in the instrument when it is fully expanded and when it is compressed. (*Hint:* Each endplate is a regular hexagonal prism and contains no air.) *(Lesson 13-1)*

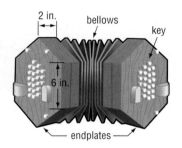

4. **ENGINEERING**   The base of an oil drilling platform is made up of 24 concrete cylindrical cells. Twenty of the cells are used for oil storage. The pillars that support the platform deck rest on the four other cells. Find the total volume of the storage cells. *(Lesson 13-1)*

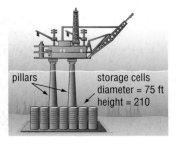

5. **HOME BUSINESS**   Jodi has a home-based business selling homemade candies. She is designing a pyramid-shaped box for the candy. The base is a square measuring 14.5 centimeters on a side. The slant height of the pyramid is 16 centimeters. Find the volume of the box. Round to the nearest cubic centimeter. *(Lesson 13-2)*

ENTERTAINMENT   For Exercises 6–10, use the following information. Some people think that the Spaceship Earth geosphere at Epcot® in Disney World resembles a golf ball. The building is a sphere measuring 165 feet in diameter. A typical golf ball has a diameter of approximately 1.5 inches.

6. Find the volume of Spaceship Earth. Round to the nearest cubic foot. *(Lesson 13-3)*

7. Find the volume of a golf ball. Round to the nearest tenth. *(Lesson 13-3)*

8. What is the scale factor that compares Spaceship Earth to a golf ball? *(Lesson 13-4)*

9. What is the ratio of the volume of Spaceship Earth to the volume of a golf ball? *(Lesson 13-4)*

10. Suppose a six-foot-tall golfer plays golf with a 1.5 inch diameter golf ball. If the ratio between golfer and ball remains the same, how tall would a golfer need to be to use Spaceship Earth as a golf ball? *(Lesson 13-4)*

ASTRONOMY   For Exercises 11 and 12, use the following information. A museum has set aside a children's room containing objects suspended from the ceiling to resemble planets and stars. Suppose an imaginary coordinate system is placed in the room with the center of the room at $(0, 0, 0)$. Three particular stars are located at $S(-10, 5, 3)$, $T(3, -8, -1)$, and $R(-7, -4, -2)$, where the coordinates represent the distance in feet from the center of the room. *(Lesson 13-5)*

11. Find the distance between each pair of stars.

12. Which star is farthest from the center of the room?

Becoming a Better Test-Taker

At some time in your life, you will have to take a standardized test. Sometimes this test may determine if you go on to the next grade or course, or even if you will graduate from high school. This section of your textbook is dedicated to making you a better test-taker.

TYPES OF TEST QUESTIONS   In the following pages, you will see examples of four types of questions commonly seen on standardized tests. A description of each type of question is shown in the table below.

Type of Question	Description	See Pages
multiple choice	Four or five possible answer choices are given from which you choose the best answer.	796–797
gridded response	You solve the problem. Then you enter the answer in a special grid and color in the corresponding circles.	798–801
short response	You solve the problem, showing your work and/or explaining your reasoning.	802–805
extended response	You solve a multi-part problem, showing your work and/or explaining your reasoning.	806–810

PRACTICE   After being introduced to each type of question, you can practice that type of question. Each set of practice questions is divided into five sections that represent the categories most commonly assessed on standardized tests.

- Number and Operations
- Algebra
- Geometry
- Measurement
- Data Analysis and Probability

USING A CALCULATOR   On some tests, you are permitted to use a calculator. You should check with your teacher to determine if calculator use is permitted on the test you will be taking, and, if so, what type of calculator can be used.

TEST-TAKING TIPS   In addition to the Test-Taking Tips like the one shown at the right, here are some additional thoughts that might help you.

- Get a good night's rest before the test. Cramming the night before does not improve your results.

- Budget your time when taking a test. Don't dwell on problems that you cannot solve. Just make sure to leave that question blank on your answer sheet.

- Watch for key words like NOT and EXCEPT. Also look for order words like LEAST, GREATEST, FIRST, and LAST.

> **Test-Taking Tip**
> If you are allowed to use a calculator, make sure you are familiar with how it works so that you won't waste time trying to figure out the calculator when taking the test.

Multiple-Choice Questions

Multiple-choice questions are the most common type of question on standardized tests. These questions are sometimes called *selected-response questions*. You are asked to choose the best answer from four or five possible answers.

To record a multiple-choice answer, you may be asked to shade in a bubble that is a circle or an oval or just to write the letter of your choice. Always make sure that your shading is dark enough and completely covers the bubble.

Incomplete shading
Ⓐ Ⓑ Ⓒ Ⓓ

Too light shading
Ⓐ Ⓑ Ⓒ Ⓓ

Correct shading
Ⓐ Ⓑ Ⓒ Ⓓ

Sometimes a question does not provide you with a figure that represents the problem. Drawing a diagram may help you to solve the problem. Once you draw the diagram, you may be able to eliminate some of the possibilities by using your knowledge of mathematics. Another answer choice might be that the correct answer is not given.

Example

Strategy

Diagrams
Draw a diagram of the playground.

A coordinate plane is superimposed on a map of a playground. Each side of each square represents 1 meter. The slide is located at (5, –7), and the climbing pole is located at (–1, 2). What is the distance between the slide and the pole?

Ⓐ $\sqrt{15}$ m   Ⓑ 6 m   Ⓒ 9 m   Ⓓ $9\sqrt{13}$ m   Ⓔ none of these

Draw a diagram of the playground on a coordinate plane. Notice that the difference in the x-coordinates is 6 meters and the difference in the y-coordinates is 9 meters.

Since the two points are two vertices of a right triangle, the distance between the two points must be greater than either of these values. So we can eliminate Choices B and C.

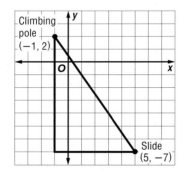

Use the Distance Formula or the Pythagorean Theorem to find the distance between the slide and the climbing pole. Let's use the Pythagorean Theorem.

$a^2 + b^2 = c^2$ Pythagorean Theorem

$6^2 + 9^2 = c^2$ Substitution

$36 + 81 = c^2$

$117 = c^2$

$3\sqrt{13} = c$ Take the square root of each side and simplify.

So, the distance between the slide and pole is $3\sqrt{13}$ meters. Since this is not listed as choice A, B, C, or D, the answer is Choice E.

If you are short on time, you can test each answer choice to find the correct answer. Sometimes you can make an educated guess about which answer choice to try first.

Multiple-Choice Practice

Choose the best answer.

Number and Operations

1. Carmen designed a rectangular banner that was 5 feet by 8 feet for a local business. The owner of the business asked her to make a larger banner measuring 10 feet by 20 feet. What was the percent increase in size from the first banner to the second banner?

Ⓐ 4%   Ⓑ 20%

Ⓒ 80%   Ⓓ 400%

2. A roller coaster casts a shadow 57 yards long. Next to the roller coaster is a 35-foot tree with a shadow that is 20 feet long at the same time of day. What is the height of the roller coaster to the nearest whole foot?

Ⓐ 98 ft   Ⓑ 100 ft

Ⓒ 299 ft   Ⓓ 388 ft

Algebra

3. At Speedy Car Rental, it costs $32 per day to rent a car and then $0.08 per mile. If y is the total cost of renting the car and x is the number of miles, which equation describes the relation between x and y?

Ⓐ $y = 32x + 0.08$ Ⓑ $y = 32x - 0.08$

Ⓒ $y = 0.08x + 32$ Ⓓ $y = 0.08x - 32$

4. Eric plotted his house, school, and the library on a coordinate plane. Each side of each square represents one mile. What is the distance from his house to the library?

Ⓐ $\sqrt{24}$ mi

Ⓑ 5 mi

Ⓒ $\sqrt{26}$ mi

Ⓓ $\sqrt{29}$ mi

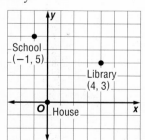

Geometry

5. The grounds outside of the Custer County Museum contain a garden shaped like a right triangle. One leg of the triangle measures 8 feet, and the area of the garden is 18 square feet. What is the length of the other leg?

Ⓐ 2.25 in.   Ⓑ 4.5 in.   Ⓒ 13.5 in.

Ⓓ 27 in.   Ⓔ 54 in.

Test-Taking Tip

Questions 2, 5 and 7
The units of measure given in the question may not be the same as those given in the answer choices. Check that your solution is in the proper unit.

6. The circumference of a circle is equal to the perimeter of a regular hexagon with sides that measure 22 inches. What is the length of the radius of the circle to the nearest inch? Use 3.14 for π.

Ⓐ 7 in.   Ⓑ 14 in.   Ⓒ 21 in.

Ⓓ 24 in.   Ⓔ 28 in.

Measurement

7. Eduardo is planning to install carpeting in a rectangular room that measures 12 feet 6 inches by 18 feet. How many square yards of carpet does he need for the project?

Ⓐ 25 yd²   Ⓑ 50 yd²

Ⓒ 225 yd²   Ⓓ 300 yd²

8. Marva is comparing two containers. One is a cylinder with diameter 14 centimeters and height 30 centimeters. The other is a cone with radius 15 centimeters and height 14 centimeters. What is the ratio of the volume of the cylinder to the volume of the cone?

Ⓐ 3 to 1   Ⓑ 2 to 1

Ⓒ 7 to 5   Ⓓ 7 to 10

Data Analysis and Probability

9. Refer to the table. Which statement is true about this set of data?

Country	Spending per Person
Japan	$8622
United States	$8098
Switzerland	$6827
Norway	$6563
Germany	$5841
Denmark	$5778

Source: *Top 10 of Everything 2003*

Ⓐ The median is less than the mean.

Ⓑ The mean is less than the median.

Ⓒ The range is 2844.

Ⓓ A and C are true.

Ⓔ B and C are true.

Gridded-Response Questions

Gridded-response questions are another type of question on standardized tests. These questions are sometimes called *student-produced response* or *grid-in*, because you must create the answer yourself, not just choose from four or five possible answers.

For gridded response, you must mark your answer on a grid printed on an answer sheet. The grid contains a row of four or five boxes at the top, two rows of ovals or circles with decimal and fraction symbols, and four or five columns of ovals, numbered 0–9. Since there is no negative symbol on the grid, answers are never negative. An example of a grid from an answer sheet is shown at the right.

How do you correctly fill in the grid?

Example 1   **In the diagram, $\triangle MPT \sim \triangle RPN$. Find PR.**

What do you need to find?

You need to find the value of x so that you can substitute it into the expression $3x + 3$ to find PR. Since the triangles are similar, write a proportion to solve for x.

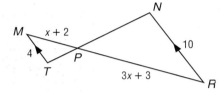

$\dfrac{MT}{RN} = \dfrac{PM}{PR}$ Definition of similar polygons

$\dfrac{4}{10} = \dfrac{x + 2}{3x + 3}$ Substitution

$4(3x + 3) = 10(x + 2)$ Cross products

$12x + 12 = 10x + 20$ Distributive Property

$2x = 8$ Subtract 12 and 10x from each side.

$x = 4$ Divide each side by 2.

Now find PR.
$PR = 3x + 3$
$= 3(4) + 3$ or 15

How do you fill in the grid for the answer?

- Write your answer in the answer boxes.
- Write only one digit or symbol in each answer box.
- Do not write any digits or symbols outside the answer boxes.
- You may write your answer with the first digit in the left answer box, or with the last digit in the right answer box. You may leave blank any boxes you do not need on the right or the left side of your answer.
- Fill in only one bubble for every answer box that you have written in. Be sure not to fill in a bubble under a blank answer box.

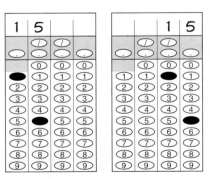

Many gridded-response questions result in an answer that is a fraction or a decimal. These values can also be filled in on the grid.

How do you grid decimals and fractions?

Example 2

A triangle has a base of length 1 inch and a height of 1 inch. What is the area of the triangle in square inches?

Use the formula $A = \frac{1}{2}bh$ to find the area of the triangle.

$$A = \frac{1}{2}bh \qquad \text{Area of a triangle}$$

$$= \frac{1}{2}(1)(1) \qquad \text{Substitution}$$

$$= \frac{1}{2} \text{ or } 0.5 \qquad \text{Simplify.}$$

How do you grid the answer?

You can either grid the fraction or the decimal. Be sure to write the decimal point or fraction bar in the answer box. The following are acceptable answer responses.

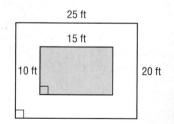

Do not leave a blank answer box in the middle of an answer.

Sometimes an answer is an improper fraction. Never change the improper fraction to a mixed number. Instead, grid either the improper fraction or the equivalent decimal.

How do you grid mixed numbers?

Example 3

The shaded region of the rectangular garden will contain roses. What is the ratio of the area of the garden to the area of the shaded region?

Strategy

Formulas
If you are unsure of a formula, check the reference sheet.

First, find the area of the garden.

$$A = \ell w$$

$$= 25(20) \text{ or } 500$$

Then find the area of the shaded region.

$$A = \ell w$$

$$= 15(10) \text{ or } 150$$

Write the ratio of the areas as a fraction.

$$\frac{\text{area of garden}}{\text{area of shaded region}} = \frac{500}{150} \text{ or } \frac{10}{3}$$

Leave the answer as the improper fraction $\frac{10}{3}$,

as there is no way to correctly grid $3\frac{1}{3}$.

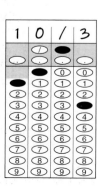

Gridded-Response Practice

Solve each problem and complete the grid.

Number and Operations

1. A large rectangular meeting room is being planned for a community center. Before building the center, the planning board decides to increase the area of the original room by 40%. When the room is finally built, budget cuts force the second plan to be reduced in area by 25%. What is the ratio of the area of the room that is built to the area of the original room?

2. Greenville has a spherical tank for the city's water supply. Due to increasing population, they plan to build another spherical water tank with a radius twice that of the current tank. How many times as great will the volume of the new tank be as the volume of the current tank?

3. In Earth's history, the Precambrian period was about 4600 million years ago. If this number of years is written in scientific notation, what is the exponent for the power of 10?

4. A virus is a type of microorganism so small it must be viewed with an electron microscope. The largest shape of virus has a length of about 0.0003 millimeter. To the nearest whole number, how many viruses would fit end to end on the head of a pin measuring 1 millimeter?

Algebra

5. Kaia has a painting that measures 10 inches by 14 inches. She wants to make her own frame that has an equal width on all sides. She wants the total area of the painting and frame to be 285 square inches. What will be the width of the frame in inches?

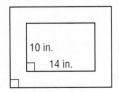

10 in.

14 in.

---

6. The diagram shows a triangle graphed on a coordinate plane. If $\overline{AB}$ is extended, what is the value of the y-intercept?

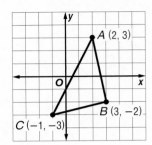

7. Tyree networks computers in homes and offices. In many cases, he needs to connect each computer to every other computer with a wire. The table shows the number of wires he needs to connect various numbers of computers. Use the table to determine how many wires are needed to connect 20 computers.

Computers	Wires	Computers	Wires
1	0	5	10
2	1	6	15
3	3	7	21
4	6	8	28

8. A line perpendicular to $9x - 10y = -10$ passes through $(-1, 4)$. Find the x-intercept of the line.

9. Find the positive solution of $6x^2 - 7x = 5$.

Geometry

10. The diagram shows $\triangle RST$ on the coordinate plane. The triangle is first rotated 90° counterclockwise about the origin and then reflected in the y-axis. What is the x-coordinate of the image of T after the two transformations?

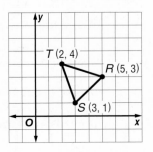

11. An octahedron is a solid with eight faces that are all equilateral triangles. How many edges does the octahedron have?

12. Find the measure of ∠A to the nearest tenth of a degree.

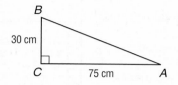

Measurement

13. The Pep Club plans to decorate some large garbage barrels for Spirit Week. They will cover only the sides of the barrels with decorated paper. How many square feet of paper will they need to cover 8 barrels like the one in the diagram? Use 3.14 for π. Round to the nearest square foot.

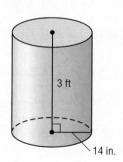

14. Kara makes decorative paperweights. One of her favorites is a hemisphere with a diameter of 4.5 centimeters. What is the surface area of the hemisphere including the bottom on which it rests? Use 3.14 for π. Round to the nearest tenth of a square centimeter.

15. The record for the fastest land speed of a car traveling for one mile is approximately 763 miles per hour. The car was powered by two jet engines. What was the speed of the car in feet per second? Round to the nearest whole number.

16. On average, a B-777 aircraft uses 5335 gallons of fuel on a 2.5-hour flight. At this rate, how much fuel will be needed for a 45-minute flight? Round to the nearest gallon.

Data Analysis and Probability

17. The table shows the heights of the tallest buildings in Kansas City, Missouri. To the nearest tenth, what is the positive difference between the median and the mean of the data?

Name	Height (m)
One Kansas City Place	193
Town Pavilion	180
Hyatt Regency	154
Power and Light Building	147
City Hall	135
1201 Walnut	130

Source: skyscrapers.com

18. A long-distance telephone service charges 40 cents per call and 5 cents per minute. If a function model is written for the graph, what is the rate of change of the function?

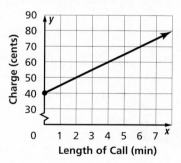

19. In a dart game, the dart must land within the innermost circle on the dartboard to win a prize. If a dart hits the board, what is the probability, as a percent, that it will hit the innermost circle?

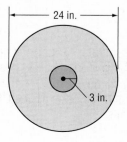

Short-Response Questions

Short-response questions require you to provide a solution to the problem, as well as any method, explanation, and/or justification you used to arrive at the solution. These are sometimes called *constructed-response, open-response, open-ended, free-response,* or *student-produced questions.* The following is a sample rubric, or scoring guide, for scoring short-response questions.

Credit	Score	Criteria
Full	2	Full credit: The answer is correct and a full explanation is provided that shows each step in arriving at the final answer.
Partial	1	Partial credit: There are two different ways to receive partial credit. • The answer is correct, but the explanation provided is incomplete or incorrect. • The answer is incorrect, but the explanation and method of solving the problem is correct.
None	0	No credit: Either an answer is not provided or the answer does not make sense.

On some standardized tests, no credit is given for a correct answer if your work is not shown.

Example

Mr. Solberg wants to buy all the lawn fertilizer he will need for this season. His front yard is a rectangle measuring 55 feet by 32 feet. His back yard is a rectangle measuring 75 feet by 54 feet. Two sizes of fertilizer are available—one that covers 5000 square feet and another covering 15,000 square feet. He needs to apply the fertilizer four times during the season. How many bags of each size should he buy to have the least amount of waste?

Full Credit Solution

Strategy

Estimation
Use estimation to check your solution.

Find the area of each part of the lawn and multiply by 4 since the fertilizer is to be applied 4 times. Each portion of the lawn is a rectangle, so $A = lw$.

$$4[(55 \times 32) + (75 \times 54)] = 23{,}240 \text{ ft}^2$$

If Mr. Solberg buys 2 bags that cover 15,000 ft², he will have too much fertilizer. If he buys 1 large bag, he will still need to cover $23{,}240 - 15{,}000$ or 8240 ft².

The steps, calculations, and reasoning are clearly stated.

Find how many small bags it takes to cover 82400 ft².

$$8240 \div 5000 = 1.648$$

Since he cannot buy a fraction of a bag, he will need to buy 2 of the bags that cover 5000 ft² each.

The solution of the problem is clearly stated.

Mr. Solberg needs to buy 1 bag that covers 15,000 square feet and 2 bags that cover 5000 square feet each.

Partial Credit Solution

In this sample solution, the answer is correct. However, there is no justification for any of the calculations.

There is not an explanation of how 23,240 was obtained.

23,240

$$23,240 - 15,000 = 8240$$

$$8240 \div 5000 = 1.648$$

Mr. Solberg needs to buy 1 large bag and 2 small bags.

Partial Credit Solution

In this sample solution, the answer is incorrect. However, after the first statement all of the calculations and reasoning are correct.

The first step of multiplying the area by 4 was left out.

First find the total number of square feet of lawn. Find the area of each part of the yard.

$$(55 \times 32) + (75 \times 54) = 5810 \text{ ft}^2$$

The area of the lawn is greater than 5000 ft², which is the amount covered by the smaller bag, but buying the bag that covers 15,000 ft² would result in too much waste.

$$5810 \div 5000 = 1.162$$

Therefore, Mr. Solberg will need to buy 2 of the smaller bags of fertilizer.

No Credit Solution

In this sample solution, the response is incorrect and incomplete.

The wrong operations are used, so the answer is incorrect. Also, there are no units of measure given with any of the calculations.

$$55 + 75 = 130$$
$$32 + 54 = 86$$
$$130 \times 86 \times 4 = 44{,}720$$
$$44{,}720 \div 15{,}000 = 2.98$$

Mr. Solberg will need 3 bags of fertilizer.

Short-Response Practice

Solve each problem. Show all your work.

Number and Operations

1. In 2000, approximately $191 billion in merchandise was sold by a popular retail chain store in the United States. The population at that time was 281,421,906. Estimate the average amount of merchandise bought from this store by each person in the U.S.

2. At a theme park, three educational movies run continuously all day long. At 9 A.M., the three shows begin. One runs for 15 minutes, the second for 18 minutes, and the third for 25 minutes. At what time will the movies all begin at the same time again?

3. Ming found a sweater on sale for 20% off the original price. However, the store was offering a special promotion, where all sale items were discounted an additional 60%. What was the total percent discount for the sweater?

4. The serial number of a DVD player consists of three letters of the alphabet followed by five digits. The first two letters can be any letter, but the third letter cannot be O. The first digit cannot be zero. How many serial numbers are possible with this system?

Algebra

5. Solve and graph $2x - 9 \le 5x + 4$.

6. Vance rents rafts for trips on the Jefferson River. You have to reserve the raft and provide a $15 deposit in advance. Then the charge is $7.50 per hour. Write an equation that can be used to find the charge for any amount of time, where y is the total charge in dollars and x is the amount of time in hours.

Test-Taking Tip (A) (B) (C) (D)

Question 4
Be sure to completely and carefully read the problem before beginning any calculations. If you read too quickly, you may miss a key piece of information.

7. Hector is working on the design for the container shown below that consists of a cylinder with a hemisphere on top. He has written the expression $\pi r^2 + 2\pi rh + 2\pi r^2$ to represent the surface area of any size container of this shape. Explain the meaning of each term of the expression.

8. Find all solutions of the equation $6x^2 + 13x = 5$.

9. In 1999, there were 2,192,070 farms in the U.S., while in 2001, there were 2,157,780 farms. Let x represent years since 1999 and y represent the total number of farms in the U.S. Suppose the number of farms continues to decrease at the same rate as from 1999 to 2001. Write an equation that models the number of farms for any year after 1999.

Geometry

10. Refer to the diagram. What is the measure of $\angle 1$?

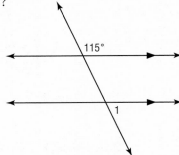

11. Quadrilateral *JKLM* is to be reflected in the line $y = x$. What are the coordinates of the vertices of the image?

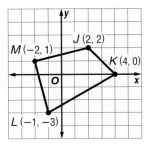

12. Write an equation in standard form for a circle that has a diameter with endpoints at $(-3, 2)$ and $(4, -5)$.

13. In the Columbia Village subdivision, an unusually shaped lot, shown below, will be used for a small park. Find the exact perimeter of the lot.

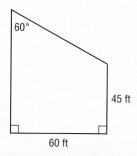

60°

45 ft

60 ft

Measurement

14. The Astronomical Unit (AU) is the distance from Earth to the Sun. It is usually rounded to 93,000,000 miles. The star Alpha Centauri is 25,556,250 million miles from Earth. What is this distance in AU?

15. Linesse handpaints unique designs on shirts and sells them. It takes her about 4.5 hours to create a design. At this rate, how many shirts can she design if she works 22 days per month for an average of 6.5 hours per day?

16. The world's largest pancake was made in England in 1994. To the nearest cubic foot, what was the volume of the pancake?

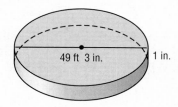

49 ft 3 in.   1 in.

17. Find the ratio of the volume of the cylinder to the volume of the pyramid.

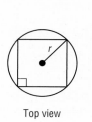

Top view

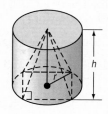

Front view

r

h

Data Analysis and Probability

18. The table shows the winning times for the Olympic men's 1000-meter speed skating event. Make a scatter plot of the data and describe the pattern in the data. Times are rounded to the nearest second.

Men's 1000-m Speed Skating Event		
Year	Country	Time(s)
1976	U.S.	79
1980	U.S.	75
1984	Canada	76
1988	USSR	73
1992	Germany	75
1994	U.S.	72
1998	Netherlands	71
2002	Netherlands	67

Source: *The World Almanac*

19. Bradley surveyed 70 people about their favorite spectator sport. If a person is chosen at random from the people surveyed, what is the probability that the person's favorite spectator sport is basketball?

Favorite Spectator Sport

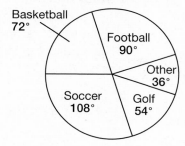

Basketball 72°
Football 90°
Other 36°
Golf 54°
Soccer 108°

20. The graph shows the altitude of a small airplane. Write a function to model the graph. Explain what the model means in terms of the altitude of the airplane.

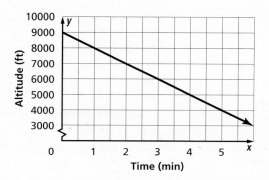

Altitude (ft)

Time (min)

Extended-Response Questions

Extended-response questions are often called *open-ended* or *constructed-response questions.* Most extended-response questions have multiple parts. You must answer all parts to receive full credit.

Extended-response questions are similar to short-response questions in that you must show all of your work in solving the problem. A rubric is also used to determine whether you receive full, partial, or no credit. The following is a sample rubric for scoring extended-response questions.

Credit	Score	Criteria
Full	4	Full credit: A correct solution is given that is supported by well-developed, accurate explanations.
Partial	3, 2, 1	Partial credit: A generally correct solution is given that may contain minor flaws in reasoning or computation or an incomplete solution. The more correct the solution, the greater the score.
None	0	No credit: An incorrect solution is given indicating no mathematical understanding of the concept, or no solution is given.

On some standardized tests, no credit is given for a correct answer if your work is not shown.

Make sure that when the problem says to *Show your work,* you show every part of your solution including figures, sketches of graphing calculator screens, or the reasoning behind your computations.

Example

Polygon *WXYZ* with vertices *W*(−3, 2), *X*(4, 4), *Y*(3, −1), and *Z*(−2, −3) is a figure represented on a coordinate plane to be used in the graphics for a video game. Various transformations will be performed on the polygon to use for the game.

a. Graph *WXYZ* and its image *W′X′Y′Z′* under a reflection in the *y*-axis. Be sure to label all of the vertices.

b. Describe how the coordinates of the vertices of *WXYZ* relate to the coordinates of the vertices of *W′X′Y′Z′*.

Strategy

Make a List

Write notes about what to include in your answer for each part of the question.

c. Another transformation is performed on *WXYZ*. This time, the vertices of the image *W′X′Y′Z′* are *W*′(2, −3), *X*′(4, 4), *Y*′(−1, 3), and *Z*′(−3, −2). Graph *WXYZ* and its image under this transformation. What transformation produced *W′X′Y′Z′*?

Full Credit Solution

Part a A complete graph includes labels for the axes and origin and labels for the vertices, including letter names and coordinates.

- The vertices of the polygon should be correctly graphed and labeled.

- The vertices of the image should be located such that the transformation shows a reflection in the *y*-axis.

- The vertices of the polygons should be connected correctly. Optionally, the polygon and its image could be graphed in two contrasting colors.

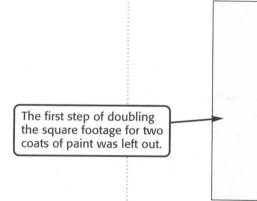

The first step of doubling the square footage for two coats of paint was left out.

Part b

The coordinates of W and W' are $(-3, 2)$ and $(3, 2)$. The x-coordinates are the opposite of each other and the y-coordinates are the same. For any point (a, b), the coordinates of the reflection in the y-axis are $(-a, b)$.

Part c

For full credit, the graph in Part C must also be accurate, which is true for this graph.

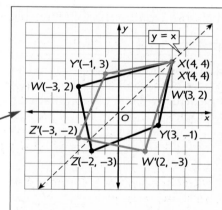

The coordinates of Z and Z' have been switched. In other words, for any point (a, b), the coordinates of the reflection in the y-axis are (b, a). Since X and X' are the same point, the polygon has been reflected in the line $y = x$.

Partial Credit Solution

Part a This sample graph includes no labels for the axes and for the vertices of the polygon and its image. Two of the image points have been incorrectly graphed.

More credit would have been given if all of the points were reflected correctly. The images for X and Y are not correct.

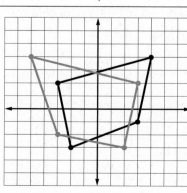

(continued on the next page)

Part b Partial credit is given because the reasoning is correct, but the reasoning was based on the incorrect graph in Part a.

> For two of the points, W and Z, the y-coordinates are the same and the x-coordinates are opposites. But, for points X and Y, there is no clear relationship.

Part c Full credit is given for Part c. The graph supplied by the student was identical to the graph shown for the full credit solution for Part c. The explanation below is correct, but slightly different from the previous answer for Part c.

> I noticed that point X and point X' were the same. I also guessed that this was a reflection, but not in either axis. I played around with my ruler until I found a line that was the line of reflection. The transformation from WXYZ to W'X'Y'Z' was a reflection in the line y = x.

This sample answer might have received a score of 2 or 1, depending on the judgment of the scorer. Had the student graphed all points correctly and gotten Part b correct, the score would probably have been a 3.

No Credit Solution

Part a The sample answer below includes no labels on the axes or the coordinates of the vertices of the polygon. The polygon *WXYZ* has three vertices graphed incorrectly. The polygon that was graphed is not reflected correctly either.

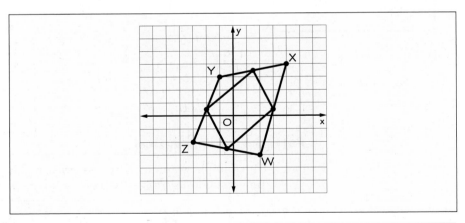

Part b

> I don't see any way that the coordinates relate.

Part c

> It is a reduction because it gets smaller.

In this sample answer, the student does not understand how to graph points on a coordinate plane and also does not understand the reflection of figures in an axis or other line.

Extended-Response Practice

Solve each problem. Show all your work.

Number and Operations

1. Refer to the table.

Population		
City	1990	2000
Phoenix, AZ	983,403	1,321,045
Austin, TX	465,622	656,562
Charlotte, NC	395,934	540,828
Mesa, AZ	288,091	396,375
Las Vegas, NV	258,295	478,434

Source: census.gov

a. For which city was the increase in population the greatest? What was the increase?

b. For which city was the percent of increase in population the greatest? What was the percent increase?

c. Suppose that the population increase of a city was 30%. If the population in 2000 was 346,668, find the population in 1990.

2. Molecules are the smallest units of a particular substance that still have the same properties as that substance. The diameter of a molecule is measured in angstroms (Å). Express each value in scientific notation.

a. An angstrom is exactly 10^{-8} centimeter. A centimeter is approximately equal to 0.3937 inch. What is the approximate measure of an angstrom in inches?

b. How many angstroms are in one inch?

c. If a molecule has a diameter of 2 angstroms, how many of these molecules placed side by side would fit on an eraser measuring $\frac{1}{4}$ inch?

Algebra

3. The Marshalls are building a rectangular in-ground pool in their backyard. The pool will be 24 feet by 29 feet. They want to build a deck of equal width all around the pool. The final area of the pool and deck will be 1800 square feet.

a. Draw and label a diagram.

b. Write an equation that can be used to find the width of the deck.

c. Find the width of the deck.

4. The depth of a reservoir was measured on the first day of each month. (Jan. = 1, Feb. = 2, and so on.)

Depth of the Reservoir

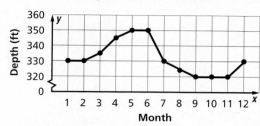

a. What is the slope of the line joining the points with x-coordinates 6 and 7? What does the slope represent?

b. Write an equation for the segment of the graph from 5 to 6. What is the slope of the line and what does this represent in terms of the reservoir?

c. What was the lowest depth of the reservoir? When was this depth first measured and recorded?

Geometry

5. The Silver City Marching Band is planning to create this formation with the members.

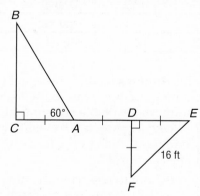

a. Find the missing side measures of $\triangle EDF$. Explain.

b. Find the missing side measures of $\triangle ABC$. Explain.

c. Find the total distance of the path: A to B to C to A to D to E to F to D.

d. The director wants to place one person at each point A, B, C, D, E, and F. He then wants to place other band members approximately one foot apart on all segments of the formation. How many people should he place on each segment of the formation? How many total people will he need?

Measurement

6. Two containers have been designed. One is a hexagonal prism, and the other is a cylinder.

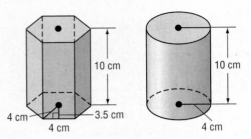

   a. What is the volume of the hexagonal prism?

   b. What is the volume of the cylinder?

   c. What is the percent of increase in volume from the prism to the cylinder?

7. Kabrena is working on a project about the solar system. The table shows the maximum distances from Earth to the other planets in millions of miles.

Distance from Earth to Other Planets

Planet	Distance	Planet	Distance
Mercury	138	Saturn	1031
Venus	162	Uranus	1962
Mars	249	Neptune	2913
Jupiter	602	Pluto	4681

Source: *The World Almanac*

   a. The maximum speed of the Apollo moon missions spacecraft was about 25,000 miles per hour. Make a table showing the time it would take a spacecraft traveling at this speed to reach each of the four closest planets.

   b. Describe how to use scientific notation to calculate the time it takes to reach any planet.

   c. Which planet would it take approximately 13.3 years to reach? Explain.

Test-Taking Tip Ⓐ Ⓑ Ⓒ Ⓓ

Question 6
While preparing to take a standardized test, familiarize yourself with the formulas for surface area and volume of common three-dimensional figures.

Data Analysis and Probability

8. The table shows the average monthly temperatures in Barrow, Alaska. The months are given numerical values from 1-12. (Jan. = 1, Feb. = 2, and so on.)

Average Monthly Temperature

Month	°F	Month	°F
1	−14	7	40
2	−16	8	39
3	−14	9	31
4	−1	10	15
5	20	11	−1
6	35	12	−11

   a. Make a scatter plot of the data. Let x be the numerical value assigned to the month and y be the temperature.

   b. Describe any trends shown in the graph.

   c. Find the mean of the temperature data.

   d. Describe any relationship between the mean of the data and the scatter plot.

9. A dart game is played using the board shown. The inner circle is pink, the next ring is blue, the next red, and the largest ring is green. A dart must land on the board during each round of play.

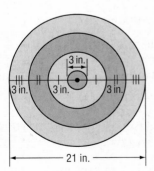

   a. What is the probability that a dart landing on the board hits the pink circle?

   b. What is the probability that the first dart thrown lands in the blue ring and the second dart lands in the green ring?

   c. Suppose players throw a dart twice. For which outcome of two darts would you award the most expensive prize? Explain your reasoning.

Postulates, Theorems, and Corollaries

Chapter 2 Reasoning and Proof

Postulate 2.1   Through any two points, there is exactly one line. (p. 89)

Postulate 2.2   Through any three points not on the same line, there is exactly one plane. (p. 89)

Postulate 2.3   A line contains at least two points. (p. 90)

Postulate 2.4   A plane contains at least three points not on the same line. (p. 90)

Postulate 2.5   If two points lie in a plane, then the entire line containing those points lies in that plane. (p. 90)

Postulate 2.6   If two lines intersect, then their intersection is exactly one point. (p. 90)

Postulate 2.7   If two planes intersect, then their intersection is a line. (p. 90)

Theorem 2.1   **Midpoint Theorem** If M is the midpoint of $\overline{AB}$, then $\overline{AM} \cong \overline{MB}$. (p. 91)

Postulate 2.8   **Ruler Postulate** The points on any line or line segment can be paired with real numbers so that, given any two points A and B on a line, A corresponds to zero, and B corresponds to a positive real number. (p. 101)

Postulate 2.9   **Segment Addition Postulate** If B is between A and C, then $AB + BC = AC$. If $AB + BC = AC$, then B is between A and C. (p. 102)

Theorem 2.2   Congruence of segments is reflexive, symmetric, and transitive. (p. 102)

Postulate 2.10   **Protractor Postulate** Given $\overrightarrow{AB}$ and a number r between 0 and 180, there is exactly one ray with endpoint A, extending on either side of $\overrightarrow{AB}$, such that the measure of the angle formed is r. (p. 107)

Postulate 2.11   **Angle Addition Postulate** If R is in the interior of $\angle PQS$, then $m\angle PQR + m\angle RQS = m\angle PQS$. If $m\angle PQR + m\angle RQS = m\angle PQS$, then R is in the interior of $\angle PQS$. (p. 107)

Theorem 2.3   **Supplement Theorem** If two angles form a linear pair, then they are supplementary angles. (p. 108)

Theorem 2.4   **Complement Theorem** If the noncommon sides of two adjacent angles form a right angle, then the angles are complementary angles. (p. 108)

Theorem 2.5   Congruence of angles is reflexive, symmetric, and transitive. (p. 108)

Theorem 2.6   Angles supplementary to the same angle or to congruent angles are congruent. (p. 109) Abbreviation: ∠s suppl. to same ∠ or ≅ ∠s are ≅.

Theorem 2.7   Angles complementary to the same angle or to congruent angles are congruent. (p. 109) Abbreviation: ∠s compl. to same ∠ or ≅ ∠s are ≅.

Theorem 2.8   **Vertical Angle Theorem** If two angles are vertical angles, then they are congruent. (p. 110)

Theorem 2.9   Perpendicular lines intersect to form four right angles. (p. 110)

Theorem 2.10   All right angles are congruent. (p. 110)

Theorem 2.11   Perpendicular lines form congruent adjacent angles. (p. 110)

Theorem 2.12   If two angles are congruent and supplementary, then each angle is a right angle. (p. 110)

Theorem 2.13   If two congruent angles form a linear pair, then they are right angles. (p. 110)

Chapter 3 Perpendicular and Parallel Lines

Postulate 3.1   **Corresponding Angles Postulate** If two parallel lines are cut by a transversal, then each pair of corresponding angles is congruent. (p. 133)

Theorem 3.1   **Alternate Interior Angles Theorem** If two parallel lines are cut by a transversal, then each pair of alternate interior angles is congruent. (p. 134)

Theorem 3.2   **Consecutive Interior Angles Theorem** If two parallel lines are cut by a transversal, then each pair of consecutive interior angles is supplementary. (p. 134)

Theorem 3.3   **Alternate Exterior Angles Theorem** If two parallel lines are cut by a transversal, then each pair of alternate exterior angles is congruent. (p. 134)

Theorem 3.4   **Perpendicular Transversal Theorem** In a plane, if a line is perpendicular to one of two parallel lines, then it is perpendicular to the other. (p. 134)

Postulate 3.2   Two nonvertical lines have the same slope if and only if they are parallel. (p. 141)

Postulate 3.3   Two nonvertical lines are perpendicular if and only if the product of their slopes is -1. (p. 141)

Postulate 3.4   If two lines in a plane are cut by a transversal so that corresponding angles are congruent, then the lines are parallel. (p. 151) Abbreviation: If corr. ∠ are ≅, lines are ∥.

Postulate 3.5   **Parallel Postulate** If there is a line and a point not on the line, then there exists exactly one line through the point that is parallel to the given line. (p. 152)

Theorem 3.5   If two lines in a plane are cut by a transversal so that a pair of alternate exterior angles is congruent, then the two lines are parallel. (p. 152) Abbreviation: If alt. ext. ∠ are ≅, then lines are ∥.

Theorem 3.6   If two lines in a plane are cut by a transversal so that a pair of consecutive interior angles is supplementary, then the lines are parallel. (p. 152) Abbreviation: If cons. int. ∠ are suppl., then lines are ∥.

Theorem 3.7   If two lines in a plane are cut by a transversal so that a pair of alternate interior angles is congruent, then the lines are parallel. (p. 152) Abbreviation: If alt. int. ∠ are ≅, then lines are ∥.

Theorem 3.8   In a plane, if two lines are perpendicular to the same line, then they are parallel. (p. 152) Abbreviation: If 2 lines are ⊥ to the same line, then lines are ∥.

Theorem 3.9   In a plane, if two lines are each equidistant from a third line, then the two lines are parallel to each other. (p. 161)

Chapter 4 Congruent Triangles

Theorem 4.1   **Angle Sum Theorem** The sum of the measures of the angles of a triangle is 180. (p. 185)

Theorem 4.2   **Third Angle Theorem** If two angles of one triangle are congruent to two angles of a second triangle, then the third angles of the triangles are congruent. (p. 186)

Theorem 4.3	**Exterior Angle Theorem** The measure of an exterior angle of a triangle is equal to the sum of the measures of the two remote interior angles. (p. 186)
Corollary 4.1	The acute angles of a right triangle are complementary. (p. 188)
Corollary 4.2	There can be at most one right or obtuse angle in a triangle. (p. 188)
Theorem 4.4	Congruence of triangles is reflexive, symmetric, and transitive. (p. 193)
Postulate 4.1	**Side-Side-Side Congruence (SSS)** If the sides of one triangle are congruent to the sides of a second triangle, then the triangles are congruent. (p. 201)
Postulate 4.2	**Side-Angle-Side Congruence (SAS)** If two sides and the included angle of one triangle are congruent to two sides and the included angle of another triangle, then the triangles are congruent. (p. 202)
Postulate 4.3	**Angle-Side-Angle Congruence (ASA)** If two angles and the included side of one triangle are congruent to two angles and the included side of another triangle, the triangles are congruent. (p. 207)
Theorem 4.5	**Angle-Angle-Side Congruence (AAS)** If two angles and a nonincluded side of one triangle are congruent to the corresponding two angles and side of a second triangle, then the two triangles are congruent. (p. 208)
Theorem 4.6	**Leg-Leg Congruence (LL)** If the legs of one right triangle are congruent to the corresponding legs of another right triangle, then the triangles are congruent. (p. 214)
Theorem 4.7	**Hypotenuse-Angle Congruence (HA)** If the hypotenuse and acute angle of one right triangle are congruent to the hypotenuse and corresponding acute angle of another right triangle, then the two triangles are congruent. (p. 215)
Theorem 4.8	**Leg-Angle Congruence (LA)** If one leg and an acute angle of one right triangle are congruent to the corresponding leg and acute angle of another right triangle, then the triangles are congruent. (p. 215)
Postulate 4.4	**Hypotenuse-Leg Congruence (HL)** If the hypotenuse and a leg of one right triangle are congruent to the hypotenuse and corresponding leg of another right triangle, then the triangles are congruent. (p. 215)
Theorem 4.9	**Isosceles Triangle Theorem** If two sides of a triangle are congruent, then the angles opposite those sides are congruent. (p. 216)
Theorem 4.10	If two angles of a triangle are congruent, then the sides opposite those angles are congruent. (p. 218) Abbreviation: Conv. of Isos. △Th.
Corollary 4.3	A triangle is equilateral if and only if it is equiangular. (p. 218)
Corollary 4.4	Each angle of an equilateral triangle measures 60°. (p. 218)

Chapter 5 Relationships in Triangles

Theorem 5.1	Any point on the perpendicular bisector of a segment is equidistant from the endpoints of the segment. (p. 238)
Theorem 5.2	Any point equidistant from the endpoints of a segment lies on the perpendicular bisector of the segment. (p. 238)

Theorem 5.3   Circumcenter Theorem The circumcenter of a triangle is equidistant from the vertices of the triangle. (p. 239)

Theorem 5.4   Any point on the angle bisector is equidistant from the sides of the angle. (p. 239)

Theorem 5.5   Any point equidistant from the sides of an angle lies on the angle bisector. (p. 239)

Theorem 5.6   Incenter Theorem The incenter of a triangle is equidistant from each side of the triangle. (p. 240)

Theorem 5.7   Centroid Theorem The centroid of a triangle is located two-thirds of the distance from a vertex to the midpoint of the side opposite the vertex on a median. (p. 240)

Theorem 5.8   Exterior Angle Inequality Theorem If an angle is an exterior angle of a triangle, then its measure is greater than the measure of either of its corresponding remote interior angles. (p. 248)

Theorem 5.9   If one side of a triangle is longer than another side, then the angle opposite the longer side has a greater measure than the angle opposite the shorter side. (p. 249)

Theorem 5.10   If one angle of a triangle has a greater measure than another angle, then the side opposite the greater angle is longer than the side opposite the lesser angle. (p. 250)

Theorem 5.11   Triangle Inequality Theorem The sum of the lengths of any two sides of a triangle is greater than the length of the third side. (p. 261)

Theorem 5.12   The perpendicular segment from a point to a line is the shortest segment from the point to the line. (p. 262)

Corollary 5.1   The perpendicular segment from a point to a plane is the shortest segment from the point to the plane. (p. 263)

Theorem 5.13   SAS Inequality/Hinge Theorem If two sides of a triangle are congruent to two sides of another triangle and the included angle in one triangle has a greater measure than the included angle in the other, then the third side of the first triangle is longer than the third side of the second triangle. (p. 267)

Theorem 5.14   SSS Inequality If two sides of a triangle are congruent to two sides of another triangle and the third side in one triangle is longer than the third side in the other, then the angle between the pair of congruent sides in the first triangle is greater than the corresponding angle in the second triangle. (p. 268)

Chapter 6 Proportions and Similarity

Postulate 6.1   Angle-Angle (AA) Similarity If the two angles of one triangle are congruent to two angles of another triangle, then the triangles are similar. (p. 298)

Theorem 6.1   Side-Side-Side (SSS) Similarity If the measures of the corresponding sides of two triangles are proportional, then the triangles are similar. (p. 299)

Theorem 6.2   Side-Angle-Side (SAS) Similarity If the measures of two sides of a triangle are proportional to the measures of two corresponding sides of another triangle and the included angles are congruent, then the triangles are similar. (p. 299)

Theorem 6.3   Similarity of triangles is reflexive, symmetric, and transitive. (p. 300)

Postulates, Theorems, and Corollaries

Theorem 6.4 **Triangle Proportionality Theorem** If a line is parallel to one side of a triangle and intersects the other two sides in two distinct points, then it separates these sides into segments of proportional lengths. (p. 307)

Theorem 6.5 **Converse of the Triangle Proportionality Theorem** If a line intersects two sides of a triangle and separates the sides into corresponding segments of proportional lengths, then the line is parallel to the third side. (p. 308)

Theorem 6.6 **Triangle Midsegment Theorem** A midsegment of a triangle is parallel to one side of the triangle, and its length is one-half the length of that side. (p. 308)

Corollary 6.1 If three or more parallel lines intersect two transversals, then they cut off the transversals proportionally. (p. 309)

Corollary 6.2 If three or more parallel lines cut off congruent segments on one transversal, then they cut off congruent segments on every transversal. (p. 309)

Theorem 6.7 **Proportional Perimeters Theorem** If two triangles are similar, then the perimeters are proportional to the measures of corresponding sides. (p. 316)

Theorem 6.8 If two triangles are similar, then the measures of the corresponding altitudes are proportional to the measures of the corresponding sides. (p. 317)
Abbreviation: ~ △s have corr. altitudes proportional to the corr. sides.

Theorem 6.9 If two triangles are similar, then the measures of the corresponding angle bisectors of the triangles are proportional to the measures of the corresponding sides. (p. 317)
Abbreviation: ~ △s have corr. ∠ bisectors proportional to the corr. sides.

Theorem 6.10 If two triangles are similar, then the measures of the corresponding medians are proportional to the measures of the corresponding sides. (p. 317)
Abbreviation: ~ △s have corr. medians proportional to the corr. sides.

Theorem 6.11 **Angle Bisector Theorem** An angle bisector in a triangle separates the opposite side into segments that have the same ratio as the other two sides. (p. 319)

Chapter 7 Right Triangles and Trigonometry

Theorem 7.1 If the altitude is drawn from the vertex of the right angle of a right triangle to its hypotenuse, then the two triangles formed are similar to the given triangle and to each other. (p. 343)

Theorem 7.2 The measure of the altitude drawn from the vertex of the right angle of a right triangle to its hypotenuse is the geometric mean between the measures of the two segments of the hypotenuse. (p. 343)

Theorem 7.3 If the altitude is drawn from the vertex of the right angle of a right triangle to its hypotenuse, then the measure of a leg of the triangle is the geometric mean between the measures of the hypotenuse and the segment of the hypotenuse adjacent to that leg. (p. 344)

Theorem 7.4 **Pythagorean Theorem** In a right triangle, the sum of the squares of the measures of the legs equals the square of the measure of the hypotenuse. (p. 350)

Theorem 7.5 **Converse of the Pythagorean Theorem** If the sum of the squares of the measures of two sides of a triangle equals the square of the measure of the longest side, then the triangle is a right triangle. (p. 351)

Theorem 7.6 In a 45°-45°-90° triangle, the length of the hypotenuse is $\sqrt{2}$ times the length of a leg. (p. 357)

Theorem 7.7   In a 30°-60°-90° triangle, the length of the hypotenuse is twice the length of the shorter leg, and the length of the longer leg is $\sqrt{3}$ times the length of the shorter leg. (p. 359)

Chapter 8 Quadrilaterals

Theorem 8.1   **Interior Angle Sum Theorem** If a convex polygon has n sides and S is the sum of the measures of its interior angles, then $S = 180(n - 2)$. (p. 404)

Theorem 8.2   **Exterior Angle Sum Theorem** If a polygon is convex, then the sum of the measures of the exterior angles, one at each vertex, is 360. (p. 406)

Theorem 8.3   Opposite sides of a parallelogram are congruent. (p. 412)
Abbreviation: Opp. sides of ▱ are ≅.

Theorem 8.4   Opposite angles of a parallelogram are congruent. (p. 412)
Abbreviation: Opp. ∠s of ▱ are ≅.

Theorem 8.5   Consecutive angles in a parallelogram are supplementary. (p. 412)
Abbreviation: Cons. ∠s in ▱ are suppl.

Theorem 8.6   If a parallelogram has one right angle, it has four right angles. (p. 412)
Abbreviation: If ▱ has 1 rt. ∠, it has 4 rt. ∠s.

Theorem 8.7   The diagonals of a parallelogram bisect each other. (p. 413)
Abbreviation: Diag. of ▱ bisect each other.

Theorem 8.8   The diagonal of a parallelogram separates the parallelogram into two congruent triangles. (p. 414) Abbreviation: Diag. of ▱ separates ▱ into 2 ≅ △s.

Theorem 8.9   If both pairs of opposite sides of a quadrilateral are congruent, then the quadrilateral is a parallelogram. (p. 418) Abbreviation: If both pairs of opp. sides are ≅ , then quad. is ▱.

Theorem 8.10   If both pairs of opposite angles of a quadrilateral are congruent, then the quadrilateral is a parallelogram. (p. 418) Abbreviation: If both pairs of opp. ∠s are ≅, then quad. is ▱.

Theorem 8.11   If the diagonals of a quadrilateral bisect each other, then the quadrilateral is a parallelogram. (p. 418) Abbreviation: If diag. bisect each other, then quad. is ▱.

Theorem 8.12   If one pair of opposite sides of a quadrilateral is both parallel and congruent, then the quadrilateral is a parallelogram. (p. 418)
Abbreviation: If one pair of opp. sides is ∥ and ≅, then the quad. is a ▱.

Theorem 8.13   If a parallelogram is a rectangle, then the diagonals are congruent. (p. 424)
Abbreviation: If ▱ is rectangle, diag. are ≅.

Theorem 8.14   If the diagonals of a parallelogram are congruent, then the parallelogram is a rectangle. (p. 426) Abbreviation: If diagonals of ▱ are ≅, ▱ is a rectangle.

Theorem 8.15   The diagonals of a rhombus are perpendicular. (p. 431)

Theorem 8.16   If the diagonals of a parallelogram are perpendicular, then the parallelogram is a rhombus. (p. 431)

Theorem 8.17   Each diagonal of a rhombus bisects a pair of opposite angles. (p. 431)

Theorem 8.18   Both pairs of base angles of an isosceles trapezoid are congruent. (p. 439)

Theorem 8.19   The diagonals of an isosceles trapezoid are congruent. (p. 439)

Theorem 8.20   The median of a trapezoid is parallel to the bases, and its measure is one-half the sum of the measures of the bases. (p. 441)

Chapter 9 Transformations

Postulate 9.1   In a given rotation, if A is the preimage, A' is the image, and P is the center of rotation, then the measure of the angle of rotation, $\angle APA'$ is twice the measure of the acute or right angle formed by the intersecting lines of reflection. (p. 477)

Corollary 9.1   Reflecting an image successively in two perpendicular lines results in a 180° rotation. (p. 477)

Theorem 9.1   If a dilation with center C and a scale factor of r transforms A to E and B to D, then $ED = |r|(AB)$. (p. 491)

Theorem 9.2   If $P(x, y)$ is the preimage of a dilation centered at the origin with a scale factor r, then the image is $P'(rx, ry)$. (p. 492)

Chapter 10 Circles

Theorem 10.1   Two arcs are congruent if and only if their corresponding central angles are congruent. (p. 530)

Postulate 10.1   Arc Addition Postulate The measure of an arc formed by two adjacent arcs is the sum of the measures of the two arcs. (p. 531)

Theorem 10.2   In a circle or in congruent circles, two minor arcs are congruent if and only if their corresponding chords are congruent. (p. 536)
Abbreviations: In $\odot$, 2 minor arcs are $\cong$, *iff* corr. chords are $\cong$.
In $\odot$, 2 chords are $\cong$, *iff* corr. minor arcs are $\cong$.

Theorem 10.3   In a circle, if a diameter (or radius) is perpendicular to a chord, then it bisects the chord and its arc. (p. 537)

Theorem 10.4   In a circle or in congruent circles, two chords are congruent if and only if they are equidistant from the center. (p. 539)

Theorem 10.5   If an angle is inscribed in a circle, then the measure of the angle equals one-half the measure of its intercepted arc (or the measure of the intercepted arc is twice the measure of the inscribed angle). (p. 544)

Theorem 10.6   If two inscribed angles of a circle (or congruent circles) intercept congruent arcs or the same arc, then the angles are congruent. (p. 546) Abbreviations: Inscribed $\angle$ of same arc are $\cong$. Inscribed $\angle$ of $\cong$ arcs are $\cong$.

Theorem 10.7   If an inscribed angle intercepts a semicircle, the angle is a right angle. (p. 547)

Theorem 10.8   If a quadrilateral is inscribed in a circle, then its opposite angles are supplementary. (p. 548)

Theorem 10.9   If a line is tangent to a circle, then it is perpendicular to the radius drawn to the point of tangency. (p. 553)

Theorem 10.10   If a line is perpendicular to a radius of a circle at its endpoint on the circle, then the line is a tangent to the circle. (p. 553)

Theorem 10.11   If two segments from the same exterior point are tangent to a circle, then they are congruent. (p. 554)

Theorem 10.12   If two secants intersect in the interior of a circle, then the measure of an angle formed is one-half the sum of the measure of the arcs intercepted by the angle and its vertical angle. (p. 561)

Theorem 10.13   If a secant and a tangent intersect at the point of tangency, then the measure of each angle formed is one-half the measure of its intercepted arc. (p. 562)

Theorem 10.14   If two secants, a secant and a tangent, or two tangents intersect in the exterior of a circle, then the measure of the angle formed is one-half the positive difference of the measures of the intercepted arcs. (p. 563)

Theorem 10.15   If two chords intersect in a circle, then the products of the measures of the segments of the chords are equal. (p. 569)

Theorem 10.16   If two secant segments are drawn to a circle from an exterior point, then the product of the measures of one secant segment and its external secant segment is equal to the product of the measures of the other secant segment and its external secant segment. (p. 570)

Theorem 10.17   If a tangent segment and a secant segment are drawn to a circle from an exterior point, then the square of the measure of the tangent segment is equal to the product of the measures of the secant segment and its external secant segment. (p. 571)

Chapter 11 Area of Polygons And Circles

Postulate 11.1   Congruent figures have equal areas. (p. 603)

Postulate 11.2   The area of a region is the sum of the areas of all of its nonoverlapping parts. (p. 619)

Chapter 13 Volume

Theorem 13.1   If two solids are similar with a scale factor of $a : b$, then the surface areas have a ratio of $a^2 : b^2$, and the volumes have a ratio of $a^3 : b^3$. (p. 709)

Glossary/Glosario

English	Español

A

acute angle (p. 30) An angle with a degree measure less than 90.

$0 < m\angle A < 90$

ángulo agudo Ángulo cuya medida en grados es menos de 90.

acute triangle (p. 178) A triangle in which all of the angles are acute angles.

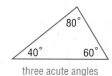

three acute angles
tres ángulos agudos

triángulo acutángulo Triángulo cuyos ángulos son todos agudos.

adjacent angles (p. 37) Two angles that lie in the same plane, have a common vertex and a common side, but no common interior points.

ángulos adyacentes Dos ángulos que yacen sobre el mismo plano, tienen el mismo vértice y un lado en común, pero ningún punto interior.

alternate exterior angles (p. 128) In the figure, transversal t intersects lines ℓ and m. $\angle 5$ and $\angle 3$, and $\angle 6$ and $\angle 4$ are alternate exterior angles.

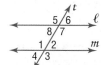

ángulos alternos externos En la figura, la transversal t interseca las rectas ℓ y m. $\angle 5$ y $\angle 3$, y $\angle 6$ y $\angle 4$ son ángulos alternos externos.

alternate interior angles (p. 128) In the figure above, transversal t intersects lines ℓ and m. $\angle 1$ and $\angle 7$, and $\angle 2$ and $\angle 8$ are alternate interior angles.

ángulos alternos internos En la figura anterior, la transversal t interseca las rectas ℓ y m. $\angle 1$ y $\angle 7$, y $\angle 2$ y $\angle 8$ son ángulos alternos internos .

altitude   1. (p. 241) In a triangle, a segment from a vertex of the triangle to the line containing the opposite side and perpendicular to that side. **2.** (pp. 649, 655) In a prism or cylinder, a segment perpendicular to the bases with an endpoint in each plane.   **3.** (pp. 660, 666) In a pyramid or cone, the segment that has the vertex as one endpoint and is perpendicular to the base.

altura   1. En un triángulo, segmento trazado desde el vértice de un triángulo hasta el lado opuesto y que es perpendicular a dicho lado.   **2.** El segmento perpendicular a las bases de prismas y cilindros que tiene un extremo en cada plano. **3.** El segmento que tiene un extremo en el vértice de pirámides y conos y que es perpendicular a la base.

ambiguous case of the Law of Sines (p. 384) Given the measures of two sides and a nonincluded angle, there exist two possible triangles.

caso ambiguo de la ley de los senos Dadas las medidas de dos lados y de un ángulo no incluido, existen dos triángulos posibles.

angle (p. 29) The intersection of two noncollinear rays at a common endpoint. The rays are called *sides* and the common endpoint is called the *vertex.*

ángulo La intersección de dos semirrectas no colineales en un punto común. Las semirrectas se llaman *lados* y el punto común se llama *vértice.*

angle bisector (p. 32) A ray that divides an angle into two congruent angles.

$\overrightarrow{PW}$ is the bisector of $\angle P$.
$\overrightarrow{PW}$ *es la bisectriz del* $\angle P$.

bisectriz de un ángulo Semirrecta que divide un ángulo en dos ángulos congruentes.

angle of depression (p. 372)   The angle between the line of sight and the horizontal when an observer looks downward.

angle of elevation (p. 371)   The angle between the line of sight and the horizontal when an observer looks upward.

angle of rotation (p. 476)   The angle through which a preimage is rotated to form the image.

apothem (p. 610)   A segment that is drawn from the center of a regular polygon perpendicular to a side of the polygon.

apothem
apotema

arc (p. 530)   A part of a circle that is defined by two endpoints.

axis   **1.** (p. 655) In a cylinder, the segment with endpoints that are the centers of the bases.
2. (p. 666)   In a cone, the segment with endpoints that are the vertex and the center of the base.

ángulo de depresión   Ángulo formado por la horizontal y la línea de visión de un observador que mira hacia abajo.

ángulo de elevación   Ángulo formado por la horizontal y la línea de visión de un observador que mira hacia arriba.

ángulo de rotación   El ángulo a través del cual se rota una preimagen para formar la imagen.

apotema   Segmento perpendicular trazado desde el centro de un polígono regular hasta uno de sus lados.

arco   Parte de un círculo definida por los dos extremos de una recta.

eje   **1.** El segmento en un cilindro cuyos extremos forman el centro de las bases.
2. El segmento en un cono cuyos extremos forman el vértice y el centro de la base.

B

between (p. 14)   For any two points A and B on a line, there is another point C between A and B if and only if A, B, and C are collinear and $AC + CB = AB$.

biconditional (p. 81)   The conjunction of a conditional statement and its converse.

ubicado entre   Para cualquier par de puntos A y B de una recta, existe un punto C ubicado entre A y B si y sólo si A, B y C son colineales y $AC + CB = AB$.

bicondicional   La conjunción entre un enunciado condicional y su recíproco.

C

center of rotation (p. 476)   A fixed point around which shapes move in a circular motion to a new position.

central angle (p. 529)   An angle that intersects a circle in two points and has its vertex at the center of the circle.

centroid (p. 240)   The point of concurrency of the medians of a triangle.

chord   **1.** (p. 522) For a given circle, a segment with endpoints that are on the circle.
2. (p. 671) For a given sphere, a segment with endpoints that are on the sphere.

circle (p. 522)   The locus of all points in a plane equidistant from a given point called the *center* of the circle.

centro de rotación   Punto fijo alrededor del cual gira una figura hasta alcanzar una posición determinada.

ángulo central   Ángulo que interseca un círculo en dos puntos y cuyo vértice se localiza en el centro del círculo.

centroide   Punto de intersección de las medianas de un triángulo.

cuerda   **1.** Segmento cuyos extremos están en un círculo.
2. Segmento cuyos extremos están en una esfera.

círculo   Lugar geométrico formado por el conjunto de puntos en un plano, equidistantes de un punto dado llamado *centro*.

P is the center of the circle.
P es el centro del círculo.

circumcenter (p. 238)   The point of concurrency of the perpendicular bisectors of a triangle.

circuncentro   Punto de intersección de las mediatrices de un triángulo.

circumference (p. 523)   The distance around a circle.

circunferencia   Distancia alrededor de un círculo.

circumscribed (p. 537)   A circle is circumscribed about a polygon if the circle contains all the vertices of the polygon.

circunscrito   Un polígono está circunscrito a un círculo si todos sus vértices están contenidos en el círculo.

⊙E is circumscribed about quadrilateral *ABCD*.
⊙E está circunscrito al cuadrilátero *ABCD*.

collinear (p. 6)   Points that lie on the same line.

colineal   Puntos que yacen en la misma recta.

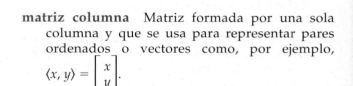

P, *Q*, and *R* are collinear.
P, *Q* y *R* son colineales.

column matrix (p. 506)   A matrix containing one column often used to represent an ordered pair or a vector, such as $\langle x, y \rangle = \begin{bmatrix} x \\ y \end{bmatrix}$.

matriz columna   Matriz formada por una sola columna y que se usa para representar pares ordenados o vectores como, por ejemplo, $\langle x, y \rangle = \begin{bmatrix} x \\ y \end{bmatrix}$.

complementary angles (p. 39)   Two angles with measures that have a sum of 90.

ángulos complementarios   Dos ángulos cuya suma es igual a 90 grados.

component form (p. 498)   A vector expressed as an ordered pair, $\langle$change in x, change in $y\rangle$.

componente   Vector representado en forma de par ordenado, $\langle$cambio en x, cambio en $y\rangle$.

composition of reflections (p. 471)   Successive reflections in parallel lines.

composición de reflexiones   Reflexiones sucesivas en rectas paralelas.

compound statement (p. 67)   A statement formed by joining two or more statements.

enunciado compuesto   Enunciado formado por la unión de dos o más enunciados.

concave polygon (p. 45)   A polygon for which there is a line containing a side of the polygon that also contains a point in the interior of the polygon.

polígono cóncavo   Polígono para el cual existe una recta que contiene un lado del polígono y un punto interior del polígono.

conclusion (p. 75)   In a conditional statement, the statement that immediately follows the word *then*.

conclusión   Parte del enunciado condicional que está escrita después de la palabra *entonces*.

concurrent lines (p. 238)   Three or more lines that intersect at a common point.

rectas concurrentes   Tres o más rectas que se intersecan en un punto común.

conditional statement (p. 75)   A statement that can be written in *if-then form*.

enunciado condicional   Enunciado escrito en la forma *si-entonces*.

cone (p. 666)   A solid with a circular base, a vertex not contained in the same plane as the base, and a lateral surface area composed of all points in the segments connecting the vertex to the edge of the base.

vertex
vértice

base
base

cono   Sólido de base circular cuyo vértice no se localiza en el mismo plano que la base y cuya superficie lateral está formada por todos los segmentos que unen el vértice con los límites de la base.

congruence transformations (p. 194) A mapping for which a geometric figure and its image are congruent.

transformación de congruencia Transformación en un plano en la que la figura geométrica y su imagen son congruentes.

congruent (p. 15) Having the same measure.

congruente Que miden lo mismo.

congruent arcs (p. 530) Arcs of the same circle or congruent circles that have the same measure.

arcos congruentes Arcos de un mismo círculo, o de círculos congruentes, que tienen la misma medida.

congruent solids (p. 707) Two solids are congruent if all of the following conditions are met.
1. The corresponding angles are congruent.
2. Corresponding edges are congruent.
3. Corresponding faces are congruent.
4. The volumes are congruent.

sólidos congruentes Dos sólidos son congruentes si cumplen todas las siguientes condiciones:
1. Los ángulos correspondientes son congruentes.
2. Las aristas correspondientes son congruentes.
3. Las caras correspondientes son congruentes.
4. Los volúmenes son congruentes.

congruent triangles (p. 192) Triangles that have their corresponding parts congruent.

triángulos congruentes Triángulos cuyas partes correspondientes son congruentes.

conjecture (p. 62) An educated guess based on known information.

conjetura Juicio basado en información conocida.

conjunction (p. 68) A compound statement formed by joining two or more statements with the word *and*.

conjunción Enunciado compuesto que se obtiene al unir dos o más enunciados con la palabra *y*.

consecutive interior angles (p. 128) In the figure, transversal *t* intersects lines ℓ and *m*. There are two pairs of consecutive interior angles: ∠8 and ∠1, and ∠7 and ∠2.

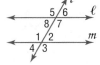

ángulos internos consecutivos En la figura, la transversal *t* interseca las rectas ℓ y *m*. La figura presenta dos pares de ángulos consecutivos internos: ∠8 y ∠1, y ∠7 y ∠2.

construction (p. 15) A method of creating geometric figures without the benefit of measuring tools. Generally, only a pencil, straightedge, and compass are used.

construcción Método para dibujar figuras geométricas sin el uso de instrumentos de medición. En general, sólo requiere de un lápiz, una regla sin escala y un compás.

contrapositive (p. 77) The statement formed by negating both the hypothesis and conclusion of the converse of a conditional statement.

antítesis Enunciado formado por la negación de la hipótesis y la conclusión del recíproco de un enunciado condicional dado.

converse (p. 77) The statement formed by exchanging the hypothesis and conclusion of a conditional statement.

recíproco Enunciado que se obtiene al intercambiar la hipótesis y la conclusión de un enunciado condicional dado.

convex polygon (p. 45) A polygon for which there is no line that contains both a side of the polygon and a point in the interior of the polygon.

polígono convexo Polígono para el cual no existe recta alguna que contenga un lado del polígono y un punto en el interior del polígono.

coordinate proof (p. 222) A proof that uses figures in the coordinate plane and algebra to prove geometric concepts.

prueba de coordenadas Demostración que usa álgebra y figuras en el plano de coordenadas para demostrar conceptos geométricos.

coplanar (p. 6) Points that lie in the same plane.

coplanar Puntos que yacen en un mismo plano.

corner view (p. 636)   The view from a corner of a three-dimensional figure, also called the *perspective view*.

corollary (p. 188)   A statement that can be easily proved using a theorem is called a corollary of that theorem.

corresponding angles (p. 128)   In the figure, transversal t intersects lines ℓ and m. There are four pairs of corresponding angles: $\angle 5$ and $\angle 1$, $\angle 8$ and $\angle 4$, $\angle 6$ and $\angle 2$, and $\angle 7$ and $\angle 3$.

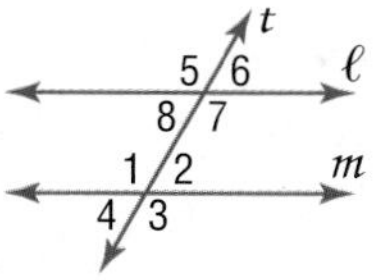

cosine (p. 364)   For an acute angle of a right triangle, the ratio of the measure of the leg adjacent to the acute angle to the measure of the hypotenuse.

counterexample (p. 63)   An example used to show that a given statement is not always true.

cross products (p. 283)   In the proportion $\frac{a}{b} = \frac{c}{d}$, where $b \neq 0$ and $d \neq 0$, the cross products are ad and bc. The proportion is true if and only if the cross products are equal.

cylinder (p. 638)   A figure with bases that are formed by congruent circles in parallel planes.

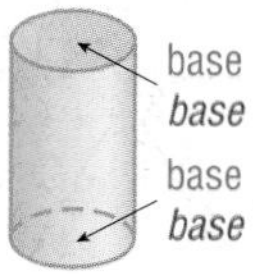

vista de esquina   Vista de una figura tridimensional desde una esquina. También se conoce como *vista de perspectiva*.

corolario   La afirmación que puede demostrarse fácilmente mediante un teorema se conoce como corolario de dicho teorema.

ángulos correspondientes   En la figura, la transversal t interseca las rectas ℓ y m. La figura muestra cuatro pares de ángulos correspondientes: $\angle 5$ y $\angle 1$, $\angle 8$ y $\angle 4$, $\angle 6$ y $\angle 2$, y $\angle 7$ y $\angle 3$.

coseno   Para un ángulo agudo de un triángulo rectángulo, la razón entre la medida del cateto adyacente al ángulo agudo y la medida de la hipotenusa de un triángulo rectángulo.

contraejemplo   Ejemplo que se usa para demostrar que un enunciado dado no siempre es verdadero.

productos cruzados   En la proporción, $\frac{a}{b} = \frac{c}{d}$, donde $b \neq 0$ y $d \neq 0$, los productos cruzados son ad y bc. La proporción es verdadera si y sólo si los productos cruzados son iguales.

cilindro   Figura cuyas bases son círculos congruentes localizados en planos paralelos.

D

deductive argument (p. 94)   A proof formed by a group of algebraic steps used to solve a problem.

deductive reasoning (p. 82)   A system of reasoning that uses facts, rules, definitions, or properties to reach logical conclusions.

degree (p. 29)   A unit of measure used in measuring angles and arcs. An arc of a circle with a measure of $1°$ is $\frac{1}{360}$ of the entire circle.

diagonal (p. 404)   In a polygon, a segment that connects nonconsecutive vertices of the polygon.

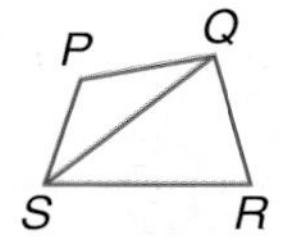

diameter   1. (p. 522) In a circle, a chord that passes through the center of the circle.   2. (p. 671) In a sphere, a segment that contains the center of the sphere, and has endpoints that are on the sphere.

argumento deductivo   Demostración que consta del conjunto de pasos algebraicos que se usan para resolver un problema.

razonamiento deductivo   Sistema de razonamiento que emplea hechos, reglas, definiciones y propiedades para obtener conclusiones lógicas.

grado   Unidad de medida que se usa para medir ángulos y arcos. El arco de un círculo que mide $1°$ equivale a $\frac{1}{360}$ del círculo completo.

diagonal   Recta que une vértices no consecutivos de un polígono.

$\overline{SQ}$ is a diagonal.
$\overline{SQ}$ *es una diagonal.*

diámetro   1. Cuerda que pasa por el centro de un círculo.   2. Segmento que incluye el centro de una esfera y cuyos extremos se localizan en la esfera.

dilation (p. 490) A transformation determined by a center point C and a scale factor k. When $k > 0$, the image P' of P is the point on $\overrightarrow{CP}$ such that $CP' = |k| \cdot CP$. When $k < 0$, the image P' of P is the point on the ray opposite $\overrightarrow{CP}$ such that $CP' = k \cdot CP$.

dilatación Transformación determinada por un punto central C y un factor de escala k. Cuando $k > 0$, la imagen P' de P es el punto en $\overrightarrow{CP}$ tal que $CP' = |k| \cdot CP$. Cuando $k < 0$, la imagen P' de P es el punto en la semirrecta opuesta $\overrightarrow{CP}$ tal que $CP' = k \cdot CP$.

direct isometry (p. 481) An isometry in which the image of a figure is found by moving the figure intact within the plane.

isometría directa Isometría en la cual se obtiene la imagen de una figura, al mover la figura intacta junto con su plano.

direction (p. 498) The measure of the angle that a vector forms with the positive x-axis or any other horizontal line.

dirección Medida del ángulo que forma un vector con el eje positivo x o con cualquier otra recta horizontal.

disjunction (p. 68) A compound statement formed by joining two or more statements with the word *or*.

disyunción Enunciado compuesto que se forma al unir dos o más enunciados con la palabra *o*.

E

equal vectors (p. 499) Vectors that have the same magnitude and direction.

vectores iguales Vectores que poseen la misma magnitud y dirección.

equiangular triangle (p. 178) A triangle with all angles congruent.

triángulo equiangular Triángulo cuyos ángulos son congruentes entre sí.

equilateral triangle (p. 179) A triangle with all sides congruent.

triángulo equilátero Triángulo cuyos lados son congruentes entre sí.

exterior (p. 29) A point is in the exterior of an angle if it is neither on the angle nor in the interior of the angle.

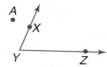

A is in the exterior of $\angle XYZ$.
A está en el exterior del $\angle XYZ$.

exterior Un punto yace en el exterior de un ángulo si no se localiza ni en el ángulo ni en el interior del ángulo.

exterior angle (p. 186) An angle formed by one side of a triangle and the extension of another side.

$\angle 1$ is an exterior angle.
$\angle 1$ es un ángulo externo.

ángulo externo Ángulo formado por un lado de un triángulo y la extensión de otro de sus lados.

extremes (p. 283) In $\frac{a}{b} = \frac{c}{d}$, the numbers a and d.

extremos Los números a y d en $\frac{a}{b} = \frac{c}{d}$.

F

flow proof (p. 187) A proof that organizes statements in logical order, starting with the given statements. Each statement is written in a box with the reason verifying the statement written below the box. Arrows are used to indicate the order of the statements.

demostración de flujo Demostración en que se ordenan los enunciados en orden lógico, empezando con los enunciados dados. Cada enunciado se escribe en una casilla y debajo de cada casilla se escribe el argumento que verifica el enunciado. El orden de los enunciados se indica mediante flechas.

fractal (p. 325)   A figure generated by repeating a special sequence of steps infinitely often. Fractals often exhibit self-similarity.

fractal   Figura que se obtiene mediante la repetición infinita de una sucesión particular de pasos. Los fractales a menudo exhiben autosemejanza.

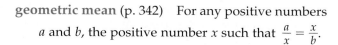

geometric mean (p. 342)   For any positive numbers a and b, the positive number x such that $\frac{a}{x} = \frac{x}{b}$.

media geométrica   Para todo número positivo a y b, existe un número positivo x tal que $\frac{a}{x} = \frac{x}{b}$.

geometric probability (p. 622)   Using the principles of length and area to find the probability of an event.

probabilidad geométrica   El uso de los principios de longitud y área para calcular la probabilidad de un evento.

glide reflection (p. 475)   A composition of a translation and a reflection in a line parallel to the direction of the translation.

reflexión de deslizamiento   Composición que consta de una traslación y una reflexión realizadas sobre una recta paralela a la dirección de la traslación.

great circle (p. 671)   For a given sphere, the intersection of the sphere and a plane that contains the center of the sphere.

círculo máximo   La intersección entre una esfera dada y un plano que contiene el centro de la esfera.

height of a parallelogram (p. 595)   The length of an altitude of a parallelogram.

h is the height of parallelogram *ABCD*.
H es la altura del paralelogramo ABCD.

altura de un paralelogramo   La longitud de la altura de un paralelogramo.

hemisphere (p. 672)   One of the two congruent parts into which a great circle separates a sphere.

hemisferio   Cada una de las dos partes congruentes en que un círculo máximo divide una esfera.

hypothesis (p. 75)   In a conditional statement, the statement that immediately follows the word *if*.

hipótesis   El enunciado escrito a continuación de la palabra *si* en un enunciado condicional.

if-then statement (p. 75)   A compound statement of the form "if A, then B", where A and B are statements.

enunciado si-entonces   Enunciado compuesto de la forma "si A, entonces B", donde A y B son enunciados.

incenter (p. 240)   The point of concurrency of the angle bisectors of a triangle.

incentro   Punto de intersección de las bisectrices interiores de un triángulo.

included angle (p. 201)   In a triangle, the angle formed by two sides is the included angle for those two sides.

ángulo incluido   En un triángulo, el ángulo formado por dos lados cualesquiera del triángulo es el ángulo incluido de esos dos lados.

included side (p. 207)   The side of a triangle that is a side of each of two angles.

lado incluido   El lado de un triángulo que es común a de sus dos ángulos.

indirect isometry (p. 481)   An isometry that cannot be performed by maintaining the orientation of the points, as in a direct isometry.

isometría indirecta   Tipo de isometría que no se puede obtener manteniendo la orientación de los puntos, como ocurre durante la isometría directa.

indirect proof (p. 255) In an indirect proof, one assumes that the statement to be proved is false. One then uses logical reasoning to deduce that a statement contradicts a postulate, theorem, or one of the assumptions. Once a contradiction is obtained, one concludes that the statement assumed false must in fact be true.

demostración indirecta En una demostración indirecta, se asume que el enunciado por demostrar es falso. Después, se deduce lógicamente que existe un enunciado que contradice un postulado, un teorema o una de las conjeturas. Una vez hallada una contradicción, se concluye que el enunciado que se suponía falso debe ser, en realidad, verdadero.

indirect reasoning (p. 255) Reasoning that assumes that the conclusion is false and then shows that this assumption leads to a contradiction of the hypothesis or some other accepted fact, like a postulate, theorem, or corollary. Then, since the assumption has been proved false, the conclusion must be true.

razonamiento indirecto Razonamiento en que primero se asume que la conclusión es falsa y, después, se demuestra que esto contradice la hipótesis o un hecho aceptado como un postulado, un teorema o un corolario. Finalmente, dado que se ha demostrado que la conjetura es falsa, entonces la conclusión debe ser verdadera.

inductive reasoning (p. 62) Reasoning that uses a number of specific examples to arrive at a plausible generalization or prediction. Conclusions arrived at by inductive reasoning lack the logical certainty of those arrived at by deductive reasoning.

razonamiento inductivo Razonamiento que usa varios ejemplos específicos para lograr una generalización o una predicción creíble. Las conclusiones obtenidas mediante el razonamiento inductivo carecen de la certidumbre lógica de aquellas obtenidas mediante el razonamiento deductivo.

inscribed (p. 537) A polygon is inscribed in a circle if each of its vertices lie on the circle.

$\triangle LMN$ is inscribed in $\odot P$.
$\triangle LMN$ está inscrito en $\odot P$.

inscrito Un polígono está inscrito en un círculo si todos sus vértices yacen en el círculo.

intercepted (p. 544) An angle intercepts an arc if and only if each of the following conditions are met.
1. The endpoints of the arc lie on the angle.
2. All points of the arc except the endpoints are in the interior of the circle.
3. Each side of the angle contains an endpoint of the arc.

intersecado Un ángulo interseca un arco si y sólo si se cumplen todas las siguientes condiciones.
1. Los extremos del arco yacen en el ángulo.
2. Todos los puntos del arco, exceptuando sus extremos, yacen en el interior del círculo.
3. Cada lado del ángulo contiene un extremo del arco.

interior (p. 29) A point is in the interior of an angle if it does not lie on the angle itself and it lies on a segment with endpoints that are on the sides of the angle.

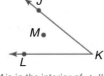

M is in the interior of $\angle JKL$.
M está en el interior del $\angle JKL$.

interior Un punto se localiza en el interior de un ángulo, si no yace en el ángulo mismo y si está en un segmento cuyos extremos yacen en los lados del ángulo.

inverse (p. 77) The statement formed by negating both the hypothesis and conclusion of a conditional statement.

inversa Enunciado que se obtiene al negar la hipótesis y la conclusión de un enunciado condicional.

irregular figure (p. 617) A figure that cannot be classified as a single polygon.

figura irregular Figura que no se puede clasificar como un solo polígono.

irregular polygon (p. 618) A polygon that is not regular.

polígono irregular Polígono que no es regular.

isometry (p. 463) A mapping for which the original figure and its image are congruent.

isometría Transformación en que la figura original y su imagen son congruentes.

isosceles trapezoid (p. 439) A trapezoid in which the legs are congruent, both pairs of base angles are congruent, and the diagonals are congruent.

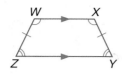

trapecio isósceles Trapecio cuyos catetos son congruentes, ambos pares de ángulos son congruentes y las diagonales son congruentes.

isosceles triangle (p. 179) A triangle with at least two sides congruent. The congruent sides are called *legs*. The angles opposite the legs are *base angles*. The angle formed by the two legs is the *vertex angle*. The side opposite the vertex angle is the *base*.

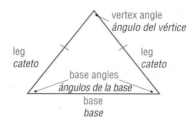

triángulo isósceles Triángulo que tiene por lo menos dos lados congruentes. Los lados congruentes se llaman *catetos*. Los ángulos opuestos a los catetos son los *ángulos de la base*. El ángulo formado por los dos catetos es el *ángulo del vértice*. Los lados opuestos al ángulo del vértice forman la *base*.

iteration (p. 325) A process of repeating the same procedure over and over again.

iteración Proceso de repetir el mismo procedimiento una y otra vez.

K

kite (p. 438) A quadrilateral with exactly two distinct pairs of adjacent congruent sides.

cometa Cuadrilátero que tiene exactamente dospares de lados congruentes adyacentes distintivos.

L

lateral area (p. 649) For prisms, pyramids, cylinders, and cones, the area of the figure, not including the bases.

área lateral En prismas, pirámides, cilindros y conos, es el área de la figura, sin incluir el área de las bases.

lateral edges **1.** (p. 649) In a prism, the intersection of two adjacent lateral faces. **2.** (p. 660) In a pyramid, lateral edges are the edges of the lateral faces that join the vertex to vertices of the base.

aristas laterales **1.** En un prisma, la intersección de dos caras laterales adyacentes. **2.** En una pirámide, las aristas de las caras laterales que unen el vértice de la pirámide con los vértices de la base.

lateral faces **1.** (p. 649) In a prism, the faces that are not bases. **2.** (p. 660) In a pyramid, faces that intersect at the vertex.

caras laterales **1.** En un prisma, las caras que no forman las bases. **2.** En una pirámide, las caras que se intersecan en el vértice.

Law of Cosines (p. 385) Let $\triangle ABC$ be any triangle with a, b, and c representing the measures of sides opposite the angles with measures A, B, and C respectively. Then the following equations are true.
$a^2 = b^2 + c^2 - 2bc \cos A$
$b^2 = a^2 + c^2 - 2ac \cos B$
$c^2 = a^2 + b^2 - 2ab \cos C$

ley de los cosenos Sea $\triangle ABC$ cualquier triángulo donde a, b y c son las medidas de los lados opuestos a los ángulos que miden A, B y C respectivamente. Entonces las siguientes ecuaciones son ciertas.
$a^2 = b^2 + c^2 - 2bc \cos A$
$b^2 = a^2 + c^2 - 2ac \cos B$
$c^2 = a^2 + b^2 - 2ab \cos C$

Law of Detachment (p. 82) If $p \rightarrow q$ is a true conditional and p is true, then q is also true.

ley de indiferencia Si $p \rightarrow q$ es un enunciado condicional verdadero y p es verdadero, entonces q es verdadero también.

Law of Sines (p. 377)   Let $\triangle ABC$ be any triangle with a, b, and c representing the measures of sides opposite the angles with measures A, B, and C respectively. Then, $\dfrac{\sin A}{a} = \dfrac{\sin B}{b} = \dfrac{\sin C}{c}$.

ley de los senos   Sea $\triangle ABC$ cualquier triángulo donde a, b y c representan las medidas de los lados opuestos a los ángulos A, B y C respectivamente. Entonces, $\dfrac{\sin A}{a} = \dfrac{\sin B}{b} = \dfrac{\sin C}{c}$.

Law of Syllogism (p. 83)   If $p \rightarrow q$ and $q \rightarrow r$ are true conditionals, then $p \rightarrow r$ is also true.

ley del silogismo   Si $p \rightarrow q$ y $q \rightarrow r$ son enunciados condicionales verdaderos, entonces $p \rightarrow r$ también es verdadero.

line (p. 6)   A basic undefined term of geometry. A line is made up of points and has no thickness or width. In a figure, a line is shown with an arrowhead at each end. Lines are usually named by lowercase script letters or by writing capital letters for two points on the line, with a double arrow over the pair of letters.

recta   Término primitivo en geometría. Una recta está formada por puntos y carece de grosor o ancho. En una figura, una recta se representa con una flecha en cada extremo. Por lo general, se designan con letras minúsculas o con las dos letras mayúsculas de dos puntos sobre la línea. Se escribe una flecha doble sobre el par de letras mayúsculas.

line of reflection (p. 463)   A line through a figure that separates the figure into two mirror images.

línea de reflexión   Línea que divide una figura en dos imágenes especulares.

line of symmetry (p. 466)   A line that can be drawn through a plane figure so that the figure on one side is the reflection image of the figure on the opposite side.

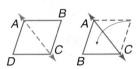

$\overleftrightarrow{AC}$ is a line of symmetry.
$\overleftrightarrow{AC}$ es un eje de simetría.

eje de simetría   Recta que se traza a través de una figura plana, de modo que un lado de la figura es la imagen reflejada del lado opuesto.

line segment (p. 13)   A measurable part of a line that consists of two points, called endpoints, and all of the points between them.

segmento de recta   Sección medible de una recta. Consta de dos puntos, llamados extremos, y todos los puntos localizados entre ellos.

linear pair (p. 37)   A pair of adjacent angles whose non-common sides are opposite rays.

$\angle PSQ$ and $\angle QSR$ are a linear pair.
$\angle PSQ$ y $\angle QSR$ forman un par lineal.

par lineal   Par de ángulos adyacentes cuyos lados no comunes forman semirrectas opuestas.

locus (p. 11)   The set of points that satisfy a given condition.

lugar geométrico   Conjunto de puntos que satisfacen una condición dada.

logically equivalent (p. 77)   Statements that have the same truth values.

equivalente lógico   Enunciados que poseen el mismo valor de verdad.

M

magnitude (p. 498)   The length of a vector.

magnitud   La longitud de un vector.

major arc (p. 530)   An arc with a measure greater than 180.

$\overset{\frown}{ACB}$ is a major arc.

arco mayor   Arco que mide más de 180°.

$\overset{\frown}{ACB}$ es un arco mayor.

matrix logic (p. 88) A method of deductive reasoning that uses a table to solve problems.

means (p. 283) In $\frac{a}{b} = \frac{c}{d}$, the numbers b and c.

median **1.** (p. 240) In a triangle, a line segment with endpoints that are a vertex of a triangle and the midpoint of the side opposite the vertex. **2.** (p. 440) In a trapezoid, the segment that joins the midpoints of the legs.

midpoint (p. 22) The point halfway between the endpoints of a segment.

midsegment (p. 308) A segment with endpoints that are the midpoints of two sides of a triangle.

minor arc (p. 530) An arc with a measure less than 180. $\overarc{AB}$ is a minor arc.

lógica matricial Método de razonamiento deductivo que utiliza una tabla para resolver problemas.

medios Los números b y c en la proporción $\frac{a}{b} = \frac{c}{d}$.

mediana **1.** Segmento de recta de un triángulo cuyos extremos son un vértice del triángulo y el punto medio del lado opuesto a dicho vértice. **2.** Segmento que une los puntos medios de los catetos de un trapecio.

punto medio Punto que es equidistante entre los extremos de un segmento.

segmento medio Segmento cuyos extremos son los puntos medios de dos lados de un triángulo.

arco menor Arco que mide menos de 180°. $\overarc{AB}$ es un arco menor.

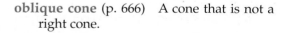

negation (p. 67) If a statement is represented by p, then *not p* is the negation of the statement.

net (p. 644) A two-dimensional figure that when folded forms the surfaces of a three-dimensional object.

n-**gon** (p. 46) A polygon with n sides.

non-Euclidean geometry (p. 165) The study of geometrical systems that are not in accordance with the Parallel Postulate of Euclidean geometry.

negación Si p representa un enunciado, entonces *no p* representa la negación del enunciado.

red Figura bidimensional que al ser plegada forma las superficies de un objeto tridimensional.

*en*ágono Polígono con n lados.

geometría no euclidiana El estudio de sistemas geométricos que no satisfacen el Postulado de las Paralelas de la geometría euclidiana.

oblique cone (p. 666) A cone that is not a right cone.

cono oblicuo Cono que no es un cono recto.

oblique cylinder (p. 655) A cylinder that is not a right cylinder.

cilindro oblicuo Cilindro que no es un cilindro recto.

oblique prism (p. 649) A prism in which the lateral edges are not perpendicular to the bases.

prisma oblicuo Prisma cuyas aristas laterales no son perpendiculares a las bases.

obtuse angle (p. 30) An angle with degree measure greater than 90 and less than 180.

$90 < m\angle A < 180$

ángulo obtuso Ángulo que mide más de 90° y menos de 180°.

obtuse triangle (p. 178) A triangle with an obtuse angle.

one obtuse angle
un ángulo obtuso

triángulo obtusángulo Triángulo que tiene un ángulo obtuso.

opposite rays (p. 29) Two rays $\overrightarrow{BA}$ and $\overrightarrow{BC}$ such that B is between A and C.

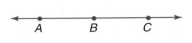

semirrectas opuestas Dos semirrectas $\overrightarrow{BA}$ y $\overrightarrow{BC}$ tales que B se localiza entre A y C.

ordered triple (p. 714) Three numbers given in a specific order used to locate points in space.

triple ordenado Tres números dados en un orden específico que sirven para ubicar puntos en el espacio.

orthocenter (p. 240) The point of concurrency of the altitudes of a triangle.

ortocentro Punto de intersección de las alturas de un triángulo.

orthogonal drawing (p. 636) The two-dimensional top view, left view, front view, and right view of a three-dimensional object.

vista ortogonal Vista bidimensional desde arriba, desde la izquierda, desde el frente o desde la derecha de un cuerpo tridimensional.

Ⓟ

paragraph proof (p. 90) An informal proof written in the form of a paragraph that explains why a conjecture for a given situation is true.

demostración de párrafo Demostración informal escrita en forma de párrafo que explica por qué una conjetura acerca de una situación dada es verdadera.

parallel lines (p. 126) Coplanar lines that do not intersect.

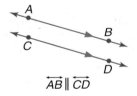

$\overleftrightarrow{AB} \parallel \overleftrightarrow{CD}$

rectas paralelas Rectas coplanares que no se intersecan.

parallel planes (p. 126) Planes that do not intersect.

planos paralelos Planos que no se intersecan.

parallel vectors (p. 499) Vectors that have the same or opposite direction.

vectores paralelos Vectores que tienen la misma dirección o la dirección opuesta.

parallelogram (p. 411) A quadrilateral with parallel opposite sides. Any side of a parallelogram may be called a *base*.

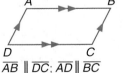

$\overline{AB} \parallel \overline{DC}; \overline{AD} \parallel \overline{BC}$

paralelogramo Cuadrilátero cuyos lados opuestos son paralelos entre sí. Cualquier lado del paralelogramo puede ser la *base*.

perimeter (p. 46) The sum of the lengths of the sides of a polygon.

perímetro La suma de la longitud de los lados de un polígono.

perpendicular bisector (p. 238) In a triangle, a line, segment, or ray that passes through the midpoint of a side and is perpendicular to that side.

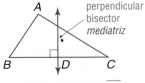

D is the midpoint of $\overline{BC}$.
D es el punto medio de $\overline{BC}$.

mediatriz Recta, segmento o semirrecta que atraviesa el punto medio del lado de un triángulo y que es perpendicular a dicho lado.

perpendicular lines (p. 40)   Lines that form right angles.

line $m \perp$ line n
recta $m \perp$ recta n

rectas perpendiculares   Rectas que forman ángulos rectos.

perspective view (p. 636)   The view of a three-dimensional figure from the corner.

vista de perspectiva   Vista de una figura tridimensional desde una de sus esquinas.

pi (π) (p. 524)   An irrational number represented by the ratio of the circumference of a circle to the diameter of the circle.

pi (π)   Número irracional representado por la razón entre la circunferencia de un círculo y su diámetro.

plane (p. 6)   A basic undefined term of geometry. A plane is a flat surface made up of points that has no depth and extends indefinitely in all directions. In a figure, a plane is often represented by a shaded, slanted 4-sided figure. Planes are usually named by a capital script letter or by three noncollinear points on the plane.

plano   Término primitivo en geometría. Es una superficie formada por puntos y sin profundidad que se extiende indefinidamente en todas direcciones. Los planos a menudo se representan con un cuadrilátero inclinado y sombreado. Los planos en general se designan con una letra mayúscula o con tres puntos no colineales del plano.

plane Euclidean geometry (p. 165)   Geometry based on Euclid's axioms dealing with a system of points, lines, and planes.

geometría del plano euclidiano   Geometría basada en los axiomas de Euclides, los que integran un sistema de puntos, rectas y planos.

Platonic Solids (p. 637)   The five regular polyhedra: tetrahedron, hexahedron, octahedron, dodecahedron, or icosahedron.

sólidos platónicos   Cualquiera de los siguientes cinco poliedros regulares: tetraedro, hexaedro, octaedro, dodecaedro e icosaedro.

point (p. 6)   A basic undefined term of geometry. A point is a location. In a figure, points are represented by a dot. Points are named by capital letters.

punto   Término primitivo en geometría. Un punto representa un lugar o localización. En una figura, se representa con una marca puntual. Los puntos se designan con letras mayúsculas.

point of concurrency (p. 238)   The point of intersection of concurrent lines.

punto de concurrencia   Punto de intersección de rectas concurrentes.

point of symmetry (p. 466)   The common point of reflection for all points of a figure.

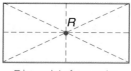

R is a point of symmetry.
R es un punto de simetría.

punto de simetría   El punto común de reflexión de todos los puntos de una figura.

point of tangency (p. 552)   For a line that intersects a circle in only one point, the point at which they intersect.

punto de tangencia   Punto de intersección de una recta que interseca un círculo en un solo punto, el punto en donde se intersecan.

point-slope form (p. 145)   An equation of the form $y - y_1 = m(x - x_1)$, where (x_1, y_1) are the coordinates of any point on the line and m is the slope of the line.

forma punto-pendiente   Ecuación de la forma $y - y_1 = m(x - x_1)$, donde (x_1, y_1) representan las coordenadas de un punto cualquiera sobre la recta y m representa la pendiente de la recta.

polygon (p. 45)   A closed figure formed by a finite number of coplanar segments called *sides* such that the following conditions are met.
   1. The sides that have a common endpoint are noncollinear.
   2. Each side intersects exactly two other sides, but only at their endpoints, called the *vertices*.

polyhedrons (p. 637)   Closed three-dimensional figures made up of flat polygonal regions. The flat regions formed by the polygons and their interiors are called *faces*. Pairs of faces intersect in segments called *edges*. Points where three or more edges intersect are called *vertices*.

postulate (p. 89)   A statement that describes a fundamental relationship between the basic terms of geometry. Postulates are accepted as true without proof.

precision (p. 14)   The precision of any measurement depends on the smallest unit available on the measuring tool.

prism (p. 637)   A solid with the following characteristics.
   1. Two faces, called *bases*, are formed by congruent polygons that lie in parallel planes.
   2. The faces that are not bases, called *lateral faces*, are formed by parallelograms.
   3. The intersections of two adjacent lateral faces are called *lateral edges* and are parallel segments.

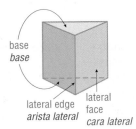

base
base

lateral edge
arista lateral

lateral face
cara lateral

triangular prism
prisma triangular

proof (p. 90)   A logical argument in which each statement you make is supported by a statement that is accepted as true.

proof by contradiction (p. 255)   An indirect proof in which one assumes that the statement to be proved is false. One then uses logical reasoning to deduce a statement that contradicts a postulate, theorem, or one of the assumptions. Once a contradiction is obtained, one concludes that the statement assumed false must in fact be true.

proportion (p. 283)   An equation of the form $\frac{a}{b} = \frac{c}{d}$ that states that two ratios are equal.

pyramid (p. 637)   A solid with the following characteristics.
   1. All of the faces, except one face, intersect at a point called the *vertex*.
   2. The face that does not contain the vertex is called the *base* and is a polygonal region.
   3. The faces meeting at the vertex are called *lateral faces* and are triangular regions.

vertex
vértice

lateral face
cara lateral

base
base

rectangular pyramid
pirámide rectangular

polígono   Figura cerrada formada por un número finito de segmentos coplanares llamados *lados*, y que satisface las siguientes condiciones:
   1. Los lados que tienen un extremo común son no colineales.
   2. Cada lado interseca exactamente dos lados, pero sólo en sus extremos, formando los *vértices*.

poliedro   Figura tridimensional cerrada formada por regiones poligonales planas. Las regiones planas definidas por un polígono y sus interiores se llaman *caras*. Cada intersección entre dos caras se llama *arista*. Los puntos donde se intersecan tres o más aristas se llaman *vértices*.

postulado   Enunciado que describe una relación fundamental entre los términos primitivos de geometría. Los postulados se aceptan como verdaderos sin necesidad de demostración.

precisión   La precisión de una medida depende de la unidad de medida más pequeña del instrumento de medición.

prisma   Sólido que posee las siguientes características:
   1. Tiene dos caras llamadas *bases*, formadas por polígonos congruentes que yacen en planos paralelos.
   2. Las caras que no son las bases, llamadas *caras laterales*, son formadas por paralelogramos.
   3. Las intersecciones de dos aristas laterales adyacentes se llaman *aristas laterales* y son segmentos paralelos.

demostración   Argumento lógico en que cada enunciado está basado en un enunciado que se acepta como verdadero.

demostración por contradicción   Demostración indirecta en que se asume que el enunciado que se va a demostrar es falso. Después, se razona lógicamente para deducir un enunciado que contradiga un postulado, un teorema o una de las conjeturas. Una vez que se obtiene una contradicción, se concluye que el enunciado que se supuso falso es, en realidad, verdadero.

proporción   Ecuación de la forma $\frac{a}{b} = \frac{c}{d}$ que establece que dos razones son iguales.

pirámide   Sólido con las siguientes características:
   1. Todas, excepto una de las caras, se intersecan en un punto llamado *vértice*.
   2. La cara que no contiene el vértice se llama *base* y es una región poligonal.
   3. Las caras que se encuentran en los vértices se llaman *caras laterales* y son regiones triangulares.

Pythagorean identity (p. 391) The identity $\cos^2\theta + \sin^2\theta = 1$.

identidad pitagórica La identidad $\cos^2\theta + \sin^2\theta = 1$.

Pythagorean triple (p. 352) A group of three whole numbers that satisfies the equation $a^2 + b^2 = c^2$, where c is the greatest number.

triplete de Pitágoras Grupo de tres números enteros que satisfacen la ecuación $a^2 + b^2 = c^2$, donde c es el número más grande.

R

radius **1.** (p. 522) In a circle, any segment with endpoints that are the center of the circle and a point on the circle. **2.** (p. 671) In a sphere, any segment with endpoints that are the center and a point on the sphere.

radio **1.** Cualquier segmento cuyos extremos están en el centro de un círculo y en un punto cualquiera del mismo. **2.** Cualquier segmento cuyos extremos forman el centro y en punto de una esfera.

rate of change (p. 140) Describes how a quantity is changing over time.

tasa de cambio Describe cómo cambia una cantidad a través del tiempo.

ratio (p. 282) A comparison of two quantities.

razón Comparación entre dos cantidades.

ray (p. 29) $\overrightarrow{PQ}$ is a ray if it is the set of points consisting of $\overline{PQ}$ and all points S for which Q is between P and S.

semirrecta $\overrightarrow{PQ}$ es una semirrecta si consta del conjunto de puntos formado por $\overline{PQ}$ y todos los S puntos S para los que Q se localiza entre P y S.

reciprocal identity (p. 391) Each of the three trigonometric ratios called *cosecant*, *secant*, and *cotangent*, that are the reciprocals of sine, cosine, and tangent, respectively.

identidad recíproca Cada una de las tres razones trigonométricas llamadas *cosecante*, *secante* y *tangente* y que son los recíprocos del seno, el coseno y la tangente, respectivamente

rectangle (p. 424) A quadrilateral with four right angles.

rectángulo Cuadrilátero que tiene cuatro ángulos rectos.

reflection (p. 463) A transformation representing a flip of the figure over a point, line, or plane.

reflexión Transformación que se obtiene cuando se "voltea" una imagen sobre un punto, una línea o un plano.

reflection matrix (p. 507) A matrix that can be multiplied by the vertex matrix of a figure to find the coordinates of the reflected image.

matriz de reflexión Matriz que al ser multiplicada por la matriz de vértices de una figura permite hallar las coordenadas de la imagen reflejada.

regular polygon (p. 46) A convex polygon in which all of the sides are congruent and all of the angles are congruent.

polígono regular Polígono convexo en el que todos los lados y todos los ángulos son congruentes entre sí.

regular pentagon
pentágono regular

regular polyhedron (p. 637) A polyhedron in which all of the faces are regular congruent polygons.

poliedro regular Poliedro cuyas caras son polígonos regulares congruentes.

regular prism (p. 637) A right prism with bases that are regular polygons.

prisma regular Prisma recto cuyas bases son polígonos regulares.

regular tessellation (p. 484)   A tessellation formed by only one type of regular polygon.

teselado regular   Teselado formado por un solo tipo de polígono regular.

related conditionals (p. 77)   Statements such as the converse, inverse, and contrapositive that are based on a given conditional statement.

enunciados condicionales relacionados   Enunciados tales como el recíproco, la inversa y la antítesis que están basados en un enunciado condicional dado.

relative error (p. 19)   The ratio of the half-unit difference in precision to the entire measure, expressed as a percent.

error relativo   La razón entre la mitad de la unidad más precisa de la medición y la medición completa, expresada en forma de porcentaje.

remote interior angles (p. 186)   The angles of a triangle that are not adjacent to a given exterior angle.

ángulos internos no adyacentes   Ángulos de un triángulo que no son adyacentes a un ángulo exterior dado.

resultant (p. 500)   The sum of two vectors.

resultante   La suma de dos vectores.

rhombus (p. 431)   A quadrilateral with all four sides congruent.

rombo   Cuadrilátero cuyos cuatro lados son congruentes.

right angle (p. 30)   An angle with a degree measure of 90.

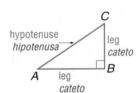
A
$m\angle A = 90$

ángulo recto   Ángulo cuya medida en grados es 90.

right cone (p. 666)   A cone with an axis that is also an altitude.

cono recto   Cono cuyo eje es también su altura.

right cylinder (p. 655)   A cylinder with an axis that is also an altitude.

cilindro recto   Cilindro cuyo eje es también su altura.

right prism (p. 649)   A prism with lateral edges that are also altitudes.

prisma recto   Prisma cuyas aristas laterales también son su altura.

right triangle (p. 178)   A triangle with a right angle. The side opposite the right angle is called the *hypotenuse*. The other two sides are called legs.

hypotenuse
hipotenusa
leg
cateto
C
B
A
leg
cateto

triángulo rectángulo   Triángulo con un ángulo recto. El lado opuesto al ángulo recto se conoce como *hipotenusa*. Los otros dos lados se llaman catetos.

rotation (p. 476)   A transformation that turns every point of a preimage through a specified angle and direction about a fixed point, called the *center of rotation*.

rotación   Transformación en que se hace girar cada punto de la preimagen a través de un ángulo y una dirección determinadas alrededor de un punto, conocido como *centro de rotación*.

rotation matrix (p. 507)   A matrix that can be multiplied by the vertex matrix of a figure to find the coordinates of the rotated image.

matriz de rotación   Matriz que al ser multiplicada por la matriz de vértices de la figura permite calcular las coordenadas de la imagen rotada.

rotational symmetry (p. 478)   If a figure can be rotated less than 360° about a point so that the image and the preimage are indistinguishable, the figure has rotational symmetry.

simetría de rotación   Si se puede rotar una imagen menos de 360° alrededor de un punto y la imagen y la preimagen son idénticas, entonces la figura presenta simetría de rotación.

scalar (p. 501)   A constant multiplied by a vector.

escalar   Una constante multiplicada por un vector.

scalar multiplication (p. 501)   Multiplication of a vector by a scalar.

multiplicación escalar   Multiplicación de un vector por una escalar.

scale factor (p. 290)   The ratio of the lengths of two corresponding sides of two similar polygons or two similar solids.

factor de escala   La razón entre las longitudes de dos lados correspondientes de dos polígonos o sólidos semejantes.

scalene triangle (p. 179)   A triangle with no two sides congruent.

triángulo escaleno   Triángulo cuyos lados no son congruentes.

secant (p. 561)   Any line that intersects a circle in exactly two points.

$\overleftrightarrow{CD}$ is a secant of $\odot P$.
$\overleftrightarrow{CD}$ es una secante de $\odot P$.

secante   Cualquier recta que interseca un círculo exactamente en dos puntos.

sector of a circle (p. 623)   A region of a circle bounded by a central angle and its intercepted arc.

The shaded region is a sector of $\odot A$.
La región sombreada es un sector de $\odot A$.

sector de un círculo   Región de un círculo que está limitada por un ángulo central y el arco que interseca.

segment (p. 13)   *See* line segment.

segmento   *Ver* segmento de recta.

segment bisector (p. 24)   A segment, line, or plane that intersects a segment at its midpoint.

bisectriz de segmento   Segmento, recta o plano que interseca un segmento en su punto medio.

segment of a circle (p. 624)   The region of a circle bounded by an arc and a chord.

The shaded region is a segment of $\odot A$.
La región sombreada es un segmento de $\odot A$.

segmento de un círculo   Región de un círculo limitada por un arco y una cuerda.

self-similar (p. 325)   If any parts of a fractal image are replicas of the entire image, the image is self-similar.

autosemejante   Si cualquier parte de una imagen fractal es una réplica de la imagen completa, entonces la imagen es autosemejante.

semicircle (p. 530)   An arc that measures 180.

semicírculo   Arco que mide 180°.

semi-regular tessellation (p. 484)   A uniform tessellation formed using two or more regular polygons.

teselado semirregular   Teselado uniforme compuesto por dos o más polígonos regulares.

similar polygons (p. 289)   Two polygons are similar if and only if their corresponding angles are congruent and the measures of their corresponding sides are proportional.

polígonos semejantes   Dos polígonos son semejantes si y sólo si sus ángulos correspondientes son congruentes y las medidas de sus lados correspondientes son proporcionales.

similar solids (p. 707)   Solids that have exactly the same shape, but not necessarily the same size.

sólidos semejantes   Sólidos que tienen exactamente la misma forma, pero no necesariamente el mismo tamaño.

similarity transformation (p. 491)   When a figure and its transformation image are similar.

transformación de semejanza   Aquélla en que la figura y su imagen transformada son semejantes.

sine (p. 364)   For an acute angle of a right triangle, the ratio of the measure of the leg opposite the acute angle to the measure of the hypotenuse.

seno   Es la razón entre la medida del cateto opuesto al ángulo agudo y la medida de la hipotenusa de un triángulo rectángulo.

skew lines (p. 127)   Lines that do not intersect and are not coplanar.

rectas alabeadas   Rectas que no se intersecan y que no son coplanares.

slope (p. 139)   For a (nonvertical) line containing two points (x_1, y_1) and (x_2, y_2), the number m given by the formula $m = \dfrac{y_2 - y_1}{x_2 - x_1}$ where $x_2 \neq x_1$.

pendiente   Para una recta (no vertical) que contiene dos puntos (x_1, y_1) y (x_2, y_2), el número m dado por la fórmula $m = \dfrac{y_2 - y_1}{x_2 - x_1}$ donde $x_2 \neq x_1$.

slope-intercept form (p. 145)   A linear equation of the form $y = mx + b$. The graph of such an equation has slope m and y-intercept b.

forma pendiente-intersección   Ecuación lineal de la forma $y = mx + b$. En la gráfica de tal ecuación, la pendiente es m y la intersección y es b.

solving a triangle (p. 378)   Finding the measures of all of the angles and sides of a triangle.

resolver un triángulo   Calcular las medidas de todos los ángulos y todos los lados de un triángulo.

space (p. 8)   A boundless three-dimensional set of all points.

espacio   Conjunto tridimensional no acotado de todos los puntos.

sphere (p. 638)   In space, the set of all points that are a given distance from a given point, called the *center*.

esfera   El conjunto de todos los puntos en el espacio que se encuentran a cierta distancia de un punto dado llamado *centro*.

C is the center of the sphere.
C es el centro de la esfera.

spherical geometry (p. 165)   The branch of geometry that deals with a system of points, greatcircles (lines), and spheres (planes).

geometría esférica   Rama de la geometría que estudia los sistemas de puntos, círculos máximos (rectas) y esferas (planos).

square (p. 432)   A quadrilateral with four right angles and four congruent sides.

cuadrado   Cuadrilátero con cuatro ángulos rectos y cuatro lados congruentes.

standard position (p. 498)   When the initial point of a vector is at the origin.

posición estándar   Ocurre cuando la posición inicial de un vector es el origen.

statement (p. 67)   Any sentence that is either true or false, but not both.

enunciado   Una oración que puede ser falsa o verdadera, pero no ambas.

strictly self-similar (p. 325)   A figure is strictly self-similar if any of its parts, no matter where they are located or what size is selected, contain the same figure as the whole.

estrictamente autosemejante   Una figura es estrictamente autosemejante si cualquiera de sus partes, sin importar su localización o su tamaño, contiene la figura completa.

supplementary angles (p. 39)   Two angles with measures that have a sum of 180.

ángulos suplementarios   Dos ángulos cuya suma es igual a 180°.

surface area (p. 644)   The sum of the areas of all faces and side surfaces of a three-dimensional figure.

área de superficie   La suma de las áreas de todas las caras y superficies laterales de una figura tridimensional.

T

tangent   **1.** (p. 364) For an acute angle of a right triangle, the ratio of the measure of the leg opposite the acute angle to the measure of the leg adjacent to the acute angle.   **2.** (p. 552) A line in the plane of a circle that intersects the circle in exactly one point. The point of intersection is called the *point of tangency*.   **3.** (p. 671) A line that intersects a sphere in exactly one point.

tangente   **1.** La razón entre la medida del cateto opuesto al ángulo agudo y la medida del cateto adyacente al ángulo agudo de un triángulo rectángulo.   **2.** La recta situada en el mismo plano de un círculo y que interseca dicho círculo en un sólo punto. El punto de intersección se conoce como *punto de tangencia*.   **3.** Recta que interseca una esfera en un sólo punto.

tessellation (p. 483)   A pattern that covers a plane by transforming the same figure or set of figures so that there are no overlapping or empty spaces.

teselado   Patrón que cubre un plano y que se obtiene transformando la misma figura o conjunto de figuras, sin que haya traslapes ni espacios vacíos.

theorem (p. 90)   A statement or conjecture that can be proven true by undefined terms, definitions, and postulates.

teorema   Enunciado o conjetura que se puede demostrar como verdadera mediante el uso de términos primitivos, definiciones y postulados.

transformation (p. 462)   In a plane, a mapping for which each point has exactly one image point and each image point has exactly one preimage point.

transformación   La relación en el plano en que cada punto tiene un único punto imagen y cada punto imagen tiene un único punto preimagen.

translation (p. 470)   A transformation that moves all points of a figure the same distance in the same direction.

traslación   Transformación en que todos los puntos de una figura se trasladan la misma distancia, en la misma dirección.

translation matrix (p. 506)   A matrix that can be added to the vertex matrix of a figure to find the coordinates of the translated image.

matriz de traslación   Matriz que al sumarse a la matriz de vértices de una figura permite calcular las coordenadas de la imagen trasladada.

transversal (p. 127)   A line that intersects two or more lines in a plane at different points.

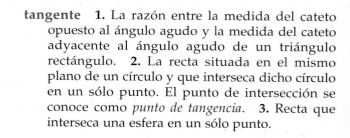

Line *t* is a transversal.
La recta t es una transversal.

transversal   Recta que interseca en diferentes puntos dos o más rectas en el mismo plano.

trapezoid (p. 439)   A quadrilateral with exactly one pair of parallel sides. The parallel sides of a trapezoid are called *bases*. The nonparallel sides are called *legs*. The pairs of angles with their vertices at the endpoints of the same base are called *base angles*.

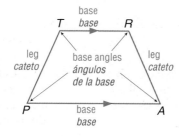

trapecio   Cuadrilátero con un sólo par de lados paralelos. Los lados paralelos del trapecio se llaman *bases*. Los lados no paralelos se llaman *catetos*. Los ángulos cuyos vértices se encuentran en los extremos de la misma base se llaman *ángulos de la base*.

trigonometric identity (p. 391) An equation involving a trigonometric ratio that is true for all values of the angle measure.

identidad trigonométrica Ecuación que contiene una razón trigonométrica que es verdadera para todos los valores de la medida del ángulo.

trigonometric ratio (p. 364) A ratio of the lengths of sides of a right triangle.

razón trigonométrica Razón de las longitudes de los lados de un triángulo rectángulo.

trigonometry (p. 364) The study of the properties of triangles and trigonometric functions and their applications.

trigonometría Estudio de las propiedades de los triángulos y de las funciones trigonométricas y sus aplicaciones.

truth table (p. 70) A table used as a convenient method for organizing the truth values of statements.

tabla verdadera Tabla que se utiliza para organizar de una manera conveniente los valores de verdad de los enunciados.

truth value (p. 67) The truth or falsity of a statement.

valor verdadero La condición de un enunciado de ser verdadero o falso.

two-column proof (p. 95) A formal proof that contains statements and reasons organized in two columns. Each step is called a *statement*, and the properties that justify each step are called *reasons*.

demostración a dos columnas Aquélla que contiene enunciados y razones organizadas en dos columnas. Cada paso se llama *enunciado* y las propiedades que lo justifican son las *razones*.

U

undefined terms (p. 7) Words, usually readily understood, that are not formally explained by means of more basic words and concepts. The basic undefined terms of geometry are point, line, and plane.

términos primitivos Palabras que por lo general se entienden fácilmente y que no se explican formalmente mediante palabras o conceptos más básicos. Los términos básicos primitivos de la geometría son el punto, la recta y el plano.

uniform tessellations (p. 484) Tessellations containing the same arrangement of shapes and angles at each vertex.

teselado uniforme Teselados que contienen el mismo patrón de formas y ángulos en cada vértice.

V

vector (p. 498) A directed segment representing a quantity that has both magnitude, or length, and direction.

vector Segmento dirigido que representa una cantidad que posee tanto magnitud, o longitud, como dirección.

vertex matrix (p. 506) A matrix that represents a polygon by placing all of the column matrices of the coordinates of the vertices into one matrix.

matriz del vértice Matriz que representa un polígono al colocar todas las matrices columna de las coordenadas de los vértices en una matriz.

vertical angles (p. 37) Two nonadjacent angles formed by two intersecting lines.

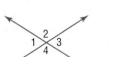

∠1 and ∠3 are vertical angles.
∠2 and ∠4 are vertical angles.
∠1 y ∠3 son ángulos opuestos por el vértice.
∠2 y ∠4 son ángulos opuestos por el vértice.

ángulos opuestos por el vértice Dos ángulos no adyacentes formados por dos rectas que se intersecan.

volume (p. 688) A measure of the amount of space enclosed by a three-dimensional figure.

volumen La medida de la cantidad de espacio dentro de una figura tridimensional.

Selected Answers

Chapter 1 Points, Lines, Planes, and Angles

Page 5   Chapter 1 Getting Started

1–4.
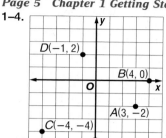

5. $1\frac{1}{8}$ **7.** $\frac{5}{16}$ **9.** -15
11. 25 **13.** 20 in.
15. 24.6 m

Pages 9–11   Lesson 1-1

1. point, line, plane **3.** Micha; the points must be noncollinear to determine a plane.
5. Sample answer:

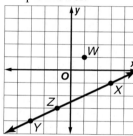

7. 6 **9.** No; A, C, and J lie in plane ABC, but D does not.
11. point **13.** n **15.** $\mathcal{R}$
17. Sample answer: $\overrightarrow{PR}$
19. (D, 9)
21.

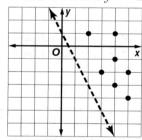

23. Sample answer:

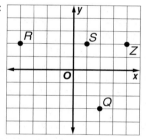

25.
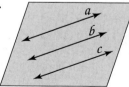
27.
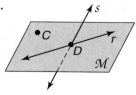

29. points that seem collinear; sample answer: $(0, -2)$, $(1, -3)$, $(2, -4)$, $(3, -5)$

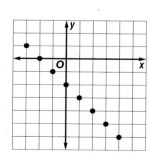

31. 1 **33.** anywhere on $\overleftrightarrow{AB}$ **35.** A, B, C, D or E, F, C, B
37. $\overrightarrow{AC}$ **39.** lines **41.** plane **43.** point **45.** point
47.

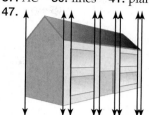

49. See students' work.

51. Sample answer:

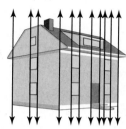

53. vertical **55.** Sample answer: Chairs wobble because all four legs do not touch the floor at the same time. Answers should include the following.
- The ends of the legs represent points. If all points lie in the same plane, the chair will not wobble.
- Because it only takes three points to determine a plane, a chair with three legs will never wobble.

57. B
59. part of the coordinate plane above the line $y = -2x + 1$.

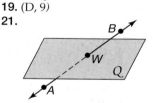

61. $=$
63. $=$
65. $<$

Pages 16–19   Lesson 1-2

1. Align the 0 point on the ruler with the leftmost endpoint of the segment. Align the edge of the ruler along the segment. Note where the rightmost endpoint falls on the scale and read the closest eighth of an inch measurement.
3. $1\frac{3}{4}$ in. **5.** 0.5 m; 14 m could be 13.5 to 14.5 m **7.** 3.7 cm
9. $x = 3$; $LM = 9$ **11.** $\overline{BC} \cong \overline{CD}$, $\overline{BE} \cong \overline{ED}$, $\overline{BA} \cong \overline{DA}$
13. 4.5 cm or 45 mm **15.** $1\frac{1}{4}$ in. **17.** 0.5 cm; 21.5 to 22.5 mm
19. 0.5 cm; 307.5 to 308.5 cm **21.** $\frac{1}{8}$ ft.; $3\frac{1}{8}$ to $3\frac{3}{8}$ ft.
23. $1\frac{1}{4}$ in. **25.** 2.8 cm **27.** $1\frac{1}{4}$ in. **29.** $x = 11$; $ST = 22$
31. $x = 2$; $ST = 4$ **33.** $y = 2$; $ST = 3$ **35.** no **37.** yes
39. yes **41.** $\overline{CF} \cong \overline{DG}$, $\overline{AB} \cong \overline{HI}$, $\overline{CE} \cong \overline{ED} \cong \overline{EF} \cong \overline{EG}$
43. 50,000 visitors **45.** No; the number of visitors to Washington state parks could be as low as 46.35 million or as high as 46.45 million. The visitors to Illinois state parks could be as low as 44.45 million or as high as 44.55 million visitors. The difference in visitors could be as high as 2.0 million.

47. 15.5 cm; Each measurement is accurate within 0.5 cm, so the greatest perimeter is 3.5 cm + 5.5 cm + 6.5 cm.
49.

51. Sample answer: Units of measure are used to differentiate between size and distance, as well as for accuracy. Answers should include the following.
- When a measurement is stated, you do not know the precision of the instrument used to make the measure. Therefore, the actual measure could be greater or less than that stated.
- You can assume equal measures when segments are shown to be congruent.

53. 1.7% **55.** 0.08% **57.** D **59.** Sample answer: planes ABC and BCD **61.** 5 **63.** 22 **65.** 1

Page 19 Practice Quiz 1
1. $\overrightarrow{PR}$ **3.** $\overrightarrow{PR}$ **5.** 8.35

Pages 25–27 Lesson 1-3
1. Sample answers: (1) Use one of the Midpoint Formulas if you know the coordinates of the endpoints. (2) Draw a segment and fold the paper so that the endpoints match to locate the middle of the segment. (3) Use a compass and straightedge to construct the bisector of the segment.
3. 8 **5.** 10 **7.** −6 **9.** (−2.5, 4) **11.** (3, 5) **13.** 2
15. 3 **17.** 11 **19.** 10 **21.** 13 **23.** 15 **25.** $\sqrt{90} \approx 9.5$
27. $\sqrt{61} \approx 7.8$ **29.** 17.3 units **31.** −3 **33.** 2.5 **35.** 1
37. (10, 3) **39.** (−10, −3) **41.** (5.6, 2.85) **43.** $R(2, 7)$
45. $T\left(\frac{8}{3}, 11\right)$ **47.** LaFayette, LA **49a.** 111.8 **49b.** 212.0
49c. 353.4 **49d.** 420.3 **49e.** 37.4 **49f.** 2092.9 **51.** ≈ 73.8
53. Sample answer: The perimeter increases by the same factor. **55.** (−1, −3) **57.** B **59.** $4\frac{1}{4}$ in.
61. Sample answer:

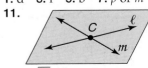

63. 10 **65.** 9
67. $\frac{13}{3}$

Pages 33–36 Lesson 1-4
1. Yes; they all have the same measure. **3.** $m\angle A = m\angle Z$
5. $\overrightarrow{BA}, \overrightarrow{BC}$ **7.** 135°, obtuse **9.** 47 **11.** $\angle 1$, right; $\angle 2$, acute; $\angle 3$, obtuse **13.** B **15.** A **17.** $\overrightarrow{AB}, \overrightarrow{AD}$ **19.** $\overrightarrow{AD}, \overrightarrow{AE}$
21. $\angle FEA, \angle 4$ **23.** $\angle AED, \angle DEA, \angle AEB, \angle BEA, \angle AEC,$ $\angle CEA$ **25.** $\angle 2$ **27.** 30, 30 **29.** 60°, acute **31.** 90°, right
33. 120°, obtuse **35.** 65 **37.** 4 **39.** 4 **41.** Sample answer: *Acute* can mean something that is sharp or having a very fine tip like a pen, a knife, or a needle. *Obtuse* means not pointed or blunt, so something that is obtuse would be wide. **43.** 31; 59 **45.** 1, 3, 6, 10, 15 **47.** 21, 45 **49.** Sample answer: A degree is $\frac{1}{360}$ of a circle. Answers should include the following.
- Place one side of the angle to coincide with 0 on the protractor and the vertex of the angle at the center point of the protractor. Observe the point at which the other side of the angle intersects the scale of the protractor.
- See students' work.

51. C **53.** $\sqrt{80} \approx 8.9$; (2, 2) **55.** $9\frac{2}{3}$ in. **57.** 13 **59.** F, L, J
61. 5 **63.** −45 **65.** 8

1. $\left(-\frac{1}{2}, 1\right)$; $\sqrt{65} \approx 8.1$ **3.** (0, 0); $\sqrt{2000} \approx 44.7$ **5.** 34; 135

Pages 41–62 Lesson 1-5
1.

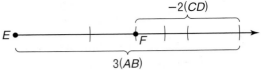

3. Sample answer: The noncommon sides of a linear pair of angles form a straight line.

5. Sample answer: $\angle ABC, \angle CBE$ **7.** $x = 24, y = -20$
9. Yes; they share a common side and vertex, so they are adjacent. Since $\overrightarrow{PR}$ falls between $\overrightarrow{PQ}$ and $\overrightarrow{PS}$, $m\angle QPR < 90$, so the two angles cannot be complementary or supplementary.
11. $\angle WUT, \angle VUX$ **13.** $\angle UWT, \angle TWY$ **15.** $\angle WTY,$ $\angle WTU$ **17.** 53, 37 **19.** 148 **21.** 84, 96 **23.** always
25. sometimes **27.** 3.75 **29.** 114 **31.** Yes; the symbol denotes that $\angle DAB$ is a right angle. **33.** Yes; their sum of their measures is $m\angle ADC$, which 90. **35.** No; we do not know $m\angle ABC$.
37. Sample answer:

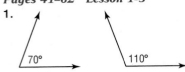

39. Because $\angle WUT$ and $\angle TUV$ are supplementary, let $m\angle WUT = x$ and $m\angle TUV = 180 - x$. A bisector creates measures that are half of the original angle, so $m\angle YUT = \frac{1}{2}m\angle WUT$ or $\frac{x}{2}$ and $m\angle TUZ = \frac{1}{2}m\angle TUV$ or $\frac{180 - x}{2}$. Then $m\angle YUZ = m\angle YUT + m\angle TUZ$ or $\frac{x}{2} + \frac{180 - x}{2}$. This sum simplifies to $\frac{180}{2}$ or 90. Because $m\angle YUZ = 90$, $\overline{YU} \perp \overline{UZ}$. **41.** A **43.** $\ell \perp \overrightarrow{AB}, m \perp \overrightarrow{AB},$ $n \perp \overrightarrow{AB}$ **45.** obtuse **47.** right **49.** obtuse **51.** 8
53. $\sqrt{173} \approx 13.2$ **55.** $\sqrt{20} \approx 4.5$ **57.** $n = 3, QR = 20$
59. 24 **61.** 40

Pages 48–50 Lesson 1-6
1. Divide the perimeter by 10. **3.** $P = 3s$ **5.** pentagon; concave; irregular **7.** 33 ft **9.** 16 units **11.** 4605 ft
13. octagon; convex; regular **15.** pentagon **17.** triangle
19. 82 ft **21.** 40 units **23.** The perimeter is tripled.
25. 125 m **27.** 30 units **29.** All are 15 cm. **31.** 13 units, 13 units, 5 units **33.** 4 in., 4 in., 17 in., 17 in. **35.** 52 units
37. Sample answer: Some toys use pieces to form polygons. Others have polygon-shaped pieces that connect together. Answers should include the following.
- triangles, quadrilaterals, pentagons
-

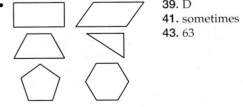

39. D
41. sometimes
43. 63

Pages 53–56 Chapter 1 Study Guide and Review
1. d **3.** f **5.** b **7.** p or m **9.** F
11.

13. $x = 6, PB = 18$
15. $s = 3, PB = 12$ **17.** yes
19. not enough information
21. $\sqrt{101} \approx 10.0$
23. $\sqrt{13} \approx 3.6$ **25.** (3, −5) **27.** (0.6, −6.35) **29.** $\overrightarrow{FE}, \overrightarrow{FG}$
31. 70°, acute **33.** 50°, acute **35.** 36 **37.** 40 **39.** $\angle TWY,$ $\angle XWY$ **41.** 9 **43.** not a polygon **45.** ≈ 22.5 units

Chapter 2 Reasoning and Proof

Page 61 Chapter 2 Getting Started
1. 10 **3.** 0 **5.** 50 **7.** 21 **9.** −9 **11.** $-\dfrac{18}{5}$ **13.** 16

Pages 63–66 Lesson 2-1
1. Sample answer: After the news is over, it's time for dinner. **3.** Sample answer: When it's cloudy, it rains. Counterexample: It is often cloudy and it does not rain.

5. 7 **7.** *A, B, C,* and *D* are noncollinear.

9. true **11.**

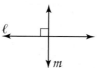

13. 32 **15.** $\dfrac{11}{3}$ **17.** 162 **19.** 30

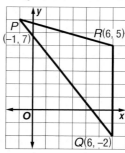

21. Lines ℓ and m form four right angles.

23. $\angle 3$ and $\angle 4$ are supplementary.

25. $\triangle PQR$ is a scalene triangle.

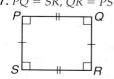

27. $PQ = SR, QR = PS$

29. false;

31. false; W X Y Z **33.** true **35.** False; *JKLM* may not have a right angle. **37.** trial and error, a process of inductive reasoning **39.** C_7H_{16} **41.** false; $n = 41$ **43.** C
45. hexagon, convex, irregular **47.** heptagon, concave, irregular **49.** No; we do not know anything about the angle measures. **51.** Yes; they form a linear pair.
53. (2, −1) **55.** (1, −12) **57.** (5.5, 2.2) **59.** 8; 56 **61.** 4; 16
63. 10; 43 **65.** 4, 5 **67.** 5, 6, 7

Pages 71–74 Lesson 2-2
1. The conjunction (p and q) is represented by the intersection of the two circles. **3.** A conjunction is a compound statement using the word *and*, while a disjunction is a compound statement using the word *or*.
5. $9 + 5 = 14$ and a square has four sides; true.
7. $9 + 5 = 14$ or February does not have 30 days; true.
9. $9 + 5 \neq 14$ or a square does not have four sides; false.

11. Sample answer:

p	q	$p \wedge q$
T	T	T
T	F	F
F	T	F
F	F	F

13. Sample answer:

p	r	$\sim p$	$\sim p \wedge r$
T	T	F	F
T	F	F	F
F	T	T	T
F	F	T	F

15. 14 **17.** 3 **19.** $\sqrt{-64} = 8$ or an equilateral triangle has three congruent sides; true. **21.** $0 < 0$ and an obtuse angle measures greater than 90° and less than 180°; false. **23.** An equilateral triangle has three congruent sides and an obtuse angle measures greater than 90° and less than 180°; true.
25. An equilateral triangle has three congruent sides and $0 < 0$; false. **27.** An obtuse angle measures greater than 90° and less than 180° or an equilateral triangle has three congruent sides; true. **29.** An obtuse angle measures greater than 90° and less than 180°, or an equilateral triangle has three congruent sides and $0 < 0$; true.

31.

p	q	$\sim p$	$\sim q$	$\sim p \wedge \sim q$
T	T	F	F	F
T	F	F	T	F
F	T	T	F	F
F	F	T	T	T

33. Sample answer:

q	r	q and r
T	T	T
T	F	F
F	T	F
F	F	F

35. Sample answer:

p	r	p or r
T	T	T
T	F	T
F	T	T
F	F	F

37. Sample answer:

q	r	$\sim r$	$q \wedge \sim r$
T	T	F	F
T	F	T	T
F	T	F	F
F	F	T	F

39. Sample answer:

p	q	r	$\sim p$	$\sim r$	$q \wedge \sim r$	$\sim p \vee (q \wedge \sim r)$
T	T	T	F	F	F	F
T	T	F	F	T	T	T
T	F	T	F	F	F	F
T	F	F	F	T	F	F
F	T	T	T	F	F	T
F	T	F	T	T	T	T
F	F	T	T	F	F	T
F	F	F	T	T	F	T

41. 42   **43.** 25

45.

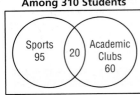

Level of Participation
Among 310 Students

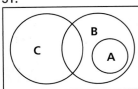

47. 135   **49.** true

51.

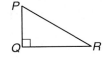

53. Sample answer: Logic can be used to eliminate false choices on a multiple choice test. Answers should include the following.
• Math is my favorite subject and drama club is my favorite activity.
• See students' work.

55. C   **57.** 81   **59.** 1   **61.** 405   **63.** 34.4   **65.** 29.5   **67.** 55°, acute   **69.** 222 feet   **71.** 44   **73.** 184

Pages 78–80   Lesson 2-3

1. Writing a conditional in if-then form is helpful so that the hypothesis and conclusion are easily recognizable.
3. In the inverse, you negate both the hypothesis and the conclusion of the conditional. In the contrapositive, you negate the hypothesis and the conclusion of the converse.
5. H: $x - 3 = 7$; C: $x = 10$ **7.** If a pitcher is a 32-ounce pitcher, then it holds a quart of liquid.   **9.** If an angle is formed by perpendicular lines, then it is a right angle.
11. true   **13.** Converse: If plants grow, then they have water; true.   Inverse: If plants do not have water, then they will not grow; true.   Contrapositive: If plants do not grow, then they do not have water. False; they may have been killed by overwatering.   **15.** Sample answer: If you are in Colorado, then aspen trees cover high areas of the mountains.   If you are in Florida, then cypress trees rise from the swamps.   If you are in Vermont, then maple trees are prevalent.   **17.** H: you are a teenager; C: you are at least 13 years old   **19.** H: three points lie on a line; C: the points are collinear   **21.** H: the measure of an is between 0 and 90; C: the angle is acute   **23.** If you are a math teacher, then you love to solve problems.   **25.** Sample answer: If two angles are adjacent, then they have a common side.
27. Sample answer: If two triangles are equiangular, then they are equilateral.   **29.** true   **31.** true   **33.** false
35. true   **37.** false   **39.** true   **41.** Converse: If you are in good shape, then you exercise regularly; true.   Inverse: If you do not exercise regularly, then you are not in good shape; true.   Contrapositive: If you are not in good shape, then you do not exercise regularly.   False; an ill person may exercise a lot, but still not be in good shape.
43. Converse: If a figure is a quadrilateral, then it is a rectangle; false, rhombus.   Inverse: If a figure is not a rectangle, then it is not a quadrilateral; false, rhombus.   Contrapositive: If a figure is not a quadrilateral, then it is not a rectangle; true.   **45.** Converse: If an angle has measure less than 90, then it is acute; true.   Inverse: If an angle is not acute, then its measure is not less than 90; true.   Contrapositive: If an angle's measure is not less than 90, then it is not acute; true.   **47.** Sample answer: In Alaska, if there are more hours of daylight than darkness, then it is summer. In Alaska, if there are more hours of darkness than daylight, then it is winter.   **49.** Conditional statements can be used to describe how to get a discount, rebate, or refund.

Sample answers should include the following.   If you are not 100% satisfied, then return the product for a full refund. Wearing a seatbelt reduces the risk of injuries.   **51.** B
53. A hexagon has five sides or $60 \times 3 = 18$.; false
55. A hexagon doesn't have five sides or $60 \times 3 = 18$.; true
57. George Washington was not the first president of the United States and $60 \times 3 \neq 18$.; false
59. The sum of the measures of the angles in a triangle is 180.   **61.** $\angle PQR$ is a right angle.

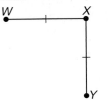

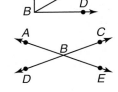

63. $\sqrt{41}$ or 6.4   **65.** $\sqrt{125}$ or 11.2
67. Multiply each side by 2.

Page 80   Practice Quiz 1
1. false   **3.** Sample answer:

W ●——|——● X

|_____|
●Y

p	q	$\sim p$	$\sim p \wedge q$
T	T	F	F
T	F	F	F
F	T	T	T
F	F	T	F

5. Converse: If two angles have a common vertex, then the angles are adjacent. False; $\angle ABD$ is not adjacent to $\angle ABC$.

Inverse: If two angles are not adjacent, then they do not have a common vertex. False; $\angle ABC$ and $\angle DBE$ have a common vertex and are not adjacent.

Contrapositive: If two angles do not have a common vertex, then they are not adjacent; true.

Pages 84–87   Lesson 2-4
1. Sample answer: a: If it is rainy, the game will be cancelled; b: It is rainy; c: The game will be cancelled.
3. Lakeisha; if you are dizzy, that does not necessarily mean that you are seasick and thus have an upset stomach.
5. Invalid; congruent angles do not have to be vertical.
7. The midpoint of a segment divides it into two segments with equal measures.   **9.** invalid   **11.** No; Terry could be a man or a woman. She could be 45 and have purchased $30,000 of life insurance.   **13.** Valid; since 5 and 7 are odd, the Law of Detachment indicates that their sum is even.
15. Invalid; the sum is even.   **17.** Invalid; E, F, and G are not necessarily noncollinear.   **19.** Valid; the vertices of a triangle are noncollinear, and therefore determine a plane.
21. If the measure of an angle is less than 90, then it is not obtuse.   **23.** no conclusion   **25.** yes; Law of Detachment
27. yes; Law of Detachment   **29.** invalid   **31.** If Catriona Le May Doan skated her second 500 meters in 37.45 seconds, then she would win the race.   **33.** Sample answer: Doctors and nurses use charts to assist in determining medications and their doses for patients. Answers should include the following.

- Doctors need to note a patient's symptoms to determine which medication to prescribe, then determine how much to prescribe based on weight, age, severity of the illness, and so on.
- Doctors use what is known to be true about diseases and when symptoms appear, then deduce that the patient has a particular illness.

35. B **37.** They are a fast, easy way to add fun to your family's menu.

39. Sample answer:

q	r	q∧r
T	T	T
T	F	F
F	T	F
F	F	F

41. Sample answer:

p	q	r	q∨r	p∧(q∨r)
T	T	T	T	T
T	T	F	T	T
T	F	T	T	T
T	F	F	F	F
F	T	T	T	F
F	T	F	T	F
F	F	T	T	F
F	F	F	F	F

43. $\angle HDG$ **45.** Sample answer: $\angle JHK$ and $\angle DHK$
47. Yes, slashes on the segments indicate that they are congruent. **49.** 10 **51.** $\sqrt{130} \approx 11.4$

53.

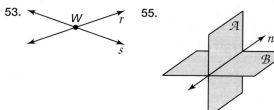

55.

57. Sample answer: $\angle 1$ and $\angle 2$ are complementary, $m\angle 1 + m\angle 2 = 90$.

Pages 91–93 Lesson 2-5

1. Deductive reasoning is used to support claims that are made in a proof. **3.** postulates, theorems, algebraic properties, definitions **5.** 15 **7.** definition of collinear
9. Through any two points, there is exactly one line.
11. 15 ribbons **13.** 10 **15.** 21 **17.** Always; if two points lie in a plane, then the entire line containing those points lies in that plane. **19.** Sometimes; the three points cannot be on the same line. **21.** Sometimes; ℓ and m could be skew so they would not lie in the same plane $\mathcal{R}$. **23.** If two points lie in a plane, then the entire line containing those points lies in that plane. **25.** If two points lie in a plane, then the entire line containing those points lies in the plane.
27. Through any three points not on the same line, there is exactly one plane. **29.** She will have 4 different planes and 6 lines. **31.** one, ten **33.** C **35.** yes; Law of Detachment
37. Converse: If $\triangle ABC$ has an angle with measure greater than 90, then $\triangle ABC$ is a right triangle. False; the triangle

would be obtuse. Inverse: If $\triangle ABC$ is not a right triangle, none of its angle measures are greater than 90. False; it could be an obtuse triangle. Contrapositive: If $\triangle ABC$ does not have an angle measure greater than 90, $\triangle ABC$ is not a right triangle. False; $m\angle ABC$ could still be 90 and $\triangle ABC$ be a right triangle. **39.** $\sqrt{17} \approx 4.1$ **41.** $\sqrt{106} \approx 10.3$
43. 25 **45.** 12 **47.** 10

Pages 97–100 Lesson 2-6

1. Sample answer: If $x = 2$ and $x + y = 6$, then $2 + y = 6$.
3. hypothesis; conclusion **5.** Multiplication Property
7. Addition Property **9a.** $5 - \frac{2}{3}x = 1$ **9b.** Mult. Prop.
9c. Dist. Prop. **9d.** $-2x = -12$ **9e.** Div. Prop.

11. Given: Rectangle $ABCD$,
$AD = 3$, $AB = 10$
Prove: $AC = BD$

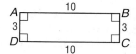

Proof:

Statement	Reasons
1. Rectangle $ABCD$, $AD = 3$, $AB = 10$	1. Given
2. Draw segments AC and DB.	2. Two points determine a line.
3. $\triangle ABC$ and $\triangle BCD$ are right triangles.	3. Def. of rt $\triangle$
4. $AC = \sqrt{3^2 + 10^2}$, $DB = \sqrt{3^2 + 10^2}$	4. Pythagorean Th.
5. $AC = BD$	5. Substitution

13. C **15.** Subt. Prop. **17.** Substitution **19.** Reflexive Property **21.** Substitution **23.** Transitive Prop.
25a. $2x - 7 = \frac{1}{3}x - 2$ **25b.** $3(2x - 7) = 3\left(\frac{1}{3}x - 2\right)$
25c. Dist. Prop. **25d.** $5x - 21 = -6$ **25e.** Add. Prop.
25f. $x = 3$

27. Given: $-2y + \frac{3}{2} = 8$
Prove: $y = -\frac{13}{4}$
Proof:

Statement	Reasons
1. $-2y + \frac{3}{2} = 8$	1. Given
2. $2\left(-2y + \frac{3}{2}\right) = 2(8)$	2. Mult. Prop.
3. $-4y + 3 = 16$	3. Dist. Prop.
4. $-4y = 13$	4. Subt. Prop.
5. $y = -\frac{13}{4}$	5. Div. Prop.

29. Given: $5 - \frac{2}{3}z = 1$
Prove: $z = 6$
Proof:

Statement	Reasons
1. $5 - \frac{2}{3}z = 1$	1. Given
2. $3\left(5 - \frac{2}{3}z\right) = 3(1)$	2. Mult. Prop.
3. $15 - 2x = 3$	3. Dist. Prop.
4. $15 - 2x - 15 = 3 - 15$	4. Subt. Prop.
5. $-2x = -12$	5. Substitution
6. $\frac{-2x}{-2} = \frac{-12}{-2}$	6. Div. Prop.
7. $x = 6$	7. Substitution

31. Given: $m\angle ACB = m\angle ABC$
Prove: $m\angle XCA = m\angle YBA$

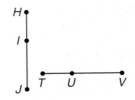

Proof:

Statement	Reasons
1. $m\angle ACB = m\angle ABC$	1. Given
2. $m\angle XCA + m\angle ACB = 180$ $m\angle YBA + m\angle ABC = 180$	2. Def. of supp. $\angle$s
3. $m\angle XCA + m\angle ACB = $ $m\angle YBA + m\angle ABC$	3. Substitution
4. $m\angle XCA + m\angle ACB = $ $m\angle YBA + m\angle ACB$	4. Substitution
5. $m\angle XCA = m\angle YBA$	5. Subt. Prop.

33. All of the angle measures would be equal. **35.** See students' work. **37.** B **39.** 6 **41.** Invalid; $27 \div 6 = 4.5$, which is not an integer. **43.** Sample answer: If people are happy, then they rarely correct their faults. **45.** Sample answer: If a person is a champion, then the person is afraid of losing. **47.** $\frac{1}{2}$ ft **49.** 0.5 in. **51.** 11 **53.** 47

Page 100   Practice Quiz 2
1. invalid **3.** If two lines intersect, then their intersection is exactly one point.

5. Given: $2(n-3) + 5 = 3(n-1)$
Prove: $n = 2$
Proof:

Statement	Reasons
1. $2(n-3) + 5 = 3(n-1)$	1. Given
2. $2n - 6 + 5 = 3n - 3$	2. Dist. Prop.
3. $2n - 1 = 3n - 3$	3. Substitution
4. $2n - 1 - 2n = 3n - 3 - 2n$	4. Subt. Prop.
5. $-1 = n - 3$	5. Substitution
6. $-1 + 3 = n - 3 + 3$	6. Add. Prop.
7. $2 = n$	7. Substitution
8. $n = 2$	8. Symmetric Prop.

Pages 103–106   Lesson 2-7
1. Sample answer: The distance from Cleveland to Chicago is the same as the distance from Cleveland to Chicago.
3. If A, B, and C are collinear and $AB + BC = AC$, then B is between A and C. **5.** Symmetric

7. Given: $\overline{PQ} \cong \overline{RS}$, $\overline{QS} \cong \overline{ST}$
Prove: $\overline{PS} \cong \overline{RT}$
Proof:

Statements	Reasons
a. $\overline{PQ} \cong \overline{RS}$, $\overline{QS} \cong \overline{ST}$	a. Given
b. $PQ = RS$, $QS = ST$	b. Def. of $\cong$ segments
c. $PS = PQ + QS$, $RT = RS + ST$	c. Segment Addition Post.
d. $PQ + QS = RS + ST$	d. Addition Property
e. $PS = RT$	e. Substitution
f. $\overline{PS} \cong \overline{RT}$	f. Def. of $\cong$ segments

9. Given: $\overline{HI} \cong \overline{TU}$, $\overline{HJ} \cong \overline{TV}$
Prove: $\overline{IJ} \cong \overline{UV}$

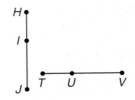

Proof:

Statements	Reasons
1. $\overline{HI} \cong \overline{TU}$, $\overline{HJ} \cong \overline{TV}$	1. Given
2. $HI = TU$, $HJ = TV$	2. Def. of $\cong$ segs.
3. $HI + IJ = HJ$	3. Seg. Add. Post.
4. $TU + IJ = TV$	4. Substitution
5. $TU + UV = TV$	5. Seg. Add. Post.
6. $TU + IJ = TU + UV$	6. Substitution
7. $TU = TU$	7. Reflexive Prop.
8. $IJ = UV$	8. Subt. Prop.
9. $\overline{IJ} \cong \overline{UV}$	9. Def. of $\cong$ segs.

11. Helena is between Missoula and Miles City.
13. Substitution **15.** Transitive **17.** Subtraction

19. Given: $\overline{XY} \cong \overline{WZ}$ and $\overline{WZ} \cong \overline{AB}$
Prove: $\overline{XY} \cong \overline{AB}$

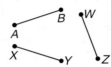

Proof:

Statements	Reasons
1. $\overline{XY} \cong \overline{WZ}$ and $\overline{WZ} \cong \overline{AB}$	1. Given
2. $XY = WZ$ and $WZ = AB$	2. Def. of $\cong$ segs.
3. $XY = AB$	3. Transitive Prop.
4. $\overline{XY} \cong \overline{AB}$	4. Def. of $\cong$ segs.

21. Given: $\overline{WY} \cong \overline{ZX}$
A is the midpoint of $\overline{WY}$.
A is the midpoint of $\overline{ZX}$.
Prove: $\overline{WA} \cong \overline{ZA}$

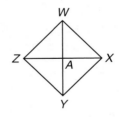

Proof:

Statements:	Reasons:
a. $\overline{WY} \cong \overline{ZX}$ A is the midpoint of $\overline{WY}$. A is the midpoint of $\overline{ZX}$.	a. Given
b. $WY = ZX$	b. Def. of $\cong$ segs.
c. $WA = AY$, $ZA = AX$	c. Definition of midpoint
d. $WY = WA + AY$, $ZX = ZA + AX$	d. Segment Addition Post.
e. $WA + AY = ZA + AX$	e. Substitution
f. $WA + WA = ZA + ZA$	f. Substitution
g. $2WA = 2ZA$	g. Substitution
h. $WA = ZA$	h. Division Property
i. $\overline{WA} \cong \overline{ZA}$	i. Def. of $\cong$ segs.

23. Given: $AB = BC$
Prove: $AC = 2BC$

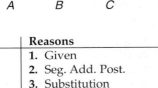

Proof:

Statements	Reasons
1. $AB = BC$	1. Given
2. $AC = AB + BC$	2. Seg. Add. Post.
3. $AC = BC + BC$	3. Substitution
4. $AC = 2BC$	4. Substitution

25. Given: $\overline{AB} \cong \overline{DE}$, C is the midpoint of $\overline{BD}$.
Prove: $\overline{AC} \cong \overline{CE}$

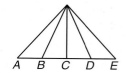

Proof:

Statements	Reasons
1. $\overline{AB} \cong \overline{DE}$, C is the midpoint of $\overline{BD}$.	1. Given
2. $BC = CD$	2. Def. of midpoint
3. $AB = DE$	3. Def. of $\cong$ segs.
4. $AB + BC = CD + DE$	4. Add. Prop.
5. $AB + BC = AC$ $CD + DE = CE$	5. Seg. Add. Post.
6. $AC = CE$	6. Substitution
7. $\overline{AC} \cong \overline{CE}$	7. Def. of $\cong$ segs.

27. Sample answers: $\overline{LN} \cong \overline{QO}$ and $\overline{LM} \cong \overline{MN} \cong \overline{RS} \cong \overline{ST} \cong \overline{QP} \cong \overline{PO}$ **29.** B **31.** Substitution **33.** Addition Property **35.** Never; the midpoint of a segment divides it into two congruent segments. **37.** Always; if two planes intersect, they intersect in a line. **39.** 3; 9 cm by 13 cm **41.** 15 **43.** 45 **45.** 25

Pages 111–114   Lesson 2-8

1. Tomas; Jacob's answer left out the part of $\angle ABC$ represented by $\angle EBF$. **3.** $m\angle 2 = 65$ **5.** $m\angle 11 = 59$, $m\angle 12 = 121$

7. Given: $\overrightarrow{VX}$ bisects $\angle WVY$. $\overrightarrow{VY}$ bisects $\angle XVZ$.
Prove: $\angle WVX \cong \angle YVZ$

Proof:

Statements	Reasons
1. $\overrightarrow{VX}$ bisects $\angle WVY$, $\overrightarrow{VY}$ bisects $\angle XVZ$.	1. Given
2. $\angle WVX \cong \angle XVY$	2. Def. of $\angle$ bisector
3. $\angle XVY \cong \angle YVZ$	3. Def. of $\angle$ bisector
4. $\angle WVX \cong \angle YVZ$	4. Trans. Prop.

9. sometimes

11. Given: $\angle ABC$ is a right angle.
Prove: $\angle 1$ and $\angle 2$ are complementary angles.

Proof:

Statements	Reasons
1. $\angle ABC$ is a right angle.	1. Given
2. $m\angle ABC = 90$	2. Def. of rt. $\angle$
3. $m\angle ABC = m\angle 1 + m\angle 2$	3. Angle Add. Post.
4. $m\angle 1 + m\angle 2 = 90$	4. Substitution
5. $\angle 1$ and $\angle 2$ are complementary angles.	5. Def. of complementary $\angle$s

13. 62 **15.** 28 **17.** $m\angle 4 = 52$ **19.** $m\angle 9 = 86$, $m\angle 10 = 94$ **21.** $m\angle 13 = 112$, $m\angle 14 = 112$ **23.** $m\angle 17 = 53$, $m\angle 18 = 53$

25. Given: $\angle A$
Prove: $\angle A \cong \angle A$

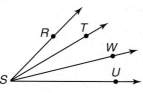

Proof:

Statements	Reasons
1. $\angle A$ is an angle.	1. Given
2. $m\angle A = m\angle A$	2. Reflexive Prop.
3. $\angle A \cong \angle A$	3. Def. of $\cong$ angles

27. sometimes   **29.** always   **31.** sometimes

33. Given: $\ell \perp m$
Prove: $\angle 2$, $\angle 3$, and $\angle 4$ are rt. $\angle$s.

Proof:

Statements	Reasons
1. $\ell \perp m$	1. Given
2. $\angle 1$ is a right angle.	2. Def. of $\perp$ lines
3. $m\angle 1 = 90$	3. Def. of rt. $\angle$s
4. $\angle 1 \cong \angle 4$	4. Vert. $\angle$s are $\cong$.
5. $m\angle 1 = m\angle 4$	5. Def. of $\cong$ $\angle$s
6. $m\angle 4 = 90$	6. Substitution
7. $\angle 1$ and $\angle 2$ form a linear pair. $\angle 3$ and $\angle 4$ form a linear pair.	7. Def. of linear pair
8. $m\angle 1 + m\angle 2 = 180$, $m\angle 4 + m\angle 3 = 180$	8. Linear pairs are supplementary.
9. $90 + m\angle 2 = 180$, $90 + m\angle 3 = 180$	9. Substitution
10. $m\angle 2 = 90$, $m\angle 3 = 90$	10. Subt. Prop.
11. $\angle 2$, $\angle 3$, and $\angle 4$ are rt. $\angle$s.	11. Def. of rt. $\angle$s (steps 6, 10)

35. Given: $\ell \perp m$
Prove: $\angle 1 \cong \angle 2$

Proof:

Statements	Reasons
1. $\ell \perp m$	1. Given
2. $\angle 1$ and $\angle 2$ rt. $\angle$s	2. $\perp$ lines intersect to form 4 rt. $\angle$s.
3. $\angle 1 \cong \angle 2$	3. All rt. $\angle$s $\cong$.

37. Given: $\angle ABD \cong \angle CBD$, $\angle ABD$ and $\angle DBC$ form a linear pair.
Prove: $\angle ABD$ and $\angle CBD$ are rt. $\angle$s.

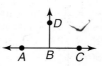

Proof:

Statements	Reasons
1. $\angle ABD \cong \angle CBD$, $\angle ABD$ and $\angle CBD$ form a linear pair.	1. Given
2. $\angle ABD$ and $\angle CBD$ are supplementary.	2. Linear pairs are supplementary.
3. $\angle ABD$ and $\angle CBD$ are rt. $\angle$s.	3. If $\angle$s are $\cong$ and suppl., they are rt. $\angle$s.

39. Given: $m\angle RSW = m\angle TSU$
Prove: $m\angle RST = m\angle WSU$

Proof:

Statements	Reasons
1. $m\angle RSW = m\angle TSU$	1. Given
2. $m\angle RSW = m\angle RST + m\angle TSW$, $m\angle TSU = m\angle TSW + m\angle WSU$	2. Angle Addition Postulate
3. $m\angle RST + m\angle TSW = m\angle TSW + m\angle WSU$	3. Substitution
4. $m\angle TSW = m\angle TSW$	4. Reflexive Prop.
5. $m\angle RST = m\angle WSU$	5. Subt. Prop.

41. Because the lines are perpendicular, the angles formed are right angles. All right angles are congruent. Therefore, $\angle 1$ is congruent to $\angle 2$. **43.** Two angles that are supplementary to the same angle are congruent. Answers should include the following.

- $\angle 1$ and $\angle 2$ are supplementary; $\angle 2$ and $\angle 3$ are supplementary.
- $\angle 1$ and $\angle 3$ are vertical angles, and are therefore congruent.
- If two angles are complementary to the same angle, then the angles are congruent. **45.** B

47. Given: X is the midpoint of $\overline{WY}$.
Prove: $WX + YZ = XZ$

Proof:

Statements	Reasons
1. X is the midpoint of $\overline{WY}$.	1. Given
2. $WX = XY$	2. Def. of midpoint
3. $XY + YZ = XZ$	3. Segment Addition Postulate
4. $WX + YZ = XZ$	4. Substitution

49. $\angle ONM, \angle MNR$ **51.** N or R **53.** obtuse
55. $\angle NML, \angle NMP, \angle NMO, \angle RNM, \angle ONM$

Pages 115–120 Chapter 2 Study Guide and Review
1. conjecture **3.** compound **5.** hypothesis **7.** Postulates
9. $m\angle A + m\angle B = 180$ **11.** $LMNO$ is a square.

13. In a right triangle with right angle C, $a^2 + b^2 = c^2$ or the sum of the measures of two supplementary angles is 180; true. **15.** $-1 > 0$, and in a right triangle with right angle C, $a^2 + b^2 = c^2$, or the sum of the measures of two supplementary angles is 180; false. **17.** In a right triangle with right angle C, $a^2 + b^2 = c^2$ and the sum of the measures of two supplementary angles is 180, and $-1 > 0$; false. **19.** Converse: If a month has 31 days, then it is March. False; July has 31 days. Inverse: If a month is not March, then it does not have 31 days. False; July has 31 days. Contrapositive: If a month does not have 31 days, then it is not March; true. **21.** true **23.** false **25.** Valid; by definition, adjacent angles have a common vertex. **27.** yes; Law of Detachment **29.** yes; Law of Syllogism **31.** Always; if P is the midpoint of $\overline{XY}$, then $\overline{XP} \cong \overline{PY}$. By definition of congruent segments, $XP = PY$. **33.** Sometimes; if the points are collinear. **35.** Sometimes; if the right angles form a linear pair. **37.** Never; adjacent angles must share a common side, and vertical angles do not. **39.** Distributive Property **41.** Subtraction Property

43. Given: $5 = 2 - \frac{1}{2}x$
Prove: $x = -6$
Proof:

Statements	Reasons
1. $5 = 2 - \frac{1}{2}x$	1. Given
2. $5 - 2 = 2 - \frac{1}{2}x - 2$	2. Subt. Prop.
3. $3 = -\frac{1}{2}x$	3. Substitution

4. $-2(3) = -2\left(-\frac{1}{2}x\right)$	4. Mult. Prop
5. $-6 = x$	5. Substitution
6. $x = -6$	6. Symmetric Prop.

45. Given: $AC = AB$, $AC = 4x + 1$,
$AB = 6x - 13$
Prove: $x = 7$

Proof:

Statements	Reasons
1. $AC = AB$, $AC = 4x + 1$, $AB = 6x - 13$	1. Given
2. $4x + 1 = 6x - 13$	2. Substitution
3. $4x + 1 - 1 = 6x - 13 - 1$	3. Subt. Prop.
4. $4x = 6x - 14$	4. Substitution
5. $4x - 6x = 6x - 14 - 6x$	5. Subt. Prop.
6. $-2x = -14$	6. Substitution
7. $\frac{-2x}{-2} = \frac{-14}{-2}$	7. Div. Prop.
8. $x = 7$	8. Substitution

47. Reflexive Property **49.** Addition Property
51. Division or Multiplication Property

53. Given: $BC = EC$, $CA = CD$
Prove: $BA = DE$

Proof:

Statements	Reasons
1. $BC = EC$, $CA = CD$	1. Given
2. $BC + CA = EC + CA$	2. Add. Prop.
3. $BC + CA = EC + CD$	3. Substitution
4. $BC + CA = BA$ $EC + CD = DE$	4. Seg. Add. Post.
5. $BA = DE$	5. Substitution

55. 145 **57.** 90

Chapter 3 Parallel and Perpendicular Lines

Page 125 Chapter 3 Getting Started
1. $\overleftrightarrow{PQ}$ **3.** $\overrightarrow{ST}$ **5.** $\angle 4, \angle 6, \angle 8$ **7.** $\angle 1, \angle 5, \angle 7$ **9.** 9 **11.** $-\frac{3}{2}$
Pages 128–131 Lesson 3-1
1. Sample answer: The bottom and top of a cylinder are contained in parallel planes.

3. Sample answer: looking down railroad tracks **5.** $\overline{AB}, \overline{JK}, \overline{LM}$ **7.** q and r, q and t, r and t **9.** p and r, p and t, r and t **11.** alternate interior **13.** consecutive interior **15.** p; consecutive interior **17.** q; alternate interior **19.** Sample answer: The roof and the floor are parallel planes. **21.** Sample answer: The top of the memorial "cuts" the pillars. **23.** $ABC, ABQ, PQR, CDS,$ APU, DET **25.** $\overline{AP}, \overline{BQ}, \overline{CR}, \overline{FU}, \overline{PU}, \overline{QR}, \overline{RS}, \overline{TU}$ **27.** $\overline{BC},$ $\overline{CD}, \overline{DE}, \overline{EF}, \overline{QR}, \overline{RS}, \overline{ST}, \overline{TU}$ **29.** a and c, a and r, r and c **31.** a and b, a and c, b and c **33.** alternate exterior **35.** corresponding **37.** alternate interior **39.** consecutive interior **41.** p; alternate interior **43.** ℓ; alternate exterior **45.** q; alternate interior **47.** m; consecutive interior **49.** $\overline{CG}, \overline{DH}, \overline{EI}$ **51.** No; plane ADE will intersect all the planes if they are extended. **53.** infinite number

55. Sample answer: Parallel lines and planes are used in architecture to make structures that will be stable. Answers should include the following.
- Opposite walls should form parallel planes; the floor may be parallel to the ceiling.
- The plane that forms a stairway will not be parallel to some of the walls.

57. 16, 20, or 28

59. Given: $\overline{PQ} \cong \overline{ZY}$, $\overline{QR} \cong \overline{XY}$
Prove: $\overline{PR} \cong \overline{XZ}$

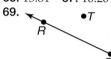

Proof: Since $\overline{PQ} \cong \overline{ZY}$ and $\overline{QR} \cong \overline{XY}$, $PQ = ZY$ and $QR = XY$ by the definition of congruent segments. By the Addition Property, $PQ + QR = ZY + XY$. Using the Segment Addition Postulate, $PR = PQ + QR$ and $XZ = XY + YZ$. By substitution, $PR = XZ$. Because the measures are equal, $\overline{PR} \cong \overline{XZ}$ by the definition of congruent segments.

61. $m\angle EFG$ is less than 90; Detachment. **63.** 8.25
65. 15.81 **67.** 10.20

69.

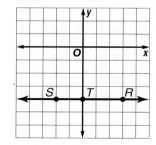

71. 90, 90 **73.** 72, 108
75. 76, 104

Pages 136–138 Lesson 3-2

1. Sometimes; if the transversal is perpendicular to the parallel lines, then $\angle 1$ and $\angle 2$ are right angles and are congruent. **3.** 1 **5.** 110 **7.** 70 **9.** 55 **11.** $x = 13$, $y = 6$ **13.** 67 **15.** 75 **17.** 105 **19.** 105 **21.** 43 **23.** 43 **25.** 137 **27.** 60 **29.** 70 **31.** 120 **33.** $x = 34$, $y = \pm 5$ **35.** 113 **37.** $x = 14$, $y = 11$, $z = 73$ **39.** (1) Given (2) Corresponding Angles Postulate (3) Vertical Angles Theorem (4) Transitive Property

41. Given: $\ell \perp m$, $m \parallel n$
Prove: $\ell \perp n$

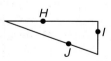

Proof: Since $\ell \perp m$, we know that $\angle 1 \cong \angle 2$, because perpendicular lines form congruent right angles. Then by the Corresponding Angles Postulate, $\angle 1 \cong \angle 3$ and $\angle 2 \cong \angle 4$. By the definition of congruent angles, $m\angle 1 = m\angle 2$, $m\angle 1 = m\angle 3$, and $m\angle 2 = m\angle 4$. By substitution, $m\angle 3 = m\angle 4$. Because $\angle 3$ and $\angle 4$ form a congruent linear pair, they are right angles. By definition, $\ell \perp n$.

43. $\angle 2$ and $\angle 6$ are consecutive interior angles for the same transversal, which makes them supplementary because $\overline{WX} \parallel \overline{YZ}$. $\angle 4$ and $\angle 6$ are not necessarily supplementary because $\overline{XY}$ may not be parallel to $\overline{WZ}$. **45.** C **47.** $\overline{FG}$ **49.** CDH **51.** $m\angle 1 = 56$ **53.** H: it rains this evening; C: I will mow the lawn tomorrow **55.** $-\dfrac{2}{3}$ **57.** $\dfrac{3}{8}$ **59.** $-\dfrac{4}{5}$

Page 138 Practice Quiz 1

1. p; alternate exterior **3.** q; alternate interior **5.** 75

Pages 142–144 Lesson 3-3

1. horizontal; vertical **3.** horizontal line, vertical line
5. $-\dfrac{1}{2}$ **7.** 2 **9.** parallel

11.

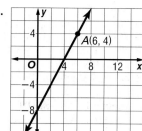

13. (1500, −120) or (−1500, −120)
15. $\dfrac{1}{7}$ **17.** −5
19. perpendicular
21. neither **23.** parallel
25. −3 **27.** 6 **29.** 6
31. undefined

33.

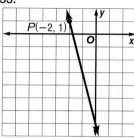

35.

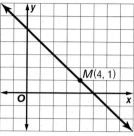

37.

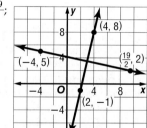

39. Sample answer: 0.24
41. 2016

43. $\dfrac{19}{2}$;

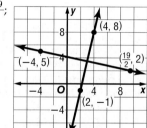

45. 2001
47. $y = \dfrac{1}{2}x - \dfrac{11}{2}$
49. C **51.** 131 **53.** 49
55. 49 **57.** ℓ; alternate exterior
59. p; alternate interior
61. m; alternate interior

63. H, I, and J are noncollinear.

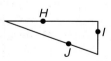

65. R, S, and T are collinear.

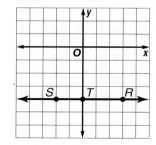

67. obtuse **69.** obtuse
71. $y = -\dfrac{1}{2}x - \dfrac{5}{4}$

Pages 147–150 Lesson 3-4

1. Sample answer: Use the point-slope form where $(x_1, y_1) = (-2, 8)$ and $m = -\dfrac{2}{5}$.

3. Sample answer: $y = x$

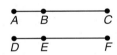

5. $y = -\frac{3}{5}x - 2$

7. $y + 1 = \frac{3}{2}(x - 4)$

9. $y - 137.5 = 1.25(x - 20)$

11. $y = -x + 2$

13. $y = 39.95$, $y = 0.95x + 4.95$

15. $y = \frac{1}{6}x - 4$

17. $y = \frac{5}{8}x - 6$

19. $y = -x - 3$

21. $y - 1 = 2(x - 3)$ **23.** $y + 5 = -\frac{4}{5}(x + 12)$
25. $y - 17.12 = 0.48(x - 5)$ **27.** $y = -3x - 2$
29. $y = 2x - 4$ **31.** $y = -x + 5$ **33.** $y = -\frac{1}{8}x$
35. $y = -3x + 5$ **37.** $y = -\frac{3}{5}x + 3$
39. $y = -\frac{1}{5}x - 4$ **41.** no slope-intercept form, $x = -6$
43. $y = \frac{2}{5}x - \frac{24}{5}$ **45.** $y = 0.05x + 750$, where $x =$ total
price of appliances sold   **47.** $y = -750x + 10,800$ **49.** in
10 days   **51.** $y = x - 180$ **53.** Sample answer: In the
equation of a line, the b value indicates the fixed rate, while
the mx value indicates charges based on usage. Answers
should include the following.

- The fee for air time can be considered the slope of the
  equation.
- We can find where the equations intersect to see where
  the plans would be equal.

55. B   **57.** undefined   **59.** 58   **61.** 75   **63.** 73

65. Given: $AC = DF$, $AB = DE$
Prove: $BC = EF$

Proof:

Statements	Reasons
1. $AC = DF$, $AB = DE$	**1.** Given
2. $AC = AB + BC$ $DF = DE + EF$	**2.** Segment Addition Postulate
3. $AB + BC = DE + EF$	**3.** Substitution Property
4. $BC = EF$	**4.** Subtraction Property

67. 26.69   **69.** $\angle 1$ and $\angle 5$, $\angle 2$ and $\angle 6$, $\angle 4$ and $\angle 8$, $\angle 3$
and $\angle 7$ **71.** $\angle 2$ and $\angle 8$, $\angle 3$ and $\angle 5$

Page 150   Practice Quiz 2
1. neither   **3.** $\frac{7}{2}$ **5.** $\frac{5}{4}$ **7.** $y = -\frac{4}{5}x + \frac{16}{5}$
9. $y + 8 = -\frac{1}{4}(x - 5)$

Pages 154–157   Lesson 3-5
1. Sample answer: Use a pair of alternate exterior $\angle$s that
are $\cong$ and cut by a transversal; show that a pair of
consecutive interior $\angle$s are suppl.; show that alternate
interior $\angle$s are $\cong$; show two lines are $\perp$ to same line; show
corresponding $\angle$s are $\cong$.   **3.** Sample answer: A basketball
court has parallel lines, as does a newspaper. The edges
should be equidistant along the entire line.   **5.** $\ell \parallel m$; $\cong$ alt.
int. $\angle$s   **7.** $p \parallel q$; $\cong$ alt. ext. $\angle$s   **9.** 11.375   **11.** The slope of $\overleftrightarrow{CD}$
is $\frac{1}{8}$, and the slope of line $\overleftrightarrow{AB}$ is $\frac{1}{7}$. The slopes are not equal,
so the lines are not parallel.   **13.** $a \parallel b$; $\cong$ alt. int. $\angle$s
15. $\ell \parallel m$; $\cong$ corr. $\angle$s   **17.** $\overleftrightarrow{AE} \parallel \overleftrightarrow{BF}$; $\cong$ corr. $\angle$s
19. $\overleftrightarrow{AC} \parallel \overleftrightarrow{EG}$; $\cong$ alt. int. $\angle$s   **21.** $\overleftrightarrow{HS} \parallel \overleftrightarrow{JT}$; $\cong$ corr. $\angle$s
23. $\overleftrightarrow{KN} \parallel \overleftrightarrow{PR}$; suppl. cons. int. $\angle$s

25. 1. Given
2. Definition of perpendicular
3. All rt. $\angle$s are $\cong$.
4. If corresponding $\angle$s are $\cong$, then lines are $\parallel$.

27. 15   **29.** -8 **31.** 21.6

33. Given: $\angle 4 \cong \angle 6$
Prove: $\ell \parallel m$

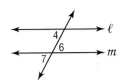

Proof: We know that $\angle 4 \cong \angle 6$. Because $\angle 6$ and $\angle 7$ are
vertical angles they are congruent. By the Transitive
Property of Congruence, $\angle 4 \cong \angle 7$. Since $\angle 4$ and $\angle 7$
are corresponding angles, and they are congruent, $\ell \parallel m$.

35. Given: $\overline{AD} \perp \overline{CD}$
$\angle 1 \cong \angle 2$
Prove: $\overline{BC} \perp \overline{CD}$

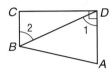

Proof:

Statements	Reasons
1. $\overline{AD} \perp \overline{CD}$, $\angle 1 \cong \angle 2$	**1.** Given
2. $\overline{AD} \parallel \overline{BC}$	**2.** If alternate interior $\angle$s are $\cong$, lines are $\parallel$.
3. $\overline{BC} \perp \overline{CD}$	**3.** Perpendicular Transversal Th.

37. Given: $\angle RSP \cong \angle PQR$
$\angle QRS$ and $\angle PQR$
are supplementary.
Prove: $\overline{PS} \parallel \overline{QR}$

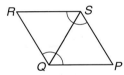

Proof:

Statements	Reasons
1. $\angle RSP \cong \angle PQR$ $\angle QRS$ and $\angle PQR$ are supplementary.	**1.** Given
2. $m\angle RSP = m\angle PQR$	**2.** Def. of $\cong$ $\angle$s
3. $m\angle QRS + m\angle PQR = 180$	**3.** Def. of suppl. $\angle$s
4. $m\angle QRS + m\angle RSP = 180$	**4.** Substitution
5. $\angle QRS$ and $\angle RSP$ are supplementary.	**5.** Def. of suppl. $\angle$s
6. $\overline{PS} \parallel \overline{QR}$	**6.** If consecutive interior $\angle$s are suppl., lines $\parallel$.

39. No, the slopes are not the same.   **41.** The 10-yard lines
will be parallel because they are all perpendicular to the
sideline and two or more lines perpendicular to the same
line are parallel.   **43.** See students' work.   **45.** B
47. $y = 0.3x - 6$ **49.** $y = -\frac{1}{2}x + \frac{19}{2}$ **51.** $-\frac{5}{4}$ **53.** 1

55. undefined

57.

p	q	p and q
T	T	T
T	F	F
F	T	F
F	F	F

59.

p	q	$\sim p$	$\sim p \wedge q$
T	T	F	F
T	F	F	F
F	T	T	T
F	F	T	F

61. complementary angles   **63.** $\sqrt{85} \approx 9.22$

Pages 162–164   Lesson 3-6
1. Construct a perpendicular line between them.
3. Sample answer: Measure distances at different parts; compare slopes; measure angles. Finding slopes is the most readily available method.

5. 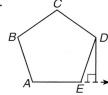   **7.** 0.9
9. 5 units;

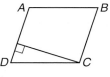

11. 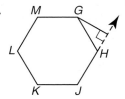   **13.**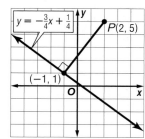

15. 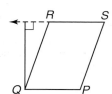   **17.** $d = 3$;

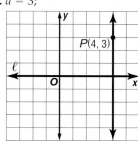

19. 4   **21.** $\sqrt{5}$ **23.** $\frac{7\sqrt{5}}{5}$

25. 1;   **27.** $\sqrt{13}$;

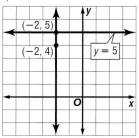   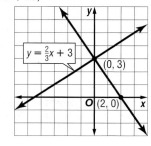

29. It is everywhere equidistant from the ceiling.   **31.** 6
33. Sample answer: We want new shelves to be parallel so they will line up. Answers should include the following.

- After marking several points, a slope can be calculated, which should be the same slope as the original brace.
- Building walls requires parallel lines.

35. D   **37.** $\overleftrightarrow{DA} \parallel \overleftrightarrow{EF}$; corresponding $\angle$s   **39.** $y = \frac{1}{2}x + 3$
41. $y = \frac{2}{3}x - 2$ **43.** $y = \frac{2}{3}x + \frac{11}{3}$

Pages 167–170   Chapter 3   Study Guide and Review
1. alternate   **3.** parallel   **5.** alternate exterior
7. consecutive   **9.** alternate exterior   **11.** corresponding
13. consecutive. interior   **15.** alternate interior   **17.** 53
19. 127   **21.** 127   **23.** neither   **25.** perpendicular

27. 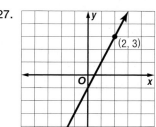   **29.** $y = 2x - 7$
31. $y = -\frac{2}{7}x + 4$
33. $y = 5x - 3$
35. $\overleftrightarrow{AL}$ and $\overleftrightarrow{BJ}$, alternate exterior $\angle$s $\cong$
37. $\overleftrightarrow{CF}$ and $\overleftrightarrow{GK}$, 2 lines $\perp$ same line
39. $\overleftrightarrow{CF}$ and $\overleftrightarrow{GK}$, consecutive interior $\angle$s suppl.   **41.** $\sqrt{5}$

Chapter 4 Congruent Triangles

Pages 177   Chapter 4   Getting Started
1. $-6\frac{1}{2}$ **3.** 1   **5.** $2\frac{3}{4}$ **7.** $\angle 2, \angle 12, \angle 15, \angle 6, \angle 9, \angle 3, \angle 13$
9. $\angle 6, \angle 9, \angle 3, \angle 13, \angle 2, \angle 8, \angle 12, \angle 15$ **11.** $\angle 11.2$
13. $\angle 14.6$

Pages 180–183   Lesson 4-1
1. Triangles are classified by sides and angles. For example, a triangle can have a right angle and have no two sides congruent.   **3.** Always; equiangular triangles have three acute angles.   **5.** obtuse   **7.** $\triangle MJK, \triangle KLM, \triangle JKN, \triangle LMN$
9. $x = 4, JM = 3, MN = 3, JN = 2$ **11.** $TW = \sqrt{125}, WZ = \sqrt{74}, TZ = \sqrt{61}$; scalene   **13.** right   **15.** acute
17. obtuse   **19.** equilateral, equiangular   **21.** isosceles, acute   **23.** $\triangle BAC, \triangle CDB$ **25.** $\triangle ABD, \triangle ACD, \triangle BAC, \triangle CDB$ **27.** $x = 5, MN = 9, MP = 9, NP = 9$
29. $x = 8, JL = 11, JK = 11, KL = 7$ **31.** Scalene; it is 184 miles from Lexington to Nashville, 265 miles from Cairo to Lexington, and 144 miles from Cairo to Nashville.
33. $AB = \sqrt{106}, BC = \sqrt{233}, AC = \sqrt{65}$; scalene
35. $AB = \sqrt{29}, BC = 4, AC = \sqrt{29}$; isosceles
37. $AB = \sqrt{124}, BC = \sqrt{124}, AC = 8$; isosceles

39. Given:
$m\angle NPM = 33$
Prove:
$\triangle RPM$ is obtuse.

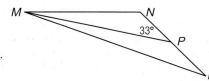

Proof: $\angle NPM$ and $\angle RPM$ form a linear pair. $\angle NPM$ and $\angle RPM$ are supplementary because if two angles form a linear pair, then they are supplementary. So, $m\angle NPM + m\angle RPM = 180$. It is given that $m\angle NPM = 33$. By substitution, $33 + m\angle RPM = 180$. Subtract to find that $m\angle RPM = 147$. $\angle RPM$ is obtuse by definition. $\triangle RPM$ is obtuse by definition.

41. $AD = \sqrt{\left(0 - \frac{a}{2}\right)^2 + (0 - b)^2}$ $CD = \sqrt{\left(a - \frac{a}{2}\right)^2 + (0 - b)^2}$

$\quad = \sqrt{\left(-\frac{a}{2}\right)^2 + (-b)^2}$ $\quad = \sqrt{\left(\frac{a}{2}\right)^2 + (-b)^2}$

$\quad = \sqrt{\frac{a^2}{4} + b^2}$ $\quad = \sqrt{\frac{a^2}{4} + b^2}$

$AD = CD$, so $\overline{AD} \cong \overline{CD}$. $\triangle ADC$ is isosceles by definition.
43. Sample answer: Triangles are used in construction as structural support. Answers should include the following.
- Triangles can be classified by sides and angles. If the measure of each angle is less than 90, the triangle is acute. If the measure of one angle is greater than 90, the triangle is obtuse. If one angle equals 90°, the triangle is right. If each angle has the same measure, the triangle is equiangular. If no two sides are congruent, the triangle is scalene. If at least two sides are congruent, it is isosceles. If all of the sides are congruent, the triangle is equilateral.
- Isosceles triangles seem to be used more often in architecture and construction.

45. B   **47.** $\sqrt{8}$;

49. 15   **51.** 44   **53.** any three: $\angle 2$ and $\angle 11$, $\angle 3$ and $\angle 6$, $\angle 4$ and $\angle 7$, $\angle 3$ and $\angle 12$, $\angle 7$ and $\angle 10$, $\angle 8$ and $\angle 11$ **55.** $\angle 6$, $\angle 9$, and $\angle 12$ **57.** $\angle 2$, $\angle 5$, and $\angle 8$

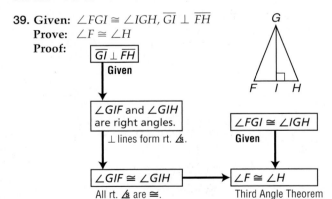

Pages 188–191   Lesson 4-2
1. Sample answer: $\angle 2$ and $\angle 3$ are the remote interior angles of exterior $\angle 1$.

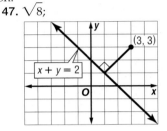

3. 43   **5.** 55   **7.** 147   **9.** 25
11. 93   **13.** 65, 65   **15.** 76
17. 49   **19.** 53   **21.** 32   **23.** 44   **25.** 123   **27.** 14   **29.** 53
31. 103   **33.** 50   **35.** 40   **37.** 129

39. Given: $\angle FGI \cong \angle IGH$, $\overline{GI} \perp \overline{FH}$
Prove: $\angle F \cong \angle H$
Proof:

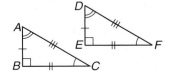

41. Given: $\triangle ABC$
Prove: $m\angle CBD = m\angle A + m\angle C$

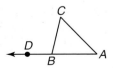

Proof:

Statements	Reasons
1. $\triangle ABC$	1. Given
2. $\angle CBD$ and $\angle ABC$ form a linear pair.	2. Def. of linear pair
3. $\angle CBD$ and $\angle ABC$ are supplementary.	3. If 2 $\angle$s form a linear pair,they are suppl.

---

4. $m\angle CBD + m\angle ABC = 180$	4. Def. of suppl.
5. $m\angle A + m\angle ABC + m\angle C = 180$	5. Angle Sum Theorem
6. $m\angle A + m\angle ABC + m\angle C = m\angle CBD + m\angle ABC$	6. Substitution
7. $m\angle A + m\angle C = m\angle CBD$	7. Subtraction Property

43. Given: $\triangle MNO$
$\quad \angle M$ is a right angle.
Prove: There can be at most one right angle in a triangle.

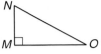

Proof:
In $\triangle MNO$, $\angle M$ is a right angle. $m\angle M + m\angle N + m\angle O = 180$. $m\angle M = 90$, so $m\angle N + m\angle O = 90$. If $\angle N$ were a right angle, then $m\angle O = 0$. But that is impossible, so there cannot be two right angles in a triangle.

Given: $\triangle PQR$
$\quad \angle P$ is obtuse.
Prove: There can be at most one obtuse angle in a triangle.

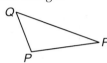

Proof:
In $\triangle PQR$, $\angle P$ is obtuse. So $m\angle P > 90$. $m\angle P + m\angle Q + m\angle R = 180$. It must be that $m\angle Q + m\angle R < 90$. So, $\angle Q$ and $\angle R$ must be acute.

45. $m\angle 1 = 48$, $m\angle 2 = 60$, $m\angle 3 = 72$ **47.** A   **49.** $\triangle AED$
51. $\triangle BEC$ **53.** $\sqrt{20}$ units   **55.** $\frac{\sqrt{117}}{13}$ units   **57.** $x = 112$, $y = 28$, $z = 22$ **59.** reflexive   **61.** symmetric   **63.** transitive

---

Pages 195–198   Lesson 4-3
1. The sides and the angles of the triangle are not affected by a congruence transformation, so congruence is preserved. **3.** $\triangle AFC \cong \triangle DFB$ **5.** $\angle W \cong \angle S$, $\angle X \cong \angle T$, $\angle Z \cong \angle J$, $\overline{WX} \cong \overline{ST}$, $\overline{XZ} \cong \overline{TJ}$, $\overline{WZ} \cong \overline{SJ}$ **7.** $QR = 5$, $Q'R' = 5$, $RT = 3$, $R'T' = 3$, $QT = \sqrt{34}$, and $Q'T' = \sqrt{34}$. Use a protractor to confirm that the corresponding angles are congruent; flip.   **9.** $\triangle CFH \cong \triangle JKL$ **11.** $\triangle WPZ \cong \triangle QVS$ **13.** $\angle T \cong \angle X$, $\angle U \cong \angle Y$, $\angle V \cong \angle Z$, $\overline{TU} \cong \overline{XY}$, $\overline{UV} \cong \overline{YZ}$, $\overline{TV} \cong \overline{XZ}$ **15.** $\angle B \cong \angle D$, $\angle C \cong \angle G$, $\angle F \cong \angle H$, $\overline{BC} \cong \overline{DG}$, $\overline{CF} \cong \overline{GH}$, $\overline{BF} \cong \overline{DH}$ **17.** $\angle 1 \cong \angle 10$, $\angle 2 \cong \angle 9$, $\angle 3 \cong \angle 8$, $\angle 4 \cong \angle 7$, $\angle 5 \cong \angle 6$ **19.** $\angle$s 1, 5, 6, and 11, $\angle$s 3, 8, 10, and 12, $\angle$s 2, 4, 7, and 9   **21.** We need to know that all of the angles are congruent and that the other corresponding sides are congruent.   **23.** Flip; $MN = 8$, $M'N' = 8$, $NP = 2$, $N'P' = 2$, $MP = \sqrt{68}$, and $M'P' = \sqrt{68}$. Use a protractor to confirm that the corresponding angles are congruent.
25. Turn; $JK = \sqrt{40}$, $J'K' = \sqrt{40}$, $KL = \sqrt{29}$, $K'L' = \sqrt{29}$, $JL = \sqrt{17}$, and $J'L' = \sqrt{17}$. Use a protractor to confirm that the corresponding angles are congruent.
27. True;

29.

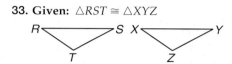

31.

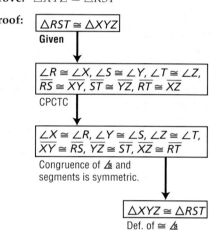

33. Given: $\triangle RST \cong \triangle XYZ$

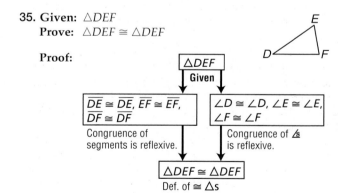

Prove: $\triangle XYZ \cong \triangle RST$

Proof:

| $\triangle RST \cong \triangle XYZ$ |
| Given |

↓

| $\angle R \cong \angle X, \angle S \cong \angle Y, \angle T \cong \angle Z,$ $\overline{RS} \cong \overline{XY}, \overline{ST} \cong \overline{YZ}, \overline{RT} \cong \overline{XZ}$ |
| CPCTC |

↓

| $\angle X \cong \angle R, \angle Y \cong \angle S, \angle Z \cong \angle T,$ $\overline{XY} \cong \overline{RS}, \overline{YZ} \cong \overline{ST}, \overline{XZ} \cong \overline{RT}$ |
| Congruence of $\angle$s and segments is symmetric. |

↓

| $\triangle XYZ \cong \triangle RST$ |
| Def. of $\cong$ $\triangle$s |

35. Given: $\triangle DEF$
Prove: $\triangle DEF \cong \triangle DEF$

Proof:

| $\triangle DEF$ |
| Given |

↓

| $\overline{DE} \cong \overline{DE}, \overline{EF} \cong \overline{EF},$ $\overline{DF} \cong \overline{DF}$ | $\angle D \cong \angle D, \angle E \cong \angle E,$ $\angle F \cong \angle F$ |
| Congruence of segments is reflexive. | Congruence of $\angle$s is reflexive. |

↓

| $\triangle DEF \cong \triangle DEF$ |
| Def. of $\cong$ $\triangle$s |

37. Sample answer: Triangles are used in bridge design for structure and support. Answers should include the following.
• The shape of the triangle does not matter.
• Some of the triangles used in the bridge supports seem to be congruent.
39. D **41.** 58 **43.** $x = 3$, $BC = 10$, $CD = 10$, $BD = 5$
45. $y = -\frac{3}{2}x + 3$ **47.** $y = -4x - 11$ **49.** $\sqrt{5}$ **51.** $\sqrt{13}$

Page 198 Chapter 4 Practice Quiz 1
1. $\triangle DFJ, \triangle GJF, \triangle HJG, \triangle DJH$ **3.** $AB = BC = AC = 7$
5. $\angle M \cong \angle J, \angle N \cong \angle K, \angle P \cong \angle L; \overline{MN} \cong \overline{JK}, \overline{NP} \cong \overline{KL},$ and $\overline{MP} \cong \overline{JL}$

Pages 203–206 Lesson 4-4
1. Sample answer: In $\triangle QRS$, $\angle R$ is the included angle of the sides $\overline{QR}$ and $\overline{RS}$.

3. $EG = 2, MP = 2, FG = 4, NP = 4, EF = \sqrt{20}$, and $MN = \sqrt{20}$. The corresponding sides have the same measure and are congruent. $\triangle EFG \cong \triangle MNP$ by SSS.
5. Given: $\overline{DE}$ and $\overline{BC}$ bisect each other
Prove: $\triangle DGB \cong \triangle EGC$
Proof:

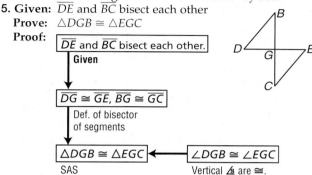

| $\overline{DE}$ and $\overline{BC}$ bisect each other. |
| Given |

↓

| $\overline{DG} \cong \overline{GE}, \overline{BG} \cong \overline{GC}$ |
| Def. of bisector of segments |

↓

| $\triangle DGB \cong \triangle EGC$ | ← | $\angle DGB \cong \angle EGC$ |
| SAS | | Vertical $\angle$s are $\cong$. |

7. SAS
9. Given: T is the midpoint of $\overline{SQ}$.
   $\overline{SR} \cong \overline{QR}$
Prove: $\triangle SRT \cong \triangle QRT$
Proof:

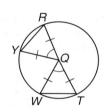

Statements	Reasons
1. T is the midpoint of $\overline{SQ}$.	1. Given
2. $\overline{ST} \cong \overline{TQ}$	2. Midpoint Theorem
3. $\overline{SR} \cong \overline{QR}$	3. Given
4. $\overline{RT} \cong \overline{RT}$	4. Reflexive Property
5. $\triangle SRT \cong \triangle QRT$	5. SSS

11. $JK = \sqrt{10}, KL = \sqrt{10}, JL = \sqrt{20}, FG = \sqrt{2}, GH = \sqrt{50},$ and $FH = 6$. The corresponding sides are not congruent so $\triangle JKL$ is not congruent to $\triangle FGH$. **13.** $JK = \sqrt{10}, KL = \sqrt{10}, JL = \sqrt{20}, FG = \sqrt{10}, GH = \sqrt{10},$ and $FH = \sqrt{20}$. Each pair of corresponding sides have the same measure so they are congruent. $\triangle JKL \cong \triangle FGH$ by SSS.

15. Given: $\overline{RQ} \cong \overline{TQ} \cong \overline{YQ} \cong \overline{WQ},$
   $\angle RQY \cong \angle WQT$
Prove: $\triangle QWT \cong \triangle QYR$

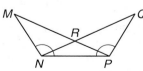

Proof:

| $\overline{RQ} \cong \overline{TQ} \cong \overline{YQ} \cong \overline{WQ}$ | $\angle RQY \cong \angle WQT$ |
| Given | Given |

↓

| $\triangle QWT \cong \triangle QYR$ |
| SAS |

17. Given: $\triangle MRN \cong \triangle QRP$
   $\angle MNP \cong \angle QPN$
Prove: $\triangle MNP \cong \triangle QPN$

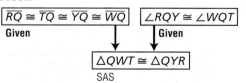

Proof:

Statement	Reason
1. $\triangle MRN \cong \triangle QRP,$ $\angle MNP \cong \angle QPN$	1. Given
2. $\overline{MN} \cong \overline{QP}$	2. CPCTC
3. $\overline{NP} \cong \overline{NP}$	3. Reflexive Property
4. $\triangle MNP \cong \triangle QPN$	4. SAS

19. Given: $\triangle GHJ \cong \triangle LKJ$
Prove: $\triangle GHL \cong \triangle LKG$

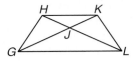

Proof:

Statement	Reason
1. $\triangle GHJ \cong \triangle LKJ$	1. Given
2. $\overline{HJ} \cong \overline{KJ}$, $\overline{GJ} \cong \overline{LJ}$, $\overline{GH} \cong \overline{LK}$,	2. CPCTC
3. $HJ = KJ$, $GJ = LJ$	3. Def. of $\cong$ segments
4. $HJ + LJ = KJ + JG$	4. Addition Property
5. $KJ + GJ = KG$; $HJ + LJ = HL$	5. Segment Addition
6. $KG = HL$	6. Substitution
7. $\overline{KG} \cong \overline{HL}$	7. Def. of $\cong$ segments
8. $\overline{GL} \cong \overline{GL}$	8. Reflexive Property
9. $\triangle GHL \cong \triangle LKG$	9. SSS

21. Given: $\overline{EF} \cong \overline{HF}$
G is the midpoint of $\overline{EH}$.
Prove: $\triangle EFG \cong \triangle HFG$

Proof:

Statements	Reasons
1. $\overline{EF} \cong \overline{HF}$; G is the midpoint of $\overline{EH}$.	1. Given
2. $\overline{EG} \cong \overline{GH}$	2. Midpoint Theorem
3. $\overline{FG} \cong \overline{FG}$	3. Reflexive Property
4. $\triangle EFG \cong \triangle HFG$	4. SSS

23. not possible **25.** SSS or SAS

27. Given: $\overline{TS} \cong \overline{SF} \cong \overline{FH} \cong \overline{HT}$
$\angle TSF$, $\angle SFH$, $\angle FHT$, and $\angle HTS$ are right angles.
Prove: $\triangle SHT \cong \triangle SHF$

Proof:

Statements	Reasons
1. $\overline{TS} \cong \overline{SF} \cong \overline{FH} \cong \overline{HT}$	1. Given
2. $\angle TSF$, $\angle SFH$, $\angle FHT$, and $\angle HTS$ are right angles.	2. Given
3. $\angle STH \cong \angle SFH$	3. All rt. $\angle$s are $\cong$.
4. $\triangle STH \cong \triangle SFH$	4. SAS
5. $\angle SHT \cong \angle SHF$	5. CPCTC

29. Sample answer: The properties of congruent triangles help land surveyors double check measurements. Answers should include the following.
• If each pair of corresponding angles and sides are congruent, the triangles are congruent by definition. If two pairs of corresponding sides and the included angle are congruent, the triangles are congruent by SAS. If each pair of corresponding sides are congruent, the triangles are congruent by SSS.
• Sample answer: Architects also use congruent triangles when designing buildings.
31. B **33.** $\triangle WXZ \cong \triangle YXZ$ **35.** 78 **37.** 68 **39.** 59
41. -1 **43.** There is a steeper rate of decline from the second quarter to the third. **45.** $\angle CBD$ **47.** $\overline{CD}$

Pages 210–213 Lesson 4-5
1. Two triangles can have corresponding congruent angles without corresponding congruent sides. $\angle A \cong \angle D$, $\angle B \cong \angle E$, and

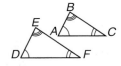

$\angle C \cong \angle F$. However, $\overline{AB} \neq \overline{DE}$, so $\triangle ABC \neq \triangle DEF$.
3. AAS can be proven using the Third Angle Theorem. Postulates are accepted as true without proof.

5. Given: $\overline{XW} \parallel \overline{YZ}$, $\angle X \cong \angle Z$
Prove: $\triangle WXY \cong \triangle YZW$

Proof:

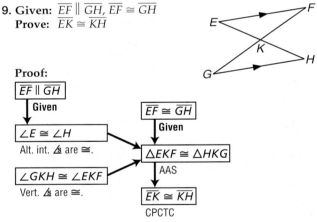

7. Given: $\angle E \cong \angle K$,
$\angle DGH \cong \angle DHG$,
$\overline{EG} \cong \overline{KH}$
Prove: $\triangle EGD \cong \triangle KHD$
Proof:
Since $\angle EGD$ and $\angle DGH$ are a linear pair, the angles are supplementary. Likewise, $\angle KHD$ and $\angle DHG$ are supplementary. We are given that $\angle DGH \cong \angle DHG$. Angles supplementary to congruent angles are congruent so $\angle EGD \cong \angle KHD$. Since we are given that $\angle E \cong \angle K$ and $\overline{EG} \cong \overline{KH}$, $\triangle EGD \cong \triangle KHD$ by ASA.

9. Given: $\overline{EF} \parallel \overline{GH}$, $\overline{EF} \cong \overline{GH}$
Prove: $\overline{EK} \cong \overline{KH}$

Proof:

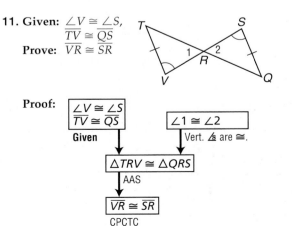

11. Given: $\angle V \cong \angle S$,
$\overline{TV} \cong \overline{QS}$
Prove: $\overline{VR} \cong \overline{SR}$

13. Given: $\overline{MN} \cong \overline{PQ}$, $\angle M \cong \angle Q$
$\angle 2 \cong \angle 3$
Prove: $\triangle MLP \cong \triangle QLN$

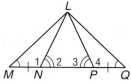

Proof:

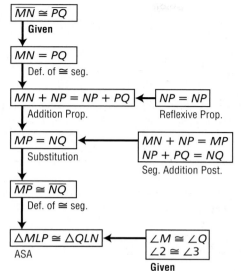

$\boxed{\overline{MN} \cong \overline{PQ}}$
↓ **Given**

$\boxed{MN = PQ}$
↓ **Def. of ≅ seg.**

$\boxed{MN + NP = NP + PQ}$ ← $\boxed{NP = NP}$
↓ **Addition Prop.**   **Reflexive Prop.**

$\boxed{MP = NQ}$ ← $\boxed{\begin{array}{l} MN + NP = MP \\ NP + PQ = NQ \end{array}}$
↓ **Substitution**   **Seg. Addition Post.**

$\boxed{\overline{MP} \cong \overline{NQ}}$
↓ **Def. of ≅ seg.**

$\boxed{\triangle MLP \cong \triangle QLN}$ ← $\boxed{\begin{array}{l} \angle M \cong \angle Q \\ \angle 2 \cong \angle 3 \end{array}}$
ASA   **Given**

15. Given: $\angle NOM \cong \angle POR$,
$\overline{NM} \perp \overline{MR}$,
$\overline{PR} \perp \overline{MR}$,
$\overline{NM} \cong \overline{PR}$

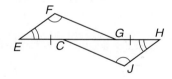

Prove: $\overline{MO} \cong \overline{OR}$
Proof: Since $\overline{NM} \perp \overline{MR}$ and $\overline{PR} \perp \overline{MR}$, $\angle M$ and $\angle R$ are right angles. $\angle M \cong \angle R$ because all right angles are congruent. We know that $\angle NOM \cong \angle POR$ and $\overline{NM} \cong \overline{PR}$. By AAS, $\triangle NMO \cong \triangle PRO$. $\overline{MO} \cong \overline{OR}$ by CPCTC.

17. Given: $\angle F \cong \angle J$,
$\angle E \cong \angle H$,
$\overline{EC} \cong \overline{GH}$
Prove: $\overline{EF} \cong \overline{HJ}$

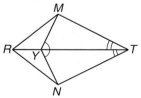

Proof: We are given that $\angle F \cong \angle J$, $\angle E \cong \angle H$, and $\overline{EC} \cong \overline{GH}$. By the Reflexive Property, $\overline{CG} \cong \overline{CG}$. Segment addition results in $EG = EC + CG$ and $CH = CG + GH$. By the definition of congruence, $EC = GH$ and $CG = CG$. Substitute to find $EG = CH$. By AAS, $\triangle EFG \cong \triangle HJC$. By CPCTC, $\overline{EF} \cong \overline{HJ}$.

19. Given: $\angle MYT \cong \angle NYT$
$\angle MTY \cong \angle NTY$
Prove: $\triangle RYM \cong \triangle RYN$

Proof:

Statement	Reason
1. $\angle MYT \cong \angle NYT$ $\angle MTY \cong \angle NTY$	1. Given
2. $\overline{YT} \cong \overline{YT}$, $\overline{RY} \cong \overline{RY}$	2. Reflexive Property
3. $\triangle MYT \cong \triangle NYT$	3. ASA
4. $\overline{MY} \cong \overline{NY}$	4. CPCTC
5. $\angle RYM$ and $\angle MYT$ are a linear pair; $\angle RYN$ and $\angle NYT$ are a linear pair	5. Def. of linear pair

6. $\angle RYM$ and $\angle MYT$ are supplementary and $\angle RYN$ and $\angle NYT$ are supplementary. — **6.** Supplement Theorem
7. $\angle RYM \cong \angle RYN$ — **7.** $\angle s$ suppl. to $\cong$ $\angle s$ are $\cong$.
8. $\triangle RYM \cong \triangle RYN$ — **8.** SAS

21. $\overline{CD} \cong \overline{GH}$, because the segments have the same measure. $\angle CFD \cong \angle HFG$ because vertical angles are congruent. Since F is the midpoint of $\overline{DG}$, $\overline{DF} \cong \overline{FG}$. It cannot be determined whether $\triangle CFD \cong \triangle HFG$. The information given does not lead to a unique triangle.

23. Since N is the midpoint of $\overline{JL}$, $\overline{JN} \cong \overline{NL}$. $\angle JNK \cong \angle LNK$ because perpendicular lines form right angles and right angles are congruent. By the Reflexive Property, $\overline{KN} \cong \overline{KN}$. $\triangle JKN \cong \triangle LKN$ by SAS. **25.** $\triangle VNR$, AAS or ASA **27.** $\triangle MIN$, SAS **29.** Since Aiko is perpendicular to the ground, two right angles are formed and right angles are congruent. The angles of sight are the same and her height is the same for each triangle. The triangles are congruent by ASA. By CPCTC, the distances are the same. The method is valid. **31.** D

33. Given: $\overline{BA} \cong \overline{DE}$,
$\overline{DA} \cong \overline{BE}$
Prove: $\triangle BEA \cong \triangle DAE$

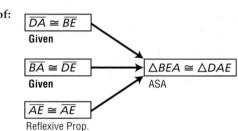

Proof:

$\boxed{\overline{DA} \cong \overline{BE}}$
Given

$\boxed{\overline{BA} \cong \overline{DE}}$ → $\boxed{\triangle BEA \cong \triangle DAE}$
Given   **ASA**

$\boxed{\overline{AE} \cong \overline{AE}}$
Reflexive Prop.

35. Turn; $RS = \sqrt{2}$, $R'S' = \sqrt{2}$, $ST = 1$, $S'T' = 1$, $RT = 1$, $R'T' = 1$. Use a protractor to confirm that the corresponding angles are congruent. **37.** If people are happy, then they rarely correct their faults. **39.** isosceles **41.** isosceles

---

Pages 219–221 Lesson 4-6
1. The measure of only one angle must be given in an isosceles triangle to determine the measures of the other two angles. **3.** Sample answer: Draw a line segment. Set your compass to the length of the line segment and draw an arc from each endpoint. Draw segments from the intersection of the arcs to each endpoint. **5.** $\overline{BH} \cong \overline{BD}$
7. Given: $\triangle CTE$ is isosceles with vertex $\angle C$.
$m\angle T = 60$
Prove: $\triangle CTE$ is equilateral.

Proof:

Statements	Reasons
1. $\triangle CTE$ is isosceles with vertex $\angle C$.	1. Given
2. $\overline{CT} \cong \overline{CE}$	2. Def. of isosceles triangle
3. $\angle E \cong \angle T$	3. Isosceles Triangle Theorem
4. $m\angle E = m\angle T$	4. Def. of $\cong$ $\angle s$

5. $m\angle T = 60$
6. $m\angle E = 60$
7. $m\angle C + m\angle E +$ $m\angle T = 180$
8. $m\angle C + 60 + 60 = 180$
9. $m\angle C = 60$
10. $\triangle CTE$ is equiangular.
11. $\triangle CTE$ is equilateral.

5. Given
6. Substitution
7. Angle Sum Theorem
8. Substitution
9. Subtraction
10. Def. of equiangular $\triangle$
11. Equiangular $\triangle$s are equilateral.

9. $\angle LTR \cong \angle LRT$ **11.** $\angle LSQ \cong \angle LQS$ **13.** $\overline{LS} \cong \overline{LR}$
15. 20 **17.** 81 **19.** 28 **21.** 56 **23.** 36.5 **25.** 38
27. $x = 3; y = 18$

29. Given: $\triangle XKF$ is equilateral.
$\overline{XJ}$ bisects $\angle KXF$.
Prove: J is the midpoint of $\overline{KF}$.

Proof:

Statements	Reasons
1. $\triangle XKF$ is equilateral.	1. Given
2. $\overline{KX} \cong \overline{FX}$	2. Definition of equilateral $\triangle$
3. $\angle 1 \cong \angle 2$	3. Isosceles Triangle Theorem
4. $\overline{XJ}$ bisects $\angle X$	4. Given
5. $\angle KXJ \cong \angle FXJ$	5. Def. of $\angle$ bisector
6. $\triangle KXJ \cong \triangle FXJ$	6. ASA
7. $\overline{KJ} \cong \overline{JF}$	7. CPCTC
8. J is the midpoint of $\overline{KF}$.	8. Def. of midpoint

31. Case I:
Given: $\triangle ABC$ is an equilateral triangle.
Prove: $\triangle ABC$ is an equiangular triangle.

Proof:

Statements	Reasons
1. $\triangle ABC$ is an equilateral triangle.	1. Given
2. $\overline{AB} \cong \overline{AC} \cong \overline{BC}$	2. Def. of equilateral $\triangle$
3. $\angle A \cong \angle B$, $\angle B \cong \angle C$, $\angle A \cong \angle C$	3. Isosceles Triangle Theorem
4. $\angle A \cong \angle B \cong \angle C$	4. Substitution
5. $\triangle ABC$ is an equiangular $\triangle$.	5. Def. of equiangular $\triangle$

Case II:
Given: $\triangle ABC$ is an equiangular triangle.
Prove: $\triangle ABC$ is an equilateral triangle.

Proof:

Statements	Reasons
1. $\triangle ABC$ is an equiangular triangle.	1. Given
2. $\angle A \cong \angle B \cong \angle C$	2. Def. of equiangular $\triangle$
3. $\overline{AB} \cong \overline{AC}$, $\overline{AB} \cong \overline{BC}$, $\overline{AC} \cong \overline{BC}$	3. Conv. of Isos. $\triangle$ Th.
4. $\overline{AB} \cong \overline{AC} \cong \overline{BC}$	4. Substitution
5. $\triangle ABC$ is an equilateral $\triangle$.	5. Def. of equilateral $\triangle$

33. Given: $\triangle ABC$
$\angle A \cong \angle C$
Prove: $\overline{AB} \cong \overline{CB}$

Proof:

Statements	Reasons
1. Let $\overrightarrow{BD}$ bisect $\angle ABC$.	1. Protractor Postulate
2. $\angle ABD \cong \angle CBD$	2. Def. of $\angle$ bisector
3. $\angle A \cong \angle C$	3. Given
4. $\overline{BD} \cong \overline{BD}$	4. Reflexive Property
5. $\triangle ABD \cong \triangle CBD$	5. AAS
6. $\overline{AB} \cong \overline{CB}$	6. CPCTC

35. 18 **37.** 30 **39.** The triangles in each set appear to be acute. **41.** Sample answer: Artists use angles, lines, and shapes to create visual images. Answers should include the following.
- Rectangle, squares, rhombi, and other polygons are used in many works of art.
- There are two rows of isosceles triangles in the painting. One row has three congruent isosceles triangles. The other row has six congruent isosceles triangles.

43. D
45. Given: $\overline{VR} \perp \overline{RS}$, $\overline{UT} \perp \overline{SU}$, $\overline{RS} \cong \overline{US}$
Prove: $\triangle VRS \cong \triangle TUS$

Proof: We are given that $\overline{VR} \perp \overline{RS}$, $\overline{UT} \perp \overline{SU}$, and $\overline{RS} \cong \overline{US}$. Perpendicular lines form four right angles so $\angle R$ and $\angle U$ are right angles. $\angle R \cong \angle U$ because all right angles are congruent. $\angle RSV \cong \angle UST$ since vertical angles are congruent. Therefore, $\triangle VRS \cong \triangle TUS$ by ASA.
47. $QR = \sqrt{52}$, $RS = \sqrt{2}$, $QS = \sqrt{34}$, $EG = \sqrt{34}$, $GH = \sqrt{10}$, and $EH = \sqrt{52}$. The corresponding sides are not congruent so $\triangle QRS$ is not congruent to $\triangle EGH$.

49.

p	q	~ p	~ q	~ p or ~ q
T	T	F	F	F
T	F	F	T	T
F	T	T	F	T
F	F	T	T	T

51.

y	z	~ y	~ y or z
T	T	F	T
T	F	F	F
F	T	T	T
F	F	T	T

53. $(-1, -3)$

Page 221   Chapter 4   Practice Quiz 2
1. $JM = \sqrt{5}$, $ML = \sqrt{26}$, $JL = 5$, $BD = \sqrt{5}$, $DG = \sqrt{26}$, and $BG = 5$. Each pair of corresponding sides have the same measure so they are congruent. $\triangle JML \cong \triangle BDG$ by SSS. **3.** 52 **5.** 26

Pages 224–226   Lesson 4-7
1. Place one vertex at the origin, place one side of the triangle on the positive x-axis. Label the coordinates with expressions that will simplify the computations.

3.

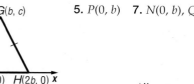

$G(b, c)$, $F(0, 0)$, $H(2b, 0)$

5. $P(0, b)$ **7.** $N(0, b)$, $Q(a, 0)$

9. Given: △ABC
Prove: △ABC is isosceles.

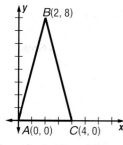

$B(2, 8)$, $A(0, 0)$, $C(4, 0)$

Proof: Use the Distance Formula to find AB and BC.
$AB = \sqrt{(2 - 0)^2 + (8 - 0)^2} = \sqrt{4 + 64}$ or $\sqrt{68}$
$BC = \sqrt{(4 - 2)^2 + (0 - 8)^2} = \sqrt{4 + 64}$ or $\sqrt{68}$
Since $AB = BC$, $\overline{AB} \cong \overline{BC}$. Since the legs are congruent, △ABC is isosceles.

11.

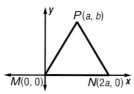

$P(a, b)$, $M(0, 0)$, $N(2a, 0)$

13.

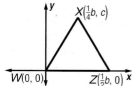

$X(\frac{1}{4}b, c)$, $W(0, 0)$, $Z(\frac{1}{2}b, 0)$

15.

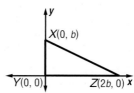

$X(0, b)$, $Y(0, 0)$, $Z(2b, 0)$

17. $Q(a, a)$, $P(a, 0)$
19. $D(2b, 0)$ **21.** $P(0, c)$, $N(2b, 0)$ **23.** $J(c, b)$

25. Given: isosceles △ABC with $\overline{AC} \cong \overline{BC}$
R and S are midpoints of legs $\overline{AC}$ and $\overline{BC}$.
Prove: $\overline{AS} \cong \overline{BR}$

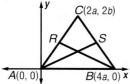

$C(2a, 2b)$, R, S, $A(0, 0)$, $B(4a, 0)$

Proof:
The coordinates of R are $\left(\frac{2a + 0}{2}, \frac{2b + 0}{2}\right)$ or (a, b).
The coordinates of S are $\left(\frac{2a + 4a}{2}, \frac{2b + 0}{2}\right)$ or $(3a, b)$.
$BR = \sqrt{(4a - a)^2 + (0 - b)^2} = \sqrt{(3a)^2 + (-b)^2}$
    or $\sqrt{9a^2 + b^2}$
$AS = \sqrt{(3a - 0)^2 + (b - 0)^2} = \sqrt{(3a)^2 + (b)^2}$
    or $\sqrt{9a^2 + b^2}$
Since $BR = AS$, $\overline{AS} \cong \overline{BR}$.

27. Given: △ABC
S is the midpoint of $\overline{AC}$.
T is the midpoint of $\overline{BC}$.
Prove: $\overline{ST} \parallel \overline{AB}$

$C(b, c)$, S, T, $A(0, 0)$, $B(a, 0)$

Proof:
Midpoint S is $\left(\frac{b + 0}{2}, \frac{c + 0}{2}\right)$ or $\left(\frac{b}{2}, \frac{c}{2}\right)$.

Midpoint T is $\left(\frac{a + b}{2}, \frac{c + 0}{2}\right)$ or $\left(\frac{a + b}{2}, \frac{c}{2}\right)$.

Slope of $\overline{ST} = \dfrac{\frac{c}{2} - \frac{c}{2}}{\frac{a + b}{2} - \frac{b}{2}} = \dfrac{0}{\frac{a}{2}}$ or 0.

Slope of $\overline{AB} = \dfrac{0 - 0}{a - 0} = \dfrac{0}{a}$ or 0.

$\overline{ST}$ and $\overline{AB}$ have the same slope so $\overline{ST} \parallel \overline{AB}$.

29. Given: △ABD, △FBD
$AF = 6$, $BD = 3$
Prove: △ABD ≅ △FBD

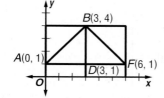

$B(3, 4)$, $A(0, 1)$, $D(3, 1)$, $F(6, 1)$, O

Proof: $\overline{BD} \cong \overline{BD}$ by the Reflexive Property.
$AD = \sqrt{(3 - 0)^2 + (1 - 1)^2} = \sqrt{9 + 0}$ or 3
$DF = \sqrt{(6 - 3)^2 + (1 - 1)^2} = \sqrt{9 + 0}$ or 3
Since $AD = DF$, $\overline{AD} \cong \overline{DF}$.
$AB = \sqrt{(3 - 0)^2 + (4 - 1)^2} = \sqrt{9 + 9}$ or $3\sqrt{2}$
$BF = \sqrt{(6 - 3)^2 + (1 - 4)^2} = \sqrt{9 + 9}$ or $3\sqrt{2}$
Since $AB = BF$, $\overline{AB} \cong \overline{BF}$.
△ABD ≅ △FBD by SSS.

31. Given: △BPR, △BAR
$PR = 800$, $BR = 800$, $RA = 800$
Prove: $\overline{PB} \cong \overline{BA}$
Proof:
$PB = \sqrt{(800 - 0)^2 + (800 - 0)^2}$ or $\sqrt{1{,}280{,}000}$
$BA = \sqrt{(800 - 1600)^2 + (800 - 0)^2}$ or $\sqrt{1{,}280{,}000}$
$PB = BA$, so $\overline{PB} \cong \overline{BA}$.

33. $\sqrt{680{,}000}$ or about 824.6 ft   **35.** $(2a, 0)$ **37.** $AB = 4a$;
$AC = \sqrt{(0 - (-2a))^2 + (2a - 0)^2} = \sqrt{4a^2 + 4a^2}$ or
$\sqrt{8a^2}$; $CB = \sqrt{(0 - 2a)^2 + (2a - 0)^2} = \sqrt{4a^2 + 4a^2}$ or
$\sqrt{8a^2}$; Slope of $\overline{AC} = \dfrac{2a - 0}{0 - (-2a)}$ or 1; slope of $\overline{CB} = \dfrac{2a - 0}{0 - 2a}$
or -1. $\overline{AC} \perp \overline{CB}$ and $\overline{AC} \cong \overline{CB}$, so △ABC is a right isosceles triangle.   **39.** C

41. Given: $\angle 3 \cong \angle 4$
Prove: $\overline{QR} \cong \overline{QS}$

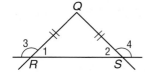

Q, 3, 1, R, 2, 4, S

Proof:

Statements	Reasons
1. $\angle 3 \cong \angle 4$	1. Given
2. $\angle 2$ and $\angle 4$ form a linear pair. $\angle 1$ and $\angle 3$ form a linear pair.	2. Def. of linear pair
3. $\angle 2$ and $\angle 4$ are supplementary. $\angle 1$ and $\angle 3$ are supplementary.	3. If 2 ⩘ form a linear pair, then they are suppl.
4. $\angle 2 \cong \angle 1$	4. Angles that are suppl. to ≅ ⩘ are ≅.
5. $\overline{QR} \cong \overline{QS}$	5. Conv. of Isos. △ Th.

43. Given: $\overline{AD} \cong \overline{CE}$,
$\overline{AD} \parallel \overline{CE}$
Prove: $\triangle ABD \cong$
$\triangle EBC$

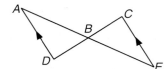

Proof:

Statements	Reasons
1. $\overline{AD} \parallel \overline{CE}$	1. Given
2. $\angle A \cong \angle E$, $\angle D \cong \angle C$	2. Alt. int. $\angle$s are $\cong$.
3. $\overline{AD} \cong \overline{CE}$	3. Given
4. $\triangle ABD \cong \triangle EBC$	4. ASA

45. $\overline{BC} \parallel \overline{AD}$; if alt. int. $\angle$s are $\cong$, lines are $\parallel$. **47.** $\ell \parallel m$; if 2 lines are $\perp$ to the same line, they are $\parallel$.

Pages 227–230 Chapter 4 Study Guide and Review
1. h **3.** d **5.** a **7.** b **9.** obtuse, isosceles
11. equiangular, equilateral **13.** 25 **15.** $\angle E \cong \angle D$, $\angle F \cong$
$\angle C$, $\angle G \cong \angle B$, $\overline{EF} \cong \overline{DC}$, $\overline{FG} \cong \overline{CB}$, $\overline{GE} \cong \overline{BD}$ **17.** $\angle KNC$
$\cong \angle RKE$, $\angle NCK \cong \angle KER$, $\angle CKN \cong \angle ERK$, $\overline{NC} \cong \overline{KE}$,
$\overline{CK} \cong \overline{ER}$, $\overline{KN} \cong \overline{RK}$ **19.** $MN = \sqrt{20}$, $NP = \sqrt{5}$, $MP = 5$,
$QR = \sqrt{20}$, $RS = \sqrt{5}$, and $QS = 5$. Each pair of
corresponding sides has the same measure. Therefore,
$\triangle MNP \cong \triangle QRS$ by SSS.

21. Given: $\triangle DGC \cong \triangle DGE$,
$\triangle GCF \cong \triangle GEF$
Proof: $\triangle DFC \cong \triangle DFE$

Proof:

Statement	Reason
1. $\triangle DGC \cong \triangle DGE$, $\triangle GCF \cong \triangle GEF$	1. Given
2. $\angle CDG \cong \angle EDG$, $\overline{CD} \cong \overline{ED}$, and $\angle CFD \cong \angle EFD$	2. CPCTC
3. $\triangle DFC \cong \triangle DFE$	3. AAS

23. 40 **25.** 80
27.

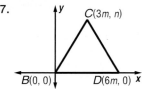

Chapter 5 Relationships in Triangles

Page 235 Chapter 5 Getting Started
1. $(-4, 5)$ **3.** $(-0.5, -5)$ **5.** 68 **7.** 40 **9.** 26 **11.** 14
13. The sum of the measures of the angles is 180.

Pages 242–245 Lesson 5-1
1. Sample answer: Both pass through the midpoint of a side. A perpendicular bisector is perpendicular to the side of a triangle, and does not necessarily pass through the vertex opposite the side, while a median does pass through the vertex and is not necessarily perpendicular to the side.
3. Sample answer: An altitude and angle bisector of a triangle are the same segment in an equilateral triangle.

5. Given: $\overline{XY} \cong \overline{XZ}$
$\overline{YM}$ and $\overline{ZN}$ are medians.
Prove: $\overline{YM} \cong \overline{ZN}$

Proof:

Statements	Reasons
1. $\overline{XY} \cong \overline{XZ}$, $\overline{YM}$ and $\overline{ZN}$ are medians.	1. Given
2. M is the midpoint of $\overline{XZ}$. N is the midpoint of $\overline{XY}$.	2. Def. of median
3. $XY = XZ$	3. Def. of $\cong$ segs.
4. $\overline{XM} \cong \overline{MZ}$, $\overline{XN} \cong \overline{NY}$	4. Def. of median
5. $XM = MZ$, $XN = NY$	5. Def. of $\cong$ segs.
6. $XM + MZ = XZ$, $XN + NY = XY$	6. Segment Addition Postulate
7. $XM + MZ = XN + NY$	7. Substitution
8. $MZ + MZ = NY + NY$	8. Substitution
9. $2MZ = 2NY$	9. Addition Property
10. $MZ = NY$	10. Division Property
11. $\overline{MZ} \cong \overline{NY}$	11. Def. of $\cong$ segs.
12. $\angle XZY \cong \angle XYZ$	12. Isosceles Triangle Theorem
13. $\overline{YZ} \cong \overline{YZ}$	13. Reflexive Property
14. $\triangle MYZ \cong \triangle NZY$	14. SAS
15. $\overline{YM} \cong \overline{ZN}$	15. CPCTC

7. $\left(\frac{2}{3}, 3\frac{1}{3}\right)$ **9.** $\left(1\frac{2}{5}, 2\frac{3}{5}\right)$

11. Given: $\triangle UVW$ is isosceles with vertex angle UVW. $\overline{YV}$ is the bisector of $\angle UVW$.
Prove: $\overline{YV}$ is a median.

Proof:

Statements	Reasons
1. $\triangle UVW$ is an isosceles triangle with vertex angle UVW, $\overline{YV}$ is the bisector of $\angle UVW$.	1. Given
2. $\overline{UV} \cong \overline{WV}$	2. Def. of isosceles $\triangle$
3. $\angle UVY \cong \angle WVY$	3. Def. of angle bisector
4. $\overline{YV} \cong \overline{YV}$	4. Reflexive Property
5. $\triangle UVY \cong \triangle WVY$	5. SAS
6. $\overline{UY} \cong \overline{WY}$	6. CPCTC
7. Y is the midpoint of $\overline{UW}$.	7. Def. of midpoint
8. $\overline{YV}$ is a median.	8. Def. of median

13. $x = 7$, $m\angle 2 = 58$ **15.** $x = 20$, $y = 4$; yes; because $m\angle WPA = 90$ **17.** always **19.** never **21.** 2 **23.** 40
25. $PR = 18$ **27.** $(0, 7)$ **29.** $-\frac{4}{3}$

31. Given: $\overline{CA} \cong \overline{CB}$, $\overline{AD} \cong \overline{BD}$
Prove: C and D are on the perpendicular bisector of $\overline{AB}$.

Proof:

Statements	Reasons
1. $\overline{CA} \cong \overline{CB}$, $\overline{AD} \cong \overline{BD}$	1. Given
2. $\overline{CD} \cong \overline{CD}$	2. Reflexive Property
3. $\triangle ACD \cong \triangle BCD$	3. SSS
4. $\angle ACD \cong \angle BCD$	4. CPCTC
5. $\overline{CE} \cong \overline{CE}$	5. Reflexive Property
6. $\triangle CEA \cong \triangle CEB$	6. SAS

7. $\overline{AE} \cong \overline{BE}$ | **7.** CPCTC
8. E is the midpoint of $\overline{AB}$. | **8.** Def. of midpoint
9. $\angle CEA \cong \angle CEB$ | **9.** CPCTC
10. $\angle CEA$ and $\angle CEB$ form a linear pair. | **10.** Def. of linear pair
11. $\angle CEA$ and $\angle CEB$ are supplementary. | **11.** Supplement Theorem
12. $m\angle CEA + m\angle CEB = 180$ | **12.** Def. of suppl. $\angle$s
13. $m\angle CEA + m\angle CEA = 180$ | **13.** Substitution
14. $2(m\angle CEA) = 180$ | **14.** Substitution
15. $m\angle CEA = 90$ | **15.** Division Property
16. $\angle CEA$ and $\angle CEB$ are rt. $\angle$s. | **16.** Def. of rt. $\angle$
17. $\overline{CD} \perp \overline{AB}$ | **17.** Def. of $\perp$
18. $\overline{CD}$ is the perpendicular bisector of $\overline{AB}$. | **18.** Def. of $\perp$ bisector
19. C and D are on the perpendicular bisector of $\overline{AB}$. | **19.** Def. of points on a line

33. Given: $\triangle ABC$, $\overrightarrow{AD}$, $\overrightarrow{BE}$, $\overrightarrow{CF}$, $\overline{KP} \perp \overline{AB}$, $\overline{KQ} \perp \overline{BC}$, $\overline{KR} \perp \overline{AC}$
Prove: $KP = KQ = KR$

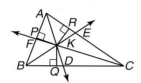

Proof:

Statements	**Reasons**
1. $\triangle ABC$, $\overrightarrow{AD}$, $\overrightarrow{BE}$, $\overrightarrow{CF}$, $\overline{KP} \perp \overline{AB}$, $\overline{KQ} \perp \overline{BC}$, $\overline{KR} \perp \overline{AC}$	**1.** Given
2. $KP = KQ$, $KQ = KR$, $KP = KR$	**2.** Any point on the $\angle$ bisector is equidistant from the sides of the angle.
3. $KP = KQ = KR$	**3.** Transitive Property

35. 4

37.

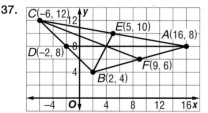

39. The altitude will be the same for both triangles, and the bases will be congruent, so the areas will be equal.   **41.** C

43. Sample answer:

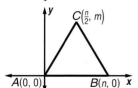

45. Sample answer:

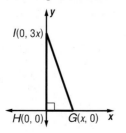

47. $\angle 5 \cong \angle 11$ **49.** $\overline{ML} \cong \overline{MN}$
51. $>$ **53.** $>$

Pages 251–254   Lesson 5-2
1. never   **3.** Grace; she placed the shorter side with the smaller angle, and the longer side with the larger angle.
5. $\angle 3$ **7.** $\angle 4$, $\angle 5$, $\angle 6$ **9.** $\angle 2$, $\angle 3$, $\angle 5$, $\angle 6$ **11.** $m\angle XZY < m\angle XYZ$ **13.** $AE < EB$ **15.** $BC = EC$ **17.** $\angle 1$ **19.** $\angle 7$
21. $\angle 7$ **23.** $\angle 2$, $\angle 7$, $\angle 8$, $\angle 10$ **25.** $\angle 3$, $\angle 5$ **27.** $\angle 8$, $\angle 7$,

$\angle 3$, $\angle 1$ **29.** $m\angle KAJ < m\angle AJK$ **31.** $m\angle SMJ > m\angle MJS$
33. $m\angle MYJ < m\angle JMY$
35. Given: $\overline{JM} \cong \overline{JL}$, $\overline{JL} \cong \overline{KL}$
Prove: $m\angle 1 > m\angle 2$

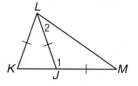

Proof:

Statements	**Reasons**
1. $\overline{JM} \cong \overline{JL}$, $\overline{JL} \cong \overline{KL}$	**1.** Given
2. $\angle LKJ \cong \angle LJK$	**2.** Isosceles $\triangle$ Theorem
3. $m\angle LKJ = m\angle LJK$	**3.** Def. of $\cong$ $\angle$s
4. $m\angle 1 > m\angle LKJ$	**4.** Ext. $\angle$ Inequality Theorem
5. $m\angle 1 > m\angle LJK$	**5.** Substitution
6. $m\angle LJK > m\angle 2$	**6.** Ext. $\angle$ Inequality Theorem
7. $m\angle 1 > m\angle 2$	**7.** Trans. Prop. of Inequality

37. $ZY > YR$ **39.** $RZ > SR$ **41.** $TY < ZY$ **43.** $\angle M$, $\angle L$, $\angle K$ **45.** Phoenix to Atlanta, Des Moines to Phoenix, Atlanta to Des Moines   **47.** 5; $\overline{PR}$, $\overline{QR}$, $\overline{PQ}$ **49.** 12; $\overline{QR}$, $\overline{PR}$, $\overline{PQ}$ **51.** $2(y + 1) > \frac{x}{3}$, $y > \frac{x - 6}{6}$ **53.** $3x + 15 > 4x + 7 > 0$, $-\frac{7}{4} < x < 8$ **55.** A   **57.** $(15, -6)$ **59.** Yes; $\frac{1}{3}(-3) = -1$, and F is the midpoint of $\overline{BD}$.   **61.** Label the midpoints of $\overline{AB}$, $\overline{BC}$, and $\overline{CA}$ as E, F, and G respectively. Then the coordinates of E, F, and G are $\left(\frac{a}{2}, 0\right)$, $\left(\frac{a + b}{2}, \frac{c}{2}\right)$, and $\left(\frac{b}{2}, \frac{c}{2}\right)$ respectively. The slope of $\overline{AF} = \frac{c}{a + b}$, and the slope of $\overline{AD} = \frac{c}{a + b}$, so D is on $\overline{AF}$. The slope of $\overline{BG} = \frac{c}{b - 2a}$ and the slope of $\overline{BD} = \frac{c}{b - 2a}$, so D is on $\overline{BG}$. The slope of $\overline{CE} = \frac{2c}{2b - a}$ and the slope of $\overline{CD} = \frac{2c}{2b - a}$, so D is on $\overline{CE}$. Since D is on $\overline{AF}$, $\overline{BG}$, and $\overline{CE}$, it is the intersection point of the three segments.   **63.** $\angle C \cong \angle R$, $\angle D \cong \angle S$, $\angle G \cong \angle W$, $\overline{CD} \cong \overline{RS}$, $\overline{DG} \cong \overline{SW}$, $\overline{CG} \cong \overline{RW}$ **65.** 9.5   **67.** false

Page 254   Practice Quiz 1
1. 5   **3.** never   **5.** sometimes   **7.** no triangle   **9.** $m\angle Q = 56$, $m\angle R = 61$, $m\angle S = 63$

Pages 257–260   Lesson 5-3
1. If a statement is shown to be false, then its opposite must be true.
3. Sample answer: $\triangle ABC$ is scalene.
Given: $\triangle ABC$; $AB \neq BC$; $BC \neq AC$; $AB \neq AC$
Prove: $\triangle ABC$ is scalene.
Proof:
Step 1: Assume $\triangle ABC$ is not scalene.
Case 1: $\triangle ABC$ is isosceles.
If $\triangle ABC$ is isosceles, then $AB = BC$, $BC = AC$, or $AB = AC$. This contradicts the given information, so $\triangle ABC$ is not isosceles.
Case 2: $\triangle ABC$ is equilateral.
In order for a triangle to be equilateral, it must also be isosceles, and Case 1 proved that $\triangle ABC$ is not isosceles. Thus, $\triangle ABC$ is not equilateral. Therefore, $\triangle ABC$ is scalene.
5. The lines are not parallel.

7. Given: $a > 0$
Prove: $\frac{1}{a} > 0$
Proof:
Step 1: Assume $\frac{1}{a} \le 0$.
Step 2: $\frac{1}{a} \le 0$; $a \cdot \frac{1}{a} \le 0 \cdot a$, $1 \le 0$
Step 3: The conclusion that $1 \le 0$ is false, so the assumption that $\frac{1}{a} \le 0$ must be false. Therefore, $\frac{1}{a} > 0$.

9. Given: $\triangle ABC$
Prove: There can be no more than one obtuse angle in $\triangle ABC$.

Proof:
Step 1: Assume that there can be more than one obtuse angle in $\triangle ABC$.
Step 2: The measure of an obtuse angle is greater than 90, $x > 90$, so the measure of two obtuse angles is greater than 180, $2x > 180$.
Step 3: The conclusion contradicts the fact that the sum of the angles of a triangle equals 180. Thus, there can be at most one obtuse angle in $\triangle ABC$.

11. Given: $\triangle ABC$ is a right triangle; $\angle C$ is a right angle.
Prove: $AB > BC$ and $AB > AC$
Proof:
Step 1: Assume that the hypotenuse of a right triangle is not the longest side. That is, $AB < BC$ or $AB < AC$.
Step 2: If $AB < BC$, then $m\angle C < m\angle A$. Since $m\angle C = 90$, $m\angle A > 90$.
So, $m\angle C + m\angle A > 180$. By the same reasoning, if $AB < BC$, then $m\angle C + m\angle B > 180$.
Step 3: Both relationships contradict the fact that the sum of the measures of the angles of a triangle equals 180. Therefore, the hypotenuse must be the longest side of a right triangle.

13. $\overline{PQ} \not\cong \overline{ST}$ **15.** A number cannot be expressed as $\frac{a}{b}$.
17. Points P, Q, and R are noncollinear.

19. Given: $\frac{1}{a} < 0$
Prove: a is negative.
Proof:
Step 1: Assume $a > 0$. $a \ne 0$ since that would make $\frac{1}{a}$ undefined.
Step 2: $\frac{1}{a} < 0$
$a\left(\frac{1}{a}\right) < 0 \cdot a$
$1 < 0$
Step 3: $1 > 0$, so the assumption must be false. Thus, a must be negative.

21. Given: $\overline{PQ} \cong \overline{PR}$
$\angle 1 \not\cong \angle 2$
Prove: $\overline{PZ}$ is not a median of $\triangle PQR$.

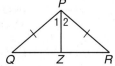

Proof:
Step 1: Assume $\overline{PZ}$ is a median of $\triangle PQR$.
Step 2: If $\overline{PZ}$ is a median of $\triangle PQR$, then Z is the midpoint of $\overline{QR}$, and $\overline{QZ} \cong \overline{RZ}$. $\overline{PZ} \cong \overline{PZ}$ by the Reflexive Property. $\triangle PZQ \cong \triangle PZR$ by SSS. $\angle 1 \cong \angle 2$ by CPCTC.

Step 3: This conclusion contradicts the given fact $\angle 1 \not\cong \angle 2$. Thus, $\overline{PZ}$ is not a median of $\triangle PQR$.

23. Given: $a > 0$, $b > 0$, and $a > b$
Prove: $\frac{a}{b} > 1$
Proof:
Step 1: Assume that $\frac{a}{b} \le 1$.
Step 2:

Case 1	Case 2
$\frac{a}{b} < 1$	$\frac{a}{b} = 1$
$a < b$	$a = b$

Step 3: The conclusion of both cases contradicts the given fact $a > b$. Thus, $\frac{a}{b} > 1$.

25. Given: $\triangle ABC$ and $\triangle ABD$ are equilateral. $\triangle ACD$ is not equilateral.
Prove: $\triangle BCD$ is not equilateral.

Proof:
Step 1: Assume that $\triangle BCD$ is an equilateral triangle.
Step 2: If $\triangle BCD$ is an equilateral triangle, then $\overline{BC} \cong \overline{CD} \cong \overline{DB}$. Since $\triangle ABC$ and $\triangle ABD$ are equilateral triangles, $\overline{AC} \cong \overline{AB} \cong \overline{BC}$ and $\overline{AD} \cong \overline{AB} \cong \overline{DB}$. By the Transitive Property, $\overline{AC} \cong \overline{AD} \cong \overline{CD}$. Therefore, $\triangle ACD$ is an equilateral triangle.
Step 3: This conclusion contradicts the given information. Thus, the assumption is false. Therefore, $\triangle BCD$ is not an equilateral triangle.

27. Use $r = \frac{d}{t}$, $t = 3$, and $d = 175$.
Proof:
Step 1: Assume that Ramon's average speed was greater than or equal to 60 miles per hour, $r \ge 60$.
Step 2:

Case 1	Case 2
$r = 60$	$r > 60$
$60 \overset{?}{=} \frac{175}{3}$	$\frac{175}{3} \overset{?}{>} 60$
$60 \ne 58.3$	$58.3 \not> 60$

Step 3: The conclusions are false, so the assumption must be false. Therefore, Ramon's average speed was less than 60 miles per hour.

29. $1500 \cdot 15\% \overset{?}{=} 225$
$1500 \cdot 0.15 \overset{?}{=} 225$
$225 = 225$

31. Yes; if you assume the client was at the scene of the crime, it is contradicted by his presence in Chicago at that time. Thus, the assumption that he was present at the crime is false.

33. Proof:
Step 1: Assume that $\sqrt{2}$ is a rational number.
Step 2: If $\sqrt{2}$ is a rational number, it can be written as $\frac{a}{b}$, where a and b are integers with no common factors, and $b \ne 0$. If $\sqrt{2} = \frac{a}{b}$, then $2 = \frac{a^2}{b^2}$, and $2b^2 = a^2$. Thus a^2 is an even number, as is a. Because a is even it can be written as $2n$.
$2b^2 = a^2$
$2b^2 = (2n)^2$
$2b^2 = 4n^2$
$b^2 = 2n^2$
Thus, b^2 is an even number. So, b is also an even number.
Step 3: Because b and a are both even numbers, they have a common factor of 2. This contradicts the definition of rational numbers. Therefore, $\sqrt{2}$ is not rational.

35. D **37.** ∠P

39. Given: $\overline{CD}$ is an angle bisector.
$\overline{CD}$ is an altitude.
Prove: △ABC is isosceles.

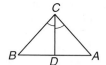

Proof:

Statements	Reasons
1. $\overline{CD}$ is an angle bisector. $\overline{CD}$ is an altitude.	1. Given
2. ∠ACD ≅ ∠BCD	2. Def. of ∠ bisector
3. $\overline{CD} \perp \overline{AB}$	3. Def. of altitude
4. ∠CDA and ∠CDB are rt. ∠s.	4. ⊥ lines form 4 rt. ∠s.
5. ∠CDA ≅ ∠CDB	5. All rt. ∠s are ≅.
6. $\overline{CD} \cong \overline{CD}$	6. Reflexive Prop.
7. △ACD ≅ △BCD	7. ASA
8. $\overline{AC} \cong \overline{BC}$	8. CPCTC
9. △ABC is isosceles.	9. Def. of isosceles △

41. Given: △ABC ≅ △DEF; $\overline{BG}$ is an angle bisector of ∠ABC. $\overline{EH}$ is an angle bisector of ∠DEF.
Prove: $\overline{BG} \cong \overline{EH}$

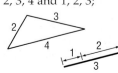

Proof:

Statements	Reasons
1. △ABC ≅ △DEF	1. Given
2. ∠A ≅ ∠D, $\overline{AB} \cong \overline{DE}$, ∠ABC ≅ ∠DEF	2. CPCTC
3. $\overline{BG}$ is an angle bisector of ∠ABC. $\overline{EH}$ is an angle bisector of ∠DEF.	3. Given
4. ∠ABG ≅ ∠GBC, ∠DEH ≅ ∠HEF	4. Def. of ∠ bisector
5. m∠ABC = m∠DEF	5. Def. of ≅ ∠s
6. m∠ABG = m∠GBC, m∠DEH = m∠HEF	6. Def. of ≅ ∠s
7. m∠ABC = m∠ABG + m∠GBC, m∠DEF = m∠DEH + m∠HEF	7. Angle Addition Property
8. m∠ABC = m∠ABG + m∠ABG, m∠DEF = m∠DEH + m∠DEH	8. Substitution
9. m∠ABG + m∠ABG = m∠DEH + m∠DEH	9. Substitution
10. 2m∠ABG = 2m∠DEH	10. Addition
11. m∠ABG = m∠DEH	11. Division
12. ∠ABG ≅ ∠DEH	12. Def. of ≅ ∠s
13. △ABG ≅ △DEH	13. ASA
14. $\overline{BG} \cong \overline{EH}$	14. CPCTC

43. $y - 3 = 2(x - 4)$ **45.** $y + 9 = 11(x + 4)$ **47.** false

Pages 263–266 Lesson 5-4
1. Sample answer: If the lines are not horizontal, then the segment connecting their y-intercepts is not perpendicular to either line. Since distance is measured along a perpendicular segment, this segment cannot be used.
3. Sample answer: 2, 3, 4 and 1, 2, 3;

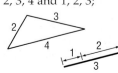

5. no; $5 + 10 \not> 15$
7. yes; $5.2 + 5.6 > 10.1$
9. $9 < n < 37$ **11.** $3 < n < 33$
13. B **15.** no; $2 + 6 \not> 11$
17. no; $13 + 16 \not> 29$ **19.** yes; $9 + 20 > 21$ **21.** yes; $17 + 30 > 30$ **23.** yes; $0.9 + 4 > 4.1$

25. no; $0.18 + 0.21 \not> 0.52$ **27.** $2 < n < 16$ **29.** $6 < n < 30$
31. $29 < n < 93$ **33.** $24 < n < 152$ **35.** $0 < n < 150$
37. $97 < n < 101$

39. Given: $\overline{HE} \cong \overline{EG}$
Prove: $HE + FG > EF$

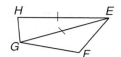

Proof:

Statements	Reasons
1. $\overline{HE} \cong \overline{EG}$	1. Given
2. HE = EG	2. Def. of ≅ segments
3. EG + FG > EF	3. Triangle Inequality
4. HE + FG > EF	4. Substitution

41. yes; $AB + BC > AC$, $AB + AC > BC$, $AC + BC > AB$
43. no; $XY + YZ = XZ$ **45.** 4 **47.** 3 **49.** $\frac{1}{2}$ **51.** Sample answer: You can use the Triangle Inequality Theorem to verify the shortest route between two locations. Answers should include the following.
- A longer route might be better if you want to collect frequent flier miles.
- A straight route might not always be available.
53. A **55.** $\overline{QR}, \overline{PQ}, \overline{PR}$ **57.** $JK = 5$, $KL = 2$, $JL = \sqrt{29}$, $PQ = 5$, $QR = 2$, and $PR = \sqrt{29}$. The corresponding sides have the same measure and are congruent. △JKL ≅ △PQR by SSS. **59.** $JK = \sqrt{113}$, $KL = \sqrt{50}$, $JL = \sqrt{65}$, $PQ = \sqrt{58}$, $QR = \sqrt{61}$, and $PR = \sqrt{65}$. The corresponding sides are not congruent, so the triangles are not congruent. **61.** $x < 6.6$

Page 266 Practice Quiz 2
1. The number 117 is not divisible by 13.
3. Step 1: Assume that $x \leq 8$.
Step 2: $7x > 56$ so $x > 8$.
Step 3: The solution of $7x > 56$ contradicts the assumption. Thus, $x \leq 8$ must be false. Therefore, $x > 8$.

5. Given: $m\angle ADC \neq m\angle ADB$
Prove: $\overline{AD}$ is not an altitude of △ABC.

Proof:

Statements	Reasons
1. $\overline{AD}$ is an altitude of △ABC.	1. Assumption
2. ∠ADC and ∠ADB are right angles.	2. Def. of altitude
3. ∠ADC ≅ ∠ADB	3. All rt ∠s are ≅.
4. m∠ADC = m∠ADB	4. Def. of ≅ angles

This contradicts the given information that $m\angle ADC \neq m\angle ADB$. Thus, $\overline{AD}$ is not an altitude of △ABC.
7. no; $25 + 35 \not> 60$ **9.** yes; $5 + 6 > 10$

Pages 270–273 Lesson 5-5
1. Sample answer: A pair of scissors illustrates the SSS inequality. As the distance between the tips of the scissors decreases, the angle between the blades decreases, allowing the blades to cut. **3.** $AB < CD$ **5.** $\frac{7}{3} < x < 6$

7. Given: $\overline{PQ} \cong \overline{SQ}$
Prove: $PR > SR$

Proof:

Statements	Reasons
1. $\overline{PQ} \cong \overline{SQ}$	1. Given
2. $\overline{QR} \cong \overline{QR}$	2. Reflexive Property
3. $m\angle PQR = m\angle PQS + m\angle SQR$	3. Angle Addition Postulate
4. $m\angle PQR > m\angle SQR$	4. Def. of inequality
5. $PR > SR$	5. SAS Inequality

9. Sample answer: The pliers are an example of the SAS inequality. As force is applied to the handles, the angle between them decreases causing the distance between the ends of the pliers to decrease. As the distance between the ends of the pliers decreases, more force is applied to a smaller area. **11.** $m\angle BDC < m\angle FDB$ **13.** $AD > DC$ **15.** $m\angle AOD > m\angle AOB$ **17.** $4 < x < 10$ **19.** $7 < x < 20$

21. Given: $\overline{PQ} \cong \overline{RS}$, $QR < PS$
Prove: $m\angle 3 < m\angle 1$

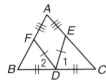

Proof:

Statements	Reasons
1. $\overline{PQ} \cong \overline{RS}$	1. Given
2. $\overline{QS} \cong \overline{QS}$	2. Reflexive Property
3. $QR < PS$	3. Given
4. $m\angle 3 < m\angle 1$	4. SSS Inequality

23. Given: $\overline{ED} \cong \overline{DF}$; $m\angle 1 > m\angle 2$; D is the midpoint of $\overline{CB}$; $\overline{AE} \cong \overline{AF}$.
Prove: $AC > AB$

Proof:

Statements	Reasons
1. $\overline{ED} \cong \overline{DF}$; D is the midpoint of $\overline{DB}$.	1. Given
2. $CD = BD$	2. Def. of midpoint
3. $\overline{CD} \cong \overline{BD}$	3. Def. of $\cong$ segments
4. $m\angle 1 > m\angle 2$	4. Given
5. $EC > FB$	5. SAS Inequality
6. $\overline{AE} \cong \overline{AF}$	6. Given
7. $AE = AF$	7. Def. of $\cong$ segments
8. $AE + EC > AE + FB$	8. Add. Prop. of Inequality
9. $AE + EC > AF + FB$	9. Substitution Prop. of Inequality
10. $AE + EC = AC$, $AF + FB = AB$	10. Segment Add. Post.
11. $AC > AB$	11. Substitution

25. As the door is opened wider, the angle formed increases and the distance from the end of the door to the door frame increases.

27. As the vertex angle increases, the base angles decrease. Thus, as the base angles decrease, the altitude of the triangle decreases.

29.

Stride (m)	Velocity (m/s)
0.25	0.07
0.50	0.22
0.75	0.43
1.00	0.70
1.25	1.01
1.50	1.37

31. Sample answer: A backhoe digs when the angle between the two arms decreases and the shovel moves through the dirt. Answers should include the following.
- As the operator digs, the angle between the arms decreases.
- The distance between the ends of the arms increases as the angle between the arms increases, and decreases as the angle decreases.

33. B **35.** yes; $16 + 6 > 19$ **37.** $\overline{AD}$ is a not median of $\triangle ABC$.

39. Given: $\overline{AD}$ bisects $\overline{BE}$; $\overline{AB} \parallel \overline{DE}$.
Prove: $\triangle ABC \cong \triangle DEC$

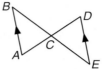

Proof:

Statements	Reasons
1. $\overline{AD}$ bisects $\overline{BE}$; $\overline{AB} \parallel \overline{DE}$.	1. Given
2. $\overline{BC} \cong \overline{EC}$	2. Def. of seg. bisector
3. $\angle B \cong \angle E$	3. Alt. int. $\angle$s Thm.
4. $\angle BCA \cong \angle ECD$	4. Vert. $\angle$s are $\cong$.
5. $\triangle ABC \cong \triangle DEC$	5. ASA

41. $EF = 5$, $FG = 50$, $EG = 5$; isosceles **43.** $EF = \sqrt{145}$, $FG = \sqrt{544}$, $EG = 35$; scalene **45.** yes, by the Law of Detachment

Pages 274–276 Chapter 5 Study Guide and Review
1. incenter **3.** Triangle Inequality Theorem **5.** angle bisector **7.** orthocenter **9.** 72 **11.** $m\angle DEF > m\angle DFE$ **13.** $m\angle DEF > m\angle FDE$ **15.** $DQ < DR$ **17.** $SR > SQ$ **19.** The triangles are not congruent. **21.** no; $7 + 5 \not> 20$ **23.** yes; $6 + 18 > 20$ **25.** $BC > MD$ **27.** $x > 7$

Chapter 6 Proportions and Similarity

Page 281 Chapter 6 Getting Started
1. 15 **3.** 10 **5.** 2 **7.** $-\dfrac{6}{5}$ **9.** yes; $\cong$ alt. int. $\angle$s **11.** 2, 4, 8, 16 **13.** 1, 7, 25, 79

Page 284–287 Lesson 6-1
1. Cross multiply and divide by 28. **3.** Suki; Madeline did not find the cross products correctly. **5.** $\dfrac{1}{12}$ **7.** 2.1275
9. 54, 48, 42 **11.** 320 **13.** 76:89 **15.** 25.3:1 **17.** 18 ft, 24 ft
19. 43.2, 64.8, 72 **21.** 18 in., 24 in., 30 in. **23.** $\dfrac{3}{2}$ **25.** 2:19
27. 16.4 lb **29.** 1.295 **31.** 14 **33.** 3 **35.** -1, $\dfrac{-2}{3}$ **37.** 36%

39. Sample answer: It appears that Tiffany used rectangles with areas that were in proportion as a background for this artwork. Answers should include the following.
- The center column pieces are to the third column from the left pieces as the pieces from the third column are to the pieces in the outside column.
- The dimensions are approximately 24 inches by 34 inches.
41. D **43.** always **45.** $15 < x < 47$ **47.** $12 < x < 34$
49.

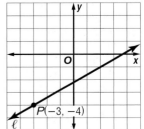

51.

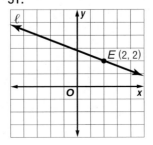

53. Yes; 100 km and 62 mi are the same length, so $AB = CD$. By the definition of congruent segments, $\overline{AB} \cong \overline{CD}$. **55.** 13.0 **57.** 1.2

Page 292–297   Lesson 6-2

1. Both students are correct. One student has inverted the ratio and reversed the order of the comparison. **3.** If two polygons are congruent, then they are similar. All of the corresponding angles are congruent, and the ratio of measures of the corresponding sides is 1. Two similar figures have congruent angles, and the sides are in proportion, but not always congruent. If the scale factor is 1, then the figures are congruent. **5.** Yes; $\angle A \cong \angle E$, $\angle B \cong \angle F$, $\angle C \cong \angle G$, $\angle D \cong \angle H$ and $\frac{AD}{EH} = \frac{DC}{HG} = \frac{CB}{GF} = \frac{BA}{FE} = \frac{2}{3}$. So $\square ABCD \sim \square EFGH$. **7.** polygon $ABCD \sim$ polygon $EFGH$; 23; 28; 20; 32; $\frac{1}{2}$ **9.** 60 m **11.** $ABCF$ is similar to $EDCF$ since they are congruent. **13.** $\triangle ABC$ is not similar to $\triangle DEF$. $\angle A \ncong \angle D$. **15.** $\frac{1}{3}$ **17.** polygon $ABCD \sim$ polygon $EFGH$; $\frac{13}{3}$; $AB = \frac{16}{3}$; $CD = \frac{10}{3}$; $\frac{2}{3}$

19. $\triangle ABE \sim \triangle ACD$; 6; $BC = 8$; $ED = 5$; $\frac{5}{9}$ **21.** about 3.9 in. by 6.25 in. **23.** $\frac{25}{16}$

25.

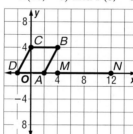

$5\frac{1}{4}$ in.
$3\frac{1}{8}$ in.
Figure not shown actual size.

27. always **29.** never **31.** sometimes **33.** always **35.** 30; 70 **37.** 27; 14 **39.** 71.05; 48.45 **41.** 7.5 **43.** 108 **45.** 73.2 **47.** $\frac{8}{5}$

49. $L(16, 8)$ and $P(8, 8)$ or $L(16, -8)$ and $P(8, -8)$

51. 18 ft by 15 ft **53.** 16:1 **55.** 16:1 **57.** 2:1; ratios are the same.

59. $\frac{a}{3a} = \frac{b}{3b} = \frac{c}{3c} = \frac{a+b+c}{3(a+b+c)} = \frac{1}{3}$

61. Sample answer: Artists use geometric shapes in patterns to create another scene or picture. The included objects have the same shape but are different sizes. Answers should include the following.
- The objects are enclosed within a circle. The objects seem to go on and on
- Each "ring" of figures has images that are approximately the same width, but vary in number and design.

63. D

65.

67. $\frac{AB}{A'B'} = \frac{AC}{A'C'} = \frac{BC}{B'C'} = \frac{1}{2}$ **69.** The sides are proportional and the angles are congruent, so the triangles are similar. **71.** -23 **73.** $OC > AO$ **75.** $m\angle ABD > m\angle ADB$ **77.** 91 **79.** $m\angle 1 = m\angle 2 = 111$ **81.** 62 **83.** 118 **85.** 62 **87.** 118

Page 301–306   Lesson 6-3

1. Sample answer: Two triangles are congruent by the SSS, SAS, and ASA Postulates and the AAS Theorem. In these triangles, corresponding parts must be congruent. Two triangles are similar by AA Similarity, SSS Similarity, and SAS Similarity. In similar triangles, the sides are proportional and the angles are congruent. Congruent triangles are always similar triangles. Similar triangles are congruent only when the scale factor for the proportional sides is 1. SSS and SAS are common relationships for both congruence and similarity. **3.** Alicia; while both have corresponding sides in a ratio, Alicia has them in proper order with the numerators from the same triangle.

5. $\triangle ABC \sim \triangle DEF$; $x = 10$; $AB = 10$; $DE = 6$ **7.** yes: $\triangle DEF \sim \triangle ACB$ by SSS Similarity **9.** 135 ft **11.** yes; $\triangle QRS \sim \triangle TVU$ by SSS Similarity **13.** yes; $\triangle RST \sim \triangle JKL$ by AA Similarity **15.** Yes; $\triangle ABC \sim \triangle JKL$ by SAS Similarity **17.** No; sides are not proportional.

19. $\triangle ABE \sim \triangle ACD$; $x = \frac{8}{5}$; $AB = 3\frac{3}{5}$; $AC = 9\frac{3}{5}$ **21.** $\triangle ABC \sim \triangle ARS$; $x = 8$; 15; 8 **23.** $\frac{3}{2}$ **25.** true **27.** $\triangle EAB \sim \triangle EFC \sim \triangle AFD$: AA Similarity **29.** $KP = 5$, $KM = 15$, $MR = 13\frac{1}{3}$, $ML = 20$, $MN = 12$, $PR = 16\frac{2}{3}$ **31.** $m\angle TUV = 43$, $m\angle R = 43$, $m\angle RSU = 47$, $m\angle SUV = 47$ **33.** $x = y$; if $\overline{BD} \parallel \overline{AE}$, then $\triangle BCD \sim \triangle ACE$ by AA Similarity and $\frac{BC}{AC} = \frac{DC}{EC}$. Thus, $\frac{2}{4} = \frac{x}{x+y}$. Cross multiply and solve for y, yielding $y = x$.

35. Given: $\overline{LP} \parallel \overline{MN}$

Prove: $\frac{LJ}{JN} = \frac{PJ}{JM}$

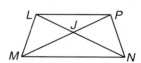

Proof:

Statements	Reasons
1. $\overline{LP} \parallel \overline{MN}$	1. Given
2. $\angle PLN \cong \angle LNM$, $\angle LPM \cong \angle PMN$	2. Alt. Int. $\angle$ Theorem
3. $\triangle LPJ \sim \triangle NMJ$	3. AA Similarity
4. $\frac{LJ}{JN} = \frac{PJ}{JM}$	4. Corr. sides of $\sim$ $\triangle$s are proportional.

37. Given: $\triangle BAC$ and $\triangle EDF$ are right triangles. $\frac{AB}{DE} = \frac{AC}{DF}$

Prove: $\triangle ABC \sim \triangle DEF$

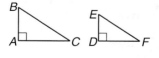

Proof:

Statements	Reasons
1. $\triangle BAC$ and $\triangle EDF$ are right triangles.	1. Given
2. $\angle BAC$ and $\angle EDF$ are right angles.	2. Def. of rt. $\triangle$
3. $\angle BAC \cong \angle EDF$	3. All rt. $\angle$s are $\cong$.
4. $\frac{AB}{DE} = \frac{AC}{DF}$	4. Given
5. $\triangle ABC \sim \triangle DEF$	5. SAS Similarity

39. 13.5 ft **41.** about 420.5 m **43.** 10.75 m

45.

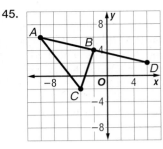

47. $\triangle ABC \sim \triangle ACD$; $\triangle ABC \sim \triangle CBD$; $\triangle ACD \sim \triangle CBD$; they are similar by AA Similarity. **49.** A **51.** $PQRS \sim ABCD$; 1.6; 1.4; 1.1; $\frac{1}{2}$

53. 5 **55.** 15 **57.** No; $\overline{AT}$ is not perpendicular to $\overline{BC}$.
59. (5.5, 13) **61.** (3.5, −2.5)

Page 306 Practice Quiz 1

1. yes; $\angle A \cong \angle E$, $\angle B \cong \angle D$, $\angle 1 \cong \angle 3$, $\angle 2 \cong \angle 4$ and $\frac{AB}{ED} = \frac{BC}{DC} = \frac{AF}{EF} = \frac{FC}{FC} = 1$ **3.** $\triangle ADE \sim \triangle CBE$; 2; 8; 4
5. 1947 mi

Page 311–315 Lesson 6-4

1. Sample answer: If a line intersects two sides of a triangle and separates sides into corresponding segments of proportional lengths, then it is parallel to the third side.
3. Given three or more parallel lines intersecting two transversals, Corollary 6.1 states that the parts of the transversals are proportional. Corollary 6.2 states that if the parts of one transversal are congruent, then the parts of every transversal are congruent. **5.** 10 **7.** The slopes of $\overline{DE}$ and $\overline{BC}$ are both 0. So $\overline{DE} \parallel \overline{BC}$. **9.** Yes; $\frac{MN}{NP} = \frac{MR}{RQ} = \frac{9}{16}$, so $\overline{RN} \parallel \overline{QP}$. **11.** $x = 2$; $y = 5$ **13.** 1100 yd **15.** 3
17. $x = 6$, ED = 9 **19.** BC = 10, FE = $13\frac{1}{3}$, CD = 9, DE = 15
21. 10 **23.** No; segments are not proportional; $\frac{PQ}{QR} = \frac{3}{7}$ and $\frac{PT}{TS} = 2$. **25.** yes **27.** $\sqrt{52}$ **29.** The endpoints of $\overline{DE}$ are $D(3, \frac{1}{2})$ and $E(\frac{3}{2}, -4)$. Both $\overline{DE}$ and $\overline{AB}$ have slope of 3. **31.** (3, 8) or (4, 4) **33.** $x = 21$, $y = 15$ **35.** 25 ft
37. 18.75 ft

39. Given: D is the midpoint of $\overline{AB}$.
E is the midpoint of $\overline{AC}$.
Prove: $\overline{DE} \parallel \overline{BC}$; DE = $\frac{1}{2}BC$
Proof:

Statements	Reasons
1. D is the midpoint of $\overline{AB}$. E is the midpoint of $\overline{AC}$.	1. Given
2. $\overline{AD} \cong \overline{DB}$, $\overline{AE} \cong \overline{EC}$	2. Midpoint Theorem
3. AD = DB, AE = EC	3. Def. of $\cong$ segments
4. AB = AD + DB, AC = AE + AC	4. Segment Addition Postulate
5. AB = AD + AD, AC = AE + AE	5. Substitution
6. AB = 2AD, AC = 2AE	6. Substitution
7. $\frac{AB}{AD} = 2$, $\frac{AC}{AE} = 2$	7. Division Prop.
8. $\frac{AB}{AD} = \frac{AC}{AE}$	8. Transitive Prop.
9. $\angle A \cong \angle A$	9. Reflexive Prop.
10. $\triangle ADE \sim \triangle ABC$	10. SAS Similarity
11. $\angle ADE \cong \angle ABC$	11. Def. of $\sim$ polygons
12. $\overline{DE} \parallel \overline{BC}$	12. If corr. $\angle$s are $\cong$, the lines are parallel.
13. $\frac{BC}{DE} = \frac{AB}{AD}$	13. Def. of $\sim$ polygons
14. $\frac{BC}{DE} = 2$	14. Substitution

15. $2DE = BC$
16. $DE = \frac{1}{2}BC$

41.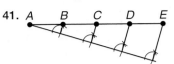

43. $u = 24$; $w = 26.4$; $x = 30$; $y = 21.6$; $z = 33.6$

45. Sample answer: City planners use maps in their work. Answers should include the following.
- City planners need to know geometry facts when developing zoning laws.
- A city planner would need to know that the shortest distance between two parallel lines is the perpendicular distance.

47. 4 **49.** yes; AA **51.** no; angles not congruent **53.** $x = 12$, $y = 6$ **55.** $m\angle ABD > m\angle BAD$ **57.** $m\angle CBD > m\angle BCD$ **59.** 18 **61.** false **63.** true **65.** $\angle R \cong \angle X$, $\angle S \cong \angle Y$, $\angle T \cong \angle Z$, $\overline{RS} \cong \overline{XY}$, $\overline{ST} \cong \overline{YZ}$, $\overline{RT} \cong \overline{XZ}$

Page 319–323 Lesson 6-5

1. $\triangle ABC \sim \triangle MNQ$ and $\overline{AD}$ and $\overline{MR}$ are altitudes, angle bisectors, or medians. **3.** 10.8 **5.** 6 **7.** 6.75 **9.** 330 cm or 3.3 m **11.** 63 **13.** 20.25 **15.** 78 **17.** Yes; the perimeters are in the same ratio as the sides, $\frac{300}{600}$ or $\frac{1}{2}$.
19. $\frac{3}{2}$ **21.** 4 **23.** $11\frac{1}{5}$ **25.** 6 **27.** 5, 13.5
29. $xy = z^2$; $\triangle ACD \sim \triangle CBD$ by AA Similarity. Thus, $\frac{CD}{BD} = \frac{AD}{CD}$ or $\frac{z}{y} = \frac{x}{z}$. The cross products yield $xy = z^2$.

31. Given: $\triangle ABC \sim \triangle RST$, $\overline{AD}$ is a median of $\triangle ABC$. $\overline{RU}$ is a median of $\triangle RST$.
Prove: $\frac{AD}{RU} = \frac{AB}{RS}$

Proof:

Statements	Reasons
1. $\triangle ABC \sim \triangle RST$ $\overline{AD}$ is a median of $\triangle ABC$. $\overline{RU}$ is a median of $\triangle RST$.	1. Given
2. CD = DB; TU = US	2. Def. of median
3. $\frac{AB}{RS} = \frac{CB}{TS}$	3. Def. of $\sim$ polygons
4. CB = CD + DB; TS = TU + US	4. Segment Addition Postulate
5. $\frac{AB}{RS} = \frac{CD + DB}{TU + US}$	5. Substitution
6. $\frac{AB}{RS} = \frac{DB + DB}{US + US}$ or $\frac{2(DB)}{2(US)}$	6. Substitution
7. $\frac{AB}{RS} = \frac{DB}{US}$	7. Substitution
8. $\angle B \cong \angle S$	8. Def. of $\sim$ polygons
9. $\triangle ABD \sim \triangle RSU$	9. SAS Similarity
10. $\frac{AD}{RU} = \frac{AB}{RS}$	10. Def. of $\sim$ polgyons

33. Given: $\triangle ABC \sim \triangle PQR$, $\overline{BD}$ is an altitude of $\triangle ABC$. $\overline{QS}$ is an altitude of $\triangle PQR$.
Prove: $\frac{QP}{BA} = \frac{QS}{BD}$

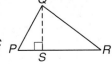

Proof:

$\angle A \cong \angle P$ because of the definition of similar polygons. Since $\overline{BD}$ and $\overline{QS}$ are perpendicular to $\overline{AC}$ and $\overline{PR}$, $\angle BDA \cong \angle QSP$. So, $\triangle ABD \sim \triangle PQS$ by AA Similarity and $\frac{QP}{BA} = \frac{QS}{BD}$ by definition of similar polygons.

35. Given: $\overline{JF}$ bisects $\angle EFG$.
$\overline{EH} \parallel \overline{FG}, \overline{EF} \parallel \overline{HG}$
Prove: $\frac{EK}{KF} = \frac{GJ}{JF}$

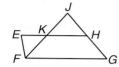

Proof:

Statements	Reasons
1. $\overline{JF}$ bisects $\angle EFG$. $\overline{EH} \parallel \overline{FG}, \overline{EF} \parallel \overline{HG}$	1. Given
2. $\angle EFK \cong \angle KFG$	2. Def. of $\angle$ bisector
3. $\angle KFG \cong \angle JKH$	3. Corresponding $\angle$s Post.
4. $\angle JKH \cong \angle EKF$	4. Vertical $\angle$s are $\cong$.
5. $\angle EFK \cong \angle EKF$	5. Transitive Prop.
6. $\angle FJH \cong \angle EFK$	6. Alternate Interior $\angle$s Th.
7. $\angle FJH \cong \angle EKF$	7. Transitive Prop.
8. $\triangle EKF \sim \triangle GJF$	8. AA Similarity
9. $\frac{EK}{KF} = \frac{GJ}{JF}$	9. Def. of $\sim \triangle$s

37. Given: $\triangle RST \sim \triangle ABC$, W and D are midpoints of $\overline{TS}$ and $\overline{CB}$, respectively.
Prove: $\triangle RWS \sim \triangle ADB$

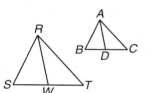

Proof:

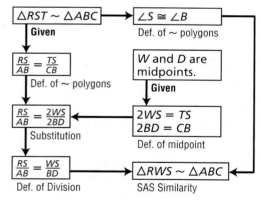

39. 12.9 **41.** no; sides not proportional **43.** yes; $\frac{LM}{MO} = \frac{LN}{NP}$
45. $\triangle PQT \sim \triangle PRS, x = 7, PQ = 15$ **47.** $y = 2x + 1$
49. 320, 640 **51.** $-27, -33$

Page 323 Practice Quiz 2
1. 20 **3.** no; sides not proportional **5.** 12.75 **7.** 10.5 **9.** 5

Page 328–331 Lesson 6-6
1. Sample answer: irregular shape formed by iteration of self-similar shapes **3.** Sample answer: icebergs, ferns, leaf veins **5.** $A_n = 2(2^n - 1)$ **7.** 1.4142...; 1.1892... **9.** Yes, the procedure is repeated over and over again.

11. 9 holes

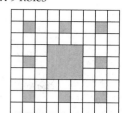

13. Yes, any part contains the same figure as the whole, 9 squares with the middle shaded. **15.** 1, 3, 6, 10, 15...; Each difference is 1 more than the preceding difference. **17.** The result is similar to a Stage 3 Sierpinski triangle. **19.** 25

21. Given: $\triangle ABC$ is equilateral. $CD = \frac{1}{3}CB$ and $CE = \frac{1}{3}CA$
Prove: $\triangle CED \sim \triangle CAB$

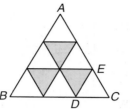

Proof:

Statements	Reasons
1. $\triangle ABC$ is equilateral. $CD = \frac{1}{3}CB, CE = \frac{1}{3}CA$	1. Given
2. $\overline{AC} \cong \overline{BC}$	2. Def. of equilateral $\triangle$
3. $AC = BC$	3. Def. of $\cong$ segments
4. $\frac{1}{3}AC = \frac{1}{3}CB$	4. Mult. Prop.
5. $CD = CE$	5. Substitution
6. $\frac{CD}{CB} = \frac{CE}{CB}$	6. Division Prop.
7. $\frac{CD}{CB} = \frac{CE}{CA}$	7. Substitution
8. $\angle C \cong \angle C$	8. Reflexive Prop.
9. $\triangle CED \sim \triangle CAB$	9. AA Similarity

23. Yes; the smaller and smaller details of the shape have the same geometric characteristics as the original form.
25. $A_n = 4^n$; 65,536 **27.** Stage 0: 3 units, Stage 1: $3 \cdot \frac{4}{3}$ or 4 units, Stage 2: $3\left(\frac{4}{3}\right)\frac{4}{3} = 3\left(\frac{4}{3}\right)^2$ or $5\frac{1}{3}$ units, Stage 3: $3\left(\frac{4}{3}\right)^3$ or $7\frac{1}{9}$ units **29.** The original triangle and the new triangles are equilateral and thus, all of the angles are equal to 60. By AA Similarity, the triangles are similar. **31.** 0.2, 5, 0.2, 5, 0.2; the numbers alternate between 0.2 and 5.0. **33.** 1, 2, 4, 16, 65,536; the numbers approach positive infinity. **35.** 0, $-5, -10$ **37.** $-6, 24, -66$ **39.** When $x = 0.00$: 0.64, 0.9216, 0.2890..., 0.8219..., 0.5854..., 0.9708..., 0.1133..., 0.4019..., 0.9615..., 0.1478...; when $x = 0.201$: 0.6423..., 0.9188..., 0.2981..., 0.8369..., 0.5458..., 0.9916..., 0.0333..., 0.1287..., 0.4487..., 0.9894.... Yes, the initial value affected the tenth value. **41.** The leaves in the tree and the branches of the trees are self-similar. These self-similar shapes are repeated throughout the painting. **43.** See students' work.

45. Sample answer: Fractal geometry can be found in the repeating patterns of nature. Answers should include the following.

- Broccoli is an example of fractal geometry because the shape of the florets is repeated throughout; one floret looks the same as the stalk.
- Sample answer: Scientists can use fractals to study the human body, rivers, and tributaries, and to model how landscapes change over time.

47. C **49.** $13\frac{3}{5}$ **51.** $\frac{7}{3}$ **53.** $16\frac{1}{4}$ **55.** Miami, Bermuda, San Juan **57.** 10 ft, 10 ft, 17 ft, 17 ft

1. true **3.** true **5.** false, iteration **7.** true **9.** false,
parallel to **11.** 12 **13.** $\frac{58}{3}$ **15.** $\frac{3}{5}$ **17.** 24 in. and 84 in.
19. Yes, these are rectangles, so all angles are congruent.
Additionally, all sides are in a $3:2$ ratio. **21.** $\triangle PQT \sim \triangle RQS$;
0; $PQ = 6$; $QS = 3$; 1 **23.** yes, $\triangle GHI \sim \triangle GJK$ by AA Similarity
25. $\triangle ABC \sim \triangle DEC$, 4 **27.** no; lengths not proportional
29. yes; $\frac{HI}{GH} = \frac{IK}{KL}$ **31.** 6 **33.** 9 **35.** 24 **37.** 36 **39.** Stage
2 is not similar to Stage 1. **41.** $-8, -20, -56$
43. $-6, -9.6, -9.96$

Chapter 7 Right Triangles and Trigonometry

1. $a = 16$ **3.** $e = 24, f = 12$ **5.** 13 **7.** 21.21 **9.** $2\sqrt{2}$
11. 15 **13.** 98 **15.** 23

1. Sample answer: 2 and 72 **3.** Ian; his proportion shows
that the altitude is the geometric mean of the two segments
of the hypotenuse. **5.** 42 **7.** $2\sqrt{3} \approx 3.5$ **9.** $4\sqrt{3} \approx 6.9$
11. $x = 6; y = 4\sqrt{3}$ **13.** $\sqrt{30} \approx 5.5$ **15.** $2\sqrt{15} \approx 7.7$
17. $\frac{\sqrt{15}}{5} \approx 0.8$ **19.** $\frac{\sqrt{5}}{3} \approx 0.7$ **21.** $3\sqrt{5} \approx 6.7$
23. $8\sqrt{2} \approx 11.3$ **25.** $\sqrt{26} \approx 5.1$ **27.** $x = 2\sqrt{15} \approx 9.4$;
$y = \sqrt{33} \approx 5.7; z = 2\sqrt{6} \approx 4.9$ **29.** $x = \frac{40}{3}; y = \frac{5}{3}$;
$z = 10\sqrt{2} \approx 14.1$ **31.** $x = 6\sqrt{6} \approx 14.7; y = 6\sqrt{42} \approx 38.9$;
$z = 36\sqrt{7} \approx 95.2$ **33.** $\frac{17}{7}$ **35.** never **37.** sometimes
39. $\triangle FGH$ is a right triangle. $\overline{OG}$ is the altitude from the
vertex of the right angle to the hypotenuse of that triangle.
So, by Theorem 7.2, OG is the geometric mean between OF
and OH, and so on. **41.** 2.4 yd **43.** yes; Indiana and
Virginia

45. Given: $\angle PQR$ is a right angle.
  $\overline{QS}$ is an altitude of
  $\triangle PQR$.
Prove: $\triangle PSQ \sim \triangle PQR$
  $\triangle PQR \sim \triangle QSR$
  $\triangle PSQ \sim \triangle QSR$

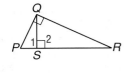

Proof:

Statements	Reasons
1. $\angle PQR$ is a right angle. $\overline{QS}$ is an altitude of $\triangle PQR$.	1. Given
2. $\overline{QS} \perp \overline{RP}$	2. Definition of altitude
3. $\angle 1$ and $\angle 2$ are right angles.	3. Definition of perpendicular lines
4. $\angle 1 \cong \angle PQR$ $\angle 2 \cong \angle PQR$	4. All right $\angle$s are $\cong$.
5. $\angle P \cong \angle P$ $\angle R \cong \angle R$	5. Congruence of angles is reflexive.
6. $\triangle PSQ \sim \triangle PQR$ $\triangle PQR \sim \triangle QSR$	6. AA Similarity Statements 4 and 5
7. $\triangle PSQ \sim \triangle QSR$	7. Similarity of triangles is transitive.

47. Given: $\angle ADC$ is a right angle. $\overline{DB}$ is an altitude of
  $\triangle ADC$.
Prove: $\dfrac{AB}{AD} = \dfrac{AD}{AC}$
  $\dfrac{BC}{DC} = \dfrac{DC}{AC}$

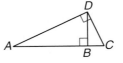

Proof:

Statements	Reasons
1. $\angle ADC$ is a right angle. $\overline{DB}$ is an altitude of $\triangle ADC$.	1. Given
2. $\triangle ADC$ is a right triangle.	2. Definition of right triangle
3. $\triangle ABD \sim \triangle ADC$ $\triangle DBC \sim \triangle ADC$	3. If the altitude is drawn from the vertex of the rt. $\angle$ to the hypotenuse of a rt. $\triangle$, then the 2 $\triangle$s formed are similar to the given $\triangle$ and to each other.
4. $\dfrac{AB}{AD} = \dfrac{AD}{AC}; \dfrac{BC}{DC} = \dfrac{DC}{AC}$	4. Definition of similar polygons

49. C **51.** 15, 18, 21 **53.** 7, 47, 2207 **55.** $8\frac{8}{9}, 11\frac{1}{9}$
57. $\angle 5, \angle 7$ **59.** $\angle 2, \angle 7, \angle 8$ **61.** $y = 4x - 8$
63. $y = -4x - 11$ **65.** 13 ft

1. Maria; Colin does not have the longest side as the value
of c.
3.

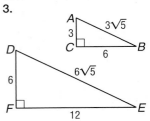

Sample answer : $\triangle ABC \sim$
$\triangle DEF$, $\angle A \cong \angle D$, $\angle B \cong \angle E$,
and $\angle C \cong \angle F$, $\overline{AB}$ corresponds
to $\overline{DE}$, $\overline{BC}$ corresponds to $\overline{EF}$,
$\overline{AC}$ corresponds to $\overline{DF}$. The scale
factor is $\frac{2}{1}$. No; the measures do
not form a Pythagorean triple
since $6\sqrt{5}$ and $3\sqrt{5}$ are not
whole numbers.
5. $\frac{3}{7}$ **7.** yes; $JK = \sqrt{17}, KL = \sqrt{17}, JL = \sqrt{34}; (\sqrt{17})^2 +$
$(\sqrt{17})^2 = (\sqrt{34})^2$ **9.** no, no **11.** about 15.1 in.
13. $4\sqrt{3} \approx 6.9$ **15.** $8\sqrt{41} \approx 51.2$ **17.** 20 **19.** no; $QR =$
5, $RS = 6, QS, = 5; 5^2 + 5^2 \neq 6^2$ **21.** yes; $QR = \sqrt{29}, RS =$
$\sqrt{29}, QS = \sqrt{58}; (\sqrt{29})^2 + (\sqrt{29})^2 = (\sqrt{58})^2$ **23.** yes, yes
25. no, no **27.** no, no **29.** yes, no **31.** 5-12-13
33. Sample answer: They consist of any number of similar
triangles. **35a.** 16-30-34; 24-45-51 **35b.** 18-80-82;
27-120-123 **35c.** 14-48-50; 21-72-75 **37.** 10.8 degrees
39. Given: $\triangle ABC$ with right angle at C, $AB = d$
  Prove: $d = \sqrt{(x_2 - x_1)^2 + (y_2 - y_1)^2}$

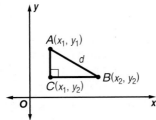

Proof:

Statements	Reasons
1. $\triangle ABC$ with right angle at C, $AB = d$	1. Given

1. Opposite sides are congruent; opposite angles are congruent; consecutive angles are supplementary; and if there is one right angle, there are four right angles.

3. Sample answer:

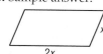

5. $\triangle VTQ$, SSS; diag. bisect each other and opp. sides of $\square$ are $\cong$.
7. 100 **9.** 80 **11.** 7

13. Given: $\square VZRQ$ and $\square WQST$
Prove: $\angle Z \cong \angle T$

Proof:

Statements	Reasons
1. $\square VZRQ$ and $\square WQST$	1. Given
2. $\angle Z \cong \angle Q$, $\angle Q \cong \angle T$	2. Opp. $\angle$s of a $\square$ are $\cong$.
3. $\angle Z \cong \angle T$	3. Transitive Prop.

15. C **17.** $\angle CDB$, alt. int. $\angle$s are $\cong$. **19.** $\overline{GD}$, diag. of $\square$ bisect each other. **21.** $\angle BAC$, alt. int. $\angle$s are $\cong$. **23.** 33
25. 109 **27.** 83 **29.** 6.45 **31.** 6.1 **33.** $y = 5$, $FH = 9$
35. $a = 6$, $b = 5$, $DB = 32$ **37.** $EQ = 5$, $QG = 5$, $HQ = \sqrt{13}$, $QF = \sqrt{13}$ **39.** Slope of $\overline{EH}$ is undefined, slope of $\overline{EF} = -\frac{1}{3}$; no, the slopes of the sides are not negative reciprocals of each other.

41. Given: $\square PQRS$
Prove: $\overline{PQ} \cong \overline{RS}$
$\overline{QR} \cong \overline{SP}$
Proof:

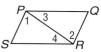

Statements	Reasons
1. $\square PQRS$	1. Given
2. Draw an auxiliary segment $\overline{PR}$ and label angles 1, 2, 3, and 4 as shown.	2. Diagonal of $\square PQRS$
3. $\overline{PQ} \parallel \overline{SR}$, $\overline{PS} \parallel \overline{QR}$	3. Opp. sides of $\square$ are $\parallel$.
4. $\angle 1 \cong \angle 2$, and $\angle 3 \cong \angle 4$	4. Alt. int. $\angle$s are $\cong$.
5. $\overline{PR} \cong \overline{PR}$	5. Reflexive Prop.
6. $\triangle QPR \cong \triangle SRP$	6. ASA
7. $\overline{PQ} \cong \overline{RS}$ and $\overline{QR} \cong \overline{SP}$	7. CPCTC

43. Given: $\square MNPQ$
$\angle M$ is a right angle.
Prove: $\angle N$, $\angle P$ and $\angle Q$ are right angles.

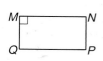

Proof:
By definition of a parallelogram, $\overline{MN} \parallel \overline{QP}$. Since $\angle M$ is a right angle, $\overline{MQ} \perp \overline{MN}$. By the Perpendicular Transversal Theorem, $\overline{MQ} \perp \overline{QP}$. $\angle Q$ is a right angle, because perpendicular lines form a right angle. $\angle N \cong \angle Q$ and $\angle M \cong \angle P$ because opposite angles in a parallelogram are congruent. $\angle P$ and $\angle N$ are right angles, since all right angles are congruent.

45. Given: $\square WXYZ$
Prove: $\triangle WXZ \cong \triangle YZX$

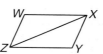

Proof:

Statements	Reasons
1. $\square WXYZ$	1. Given
2. $\overline{WX} \cong \overline{ZY}$, $\overline{WZ} \cong \overline{XY}$	2. Opp. sides of $\square$ are $\cong$.
3. $\angle ZWX \cong \angle XYZ$	3. Opp. $\angle$s of $\square$ are $\cong$.
4. $\triangle WXZ \cong \triangle YZX$	4. SAS

47. Given: $\square BCGH$, $\overline{HD} \cong \overline{FD}$
Prove: $\angle F \cong \angle GCB$

Proof:

Statements	Reasons
1. $\square BCGH$, $\overline{HD} \cong \overline{FD}$	1. Given
2. $\angle F \cong \angle H$	2. Isosceles Triangle Th.
3. $\angle H \cong \angle GCB$	3. Opp. $\angle$s of $\square$ are $\cong$.
4. $\angle F \cong \angle GCB$	4. Congruence of angles is transitive.

49. The graphic uses the illustration of wedges shaped like parallelograms to display the data. Answers should include the following.
- The opposite sides are parallel and congruent, the opposite angles are congruent, and the consecutive angles are supplementary.
- Sample answer:

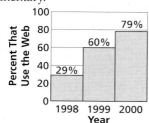

51. B **53.** 3600 **55.** 6120 **57.** Sines; $m\angle C \approx 69.9$, $m\angle A \approx 53.1$, $a \approx 11.9$ **59.** 30 **61.** side, $\frac{7}{3}$ **63.** side, $\frac{7}{3}$

Pages 420–423 Lesson 8-3
1. Both pairs of opposite sides are congruent; both pairs of opposite angles are congruent; diagonals bisect each other; one pair of opposite sides is parallel and congruent.
3. Shaniqua; Carter's description could result in a shape that is not a parallelogram. **5.** Yes; each pair of opp. $\angle$s is $\cong$. **7.** $x = 41$, $y = 16$ **9.** yes

11. Given: $\overline{PT} \cong \overline{TR}$
$\angle TSP \cong \angle TQR$
Prove: $PQRS$ is a parallelogram.

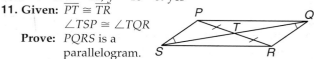

Proof:

Statements	Reasons
1. $\overline{PT} \cong \overline{TR}$, $\angle TSP \cong \angle TQR$	1. Given
2. $\angle PTS \cong \angle RTQ$	2. Vertical $\angle$s are $\cong$.
3. $\triangle PTS \cong \triangle RTQ$	3. AAS
4. $\overline{PS} \cong \overline{QR}$	4. CPCTC
5. $\overline{PS} \parallel \overline{QR}$	5. If alt. int. $\angle$s are $\cong$, lines are $\parallel$.
6. $PQRS$ is a parallelogram.	6. If one pair of opp. sides is $\parallel$ and $\cong$, then the quad. is a $\square$.

13. Yes; each pair of opposite angles is congruent. **15.** Yes; opposite angles are congruent. **17.** Yes; one pair of opposite sides is parallel and congruent. **19.** $x = 6$, $y = 24$ **21.** $x = 1$, $y = 2$ **23.** $x = 34$, $y = 44$ **25.** yes **27.** yes **29.** no **31.** yes **33.** Move M to $(-4, 1)$, N to $(-3, 4)$, P to $(0, -9)$, or R to $(-7, 3)$. **35.** $(-2, -2)$, $(4, 10)$, or $(10, 0)$ **37.** Parallelogram; $\overline{KM}$ and $\overline{JL}$ are diagonals that bisect each other.

39. Given: $\overline{AD} \cong \overline{BC}$
$\overline{AB} \cong \overline{DC}$
Prove: $ABCD$ is a parallelogram.

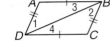

Proof:

Statements	Reasons
1. $\overline{AD} \cong \overline{BC}$, $\overline{AB} \cong \overline{DC}$	1. Given
2. Draw $\overline{DB}$.	2. Two points determine a line.
3. $\overline{DB} \cong \overline{DB}$	3. Reflexive Property
4. $\triangle ABD \cong \triangle CDB$	4. SSS
5. $\angle 1 \cong \angle 2$, $\angle 3 \cong \angle 4$	5. CPCTC
6. $\overline{AD} \parallel \overline{BC}$, $\overline{AB} \parallel \overline{DC}$	6. If alt. int. $\angle$s are $\cong$, lines are $\parallel$.
7. $ABCD$ is a parallelogram.	7. Definition of parallelogram

41. Given: $\overline{AB} \cong \overline{DC}$
$\overline{AB} \parallel \overline{DC}$
Prove: $ABCD$ is a parallelogram.
Proof:

Statements	Reasons
1. $\overline{AB} \cong \overline{DC}$, $\overline{AB} \parallel \overline{DC}$	1. Given
2. Draw $\overline{AC}$	2. Two points determine a line.
3. $\angle 1 \cong \angle 2$	3. Alternate Interior Angles Theorem
4. $\overline{AC} \cong \overline{AC}$	4. Reflexive Property
5. $\triangle ABC \cong \triangle CDA$	5. SAS
6. $\overline{AD} \cong \overline{BC}$	6. CPCTC
7. $ABCD$ is a parallelogram.	7. If both pairs of opp. sides are $\cong$, then the quad. is $\square$.

43. Given: $ABCDEF$ is a regular hexagon.
Prove: $FDCA$ is a parallelogram.

Proof:

Statements	Reasons
1. $ABCDEF$ is a regular hexagon.	1. Given
2. $\overline{AB} \cong \overline{DE}$, $\overline{BC} \cong \overline{EF}$ $\angle E \cong \angle B$, $\overline{FA} \cong \overline{CD}$	2. Def. of regular hexagon
3. $\triangle ABC \cong \triangle DEF$	3. SAS
4. $\overline{AC} \cong \overline{DF}$	4. CPCTC
5. $FDCA$ is a $\square$.	5. If both pairs of opp. sides are $\cong$, then the quad. is $\square$.

45. B **47.** 12 **49.** 14 units **51.** 8 **53.** 30 **55.** 72 **57.** 45, $12\sqrt{2}$ **59.** $16\sqrt{3}$, 16 **61.** $5, -\frac{3}{2}$; not $\perp$ **63.** $\frac{2}{3}, -\frac{3}{2}$; $\perp$

---

1. 11 **3.** 66 **5.** $x = 8$, $y = 6$

Pages 427–430 Lesson 8-4
1. If consecutive sides are perpendicular or diagonals are congruent, then the parallelogram is a rectangle.
3. McKenna; Consuelo's definition is correct if one pair of opposite sides is parallel and congruent. **5.** 40 **7.** 52 or 10
9. Make sure that the angles measure 90 or that the diagonals are congruent. **11.** 11 **13.** $29\frac{1}{3}$ **15.** 4 **17.** 60
19. 30 **21.** 60 **23.** 30 **25.** Measure the opposite sides and the diagonals to make sure they are congruent. **27.** No; $\overline{DH}$ and $\overline{FG}$ are not parallel. **29.** Yes; opp. sides are $\parallel$, diag. are $\cong$. **31.** $\left(\frac{1}{2}, -\frac{3}{2}\right), \left(\frac{7}{2}, \frac{3}{2}\right)$ **33.** Yes; consec. sides are $\perp$.
35. Move L and K until the length of the diagonals is the same. **37.** See students' work.
39. Sample answer:
$\overline{AC} \cong \overline{BD}$ but $ABCD$ is not a rectangle

41. Given: $\square WXYZ$ and $\overline{WY} \cong \overline{XZ}$
Prove: $WXYZ$ is a rectangle.

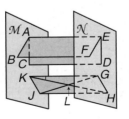

Proof:

Statements	Reasons
1. $\square WXYZ$ and $\overline{WY} \cong \overline{XZ}$	1. Given
2. $\overline{XY} \cong \overline{WZ}$	2. Opp. sides of $\square$ are $\cong$.
3. $\overline{WX} \cong \overline{WX}$	3. Reflexive Property
4. $\triangle WZX \cong \triangle XYW$	4. SSS
5. $\angle ZWX \cong \angle YXW$	5. CPCTC
6. $\angle ZWX$ and $\angle YXW$ are supplementary.	6. Consec. $\angle$s of $\square$ are suppl.
7. $\angle ZWX$ and $\angle YXW$ are right angles.	7. If 2 $\angle$s are $\cong$ and suppl, each $\angle$ is a rt. $\angle$.
8. $\angle WZY$ and $\angle XYZ$ are right angles.	8. If $\square$ has 1 rt. $\angle$, it has 4 rt. $\angle$s.
9. $WXYZ$ is a rectangle.	9. Def. of rectangle

43. Given: $DEAC$ and $FEAB$ are rectangles.
$\angle GKH \cong \angle JHK$; $\overline{GJ}$ and $\overline{HK}$ intersect at L.
Prove: $GHJK$ is a parallelogram.

Proof:

Statements	Reasons
1. $DEAC$ and $FEAB$ are rectangles. $\angle GKH \cong \angle JHK$ $\overline{GJ}$ and $\overline{HK}$ intersect at L.	1. Given
2. $\overline{DE} \parallel \overline{AC}$ and $\overline{FE} \parallel \overline{AB}$	2. Def. of parallelogram
3. plane $\mathcal{N} \parallel$ plane $\mathcal{M}$	3. Def. of parallel planes
4. G, J, H, K, L are in the same plane.	4. Def. of intersecting lines
5. $\overline{GH} \parallel \overline{KJ}$	5. Def. of parallel lines
6. $\overline{GK} \parallel \overline{HJ}$	6. If alt. int. $\angle$s are $\cong$, lines are $\parallel$.
7. $GHJK$ is a parallelogram.	7. Def. of parallelogram

45. No; there are no parallel lines in spherical geometry.
47. No; the sides are <u>not</u> parallel. **49.** A **51.** 31 **53.** 43
55. 49 **57.** 5 **59.** $\sqrt{297} \approx 17.2$ **61.** 5 **63.** 29

Pages 434–437 Lesson 8-5

1. Sample answer:

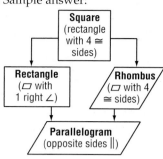

3. A square is a rectangle with all sides congruent.
5. 5 **7.** 96.8 **9.** None; the diagonals are not congruent or perpendicular. **11.** If the measure of each angle is 90 or if the diagonals are congruent, then the floor is a square. **13.** 120 **15.** 30

17. 53 **19.** 5 **21.** Rhombus; the diagonals are perpendicular. **23.** None; the diagonals are not congruent or perpendicular.
25. Sample answer:

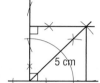

27. always **29.** sometimes
31. always **33.** 40 cm

35. Given: $ABCD$ is a parallelogram.
$\overline{AC} \perp \overline{BD}$
Prove: $ABCD$ is a rhombus.

Proof: We are given that $ABCD$ is a parallelogram. The diagonals of a parallelogram bisect each other, so $\overline{AE} \cong \overline{EC}$. $\overline{BE} \cong \overline{BE}$ because congruence of segments is reflexive. We are also given that $\overline{AC} \perp \overline{BD}$. Thus, $\angle AEB$ and $\angle BEC$ are right angles by the definition of perpendicular lines. Then $\angle AEB \cong \angle BEC$ because all right angles are congruent. Therefore, $\triangle AEB \cong \triangle CEB$ by SAS. $\overline{AB} \cong \overline{CB}$ by CPCTC. Opposite sides of parallelograms are congruent, so $\overline{AB} \cong \overline{CD}$ and $\overline{BC} \cong \overline{AD}$. Then since congruence of segments is transitive, $\overline{AB} \cong \overline{CD} \cong \overline{CB} \cong \overline{AD}$. All four sides of $ABCD$ are congruent, so $ABCD$ is a rhombus by definition.
37. No; it is about 11,662.9 mm. **39.** The flag of Denmark contains four red rectangles. The flag of St. Vincent and the Grenadines contains a blue rectangle, a green rectangle, a yellow rectangle, a blue and yellow rectangle, a yellow and green rectangle, and three green rhombi. The flag of Trinidad and Tobago contains two white parallelograms and one black parallelogram.

41. Given: $\triangle TPX \cong \triangle QPX \cong$
$\triangle QRX \cong \triangle TRX$
Prove: $TPQR$ is a rhombus.

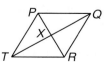

Proof:

Statements	Reasons
1. $\triangle TPX \cong \triangle QPX \cong$ $\triangle QRX \cong \triangle TRX$	1. Given
2. $\overline{TP} \cong \overline{PQ} \cong \overline{QR} \cong \overline{TR}$	2. CPCTC
3. $TPQR$ is a rhombus.	3. Def. of rhombus

43. Given: $QRST$ and $QRTV$ are rhombi.
Prove: $\triangle QRT$ is equilateral.

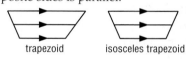

Proof:

Statements	Reasons
1. $QRST$ and $QRTV$ are rhombi.	1. Given
2. $\overline{QV} \cong \overline{VT} \cong \overline{TR} \cong \overline{QR}$, $\overline{QT} \cong \overline{TS} \cong \overline{RS} \cong \overline{QR}$	2. Def. of rhombus
3. $\overline{QT} \cong \overline{TR} \cong \overline{QR}$	3. Substitution Property
4. $\triangle QRT$ is equilateral.	4. Def. of equilateral triangle

45. Sample answer: You can ride a bicycle with square wheels over a curved road. Answers should include the following.
- Rhombi and squares both have all four sides congruent, but the diagonals of a square are congruent. A square has four right angles and rhombi have each pair of opposite angles congruent, but not all angles are necessarily congruent.
- Sample answer: Since the angles of a rhombus are not all congruent, riding over the same road would not be smooth.

47. C **49.** 140 **51.** $x = 2, y = 3$ **53.** yes **55.** no
57. 13.5 **59.** 20 **61.** $\angle AJH \cong \angle AHJ$ **63.** $\overline{AK} \cong \overline{AB}$
65. 2.4 **67.** 5

Pages 442–445 Lesson 8-6

1. Exactly one pair of opposite sides is parallel.
3. Sample answer: The median of a trapezoid is parallel to both bases.

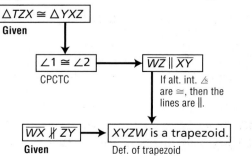

trapezoid   isosceles trapezoid

5. isosceles, $QR = \sqrt{20}$, $ST = \sqrt{20}$ **7.** 4 **9a.** $\overline{AD} \parallel \overline{BC}$, $\overline{CD} \parallel \overline{AB}$ **9b.** not isosceles, $AB = \sqrt{17}$ and $CD = 5$
11a. $\overline{DC} \parallel \overline{FE}$, $\overline{DE} \parallel \overline{FC}$ **11b.** isosceles, $DE = \sqrt{50}$, $CF = \sqrt{50}$ **13.** 8 **15.** 14, 110, 110 **17.** 62 **19.** 15
21. Sample answer: triangles, quadrilaterals, trapezoids, hexagons **23.** trapezoid, exactly one pair opp. sides $\parallel$
25. square, all sides $\cong$, consecutive sides $\perp$ **27.** $A(-2, 3.5)$, $B(4, -1)$ **29.** $\overline{DG} \parallel \overline{EF}$, not isosceles, $DE \neq GF$, $\overline{DE} \parallel GF$
31. $WV = 6$

33. Given: $\triangle TZX \cong \triangle YXZ$, $\overline{WX} \parallel\!\!\!/ \ \overline{ZY}$
Prove: $XYZW$ is a trapezoid.

Proof:

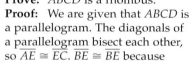

35. Given: *E* and *C* are midpoints of $\overline{AD}$ and $\overline{DB}$;
$\overline{AD} \cong \overline{DB}$

Prove: *ABCE* is an isosceles trapezoid.

Proof:

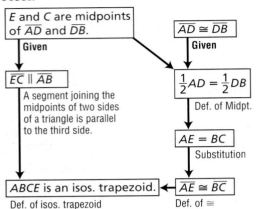

| E and C are midpoints of $\overline{AD}$ and $\overline{DB}$. | $\overline{AD} \cong \overline{DB}$ |

Given → Given

$\overline{EC} \parallel \overline{AB}$ — A segment joining the midpoints of two sides of a triangle is parallel to the third side.

$\frac{1}{2}AD = \frac{1}{2}DB$ — Def. of Midpt.

$AE = BC$ — Substitution

ABCE is an isos. trapezoid. ← $\overline{AE} \cong \overline{BC}$

Def. of isos. trapezoid   Def. of $\cong$

37. Sample answer: 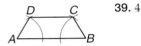   **39.** 4

41. Sample answer: Trapezoids are used in monuments as well as other buildings. Answers should include the following.
- Trapezoids have exactly one pair of opposite sides parallel.
- Trapezoids can be used as window panes.

43. B   **45.** 10   **47.** 70   **49.** $RS = 7\sqrt{2}$, $TV = \sqrt{113}$
51. No; opposite sides are not congruent and the diagonals do not bisect each other.   **53.** $\frac{17}{5}$ **55.** $\frac{13}{2}$ **57.** 0   **59.** $\frac{2b}{a}$
61. $\frac{c}{b}$

Page 445   Chapter 8   Practice Quiz 2
1. 12   **3.** rhombus, opp. sides $\parallel$, diag. $\perp$, consec. sides not $\perp$ **5.** 18

Pages 449–451   Lesson 8-7
1. Place one vertex at the origin and position the figure so another vertex lies on the positive *x*-axis.
3. 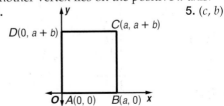   **5.** (c, b)

7.   Given: *ABCD* is a square.
Prove: $\overline{AC} \perp \overline{DB}$

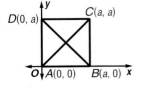

Proof:
Slope of $\overline{DB} = \dfrac{0 - a}{a - 0}$ or -1
Slope of $\overline{AC} = \dfrac{0 - a}{0 - a}$ or 1
The slope of $\overline{AC}$ is the negative reciprocal of the slope of $\overline{DB}$, so they are perpendicular.

9.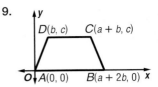

11. $B(-b, c)$
13. $G(a, 0)$, $E(-b, c)$
15. $T(-2a, c)$, $W(-2a, -c)$

17.   Given: *ABCD* is a rectangle.
Prove: $\overline{AC} \cong \overline{DB}$
Proof:
Use the Distance Formula to find $AC = \sqrt{a^2 + b^2}$ and $BD = \sqrt{a^2 + b^2}$. $\overline{AC}$ and $\overline{BC}$ have the same length, so they are congruent.

19.   Given: isosceles trapezoid *ABCD* with $\overline{AD} \cong \overline{BC}$
Prove: $\overline{BD} \cong \overline{AC}$
Proof:
$BD = \sqrt{(a - b)^2 + (0 - c)^2} = \sqrt{(a - b)^2 + c^2}$
$AC = \sqrt{((a - b) - 0)^2 + (c - 0)^2} = \sqrt{(a - b)^2 + c^2}$
$BD = AC$ and $\overline{BD} \cong \overline{AC}$

21.   Given: *ABCD* is a rectangle. *Q*, *R*, *S*, and *T* are midpoints of their respective sides.

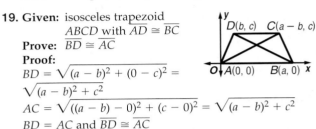

Prove: *QRST* is a rhombus.
Proof:
Midpoint *Q* is $\left(\dfrac{0 + 0}{2}, \dfrac{b + 0}{2}\right)$ or $\left(0, \dfrac{b}{2}\right)$.
Midpoint *R* is $\left(\dfrac{a + 0}{2}, \dfrac{b + b}{2}\right)$ or $\left(\dfrac{a}{2}, \dfrac{2b}{2}\right)$ or $\left(\dfrac{a}{2}, b\right)$
Midpoint *S* is $\left(\dfrac{a + a}{2}, \dfrac{b + 0}{2}\right)$ or $\left(\dfrac{2a}{2}, \dfrac{b}{2}\right)$ or $\left(a, \dfrac{b}{2}\right)$.
Midpoint *T* is $\left(\dfrac{a + 0}{2}, \dfrac{0 + 0}{2}\right)$ or $\left(\dfrac{a}{2}, 0\right)$.

$QR = \sqrt{\left(\dfrac{a}{2} - 0\right)^2 + \left(b - \dfrac{b}{2}\right)^2} = \sqrt{\left(\dfrac{a}{2}\right)^2 + \left(\dfrac{b}{2}\right)^2}$

$RS = \sqrt{\left(a - \dfrac{a}{2}\right)^2 + \left(\dfrac{b}{2} - b\right)^2} = \sqrt{\left(\dfrac{a}{2}\right)^2 + \left(-\dfrac{b}{2}\right)^2}$ or $\sqrt{\left(\dfrac{a}{2}\right)^2 + \left(\dfrac{b}{2}\right)^2}$

$ST = \sqrt{\left(a - \dfrac{a}{2}\right)^2 + \left(\dfrac{b}{2} - 0\right)^2} = \sqrt{\left(\dfrac{a}{2}\right)^2 + \left(\dfrac{b}{2}\right)^2}$

$QT = \sqrt{\left(\dfrac{a}{2} - 0\right)^2 + \left(0 - \dfrac{b}{2}\right)^2} = \sqrt{\left(\dfrac{a}{2}\right)^2 + \left(-\dfrac{b}{2}\right)^2}$ or $\sqrt{\left(\dfrac{a}{2}\right)^2 + \left(\dfrac{b}{2}\right)^2}$

$QR = RS = ST = QT$ so $\overline{QR} \cong \overline{RS} \cong \overline{ST} \cong \overline{QT}$. *QRST* is a rhombus.

23. Sample answer: $C(a + c, b)$, $D(2a + c, 0)$ **25.** No, there is not enough information given to prove that the sides of the tower are parallel.   **27.** Sample answer: The coordinate plane is used in coordinate proofs. The Distance Formula, Midpoint Formula and Slope Formula are used to prove theorems. Answers should include the following.
- Place the figure so one of the vertices is at the origin. Place at least one side of the figure on the positive *x*-axis. Keep the figure in the first quadrant if possible and use coordinates that will simplify calculations.
- Sample answer: Theorem 8.3 Opposite sides of a parallelogram are congruent.

29. A   **31.** 55   **33.** 160   **35.** $\sqrt{60} \approx 7.7$ **37.** $m\angle XVZ = m\angle VXZ$ **39.** $m\angle XZY > m\angle ZXY$

1. true **3.** false, rectangle **5.** false, trapezoid **7.** true
9. 120 **11.** 90 **13.** $m\angle W = 62$, $m\angle X = 108$, $m\angle Y = 80$,
$m\angle Z = 110$ **15.** 52 **17.** 87.9 **19.** 6 **21.** no **23.** yes
25. 52 **27.** 28 **29.** Yes, opp. sides are parallel and diag.
are congruent **31.** 7.5 **33.** 102

35. Given: $ABCD$ is a square.
Prove: $\overline{AC} \perp \overline{BD}$

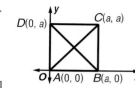

Proof:
Slope of $\overline{AC} = \dfrac{a - 0}{a - 0}$ or 1
Slope of $\overline{BD} = \dfrac{a - 0}{0 - a}$ or -1
The slope of $\overline{AC}$ is the negative reciprocal of the slope
of $\overline{BD}$. Therefore, $\overline{AC} \perp \overline{BD}$.
37. $P(3a, c)$

Chapter 9 Transformations

Page 461 Chapter 9 Getting Started

1.

3.
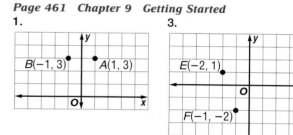

5.
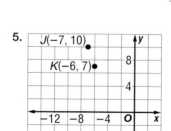

7. 36.9 **9.** 41.8 **11.** 41.4
13. $\begin{bmatrix} -5 & -1 \\ 10 & 5 \end{bmatrix}$

15. $\begin{bmatrix} -2 & -5 & 1 \\ 3 & -4 & -5 \end{bmatrix}$

1. Sample Answer: The centroid of an equilateral triangle is
not a point of symmetry. **3.** angle measure, betweenness
of points, collinearity, distance **5.** 4; yes **7.** 6; yes

9.
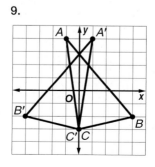

11.
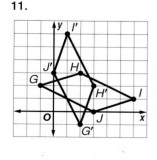

13. 4, yes **15.** $\overline{YX}$ **17.** $\angle XZW$ **19.** $\overline{UV}$ **21.** T
23. $\triangle WTZ$

25.
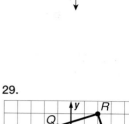

27.
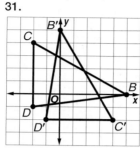

29.
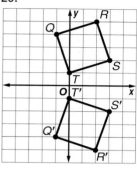

31.
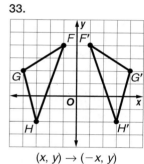

33.
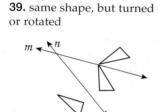

$(x, y) \rightarrow (-x, y)$

35. 2; yes **37.** 1; no
39. same shape, but turned
or rotated

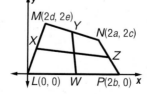

41. $A(4, 7)$, $B(10, -3)$, and $C(-6, -8)$ **43.** Consider point
(a, b). Upon reflection in the origin, its image is $(-a, -b)$.
Upon reflection in the x-axis and then the y-axis, its image
is $(a, -b)$ and then $(-a, -b)$. The images are the same.
45. vertical line of symmetry **47.** vertical, horizontal lines
of symmetry; point of symmetry at the center **49.** D

51. Given: Quadrilateral $LMNP$; X, Y, Z, and W are
midpoints of their respective sides.
Prove: $\overline{YW}$ and $\overline{XZ}$
bisect each other.
Proof:
Midpoint Y of $\overline{MN}$ is
$\left(\dfrac{2d + 2a}{2}, \dfrac{2e + 2c}{2}\right)$ or
$(d + a, e + c)$.
Midpoint Z of $\overline{NP}$ is
$\left(\dfrac{2a + 2b}{2}, \dfrac{2c + 0}{2}\right)$
or $(a + b, c)$. Midpoint W of $\overline{PL}$ is $\left(\dfrac{0 + 2b}{2}, \dfrac{0 + 0}{2}\right)$ or $(b, 0)$.
Midpoint X of $\overline{LM}$ is $\left(\dfrac{0 + 2d}{2}, \dfrac{0 + 2e}{2}\right)$ or (d, e). Midpoint
of $\overline{WY}$ is $\left(\dfrac{d + a + b}{2}, \dfrac{e + c + 0}{2}\right)$ or $\left(\dfrac{a + b + d}{2}, \dfrac{c + e}{2}\right)$.
Midpoint of $\overline{XZ}$ is $\left(\dfrac{d + a + b}{2}, \dfrac{e + c}{2}\right)$ or $\left(\dfrac{a + b + d}{2}, \dfrac{c + e}{2}\right)$.

The midpoints of $\overline{XZ}$ and $\overline{WY}$ are the same, so $\overline{XZ}$ and
$\overline{WY}$ bisect each other.

53. 40 **55.** 36 **57.** $f \approx 25.5$, $m\angle H = 76$, $h \approx 28.8$ **59.** $\sqrt{2}$
61. $\sqrt{5}$

Pages 470–475 Lesson 9-2
1. Sample answer: $A(3, 5)$ and $B(-4, 7)$; start at 3, count to the left to -4, which is 7 units to the left or -7. Then count up 2 units from 5 to 7 or $+2$. The translation from A to B is $(x, y) \rightarrow (x - 7, y + 2)$. **3.** Allie; counting from the point $(-2, 1)$ to $(1, -1)$ is right 3 and down 2 to the image. The reflections would be too far to the right. The image would be reversed as well. **5.** No; quadrilateral $WXYZ$ is oriented differently than quadrilateral $NPQR$.

7.
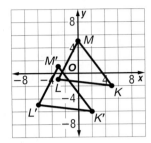

9. Yes; it is one reflection after another with respect to the two parallel lines.
11. No; it is a reflection followed a rotation.
13. Yes; it is one reflection after another with respect to the two parallel lines.

15.
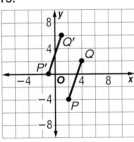

17.
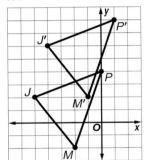

19.
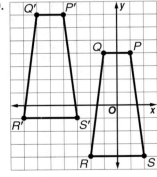

21. left 3 squares and down 7 squares
23. 48 in. right
25. 72 in. right, $24\sqrt{3}$ in. down

27.
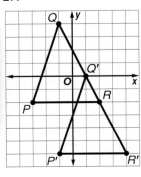

29.
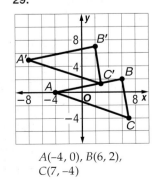

$A(-4, 0)$, $B(6, 2)$, $C(7, -4)$

31. more brains; more free time **33.** No; the percent per figure is different in each category. **35.** Translations and reflections preserve the congruences of segments and angles. The composition of the two transformations will preserve both congruences. Therefore, a glide reflection is an isometry.

37.
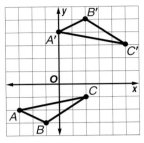

39. A
41.
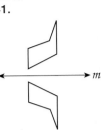

43. $Q(a - b, c)$, $T(0, 0)$ **45.** 23 ft **47.** You did not fill out an application. **49.** The two lines are not parallel. **51.** 5
53. $3\sqrt{2}$
55.

45°
57.

60°

59.

150°

Pages 476–482 Lesson 9-3
1. clockwise $(x, y) \rightarrow (y, -x)$; counterclockwise $(x, y) \rightarrow (-y, x)$

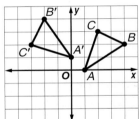

3. Both translations and rotations are made up of two reflections. The difference is that translations reflect across parallel lines and rotations reflect across intersecting lines.

5.

7.

$X(-5, 8)$ $X'\left(\dfrac{3\sqrt{2}}{2}, \dfrac{13\sqrt{2}}{2}\right)$

$Y(0, 3)$ $Y'\left(\dfrac{3\sqrt{2}}{2}, \dfrac{3\sqrt{2}}{2}\right)$

9. order 6; magnitude 60°
11. order 5 and magnitude 72°; order 4 and magnitude 90°; order 3 and magnitude 120°

13.

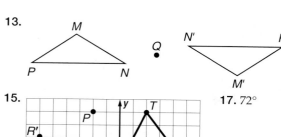

15.

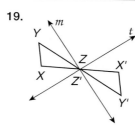

17. 72°

19.

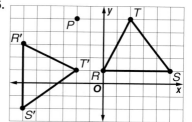

21.

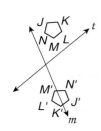

23. $K''(0, -5)$, $L''(4, -2)$, and $M''(4, 2)$; 90° clockwise

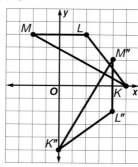

25. $(\sqrt{3}, 1)$ **27.** Yes; it is a proper successive reflection with respect to the two intersecting lines. **29.** yes **31.** no **33.** 9 **35.** $(x, y) \rightarrow (y, -x)$ **37.** any point on the line of reflection **39.** no invariant points **41.** B

43.

Transformation	angle measure	betweenness of points	orientation	collinearity	distance measure
reflection	yes	yes	no	yes	yes
translation	yes	yes	yes	yes	yes
rotation	yes	yes	yes	yes	yes

45. direct **47.** Yes; it is one reflection after another with respect to the two parallel lines. **49.** Yes; it is one reflection after another with respect to the two parallel lines. **51.** C **53.** $\angle AGF$ **55.** $\overline{TR}$; diagonals bisect each other **57.** $\angle QRS$; opp. $\angle s \cong$ **59.** no **61.** yes **63.** (0, 4), (1, 2), (2, 0) **65.** (0, 12), (1, 8), (2, 4), (3, 0) **67.** (0, 12), (1, 6), (2, 0)

Page 482 Chapter 9 Practice Quiz 1

1.

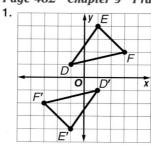

3.

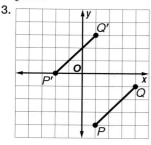

5. order 36; magnitude 10°

Pages 483–488 Lesson 9-4

1. Semi-regular tessellations contain two or more regular polygons, but uniform tessellations can be any combination of shapes. **3.** The figure used in the tesselation appears to be a trapezoid, which is not a regular polygon. Thus, the tessellation cannot be regular. **5.** no; measure of interior angle = 168 **7.** yes **9.** yes; not uniform **11.** no; measure of interior angle = 140 **13.** yes; measure of interior angle = 60 **15.** no; measure of interior angle ≈ 164.3 **17.** no **19.** yes **21.** yes; uniform **23.** yes; not uniform **25.** yes; not uniform **27.** yes; uniform, regular **29.** semi-regular, uniform **31.** Never; semi-regular tessellations have the same combination of shapes and angles at each vertex like uniform tessellations. The shapes for semi-regular tessellations are just regular. **33.** Always; the sum of the measures of the angles of a quadrilateral is 360°. So if each angle of the quadrilateral is rotated at the vertex, then that equals 360° and the tessellation is possible. **35.** yes **37.** uniform, regular **39.** Sample answer: Tessellations can be used in art to create abstract art. Answers should include the following.

- The equilateral triangles are arranged to form hexagons, which are arranged adjacent to one another.
- Sample answers: kites, trapezoids, isosceles triangles

41. A

43.

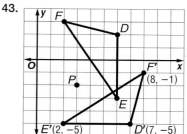

45.

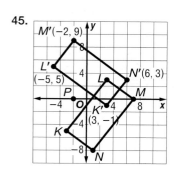

47. $x = 4$, $y = 1$
49. $x = 56$, $y = 12$
51. no, no **53.** yes, no
55. no, no **57.** $AB = 7$, $BC = 10$, $AC = 9$
59. $1(-1) = -1$ and $-1(1) = -1$ **61.** square
63. 15 **65.** 22.5

Pages 490–497 Lesson 9-5

1. Dilations only preserve length if the scale factor is 1 or −1. So for any other scale factor, length is not preserved and the dilation is not an isometry. **3.** Trey; Desiree found the image using a positive scale factor.

5.

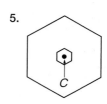

7. $A'B' = 12$

9.

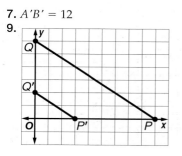

11. $r = 2$; enlargement   **13.** C

15.
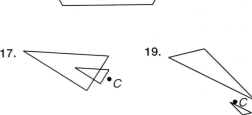

17.
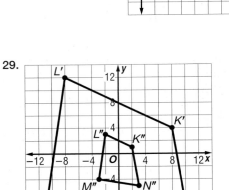

19.

21. $S'T' = \frac{3}{5}$
23. $ST = 4$
25. $S'T' = 0.9$

27.
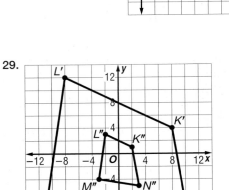

29.
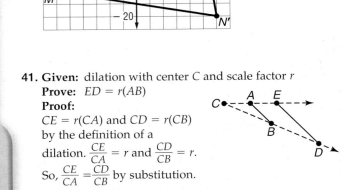

31. $\frac{1}{2}$; reduction
33. $\frac{1}{3}$; reduction
35. -2; enlargement
37. 7.5 by 10.5
39. The perimeter is four times the original perimeter.

41. Given: dilation with center C and scale factor r
Prove: $ED = r(AB)$
Proof:
$CE = r(CA)$ and $CD = r(CB)$ by the definition of a dilation. $\frac{CE}{CA} = r$ and $\frac{CD}{CB} = r$.
So, $\frac{CE}{CA} = \frac{CD}{CB}$ by substitution.
$\angle ACB \cong \angle ECD$, since congruence of angles is reflexive. Therefore, by SAS Similarity, $\triangle ACB$ is similar to $\triangle ECD$. The corresponding sides of similar triangles are proportional, so $\frac{ED}{AB} = \frac{CE}{CA}$. We know that $\frac{CE}{CA} = r$, so $\frac{ED}{AB} = r$ by substitution. Therefore, $ED = r(AB)$ by the Multiplication Property of Equality.

43. 2   **45.** $\frac{1}{20}$ **47.** 60%   **49.**
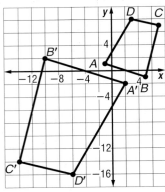

51.
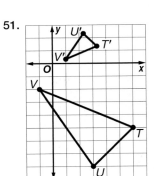

53. Sample answer: Yes; a cut and paste produces an image congruent to the original. Answers should include the following.
- Congruent figures are similar, so cutting and pasting is a similarity transformation.
- If you scale both horizontally and vertically by the same factor, you are creating a dilation.

55. A   **57.** no   **59.** no

61.
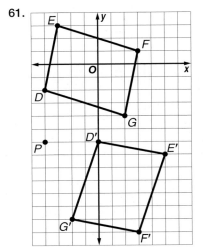

63. Given: $\angle J \cong \angle L$ B is the midpoint of $\overline{JL}$.
Prove: $\triangle JHB \cong \triangle LCB$
Proof: It is known that $\angle J \cong \angle L$. Since B is the midpoint of $\overline{JL}$, $\overline{JB} \cong \overline{LB}$ by the Midpoint Theorem.
$\angle JBH \cong \angle LBC$ because vertical angles are congruent. Thus, $\triangle JHB \cong \triangle LCB$ by ASA.   **65.** 76.0

1. yes; uniform; semi-regular   **3.**

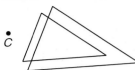

5. $A'(-5, -1)$, $B'\left(-\frac{1}{2}, -3\right)$, $C'(2, -2)$

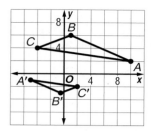

Pages 498–505   Lesson 9-6

1. Sample answer; $\langle 7, 7 \rangle$

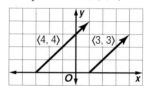

3. Sample answer: Using a vector to translate a figure is the same as using an ordered pair because a vector has horizontal and vertical components, each of which can be represented by one coordinate of an ordered pair.

5. $\langle 4, -3 \rangle$

7. $2\sqrt{13} \approx 7.2$, $\approx 213.7°$

9.

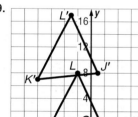

11.

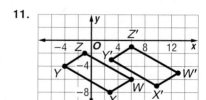

13. $6\sqrt{13} \approx 21.6$, $303.7°$ **15.** $\langle 2, 6 \rangle$ **17.** $\langle -7, -4 \rangle$
19. $\langle -3, 5 \rangle$ **21.** $5, 0°$ **23.** $2\sqrt{5} \approx 4.5$, $296.6°$ **25.** $7\sqrt{5} \approx$
15.7, $26.6°$ **27.** 25, $\approx 73.7°$ **29.** $5\sqrt{41} \approx 32.0$, $\approx 218.7°$
31. $6\sqrt{2} \approx 8.5$, $135.0°$ **33.** $4\sqrt{10} \approx 12.6$, $198.4°$
35. $2\sqrt{122} \approx 22.1$, $275.2°$
37.   **39.**

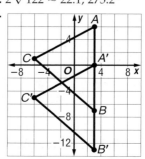

41.

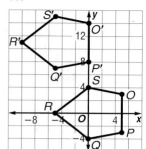

43.

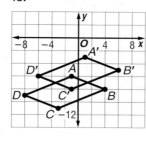

45.

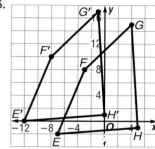

47. 13, $\approx 67.4°$
49. 5, $\approx 306.9°$
51. $2\sqrt{5} \approx 4.5$, $\approx 26.6°$
53. about 44.8 mi; about $38.7°$ south of due east
55. $\langle -350, 450 \rangle$ mph
57. $52.1°$ north of due west

59. Sample answer: Quantities such as velocity are vectors. The velocity of the wind and the velocity of the plane together factor into the overall flight plan. Answers should include the following.
• A wind from the west would add to the velocity contributed by the plane resulting in an overall velocity with a larger magnitude.
• When traveling east, the prevailing winds add to the velocity of the plane. When traveling west, they detract from it.
61. D   **63.** $A'B' = 6$ **65.** $AB = 48$ **67.** yes; not uniform
69. 12   **71.** 30
73. $\begin{bmatrix} -4 & -3 \\ -10 & 4 \end{bmatrix}$ **75.** $\begin{bmatrix} -27 & -15 & -3 \\ 27 & 3 & 15 \end{bmatrix}$ **77.** $\begin{bmatrix} 12 & 4 \\ -4 & -12 \end{bmatrix}$

Pages 506–511   Lesson 9-7

1. $\begin{bmatrix} 0 & 1 \\ 1 & 0 \end{bmatrix}$ **3.** Sample answer: $\begin{bmatrix} -2 & -2 & -2 & -2 \\ -1 & -1 & -1 & -1 \end{bmatrix}$

5. $D'(-1, 9)$, $E'(5, 9)$, $F'(3, 6)$, $G'(-3, 6)$ **7.** $A'\left(-\frac{1}{4}, -\frac{1}{2}\right)$,
$B'\left(-\frac{3}{4}, -\frac{3}{4}\right)$, $C'\left(-\frac{3}{4}, -\frac{5}{4}\right)$, $D'\left(-\frac{1}{4}, -1\right)$ **9.** $H'(5, 4)$, $I'(1, -1)$,
$J'(3, -6)$, $K'(7, -3)$ **11.** $P'(3, -6)$, $Q'(7, -6)$, $R'(7, -2)$
13. $(1.5, -0.5)$, $(3.5, -1.5)$, $(2.5, -3.5)$, $(0.5, -2.5)$
15. $E'(-6, 6)$, $F'(-3, 8)$ **17.** $M'(1, 1)$, $N'(5, 3)$, $O'(5, 1)$,
$P'(1, -1)$ **19.** $A'(12, 10)$, $B'(8, 10)$, $C'(6, 14)$ **21.** $G'(-2, -1)$,
$H'(2, -3)$, $I'(3, 4)$, $J'(-3, 5)$ **23.** $X'(-2, 2)$, $Y'(-4, -1)$
25. $D'(-4, -5)$, $E'(2, -6)$, $F'(3, -1)$, $G'(-3, 4)$
27. $V'(-2, 2)$, $W'\left(\frac{2}{3}, 2\right)$, $X'\left(2, -\frac{4}{3}\right)$ **29.** $V'(-3, -3)$,
$W'(-3, 1)$, $X'(2, 3)$ **31.** $P'(2, -3)$, $Q'(-1, -1)$, $R'(1, 2)$,
$S'(3, 2)$, $T'(5, -1)$ **33.** $P'(1, -1)$, $Q'(4, 1)$, $R'(2, 4)$, $S'(0, 4)$,
$T'(-2, 1)$ **35.** $M'(-1, 12)$, $N'(-10, -3)$ **37.** $S'(-1, 2)$,
$T'(-1, 6)$, $U'(3, 5)$, $V'(3, 1)$ **39.** $A'\left(-1, -\frac{1}{3}\right)$, $B'\left(-\frac{2}{3}, -\frac{4}{3}\right)$,
$C'\left(\frac{2}{3}, -\frac{4}{3}\right)$, $D'\left(1, -\frac{1}{3}\right)$, $E'\left(\frac{2}{3}, \frac{2}{3}\right)$, $F'\left(-\frac{2}{3}, \frac{2}{3}\right)$ **41.** $A'(2, 1)$,
$B'(5, 2)$, $C'(5, 6)$, $D'(2, 7)$, $E'(-1, 6)$, $F'(-1, 2)$ **43.** Each footprint is reflected in the y-axis, then translated up two units.
45. $\begin{bmatrix} -1 & 0 \\ 0 & 1 \end{bmatrix}$ **47.** $\begin{bmatrix} 0 & -1 \\ -1 & 0 \end{bmatrix}$ **49.** $\begin{bmatrix} 0 & 1 \\ -1 & 0 \end{bmatrix}$

51.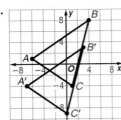

53. $-\frac{1}{2}$; reduction
55. 60, 120   **57.** 36, 144

diameter, but $2r$ is the measure of the diameter. So the diameter has to be longer than any other chord of the circle.
5. $\overline{EA}$, $\overline{EB}$, $\overline{EC}$, or $\overline{ED}$ **7.** $\overline{AC}$ or $\overline{BD}$ **9.** 10.4 in.   **11.** 6
13. 10 m, 31.42 m   **15.** B   **17.** $\overline{FA}$, $\overline{FB}$, or $\overline{FE}$ **19.** $\overline{BE}$
21. $\odot R$ **23.** $\overline{ZV}$, $\overline{TX}$, or $\overline{WZ}$ **25.** $\overline{RU}$, $\overline{RV}$ **27.** 2.5 ft
29. 64 in. or 5 ft 4 in.   **31.** 0.6 m   **33.** 3   **35.** 12   **37.** 34
39. 20   **41.** 5   **43.** 2.5   **45.** 13.4 cm, 84.19 cm
47. 24.32 m, 12.16 m   **49.** $13\frac{1}{2}$ in., 42.41 in.   **51.** 0.33a, 1.05a
53. 5π ft   **55.** 8π cm   **57.** 0; The longest chord of a circle is the diameter, which contains the center.   **59.** 500–600 ft
61. 24π units   **63.** 27   **65.** $10\pi, 20\pi, 30\pi$ **67.** 9.8; 66°
69. 44.7; 27°   **71.** 24

Pages 512–516   Chapter 9   Study Guide and Review
1. false, center   **3.** false, component form   **5.** false, center of rotation   **7.** false, scale factor
9. 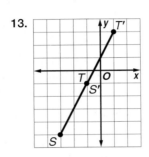   **11.**

13. 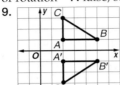   **15.** $B'(3, -5)$, $C'(3, -3)$, $D'(5, -3)$; 180°

17. $L'(-2, 2)$, $M'(-3, 5)$, N' $(-6, 3)$; 90° counterclockwise

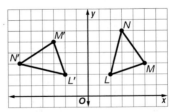

19. 200°   **21.** yes; not uniform   **23.** yes; uniform
25. Yes; the measure of an interior angle is 60, which is a factor of 360.
27. $C'D' = 24$
29. $CD = 4$

31. $C'D' = 10$ **33.** $P'(2, -6)$, $Q'(-4, -4)$, $R'(-2, 2)$
35. $\langle 3, 4 \rangle$ **37.** $\langle 0, 8 \rangle$ **39.** ≈ 14.8, $\approx 208.3°$ **41.** ≈ 72.9, $\approx 213.3°$ **43.** $D'\left(-\frac{12}{5}, -\frac{8}{5}\right)$, $E'(0, 4)$, $F'\left(\frac{8}{5}, -\frac{16}{5}\right)$
45. $D'(-2, 3)$, $E'(5, 0)$, $F'(-4, -2)$ **47.** $W'(-16, 2)$, $X'(-4, 6)$, $Y'(-2, 0)$, $Z'(-12, -6)$

Chapter 10   Circles

Pages 521   Chapter 10   Getting Started
1. 162   **3.** 2.4   **5.** $r = \frac{C}{2p}$ **7.** 15   **9.** 17.0
11. 1.5, −0.9   **13.** 2.5, −3

Pages 522–528   Lesson 10-1
1. Sample answer: The value of π is calculated by dividing the circumference of a circle by the diameter.   **3.** Except for a diameter, two radii and a chord of a circle can form a triangle. The Triangle Inequality Theorem states that the sum of two sides has to be greater than the third. So, $2r$ has to be greater than the measure of any chord that is not a

73. Given: $\overline{RQ}$ bisects $\angle SRT$.
   Prove: $m\angle SQR > m\angle SRQ$

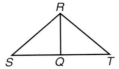

Proof:

Statements	Reasons
1. $\overline{RQ}$ bisects $\angle SRT$.	1. Given
2. $\angle SRQ \cong \angle QRT$	2. Def. of $\angle$ bisector
3. $m\angle SRQ = m\angle QRT$	3. Def. of $\cong$ $\angle s$
4. $m\angle SQR = m\angle T + m\angle QRT$	4. Exterior Angle Theorem
5. $m\angle SQR > m\angle QRT$	5. Def. of Inequality
6. $m\angle SQR > m\angle SRQ$	6. Substitution

75. 60   **77.** 30   **79.** 30

Pages 529–535   Lesson 10-2
1. Sample answer:
$\widehat{AB}$, $\widehat{BC}$, $\widehat{AC}$, $\widehat{ABC}$, $\widehat{BCA}$, $\widehat{CAB}$; $m\widehat{AB} = 110$, $m\widehat{BC} = 160$, $m\widehat{AC} = 90$, $m\widehat{ABC} = 270$, $m\widehat{BCA} = 250$, $m\widehat{CAB} = 200$ **3.** Sample answer: Concentric circles have the same center, but different radius measures; congruent circles usually have different centers but the same radius measure.   **5.** 137   **7.** 103   **9.** 180   **11.** 138
13. Sample answer: 25% = 90°, 23% = 83°, 28% = 101°, 22% = 79°, 2% = 7°   **15.** 60   **17.** 30   **19.** 120   **21.** 115
23. 65   **25.** 90   **27.** 90   **29.** 135   **31.** 270   **33.** 76   **35.** 52
37. 256   **39.** 308   **41.** $24\pi \approx 75.40$ units   **43.** $4\pi \approx 12.57$ units   **45.** The first category is a major arc, and the other three categories are minor arcs.   **47.** always   **49.** never
51. $m\angle 1 = 80$, $m\angle 2 = 120$, $m\angle 3 = 160$ **53.** 56.5 ft
55. No; the radii are not equal, so the proportional part of the circumferences would not be the same. Thus, the arcs would not be congruent.   **57.** B   **59.** 20; 62.83
61. 28; 14   **63.** 84.9 newtons, 32° north of due east
65. 36.68   **67.** $\sqrt{24.5}$ **69.** If ABC has three sides, then ABC is a triangle.   **71.** 42   **73.** 100   **75.** 36

Pages 536–543   Lesson 10-3
1. Sample answer: An inscribed polygon has all vertices on the circle. A circumscribed circle means the circle is drawn around so that the polygon lies in its interior and all vertices lie on the circle.   **3.** Tokei; to bisect the chord, it must be a diameter and be perpendicular.   **5.** 30
7. $5\sqrt{3}$ **9.** $10\sqrt{5} \approx 22.36$ **11.** 15   **13.** 15   **15.** 40
17. 80   **19.** 4   **21.** 5   **23.** $m\widehat{AB} = m\widehat{BC} = m\widehat{CD} = m\widehat{DE} = m\widehat{EF} = m\widehat{FG} = m\widehat{GH} = m\widehat{HA} = 45$ **25.** $m\widehat{NP} = m\widehat{RQ} = 120$; $m\widehat{NR} = m\widehat{PQ} = 60$ **27.** 30   **29.** 15   **31.** 16   **33.** 6
35. $\sqrt{2} \approx 1.41$

37. Given: $\odot O, \overline{OS} \perp \overline{RT}, \overline{OV} \perp \overline{UW}, \overline{OS} \cong \overline{OV}$
Prove: $\overline{RT} \cong \overline{UW}$

Proof:

Statements	Reasons
1. $\overline{OT} \cong \overline{OW}$	1. All radii of a $\odot$ are $\cong$.
2. $\overline{OS} \perp \overline{RT}, \overline{OV} \perp \overline{VW},$ $\overline{OS} \cong \overline{OV}$	2. Given
3. $\angle OST, \angle OVW$ are right angles.	3. Definition of $\perp$ lines
4. $\triangle STO \cong \triangle VWO$	4. HL
5. $\overline{ST} \cong \overline{VW}$	5. CPCTC
6. $ST = VW$	6. Definition of $\cong$ segments
7. $2(ST) = 2(VW)$	7. Multiplication Property
8. $\overline{OS}$ bisects $\overline{RT}$; $\overline{OV}$ bisects $\overline{UW}$.	8. Radius $\perp$ to a chord bisects the chord.
9. $RT = 2(ST), UW = 2(VW)$	9. Definition of segment bisector
10. $RT = UW$	10. Substitution
11. $\overline{RT} \cong \overline{UW}$	11. Definition of $\cong$ segments

39. 2.82 in.
41. 18 inches

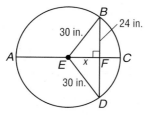

43. $2\sqrt{135} \approx 23.24$ yd

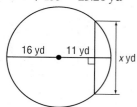

45. Let r be the radius of $\odot P$. Draw radii to points D and E to create triangles. The length DE is $r\sqrt{3}$ and $AB = 2r$; $r\sqrt{3} \neq \frac{1}{2(2r)}$. **47.** Inscribed equilateral triangle; the six arcs making up the circle are congruent because the chords intercepting them were congruent by construction. Each of the three chords drawn intercept two of the congruent chords. Thus, the three larger arcs are congruent. So, the three chords are congruent, making this an equilateral triangle.
49. No; congruent arcs are must be in the same circle, but these are in concentric circles. **51.** Sample answer: The grooves of a waffle iron are chords of the circle. The ones that pass horizontally and vertically through the center are diameters. Answers should include the following.

- If you know the measure of the radius and the distance the chord is from the center, you can use the Pythagorean Theorem to find the length of half of the chord and then multiply by 2.

- There are four grooves on either side of the diameter, so each groove is about 1 in. from the center. In the figure, $EF = 2$ and $EB = 4$ because the radius is half the diameter. Using the Pythagorean Theorem, you find that $FB \approx 3.464$ in. so $AB \approx 6.93$ in. Approximate lengths for

other chords are 5.29 in. and 7.75 in., but exactly 8 in. for the diameter.
53. 14,400 **55.** 180 **57.** $\overline{SU}$ **59.** $\overline{RM}, \overline{AM}, \overline{DM}, \overline{IM}$
61. 50 **63.** 10 **65.** 20

Page 543 Chapter 10 Practice Quiz 1
1. $\overline{BC}, \overline{BD}, \overline{BA}$ **3.** 95 **5.** 9 **7.** 28 **9.** 21

Page 544–551 Lesson 10-4
1. Sample answer:

3. $m\angle 1 = 30, m\angle 2 = 60, m\angle 3 = 60,$ $m\angle 4 = 30, m\angle 5 = 30, m\angle 6 = 60,$ $m\angle 7 = 60, m\angle 8 = 30$ **5.** $m\angle 1 = 35,$ $m\angle 2 = 55, m\angle 3 = 39, m\angle 4 = 39$
7. 1 **9.** $m\angle 1 = m\angle 2 = 30, m\angle 3 = 25$

11. Given: $\widehat{AB} \cong \widehat{DE}, \widehat{AC} \cong \widehat{CE}$
Prove: $\triangle ABC \cong \triangle EDC$

Proof:

Statements	Reasons
1. $\widehat{AB} \cong \widehat{DE}, \widehat{AC} \cong \widehat{CE}$	1. Given
2. $m\widehat{AB} = m\widehat{DE},$ $m\widehat{AC} = m\widehat{CE}$	2. Def. of $\cong$ arcs
3. $\frac{1}{2}m\widehat{AB} = \frac{1}{2}m\widehat{DE}$ $\frac{1}{2}m\widehat{AC} = \frac{1}{2}m\widehat{CE}$	3. Mult. Prop.
4. $m\angle ACB = \frac{1}{2}m\widehat{AB},$ $m\angle ECD = \frac{1}{2}m\widehat{DE},$ $m\angle 1 = \frac{1}{2}m\widehat{AC},$ $m\angle 2 = \frac{1}{2}m\widehat{CE}$	4. Inscribed Angle Theorem
5. $m\angle ACB = m\angle ECD,$ $m\angle 1 = m\angle 2$	5. Substitution
6. $\angle ACB \cong \angle ECD,$ $\angle 1 \cong \angle 2$	6. Def. of $\cong \angle$s
7. $\overline{AB} \cong \overline{DE}$	7. $\cong$ arcs have $\cong$ chords.
8. $\triangle ABC \cong \triangle EDC$	8. AAS

13. $m\angle 1 = m\angle 2 = 13$ **15.** $m\angle 1 = 51, m\angle 2 = 90, m\angle 3 = 39$ **17.** 45, 30, 120 **19.** $m\angle B = 120, m\angle C = 120, m\angle D = 60$ **21.** Sample answer: $\overline{EF}$ is a diameter of the circle and a diagonal and angle bisector of $EDFG$. **23.** 72 **25.** 144
27. 162 **29.** 9 **31.** $\frac{8}{9}$ **33.** 1

35. Given: T lies inside $\angle PRQ$. $\overline{RK}$ is a diameter of $\odot T$.
Prove: $m\angle PRQ = \frac{1}{2}m\widehat{PKQ}$

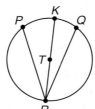

Proof:

Statements	Reasons
1. $m\angle PRQ = m\angle PRK + m\angle KRQ$	1. Angle Addition Theorem
2. $m\widehat{PKQ} = m\widehat{PK} + m\widehat{KQ}$	2. Arc Addition Theorem
3. $\frac{1}{2}m\widehat{PKQ} = \frac{1}{2}m\widehat{PK} + \frac{1}{2}m\widehat{KQ}$	3. Multiplication Property

4. $m\angle PRK = \frac{1}{2}m\widehat{PK}$, | 4. The measure of an inscribed angle whose side is a diameter is half the measure of the intercepted arc (Case 1).
$m\angle KRQ = \frac{1}{2}m\widehat{KQ}$ |

5. $\frac{1}{2}m\widehat{PKQ} = m\angle PRK + m\angle KRQ$ | 5. Substitution (Steps 3, 4)

6. $\frac{1}{2}m\widehat{PKQ} = m\angle PRQ$ | 6. Substitution (Steps 5, 1)

37. Given: inscribed $\angle MLN$ and $\angle CED$, $\overline{CD} \cong \overline{MN}$
Prove: $\angle CED \cong \angle MLN$

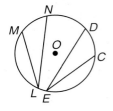

Proof:

Statements	Reasons
1. $\angle MLN$ and $\angle CED$ are inscribed; $\overline{CD} \cong \overline{MN}$	1. Given
2. $m\angle MLN = \frac{1}{2}m\widehat{MN}$; $m\angle CED = \frac{1}{2}m\widehat{CD}$	2. Measure of an inscribed $\angle$ = half measure of intercepted arc.
3. $m\widehat{CD} = m\widehat{MN}$	3. Def. of $\cong$ arcs
4. $\frac{1}{2}m\widehat{CD} = \frac{1}{2}m\widehat{MN}$	4. Mult. Prop.
5. $m\angle CED = m\angle MLN$	5. Substitution
6. $\angle CED \cong \angle MLN$	6. Def. of $\cong$ $\angle$s

39. Given: quadrilateral $ABCD$ inscribed in $\odot O$
Prove: $\angle A$ and $\angle C$ are supplementary. $\angle B$ and $\angle D$ are supplementary.

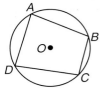

Proof: By arc addition and the definitions of arc measure and the sum of central angles, $m\widehat{DCB} + m\widehat{DAB} = 360$. Since $m\angle C = \frac{1}{2}m\widehat{DAB}$ and $m\angle A = \frac{1}{2}m\widehat{DCB}$, $m\angle C + m\angle A = \frac{1}{2}(m\widehat{DCB} + m\widehat{DAB})$, but $m\widehat{DCB} + m\widehat{DAB} = 360$, so $m\angle C + m\angle A = \frac{1}{2}(360)$ or 80. This makes $\angle C$ and $\angle A$ supplementary. Because the sum of the measures of the interior angles of a quadrilateral is 360, $m\angle A + m\angle C + m\angle B + m\angle D = 360$. But $m\angle A + m\angle C = 180$, so $m\angle B + m\angle D = 180$, making them supplementary also.

41. Isosceles right triangle because sides are congruent radii making it isosceles and $\angle AOC$ is a central angle for an arc of 90°, making it a right angle. **43.** Square because each angle intercepts a semicircle, making them 90° angles. Each side is a chord of congruent arcs, so the chords are congruent.

45. Sample answer: The socket is similar to an inscribed polygon because the vertices of the hexagon can be placed on a circle that is concentric with the outer circle of the socket. Answers should include the following.
- An inscribed polygon is one in which all of its vertices are points on a circle.
- The side of the regular hexagon inscribed in a circle $\frac{3}{4}$ inch wide is $\frac{3}{8}$ inch.

47. 234 **49.** $\sqrt{135} \approx 11.62$ **51.** 4π units **53.** always **55.** sometimes **57.** no

Page 552–558 Lesson 10-5

1a. Two; from any point outside the circle, you can draw only two tangents. **1b.** None; a line containing a point inside the circle would intersect the circle in two points. A tangent can only intersect a circle in one point. **1c.** One; since a tangent intersects a circle in exactly one point, there is one tangent containing a point on the circle.
3. Sample answer:

polygon circumscribed about a circle

polygon inscribed in a circle

5. Yes; $5^2 + 12^2 = 13^2$ **7.** 576 ft **9.** no **11.** yes **13.** 16
15. 12 **17.** 3 **19.** 30 **21.** See students' work. **23.** 60 units **25.** $15\sqrt{3}$ units
27. Given: $\overline{AB}$ is tangent to $\odot X$ at B. $\overline{AC}$ is tangent to $\odot X$ at C.
Prove: $\overline{AB} \cong \overline{AC}$

Proof:

Statements	Reasons
1. $\overline{AB}$ is tangent to $\odot X$ at B. $\overline{AC}$ is tangent to $\odot X$ at C.	1. Given
2. Draw $\overline{BX}$, $\overline{CX}$, and $\overline{AX}$.	2. Through any two points, there is one line.
3. $\overline{AB} \perp \overline{BX}$, $\overline{AC} \perp \overline{CX}$	3. Line tangent to a circle is $\perp$ to the radius at the pt. of tangency.
4. $\angle ABX$ and $\angle ACX$ are right angles.	4. Def. of $\perp$ lines
5. $\overline{BX} \cong \overline{CX}$	5. All radii of a circle are $\cong$.
6. $\overline{AX} \cong \overline{AX}$	6. Reflexive Prop.
7. $\triangle ABX \cong \triangle ACX$	7. HL
8. $\overline{AB} \cong \overline{AC}$	8. CPCTC

29. $\overline{AE}$ and $\overline{BF}$
31. 12; Draw $\overline{PG}$, $\overline{NL}$, and $\overline{PL}$. Construct $\overline{LQ} \perp \overline{GP}$, thus $LQGN$ is a rectangle. $GQ = NL = 4$, so $QP = 5$. Using the Pythagorean Theorem, $(QP)^2 + (QL)^2 = (PL)^2$. So, $QL = 12$. Since $GN = QL$, $GN = 12$.

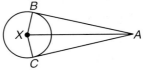

33. 27 **35.** $\overline{AD}$ and $\overline{BC}$ **37.** 45, 45 **39.** 4
41. Sample answer:
Given: $ABCD$ is a rectangle. E is the midpoint of $\overline{AB}$.
Prove: $\triangle CED$ is isosceles.

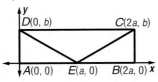

Proof: Let the coordinates of E be $(a, 0)$. Since E is the midpoint and is halfway between A and B, the coordinates of B will be $(2a, 0)$. Let the coordinates of D be $(0, b)$. The coordinates of C will be $(2a, b)$ because it is on the same horizontal as D and the same vertical as B.

$$ED = \sqrt{(a-0)^2 + (0-b)^2} \quad EC = \sqrt{(a-2a)^2 + (0-b)^2}$$
$$= \sqrt{a^2 + b^2} \qquad\qquad = \sqrt{a^2 + b^2}$$

Since $ED = EC$, $\overline{ED} \cong \overline{EC}$. $\triangle DEC$ has two congruent sides, so it is isosceles.

43. 6 **45.** 20.5

Page 561–568 Lesson 10-6

1. Sample answer: A tangent intersects the circle in only one point and no part of the tangent is in the interior of the circle. A secant intersects the circle in two points and some of its points do lie in the interior of the circle. **3.** 138
5. 20 **7.** 235 **9.** 55 **11.** 110 **13.** 60 **15.** 110 **17.** 90
19. 50 **21.** 30 **23.** 8 **25.** 4 **27.** 25 **29.** 130 **31.** 10
33. 141 **35.** 44 **37.** 118 **39.** about 103 ft **41.** 4.6 cm

43a. Given: $\overleftrightarrow{AB}$ is a tangent to $\odot O$. $\overrightarrow{AC}$ is a secant to $\odot O$. $\angle CAB$ is acute.
Prove: $m\angle CAB = \frac{1}{2}m\widehat{CA}$

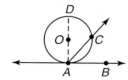

Proof: $\angle DAB$ is a right $\angle$ with measure 90, and $\widehat{DCA}$ is a semicircle with measure 180, since if a line is tangent to a $\odot$, it is $\perp$ to the radius at the point of tangency. Since $\angle CAB$ is acute, C is in the interior of $\angle DAB$, so by the Angle and Arc Addition Postulates, $m\angle DAB = m\angle DAC + m\angle CAB$ and $m\widehat{DCA} = m\widehat{DC} + m\widehat{CA}$. By substitution, $90 = m\angle DAC + m\angle CAB$ and $180 = m\widehat{DC} + m\widehat{CA}$. So, $90 = \frac{1}{2}m\widehat{DC} + \frac{1}{2}m\widehat{CA}$ by Division Prop., and $m\angle DAC + m\angle CAB = \frac{1}{2}m\widehat{DC} + \frac{1}{2}m\widehat{CA}$ by substitution. $m\angle DAC = \frac{1}{2}m\widehat{DC}$ since $\angle DAC$ is inscribed, so substitution yields $\frac{1}{2}m\widehat{DC} + m\angle CAB = \frac{1}{2}m\widehat{DC} + \frac{1}{2}m\widehat{CA}$. By Subtraction Prop., $m\angle CAB = \frac{1}{2}m\widehat{CA}$.

43b. Given: $\overleftrightarrow{AB}$ is a tangent to $\odot O$. $\overrightarrow{AC}$ is a secant to $\odot O$. $\angle CAB$ is obtuse.
Prove: $m\angle CAB = \frac{1}{2}m\widehat{CDA}$

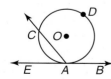

Proof: $\angle CAB$ and $\angle CAE$ form a linear pair, so $m\angle CAB + m\angle CAE = 180$. Since $\angle CAB$ is obtuse, $\angle CAE$ is acute and Case 1 applies, so $m\angle CAE = \frac{1}{2}m\widehat{CA}$. $m\widehat{CA} + m\widehat{CDA} = 360$, so $\frac{1}{2}m\widehat{CA} + \frac{1}{2}m\widehat{CDA} = 180$ by Division Prop., and $m\angle CAE + \frac{1}{2}m\widehat{CDA} = 180$ by substitution. By the Transitive Prop., $m\angle CAB + m\angle CAE = m\angle CAE + \frac{1}{2}m\widehat{CDA}$, so by Subtraction Prop., $m\angle CAB = \frac{1}{2}m\widehat{CDA}$.

45. $\angle 3$, $\angle 1$, $\angle 2$; $m\angle 3 = m\widehat{RQ}$, $m\angle 1 = \frac{1}{2}m\widehat{RQ}$ so $m\angle 3 > m\angle 1$, $m\angle 2 = \frac{1}{2}(m\widehat{RQ} - m\widehat{TP}) = \frac{1}{2}m\widehat{RQ} - \frac{1}{2}m\widehat{TP}$, which is less than $\frac{1}{2}m\widehat{RQ}$, so $m\angle 2 < m\angle 1$. **47.** A **49.** 16
51. 33 **53.** 44.5 **55.** 30 in. **57.** 4, -10 **59.** 3, 5

Page 568 Chapter 10 Practice Quiz 2
1. 67.5 **3.** 12 **5.** 115.5

Page 569–574 Lesson 10-7
1. Sample answer: The product equation for secant segments equates the product of exterior segment measure and the whole segment measure for each secant. In the case of secant-tangent, the product involving the tangent segment becomes (measure of tangent segment)^2 because the exterior segment and the whole segment are the same segment.
3. Sample answer:

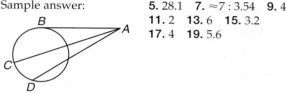

5. 28.1 **7.** $\approx 7 : 3.54$ **9.** 4
11. 2 **13.** 6 **15.** 3.2
17. 4 **19.** 5.6

21. Given: $\overline{WY}$ and $\overline{ZX}$ intersect at T.
Prove: $WT \cdot TY = ZT \cdot TX$

Proof:

Statements	Reasons
a. $\angle W \cong \angle Z$, $\angle X \cong \angle Y$	**a.** Inscribed angles that intercept the same arc are congruent.
b. $\triangle WXT \sim \triangle ZYT$	**b.** AA Similarity
c. $\dfrac{WT}{ZT} = \dfrac{TX}{TY}$	**c.** Definition of similar triangles
d. $WT \cdot TY = ZT \cdot TX$	**d.** Cross products

23. 4 **25.** 11 **27.** 14.3 **29.** $113.\overline{3}$ cm

31. Given: tangent $\overline{RS}$ and secant $\overline{US}$
Prove: $(RS)^2 = US \cdot TS$

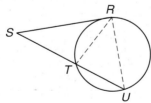

Proof:

Statements	Reasons
1. tangent $\overline{RS}$ and secant $\overline{US}$	**1.** Given
2. $m\angle RUT = \frac{1}{2}m\widehat{RT}$	**2.** The measure of an inscribed angle equals half the measure of its intercepted arc.
3. $m\angle SRT = \frac{1}{2}m\widehat{RT}$	**3.** The measure of an angle formed by a secant and a tangent equals half the measure of its intercepted arc.
4. $m\angle RUT = m\angle SRT$	**4.** Substitution

5. $\angle RUT \cong \angle SRT$ **5.** Definition of $\cong \angle s$
6. $\angle S \cong \angle S$ **6.** Reflexive Prop.
7. $\triangle SUR \sim \triangle SRT$ **7.** AA Similarity
8. $\frac{RS}{US} = \frac{TS}{RS}$ **8.** Definition of $\sim \triangle s$
9. $(RS)^2 = US \cdot TS$ **9.** Cross products

33. Sample answer: The product of the parts of one intersecting chord equals the product of the parts of the other chord. Answers should include the following.
- $\overline{AF}, \overline{FD}, \overline{EF}, \overline{FB}$
- $AF \cdot FD = EF \cdot FB$

35. C **37.** 157.5 **39.** 7 **41.** 36 **43.** scalene, obtuse
45. equilateral, acute or equiangular **47.** $\sqrt{13}$

Pages 575–580 Lesson 10-8
1. Sample answer:

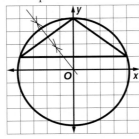

3. $(x + 3)^2 + (y - 5)^2 = 100$
5. $(x + 2)^2 + (y - 11)^2 = 32$
7.

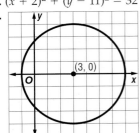

9. $x^2 + y^2 = 1600$ **11.** $(x + 2)^2 + (y + 8)^2 = 25$
13. $x^2 + y^2 = 36$ **15.** $x^2 + (y - 5)^2 = 100$
17. $(x + 3)^2 + (y + 10)^2 = 144$ **19.** $x^2 + y^2 = 8$
21. $(x + 2)^2 + (y - 1)^2 = 10$ **23.** $(x - 7)^2 + (y - 8)^2 = 25$

25.

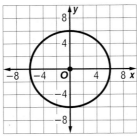

27.

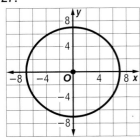

29.

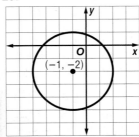

31. $(x + 3)^2 + y^2 = 9$ **33.** 2
35. $x^2 + y^2 = 49$ **37.** 13
39. $(2, -4); r = 6$ **41.** See students' work **43a.** $(0, 3)$ or $(-3, 0)$ **43b.** none
43c. $(0, 0)$ **45.** B **47.** 24
49. 18 **51.** 59 **53.** 20
55. $(3, 2), (-4, -1), (0, -4)$

Pages 581–586 Chapter 10 Study Guide and Review
1. a **3.** h **5.** b **7.** d **9.** c **11.** 7.5 in.; 47.12 in.
13. 10.82 yd; 21.65 yd **15.** 21.96 ft; 43.93 ft **17.** 60
19. 117 **21.** 30 **23.** 30 **25.** 150 **27.** $\frac{22}{5}\pi$ **29.** 10 **31.** 10

33. 45 **35.** 48 **37.** 32 **39.** $m\angle 1 = m\angle 3 = 30, m\angle 2 = 60$
41. 9 **43.** 18 **45.** 37 **47.** 17.1 **49.** 7.2 **51.** $(x + 4)^2 = (y - 8)^2 = 9$ **53.** $(x + 1)^2 + (y - 4)^2 = 4$
55.

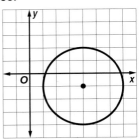

57.

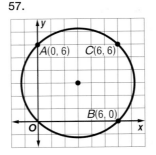

Chapter 11 Areas of Polygons and Circles

Page 593 Chapter 11 Getting Started
1. 10 **3.** 4.6 **5.** 18 **7.** 54 **9.** 13 **11.** 9 **13.** $6\sqrt{3}$
15. $\frac{15\sqrt{2}}{2}$

Pages 598–600 Lesson 11-1
1. The area of a rectangle is the product of the length and the width. The area of a parallelogram is the product of the base and the height. For both quadrilaterals, the measure of the length of one side is multiplied by the length of the altitude. **3.** 28 ft; 39.0 ft² **5.** 12.8 m; 10.2 m² **7.** rectangle, 170 units² **9.** 80 in.; 259.8 in² **11.** 21.6 cm; 29.2 cm²
13. 44 m; 103.9 m² **15.** 45.7 mm² **17.** 108.5 m **19.** $h = 40$ units, $b = 50$ units **21.** parallelogram, 56 units²
23. parallelogram, 64 units² **25.** square, 13 units²
27. 150 units² **29.** Yes; the dimensions are 32 in. by 18 in.
31. ≈ 13.9 ft **33.** The perimeter is 19 m, half of 38 m. The area is 20 m². **35.** 5 in., 7 in. **37.** C **39.** $(5, 2), r = 7$
41. $\left(-\frac{2}{3}, \frac{1}{9}\right), r = \frac{2}{3}$ **43.** 32 **45.** 21 **47.** $F''(-4, 0),$
$G''(-2, -2), H''(-2, 2); 90°$ counterclockwise **49.** 13 ft
51. 16 **53.** 20

Pages 605–609 Lesson 11-2
1. Sample answer:

3. Sometimes; two rhombi can have different corresponding diagonal lengths and have the same area. **5.** 499.5 in²

7. 21 units² **9.** 4 units² **11.** 45 m **13.** 12.4 cm²
15. 95 km² **17.** 1200 ft² **19.** 50 m² **21.** 129.9 mm²
23. 55 units² **25.** 22.5 units² **27.** 20 units² **29.** 16 units²
31. ≈ 26.8 ft **33.** ≈ 22.6 m **35.** 20 cm **37.** about 8.7 ft
39. 13,326 ft² **41.** 120 in² **43.** ≈ 10.8 in² **45.** 21 ft²
47. False; sample answer: the area for each of these right triangles is 6 square units. The perimeter of one triangle is 12 and the perimeter of the other is $8 + \sqrt{40}$ or about 14.3.
49. area = 12, area = 3; perimeter = $8\sqrt{13}$, perimeter = $4\sqrt{13}$; scale factor and ratio of perimeters = $\frac{1}{2}$, ratio of areas = $\left(\frac{1}{2}\right)^2$ **51.** $\frac{2}{1}$ **53.** The ratio is the same.
55. 4 : 1; The ratio of the areas is the square of the scale factor. **57.** 45 ft²; The ratio of the areas is 5 : 9. **59.** B
61. area = $\frac{1}{2}ab \sin C$ **63.** 6.02 cm² **65.** 374 cm²

67. 231 ft² **69.** $(x + 4)^2 + \left(y - \frac{1}{2}\right)^2 = \frac{121}{4}$ **71.** 275 in.
73. ⟨172.4, 220.6⟩ **75.** 20.1

Page 609 Practice Quiz 1
1. square **3.** 54 units² **5.** 42 yd

Pages 613–616 Lesson 11-3
1. Sample answer: Separate a hexagon inscribed in a circle into six congruent nonoverlapping isosceles triangles. The area of one triangle is one-half the product of one side of the hexagon and the apothem of the hexagon. The area of the hexagon is $6\left(\frac{1}{2}sa\right)$. The perimeter of the hexagon is $6s$, so the formula is $\frac{1}{2}Pa$. **3.** 127.3 yd² **5.** 10.6 cm² **7.** about 3.6 yd² **9.** 882 m² **11.** 1995.3 in² **13.** 482.8 km² **15.** 30.4 units² **17.** 26.6 units² **19.** 4.1 units² **21.** 271.2 units² **23.** 2 : 1 **25.** One 16-inch pizza; the area of the 16-inch pizza is greater than the area of two 8-inch pizzas, so you get more pizza for the same price. **27.** 83.1 units² **29.** 48.2 units² **31.** 227.0 units² **33.** 664.8 units² **35.** triangles; 629 tiles **37.** ≈ 380.1 in² **39.** 34.6 units² **41.** 157.1 units² **43.** 471.2 units² **45.** 54,677.8 ft²; 899.8 ft **47.** 225π ≈ 706.9 ft² **49.** 2 : 3 **51.** The ratio is the same. **53.** The ratio of the areas is the square of the scale factor. **55.** 3 to 4 **57.** B **59.** 260 cm² **61.** ≈ 2829.0 yd² **63.** square; 36 units² **65.** rectangle; 30 units² **67.** 42 **69.** 6 **71.** $4\sqrt{2}$

Pages 619–621 Lesson 11-4
1. Sample answer: ≈ 18.3 units² **3.** 53.4 units² **5.** 24 units²

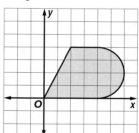

7. ≈ 1247.4 in² **9.** 70.9 units² **11.** 4185 units² **13.** 154.1 units² **15.** ≈ 2236.9 in² **17.** 23.1 units² **19.** 21 units² **21.** 33 units² **23.** Sample answer: 57,500 mi² **25.** 462 **27.** Sample answer: Reduce the width of each rectangle.

29. Sample answer: Windsurfers use the area of the sail to catch the wind and stay afloat on the water. Answers should include the following.
• To find the area of the sail, separate it into shapes. Then find the area of each shape. The sum of areas is the area of the sail.
• Sample answer: Surfboards and sailboards are also irregular figures.
31. C **33.** 154.2 units² **35.** 156.3 ft² **37.** ≈ 429.0 m² **39.** 0.63 **41.** 0.19

Page 621 Practice Quiz 2
1. 679.0 mm² **3.** 1208.1 units² **5.** 44.5 units²

Pages 625–627 Lesson 11-5
1. Multiply the measure of the central angle of the sector by the area of the circle and then divide the product by 360°.
3. Rachel; Taimi did not multiply $\frac{62}{360}$ by the area of the circle. **5.** ≈ 114.2 units², ≈ 0.36 **7.** 0.60 **9.** 0.54 **11.** ≈ 58.9 units², 0.$\overline{3}$ **13.** ≈ 19.6 units², 0.$\overline{1}$ **15.** 74.6 units², 0.42 **17.** ≈ 3.3 units², ≈ 0.03 **19.** ≈ 25.8 units², ≈ 0.15 **21.** 0.68 **23.** 0.68 **25.** 0.19 **27.** ≈ 0.29 **29.** The chances of landing on a black or white sector are the same, so they should have the same point value. **31a.** No; each colored sector

has a different central angle. **31b.** No; there is not an equal chance of landing on each color. **33.** C **35.** 1050 units² **37.** 110.9 ft² **39.** 221.7 in² **41.** 123 **43.** 165 **45.** $g = 21.5$

Pages 628–630 Chapter 11 Study Guide and Review
1. c **3.** a **5.** b **7.** 78 ft, ≈ 318.7 ft² **9.** square; 49 units² **11.** parallelogram; 20 units² **13.** 28 in. **15.** 688.2 in² **17.** 31.1 units² **19.** 0.$\overline{3}$

Chapter 12 Surface Area

Page 635 Chapter 12 Getting Started
1. true **3.** cannot be determined **5.** 384 ft² **7.** 1.8 m² **9.** 7.1 yd²

Pages 639–642 Lesson 12-1
1. The Platonic solids are the five regular polyhedra. All of the faces are congruent, regular polygons. In other polyhedra, the bases are congruent parallel polygons, but the faces are not necessarily congruent.
3. Sample answer:

5. Hexagonal pyramid; base: $ABCDEF$; faces: $ABCDEF$, $\triangle AGF$, $\triangle FGE$, $\triangle EGD$, $\triangle DGC$, $\triangle CGB$, $\triangle BGA$; edges: $\overline{AF}$, $\overline{FE}$, $\overline{ED}$, $\overline{DC}$, $\overline{CB}$, $\overline{BA}$, $\overline{AG}$, $\overline{FG}$, $\overline{EG}$, $\overline{DG}$, $\overline{CG}$, and $\overline{BG}$; vertices: $A, B, C, D, E, F,$ and G **7.** cylinder; bases: circles P and Q

9.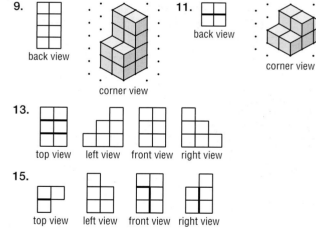

11. back view, corner view

13. top view, left view, front view, right view

15. top view, left view, front view, right view

17. rectangular pyramid; base: $\square DEFG$; faces: $\square DEFG$, $\triangle DHG$, $\triangle GHF$, $\triangle FHE$, $\triangle DHE$; edges: $\overline{DG}$, $\overline{GF}$, $\overline{FE}$, $\overline{ED}$, $\overline{DH}$, $\overline{EH}$, $\overline{FH}$, and $\overline{GH}$; vertices: $D, E, F, G,$ and H
19. cylinder: bases: circles S and T **21.** cone; base: circle B; vertex A **23.** No, not enough information is provided by the top and front views to determine the shape.
25. parabola **27.** circle **29.** rectangle

31. intersecting three faces and parallel to base;

33. intersecting all four faces, not parallel to any face;

35. cylinder **37.** rectangles, triangles, quadrilaterals

39a. triangular **39b.** cube, rectangular, or hexahedron
39c. pentagonal **39d.** hexagonal **39e.** hexagonal
41. No; the number of faces is not enough information to classify a polyhedron. A polyhedron with 6 faces could be a cube, rectangular prism, hexahedron, or a pentagonal pyramid. More information is needed to classify a polyhedron. **43.** Sample answer: Archaeologists use two dimensional drawings to learn more about the structure they are studying. Egyptologists can compare two-dimensional drawings to learn more about the structure they are studying. Egyptologists can compare two-dimensional drawings of the pyramids and note similarities and any differences. Answers should include the following.
- Viewpoint drawings and corner views are types of two-dimensional drawings that show three dimensions.
- To show three dimensions in a drawing, you need to know the views from the front, top, and each side.

45. D **47.** infinite **49.** 0.242 **51.** 0.611 **53.** 21 units²
55. 11 units² **57.** 90 ft, 433.0 ft² **59.** 300 cm² **61.** 4320 in²

Pages 645–648 Lesson 12-2

1. Sample answer: **3.**

5. 188 in²;

7. 64 cm²;

9. **11.**

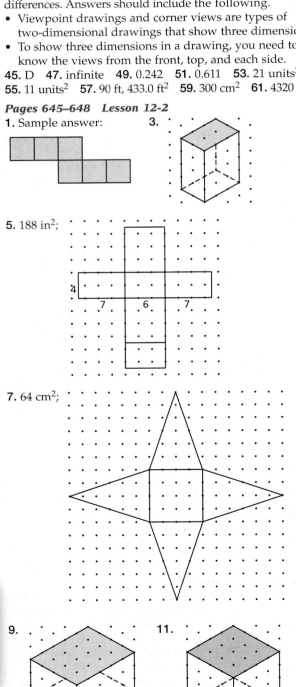

13.

15. 66 units²;

17. 56 units²;

19. 121.5 units²;

21. 116.3 units²;

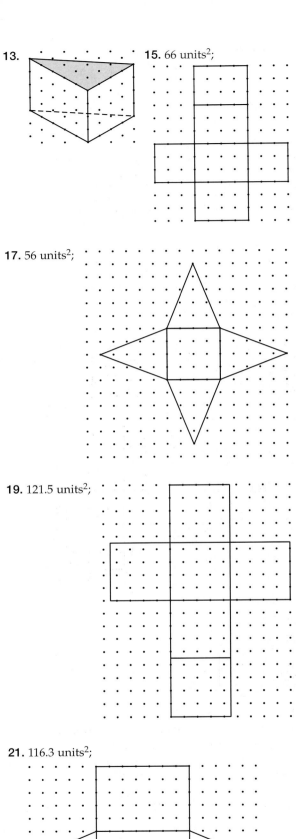
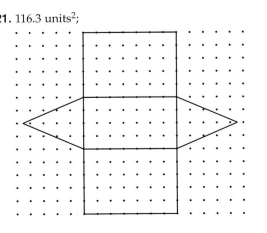

23. 108.2 units²;

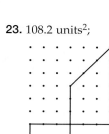

25. 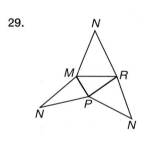   **27.**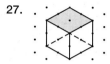

29. 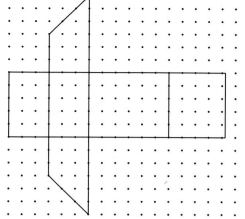   **31.**

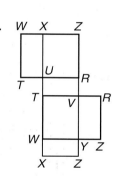

33.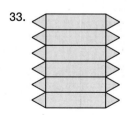

35. A 6 units²;   **B** $\left(9 + \frac{\sqrt{3}}{2}\right)$ = 9.87 units²;

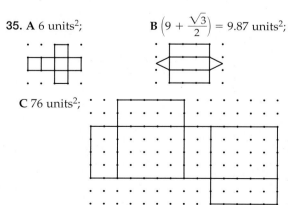

C 76 units²;

37. The surface area quadruples when the dimensions are doubled. For example, the surface area of the cube is 6(1²)

or 6 square units. When the dimensions are doubled the surface area is 6(2²) or 24 square units.   **39.** No; 5 and 3 are opposite faces; the sum is 8.   **41.** C   **43.** rectangle **45.** rectangle   **47.** 90   **49.** 120   **51.** 63 cm²   **53.** 110 cm²

Pages 651–654   Lesson 12-3
1. In a right prism a lateral edge is also an altitude. In an oblique prism, the lateral edges are not perpendicular to the bases.   **3.** 840 units², 960 units²   **5.** 1140 ft²   **7.** 128 units²   **9.** 162 units²   **11.** 160 units² (square base), 126 units² (rectangular base)   **13.** 16 cm   **15.** The perimeter of the base must be 24 meters. There are six rectangles with integer values for the dimensions that have a perimeter of 24. The dimensions of the base could be 1 × 11, 2 × 10, 3 × 9, 4 × 8, 5 × 7, or 6 × 6.   **17.** 114 units²   **19.** 522 units² **21.** 454.0 units²   **23.** 3 gallons for 2 coats   **25.** 44,550 ft² **27.** The actual amount needed will be higher because the area of the curved architectural element appears to be greater than the area of the doors.   **29.** base of A ≅ base of C; base of A ~ base of B; base of C ~ base of B   **31.** A : B = 1 : 4, B : C = 4 : 1, A : C = 1 : 1   **33.** A : B, because the heights of A and B are in the same ratio as perimeters of bases   **35.** No, the surface area of the finished product will be the sum of the lateral areas of each prism plus the area of the bases of the TV and DVD prisms. It will also include the area of the overhang between each prism, but not the area of the overlapping prisms.   **37.** 198 cm²   **39.** B   **41.** L = 1416 cm², T = 2056 cm² **43.** See students' work.

45. 108 units²;

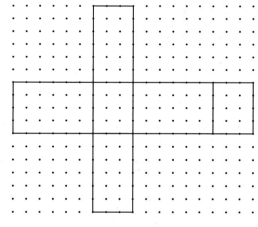

47.   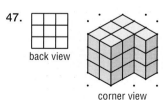   **49.** 43
back view   corner view   **51.** 35   **53.** $\frac{1}{72}$
55. 1963.50 in²
57. 21,124.07 mm²

Pages 657–659   Lesson 12-4
1. Multiply the circumference of the base by the height and add the area of each base.   **3.** Jamie; since the cylinder has one base removed, the surface area will be the sum of the lateral area and one base.   **5.** 1520.5 m²   **7.** 5 ft   **9.** 2352.4 m² **11.** 517.5 in²   **13.** 251.3 ft²   **15.** 30.0 cm²   **17.** 3 cm   **19.** 8 m **21.** The lateral areas will be in the ratio 3 : 2 : 1; 45π in², 30π in², 15π in².   **23.** The lateral area is tripled. The surface area is increased, but not tripled.   **25.** 1.25 m   **27.** Sample answer: Extreme sports participants use a semicylinder for a ramp. Answers should include the following.

- To find the lateral area of a semicylinder like the half-pipe, multiply the height by the circumference of the base and then divide by 2.
- A half-pipe ramp is half of a cylinder if the ramp is an equal distance from the axis of the cylinder.

29. C

31. a plane perpendicular to the line containing the opposite vertices of the face of the cube

33. 300 units^2

35.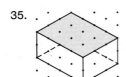

37. 27 **39.** 8
41. $m\angle A = 64$, $b \approx 12.2$, $c \approx 15.6$
43. 54 cm^2

Page 659 Practice Quiz 1

1.

corner view

3. 231.5 m^2 **5.** 5.4 ft

Pages 663–665 Lesson 12-5

1. Sample answer:

square base   rectangular base
(regular)   (not regular)

3. 74.2 ft^2
5. 340 cm^2
7. 119 cm^2
9. 147.7 ft^2
11. 173.2 yd^2
13. 326.9 in^2

15. 27.7 ft^2 **17.** $\approx$ 2.3 inches on each side **19.** $\approx$ 615,335.3 ft^2
21. 20 ft **23.** 960 ft^2 **25.** The surface area of the original cube is 6 square inches. The surface area of the truncated cube is approximately 5.37 square inches. Truncating the corner of the cube reduces the surface area by about 0.63 square inch. **27.** D **29.** 967.6 m^2 **31.** 1809.6 yd^2 **33.** 74 ft, 285.8 ft^2 **35.** 98 m, 366 m^2 **37.** $\overline{GF}$ **39.** $\overline{JM}$ **41.** True; each pair of opposite sides are congruent. **43.** 21.3 m

Pages 668–670 Lesson 12-6

1. Sample answer:

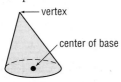

vertex
center of base

3. 848.2 cm^2 **5.** 485.4 in^2
7. 282.7 cm^2 **9.** 614.3 in^2
11. 628.8 m^2 **13.** 679.9 in^2
15. 7.9 m **17.** 5.6 ft
19. 475.2 in^2 **21.** 1509.8 m^2
23. 1613.7 in^2 **25.** $\approx$ 12 ft
27. 8.1 in.; 101.7876 in^2

29. Using the store feature on the calculator is the most accurate technique to find the lateral area. Rounding the slant height to either the tenths place or hundredths place changes the value of the slant height, which affects the final computation of the lateral area. **31.** Sometimes; only when the heights are in the same ratio as the radii of the bases. **33.** Sample answer: Tepees are conical shaped structures. Lateral area is used because the ground may not always be covered in circular canvas. Answers should include the following.
- We need to know the circumference of the base or the radius of the base and the slant height of the cone.
- The open top reduces the lateral area of canvas needed to cover the sides. To find the actual lateral area, subtract

the lateral area of the conical opening from the lateral area of the structure.

35. D **37.** 5.8 ft **39.** 6.0 yd **41.** 48 **43.** 24 **45.** 45
47. 21 **49.** $8\sqrt{11} \approx 26.5$ **51.** 25.1 **53.** 51.5 **55.** 25.8

Page 670 Practice Quiz 2

1. 423.9 cm^2 **3.** 144.9 ft^2 **5.** 3.9 in.

Pages 674–676 Lesson 12-7

1. Sample answer:

3. 15 **5.** 18 **7.** 150.8 cm^2 **9.** $\approx$ 283.5 in^2
11. $\approx$ 8.5 **13.** 8 **15.** 12.8 **17.** 7854.0 in^2
19. 636,172.5 m^2 **21.** 397.4 in^2
23. 3257.2 m^2 **25.** true **27.** true
29. true **31.** $\approx$ 206,788,161.4 mi^2
33. 398.2 ft^2

35. $\dfrac{\sqrt{2}}{2}$: 1 **37.** The surface area can range from about 452.4 to about 1256.6 mi^2. **39.** The radius of the sphere is half the side of the cube. **41.** None; every line (great circle) that passes through X will also intersect g. All great circles intersect. **43.** A **45.** 1430.3 in^2 **47.** 254.7 cm^2 **49.** 969 yd^2
51. 649 cm^2 **53.** $(x + 2)^2 + (y - 7)^2 = 50$

Pages 678–682 Chapter 12 Study Guide and Review

1. d **3.** b **5.** a **7.** e **9.** c **11.** cylinder; bases: $\odot F$ and $\odot G$ **13.** triangular prism; base: $\triangle BCD$; faces: $\triangle ABC$, $\triangle ABD$, $\triangle ACD$, and $\triangle BCD$; edges: $\overline{AB}$, $\overline{BC}$, $\overline{AC}$, $\overline{AD}$, $\overline{BD}$, $\overline{CD}$; vertices: A, B, C, and D
15. 340 units^2;

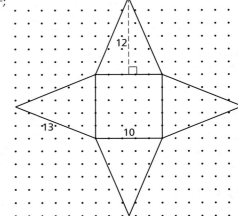

17. $\approx$ 133.7 units^2;

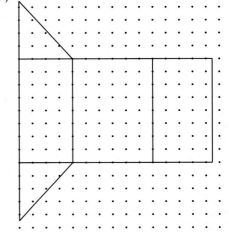

19. 228 units^2;

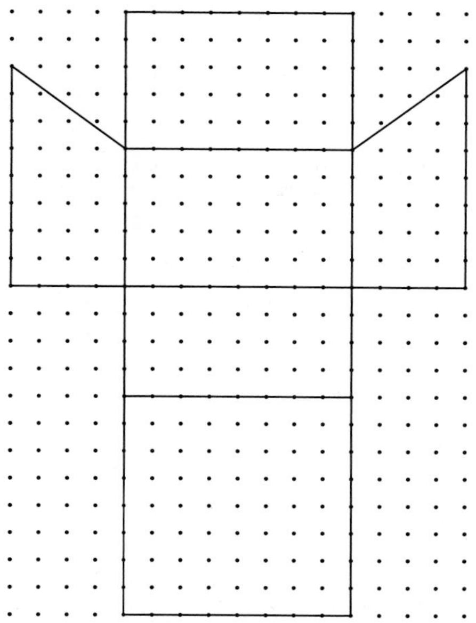

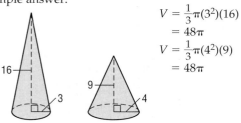

21. 72 units^2 **23.** 175.9 in^2 **25.** 1558.2 mm^2 **27.** 304 units^2 **29.** 33.3 units^2 **31.** 75.4 yd^2 **33.** 1040.6 ft^2 **35.** 363 mm^2 **37.** 2412.7 ft^2 **39.** 880 ft^2

Chapter 13 Volume

Page 687 Chapter 13 Getting Started
1. ±5 **3.** ±3 **5.** ± $\sqrt{305}$ **7.** 134.7 cm^2 **9.** 867.0 mm^2
11. 25b^2 **13.** $\frac{9x^2}{16y^2}$ **15.** $W(-2.5, 1.5)$ **17.** $B(19, 21)$

Pages 691–694 Lesson 13-1
1. Sample answers: cans, roll of paper towels, and chalk; boxes, crystals, and buildings **3.** 288 cm^3 **5.** 3180.9 mm^3
7. 763.4 cm^3 **9.** 267.0 cm^3 **11.** 750 in^3 **13.** 28 ft^3
15. 15,108.0 mm^3 **17.** ≈ 14 m **19.** 24 units^3 **21.** 48.5 mm^3 **23.** 173.6 in^3 **25.** ≈ 304.1 cm^3 **27.** about 19.2 ft
29. ≈ 104,411.5 mm^3 **31.** ≈ 137.6 ft^3 **33.** A **35.** 452.4 ft^2
37. 1017.9 m^2 **39.** 320.4 m^2 **41.** 282.7 in^2 **43.** ≈ 0.42
45. 186 m^2 **47.** 8.8 **49.** 21.22 in^2 **51.** 61.94 m^2

Pages 698–701 Lesson 13-2
1. Each volume is 8 times as large as the original.
3. Sample answer:

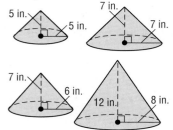

$$V = \frac{1}{3}\pi(3^2)(16)$$
$$= 48\pi$$
$$V = \frac{1}{3}\pi(4^2)(9)$$
$$= 48\pi$$

5. 603.2 mm^3 **7.** 975,333.3 ft^3 **9.** 1561.2 ft^3
11. 8143.0 mm^3 **13.** 2567.8 m^3 **15.** 188.5 cm^3
17. 1982.0 mm^3 **19.** 7640.4 cm^3 **21.** ≈ 2247.5 km^3
23. ≈ 158.8 km^3 **25.** ≈ 91,394,008.3 ft^3 **27.** ≈ 6,080,266.7 ft^3
29. ≈ 522.3 units^3 **31.** ≈ 203.6 in^3 **33.** B **35.** 1008 in^3
37. 1140 ft^3 **39.** 258 yd^2 **41.** 145.27 **43.** 1809.56

Page 701 Practice Quiz 1
1. 125.7 in^3 **3.** 935.3 cm^3 **5.** 42.3 in^3

Pages 704–706 Lesson 13-3
1. The volume of a sphere was generated by adding the volumes of an infinite number of small pyramids. Each pyramid has its base on the surface of the sphere and its height from the base to the center of the sphere.
3. 9202.8 in^3 **5.** 268.1 in^3 **7.** 155.2 m^3 **9.** 1853.3 m^3
11. 3261.8 ft^3 **13.** 233.4 in^3 **15.** 68.6 m^3 **17.** 7238.2 in^3
19. ≈ 21,990,642,871 km^3 **21.** No, the volume of the cone is 41.9 cm^3; the volume of the ice cream is about 33.5 cm^3.
23. ≈ 20,579.5 mm^3 **25.** ≈ 1162.1 mm^2 **27.** $\frac{2}{3}$
29. ≈ 587.7 in^3 **31.** 32.7 m^3 **33.** about 184 mm^3
35. See students' work. **37.** A **39.** 412.3 m^3
41. $(x-2)^2 + (y+1)^2 = 64$ **43.** $(x-2)^2 + (y-1)^2 = 34$
45. 27x^3 **47.** $\frac{8k^3}{125}$

Pages 710–713 Lesson 13-4
1. Sample answer:

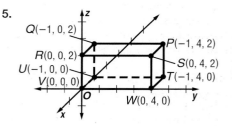

3. congruent **5.** $\frac{4}{3}$
7. $\frac{64}{27}$ **9.** 1:64
11. neither
13. congruent
15. neither
17. 130 m high, 245 m wide, and 465 m long
19. Always; congruent solids have equal dimensions.
21. Never; different types of solids cannot be similar.
23. Sometimes; solids that are not similar can have the same surface area. **25.** 1,000,000x cm^2 **27.** $\frac{2}{5}$ **29.** $\frac{8}{125}$
31. 18 cm **33.** $\frac{29}{30}$ **35.** $\frac{24,389}{27,000}$ **37.** ≈ 0.004 in^3 **39.** 3:4; 3:1
41. The volume of the cone on the right is equal to the sum of the volumes of the cones inside the cylinder. Justification: Call h the height of both solids. The volume of the cone on the right is $\frac{1}{3}\pi r^2 h$. If the height of one cone inside the cylinder is c, then the height of the other one is $h - c$. Therefore, the sum of the volumes of the two cones is: $\frac{1}{3}\pi r^2 c + \frac{1}{3}\pi r^2(h - c)$ or $\frac{1}{3}\pi r^2(c + h - c)$ or $\frac{1}{3}\pi r^2 h$. **43.** C **45.** 268.1 ft^3
47. 14,421.8 cm^3 **49.** 323.3 in^3 **51.** 2741.8 ft^3 **53.** 2.8 yd
55. 36 ft^2 **57.** yes **59.** no

Page 713 Practice Quiz 2
1. 67,834.4 ft^3 **3.** $\frac{7}{5}$ **5.** $\frac{343}{125}$

Pages 717–719 Lesson 13-5
1. The coordinate plane has 4 regions or quadrants with 4 possible combinations of signs for the ordered pairs. Three-dimensional space is the intersection of 3 planes that create 8 regions with 8 possible combinations of signs for the ordered triples. **3.** A dilation of a rectangular prism will provide a similar figure, but not a congruent one unless $r = 1$ or $r = -1$.

5.

Selected Answers (side tab)

7. $\sqrt{186}$; $\left(1, -\frac{7}{2}, \frac{1}{2}\right)$ **9.** (12, 8, 8), (12, 0, 8), (0, 0, 8), (0, 8, 8), (12, 8, 0), (12, 0, 0), (0, 0, 0), and (0, 8, 0); (−36, 8, 24), (−36, 0, 24), (−48, 0, 24), (−48, 8, 24) (−36, 8, 16), (−36, 0, 16), (−48, 0, 16), and (−48, 8, 16)

11.

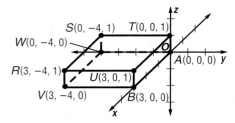

13.

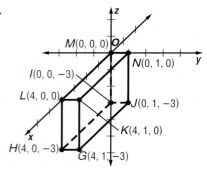

15.

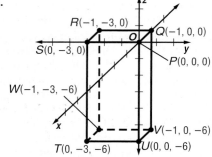

17. $PQ = \sqrt{115}$; $\left(\frac{1}{2}, -\frac{7}{2}, \frac{7}{2}\right)$ **19.** $GH = \sqrt{17}$; $\left(\frac{3}{5}, -\frac{7}{10}, 4\right)$

21. $BC = \sqrt{39}$; $\left(-\frac{\sqrt{3}}{2}, 3, 3\sqrt{2}\right)$

23.

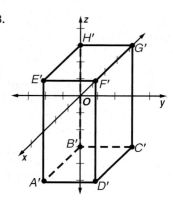

25. $P'(0, 2, -2)$, $Q'(0, 5, -2)$, $R'(2, 5, -2)$, $S'(2, 2, -2)$ $T'(0, 5, -5)$, $U'(0, 2, -5)$, $V'(2, 2, -5)$, and $W'(2, 5, -5)$

27. $A'(4, 5, 1)$, $B'(4, 2, 1)$, $C'(1, 2, 1)$, $D'(1, 5, 1)$ $E'(4, 5, -2)$, $F'(4, 2, -2)$, $G'(1, 2, -2)$, and $H'(1, 5, -2)$;

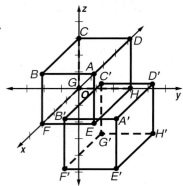

29. $A'(6, 6, 6)$, $B'(6, 0, 6)$, $C'(0, 0, 6)$, $D'(0, 6, 6)$, $E'(6, 6, 0)$, $F'(6, 0, 0)$, $G'(0, 0, 0)$, and $H'(0, 6, 0)$; $V = 216$ units^3;

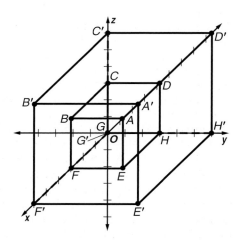

31. 8.2 mi   **33.** (0, −14, 14)   **35.** $(x, y, z) \rightarrow (x + 2, y + 3, z - 5)$ **37.** Sample answer: Three-dimensional graphing is used in computer animation to render images and allow them to move realistically. Answers should include the following.
- Ordered triples are a method of locating and naming points in space. An ordered triple is unique to one point.
- Applying transformations to points in space would allow an animator to create realistic movement in animation.

39. B   **41.** The locus of points in space with coordinates that satisfy the equation of $x + z = 4$ is a plane perpendicular to the xz-plane whose intersection with the xz-plane is the graph of $z = -x + 4$ in the xz-plane.

43. similar   **45.** 1150.3 yd^3 **47.** 12,770.1 ft^3

Pages 720–722   Chapter 13   Study Guide and Review
1. pyramid   **3.** an ordered triple   **5.** similar   **7.** the Distance Formula in Space   **9.** Cavalieri's Principle
11. 504 in^3 **13.** 749.5 ft^3 **15.** 1466.4 ft^3 **17.** 33.5 ft^3
19. 4637.6 mm^3 **21.** 523.6 units^3 **23.** similar   **25.** $CD = \sqrt{58}$; (−9, 5.5, 5.5)   **27.** $FG = \sqrt{422}$; $\left(1.5\sqrt{2}, 3\sqrt{7}, -3\right)$

Photo Credits

About the Cover: This photo of the financial district of Hong Kong illustrates a variety of geometrical shapes. The building on the right is called Jardine House. Because the circular windows resemble holes in the rectangular blocks, this building was given the nickname "House of a Thousand Orifices." The other building is one of the three towers that comprise the Exchange Square complex, home to the Hong Kong Stock Exchange. These towers appear to be a combination of large rectangular prisms and cylinders.

Acheson Wallace Gift, 1993 (1993.303a–f), (r)courtesy Dorothea Rockburne and Artists Rights Society; **439** Bill Bachmann/PhotoEdit; **440** (l)Bernard Gotfryd/Woodfin Camp & Associates, (r)San Francisco Museum of Modern Art. Purchased through a gift of Phyllis Wattis/©Barnett Newman Foundation/Artists Rights Society, New York; **442** Tim Hall/PhotoDisc; **451** Paul Trummer/Getty Images; **460–461** William A. Bake/CORBIS; **463** Robert Glusic/PhotoDisc; **467** (l)Siede Pries/PhotoDisc, (c)Spike Mafford/PhotoDisc, (r)Lynn Stone; **468** Hulton Archive; **469** Phillip Hayson/Photo Researchers; **470** James L. Amos/CORBIS; **476** Sellner Manufacturing Company; **478** Courtesy Judy Mathieson; **479** (l)Matt Meadows, (c)Nick Carter/Elizabeth Whiting & Associates/CORBIS, (r)Massimo Listri/CORBIS; **480** (t)Sony Electronics/AP/Wide World Photos, (bl)Jim Corwin/Stock Boston, (bc)Spencer Grant/PhotoEdit, (br)Aaron Haupt; **483** *Symmetry Drawing E103*. M.C. Escher. ©2002 Cordon Art, Baarn, Holland. All rights reserved; **486** Smithsonian American Art Museum, Washington DC/Art Resource, NY; **487** (tl)Sue Klemens/Stock Boston, (tr)Aaron Haupt, (b)Digital Vision; **495** Phillip Wallick/CORBIS; **501** CORBIS; **504** Georg Gerster/Photo Researchers; **506** Rob McEwan/TriStar/Columbia/Motion Picture & Television Photo Archive; **520–521** Michael Dunning/Getty Images; **522** Courtesy The House on The Rock, Spring Green WI; **524** Aaron Haupt; **529** Carl Purcell/Photo Researchers; **534** Craig Aurness/CORBIS; **536** KS Studios; **541** (l)Hulton Archive/Getty Images, (r)Aaron Haupt; **543** Profolio/Index Stock; **544 550** Aaron Haupt; **552** Andy Lyons/Getty Images; **557** Ray Massey/Getty Images; **558** Aaron Haupt; **566** file photo; **569** Matt Meadows; **572** Doug Martin; **573** David Young-Wolff/PhotoEdit; **575** Pete Turner/Getty Images; **578** NOAA; **579** NASA; **590** Courtesy National World War II Memorial; **590–591** Rob Crandall/Stock Boston; **592–593** Ken Fisher/Getty Images; **595** Michael S. Yamashita/CORBIS; **599** (l)State Hermitage Museum, St. Petersburg, Russia/CORBIS, (r)Bridgeman Art Library; **601** (t)Paul Baron/CORBIS, (b)Matt Meadows; **607** Chuck Savage/CORBIS;

610 R. Gilbert/H. Armstrong Roberts; **613** Christie's Images; **615** Sakamoto Photo Research Laboratory/CORBIS; **617** Peter Stirling/CORBIS; **620** Mark S. Wexler/Woodfin Camp & Associates; **622** C Squared Studios/PhotoDisc; **626** Stu Forster/Getty Images; **634–635** Getty Images; **636** (t)Steven Studd/Getty Images, (b)Collection Museum of Contemporary Art, Chicago, gift of Lannan Foundation. Photo by James Isberner; **637** Aaron Haupt; **638** Scala/Art Resource, NY; **641** (l)Charles O'Rear/CORBIS, (c)Zefa/Index Stock, (r)V. Fleming/Photo Researchers; **643** (t)Image Port/Index Stock, (b)Chris Alan Wilton/Getty Images; **647** (t)Doug Martin, (b)CORBIS; **649** Lon C. Diehl/PhotoEdit; **652** G. Ryan & S. Beyer/Getty Images; **655** Paul A. Souders/CORBIS; **658** Michael Newman/PhotoEdit; **660** First Image; **664** (tl)Elaine Rebman/Photo Researchers, (tr)Dan Callister/Online USA/Getty Images, (b)Massimo Listri/CORBIS; **666** EyeWire; **668** CORBIS; **669** Courtesy Tourism Medicine Hat. Photo by Royce Hopkins; **671** StudiOhio; **672** Aaron Haupt; **673** Don Tremain/PhotoDisc; **675** (l)David Rosenberg/Getty Images, (r)StockTrek/PhotoDisc; **686–687** Ron Watts/CORBIS; **688** (t)Tribune Media Services, Inc. All Rights Reserved. Reprinted with permission., (b)Matt Meadows; **690** Aaron Haupt; **693** (l)Peter Vadnai/CORBIS, (r)CORBIS; **696** (t)Lightwave Photo, (b)Matt Meadows; **699** Courtesy American Heritage Center; **700** Roger Ressmeyer/CORBIS; **702** Dominic Oldershaw; **705** Yang Liu/CORBIS; **706** Brian Lawrence/SuperStock; **707** Matt Meadows; **709** Aaron Haupt; **711** Courtesy Denso Corp.; **712** (l)Doug Pensinger/Getty Images, (r)AP/Wide World Photos; **714** Rein/CORBIS SYGMA; **717** Gianni Dagli Orti/CORBIS; **727** Grant V. Faint/Getty Images; **782** (t)Walter Bibikow/Stock Boston, (b)Serge Attal/TimePix; **784** (l)Carl & Ann Purcell/CORBIS, (r)Doug Martin; **789** John D. Norman/CORBIS; **790** Stella Snead/Bruce Coleman, Inc.; **793** (t)Yann Arthus-Bertrand/CORBIS, (c)courtesy M-K Distributors, Conrad MT, (b)Aaron Haupt; **794** F. Stuart Westmorland/Photo Researchers.

Index

Index

Index

Index

Index

Index